Tribology in Metalworking

Friction, Lubrication and Wear

Tribology in Metalworking

Friction, Lubrication and Wear

JOHN A. SCHEY

Professor
Department of Mechanical Engineering
University of Waterloo
Ontario, Canada

AMERICAN SOCIETY FOR METALS
Metals Park, Ohio 44073

Second printing, June 1984

Library of Congress Catalog Card No.: 82-73612

ISBN: 0-87170-155-3

SAN: 204-7586

Editorial and production coordination by
Carnes Publication Services, Inc.

PRINTED IN THE UNITED STATES OF AMERICA

Preface

Some thirteen years ago, Professor S. Kalpakjian and Drs. J.A. Newnham and C.H. Riesz cooperated with me in producing a reference book on "Metal Deformation Processes: Friction and Lubrication," published by Marcel Dekker, Inc., New York (1970). The intervening years have seen a veritable explosion of information in the field that meanwhile has come to be described as "tribology," a term encompassing the science and technology of friction, lubrication, and wear. Knowledge pertaining to metalworking tribology has also increased greatly. Contacts with colleagues in industry and research have impressed upon me the need for a book that would serve better as a self-learning text. Therefore, when the American Society for Metals expressed interest in an updated version, I decided to write an essentially new research monograph, which, hopefully, will be found more useful as an introduction to the subject while serving also as a reference source. In view of its importance, the tribology of machining processes is now also included.

I could have chosen to limit the scope of this volume strictly to metalworking, and to direct the reader to the vast literature from which the necessary background could be acquired. Such an approach would have been patently unfair to the majority of readers, who may be specialists in one field but not in others, or who may be newcomers to the entire subject. Tribology itself is a truly interdisciplinary discipline, drawing on various aspects of physics, chemistry, mechanical engineering, and materials science. Metalworking tribology is yet further broadened by the inclusion of the mechanics and metallurgy of metalworking processes. Therefore, I have endeavored to give a brief but comprehensive account of the essential background material, so that the volume can stand on its own, and so that the reader need delve into the literature only if interested in following up specific aspects.

The material relevant to metalworking tribology would easily fill several monographs of this size. Complete coverage thus not only was an unrealistic goal but also would have made the volume virtually indigestible, with a quite unreasonable number of references. Therefore, some selection had to be exercised, and this necessarily involved personal judgments of what is most important and what can be readily followed up on elsewhere. In making such judgments, I have endeavored to keep in mind the varied background of the readership: what is trivial to a chemist may be new to a mechanical engineer, and *vice versa*. The emphasis here is on physical observations and their interpretation; details of analysis, for the interested reader, can be found in the references given. I hope that this volume, as a result of the selections made, will provide adequate information for the novice without boring the expert.

In writing the text I had to venture into areas beyond my immediate expertise. To ensure accuracy of treatment, I imposed on friends for advice. I am greatly indebted to

Donald H. Buckley, Bernard J. Hamrock, John Hickman, J. Brian Peace, Clyde A. Sluhan, Joseph L. Tevaarwerk, Joseph Wright, and several graduate students for their valuable remarks on Chapters 2 to 5; to Ranga Komanduri, Milton C. Shaw, and Paul K. Wright for their suggestions regarding Chapter 11; and to George E. Dieter for the enormous task of reviewing all but Chapter 11. For any remaining errors, I alone am responsible.

An operating grant from the Natural Sciences and Engineering Research Council of Canada allowed me to explore topics that were in need of clarification, and I made extensive use of the library facilities and services of the University of Waterloo.

There are many friends and colleagues without whose help this undertaking would have been impossible. Kurt Lange provided me with the opportunity to search the German literature in his Institut für Umformtechnik, and his co-worker Th. Gräbener made available his collection of references. Bruce Uttley and Marjorie Kohli of the University of Waterloo found solutions to the many problems of transferring the text from word processor to phototypesetter; Beryl Hultin, together with others, helped in correcting the text on the word processor; Evie Hill of Historical Reflections took care of readying the text for phototypesetting. I enjoyed working with Timothy L. Gall of ASM and with William J. Carnes in the preparation of this volume. The brunt of the task was carried by my wife, Gitta, who for three years spent endless hours working on references, typing drafts, and typing the text for the OCR transfer; to her I dedicate this volume.

It is neither likely nor necessary that anyone should read this volume from cover to cover as a continuum. Cross references are included to facilitate reading on individual topics and processes. I should be most grateful to have readers' reactions and suggestions regarding the treatment and any errors that may be found.

John A. Schey

Waterloo, Ontario
February, 1983

Contents

1

Background and Method of Treatment

The history of tribology, set against the broader background of industrial and social development, is the subject of a recent and thoroughly enjoyable book by Dowson [1]. Regrettably, it contains but few references to metalworking; therefore, a brief account based on the earlier review of Schey [2] is given here, followed by some remarks on the method of treatment adopted in the present book.

1.1 Historical Background

Metalworking is probably the earliest technological occupation known to mankind; native metals were forged and shaped more than 7000 years ago. Considering the importance of lubricants in deformation processes, it is rather amazing that no account of their use can be found until relatively recent times.

There are, perhaps, three main reasons for this anomaly. First, the composition, manufacture, and use of lubricants were—and to some extent, still are—the most closely guarded secrets of the whole operation. Second, nature very kindly provided some of the best lubricants known today, so that the artisan engaged in metalworking did not have to rely on the skills and knowledge of other crafts, arts, or sciences; he was not compelled to reveal his practices, and the nonexpert observers reporting on metalworking processes failed to grasp the significance of the little they may have been allowed to see. Third, lubricants assumed a vital role only at a relatively late stage of development.

1.1.1 Forging

There is little doubt that forging was the first, and for a long time the only, bulk deformation technique. Native gold, silver, and copper were hammered into thin sheets and then shaped into jewelry and household utensils as early as 5000 B.C. [3]. These metals, and, later, copper reduced from ores, were readily cold worked without a lubricant. Some unintentional lubrication may have occurred, for instance, when sheet was driven into asphalt for shaping, or was placed between animal skins for thinning down to metal (gold) leaf. Copper was annealed well before 4000 B.C., and hot forging must have soon followed. Copper oxide is a good parting agent, as is the oxide of iron, a metal which was worked with great skill under the Hittites in the 13th century B.C.

Iron became the dominant material for weapons, tools, and agricultural implements from 800 B.C. on, replacing the cast and also wrought bronze that dominated earlier centuries [4]. This marks the ascendancy of the blacksmith, with his Greek god Hephaistos [5]. The cold forming of metals, nevertheless, continued uninterrupted. Cast bronze forms were used for hammering decorative parts from sheet, and the first known coin was driven into a die with several punch strokes in the 7th century B.C. Again, no reference to lubricants is known and it is probable that natural contaminants, even if only from greasy fingers, provided all that was needed.

One must not forget, though, that seed oils, animal oils, tallow, and waxes were available in early antiquity [6]. In the 5th century B.C. Herodotus wrote about the extraction of bitumen and light oil from petroleum, and in about 60 A.D. Pliny the elder described the manufacture of soap [7]. Thus, at whatever point of development the first lubricant may have been used, a variety of eminently suitable substances certainly must have been available.

Forging continued to be the dominant metalworking process for many centuries. Coin pressing achieved high standards under the Romans, who drove the blank with a hand punch of suitable profile into a counterforming die [4]. The first forgings made in closed dies were produced in the northern Germanic areas, where local iron ores yielded higher-carbon iron and, therefore, better weapons. Chain mail was probably first made there and, while the starting material would ideally have been drawn wire, it is now generally conceded that this wire was actually forged in swaging dies. Cast bronze anvils survive from the Bronze Age with indentations that probably served for the forging of needles, and the goldsmiths and silversmiths had no problem in shaping their precious metals into decorative wire.

Forging is, nevertheless, the only hot working process that can claim a long history of lubrication. Steel rifle parts were forged in die impressions for interchangeability as early as the 18th century [4]; lubrication practices were probably not too different from those found some fifty years ago, with sawdust, a thin smear of a heavy oil, or oil mixed with graphite serving as a lubricant.

1.1.2 Wiredrawing

The argument on the earliest date of wiredrawing is still not settled. Theophilus, the German monk writing in the early 12th century, treats wiredrawing as an established art [8]. He refers to steel dies used for drawing a tin-lead alloy, and to oak dies for smoothing hammered gold and silver wire. Iron and steel plates bearing a number of holes, connected with a surface channel, have been identified by many researchers as drawing dies, on the assumption that the surface channel served for retaining the lubricant [9]. The earliest such finds date back to the first century A.D. and, if the interpretation of the channels is correct, these would provide circumstantial evidence of the use of metalworking lubricants for two millenia.

Guilds of wiredrawers were registered and regulated as early as the 13th century [10], and water power was introduced in the 14th [11]. Yet, the first written reference to wiredrawing lubrication is of much later date. Biringuccio [12], writing of the drawing of gold and silver wire, admonishes to "remember that while you are working it you must always keep it greased with new wax, for besides easing its passage through the holes, this also keeps its color yellow and beautiful." Rather casual references to oil-soaked rags, applied to the iron wire before it entered the drawing die, were made by early 18th century observers [8]. It is likely that the lubricant was a locally available product—probably lard oil in the north, and vegetable oils in the south, of Europe [6]. Records show that olive oil was imported into Altena, the center of the German steel wiredrawing industry, far in excess of any conceivable household needs, and it is reasonable to assume that this was used for drawing the highest grades of iron and, later, steel wires [10].

It was in Altena that the first surface treatment was discovered. For many centuries the wire had been prepared for drawing by hand scouring it with bricks or sand,

and it was only in the 17th century that bundles were mechanically threshed with a waterwheel, with frequent dashes of water and sand [10]. With the usual grease or oil lubrication, it was possible to draw the clean rod, but only if made of the softest, best quality Osmund iron. Steel, while known to be superior for many applications, including needles, fish hooks, and the like, could not be drawn, presumably because high friction caused the wire to break. It remained for one Johann Gerdes to discover the value of a surface treatment around 1650, the story being recorded by a wandering minstrel only a few years later. As related by Lewis [10], after an unsuccessful effort, Gerdes threw the steel rods out of the window, into the area where "men came to cast their water." After a while, he retrieved the rods and—without bothering to remove the soft, brown film that had formed on them—found that they now drew with great ease. This original sull-coating technique was used for almost 150 years. Then dilute sour beer was found just as effective, and in another 50 years it was recognized that plain water would serve equally well, although the original technique has survived in some places until this century.

Prior to the introduction of sulfuric and hydrochloric acids, light scale was removed or softened on the wire surface by lengthy immersion in weak solutions of tartaric acid, brewers' yeast, or similar organic liquids. Acid cleaning was first used for iron sheet in the early 19th century. Pickling, lime coating, and baking were developed for steel wire in the middle of the 19th century when large quantities of signal wire were demanded by the expanding railroad networks.

Wet drawing practices employing oils, emulsions, or greases as lubricants remained dominant in Europe until recent years, and were often aided by the deposition of a thin copper or copper-tin coat on the steel wire by drawing from baths containing copper sulfate. Like so many innovations, this too was a result of accident; the coating was observed to develop when hot brass ingots were used for warming up an acid pickling bath [13]. While thin steel wire was drawn wet with a copper-tin coat, dry drawing with a soap was the main production technique for heavier gages in the U.S. [8] but found general acceptance in Europe only in the last few decades. The latest change affecting lubrication in wiredrawing came with the introduction of the sintered tungsten carbide die in 1923 [11] and phosphate coating of steel in 1934 [14].

1.1.3 Rolling

Rolling lubricants, which now occupy such an important position, have a much shorter history. Rolling itself was first applied to metals in the cold working of lead and gold in the 15th century [5], and for narrow strips of coinage alloys in the 16th century [15]. The first sketch of a mill is due to Leonardo da Vinci. The 17th century saw the appearance of slitting mills in which forged flats were split with collared rolls, then rounded by forging, and finally drawn into wire. Grooved rolls were used in 1728 in France by M. Fleuer some 60 years before they were patented by Henry Cort in England. To quote Lewis [16], "these early ironmasters never scrupled to make use of a process many years before it had been invented." Nevertheless, it was another hundred years before wire rod was regularly rolled, and Bedson's continuous mill appeared in Manchester in 1862 [17]. Rounds and sections, however, were and still are often rolled without any lubricant, and we must look to the rolling of sheet to trace the development of lubricants.

Sheet was traditionally hammered from a cast ingot, with up to 50 layers forged in a pack for tinplate [5]; rolling had to wait until rolls could be turned accurately. Wide lead sheet was rolled in the 18th century, and nonferrous metal sheets were soon cold rolled in gradually increasing widths. However, lubrication requirements of copper and copper-base alloys were relatively modest and, until well into this century, an occasional smear with a lubricant of jealously guarded composition was sufficient. The lubricant was usually based on mineral oils (available in quantity since 1860) and compounded with animal and vegetable fats and oils. The cold rolling of aluminum sheet, gaining prominence in the 20th century, presented probably the first challenge for developing lubricants on a more systematic basis. Light mineral oils chosen for their low staining propensity, sometimes compounded with small quantities of additives such as lanolin, were soon found to be adequate for the low speeds then employed.

Until the same period, steel was rolled cold in relatively narrow widths at low speeds, for which water cooling and a rather marginal lubrication with some heavy and possibly contaminated mineral oil sufficed. Wider plate was hot rolled in the mid 18th century, while thin sheets were hot rolled in packs and thus needed no lubricant at all [5]. The first continuous wide strip mill, which was built in Teplitz (at that time, Austria) in 1892, made little impact, but the hot rolling of wide strip became general practice after the success of Townsend and Naugle's plant at Columbia Steel Company, Pennsylvania, in 1926 [18]. In the same year, the Revere Copper and Brass Company of America introduced the tandem cold mill, soon followed by similar mills for steel and some years later for aluminum.

Tinplate had its beginnings in the early 14th century in Germany [19]. As with all sheet, it was first forged (hammered) and, in the 18th century, rolled. Finally, hot rolling in packs was established around 1790. The increasing consumption of tinplate promoted the installation of wide strip mills for cold rolling. These first presented a serious problem, because the usual mineral-oil-base lubricants proved completely inadequate. An inspired guess by A.J. Castle [20] in 1930 resulted in the immediate acceptance of palm oil—previously used on tinning lines—as a rolling lubricant. It spread from the U.S. to all parts of the world and is still a lubricant of recognized quality. High rolling speeds also created a substantial problem of roll heating, and water—applied either separately or in the form of an aqueous dispersion or emulsion—gained in importance.

1.1.4 Extrusion and Other Processes

Lubrication techniques in other cold metalworking operations such as tube drawing, deep drawing, and cold heading usually drew upon experience gained in wiredrawing and rolling, although a good deal of interaction must have existed. By and large, highly viscous oils, greases, fats, and soaps, often fortified with filler materials, were chosen for low-speed deformation at high reductions, and lower-viscosity or aqueous lubricants for lighter duties.

Extrusion was a relative latecomer among the metalworking processes. Like so many others, it was first practiced on lead [19]. Joseph Bramah patented a tube extrusion press in 1797; the hot extrusion of copper and its alloys had to wait until 1894, when Dick successfully extruded brass. Lubrication was and is important, but simple oils and graphited mixtures were found acceptable. A complete absence of lubricants was found to ensure the best-quality extrusions in aluminum. A significant development took place during World War II when Sejournet developed glass pad lubrication for the hot extrusion of steel (Chapter 8).

Cold extrusion is relatively recent; the reverse extrusion of cans originated in France around 1886, while the forward extrusion of tubes was patented by Hooker in the U.S. in 1909. The processes were limited to soft metals such as lead, tin, zinc, aluminum, and copper until the adoption of the phosphate conversion coating as a lubricant carrier for the cold extrusion of steel by Singer [14]. After extensive use and development during World War II in Germany, the process spread to change the balance between machined and cold-forged parts.

1.1.5 Metal Cutting

Abrasive machining is a controlled wear process and as such can be regarded as a part of tribology in the broader interpretation of the subject. In this sense, metal removal processes can claim great antiquity, because hammered or cast objects were often finished by polishing with a stone. Similarly, the bow drill was used with abrasive powders for the making of holes.

Metal cutting was a much later development. The lathe has been used for many centuries for the turning of wood, with the workpiece rotated by hand or, from about 1400 onward, with the help of a treadle [21,22]. Metal cutting was probably a development of the Middle Ages, spurred on by the need for making smoother gun bores. Biringuccio [12] shows a horizontal boring mill for cleaning out precast bores; more powerful drills, driven by water, were later used for cutting holes into solid castings. Meanwhile, red-metal (copper and brass) turners developed smaller lathes, at first operated by a bow, for making precision clock components. Exact screws could be cut by 1770.

The rapid development of machining was, however, linked to the industrial revolution; from the late 18th century onward, the steam engine placed new demands on machining accuracy and also provided the motive power for new machine tools, constructed of iron and steel, on which heavier cuts could be taken. The steel tools then available limited the cutting speeds, and lubrication with some heavy oil or fat gave immediately observable improvements [23]. Higher cutting speeds, made possible with the advent of high speed steels at the turn of the century, and the growth of grinding made the development of coolants necessary.

1.1.6 Theory

The reader will note that until 50 years ago most of the noteworthy developments in metalworking lubrication came about by accident, by inspiration, or, at best, as a result of persistent experimentation. Nevertheless, some theoretical background for a more systematic approach has been available for a considerable time [1]. The basic laws of friction, recognized by Leonardo da Vinci in 1508, were rediscovered by Amontons in 1699. Since then, the concept of a coefficient of friction has been adopted, but almost a century passed before Coulomb developed a theory postulating that friction was due to both surface roughness and adhesion. Another hundred years later, at the end of the 19th century, the principles of the hydrodynamic theory of lubrication were developed by a number of scientists, all working within a time span of 20 years. Boundary lubrication (Sec.3.5), which plays such an important role in metalworking processes, was closely investigated by Hardy between 1919 and 1933. Development of theories relating to dry and lubricated friction had to await the forties, when research centers devoted entirely to this subject formed around leading scientists in a number of countries.

1.1.7 Tribology

The subject of friction, lubrication, and wear cuts through many traditional scientific disciplines; it is a truly interdisciplinary science even though its practitioners are usually trained in one of the related classical disciplines. It was not until 1966 that, at the instigation of the Jost Committee in England, a new word, derived from the Greek "tribos" (rubbing), was adopted [1]. Since then the term "tribology" (defined as the science and practice of interacting surfaces in relative motion and the practices related thereto) has gained wide acceptance. It has also been recognized that a better understanding of this newly named field could yield tremendous economic benefits and that, as shown by a study under the guidance of the ASME [24] and by Jost and Schofield [25], tribology can make significant contributions to energy conservation, too. The subject is no less important for metalworking processes, since tribological considerations are often decisive in determining the very feasibility of the process itself and greatly affect the economy of operation and the quality of the issuing product.

1.1.8 Summary

We have seen that natural lubricants (mostly fats and fatty oils), or their modifications obtained through design or accident, provided satisfactory lubricants for most purposes until recent times. On prolonged standing or in continued use, many fatty substances develop free fatty acids, later recognized as being among the most useful boundary lubricants. Favorite lubricants or lubricant combinations were often made to work by sheer persistence and by adjusting process conditions until optimum results were obtained. However, the need for a more systematic approach became increasingly evident from the 1940's on. More powerful drives permitted increased production rates, heavier reductions, and higher speeds, and imposed more severe conditions on the lubricant. New metals that were introduced to satisfy the needs of the developing aerospace, chemical, and electronic industries often had properties quite different from those of the more common metals previously used, and have presented some of the most difficult lubrication problems. This resulted in an unprecedented upsurge in interest, experimentation, and analysis, and this is the subject of the present book.

1.2 Method of Treatment

This book is essentially presented in two parts. The first five chapters bring together what is general and can be applied to all processes.

a. Metalworking in all its forms is based on the plastic flow of metals, and no discussion of metalworking tribology is possible without understanding some basic concepts. These are covered in Chapter 2.

b. Much of what we know about tribology has been learned under nonmetalworking conditions, yet much of it can be applied to metalworking too, and Chapter 3 brings together the relevant knowledge, always with an eye on its metalworking application. Chapter 3 is thus, in many ways, a cornerstone of the entire line of argument, and should be studied by the novice and at least skimmed by the practicing tribologist.

c. Metalworking lubricants are often quite different from machinery lubricants, and Chapter 4 gives a brief introduction to them, primarily for the benefit of the nonchemist.

d. Laboratory and industrial observations can be quantified only when reliable and relevant measurements are made; appropriate techniques are discussed in Chapter 5, including techniques that could be but have not yet been applied to metalworking.

The second part of the book deals with individual processes. Here too, general knowledge is emphasized and is discussed first. The sequence of treatment within individual chapters is always the same. The process and the effects of friction on forces, power requirements, material flow, and defects are described first, and the best methods available for the quantification of observations are interwoven. Then, lubricating mechanisms are discussed; because of their importance, die materials are treated as an integral part of the system. Only what is truly specific is then dealt with by workpiece material.

Numerical data are generally presented in S.I. units, with U.S. customary units given in parentheses only when absolutely required. Because of the uncertainty attached to converting viscosity data, these are often quoted in the units given in the original reference. A conversion graph will be found in the Appendix.

1.3 Literature Sources

The literature of tribology has grown exponentially and now encompasses some 10,000 titles every year.

1.3.1 Sources

The major English-language publications are given below, together with the abbreviations used in this book.

Periodicals

American Society of Lubrication Engineers Transactions (ASLE Trans.) from 1958 onwards.

Lubrication Engineering (Lubric. Eng.), also of ASLE, since 1945.

Transactions of the American Society of Mechanical Engineers, Series F, Journal of Lubrication Technology (Trans. ASME, Ser.F, J. Lub. Tech.) from 1967 (Volume 89) on.

Tribology (Tribology) from 1968, continued as Tribology International (Trib. Int.) from 1974.

Wear (Wear), from 1958, now in more than one volume annually.

Journal of the Japan Society of Lubrication Engineers, International Edition (J. JSLE, Int. Ed.), since 1980, contains English condensations of articles appearing in the Japanese-language journal (J. JSLE) of the Society.

Many articles of importance are published in the Japanese-language journal of the Japan Society for Technology of Plasticity (J. JSTP); since 1980, one-page English condensations are available in the Annual Report (Ann. Rep.).

Bibliographies and Abstract Journals

Documentation Tribology (published annually by the Bundesanstalt für Materialprüfung, Berlin) is the most comprehensive collection of titles from all fields of tribology.

Tribos, from 1968, is also of general coverage but contains selected abstracts.

Metalworking Interfaces (MW Interf.) from 1976, contains review articles and patent surveys in addition to extended abstracts limited to metalworking.

Conference Publications

International Machine Tool Design and Research Conference (Proc. n-th Int. MTDR Conf.), since 1959, continues with changes in publishers and editors.

North American Metalworking (now: Manufacturing) Research Conference (Proc. n-th NAMRC), since 1973, published by McMaster University, 1973, by the University of Wisconsin, Madison, 1974, and by the Society of Manufacturing Engineers (SME) since then. Combined with Manufacturing Engineering Transactions (Manuf. Eng. Trans.) since 1977.

The tribology centers of Leeds and Lyon publish their annual symposia, since 1974, through Mechanical Engineering Publications Ltd, London.

1.3.2 Selection and Styling of References

The number of relevant publications would have exceeded 10,000 and, in order to save space, some radical measures had to be adopted.

1. In general, references are given in the shortest form that still ensures unequivocal identification.

2. References are collected at the end of each chapter and are numbered individually by chapter; if used repeatedly, in later chapters they are identified by chapter number and reference number—e.g., [3.55].

3. Books and conference proceedings containing several articles of interest are fully referenced only in Chapters 2 and 3. In later chapters the reference is given by identifying the earlier chapter and reference number.

4. If the work of a researcher or group of researchers continued through a series of publications and a summary treatment is available, only this summary is referenced.

5. If a paper was published in several versions or languages, only the most readily accessible English version is cited.

6. Multiauthored papers (five or more authors) are referenced with the name of the first author, followed by "et al".

7. In the general field of tribology, recent review papers are cited whenever possible; this has the unfortunate consequence that the names of researchers previously active in the field can be ascertained only by consulting the review article.

8. To retain the value of this book as a reference source, all publications unearthed in the field of metalworking tribology are included irrespective of age. However, excluded are papers in which tribology plays only a subordinate role (such as theoretical or experimental papers in which friction or lubrication is one of the constants, practical papers in which plant or process descriptions include well-established lubrication practices, or papers in which lubricants are insufficiently identified).

9. Patents and internal or government reports are referenced only exceptionally, on the assumption that the most salient points will have found their way into the open literature. Professional society technical reports are referenced only when no later open publication could be located.

References

[1] D. Dowson, *History of Tribology,* Longman, London, 1979.

[2] J.A. Schey (ed.), *Metal Deformation Processes: Friction and Lubrication,* Dekker, New York, 1970, pp.1-13.

[3] C. Singer, E.J. Holmyard, and A.R. Hall (ed.), *A History of Technology,* Vol. 1, Oxford at the Clarendon Press, 1954, pp. 573-662.

[4] E. von Wedel, *Die Geschichtliche Entwicklung des Umformens in Gesenken,* VDI Verlag, Düsseldorf, 1960.

[5] O. Johannsen, *Geschichte des Eisens,* Verlag Stahleisen, Düsseldorf, 1953.

[6] W.F. Parish, Mill & Factory, 1935(3), 27-30, 86-87.

[7] C.St.C. Davison, Wear, *1,* 1957/1958, 155-157.

[8] K.B. Lewis, *Steel Wire in America,* The Wire Association, Stamford, CT, 1952, pp.1-41.

[9] P. Rump, Stahl u. Eisen, *88,* 1968, 53-57.

[10] K.B. Lewis, Wire Ind., Jan. 1936, 4-8; Feb. 1936, 49-55.

[11] A. Pomp, *The Manufacture and Properties of Steel Wire* (Engl. ed.), Wire Industry, Ltd., London, 1954.

[12] V. Biringuccio, *The Pirotechnia of Vannoccio Biringuccio* (1540), transl. by C.S. Smith and M.T. Grudi, AIME, New York, 1943.

[13] F.C. Thompson, Wire Ind., May 1935, 159-162; Feb. 1936, 58.

[14] F. Singer, German Patent, D.R.P. 673,405, 1934.

[15] C.S. Smith, in "Metallurgy," *Encyclopaedia Britannica,* Vol.15, 1967, p. 237.

[16] K.B. Lewis, Wire Wire Prod., *17,* 1942, 16-27, 56-60.

[17] H.R. Schuberth, *History of the British Iron and Steel Industry,* Rutledge & Kegan Paul, London, 1957.

[18] W.H. Dennis, *A Hundred Years of Metallurgy,* Aldine Publishing Co., Chicago, 1963.

[19] L. Aitchison, *A History of Metals,* Interscience Publishers, New York, 1960.

[20] R.J. Nekervis, in discussion to R.C. Williams, Iron Steel Eng., *30* (8), 1953, 65-68.

[21] T.K. Derry and T.I. Williams, *A Short History of Technology,* Oxford University Press, 1961.
[22] F. Klemm, *A History of Western Technology,* MIT Press, Cambridge, 1964.
[23] W.H. Northcott, *A Treatise on Lathes and Turning,* Longman, Green & Co., London, 1868.
[24] O. Pinkus and D.F. Wilcock, *Energy Conservation through Tribology,* ASME, New York, 1977.
[25] H.P. Jost and J. Schofield, Proc. Inst. Mech. Eng., *195*, 1981, 151-173.

2

Friction In Metalworking

With very few exceptions, metalworking processes entail contact between two bodies: the tool (die) and the workpiece. Friction is "the resisting force tangential to the common boundary between two bodies when, under the action of an external force, one body moves or tends to move relative to the surface of the other" [1]. Since forces can be relatively easily measured, studies of the effects of friction have become a central point of metalworking research. Some critics regard this concentration as an obsession, pointing out that wear or surface finish may be far more important in practical applications. While basically this criticism is valid, one cannot deny that friction does have importance and that, because it is measurable, it allows quantification of processes that would otherwise become too vague and elusive.

2.1 Plastic Deformation

The effects of friction in metalworking cannot be understood without a nodding acquaintance with the pertinent aspects of plasticity theory. More detailed treatments can be found in books dealing with metalworking in general [2-11] and with applied plasticity theory in particular [12-27]. Experimental aspects are discussed in [28].

2.1.1 Plastic Flow

Metalworking involves, by definition, the plastic flow of metals, allowing shape change by deforming the body without fracture (plastic deformation processes) or creating a new shape by removing excess material in a controlled manner (machining processes). The ability of metals to undergo shape change without fracture is called ductility. In some metals it is very limited, and fracture may even set in without perceptible plastic deformation (brittle materials). Ductility is a material property which is greatly affected by the imposed stress state. Thus, a brittle material may be made to behave in a ductile manner by imposing high compressive stresses from all directions (hydrostatic pressure, σ_H).

2.1.2 Yield Criteria

To deform a metallic body plastically, appropriate external forces must be applied. These forces generate stresses, and the combination of stresses must satisfy certain conditions for plastic flow to occur. The stress state is, in the general case, triaxial—that is, stresses act in all directions. To simplify matters, the stresses are referenced to a coordinate system chosen in such a way that shear stresses disappear and only three normal stresses (principal stresses) act (Fig.2.1a). The condition that leads to

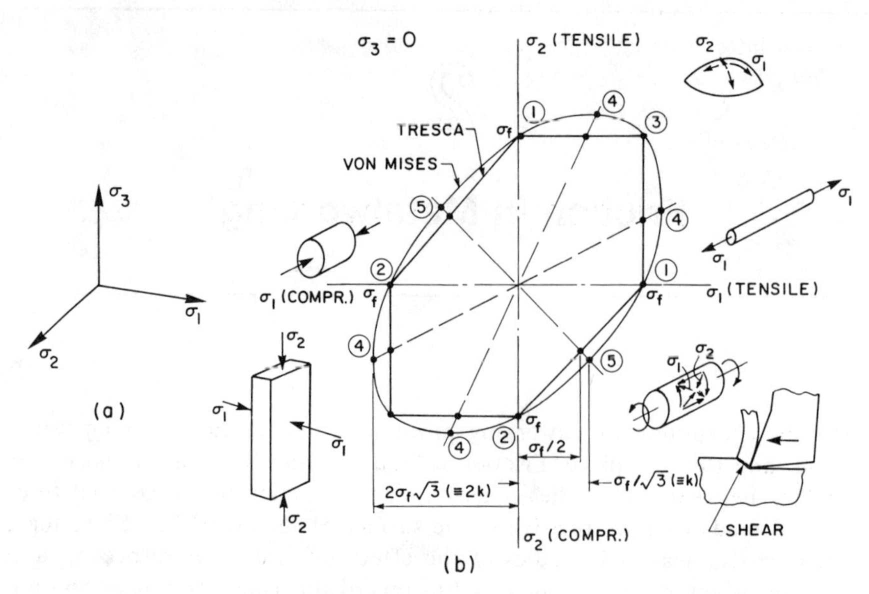

Fig.2.1. Directions of principal stresses (a) and yield criteria with some typical stress states (b)

plastic deformation can be expressed by appropriate yield criteria. For metals, two criteria are frequently used. For our purpose, the one due to Tresca can be written as

$$\frac{\sigma_{max} - \sigma_{min}}{2} = \frac{\sigma_f}{2} \tag{2.1}$$

and the one due to von Mises as

$$(\sigma_1 - \sigma_2)^2 + (\sigma_2 - \sigma_3)^2 + (\sigma_3 - \sigma_1)^2 = 2\sigma_f^2 \tag{2.2}$$

where σ_1, σ_2 and σ_3 are the three principal stresses and σ_f is the uniaxial flow stress of the material.

The significance of yield criteria is best illustrated by examining a simplified stress state in which $\sigma_3 = 0$ (plane stress). Then, the Tresca yield criterion defines a hexagon, and the von Mises criterion an ellipse (Fig.2.1b). It is seen that yielding (plastic flow) can be initiated in many ways, some of which are readily visualized:

1. In pure tension, flow occurs at the flow stress σ_f (points 1, corresponding to two directions in the plane of a sheet).

2. In pure compression, the material yields at the compressive flow stress which in ductile materials is usually equal to the tensile flow stress σ_f (points 2).

3. When a sheet is bulged by a punch or a pressurized medium (as a balloon is blown up by air), the two principal stresses in the surface of the sheet are equal (balanced biaxial tension) and must reach σ_f (points 3).

4. A technically very important condition is reached when deformation of the workpiece is prevented in one of the principal directions for one of two reasons: a die element keeps one dimension constant (Fig.2.2a), or only one part of the workpiece is deformed and adjacent nondeforming portions exert a restraining influence (Fig.2.2b). In either case, the restraint creates a stress in that principal direction; the stress is the average of the two other principal stresses (corresponding to points 4 in Fig.2.1b). The stress required for deformation is still σ_f according to Tresca but is $1.15\sigma_f$ according to

von Mises. The latter is usually the one regarded as the plane-strain flow stress of the material, denoted as $2k$ (see below). It is sometimes called the constrained flow stress.

5. Another important stress state is pure shear, in which the two principal stresses are of equal magnitude but of opposing sign (points 5 in Fig.2.1b). Flow now occurs at the shear flow stress τ_f, which is equal to $0.5\sigma_f$ according to Tresca and $0.577\sigma_f$ according to von Mises. The shear flow stress according to von Mises is often denoted as k. This stress state is of special importance for metal cutting.

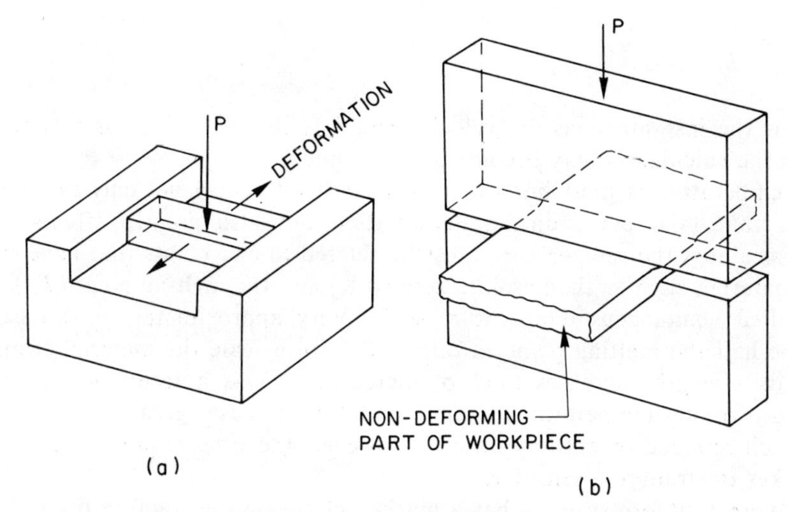

Fig.2.2. Plane-strain condition created by the restraint imposed (a) by a die and (b) by a non-deforming portion of the workpiece

2.1.3 The Flow Stress

It is clear from the above discussion that the flow stress σ_f is a most important value if numerical calculations are to be made. For convenience, it is often taken as a constant, but in reality it is a function of a great many variables. Among these are:

1. Strain—that is, the degree of deformation. For everyday use it is expressed as engineering strain, defined as the change in dimension relative to the original dimension. For tensile deformation

$$e_t = \frac{l_1 - l_0}{l_0} \tag{2.3}$$

where l_0 is the original and l_1 the final length of the workpiece. To avoid a negative sign, for compressive deformation it is usual to take

$$e_c = \frac{h_0 - h_1}{h_0} \tag{2.4}$$

where h_0 is the original and h_1 the final thickness or height of the workpiece. For purposes of calculation, the true strain (also referred to as natural or logarithmic strain) is much more convenient:

$$\varepsilon = \ln \frac{l_1}{l_0} \quad \text{or} \quad \varepsilon = \ln \frac{h_0}{h_1} \tag{2.5}$$

2. Strain rate, which must not be confused with velocity. It is the rate at which deformation proceeds:

$$\dot{\varepsilon} = \frac{d\varepsilon}{dt} \tag{2.6}$$

It can be calculated for simple processes, such as upsetting, from

$$\dot{\varepsilon} = \frac{v}{h} \tag{2.7}$$

where v is the instantaneous die velocity and h is the instantaneous height. For other processes the calculation may become quite complex.

3. Temperature is most important because it influences not only the metallurgical processes that take place during deformation but also greatly affects lubrication. Strictly speaking, the temperature must be related to end points that have significance for the material, such as the absolute zero (0 K) and the melting point (T_m), leading to the so-called homologous temperature scale. Very approximately it can be said that below one half the melting point on this temperature scale the material strain hardens, that is, its strength increases (and σ_f increases too) as a result of deformation. At higher hot working temperatures, atomic mobility increases greatly, and strain hardening is much reduced or entirely absent. However, the time required for atomic movement makes σ_f strain-rate sensitive.

4. Hydrostatic pressure σ_H has a marked effect on the fracture limit of metals, as mentioned earlier. However, from evidence available, it does not significantly affect the flow stress σ_f (and shear flow stress k) of metals (as opposed to polymers and many lubricants).

Metalworking processes present an almost infinite variety of process conditions which affect not only the strain, strain rate, temperature, and hydrostatic pressure acting on the body, but also lead to changes of individual variables within the deforming body itself. One of the most important factors that introduces such internal variations is the presence of friction at the interface between die and workpiece.

2.2 The Die-Workpiece Interface

Plasticity theory takes, in most instances, a highly simplified view of the interface. The die (or tool; we will use the terms interchangeably) and workpiece are assumed to have well-defined geometries, and the interface is regarded as a continuous substance of a known shear strength τ_i. However, for convenience of calculation, it is usually preferable that the interface be described by a nondimensional factor.

2.2.1 Coefficient of Friction

One possibility is to adopt from mechanical engineering the definition of the coefficient of friction (Fig.2.3a):

$$\mu = \frac{F}{P} = \frac{\tau_i}{p} \tag{2.8}$$

where F is the force required to move the body, P is the normal force, τ_i is the average

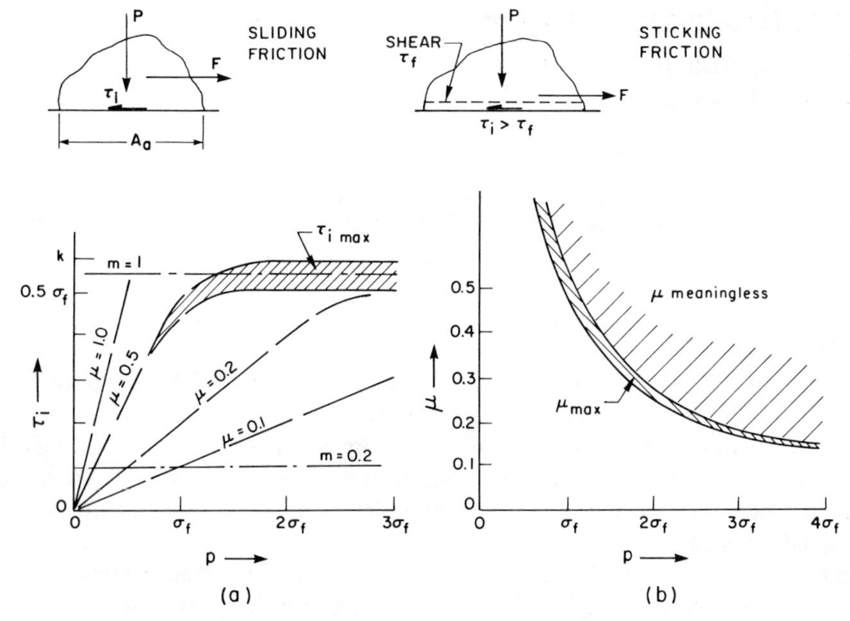

Fig.2.3. Shear stress (a) and maximum coefficient of friction (b) in sliding at various interface pressures

frictional shear stress, and p is the normal pressure; both τ_i and p are obtained by dividing the forces by the apparent area of contact A_a between the two bodies (Fig.2.3a). The definition embodies Amonton's two basic laws of friction: the frictional force is proportional to normal force, and it is independent of the size of the apparent contact area. For a constant μ, the interface shear stress τ_i must increase at the same rate as the interface pressure p. It is usual to speak of Coulomb friction when this condition is satisfied (broken lines in Fig.2.3a).

A look at Fig.2.1 will show that this is not necessarily realistic for metalworking. When τ_i reaches the value of k, the material has a further option: instead of sliding against the die surface, it will take less energy for the material to shear inside the body of the workpiece, while the surface remains immobile (Fig.2.3b). This is described as sticking friction, although no actual sticking to the die surface needs to take place. From Eq.2.8, the condition of sticking is

$$\tau_i = \mu p > k \tag{2.9}$$

Since $k = 0.577\sigma_f$ according to von Mises, it is sometimes said that $\mu_{max} = 0.577$, but this is really true only when full surface conformity is reached at $p = \sigma_f$. In many metalworking processes the interface pressure p reaches a multiple of the flow stress and, because for metals k remains constant with increasing hydrostatic pressure, the calculated μ actually drops (Fig.2.3b). It is, therefore, much better to say that the coefficient of friction becomes meaningless when $\mu p > k$ since there is no relative sliding at the interface. This limitation has been frequently overlooked, and quite unrealistically low μ values have sometimes been reported. Conversely, the interface pressure may drop below σ_f if large tensile stresses are imposed on the workpiece (as in deep drawing, wiredrawing, or thin strip rolling), and then μ can become quite high.

2.2.2 Interface Shear Factor

The difficulties with μ prompted several researchers to propose that the interface should be described by

$$\tau_i = mk \tag{2.10}$$

where m is the frictional shear factor, which has a value $m = 0$ for a frictionless interface and $m = 1$ for sticking friction (Fig. 2.3a). This has great mathematical convenience because τ_i is now defined with the aid of k, the value of which is known from the outset. In contrast, the use of μ can lead to complications because the value of p (which in itself is dependent on μ) has to be found.

The convenience of using m is, however, bought at a price. The interface is usually of some material quite different from the workpiece material, yet it is now assumed to have properties closely related to those of the workpiece material. In discussing lubricating mechanisms in Chapter 3, we will examine the possibility of such occurrence. Suffice it to say here that finding a lubricant that responds to imposed stresses and other process conditions in the same way as the workpiece material does is most unlikely. Furthermore, the practice of regarding k as a known and well-defined property is oversimplified; as pointed out by Shaw [29], severe working associated with interface sliding substantially changes the properties of the near-surface layer, making calculations of m on the basis of bulk properties quite unreliable.

Combining Eq.2.8 and 2.10 gives

$$\tau_i = \mu p = mk \simeq m\sigma_f/2 \tag{2.10a}$$

From this, it can be seen that the difference in the two treatments increases with increasing interface pressure:

p	σ_f	$2\sigma_f$	$4\sigma_f$
m	2μ	4μ	8μ

Numerical techniques of calculation, developed in the last few years, allow other methods of interface characterization to be used. For example, Hartley et al [30] regard the interface as a layer of elements whose stiffness is modified by a function of m. Yamada et al [31] use a "modulus of interface friction" which can be freely chosen. Oh et al [32] allow interface strength to assume any value.

2.2.3 Variable Friction

Because interface pressures, sliding velocities, temperatures, and other conditions affecting lubricant behavior tend to vary within the die/workpiece contact zone itself, it is to be expected that μ or m should vary too. An exact description of the interface can then be made only with the local value of τ_i although, for convenience, an average μ or m is most often assumed in calculations. This is permissible for force calculations but can lead to errors in the calculation of strain distributions (localized variations of deformation within the body) [33].

2.3 Effects of Friction

The effects of friction in metalworking are often intertwined with those of other factors. They must be considered together, because an artificial separation of friction effects could invalidate all conclusions.

2.3.1 Pressures and Forces

In the most general form, the stress required for deformation has three components:

$$p = f_1(\sigma_f) \cdot f_2(\tau_i) \cdot f_3\,(\text{process geometry}) \qquad (2.11)$$

The first component is due to the work of pure deformation: it is the stress required for the frictionless deformation of a material of flow stress σ_f, in a given stress state, to a given strain ε, at a given strain rate $\dot{\varepsilon}$, homogeneously—that is, with each particle of the workpiece subjected to the same deformation. Since σ_f is a material property subject to a number of influences (Sec.2.1) and in a real material cannot be reproduced to better than $\pm 5\%$, the value of the function $f_1(\sigma_f)$ will also be subject to some uncertainty.

The second component represents the effects of friction, and will be the main subject of our attention. This does not mean that it is necessarily the most significant contributor to the deformation stress; indeed, in well-lubricated cold rolling or wiredrawing it can be less than 5% of the pure deformation stress, and can be easily lost in the uncertainty of the σ_f value.

The third component reflects the difficulty of achieving homogeneous deformation. The degree of inhomogeneity is governed by process geometry, often in combination with the effects of friction. It is this interaction that accounts for most of the difficulties in finding valid theoretical solutions for metalworking processes. The work expended in introducing inhomogeneity of deformation is often called redundant work, implying that it is always additional to the work of pure deformation and friction. There are, however, examples where inhomogeneity actually reduces the effects of friction, and it is better to regard it as a function of process geometry. It can cause significant increases (100% or greater) in stress and can swamp friction effects. We shall take up this point again here and in discussing individual processes (Chapters 6 to 11).

For the choice of proper equipment, it is important to know the magnitude of the deforming force P, and this is obtained by multiplying the average deformation stress by the area A over which it acts:

$$P = Ap \qquad (2.12)$$

For dimensioning the drive units of metalworking equipment, the work W required to perform the deformation needs to be known. In general, this is

$$W = Ps \qquad (2.13)$$

where s is the distance over which the force P acts. The power is then readily obtained by taking the time required for deformation, or, often more conveniently, the power is calculated from force P and deforming velocity v:

$$\text{Power} = Pv \qquad (2.14)$$

Friction is not always a villain; occasionally it may even be desirable, as when it transfers stresses from the workpiece to the punch in deep drawing or ironing (Fig.2.4a) and thus allows greater reductions, or when it draws the workpiece into the roll gap (Fig.2.4b) and ensures its continuing passage through the rolls (Fig.2.4c).

2.3.2 The Effects of Friction

The presence of a frictional force is readily appreciated when the workpiece is bodily moved relative to the die. Such sliding is imposed when a billet is moved along the con-

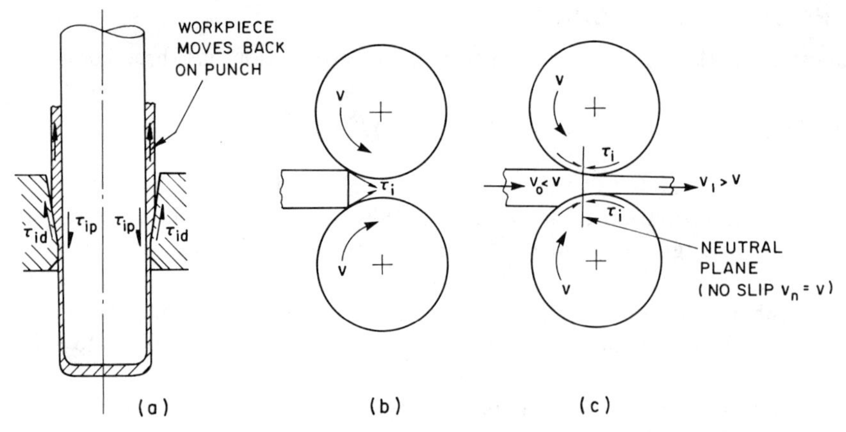

Fig.2.4. Examples of the beneficial effects of friction (a) on the punch in ironing, (b) at the start of rolling, and (c) on the roll surface during steady-state rolling

tainer in forward extrusion (Fig.2.5a), when a wire is drawn through a die (Fig.2.5b), or when a chip flows up against the rake face or the new surface moves past the flank face of the tool in machining (Fig.2.5c). The origin of sliding motion is not so obvious when sliding arises from the deformation of the workpiece itself, as it does in the compression (axial upsetting) of a cylinder (Fig.2.6). We are going to use this example to introduce the effects of friction on die pressures and, later, on strain distribution.

The volume of metals remains constant in the course of plastic deformation: therefore, the diameter of a cylinder must increase as its height is reduced (Fig.2.6a). The expanding surface moves radially outward, sliding over the die surface. A frictional stress arises which opposes free expansion of the end face of the cylinder, and the interface pressure (die pressure) must increase to overcome this resistance. As the distance from the edge increases, the frictional force and thus the die pressure must also increase, leading to the development of the so-called friction hill (Fig.2.6b). In consequence, die pressure can be very much higher than would be expected from the σ_f of the workpiece material. One of the aims of the theory of metalworking is to determine

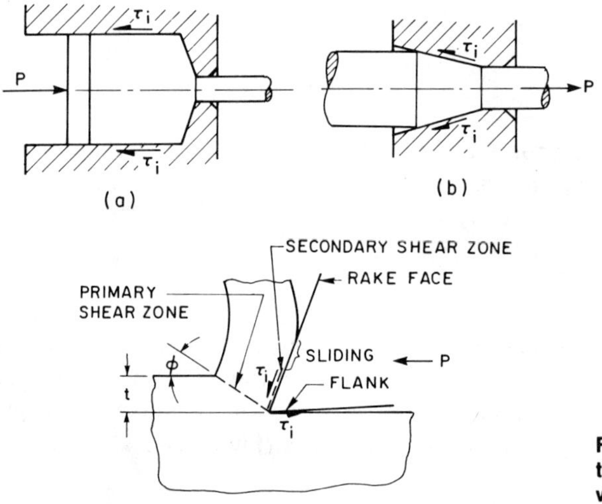

Fig.2.5. Examples of friction contributing to an increase in forces in (a) extrusion, (b) wiredrawing, and (c) machining (chip formation)

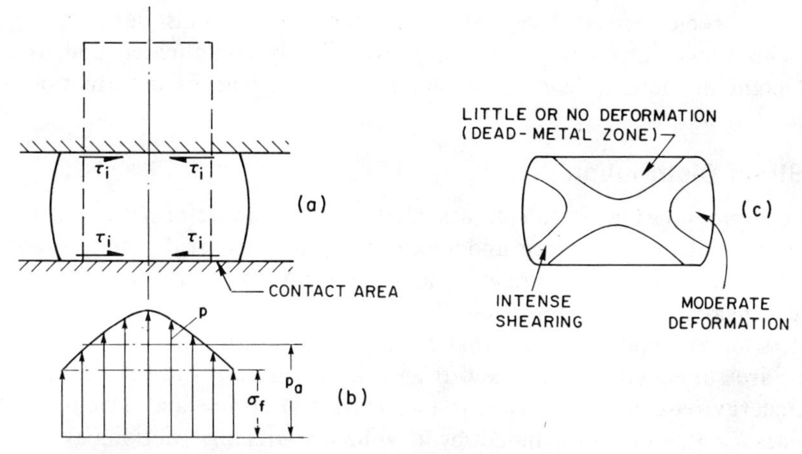

Fig.2.6. The consequences of friction illustrated in the upsetting of a cylinder. (a) Directions of shear stresses. (b) Consequent rise in interface pressure. (c) Inhomogeneity of deformation.

the pressure distribution in the die/workpiece contact zone. However, for many purposes it is sufficient to know the average die pressure p_a; it can be shown, and one can intuitively feel, that this must increase with increasing friction and diameter-to-height ratio of the workpiece. Thus, we have all three components of Eq.2.11, which will therefore affect forces (Eq.2.12) and power requirements (Eq.2.14). In the case of plastic deformation, a lot of energy is expended in a small space, and the consequences of this must be followed up before further effects are considered.

2.3.3 Temperature Rise

Most of the work of deformation (Eq.2.13) is transformed into heat. If there are no heat losses whatsoever (adiabatic conditions), the temperature rise ΔT is simply

$$\Delta T = \frac{W}{V\rho c} \tag{2.15}$$

where V is the volume of the deformation zone, and ρ is the density and c the specific heat of the workpiece material. This temperature rise can be several hundred degrees C. In most practical situations heat is lost to the dies, equipment, surrounding atmosphere, and nondeforming parts of the workpiece. Nevertheless, in plastic deformation or metal cutting at high strain rates the temperature rise can be significant, especially when conduction of heat from the deformation zone is poor.

Friction affects the temperature rise in two ways:

1. Under conditions of sliding friction (Fig.2.3a), the work required to overcome the frictional force is transformed into heat right at the interface. Because heat is generated in a generally very thin film, the temperature increase can be, at least instantaneously, quite substantial (sufficient even to cause surface melting under the special conditions prevailing in grinding). It almost always changes the properties of the interface, and one of the greatest challenges of tribology is to understand and predict the nature of these changes. Mechanical properties are affected, lubricant films may be destroyed or activated, and reactions between die and workpiece materials may be initiated.

2. Under conditions of sticking friction, changes in interface temperature are no greater than changes in bulk temperature but, because deformation now proceeds by

shear in the workpiece itself, localization of deformation can also lead to localization of heating. This makes analysis of processes exceedingly complicated and, from a more practical point of view, it leads to complex, nonhomogeneous deformation within the workpiece.

2.3.4 Strain Distribution

To a first approximation, it can be assumed that plastic deformation is homogeneous (with each part of the workpiece undergoing the same strain). In many processes (such as upset forging, rolling, wiredrawing, and extrusion), two factors, often inextricably interwoven, invalidate this assumption:

First, friction imposes an external restraint on deformation. Surface layers of the workpiece are subjected to shear, and the workpiece deforms in such a way as to minimize the energy required. This leads to localization of strains and development of dead-metal zones in the body, bounded by a zone of sticking friction, as shown in the example of forging in Fig.2.6c. Combined with heating and cooling effects, deformation inside the workpiece may be quite different from that deduced from outside appearances. Furthermore, material in the heavily barreled free surface is subjected to tensile stresses which may lead to fracture in less-ductile materials. Because these stresses are induced by the deformation of the workpiece rather than imposed externally, they are usually referred to as secondary tensile stresses.

Secondly, the geometry of the process itself may induce inhomogeneity. This is quite obviously so in metal cutting (Fig.2.5c) where deformation is concentrated in a narrow primary shear zone, whereas the remaining workpiece itself is strain hardened only in a thin surface layer (sometimes termed the tertiary shear zone). Friction on the flank face of the tool contributes to this strain hardening. More importantly, the angle of shear ϕ decreases with increasing friction. For a given undeformed chip thickness t, a higher τ_i results in a longer shear zone and thus an increased cutting force. In the limit, the condition of sticking friction may be reached on the rake face, and then the chip flows by internal shear in the secondary shear zone. Friction also increases the normal (hydrostatic) pressure acting on the shear zone and delays separation, clearly an undesirable consequence.

Not so obvious is the inhomogeneity induced by process geometry in bulk metalworking processes in which the entire thickness of the workpiece is deformed. The problem may be approached by first looking at the extreme example of hardness testing (or indentation of a semi-infinite workpiece with a small punch, Fig.2.7a). A dead-metal zone forming on the punch nose behaves like a part of the punch and displaces material by pushing it sideways and up. The rest of the workpiece is only elastically loaded, and localized deformation must be attained against the restraining influence of this nondeforming zone. Correspondingly, a rather large pressure $p = 3\sigma_f$ is needed for permanent indentation, irrespective of whether the punch is lubricated or not (hence the hardness of materials is $H = 3\sigma_f$ if expressed in consistent units of stress). In other processes deformation proceeds by contact with two tools (forging, rolling) or with a die that surrounds the workpiece (extrusion, wiredrawing). If the contact length is smaller than the mean thickness (Fig.2.7b) or diameter (Fig.2.7c), deformation will be inhomogeneous, with the center deforming less than would be expected. A detailed treatment is given by Blazynski [16]. In general, there are two consequences:

a. Because deformation proceeds against the restraining effect of the less-deformed center, deformation stresses, forces, and power requirements increase (f_3 in Eq.2.11). Just as in indentation, friction has a minor effect on strain distribution when the h/L ratio is large, and lubrication does not help much in reducing forces.

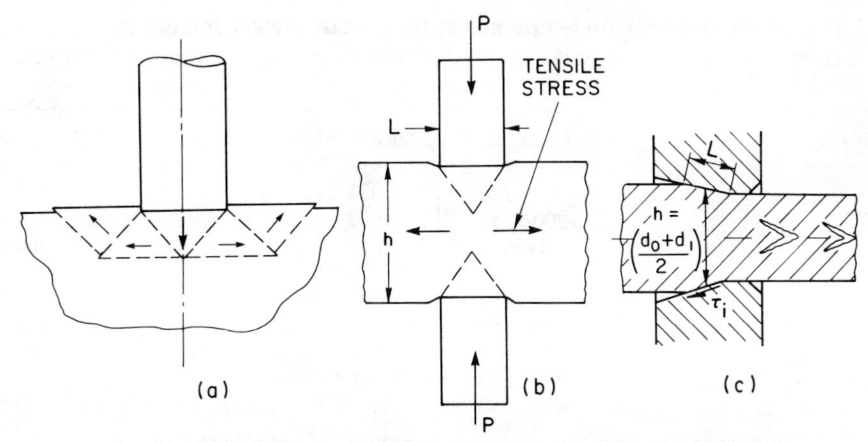

Fig.2.7. Examples of inhomogeneous deformation in (a) indenting of a semi-infinite body, (b) forging between opposing anvils, and (c) drawing of a bar

b. Usually, a more important consequence of inhomogeneous deformation is the generation of tensile stresses in the less-deformed parts of the workpiece. If these secondary tensile stresses reach high values, a workpiece material of low ductility may suffer internal fracture (arrowhead defect, Fig.2.7c). Even if there is no failure during deformation, residual stresses may remain in the workpiece, which may lead to subsequent distortion or to delayed failure due to stress corrosion.

Any inhomogeneously deformed workpiece undergoes differential strain hardening and, if subsequently annealed, will show large grain-size and property variations.

In many processes, friction effects and inhomogeneity induced by process geometry occur simultaneously, sometimes reinforcing and sometimes opposing each other. Thus, die friction increases the compressive stresses in the deformation zone and thus delays fracture in extrusion, whereas die friction combines with inhomogeneity to increase tensile stresses and thus also the danger of fracture in wiredrawing (Fig.2.7c). The important point is that the effects of friction cannot be artificially divorced from the effects of inhomogeneous deformation.

2.4 Classification of Processes

Metalworking processes are described in the literature [2-11a] and here it will suffice to offer only brief definitions essential for following the arguments presented in this book. Classification can be made according to several criteria:

1. Temperature of Deformation. We have seen that the term "hot working" has a very specific meaning in relation to material behavior. However, in everyday industrial usage it simply refers to the working of a preheated material. Usual preheating temperatures are between 70 and 90% of the melting point on the homologous temperature scale. Numerical values are given for the most important materials in Table 2.1. Special benefits are sometimes obtained by working at an elevated temperature below the hot working temperature; this is referred to as warm working. Cold working is performed on material that is not intentionally preheated, although the temperature may rise above ambient because of heat generated during deformation. When the die (tool) is preheated and held at the same temperature as the workpiece, one speaks of isothermal working. In the more common hot working operations the die is colder, although not necessarily at room temperature. Very generally, hot working has the advantage of

Table 2.1. Typical hot working temperatures for various metals and alloys

Compiled primarily from *Metals Handbook*, 9th Ed., Vol. 1 (1978), Vol. 2 (1979) and Vol. 3 (1980), ASM, Metals Park, OH.

Metal or alloy	Temperature, C	Metal or alloy	Temperature, C
Steels:		2024	360-450
1008	900-1250	5052	300-510
1045	<1150	7075	260-450
Tool steels:		Mg-base:	
D2 (1.5C-12Cr-1Mo)	900-1080	Mg-1Mn	290-540
H13 (0.4C-5Cr-1.5Mo-1V)	900-1150	Be	650-1100
Stainless steels:		Zr	600-1000
302 (18Cr-9Ni)	930-1200	Ti and Ti-base:	
410 (13Cr)	870-1150	Ti (99%)	760-980
Ni and Ni-base:		Ti-6Al-4V	920-1100
Ni	650-1230	Sn	20-100
Hastelloy X	1000-1150	Pb	20-200
Cu and Cu-base:		Zn	120-275
Cu	750-950	W	1600-1900
Cartridge brass (30Zn)	725-850	Mo-base:	
Muntz metal (40Zn)	625-800	TZM (0.5Ti)	1300-1800
Leaded brass (39Zn-1Pb)	625-800	Nb	850-1200
Phosphor bronze (5Sn)	540-560	Ta	500-1100
Al bronze (5Al)	815-870	U	750-875
Al and Al-base:			
1100 Al	250-550		

lower flow stress, permits heavy deformations even on large workpieces, and offers numerous ways of controlling the properties of the issuing product. Cold working generally yields products of closer dimensional tolerances and better surface finish and, if desired, the strength acquired through strain hardening may be retained in the part.

2. Deformation vs. Machining. One of the purposes of metalworking is always that of creating a new shape. When this is accomplished largely by taking advantage of the ductility of the material, one speaks of plastic deformation. When the new shape is obtained by removing material in the form of chips, one speaks of machining (Chapter 11). Within the plastic deformation processes, it is useful to distinguish between (a) bulk deformation processes, in which at least nominally the entire thickness of the workpiece is in the plastic state in the deformation zone and thickness changes are substantial (Chapters 6 to 9), and (b) sheet metalworking processes in which thickness changes are only incidental (Chapter 10).

3. Steady-state vs. Non-steady-state. In plastic deformation, especially bulk deformation processes, it is useful to make a distinction based on the mode of deformation (Fig.2.8) [11]. In steady-state processes all parts of the workpiece are subjected to the same mode of deformation, whereas in non-steady-state processes the geometry changes continuously. The distinction is highly significant from the lubrication point of view. In steady-state processes the surface of the workpiece arriving at the deformation zone can be coated with a fresh lubricant film. In non-steady-state processes the preapplied film has to suffice in the course of deformation, and reapplication is usually not feasible unless the process is interrupted. Some processes have a transitionary character: for example, deformation is non-steady-state at the beginning and end of extrusion but acquires steady-state characteristics while the greater part of a long billet is extruded. Individual bulk deformation processes will be treated in the sequence given by this

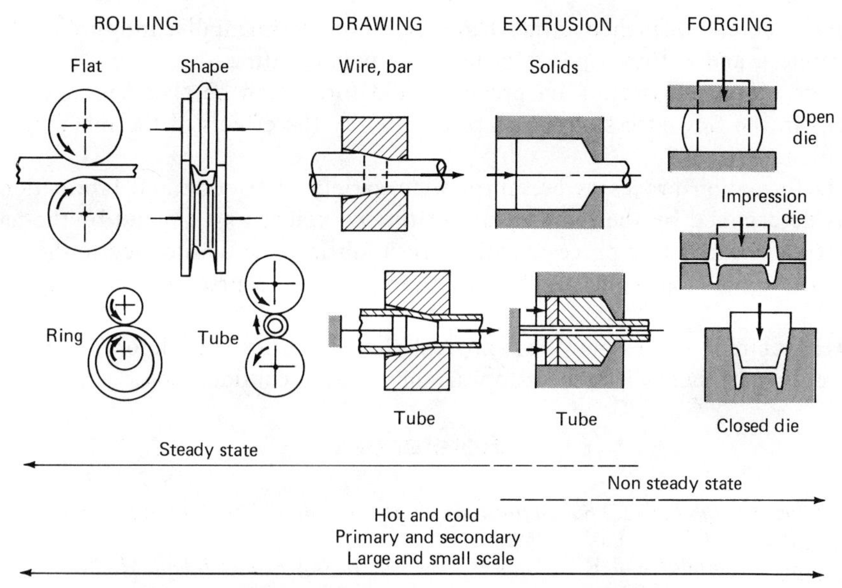

Fig.2.8. Classification of bulk deformation processes [11]

classification. The steady-state processes of rolling and drawing will be discussed first, then the extrusion of long products (which has many steady-state features), and finally the non-steady-state processes of cold and hot forging (which also include the essentially non-steady-state versions of extrusion).

2.5 Summary

1. Friction in metalworking results in a tangential (shear) force that acts at the interface between die and workpiece and resists movement of the two relative to each other.

2. The magnitude of friction can be described by an average interface shear strength $\bar{\tau}_i$.

3. For convenience in calculation, friction can be related to the interface (die) pressure p by the coefficient of friction $\mu = \tau_i/p$. A constant μ implies that τ_i increases linearly with p. There is, however, a limit to this. When the product of μp reaches the flow strength of the workpiece material in shear k, it becomes energetically more favorable for deformation to proceed by shear within the body of the workpiece, leaving the interface immobile (so-called sticking friction). The maximum value of μ is 0.5 when $p = \sigma_f$, and yet is smaller if interface pressures p reach a multiple of σ_f. Under conditions of sticking friction ($\mu p > k$) the concept of μ is meaningless.

4. Alternatively, the interface can be described by the frictional shear factor m which assumes that $\tau_i = mk$, implying that the properties of the interface respond to process variables in the same way as do those of the workpiece.

5. The presence of friction requires the exertion of force to maintain movement between die and workpiece, and this leads to increased interface pressures, higher force, work, and energy requirements, and greater heating. It can thus set a limit on metalworking by creating practically intolerable die pressures and uneconomically high forces or power requirements.

6. Occasionally, friction can be desirable and even essential, as when the workpiece has to be gripped by the die.

7. Friction can influence, either beneficially or detrimentally, fracture of the workpiece material, and is thus a powerful factor in metal cutting.

8. The effects of friction on pressures and forces can be overshadowed by minor variations in the flow stress of the material, and by the effects of the inhomogeneity of deformation.

9. Deformation processes present a wide variety of friction and lubrication conditions, as determined by the temperature of the workpiece and die and by the nature of deformation. Steady-state processes allow fresh lubricant to be supplied to the interface, whereas the initial lubricant supply must suffice in the course of non-steady-state processes.

10. Machining shares many tribological characteristics with deformation processes, but there are sufficient differences to warrant separate consideration.

References

[1] *Glossary of Terms and Definitions in the Field of Friction, Wear and Lubrication: Tribology,* OECD, Paris, 1969.

[2] J.M. Alexander and R.C. Brewer, *Manufacturing Properties of Materials,* Van Nostrand, London, 1963.

[3] B. Avitzur, *Metal-Forming Processes,* Wiley-Interscience, New York, 1981.

[4] N.H. Cook, *Manufacturing Analysis,* Addison-Wesley, Reading, MA, 1966.

[5] F.A.A. Crane, *Mechanical Working of Metals,* Macmillan, London, 1964.

[6] G.E. Dieter, Jr., *Mechanical Metallurgy,* 2nd Ed., McGraw-Hill, New York, 1974.

[7] A. Geleji, *Forge Equipment, Rolling Mills and Accessories,* Akadémiai Kiado, Budapest, 1967.

[8] S. Kalpakjian, *Mechanical Processing of Materials,* Van Nostrand, Princeton, NJ, 1967.

[9] K. Lange (ed.), *Umformtechnik* (3 volumes), Springer, Berlin, 1972-1975; also: *Handbook of Metalworking,* McGraw-Hill, New York, to be published.

[10] R.N. Parkins, *Mechanical Treatment of Metals,* George Allen and Unwin, London, 1958.

[11] J.A. Schey, *Introduction to Manufacturing Processes,* McGraw-Hill, New York, 1977.

[11a] T. Altan, *Metal Forming: Fundamentals and Applications,* ASM, Metals Park, 1983.

[12] B. Avitzur, *Metal Forming: Processes and Analysis,* McGraw-Hill, New York, 1968; 2nd Ed.: Krieger, Huntington, NY, 1979.

[13] B. Avitzur, *Metal Forming, the Application of Limit Analysis,* Dekker, New York, 1980.

[14] W.A. Backofen, *Deformation Processing,* Addison-Wesley, Reading, MA, 1972.

[15] P. Baqué, E. Felder, J. Hyafil, and Y. Descatha, *Mise en Forme des Métaux,* Dunod, Paris, 1973.

[16] T.Z. Blazynski, *Metal Forming, Tool Profiles and Flow,* Halstead Press, New York, 1976.

[17] H. Ford and J.M. Alexander, *Advanced Mechanics of Materials,* 2nd Ed., Halstead Press, New York, 1977.

[18] A. Geleji, *Bildsame Formgebung der Metalle,* Akademie Verlag, Berlin, 1967.

[19] R. Hill, *The Mathematical Theory of Plasticity,* Clarendon Press, Oxford, 1950.

[20] O. Hoffman and G. Sachs, *Introduction to the Theory of Plasticity for Engineers,* McGraw-Hill, New York, 1953.

[21] W. Johnson and P.B. Mellor, *Engineering Plasticity,* Van Nostrand, London, 1973.

[22] G.W. Rowe, *Elements of Metalworking Theory,* Edward Arnold, London, 1979.

[23] G.W. Rowe, *Principles of Industrial Metalworking Processes,* Edward Arnold, London, 1977.

[24] R.A. Slater, *Engineering Plasticity, Theory and Its Application to Metal Forming Processes,* Halstead Press, New York, 1974.

[25] E.G. Thomsen, C.T. Yang, and S. Kobayashi, *Mechanics of Plastic Deformation in Metal Processing,* Macmillan, New York, 1965.

[26] E.P. Unksov, *An Engineering Theory of Plasticity,* Butterworths, London, 1961.

[27] W. Johnson, R. Sowerby, and J.B. Haddow, *Plane-Strain Slip Line Fields for Metal Deformation Processes,* Pergamon, New York, 1982.

[28] J.A. Schey, in *Techniques of Metals Research,* Vol.1, Pt.3, R.F. Bunshah (ed.), Interscience, New York, 1968, pp.1409-1538.

[29] M.C. Shaw, in *Fundamentals of Deformation Processing,* Syracuse University Press, 1964, pp. 107-130.

[30] P. Hartley, C.E.N. Sturgess, and G.W. Rowe, Int. J. Mech. Sci., *21,* 1979, 301-311.

[31] Y. Yamada, A.S. Wifi, and T. Hirakawa, in *Metal Forming Plasticity,* H. Lippmann (ed.), Springer, Berlin, 1979, pp. 158-176.

[32] S.I. Oh, N. Rebelo, and S. Kobayashi, ibid., pp. 273-291.

[33] R. Bartram, M. Lung, and O. Mahrenholz, Arch. Eisenhüttenw., *41,* 1970, 969-973.

3

Fundamentals of Tribology in Metalworking

This chapter deals with friction, lubrication, and wear mechanisms as they pertain to metalworking. The aim is to develop a feel for general principles applicable to the broadest range of processes, allowing a more systematic view of the often isolated observations to be presented in Chapters 6 to 11. Much of the material dealt with in the preceding book [1] by Riesz [2] is included and considerably expanded.

Most of the basic information has always been generated in nonmetalworking fields, yet much of it is essential for understanding the phenomena that govern the success or failure of metalworking lubrication. There is no lack of publications on tribology; indeed, there is an embarrassment of riches, from introductory [3-16] and advanced texts [17-31], through handbooks of broad coverage [32-34], general tribological [35-49b] and metalworking-oriented [50-56] conferences and symposia, to countless research publications. In addition to a review of tribology [57], several reviews of metalworking tribology, of varying coverage, have also appeared [58-69].

To make the task of the nonspecialist reader easier, the present survey covers all relevant topics but to varying depths, depending on the importance of the topic and on the ready availability of convenient summary treatments. Applicability to metalworking is shown, and results of the more basic metalworking lubrication research are interwoven whenever possible. When necessary, general principles are illustrated by examples taken from metalworking research; further examples will be given in the chapters dealing with individual processes. Even though metal cutting embodies many of the general principles, it also presents unique problems, and, to avoid discontinuity of treatment, these will be covered in Chapter 11.

It should be noted that aspects relating to the mechanisms by which lubricants act are discussed in this chapter, whereas lubricant formulations are covered in Chapter 4.

3.1 Real Interfaces

In the purely mechanical view taken in Chapter 2, we looked upon the interface as a continuous film of τ_i shear strength, interposed between a rigid die and a deforming workpiece of σ_f (or $2k$) flow stress (Fig.3.1a). We did not inquire what that interface substance was, and we assumed that whatever changes may occur in the interface through the influence of process variables can be taken into account simply by choosing an appropriate value of τ_i (or μ or m). This simplified view is often quite acceptable for calculations of pressures, forces, and power requirements, but it is totally inadequate if the sources of friction and the mechanisms of lubrication and wear are to be understood.

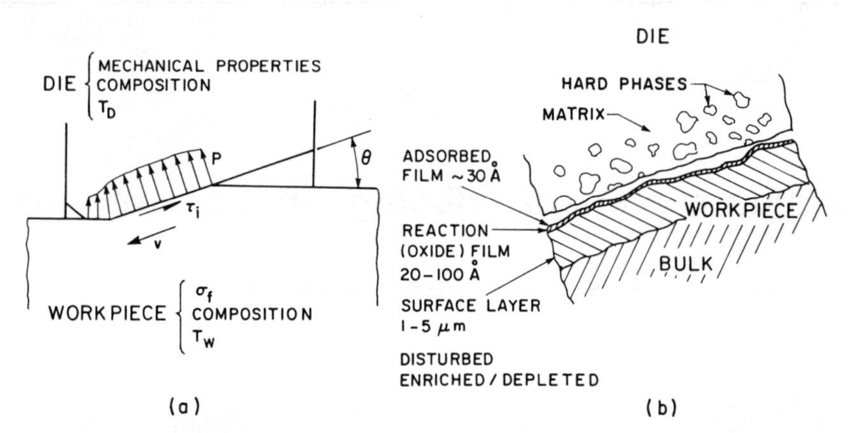

Fig.3.1. The die-workpiece interface on the (a) macro and (b) micro scales

3.1.1 Macroscopic View

On the macroscopic scale (Fig.3.1a), we have to recognize that the die material is not rigid but has a finite elastic modulus and yield strength; thus the nominal die configuration is always deformed through elastic deflections and, if some critical loading is exceeded, also by plastic flow or perhaps catastrophic failure.

Die and workpiece are not simply mechanical components, but also have more or less well-defined compositions which govern the consequences of intimate contact; if the pairing is so constituted, mutual attraction or adhesion may occur with vast consequences in terms of friction and wear.

The die temperature T_D and workpiece temperature T_W may be intentionally preset at the beginning of the process but will change in the course of metalworking itself. In cold working in the everyday sense, both T_D and T_W are at room temperature initially; the work of deformation transformed into heat and the work of friction cause a temperature rise, with important consequences for lubrication and the properties of the product. In hot working, both T_D and T_W are above room temperature, creating problems of lubrication and heat transfer.

3.1.2 Microscopic View

Even this more detailed macroscopic view fails to account for many tribological phenomena, and the interface must be viewed on a microscopic scale (Fig.3.1b):

a. The die and workpiece surfaces now show minute peaks (asperities) and valleys [70,71]. The magnitude, spacing, detailed geometry, and directionality of these microgeometrical features all play important roles not only in creating friction but also in establishing and sustaining a lubricant film designed to mitigate friction and wear.

b. A metallic die material of given composition will now appear as a multiphase structure in which hard, wear-resistant particles (usually intermetallic compounds) are embedded in a somewhat softer and more ductile matrix. The surface of the die may be different from the bulk, either because of intentional surface treatment (Sec.3.11) or because of depletion or enrichment of the surface in alloying elements through diffusion processes that occur during heat treatment or during the frictional contact itself.

c. The workpiece material may be of a single phase (as are pure metals and solid-solution alloys) or may have a multiphase structure (alloys containing more than one solid-solution phase and/or intermetallic compound). Because of friction (Sec.2.2) and nonhomogeneous deformation (Sec.2.3.4) in prior processing steps, the surface of the workpiece is very often worked to a greater degree than its bulk. This is especially noticeable in machined surfaces [72]. The frictional process itself can result in very heavy working of a thin, "micronized" surface layer [73]. The process of deformation may even lead to metallurgical changes in the material (Sec.5.5.5). The composition may be different because a surface film has been intentionally deposited (for example, copper or tin on the surface of stainless steel or steel) or because alloying elements have diffused to enrich the surface [74,75]. Some surface segregation is irreversible (Al in Cu), and some disappears upon unloading (Si in Fe). In general, segregation will occur if this leads to a lowering of surface energy [76]. The surfaces of technically processed metals may be of substantially different composition from the bulk [77].

d. Pure metal surfaces (in the language of tribology, "virgin surfaces") seldom exist. Technical surfaces are covered with reaction products formed through exposure to air, its humidity, and lubricants [74]. Often, films resulting from different reactions are superimposed—for example, a naturally formed oxide film may be covered with an absorbed lubricant film which may also enter into chemical reactions with the oxide and/or substrate.

e. Intentionally interposed substances (commonly called lubricants) are not simply inert films. True, they do have measurable shear strengths, but this is usually very much dependent on film thickness, temperature, pressure, and shear rate; such response is the subject of study in rheology of lubricants. Most importantly, though, most lubricants also have chemical properties which depend not just on elemental composition but also on the molecular forms in which various elements are present. More than one molecular species is usually included, and the surface activities and reactivities of various constituents become most important. The application, distribution, and resupply of the lubricant to the interface may affect its performance as much as its composition or rheology.

3.1.3 Process Environment

The process itself creates an environment within which the lubricant must fulfill its function:

a. Metalworking differs from other tribological processes in that deformation of the workpiece results in extension and break-up of the original surface. Entirely new, virgin surfaces are created, ranging in area from a few percent in sheet metalworking to one hundred percent in machining. Freshly generated surfaces emit exo-electrons [78], indicating that they are in a temporary energy imbalance. They are thus highly reactive and capable of chemical reactions with lubricants. Unprotected by contaminant films, the new surfaces form adhesive bonds much more readily than in the usual machinery contacts; indeed, the bond is strong enough to be exploited in processes such as roll bonding.

b. The temperature of the interface may rise above the die or workpiece temperature, and this may promote adhesion and destruction of lubricant films; but it also may promote reactions with reactive lubricants or the atmosphere, provided that access to the newly formed surfaces is allowed. Reactions are usually time- and temperature-dependent, and therefore residence time in the deformation zone is important. Asperity contact may dissipate a large amount of energy in a small volume and in a very short time, and thus local flash temperatures may be very much higher than the bulk temperature [79]. Heat generated in the interface is transferred to the die and workpiece, and hence contact geometry as well as properties of the contacting material pair dictate the flash temperature.

c. Pressures at the interface may reach multiples of the flow stress of the workpiece material, and may be as high as 3 GPa. Such pressures affect not only the response of lubricants to stresses but also the reactions that may take place within the lubricant and between the lubricant and the die and workpiece materials.

d. The very process of rubbing initiates and accelerates chemical reactions that otherwise would proceed only at much higher temperatures or not at all. The facts have long been observed, but explanations are still controversial. The heavily deformed asperities are sites of high-energy concentration with a very high density of lattice defects, and these have been suggested as sources of high catalytic and adsorption activity. The enhanced surface activity may thus be expressed [80] as follows:

$$\text{Enhanced reactivity} = \text{exo-electron} + \text{catalytic factors} + \text{elevated temperature} + \text{high pressure} \tag{3.1}$$

These phenomena constitute the subject of the rapidly growing branch of science known as tribochemistry or mechanochemistry [75,80-84]. A review of surface effects is also available [85].

Multiple interactions of the above variables determine the type of lubricating mechanisms that can be maintained; this in turn controls the magnitude of friction, the occurrence and rate of wear, and, most importantly, the quality of the issuing product. The most influential variables and some of their modes of interaction are shown schematically in Fig.3.2 [68]. This figure gives some feel for the complexity of the situation and can serve as a useful road map in exploring lubrication mechanisms. A more formal application of the systems approach [22] may also prove beneficial.

3.2 Surface Contact and Friction Theories

Contact between real surfaces is much too complex to be analyzed in its entirety. A useful first approximation can be made by observing what happens at an interface to which no lubricant has been intentionally applied, a condition usually described as dry friction. The contacting material pair is assumed to have technical surfaces of the typical microgeometry and surface contaminants (Fig.3.1b). Experiments of Wheeler [85a] showed that even partial monolayers of oxygen or chlorine are effective in reducing interfacial shear strength; therefore, dry friction is really a very ill-defined condition. Nevertheless, it is useful for discussion purposes because complications arising from modeling the real lubricant behavior are avoided and because the contact between rough surfaces can be analyzed. Much progress has been made in this area in the last two decades; good introductions are to be found in Halling [23], Ling [86], and Thomas and King [87]. Conferences have been devoted to the subject [70,88,89], and reviews are also available [90,91].

3.2.1 Contact Area

Before discussing some results of contact mechanics, it is necessary to understand the wide range of conditions that may exist in metalworking. For ease of analysis, the die is often assumed to be perfectly smooth, and the workpiece to have some regular roughness features. A more realistic view admits roughness on the die, and not necessarily regular roughness on the workpiece.

In a three-dimensional view (Fig.3.3), the surfaces make first contact only at the highest points (summits). The actual or real area of contact A_r is but a small fraction of the over-all or apparent area of contact A_a. For practical reasons, a cut perpendicular to the interface is taken and then a view similar to that in Fig.3.1b is observed;

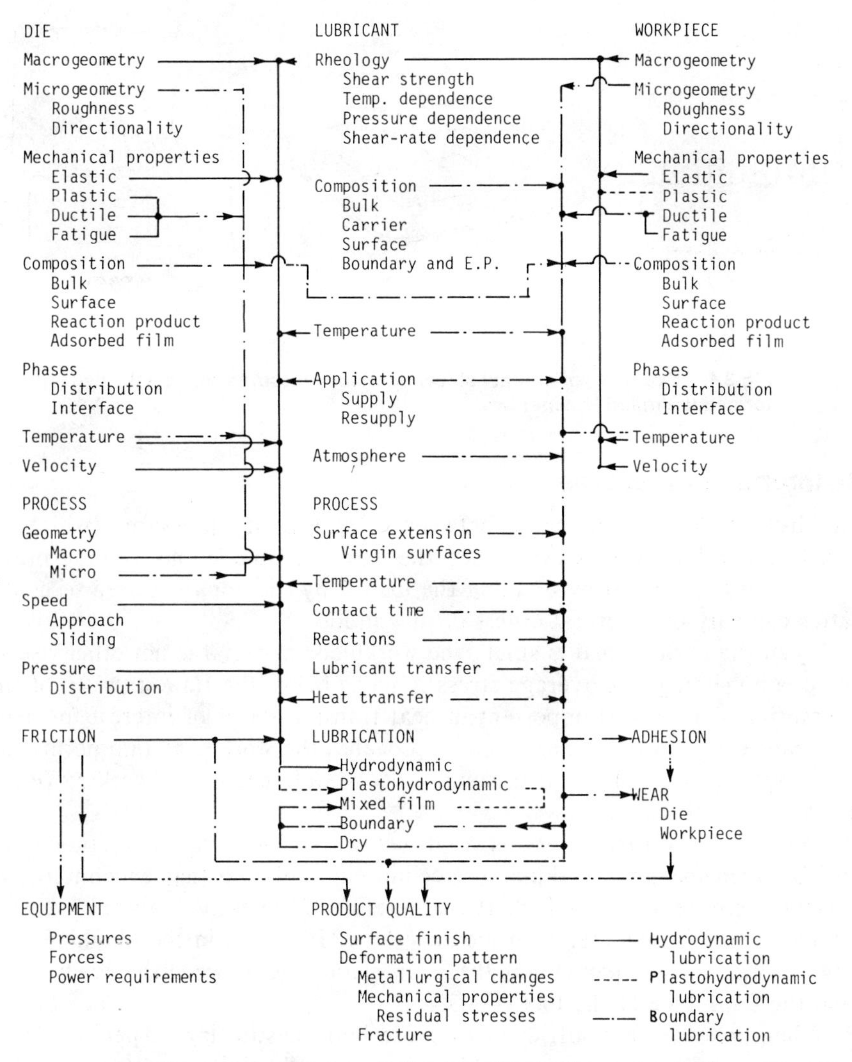

Fig.3.2. Elements of a metalworking lubrication system [68]

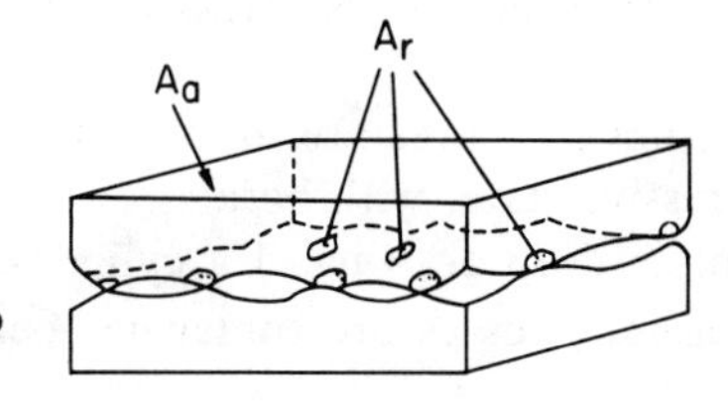

Fig.3.3. Contact points between two rough surfaces

peaks that show up in the surface profile may or may not be summits, and are called asperities. Asperities usually have rather gentle (3 to 20°) average slopes and generous radii, even though they may appear quite ragged on tracings which are intentionally distorted to make the roughness features more prominent (Sec.5.7.1). Superimposed on these asperities are yet smaller-scale features, called microasperities or asperities of second and higher orders. For a given surface configuration the magnitude of A_r depends on the magnitude of the imposed load P and, most importantly, on whether there is relative sliding of the surfaces.

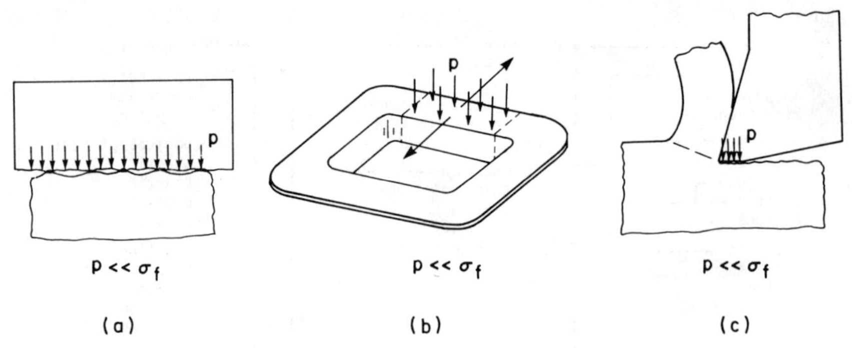

Fig.3.4. Examples of contact at low interface pressures with plastic deformation limited to asperities

3.2.2 Interface Pressures

Careful distinction must be made between local asperity pressures (which, if A_r is small, may reach high values even when the load P is small) and average pressures or normal stresses (calculated by dividing the load P by the apparent area of contact A_a). The latter can vary over several orders of magnitude:

1. When the normal load is small, the workpiece material is not otherwise stressed, and there is no sliding, the average stress remains below the flow stress σ_f of the workpiece material. This case is important in heat transfer and is of interest for metalworking as a transient condition when a tool approaches the workpiece in a normal direction (e.g., in upsetting; Fig.3.4a) and in some processes of sheet metalworking (e.g., on the punch nose in bending).

2. When the average stress is kept below σ_f, but relative motion (sliding) is imposed between tool and workpiece, a condition similar to that encountered between mechanical components is reached. It is, therefore, frequently analyzed even though it is only occasionally encountered in metalworking. It occurs in the blankholder zone in drawing of a rectangular pan (Fig.3.4b), or in machining between the newly formed surface and the flank face of the tool (Fig.3.4c).

3. When pressures are sufficient to cause bulk plastic flow, asperity contact takes place against a deforming substrate. There may be only normal contact, as when sticking friction is attained, or sliding may be superimposed. Because the yield criterion may be satisfied by a number of stress combinations (Fig.2.1), interface pressures may vary widely:

a. Normal pressures are much below σ_f when yielding occurs in a combined tensile and compressive state, with both stresses acting parallel to the interface, as in the blankholder zone in deep drawing of a cylindrical cup (Fig.3.5a).

b. Normal stresses are somewhat below σ_f when the tensile stress acts parallel to

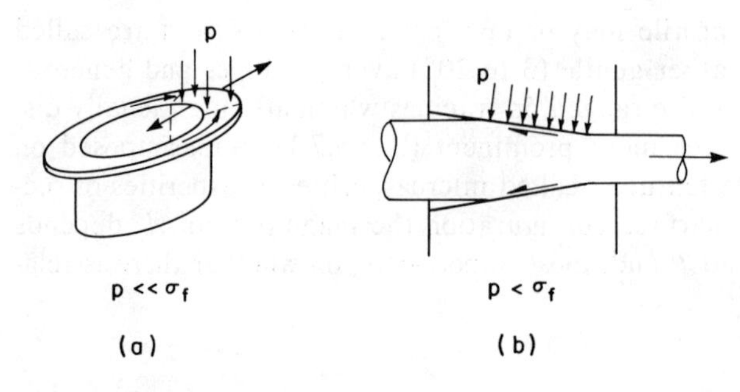

Fig.3.5. Examples of low interface pressure resulting from yielding in a combined stress state

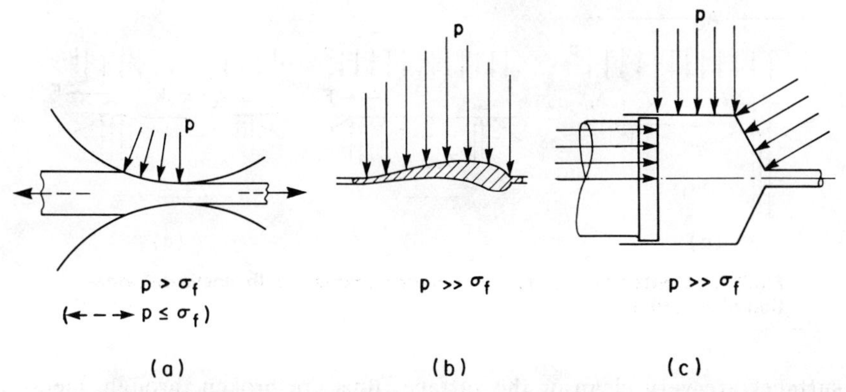

Fig.3.6. Examples of high interface pressure resulting from yielding in a compressive stress state

the interface and the compressive stress normal to it, as in wiredrawing (Fig.3.5b) or in strip rolling with front and back tensions (Fig.3.6a).

c. Normal stresses become higher than σ_f when all stresses are compressive, as in rolling without tensions (Fig.3.6a), in forging (Fig.3.6b), and in extrusion (Fig.3.6c). Stresses are limited only by the strength of the tooling, and may reach a multiple of σ_f (it is here that expressing friction by a coefficient of friction can become misleading; Fig.2.3b).

3.2.3 Elastically Loaded Workpiece, Normal Contact

From a macroscopic point of view, an asperity pressed on a surface is assumed to suffer only elastic deformation, and the contact area can be calculated from the equations developed by Hertz (hence the term "Hertzian contact"). A closer look reveals, however, that even when interface pressures are insufficient to cause bulk plastic deformation, some asperities of the workpiece deform plastically; a useful guide is the plasticity index [92,93]

$$\psi = \left[\frac{E^*}{H}\right] \left[\frac{\sigma^*}{\beta}\right]^n \tag{3.2}$$

where H is hardness, σ^* is the mean deviation of asperity height, β is the mean asperity tip radius [92] or asperity spacing [93], n = 1/2 [92] or unity [93], and E^* is the composite elastic modulus for surfaces 1 and 2, which may be derived from

$$\frac{1}{E^*} = \frac{1 - \nu_1^2}{E_1} + \frac{1 - \nu_2^2}{E_2} \tag{3.3}$$

where ν is Poisson's ratio (for reviews, see [93,94]). For most practical metalworking situations, $\psi > 1$ and plastic deformation of some asperities is unavoidable.

The bulk, which is only elastically loaded, imposes a restraint on asperity deformation, and thus a pressure higher than σ_f is needed locally. The problem is the inverse of Fig.2.7; the required pressure increases with decreasing asperity slope and reaches p = $3\sigma_f$ in the limit (although localized strain hardening may raise the pressure even higher). Then, the real contact area A_r must increase until it can support the load P (Fig.3.7a):

$$P = A_r H \tag{3.4}$$

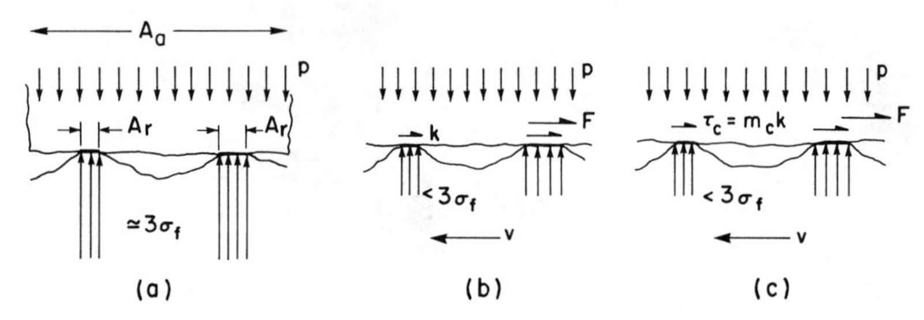

Fig.3.7. Contact between die and rough workpiece with plastic deformation of asperities

If the surfaces are very clean or the surface films are broken through, metal-to-metal contact occurs. Atoms approach each other so that interatomic forces come into play, and, if the die and workpiece are to be separated at this point, a force is needed for such separation. We say that adhesion has taken place. An examination of the surface would show that localized pressure welding (cold welding) has occurred at the asperity.

3.2.4 Sliding and the Sources of Friction

Adhesion Theory. The shear strength of the junction formed between very clean surfaces would roughly equal the shear strength k of the workpiece material; therefore, if the workpiece is now moved relative to the die (Fig.3.7b), the junctions have to be sheared with a force

$$F = A_r k \tag{3.5}$$

and a coefficient of friction can be defined:

$$\mu = \frac{F}{P} = \frac{A_r k}{A_r H} = \frac{k}{H} \tag{3.6}$$

This simple adhesion theory of friction, due to Bowden and Tabor [18], accounts for the first two empirical laws of friction (Sec.2.2.1). We have seen (Fig.2.1) that $k = 0.577\sigma_f$, and thus the maximum value of μ in Eq.3.6 would be around 1/6.

To account for the much higher friction values normally observed with clean surfaces, one has to consider that, at the point of sliding initiation, the junction is subjected to a combined stress state consisting of a normal pressure and a shear stress (Fig.3.7b). We have already seen in Fig.2.1 that the pressure required for plastic flow drops if a combined stress state is imposed. In a similar manner, it can be shown that the junction will now yield at a much lower normal pressure than $3\sigma_f$, and therefore the real contact area must grow until the load can again be carried. Such junction growth increases the junction area to be sheared and thus the frictional force, leading to high values of μ [18].

Contaminant Films. In the presence of a contaminant film, welding of asperities is prevented by a contaminant layer of shear strength τ_c (Fig.3.7c). By analogy to Eq.2.10, this can be expressed as some m_c fraction of k (to make it clear that we are now looking only at the real area of contact A_r, we use m_c rather than m, which refers to the apparent area of contact A_a):

$$\tau_c = m_c k \tag{3.7}$$

From the simple arguments of the junction-growth theory [95] it can then be shown that

$$\mu = \left[\frac{m_c^2}{\alpha(1 - m_c^2)} \right]^{1/2} \tag{3.8a}$$

where $\alpha = (H/k)^2$ for a three-dimensional junction and is taken to be around 9 from experimental evidence (more recent work [96] found it to range from 4 for brass to 20 for mild steel). On very clean surfaces $m_c = 1$; thus μ can become very high, but falls rapidly as m_c decreases, as is indeed found for contaminated surfaces. For $m_c < 0.2$, Eq.3.8a becomes

$$\mu = \left[\frac{m_c^2}{\alpha} \right]^{1/2} = \left[\frac{m_c^2 k^2}{H^2} \right]^{1/2} = \frac{\tau_c}{H} \tag{3.8b}$$

which takes us back to Eq.3.4 but shows that cold welding is not essential to explain friction. Neither is junction growth necessary when m_c is small, because friction is governed by the shear strength of the contaminant film.

Asperity Interaction. In the above discussion it is assumed that the junction between die and workpiece is sheared in a plane parallel to the direction of sliding (Fig.3.7b and c). In reality, asperities may also encounter each other in the course of sliding (Fig.3.8a), and then plastic deformation of the softer (workpiece) asperity will occur. By application of plasticity theory, the shear force can be calculated both for clean (virgin) surfaces and for asperities coated with a contaminant film of shear strength τ_c [23]. When the asperity slope becomes zero, the solution gives the same result as the junction-growth theory, and thus the latter can be regarded as a limiting case of the more generalized asperity encounter. A large asperity on the die surface (such as would develop when cold welding leads to pickup) plows through the workpiece material (Fig.3.8b) and contributes to the frictional force [18]. Indeed, deformation of a thin,

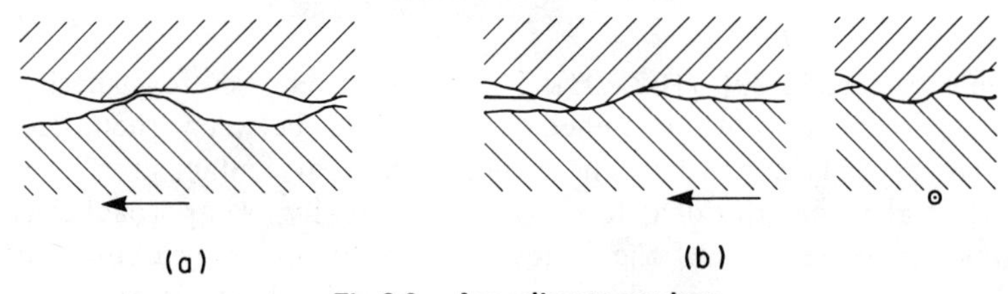

Fig.3.8. Asperity encounters

very heavily worked surface layer may be the only source of friction when adhesion is negligible [73], and theories based on this premise have been reviewed and further developed by Heilmann and Rigney [97]. The effects of crystal structure and the role of dislocation concepts have been considered by Kuhlmann-Wilsdorf [97a].

Molecular Theory and Surface Energy. There are yet further possible sources of friction. The intimate contact brings the die and workpiece close enough for interatomic forces to come into play (these are, after all, sources of adhesion too). Thus, it is not necessary that cold welding actually take place; proximity is enough to give rise to the frictional force. This "molecular theory of friction" is given great emphasis in Soviet work [26,98]. Kragelski [26] considers friction to have both mechanical and molecular sources, with elastic and plastic surface deformation, asperity encounter (shearing, plowing), and molecular attraction playing important roles.

A further view of friction is given by Rabinowicz [30], who regards the surface energy of adhesion as the source of friction. An effective lubricant has a low energy of cohesion, and low friction is obtained if the cohesive energy is small relative to the adhesive energy of the lubricant to the solid surface. The problem with this concept is that the surface energy of adhesion is difficult to determine for solids [75], and the ratio of cohesive to adhesive energy becomes an alternative description of the ratio of shear strength to hardness (Eq.3.6). Nevertheless, surface energy considerations are useful, especially in lubricated sliding [99] and adsorption phenomena [80].

A critical review of various friction theories has been given by Brown, Owens and Booser [100]. The topic has always been controversial and has led to some of the more spirited discussions in tribology [101]; for our purpose the above simple views will suffice.

3.2.5 Plastically Deforming Workpiece

As pressures increase, more and more asperities make contact until their deformation cannot be viewed in isolation; instead, interactions between the deformation zones must be taken into account. Several aspects of this problem have been dealt with by Wanheim, Bay and co-workers; their work is summarized in [102].

We have seen that an isolated asperity yields at about $3\sigma_f$ because the nondeforming substrate prevents its free deformation. As the number of asperity contacts increases, the stress fields acting on them begin to interact (Fig.3.9), and plastic flow

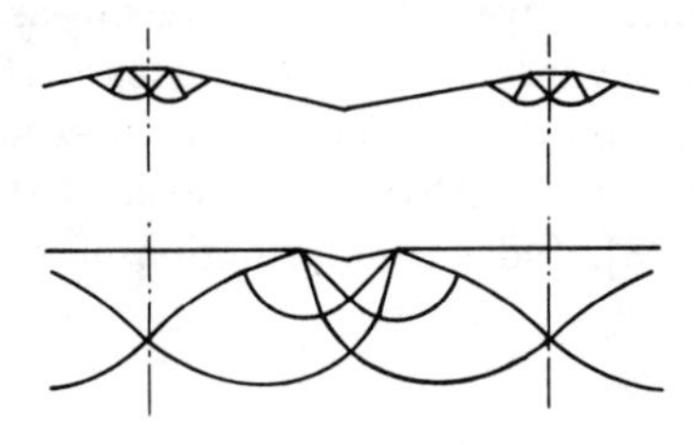

Fig.3.9. Slip-line fields for normal contact with (top) asperity deformation and (bottom) substrate deformation [102]

begins in the substrate [2.29], [102,103]. The pressure required for asperity deformation drops, the asperities become flattened, and valleys between them rise, while the material of the asperity strain hardens heavily. The real contact area A_r increases rapidly and with it also the frictional force (Eq.3.5). Finally, A_r approaches A_a asymptotically, although some vestiges of the valleys survive to very high deformations. When a nominal 100% contact area is reached and an absolutely clean surface pressure welds to the die, sliding must be accommodated by shearing in the bulk of the workpiece at a stress k. At this point,

$$\mu = k/p \tag{3.9}$$

and we are back to Eq.2.8 and the macroscopic picture presented in Fig.2.3, with μ = 0.577 at $p = \sigma_f$. At yet higher p the coefficient of friction drops (Fig.2.3b), as shown by Shaw, Ber and Mamin [104] and by others [105-107].

It should be noted that the pressure required for asperity deformation drops and A_r increases rapidly also when the workpiece is made to yield by tensile stresses [107], and this can account for the high μ often found in sheet metalworking, particularly on the punch nose.

For our purpose, most important is the case where sliding is superimposed on normal contact [102,108-113]. Assuming that the workpiece surface consists of identical wedge-shaped asperities which are deformed by contact with a flat tool surface, Wanheim and Bay [102] showed from slip-line field theory that the real contact area (expressed as a fraction of the apparent contact area $c = A_r/A_a$) depends on both the interface pressure and the shear strength in the real contact zone m_c. If Eq.3.7 holds, $\tau_c = m_c k$, and the average frictional stress over the entire apparent area of contact is

$$\tau_i = m_c ck \tag{3.10}$$

Initially, the calculated contact area increases linearly with interface pressure p, and rises more steeply for a higher m_c (Fig. 3.10a). Since τ_i also increases linearly with p,

$$\mu = \frac{\tau_i}{p} = \frac{m_c ck}{p} \tag{3.11}$$

we find that μ is also a constant, as is typical of Coulomb friction. At higher pressures ($p/2k > 1.3$) the growth of c is not linear and the limiting value of $c = 1$ is reached only gradually, at higher pressures for lower m_c values. Correspondingly, the interface shear strength τ_i also reaches a limiting value equal to m_c (Fig.3.10b). Because of this, the coefficient of friction becomes pressure dependent, as previously described for clean surfaces (Fig.2.3). However, with very good lubrication where $m_c < 0.2$, the τ_i value increases approximately linearly with p (Fig.3.10b) and Coulomb friction still prevails. The analysis thus yields results similar to the junction-growth theory except that interface pressure (Eq.3.11) rather than hardness (Eq.3.8b) is now in the denominator.

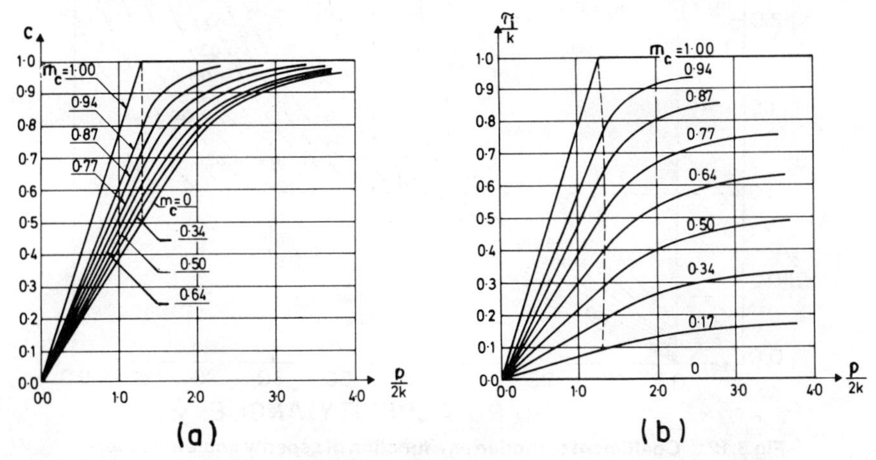

Fig.3.10. Effects of interface pressure on (a) true contact area and (b) interface shear strength [102]

Of course, real asperities are not of the pure wedge shape assumed in the theory, but have smaller microasperities (higher-order asperities) superimposed on them. When these are taken into account, it is found that the real area of contact increases much more rapidly with interface pressure, and the interesting result is that, irrespective of the value of m_c, $c = 0.86$ and higher even at low interface pressures [102]. A further

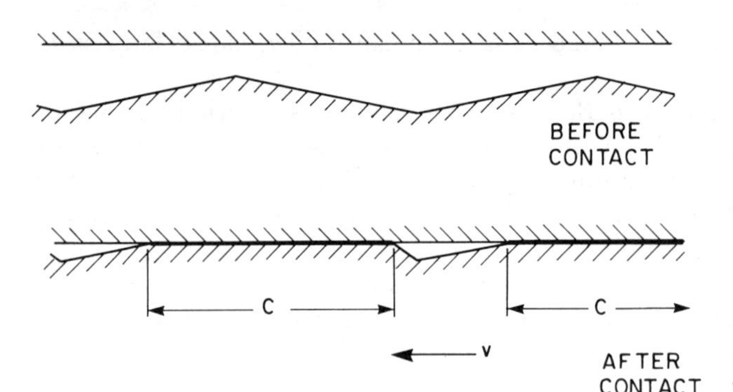

Fig.3.11. Distortion of asperities on sliding

complication that has not yet been solved theoretically is that at high m_c values the asperities tilt on sliding (Fig.3.11), and this gives rise to a more rapid increase in c than that predicted from the assumption of a constant-asperity slope.

Many die surfaces are, intentionally or through wear, roughened to the point that they can no longer be considered flat. Then asperities of the harder die material encounter and plastically deform the asperities of the workpiece, giving rise to friction, penetration of surface films, and possibly adhesion and wear. Experimentally it is often found that friction increases with surface roughness [106], at least under conditions of low adhesion. Several authors [86,110-112] have analyzed the mechanics of this situation. Challen and Oxley [112] show (Fig.3.12) that for a given die asperity slope (hard asperity angle γ) the workpiece asperity may survive by plastic deformation (rubbing model), may be removed after deformation (wear model), or may be machined away (cutting model). The shear factor in the contact area m_c (Fig.3.12) has a substantial influence

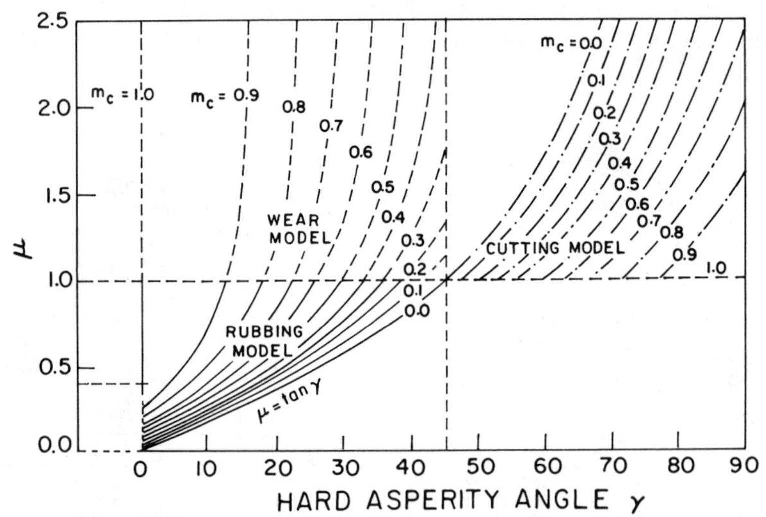

Fig.3.12. Coefficient of friction as a function of asperity angle and interface shear factor [112]

on the calculated μ which covers the wide ranges found in practice. In a later work, Oxley [113] considered the development of an intense plastic shear zone on the top of an asperity of a strain-rate-sensitive material. The deformation of a smooth workpiece with a rough tool in normal contact has also been investigated [102].

One difficulty with all theoretical analyses is that the strength of the contact film at a "weak junction" has to be described by some constant factor such as m (at total conformity, $m_c = m$, from Eq.3.10 and 2.10). The probability is very slim indeed that any surface film would survive intact the severe contact conditions existing in metalworking. Thus, the surface film is damaged at least in some places (Fig.3.13) and then a higher shear strength (or m_c value) applies locally over a fraction of A_a. If the workpiece tends to adhere to the die, the local shear stress at this "strong junction" reaches k. The average shear stress τ_i taken over the whole contact area is then

$$\tau_i = ak + cm_c k = \left(a + cm_c\right)k \tag{3.12}$$

and, if we assume that interface pressure is constant over the entire apparent area of contact, the coefficient of friction is

$$\mu = \frac{\tau_i}{p} = \left(a + cm_c\right)\frac{k}{p} \tag{3.13}$$

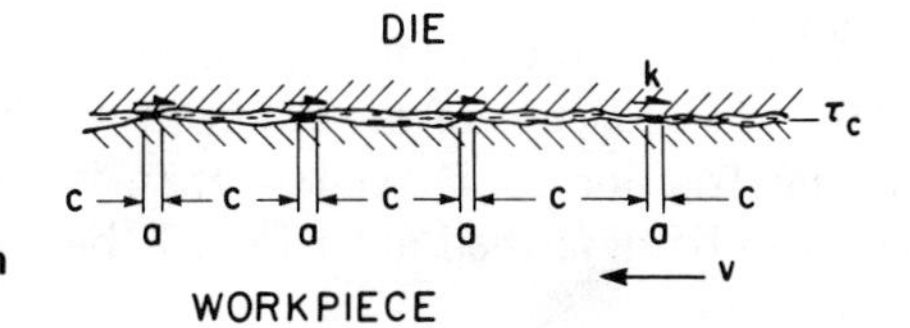

Fig.3.13. Fully conforming dry interface, with metallic contact (a) and contaminant films (c)

If the surface is only partially conforming, A_r is an α = a + c fraction of A_a. As it is highly unlikely that τ_i would change linearly with p, a constant μ (sometimes found in hot working) must be fortuitous. It is conceivable that changes in the a and c ratios with temperature contribute to this result.

Contact with large die asperities and with pickup formed on the die leads to plowing of subsequently contacted workpiece material. Plowing is included in some models, but in practical situations not much of this can be allowed because the surface finish of the product may become unacceptably damaged, and pickup could terminate the process. If m_c is fairly high, a small increase due to localized metallic contact will pass unnoticed and modeling by Eq.3.10 is still realistic. If, however, m_c is low, the force required to shear the metallic contact areas may become large enough to necessitate modeling with Eq.3.12. Of course, metallic contact can be reduced by application of proper lubricants, and this leads us to lubrication with the aid of solid films of low shear strength.

Some of the interactions in dry contact can be summarized by following the dash-dot lines in Fig.3.2.

3.3 Solid-Film Lubrication

In the presence of a solid-film lubricant, the interface looks similar to that in dry friction, except that the contaminant film is now intentionally introduced. It may act simply as a separating or parting agent, preventing metal-to-metal contact, or it may also possess a low shear strength. From a purely mechanistic point of view, it does not matter whether the film is attached to the workpiece surface, the die surface, or both, or whether this attachment is purely mechanical (by engagement in asperities) or enhanced by physical adsorption and chemical reaction. When attached to the die surface, the film may lubricate by controlled wear; when attached to the workpiece surface, attachment to the die surface must be less so as to prevent removal of the lubricant from the workpiece. Sliding is assumed to take place by shear at a stress τ_s. If

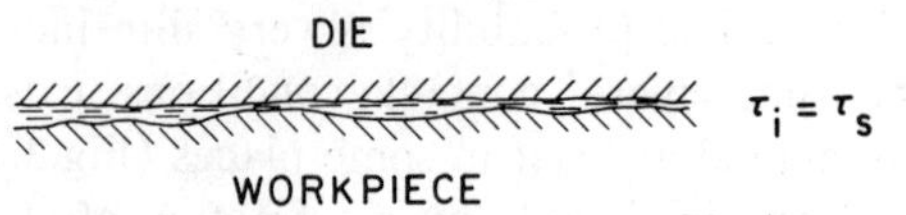

DIE

$\tau_i = \tau_s$

WORKPIECE

Fig.3.14. Solid-lubricated interface

there is full conformity between die and workpiece surface (Fig.3.14), the interface can be described either by a coefficient of friction:

$$\mu = \frac{\tau_s}{p} \tag{3.14}$$

or with the aid of an interface shear factor m_s:

$$\tau_s = m_s k \tag{3.15}$$

Obviously, the former assumes that τ_s increases linearly with imposed pressure, whereas the concept of m_s implies that the interface shear strength somehow remains a constant fraction of k. At less than full conformity, the shear strength in the contact zones can be described as in Eq.3.7, but now

$$\tau_{cs} = m_{cs} k \tag{3.7a}$$

and the average shear strength of the interface is

$$\tau_i = m_{cs} ck \tag{3.10a}$$

Nothing can be said about the validity of any of the above approaches without inquiring into the physical and chemical nature of solid-film lubrication [114,115].

3.3.1 Oxide Films

We have seen (Fig.3.1b) that oxide films are normally present on technical surfaces. Together with adsorbed films, they are important constituents of the contaminant film that permits limited deformation with dry surfaces. The oxidation of metals and alloys depends on the atmosphere and on the composition (and often minor elements) of the metal [116-119]. To act as a lubricant, an oxide film must fulfill a number of requirements, depending also on whether the oxide is on the die or on the workpiece:

1. The film must be continuous and thick enough to resist easy penetration by asperities and ensure reliable separation of the surfaces. Rabinowicz [120] suggests a minimum of 0.01 mm, but there is no confirmation of such a minimum for metalworking. An excessively thick film has poor wear resistance, and it may spall off the die and produce an unacceptably rough surface on the workpiece.

2. If attached to the workpiece, the film must have sufficient ductility to follow surface extension. This condition is seldom fulfilled even when only the asperities are deformed, as shown by electrical resistance measurements.

3. If the film is attached to the workpiece, its shear strength τ_s should be lower than k; if it is attached to the die, τ_s should be higher than k. Usually only hardness data are available [120,121]. At room temperature only a few oxides can be regarded as lubricants for the workpiece (Fig.3.15). At elevated temperatures some oxides soften quite rapidly and act as solid-film lubricants. Transformations in metals may contribute too; thus, FeO becomes softer than iron above 900 C where Fe transforms from ferrite

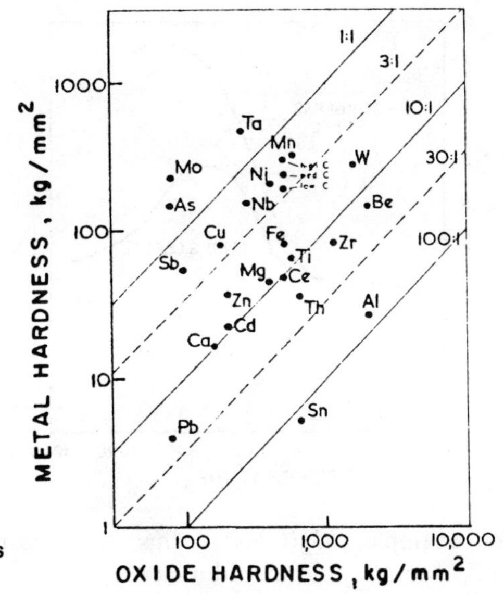

Fig.3.15. Hardnesses of metals and their oxides (the higher oxide where it occurs) [121]

into the stronger austenite. Mrozek [122] reports on experiments in which iron scale was hot extruded. As expected, extrusion pressure decreased with temperature above 700 C, but it also increased with rate of deformation.

4. If the film is damaged, it should reform rapidly. This depends on the rate of oxidation and thus temperature, and also on oxygen access and thus process geometry. It is, therefore, possible to hot roll or hammer forge steel without a lubricant because reoxidation between passes or blows is permitted, but total failure would occur in extrusion where access of oxygen to the interface is denied.

5. Some oxides are friable—that is, they break up into small, loose particles. These are often found to lubricate in nonmetalworking contact situations, and are frequently incorporated into high-temperature lubricants; especially when bonded to the surface by a suitable binder, they can provide a very useful separating, parting function, even though their shear strength may not be low. They cannot, however, follow the extension of a workpiece surface, and therefore they are seldom used in metalworking except as die (tool) coatings formed from the very hard and temperature-resistant oxides (Sec.3.11.5).

The oxides of only a few metals fulfill the requirements of a workpiece lubricant, and then only imperfectly. Luckily, iron is among them. Scale formed on iron at high temperatures is of a layered structure; the outermost oxygen-rich Fe_2O_3 layer and the intermediate layer of Fe_3O_4 are rather brittle, but the FeO layer immediately adjacent to the metal is capable of some limited deformation (Fig.3.16) [123]. Similarly, copper oxide softens above 500 C [118]. Oxides of nickel, aluminum, titanium, lead and zinc are all harder than the substrate and are also brittle; however, magnesium oxide is friable and can act as a lubricant [124]. No generalizations can be made regarding alloys. On steel, some alloying elements migrate to enter into the oxide (e.g., Si forms SiO_2), and others form more complex spinels (e.g., $NiFe_2O_4$). On brass the oxide is mostly ZnO, which is covered by the deformable CuO on low-Zn brasses. When the CuO layer is penetrated, friction rises [124]. The oxide of aluminum is very hard and brittle; of the alloying elements, copper has no effect, manganese increases resistance to damage when present as $MnAl_6$, and magnesium forms MgO at the oxide/air interface, resisting damage [125]. Reactions with the atmosphere may further change the oxide film.

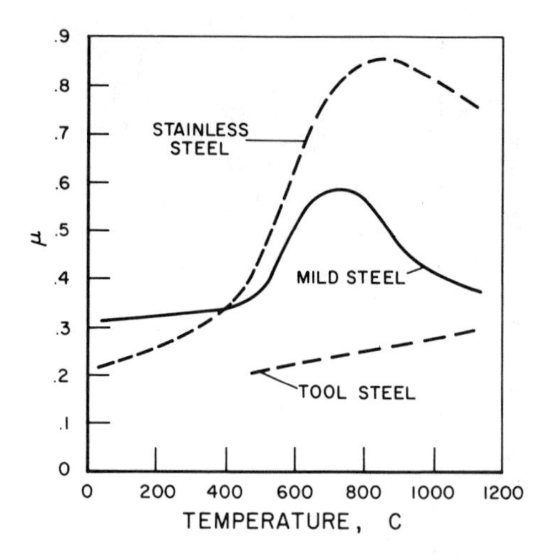

Fig.3.16. Coefficient of friction in twist compression of mild steel against three anvil materials [123]

For example, $Mg(OH_2)$ forms on exposing the mixed Al_2O_3-MgO oxide film of Al-Zn-Mg alloys to humid air [83a].

A truly lubricating, low-shear-strength film is obtained only when the oxide melts at the interface temperature, as for example the oxides of molybdenum and tungsten do; unfortunately, they also evaporate at relatively low temperatures, and the weight loss becomes intolerable. Molten lubricants should really be classified under hydrodynamic agents (Sec.3.6). Oxides chosen for their low shear strength or favorable melting range may be used as additives to other lubricants; some minerals also fall into the same category. In a sliding situation the key point is always the need for $\tau_s < k$; a harder oxide behaves as an abrasive.

In machining, the task is to provide lubrication on the tool faces which contact the virgin surfaces generated by the cutting process (Fig.2.4c). This can be most effectively done by incorporating the lubricant into the workpiece material. Some steels contain calcium-alumino-silicate inclusions which affect flow in the secondary shear zone (Chapter 11).

Apart from providing some lubricating functions of their own, oxides are of utmost importance in the lubrication system because they can aid the action of other lubricants, either by simple mechanical entrapment of liquids into microscopic features of the oxide film or by reaction with lubricants that otherwise would be rather ineffective on pure metal surfaces.

3.3.2 Metal Films

Coating of a die or workpiece with a continuous film of another metal can fulfill a number of lubricating functions:

1. Adhesion between die and workpiece can be reduced; this is the prime purpose of die coatings, and one of the purposes of workpiece coatings.

2. In the absence of other lubricants, a metal film of low shear strength τ_s deposited on the workpiece surface will reduce friction (Eq.3.14 and 3.15) as long as τ_s remains lower than k, even under the pressures and shear rates prevailing in the process. We have mentioned (Sec.2.1.3) that the flow stress σ_f and thus the shear strength k of bulk metals is not affected by hydrostatic (normal) pressure. There is some argument as to whether this is also true of thin films. Careful experiments by Riecker and Towle [126] seem to indicate some slight pressure dependence and a much greater—and undisputed—temperature dependence following a law of the type

$$\tau_s = \tau_0 \exp\left[bT/T_{m(P)}\right] \tag{3.16}$$

where τ_0 and b are material parameters and where the pressure effect comes in through its influence on the melting point $T\, m(P)$. For copper, silver, and gold films on steel, μ = 0.1 at room temperature; whether this is generally valid is open to doubt. A cursory examination of the situation might suggest that a constant m_s (Eq.3.15) should be more attractive, but this can be true only if strains, strain rates, strain hardening, and strain-rate sensitivity are the same in the coating as in the workpiece material. These conditions can hardly be satisfied, as shown by the greatly varying m_s value found by Jovane and Ludovico [127] in forging aluminum with a superplastic Sn-Pb sheet. Therefore, the interface is best modeled simply by τ_s. A difficulty might be that interface temperatures rise more than bulk temperatures, and thus τ_s may not be known with sufficient accuracy, although the drop in τ_s with temperature may be counterbalanced by an increase due to higher strain rates.

3. The metal film may improve the performance of another (usually liquid) lubricant either by providing better entrapment through favorable surface configuration or by allowing chemical reactions. The latter aspect is particularly important when the workpiece material itself is nonreactive (e.g., stainless steel or titanium).

4. The metal film can be effective only if it is well bonded to the surface. Bond strength depends on the deposition method and is enhanced by diffusion, although the formation of brittle intermetallics is harmful in a workpiece coating.

5. The film must be thick enough to ensure a continuous coating, although coatings thicker than 1 μm give lower wear resistance in nonmetalworking situations; they may also increase friction by allowing A_r to grow. In metalworking, thick films may be scraped off by the tool. Film thickness in the deformation zone is controlled by the thickness of metal deposited on the workpiece, and by surface extension. Asperities of a rough tool may pierce through the metal coating and then a situation similar to Fig.3.13 arises, followed possibly by pickup and scoring. The case of hard asperities penetrating a soft metal film deposited on a hard substrate has been investigated by Halling [128].

As mentioned earlier, internal lubrication is desired for machining, and metals such as lead and bismuth are among the oldest free-machining additives. The mechanism by which they lubricate will be discussed in Chapter 11.

3.3.3 Polymer Films

Most polymers are weaker than metals, and many of them have sufficient ductility to deform together with the workpiece surface. Thus they may be interposed either as a separate film, or they may be deposited on the workpiece surface. A number of points must be observed:

1. Strong adhesion between the polymer and the metal substrate can be established only by rubbing, melting on, fusing of dry powder, or application from a dispersion or solution. Survival of a separately interposed film depends entirely on geometric conditions. Thus, when a polymer sheet is interposed in the upsetting of a cylinder, the sheet is cut through at the edge, and then deformation proceeds with very little expansion of the polymer film (Fig.3.17a). If, on the other hand, a polymer is deposited on the die surface from a solution, it may well survive contact with the expanding cylinder end face. Similarly, a coating deposited on a bar or strip withstands drawing or rolling (Fig.3.17b), whereas a polymer sheet wrinkles or breaks.

2. Suitable polymers are thermoplastic, and operate above their glass-transition temperature T_g. Below the glass-transition temperature they behave like elastic-brittle

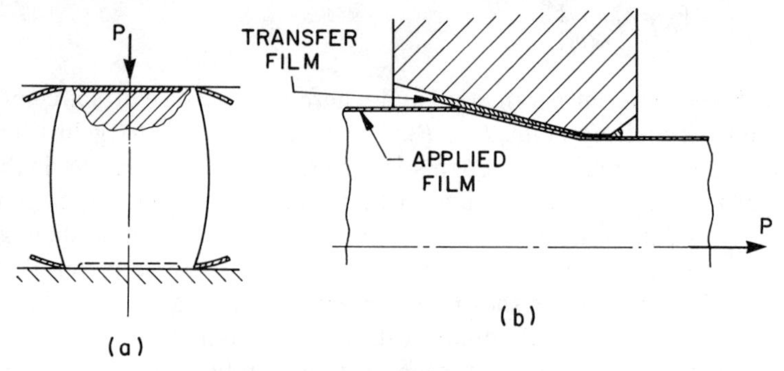

Fig.3.17. Polymer films, illustrating (a) failure in upsetting and (b) formation of transfer film in drawing

solids and are unable to follow surface deformation; they allow exposure of virgin surfaces, and die pickup usually ensues. Above the glass-transition temperature polymer films behave as viscoelastic solids but have no spreading ability (they can spread above their melting point but then they cease to be solid films). Thus, a defect in the coating can lead to die pickup, subsequent scoring, and severe surface damage.

Hydrostatic pressure σ_H increases the temperature of glass transition T_g (Fig.3.18), as does fast cooling. It is therefore important that the polymer chosen should remain above T_g under the conditions prevailing in the interface.

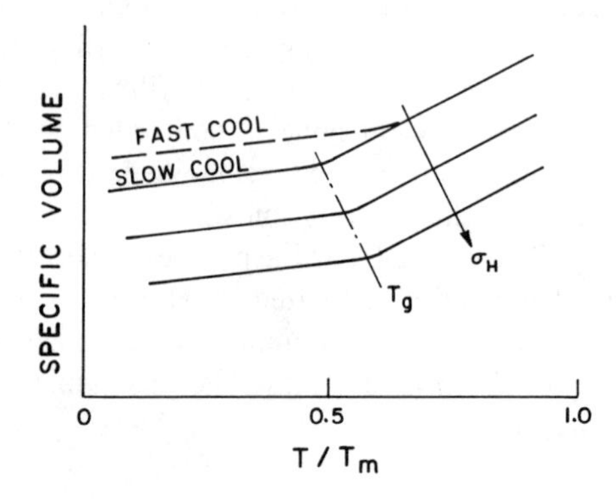

Fig.3.18. Effect of hydrostatic pressure on glass-transition temperature

3. When the polymer has a simple molecular structure (essentially straight chain) and the intermolecular forces are low, a polymer film which has been deposited on one surface can be transferred from the original surface to the contacting surface, and the films become oriented [129-132]. Shearing is concentrated between the original film and the transfer film (Fig.3.17b) rather than being equally distributed throughout the entire film thickness. If the process geometry is favorable, the transfer film can provide some protection against metal-to-metal contact. Film thickness is governed primarily by the thickness of film deposited on the workpiece. Surface roughening in the course of deformation is helpful because, together with the original roughness of die and workpiece, it helps to key the polymer to the surface.

4. In contrast to metals, the flow stress of polymers is greatly affected by hydrostatic pressure as well as temperature and strain rate. Thus, the rheology of polymers can become exceedingly complex [133]. For our purpose the most important observation is that the shear strength τ_s of thin, well-worked polymer films interposed between metal surfaces is a function of hydrostatic pressure [131,132,134,135]:

$$\tau_s = \tau_0 + \alpha p \tag{3.17}$$

where τ_0 is the shear strength at zero pressure; this component is important only at low interface pressures, and it becomes negligible at high interface pressures. Then

$$\tau_s = \alpha p \tag{3.18}$$

Comparing Eq.3.18 with Eq.3.14, it is obvious that α is equal to μ. Thus, an interface lubricated with a continuous polymer film can be approximately modeled by a constant μ usually on the order of 0.05 to 0.1 (Fig.3.19 [115]), although exceptions abound; μ increases with p in nylon at low pressures, and a phase change in PTFE at around 200 N/mm^2 results in an increase of μ (Fig.3.20 [131]). A possible explanation for a constant μ might be found by considering that sliding takes place between the original and

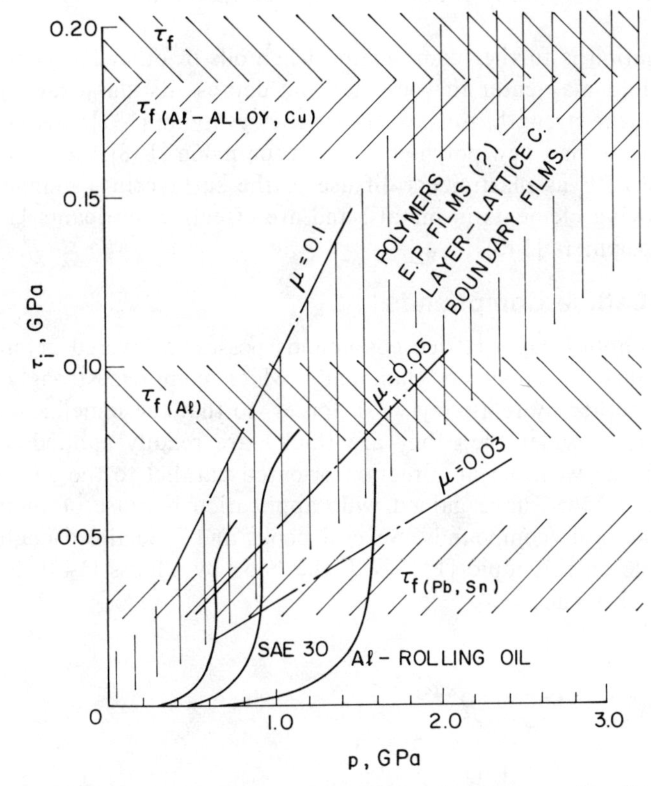

Fig.3.19. Shear strengths of various lubricant classes as a function of interface (hydrostatic) pressure [115]

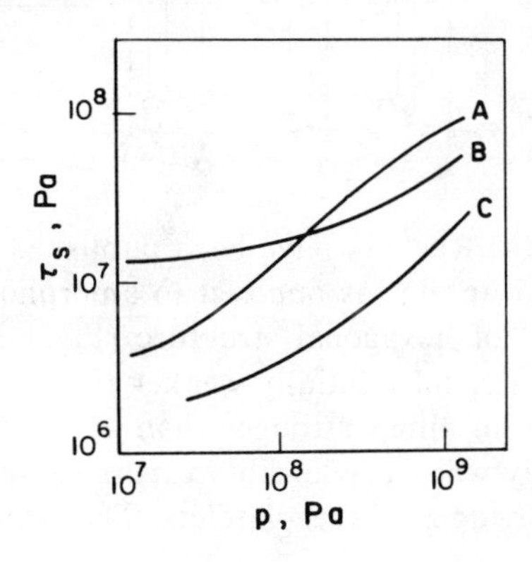

Fig.3.20. Pressure dependence of shear strengths of various polymers [131]

transfer films. Alternatively, an Eyring model [136] for viscous flow could be assumed to operate [131]. These hypotheses gain support from the observed temperature dependence of τ_s

$$\tau_s = \tau_0 \exp(-Q/RT) \tag{3.19}$$

where Q is activation energy (a negative quantity) and R is the universal gas constant. Significantly, Q is the same for such diverse polymers as polyethylene (PE), polypropylene, and polytetrafluoroethylene (PTFE). There is also some evidence that τ_s is not greatly dependent on shear strain rate, which can be possible only if shear takes place between two films rather than in the film itself. Obviously, τ_s has no relation whatsoever to the shear flow strength of the workpiece material k, and thus no constant m value can exist.

5. The situation is further complicated when oils or other fluids are applied to the polymer. The fluid may enter the polymer and depress its glass-transition temperature and reduce its shear strength; thus, water plasticizes nylon [137], oils plasticize PE, and siloxanes seem to form a new polymer of low τ_s on nylon [138]. Amides which are often incorporated into PE as plasticizers diffuse to the surface in a manner similar to the diffusion of alloying elements in metals, and are effective lubricants [139]. Fatty acids also adsorb on polymers [140].

3.3.4 Layer-Lattice Compounds

As their name implies, layer-lattice compounds possess a layered crystal structure. To serve as lubricants, the platelets (lamellae), which comprise strongly bonded atoms, must be held together by relatively weak forces, so that the lamellae—which are some ten to twenty times wider then they are thick—are readily applied to the workpiece and/or die with the weak shear direction oriented parallel to the surface. In practice, only graphite and MoS_2 have gained wide application because of their low cost compared to other suitable compounds. Several books and specialized conference proceedings are available on this topic [141-145]; the book by Clauss [143] deals with several classes of solid lubricants.

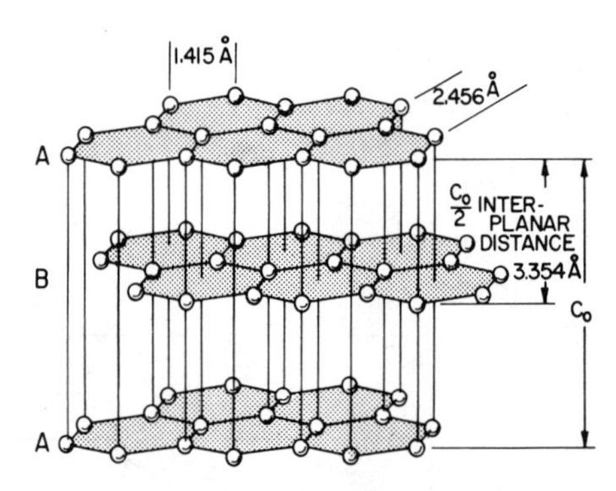

Fig.3.21. Structure of graphite crystals

Graphite. Reviews have been published by Peace [146], Feneberger [147], and Mrozek [148]. Graphite (as opposed to amorphous carbons such as lamp black) is a crystalline material of hexagonal structure (Fig.3.21), irrespective of its origin. Bonding in the c direction is substantially weaker than in the basal plane. Nevertheless, the bond is still two to four times stronger than a van der Waals bond, and graphite becomes a lubricant only when condensible vapors (gases, water vapor, or organic vapors) are adsorbed onto the edges of the platelets. The bond in the c direction is then weakened, and slid-

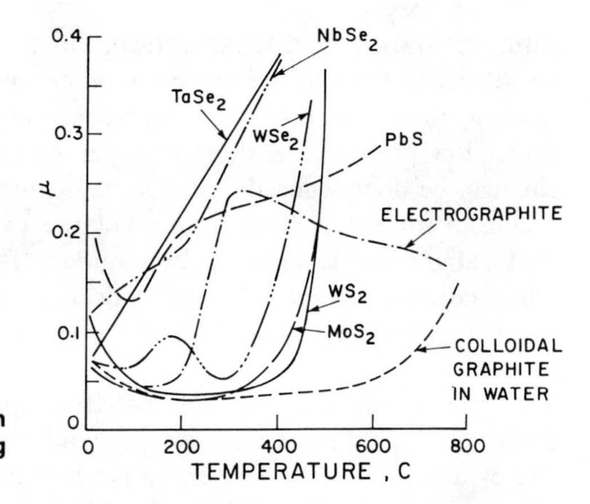

Fig.3.22. Variation of coefficient of friction with temperature for various solid lubricants in drawing of steel sheet with 20% reduction [149]

ing between the basal planes can proceed. Alignment of platelets is facilitated by milling the natural or man-made graphite to a fine particle size (0.5 to 2.0 μm), making such colloidal graphite a good lubricant (Fig.3.22), whereas randomly oriented graphite agglomerates are less effective (electrographite in Fig.3.22 [149]). Special milling techniques produce very thin flakes with large ratios of basal plane to edge surface area; such powders adsorb hydrocarbons (they are oleophilic) and are reported to give better alignment on surfaces [150]. Furthermore, they make the graphite-coated surface more receptive to any oil lubricants that may be superimposed [151].

Molybdenum Disulfide. Reviews have been published by Farr [152], Barry [153], and Mrozek [154]. In contrast to graphite, no condensible vapors are needed for MoS_2 lubrication [74,75]. The favorable frictional characteristics derive from the structure itself. Strong covalent bonding prevails within lamellae, with a high polarization of the sulfur atoms. Adjacent platelets are held only by relatively weak forces, but the presence of charge centers ensures that adjacent layers are in a definite relationship to each other (Fig.3.23); thus layers are readily deposited on top of each other in the course of film formation. Adsorbates increase the friction and wear rates of films by increasing adhesion between lamellae.

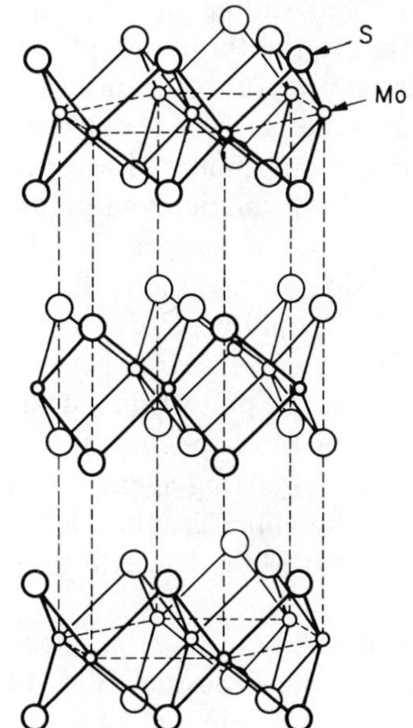

Fig.3.23. Structure of molybdenum disulfide

Film Formation and Destruction. For both MoS_2 and graphite, alignment of platelets parallel with the surface can be promoted by rubbing (burnishing) or by running the surfaces against each other in the presence of the lubricant for a prolonged period of time under low pressure, as directly observed under the microscope [155]. In metalworking this can be done when dies are to be coated, but it is less practical for coating the workpiece. In films deposited from a volatile carrier, such as water or a solvent, an approximate alignment is induced by capillary forces. Some running-in actually takes place when contact between die and workpiece is of the continuing or repetitive type. The critical period is the early phase of contact in which the platelets are not yet fully aligned.

Film formation is easier with finer particles [156] and is greatly accelerated on materials capable of forming compounds. This has been observed for graphite, although the evidence is contradictory because an increase in friction with carbide formation has also been found. The evidence for MoS_2 is unequivocal. Sulfide formation was observed with MoS_2 greases on steel [157]. In twist-compression tests, Reid and Schey [158] found that films developed quickly on sulfide-forming iron and copper whereas on aluminum and titanium the films formed more slowly, by purely mechanical means, and, if metallic contact occurred during the film-forming phase, surface damage was inevitable. Films developed more rapidly and were more durable on heat treated than on annealed die steel, presumably because higher asperity temperatures were generated on the harder steel. A further contribution to film formation may be made through oxides such as Fe_2O_3, which have a polarized bonding similar to MoS_2 as opposed to the weak attraction to the less strongly polarized Al_2O_3 [159].

The possibility of reaction is extremely important also in enabling the film to resist the severe wiping action typical of many metalworking processes. Thus, even well-formed films have limited lives on aluminum and titanium surfaces and, once worn off, allow metal-to-metal contact with consequent—and usually catastrophic—pickup [158]; the very hard and abrasive oxide of aluminum makes things worse. In contrast, films on iron and copper surfaces have much extended lives, although soft copper surfaces may break up, allowing metal transfer and thus limiting film life (Fig.3.24).

Reactions are possible also between graphite and ceramics. Thus, a drop in friction on prolonged sliding in vacuum at elevated temperatures on Al_2O_3 surfaces [160] is attributed to the formation of oxicarbides and carbides of aluminum, and this could be a potentially useful mechanism in hot extrusion or metal cutting with ceramic dies.

Favorable surface roughness of one or both of the contacting materials helps the mechanical entrapment of layer-lattice compounds. It is, therefore, extremely important for both film formation and durability (Sec.3.8.5 and 3.8.6).

Layer-lattice compounds suffer from oxidation at elevated temperatures [161]; the rate of oxidation is greater for finer powder. Thus, very finely ground (600 m^2/g surface area) graphite powder begins to oxidize above 450 C with the formation of CO and CO_2, whereas coarse powder (3 m^2/g) survives to 600 C [141]. In metalworking applications, graphite trapped at the interface survives to much higher temperatures. Oxidation of MoS_2 begins at 300 C. The oxidation product MoO_3 is not a lubricant; it is an abrasive parting agent which melts at about 800 C and subsequently sublimes. On heating in vacuum, graphite loses the adsorbed condensible vapors and becomes a poor lubricant, but MoS_2 is useful to about 730 C, where it dissociates [20].

Film Properties. Shearing does not take place in a uniformly distributed manner throughout the deposited film. Instead, a transfer film forms if only the workpiece or die is initially lubricated, and sliding takes place (as with polymers; Fig.3.17b) between

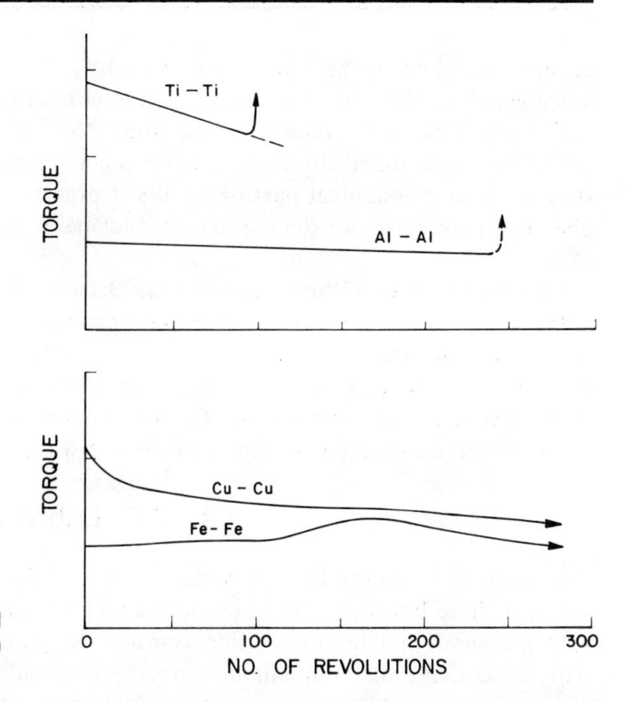

Fig.3.24. Interface friction (torque) and durability of MoS_2 in twist compression (after [158])

the original and transfer films. It is perhaps this similarity of operating mechanisms which leads to an observed pressure dependence of the shear strength of graphite and MoS_2, as in Eq.3.18 [114,162]. Even the typical coefficient of friction is of similar magnitude, around 0.05, although it depends markedly on the presence of humidity or other condensible vapors. Much higher friction can, of course, be measured on imperfectly formed films or on films that have been partially worn. Thus, even though μ is constant for both physically and chemically bonded films, the duration for which such low friction can be maintained is very much a function of factors discussed in the preceding subsection. It is conceivable that under certain conditions other mechanisms of action may also occur—for example, generation of roller-like agglomerations at the interface with perhaps a further lowering of μ. Under any circumstances, the shear strength of the films has no relation to that of the substrate, and therefore m has no application.

Other Substances. There is a great abundance of layer-lattice materials, not all of which are lubricants. Disulfides, diselenides, and ditellurides have lubricating properties if their structures are close to that of MoS_2. Jamison [163] based a review of criteria on chemical bonding and electronic structure, while Holinski [159] emphasized the importance of the polarization of the sulfur atom. Iodides are also potential lubricants [164], as are man-made modifications of other substances such as ferric chloride intercalated with graphite [165], MoS_2 and graphite with polymer chains grafted [166], graphite fluoride (CF_x) [167], and fluorinated mica [168]. The latter two have the advantage of being white, as is boron nitride, which had appeared very promising on the basis of its crystal structure yet turned out to be a poor lubricant [169] and now serves primarily as a parting agent that is stable to temperatures over 1000 C. The basic problem is adherence to the metal surfaces. Amorphous chalcogenides such as $SbSbS_4$ are promising solid additives [170].

A number of minerals also have layered structures but are not good lubricants (in fact, they are often abrasive) in the dry form. This can be attributed to mixed ionic-covalent bonding in the layers, aggravated by strong hydrogen bonds between layers of silicates. Once sheared, such layers cannot adhere to each other or to the metal surfaces and simply flake off without forming films. Nevertheless, many inorganic substances are found in compounded lubricants. Their mechanisms of action are obscure; most likely, they serve as mechanical parting agents if present in fine enough distribution, although chemical reactions with the surfaces cannot be excluded for some.

Interactions With Other Lubricants. Because film formation is aided by reactions with the metal surfaces, layer-lattice compounds are in competition with each other and with other surface-active lubricants for available surface sites. This will, in general, impair the lubricating performance of the layer-lattice compound when the other species is more reactive and the layer-lattice compound cannot attach itself to the new surface layer. However, synergistic effects are found with some E.P. additives [171].

3.4 E.P. Lubrication

The term E.P. stands for "extreme pressure" lubrication and is, as such, a misnomer, because these lubricants were developed for steel machinery elements in which high contact pressures and intense sliding combine to generate high localized temperatures on asperity contact, allowing chemical reactions to take place, presumably with the formation of organometallic compounds of low shear strength. Thus, E.P. lubrication is a version of solid-film lubrication with the difference that the low-shear-strength film is now largely limited to contact points. Quite often the shear strength of the film is not really low enough to justify its description as a lubricant (and occasionally it may even increase friction [172]), and its most important action may actually be that of wear prevention or mitigation. In industrial usage it is, therefore, quite common to distinguish between antiwear additives and E.P. lubricants.

3.4.1 Mechanism of Action

Organic E.P. compounds usually contain phosphorus, chlorine, sulfur, or combinations thereof, applied in a carrier such as mineral oil. The mechanisms of their action have long been investigated (as reviewed by Forbes [173] and by Kapsa and Martin [174]) but are not fully understood; the subject continues to raise more controversy as analytical techniques improve. The difficulty is that the surface films are not continuous and are thus not easily analyzed even by the most modern techniques [175]. Furthermore, interactions between lubricants, surfaces, and the environment prevailing in the test may change the mechanism itself; many investigations on reaction mechanisms are made in a nontribological environment where reactions may be different.

In the most general sense it is now known that earlier, simple explanations based exclusively on formation of metal phosphides, chlorides, and sulfides are oversimplified and even wrong. E.P. action proceeds by some (but not necessarily all) of the following steps:

1. Interactions of the additive with the environment (oxygen, water, other additives, and carrier fluid) produce a more reactive species.

2. The additive and/or the reactive species adsorbs on the metal surface.

3. Under the intense pressure and temperature conditions of contact, further reactive species may be formed, and bonds in the active molecule are broken to form polymeric films which usually show an increased concentration of the active element in the form of organic or organometallic compounds, often together with oxides. Remaining

organic radicals enter the bulk lubricant and may contribute to reactions under step 1. Rubbing promotes reaction [75] but also leads to the next step:

4. The reaction product is worn away by sliding, and also by chemical dissolution, and either is lost on the exiting workpiece or re-enters the bulk lubricant, changing its rheology and reactivity.

5. The reaction product is re-established by steps 1 to 3 above. Success of lubrication depends on maintaining a balance between removal and regeneration of the reaction product.

One thing is certain: the formation of an effective film takes time even at an elevated temperature, just as any other chemical reaction would do. Increasing the reactivity of the additive would shorten this time, but would also increase the danger of corrosive action on machine elements even on exposure at ambient temperature. Rubbing action and the new surfaces exposed in the course of asperity interaction are, however, enormously helpful. Even so, the time available for reaction on the workpiece surface is almost certainly insufficient in most metalworking processes. The die itself, however, makes contact for a long enough time. It is thus likely that in many metalworking operations it is the die or cutting tool, and not the workpiece surface, that forms a reaction product. A possible exception is drilling in which the hole surface is exposed to continued rubbing; the conditions are favorable enough to make drilling a useful method for E.P. studies.

Phosphorus-Containing Additives. There are innumerable organic phosphorus compounds that show powerful antiwear and more modest E.P. activity. For the most frequently investigated substance, tricresyl phosphate (TCP), it was believed that phosphoric acid was formed through hydrolysis and then reacted to form iron phosphate films. More recent work [176] indicates that the phosphate is mixed with iron oxides, and that the phosphorus is more likely present as an organic phosphate compound. It appears that neutral phosphate esters (including TCP) first adsorb on the surface, hydrolyze and/or decompose on contact to give some acid phosphates which then form the reaction product. The process is accelerated by moisture, by polar acid impurities, and by reaction products. The antiwear properties of phosphonates are similar, and hydrolysis produces a strongly adsorbed film (Fig.3.25 [177]). In contrast, weak phosphinic and phosphonic acid derivatives appear to be ineffective, and thus the presence of phosphorus does not in itself impart E.P. properties [173]. The nature of the base also affects performance: trioctyl phosphate is more reactive and thus more suitable for mild conditions than is TCP [176].

Test conditions greatly affect performance [178]; water is detrimental perhaps because uncontrolled hydrolysis produces excessive amounts of corrosive acids. Even the base oil has an effect: at low additive concentrations a naphthenic or aromatic oil is less favorable than a paraffinic one, most likely because the oil molecules compete for adsorption sites with the additive molecules [179].

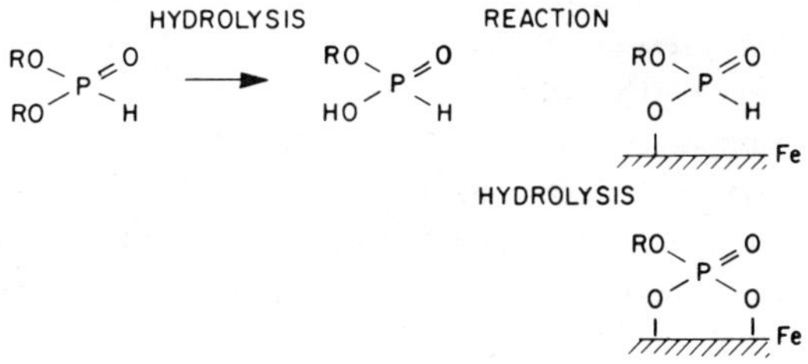

Fig.3.25. Possible mode of action of dialkylphosphonates (after [177])

When active products are present, the temperature for forming metal organophosphorus compounds can be quite low, and reactions are no doubt greatly accelerated under tribochemical conditions. The film is, however, worn away and/or dissolved at relatively low temperatures, around 350 C, and additives containing chlorine or sulfur are needed for higher temperatures.

Halogen-Containing Compounds. Halogen elements can be joined to organic molecules, yielding a large group of potential E.P. additives. Presumably, these additives react with the metal surface to form (in addition to other, less well-defined products) metal halides, many of which possess layered structures and have inherently low shear strengths. The halides form by tribochemical reactions during asperity encounters, and thus E.P. activity is directly related to the reactivity (and thus corrosivity) of the compounds. Reactivity is a function of the activity of the halogen, which can be judged from the bond strength between the carbon and halogen atoms. Within a given group of additives, those containing two or more substituted halogen atoms are generally more effective than monohalides [173].

In general, iodine compounds are the most reactive, and lubricants based on them have shown very good lubricating performance but also have resulted in serious corrosion problems. Reactivity is inversely proportional to the C-I bond strength, and elemental iodine is the most reactive [180]. It is a pity that corrosion is such a problem, because very small amounts of iodine can be effective even on metals such as titanium [181]. Iodides can be formed also directly by vapor-phase reaction [165], as can be other metal halides.

Next in activity are bromine compounds. The ability of these compounds to form E.P. films has been demonstrated [182] but has not been exploited, because the still adequately reactive chlorine compounds are much cheaper.

Chlorinated paraffins and aromatics are among the most widely used metalworking additives. They suffer hydrolysis in the presence of water, although it is not clear whether an intermediate acid is necessary to form the low-shear-strength film of metal chloride (or perhaps also a polymeric organometallic compound, a so-called friction polymer). Certainly, reactivity and corrosivity go hand in hand and decrease as the C-Cl bond strength increases in the order benzyl chloride < tertiary chloride < secondary chloride < primary chloride < aryl chloride [183]. Excessive reactivity could lead to serious corrosion problems, and therefore compounds with high hydrolytic stability have been advocated [184].

Sulfur-Containing Compounds. Elemental sulfur is a powerful additive. Presumably, on a steel surface some iron sulfide is formed; even though it has neither a lamellar structure nor a particularly low shear strength, it does have the advantage of not melting below 1100 C. With organo-sulfur compounds (such as sulfurized fats, olefins, and terpenes) there is considerable doubt as to whether the E.P. action should be attributed to iron sulfide or to some metal organo-sulfur compounds, or to both. Analysis reveals the presence of large proportions of iron oxide, conceivably formed as a result of oxidation of the sulfur reaction products. As with chlorine, the effectiveness of sulfur compounds is proportional to their reactivity [173,179], which in turn is determined by the strength of the sulfur-carbon bond. Monosulfides are less effective than disulfides, and reactivity increases in the series diphenyl < di-n-octyl < di-n-butyl < di-tert-butyl < di-benzyl [185]. The reaction mechanism is thought to involve adsorption on the steel surface, breaking of the S-S bond, and—at least at elevated temperatures—reaction of sulfur with the metal (Fig.3.26), with some of the freed thio radicals either entering the

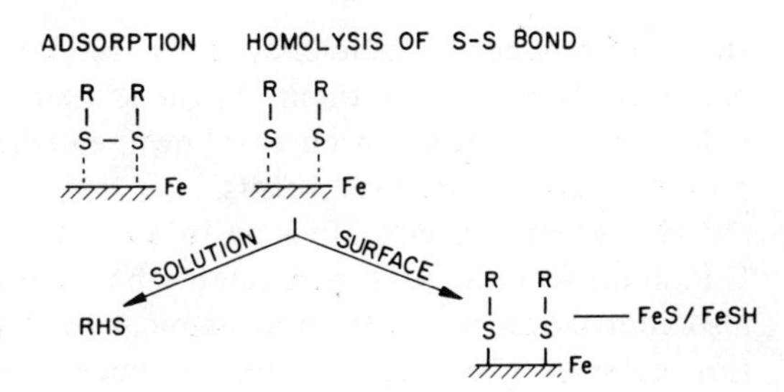

Fig.3.26. Possible mode of action of a disulfide (after [185])

bulk lubricant or reacting with the metal [185]. Polysulfides are more reactive yet [186].

Reactions are possible on other metals, too. With copper, effective lubrication is ensured only when dibenzyl disulfide is adsorbed as well as reacted [187]. On austenitic stainless steel the reaction product is oxide and, in severe wear, sulfide [188].

As with all lubricants, test conditions affect performance. The viscosity of the base oil has some effect even under ostensibly E.P. conditions, and the interface configuration and sliding speed affect the rate of replenishment [189].

Somewhat related are the borate ester additives. The effective film seems to be an organic carbon-base matrix in which iron, boron, and oxygen are dispersed, although recent work casts doubt on their effectiveness [190].

3.4.2 Complex Additives and Additive Interactions

Many E.P. additives contain more than one reactive element, and fully formulated lubricants also contain dispersants, detergents, antioxidants, and other potentially surface-active substances which can compete with the E.P. additives. The interactions are exceedingly complex and can be detrimental or beneficial, even for the same composition, depending on interface conditions [191,192].

1. The widely used and investigated additive zinc dialkyl dithiophosphate (ZDTP) forms complex surface films in which organic and inorganic phosphorus, sulfur, and oxygen compounds, including thiophosphate and ferrous sulfide films, are identifiable [193-195]. The films form on all surfaces and their composition varies with temperature [193]. The pasty consistency of these products must be a contributing factor in their effectiveness [194]. In static experiments, reaction with ferrous metal surfaces leads to the formation of sulfides and phosphides above 600 C [193]. The thermal stability and thus E.P. activity of metal dithiophosphates can be adjusted by changing either the metallic or the organic part of the compound.

2. Reaction products formed by various E.P. compounds break down at different temperatures, and therefore more than one additive is frequently used. A combination of sulfur and chlorine additives often produces better results than either of them alone; their synergism is ascribed to chemical interaction in the fluid which results in more rapid action on the metal [179]. Nevertheless, these two elements compete for surface sites [191,192,196]. Sulfur penetrates deeper than chlorine, and this may contribute to the better performance of sulfur under severe conditions.

In general, the efficiency of phosphorus compounds is impaired by the presence of chlorine compounds, perhaps because HCl is liberated and attacks the phosphorus surface product. Additives that compete for surface sites often reduce the efficiency of E.P. compounds, as was found with detergents, dispersants, boundary agents [197], and antioxidants [198]. The situation is further complicated by the interactions between reaction products and bulk fluid.

3. The base oil itself also has an effect. Aromatic molecules compete for active

sites and reduce the efficiency of the milder additives [199], whereas highly paraffinic base oils often enhance them. At the same time it must be noted that aromatic oils provide some E.P. function on their own. Furthermore, many base oils inherently contain powerful minor E.P. constituents.

4. The presence of air is an important factor. Oxygen itself may be regarded as an E.P. additive [200,201], presumably because it helps to form protective oxide films and also contributes to other tribochemical reactions and thus to the formation of the friction polymer [202,203]. Oxygen is necessary also for some additives (such as metal dithiophosphates), while others (such as dibenzyl disulfide) are effective even in an argon atmosphere [204].

5. Similarly, water (or humidity) has an important but often unpredictable influence on additive and base oil performance [191,192].

6. Most E.P. work has been carried out with iron and steel surfaces. Metal composition must have an effect [196,205], although no general principles have been established yet.

7. Surface roughness of the die affects local pressures and temperatures, and it should have a substantial influence on reactions that take place on asperities, while valleys provide a means of entrapping reaction products (Sec.3.8.6).

3.4.3 Modeling of E.P. Films

Little is known about the rheology of E.P. films. Extrapolation from experience with layer-lattice compounds could lead one to assume that shear strength should be pressure sensitive (Fig.3.19). Indeed, $\mu = 0.2$ has been reported [206] for various E.P. reaction products formed above 500 C under static conditions. No measurements seem to have been made on films formed by tribochemical reactions. In practice, only part of the surface is covered with E.P. films; troughs are filled with reaction products worn away from asperities and with the bulk lubricant used as a carrier. Thus, the interface really operates under mixed-film conditions (Sec.3.7). A model of E.P. film formation based on thermodynamic considerations has been offered [207].

3.5 Boundary Lubrication

The term "boundary lubrication" is often used to describe lubrication by any thin film, including E.P. lubricants and even layer-lattice compounds, but here we will use it in the narrower sense to mean lubrication by thin organic films physically adsorbed or chemisorbed on the metal surface. Such a film is thin enough to allow substantial asperity interaction. As shown in Fig.3.2 by dash-dot lines, the properties of the bulk lubricant (such as the viscosity of the boundary lubricant or the carrier fluid) play virtually no role. Chemical reactions with the surface may occur but are not essential as they are in E.P. lubrication. The mechanism of action is reasonably well understood. Earlier work is reviewed in [191,208-211], and the energetics of adsorption in [212-216].

3.5.1 Lubrication Mechanisms

Molecules of nonpolar fluids (such as pure mineral oils) attach themselves to metal surfaces by weak van der Waals forces. Such physical adsorption is sufficient to transmit shear stresses into the bulk fluid and is important for the development of hydrodynamic lubrication (Sec.3.6). The weak attachment means, however, that the film is readily wiped off and offers no real protection when asperities encounter each other (although unsaturated hydrocarbons afford better wear protection, indicating that carbon-metal bonds are formed). However, the low friction typical of true boundary lubrication is

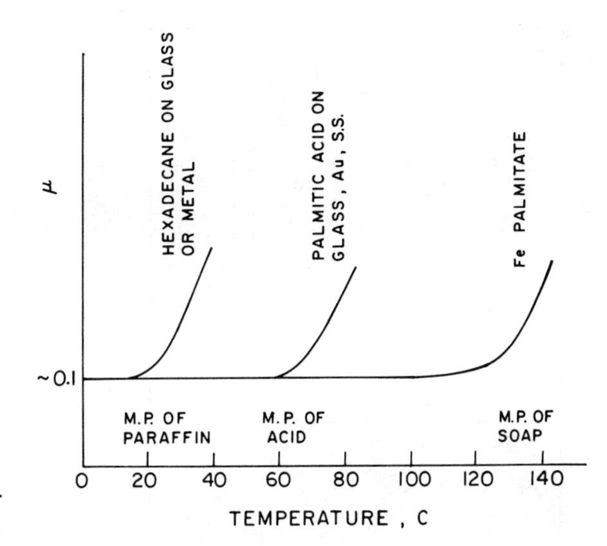

Fig.3.27. Critical (breakdown) temperatures

achieved only below the melting point of the hydrocarbon, and a rapid increase in friction takes place when the temperature characteristic of the lubricant (the transition temperature, in this instance the melting point) is reached (Fig.3.27).

The classical phenomena of boundary lubrication are exhibited by substances that contain molecules with permanent dipole moments, primarily derivatives of fatty oils, such as fatty acids, alcohols, and amines. They are characterized by a long hydrocarbon chain and a polar end (Fig.3.28). From the pioneering work of Hardy [217], Bowden and Tabor [18], and Rehbinder [218-220] it is known that their action follows several steps; these steps will now be discussed.

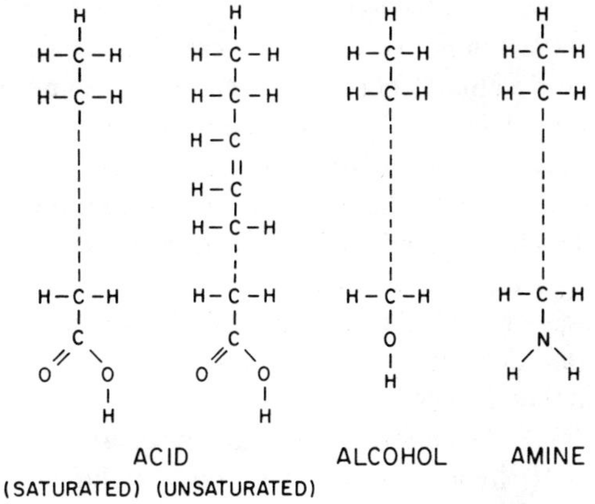

ACID (SATURATED) (UNSATURATED) ALCOHOL AMINE

Fig.3.28. Structures of fatty-oil derivatives

Adsorption.

1. Molecules of the polar liquid adsorb on the surface, forming an oriented layer; attraction between the polar group and active sites on the surface makes for strong physical adsorption (Fig.3.29) and lateral interaction between the molecules develops: a thin solid film forms and covers the surface. These aspects are discussed in great detail by Akhmatov [98]. Adsorption is aided by an oxide film and by the presence of water, possibly because in a dry substance the molecules of acid exist as linked pairs (dimers) which are then broken up by water. The adsorbed film exhibits a reasonably low shear strength, resists removal, limits junction growth, and protects against metal-to-metal contact. Controlled layers deposited by special techniques (such as retraction of monolayers

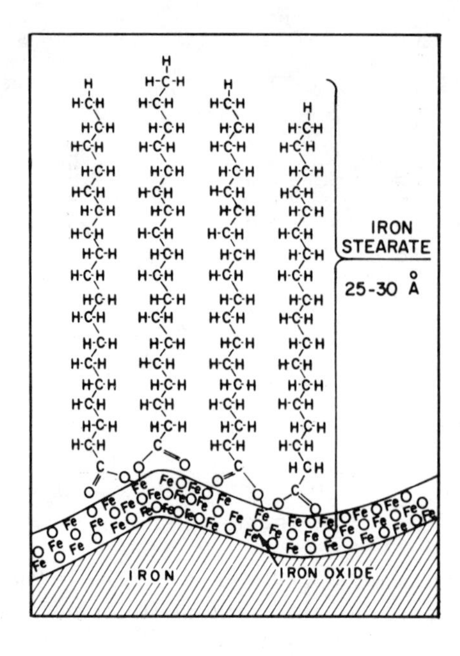

Fig.3.29. Adsorption of stearic acid on iron [208]

from the surface of water) show that a monomolecular film cannot resist removal indefinitely; multilayer films are much more durable. There is evidence that under practical conditions multimolecular layers build up, but with increasing disorientation on moving from the surface monolayer into the bulk of the fluid. On sliding, the adsorbed layers may align themselves as in a polymer [131] but remain attached to the surface, thus ensuring a low shear strength.

2. An alternative explanation offered by Rehbinder [218] suggests that adsorption of the boundary agent lowers the surface free energy and thus the flow stress of the metal. This effect is observed when some metal single crystals are deformed while submerged in a polar fluid such as oleic acid. It does not occur with all metals, and it is likely that the nature of the surface oxides has an effect. A strong oxide prevents the emergence of dislocations on the surface and increases the flow strength of the metal (Roscoe effect [221]), and an apparent softening occurs when these surface layers are removed by chemical polishing (Joffe effect [222]). An adsorbed fatty acid film could then change the strength of the oxide film by soap formation (see next section) and thus lower the observed flow strength or, in another sense, increase the plasticity of the surface [75,84].

3. Yet another mechanism of action has been suggested by Shaw [223] and Kohn [224]. In low-speed metal cutting the efficiency of boundary agents (and of some E.P. agents such as CCl_4) seems to be linked to a stabilization of microcracks in the chip, affecting the shear behavior of the metal. These phenomena could be regarded as specific examples of environment-sensitive fracture, reviewed by Westwood [225].

4. Effectiveness generally increases in the series ester-amine-alcohol-acid. Smaller concentrations of an acid are sufficient if the additive is used in a carrier (Fig.3.30) [226], suggesting again that adsorption is an essential feature. The work of adsorption has been repeatedly correlated with the effectiveness of lubricants [227-229]. However, the work of adsorption is difficult to determine, and mathematical models of boundary lubrication, reviewed by Beerbower [230], are less successful for friction than for wear.

The above explanations are not necessarily mutually exclusive; it is quite possible that in the complex events of boundary lubrication each of several mechanisms has a role. A quite unusual class of boundary lubricants comprises molten metals which have

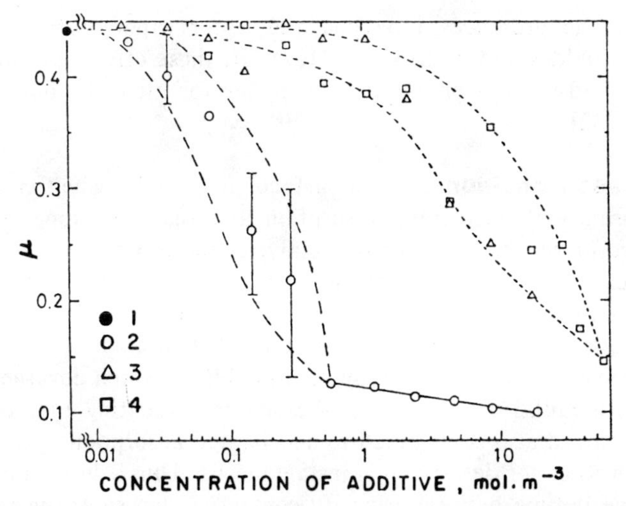

Fig.3.30. Effect of additive concentration at 30 C [226]. 1 - n-hexadecane; 2 - palmitic acid; 3 - cetyl amine; 4 - cetyl alcohol.

been found effective in drilling (Chapter 11), presumably by a controlled liquid-embrittlement process.

Desorption. When the temperature of the system is raised, the adsorbed layer becomes increasingly disoriented, and friction rises suddenly at a critical temperature. On a nonreactive surface (such as glass or gold) this is the melting point of the boundary agent (Fig.3.27). At yet higher temperatures, molecules desorb and enter into the solution, and the surface is unprotected.

If the metal surface is capable of chemical reaction with the fatty acid to form a metal soap, the transition temperature shifts to the melting point of the soap, indicating that reaction has indeed occurred (Fig.3.27 and 3.31). The reactivity of metal to fatty acid diminishes [231] in the order Pb > Cu > Zn > Cd > Ag > Fe,Ni,Al. Reaction

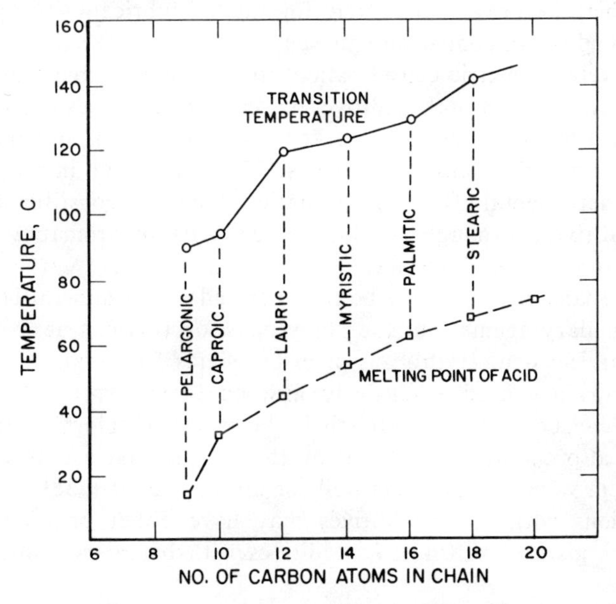

Fig.3.31. Critical (transition) temperatures for fatty acids on copper [18]

is accelerated by the presence of oxygen and water vapor, perhaps because of the greater activity of oxides and hydroxides. However, these effects are quite specific; for example, heats of adsorption on Fe_2O_3 are higher for alcohols, but on iron they are higher for acids [232].

Effects of Process Conditions. Fresh surfaces (especially when created by rubbing and plastic deformation) accelerate adsorption and reaction, apparently even in the absence of oxygen or water, as shown by catalytic studies on clean, fresh (but not rubbed) iron surfaces [233] and on freshly machined copper surfaces [234]. That plastic deformation accelerates adsorption was also shown in compression tests on zinc-plated steel specimens lubricated with radioactively tagged stearic acid [235]. The quantity of acid adsorbed increased linearly with reduction in height (which governed surface expansion) and also with radial distance from the center, indicating that the distance over which rubbing occurred also contributed to accelerated adsorption.

Neither adsorption nor desorption is instantaneous. Thus a lubricant may fail to protect the workpiece if time between asperity contacts is too short; on the other hand, a film formed prior to contact may survive even above the transition temperature [236]. This may explain the contradictory findings sometimes encountered with the same lubricants and materials in different processes, and it also points to the importance of lubricant application and carryover from previous production stages.

When a hard asperity plows through the boundary film previously deposited on a softer surface, the underlying metal is exposed [237]. Such a mechanism may account for the difficulties sometimes encountered under otherwise relatively mild sheet metalworking conditions.

3.5.2 Lubricant Interactions

Boundary lubricants are most often used in a carrier fluid, typically a mineral oil. Depending on the species, concentrations ranging from 0.1 to 5% of the active agent are usually sufficient to ensure effective boundary protection; the minimum effective concentration is lower for acids of longer chain length. This means, of course, that the polar molecules must compete for adsorption sites with the carrier fluid and any other molecules that may be present in a fully formulated lubricant. The properties of such mixed films depend on all components present:

1. From purely geometric considerations, it is to be expected that well-oriented surface films are formed when both additive and carrier molecules are straight-chain molecules of identical length. Indeed, a marked improvement in boundary lubrication has been observed for matched-length molecules [236]. Mismatch in shape may contribute to the observed detrimental effects of aromatic mineral oil constituents on the lubricity of boundary additives, although the higher reactivity of aromatics may also lead to increased competition for surface sites.

2. Neutral soaps or esters may be incorporated into a mineral oil in place of some or all of the boundary agents, but then only physisorption can be expected, and larger concentrations are needed. Hydrolysis of esters yields free acids, thus improving their attachment. Soaps are used extensively also on their own; at least some of their effectiveness is to be attributed to their favorable rheology, but some boundary attraction must also occur, as indicated by the finding that a soap completely fails in drawing of steel in vacuum but works well for drawing in air [238].

3. Interactions with other additives may have either beneficial or detrimental effects. The synergism of sulfur E.P. additives with boundary additives is well docu-

mented [239,240]; adsorption of the fatty acid in the presence of oxygen leads to much lower wear. The difficulty of generalizing on such observations is shown by the unfavorable interference between dibenzyl disulfide and surfactants [236]. Amines enhance lubrication with ZDTP at low concentrations but impair it at high concentrations [241]. Additives of combined E.P.-boundary character, such as chlorinated esters, alcohols, and acids, also show promise [242], with increasing effectiveness in the order presented. Alkyl- and alkylarylpolyethyleneoxyphosphate ester acids can be made oil or water soluble and also are of dual character [243].

4. The presence of a carrier fluid means that lubrication cannot be pure boundary lubrication but must be of the mixed-film type (Sec.3.7). The bulk viscosity of the lubricant helps to build a load-carrying film in which the influence of boundary additives becomes less significant, rendering boundary additives relatively ineffective in metalworking when added to high-viscosity base oils. Nevertheless, polar molecules help to build up and maintain an effective film [244].

5. A further complication arises from the potential interactions between the produced boundary film and bulk fluid. If the boundary additive or soap film is soluble in the carrier, its effectiveness is greatly reduced or eliminated [245]. The remarkable performance of the longer-chain fatty acid derivatives may well be attributable to the very low solubility of iron and aluminum soaps formed *in situ.* Oleic acid is, however, corrosive to copper and brass because of the high solubility of Cu-oleate in oil.

6. Tribochemical reactions in the presence of polar substances often lead to the formation of the so-called friction polymers [202,243,246]. They were first noted on electrical contacts and are often identifiable on boundary (and sometimes E.P.) lubricated machinery elements and metalworking dies. The exact mechanism of formation is not clear but appears to involve oxidation of the metal, which catalyzes the polymerization of the lubricant constituents under the intense heat and pressure [230] prevailing at the interface. The friction polymer has a very low shear strength and can actually cause difficulties in metalworking when some minimum friction is to be maintained, as in rolling. It may also account for some puzzling effects: with aluminum, for example, the strength of bond (heat of adsorption) which should contribute to lubricant effectiveness decreases in the order amine $>$ alcohol $>$ acid for a given chain length. Yet friction and wear diminish also in the same order, and acids are in practice the most powerful additives. An explanation suggested by Tripathi [247] invokes tribochemical reactions: the oxide-covered aluminum surface interacts with the mineral base oil to form a friction polymer, and the acid molecules then attach to this film. In the sliding of copper against steel, a porous, lubricant-impregnated transfer film is formed, which accounts for the very low friction often observed [248] even though the copper color of the steel surface indicates metal pickup.

7. Adsorption, reaction, and interaction between additive, bulk fluid, and reaction products make the metal surface a vital element in the success of boundary lubrication. Much depends also on the reactivity of the oxides, as has long been recognized for bearing materials. The SnO first formed on a Pb-Sn-Sb bearing metal is not reactive, but it is brittle and likely to break up, exposing the lead-rich substrate. The oxide of lead forms very effective boundary films which are, however, soluble in oil, and corrosion may ensue. The behavior of oxides offers a qualitative explanation of results for binary copper alloys [249]: the critical temperature drops greatly when the copper oxide changes to predominantly ZnO, but rises when additions of aluminum, silicon, or tin result in the formation of a harder oxide. Of course, reduced adhesion between these alloys and steel may also have an effect.

8. A basic problem with the study of boundary lubrication is that asperity tips may be lubricated by more than one mechanism, most notably by squeeze films (Sec.3.6) which then affect transition temperatures and can make lubricant performance sensitive also to surface roughness effects (Sec.3.8). The response of the lubricant to pressure then becomes also important, and in many applications the transition between boundary and EHD lubrication is quite gradual.

There have been many investigations of boundary lubricants in metalworking, and these will be discussed in Chapters 6 to 11.

3.5.3 Shear Properties of Boundary Films

The few available measurements that have been made of the shear strength of one or more monolayers of boundary additives [131,132] have shown shear strength to be a function of interface pressure (Fig.3.19). Thus, they can be modeled by a constant μ on the order of 0.05 to 0.1. Sliding takes place between the boundary films adsorbed on the contacting surfaces, rather like the sliding that occurs between the applied and transfer films found in lubrication with layer-lattice compounds and polymers. Thus, the similarity in rheological response is perhaps not surprising. Theoretical considerations based on interactions between molecular films [250] also indicate a constant μ, at least at high pressures. In other studies μ was found to increase with imposed pressure, but this was almost certainly due to metallic contact under the particular test conditions. At a given contact pressure, the shear strength of a soap film is almost independent of the number of monolayers, whereas the shear strength of stearic acid approaches that of the soap only when at least five monolayers are present. In such thicker films the initial orientation is rapidly destroyed and the chains realign on sliding, much the same as in thin polymer films. The temperature dependence of shear strength is also the same and follows Eq.3.19.

Instead of being formed *in situ,* metal soaps may be prepared as separate lubricants and be introduced to the interface. Their reactivity with the substrate then determines whether they will act as true boundary lubricants or simply as semisolid films. In either case, one possibility for modeling their behavior is to regard them as obeying Eq.3.18.

Since the shear strength of boundary films does not seem to be strain-rate sensitive, it is generally found that stick-slip phenomena (Sec.3.9.1) are also absent.

One is inclined to think of asperity contact always taking place in pure shear (Fig.3.7c). Yet in many instances die and workpiece asperities suffer a glancing encounter, and then the low elastic modulus of boundary films should help in spreading the load over a larger area and in cushioning the severity of the encounter [191].

3.6 Fluid-Film Lubrication

We have repeatedly mentioned that many metalworking lubricants are fluids which, under appropriate conditions, provide separation of the tool and die surfaces over some of the contact area while allowing boundary contact at other places. Before such mixed-film lubrication can be discussed, it is necessary to examine the extreme case where the two surfaces are fully separated by a fluid film. Conditions in the interface are then governed by the bulk properties (such as viscosity) of the lubricant; the chemical nature and reactivity of the lubricant and of the surfaces are immaterial. Factors entering into consideration are shown in Fig.3.2 (solid lines). This subject is well covered in specialized texts [251-253] and books [17,21-24,29-34] on tribology, and needs to be discussed here only very briefly.

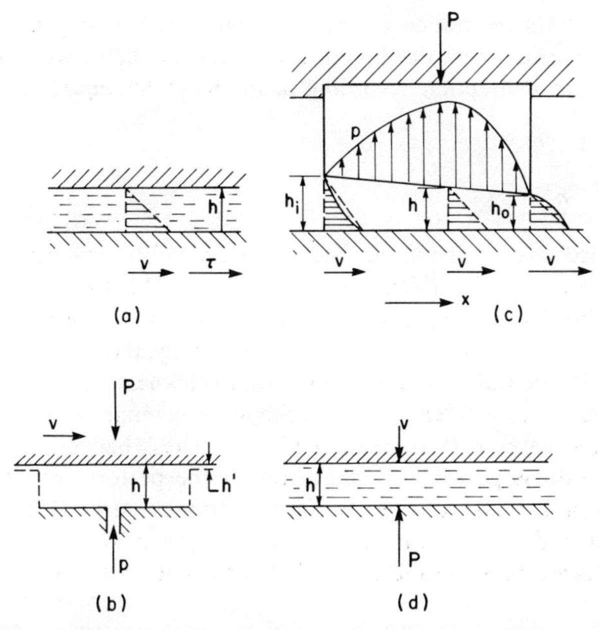

Fig.3.32. Hydrodynamic lubrication, illustrating (a) measurement of viscosity, (b) hydrostatic lubrication, (c) converging gap, and (d) approaching surfaces

3.6.1 Thick-Film (Full-Fluid-Film) Lubrication

In the simplest case, the surfaces of the two bodies (say a journal and a bearing) are separated by a film thick enough for surface roughness effects to be ignored, and the film geometry is simply defined by regarding the components as nondeforming, rigid bodies.

When the two surfaces are parallel and one is moved against the other at a relative velocity v, the fluid is dragged at the moving surface at the same velocity, and is stationary at the nonmoving surface (Fig.3.32a). If the fluid is Newtonian, the shear stress necessary for movement is

$$\tau = \eta \frac{dv}{dh} = \eta \dot{\gamma} \tag{3.20}$$

where h is local film thickness, $\dot{\gamma}$ is shear strain rate, and η is dynamic viscosity in units of $N \cdot s/m^2$ (= Pa·s). Dividing by density gives kinematic viscosity in units of m^2/s. For other viscosity units and conversions, see Appendix A.

If a load normal to the lubricant film were now applied, the film would collapse. To make the film load-bearing, one of two measures may be taken:

1. The lubricant may be supplied under sufficient pressure p to balance the applied load P over the bearing area (Fig.3.32b). This is called hydrostatic lubrication; because of heating, load-bearing capacity decreases with an imposed velocity v. The leakage rate of the lubricant is a function of η and gap thickness, hence a hydrostatic bearing usually has a much-reduced gap and a recess to which the fluid is supplied.

2. The pressure may be generated in the lubricant film itself by creation of a converging gap in the direction of movement (Fig.3.32c). The moving surface drags the lubricant into the gap at velocity v. If the fluid is incompressible and cannot leak out at

the sides, it encounters resistance in the converging gap, and generates an increasing pressure from h_i to $\bar{h}$; at the exit the oil is expelled as the pressure drops to ambient. Pressure varies in the x direction according to the Reynolds equation

$$\frac{dp}{dx} = 6\eta v \frac{h - \bar{h}}{h^3} \tag{3.21}$$

where η is the viscosity of the fluid under the prevailing conditions of pressure and temperature, and h is the film thickness at location x. This shows the essential features of hydrodynamic lubrication: for any given geometry, viscosity and sliding speed govern the load-bearing capacity, and the two are interchangeable in terms of their effects. When the system is allowed to select its own film thickness, it is found that a larger ηv or a lower load generates a thicker film. Although not immediately obvious, it can also be shown that μ increases with increasing film thickness because of increasing viscous drag. These observations permit rationalization of the performance of journal bearings in terms of a diagram, often referred to as the Stribeck curve (Fig.3.33), which shows the variation of μ and h as a function of the parameter $\eta v/p$ (or, from the German form, ZN/P), referred to also (and somewhat incorrectly) as the Sommerfeld parameter.

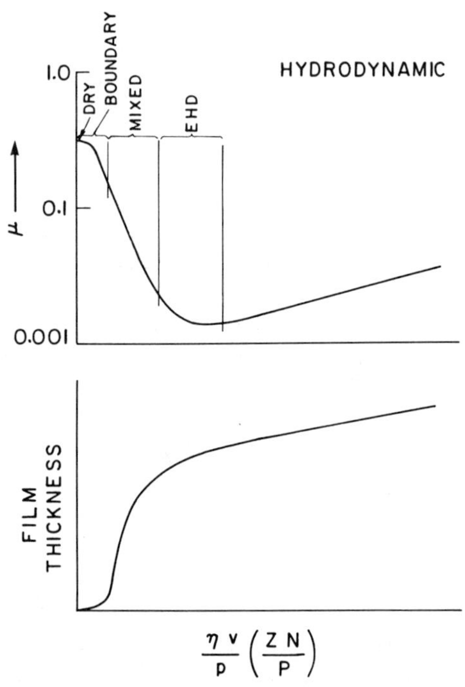

Fig.3.33. Stribeck curve

The converging gap necessary for generating a hydrodynamic load-bearing film, a so-called squeeze film, can be created also by two flat surfaces that approach each other at a given velocity v (Fig.3.32d). In many instances, the hydrodynamic effects created by relative sliding and process geometry are combined with the squeeze-film effect produced by the normal approach of surfaces.

Hydrodynamic lubrication is amenable to analysis and is, in many ways, the quantitatively most developed field of tribology. Recent advances even allow consideration of surface roughness effects [254]. In all instances, however, the film is thick enough to prevent asperity interaction; therefore, friction is low and, for clean fluids, wear is absent. The critical condition is that the minimum lubricant film thickness h_0 calculated for smooth surfaces should be much greater than the combined roughness of the two surfaces. It is usual to refer to the film thickness ratio

$$\lambda = \frac{h}{\Sigma} \tag{3.22}$$

where Σ is the composite surface roughness. Unfortunately, there is no agreement on its definition: it is taken variously as the sum of the AA (or RMS) values for the two surfaces, as the sum of the peak-to-valley heights, or as

$$\Sigma_c = \left(\sigma_1^2 + \sigma_2^2\right)^{1/2} \tag{3.23}$$

where σ_1 and σ_2 are the standard deviations of height of the two surfaces. Because individual asperities are much (typically 5 to 10 times) larger than the average roughness, asperity interaction would take place around $\lambda_c < 3$, and friction—and, even more so, wear—should rapidly increase. Yet many machine elements such as gears and roller bearings operate at $0.6 < \lambda_c < 3$ without excessive wear, even though μ does increase (Fig.3.33). Obviously, something must happen that makes actual film thickness or, rather, film thickness ratio, greater than that calculated from simple hydrodynamic theory.

3.6.2 EHD Lubrication

Two major factors combine to maintain thicker films:

1. Practical lubricants are not isoviscous fluids but exhibit increased viscosity (and even solidification) at high pressures.

2. No material surface is truly rigid, but deforms elastically to change the contact configuration.

The effects of these phenomena on film thickness and friction in concentrated contacts (also called conjunctions) are the subject of elastohydrodynamic (EHD) theory, which has seen vast development in the last 20 years [255-260]. Depending on whether elastic deformation or pressure-viscosity effects are more important, various theoretical solutions are applicable, as expressed conveniently by Johnson's map [261]. For our purpose, most important is the general principle that lubricant film thickness and traction (frictional force) can be calculated by simultaneously solving the Reynolds equation and the equations pertaining to the elastic deformation of a smooth, elastic surface, for a piezoviscous fluid (a liquid of pressure-dependent viscosity). For line contact such as that existing between gears, the results can be expressed with the aid of nondimensional groups. One of the possible forms [255] is

$$\frac{h_0}{r} = 1.6\left[\frac{\eta_0 v}{Er}\right]^{0.7}\left[\alpha E\right]^{0.6}\left[\frac{P/L}{Er}\right]^{-0.13} \tag{3.24}$$

where η_0 is viscosity at atmospheric pressure, r is equivalent radius, v is rolling velocity, E is equivalent elastic modulus (Eq.3.3), α is the pressure-viscosity exponent, and P/L is the load per unit width of contact. The most surprising result is that, under EHD conditions, the average film thickness is only slightly sensitive to pressure and, above $\lambda_c > 1$, is not too sensitive to surface roughness [94].

Theoretically more difficult but for our purpose even more important is lubrication at asperity tips, which, following Fowles [262], is referred to as micro-EHD lubrication. As in hydrodynamic lubrication, the fluid film builds up in response to two factors: normal approach of asperities creates localized squeeze films which cannot escape during the typically very brief contact time, and each asperity presents a separate converging gap in which the lubricant generates a load-bearing film. Thus, a moderately rough (say, $R = 0.3\ \mu m$) surface is better than either a smoother or a rougher one. The rheological properties of adsorbed layers make this a particularly effective microrheodynamic mechanism when boundary additives are present [209]. Qualitatively it is not difficult to see that not just the magnitude of surface roughness, but also its texture, should be important. A directional finish allows lubricant to escape in the direction of the valleys. When sliding takes place across ridges produced by directional surface preparation (as by surface grinding), the wedge effect is more powerful (and micro-EHD more effective) than when sliding is parallel to the grinding marks. Recent theoretical solutions correctly predict the effects of roughness direction [254,260].

It appears relatively simple to predict EHD film thicknesses (including thermal corrections [259]), partly because they depend so little on imposed load (Eq.3.24). The real difficulty lies in calculating the traction (frictional force), which depends very much on the shear properties of the lubricant under the particular conditions imposed by the sliding/rolling contact. The situation is complicated by the need for calculating the temperatures prevailing on the contact surface and in the film, and the effects of heat generation and conduction on these temperatures [255,259].

3.6.3 Lubricant Rheology

It is obvious from the preceding discussion that valid solutions to many lubrication problems can be obtained only if the response of the lubricant to the imposed high pressures, shear rates, and temperatures is known. This is the subject of lubricant rheology [263-266]. The mathematical expression of the true lubricant response can become almost hopelessly complex, but for our purpose it will be sufficient to understand the effect of the most important variables, some of which can be rationalized and sometimes related to the structure of fluids through various theories [136,267].

Pressure. The vast majority of data on the viscosity of specific lubricants is generated by measurements at atmospheric pressure. We have already seen that the shear strength of most solid organic lubricants increases with imposed pressure (Sec.3.3.3); in view of the more open molecular structure of fluids, it is not surprising that their viscosity should increase even more steeply as they are subjected to pressure. Of the various formulae proposed to fit the experimentally observed variations, the most convenient for our purpose is a power function (the Barus formula):

$$\eta = \eta_0 \exp(\alpha p) \tag{3.25}$$

where η is viscosity at pressure p, η_0 is viscosity at atmospheric pressure, and α is the pressure coefficient of viscosity (for selected values, see Sec.4.2.2). In general, a fluid consisting of uncomplicated straight-chain molecules is more densely packed, is less compressible (has a higher bulk modulus), and shows a lesser increase of viscosity with pressure than a substance of highly branched molecules or a fused ring structure. One can intuitively feel that viscosity cannot rise indefinitely; indeed, analysis based on fluid behavior ceases to be valid above $\eta \simeq 10^5$ Pa · s [268].

Data derived from experiments in high-pressure viscometers and, more recently, machines with EHD contact conditions indicate the gradual solidification of lubricating fluids on reaching some critical pressure. Both direct observation [269] and interpretation of experimental results [268,270,271] indicate that at this pressure the lubricant molecules become aligned, essentially as though they were solid-like polymers (alternatively they have also been visualized as granular substances [272]). In view of Eq.3.18 it is then not surprising that a pressure-dependent shear strength is found, which reduces to $\mu \simeq 0.1$ (Fig.3.19). This is also the typical value for polymers and boundary lubricants. Thus the lubricant at the tip of an asperity, although it may have been described as a boundary film in the past, may in reality be a micro-EHD film; from their shear properties, the two are indistinguishable, and the two regimes blend imperceptibly [209,273]. Thus, distinction between EHD and boundary lubrication may have to be made on the basis of wear rather than shear behavior.

Temperature. Viscosity has its roots in intermolecular attraction and is thus very sensitive to temperature. With increasing temperature, viscosity drops, increasingly so in lubricants of higher viscosities. Again, a large number of equations have been proposed. In addition, a very practically oriented system still used today compares the viscosities of the oil at two temperatures, 100 F (37.8 C) and 210 F (98.9 C), to the viscosities of two reference oils; a highly aromatic oil (which suffers a great viscosity loss) was arbitrarily assigned a viscosity index (V.I.) of 0, and a highly paraffinic oil (with a much smaller viscosity loss) was assigned a V.I. of 100. The properties of the oil in question are then expressed by a relative V.I. which may actually range outside the original boundaries of 0 and 100, especially for synthetic oils.

Temperature and viscosity have opposing effects, and their combined influence is most important. Again, many formulae exist, of which one in the form of a power law [66] is quite convenient:

$$\eta = \eta_0 \exp\left[\alpha p - \beta\left(T - T_0\right)\right] \tag{3.26}$$

where β is the temperature coefficient of viscosity, T is the interface temperature, and T_0 is the reference temperature (temperature at entry to the interface).

Non-Newtonian Flow. The definition of a Newtonian fluid (Eq.3.20) implies that viscosity is independent of shear strain rate (because the term dv/dh is actually strain rate) (Fig.3.34a, line A). Many fluids exhibit a more complex response, which can be conveniently (but not very enlighteningly) characterized by an apparent viscosity, which is simply viscosity at a specified shear rate. Non-Newtonian behavior may take many forms (Fig.3.34):

1. Dilatant fluids show an increase in viscosity with shear rate (Fig.3.34a, line B). This is typical of lubricants that are high in solid content. With the lubricant at rest,

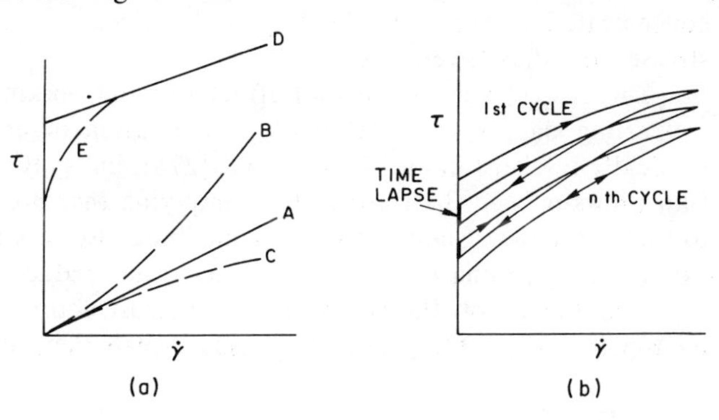

Fig.3.34. Non-Newtonian flow (a) and thixotropy (b)

the solid particles are closely packed and the fluid fills the interstices, giving a low initial shear strength. As shearing proceeds, the liquid is unable to fill all gaps, and particles interfere with each other, thus increasing the apparent viscosity.

2. Pseudoplastic substances suffer shear thinning (a loss of viscosity on shearing; Fig.3.34a, line C). Substances with randomly oriented long-chain molecules, including molten polymers, exhibit this trend, because flow is easier when the molecules become aligned on sliding. The trend is also observed in EHD contacts prior to the solid-like transition.

3. Bingham solids are stiffened by a structure that has to be broken down before flow can begin; a definite yield strength is first measured beyond which the stress is again a function of strain rate, as in a viscous fluid (Fig.3.34a, line D). Such substances are exemplified by greases in which the soap structure must be broken down before the properties of the base oil become evident. This helps a grease to build up a thicker initial film even in an EHD contact [274], especially at low speeds—a property that is valuable also in metalworking. On prolonged shearing and at increased shear rates, the properties of the base oil dominate, and the film may become even thinner than it would be with the oil alone because of lubricant starvation at the inlet. In a fully flooded contact, however, some greases maintain a thicker film [275]. Grease behavior can be very complex indeed, as when initial yielding is followed by pseudoplastic and then viscous flow (Fig.3.34a, line E).

4. Thixotropic substances suffer a temporary loss of viscosity on shearing. Their structure is broken on shear but reforms soon afterwards: viscosity attains a steady value if breakdown and rebuilding of the structure progress at a balanced rate (Fig.3.34b). A permanent loss of viscosity is experienced when the original structure is not rebuilt upon cessation of shear. This behavior is often found in polymer-thickened oils.

Wall Effect. The viscosity considered hitherto is bulk viscosity; this is not necessarily the viscosity of the fluid layer adsorbed on a metal surface. There is, indeed, mounting evidence that the effective viscosity increases as the film thickness diminishes and approaches molecular dimensions. Surprisingly enough, this effect has been found even with nonpolar lubricants; if polar additives are present, the multiple layers formed on the surface (Sec.3.5.1) provide a reasonable explanation for the relatively thick (around 0.1 μm) film observed when the bulk lubricant is blown off or a ball approaches a plate [236,276]. This is of particular importance because it would indicate that a semisolid film could exist even at pressures insufficient to cause solidification of the bulk lubricant, and may explain why many lubricants with boundary additives are so successful in reducing friction in metalworking even when the metal is not reactive. The oriented structure of boundary films makes them more viscous than the bulk fluid; shear is then concentrated in the bulk, leading to localized viscous heating which reduces shear stresses and thus lowers friction [277].

Direct evidence of the wall effect in metalworking is not available, but there are supporting data from EHD work. On addition of an organic phosphonate, the film gradually thickens in an EHD contact [278], but only at temperatures above 70 C. The film grows only in the rolling track, implying that pressure and/or shearing contribute to film formation. Such films may contribute also to a micro-EHD mechanism and thus extend the operating range of the lubricant and reduce wear.

The theory of fluids possessing microstructures (microfluids and, particularly, micropolar fluids [279]) may offer some explanations of the wall effect.

Viscoelasticity. In viscous flow, whether Newtonian or non-Newtonian, all deformation is irreversible; after the shear force is removed, the lubricant retains its new shape. There are, however, many materials that partially recover from the shape change, thus exhibiting a combined viscous and elastic behavior [266]. They usually can be modeled by various combinations of dashpots (for the viscous component) and springs (for the elastic component). One of the simplest is the Maxwell model (a spring and dashpot in series), in which a relaxation time t_R (in seconds) can be defined as

$$t_r = \frac{\eta}{G} \tag{3.27}$$

where G is the shear modulus of the substance. This has a clear physical meaning: when a shear strain is suddenly imposed on the lubricant, the stress thus generated takes t_R seconds to fall to $1/e$ of its original value (where e is the base of natural logarithm, 2.7183).

The relaxation time in liquid mineral oils is around 10^{-4} to 10^{-6} s. When the contact time in the deformation zone is of this order of magnitude, viscoelastic properties become significant. According to Johnson and Tevaarwerk [280], the critical quantity is the Deborah number

$$D = \left(\frac{\eta}{G}\right) \Big/ \left(\frac{a}{v}\right) \tag{3.28}$$

where a is the length of the contact zone and v is the sliding velocity. If the pressure and D are low, the lubricant is fluid and behaves in a viscous manner (Fig.3.35a); at high values of D elastic behavior dominates (Fig.3.35b). If, however, the pressure is high enough to make the lubricant behave like a solid, η/G reaches very high values (on the order of 15 min [268]), and a constant but pressure-dependent shear strength corresponding to $\mu = 0.1$ is found irrespective of contact time. Thus, full modeling of EHD films must take all these possibilities into account [280-282], including softening of the solid through heating (Fig.3.35c, broken line) as well as the influence of strain rate.

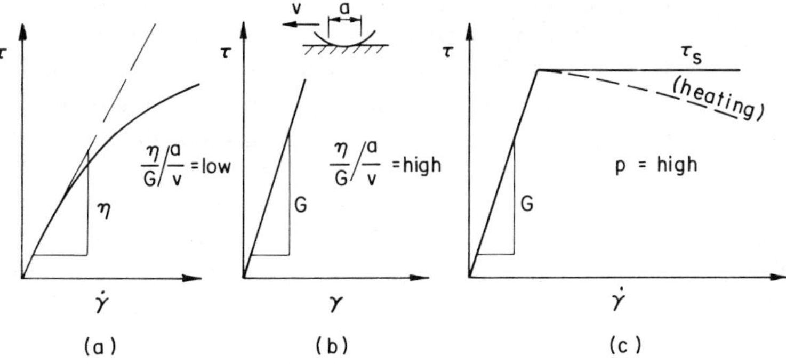

Fig.3.35. Modeling of lubricants in concentrated contacts [287]

The question has long been posed as to whether the viscosity increase and solidification due to pressure are instantaneous. Observations of density changes suggest that viscosity initially rises because of a squeezing out of free volume, but that molecular rearrangement—a time-dependent process—occurs on further pressure increase [283]. In contrast, time delay in response has been judged minimal or nonexistent from EHD work [284]. Evidence for solidification indicates a more gradual change but no substantial delay [270,285]. The question is still debated and is of considerable importance for metalworking, because contact times are short enough (especially in high-speed drawing or rolling) to be commensurate with possible time delays in response.

Phase Change. There is now little doubt that, under the influence of pressure, a transition from the liquid to a solid-like state takes place in many fluids. Less frequently, the reverse is encountered. Melting could be caused by pressure itself in a few substances such as ice, but a more likely source of melting is heat, either from the friction process itself [286] or from contact with a hot surface, as in glass-lubricated extrusion of metals (Sec.8.5.3). In the case of ice and glass the melt phase is a Newtonian fluid, but liquid phases of very complex rheology may form when a wax or soap melts.

Prior Loading (Shearing) History. Because the response of lubricants to stresses reflects their structure, the prior history of the lubricant may be important. Accurate description of properties may become too complex, and technological properties are then used to characterize them. Thus, it is usual to specify the "consistency" of a grease, which is essentially its resistance to deformation under the application of a force. It is expressed as the depth of penetration (in units of 0.1 mm) of a cone of standard shape under the application of a given weight for 5 s at 25 C (77 F). Penetrations are usually quoted for a grease that has been allowed to stand and for one that has been worked (sheared) in a standardized fashion [263].

Often it is difficult to decide what model fits best a given lubricant, partly because rheological response depends also on the particular test method employed; thus the model may be limited to conditions closely duplicating those of the test procedure. Nevertheless, modeling has seen considerable advances, covering a wide range of pressures, temperatures, and shear rates, and the many possible states (viscous, elastic, and plastic) of the same materials, and has led to successful predictions of traction in EHD contacts [287].

3.6.4 Plastohydrodynamic Lubrication

Under some very specific conditions, it is possible to maintain a full fluid film between die and workpiece materials in plastic deformation processes. The theoretical treatment of such plastohydrodynamic (PHD) lubrication generally involves the simultaneous solution of the equations governing hydrodynamic lubrication and plastic deformation. Metallurgical and chemical effects are ignored, and only variables of the lubricant film linked by solid lines in Fig.3.2 are considered. Because of the many similarities to EHD lubrication, much of the early theoretical work came from researchers active in that field. The treatment given here leans, to some extent, on summaries by Wilson [66,288]. Details are discussed in Chapters 6 to 10; at this point only general principles will be presented.

Lubricant Entrainment in Steady-State Processes. In the simplest case, the die and workpiece are assumed to be rigid prior to entry into the deformation zone, and then the entrained film thickness is

$$h_0 = \frac{6\eta v}{\sigma_f \tan\theta} \tag{3.29}$$

where v is the mean surface speed, σ_f is the flow strength of the workpiece material, and θ is the angle between the converging surfaces (Fig.3.36a). This shows that the entrained film thickness is larger for a more viscous lubricant, for a softer workpiece,

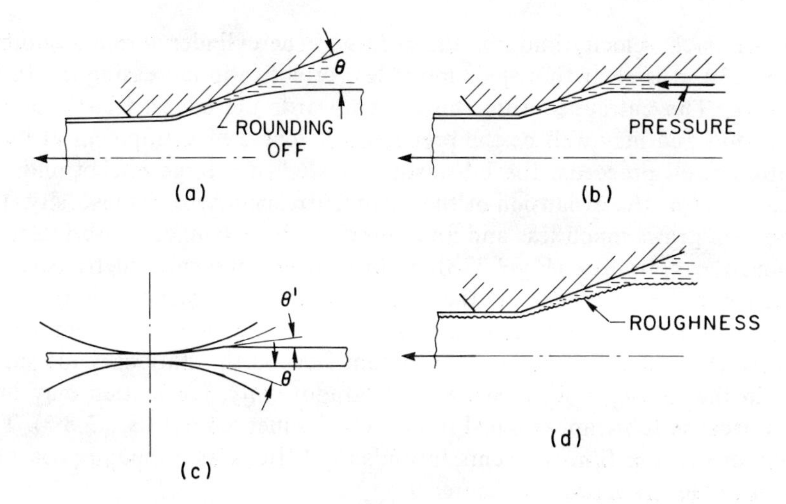

Fig.3.36. Factors promoting plastohydrodynamic lubrication

and for more gently converging surfaces, as would be expected from general considerations. The film thickness increases in steady-state processes if:

a. The fluid pressure at the die entry is increased (as in wire-drawing, hydrostatic extrusion, or deep drawing; Fig.3.36b),

b. The angle θ is reduced by elastic deflections of the die (as in rolling; Fig.3.36c),

c. The angle θ or the entire entry geometry changes by deformation (rounding) of the workpiece just before entry into the die (as in wire-drawing, extrusion, etc.; Fig.3.36a), or

d. The workpiece surface roughness helps to carry lubricant into the deformation zone (Fig.3.36d).

The entrained film cannot, in general, escape from the deformation zone, and thus thins out in proportion to the surface extension during deformation. There is, however, nothing to prevent lubricant escape at a free edge (as in sheet rolling) and film thickness drops quite steeply there. Substantial advances in analysis have been made and will be discussed in Chapters 6 to 8.

Non-Steady-State Processes. In other processes, exemplified by upsetting, a PHD squeeze film builds up as a result of normal approach. The thickest part of the film is in the center of the billet (Fig.3.37a):

$$h_{max} = \left[\frac{3\eta v r^2}{\sigma_f}\right]^{1/3} \tag{3.30}$$

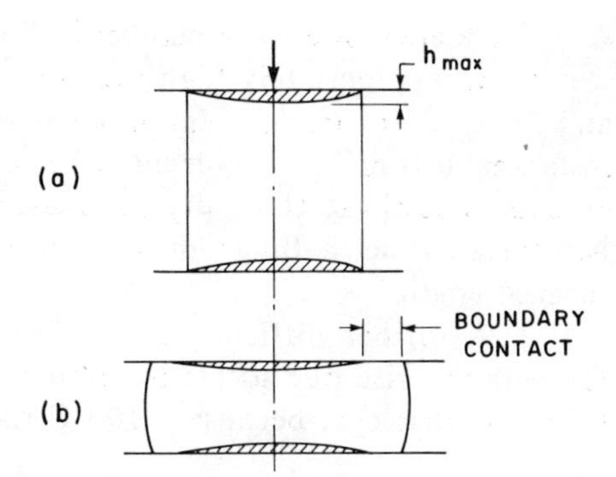

Fig.3.37. Squeeze-film development and breakdown in upsetting

where v is approach velocity and r is the radius of the cylinder. From a comparison of Eq.3.29 and 3.30, it is clear that ηv is much less effective in increasing the thickness of a squeeze film. The entrapped film thins out towards the edges. Elastic deflection of the die is helpful and may well be the predominant source of entrapment at low speeds.

As deformation proceeds, the lubricant is sealed in a large pocket and should, in the ideal case, follow the expansion of the surface irrespective of its viscosity. In reality, the squeeze film grows much less and an unlubricated (or boundary-lubricated) annulus forms around the periphery (Fig.3.37b). Lubricant entrapment and transport are promoted by a controlled roughness of the surface. Lubricant transport is reduced by the development of surface pockets during deformation (because they accommodate much of the lubricant); by the retardation of lubricant flow on the stationary die surface; and by heating in the lubricant film. Somewhat paradoxically, lubrication may improve in later stages because lubricant trapped in pockets is squeezed out (Sec.3.8.5). The realistic modeling of squeeze films presents formidable difficulties although good beginnings have been made, as discussed in Chapter 9.

By inspecting Eq.3.29 and 3.30 we find that, of the influencing factors, σ_f varies by about an order of magnitude for all technical metals, whereas η and v can vary by several orders of magnitude. Thus it should be possible to develop PHD films with any material, yet in practice one finds that this is almost impossible for harder alloys. There may be several reasons for this, linked to lubricant rheology and the effects of surface roughening.

Effect of Lubricant Rheology. The viscosity η in Eq.3.29 and 3.30 must be taken to reflect conditions prevailing in the process. Thus both viscosity and pressure response must be accounted for; Wilson [66] uses Eq.3.25 and 3.26, and Dowson and co-workers [289] use a more complex empirical formula.

Because of the marked effect of viscosity, the magnitude of the viscosity-pressure coefficient becomes a most powerful factor in determining film thickness. For lubricant entrainment by the wedge effect,

$$h_0 = \frac{6\eta_0 \alpha v}{\tan\theta\left[1 - \exp\left(-\alpha\sigma_f\right)\right]} \tag{3.29a}$$

where η_0 is viscosity at atmospheric pressure and α is the pressure coefficient of viscosity. For squeeze films,

$$h_{max} = \left[\frac{3\eta_0 \alpha v r^2}{1 - \exp\left(-\alpha\sigma_f\right)}\right]^{1/3} \tag{3.30a}$$

There are, however, a number of limitations:

1. At appropriately high pressures the lubricant solidifies, invalidating Eq.3.29a and 3.30a. A solution for entrainment with such a lubricant is available [66]. In many instances, it is difficult to decide what model best fits the lubricant under the particular process conditions; thus, dry soap used in wiredrawing could be modeled as a non-Newtonian fluid, a Bingham solid, or a solid of temperature- and pressure-dependent shear strength.

2. A further difficulty arises from the very marked effect of temperature. The temperature rise due to plastic deformation (Sec.2.3) is highly significant in terms of lubricant rheology, because a 100 C rise is not unusual at all, and this is sufficient to

change many semisolids into liquids (although pressure may change them back into solids). Temperature rises more at higher speeds of deformation, and therefore speed has a complex effect (Fig.3.38) [288]. At moderate speeds lubricant-film thickness increases according to Eq.3.29a. With a further increase in speed, the increasing temperature causes a drop in viscosity and, after reaching a maximum, film thickness diminishes until asperity contact occurs and the mixed-film regime is entered. Numerical [290,291] and analytical [292] solutions of thermal effects are in good agreement with experimental observations.

3. Shear heating of the lubricant film brings further complications. Because viscosity decreases with temperature, viscous heating tends to localize shear into a thin layer inside the thick film. In isothermal working the intense shear zone is in the middle of the film, but it moves towards the hotter surface (usually the workpiece) in nonisothermal operations [293].

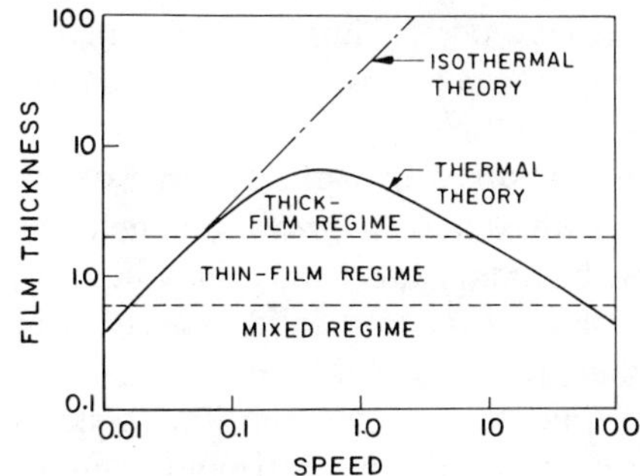

Fig.3.38. Effect of heating on plastohydrodynamic film thickness [288]

Surface Deformation. Even if the workpiece and die had perfectly smooth surfaces, a continuous film of liquid lubricant entrained or trapped at the interface would result in a roughening of the workpiece surface, because the workpiece deforms through the cushion of lubricant which acts as a pressure-transmitting medium. If the workpiece were amorphous and homogeneous, the lubricant film would—at least in theory—thin out in proportion to surface extension, as stated earlier. However, all real metals have a crystalline structure and this, combined with thick-film effects, leads to a marked overall roughening of the surface, as described in Sec.3.8.4. Excessive roughening is one of the reasons that true PHD lubrication is seldom desirable.

Applicability of Theory. The relevance of EHD research to plastohydrodynamic lubrication is obvious, and it is to be hoped that cross-fertilization will occur to a yet greater extent. The complexity of the problem often calls for numerical methods of solution, and the general approach is shared by both fields.

Theory has important contributions to make in two respects: first, the prediction of film thickness, which is so important for surface finish; and second, the prediction of frictional shear stress, which affects the mechanics of the process.

1. One can, with some difficulty, determine experimentally the lubricant throughput and thus check the average film thickness. The evidence is contradictory: some predictions seem to be fairly good (partly because of the insensitivity of film thickness to pressure), while others are off by almost an order of magnitude (partly because surface roughening is bound to make predictions based on uniform film thickness inaccurate).

2. The influence of lubrication on friction in the deformation zone is more difficult to check. The problem is that the friction contribution is usually less than 5% of the deformation force in a well-lubricated process and thus gets lost in the uncertainty of σ_f determination (Sec.2.1.3). For this reason, it would be quite a hopeless task to back-

calculate the magnitude of friction from measured forces. Only the unique and very delicate friction balance existing in rolling with negative forward slip offers a way of testing the validity of plastohydrodynamic theory. Experiments carried out under these conditions [294] showed that a theoretical prediction of issuing strip speed was, at that time, not yet possible. The difficulty most likely had two sources: first, thermal effects are extremely important, as suggested by subsequent analysis (Sec.6.3.1); and secondly, electrical resistance measurements always show some metal-to-metal contact, indicating that lubrication is actually bordering on the mixed-film regime, even though much of the surface may in fact be separated by a full fluid film. Traction must, obviously, be very sensitive to the smallest boundary contact contribution. The problem is likely to be universal in metalworking processes.

Despite these difficulties, theory can be valuable in identifying which factors have significant influences on the lubricant film and the process, and which directions these influences take. Thus, theory can contribute to the rational design of processes and operating conditions and, in the longer term, also to lubricant development.

Lubrication with Solids and Semisolids. The lubricant may be solid to begin with, and it may have a pressure-dependent shear flow stress (layer-lattice compound, polymer, boundary lubricant), may have a pressure-independent shear flow stress (metal), or may be a Bingham solid (grease, soap). Analysis of the problem follows some of the techniques used in PHD studies, and surface roughening is similar to that caused by a full fluid film. However, film thickness is now governed by the applied film thickness, the physical (mechanical) and chemical attachment of lubricant, and the process conditions:

a. In steady-state processes a gently converging entry (small θ in Fig.3.36a) allows a thicker film to enter, whereas a steep entry may cause the lubricant to be scraped off, and then firm attachment to the metal is vital (as is given for soap by a phosphate coating on steel). Once the lubricant enters the deformation zone, its average thickness is reduced in proportion to surface extension. In many instances, transfer films form (Fig.3.17b), shearing takes place between two films, and film thickness is less important as long as no metal-to-metal contact is allowed.

b. In a squeeze-film situation, a film of excessive thickness is reduced essentially by forging it between the rigid-elastic die and the initially only elastically loaded workpiece. Excess lubricant is squeezed out until the pressure required for its further deformation satisfies the yield criterion in the workpiece material. At this time bulk deformation of the workpiece begins; if the lubricant is firmly attached to the surface (as a metal film is by diffusion bonds, or a conversion coating is by virtue of its growth from the surface), it expands together with the workpiece surface. If attachment is weak, a nonlubricated band is formed at the edges (Fig.3.37b), the width of which depends on the spreading ability of the lubricant and its response to surface roughening during deformation. A film of less than equilibrium thickness can also support the pressures and thins out if strongly attached. Most of it may, however, become entrapped in a roughened surface, and this is often seen with an imperfectly formed layer-lattice film or with an unduly thin polymer film; expansion is then minimal. No general theoretical solution is possible to cover all possibilities, but specific cases have been tackled [295].

3.7 Mixed-Film Lubrication

In most metal deformation processes conducted with a liquid lubricant, film thickness is insufficient to provide a complete separation of surfaces. The situation has been variously described as quasihydrodynamic or thin-film lubrication, but the physical reality is best conveyed by the term "mixed-film lubrication." This term expresses the essence of a vast body of observations, according to which asperities are in boundary or micro-PHD/EHD contact whereas troughs (valleys) are filled with a liquid. The mixed-film regime is important also in nonmetalworking applications, and some beginnings have been made in modeling it, but plastic deformation of one contacting partner makes the case of metalworking substantially different. Because both PHD and boundary lubrication make contributions, all variables shown in Fig.3.2 are of importance.

3.7.1 Lubrication Mechanisms

Qualitatively, the mixed-film lubrication mechanism is fairly well understood.

1. A viscous fluid, compounded with boundary and/or E.P. additives, enters the deformation zone. The film thickness established as a result of initial conditions (Eq.3.29 and 3.30) is insufficient to maintain full die-workpiece separation, especially if favorable factors (Fig.3.36) are absent. Film thickness may even drop below calculated values under unfavorable conditions, such as when surface roughness is of the wrong kind or when a bulge forms ahead of the die at light reductions with high friction.

2. Differential yielding of the workpiece material results in local entrapment of a thick fluid film in surface pockets (Fig.3.39a, area *h*). Such pockets are visible on all surfaces deformed under mixed-film conditions (Fig.3.39b); their origin lies in hydrodynamic effects, and thus they are commonly called hydrodynamic pockets. As deformation proceeds, lubricant trapped in the pockets fails to keep up with surface extension, and thus an increasing proportion of the surface makes boundary contact (Fig.3.39a, area *b*).

3. Lubricant trapped in individual pockets behaves like a hydrostatic medium; at this stage, therefore, one can speak also of hydrostatic pockets. Upsetting experiments on grooved specimens (Sec.9.5.1) have shown that in the course of further deformation some of the lubricant feeds out into the adjacent boundary contact zones; at these zones hydrodynamic lubrication occurs on a very fine scale and could be termed micro-PHD lubrication of asperities. This may be induced also by sliding over previously established pockets, as shown in experiments with a special plane-strain compression test in which

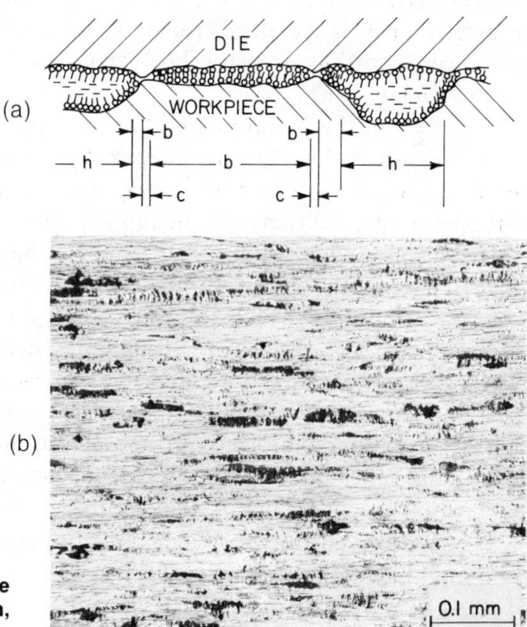

Fig.3.39. Mixed-film lubrication. (a) Interface [68]. (b) Rolled strip surface (magnification, 100×; shown here at 55%).

the top anvil can be moved laterally in the impression after compression has been completed [296]. During this subsequent sliding, the results with a wide variety of hydrocarbon oils and boundary lubricants can be rationalized in terms of τ_i versus ηv (with η taken at the operating pressure). This gives an ascending trend (Fig.3.40), indicating a full-fluid-film mechanism (compare Fig.3.33). This was confirmed by SEM photographs which, after sliding, showed the development of secondary hydrodynamic ridges on the plateaus that had been formed during the compression stage.

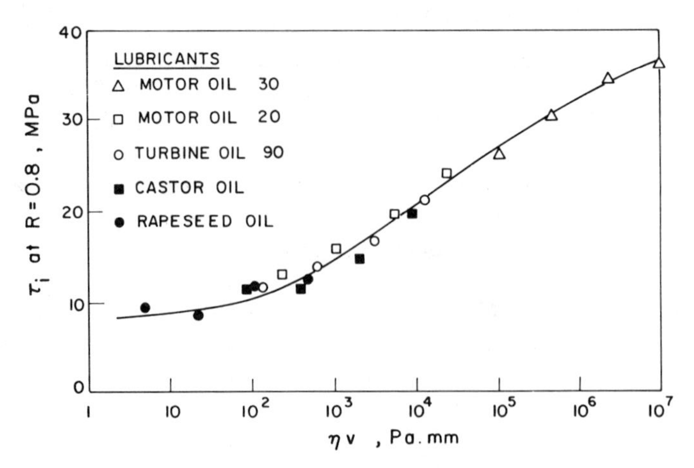

Fig.3.40. Interface shear strength with various lubricants on aluminum after 80% reduction in height [296]

4. The concept of asperities as defined for machinery elements is somewhat misleading for plastic deformation. A real interface looks more like that in Fig.3.39a (where the vertical scale is exaggerated). Asperity slope and radius have little meaning. Much of the interface (fraction b) is actually in boundary contact, so that features of the die surface finish are imprinted on the plateaus of the workpiece surface. The lubricant pockets (fraction h) show evidence of their formation and subsequent deformation. Electrical resistance measurements indicate metal-to-metal contact in all but a few instances, and thus even the boundary film is damaged at least over a small c fraction of the contact zone.

5. Few of the metal-to-metal contact points result in adhesion and the formation of cold pressure welds, because die/workpiece combinations are chosen for low adhesion. However, when welding does occur, some die pickup is unavoidable. Under many practical operating conditions the pickup is removed on subsequent contact, and the resulting wear particles enter the lubricant and can be filtered out. The site of the prior pickup is recoated with the lubricant, and a dynamic balance between pickup formation and removal is attained which results in some workpiece wear but still gives acceptable surface appearance. If, however, pickup is heavy and localized, subsequently contacted workpiece surfaces are scored, and the process may have to be terminated because of surface damage. In yet other instances, pickup on the die cannot be avoided; the aim then is to ensure uniform coverage of the die with a coating consisting of transferred metal and other products, and deformation proceeds by lubrication of this modified die surface (as in hot rolling of aluminum; see Sec.6.5.1).

6. Vast quantities of aqueous metalworking fluids are used in both plastic deformation and machining processes. Their mechanism of action is imperfectly understood but is most likely of the mixed-film type. Electrochemical effects, reviewed by Waterhouse [297], assume great importance. Some fluids are electrolytes, and a virgin metal surface

submerged in them develops an electrical double layer (Fig.3.41a): positively charged metal ions enter into solution by adsorption of water molecules, leaving behind a negatively charged metal surface onto which the aqua-cations are back-adsorbed. Hydrolysis leads to the formation of oxide/hydroxide films, and rubbing with its concomitant generation of new surfaces accelerates these processes, which could be regarded as a form of E.P. lubrication except that the presence of a liquid phase (even one of very low viscosity) may also promote some squeeze-film or hydrodynamic contribution. This is particularly true of oil-in-water emulsions, in which an oil is dispersed in water with the aid of emulsifying agents (Sec.4.6.2); such organic molecules have water-soluble and oil-soluble ends. Anionic emulsifiers, especially in conjunction with an oil-soluble, less-polar molecule such as a long-chain alcohol, form a strongly polarized surface film (a penetrated monolayer) on the oil droplet to make it adsorb (or, as it is often called in practice, plate out) on the metal surface (Fig.3.41b). This is a powerful mechanism, making some emulsions almost equivalent to the neat dispersed phase in their effect of lowering friction. The effect on film thickness is controversial. Concentrations of less than 5% gave a measurably increased EHD film thickness relative to plain water in the work of Dow [298], whereas Hamaguchi, Spikes and Cameron [299] found only negligible EHD films with oil-in-water emulsions. Most likely, the films are thinner and less continuous with an emulsion than they would be with the oily phase on its own; however, the greatly increasing cooling capacity of water often justifies the slight sacrifice in friction and wear reduction. Water-in-oil emulsions (Sec.4.6.2) have properties closer to those of the oily phase, and result in EHD films that are as thick as those resulting from use of the neat oil [298].

The breakdown of lubrication in metalworking is analogous to film failure in concentrated contacts, as in four-ball or pin-on-ring tests. Czichos [22] has shown that failure (manifested in suddenly increasing friction and wear) occurs when a critical combination of pressure p, velocity v, time t, and temperature T is reached:

$$p^a v^b t^c T = C \qquad (3.31)$$

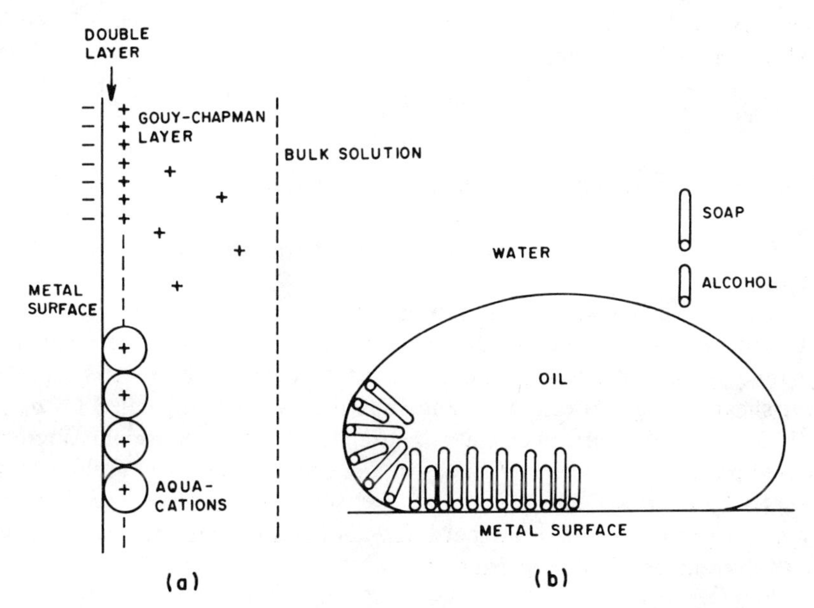

Fig.3.41. Aqueous lubricants [297]. (a) Development of double layer. (b) Adsorption of emulsion particle.

where a, b, c, and C are empirical constants, and where lubricant viscosity enters through its effect on b. This relation defines a failure surface and, with some modifications, could find application to metalworking.

3.7.2 Modeling of Mixed Films

The mathematical description of a surface lubricated by a mixed film represents formidable difficulties. The problem is not too great as far as load-bearing capacity is concerned, because liquid in the hydrodynamic pockets carries the load as well as do the boundary contact areas. The real problem lies in calculating the average shear strength τ_i of the interface.

1. First of all, the surface areas lubricated by fluid films h (or the fraction $b + c$) has to be predicted (Fig.3.39a). For simple asperity profiles, this can be done [300,301]. In the presence of a liquid lubricant, the change in boundary contact area is governed mostly by the bulk modulus of the lubricant and is rather insensitive to the shape of the individual asperity or to the shear strength of the boundary film. Such analysis, however, has to resort to asperities of idealized shape.

Other solutions follow a different approach. Starting with the roughness height distribution of the workpiece surface, a mean lubricant film thickness can be calculated from thick-film theory, and then the intercept between the roughness profile and the thus calculated tool position can be found to give the fraction of direct contact surface, $b + c$ [301]. Alternatively, one can proceed by assuming that when the film thickness diminishes sufficiently for boundary contact to occur, the mean film thickness is equal to the combined roughness of the die and workpiece surfaces. From thick-film theory, the pressure carried in the lubricant pockets can be calculated, and, from a knowledge of the required pressure to be supported, the h fraction can be found [66]. A difficulty with all these approaches is that the b fraction changes during the process and, indeed, roughness develops in response to the lubricated deformation process itself. At this time, only experimental methods of determining $b + c$ can yield accurate values.

2. Even if the $b + c$ area is found, difficulties are not over. To calculate the average interface shear strength τ_i of the mixed film, the appropriate shear strengths existing in the areas h, b, and c must be known. Following the argument of Sec.3.2.5, the average shear stress in the interface is

$$\tau_i = h\tau_h + b\tau_b + c\tau_c \tag{3.32}$$

where τ_h is the shear strength of the hydrodynamic film, τ_b is the shear strength of the boundary film, and τ_c is the shear strength of the areas in nominal dry contact (i.e., not covered by a boundary film of known shear strength, but not so clean that welding takes place).

The expression is not very useful in this form, because it does not relate to physical reality. Friction in the hydrodynamic pockets is best expressed by μ_h, calculated from hydrodynamic theory. We have seen that most boundary lubricants have a pressure-dependent shear strength (Sec.3.5.3), and thus they are best modeled by μ_b (typically 0.05 to 0.15). The greatest uncertainty is attached to τ_c, because this factor is influenced by the presence of imperfect films. In any event, it has little to do with the shear flow strength k of the workpiece material, and thus some arbitrary coefficient of friction may be assigned to it. Then the behavior of the entire interface can be described in terms of an average coefficient of friction:

$$\mu_{\text{mix}} = h\mu_h + b\mu_b + c\mu_c \tag{3.33}$$

If, over some a fraction of the surface, contact results in actual metal pickup, $\tau_a = k$ and

$$\mu_{mix} = h\mu_h + b\mu_b + c\mu_c + a(k/p) \quad (3.34)$$

Even though a may be small, its contribution can be quite significant, especially at low p. This may well account for the rather high μ_{mix} (and wear) occasionally encountered under relatively mild interface conditions.

If the lubricant entrapped in the h area solidifies under pressure, its shear strength becomes indistinguishable from those of boundary lubricants (Sec.3.6.3), and the entire surface behaves as though it were covered with a film of pressure-dependent shear strength, expressed by $\mu = 0.05 - 0.1$. This may be the reason for the surprisingly large number of observations in which μ was of this order of magnitude even under ostensibly widely varying conditions. If, on the other hand, pressures are insufficient to cause solidification, μ_h becomes a vanishingly small contribution to the total, and μ is governed by the $b + c$ "real area of contact," often denoted simply as b. Then

$$\mu_{mix} = b\mu_b + (1 - b)\mu_h \quad (3.34a)$$

—a form often used in simplified analyses, and originally proposed by Vogelpohl [302]. When the pressures p_b and p_h are not equal, the average μ becomes

$$\mu_{mix} = \mu_b \left[\frac{1}{1 + \left(\frac{1 - b}{b}\right) \frac{p_h}{p_b}} \right] \quad (3.34b)$$

3.7.3 Regimes of Lubrication Mechanisms

The preceding discussion showed that lubricant film thickness and coefficient of friction respond to variables which can be conveniently summarized with the help of a Stribeck curve.

For nonmetalworking situations, hydrodynamic and EHD lubrication covers much of the $\eta v/p$ range (Fig.3.33), and the boundary and EHD regimes merge through a relatively small zone of mixed-film lubrication. The concept of an average film thickness is meaningful because even EHD films are thick enough for surface separation.

The regimes of lubrication mechanisms have been rationalized by Schey [115] for metalworking with the aid of a modified Stribeck diagram in which due allowance is made for the plastic deformation of the workpiece. A map of lubrication regimes can be constructed for an interface pressure $p = \sigma_f$ (Fig.3.42); remembering that the calculated coefficient of friction drops with increasing interface pressure (Sec.2.2.1), a three-dimensional diagram would have to be constructed for other pressures. In Fig.3.42 a single cut (A-A) is shown at the point where the coefficient of friction is minimum; to compress the curves, the interface pressure is plotted on a logarithmic scale. Similarly, the wide range of μ values is accommodated on a logarithmic ordinate.

a. Beginning on the left-hand side, with low η and/or v, hydrodynamic effects are discouraged, the entrapped film thickness is insufficient to generate hydrodynamic pockets, and only boundary lubricant films separate the surfaces. The tool or die surface finish is faithfully reproduced (Fig.3.43a), although roughening may occur if weak junctions are formed and the surface is smeared (Fig.3.43b) [303].

b. With increasing ηv, the entrapped film thickness increases, some hydrodynamic pockets form, and we enter into the mixed-film regime (Fig.3.42). One cannot really

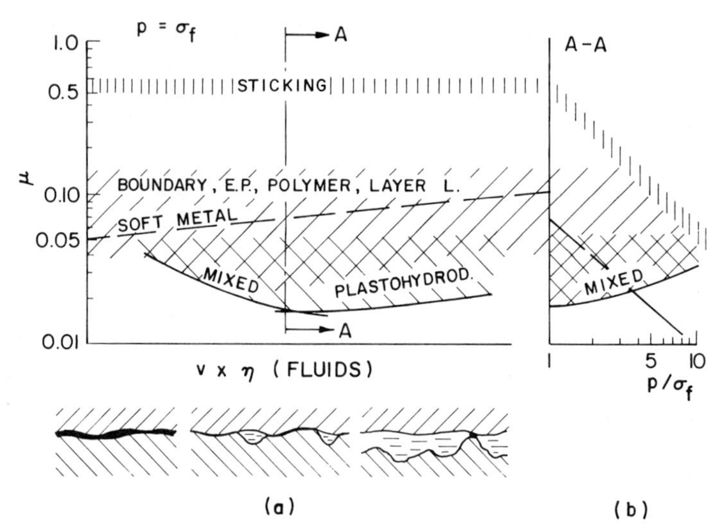

Fig.3.42. Modified Stribeck diagram for plastic deformation [115]

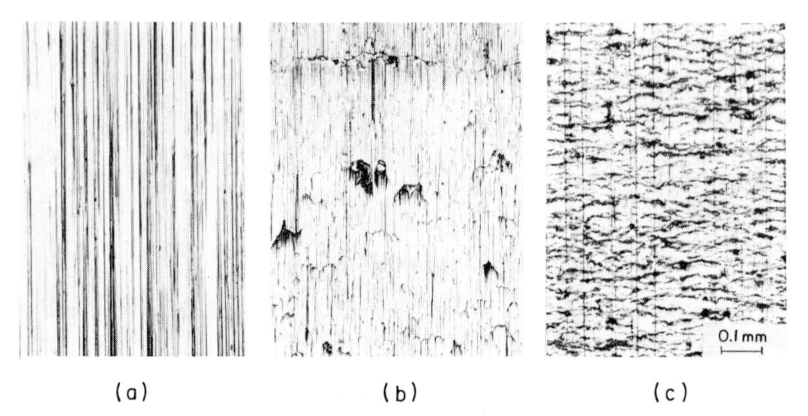

Fig.3.43. Appearances of deformed surfaces [303]. (a) Boundary lubrication. (b) Smearing due to lubricant breakdown. (c) Hydrodynamic lubrication.

speak of an average film thickness because of the very large variations from boundary-lubricated contact zones to deep (typically 1 to 5 μm) pockets (Fig.3.39). The coefficient of friction drops rapidly, in response to the decreasing b portion.

c. Once the entire surface is covered with a hydrodynamic film, the minimum coefficient of friction is reached, but the surface will now be substantially roughened (Fig.3.43c [303]). Upon further increase of ηv, increasing drag in the hydrodynamic film causes a slight increase in μ just as it does in a journal bearing. This increase may be cancelled by the drop in viscosity resulting from heating at higher rates of shear. In practice, PHD lubrication is rare, and most processes are lubricated by a mixed film which may change from predominantly boundary to predominantly hydrodynamic as ηv increases, with a corresponding drop in μ.

d. The effect of pressure on the lubrication mechanism is complex. For any given value of ηv, increasing interface pressure reduces the entrained film thickness (Eq.3.29 and 3.30). Furthermore, increased pressure is often the consequence of heavier deformation and thus larger surface extension; therefore, the b area increases, and the lubrication mechanism shifts yet further to predominantly boundary. Thus, an increasing pressure has the opposite effect of increasing ηv, and the coefficient of friction usually increases (Fig.3.42b), because b increases more rapidly than μ would decrease as a result of calculating with a higher interface pressure (Eq.2.8). In principle, however, μ_{mix} could increase, stay constant, or even decrease. Once friction is high enough to immobilize the interface (sticking friction), μ_{max} drops, as discussed in Sec.2.2.1 and Fig.2.3. Thus, it could be extremely misleading to cite any particular value of μ (measured or calculated) as evidence of the prevalence of a given lubricating mechanism; a low μ may simply mean that interface pressures were high.

e. Until now we have concentrated on situations where $p > \sigma_f$. There are, however, many instances in which $p < \sigma_f$ even though the workpiece deforms plastically (Fig.3.4 to 3.6). At very light pressures, contact is established over asperity tips only; any fluid present moves freely around and does not share in supporting the load (Fig.3.44). Thus, the system behaves as though it were a case of dry (or boundary-lubricated and dry) friction: as the pressure increases, so does the real contact area $(b + c)$ and, according to Eq.3.1, the average μ remains constant, as recognized by Kasuga and Yamaguchi [304] and by Kloos [72]. Gradually, the fluid becomes pressurized and the entrapped lubricant limits the growth of the contact area $(b + c)$ with increasing pressure. Since at low interface pressures the average μ_h in the hydrostatic pockets is negligible compared with μ_b and μ_c, and because $b + c$ does not change, the average μ will actually drop (Fig.3.42b) (Eq.3.34b). If, however, pressure is further increased and the stress state induces bulk plastic flow with surface expansion, $b + c$ increases too and μ_{mix} either remains constant or, if the friction mechanism shifts to even greater boundary contribution, rises, as in Fig.3.42a. Even relatively small changes in the lubrication mechanism may cause μ_{mix} to rise or drop with pressure, and it is for this reason that apparently contradictory trends are often reported in the literature.

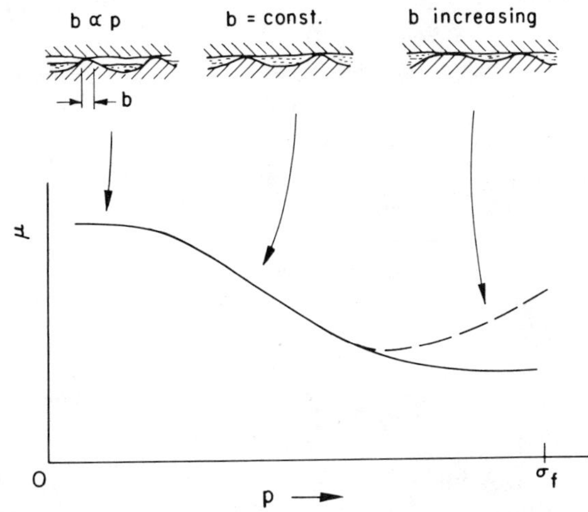

Fig.3.44. Variation of coefficient of friction at low interface pressures

The only reliable guide to the lubrication mechanism is the evidence offered by the surface itself. A uniformly roughened surface must have been deformed by a hydrodynamic or solid (polymer or metal) film capable of following the deformation of individual crystals. A surface that bears the imprints of the die surface was deformed through an effective boundary film. Localized pockets are evidence of mixed-film lubrication, whereas localized smeared or torn features indicate local lubrication breakdown and metal-to-metal contact with subsequent shearing and, if the process allows, smearing of the junctions.

3.8 Surface Roughness Effects

We have seen that the metalworking process contributes to changes in the surface configuration of the workpiece in both plastic deformation (for which the lubrication mechanism is the controlling factor) and machining (where the process itself is the primary determinant of the resulting roughness, and lubrication enters into it as a second-order effect). All workpieces are, of course, prepared by one of these techniques, and dies too are produced by one or several of the metalworking processes. Thus, in a real interface, the die and workpiece roughnesses reflect their processing history, and the lubrication mechanism will be influenced by and will interact with these microgeometrical factors in an often decisive way for the success of the operation. Before reviewing these interactions on the basis of an article by Schey [305], we must introduce some concepts dealing with characterization of the roughness of a surface.

3.8.1 Characterization of Surface Roughness

Even the most carefully manufactured die or workpiece shows deviations from its nominal dimensions and shape, even on a macro scale, but these large-scale effects seldom enter into lubrication considerations. Much more important is that, on the finer scale, the surface exhibits waviness and roughness. The method of measuring these will be briefly discussed in Sec.5.7.1. Here we will limit our attention to the results of measurement by a stylus instrument, because they are most suitable for quantitative analysis. In this measurement, an arm with a reference rest (skid) is drawn across the surface, while a stylus follows the finer surface details (Fig.3.45 [2.11]). The surface profile can then be recorded after amplification, usually distorted (with a larger gain on the vertical axis), or the signal may be electronically processed to give numbers that characterize the surface, or it may be digitized for further manipulation in a computer. The surface profile defines a number of characteristics [306].

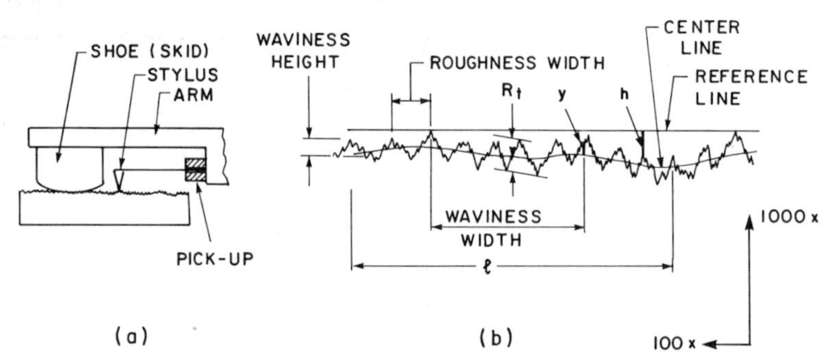

Fig.3.45. Measurement (a) and characterization (b) of surface roughness [2.11]

a. A line, drawn in such a way that the area filled with material equals the area of unfilled portions, defines the centerline or mean surface. The average deviation from this mean surface is called the centerline average (CLA) or arithmetical average (AA), denoted also as R_a:

$$R_a = \frac{1}{l} \int_0^l |y| \, dl \qquad (3.35)$$

The root mean square (RMS), also denoted as Rq, used to be preferred in practice and is now again gaining favor (because of its application in Eq.3.23):

$$R_q = \left[\frac{1}{l}\int_0^l y^2\, dl\right]^{1/2} \tag{3.36}$$

Instruments measuring average values integrate over a predetermined travel distance known as the cutoff length, and CLA or RMS values can be meaningful only if the cutoff length is large enough compared to surface features. As a rule of thumb, at least 15 to 20 waves should be traversed so that the effect of waviness is included in the averages (this calls for a cutoff of about 2.5 mm on rougher surfaces).

b. The peak-to-valley height R_t is often of greater interest, especially if there are localized asperities or troughs that fail to show up in the averages yet may pierce through the lubricant film or provide channels for the lubricant to leak away. Often a more meaningful figure is obtained by taking the average height difference between the five highest peaks and the five deepest valleys within the sampling length (ten-point height R_z).

c. Alternatively, a reference surface may be established, essentially by taking the path that the skid surface travels in the course of surface roughness measurement (Fig.3.45) or, if no skid is used, by taking the path through the highest peaks. The mean distance of the actual surface from this reference surface then defines the leveling or smoothing depth R_p, a concept widely used in Germany:

$$R_p = \frac{1}{l}\int_0^l h_i\, dl \tag{3.37}$$

This has the benefit that waviness is automatically included in the integrated value.

d. Height values in themselves convey little information. From the tribological point of view, however, their distribution is important. For a given peak-to-valley height, the profile may have quite a different character (Fig.3.46 [307]), depending on asperity radius and slope and on roughness width. Waviness can cause local variations in the entry angle and thus affect lubricant film thickness (Fig.3.36a).

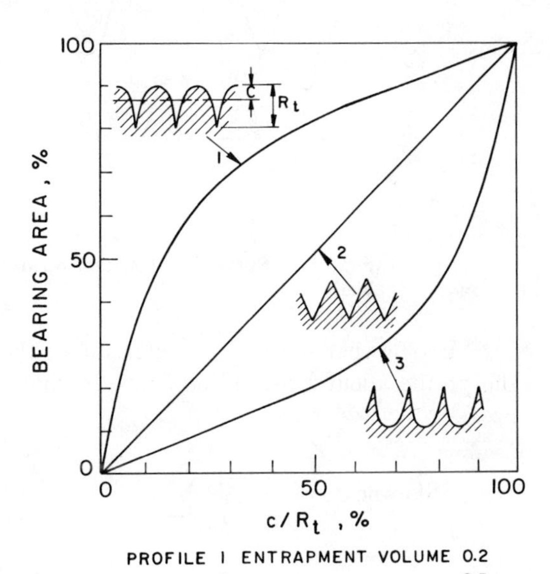

Fig.3.46. Bearing area for constant peak-to-valley roughness (after [307])

e. The surface roughness may be of random orientation or may have a marked directionality [2.11,308-310]. Some finishes have a unidirectional lay—for example, those produced by grinding, turning, side milling, or rolling on ground rolls. Others have a two-directional lay, such as those produced by Blanchard grinding or face milling. Still others are random, such as those produced by lapping, shot blasting, EDM machining, or rolling on shot-blasted rolls. Cast and forged surfaces are usually random but tend to have more uneven asperity distributions than machined surfaces.

f. For a given roughness profile, it is possible to derive numbers of immediate physical significance. First, by taking a line at c distance from the reference surface (Fig.3.46), the bearing area—on which the load would be carried if asperities were squashed or worn—can be determined (provided, of course, that the asperity is a linear ridge). The influence of asperity radius and slope and of roughness width is then clearly visible (Fig.3.46). The cumulative bearing area curve or Abbot curve (Fig.3.47) is of limited utility for our purpose. Secondly, the fullness of the profile can be determined. This measure is extremely useful in that it gives direct information on the ability of the surface to entrap lubricants or accommodate wear particles. A profile of greater fullness provides less entrapment volume, as shown for idealized unidirectional finishes in Fig.3.46. If asperities have a limited length perpendicular to the plane of the paper, entrapment volume increases. Thus ground surfaces, and, especially, shot-blasted surfaces, entrap much more lubricant than a turned surface of the same roughness. To determine the spatial free volume (entrapment volume), at least two traces must be obtained (in the direction of lay and transverse to it); only on a statistically random surface will a single trace contain all the information. Five nonparallel traces are needed to define a nonisotropic surface.

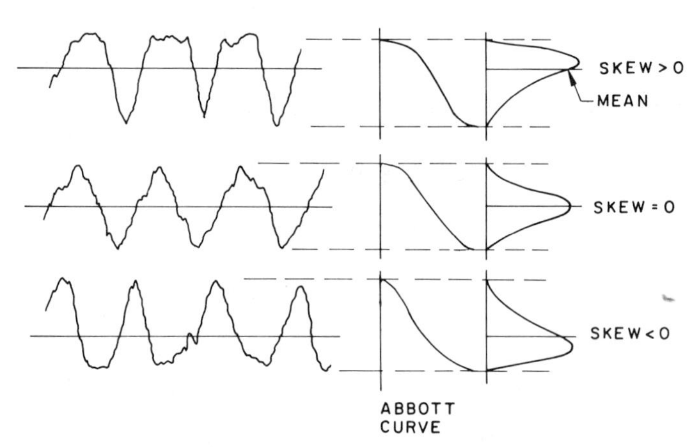

Fig.3.47. Surface profile, bearing area, and symmetry about mean [after 305]

A somewhat similar characteristic of the profile is skewness—i.e., the symmetry of the profile about a mean line (thus, in contrast to fullness, it does not encompass waviness).

$$\text{Skewness} = \frac{1}{R_q^3}\,\frac{1}{n}\sum_{1}^{n} y^3 \tag{3.38}$$

where n is the number of coodinates (y values) chosen in the record length. A perfectly random height distribution has zero skewness. Positive skewness indicates flat or large-radius asperities, whereas negative skewness is typical of surfaces with much free space (Fig.3.47 [306]).

Most stylus-type instruments used in production plants give readings only for R_a values; if the number of peaks within a specified distance is also given, the information becomes much more meaningful.

3.8.2 Types of Surfaces

It would be of considerable help if the enormous variety of roughness profiles encountered could be described by numerical values. However, progress in applying such techniques to manufacturing has been slow, partly because there is a gradual change of surface roughness in the course of the deformation or machining sequence itself.

The starting material is very often a cast, forged, extruded, rolled, or drawn workpiece which may have a highly irregular surface roughness distribution. Nonperiodic tool and score marks can make any average roughness designation meaningless.

When the workpiece surface is prepared by machining or etching that produces a more or less Gaussian distribution, the cumulative distribution of roughness heights plots as a straight line (Fig.3.48 [309]). This, however, is not likely to survive long, because deformation of the workpiece, coupled with the presence of a mixed lubricant film, rapidly changes the height distribution, as shown by the curving of lines in Fig.3.48.

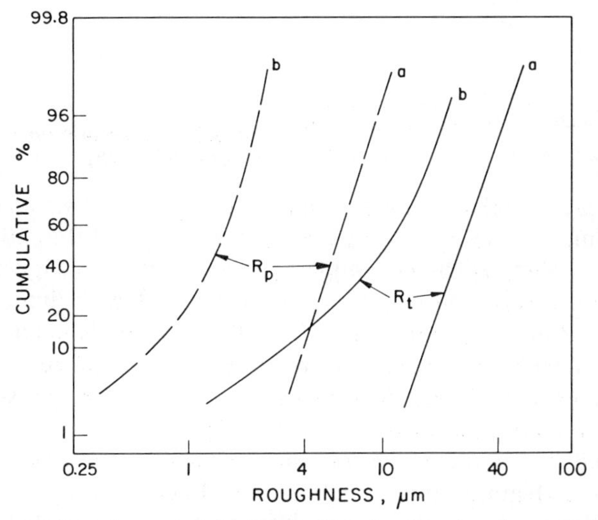

Fig.3.48. Roughness-height distributions for (a) annealed and pickled bar and (b) drawn bar [309], [7.255].

Sometimes the workpiece surface is deformed by a tool which itself shows a nearly Gaussian distribution. If the lubricant film is very thin, this distribution may be preserved on the workpiece. However, even rolling mill rolls (which are ground with exceptional care) show wide deviations from the Gaussian distribution, even if ground with the same type of wheel but by different operators on different machines [311].

There are some surfaces for which the only description is the actual surface profile (or its digital version). Others, however, are sufficiently random to be treated by statistical techniques, even though random and periodic components may have to be separated. The classification of surfaces through such characteristics has already shown promise for machined surfaces [312]. It must be remembered, though, that the surface profile obtained by a stylus instrument does not contain all the details of the actual surface, and this may have special significance for metalworking.

3.8.3 Roughening of Free Surfaces

In some deformation processes, parts of the workpiece surface deform without contacting a tool or die. The free surface roughens in a mode that is qualitatively and quantitatively similar to that observed on a lubricated surface of a deforming specimen and is, therefore, of basic interest. Roughening depends on the deformation mode (Fig.3.49) [309,313,314]:

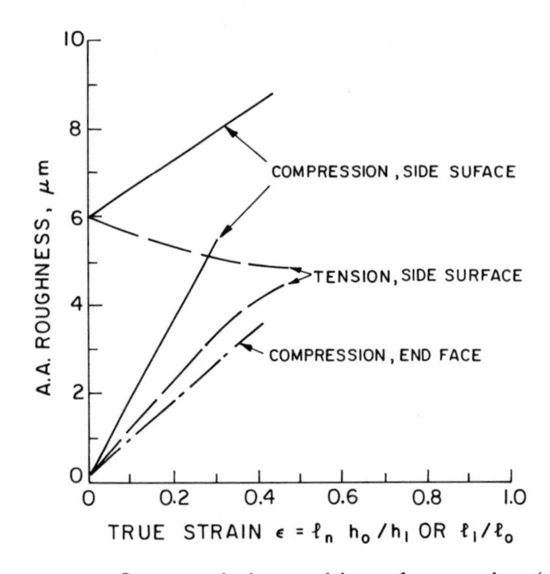

Fig.3.49. Change in roughness of surfaces with deformation [305]

a. On a workpiece subjected to tension (stretching), initially smooth surfaces (those with R_a = 0.1 μm, for example) roughen because of the differential deformation of individual grains, as dictated by crystallographic orientation. Roughening is greater for larger strains, for coarser-grain materials, and for metals with limited numbers of slip systems. Thus, HCP metals roughen most, and FCC materials much less; BCC materials are somewhat smoother yet because of their very large numbers of operative slip systems [313]. However, initially rough surfaces (e.g., R_a = 6 μm) get somewhat smoother because the surface features are stretched out [308].

b. In upsetting of a cylinder, the free surface roughens irrespective of initial roughness, because surface features are pushed closer and become deeper [314]. When differences in surface area change are taken into account, roughening for all modes of deformation becomes a function of natural strain ε [313].

c. In some processes a combination of extension and compression may occur. Thus, in bending of sheet, the outer surface changes as in tension, whereas the inner, compressed surface changes as in upsetting.

d. The human eye is a very critical judge of surface finish, and localized roughening may lead to a visible contrast, especially on sheet products (Sec.6.3.5), in the form of roping, streaks, bands, or, when the deformation mode changes direction, curved patterns (as, for example, parabolic markings on deep-drawn cups). They can usually be traced back to structural features.

3.8.4 Roughening of Lubricated Surfaces

In discussing PHD lubrication (Sec.3.6.4) we saw that deformation through a lubricant cushion results in surface roughening. In principle, the effect is the same whether the lubricant is liquid, semisolid, or solid, as long as it can deform as an amorphous substance. Roughening involves the following steps:

a. At very light strains, slip planes develop and their termination at the interface is visible as a fine structure of parallel lines, especially on single crystals and annealed polycrystalline materials [315-317].

b. If a hard oxide film is present (as on aluminum alloys), the lubricant film causes it to break up; with increasing ηv and deformation the breakup spreads over a larger area [317].

c. Even though during homogeneous deformation all parts of the entire body must undergo the same shape change, individual grains deform differentially, as dictated by their crystallographic orientations and the constraint imposed by surrounding grains. Grains in which easy shear directions are favorably oriented deform somewhat earlier; pockets of lubricant form, and the surface roughens. An analysis of the roughening of an initially smooth surface has been made by Wilson [318].

d. When the liquid film is thick enough to ensure complete die-workpiece separation, the surface roughness is the same as it would be in free deformation. Annealed materials with equiaxed grain structures develop a nondirectional roughness. Tokizawa [315] observed that individual pockets correspond to single grains in FCC and BCC materials (such as Cu, Al, and Fe) with their many slip systems, but a pocket extends over several grains in HCP materials (such as Mg and Zn). In two-phase materials, the pockets first develop in the softer phase or in softer zones between hard precipitate particles [319]. The pockets remain equiaxed in a squeeze-film situation (upsetting [320-323]) but are elongated—and, on heavy reduction, broken up—in proportion to the elongation of the workpiece in steady-state processes.

e. Strain-hardened materials have elongated grain structures in which crystallographic alignment (texture or directionality of properties) is often present. Individual grains are strain hardened and yield at about the same pressure. In a squeeze-film situation the pockets reflect structural features; when the interface is perpendicular to the alignment, pockets will again be roughly equiaxed, but when the interface is parallel to the alignment, pockets too are elongated. For the same lubrication conditions, pockets are smaller and more broken up than on an annealed material. In steady-state processes the pockets are affected by process geometry and grain alignment. In general, they are elongated in the direction of major deformation (Fig. 3.39b), and within the pockets transverse fissures—which seem to have their origins in slip bands [316]—develop. On heavier deformation the transverse features become dominant [303,316,323].

f. In principle, it does not matter whether the continuous film was established through PHD effects or interposed as a polymer or metal film. In all instances, if the film is too thick, the surface may roughen beyond acceptable limits. Lubrication instabilities may also develop and result in objectionable variations in gage or surface finish.

g. Lubricant trapped in the depressions of the surface is practically incompressible and survives intact to very heavy deformations. The pockets themselves are remarkably stable even after removal of lubricant residues and can be eliminated only by taking substantial further deformation in a boundary-lubricated process.

h. Roughening on the scale of the grain size, combined with the superimposed finer roughness stemming from the termination of slip planes on the surface, make for a surface with many light-reflecting facets. The dull, matte appearance of the surface can be quite attractive, but is seldom exploited in practice because it is difficult to reproduce.

All the above observations apply to originally smooth workpieces deformed in contact with smooth dies, and represent an idealization. In reality, the surface roughness generated in one process step affects lubrication in the follow-up processes. This is true not only of operations performed in several successive passes, such as rolling, or drawing of wire and tube, but also of sequences of different processes. An obvious example is the effect of sheet surface finish on subsequent performance in deep drawing and pressing operations.

3.8.5 Effects of Workpiece Surface Roughness

Roughness features of a workpiece surface have a powerful influence on the process and on the issuing product.

Effect on Lubrication Mechanism.

a. Surface roughness helps the mechanical entrapment of lubricants [238]. The quantity entrapped depends on the volume of free space available in the troughs; thus, for a given peak-to-valley roughness, a less full profile (Fig.3.46) is preferable. This has led to the specification of the number of peaks per unit length in addition to average roughness for sheet metal to be used in automotive pressing (Sec.10.4.3). Entrapment is a powerful mechanism for solid and liquid lubricants alike, because asperities provide a mechanical keying action which may be all-important when the lubricant or surface is nonreactive.

b. Surface roughness increases the lubricant intake into the work zone, and thus for a given liquid lubricant and ηv the mechanism shifts from predominantly boundary to predominantly hydrodynamic. For a solid lubricant, the average film thickness retained increases.

c. Within the deformation zone, the lubricant entrapped becomes pressurized, and roughness gives rise to micro-PHD lubrication; this effect becomes most important when process configuration and local sliding conditions would wipe off the lubricant from a very smooth surface. Roughness also helps to retain solid and semisolid lubricants in the deformation zone, and aids the transport of solid lubricants [324].

d. A rougher finish is more tolerant of variations in production conditions because it can accommodate wear particles.

e. The above effects are greatly influenced by the lay of surface roughness. Surfaces with a lay parallel to the sliding direction allow escape of the lubricant much more than a lay perpendicular to sliding. Therefore, lubricant intake, lubricant retention, and the micro-PHD lubrication mechanism are all more powerful if roughness features are oriented perpendicular to the sliding direction. Modeling by Sargent and Tsao [311] shows that the film thickness h_0 calculated from isothermal theory has to be adjusted by a factor which is a function of the angle θ between the roughness lay and rolling direction and of the film thickness relative to the roughness of the surface (Fig.3.50). Film thickness is greater when the lay is perpendicular to the sliding direction, but it is lower when the lay is parallel. This was indeed found experimentally in rolling with negative forward slip (Sec.6.3.1) [294]. When the direction of sliding changes in the course of deformation, as it often does in sheet metalworking, the best results are obtained with a random surface which, even though it represents a compromise between parallel and transverse lay, ensures a nondirectionality of response.

Lubricant entrapment generally results in reduced friction; therefore, moderate workpiece surface roughness tends to reduce forces and allows greater reductions to be taken. Numerous examples are found in rolling and in wire and bar drawing.

Detrimental Effects. We must emphasize that surface roughness is not universally beneficial, for a number of reasons:

a. For a given film thickness, a rougher surface will give more boundary contact,

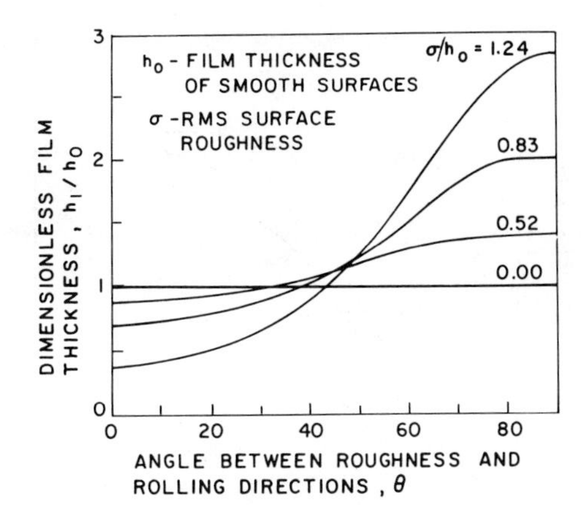

Fig.3.50. Effect of roughness orientation on inlet oil film thickness in rolling [311]

and the benefit of lubricant entrapment may be lost by asperities piercing through the film.

b. The effect of roughness orientation can be reversed too; a perpendicular lay allows more asperity contact under full fluid conditions than a parallel lay [294].

c. Transverse roughness may lead to a folding over of asperity peaks [325], resulting in surface defects which may be difficult to detect but will unfavorably influence the performance of the workpiece in service. Parallel-oriented surface features also may close up if the lubricant has a chance to escape; asperity ridges lose their support, become deformed by upsetting, and are finally folded into the surface.

d. Reduced friction may mean a loss of desirable process control features. Thus, it is sometimes found that an otherwise stable rolling process becomes unstable (the strip may not enter the roll gap, or may skid sideways) when the incoming strip surface becomes too rough. Similarly, bottom fracture may occur in deep drawing if friction on the punch drops below minimum.

e. An excessively thick fluid or semisolid film may cause instabilities and consequent objectionable variations in gage or surface finish. A too-rough surface may give an unacceptable appearance, and leads to high polishing costs.

Effect on Issuing Roughness. The interaction of the initial roughness of the workpiece with the lubrication mechanism may cause the roughness to increase, decrease, or remain unchanged, depending on a number of factors:

a. If deformation proceeds without substantial change in surface area, sliding over the tool surface under pressure flattens the asperities, and the surface roughness decreases while the bearing area increases; the rate of change is less for semisolids than for liquid lubricants.

b. If deformation results in the formation of a thick film of liquids, polymers, or semisolids (waxes, soaps, etc.), the natural surface roughness (equivalent to that of a freely deforming surface) develops. However, metal films tend to make the surface slightly smoother, as do graphite or MoS_2 films (Fig.3.51) [320,326].

c. If the combination of lubricant properties and process conditions is not successful in maintaining a film thick enough to fill the roughness features completely, the pressure to support deformation cannot build up, and surface roughness is much reduced.

d. The initial surface roughness is a particularly powerful factor in mixed-film lubrication. With moderate initial roughness (typically on the order of 0.5 μm AA), the better lubricant entrapment and consequent shift to more hydrodynamic conditions

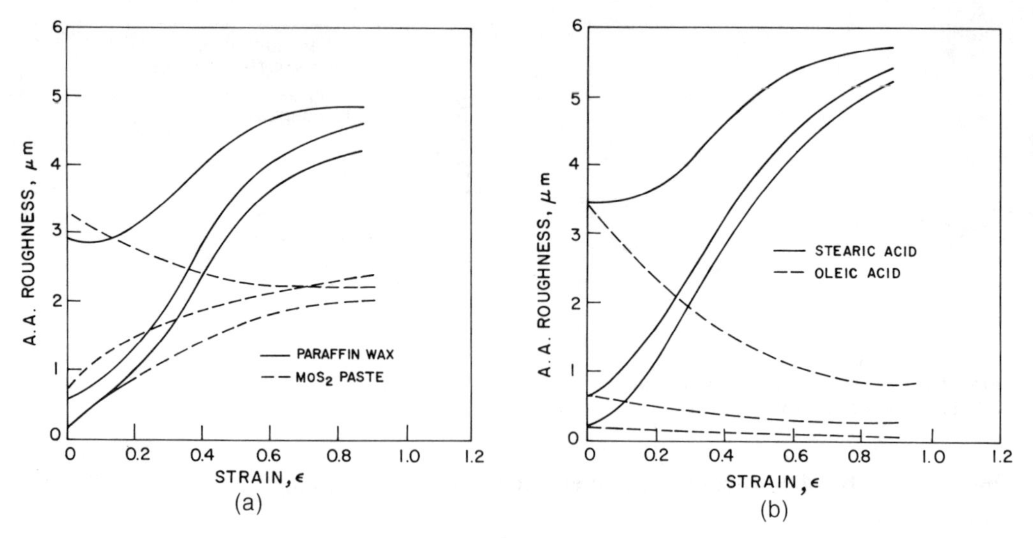

Fig.3.51. Effects of lubricants on change in roughness of end face in compression (after [320])

contribute to a further roughening of the surface [320,327]. On a strain-hardened material transverse features become more prominent. If the conditions allow more boundary contact to develop, and/or if the starting roughness is too great, the surface becomes smoother (Fig.3.51), but failure of the boundary or E.P. additives may allow severe asperity deformation which, on a more adhesion-prone material such as aluminum, results in a smeared appearance (Fig.3.43c). When this becomes excessive, the surface roughens even though no hydrostatic pockets are present at all. Damage to the surface is, however, limited to a thin surface layer, and the very high reflectivity may actually be desirable.

e. Hydrostatic pockets are extremely difficult to eliminate completely, even if in later processing stages only boundary lubrication is used with very smooth dies. Ghost images (a mottled surface appearance) may remain even when the surface is smooth as measured on a stylus instrument. Therefore, if a bright surface is desired, hydrodynamic effects must be discouraged throughout, and the initial, starting surface must be smooth, too. The structure of the metal should be free of features (inclusions, grain clusters) that could lead to visible surface markings (such as parabolic markings) accentuated in the presence of lubricant pockets.

3.8.6 Effects of Die Surface Roughness

One might easily believe that die surface roughness should be undesirable under any conditions because hard asperities pierce through the lubricant film. Indeed, the statement is sometimes made that the dies and tools should be as smooth as possible, and much time and money are often spent on hand polishing of dies. Yet the smoothest surface is not necessarily the best; there are desirable as well as undesirable types of die surface roughnesses in metalworking, depending on lubrication mechanisms and process conditions, although relatively few systematic investigations have been carried out.

Solid Lubricants. A solid lubricant film applied to the die is generally mechanically keyed by surface roughness. Roughness is thus beneficial if the combined asperity and waviness height is less than the lubricant film thickness, but only if the roughness lay is perpendicular to the sliding direction or is random. The effect is most marked with layer-lattice compounds; in the work of Schey and Lonn [328], with aluminum sliding against steel, the durability of graphite films increased when both die and workpiece surfaces were shot blasted. Film durability was much lower when either the aluminum or the steel surface was finished in the direction of sliding, and was lowest for smooth surfaces (Fig.3.52).

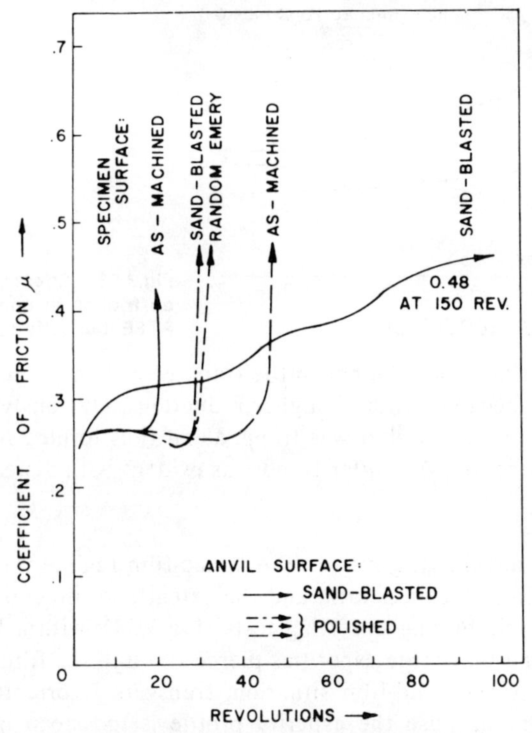

Fig.3.52. Durability of graphite films as a function of surface roughness in twist compression [328]

Under favorable conditions, a solid lubricant may be fed out from the reservoirs formed in valleys. This effect was shown by Schey and Myslivy [329] in upsetting smooth rings between anvils finished to three levels of random roughness and one intermediate level of directional roughness.

	Roughness, μm		
	AA	10-point avg.	Peak-to-valley
Lapped	0.12-0.15	0.25	0.5
Fine shot blasted	0.5 -0.7	0.8	2.5
Rough shot blasted	2.0 -2.5	4.0	11.0
Ground (perpendicular)	0.5 -0.7	1.0	3.0
(parallel)	0.18-0.2	0.4	1.0

A PTFE film deposited from a solvent was thick enough (10 μm) to prevent workpiece-to-anvil contact at low reductions. At heavy reductions the film thinned out, yet friction dropped with all surface finishes, indicating that lubricant was fed out from valleys of the surface (Fig.3.53). This interpretation is supported by the crossover of friction on

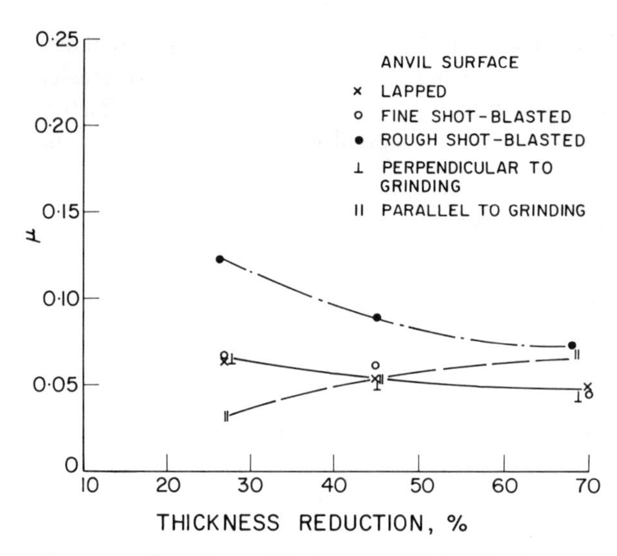

Fig.3.53. Effect of die-surface roughness on friction in compression of rings with PTFE lubrication [329]

the ground surface; friction was higher in the transverse than in the parallel direction at lower reductions, but became lower at higher reductions. Obviously, the lubricant could escape in the grinding direction, but was trapped and thus formed reservoirs perpendicular to the grinding direction. A similar trend was evident with a stearic acid film 10 μm thick.

Mixed-Film and Boundary Regime. In the mixed-film regime the effect of die roughness depends on the film thickness ratio and the extent and direction of relative sliding.

This is again visible in ring upsetting tests (Fig.3.54); with a lower-viscosity lubricant it is not surprising that large asperities pierce through the film and result in higher friction. In contrast to the solid-film situation, transversely oriented grinding features are harmful, no doubt because the asperity profile is too steep in that direction and because conditions are unfavorable for developing a micro-PHD mechanism. Even though the lubricant escapes more readily in the direction of grinding, friction is still lower because the more gently rising ridges are less damaging (Fig.3.54a).

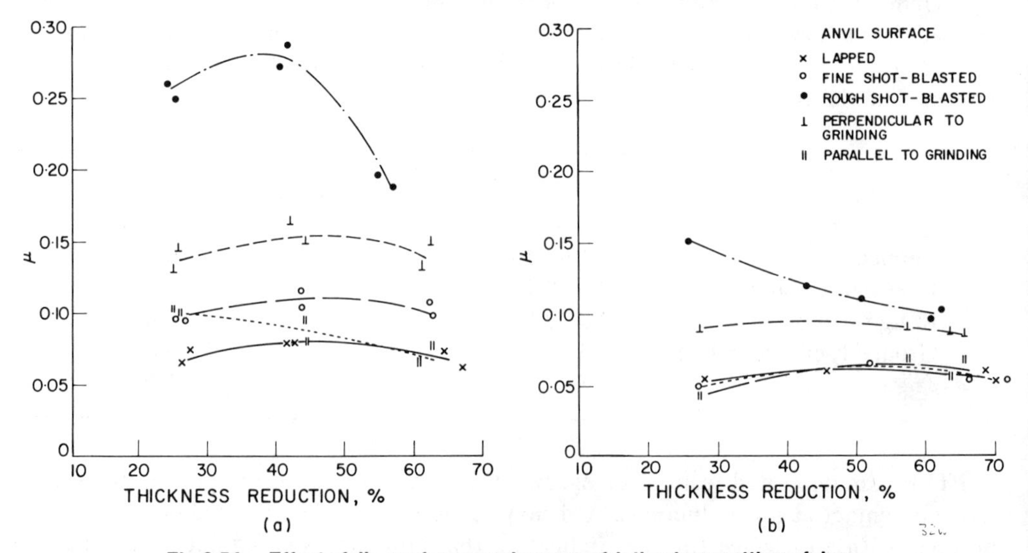

Fig.3.54. Effect of die-surface roughness on friction in upsetting of rings with mineral oil of (a) low (SAE 5) and (b) high (SAE 60) viscosity [329]

With the thicker film produced by SAE 60 oil, friction is generally lower and, except for the very rough shot-blasted and the transversely oriented ground surfaces, all the other finishes give identical friction, because the lubricant film is thick enough to mask their effect (Fig.3.54b). Similar trends showed in the work of Mizuno and Hasegawa [330] in a plane-strain compression test where the anvil was slid on the previously compressed surface. The frictional force decreased with increasing film thickness ratio (i.e., diminishing boundary contact area), with the addition of a boundary agent (i.e., reduced shear strength on asperities), and in the grinding direction (i.e., lower asperity slope). At higher sliding speeds the effect of directionality was obscured, no doubt because the film was too thick.

The difference in the role of valleys can also be seen by observing the occurrence of die pickup. With a solid lubricant, pickup is greater when sliding is parallel to the grinding direction, whereas with a mixed film pickup is greater perpendicular to the grinding direction.

An important exception to the above rule occurs when the die surface features are allowed to become filled up with byproducts (wear particles and friction polymers) formed on rubbing. The transverse troughs or valleys then serve as reservoirs even under mixed-film conditions, and friction drops in that direction. This may account for the gradual drop in measured friction that occurs when the die surface is not cleaned between successive tests in ring upsetting and, particularly, in plane-strain compression (Sec.9.5.1). Thus, die surface roughness is beneficial if the valleys can fill up with reaction and wear products before asperity contact can result in pickup. Therefore, the first few encounters are crucial and can determine the success of lubrication.

In the boundary-lubrication regime the hard die asperities are, in general, harmful because they pierce through the lubricant film and, if the die-workpiece material combination is prone to adhesion, cold welds and die pickup may form. Thus, die roughness is generally undesirable, and care is taken to finish the dies with the finishing marks in the direction of sliding, thus presenting asperities of lower slope in that direction.

Effect on Product Finish. The surface finish of the issuing product shows the effects of the lubrication mechanism superimposed on workpiece and die finish (Fig.3.55). With

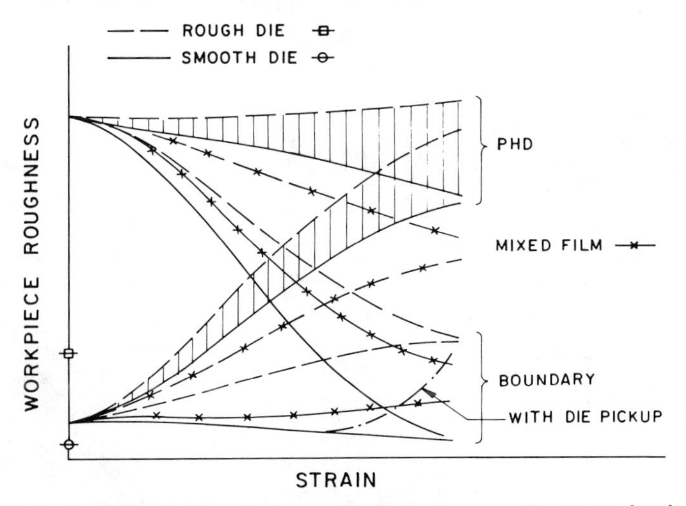

Fig.3.55. Effects of workpiece- and die-surface roughness on issuing roughness with various lubrication mechanisms

full fluid or solid-film lubrication the workpiece finish is dictated by workpiece roughness and lubricant. In the mixed-film regime, some of the die surface features are reproduced on the workpiece, but hydrostatic pockets scattered on the surface make the product rougher than the die. If, however, valleys in the die finish become filled with wear products, the workpiece surface becomes smoother. In the boundary-lubricated regime the die finish is faithfully reproduced on normal contact, and reproduced but modified by sliding in sliding contact. When film breakdown occurs, asperity contact, deformation, and smearing lead to greater roughening when the die is rough.

3.9 Frictional Instabilities

In our previous discussions we assumed that all changes in lubrication mechanisms were gradual and that the frictional force, even though subject to variations, was certainly not varying in a periodic manner. Yet there are a number of instances when just such phenomena appear, with undesirable consequences.

3.9.1 Stick-Slip Motion

The phenomenon of stick-slip motion is most readily observed in a pin-on-disk friction tester or in bar (or tube) drawing.

a. Sliding motion is imparted to the workpiece through some mechanical device, the properties of which can be modeled, in the simplest case, as a system consisting of a mass, spring, and damper (Fig.3.56a). When the draw is started by pulling at a speed v, the frictional stress τ_i rises steeply, asperities are stressed, microdeformation of asperities begins, and, finally, after reaching some critical F value, the workpiece begins to move bodily at speed v_m. The mechanical elements of the system are elastically stressed, and events are controlled by the properties of both the interface and the mechanical system.

b. If τ_i is constant irrespective of sliding speed, as it is indeed for boundary lubricants below their transition temperature, the force F will be steady (Fig.3.56b). The same applies for full fluid lubrication, in which F must actually rise with increasing v, and thus stability is ensured.

c. In the mixed-film regime the interface shear stress τ_i drops with increasing velocity because of the larger contribution of PHD components. Sometimes, it takes a small

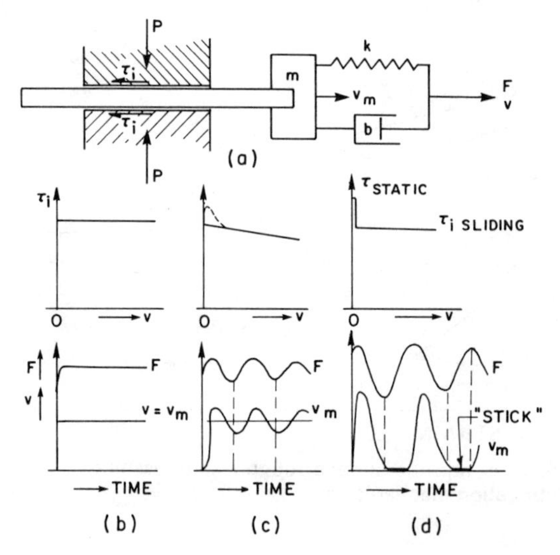

Fig.3.56. Frictional instabilities

finite speed for the PHD components to become active, and then the initial τ_i will be high, subsequently dropping steeply to its typical value (Fig.3.56c). This results in a sudden acceleration; the velocity of mass v_m may exceed the imposed velocity v, and the spring force is reduced. The velocity of m decreases at a rate determined by the damping b of the system until the velocity v_m drops below the imposed velocity v and the force F again starts to accelerate the mass. Depending on the magnitude of damping, the force and velocity variations may persist or may be damped out.

d. The extreme case is encountered when τ_i is large prior to commencement of gross sliding. This is usually referred to as static friction or stiction (not to be confused with sticking friction; Sec.2.2.1). Its origin is obscure and subject to debate [331]; nevertheless, it is often observed in systems that include a relatively soft, elastically deforming member. It is associated with the very first phase of sliding, when asperities deform without breaking away. It is particularly noticeable when the friction pair is allowed to stay in contact under load for some time, so that asperities can break through weak boundary films and establish junctions. Once sliding begins, a new and lower τ_i is reached very suddenly. Thus, the sequence of events described above occurs, except that the mass m and thus the workpiece may come to a temporary stop; hence the term "stick-slip." There are a number of analyses available [23,332,333].

Periodic variations in velocity result in corresponding variations in the lubrication mechanism, which in turn cause visible and objectionable variations in surface finish. If force variations are large enough, they also are reflected in a change in product thickness in deformation processes and in waviness in machining, depending on the stiffness of the tooling and its support. The dynamic variation of force can also cause premature fatigue of some metalworking machine elements. Examples of this will be discussed in Sec.6.3.4, 7.3.5, and 8.4.

3.9.2 Ultrasonic Vibration

Paradoxically, one of the most effective ways of combatting uncontrolled periodic fluctuations (such as chatter in wiredrawing or in drawing a tube on a plug) is the superimposition of controlled vibration of a higher frequency.

Over the last 25 years, vast efforts have gone into exploiting the potential of ultrasonic vibration in metalworking. Several reviews [334,335] are available, including books on the extensive Soviet work [336,337], and thus it will suffice here to review only the basic features. Originally, interest in the technique was awakened because it appeared that deformation forces can be reduced if the die or workpiece is vibrated at ultrasonic (over 20 kHz) frequencies. The effect appeared to be twofold:

1. It was believed that the flow strength of the material is reduced by some mechanism involving the interference of ultrasonic energy with dislocations. Critical work by Kristoffy [338] and others proved that what really happens is simply a matter of replacing the static force application with another, more expensive source of energy. The observed drop in deformation force is equal to the amplitude of the applied vibrational force. At very high levels of energy input, heating of the workpiece lowers its flow strength, but not more than heating by conventional (and more economical) means.

2. The friction mechanism at the die-workpiece interface may be affected.

a. In non-steady-state processes, such as forging, the vibration amplitude and frequency may be chosen so that the anvil is lifted away from the workpiece periodically, thus allowing liquid to flow into the interface and the squeeze-film to be replenished.

b. In steady-state processes such as wiredrawing, axial vibration of the die induces higher instantaneous velocities. In the mixed-film regime this helps to shift the mechanism towards the more favorable PHD regime. Therefore, benefits are lost when the mechanism is predominantly PHD even without vibration.

c. When process conditions are unfavorable so that not even a mixed film can be maintained, ultrasonic vibration can be helpful in that incipient pickup points are neutralized: the junction is sheared before it can grow and form a pickup particle large enough to damage the subsequently contacted lubricant film. However, the heat generated by ultrasonic energy can also cause the lubricant to fail, as was found by Gurskii et al [339] in compressing aluminum with oleic acid. With some exceptions, clear improvements can usually be demonstrated only if the lubricant chosen is poorer than the best industrially accepted lubricant. In many ways, the improvements achievable by ultrasonic vibration can often be secured at a much lower cost by measures such as controlling the surface finish of the entering workpiece or applying one of the standard lubricant carriers.

d. In a process where one of the machine elements has a low stiffness (e.g., the long supporting bar in tube drawing on a plug), ultrasonic vibration offers a means of eliminating stick-slip (Sec.7.3.5).

e. A different mechanism becomes active when a draw die or ironing die is oscillated in a circumferential direction (torsional oscillation). This results in a change in the friction vector; instead of directly opposing the deforming force, it now acts at an angle, and only the component pointing in the direction of deformation has to be overcome by the deforming force. An additional benefit may well be that incipient pickup points are again sheared and neutralized before they can grow. The same benefit can of course be obtained (and at a lower cost) by rotating the die (Sec.7.3.5).

3.10 Wear

Deformation would proceed without wear if the die and workpiece were separated by a thick film of a nonreactive lubricant and no foreign particles were present. We have seen, however, that pure PHD lubrication is seldom achieved; lubrication is mostly of the mixed-film or boundary type. Thus reactions with the metal surface are not only tolerated but encouraged. In the course of sliding, some of the reaction products are removed and a progressive loss of material (wear) is inevitable. Such mild wear of the workpiece material is, therefore, regarded as normal. Wear of the die, or severe wear of the workpiece—perhaps coupled with scoring of the workpiece surface—is obviously undesirable, and one of the major purposes of lubrication is their prevention. In some machining processes, such as grinding, lapping, and honing, the purpose is different: the lubricant/coolant is applied to increase workpiece material removal rate. In either case, a better understanding of underlying factors and events is needed for rational choices of dies (tools), lubricants, and process conditions.

Considering the importance of the subject, the paucity of information on wear in metalworking (other than metal removal) is most regrettable but, in view of the complexity of the subject, not surprising. Wear is poorly understood even in machinery where repeated contact occurs over the same surfaces, and some kind of accelerated testing is possible even if not necessarily always relevant. In metalworking, only one of the contacting partners (the die or tool) is repeatedly or continuously contacted, and the other partner (the workpiece) moves out of the contact zone. Even if the workpiece surface is repeatedly contacted in the course of a process or pass sequence, the surface moving into the contact zone is highly variable, ranging from the technically clean surface of a pickled workpiece to the mangled, chemically transformed surface of a previously worked part. Gross plastic deformation and extension of the surface aggravate the potential problems of adhesion and also affect the influence of any wear particles that may be present; particles thicker than the oil film may be pressed into the surface and either become neutralized or, if they are sharp and hard, stick out at an unfavorable angle and act as cutting tools embedded in a plastic substrate.

It is very difficult if not impossible to accelerate the wear processes prevailing in a well-controlled production situation without affecting the wear mode, and thus laboratory simulation is also difficult. Full-scale production observations are influenced by incidental variables and can be misleading. Therefore, rationalization of wear in metalworking processes leans heavily on knowledge obtained from machinery situations, the subject of several books [340-345] and reviews [346-357]. This is the course we will follow here, with due allowance for differences introduced by the plastic deformation of the workpiece.

The very complex nature of wear has long prompted attempts to bring order into chaos by various schemes of classification. No such scheme can be perfect, but these schemes do provide at least a frame of reference. One useful approach offered by Burwell [356] assumes that, at least in some situations, a single kind of wear mechanism prevails. Another system is based on the morphology of wear particles [356a].

3.10.1 Adhesive Wear

In metalworking, the die and workpiece surfaces are always contaminated and often lubricated, yet virgin surfaces are exposed partly because of the generation of new surfaces and partly because asperities plow through the surface films. Depending on the material pairing, adhesion may then lead to cold welding (solid-phase welding); further relative movement (sliding or normal separation of the surfaces) destroys the junction by one of several mechanisms:

1. If the junction is as strong as or stronger than the die material (for example, through the formation of intermetallic compounds), the junction separates either in the workpiece or in the die. Die wear rates become unacceptably high, particularly when high interface temperatures promote diffusion of alloying elements from the die material into the workpiece, thus reducing the strength of the die. This is particularly noted on the tools in some machining processes, but is also responsible for some part of die wear in rolling and wiredrawing.

2. If the junction is stronger than the workpiece, separation occurs in the workpiece with incipient pickup on the die. On subsequent contact with the workpiece surface, several possibilities exist:

a. The pickup grows by accumulating more material. It also becomes strain hardened. Once it reaches some critical size, it is subjected to periodic loading which, after a number of cycles, leads to fatigue separation and the formation of a wear particle [26,357].

b. The separated particle may be transferred onto the workpiece; in repeated contact it may back-transfer onto the die, then again onto the workpiece until finally the fragment leaves the deformation zone on the exiting workpiece surface or with the lubricant.

c. The pickup site may be recoated with the lubricant, and then dynamic equilibrium between metal transfer and removal can result in a low wear rate with a gradual buildup of wear debris in the lubricant [358]. This is typical of, for example, the cold rolling of aluminum.

d. If pickup is not removed by any of the above processes, the transfer particle becomes a new and, because of work hardening and perhaps also chemical changes, very hard asperity, which plows through the surface film on a subsequently contacted material and grows by further metal transfer. The process becomes self-accelerating, and working must be terminated because of excessive friction and/or unacceptable workpiece finish.

3. If the junction is weaker than the workpiece material, separation occurs at the interface. Wear, if it occurs at all, is due to other causes.

It should be noted that adhesive wear can take place even when the die is very smooth. The very process of sliding [359] and plastic deformation of the workpiece results in roughening of originally smooth surfaces, and this in itself may sufficiently reduce the thickness of protective lubricants at some locations to allow junction formation. Experiments [360] have shown that shearing takes place in a direction slightly inclined to the surface (Fig.3.57), creating wedges or prows of workpiece material which are then transformed into wear debris.

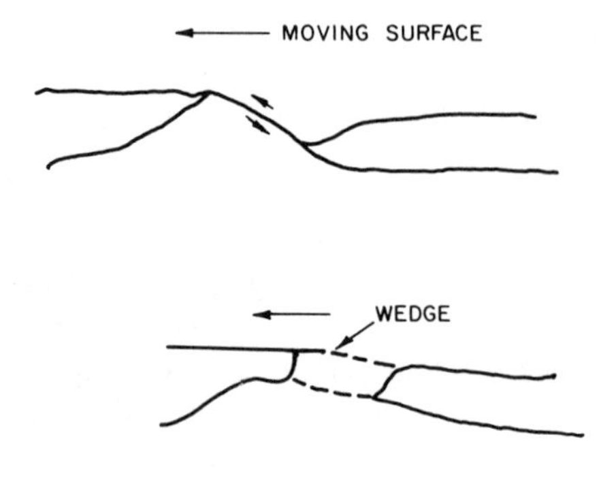

Fig.3.57. Generation of wear debris by wedge formation [360]

3.10.2 Abrasive Wear

Abrasive wear is the removal of material by a hard asperity (two-body wear) or a particle interposed between the two surfaces (three-body wear); for reviews see [361,362]. For an encounter to produce a wear particle, a number of conditions must be fulfilled:

1. Experiments have shown that the Vickers hardness of the asperity or particle must be at least 1.5 times greater than that of the surface to be abraded [363]. In metalworking, the die almost invariably satisfies this criterion. This does not mean, however, that only the workpiece wears. In contrast, the major concern is abrasive wear of the die. There are several sources of hard particles. Oxides (Fig.3.15) may be embedded in the surface of the workpiece or loosely carried into the deformation zone to become lodged in the softer workpiece material. Many workpiece materials, particularly precipitation-hardening alloys, contain hard intermetallics as part of their structure. Intermetallics may also form by diffusion.

2. The damaging effect of a hard particle depends on particle size, distribution, and relative orientation to the interface:

a. When the hard particle is round or flat but oriented at a small attack angle (angle of inclination to the surface), and the depth of penetration is small, only elastic deformation of the softer material occurs. This results in the expenditure of energy and therefore heating, which may lead to a breakdown of lubrication but in itself does not cause wear yet.

b. With a greater depth of penetration, the same particles exert a plowing action. If the material is of inherently low ductility, or is strain hardened heavily by repeated contact, the displaced material is removed as a wear particle. If the material is ductile, the displaced material is pushed sideways into ridges, which may then be removed on subsequent contact.

c. When the attack angle is rather steep, a cutting action results in instant removal of material. The angle at which transition to this wear mode occurs can be predicted (see Fig.3.12).

d. At high sliding velocities, the energy expended in the very small contact zone may lead to localized melting, and material is then ejected in the form of molten globules.

e. Long-term effects depend also on process geometry. If wear debris is allowed to accumulate (as on a piercing punch, or at the entry to a drawing die), three-body wear occurs. If, however, the wear particles leave the deformation zone and are filtered out of the lubricant, only two-body wear takes place.

3. Oxides and intermetallic compounds can be as hard as the constituents of the deforming or cutting tool; therefore, abrasive wear becomes a competition between die and workpiece whenever the lubricant fails to ensure complete separation of the interface. Particles embedded in the workpiece material become particularly damaging when they receive strong support from a matrix that is almost as hard as the particle itself [364]. Blunting of particles reduces their effectiveness, whereas fracture retains their damaging action because new sharp edges are formed.

4. Because a given surface is scored only if it is softer than the abrasive, it is not surprising that resistance to abrasive wear is a function of hardness (Fig.3.58 [365]) or, alternatively, cohesive energy. The hardness must be taken for the conditions prevailing at the interface, and therefore metallurgical phase changes [366,367] occurring during sliding are helpful. The transformation of retained austenite into martensite is particularly effective in steels (an outstanding example is Hadfield steel). Slight heating on sliding strengthens the martensite by carbon diffusion to dislocation sites [367]. For technically pure metals, cold working brings no improvement (after all, the surface hardens during frictional contact anyway), but heat treatment of alloys, to increase matrix hardness and precipitate population, does help. Conversely, diffusion of alloying elements from the die into the workpiece weakens the die and results in rapid wear (sometimes denoted specifically as diffusive wear), which is one of the main causes of high speed steel tool failure. Failure is accelerated by the presence of a magnetic field [368].

5. For the same hardness, ceramics have lower wear resistance than metals or alloys because of their lower shear strength and greater brittleness. Nevertheless, they are much harder than metals, and they remain indispensable for machining by abrasive techniques.

6. Erosion is a form of two-body abrasive wear in which the abrasive particle

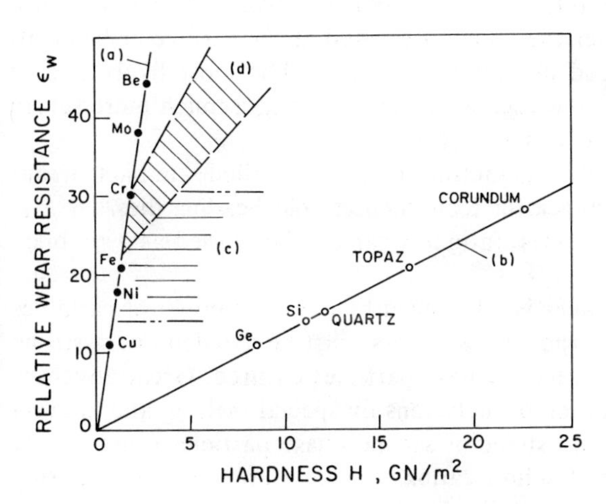

Fig.3.58. Wear resistance of (a) pure metals, (b) ceramics, (c) strain-hardened metals, and (d) heat treated steels [365]

receives its energy from the velocity of the fluid vehicle [369]. Its major application in metalworking is to processes such as liquid honing, and therefore this too will be discussed in Chapter 11.

7. The term "attritive wear" is sometimes used to describe tool wear by the removal of small surface particles and, in particular, the removal of carbide particles after the binder has been lost.

3.10.3 Fatigue Wear

This type of wear is predominant in cyclically loaded machine elements such as rolling-element bearings and gears. Because of its importance, the subject of contact fatigue has been extensively investigated, as shown by several reviews [370-372]. Much of this work has direct relevance to the wear of periodically loaded dies and tools, such as rolls, forging dies, and tools for intermittent cutting.

Details of the fatigue-wear mechanism are still debated, but in general it involves the following steps:

1. When a well-lubricated machine element, usually of hardened steel (or some other heavily alloyed material which contains large populations of second-phase particles), is subjected to elastic compressive loading over part of its surface, the surface is in compression (Fig.2.7a), but it can be shown that shear stresses are generated below the surface.

2. Repeated loading causes the generation of a microcrack, usually below the surface, at the site of a pre-existing point of weakness, such as an inclusion or second-phase particle.

3. On subsequent loading and unloading, the microcrack propagates, voids coalesce, and the cracks branch at a rate controlled by the stress state and lubricant environment [373]. Once the crack reaches a critical size, it changes direction to emerge at the surface, and a flat sheetlike particle is detached.

4. The process of forming a flat wear particle by cyclic shear deformation due to surface traction has been named "delamination wear" by Suh [374] and has been the subject of intensive studies. Severe deformation in the subsurface layer, the formation of a cell structure, and crystallographic factors [375] all contribute. This mechanism, and fatigue effects in general [375a], are of importance in processes where the workpiece is subjected to a sequence of contacts (multipass rolling, drawing, grinding).

5. When normal loading is combined with sliding, the location of maximum shear stress moves towards the surface, and fatigue cracks may originate from surface defects. When the detached particle is relatively large and sheetlike, the process is also called "spalling," irrespective of whether the crack originated at the surface or below it. Small-scale surface fatigue is described as "micropitting," and is most likely due to fatigue of asperities. When these pits allow escape of the lubricant, a much more severe form of wear, called "fatigue scoring," may set in [376].

6. Fatigue wear between normally contacting bodies is called "impact wear" [377,378], a mechanism of great importance for tools such as cold heading dies.

7. Because of the similarity of processes, fatigue wear can be viewed as a problem in fracture mechanics [379,379a].

As are all wear processes, fatigue wear is influenced by a large number of variables [372]. Obviously, both external and internal stress raisers must be avoided and a strong interface (wetting) between matrix and second-phase particles ensured. Great advances have been made in reduction or elimination of inclusions by special melting and pouring techniques, and in control of the size and shape of second-phase particles (particularly carbides in die steels) through control of solidification, breaking up of particles during hot working, and manufacture by powder metallurgy techniques. Control of surface finish [380] and of residual stresses is also important, and the nature of the lubricant has a marked effect, too [381]; for example, water in a lubricant accelerates fatigue, presumably through hydrogen embrittlement.

A further complication arises in hot working. On contact with the hot workpiece, sudden heating of the surface layer results in surface expansion and the generation of stresses between substrate and surface. After contact, the cool substrate quenches the surface material, again inducing stresses and, if temperatures were high enough, also metallurgical changes. In combination with stresses due to loading, a rapid and very destructive thermal fatigue process may set in, resulting in a mosaic-like network of cracks, described as "crazing" or "firecracking."

Fatigue may also cause sudden catastrophic failure of the tooling, as in chipping of cutting tools or in complete fracture of rolls or forging and extrusion dies. Tribological processes may contribute to this through the initiation of fracture on the tool surface, although macroscopic stress raisers (tight radii) may be the primary culprits.

3.10.4 Chemical Wear

What is meant here by "chemical wear" is not material loss due to straight chemical attack (corrosion or chemical milling) but rather loss due to tribochemical reactions. The process consists of well-identifiable stages:

1. A film forms on the surface of the die and/or workpiece through chemical reaction (e.g., oxidation) or tribochemical reaction (e.g., E.P. or soap-film formation on reactive surfaces).

2. The reaction product is mechanically removed in the deformation zone and enters the bulk lubricant or exits on the workpiece.

3. The highly reactive surface thus produced is recoated with a reaction product if the reactivity of the lubricant or environment is adequate. As are all chemical reactions, this too is greatly accelerated by temperature and pressure. Environmental effects in wear are discussed by Bentley and Duquette [381a].

When lubrication is successful, wear is minimal. To some extent, this is sacrificial wear; protection from the more severe forms of adhesive wear is bought at the expense of some slight material loss through tribochemical action. It is generally preferred that the workpiece should wear, although in the case of nonreactive materials (such as stainless steel or titanium) die life may have to be sacrificed if complete separation of die and workpiece cannot be achieved. Under relatively mild conditions, with partial PHD lubrication, oxidation may be sufficient to maintain reasonably low wear rates just as in EHD lubrication [382]; this may explain, for example, why almost any lubricant serves satisfactorily for relatively light-duty sheet metalworking.

A contributor to tribochemical wear is the electrical potential generated in the course of sliding [84]. There are several possible sources of such electrochemical wear: a thermal emf may be set up between two dissimilar materials [383], electron (charge) transfer between metal and fluid medium may take place, and oxidation may also give rise to a charge [384]. The current path is closed from workpiece to die through the equipment and through the lubricant. The accelerating effect can be neutralized by isolation or by the imposition of a reverse voltage (see Chapter 11). In aqueous solutions and oil-in-water emulsions these effects are very marked. Making the die negative offers cathodic protection and practically eliminates corrosive wear.

3.10.5 Mixed Wear Mechanisms

In many instances several of the above wear mechanisms operate simultaneously. Indeed, the separation of wear modes may be extremely difficult, and specific terms have been adopted in practice to describe some of the more common mixed mechanisms. Thus, "scuffing" is a term widely adopted in the U.K. for a form of adhesive wear observed on machine elements and metalworking dies; in the U.S. the term also embraces abrasion. "Scoring" is a synonym in American usage. Removal of protective films, combined perhaps with plastic deformation of asperities, and subsequent formation of adhesive junctions are characteristic of this failure mode [385].

Adhesive and abrasive wear at times go hand-in-hand—indeed, sometimes it is difficult to separate them at all—and therefore the term "penetrative wear" has been suggested [386].

Tribochemical wear also tends to appear mixed with other modes, as reviewed by Rowe [387]. Thus, oxidation and the mechanical characteristics of the oxide film determine to what extent adhesive and abrasive wear will take place [388]. The presence of oxygen and particularly of water affect wear even in lubricated sliding [389]. More complex phenomena, in part attributable to changes in lubricant composition and rheology, occur on prolonged exposure. Further complications arise from reactions with metals, some of which yield small amounts of hydrogen even under ostensibly abrasive wear conditions. Interactions between contacting materials and lubricants greatly affect wear rates, so as to make hardness of secondary importance [390]. An active chlorine lubricant affects both tungsten carbide and cobalt, thus increasing the wear of carbide tooling; oleic acid attacks only cobalt but still loosens the tungsten carbide particles.

Much depends also on the relative geometry of the interface. In machinery situations, small-amplitude oscillating contact results in fretting, a form of wear in which the debris remains trapped in the contact face. This has only limited application to metalworking, but illustrates the complexity of wear processes [121].

In view of the many possible interactions, it is not surprising that no basic laws of wear have yet emerged. Only some trends can be discerned, which can at least be systematized by regarding wear as a tribosystem [22,391].

3.10.6 Quantification of Wear

In a given lubrication regime, wear often reaches a steady-state condition, and then it is found that wear volume V is proportional to the distance l or time t of sliding. In some but not all cases, wear is proportional also to normal load P, inversely proportional to the hardness of the softer material H, and independent of the apparent area of contact A_a. Thus

$$\frac{V}{l} = K\frac{P}{H} \tag{3.39}$$

where K is some constant typical of the wear situation. Its value can range over several orders of magnitude, from 10^{-9} in well-lubricated sliding to almost unity in dry friction (compilations are given by Rowe [387] and Rabinowicz [392]). Minor variations in interface conditions have a much more marked effect on wear rate than on friction and, even under ostensibly identical conditions, K may range over one or two orders of magnitude. The chief difficulty of all wear theories is the interpretation and prediction of K for various wear mechanisms.

Adhesive Wear. In the simplest case of an elastically loaded workpiece in normal contact, the load is carried on a real contact area A_r determined by the workpiece indentation hardness H:

$$P = A_rH \tag{3.4}$$

Substituting into Eq.3.39, the wear rate is

$$\frac{V}{l} = KA_r \qquad (3.40)$$

which immediately suggests that wear must be due to something like adhesion over the real contact area. The trouble is that if the wear particles are then assumed to have some reasonably flat (say hemispherical) shape, calculated wear rates become orders of magnitude greater than those observed in practice. Therefore, K must represent the probability that an encounter will actually result in wear-particle formation [355]. The low probability could be due to: low adhesion between die and workpiece; back transfer; the presence of contaminant films; relatively mild asperity encounters; the need to accumulate sufficient energy in a transferred particle before it becomes detached [30]; or the number of encounters needed for fatiguing the pickup particles. We have seen that under lubricated conditions metallic contact is limited to a small fraction of the interface, much smaller even than the real contact area A_r (Fig.3.39). Adhesion then can be only a small contributor, and a model based on fundamental properties, such as heat of adsorption of lubricants, may be more relevant [227]. A theory of diffusion wear has also been proposed [393].

Abrasive Wear. The relationship is again that given in Eq.3.39, but wear debris is now formed by plowing out material of a given volume. Since the load is borne by the leading half of a nominally conical asperity, the cone half angle α (Fig.3.59) comes into the expression

$$\frac{V}{l} = K\frac{2P \cot \alpha}{H} \qquad (3.41)$$

The value of K then depends on the ductility of the material, its fracture properties, and the angle α itself, because it determines events (Sec.3.10.2). The benefits of high hardness H are obvious [355].

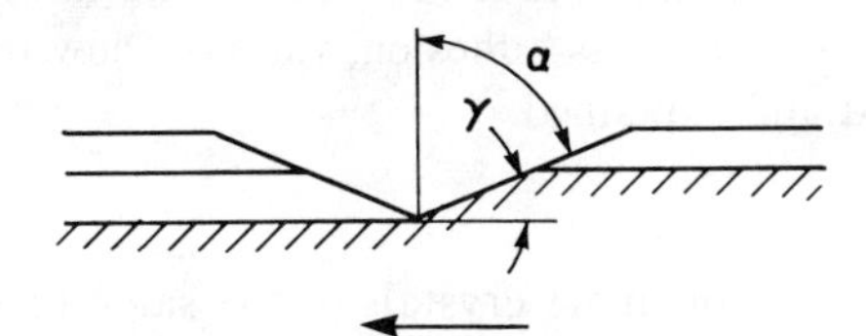

Fig.3.59. Attack angle in abrasive wear

Fatigue Wear. Equation 3.39 is again valid, but $1/K$ now represents the number of passes required to generate a wear particle by fatigue. Wear under impact situations has also been rationalized [394].

Chemical Wear. Analysis of wear rates is possible by assuming that reactions proceed according to a usually parabolic rate law [395] and that a critical film thickness has to be built up before it can be worn away.

Many of the attempts at formulating quantitative wear models have been reviewed by Beerbower [230]. Because of the great number of variables, it appears that there will never be a general law of wear, but transitions from mild to severe wear in well-defined situations may be predictable. Life prediction, based on zero-wear conditions below a critical interface shear strength and the onset of fatigue wear above this value, has shown promise [396]. A more global view is possible by regarding wear as an energy transformation process [22,397]. Dimensional analysis can be helpful in identifying parameters of importance [398,399]. Of course, before any analysis of wear can be meaningfully undertaken, one has to know what mechanism was operative. As with lubrication, the appearance and chemical analysis of the surface and of wear particles are the most reliable guides (Sec.5.9).

It will be noted that nowhere in this discussion of wear has friction or the coefficient of friction been mentioned. Indeed, there is no generally valid relationship between the two. It is true that wear is very much a function of the lubrication mechanism; there are, however, instances when within the same mechanism (say E.P. lubrication) friction is low but wear fairly high (because of the aggressiveness of the friction-reducing additives) or friction is high and wear low (because effective separating films of relatively high shear strength are present).

Systematic investigations of wear in metalworking are much rarer than those relating to general tribological situations. The reasons for this are manyfold. There are plentiful industrial observations on wear and its progression, but quantification and rationalization of these data are usually difficult because the number of operating variables is large, many of them are not quantified, and their interactions and roles are debatable. Laboratory tests are difficult to conduct because lengthy runs are needed to reveal wear trends. When operating conditions are changed to accelerate wear, conditions may become irrelevant. This is a problem of simulation testing to be discussed in Sec.5.1. One can only hope that as more systematic views of wear emerge in general tribology, the resulting hypotheses can be drawn upon to help the interpretation of metalworking wear. This is happening already, as will be seen in Chapters 6 to 11.

3.11 Material Effects

From the discussion of lubrication and wear mechanisms it is clear that in metalworking tribology the composition and structure of tool and workpiece materials, and their interactions with each other and with the lubricant, are of vital importance (Fig.3.2). Note that the term "tool" is used here in the broadest sense, denoting tools and dies for both plastic deformation and cutting processes. Throughout this chapter we have referred to the interaction between tool and die materials as adhesion; the task now remains to see if this phenomenon can be quantified and explained.

3.11.1 Adhesion of Similar Metals

It is not difficult to see that if two perfectly clean single crystals of the same metal are brought into normal contact to ensure total surface conformity, the two crystals will be joined by interatomic forces characteristic of that metal and crystal orientation. In essence, a grain boundary is created, and a large stress would be needed to separate the two crystals again. Such an experiment can indeed be performed by breaking and rejoining a specimen inside an airtight ductile envelope or in high vacuum, or by sputter cleaning the metal surfaces and joining them in high vacuum. Extensive investigations along these lines are summarized by Buckley [74,75], whose book [75] contains the most up-to-date, comprehensive treatment of adhesion in general. Adhesion under space conditions has also been the subject of a conference [400].

Tool and workpiece materials are usually chosen to be different; nevertheless, adhesion between similar metals is of interest because, once metal transfer (tool pickup) occurs, further contact between the transferred particles and the workpiece involves similar materials. Even if deformation takes place in an oxidizing atmosphere and in the presence of a lubricant, virgin surfaces exhibit the same characteristics as do the perfectly clean crystals discussed above. However, establishment of the bond is only one aspect; equally or more important is what happens when that bond is broken by normal separation or sliding, as emphasized by Tabor [351] in his critical review. These aspects will determine whether incipient pickup is removed from the die or whether it grows by accretion.

Adhesion is usually quantified by the coefficient of adhesion, which is simply the ratio of the normal separating force to the normal force applied when the contact was established. Because the real area of contact is proportional to the applied force (Eq.3.4), the coefficient of adhesion is independent of applied force and apparent area of contact. However, even minor surface contamination can significantly change adhesion, and therefore most experiments are conducted by twisting or sliding the specimen after normal contact has been made. Asperities then break through surface films and a higher coefficient of adhesion is measured, the magnitude of which depends on a number of factors:

1. At a given absolute temperature, adhesion decreases with increasing hardness of the pure metal (Fig.3.60), as shown in twist compression by Sikorski [401], presumably because hardness is a manifestation of interatomic bonding and is thus related also to the melting point. For a material of higher melting point, a constant absolute temperature corresponds to a lower temperature on the homologous temperature scale and is, therefore, further removed from the temperature at which full bond strength could be established. The same result is obtained if boiling temperature, recrystallization temperature, interfacial energy [402], or ratio of interfacial energy to hardness [30] is used as the reference.

2. For platinum, nickel, gold, silver, and lead, adhesion increases rapidly as temperature reaches about $0.4T_m$ [403], indicating that adhesion at constant homologous temperatures may be similar for various metals.

3. Diffusion increases bond strength; therefore, prolonged contact prior to separation increases the separating force, especially at high temperatures (above $0.5T_m$)

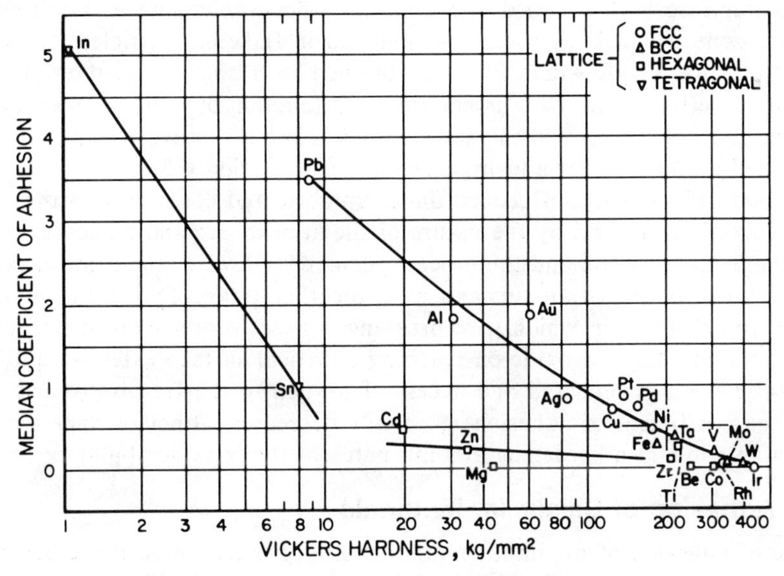

Fig.3.60. Influences of hardness and crystal structure on adhesion of pure metals [401]

where diffusion is accelerated. A high static friction is then measured, which drops to a lower value typical of that for the dynamic situation.

4. Adhesion depends on crystal structure and is lower in HCP metals. In the latter, adhesion decreases with increasing c/a ratio [75]; slip is limited to the basal plane, asperity deformation is less, separation along the basal plane takes a relatively low force, and wear rates are also low (as with zinc and cadmium). At low c/a ratios, cross slip is necessary to accommodate deformation; therefore, adhesion and wear rates are high (as with titanium and zirconium).

5. Atomic registry at the interface increases bond strength [75], whereas lattice mismatch decreases it. High atomic density planes and directions in which slip is easiest give lowest friction and adhesion; polycrystalline materials of random orientation exhibit greater adhesion than high-density planes of single crystals.

6. A metal of low elastic modulus suffers greater elastic deflection on loading, and the resultant elastic recovery on unloading helps to break the junction. Similarly, a metal of lower fracture toughness should give lower coefficients of adhesion and friction, and also lower wear.

7. High rates of strain hardening cause the junction to become harder than the bulk metal, and thus shearing requires higher stresses or may take place adjacent to the junction over a large area. Thus, the coefficients of adhesion and friction increase, but much less than does wear, which may increase by orders of magnitude.

8. Adhesion of alloys against themselves follows, in general, the above rules. Thus increasing hardness by alloying (whether due to solution effects or ordering) reduces adhesion [404], as does an increasing c/a ratio in HCP metals [75], as long as no oxides are present. There are, however, anomalies: in Cu-Ni alloys with 0 to 40 at. % Cu, the coefficient of adhesion increases with nickel content and thus with increasing hardness. Bailey and Sikorski [404] attributed this to the decrease in stacking-fault energy (SFE) which results in greater strain hardening and strengthening of the junctions, overwhelming the reduced ductility normally associated with greater hardness. This is in contradiction to Rigney and Glaeser [375] and Bhansali and Miller [405], who found in sliding tests a decreased wear resistance with higher SFE, a not unreasonable result in view of the greater ease with which dislocation climb occurs, leading to a cell structure in the surface zone which is then readily detached in flakelike particles. The question is whether the same result would also be obtained in a single encounter. It should be noted that the effect of alloying is not simply a function of bulk composition; segregation of the alloying element to the surface may make a relatively lean alloy behave as though it had a much higher alloying element concentration [75].

9. Adhesion is greatly influenced under unlubricated conditions by the presence of preformed oxide films, and by the nature of the atmosphere and its access to the interface. If the oxide is brittle and the process geometry allows it, the film breaks as ice on water and the exposed virgin surfaces weld together [406]. This is the reason for the success of roll bonding in which wire brushing is used to create strain-hardened ridges that allow the breakage of oxide on reaching a critical surface extension [407]. In general, if the contact is repeated and access of oxygen to the freshly formed surfaces is allowed, the coefficient of adhesion gradually decreases, although some decrease may be attributable to strain hardening and mismatch at the interface [408].

3.11.2 Adhesion of Dissimilar Materials

The topic of adhesion of dissimilar materials is of greatest importance because it may shed some light on the question of whether workpiece metal will pick up on the tool at all. Unfortunately, there is a general lack of understanding despite extensive work in the field.

Dissimilar Metal Pairs. There is controversy even about the factors that control adhesion between two dissimilar elemental metals.

1. In the most general sense, some of the rules developed for similar material pairs apply here also: adhesion increases with temperature and contact time, with better lattice matching, and, if one of the contacting pairs is of HCP structure, with decreasing c/a ratio.

2. There is no unequivocal criterion by which adhesion in a given material pair can be judged. Mutual solid solubility may be taken as a rough guide, even though under high vacuum mutually insoluble metals such as iron and silver will bond [75]. The concept has been refined by successive investigators; reviews are given by Merchant [409], deGee [410], and Habig [349]. It was investigated in its most developed form by Rabinowicz [411], who judged compatibility on the basis of binary phase diagrams, allowing for both liquid and solid solubility (Fig.3.61). A highly compatible pair would give high adhesion and friction, subject to restraints imposed by crystal structure. The correlation is bound to fail when intermetallic compounds, which are generally strong but brittle, form [230,410]. Beerbower [230] found good agreement between past experiments and miscibility temperatures calculated from molar volumes and solubility parameters; thus, quantification may be possible. The solubility criterion has many critics (see Keller in discussion to [411]); still, it is the only one to give any guidelines at all, and it has been generally successful even in metalcutting (Chapter 11) where the tool generates an all-virgin-metal surface.

2. Adhesion implies the establishment of an interatomic bond, and in principle it should be possible to predict adhesion from the electronic structure of metals. This line of argument, pursued by Buckley [75], Czichos [22], and others, and reviewed in [351]

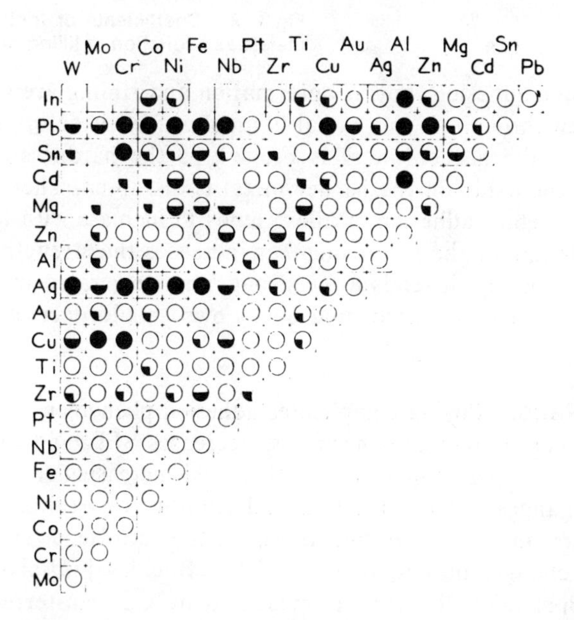

● —Two liquid phases, solid solution less than 0.1%.

◒ —Two liquid phases, solid solution greater than 0.1%, or one liquid phase, solid solution less than 0.1%.

◐ —One liquid phase, solid solution between 0.1% and 1%.

○ —One liquid phase, solid solution over 1%.

Blank boxes indicate insufficient information.

Fig.3.61. Compatibilities of dissimilar material pairs [411]

and [410], has been successful in rationalizing some results between similar metals. Ohmae et al [412] show that friction (attributed to adhesion) of 3d transition metals against copper increases as filling of the *d* bands increases, and that agreement is better than from the solubility criterion (Fig.3.62). The bond character determines also the chemical activity of metals, and Buckley [75] showed that adhesion increases with increasing chemical activity; however, surface energy should have the same effect, yet in the examples studied correlation with adhesion is poor [351]. Ferrante et al [412a] calculated binding energies resulting from electron sharing, and discovered scaling relations which allow mapping of adhesive energies between dissimilar metals onto a universal binding energy curve. Their approach may open the way to a fundamental treatment of adhesion.

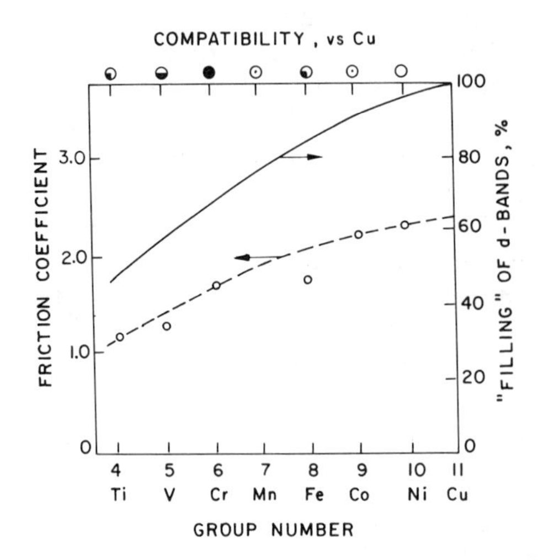

Fig.3.62. Coefficients of friction of 3d transition metals as a function of filling of d bands [412]

3. Oxides reduce adhesion, but, if deformation and sliding are sufficient, the oxide films will be broken through. If one metal is much softer, it deforms and adheres even before the oxide on the harder metal is broken; as the hardness of the softer metal increases, more of the oxide on the harder metal is disrupted. Therefore, in contrast to adhesion of similar metals, adhesion of alloys tends to increase with hardness [413].

4. Strain hardening of the softer member increases the strength of the junction. If an adhesive joint is formed, it tends to be strong, and wear rates are also high. It is not clear, however, whether strain hardening on its own should decrease or increase adhesion.

Dissimilar Alloy Pairs. Alloying complicates adhesion phenomena:

1. Alloying generally increases hardness, decreases SFE, and thus increases strain-hardening rates, thus strengthening the junctions and increasing friction.

2. Alloying changes the composition and stability of oxides. The effect is often quite dramatic even in small concentrations, partly because of surface diffusion and partly because of changes in oxidation chemistry. Buckley [75] showed that the adhesion of binary copper alloys to iron (as measured by Cu transferred to Fe) decreased with decreasing stability (free energy of formation) of their oxides. Temperature and availability of oxygen are, of course, most important in allowing re-formation of broken oxide films.

Metal-Ceramic Pairs. Ceramics possess ionic and covalent bonds, and adhesion between them and metals may involve a great variety of bonds ranging from van der Waals to covalent bonds. Correspondingly, adhesion can range widely, too.

1. The strength of an adhesive bond increases with decreasing interfacial energy. Therefore, thermodynamic calculations and wetting (sessile drop) experiments serve as useful guides [414]. More active metals (such as titanium) adhere more.

2. Because the ceramics (such as SiC, Al_2O_3, and diamond) in metal-ceramic pairs are more strongly bonded than the metals, it is the metal that transfers to the ceramic. Adhesion is a function of both ceramic and alloy. Thus, Miyoshi and Buckley [415] found with binary iron alloys that metal transfer increased with an increasing difference between solute and solvent atomic radii.

3. Atomic misfit reduces adhesion. Generally, the lattice parameters of oxides are quite different from those of their parent metals, hence the interface is weak. However, totally matched (epitaxial) growth of the oxide on aluminum makes for a very strong interface [414] and explains why aluminum picks up heavily on Al_2O_3 surfaces.

4. Even though the ceramic is stronger than the metal, it is also more brittle, and the combination of shear and normal stresses may cause fracture (spalling) of the ceramic. Particles embedded in the metal act as abrasives, and very high wear rates (and surface damage) may follow.

Workpiece-Tool Pairs. There are relatively few studies relating to realistic tool-workpiece material combinations. The complexity of tool materials make predictions extremely hazardous, and general guidelines are not available. Even the test method creates problems: in laboratory studies of adhesion, it must be ensured that any oxide films present will be broken through. Thus, twist compression, strip or wiredrawing between stationary tools, and machining are suitable, whereas surface expansion in ring upsetting or rolling at low reductions is insufficient [416] to reveal trends.

1. The availability of oxygen is vital, because it reduces adhesion even in operations such as twist compression and machining, where the only access is at the edges of the interface. As might be expected, the effect is more pronounced when both die and workpiece are metallic (Fig.3.63); with oxide tools such as Al_2O_3 the effect is only minor. The nature of the oxide is also important [416]: 4140 steel and TZM (a molybdenum alloy) benefit from the low shear strengths of their oxides; titanium dissolves its own oxide at high temperatures and thus neutralizes it; Al_2O_3 is hard, brittle, and of only minor benefit; the spinel $NiO \cdot Cr_2O_3$ (Buckley in discussion to [416]) on the nickel-base alloy Hastelloy X must be rather hard and is ineffective (Fig.3.63). Although not shown in Fig.3.63, pickup went hand-in-hand with friction in these experiments. A preformed oxide film is often found beneficial on forging dies and also in metalcutting (Chapters 9 and 11). Oxidation may explain at least in part the nonadhesion of molybdenum spray coatings [417], although the relatively low hardness of molybdenum makes it suitable only for low-load conditions.

2. Some adhesive combinations can be rationalized from the characteristics of the predominant metallurgical phase in the die and workpiece surfaces; however, information is lacking on the basic adhesive characteristics of many important die materials, including diffusion coatings. Thus, Wiegand and Kloos (reported in [59]) found the greatest draw force and metal transfer between mild steel and tool steel, and between copper and aluminum bronze (Fig.3.64), as might be expected from the solubility criterion; otherwise, adhesion generally decreased with increasing die hardness. A coating of titanium carbide (TiC) was best for both steel and copper, and the low adhesion of aluminum bronze to steel was in line with other observations on austenitic stainless steel [418].

This latter example shows that hardness alone is not a good guide in itself, although there is some correlation with it if the nature of the die material is not changed, as shown by Oyane et al [419] for copper sliding against tool steel.

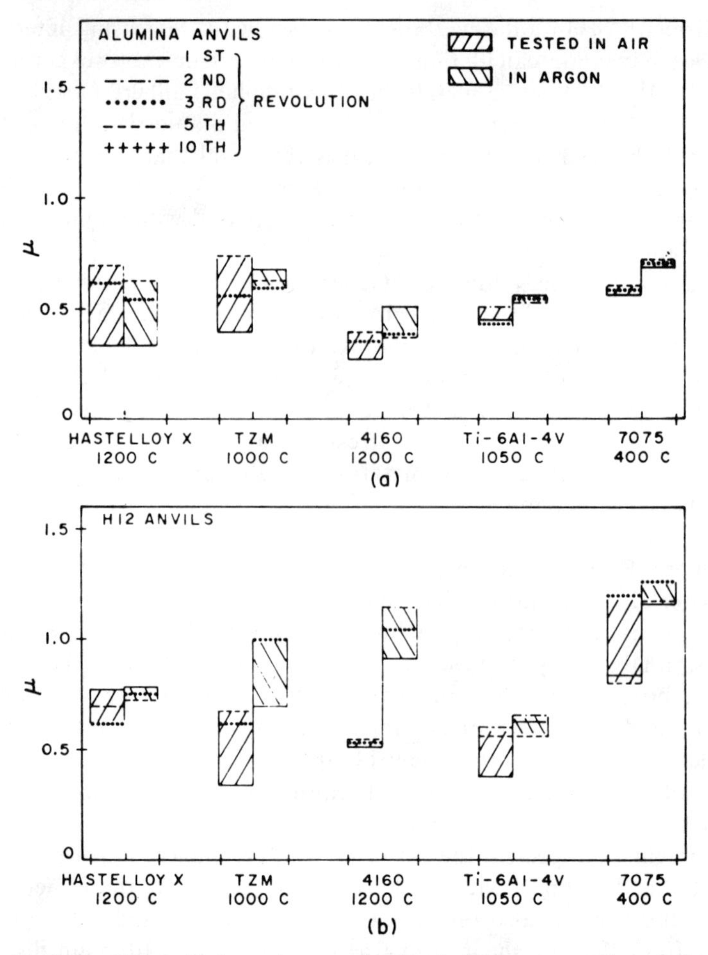

Fig.3.63. Effect of atmosphere on friction in twist compression at elevated temperatures (interface pressure = yield strength) [416]

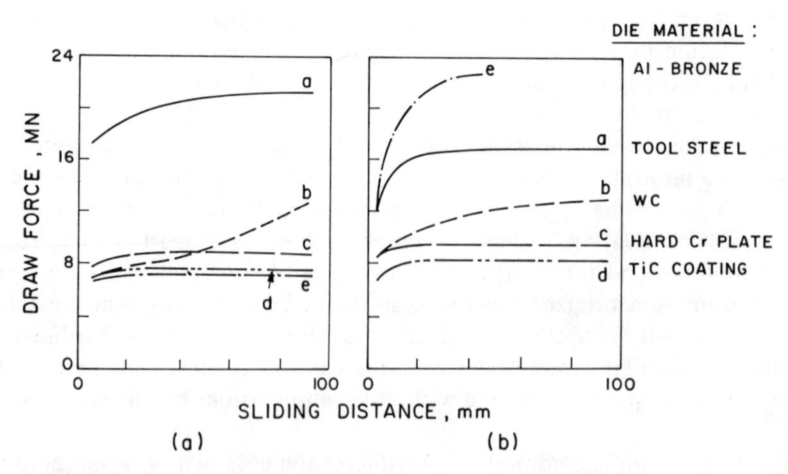

Fig.3.64. Effects of die material on draw force in drawing (a) low-carbon steel and (b) copper strip [59]

A high carbide content is generally beneficial, but the nature of the carbide is also important. In the extensive experiments of Newnham and Schey [416], TiC was generally slightly better than a chromium carbide layer (Chr) formed by diffusion, and both were greatly superior to a hard chromium plate (C.P.) despite the high hardness of the latter (Fig.3.65). A carbide (WC) bonded with 6% Co performed well in some but not all instances; Oyane et al [419] found it superior in sliding against copper. A boronized surface (Bor) consisting essentially of iron borides performed well, especially against steel and titanium at higher temperatures [416]. A ceramic such as Al_2O_3 is good from the adhesive point of view but spalls off unless supported mechanically [416].

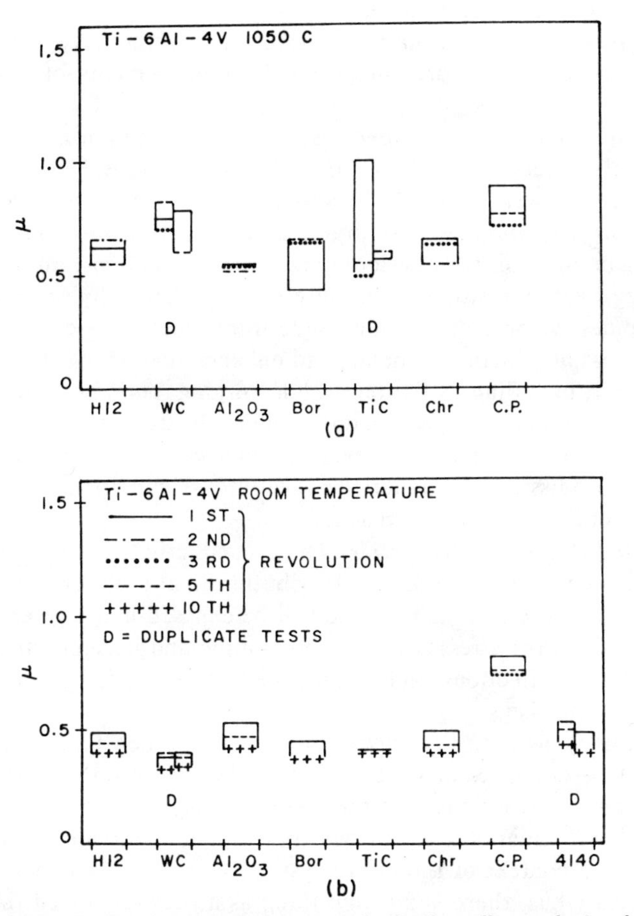

Fig.3.65. Friction in twist compression of a titanium alloy against various anvil materials at (a) 1050 C and (b) room temperature [416]

3. Because adhesion or metal transfer is difficult to quantify, the magnitude of friction is often reported, with the expressed or tacit assumption that friction is due purely to an adhesive mechanism. The danger is that other factors may have equally powerful effects. For example, Peterson and Ling [106] found good correlation between surface roughness and friction in sliding of tin against a great variety of metals and ceramics; obviously, the plowing (or plastic deformation component) overwhelmed the adhesion effect.

There are numerous examples of investigations on friction and wear under conditions conducive to adhesion, and some of these will be discussed in Chapters 6 to 11.

3.11.3 Attributes of Tool Materials

The conditions imposed by the process and the interface are varied and call for a number of sometimes contradictory tool requirements:

1. *Controlled Mechanical Properties.* The tool must maintain its geometry under the high pressures and temperatures prevailing in the process and therefore must have sufficient tensile or compressive strength to resist the deforming or cutting force; ductility (toughness) to survive impact loading; and fatigue strength to cope with repeated loading. Very generally, strength or hardness can be increased only at the expense of toughness; therefore, a compromise has to be made and then, in the best tradition of the systems approach, the process or the tool design will have to be modified to compensate for weaknesses. Many hard materials have high compressive but low tensile (flexural) strengths and are, therefore, supported by tool elements of lower hardness but greater toughness.

2. *Low adhesion to the workpiece* is required to minimize the harmful consequences of metal-to-metal contact. In the absence of adhesion, the frictional force component due to cold welding would be absent, and so would adhesive wear.

3. *High Wear Resistance.* Most practical situations involve more than one wear mechanism and, in addition to adhesion, abrasive wear is almost always a danger. High temperatures aggravate the situation because they mechanically soften the die, increase the adhesion tendency, promote usually undesirable metallurgical changes, accelerate diffusion with a loss of alloying elements, and enhance reactions that lead to accelerated tribochemical wear, including oxidation. In hot working, periodic contact with the workpiece followed by contact with the cooling medium leads to thermal fatigue. Interface temperatures can become very high, even in cold working, depending on the rate of heat generation and dissipation; under the very intense conditions prevailing in cutting, hot working temperatures are easily reached.

4. *Controlled Physical Properties.* Heat conductivity is of utmost importance because it determines the temperature distribution in the die and thus also the maximum surface temperature to which the die will be exposed under given conditions. Thermal expansion controls the stresses set up between die and its supporting structure. Elastic properties affect deflections under load, and influence die geometry and, through this, lubricating mechanisms.

5. *Compatibility with the lubricant* is important when lubricant action relies on surface-lubricant reactions. Sometimes it is sufficient if only the workpiece enters into reaction; at other times reactions with the die are essential.

From the above remarks it is obvious that any of the enumerated attributes have meaning only in the context of the entire system. One chooses a tool material not simply for the process (thus, there is no such thing as a draw-die or cutting-tool material), but also for the tribosystem created by the workpiece, the process, and the lubricant [420].

3.11.4 Tool Materials

The rapid increases in labor and machine costs have put renewed pressure on the tool industry to develop materials and processing techniques that will result in tools that allow high production rates. The rapid development that has ensued is attested to by a number of specialized conferences [421-424]; there are also several excellent compilations and reviews available [425-427], including an overview of historical development by Shaw [428]. Therefore, a brief outline of various tool material classes will suffice here.

1. *Carbon steels* derive their hardness from the martensitic transformation. Temper-

ing of carbon steels occurs above 250 C, and thus they are suitable only for low-speed (low-temperature) tools such as hand reamers and chopping knives.

2. *Alloy tool steels* [425,426,429] are characterized by the presence of hard (Table 3.1), temperature-resistant carbides. Because they have two-phase or multiphase structures, their properties are governed by the general rules relating to such structures. Thus, a high bond strength between matrix and particle is essential: the strength of the structure increases with increasing matrix strength, increasing carbide volume fraction, decreasing carbide particle size, and the absence of stress raisers, such as large angular carbides or inclusions. For these reasons, the manufacturing technique is as important as composition itself [430-434]. Great advances have been made by use of improved melting, casting, and thermomechanical processing techniques, and also by use of powder metallurgy [435], which ensures fine grain size coupled with very fine carbide distribution. The alloys range from low-alloy steels for forging dies, through medium-alloy steels specially formulated for cold or hot working [426,436], to high speed steels [437] designed for metalcutting applications but occasionally used also in plastic deformation processes. From the manufacturing point of view these steels have the advantage that they can be hot rolled or forged to dimensions from which the tool can be readily manufactured by various plastic working and machining techniques, and acquire their desirable properties in a final heat treatment. The wear of steels is discussed by Vingsbo and Hogmark [437a].

Table 3.1. Hardness of typical tool materials or of their constituents (from various sources)

Material or Constituent	Hardness, VHN	Material or Constituent	Hardness, VHN
Martensitic steel	500-1000	Mo_2C	1500
Nitrided steel	950	VC	2800
Cementite (Fe_3C)	850-1100	TiC	3200
Hard chrome coating	1200	TiN	3000
Alumina	2100-2400	B_4C	3700
WC (Co-bonded)	1800-2200	SiC	2600
WC	2600	Cubic boron nitride	6500
W_2C	2200	Polycrystalline diamond/WC	5500-8000
$(Fe, Cr)_7C_3$	1200-1600	Diamond	8000-12,000

3. *Cast Carbides.* When carbides reach very high proportions, the material is not hot workable any more and must be cast to shape. The matrix is usually a cobalt alloy, the high cost of which limits the application of solid cast carbides but has not prevented their development as hard facing alloys [426]. Abrasive wear resistance increases with increasing carbide particle size [438].

4. *Cemented carbides (sintered carbides)*, reviewed by Exner [439], are again two-phase structures, but this time the carbide is in powder form, bonded with usually 5 to 30% cobalt [426,427]. Room-temperature strength and wear resistance increase with decreasing cobalt content and particle size. For conditions involving high temperatures, coarser carbides are preferable [440]. Toughness and transverse rupture strength increase with increasing cobalt content. To increase the matrix strength, heat treatable iron- [441], nickel-, or cobalt-base superalloys or refractory metal alloys can be used.

Tribological properties are controlled also by the carbide species. The most widespread are tungsten carbide (WC) tools, which wear by gradual loss of both WC and cobalt in working of nonferrous metals [442]. In contrast, the rapid wear observed in contact with steels at high temperatures is attributable to diffusion of tungsten and carbon into the steel [443] and consequent loss of strength. It is then necessary to mix WC with titanium carbide (TiC), which forms a carbide-rich interface [444,445] that resists diffusion and softening. These mixed WC-TiC tools may be regarded as transitions to the broader class of cermets, materials composed of carbides with a small quantity of metallic bonding phase. To date, cermets based on TiC have achieved the greatest significance for cutting tools because of their greater hardness and temperature resistance [427,446]. The addition of nitrides or carbonitrides has increased shock and thermal fatigue resistance, allowing not only higher but also periodic loading of the tools. All carbides seem to rely on adsorbed films for low friction, which is necessary for prevention of premature cracking. Silicon carbide (SiC) has many desirable tribological properties [447] but is used principally for abrasive machining.

5. *Ceramics* are compounds of metallic and nonmetallic elements. By definition, carbides are ceramics too, but in tool technology the term is usually reserved for oxides, principally Al_2O_3 with the addition of ZrO_2 for higher strength. Fracture characteristics depend on microstructure and method of manufacture [448], and wear rates are again reduced in the presence of moisture [449]. Recent developments aim at improving the toughness of ceramics by the incorporation of additional elements and compounds such as TiC [427]. Furthermore, compositions such as $TaN\text{-}ZrB_2$ have been hot pressed into solid tool bits. Cubic boron nitride (CBN), obtained from the hexagonal form by diamondmaking technologies, is the hardest of the ceramics (Table 3.1) but is also brittle and must be supported on a substrate or sintered into a compact.

6. *Diamond* is the hardest material, but in its natural form suffers from a directionality of properties and has very low impact strength. In the man-made form, diamond can be used for making polycrystalline tools for both deformation and cutting applications [450,451]. At temperatures over 600 C, diamond converts to graphite. In contact with iron, it wears by a diffusion process.

The brittleness of many tool materials makes them prone to catastrophic failure, and the determination of relevant mechanical properties calls for special test methods such as the one developed by Berry et al [452].

The changes in hardness of selected tool materials with temperature are given in Fig.3.66. A more extensive treatment of tool materials will be found in Chapter 11.

Miscellaneous Die Materials. There are some materials that do not possess the high hardness normally associated with tools, but which still fulfill very important functions.

An important group of die materials is cast irons, in which the presence of carbon provides some built-in anti-adhesion and lubrication qualities, and thus makes them suitable for large pressworking dies used in the automotive industry. The presence of graphite is beneficial.

Among nonferrous metals, there are some that show remarkably low adhesion to some workpieces. Prime examples are the multiphase, age-hardenable aluminum bronzes, which exhibit very low adhesion to stainless steel—probably because of diffusion of aluminum to the surface, which then reduces adhesion as long as the surface is protected by a boundary lubricant [453]. If greater strength is required, these bronzes can also be applied as coatings on die steel surfaces [454].

Hard, castable zinc-base alloys are suitable for die-tryout, prototype, and low-volume pressworking.

Polymeric materials (plastics), especially when filled with hard, abrasion-resistant substances, have been successfully employed for low-volume sheet metalworking production dies. Elastomeric foams, particularly polyurethane, have remarkable resistance to

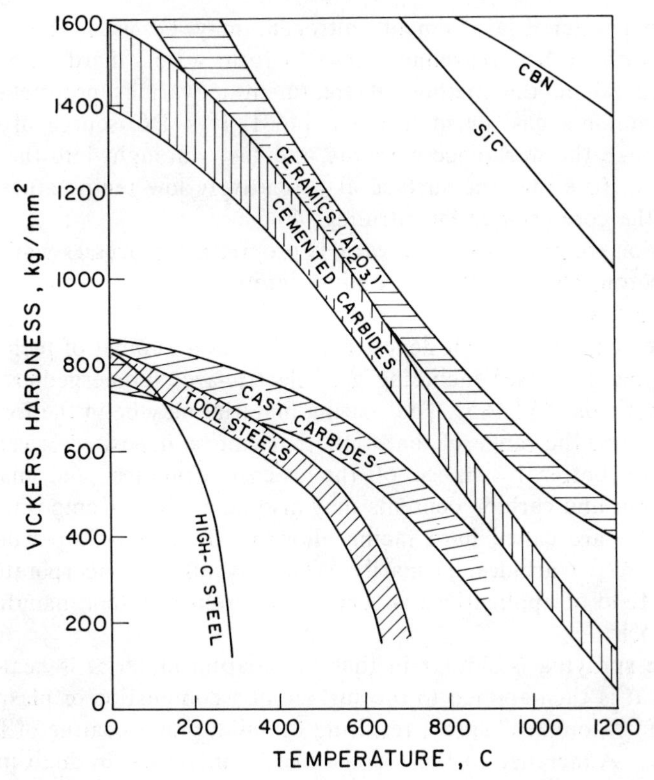

Fig.3.66. Change in hardness with temperature for some typical die materials

abrasion under moderate sliding conditions and can be used as die elements in sheet blanking, bending, and drawing operations.

3.11.5 Surface Coatings for Tools

Tribochemical and tribomechanical reactions are restricted to the surface, and therefore better tool performance can often be achieved if a material with a desirable combination of strength and toughness is used as the bulk of the tool, and the tool surface is converted or a surface layer is deposited to provide the requisite hardness and temperature resistance [455-459]. The choice of the coating is dictated by tribological considerations [22,460]; good bonding to the substrate, absence of harmful residual stresses, and matched thermal expansion are essential.

1. Surface hardening is possible in steels of sufficiently high carbon content to produce martensite on the surface by rapid quenching, while retaining a pearlitic structure in the core. Whether the effect is achieved by controlled quenching of the entire tool from the austenitization temperature, or by surface heating by means of flame or induction heating, it is important that residual stresses should not lead to cracks on subsequent grinding or use. Laser hardening [461] may be attractive. The hard case formed by spark hardening [462] benefits from the transfer of WC from the electrode.

2. Case hardening differs from the surface hardening discussed above in that the bulk of the tool has a lower carbon content and the surface is enriched in carbon to provide the requisite martensite hardness [462]. The structure of the surface layer again affects performance; residual austenite is not necessarily harmful because it can accommodate stresses and arrest cracks.

3. Another interstitial element, nitrogen, may be used alone (nitriding) or in combination with carbon (carbonitriding) to form a very hard case. Case depth and properties depend on the method of treatment in various proprietary salt baths, in dissociated ammonia gas, or in nitrogen [463]. If a DC source of up to 1500 V is attached to make the workpiece cathodic, the gas is brought into the plasma state and nitrogen ions diffuse into the surface at a relatively low temperature, permitting heat treatment of the core prior to ion-nitriding [464].

4. Diffusion treatments involve various proprietary processes which incorporate sulfur, silicon, boron, aluminum, or chromium into the surface, to form intermetallic compounds or carbides.

5. Surface coating by melt deposition embraces a number of techniques. One group employs welding processes including gas, electric-arc, submerged-arc, electroslag, and plasma-arc methods [457,465]. The coating material may be in the form of wire or powder, and therefore the range of materials that can be deposited is very wide: tool steel materials offer better resistance of the as-cast structure, and materials of higher alloying element and carbide contents may also be applied. Compositions corresponding to cast carbides are called hard facing alloys [466]. Fe-Cr-Co powders are melted on the surface [467]. Carbides, primarily WC, may also be incorporated into the fused coating. The field of applications is very broad, from new tool manufacture to reclaiming of worn tools.

6. Flame spraying is similar in that the coating material is heated to melting (or close to it), but is then applied to the surface in a combustion or plasma flame [465] in the form of an "atomized" spray, resulting in buildup of a coating of flattened particles fused together. Adherence to the surface can be increased by high impact velocity, as provided by the detonation gun technique [468]. Ceramics as well as metals and cermets may be deposited. The porosity of the coating is actually beneficial in lubricated processes, but, if porosity is undesirable, the coating may be fused after spraying. Boron and silicon added to the coating material act as a flux (by forming a glassy phase) during fusing. A somewhat related process is the deposition of coatings, including WC, by spark-hardening [469], exploiting the high temperatures generated by spark discharge.

7. In the technique of powder coating, the material (usually a carbide/metal composite) is applied as a slurry or a nonwoven cloth and then fused in a vacuum or hydrogen atmosphere [470].

8. Electroplating is very versatile in applying coatings, from hard chromium plating to Co-W and Co-Mo alloys. Electroless coatings, usually of nickel with SiC [471], CBN [472], or boron, also rely on physical adhesion, although subsequent heat treatment establishes some metallurgical bond and, in the case of Ni-B coatings, the formation of nickel borides of high hardness.

9. More recent techniques are treated in a comprehensive review by Ramalingam [473] and by Peterson and Ramalingam [473a]. Chemical vapor deposition (CVD), reviewed also by Yee [474] and a subject of biennial conferences [474a], involves the thermal decomposition of an organic metal compound or the hydrogen reduction of a metal halide. The metal (usually Ti, Cr, or W) then forms a carbide with carbon from the underlying steel. In common with other ceramics, wear of the coating is severe in vacuum but is low in a humid atmosphere or in the presence of a lubricant [475]. By admixing appropriate gases into the atmosphere, nitrides, oxides, borides, etc. [476,477], or mixed ceramics such as Ti(N,C,O) [478] and Ti(B,N) [479] of excellent tribological characteristics, are formed. Several layers are sometimes deposited on top of each other to take advantage of the specific properties of various coating types.

10. Physical vapor deposition (PVD) works on the principle of the electron tube

(diode). The material to be deposited is melted in high vacuum, by means of a filament, electron-beam melting, or induction melting, and the emerging negatively charged ions are accelerated by a probe connected to the positive terminal and thus hurled onto the substrate (tool) surface. For good bonding to the surface, the substrate is heated [473].

11. Ion plating (reviewed also by Teer [480]) is a plasma physic technique wherein a high voltage supply (500 to 5000 V) causes the substrate to be enveloped by a glow-discharge plasma. Atoms of the material are sputtered from the substrate, ensuring continuous cleaning of both substrate and coating. Thus, the coating builds up at a rate dictated by the difference between deposition and sputtering rates, and adhesion of the films is excellent [481,482]. The film structure is columnar with microporosity [483].

12. When a radiofrequency (RF) field is used in sputter deposition, charge accumulation is prevented and nonconducting materials, such as ceramics [484] and solid (layer-lattice) lubricants [485], can also be deposited. This is a very rapidly developing field, with great potential for coating of tools—particularly with TiC, TiN, and TiB. Adhesion of coatings can be improved by various techniques [486].

13. Ion implantation is a spin-off from solid-state technology. It differs from the above techniques in that the ions, accelerated in a field of 100 to 200 keV, possess sufficient energy to penetrate to a depth of 0.1 μm. In contrast to ion plating and ion nitriding, in which only a few particles are ionized and throwing power is considerable, ion implantation injects the ions in the form of a high-velocity directed beam. No surface coating is formed, but rather the surface composition is changed. Both nonmetallic (principally nitrogen) and metallic ions can be implanted [473,487,488]. The obvious application of this technique is to tool materials, although it can be applied equally well to workpieces.

There are, of course, a number of coating techniques for workpiece surfaces. Their principal aim is to provide a more favorable lubrication system, and therefore they will be discussed in Sec.4.7.1.

3.11.6 Material-Lubricant Interactions

The importance of the tool material is well recognized in metalworking tribology, and many attempts have been made to find materials that will minimize damage in the event of lubricant failure. In many ways, this desire gave the impetus to the application of surface coatings to dies and tools. An unintentional consequence of such measures may, however, be a loss of reactivity with the lubricant, negating whatever gains may have been made [489]. In view of the great importance of the subject, it is surprising that only a few investigations have been made in which die-workpiece-lubricant interactions were systematically studied. The impact of such interactions varies according to the lubrication mechanism operative in the process:

1. If the die and workpiece surfaces are effectively separated by a continuous, inert, solid, or PHD film, the real area of contact is vanishingly small and makes too little a contribution to the frictional force to reveal any differences in material pairings. There may still be differences in the tendency toward die pickup, but laboratory tests are usually of insufficient duration to reveal them. Thus, Wiegand et al [418] found in the drawing of austenitic stainless steel sheet between static rollers that lithium stearate gave a steady, low draw force and prevented pickup irrespective of tool material. However, under more severe conditions, a lubricant carrier or conversion coating may be required on the workpiece to ensure continuity and bonding of the lubricant film.

2. Any discontinuity in the lubricant film (e.g., localized damage to a solid film, or a mixed-film mechanism) makes the effect of the die-workpiece material pairing immediately felt. Thus, Wiegand et al [418] found that an aluminum bronze die was

vastly superior to a tool steel die when an essentially neutral petrolatum was used as a lubricant on stainless steel; the draw force was lower and steadier, and die pickup was absent. Very often, the real contact area is still small enough to make the magnitude of friction in itself quite unsuitable for judging the lubricant. Thus, in the work of Schlosser [490] there was little difference in friction in drawing of stainless steel with sodium palmitate. In contrast, pickup varied greatly: it was quite substantial with plain and boronized die steels and Ferro-TiC, whereas ionitrided steel, oxidized Ferro-TiC, and WC and chromium carbide coatings prevented pickup. A poor lubricant such as plain mineral oil cannot prevent pickup with any of these dies. Such differences may be observed also when the lubricant is bonded. For example, Kudo et al [491] found, in their scratch tests on phosphate-soap lubricated steel, an essentially constant friction, yet wear rate was high with die steels but low with WC and a die steel coated with WC.

3. Solid-film lubricants capable of reaction are very sensitive to the substrate material (Sec.3.3.4). Both die and workpiece may enter into reaction, as shown by Schey and Newnham [492] in twist-compression tests in which the coefficient of friction was measured and metal transfer (pickup on the anvil) was observed.

When the workpiece material is essentially nonreactive with MoS_2 (such as aluminum alloy 7075 or titanium alloy Ti-6Al-4V), the reactivity of the die and the magnitude of interface pressure determine whether a viable film can be established (Fig.3.67

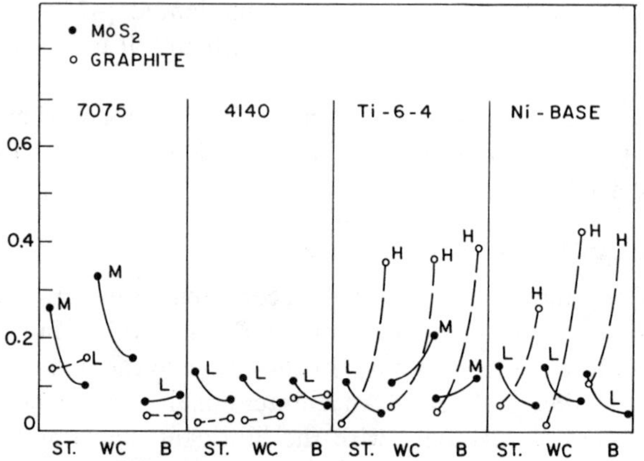

Fig.3.67. Variation of friction in first ten revolutions in twist compression at room temperature (from data in [492])

and 3.68). Thus, with aluminum alloy 7075, a boronized anvil (B), which is capable of sulfide formation, gave best results at both cold and hot working temperatures, followed by the reactive but not so hard 4140 steel (ST), whereas the virtually nonreactive carbide (WC) was the worst despite its high hardness. The same trends were evident with the harder Ti-6Al-4V, except that the higher interface pressures seemed to accelerate MoS_2 film formation on the 4140 anvils. When the workpiece material is capable of reaction, the die composition effect is washed out (see results with annealed 4140; Hastelloy X, a nickel-base superalloy; and TZM, a molybdenum alloy).

It would appear that film formation with graphite was aided by a slightly elevated temperature and a reactive die and/or workpiece (Fig.3.68). Graphite was quite consistently poor against a boronized die at elevated temperatures or higher pressures, even though a very stable boron carbide (BC) exists; one can only surmise that the severe conditions imposed on the graphite film caused a breakdown before a beneficial reac-

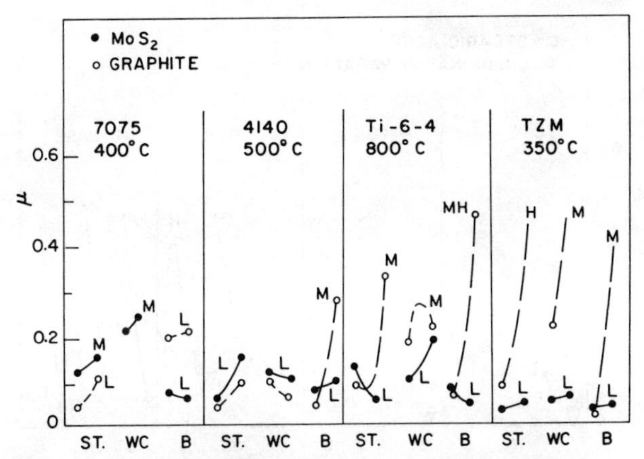

Fig.3.68. Variation of friction in first five revolutions in twist compression at elevated temperatures (from data in [492])

tion could take place. Graphite was an unqualified success only with the 4140 steel workpieces; with aluminum alloy 7075 it was better at hot working temperatures, but it was quite poor with titanium and molybdenum alloys even against a steel die. This is somewhat surprising in view of the widespread practical application of graphite but shows most clearly that a successful film can form only under favorable contact conditions.

This point was further emphasized by the results of ring upsetting tests (Fig.3.69), in which there is little of the rubbing so essential to establishment of oriented and reaction films. It is then not surprising that neither die nor workpiece material (or, for that matter, elevated temperature, not shown here) makes much difference [492]. The lower friction observed with graphite was most likely due to residual moisture; the MoS_2 film deposited from an aerosol was completely dry.

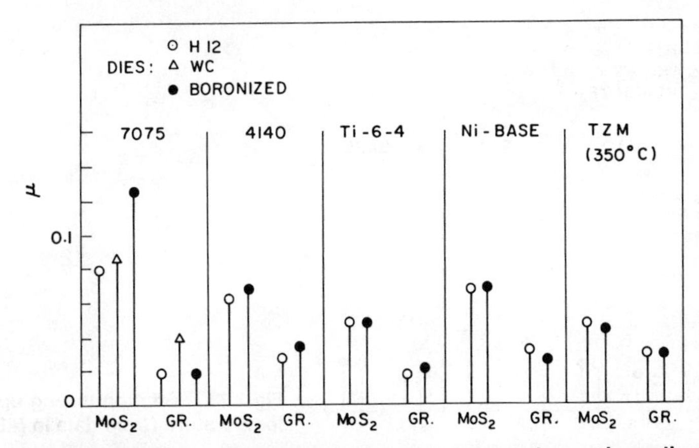

Fig.3.69. Friction in ring upsetting (at room temperature unless otherwise noted) (from data in [492])

4. E.P. lubricants also rely on reactions for their effectiveness (Sec.3.4), and either the die or the workpiece must be capable of reaction. From twist-compression experiments [492] it again appears that reactivity with an iron- or nickel-base alloy is essential for a chlorinated compound, as shown by the relatively poor performance of the boronized anvils against aluminum and titanium alloys (Fig.3.70). Friction and surface damage are lowest when both workpiece and die react (4140 anvil against 4140 steel or Ni-

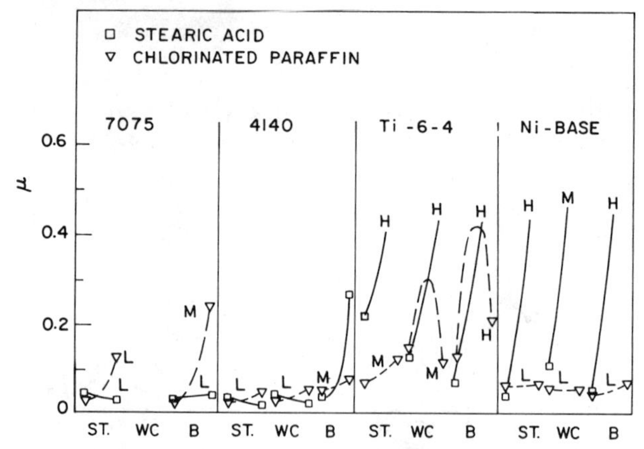

Fig.3.70. Variation of friction in first ten revolutions in twist compression at room temperature (from data in [492])

base superalloy). With a nonreactive workpiece material (Ti), even the high hardness of WC cannot prevent initial high friction and some surface damage, but the cobalt binder is capable of reaction, and friction settles to a lower final value. However, excessive reactivity with the cobalt binder could lead to damage of WC dies. The need for reaction accelerated by rubbing is shown by the insensitivity of the ring compression test (Fig.3.71); obviously the viscous properties of the chlorinated paraffin (Cl) ensured predominantly hydrodynamic lubrication and neutralized whatever die effects may have been present. This is in line with the observation of Begelinger et al [204] that a full EHD film was insensitive to a change in steel, but the load carrying capacity decreased when an oil with sulfur additives was run under thin-film conditions and the chromium content in the steel increased from 1.5 to 5.6%.

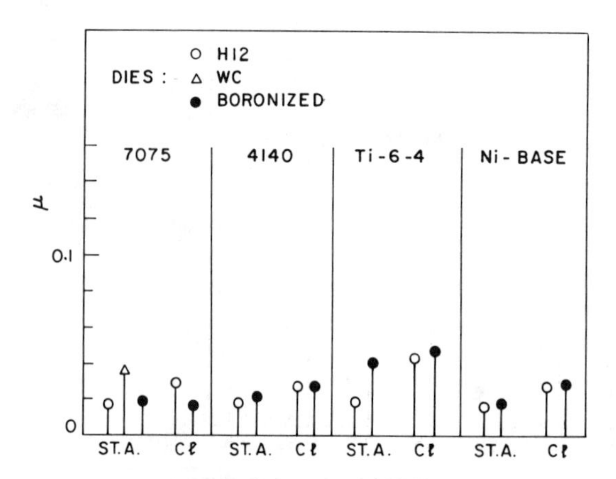

Fig.3.71. Friction in ring upsetting at room temperature (from data in [492])

5. Superior bonding of boundary additives on reactive surfaces (Sec.3.5.1) improves the performance of compounded lubricants. This becomes clearly visible only when hydrodynamic effects are completely absent, as in twist-compression tests [492]. It is preferable that the workpiece, on which virgin surfaces are exposed, be reactive. Thus, aluminum alloy 7075 is reactive enough with stearic acid to be protected against any die material (Fig.3.70), but 4140 steel was effectively lubricated only against steel (ST) and WC dies, whereas breakdown and pickup were evident with the boronized (B) die.

3.12 Summary

1. A valid approach to the tribology of metalworking must consider the die-workpiece interface in its realistic complexity. Process geometry, microtopography, the physical and chemical properties of tool, workpiece, and lubricant, and their interactions under the influence of the pressures, temperatures, sliding velocities, and atmosphere prevailing in the process must be taken into account. Of special significance is the generation of large portions of new surfaces and the consequent enhancement of tribochemical reactions.

2. In all processes, some plastic deformation is inevitable and indeed desirable. Thus, the real area of contact ranges from a few to one hundred percent of the apparent area of contact. Sliding of the workpiece over the tool is resisted by adhesive joints, asperity deformation, plowing, and molecular forces, leading to a high interface shear strength which is then reduced by lubricants. The lubricants also fulfill the vital function of separating the die from the workpiece surface and thus prevent metal transfer (tool pickup).

3. Solid films, whether of oxides, metal or polymer films, or layer-lattice compounds, have no self-healing capacity and are truly effective only when a continuous film can be established and maintained. Attachment to the workpiece surface by reaction is helpful, especially in film formation and durability.

4. Reactions are vital for the success of E.P. lubrication. Even so, the films are patchy and ensure adequate protection only if worn parts are replaced fast enough. They are relatively ineffective in reducing friction but are often indispensable in limiting metal transfer and wear.

5. Boundary lubricants are of great value because the oriented multimolecular films, adsorbed to and sometimes reacted with the surface, present a barrier to asperity penetration and endow the lubricant with greatly improved rheological properties, sometimes through tribochemical reactions leading to the formation of a friction polymer.

6. Many lubricants and additive carriers are viscous liquids. If the moving, deforming material forms a converging gap with the die, a plastohydrodynamic (PHD) film may form; the normal approach of the die and workpiece may lead to the development of a squeeze film. Even though the film is seldom thick enough to ensure complete separation of die and workpiece, the average film thickness is large enough to make PHD analysis useful. Friction and wear drop to very low levels, and the magnitude of friction can be correctly predicted only if the rheology of the lubricant is known. Conditions typical of EHD lubrication are often obtained, and the lubricant may behave as a viscous or viscoelastic liquid or as a solid-like substance of pressure-dependent shear strength.

7. In the vast majority of liquid-lubricated operations only part of the interface is separated by a fluid film; other parts are in boundary contact. Such mixed-film lubrication is highly responsive to operating conditions, as can be described by a modified Stribeck diagram. With increasing viscosity and/or sliding velocity, the mechanism shifts towards predominantly PHD lubrication and the coefficient of friction drops. Once predominantly PHD lubrication is attained, traction in the fluid film increases with viscosity and/or sliding velocity. Nevertheless, protection of asperities by boundary and/or E.P. additives is mandatory in practically all situations.

8. The tribology of metalworking is greatly affected by the roughness of contacting surfaces. The roughness of the workpiece changes in the course of deformation; nonuniform yielding of individual grains results in roughening with solid films and in the development of hydrodynamic or hydrostatic pockets in the presence of liquid lubri-

cants. Conformity to the die is typical of boundary lubrication. A workpiece with an initially rough surface carries more lubricant into the deformation zone, especially if roughness is oriented perpendicular to the sliding direction or is random; it also retains the lubricant better. Die roughness is harmful under PHD or mixed-film conditions but provides lubricant reservoirs for solid lubrication. Both die and workpiece roughness may contribute to a micro-PHD mechanism.

9. Elastic deflection of tool or machine elements, combined with a mixed-film mechanism, may give instabilities that show up as surface finish or gage variations on the product. Ultrasonic vibration of the tool may stabilize the conditions under specific circumstances and may also reduce problems of lubrication.

10. All forms of wear are encountered in metalworking. Adhesive wear leads to transfer of material to the tool surface (pickup), and to wear of the workpiece and sometimes also of the die. Detached pickup and oxide particles, together with hard constituents in the workpiece material, cause abrasive wear of the tool. Repeated loading and stress reversal induce tensile stresses which lead to fatigue or delamination of surface layers; when these stresses are combined with periodic heating, thermal fatigue may occur. Chemical reactions with oxygen and lubricant additives result in corrosive wear, sometimes accelerated by electrochemical effects.

11. Tool and workpiece materials are generally chosen for minimum mutual adhesion. Prediction of adhesion is still not possible, but the solubility criterion can give some guidance. Desirable combinations of bulk and surface properties can often be attained by coating a tougher base with a layer chosen for low adhesion and high wear resistance. Benefits may be lost if the tool material is thus rendered incapable of entering into those reactions that are essential for the functioning of the lubricant.

12. Modeling of friction, lubrication, and wear phenomena still presents formidable difficulties. The shear strength of the interface τ_i is the most reliable descriptor, but for ease of calculation the coefficient of friction μ or the friction shear factor m is preferable. As summarized in Table 3.2, the most realistic modeling uses τ_f or m for sticking friction; τ_s for metal films; μ for polymers, layer-lattice compounds, and boundary lubricants; a constant or variable μ for mixed films; and μ for PHD films.

Table 3.2. Modeling of friction in metalworking

Model	Sticking friction ($\tau_i=\tau_f$)	Solid film (metal) ($\tau_i=\tau_s\simeq$ const.)	Solid and boundary films ($\tau_i=f(p)$)	Mixed film ($\tau_i=$variable)	Plastohydrodynamic ($\tau_i=f(\eta v)$)
$\mu=\frac{\tau_i}{p}$	$\frac{\tau_f}{p}$	$\frac{\tau_s}{p}$	$\frac{\tau_i}{p}$	$\frac{\tau_i}{p}$	$f\left(\frac{\eta v}{p}\right)$
$\tau_i=f(p)$	($p\uparrow\mu\downarrow$)	($p\uparrow\mu\downarrow$)	($\mu\simeq$ constant)	($\mu\simeq$ constant or variable)	($\mu\simeq$ constant)
$m=\frac{\tau_i}{\tau_f}$	$\frac{\tau_f}{\tau_f}$	$\frac{\tau_s}{\tau_f}$	$\frac{\tau_s}{\tau_f}$	$\frac{\tau_i}{\tau_f}$	$f\left(\frac{\eta v}{\tau_f}\right)$
$\tau_i\neq f(p)$	($m=1$)	($\tau_f\uparrow m\downarrow$)	(m variable)	(m variable)	(m variable)

References

[1] J.A. Schey (ed.), *Metal Deformation Processes: Friction and Lubrication*, Dekker, New York, 1970.
[2] C.H. Riesz, in [1], pp.83-239.
[3] C.A. Bailey and J.S. Aarons (ed.), *The Lubrication Engineers' Manual*, U.S. Steel Corp., New York, 1971.

[4] M. Billet, *Industrial Lubrication*, Pergamon, Oxford, 1979.
[5] F.P. Bowden and D. Tabor, *Friction: An Introduction to Tribology*, Heinemann, London, 1974.
[6] F.P. Bowden and D. Tabor, *Friction and Lubrication*, Methuen, New York, 1967.
[7] A.F. Brewer, *Effective Lubrication*, Krieger, New York, 1974.
[8] British Petroleum Co., *Lubrication Theory and Its Application*, BP Trading, London, 1969.
[9] E.G. Ellis, *Fundamentals of Lubrication*, Scientific Publ., Broseley, Shropshire, England, 1970.
[10] G.G. Evans, V.M. Galvin, W.S. Robertson, and W.F. Waller (ed.), *Lubrication in Practice*, Macmillan, London, 1972.
[11] R.C. Gunther, *Lubrication*, Chilton, Philadelphia, 1971.
[12] J. Halling, *Introduction to Tribology*, Wykeham, London, 1976.
[13] E.D. Hondros, *Tribology*, Mills and Boon, 1971.
[14] B. Pugh, *Friction and Wear; A Tribology Text for Students*, Newnes-Butterworth, London, 1973.
[15] J.D. Summers-Smith, *An Introduction to Tribology in Industry*, Machinery Publ. Co., London, 1969.
[16] J.G. Wills, *Lubrication Fundamentals*, Dekker, New York, 1980.
[17] F.T. Barwell, *Bearing Systems-Principles and Practice*, Oxford University Press, 1979.
[18] F.P. Bowden and D. Tabor, *The Friction and Lubrication of Solids*, Clarendon Press, Oxford, Pt.I, 1950, and Pt.II, 1964.
[19] E.R. Braithwaite (ed.), *Lubrication and Lubricants*, Elsevier, Amsterdam, 1967.
[20] D.H. Buckley, *Friction, Wear and Lubrication in Vacuum*, NASA-SP 277, Washington, 1971.
[21] A. Cameron, *The Principles of Lubrication*, 2nd Ed., Longman, London, 1980.
[22] H. Czichos, *Tribology: A Systems Approach*, Elsevier, Amsterdam, 1978.
[23] J. Halling, *Principles of Tribology*, Macmillan, London, 1975.
[24] M.D. Hersey, *Theory and Research in Lubrication*, Wiley, New York, 1966.
[25] I. Iliuc, *Tribology of Thin Layers*, Elsevier, Amsterdam, 1980.
[26] I.V. Kragelski, *Friction and Wear*, Butterworth, Washington, 1965.
[27] I.V. Kragelski, M.N. Dobychin, and V.S. Kombalov, *Friction and Wear: Calculation Methods*, Pergamon, Oxford, 1981.
[28] F.F. Ling, E.F. Klaus, and R.S. Fein (ed.), *Boundary Lubrication: An Appraisal of World Literature*, ASME, New York, 1969.
[29] D.F. Moore, *Principles and Applictions of Tribology*, Pergamon, Oxford, 1975.
[30] E. Rabinowicz, *Friction and Wear of Materials*, Wiley, New York, 1965.
[31] A.Z. Szeri (ed.), *Tribology: Friction, Lubrication and Wear*, McGraw-Hill, London, 1980.
[32] J.J. O'Connor, J. Boyd, and E.A. Avallone (ed.), *Standard Handbook of Lubrication Engineering*, McGraw-Hill, New York, 1968.
[33] M.J. Neale, *Tribology Handbook*, Butterworth, London, 1973.
[34] E.S. Booser (ed.), CRC *Handbook of Lubrication (Theory and Practice of Tribology)* (ASLE), CRC Press, Boca Raton, FL, 1983.
[35] *Proc. Conf. Lubrication and Wear*, Inst. Mech. Eng., London, 1957.
[36] *Lubrication and Wear, Third Convention*, Proc. Inst. Mech. Eng., *179*, Pt.3D, 1964-65.
[37] *Lubrication and Wear, Fifth Convention*, Proc. Inst. Mech. Eng., *181*, Pt.30, 1966-67.
[38] *Lubrication and Wear: Fundamentals and Application to Design*, Proc. Inst. Mech. Eng., *182*, Pt.3A, 1967-68.
[39] P.M. Ku (ed.), *Interdisciplinary Approach to Friction and Wear*, NASA SP-181, Washington, 1968.
[40] P.M. Ku (ed.), *Interdisciplinary Approach to the Lubrication of Concentrated Contacts*, NASA SP-237, Washington, 1970.
[41] B.D. McConnell (ed.), *Assessment of Lubricant Technology (1972 Spring Lubrication Symposium)*, ASME, New York, 1972.
[42] *Tribology Convention 1972*, Inst. Mech. Eng., London, 1973.
[43] P.M. Ku (ed.), *Interdisciplinary Approach to Liquid Lubricant Technology*, NASA SP-318, Washington, 1973.
[44] F.F. Ling (ed.), *Proc. Tribology Workshop*, National Science Foundation, Washington, 1974.
[45] *First European Tribology Congress*, Inst. Mech. Eng., London, 1975.

[46] T. Sakurai (ed.), *Proc. JSLE-ASLE Int. Lubrication Conf. Tokyo 1975*, Elsevier, New York, 1976.
[47] *Eurotrib 77 (2nd European Tribology Congress)* (3 volumes), Ges. Tribol., Duisburg-Homberg, Germany, 1977.
[47a] M. Hebda and C. Kajdas (ed.), *Eurotrib 81* (8 volumes), Elsevier, Amsterdam, 1982.
[48] *Tribology 1978: Materials Performance and Conservation*, Mechanical Engineering Publs., London, 1978.
[49] N.P. Suh and N. Saka (ed.), *Fundamentals of Tribology*, MIT Press, Cambridge, MA, 1980.
[49a] D.A. Rigney (ed.), *Fundamentals of Friction and Wear of Materils*, ASM, Metals Park, 1981.
[49b] J.M. Georges (ed.), *Microscopic Aspects of Adhesion and Lubrication*, Elsevier, Amsterdam, 1982.
[50] F.F. Ling, R.L. Whitely, P.M. Ku, and M.B. Peterson (eds.), *Friction and Lubrication in Metal Processing*, ASME, New York, 1966.
[51] *Tribology in Iron and Steel Works*, ISI Publ. 125, Iron and Steel Institute, London, 1970.
[52] *Proc. 1st Int. Conf. Lubrication Challenges in Metalworking and Processing*, IIT Res. Inst. Chicago, 1978.
[53] *Proc. 2nd Int. Conf. Lubrication Challenges in Metalworking and Processing*, IIT Res. Inst. Chicago, 1979.
[54] *Tribology in Metal Working — New Developments*, Inst. Mech. Eng., London, 1980.
[55] S. Kalpakjian and S.C. Jain (ed.), *Metalworking Lubrication*, ASME, New York, 1980.
[56] W.J. Bartz and J. Wolff (ed.), *Lubrication in Metalworking* (3rd Int. Colloquium), Technische Akademie Esslingen, Germany, 1982.
[57] R.J. Wakelin, Ann. Rev. Mater. Sci., *4*, 1974, 221-253.
[58] M.G. Cockroft, in [19], pp.472-553.
[59] K.H. Kloos, in *Mechanische Umformtechnik*, O. Kienzle (ed.), Springer, Berlin, 1968, pp.293-342.
[60] T. Mang and W. Neumann, MM-Industriej., *77*, 1971, 405-408, 445-448 (German).
[61] E. Dannenmann, G. Schmitt and R. Geiger, in *Lehrbuch der Umformtechnik*, Vol.1, K. Lange (ed.), Springer, Berlin, 1972, pp.213-269.
[62] R.L. Jentgen, Tech. Rep. MFR 72-01, SME, Dearborn, 1972.
[63] J.A. Schey, in [44], pp.428-451.
[64] O. Pawelski, Schmiertech. Trib., *25*, 1978, 137-140 (German).
[65] W.J. Bartz, Wear, *49*, 1978, 1-19.
[66] W.R.D. Wilson, J. Appl. Metalwork., *1* (1), 1979, 7-19.
[67] F.T. Barwell, in [54], pp.51-63; in [56], Paper No.2.
[68] J.A.Schey, in *Proc. 4th Int. Conf. Production Engineering*, Jpn. Soc. Prec. Eng., Tokyo, 1980, pp.102-115.
[69] J.A. Schey, in [34], Vol.2, chap.19, *Metal Processing—Deformation.*
[70] *Properties and Metrology of Surfaces*, Proc. Inst. Mech. Eng., *182*, Pt.3K, 1967-1968.
[71] T.R. Thomas (ed.), *Rough Surfaces*, Longman, Harlow, England, 1982.
[72] K.H. Kloos, Fortschr.-Ber. VDI-Z, Ser.2, No.25, 1972 (German).
[73] Y. Tsuja, *Microstructures of Wear, Friction and Solid Lubrication*, Tech. Rep. No.81, Mechanical Engineering Laboratory, Tokyo, 1976.
[74] D.H. Buckley, Progr. Surface Sci., *12*, 1982, 1-154.
[75] D.H. Buckley, *Surface Effects in Adhesion, Friction, Wear and Lubrication*, Elsevier, Amsterdam, 1981.
[76] J. Oudar, Int. Met. Rev., *23*, 1978, 57-73.
[77] C.T.H. Stoddart and C.P. Hunt, Metals Techn., *8*, 1981, 205-212.
[78] J. Krämer, Z. Physik, *128*, 1950, 538 (German).
[79] H. Blok, in [40], pp.153-248.
[80] C.N. Rowe and W.R. Murphy, in [44], pp.327-401.
[81] P.A. Thiessen, K. Meyer, and G. Heinecke, *Grundlagen der Tribochemie*, Akademie Verlag, Berlin, 1967.
[82] P.G. Fox, J. Mat. Sci., *10*, 1975, 340-360.
[83] M. Ciftan and E. Saibel, Wear, *53*, 1979, 201-209.
[83a] A.R.C. Westwood and F.E. Lockwood, in [49b], pp.421-447.
[84] S.N. Postnikov, Electrophysical and Electrochemical Phenomena in Friction, Cutting and Lubrication, Van Nostrand Reinhold, New York, 1978.

[85] R.M. Latanision, in [49], pp.255-294.
[85a] D.R. Wheeler, J. Appl. Phys., *47,* 1976, 1123-1130.
[86] F.F. Ling, *Surface Mechanics,* Wiley, New York, 1973.
[87] T.R. Thomas and M. King, *Surface Topography in Engineering,* B.H.R.A., Bedford, 1980.
[88] Int. Conf. Metrology and Properties of Engineering Surfaces, Wear, *57,* 1979, 1-384.
[89] A.D. de Pater and J.J. Kalker (ed.), *The Mechanics of Contact between Deformable Bodies,* Delft University Press, 1975.
[90] D.J. Whitehouse, in [49], pp.17-52.
[91] A.H. Uppal and S.D. Probert, Wear, *23,* 1973, 173-184.
[92] J.A. Greenwood and J.B.P. Williamson, Proc. Roy. Soc. (London), *A295,* 1966, 300-319.
[93] J.F. Archard, Trib. Int., *7,* 1974, 213-220.
[94] K.L. Johnson, J.A. Greenwood, and S.Y. Poon, Wear, *19,* 1972, 91-108.
[95] D. Tabor, Proc. Roy. Soc. (London), *A251,* 1959, 378-393.
[96] T. Kayaba and K. Kato, Wear, *51,* 1978, 105-116.
[97] P. Heilmann and D.A. Rigney, Wear, *72,* 1981, 195-217.
[97a] D. Kuhlmann-Wilsdorf, in [49a], pp.119-186.
[98] A.S. Akhmatov, *Molecular Physics of Boundary Friction,* Israel Program for Scientific Translations, Jerusalem, 1966.
[99] A.J. Haltner, in [28], pp.39-60.
[100] E.D. Brown, R.S. Owens, and E.R. Booser, in [28], pp.7-18.
[101] J.J. Bikerman, Trans. ASME, Ser.F, J. Lub. Tech., *92,* 1970, 243-247.
[102] T. Wanheim and N. Bay, CIRP, *27,* 1978, 189-194.
[103] T. Wanheim, Wear, *25,* 1973, 225-244.
[104] M.C. Shaw, A. Ber, and P.A. Mamin, Trans. ASME, Ser.D, J. Basic Eng., *82,* 1960, 342-346.
[105] E.G. Thomsen, CIRP, *17,* 1969, 187-193.
[106] M.B. Peterson and F.F. Ling, Trans. ASME, Ser.F, J. Lub. Tech., *92,* 1970, 535-542.
[107] B. Fogg, in [70], pp.152-161.
[108] I.F. Collins, Int. J. Mech. Sci., *22,* 1980, 743-753.
[109] S. Ishizuka, J. JSLE (Int. Ed.), *1,* 1980, 33-38.
[110] P.K. Gupta and N.H. Cook, Wear, *20,* 1972, 73-87.
[111] J.E. Williams, Wear, *56,* 1979, 363-375.
[112] J.M. Challen and P.L.B. Oxley, Wear, *53,* 1979, 229-243.
[113] P.L.B. Oxley, Wear, *65,* 1980, 227-241.
[114] J.A. Schey, in *Proc. 4th NAMRC,* SME, Dearborn, 1976, pp.108-114.
[115] J.A. Schey, in *Metal Forming Plasticity,* H. Lippmann (ed.), Springer, Berlin, 1979, pp.336-348.
[116] O. Kubaschewski and B. Hopkins, *Oxidation of Metals and Alloys,* Butterworth, London, 1962.
[117] G.C. Wood, Brit. Corrosion J., *4,* 1969, 244-248.
[118] R.F. Tylecote, J. Iron Steel Inst., *196,* 1960, 135-141.
[119] J. Benard, Met. Rev., *9,* 1964, 473-503.
[120] E. Rabinowicz, ASLE Trans., *10,* 1967, 400-407.
[121] P.L. Hurricks, Wear, *15,* 1970, 389-409.
[122] M. Mrozek, MW Interf., *1* (4), 1976, 21-27.
[123] J.A. Schey, in [50], pp.20-38.
[124] C.F. Hinsley, A.T. Male, and G.W. Rowe, Wear, *11,* 1968, 233-238.
[125] W.T. Edwards et al, in [56], Paper No.6.
[126] R.E. Riecker and L.C. Towle, J. Appl. Phys., *38,* 1967, 5189-5194.
[127] F. Jovane and A. Ludovico, in *Proc. 6th NAMRC,* SME, Dearborn, 1978, pp.205-211.
[128] J. Halling, Trib. Int., *12,* 1979, 203-208.
[129] S.H. Rhee and K.C. Ludema, Wear, *46,* 1978, 231-240.
[130] N.S. Eiss, Jr., K.C. Wood, J.A. Herold, and K.A. Smyth, Trans. ASME, Ser.F, J. Lub. Tech., *101,* 1979, 212-219.
[131] B.J. Briscoe and A.C. Smith, Rev. Deformation Behavior Mater., *3,* 1980, 151-191.
[132] J.K.A. Amazu, B.J. Briscoe, and D. Tabor, ASLE Trans., *20,* 1977, 354-358.
[133] I.M. Ward, in *Polymer Science Symposium,* Wiley, *32,* 1971, 95-217.
[134] R.C. Bowers, J. Appl. Phys., *42,* 1971, 4961-4970.
[135] L.C. Towle, J. Appl. Phys., *42,* 1971, 2368-2376.
[136] H. Eyring and M.S. Jhon, in [40], pp.249-278.

[137] S.C. Cohen and D. Tabor, Proc. Roy. Soc. (London), *A291*, 1966, 186-207.
[138] A.A. Koutkov and D. Tabor, Tribology, *3*, 1970, 163-164.
[139] B.J. Briscoe, V. Mustafaev, and D. Tabor, Wear, *19*, 1972, 399-414.
[140] J.M. Senior and G.H. West, Wear, *18*, 1971, 311-323.
[141] E.R. Braithwaite, *Solid Lubricants and Surfaces,* Pergamon, New York, 1964.
[142] P.J. Bryant, M. Lavik, and G. Salomon (ed.), *Mechanisms of Solid Friction,* Elsevier, Amsterdam, 1964.
[143] F.J. Clauss, *Solid Lubricants and Self-Lubricating Solids,* Academic Press, New York, 1972.
[144] *Proc. Int. Conf. on Solid Lubrication,* ASLE, Park Ridge, IL, 1971.
[145] *Proc. 2nd Int. Conf. on Solid Lubrication,* ASLE SP-6, Park Ridge, IL, 1978.
[146] J.B. Peace, in [141], pp.67-118.
[147] K. Feneberger, Lubric. Eng., *31*, 1975, 456-460.
[148] M. Mrozek, MW Interf., *4* (1), 1979, 13-22.
[149] O. Pawelski, Schmiertechnik, *15*, 1968, 129-138 (German).
[150] A.J. Groszek and R.E. Whiteridge, ASLE Trans., *14*, 1971, 254-266.
[151] E.R. Braithwaite and A.B. Greene, Wear, *37*, 1976, 251-264.
[152] J.P.G. Farr, Wear, *35*, 1975, 1-22.
[153] H.F. Barry, Lubric. Eng., *33*, 1977, 475-480.
[154] M. Mrozek, MW Interf., *4* (2), 1979, 10-25.
[155] H.E. Sliney, ASLE Trans., *21*, 1978, 109-117.
[156] W.J. Bartz, in [144], pp.335-349.
[157] J. Gansheimer and R. Holinski, Wear, *19*, 1972, 439-449.
[158] J.V. Reid and J.A. Schey, in [145], pp.22-29.
[159] R. Holinski, in [144], pp.41-58.
[160] A.P. Semenov and A.A. Kazura, in [145], pp.38-40.
[161] J.B. Peace, in [144], pp.290-299.
[162] P.W. Bridgman, *The Physics of High Pressure,* Dover, New York, 1970.
[163] W.E. Jamison, ASLE Trans., *15*, 296-305.
[164] G.W. Rowe, in [35], pp.333-338.
[165] R.I. Hughes and S.G. Daniel, Proc. Inst. Mech. Eng., *178*, Pt.3N, 1963-64, 273-282.
[166] M.C. Brendle, Wear, *43*, 1977, 127-140.
[167] R.L. Fusaro, Wear, *53*, 1979, 303-323.
[168] Y. Tamai, in [55], pp.63-65.
[169] D.H. Buckley, ASLE Trans., *21*, 1978, 118-124.
[170] J.P. King and Y. Asmerom, ASLE Trans., *24*, 1981, 497-504.
[171] W.J. Bartz, Lubric. Eng., *36*, 1980, 579-585.
[172] R. Holinski, Wear, *56*, 1979, 147-154.
[173] E.S. Forbes, Tribology, *3*, 1970, 145-152.
[174] Ph. Kapsa and J.M. Martin, Trib. Int., *15*, 1982, 37-42.
[175] T.P. Debies and W.G. Johnston, ASLE Trans., *23*, 1980, 289-297.
[176] A. Gauthier, H. Montes, and J.M. Georges, ASLE Trans., *25*, 1982, 445-455.
[177] E.S. Forbes and J. Battersby, ASLE Trans., *17*, 1974, 263-269.
[178] I.L. Goldblatt and J.K. Appeldoorn, ASLE Trans., *13*, 1970, 203-214.
[179] J.J. McCarroll et al, in [39], pp.23-33.
[180] H.B. Silver, Wear, *27*, 1974, 267-271.
[181] J.P. Giltrow, Tribology, *3*, 1970, 219-224.
[182] R.L. Johnson, M.A. Swikert, and D.H. Buckley, Corrosion, *16*, 1960, 395-398.
[183] R.W. Mould, H.B. Silver, and R.J. Syrett, Wear, *19*, 1972, 67-80; *22*, 1972, 269-286; *26*, 1973, 27-37.
[184] E.J. Latos and R.H. Rosenwald, Lubric. Eng., *25*, 1969, 401-411.
[185] E.S. Forbes and A.J.D. Reid, ASLE Trans., *16*, 1973, 50-60.
[186] R.W. Hiley, H.A. Spikes, and A. Cameron, Lubric. Eng., *37*, 1981, 732-737.
[187] H. Okabe, H. Nishio, and M. Masuko, ASLE Trans., *22*, 1979, 65-70.
[188] D.R. Wheeler, Wear, *47*, 1978, 243-254.
[189] H. Okabe, M. Masuko, and H. Oshino, ASLE Trans., *25*, 1982, 39-43.
[190] B.A. Baldwin, Wear, *45*, 1977, 345-353.
[191] R.S. Fein, in [40], pp.489-527.
[192] E.E. Klaus, in [258], pp.227-236.

[193] F.T. Barcroft, R.J. Bird, J.F. Hutton, and D. Park, Wear, *77*, 1982, 355-384.
[194] J.M. Georges et al, Wear, *53*, 1979, 9-34.
[195] H. Spedding and R.C. Watkins, Trib. Int., *15*, 1982, 9-15.
[196] W.J.S. Grew and A. Cameron, Proc. Roy. Soc. (London), *A327*, 1972, 47-59.
[197] M. Hirata, A. Masuko, and H. Watanabe, Wear, *46*, 1978, 367-376.
[198] P.A. Willermet, L.R. Mahoney, and C.M. Bishop, ASLE Trans., *23*, 1980, 217-224.
[199] A.K. Misra, A.K. Mehrotra, and R.D. Srivastava, Wear, *31*, 1975, 345-357.
[200] G.V. Vinogradov and Yu.Ya. Podolsky, in [38], pp.428-430.
[201] S.M. Hsu and E.E. Klaus, ASLE Trans., *22*, 1979, 135-145.
[202] R.S. Fein and K.L. Kreuz, ASLE Trans., *10*, 1965, 29-38.
[203] Yu.S. Zaslavsky et al, Wear, *30*, 1974, 267-273.
[204] A. Begelinger, A.W.J. deGee, and G. Salomon, ASLE Trans., *23*, 1980, 23-34.
[205] F.G. Rounds, ASLE Trans., *15*, 1972, 52-66, discussion 235-238.
[206] M. Kawamura, K. Fujita, and K. Ninomiya, J. JSLE (Int. Ed.), *2*, 1981, 157-162.
[207] R.M. Matveesky and I.A. Bujanovsky, in [56], Paper No.7.
[208] D. Godfrey, in [39], pp.335-353.
[209] R.S. Fein and K.L. Kreuz, in [39], pp.358-376.
[210] C.N. Rowe, in [43], pp.527-568.
[211] W.E. Campbell, in [28], pp.87-117.
[212] A.J. Haltner, in [40], pp.463-488.
[213] G.A. Somorjai (ed.), *The Structure and Chemistry of Solid Surfaces*, Wiley, New York, 1969.
[214] J.M. Blakely (ed.), *Surface Physics of Materials* (2 volumes), Academic Press, New York, 1975.
[215] R. Aveyard and D.A. Haydon, *An Introduction to the Principles of Surface Chemistry*, Cambridge University Press, 973.
[216] N. Eustathopoulos and J.C. Joud, in *Current Topics in Materials Science*, Vol.4, E. Kaldis (ed.), North-Holland Publ. Co., 1980, pp.281-360.
[217] W.B. Hardy, *Collected Works*, Cambridge University Press, Cambridge, England, 1936.
[218] P.A. Rehbinder and E. Shchukin, Progr. Surface Sci., *3*, 1972-73, 97-188.
[219] V.I. Likhtman, E.D. Shchukin, and P.A. Rehbinder, *Physicochemical Mechanics of Metals*, Israel Program for Scientific Translations, Jerusalem, 1964.
[220] E.D. Shchukin, Sov. Mat. Sci., *12*, 1976, 1-15.
[221] R. Roscoe, Phil. Mag., *21*, 1936, 399-406.
[222] A. Joffe, A. Kirpitschewa, and Z. Lewitsky, Z. Phys., *22*, 1924, 286-302.
[223] M.C. Shaw, Int. J. Mech. Sci., *22*, 1980, 673-686.
[224] E.H. Kohn, Wear, *8*, 1965, 43-59.
[225] A.R.C. Westwood, J. Mat. Sci., *9*, 1974, 1871-1895.
[226] H. Okabe, M. Masuko, and K. Sakurai, ASLE Trans., *24*, 1981, 467-473.
[227] C.N. Rowe, ASLE Trans., *13*, 1970, 179-188.
[228] J.P. Sharma and A. Cameron, ASLE Trans., *16*, 1973, 258-266.
[229] R.M. Matveesky, A.A. Markov, and I.A. Buyanovsky, ASLE Trans., *16*, 1973, 16-21.
[230] A. Beerbower, ASLE Trans., *14*, 1973, 90-104.
[231] H. Dunken and O. Wallbraun, Schmierstoffe u. Schmiervorgangstechn., *35*, 1969, 42-51.
[232] A.J. Groszek, ASLE Trans., *13*, 1970, 278-287.
[233] D.W. Morecroft, Wear, *18*, 1971, 333-339.
[234] H.A. Smith and R.M. McGill, J. Phys. Chem., *61*, 1957, 1025.
[235] H. Iwasaki and J. Saga, J. JSTP, *12*, 1971, 776-780 (Japanese).
[236] A. Cameron, ASLE Trans., *23*, 1980, 388-392.
[237] R.M. Matveesky, Wear, *4*, 1961, 292-299.
[238] P.R. Lancaster and G.W. Rowe, Wear, *2*, 1959, 428-437.
[239] M. Kagami et al, ASLE Trans., *24*, 1981,, 517-525.
[240] T.N. Mills and A. Cameron, ASLE Trans., *25*, 1982, 117-124.
[241] F.G. Rounds, ASLE Trans., *24*, 1981, 431-440.
[242] P. Studt, Erdöl u. Kohle, *21*, 1968, 784-785 (German).
[243] W. Katzenstein, in [52], pp.99-105.
[244] H. Okabe and T. Kanno, ASLE Trans. *24*, 1981, 459-466.
[245] K.L. Kreuz, R.S. Fein, and S.J. Rand, Wear, *23*, 1973, 393-407.
[246] J.M. Georges et al, Wear, *42*, 1977, 217-228.

[247] K.C. Tripathi, Trib. Int., *8*, 1975, 146-152.
[248] I.V. Kragelskii et al, Wear, *47*, 1978, 133-138.
[249] R.M. Matveesky and O.V. Lozovskaya, Wear, *11*, 1968, 69-75.
[250] M.J. Sutcliffe, S.R. Taylor, and A. Cameron, Wear, *51*, 1978, 181-192.
[251] A. Cameron, *Basic Lubrication Theory,* 3rd Ed., Wiley, New York, 1981.
[252] W.A. Gross et al. (eds.), *Fluid Film Lubrication,* Wiley, New York, 1980.
[253] J.A. Walowit and J.N. Anno, *Modern Developments in Lubrication Mechanics,* Applied Science Publ., London, 1975.
[254] D. Dowson, C.M. Taylor, M. Godet, and D. Berthe (ed.), *Surface Roughness Effects in Lubrication (Proc. 4th Leeds-Lyon Symp. on Tribology),* Mechanical Engineering Publs., London, 1978.
[255] D. Dowson and G.R. Higginson, *Elasto-Hydrodynamic Lubrication,* Pergamon, New York, 1965; SI edition, 1977.
[256] B.J. Hamrock and D. Dowson, *Ball Bearing Lubrication,* Wiley, New York, 1981.
[257] *Elastohydrodynamic Lubrication: 1972 Symposium*, Inst. Mech. Eng., London, 1972.
[258] D. Dowson, C.M. Taylor, M. Godet, and D. Berthe (ed.), *Elastohydrodynamics and Related Topics (Proc. 5th Leeds-Lyon Symp. on Tribology),* Mechanical Engineering Publs., London, 1979.
[259] D. Dowson, C.M. Taylor, M. Godet, and D. Berthe (ed.), *Thermal Effects in Tribology (Proc. 6th Leeds-Lyon Symp. on Tribology),* Mechanical Engineering Publs., London, 1980.
[260] H.S. Cheng, in [49], pp.1009-1048.
[261] K.L. Johnson, J. Mech. Eng. Sci., *12*, 1970, 9-16.
[262] P.E. Fowles, Trans. ASME, Ser.F, J. Lub. Tech., *93*, 1971, 383-397.
[263] T.C. Davenport (ed.), *The Rheology of Lubricants,* Applied Science Publishers, Barking, Essex, 1973.
[264] J. Harris, *Rheology and Non-Newtonian Flow,* Longman, New York, 1977.
[265] N. Naylor, in [40], pp.279-307.
[266] J. Lamb, in [38], pp.293-310.
[267] S. Bair and W.O. Winer, in [52], pp.26-31.
[268] K.L. Johnson, in [48], pp.155-161.
[269] J.L. Lauer and M.E. Peterkin, ASLE Trans., *21*, 1978, 250-256.
[270] R.S. Miller, ASLE Trans., *19*, 1976, 1-16.
[271] S. Bair and W.O. Winer, Trans. ASME, Ser.F, J. Lub. Tech., *101*, 1979, 258-265.
[272] C.R. Gentle, G.R. Paul and A. Cameron, ASLE Trans., *23*, 1980, 155-162.
[273] D. Dowson, in [28], pp.229-240.
[274] S. Aihara and D. Dowson, in [258], pp.104-110.
[275] J.M. Palacios, A. Cameron, and L. Arizmendi, ASLE Trans., *24*, 1981, 474-478.
[276] G.J. Fuks, in *Research in Surface Forces,* B.V. Deryagin (ed.), Consultants Bureau, New York, 1962, pp.79-88.
[277] R.S. Fein, Lubric. Eng., *27*, 1971, 190-195.
[278] P.E. Fowles, A. Jackson, and W.R. Murphy, ASLE Trans., *24*, 1981, 107-118.
[279] J. Prakash, K. Tonder, and H. Christensen, Trans. ASME, Ser.F., J. Lub. Tech., *102*, 1980, 368-373.
[280] K.L. Johnson and J.L. Tevaarwerk, Proc. Roy. Soc. (London), *A356*, 1977, 215-236.
[281] W. Hirst and A.J. Moore, Phil. Trans. Roy. Soc. (London), *A298*, 1980, 183-208.
[282] J.L. Lauer, Trans. ASME, Ser.F, J. Lub. Tech., *101*, 1979, 67-73.
[283] G. Paul, Wear, *49*, 1978, 79-83.
[284] D.R. Adams and W. Hirst, Proc. Roy. Soc. (London), *A332*, 1973, 505-525.
[285] B. Jacobson, ASLE Trans., *17*, 1974, 290-294.
[286] W.R.D. Wilson, Trans. ASME, Ser.F., J. Lub. Tech., *98*, 1976, 22-26.
[287] J.L. Tevaarwerk, in [49], pp.1129-1147; also in *Friction and Traction* (Proc. 7th Leeds-Lyon Symp. on Tribology), Mechanical Engineering Publs., London, 1981, pp.302-309.
[288] W.R.D. Wilson, in *Mechanics of Sheet Metal Forming,* D.P. Koistinen and N.M. Wang (ed.), Plenum, New York, 1978, pp.157-177.
[289] M.S. Bloor, D. Dowson, and B. Parsons, in [257], pp.45-47.
[290] B.B. Aggarwal and W.R.D. Wilson, in [259], pp.152-161.
[291] H.B.D. Murthy, in [259], pp.273-280.
[292] E. Felder, in [259], pp.147-151.

[293] S.M. Mahdavian and W.R.D. Wilson, Trans. ASME, Ser.F., J. Lub. Tech., *98*, 1976, 16-21.
[294] J.V. Reid and J.A. Schey, ASLE Trans., *21*, 1978, 191-200.
[295] W.R.D. Wilson, in [52], pp.44-51.
[296] T. Mizuno and M. Okamoto, Trans. ASME, Ser.F, J. Lub. Tech., *104*, 1982, 53-59.
[297] R.B. Waterhouse, Tribology, *3*, 1970, 158-162.
[298] T.A. Dow, Paper No. MS77-339, SME, Dearborn, 1977.
[299] H. Hamaguchi, H.A. Spikes, and A. Cameron, Wear, *43*, 1977, 17-24.
[300] J. Kudo, Int. J. Mech. Sci., *7*, 1965, 383-388.
[301] Y.H. Tsao and L.B. Sargent, ASLE Trans., *20*, 1977, 55-63.
[302] G. Vogelpohl, Öl u. Kohle, *12*, 1936, 943-947 (German).
[303] J.A. Schey, J. Inst. Metals, *91*, 1962-63, 360-368.
[304] Y. Kasuga and K. Yamaguchi, Bull. JSME, *11*, 1968, 344-353.
[305] J.A. Schey, Lubric. Eng., *39*, 1983, 376-382.
[306] American National Standard, ANSI B46.1-1978: *Surface Texture*, ASME, New York, 1978.
[307] N. Hansen, Werkstattstechnik, *57*, 1967, 379-383 (German).
[308] O. Kienzle and K. Mietzner, *Atlas umgeformter metallischer Oberflachen*, Springer, Berlin, 1967.
[309] O. Kienzle and K. Mietzner, *Grundlagen einer Typologie umgeformter metallischer Oberflächen mittels Verfahrensanalyse*, Springer, Berlin, 1965.
[310] G. Noppen et al, *Technische Oberflächen*, Deutsches Institut für Normung, Berlin, 1981.
[311] L.B. Sargent and Y.H. Tsao, ASLE Trans., *23*, 1980, 70-76.
[312] J. Peters, P. Vanherck and M. Sastrodinoto, CIRP, *28*, 1979, 539-554.
[313] K. Osakada and M. Oyane, Bull. JSME, *14*, 1971, 171-177.
[314] E. Dannenmann, Techn. Mitt., *73*, 1980, 893-901 (German).
[315] M. Tokizawa, J. Jpn. Inst. Metals, *37*, 1973, 19-25 (Japanese).
[316] P.F. Thomson, J. Australian Inst. Metals, *13*, 1968, 169-178.
[317] D.D. Ratnagar, H.S. Cheng and J.A. Schey, Trans. ASME, Ser.F, J. Lub. Tech., *96*, 1974, 591-594.
[318] W.R.D. Wilson, Trans. ASME, Ser.F, J. Lub. Tech., *99*, 1977, 10-14.
[319] M. Tokizawa, J. Jpn. Soc. Prec. Eng., *36*, 1970, 808-813 (Japanese).
[320] H. Wiegand and K.H. Kloos, Stahl u. Eisen, *83*, 1963, 406-415 (German).
[321] H. Wiegand and K.H. Kloos, Werkstattstechnik, *56*, 1966, 129-137 (German).
[322] L.H. Butler, J. Inst. Metals, *88*, 1959-60, 337-343.
[323] L.H. Butler, J. Inst. Metals, *89*, 1960-61, 116-123.
[324] S. Lak and W.R.D. Wilson, in [254], pp.301-307.
[325] W. Dehne, reported in [64].
[326] J. Saga, H. Nojima, and K. Arita, J. JSTP, *16*, 1975, 398-404 (Japanese).
[327] M. Tokizawa and K. Yoshikawa, Bull. Jpn. Soc. Prec. Eng., 4, 1970, 105-106.
[328] J.A. Schey and A.H. Lonn, Trans. ASME, Ser.F, J. Lub. Tech., *97*, 1975, 289-295.
[329] J.A. Schey and R.E. Myslivy, in [47], Paper No.67.
[330] T. Mizuno and K. Hasegawa, Trans. ASME, Ser.F, J. Lub. Tech., *104*, 1982, 23-28.
[331] L.B. Sargent Jr., ASLE Trans., *17*, 1974, 79-83.
[332] C.A. Brockley and P.L. Ko, Trans. ASME, Ser.F, J. Lub. Tech., *92*, 1970, 550-556.
[333] J. Korycki, Wear, *55*, 1979, 261-263.
[334] G.R. Dawson, C.E. Winsper, and D.H. Sansome, Metal Forming, *37*, 1970, 234-238, 254-261.
[335] A.E. Eaves et al, Ultrasonics, *13*, 1975, 162-170.
[336] V.P. Severdenko, V.V. Klubovich, and A.V. Stepanenko, *Ultrasonic Rolling and Drawing of Metals*, Plenum, New York, 1972.
[337] L.D. Rozenberg (ed.), *Physical Principles of Ultrasonic Technology*, Vol.1 and 2, Plenum, New York, 1973.
[338] I.L. Kristoffy, Trans. ASME, Ser.B, J. Eng. Ind., *91*, 1969, 1168-1174.
[339] L.J. Gurskii, V.V. Petrenko, S.I. Petrenko, and V.P. Severdenko, Sov. Mater. Sci., *1975* (7), 94-95.
[340] E.N. Simons, *Metal Wear, A Brief Outline*, Muller, London, 1972.
[341] A.D. Sarkar, *Wear of Metals*, Pergamon Press, New York, 1976.

[342] D. Scott (ed.), *Wear*, Academic Press, New York, 1979.
[343] W.A. Glaeser, K.C. Ludema, and S.K. Rhee (ed.), *Wear of Materials*, ASME, New York, 1977.
[344] M.B. Peterson and W.O. Winer (ed.), *Wear Control Handbook*, ASME, New York, 1980.
[345] S.K. Rhee, A.W. Ruff, and K.C. Ludema (ed.), *The Wear of Materials 1981*, ASME, New York, 1981.
[346] J.F. Archard, in [39], pp.267-333.
[347] M.B. Peterson, in [28], pp.19-37.
[348] M.C. Shaw, CIRP, *19*, 1971, 533-543.
[349] K.H. Habig, Z. Werkstofftechn., *4*, 1973, 33-40 (German).
[350] T.S. Eyre, Trib. Int., *9*, 1976, 203-212.
[351] D. Tabor, Trans. ASME, Ser.F, J. Lub. Tech., *99*, 1977, 387-395.
[352] T.H.C. Childs, Trib. Int., *13*, 1980, 285-293.
[353] F.T. Barwell, in [49], pp.401-441.
[354] M.C. Shaw, in [49], pp.643-661.
[355] J.F. Archard, in [344], pp.35-80.
[356] J.T. Burwell, Wear, *1*, 1957, 119-141.
[356a] L.E. Samuels, E.D. Doyle, and D.M. Turley, in [49a], pp.13-41.
[357] E.F. Finkin, Wear, *47*, 1978, 107-117.
[358] Z. Lisowski and T.A. Stolarski, Wear, *68*, 1981, 333-345.
[359] I.-M. Feng, J. App. Phys., *23*, 1952, 1011-1019.
[360] M. Cocks, Wear, *9*, 1966, 320-328.
[361] A. Misra and I. Finnie, Trans. ASME, Ser.H, J. Eng. Mat. Tech., *104*, 1982, 94-101.
[362] M.A. Moore, in [342], pp.217-257; in [49a], pp.73-118.
[363] R.C.D. Richardson, Wear, *10*, 1969, 291-309.
[364] F. Hirano and A. Ura, in [51], pp.163-167.
[365] M.M. Khruschov, Wear, *28*, 1974, 69-88.
[366] J.F. Archard and R.A. Rowntree, in [259], pp.285-297.
[367] E. Hornbogen, Metall, *34*, 1980, 1079-1086.
[368] M.K. Muju and A. Ghosh, Wear, *58*, 1980, 137-145.
[369] C. Preece (ed.), *Erosion*, Academic Press, New York, 1979.
[370] R. Tourett and E.P. Wright, *Rolling Contact Fatigue: Performance and Testing of Lubricants*, Heyden and Son, London, 1977.
[371] L.B. Sibley, in [344], pp.699-726.
[372] W.E. Littmann, in [40], pp.309-377.
[373] C.J. Polk and C.N. Rowe, ASLE Trans., *19*, 1976, 23-32.
[374] N.P. Suh et al, Wear, *44*, 1977, 1-162; in [49a], pp.43-71.
[375] D.A. Rigney and W.A. Glaeser, Wear, *46*, 1978, 241-250.
[375a] Y. Kimura, in [49a], pp. 187-219.
[376] P.B. MacPherson and A. Cameron, ASLE Trans., *16*, 1973, 68-72.
[377] P.A. Engel, *Impact Wear of Materials*, Elsevier, Amsterdam, 1976.
[378] P.A. Engel, in [344], pp.1103-1141.
[379] D.A. Hills and D.W. Ashelby, Wear, *54*, 1979, 321-330.
[379a] A.R. Rosenfield, in [49a], pp.221-234.
[380] C.N. Rowe, ASLE Trans., *24*, 1981, 423-430.
[381] C.N. Rowe and E.L. Armstrong, Lubric. Eng., *38*, 1982, 23-30, 39-40.
[381a] R.M. Bentley and D.J. Duquette, in [49a], pp.291-329.
[382] A. Begelinger and A.W.J. deGee, Trans. ASME, Ser.F, J. Lub. Tech., *98*, 1976, 575-580.
[383] W. Reichel, Mineralöltechnik, *17* (3/4), 1972, 1-37; *17* (5/6), 41-83 (German).
[384] I.L. Goldblatt, in [49], pp.981-1005.
[385] A. Dyson, Trib. Int., *8*, 1975, 77-87, 117-122.
[386] T.R. Bates, Jr., and K.C. Ludema, Wear, *30*, 1974, 365-375.
[387] C.N. Rowe, in [344], pp.143-160.
[388] J. Molgaard, Wear, *40*, 1976, 277-291.
[389] P. Schatzberg, Lubric. Eng., *26*, 1970, 301-305, 462-463.
[390] R.S. Montgomery, Wear, *15*, 1970, 373-387.
[391] H.J. Verbeek, Wear, *56*, 1979, 81-92.
[392] E. Rabinowicz, in [344], pp.475-506.

[393] T.N. Loladze, CIRP, *30*, 1981, 71-76.
[394] P.A. Engel, Trans. ASME, Ser.F, J. Lub. Tech., *99*, 1977, 236-245.
[395] T.F.J. Quinn, Wear, *18*, 1971, 413-419.
[396] R.G. Bayer, Wear, *11*, 1968, 319-332.
[397] H. Uetz and J. Fohl, Wear, *49*, 1978, 253-264.
[398] D.C. Drucker, Wear, *40*, 1976, 129-133.
[399] M.C. Shaw, Wear, *43*, 1977, 263-266.
[400] *Adhesion or Cold Welding of Materials in Space Environments,* STP 431, ASTM, Philadelphia, 1967.
[400a] L.B. Sargent, Jr., ASLE Trans., *21*, 1978, 285-290.
[401] M.E. Sikorski, Wear, *7*, 1964, 144-162.
[402] M.O.A. Mokhtar, M. Zaki and G.S.A. Shawki, Wear, *65*, 1980, 29-34.
[403] F.P. Bowden and G.W. Rowe, Proc. Roy. Soc. (London), *A233*, 1956, 429-442.
[404] J.A. Bailey and M.E. Sikorski, Wear, *14*, 1969, 181-192.
[405] K.J. Bhansali and A.E. Miller, Wear, *75*, 1982, 241-252.
[406] C. Dayson and J. Lowe, Wear, *21*, 1972, 263-288.
[407] N. Bay, Trans. ASME, Ser.B., J. Eng. Ind., *101*, 1979, 121-127.
[408] G.P. Upit and J.J. Manik, Wear, *11*, 1968, 333-340.
[409] M.E. Merchant, in [39], pp.181-265.
[410] A.W.J. deGee, Int. Met. Rev., *24*, 1979, 57-67.
[411] E. Rabinowicz, ASLE Trans., *14*, 1971, 198-205, 206-212.
[412] N. Ohmae, T. Okuyama, and T. Tsukizoe, Trib. Int., *13*, 1980, 177-180.
[412a] J. Ferrante, J.R. Smith, and J.H. Rose, in [49b], pp.19-30.
[413] B. Bethune and R.B. Waterhouse, Wear, *12*, 1968, 289-296, 369-374.
[414] L.H. Van Vlack, Metals Eng. Quart., *5*, 1965, 7-12.
[415] K. Miyoshi and D.H. Buckley, Trans. ASME, Ser.F., J. Lub. Tech., *103*, 1981, 180-187.
[416] J.A. Newnham and J.A. Schey, Trans. ASME, Ser.F., J. Lub. Tech., *91*, 1969, 351-359.
[417] E.A. Brandes and F.A. Collins, Metals Mat., *3*, 1969, 473-477.
[418] H. Wiegand, K.H. Kloos, and K. Müller, Stahl u. Eisen, *81*, 1961, 924-933 (German).
[419] M. Oyane et al, J. JSTP, *20*, 1979, 644-651 (Japanese), (Ann. Rep., *1*, 1980, 59).
[420] K.H. Kloos, Wear, *34*, 1975, 95-107.
[421] R. Komanduri (ed.), *Advances in Hard Material Tool Technology,* Carnegie Press, Pittsburgh, 1976.
[422] *Tools and Dies for Industry,* Metals Society, London, 1976.
[423] *New Developments in Tool Materials and Applications (Proc. 1977 Symp.),* Illinois Institute of Technology, Chicago, 1977.
[424] *Materials for Metal Cutting,* ISI Publ. 126, Iron and Steel Inst., London, 1970.
[425] G.A. Roberts and R.A. Cary, *Tool Steels* (4th Ed.), ASM, Metals Park, 1980.
[426] *Metals Handbook,* 9th Ed., Vol.3, *Properties and Selection: Stainless Steels, Tool Materials and Special-Purpose Metals,* ASM, Metals Park, 1980, pp.421-567.
[427] Various, Ind. Anz., *102* (46), 1980, 21-44 (German).
[428] M.C. Shaw, in *Proc. 4th Int. Conf. Production Engineering,* Jpn. Soc. Prec. Eng., Tokyo, 1980, pp.492-511.
[429] K.G. Budinski, in [344], pp.931-985.
[430] F. Borik, in [344], pp.327-342.
[431] K.G. Budinski, in [343], pp.100-109.
[432] T.N. Loladze, G.V. Bokuchera, and G.E. Davidova, in *The Science of Hardness Testing,* ASM, Metals Park, 1973, pp.251-257.
[433] P.L. Hurricks, Wear, *26*, 1973, 285-304.
[434] F.A. Kirk, Metals Techn., *4*, 1977, 233-239.
[435] Various, Manuf. Eng., *86* (1), 1981, 66-78.
[436] P. Gumpel, Stahl u. Eisen, *100*, 1980, 905-910 (German).
[437] G. Hoyle, Metall. Rev., *9*, 1964, 49-91.
[437a] O. Vingsbo and S. Hogmark, in [49a], pp.373-408.
[438] B.J. Bhansali and W.L. Silence, Metal Progr., *112* (3), 1977, 39-43.
[439] H.E. Exner, Int. Met. Rev., *24*, 1979, 149-173.
[440] R.A. Moll and J.D. Wood, in *Proc. 4th NAMRC,* SME, Dearborn, 1976, pp.13-18.
[441] M.K. Mal and S.E. Tarcan, in [343], pp.140-147.
[442] J. Golden and G.W. Rowe, Br. J. Appl. Phys., *11*, 1960, 517-520.
[443] G. Lunde and P.B. Anderson, Int. J. Mach. Tool Des. Res., *10*, 1970, 79-93.

[444] B. Gregory, Metallurgia, *82* (8), 1970, 55-59.
[445] N. Narutaki and Y. Yamane, Bull. Jpn. Soc. Prec. Eng., *10*, 1976, 95-100.
[446] J.E. Mayer, D. Moskowitz, and M. Humenik, in [424], pp.143-151.
[447] M. Miyoshi and D.H. Buckley, in [55], pp.205-215.
[448] R.E. Shepler and E.D. Whitney, in [343], pp.468-474.
[449] H. Shimura and Y. Tsuya, in [343], pp.452-461.
[450] K.H. O'Donovan, CIRP, *24*, 1975, 265-270.
[451] H.C. Miller and R.H. Wentor in *Kirk-Othmer Encyclopedia of Chemical Technology*, 3rd Ed., Vol.4, Wiley-Interscience, New York, 1978, pp.666-688.
[452] A.A. Tseng, C. Ten Haagen, and J.F. Berry, in *Proc. 6th NAMRC*, SME, Dearborn, 1979, pp.258-263.
[453] W. Poole and J.L. Sullivan, ASLE Trans., *23*, 1980, 401-408.
[454] R.G. Lockwood, Sheet Metal Ind., *48*, 1971, 271, 273.
[455] J.C. Gregory, Tribology, *3*, 1970, 73-83.
[456] P.L. Hurricks, Wear, *22*, 1972, 291-320.
[457] Various, Tribology, *5*, 1972, 205-224.
[458] R.W. Wilson, in [45], pp.165-175.
[459] A. Oldewurtel, Draht, *31*, 1980, 78-81 (German).
[460] K.H. Kloos, Z. Werkstofftech., *10*, 1979, 456-466.
[461] K. Stanford, Metallurgia, *47* (3), 1980, 109-111, 113, 114.
[462] M.K. Gabel and D.M. Donovan, in [344], pp.343-371.
[463] *Source Book on Nitriding*, ASM, Metals Park, 1977.
[464] M. Mrozek, MW Interf., *3* (5), 1978, 13-22.
[465] *Metals Handbook*, 8th Ed., Vol.6, *Welding and Brazing*, ASM, Metals Park, 1971, pp.63-65, 152-166, 587-592.
[466] M. Riddihough, Tribology, *3*, 1970, 211-215.
[467] J.A. Catherall, R.F. Smart, and J.A. Reynolds, in [45], pp.215-221.
[468] E.P. Cashon, Trib. Int., *8*, 1975, 111-115.
[469] S. Vaidyanathan and V.C. Venkatesh, Trib. Int., *7*, 1974, 54-58.
[470] J.D. Wolf, in [343], pp.326-330.
[471] J.M. Sale, Metal Progr., *115* (4), 1979, 44, 45, 52-55.
[472] H. Wapler, T.A. Spooner, and A.M. Balfour, Trib. Int., *13*, 1980, 21-24.
[473] S. Ramalingam, in [344], pp.385-411.
[473a] M.B. Peterson and S. Ramalingam, in [49a], pp.331-372.
[474] K.K. Yee, Int. Met. Rev., *23*, 1978, 19-42.
[474a] *Chemical Vapor Deposition* (Proc. Conf.), Electrochemical Society, Princeton, NJ.
[475] H.E. Hintermann and F. Aubert, in [45], pp.207-213.
[476] H.E. Hintermann, H. Boving, and W. Hanni, Wear, *48*, 1978, 225-236.
[477] K.H. Habig, W. Evers, and H.E. Hintermann, Z.Werkstofftech., *11*, 1980, 182-190 (German).
[478] T. Sadahiro, S. Yamaga, K. Shibuki, and N. Ujiie, Wear, *48*, 1978, 291-299.
[479] J.L. Peytavy, A. Lebugle, G. Montel, and H. Pastor, Wear, *52*, 1979, 89-94.
[480] D.G. Teer, Trib. Int., *8*, 1975, 247-251.
[481] T. Spalvins, Lubric. Eng., *27*, 1971, 40-46.
[482] D.G. Teer and J. Halling, CIRP, *27*, 1978, 517-522.
[483] Y. Enomoto and K. Matsubara, J. Vac. Sci. Technol., *12*, 1975, 827-829.
[484] T. Spalvins, in [343], pp.358-364.
[485] T. Spalvins, in [145], pp.109-117.
[486] D.R. Wheeler and W.A. Brainard, Wear, *58*, 1980, 341-358.
[487] N.E.W. Hartley, Trib. Int., *8*, 1975, 65-72.
[488] G. Dearnaley, Metall. Mat. Techn., *12*, 1980, 129-135, 178-183.
[489] J.A. Schey, J. Metals, *32* (1), 1980, 28-33.
[490] D. Schlosser, Bänder Bleche Rohre, *16*, 1975, 302-306, 378-381; *17*, 1976, 97-102 (German).
[491] H. Kudo, M. Tsubouchi, Y. Fukahara, and Y. Ito, CIRP, *27*, 1979, 159-163.
[492] J.A. Schey and J.A. Newnham, Lubric. Eng., *26*, 1970, 129-137.

Metalworking Lubricants

The mechanisms of lubricant action were explored in Chapter 3; it now remains to discuss the various classes of metalworking lubricants. Most lubricants are of rather complex composition and, depending on interface conditions, may function by several mechanisms. Therefore, they will be discussed in groups corresponding to their basic formulation and physical appearance, and not to their operative mechanism. A comprehensive coverage is beyond the scope of this book, and the aim is to provide the minimum background essential for understanding later chapters. Subject material previously covered by Riesz [1] is included, as are topics from other general treatments [3.43],[2-15]. Previous attempts at classification [16-18] are taken into account.

Some indication of the economic significance of lubricants may be gained from the statistics quoted by Tamai [3.168]. In 1978, Japan consumed 21 ML of lubricants for rolling and other deformation processes and 61 ML of cutting lubricants. Consumption in other industrially developed nations should be proportional. The economic role of lubricants is, of course, far greater than these numbers suggest, because lubricants affect both production and quality of a much larger quantity of products.

Before discussing the various lubricant classes, it will be useful to recapitulate the desirable lubricant attributes.

4.1 Attributes of Metalworking Lubricants

In practice, a lubricant is expected to fulfill a number of different functions. Some of the requirements may be contradictory or mutually exclusive, and thus the choice of lubricant may become a matter of best compromise and be directed by specific needs. There are, however, some attributes which are generally valid for the majority of applications, and these, based on an evaluation by Schey [19], are as follows:

1. Controlled Friction. Low friction reduces forces and power requirements and results in more homogeneous deformation. Nevertheless, zero friction is seldom approached and is not always desirable (Sec.2.3).

2. Separation of Surfaces. Ideally, the lubricant should act as a perfect parting agent, separating the die and workpiece surfaces and preventing metal-to-metal contact.

3. Control of Metal Pickup on Tool Surface. Even if continuity of the parting lubricant film is locally lost, pickup on the die surface (Sec.3.11.2) should not become self-accelerating; even in the changed environment, the lubricant should be capable of retaining its effectiveness in preventing growth of incipient pickup points.

4. Reduced Wear. Moderate wear of the workpiece material is usually acceptable, but all forms of die wear are objectionable (Sec.3.10). The lubricant should reduce wear of the die while limiting wear of the workpiece material to tolerable proportions. The wear products themselves should be nonabrasive.

5. Protection of Old and New Surfaces. The most important phenomenon separating metalworking from all other friction problems is the continuous generation of new surfaces. The lubricant is called upon to cover both old and new surfaces efficiently even though the chemical activities of these surfaces may be radically different (Sec.3.1). The lubricant should have wetting and spreading characteristics that permit it to follow the extension of the metal surface.

6. Adaptability to Varied Working Conditions. The lubricant must perform its functions at the pressures, temperatures, and relative sliding velocities prevailing in the process (Sec.3.1.3), even when these factors change within the contact zone itself.

7. Compatibility With Die and Workpiece Materials. The lubricant must perform its function in the system defined by the die and tool surfaces, the old and new metal surfaces, and the surrounding atmosphere.

8. Rapid Response. The lubricant must exert its influence in the short time (on the order of a few milliseconds) available during the actual metalworking process. Even when the surface of the workpiece is conditioned beforehand, lubricant efficiency depends on whether the newly developed surfaces can be protected during their formation. When the lubricant functions by sacrificial action, the rate of lubricant film formation must keep pace with the rate of loss.

9. Durability of Lubricant Film. The lubricant film formed on the surface of the die must be capable of withstanding continued or repeated encounters. Even if the lubricant acts primarily by deposition on the workpiece surface, some transfer onto the die surface should also occur.

10. Controlled Surface Finish. The lubrication mechanism should produce, in conjunction with initial roughness and process conditions (Sec.3.8.4), a finish suited to subsequent applications.

11. Thermal Insulation. In hot working operations the lubricant should provide thermal insulation between workpiece and die surfaces, partly to reduce heat loss from the hot stock and partly to protect the die from excess heat. Additionally, the lubricant should protect the workpiece against oxidation or gas pickup during heating and deformation. Therefore, it should form a continuous film of high tenacity, adherence, controlled viscosity, and good self-healing capacity.

12. Cooling. In working of metals at high rates, the lubricant should cool the dies and/or workpiece to prevent lubricant breakdown and avoid catastrophic wear.

13. Controlled Stability. The lubricant should remain unchanged with time and under the influence of repeated encounters. It should be stable and should be unaffected by temperature, oxidation, bacteriological attack, and the contamination often unavoidable in industrial practice. If changes occur, they should be beneficial or at least harmless. Lubricant constituents, if used up, should be amenable to replenishment. Neither the lubricant nor its breakdown or reaction products should be abrasive. Gases that are liberated should contribute to or at least not interfere with the lubrication process.

14. Controlled Reactivity. Even though efficient lubrication frequently requires a certain degree of reactivity, the lubricant should not be corrosive to the product, the die, or the metalworking machine tool.

15. Harmless Residues. Neither the lubricant nor its residues should accumulate in recesses of dies. Lubricant residues left on the workpiece should not cause discoloration (staining) of the product on subsequent annealing or storage, and should not cause chemical or metallurgical changes in the product. They should not interfere with finishing operations such as painting, enameling, printing, welding, or electrodeposition. Residues left on products used in the preparation or storage of food should be harmless.

16. Application and Removal. The lubricant should have a long shelf life and should be easy to apply to the workpiece and/or die (and even to a hot die) in a controlled manner and, if necessary, the lubricant and its residues should be readily removable.

17. Disposal. Preferably, it should be possible to reclaim the lubricant and, after appropriate treatment, return it to the process. Otherwise, it should be possible to reclaim the oily constituents, and the effluents resulting from the treatment should be environmentally acceptable.

18. Handling and Safety. Some lubricant attributes which do not have direct bearing on the actual lubrication function have assumed increased importance in recent years. Examples are absence of skin irritation, toxicity, carcinogenic effects, odor, and fire hazard.

19. Cost. The lubricant that conforms to all of the above requirements should also be relatively inexpensive. This is an important consideration when the cost of the workpiece material is low and processing contributes the larger portion of the total cost. Nevertheless, it would be shortsighted to judge the cost of tribological control purely by the cost of the lubricant. A more costly lubricant may well be justifiable if the savings in energy consumption, processing steps, surface preparation, and lubricant removal and disposal are considered. This aspect is discussed in [20] and by Mang [21,22].

20. Integrated Approach. The metalworking lubricant should be regarded not in isolation but rather as a part of a larger lubrication planning and management system which encompasses all lubricants and lubrication practices used in the entire production plant. This broader activity, often described as terotechnology, has received much attention in recent years [23-27].

The development of a lubricant represents substantial effort, involving laboratory evaluation (Sec.5.4) of candidate base lubricants and additives, testing for toxicity and other nontribological properties, field testing, and establishment of a lubricant manufacturing process. As illustrated by Davis et al [28] in an example of cutting fluid development, commercialization may take 3 to 6 years. The great variety of considerations that enter into lubricant development is illustrated in the examples of profile drawing and hot die forging lubricants presented by Schroeder and Schey [29].

4.2 Mineral Oils

Even though mineral oils became available in quantity only in the middle of the 19th century, they soon came to occupy an important position as metalworking lubricants. Mineral oils are hydrocarbons, obtained mostly from crude oil by distillation. Their properties depend on chain length, structure, and degree of refining.

4.2.1 Effects of Chain Length

Metalworking oils contain 10 to 70 carbon atoms per molecule. As chain length increases, viscosity, flash point, fire point, and boiling point increase (for definitions, see Appendix). Viscosity increases with pressure and decreases with temperature. Temperature dependence can be described by Eq.3.26 but, in general, a much better fit is given by the semi-empirical Walther relationship. This is the basis of the ASTM D-341 chart, which is the most convenient means of finding the viscosity of a lubricant at any temperature intermediate between two experimentally determined values. Figure 4.1 presents data for three mineral oils and gives data for various engine oils for reference.

Oily residues on worked metal surfaces undergo distillation on subsequent heating (see Sec.5.10) and form brown stains. For a given temperature, staining intensity increases with chain length. For a given viscosity, an oil distilled to a narrower molecu-

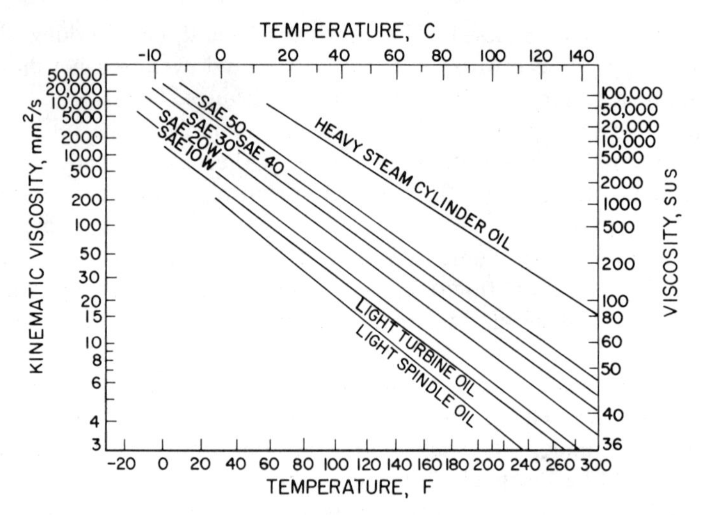

Fig.4.1. Variation of viscosity with temperature for mineral oils [15]

lar weight distribution has fewer or none of the longer-chain molecules that lead to staining; therefore, narrow-cut oils are preferred when staining is a problem.

4.2.2 Effects of Structure

Depending on origin, hydrocarbons can occur in a variety of structures, as shown in Fig.4.2 (the hydrogen atoms are omitted in this simplified representation).

For a given molecular weight, straight-chain saturated hydrocarbons (paraffins or alkanes) have relatively low densities and viscosities, but lose less of their viscosity with increasing temperature and thus have high V.I. values (Sec.3.6.3); they are also less compressible and have lower pressure-viscosity exponents (Eq.3.25). Thus, their viscosity increases with pressure less rapidly (Fig.4.3), but they solidify at lower pressures. They have relatively high flash points and cloud points (temperatures at which a solid phase appears). Their oxidation is delayed by the formation of corrosive volatile acids and can be readily improved with additives.

With increasing branching, isoparaffins behave more like naphthenes (Fig.4.2), which can be regarded as paraffins folded into closed loops, without having the regular structure of the aromatics based on the benzene ring. Aromatics oxidize heavily, but

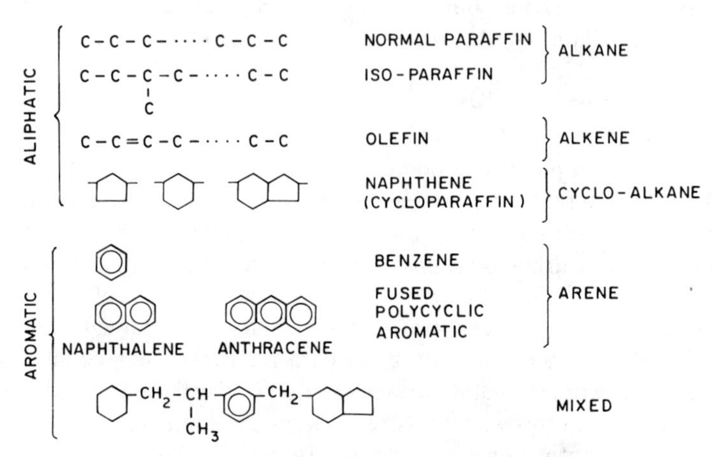

Fig.4.2. Structures of selected hydrocarbons

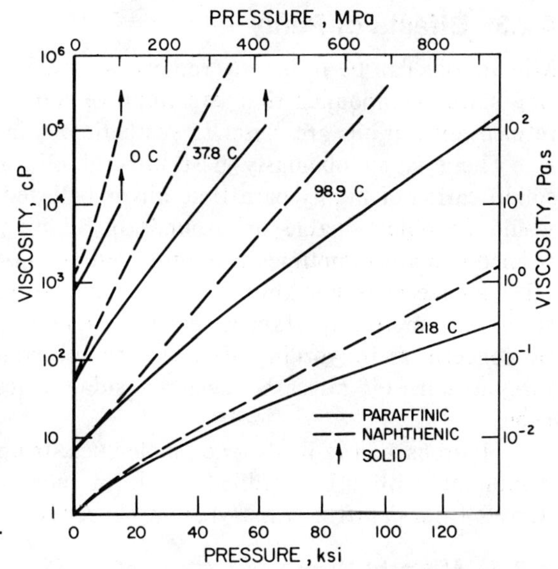

Fig.4.3. Increase in viscosity and solidification of mineral oils with pressure [15]

can serve as oxidation inhibitors if present in small quantities in other hydrocarbons. Aromatics compete with E.P. additives for surface sites, and the small quantities often found in paraffinic oils are the most probable sources of improved wear resistance and load-carrying capacity.

Mineral oils usually consist of several kinds of hydrocarbons; predominantly paraffinic oils tend to contain substantial amounts of naphthenes. With the spread of modern analytical techniques, the composition of oils can be defined in great detail. Some progress has been made in predicting viscosity from composition [30,31]; examples illustrating differences between paraffinic and naphthenic oils are given in Table 4.1. Data for a wide variety of lubricants have been compiled by Hersey and Hopkins [32]. The response of non-Newtonian fluids to high pressures is discussed by Kuss [33], and the compressibility of metalworking fluids has been studied by Goldman et al [34]. Pywell [35] and Johnston [36] suggest that the pressure viscosity coefficient can be predicted from atmospheric viscosity measured at two temperatures. A very approximate but useful rule of thumb states that a 4 MPa increase in pressure offsets a 1 C increase in temperature (38 atm for 1 C, or 500 psi for 2 F) [3.61].

Table 4.1. Viscosity-pressure coefficients for selected petroleum oils (data compiled from [32])

Molecular weight	Viscosity at 1 atm, cP (a) 37.8 C	98.9 C	Viscosity index	Average value of $\alpha \times 10^4$ at a temperature of: 0 C	25 C	37.8 C	98.9 C	218 C
Paraffinic Oils								
450	48.0	5.9	96	1.99	1.45	1.29	0.87	0.57
576	144.3	12.2	99	1.97	1.47	1.36	0.95	0.64
700	523	27.6	91	2.65	1.84	1.61	1.09	0.67
Naphthenic Oils								
349	49.9	5.1	23	2.92	2.14	1.87	1.11	0.59
383	131.4	8.3	8	3.4	2.58	2.18	1.31	0.71
400	485	15.5	−42	4.8	3.23	2.83	1.54	0.81

(a) 1 cP = 1 mPa·s

4.2.3 Effects of Purity

Mineral oils can be refined to remove waxes, aromatic compounds, and naturally occurring sulfur compounds that can improve lubrication but also lead to staining. Superrefined mineral oils are closer to synthetic oils in purity.

There is no obviously best mineral oil for metalworking purposes. The earlier solidification of highly paraffinic oils is believed to be helpful when a high-viscosity oil would be objectionable on grounds of staining (as in rolling of aluminum). Highly refined oils are sometimes recommended, again for lower staining, although experimental evidence does not always support the assumption that a more refined oil actually stains less; more important is that the oil should be narrow cut. When staining is of little concern, as in working of steel with subsequent annealing at high enough temperatures to burn off all carbonaceous residue, a less-refined oil may actually be advantageous.

Hydrocarbon radicals—i.e., molecules stripped of a hydrogen atom—are capable of combining with other radicals to form more complex compounds. An alkyl radical derives from paraffins, an aryl from benzenes.

4.2.4 Mineral Waxes

There are a number of hydrocarbons of mineral origin which are solid at room temperature. Their terminology is somewhat diffuse but for our purpose the classification accepted in everyday life will suffice. Detailed treatments are given by Bennett [37] and McLoud [38].

Most paraffin waxes are separated by pressing from a paraffin distillate of a crude oil. They consist chiefly of straight-chain C_{18}-C_{40} paraffins with melting points ranging from 27 to 80 C. On solidification they form platelike or needlelike crystals, and they are relatively brittle.

Microcrystalline waxes cannot be distilled without decomposition and are produced by solvent separation from crude oil distillation residues. They contain large quantities of branched-chain and cyclic paraffins of C_{41} to C_{50} length. On solidification they form only small, irregular crystals, and they develop no well-formed crystals at all when deposited from solvents; they are relatively ductile.

Waxes are tailor-made for special applications. They do not possess any boundary properties and are thus less frequently used for metalworking than the natural waxes. This is true also of other mineral waxes gained by solvent extraction from the various forms of coal, such as Montan wax produced from lignite.

4.3 Natural Oils, Fats, and Derivatives

Oils and fats are water-insoluble substances derived from vegetable and animal sources. They were, undoubtedly, the first metalworking lubricants (Sec.1.1), and they have retained their importance over the years. The distinction between oils and fats is rather arbitrary: oils are liquid, fats are semisolid at room temperature. For detailed treatments, see [39,40].

4.3.1 Natural Fats

Oils and fats are triglycerides (Fig.4.4)—i.e., esters of the trifunctional alcohol glycerol (glycerine) and of fatty acids (Fig.3.28). Thus, they can be decomposed by hydrolysis to yield glycerol and fatty acids in which the radicals R1, R2, and R3 need not be the same. Fats derived from various vegetable or animal sources are named according to their origin, and usually show characteristic compositions, although substantial varia-

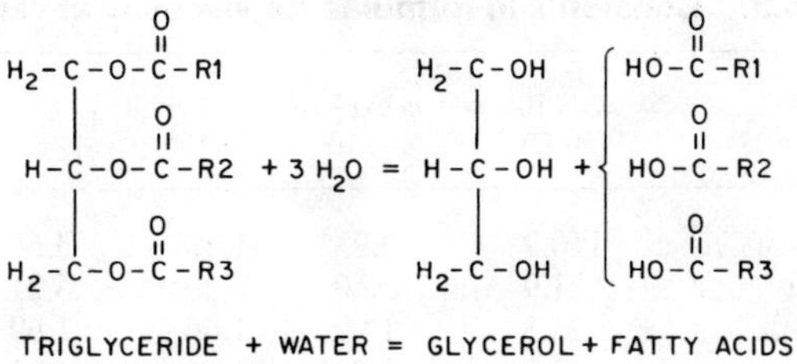

$$\begin{array}{l} H_2-C-O-\overset{O}{\overset{\|}{C}}-R1 \\ \quad | \\ H-C-O-\overset{O}{\overset{\|}{C}}-R2 \\ \quad | \\ H_2-C-O-\overset{O}{\overset{\|}{C}}-R3 \end{array} + 3\,H_2O = \begin{array}{l} H_2-C-OH \\ \quad | \\ H-C-OH \\ \quad | \\ H_2-C-OH \end{array} + \left\{ \begin{array}{l} HO-\overset{O}{\overset{\|}{C}}-R1 \\ HO-\overset{O}{\overset{\|}{C}}-R2 \\ HO-\overset{O}{\overset{\|}{C}}-R3 \end{array} \right.$$

Fig.4.4. Hydrolysis of fatty oils TRIGLYCERIDE + WATER = GLYCEROL + FATTY ACIDS

tions due to geographical or genetical differences may exist. Some examples are given in Table 4.2.

Some oils and fats, especially those of marine origin, contain not only triglycerides but also esters of higher single-valued (monohydric) alcohols. Castor oil is different from other vegetable oils in that it contains 85 to 90% of ricinoleic acid in which a hydroxyl group is attached to the carbon chain $CH_3(CH_2)_5CHOHCH_2CH{=}CH(CH_2)_7COOH$.

As shown in the Appendix, there are a number of readily determined values which have been traditionally used for characterizing fatty oils. The composition of the oil or fat affects both viscosity and boundary-lubrication properties. A most useful set of data was generated by Takatsuka et al [41] for frequently used metalworking lubricants. For widely varying initial viscosities at 50 C (Table 4.3), the relative viscosity increase is remarkably similar (Fig. 4.5), with the exception of the naphthenic oil. Equation 3.26 gives a rather poor fit, and a more complex, multiparameter formula (Table 4.3) is much better. The benefits of increased viscosity are partially lost at elevated temperatures (Fig.4.6), but under the high pressures prevailing in many processes the oils revert to solids (Fig.4.7). Highly viscous oils and semisolid fats may be applied to the workpiece surface in the molten state and allowed to solidify before entering the interface, or a volatile organic solvent may be used.

Table 4.2. Fatty acid compositions (wt %) of various fats and oils (data compiled from [40])

No. of carbon atoms	Name of acid	Coconut oil	Palm kernel oil	Palm oil	Cotton-seed oil	Rape-seed oil	Tallow (beef)	Tallow (mutton)	Lard
Saturated Acids									
12	Lauric	44-51	47-52	...	...	...	...	...	...
14	Myristic ...	13-19	14-18	0.5-3	1	1	3-6	4-10	1-2
16	Palmitic ...	8-11	7-9	32-45	26-31	1-5	25-37	24-38	22-31
18	Stearic	1-3	1-3	4-7	3-5	1-3	14-29	15-30	16-24
Total (typical)		92	81	40-50	28-35	3-8	50-55	52-57	35-40
Unsaturated Acids									
Monounsaturated									
18	Oleic	5-8	11-19	38-53	19-26	14-38	26-50	38-48	38-44
Diunsaturated									
18	Linoleic ...	1.0-2.5	0.5-2	6-12	37-50	10-22	1-3	...	4-9
22	Erucic	...	...	...	...	40-64	...	...	...
Triunsaturated									
18	Linolenic ..	...	...	...	...	8-12	...	...	1-2
Total (typical)		8	19	50-60	60-70	92-97	44-52	40-50	60-65

Table 4.3. Constants in formulae for viscosity at various pressures and temperatures [41]

Oil	$\eta_0 \times 10^3$ (50 C)	$\eta = \eta_0 \exp[\alpha P - \gamma (T-50)]$ $\alpha \times 10^2$	$\gamma \times 10^2$	$\eta = \eta_0 \exp(a + bT + cP + dT^2 + eTP + fP^2)$ a	$b \times 10^2$	$c \times 10^2$	$d \times 10^4$	$e \times 10^5$	$f \times 10^6$
Castor oil	110.2	3.23	1.10	3.60	−8.83	1.64	3.28	−7.45	−3.62
Palm oil	21.9	2.60	1.20	5.21	−5.24	1.37	1.73	−3.06	−6.83
Tallow	23.5	1.99	1.46	1.69	−3.46	1.19	0.74	−1.60	−6.69
Lard	23.3	2.94	1.13	1.90	−4.15	1.22	0.79	−2.89	−4.17
Oleic acid	11.2	2.52	1.21	2.84	−6.39	1.16	2.15	−3.82	−2.94
Paraffinic mineral oil ..	2.82	1.75	1.09	1.39	−3.39	1.45	1.21	−5.40	−1.80
Naphthenic mineral oil ..	22.3	3.75	2.46	4.59	−11.3	3.09	4.47	−1.34	−2.90

Note: η, η_0 [Pa·s], T [°C], P [MPa]

Another source of high performance is to be sought in composition. Many oils and fats contain free fatty acids, and decomposition during storage or under the conditions existing at the tool-workpiece interface liberates further free fatty acids with their remarkable boundary-lubrication properties. Small (2 to 8%) concentrations of free fatty acids are believed to be important in the performance of palm oil and lanolin. Nevertheless, the benefits are sometimes contradicted; no doubt, free fatty acids are less important when the process encourages PHD lubrication. Because most oils operate in the mixed-film regime, it is often difficult to determine whether or not the presence of a longer-chain fatty acid constituent improves boundary lubrication per se. Unsaturated compounds containing double or triple bonds (drying oils) are more soluble but are also prone to oxidation, polymerization, and the formation of gummy deposits on storage. They also contribute to brown staining on heating, although some staining is found with

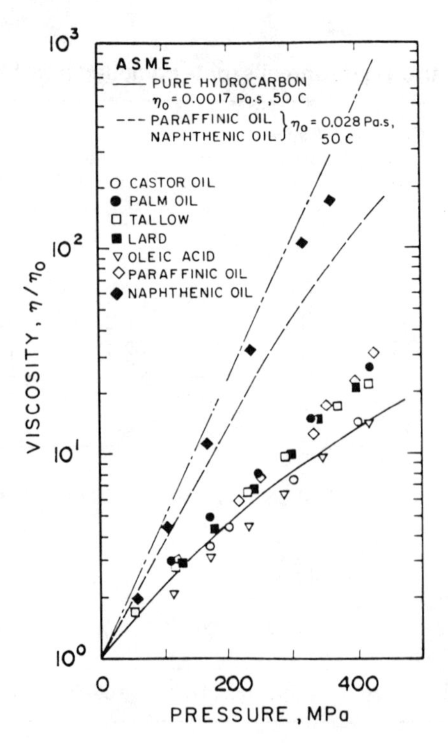

Fig.4.5. Increase in viscosity of some fatty oils with pressure at 100 C [41]

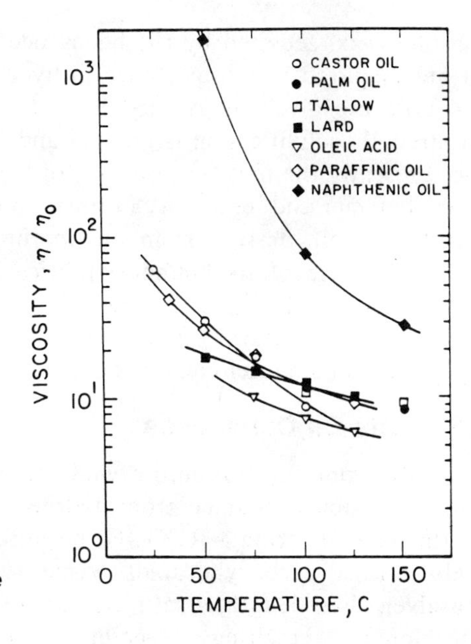

Fig.4.6. Drop in viscosity with temperature at a pressure of 300 MPa [41]

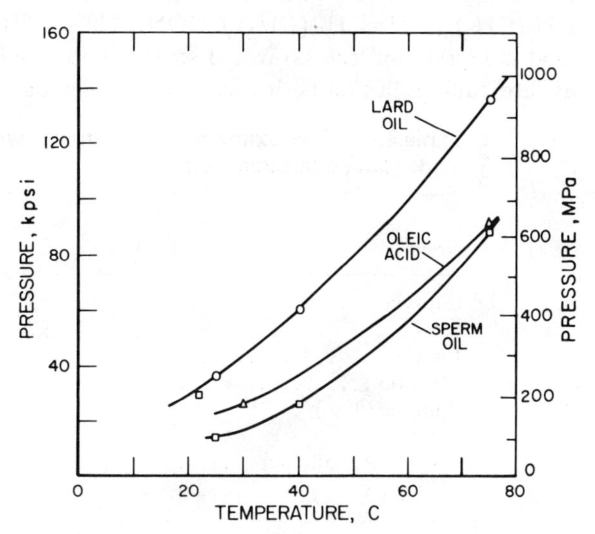

Fig.4.7. Solidification of fatty oils as a function of pressure and temperature [32]

all fats and oils. Oxidation and bacterial breakdown lead to rancidity and its unpleasant odor.

The properties of fatty oils can be modified by a number of treatments. Unsaturation of natural oils (especially that of the drying oils) is reduced by hydrogenation—i.e., the breaking of the double or triple bond with the addition of hydrogen in the presence of a catalyst. Chlorination and sulfonation (addition of the group -SO_2OH) impart useful E.P. properties.

4.3.2 Natural Waxes

Natural waxes are of vegetable or animal origin [37,38] and are esters of long-chain fatty acids and monohydric alcohols:

a. Carnauba wax, which is exuded by the leaves of a Brazilian palm, is amorphous and has a high melting point (approximately 85 C).

b. Beeswax, secreted by the honey bee, is almost amorphous, and melts around 64 C. It has acquired new importance in hydrostatic extrusion. Properties at pressures up to 1.4 GPa are given by Ahmed [42]. It has a pressure- and temperature-dependent shear strength; solidification is gradual and takes place around 0.4 GPa.

c. Presently the most important product is lanolin or refined wool wax. It is semi-solid in character and represents a transition to the higher fats and fatty acids.

d. Sperm oil, derived from the sperm whale, is actually a liquid wax which had wide use because of its boundary-lubrication properties and resistance to gumming. Vegetable oils (jojoba oil) and some waxes are potential substitutes [43].

The value of natural waxes and waxlike alcohols, etc., is well demonstrated, especially for aluminum-steel combinations.

4.3.3 Fatty-Oil Derivatives

For metalworking lubrication, fatty-oil derivatives are of greatest importance:

a. Fatty acids obtained from hydrolysis of fats are shown in Table 4.4 [44]. They have the general formula RCOOH; because of the presence of the -COOH group, they are also termed carboxylic acids. Viscosity increases with chain length (Fig.4.8) [45]. For a given chain length, a saturated acid $CH_3(CH_2)_n(COOH)$ is more viscous and has a higher melting point than its unsaturated counterpart $CH_3(CH_2)_n(\text{-}CH{=}CH\text{-})(CH_2)_nCOOH$. Thus, of the C_{18} acids, the unsaturated oleic acid is liquid, and the saturated stearic acid is solid, at room temperature. Their value as lubricants relies on both viscosity and boundary activity; because of the higher melt-

Table 4.4. Approximate compositions (wt %) of commercial saturated acids (data compiled from [44])

Acids	Saturated acids C_6	C_8	C_{10}	C_{12}	C_{14}	C_{16}	C_{18}	Unsaturated acids
Caprylic	5	92	3					
Capric		1	97	2				
Lauric			1	97	2			
Myristic				1	98			
Palmitic (90%)					1	92	5	
Stearic:								
Commercially pure						7	90	
Single pressed					2	52	38	5
Hydrogenated tallow					2	30	66	

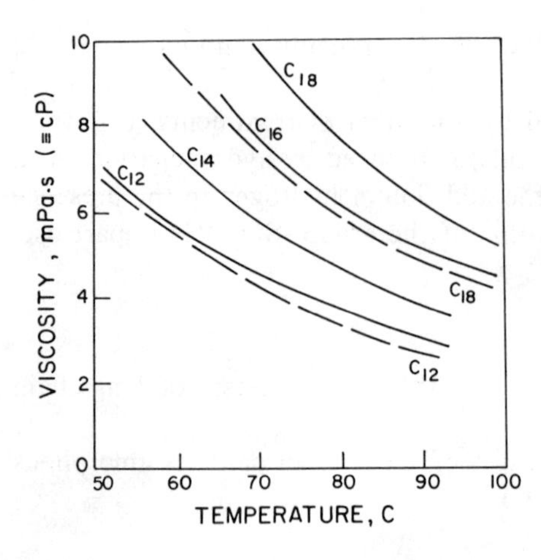

Fig.4.8. Variation of viscosity with temperature for selected fatty acids (solid lines) [45] and fatty alcohols (broken lines) [46]

ing point and thus higher transition temperature of a longer-chain acid (or its soap formed *in situ* on a reactive metal surface; Fig.3.31), longer-chain acids are generally better boundary lubricants. Whether or not unsaturation improves boundary lubricity is debated, and firm evidence seems to be lacking. Unsaturation does, however, reduce oxidative stability, allow polymerization, and promote staining.

b. Fatty alcohols of the general structure RCH_2OH are derived mostly from saturated acids or esters. Their viscosity is highly temperature dependent (Fig.4.8)[46]. Like the saturated acids, fatty alcohols of higher molecular weight turn into solid, waxlike substances. This is true also of amines (RCH_2NH_2; Fig.3.28) and amides ($RCONH_2$).

c. Esters are the reaction products of acids and alcohols, and will be discussed in Sec.4.4.2.

4.3.4 Soaps

Of great significance in metalworking lubrication, soaps are the products of reaction between a fat and a metal hydroxide, yielding a metal (M) salt of the fatty acids. Thus, they have the general structure RCOOM. When formed *in situ* (Fig.3.28), they are firmly bonded to the metal surface. When pre-reacted, usually from animal or plant fats, they are highly viscous or solid substances whose value derives mostly from their rheological properties. Soaps of the alkaline metals sodium and potassium are water soluble; those of other metals are insoluble. Of great importance are the solid calcium and zinc stearates.

The melting point of any soap generally increases with the chain length and saturation of the fatty acid radical (Table 4.5) [47]. The rheology of sodium stearate, calcium stearate, and a mixed Na-Ca stearate, extensively applied in wiredrawing, was studied by Pawelski, Rasp, and Hirouchi [48] by the back and forward extrusion of the soaps themselves. They were found to behave as a Bingham body (Fig.3.34a, line D), with the initial shear strength τ_0 and the viscosity η_B shown as a function of temperature in Fig.4.9. At high temperatures τ_0 is small and Newtonian flow is approached. Most information on their effectiveness comes from experience with metalworking processes such as wiredrawing (Sec.7.3).

Table 4.5. Approximate melting points of fatty acid soaps (data compiled from [47])

Cation	Melting point of salt, C				
	Laurate (C_{12})	Myristate (C_{14})	Palmitate (C_{16})	Stearate (C_{18})	Oleate (C_{18})
Ammonium (neutral)	75	75-90	...	...	...
Ammonium (acid)	77	84	89	93	78(a)
Potassium (acid)	80-150	95-160	100-160	100-160	...
Lithium	229.5	224	224	221	...
Barium	260	...	(a)	...	100(a)
Calcium	182	...	154	152	83
Magnesium	150	132	121	132	...
Lead	104.5	108.5	112	115.5	50
Silver	212	211	209	205	...
Zinc	128	...	129	130	70
Copper (ic)	112	...	115-120	125	100
Nickel (ic)	44	...	80	80-86	18-20
Cobalt (ic)	...	...	70-75	73-75	...

(a) Decomposes.

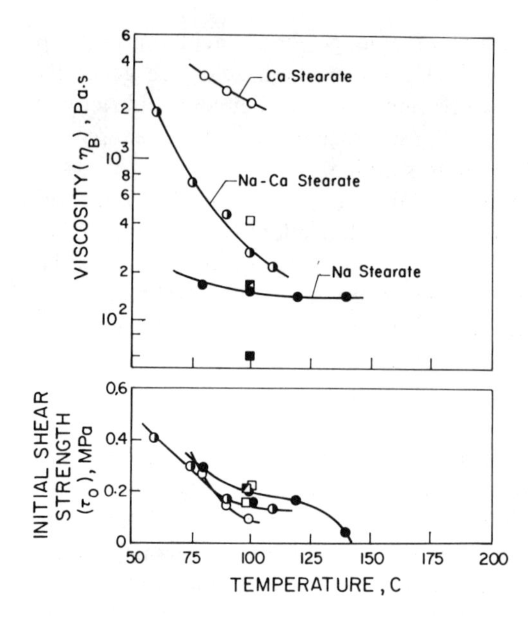

Fig.4.9. Apparent viscosity and initial yield strength vs temperature for various soaps [48]

Acid salts of the general structure RCOOM·RCOOH (where M = Na, K, or NH_2) and basic salts of multivalent metals—e.g., $Al(OH)(RCOO)_2$—are more important in grease manufacture than on their own.

4.4 Synthetic Fluids

The desire to create lubricants with improved properties has led to the development of a vast array of synthetic lubricants [49-51]. Some are the synthetic equivalents of naturally occurring oils; others are not found in nature.

4.4.1 Synthetic Mineral Oils and Derivatives

The double bond of olefins (Fig.4.2) allows polymerization to give the equivalent of saturated paraffins. Polybutene is shown in Fig.4.10. Polyisobutylene (polyisobutene) has achieved great commercial significance. Chain length and thus viscosity can be closely controlled [13]. The oil distills off without leaving a residue, and brown staining is avoided.

When the chain length is C_{40} or higher, synthetic waxes, superior in many aspects to paraffin wax, are obtained with melting points ranging from 90 to 120 C [37].

Hydrogen may be replaced to give chlorinated hydrocarbons. Of special interest are chlorinated paraffins with chlorine contents of 40 to 70%, ranging from liquids to solids of substantial E.P. activity (Sec.3.4).

All hydrogen is replaced by chlorine and fluorine in chlorotrifluoroethylene, $CF_2Cl\text{-}CFCl_2$ (Fig.4.10). When polymerized to a low molecular weight, liquid chlorofluorocarbons exhibit an E.P. activity.

Hydrogen is replaced by fluorine in the fluorocarbons, which are inert, nonflammable, thermally and oxidatively stable liquids or solids.

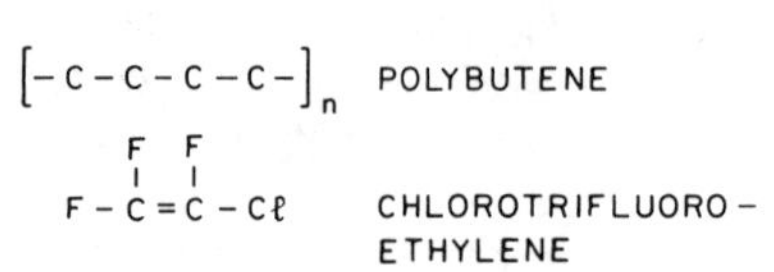

$[-C-C-C-C-]_n$ POLYBUTENE

F-C(F)=C(F)-Cℓ CHLOROTRIFLUORO-ETHYLENE

Fig.4.10. Structures of synthetic oils

4.4.2 Synthetic Esters

Esters are the reaction products of organic acids and alcohols (Fig.4.11). The ester linkage -C-O-R is very stable and the resulting lubricants have high temperature stability. Because organic acids may contain one or two -COOH groups (mono- and dibasic acids), and alcohols may have one to three -OH ends (mono-, bi-, and trihydric alcohols), a vast variety of esters is available. Many of them can be classified as fat-base synthetics [52].

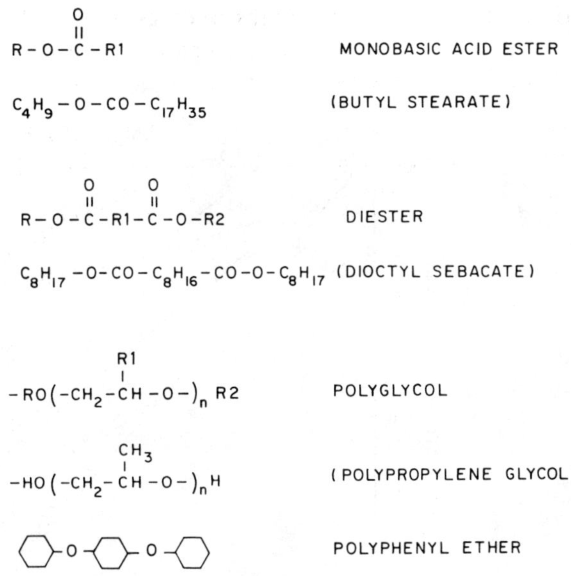

$R-O-\overset{\overset{O}{\|}}{C}-R1$ MONOBASIC ACID ESTER

$C_4H_9-O-CO-C_{17}H_{35}$ (BUTYL STEARATE)

$R-O-\overset{\overset{O}{\|}}{C}-R1-\overset{\overset{O}{\|}}{C}-O-R2$ DIESTER

$C_8H_{17}-O-CO-C_8H_{16}-CO-O-C_8H_{17}$ (DIOCTYL SEBACATE)

$-RO(-CH_2-\overset{\overset{R1}{|}}{C}H-O-)_n R2$ POLYGLYCOL

$-HO(-CH_2-\overset{\overset{CH_3}{|}}{C}H-O-)_n H$ (POLYPROPYLENE GLYCOL)

⬡-O-⬡-O-⬡ POLYPHENYL ETHER

Fig.4.11. Structures of synthetic esters

Monobasic acid esters are the reaction products of monobasic acids and monohydric alcohols. Examples of importance to metalworking lubrication include methyl stearate, butyl stearate (Fig.4.11), and hexyl laurate.

Dibasic acid esters, also called diesters, are formed from dibasic acids and primary alcohols, or from long-chain monobasic acids reacted with dihydric alcohols (glycols). Each molecule thus contains two ester groups; a typical example is given in Fig.4.11. No breakdown occurs until around 260 C, making them suitable for gas-turbine applications.

Somewhat related are neopentyl polyol esters formed from organic acids and polyfunctional alcohols, which have breakdown temperatures in excess of 300 C.

When the alcohol is replaced with a fluoroalcohol (in which fluorine substitutes for hydrogen), fluoroesters are obtained which resist hydrolysis and also oxidation to 320 C.

Synthetic fats are mono-, di-, and triglycerides, often obtained from the esterification of fatty acids and alcohols derived from natural sources, with the purpose of controlling properties, replacing fats from a scarce source, and creating fats particularly well-suited for a specific purpose. An example is synthetic palm oil produced from animal fat sources.

Polyglycols are based on dihydric alcohols or glycols, having the general formula $C_nH_{2n}(OH)_2$. The polymers have the structure shown in Fig.4.11. In polyethylene glycol, all three groups—R, R1, and R2—are hydrogen. In polypropylene glycol, R1 is a methyl (-CH_3) group (Fig.4.11). In polyether, R and R2 are alkyl groups, whereas in polyesters they are ester groups. The great advantage of polyglycols is that they can be made either oil soluble or water soluble [53]. Of importance for metalworking is that they may exhibit inverse solubility—i.e., they separate from water on reaching some characteristic temperature, ranging from 35 to 100 C depending on composition. Thus the properties of the viscous phase can be brought out at the tool-workpiece interface, and the solution is re-established once the lubricant leaves the interface. Polyglycols act primarily as hydrodynamic substances, but can be modified to impart some E.P. properties to them. At chain lengths in excess of C_{70} they become water-soluble waxes.

The temperature-viscosity characteristics of some synthetics are shown in Fig.4.12 [15] . Pressure-viscosity-temperature data for several synthetics have been given [54,55].

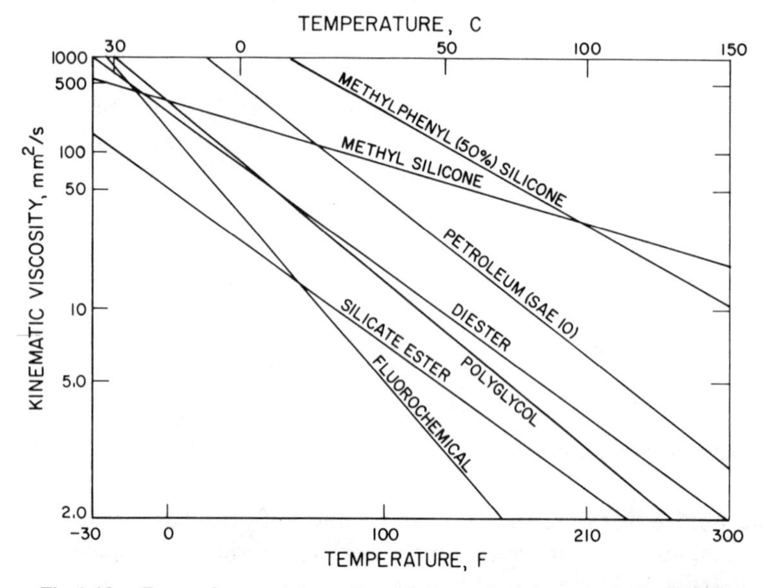

Fig.4.12. Dependence of viscosity on temperature for selected synthetic lubricants [15]

Phosphate esters differ from the above esters in that the acid is the inorganic phosphoric acid. Thus the characteristic -C-O-R linkage is replaced by the -P-O-R grouping (Fig.4.13). If two of the R groups are hydrogen and the third an alkyl or aryl radical, a primary phosphate ester is produced. Two alkyl or aryl radicals produce a secondary phosphate ester, and three produce a tertiary phosphate ester. The best known example of the latter is tricresyl phosphate (Fig.4.13), which is used as an E.P. (antiwear) additive.

By proper choice of the R radicals, substances ranging from water-soluble, low-boiling liquids to water-insoluble, viscous fluids or waxlike solids can be obtained. In resistance to thermal degradation they are comparable to mineral oils, but, in contrast to them, they are fire resistant. Their viscosity drops little with temperature, which makes them useful hydrodynamic lubricants with built-in wear protection.

When at least one of the single-bonded oxygen atoms is replaced by sulfur, thiophosphates are obtained which are frequently used as lubricant additives.

4.4.3 Silicon Compounds

Many organic compounds have their counterparts in which the carbon atom is replaced by silicon. When these are combined with organic radicals, useful lubricants result.

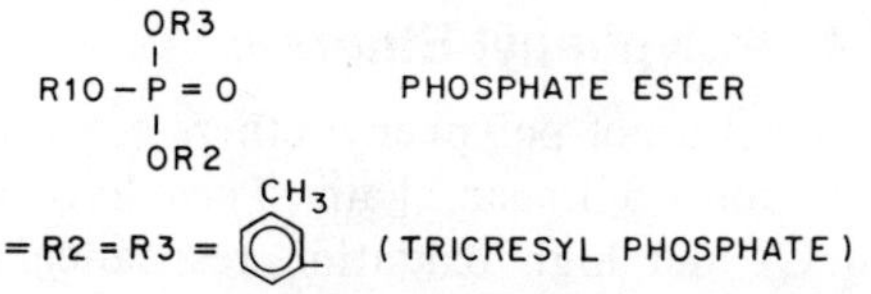

```
        OR3
         |
  R1O - P = O        PHOSPHATE ESTER
         |
        OR2
                  CH3
R1 = R2 = R3 = (benzene ring)   (TRICRESYL PHOSPHATE)
```

Fig.4.13. Structures of phosphate esters

Silicate esters. In silicate esters the organic ester linkage -C-O-R is replaced by -Si-O-R. These compounds are based on the orthosilicate structure (Fig.4.14) in which R may be similar or different alkyl or aryl groups. Disiloxanes (Fig.4.14) have similar properties. Depending on molecular weight and on the length and nature of the R radicals, these compounds range from low-viscosity liquids to high-melting solids. Their viscosity-temperature characteristics (Fig.4.12) make them attractive as hydrodynamic lubricants, but they cannot protect surfaces under boundary conditions unless suitable additives are incorporated.

Silicones are the inorganic equivalents of hydrocarbons, with SiO_2 forming the backbone of the molecule (hence the name polysiloxane, referring to the Si-O bond), and with organic R radicals attached by Si-C bonds (Fig.4.14). For a review, see Meals [56]. The R radical is mostly a methyl ($-CH_3$) group or a mixture of methyl and phenyl ($-C_6H_5$) groups. The linear dimethyl silicones have a remarkably low temperature-viscosity coefficient (Fig.4.12). Branching reduces viscosity, whereas networks formed by crosslinking impart rubberlike behavior. They are temperature resistant to 200 C in air, and to 360 C in the absence of oxygen.

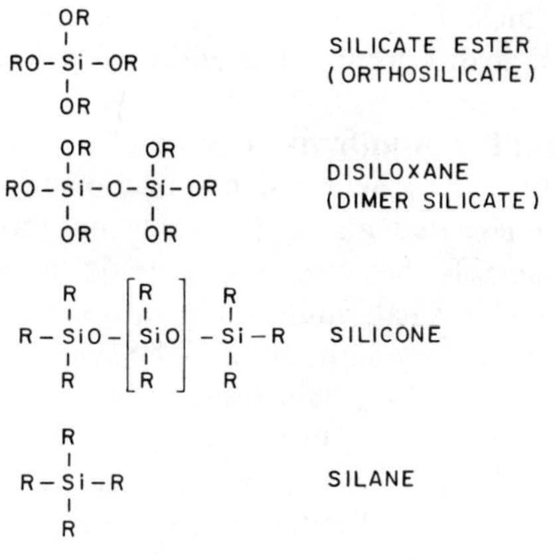

```
      OR
      |
RO - Si - OR                    SILICATE ESTER
      |                         (ORTHOSILICATE)
      OR

      OR     OR
      |      |
RO - Si - O - Si - OR           DISILOXANE
      |      |                  (DIMER SILICATE)
      OR     OR

     R     [R ]     R
     |     [| ]     |
R - SiO - [SiO] - Si - R        SILICONE
     |     [| ]     |
     R     [R ]     R

     R
     |
R - Si - R                      SILANE
     |
     R
```

Fig.4.14. Structures of silicon compounds

Silicones owe their lubricating power to viscosity alone; they are relatively poor lubricants because of their low viscosity-pressure coefficient and absence of boundary-lubrication action. Some E.P. activity can be imparted by chlorination or fluorination. At elevated temperatures some wear protection is observed, and this is attributed to the formation of a surface film from the oxidative breakdown and crosslinking of the lubricant, or to interaction of the Si-O bond with the metal [57]. High-molecular-weight silicone waxes were successful for aluminum when deposited from a chlorinated solvent [58].

Silanes are based entirely on the Si-C bond (Fig.4.14). Depending on the organic R radical, liquids of low oxidative but high thermal stability are formed which can be improved by appropriate additives.

4.4.4 Polyphenyl Ethers

The structure of polyphenyl ethers consists of three or more benzene rings linked by oxygen to form a linear chain. They have a very high thermal degradation temperature (400 C) and high oxidation resistance. Fluorinated polyethers have good boundary-lubrication properties and are thermally stable to 443 C [59].

4.5 Compounded Lubricants

With the exception of natural fats, few of the above discussed substances fulfill all the requirements of a metalworking lubricant. The predominance of the mixed-film lubrication mechanism demands that boundary and E.P. lubricants be added to substances possessing only viscous properties. Almost all lubricants require further additives to impart other characteristics of a nontribological nature such as oxidation resistance, corrosion protection, and detergency. Thus, most metalworking lubricants are compounded with one or more additives, and the industrial oil chemist has the task of selecting base stocks and additives which produce no harmful interactions and which preferably are synergistic. Additives used in lubricant formulation have been discussed by Jentgen [3.62], Molyneux [60], Rowe [61], and Smalheer [62]. Greases may be regarded as a special class of compounded lubricants and will be discussed here. Because of their importance, water-base (aqueous) lubricants will be discussed separately (Sec.4.6), as will be lubricants applied to the workpiece as a coating (Sec.4.7).

4.5.1 Compounded Oils

The most frequently encountered compounded oils are based on mineral oils or synthetic hydrocarbons. The additive package may contain the following:

Boundary Additives. Choice of additives is governed by workpiece material. As little as 0.1% fatty acid reduces friction in working of aluminum, and no further improvement is noted above 1% acid concentration. Larger percentages are usual for steel, presumably because a change in lubricant rheology is also needed (Sec.3.6.3). In particular, small additions of fatty acids cause earlier solidification of mineral oils [63].

Boundary additives may be synthesized, too; two oxygen-containing groups attached to a linear alkyl chain resulted in good aluminum-lubricant additives [64].

Fatty esters, fatty oils, or soap additions are needed in larger quantities, on the order of 1 to 10%, perhaps again because of their effect on lubricant rheology (Sec.3.5.3). A limitation on additive chain length and concentration is imposed by brown staining.

A special virtue of many fatty oils and derivatives is that they are edible, and thus their residues can be tolerated on products used for packaging. Of course, substances such as lead soaps are an exception.

Polymers are often added to engine oils as viscosity-index improvers In metalworking lubrication they serve to impart special properties such as thixotropy Finely dispersed powders of polymers such as PE are believed to function by the attachment of radicals formed by the thermomechanical degradation of the polymer in the deformation zone [65]. Adsorption to the metal surface is reported to have been increased by gamma irradiation of the polyethylene and by the addition of distillation residues of fats [65a].

E.P. Additives. Depending on the active element and the bond strength of the compound (Sec.3.4), additives can be more or less aggressive. The aim is generally to provide protection in the temperature range encountered over asperities, as shown schematically in Fig.4.15 for the mixed-film lubrication regime in the sliding of iron. Sulfonated, chlorinated, or sulfochlorinated fatty oils, which themselves can be regarded as compounded lubricants, may serve as dual-purpose additives, although separate boundary and E.P. additives are often employed. Some well-established additives may need replacement because of health or environmental considerations or because the source of the additive (sperm oil for sulfurizing) has disappeared [66].

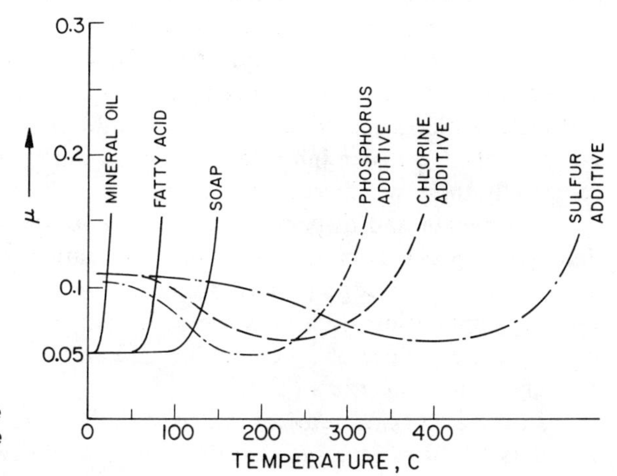

Fig.4.15. Schematic illustration of the mode of action of a compounded oil-base lubricant

E.P. additives are mandatory when boundary additives would be totally ineffective (as on stainless steel and other high-chromium steels, and on titanium) or would break down at the higher temperatures developed in the process (as in the more severe working of steels). The choice of additives depends on, among other factors, workpiece material. Thus, chlorine is the only effective element for stainless steel. Sulfur is effective on steel, but is avoided on high-nickel alloys because of the danger of forming a low-melting eutectic, and on copper alloys because of the formation of a dark sulfide stain. Staining is proportional to the activity of the additive; additives that are nonstaining at room temperature can be effective at the high temperatures that prevail in metalcutting. A nonactive additive does not react with copper below 100 C. Phosphorus, usually in the form of phosphate esters or ZDTP, is effective more as an antiwear agent.

Solid Additives. When a process fails to produce the proper conditions necessary for the activation of E.P. lubricants (e.g., in deformation processes at low speeds, with nonrepetitive contact), or for severe conditions involving higher temperatures, solid lubricants (primarily graphite and/or MoS_2) are added in various concentrations. They act by forming surface films [67] and may interfere with some additives, as graphite does with phosphates, or give synergistic effects, as with calcium sulfonate [68].

Graphite and MoS_2 are sometimes objectionable because they tend to blacken the machines, the shop, and the operator's clothing, and are difficult to remove. This has given rise to the so-called white lubricants, which are oil-base pastes or diluted suspensions of sulfides (e.g., zinc sulfide), phosphorus compounds (zinc pyrophosphate), or metal hydroxides (calcium hydroxide) [69-71]. Used at the same concentration as MoS_2, they lower friction sufficiently to allow moderately severe deformation, and they can be removed by the usual degreasing techniques. Calcium carbonate [3.169] falls into this category, too. Inert fillers are sometimes added to obtain a pastelike consistency. Some oxides such as CaO and PbO have lubricating value [72]. Solids formed by the reaction of caprolactam with a metal hydroxide are reported to have some boundary properties [73]. One point to consider is whether or not solid additives may make the removal of lubricant residues more difficult.

General Additives. Many oils have insufficient oxidative stability on their own and form acids, gums, and insoluble carbon deposits. Oxidation inhibitors act by adsorbing on the metal surface, reducing its catalytic activity, or by decomposing the organic peroxides which are the intermediate oxidation products. Many antioxidants also have E.P. properties, as have tributyl phosphite, disulfides, and ZDTP. The latter also acts as a corrosion inhibitor, an important function if the lubricant resides for a long time in the metalworking machinery or in the lubricant application system and could attack components. Corrosion inhibitors are particularly important when aggressive E.P. agents (especially those of chlorine) are used.

Detergents and dispersants are of prime importance in automotive engines, where detergents prevent high-temperature formation of deposits and dispersants prevent accumulation of sludge in a cold engine. By adsorbing on solid particles, they prevent their agglomeration. A wide variety of compounds is used, all of which are characterized by the presence of an oleophilic and an oleophobic group in the molecule and thus are related to emulsifiers (Sec.4.6.2).

In metalworking lubricants it is usually more desirable to remove various reaction products by filtration, because detergents could result in foaming and in formation of emulsions.

Small concentrations of methyl silicone polymers form minute droplets of low surface tension, which break the foam bubbles and thus serve as defoaming agents.

Some lubricants provide nutrients for microorganisms, especially in the presence of water, and are compounded with biocides (see also Sec.4.6.5).

4.5.2 Compounded Fats, Waxes, and Soaps

Lubricants of this class have in common a semisolid or solid consistency, making them especially suitable for low-speed operation where a mixed film could not be built up with a liquid lubricant.

We have already mentioned that fats can be sulfonated and chlorinated. They can also be compounded with separate E.P. agents, layer-lattice compounds, or solid fillers. The options are inexhaustible, but many practically used combinations have questionable value and often represent overkill. Only carefully controlled experiments can determine if a particular additive is indeed necessary.

Waxes and soaps may be compounded, too, often with layer-lattice compounds (Sec.7.3.1).

4.5.3 Greases

Greases differ from pastes in that the liquid phase is uniformly dispersed within a thickener or jelling phase which imparts definite rheological characteristics to the grease. Reviews have been published by Harris [74], McCarthy [75], and Polishuk [76].

The liquid phase may be a mineral oil, synthetic oil, ester, or silicone. It is often compounded with oxidation and corrosion inhibitors, and sometimes with boundary, antiwear, and E.P. additives. Thickening agents fall into three general categories:

1. Fatty-acid soaps are soluble in mineral oil at elevated temperatures, but on cooling form a gel which then turns into a fibrous skeleton, trapping the oily phase. Greases based on calcium-(lime) soaps often have an oily phase compounded by fats or fatty

acids; stabilized by water, they retain their consistency even in an aqueous environment, but become liquid around 100 C. Greases based on sodium-(soda) soaps can stand somewhat higher temperatures but are soluble in water and are sometimes termed water-wettable. Greases based on lithium soaps have gained prominence in the last decade because of their higher liquification temperature (dropping point of 185 C and over) and good water resistance. They also permit the use of synthetic oils, esters, and silicones as the liquid phase.

2. Complex soaps have a fiber structure modified by the co-crystallization of two compounds, a normal soap (usually lithium hydroxystearate) and the complexing agent (a salt usually of the same metal) [77]. They have a higher service temperature. Aluminum complex greases have found use in steel mills [78].

3. Non-soap-base greases contain gelling agents which are insoluble in the liquid phase at all temperatures. Thus they do not suffer a change in consistency at some critical temperature, and are useful in elevated-temperature applications. The thickener may be an inorganic substance such as carbon black, silica (below 50 μm particle size) or, more importantly, clay (montmorillonite) modified to make it oleophilic. The number of possible organic thickeners is very large. Of commercial significance are polymers such as PTFE [79], aryl-substituted urea, pigments such as indanthrene blue, and other substances such as copper phthalocyanine.

Most greases are Bingham solids (Sec.3.6.3), and the presence of their initial yield point makes for a high starting torque in bearing applications; this is of importance in metalworking only if the transient high friction and force would result in fracture, as in drawing. Once the thickener structure is crushed, the properties of the liquid phase dominate, although properties are modified by poorly understood interactions with the thickener. Worked penetration is the basis of the consistency classification of the National Lubricating Grease Institute (NLGI). Greases release oil on standing, and such bleeding and, in general, long-term stability are of great concern in bearing applications but, apart from storage considerations, of minor importance in metalworking. The liquid phase is often chosen for inertness, as in many of the chlorinated or fluorinated synthetic greases; it is not clear whether this is of any benefit for metalworking applications, and most greases used are still of the mineral oil - soap type.

For severe duties, layer-lattice compounds, especially MoS_2 and/or graphite, are often incorporated [67]. They seem to have little or no effect on the consistency of the grease, in contrast to antiwear and E.P. additives and oxidation and corrosion inhibitors [80].

4.6 Aqueous Lubricants

Water is inexpensive and possesses great cooling power. It is, however, a very poor lubricant. The desire to endow water with the necessary lubricating qualities has led to the development of a large number of fluids, used in vast quantities for both metal deformation and metalcutting processes. Over the years, terminology has grown in a haphazard fashion and is still confusing today.

4.6.1 Classification

The following classification, originally developed for cutting fluids [20,81], is acceptable for general purposes:

1. Emulsions (sometimes, misleadingly, also called "soluble oils") contain a lubricant, such as a mineral or compounded oil in the form of suspended droplets, dispersed with the aid of special chemical agents called emulsifiers. The emulsified oil droplets

are large enough to make the made-up lubricant milky (or sometimes translucent) in appearance. The action of emulsions as lubricants can be close to that of the dispersed phase (Sec.3.7.1).

2. Semisynthetic fluids (also called semichemical fluids or chemical emulsions) are lean emulsions; the oily phase is only 5 to 30% of the concentrate, the remainder being comprised of emulsifiers, water-soluble corrosion inhibitors, wetting agents, organic and inorganic salts and, sometimes, E.P. agents. The large proportion of emulsifiers reduces particle size to the point that the made-up lubricant is translucent or transparent. These lubricants combine reasonable lubricating qualities with excellent cooling power, making them useful in metalcutting.

3. Synthetic fluids (also called chemical fluids) have nothing to do with the synthetic lubricants discussed in Sec.4.4. They contain no oil, only water-soluble wetting agents, corrosion inhibitors, salts and, sometimes, E.P. agents. Some of the constituents may form aggregates (micelles) but are essentially water-soluble, and therefore the made-up fluid is transparent. These fluids combine limited lubricating power with excellent cooling, and are useful in grinding and less-demanding cutting.

4. A further distinction is sometimes made: synthetic fluids that contain only corrosion inhibitors are sometimes called "true solutions." Their lubricating qualities are little better than those of water itself, but they often make good grinding fluids and coolants.

Some general discussions of this topic are available [20,21,81-87]; information on specific fluids can be found in Chapters 6 to 11.

4.6.2 Emulsions

Oil can be broken into droplets and dispersed in water by intense mechanical agitation (shearing) alone. Much of the mechanical energy goes into the necessarily substantial increase in interfacial area; on standing, the system tends to return to equilibrium rapidly by coagulation of the oily phase. Thus, a mechanical dispersion is too unstable to be practically useful. In order to stabilize the dispersion, the interfacial energy must be reduced by the addition of surface-active agents (emulsifiers). Emulsion technology is complex and has been the subject of several detailed publications [88-92].

Most metalworking emulsions are oil-in-water (O/W) systems: oil is the dispersed (internal) phase, water the continuous (external) phase. Water-in-oil (W/O) emulsions find only limited use. Emulsions may be regarded as having three principal components: the oily phase, the emulsifier, and water.

The Oily Phase. Depending on the oily phase, emulsions are sometimes denoted by distinctive names:

1. For lightest duties, an uncompounded mineral oil is used. The term "soluble oil" is sometimes applied.

2. For more severe duties, mostly in the working of nonferrous metals, animal or vegetable fats or oils are added to the mineral oil, yielding what are sometimes termed "superfatted emulsions."

3. For more severe duties on ferrous metals, E.P. additives are needed. Because water-soluble additives are ineffective [93], E.P. additives are incorporated in the mineral oil, leading to so-called "heavy-duty emulsions."

4. Fatty oils or fats (such as tallow), incorporated into a soap (usually potassium stearate) in relatively high proportions (up to 50%), can be diluted with 25 to 40% water to give pastes or, if further diluted, liquids Free fatty acid (such as stearic acid) is often added. By definition these products are emulsions, but, because of their consistency, they are often termed "soap-fat pastes."

Emulsifiers are surface-active substances (surfactants). By definition, their equilibrium concentration on the oil-water interface is higher than in the bulk. An effective emulsifier has some solubility in both phases, and the molecule has a water-soluble (hydrophilic) part and an oil-soluble (hydrophobic or, better, lipophilic) part. The relative weight percentages of the two parts give the HLB (hydrophilic-lipophilic balance) of a surfactant, and form the basis of a semi-empirical emulsion formulation system devised by Griffin [94].

Highly water-soluble emulsifiers have high HLB values. Most O/W emulsions require an HLB of 10 to 13, whereas a low HLB of 3 to 6 is typical of O/W emulsions. The HLB value of a particular surfactant can be calculated [89] or experimentally determined (Table 4.6). Similarly, the HLB values required for emulsifying specific oily phases can be determined (Table 4.7). As a first attempt, an emulsifier (or, more frequently, a blend of two emulsifiers) of the appropriate HLB value is mixed with the oil and, if an emulsion forms on severe shearing, other emulsifiers of similar HLB can be used. The final choice is dictated by cost, odor, stability, and other considerations, not least of which is efficiency—i.e., the concentration required to ensure stability.

Table 4.6. HLB values of selected emulsifiers (data compiled from [92])

Surfactant	HLB value	Surfactant	HLB value
Sodium oleate	18	Sorbitan monolaurate	8.6
Sorbitan monooleate +		Sorbitan monostearate	6
20 (CH_2-CH_2-O) groups	15	Glycerol monostearate	3.8
10 (CH_2-CH_2-O) groups	13.5	Cetyl alcohol	1
5 (CH_2-CH_2-O) groups	10	Oleic acid	1

Table 4.7. HLB values required for emulsification (data compiled from [92])

Oily phase	O/W emulsion	W/O emulsion
Acid (oleic, stearic)	17	. .
Alcohol:		
Stearyl	15	7
Lauryl	14	. .
Fatty oil:		
Castor	14	. .
Cottonseed, tallow	6	. .
Palm, rapeseed	7	. .
Lanolin (anhydrous)	12	8
Butyl stearate	11	. .
Wax:		
Beeswax	9	5
Carnauba	15	. .
Paraffin, microcrystalline	10	4
Mineral oil:		
Paraffinic	10	4
Aromatic	12	4
Kerosene	12	6
Chlorinated paraffin	12-14	. .
Tricresyl phosphate	17	. .
Methyl phenyl silicone	7	. .

Most emulsions of importance in metalworking owe their stability to the development of an electrical double layer at the interface; repulsive forces between like-charged droplets prevent their coalescence. Therefore, emulsifiers are usually classified according to the charge they impart to the emulsified droplet.

1. In anionic agents the hydrophilic part of the molecule carries a negative charge. Classical examples, still employed, are the fatty-acid soaps. In the sodium soap of a mixed acid, the long-chain hydrocarbon tail is oil-soluble, and the hydrophilic -COONa heads form an oriented layer on the surfaces of the oil droplets (Fig.4.16a). The soap ionizes to a large extent, and the sodium ion enters the water phase, leaving behind a negatively charged surface. Coupling agents are often used to increase the solubility of the soap in oil. Soaps have several disadvantages. The emulsion must be formulated to tolerate their high alkalinity (pH = 10). In hard water, insoluble bivalent calcium and magnesium soaps form which deplete the emulsifier concentration and favor the formation of W/O emulsions (Fig.4.16b), thus reducing the stability of the emulsion. This is true also, but to a slightly lesser extent, of amine soaps. More resistant to calcium are sulfated agents with the -O-SO_3Na group bonded by the somewhat stronger C-O link. Classical examples are sulfated castor oil (Turkey red oil) and sulfated oleic acid (sulfonated red oil). Most resistant to calcium and most extensively used are the sulfonates, in which the hydrophilic group -SO_3Na is attached by the strong C-S linkage, which is not subject to hydrolysis or oxidation.

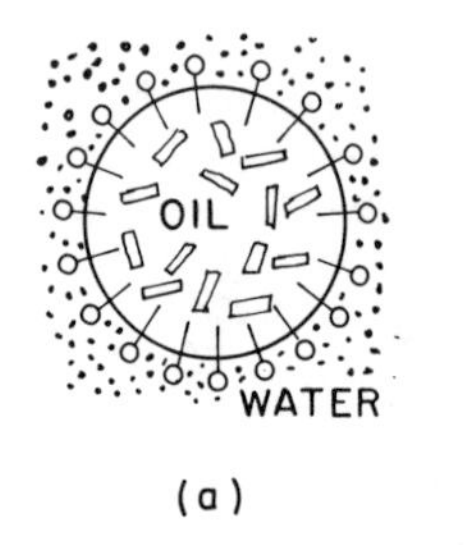

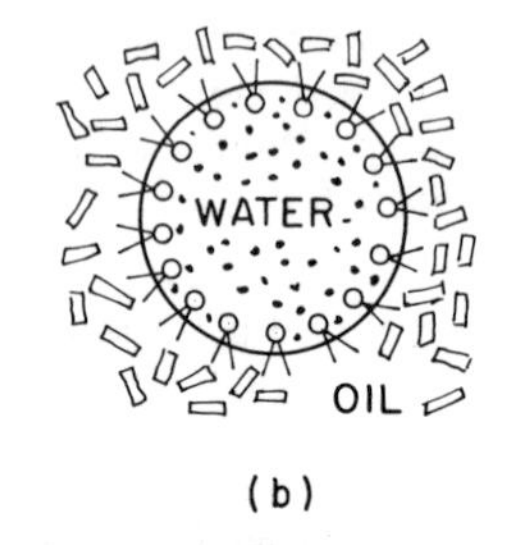

Fig.4.16. Oil-in-water emulsion stabilized with monovalent soap (a) and water-in-oil droplet stabilized with bivalent soap (b)

2. Cationic emulsifiers, most frequently quaternary amine salts, show reverse ionization relative to soaps. The long-chain lipophilic group is the cation, and the anion is usually a chloride or bromide. They are insensitive to calcium; their main use is in O/W systems, or when the emulsion has to be neutral or acidic.

3. Nonionic agents are usually esters formed by the reaction of a lipophilic long-chain fatty acid with a hydrophilic polyhydric alcohol, or of fatty acids with polymerized ethylene oxide (-$(CH_2)_2$-O-). Since the relative chain lengths of hydrophilic and lipophilic groups can be varied, emulsifiers of a broad HLB range (Table 4.6), insensitive to water hardness and emulsion pH, can be made.

4. Less frequently used are the ampholytic (amphoteric) emulsifiers, such as long-chain amino acids, which show cationic behavior at low pH and anionic behavior at high pH.

5. Naturally occurring emulsifiers, such as water-soluble vegetable gums, lanolin, cholesterol, etc., are useful in conjunction with other emulsifiers.

6. Solids, if divided into particles much finer than the oily globules, contribute to emulsion stability. Clay (bentonite), carbon black, and hydrophilic graphite are examples.

Water. It is clear from the above discussion that water has a marked affect on emulsion stability. Both the chemical and biological quality of the water has to be chosen to complement the emulsion system (or vice versa), or the water has to be subjected to an appropriate treatment, as discussed by Bennett [95] and Sluhan [96].

Water treatment follows one of three directions:

1. Water hardness is reduced by removal of the calcium and magnesium ions. This can be done by adding washing soda ($Na_2CO_3 \cdot 10H_2O$) to form water-insoluble carbonates, or by processing the water through a zeolite ion exchanger. In either case, the increased sodium concentration in the water could lead to foaming.

2. All ions are removed by chemical adsorption, making deionized water equivalent to distilled water or rainwater.

3. Where available, boiler condensate can be used.

Deionized and condensate waters are free of chlorides, sulfates, and other inorganic substances which, even though of only minor effect on emulsion stability, contribute to corrosion and bacterial deterioration. It is possible to compensate for moderate water hardness by emulsion formulation, but the cost of replenishment and the required compromises regarding emulsion performance make water treatment preferable. When water hardness exceeds 100 ppm (or, to use the still lingering system, 7 grains), treatment becomes mandatory.

Emulsion formulation is a complex blend of art and science [87,90] and is governed by several considerations:

1. Stability is a controversial issue. According to some views, less stability is desirable when maximum lubrication efficiency is sought; the oily phase is believed to separate out more readily on the metal surface (plating out). Very coarse emulsions may need continuous agitation to maintain dispersion, and a more finely distributed emulsion has, in general, greater stability. Thus, stability is often judged by appearance, which in turn is affected by emulsion particle size (Table 4.8).

Particles of the same order of magnitude as the wavelength of white light (0.4 to 0.7 μm) reflect light and appear white. Light incident on smaller particles is scattered and polarized: particles appear blue in reflected light and orange to red in transmitted light. This is the Tyndall effect, which gives color to the sky, and is useful as a quick method of estimating particle size.

Translucent emulsions are termed microemulsions [97]. They differ from macroemulsions in that the interfacial film, usually a mixed soap/fatty alcohol film (also called a penetrated monolayer; Fig.3.41b), reduces interfacial tension to zero, and the emulsion remains stable for years. So-called cutting oils often fall into this category.

To control particle size, some 5 to 20% of the emulsifier is added to the oil, on its own or with a coupling agent such as a complex alcohol or a nonionic wetting agent. In addition to emulsifier type and concentration, the factors of temperature, method of preparation, viscosity difference between phases, and energy input (agitation) all contribute to the definition of emulsion stability [98].

Table 4.8. Guidelines for visual estimation of droplet size (data compiled from [97])

Color of dilute emulsion	Tyndall effect to eye: Reflected	Tyndall effect to eye: Transmitted	Droplet diameter: μm	Droplet diameter: Å
Dead white	None	None	>0.5	>5000
White to gray	Weak blue	Weak red	0.3-0.1	3000-1000
Gray to translucent	Intense blue	Intense red	0.14-0.01	1400-100
Clear transparent	None	None	<0.01	<100

For economy in transportation, a concentrate is made up, usually in the form of a W/O emulsion containing just enough water to ensure the formation of a stable O/W emulsion by inversion when the concentrate is added to water.

2. Another controversial issue is wetting. Emulsifiers are, by definition, surfactants and thus promote wetting of the surface. They also are detergents, and thus remove dirt, and are dispersants, and thus maintain solid particles in a dispersed form. However, the best emulsifier is seldom the optimum agent for other functions and, if desired, wetting agents, etc., are added. While some researchers regard wetting to be of paramount importance, others disagree, and no unequivocal evidence seems to be available.

3. Because emulsifiers reduce surface tension, they also promote foaming [99]. This is undesirable because, in effect, an air-lubricant mixture would be applied, and foaming tendency must be suppressed, especially when the process itself (such as grinding) promotes it. Small quantities of foam suppressants, usually silicones (particularly dimethyl siloxane), are sometimes added, although it is preferable to minimize foam formation by a suitable choice of emulsifiers.

4. In principle, synthetic polymers can be emulsified just as well as can mineral or fatty oils. Thus, Proskuryakov and Isaev [100] report on the success of lean (0.01 to 0.025%) polyacrylamide emulsions in finish machining, and Tamai et al [101] on the use of 20% polyalkylmethacrylate to replace the fatty oil in the mineral oil phase of a rolling emulsion.

5. Special emulsion formulations are often developed for specific applications. Thus, the addition of a silicone emulsion to an aqueous graphite dispersion was found to reduce friction and die pickup in hot forging of aluminum [58]. Yet further additives are often incorporated; for example, the emulsion mentioned was further improved by a lead additive. Some additives affect the basic mode of action of emulsions. Thus, Maurer [102] describes metallic chelates, such as a 1:1 molar Cu (II):citrate complex, which release metal ions (in this example, cupric ions), which in turn deposit on the metal surface, resulting in lower lubricant consumption and better lubricant-system control.

Emulsion Breakdown. The stability of emulsions is a function of many variables [98], and any shift in this delicate balance can reduce emulsion stability (Fig.4.17):

a. Slow gravitational segregation of the emulsified phase results in creaming: the emulsion separates into an oil-rich phase and a water-rich phase. On agitation the emulsion is re-established.

b. When the surfactant cannot maintain a continuous film, or the surface charge is lost by depletion of emulsifier, the particles of the dispersed phase coagulate and separate as a layer. Such breaking of the emulsion may be irreversible.

c. Breaking is sometimes preceded by flocculation, a clustering of droplets without coalescence. At this early stage of breakdown the emulsion can be re-established.

d. When the proportion of the dispersed phase increases (e.g., by addition of oil to

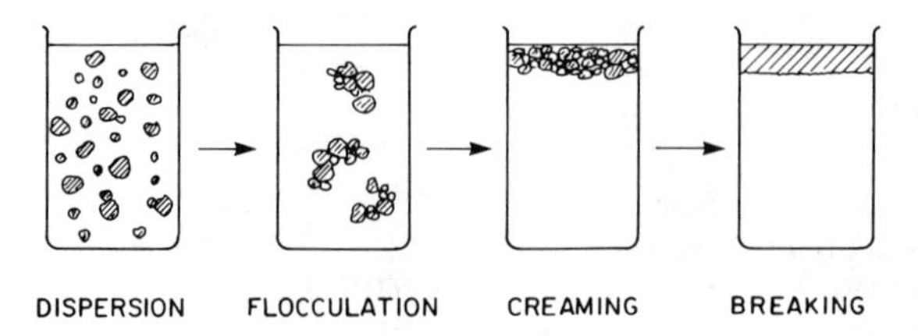

Fig.4.17. Instability in emulsions

or removal of water from an O/W emulsion), the oil droplets unite, trapping water as the dispersed phase. Thus a W/O emulsion is formed by the process of inversion.

Emulsion Rheology. At a first approximation, emulsions can be regarded as two-phase systems whose flow properties are controlled by the properties of the continuous and dispersed phases and by electrical interactions. The emulsifier layer modifies the properties of the dispersed phase. For a review, see Sherman [89]. The most important factors are the quantity and size distribution of the dispersed phase, as discussed by Lissant [90]:

a. If the internal phase is uniformly dispersed and of a low volume ratio (e.g., below 5%), it will not appreciably affect the flow of the continuous (external) phase. Viscosity rises only very slowly with increasing volume ratio. Only at volume ratios greater than 10 to 20% will the interaction between droplets cause some deviation from Newtonian behavior, usually in the form of mild thixotropy or pseudoplasticity (Sec.3.6.3; Fig.3.34). This should be true of most metalworking emulsions. Only if the droplets flocculate, especially in the form of tangled chains, will non-Newtonian behavior become pronounced. Even gelling may take place.

b. At medium internal phase ratios (e.g., from 30 to 70%), interference between emulsion droplets is severe, and at around 50% concentration (corresponding to a cubic array of uniform spheres) the emulsion turns more viscous than its internal phase. Also, if the emulsifier does not form an effective surface film, inversion may occur. This ratio is avoided in metalworking.

c. At the highest internal phase ratios, droplets must actually deform to fit without breaking through the continuous phase, or the emulsion must change into a gel in which the internal phase takes the form of fibers and droplets embedded in the continuous phase. It is likely that some drawing compounds may fall into this category, with rheological properties similar to those of greases.

Gradual breakdown of an emulsion can take it through the entire range of rheological behavior, with consequent difficulties in terms of the lubricating function, circulation, and application.

4.6.3 Synthetics

As mentioned already, semisynthetics are closest to emulsions in their characteristics. They could be compared to microemulsions with additional water-soluble components which either are truly dissolved as individual molecules or form oriented clusters (micelles) in the fluid.

True synthetics have no oil in them. The earliest versions contained simply soda ash, which tended to deposit on machineways and workpieces [103]. Most later versions were based on organic amines (such as triethanolamine) with sodium nitrite ($NaNO_2$) as the oxidation inhibitor. The nitrite is cheap and effective by sacrificial action (it oxidizes to nitrate), but the potential for the formation of nitrosamines (see Sec.4.9.2) has led to the development of other proprietary chemical solutions. Nitrites are replaced by some other compound, such as sodium molybdate [104], a carboxylic acid derivative [105], a benzylamine [106], or disodium phosphate [107]; this is still a most active field of research.

Among synthetics are special lubricants such as those based on polyacrylonitrile, which yields water-soluble polymers with $-CONH_2$, -CN, -COONa, and -COHNH polar groups that form stable complexes with metal ions. They are reported to give good wetting and low surface tension, and are presumed to form bonds which help to remove chips from the cutting zone [108]. Some solutions are improved by additives.

4.6.4 Corrosion and Equipment Damage

A water-base lubricant which contains a number of chemically active compounds presents a clear danger of attack on the workpiece and machine tool [81,109,110].

Corrosion.

a. Boundary additives react to form soaps with many metals. Zinc soaps formed on galvanized iron are insoluble in oil but can change the stability of emulsions. However, copper piping is attacked because the soaps are oil-soluble. Furthermore, copper is a catalyst for the oxidation of the oily phase, and it can affect emulsion stability. On the workpiece, a characteristic green oxide film may form on storage.

b. In order to function, E.P. additives must react with metal surfaces. Thus, chlorine is essential for the lubrication of stainless steel, some nickel-base superalloys, and titanium; residues could cause corrosion and, at elevated temperatures, stress-corrosion cracking, although the evidence is not clear whether or not small quantities remaining after machining are indeed responsible for such attack. Any chlorine that is liberated but not reacted with the workpiece can form HCl, which then corrodes the machine tool. As mentioned, problems exist with sulfur compounds in conjunction with high-nickel alloys and with the discoloration of copper alloys. Sources of sulfur may exist not only in the E.P. additives but also in the sulfates and sulfonates used as emulsifiers.

c. Oxidation (rusting) of iron in aqueous environments is a perennial problem, and is greatly accelerated when HCL is formed. Corrosion of the equipment is accelerated when chips (swarf) are allowed to accumulate because of poor housekeeping. Cast iron rusts particularly rapidly. Because it breaks into small chips on machining, it also retains large quantities of cutting fluid. Corrosion is highly dependent on the pH of the fluid [111]. To limit the corrosion of steel and cast iron machine elements, emulsions are kept alkaline, with pH values of 7 or, more typically, 8 to 10. Excessive alkalinity cannot be tolerated, partly because of irritation to the human skin and partly because aluminum and magnesium are attacked. Corrosion inhibitors are also added.

d. Oxidation of magnesium is so rapid that chips and fines ignite unless they are kept totally submerged in the lubricants.

e. The ethanolamines frequently found in synthetics form blue cupramines with copper-base metals [109].

Damage to Machine Elements.

a. We have already mentioned (Sec.3.10.3) that fatigue wear of highly stressed components is accelerated by the presence of water. If seals on the machine tool fail to keep the aqueous lubricant out, premature failure of hardened components such as roller bearings and shafts occurs [112,113]. The rate of attack is influenced by even minor changes in composition or environmental conditions. Thus, fatigue pitting of bearing balls was affected by the addition of a biocide [112].

b. Emulsions are able to emulsify bearing and gear-lubricating oils and water-wettable greases and thus either wash them out or form highly viscous phases (often W/O emulsions) which block lubricating passages. Either way, they contribute to failure due to lubricant starvation. It is essential, therefore, that machines be constructed with specific attention to keeping out aqueous lubricants and that seals be kept in good condition.

c. In a poorly controlled system, emulsion residues may accumulate and then cause sticking, staining, and bacterial attack (odor).

d. Special problems arise with synthetics. They usually provide adequate corrosion protection, but also act as detergents and remove whatever oil or grease may have been applied to machine elements. Hydrostatic bearings become incapacitated. It is imperative that seals be designed and maintained with particular care, and that sliding surfaces exposed to the synthetic be provided with automatic lubrication.

e. Emulsifiers and synthetics wet the paint-metal interface; they can also soften the paint, leading to blistering and even to total lift-off (pealing) of conventional paints, such as the air-drying cellulosic paints often preferred for their ease of application. Practically emulsion-resistant are nylon 0.2 to 0.3 mm thick, and epoxy and polyester powder coatings 0.05 mm thick. Epoxy and polyester paints are acceptable although they have less resistance to impact and suffer chipping [114].

4.6.5 Biological Control

The aqueous environment of emulsions, combined with the availability of nutrients, provides an ideal breeding ground for microorganisms such as bacteria, molds (fungi), and yeasts. The problems arising from the growth of such organisms comprise the subject of a vast and rapidly growing literature. The biology of metalworking fluids has been repeatedly reviewed [20,115-122].

Microbial Growth. Emulsion concentrates are practically free of microorganisms and can be stored for years. After dilution with water, the growth of organisms (introduced by the makeup water, or by contact with air, human skin, animals, floor sweepings, etc.) becomes rapid. In the early phases of attack, tiny (below 2 μm) aerobic bacteria (mostly of the Pseudomonas genus) grow at a fast rate, doubling every 20 to 30 minutes. They cause a breakdown of fatty acids and the formation of organic acids, with a consequent drop in pH and a destabilization of the emulsion. The oily phase rises to the surface and, together with any contaminant (tramp) oil present, reduces air access. Aerobic bacteria are facultative (they can change to a state in which they can grow in the absence of oxygen), and different species thrive at various times until conditions become favorable for the growth of anaerobic bacteria which, as their name (Desulfovibrio) suggests, attack sulfur in the floating oil, generating the foul-smelling H_2S gas. Their growth is slower, and the smell becomes noticeable mostly after a shutdown, hence "Monday morning odor"; they turn the fluid black in the presence of iron.

Fungi grow in the form of slimy masses consisting of tangled fibers up to 15 μm in diameter [121], whereas yeasts form oval cells 8 to 12 μm in diameter. Sometimes groups resembling micelles are seen. They seldom grow under conditions favorable for bacterial growth, and are found more often in chemical coolants than in emulsions.

Controlling Factors. Based on the above observations, one might expect that the rate and kind of attack are controlled by a multitude of factors:

a. Emulsion concentration is most significant. A 5:1 dilution is practically immune to attack; a 10:1 ratio allows some growth; and the rate of growth is maximum at ratios of 25:1 to 50:1. Water quality is important in two respects [95]: water is a source of infection, and water hardness and other inorganic salts accelerate microbial growth as well as reduce emulsion stability. In contrast, hardness inhibits the growth of molds in synthetics.

b. The emulsion composition has a marked effect: a high pH (over 9.5) inhibits attack [111], whereas sulfur content increases it. Fatty acids and their derivatives are nutrients and so are some of the emulsifiers, whereas amphoteric emulsifiers and oxyethylene ethers are resistant to attack. Ethylene glycol is quite resistant [123].

c. Growth rate increases with temperature to about 40 C; at higher temperatures the growth of molds is accelerated.

Consequences of attack are usually discussed with reference to cutting fluids, but the problems are similar in deformation processing applications such as rolling of aluminum. Because microorganisms are numerous (around 10 million organisms per millilitre even in well-controlled emulsions), they may have serious effects on operation.

a. The lubricating qualities of the emulsion change by loss of stability (increasing droplet size) and loss of boundary additives and sulfur compounds.

b. The work environment deteriorates, with objectionable odors, slimy deposits, discoloration of the fluid, and staining of the equipment.

c. Because microbes and fine metal debris are of comparable sizes, filters become incapacitated [124], lubricant passages become clogged, and, in zones of poor circulation, deposits build up.

d. Organic acids and H_2S formed as a byproduct of bacterial attack increase corrosion. Corrosion inhibitors may actually contribute to microbial growth.

e. Hydrogen sulfide formed by microbial growth eventually dissolves the surfaces of concrete sumps, leaving behind highly porous surfaces which then become inhabited by microbes, thus providing perpetual sources of contamination.

Control of Attack. Several approaches are viable [20,125,126].

a. Ideally, the emulsion itself should be formulated so as to make it less susceptible to attack. This is possible in some cases [122], although performance criteria may demand that emulsion components be used which serve as nutrients for microbes (Fig.4.18) [20].

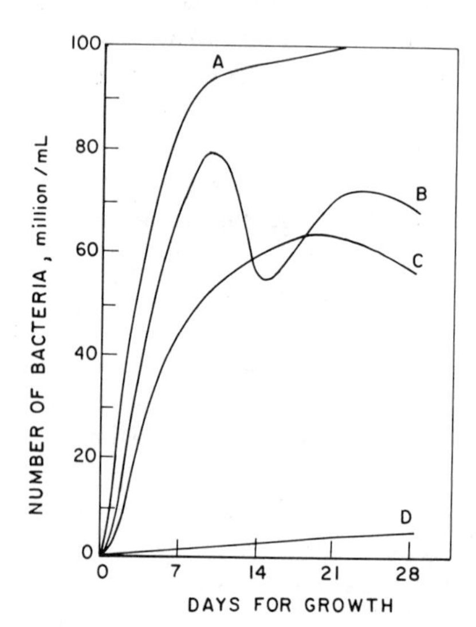

Fig.4.18. Rate of bacterial growth as a function of available nutrients in aqueous lubricants [20]. A - lard oil; B - sulfurized oil; C - paraffin oil; D - synthetic coolant.

b. Preservatives (germicides) are often incorporated into the concentrate. The problem is that they are rather short-lived, especially if the made-up emulsion is allowed to become contaminated [127,128]. Once infection occurs, it is the germicide that is first consumed.

c. Effective control requires the addition of relatively high dosages of biocides at regular intervals. Biocides usually belong to the following families: phenols, formaldehyde and formaldehyde-releasing compounds, isothiocyanates, ethanolamines, and others [125,129,130]. Mixed biocides may show synergistic effects [131]. Control is essentially the same in cutting fluids and in rolling and canmaking emulsions. An important point is that biocides should not create health hazards. Physical methods of control [132,133], including pasteurization [134], may become more important.

d. A basic problem is that ostensibly minor changes in production conditions can change the character of infection; therefore, effective control by biocides requires regular testing of the fluids. Under-dosing actually stimulates growth and can lead to the development of biocide-resistant species.

e. In all instances, cleanness and the prevention of contamination with microorganisms and oil is an essential part of control. Small lubricant systems of individual machines seldom pose problems, because makeup with fresh concentrate replenishes the system. If contamination occurs, it may be economical to dump the used fluid and refill the system after a thorough cleaning. The problem is greatest with large-capacity systems in which the lubricant is expected to last for months and makeup is relatively small.

4.6.6 Lubricant Control

Aqueous lubricants are, in general, more demanding of user knowledge than many of the oil-base lubricants. The best lubricant may totally fail or give inadequate service if it is improperly handled, or if tests which would give information on the condition of the lubricant are not carried out and proper corrective actions are not taken in time. Aspects of control [135-140] and specific test methods, often with emphasis on control of large systems [81,141-146], have been discussed. The main points to be observed are as follows:

1. Concentrates contain at least some water; therefore, they must be protected from freezing and from excessively high temperatures which could lead to loss of water and inversion of the emulsion. The mixing vessel should have built-in heaters with thermostatic control, or be connected to a warm-water supply. Means for moderate agitation should be available—usually in the form of an impeller rather than an air or steam jet, which would induce foaming. The concentrate is introduced slowly into the agitated water of controlled quality. Alternatively, commercial water/concentrate mixing units may be used [138].

2. Water quality must be periodically checked. This includes hardness, chloride and sulfate concentrations, evaporation residues, and microbial count. If municipal water is used, this information is usually available.

3. The concentration of the lubricant must be regularly checked, because oil carry-out on the workpiece, misting, leakage from the system, and water evaporation all contribute to gradual changes. In general, emulsions tend to lose the oily phase (become leaner) whereas synthetics tend to become more concentrated (richer). In larger systems, with their large surface areas, water loss dominates (and, if present, water hardness builds up). For determination of concentration, laboratory analysis (based on breaking of the emulsion) is customary [137]. However, some components of the emulsion remain in the water phase and a correction factor obtained from the supplier must be used to convert results into concentrate equivalent [147]. Analytical methods, usually based on titration, may be beyond the capability of the average user of synthetics. Specific gravity, even when determined with sensitive hydrometers, can give only approximate indications of concentration. Pocket refractometers are likely to give much better results when used in conjunction with calibration charts prepared by the synthetic lubricant supplier. This method is usable also for semisynthetics and translucent emulsions [137,148].

4. Emulsion particle size and distribution are often taken as indicators of lubricant quality, with tighter (smaller-particle-size) emulsions associated with greater stability

and lesser lubricating power. The most direct technique is observation, measurement, and counting under the microscope, usually from microphotos. In-plant, on-line monitoring is possible with the Coulter counter, which measures the change in conductivity of the continuous phase as each particle of the dispersed phase passes through a minute orifice. The problem is that very small particles will pass undetected (for example, a counter with a 50-μm orifice will miss particles below 1 μm), yet commercial instruments usually record the count out of 100%. It is, therefore, necessary to apply an empirical correction factor based on independent determination of oil concentration [147,149]. More recently, instruments based on the scattering of a laser beam passing through a thin stream of the emulsion have become available. In addition to particle-size distribution, total oily-phase volume can be derived. These readings, too, need to be corrected by empirical factors [149].

5. The pH of the emulsion is routinely checked, either by a simple litmus-paper test or by electrical techniques.

6. There is no totally adequate corrosion test. In general, metal strips thoroughly cleaned and abraded are placed in an emulsion of controlled make-up, and corrosion is assessed after various times of exposure and, if desired, after exposures at elevated temperatures. One of the most frequently used tests assesses the corrosion of cast iron, which is the material used for many machine tool castings. Steel chips, mixed with the emulsion, are placed in small heaps on specially prepared cast iron plates, and corrosion is assessed visually. Alternatively, clean cast iron chips are placed on a filter paper in a plastic petri dish. Test fluid is poured over the chips and decanted after 30 min, and corrosion is assessed after 24 h by inspection of the chips and the filter paper [20].

7. Stability measurements involve the observation of an emulsion made up under controlled conditions and held in special flasks for a specified time, usually at slightly elevated temperatures. Flocculation, phase separation (creaming), and sedimentation are noted.

8. Residues of a lubricant-coolant could interfere with machine operation, and the tendency to form residues is checked by allowing a measured quantity to stand for a prolonged time, or by driving the water off by slow boiling.

9. Foaming tendency is judged by subjecting samples to controlled shaking, or to pumping through a shower head with the stream hitting the surface of the bath.

10. The importance of microbial contamination has led to the development of a number of tests for laboratory [150] and in-plant [151] testing of metalworking fluids. Commercial test kits consist of glass slides coated with nutrients specific for various bacteria or fungi. The slides are dipped in the emulsion and incubated for a specified period of time, and the number of organisms is visually compared with calibration slides [152,153]. This technique has been shown to be quite reliable (Fig.4.19). Indirect techniques, for example, based on enzyme activity [154], have also been developed. A summary of biological control is given by Bennett [126].

Some tests are subject to national standards, whereas others are widely accepted but not standardized. American practices are summarized in [203], British practices in [82] and [141], and German practices in [139] and [146]. Relevant ASTM standards are listed in the Appendix. An important part of any fluid maintenance program is formalized recordkeeping [155], so that trends can be spotted and corrective actions taken. These include the following:

1. Addition of water, emulsion concentrate, or specifically formulated additive packages designed to replenish the constituents that are subject to selective loss. Thus the oil concentration, pH, emulsion particle size, and corrosion inhibitor, antioxidant, and foam suppressant levels are restored.

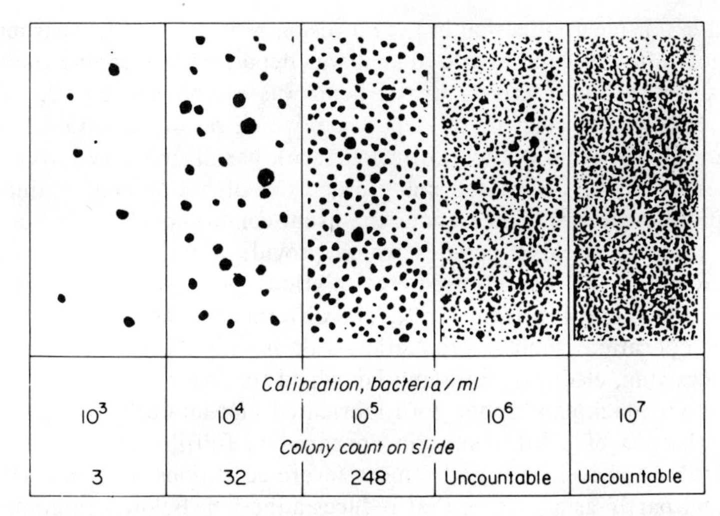

Fig.4.19. Calibration slides for estimating extent of microbial infection [153]

2. Germicides are added according to need; in many ways this is among the most important measures to be taken for emulsion systems that have an expected life of several months.

4.7 Coatings and Carriers

All lubricants discussed hitherto function by boundary, mixed-film, or PHD mechanisms and are applied, in principle, to the interface. Very often an excess quantity is supplied and film thickness is controlled by the lubricating mechanism itself.

Another class of lubricant systems relies on deposition of a lubricant or lubricant carrier onto the workpiece surface in a controlled thickness by a technique that ensures firm attachment. Thus film thickness is controlled primarily by deposition, although process conditions are influential in determining the thinning and, indeed, survival of the predeposited film. The film itself may possess lubricating qualities, or it may serve primarily as a carrier for superimposed lubricants ranging from solids to liquids. Lubricants of a special class are deposited as solids but become liquid on preheating to the temperature of deformation.

4.7.1 Metal Coatings

Even though thin metal coatings are customarily regarded as excellent lubricants in themselves (Sec.3.3.2), they are almost always used in conjunction with a liquid or semisolid lubricant. Thus, more correctly, they serve as lubricant carriers, with the lubricant chosen for the coating rather than the substrate metal. They are employed when the workpiece material is difficult to lubricate, and are selected to minimize adhesion to the die in the event of lubricant breakdown. Others are applied primarily for corrosion protection, and the lubrication system is adjusted correspondingly. Details of application techniques are given by Gabe [156] and in Metals Handbook [157]. Properties are reviewed as a function of deposition technique by Safranck [158].

The efficiency of treatment depends on the metal, the strength of the bond between the base and the coating, and the microtopography of the surface. Adhesive bond strength depends on surface cleanness and is improved by interdiffusion (alloying), but this can often lead to the formation of intermetallic compounds of low ductility.

Tin has found extensive application in the corrosion protection of low-carbon steel sheet (tinplate). Prior to the Second World War it was deposited by dipping the steel in molten tin, and a strong but relatively brittle bond was established by alloy formation at the interface. Since then, electrolytic deposition, which results in close control of thickness (ranging typically from 0.2 to 1.5 μm/surface), has all but taken over. The coating may be melted (reflowed) to match many properties of hot-dipped coatings, or it may be left as-deposited (matte). Since under the usual conditions of deposition the electrocoat consists of tiny touching nodules which provide a favorable microgeometry for lubricant entrapment, the non-reflowed electrolytic tinplate gives better performance in conjunction with a liquid lubricant [159]. The difference in performance shows up only in more severe operations such as production-scale ironing. Under the less severe conditions of deep drawing, electrolytic tin and hot-wiped tin coatings gave the same improvement relative to a blackplate under both lubricated and unlubricated conditions [160]. Thus, in the absence of a lubricant, tin appeared to fulfill some lubricant functions. There is no doubt, however, that under more severe conditions tin acts partly as a lubricant carrier and partly as a coating that reduces adhesion. Below a minimum thickness, ironing loads rise sharply; with adequate thickness, tin is redistributed on the surface, and little remains on the asperities, yet adequate protection against die pickup is ensured [161].

Tin has the advantage that die pickup does not build up rapidly. It protects steel from corrosion and is nontoxic, hence its extènsive use for food containers.

Zinc coatings formed on steel (galvanized iron) by dipping in a melt or, more frequently, by electrodeposition, offer corrosion protection by sacrificial action. Zinc is most often applied for nontribological reasons; nevertheless, it may make lubrication somewhat easier, especially if it is electrolytically deposited to provide lubricant entrapment through favorable microgeometry (matt zinc coating). Die pickup tends to be cumulative.

Shepard [162] reported that when zinc plating was controlled to give a ductile, well-adhering coating 12 to 25 μm thick, it also provided a porous base for the soap lubricant used in drawing. The importance of microgeometry was shown by Apel and Nünninghoff [163], who compared hot-dipped and electrodeposited zinc in the drawing of high-strength, patented, 0.8% carbon steel wire. On drawing from 2 mm to 0.6 mm in diameter in 17 dies with a calcium soap, much of the hot-dipped coating was lost, and the loss increased when the line was set to produce less of the hard Fe-Zn alloy normally found at the interface (Table 4.9). In electrolytic coating, no such alloy forms, yet zinc loss was minimum. The reason was found in the microgeometry of the coatings. Hot-dipped zinc had a roughness R_t of 5 μm and exhibited the locally smooth, fully dense surface characteristic of a solidified melt (Fig.4.20a), whereas the electrodeposited zinc had a roughness R_t of 22 μm and a nodular structure (Fig.4.20b) capable of

Table 4.9. Drawing of zinc-coated steel wire (data compiled from [163])

Zn coating	Coating thickness, μm Hard Zn	Soft Zn	Zn loss, %	Eccentricity of Zn layer
Hot dip 1	5.6	25.6	30.6	16.5
2	12.2	23.4	25.0	13.4
3	21.6	13.8	13.3	13.2
Electrodeposited	. . .	40.0	5.7	11.0

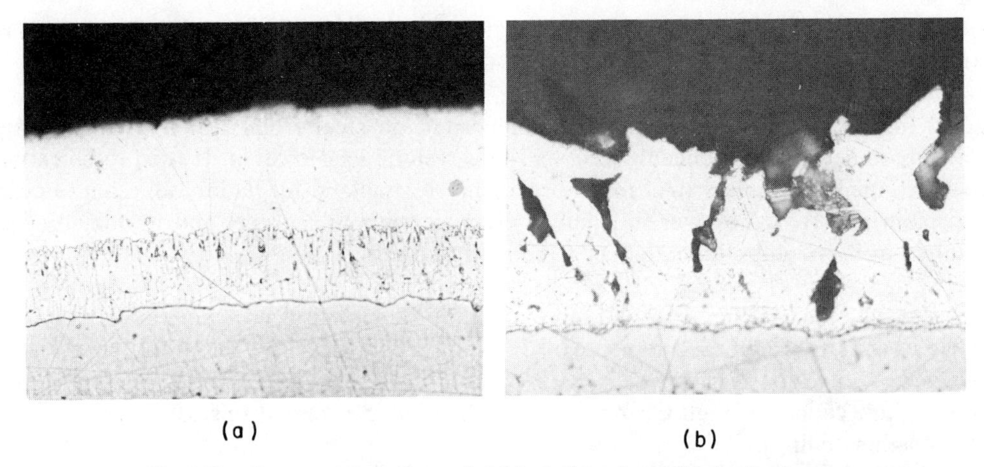

Fig.4.20. Cross sections through (a) hot-dipped and (b) electrodeposited zinc coatings [163]. (Courtesy of Dr. G. Apel, University of Wuppertal, W. Germany)

entrapping the lubricant. The increased lubricant intake prevented metal loss until the last draws, when only vestiges of the nodular structure remained. Even then, lubricant residues must have been present, because the finished surface was smooth, as opposed to the torn surface of the hot-dipped coating (Fig.4.21). The substrate steel remained smooth, whereas the greater hardness of the hot-dipped coating caused roughening of the steel. Surprisingly, the zinc layer did not retain the original uniform distribution

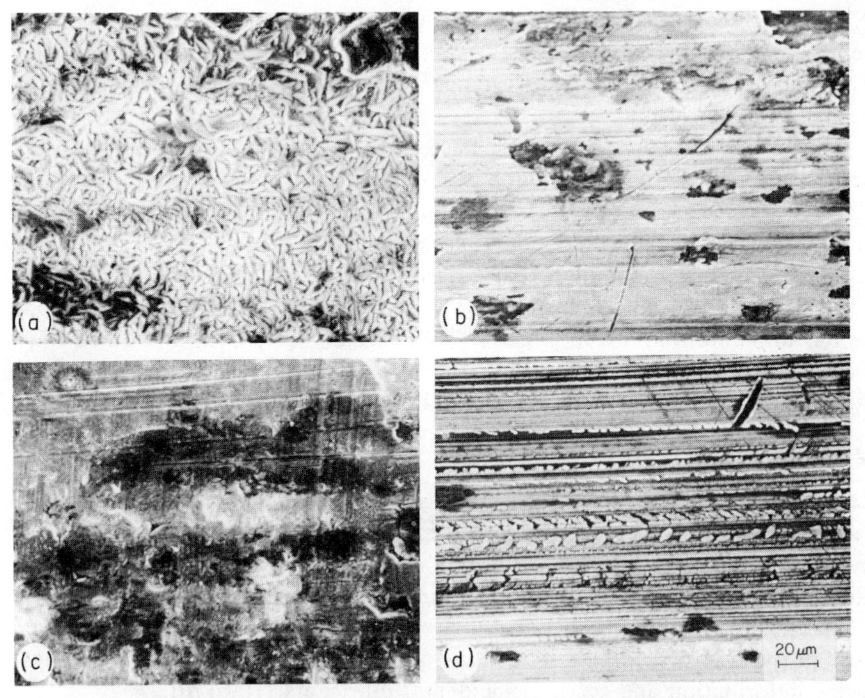

Fig.4.21. Surfaces of zinc-coated wires [163]. (a) Electrolytic zinc as-coated. (b) Electrolytic zinc after 84% reduction. (c) Hot-dipped zinc as-coated. (d) Hot-dipped zinc after 84% reduction. (Courtesy of Dr. G. Apel, University of Wuppertal, W. Germany)

around the wire circumference, showing that circumferential flow was induced by entering the wire at an angle into the draw die.

Lead (or, rather, a low-tin lead alloy) deposited on steel from a molten bath (terne plate) is sometimes still encountered, as is electrolytic lead coating. It used to be extensively applied to stainless steel for improved lubrication and for its low adhesion to most die materials. It is, however, difficult to remove without a trace, and its toxicity has resulted in its virtual elimination as a lubricating coating.

Copper, as noted in Sec.3.5.2, is unusual in that die pickup is not cumulative. It is also fairly easy to lubricate and is nontoxic in many media and has, therefore, been extensively used in various forms.

a. Pure copper or high Cu-Zn alloy cladding can be applied to steel by hot bonding (such as hot rolling).

b. Copper may be electrodeposited on steel, stainless steel, titanium, and other metals, provided that it is not harmful in the application of the finished part or, if harmful, can be removed by chemical means without attacking the base metal. Brass is electrodeposited on steel if good bonding to rubber is needed, as in steel-belted tires.

c. Electroless deposition by cementation is possible on less noble metals: freshly pickled steel is drawn through a bath of a copper salt (such as $CuSO_4$, possibly with 1 to 3% $FeSO_4$ and some H_2SO_4). The thin coating thus obtained gives a pleasing color to the finished part. To increase the adherence of the coating and the deposition rates, proprietary additives have been introduced [164,165]. The addition of Na_2SO_4 and NaCl results in a coating with E.P. characteristics [166]. Very fast deposition rates are achieved with HCl-base baths to which a $CuSO_4$-surfactant-inhibitor mixture is added [165]. The color may be made yellow by the addition of tin or zinc salts, and white by replacing copper with $NiSO_4$. These techniques have a long history of application in wiredrawing (Sec.7.3.3). Electroless copper coatings have been developed also for the drawing of titanium [167].

Other Metals. Silver and, less frequently, gold have found limited application on more expensive workpiece materials such as titanium. They may be electrodeposited or, for better adhesion, vapor deposited or ion plated [3.481].

In general, other techniques discussed in conjunction with the coating of dies (Sec.3.11.5) could also be used, but are usually too expensive. An unusual proposition is to use ion implantation. Yoshida et al [168] found that copper increased but nitrogen reduced friction when implanted at higher concentrations in mild steel. It is unlikely that such techniques would find commercial application.

4.7.2 Polymer Coatings

Solid thermoplastic polymer films, discussed in Sec.3.3.3, are sometimes interposed between the die and workpiece as loose films, but this technique is so cumbersome that it finds little application outside the laboratory. More practically, the polymer film is deposited on or bonded to the workpiece surface.

1. Large quantities of sheet metal are coated (coil-coated strip, prepainted strip) with the purpose of providing a decorative and protective finish. Many coatings can sustain some deformation and serve as incidental lubricants [169,170].

2. The polymers may be applied with the express purpose of lubrication, with the added benefit of protection in handling and storage. They are removed from the finished product:

a. Peelable films may be tapes bonded with an adhesive, or deposited in films with a minimum thickness of 3 μm. Deposition is from solvent solutions, aqueous dispersions, or plastisols (polymer dispersions in plasticizers, baked onto the strip).

b. Thin, lacquer-type coatings are dissolved after forming. Depending on the severity of deformation, thicknesses range from 0.8 (or, more typically, from 2.5) to 7.5 μm.

Polymer coatings may provide lubrication on their own or in conjunction with a liquid lubricant which may also function as a plasticizer.

Polymer Types. Numerous polymers are useful in bearing applications [3.143], but for metalworking purposes only thermoplastic polymers of the appropriate glass-transition temperature are of interest (Table 4.10):

a. Polyethylene is readily available in sheet form and possesses substantial elongation (100 to 300%). In combination with mineral oil, sheets 2.5 to 7.5 μm thick provide excellent lubrication. Polypropylene has similar properties.

Table 4.10. Thermoplastic polymers and their transition temperatures (C)

Polymer	Repeat Unit	T_g	T_m
Polyethylene (PE)	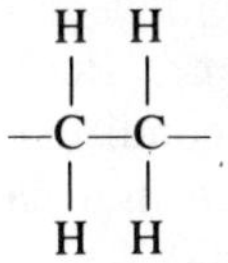 H H / —C—C— / H H	−120	137
Polypropylene (PP)	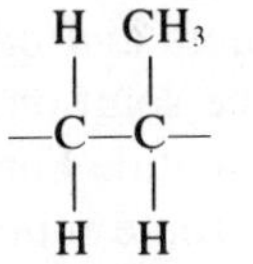 H CH_3 / —C—C— / H H	−18	176
Polytetrafluoroethylene (PTFE)	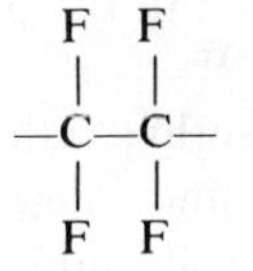 F F / —C—C— / F F	−50	327
Polyvinylchloride (PVC)	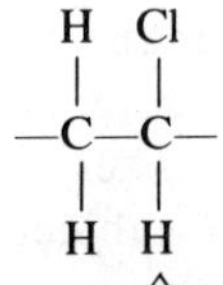 H Cl / —C—C— / H H	87	212
Polystyrene	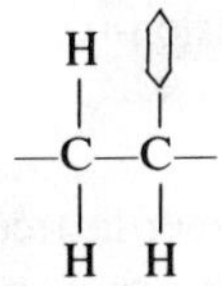 H (benzene ring) / —C—C— / H H	105	240
Polymethylmethacrylate (PMMA)	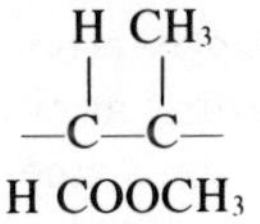 H CH_3 / —C—C— / H $COOCH_3$	105	160
Polyimide	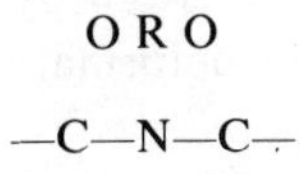 O R O / —C—N—C—	220 370	none

b. PTFE sheets have limited ductility, but well-adhering films of controlled thickness and good extendability can be deposited from a tricholorotrifluoroethylene dispersion of telomers (which have aliphatic end chains grafted onto the PTFE). Sputtering is possible, too [3.485].

c. PVC is available both in sheet form and as a deposited coating.

d. PMMA, and acrylics in general, have the advantage that they are readily deposited from either solvent solutions or emulsions, and they can be modified so as to endow them with boundary-lubrication properties.

e. Acetal resins contain the linkage -CH_2O-CH_2O- and have found some limited application.

f. Polyimides formed with aromatic radicals have outstanding temperature stability.

Lubricant Systems. As mentioned in Sec.3.3.3, friction properties can be modified by the application of a lubricant on top of the polymer film. Several patented systems have been reviewed by Wojtowicz [171] and by Sidey [172]:

a. Chlorinated rubber, PMMA, polyvinyl acetate, chlorinated polyethylene, or some other film of low melting point (37 to 110 C) is deposited and a surface modifier such as oleic acid amide is applied. Alternatively, the polymer and amide are co-deposited; subsequently, the amide diffuses to the surface.

b. An acrylic film is deposited on the metal surface. This has a rather high friction, facilitating handling and feeding of the strip; just before forming, a lubricant or "resin activator" (such as a ketone, alcohol, glycol, ester, or ether) is applied which plasticizes the surface under the conditions prevailing during working, and ensures low friction. The activators are applied in a carrier, such as mineral oil or water, which in itself does not plasticize the coating.

c. A copolymer of acrylic esters and minor amounts of acrylic or metacrylic acid is deposited from a trichloroethylene solution. The carboxylic group of the acid increases the adhesion of the film to the metal. For soft metals such as aluminum and copper, the film alone suffices; for steel, some mineral oil or chlorinated paraffin is added and, presumably, becomes encapsulated in the polymer matrix and is exuded on contact with the die.

d. A cellulose ether such as ethyl cellulose is deposited from a solvent system with the addition of softening agents and inorganic fillers, such as graphite, talc, etc. This represents a transition to bonded solid lubricant films.

A common requirement is that the film should be readily removed after deformation, to prevent interference with finishing operations or discoloration or carbon pickup during annealing. Some must be dissolved in organic solvents; others can be lifted off with hot alkaline cleaners. Films subject to the latter treatment are susceptible to deterioration on contact with emulsion-type lubricants.

4.7.3 Bonded Coatings

Many solid lubricants, including layer-lattice structures (Sec.3.3.4), function only if present in a continuous film. When the process fails to develop such a film, the lubricant may be held in place by an appropriate bonding agent, most often a polymer. Such systems may, therefore, also be regarded as filled polymers. A vast effort has been put into their development, primarily for aerospace applications, and some of the resulting systems have found use in general tribology [3.143],[173]. They are applied also in metalworking [174,175], often for deformation at elevated temperatures. There are several challenges to be met:

a. Adhesion to the metal surface is essential and is aided by mechanical attachment. For this reason, the surface is frequently roughened by shot blasting or etching. A porous surface, such as one produced by phosphating, is helpful. Firm bonding is particularly important when the process geometry would tend to scrape off the lubricant. This was shown by Oyane et al [176] in a wedge indentation test. Solid lubricants deposited loosely from an acetone dispersion were wiped off, whereas those bonded with a copolymer of an acrylate and polyethylene, or a polyamide-imide, performed well. The problem is acute in drawing operations when the die angle is large (Sec.7.3.3).

b. The bonded film, if applied to the billet, must be capable of following the extension of the workpiece surface. Therefore, only thermoplastic polymers (such as acrylics, polyethylene, etc.) and copolymers (such as ethylene-acrylic acid [177]) can be considered. Heat curing (thermosetting) resins, widely used in bearing applications, are brittle and suitable only for application to the die. The coating thickness ranges from a fraction of a micrometre to 3 μm, and even thicker for heavier duties, although surface roughening and scraping-off set limits.

c. Stability and ease of application are very practical but important requirements. Commercial preparations are available as organic solvents, as aqueous dispersions, or as aerosols. Powder coating, widely used for painting in mass production [178], should be attractive. The disposal problems created by large volumes of organic solvents have prompted a shift toward water-base formulations, among others, for resin-bonded graphite and MoS_2 films [179].

d. The dispersed phase is usually MoS_2 or graphite, which must be protected from premature oxidation, either by choosing a polymer of high temperature resistance (such as a polyimide) [173] or by incorporating an oxidation inhibitor [180]. Among the latter are metal oxides, such as Sb_2O_3, which have some lubricating power on their own and also delay the oxidation of the layer lattice compound by oxidizing into a higher state, as shown by Harmer and Pantano, quoted in [175]. In bearing applications it has been observed that fillers such as Pb_3O_4 and CuO reduce the wear rate of polymers operating in the softening range, apparently by the formation of a strongly bonded transfer film during running-in. This might be a mechanism active in metalworking, too. Some commercial formulations contain graphite with BN suspended in a water-soluble amine-amide complex. Decomposition of the latter slows the oxidation of steel.

e. For the most severe duties, a variety of other additives are used. Metal (Pb, Al) powders are incorporated, presumably for reducing adhesion to the die and for forming a partially liquid film. Lead powder coated with graphite by vibratory milling [181] could conceivably be used, although the toxicity of lead is always a problem.

4.7.4 Lubricants Deposited In Situ

In principle, a lubricant should adhere well if formed on the surface by reaction. Several possibilities exist:

a. The film is formed by reaction between two compounds. Thus, Feng et al [182] reacted a molybdenum complex and a mixture of dialkylphosphorodithioate to form MoS_2.

b. A film, deposited on the surface, is converted into a lubricating compound. Thus, Di Sapio and Maloney [183] electrodeposited molybdic oxide on steel, stainless steel, and titanium, and then converted it in H_2S gas, at elevated temperature and pressure, to MoS_2. Alternatively, the oxide may be applied by sputtering, as reviewed by Nishimura et al [184].

c. A reaction film is formed between a gas and the metal surface. Baldwin and Rowe [185] heated the specimen in vacuum to evaporate contaminant films, and then admitted a halogen gas (Cl, I) or H_2S. Useful parting layers were formed on many metals, and chlorides on chromium, tungsten and stainless steel, iodides on titanium and niobium, and sulfide on molybdenum gave low friction.

d. A reaction film is formed in a molten electrolyte. Rowe and his co-workers formed useful chloride films on titanium by electrodeposition from a molten KCl-LiCl bath and also from a commercial heat treating salt [186]. The film retained low friction to 450 C. By adding Na_2S to the salt, MoS_2 was formed on molybdenum. These techniques have the attraction that the lubricant can be selectively deposited where it is most needed [187].

4.7.5 Fusible Coatings

Lubrication at high temperatures presents exceptional challenges. The viscosity of metals is too low in the liquid state to provide lubrication. Some polymers retain a useful viscosity but are effective only if the process geometry contributes to their retention at the interface (as in extrusion) or if they are filled with solid lubricants; in the latter case, the polymer serves as a temporary binder. There are, however, a number of inorganic substances that retain a useful viscosity on melting, and resist thermal breakdown long enough to act as a lubricant and/or binding agent. On cooling they solidify and become brittle; they may remain useful bearing lubricants, especially as binders for layer-lattice compounds and oxides [188]. In metalworking they are used only in the molten or half-molten form, in warm and hot deformation processes.

Metal Oxides. Only a few oxides melt in a useful temperature range. Foremost is PbO, which converts to the nonlubricating Pb_3O_4 at 370 C but reconverts to PbO at 480 C. Saga and Arita [189] found it effective in the extrusion of stainless steel at 300 to 700 C; the lack of difficulty in the Pb_3O_4 temperature range was attributed to the formation of duplex oxides with iron and chromium from the workpiece. A small (4%) addition of SiO_2 or V_2O_5 had little effect even though the duplex oxide thus formed has a lower melting point. Addition of B_2O_3 resulted in a smooth, glassy coating. Coverage on carbon or low-alloy steel was inadequate, but addition of 5% CrO_3 ensured reaction with the workpiece and resulted in a smooth coating. The combined effects of pressure, shear, and temperature contribute to phase changes, reviewed by Klemgard [190], some of which may be beneficial.

The lubricating properties of many oxides can be improved by mixing with graphite, as shown in extensive screening tests on potential hot working lubricants by Haverstraw [191] and Cook [192].

Inorganic Salts. A vast number of salts have potentially useful melt properties. Their water solubility makes for easy application (even to a hot die, on which the water flashes off to leave a uniform salt film) and removal.

Common salt (NaCl) melts around 800 C and has been extensively used, primarily in nonisothermal hot forging. Chlorides of other metals (Ba, Zn, Mg, K, Li) and eutectic mixtures such as $ZnCl_2$+$ZnSO_4$ have been employed at least experimentally [193,194]. The temperature range of their application can be extended by the addition of 10 to 20% of a solid such as slate [193]. Various fluorides, chromates, carbonates, nitrates and phosphates have also been explored [191,192] but have not gained commercial importance.

A commercially available eutectic mixture of potassium metaphosphate and sodium tetraborate [195,196] forms a relatively low-viscosity melt (7.7 Pa · s at 700 C; 1.5 Pa · s at 1000 C). It reacts with carbon steel and its scale to form a mixture of salt, iron oxide, and iron phosphide, and thus behaves as a viscous liquid with boundary properties, giving relatively low friction at temperatures from 300 to 700 C. By changing the proportions of phosphate and borate, viscosity can be adjusted.

Glassy Films. Glasses proper will be considered under a separate heading. There are some substances which are essentially glass formers and which, in conjunction with other inorganic substances, form continuous coatings. Foremost of these is B_2O_3, which has been used, in quantities below 12 wt %, to bond MoS_2 films to steel [197].

Saga et al [198] used mixtures of anhydrous borax ($Na_2B_4O_7$) with 10% Bi_2O_3 or PbO for the extrusion of steels and stainless steels in the temperature range 300 to 1000 C. Through a series of reactions, mixed solid-liquid films are formed, giving much lower friction at 500 to 700 C than either Bi_2O_3 or $Na_2B_4O_7$ alone (Fig.4.22). The mixture can be applied to the cold workpiece or to a workpiece heated to the operating temperature, when it melts and wets the surface immediately. The lead-free compositions overcome the problem of toxicity.

4.7.6 Glasses

Glasses may be regarded as inorganic thermoplastic polymers of a spatial network structure. On cooling from a high temperature at the usual rates, their viscosity increases sufficiently for them to behave like solids, yet without acquiring a crystalline structure. On prolonged holding at an elevated temperature they may crystallize (devitrify), a change that would be undesirable during metalworking but which can be helpful in removing the glass after working.

The composition, structure, and technology of glasses are discussed in several monographs [199-203]. The monograph of Manegin and Anisimova [204] is devoted to the application of glasses in metalworking but, unfortunately, could not be consulted. Typical glass compositions used in various countries have been reviewed [205-207], and the compositions of German [208] and U.S. [203] proprietary glasses have been published.

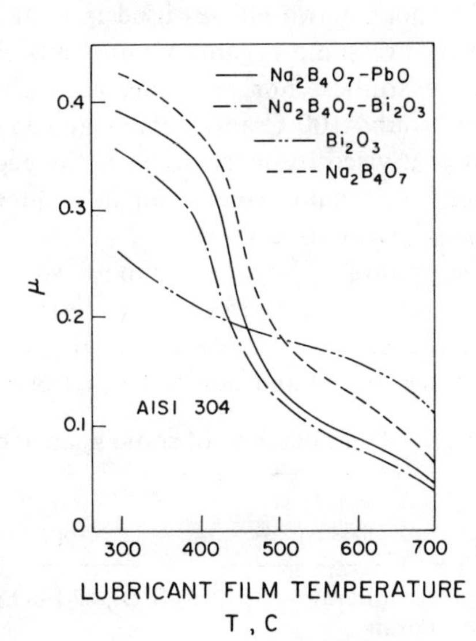

Fig.4.22. Sliding friction of various borax-type lubricants on austenitic stainless steel [198]

Composition. There are many glass-forming systems, but only those based on SiO_2, B_2O_3, and P_2O_5 are important for metalworking purposes. The network formed by the glass formers is depolymerized, and is made more ionic and also much less viscous, by the incorporation of network modifiers such as Na_2O, K_2O, Li_2O, CaO, and MgO.

Some oxides, primarily Al_2O_3 and, to a lesser extent, PbO, act as both network formers and modifiers. Some guidelines to formulation, based on the ionic potential of the atomic species, are available, and a vast variety of glasses have been developed for specific purposes.

Metalworking imposes some specific limitations. Thus, PbO is toxic, and contaminates waste waters if removed by etching; therefore, it is avoided in most countries [209]. If allowed to remain on scrap, it would contaminate many materials [210]. Pardoe [211] indicates that in the extrusion of nuclear reactor materials glasses should be free of boron or other elements of high neutron-capture cross section, although in principle no glass residue should remain anyway.

Viscosity. Glasses have no boundary-lubrication properties, and therefore their most important property is viscosity. Glasses are essentially Newtonian liquids and, to a first approximation, their viscosity decreases with temperature according to an exponential law:

$$\eta = A \exp (E/RT) \tag{4.1}$$

where A is a constant, E is activation energy, and R is the universal gas constant. In general, silicate glasses are used for the highest, borate glasses for intermediate, and phosphate glasses for the lowest temperatures. The slope of the viscosity-temperature curve can be modified by a judicious choice of network-modifying elements. Compositions of some glasses are given in Table 4.11, and viscosities are presented in Fig.4.23; viscosities associated with conventional glass technology are also marked. Extensive compilations of glass properties are given by Morey [212] and by Volf [213].

In view of the open-network structure of glasses, one would expect viscosity to increase with increasing pressure. Indeed, a slight effect was found for borate glasses [214] but not for two glasses used in metalworking [215].

There are some organic compounds that form glasses, and these can be useful for simulation studies. For example, both glucose and abietic acid have viscosities of 10 Pa · s at around 100 C and 1 Pa · s at 125 C. Their drawback is that they are not stable and suffer a viscosity increase on prolonged holding at temperature [216].

Non-Newtonian effects can be induced by partial crystallization, and also by the addition of particulate solids.

The optimum viscosity for metalworking purposes must, obviously, be a function of interface pressure and process geometry. Widely varying recommendations are found in the literature, partly because of the confusion created by a failure to distinguish between isothermal and nonisothermal operations:

Table 4.11. Compositions of some special glasses (data from Corning Glass Co.)

Corning No.	Glass type	Approximate composition, %	Suggested temperature range, C
8363	Lead borate	10 B_2O_3, 82 PbO, 5 SiO_2, 3 Al_2O_3	530
9772	Borate	. . .	870
8871	Potash lead	35 SiO_2, 7.2 K_2O, 58 PbO	870-1090
0010	Potash-soda-lead	63 SiO_2, 7.6 Na_2O, 6 K_2O, 0.3 CaO, 3.6 MgO, 21 PbO, 1 Al_2O_3	1090-1430
7052	Borosilicate	70 SiO_2, 0.5 K_2O, 1.2 PbO, 28 B_2O_3, 1.1 Al_2O_3	1260-1730
7740	Borosilicate	81 SiO_2, 4 Na_2O, 0.5 K_2O, 13 B_2O_3, 2 Al_2O_3	1540-2100
7900	Silica	96+ SiO_2	2210

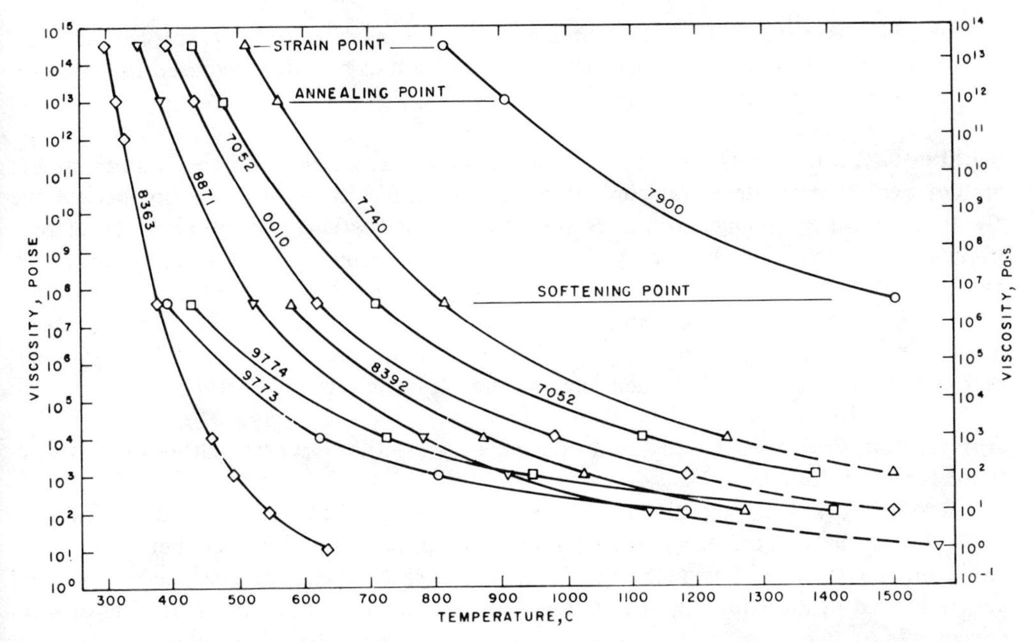

Fig.4.23. Variation of viscosity with temperature for various glasses [217]

a. In nonisothermal working, chilling on the colder die results in a higher effective viscosity. Indeed, Schey et al [217] found that in plane-strain compression the relevant value is viscosity at the mean of the die and workpiece temperatures. This was observed also by Pawelski et al [196] in pushing slugs through a converging die. Optimum viscosities range around 10^5 to 10^7 Pa · s [217] measured at the mean temperature. Spiegelberg [218] defines a window between 10 and 10^3 Pa · s for forging, whereas Manegin and Lenyashin [214] found 20 to 30 Pa · s to be optimum for low-carbon steel, 80 to 120 Pa · s for austenitic stainless steel, and 200 to 300 Pa · s for a nickel-base superalloy. The quoted viscosities are taken at the workpiece temperature.

b. Only in isothermal working can the viscosity be unambiguously specified. Leipold et al [219] recommended that viscosity should be 10^{-4} to 10^{-2} of the flow stress of the material in forging at a very low speed (0.25 mm/s). Presumably, even lower viscosities would be sufficient at higher velocities. In a process such as hot extrusion with a glass pad (Sec.8.5.3), the film gradually melts off; therefore, not only viscosity but also viscosity-temperature slope become important. A general discussion is given by Scheidler [220].

Adhesion. Because glasses have no boundary-lubrication properties, the film must be fully continuous. For this to occur, the glass must adhere to the workpiece surface: complete wetting of the workpiece by the glass is desirable, but without attack on the die material.

Surface energy of glasses can be measured but is, in itself, not a useful guide, because metal oxides on the workpiece surface can enter into the glass as network modifiers and thus change their composition and wetting properties [221]. Indeed, oxides are sometimes added for the purpose of improving wetting. A more practical way of improving wetting is by surface preparation, especially shot blasting [217].

Because adhesion and attack cannot be predicted from first principles, experiments must be conducted under conditions typical of intended applications. A measured quantity of glass is usually heated on plates representing die and workpiece materials [217,218,222]. Holding times must be long, because dies are often exposed to the lubricant for many hours [223].

Additives. Rupture of the glass film is sometimes unavoidable, and then additives that protect against metal-to-metal contact are incorporated in the glass, or applied to the die as a separate coating. Additives tend to change the rheology of glass, often to a Bingham-type behavior. They may also impart special properties. Thus, graphite reacts, and the formation of CO makes a foam. Molybdenum disulfide and BN have also found use. Their virtue is that they protect a bare steel die, as shown by twist-compression tests (Fig.4.24) [217]. In some applications the die is protected with a refractory coating, and additions to the glass then have a much lesser effect (Fig.4.24).

Application Methods. To ensure a continuous glass film, several methods of application have been developed:

a. The workpiece is preheated to operating temperature and dipped or rolled in glass powder or fibers. A smooth film forms by melting, provided that there is no loose oxide on the surface. The technique is thus useful for most metals, except for steel slowly heated in an oxidizing atmosphere, and is the basis of the hot extrusion of steel (Sec.8.5.2).

b. In modification of the dip-coating technique, excess oxidation of the metal is prevented by applying the coating at an intermediate preheating temperature and then reapplying the coating once the operating temperature is reached [217].

c. Room-temperature application is possible if temporary bonding is provided. Thus, slurries of glass powder and a polymeric binder may be made up [224] in organic solvents or in a water base. They may be applied by spraying, painting on, or dipping. During heating, the polymer film cures or melts and provides a smooth coating. Once the temperature is high enough for the polymer to burn off, the glass film forms. Thus, continuous protection against oxidation is ensured. Indeed, many of these slurries were originally developed as protective coatings for heat treatment.

d. Heating in molten glass protects and lubricates the workpiece [225].

The heat-expansion coefficient of glass is usually an order of magnitude smaller than that of metals. This mismatch creates no problem during heating, when the glass is

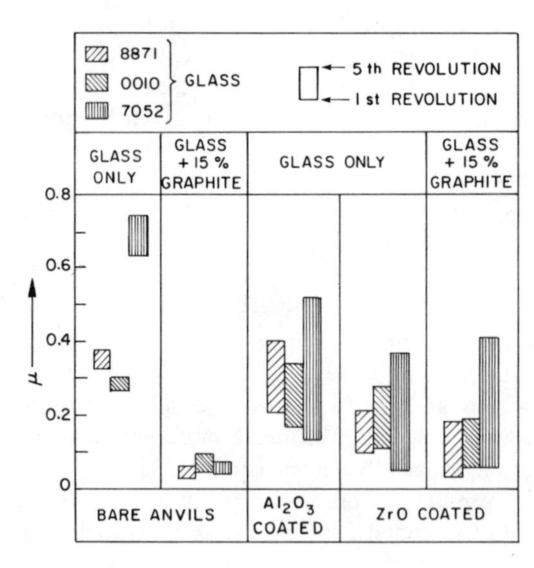

Fig.4.24. Effects of graphite additions in non-isothermal twist compression [217]

in the viscous state, and it is helpful after cooling because it facilitates removal of the glass [220].

Functional Testing. The success of glass lubrication depends critically on the choice of the right composition, and therefore much effort has gone into development of specific tests for lubricant selection [218,223]. Among these are:

a. Stability of the dispersion at elevated temperatures in the presence of a furnace atmosphere

b. Continuity of the films formed, as assessed by visual examination

c. Adhesion in shear, as determined by a suitable test involving no plastic deformation

d. Partition between die and workpiece material, and the force required to separate the two surfaces

e. Tendency for accumulation in deep-lying portions of the die

f. Friction as determined by a test relevant to the particular application

g. Heat insulating properties.

No glass is likely to have optimum ratings in all respects, and a multifactor desirability analysis may have to be performed [218].

Removal. Almost without exception, glass is objectionable on the surface of the finished product and must be removed:

a. The simplest and environmentally most acceptable method is quenching of the hot workpiece in water, whereupon much—if not all—of the glass cracks and is thrown off.

b. Also acceptable is shot blasting, which has the advantage of producing a good surface finish.

c. Pickling requires the use of a 5 to 10% HF solution with appropriate fume hoods and a method of spent liquor disposal.

d. Hot molten sodium hydride also removes glass, but at a higher installation cost.

4.7.7 Conversion Coatings

By definition, conversion coating involves the reaction of the surface atomic layers of the workpiece material with anions of a selected medium [226]. It is thus similar to E.P. lubrication in that a controlled corrosion process produces a firmly bonded surface layer. The difference is that a conversion coating in itself does not lubricate, but acts only as a parting agent and lubricant carrier. A superimposed and sometimes reacted lubricant layer is necessary to complete the lubrication system. Because of their importance, phosphate coatings will be discussed in a subsequent section.

Sull coating is perhaps the oldest type of intentional conversion coating (Sec.1.1.2). It is practiced mostly on steel wire but could, in principle, be applied to other forms of steel. The finished coating is really a three-component system:

1. The Conversion Layer. First, a freshly pickled and rinsed wire is dipped in 1% hydrochloric acid to initiate corrosion, and is then kept under fine sprays of water for some 2 h until a fine rust layer forms. The porous layer, consisting of hydrated ferrous-ferric oxide, is the base of the carrier coating [227].

2. The Carrier Film. The carrier itself is an inorganic layer, deposited on the sull coating, but more frequently on the bare wire. Three coating systems exist:

1. The wire is dipped in a lime suspension which neutralizes any residual acid, arrests further corrosion, and prevents the conversion of the rust layer into an abrasive form.

Flash-baked at 180 to 200 C in an oven for 10 to 20 min, the coating is ready to accept the lubricant. The efficiency of treatment depends on a number of factors:

a. Coating adherence is better on a rougher wire surface, and coating thickness can be increased by bath control or by repeated immersion if the wire is heavy or is to be drawn in many passes [228].

b. Both physical and chemical properties of the lime are important. Lueg and Treptow [229] find lime with a low magnesium content to be better because of its physical characteristics, and Miller [230] recommends quicklime rather than commercial hydrated lime. Such soft-burned lime generates heat at a high rate on reacting with water during slaking, and the hydroxide has the desirable high (over 5 m^2/g) specific surface and contains no harmful crystals. With a cold lime bath, a more durable coating is formed [232] but drying is slower. Bastian [233] recommends 60 C, whereas others recommend a hot but not boiling bath. Prolonged exposure to over 100 C results in crystallization and loss of adherence. A detailed treatment of the process is given in [231].

c. The lime coating must be uniform; otherwise, die contact with the sull coat or bare wire causes wear.

Lime has disadvantages in that it creates dusty conditions and must be removed from products that are to be welded or painted.

2. As an alternative to liming, a borax coating is deposited [229,234-238]. Borax is normally available in the hydrated form, $Na_2B_4O_7 \cdot nH_2O$, where n is usually 10 (decahydrate). The quality of the coating is controlled by several factors:

a. Heating above 60.6 C converts borax into the hexagonal pentahydrate (n = 5) and heating above 88 C converts it to the amorphous dihydrate (n = 2) [239]. There is no agreement as to which form is best, although the majority of opinion favors the pentahydrate. Conversion is quite rapid. In the studies of Delille [238], heavier wires (over 3.8 mm) were heated through in only 5 to 8 s in a bath at 80 to 90 C and retained heat to effect the conversion; thinner ones heated more rapidly but had to be baked in warm air to ensure transformation.

b. Coating thickness is controlled by concentration. Between 5 and 30% (120 and 350 g/L) borax decahydrate is usual; the higher values are for the thicker coatings needed on harder (higher-C) steels and for heavier draws.

c. Borax is the salt of a strong base and a weak acid, and hydrolysis gives an alkaline bath with a pH of 8.4 to 10 [236]. It neutralizes minor acid carry-over from pickling or sull coating, but good rinsing is advisable to prevent excess consumption. Water hardness causes separation of insoluble compounds, and soft water is recommended. An advantage of borax is that the wire can be stored, in dry air, without danger of rusting, and it interferes less with subsequent galvanizing.

3. The third carrier, water glass, was used only experimentally [239]. It was diluted with 7 parts water at 70 C, and caustic soda was added to bring the SiO_2/Na_2O ratio to 2:1 and thus prevent precipitation of abrasive silica. Lubrication was adequate; however, build-up of water glass on the die and the danger of SiO_2 formation have prevented widespread application of this carrier.

3. The Lubricant. The third part of the coating system is the lubricant. Most of the time it is a soap of either the water-soluble or the insoluble variety. Lime (sometimes borax or soda) is added or, rather, encapsulated in the soap to increase melting point and viscosity. The equipment and techniques employed in the compounding of soaps are described by Marr and Eckard [240]. Paste or grease is also used [236]. When ease of removal is of concern, water-soluble sodium soaps are preferred in conjunction with a borax coating, because the entire system can be removed readily, whereas a lime coating requires pickling.

Oxide Coatings. Besides their parting functions, oxides can serve also as lubricant carriers, especially at elevated temperatures, a point to which we shall return in Sec.9.5.3. In cold working their utility is limited by their brittleness; nevertheless, when lubrication difficulties are great, even the small advantage gained from oxides may be of value.

Anodizing, in which the workpiece is submerged in an electrolyte and is made the anode, is extensively used for creating decorative and protective coatings on aluminum and, to a much lesser extent, on other metals [156,157,226]. The porosity of the coating would make it a good lubricant carrier, but its brittleness precludes widespread use. Anodizing of titanium in a sodium hydroxide solution [241] has similarly limited application.

Controlled oxidation in air is more practicable (for example, in the hot extrusion of copper; see Sec.8.5). Oxide films grown on titanium by exposure to air at 630 C for 15 min are of some value in cold working [242]. Oxidizing improved lubrication in the drawing of nickel wire [243].

Oxide films can act as incidental, reactive carriers. Thus, Pikaeva et al [244] found the oxide of copper particularly beneficial in the presence of a boundary lubricant such as caprylic acid.

Sulfuration. Conversion coatings which enrich the surface in sulfur act, by their nature, as both E.P. agents and lubricant carriers.

Courtel et al [245] found that sulfuration of a 0.1% C steel in a melted salt bath resulted in communicating pores 0.2 to 2 μm in size. Impregnation with PTFE gave excellent lubrication and a deformable film. The very difficult problem of lubricating titanium is also eased by sulfuration in a salt bath composed of sulfur and cyanide compounds [242].

Chromating involves the formation of mixed metal-chrome oxide films. Originally applied to nonferrous metals, it had little influence on metalworking practice. Only with the introduction of the so-called tin-free steel [156,246,247] have chrome coatings acquired importance. The technique is different from those employed on nonferrous metals, in that chromium and its oxide are co-deposited, or a thin metallic chromium layer is electroplated and then subjected to after-oxidation by electrochemical treatment [246]. The layers are very thin (5 to 8 nm Cr; 7 to 25 nm oxide), and the system is completed by an oil film 2 to 5 nm thick [247]. The coating ensures corrosion protection and serves as a substrate for subsequently applied lacquers. It should, in principle, also give good lubricant adhesion, but no work in this direction appears to have been reported.

4.7.8 Phosphate Coatings

Phosphate coatings had been used as paint bases well before Singer [1.14] realized their value as lubricant carriers. In Germany they were applied extensively during the Second World War for wire and tube drawing and for cold extrusion. Even though some of this work became known from translations of German articles [248], it was only after the end of the war [227] that their application began to spread. Since then they have taken a dominant position around the world, and have made it possible to replace machined parts with plastically deformed ones. The process and the resultant properties of the coatings have been extensively investigated. Details may be found in the classic work of Machu [249], in books by Wiederholt [250], Rausch [251], Gabe [156], and Biestek and Weber [226], and in an article by Waterhouse [252]. There are innumerable articles dealing with phosphating for metalworking [157,165,166,174]; recent ones have been published by Darkins [253], Schmoeckel [254], and Haupt [255]. The lubricant system consists of the coating and a suitable superimposed lubricant.

The Phosphating Process. Phosphating for metalworking is really a pseudoconversion coating because cations of heavy metals such as zinc enter into the coating. The phosphating bath is an aqueous solution of primary phosphates $Me^{++}(H_2PO_4)_2$ where Me^{++} can be iron, zinc, or manganese, although for metalworking zinc is of importance. Free phosphoric acid H_3PO_4 is added to bring the pH to a value between 2 and 4.

On immersion in the acid solution, the free H^+ ions attack the workpiece surface as in corrosion:

$$Fe + 2H^+ \rightarrow Fe^{++} + H_2 \tag{4.2}$$

Removal of H^+ ions in the form of H_2 gas raises the pH in a thin layer of the bath adjacent to the workpiece surface, whereupon the equilibrium between various forms of the heavy-metal salts shifts to the right:

$$Zn(H_2PO_4)_2 \rightarrow ZnHPO_4 + H_3PO_4 \tag{4.3}$$

$$3Zn(H_2PO_4)_2 \rightarrow Zn_3(PO_4)_2 + 4H_3PO_4 \tag{4.4}$$

As shown by the ionic form of the over-all reaction

$$4Zn^{++} + 3H_2PO_4 \rightarrow ZnHPO_4 + Zn_3(PO_4)_2 + 5H^+ \tag{4.5}$$

localized depletion in the H^+ ion concentration causes a shift toward the formation of the sparsely soluble secondary and totally insoluble tertiary zinc phosphates which deposit on the steel surface in a crystalline form.

Deposition is accelerated by the addition of oxidizing agents that remove the hydrogen, formed in reaction (4.2), in the form of water, and also oxidize the Fe^{++} ions into Fe which then forms an insoluble sludge:

$$2Fe + 2Fe(H_2PO_4)_2 + 4O \rightarrow 4FePO_4 + 4H_2O \tag{4.6}$$

Different accelerators give different deposition rates and film weights (Fig.4.25) [254]. Nitrates are moderately effective in baths at 80 to 95 C but give the heaviest deposits; the iron content of the coating remains rather high. In contrast, nitrate/nitrite accelerators prevent the deposition of iron phosphates although they give lighter coatings at typically 60 to 70 C. The nitrite content may be objectionable, and similar results are obtained with chlorate accelerators at 50 to 80 C except that sludge formation is greater.

Growth and Properties of the Phosphate Layer. According to Machu [249], phosphating is a topochemical reaction in which the corrosive attack of the phosphoric

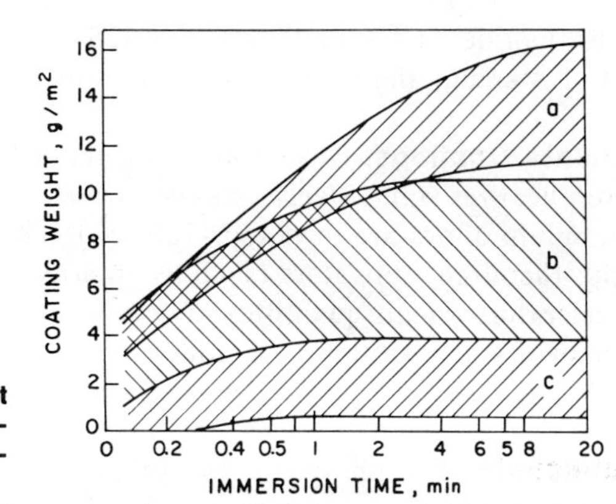

Fig.4.25. Effects of accelerators on weight of phosphate film formed [254]. a - nitrate; b - nitrate/nitrite or chlorate; c - with deposit-refining additives.

acid (Eq.4.2) takes place at sites that can be regarded as microanodes, and the formation of tertiary phosphates (Eq.4.5) at microcathodes. The secondary and tertiary phosphates are deposited in the form of crystallites which grow epitaxially on the substrate (in the same direction as the surface crystals in the metal). Individual crystals are thin platelets (and sometimes also needles or grains) 10 to 100 μm in size which form a layer somewhat resembling fallen leaves. The film thickness is 1 to 10 μm; pores amounting to several percent of the volume become the lubricant reservoirs. Crystal density and size can be controlled to give fine- or coarse-crystalline, thin or thick coatings. The effects can be rationalized by the theory of Machu [249] which states that nucleation and growth depend on the number and energy content of active sites:

a. Residual oil films prevent reaction with the surface, resulting in patchy coatings. Therefore, it is essential that the surface be clean. Degreasing in organic solvents or emulsions promotes a fine-grain coating, perhaps because a residual monolayer of oil prevents large cathodic sites from developing. In contrast, strongly alkaline degreasing results in coarse coatings, as does pickling in strong acids.

b. Mechanical cleaning, especially by shot blasting, promotes nucleation at many sites and thus a fine-grain coating, as does wire brushing of a chemically clean surface.

c. The number of nucleation sites can be increased by activation of the clean metal surface through immersion in a 1% solution of copper sulfate or nickel sulfate. The more noble metals form electrochemical deposits which provide many sites. Similar results are obtained with a 1 to 2% disodiumphosphate solution to which 0.1 to 100 mg/L of a titanium compound (e.g., orthophosphate) has been added.

d. The growth of the layer is accelerated by factors that favor the formation of cathodic areas. The chemical accelerators mentioned earlier act by maintaining cathodic areas and converting anodic ones. The same effect is ensured by making the workpiece the cathode in an electric circuit. Because anodic action is less intense, an alternating current has the same effect.

e. Mechanical acceleration can be done by spraying with a stream of the solution, thus accelerating the diffusion of phosphating constituents.

Modified Treatments. Phosphating was originally developed for steel; modifications for other metals are available:

a. Zinc and its alloys can be phosphated in solutions enriched in iron, or in proprietary solutions. This is of importance in the working of zinc-plated steel.

b. Aluminum is phosphated mostly in proprietary solutions containing chromates and fluorides in addition to phosphates [256].

c. Titanium can be treated in a solution containing sodium orthophosphate, potassium fluoride, and hydrofluoric acid [257]. The dark adherent coating is believed to consist of titanium and potassium fluorides and phosphates.

Oxalate Coatings. Steel containing more than 5% Cr cannot be phosphated. Therefore, heat-resistant steels, stainless steels, nickel-base superalloys, and other high-chromium alloys are oxalated [166,258]. The chromium and nickel oxides resist oxalic acid; therefore, activators (such as chlorides, bromides, fluorides, rhodanates, etc.) and accelerators (small quantities of tetravalent sulfur compounds such as sodium thiosulfate or titanium oxalate) are added which depassivate the surface. According to Machu [249], the oxalate layer is deposited on a film of iron and nickel sulfates.

Lubricants. The phosphate or oxalate coating does not in itself lubricate, but rather provides a key for a superimposed lubricant which is chosen according to the severity of deformation [254].

a. For lightest duties the surface is impregnated with an emulsion or oil which is often compounded. Absorption is independent of viscosity but is aided by surface-active (polar) additives.

b. Most workpieces are treated in a hot soap (sodium stearate or potassium stearate) solution. A neutral soap film forms a surface layer and penetrates the pores. Adsorption is of a chemical nature: the soap is ionized, and some of the phosphate is converted into a sodium phosphate, thus increasing the breakdown temperature of the soap coating. Use of formulations containing unsaturated additives or borax, or rinsing in borax prior to dipping in soap, results in so-called reactive soaps which form coatings suitable for operation up to 250 C. The soap must be perfectly dry before working; otherwise, breakdown occurs. A heavier phosphate coating adsorbs more soap and is thus suitable for heavy reductions, while the better adhesion of finer coatings is preferred when the process involves a severe wiping action. The low iron content of nitrate-nitrite accelerated fine coatings also contributes to better soap conversion. After deformation the film appears glazed and can withstand further deformation, although resoaping is sometimes necessary.

c. Layer-lattice lubricants are suitable for yet heavier duties. Deposition by gentle barreling is possible, but some of the phosphate is rubbed off in the process. A suspension of 5 to 15% MoS_2 in water can be applied by hot (80 C) dipping followed by drying. A polymeric binder (often a cellulosic derivative) ensures good adherence [180], and wetting agents such as polyglycol ethers give good coverage. Continuous operation is possible in a rotating drum [259]. Deposition from an organic solvent dispersion, with a polymeric binder, is also feasible.

d. For the most severe duties, a MoS_2 layer can be applied on top of the soap film, either by barreling or from a suspension [180,254,259].

Application and Removal. The sequence of phosphating treatments can be carried out by several techniques [260]:

a. Immersion treatment involves degreasing, pickling (if the steel is rusted), phosphating, and drying, with cold and/or hot water rinses between steps to prevent contamination of the baths. The last hot rinse is often in a passivating solution of borax or nitrites. Strict control of temperature and concentration is essential, and suppliers of coatings provide the necessary test kits. The lubricant is applied in a separate bath, and the workpiece is dried. Strip or wire may be guided through the baths, whereas slugs are often processed in barrels. To avoid excessive attack of the metal, the pH is maintained between 1.8 and 2.4.

b. Spray coating is a useful alternative, especially for workpieces of awkward shapes. The pH can rise as high as 3.

c. In a technique described by Darkins [253], degreasing, descaling, cleaning, and phosphating take place in one step during blasting with a slurry of abrasives in the phosphating solution. The results are fine-grain coatings and (especially for lead phosphate coatings) good performance. For reasons not yet understood, the glazing typical of zinc phosphates does not occur with this method of deposition.

d. Simultaneous phosphating and oiling is possible in processes developed for drawing applications [261,262]. The finish is smoother than with zinc phosphate plus soap.

e. Simultaneous phosphating with deposition of MoS_2 [263] is favorable in that the lubricant particles are embedded in the conversion coating, but bath control is difficult.

Removal of coatings presents no particular difficulties, but is essential for some applications [264,265]. On heat treating the coating burns off above 300 C; otherwise, it can be removed by appropriate stripping solutions. Sulfuric acid, caustic soda, or ammonium hydroxide is used on steel, and nitric acid is used on aluminum.

4.8 Lubricant Application, Maintenance, and Removal

The best lubricant will fail if it is improperly applied or allowed to deteriorate. Some techniques limited to specific applications have already been covered in conjunction with glasses and coatings; here we will review, very briefly, the general aspects of surface preparation and of the application, maintenance, and removal of liquid lubricants.

4.8.1 Surface Preparation

The purpose of surface preparation is twofold: first, to remove surface layers that may interfere with lubrication or cause wear; and second, to provide the optimum surface roughness and chemical condition for the lubricant. Processes are described in detail in Metals Handbook [157], and updated information is available for stainless steels [266].

We have seen that well-controlled oxide films can contribute to lubrication but that thick, uneven, discontinuous films can act as abrasives and lead to surface defects. Removal is possible by several techniques:

a. In mechanical descaling, slight but uniform deformation of the surface causes the scale to fall off. For this, a sheet or wire can be repeatedly bent, or discreet lengths of wire, rod, or tube can be reeled. An experimental study of the descaling of wire rods by bending is given by Juretzek and Rohland [267], and a theory has been developed by Sudo et al [268]. Discrete parts are tumbled in barrels with or without an abrasive medium. The most frequently used technique is based on blasting the surface with a high-velocity stream of abrasive or metal shot. The source of energy may be compressed air or a rotating wheel. Properly controlled blasting results in a uniform, nondirectional surface finish which is very favorable for lubricant entrapment (Sec.3.8.5). Excessive blasting can lead to the development of surface defects. The effects of several variables have been studied by Jones and Gardos [269].

b. High-pressure water descaling is applicable to hot steel only. The jet is believed to act both by mechanical action due to its high pressure and by the generation of steam at the steel-scale interface. Dimensioning of a descaling installation is discussed by Herold [270].

c. In pickling, the oxide is removed in mineral acids. Concentration, temperature, and time of immersion are controlled. The choice of the treatment is influenced by ecological considerations, because some spent pickle liquors are difficult to treat and dispose of in an environmentally acceptable way. Recovery of the acid is possible and reduces cost [271]. The removal of a thick surface layer is called chemical milling and has the advantage of removing surface defects inherited from prior processing stages.

d. Alkaline solutions (primarily sodium hydroxide) are suitable for aluminum and also for some steels.

e. Salt baths operated at the melt temperatures of salts or salt mixtures are often preferred for high-nickel alloys, stainless steels, and other alloys with tenacious and complex oxides.

4.8.2 Lubricant Storage and Application

Lubricants are subject to degradation. We have already mentioned that emulsions must not be allowed to freeze, nor must they be exposed to high temperatures that would break the emulsion. Futhermore, fatty additives in aqueous and oil-base lubricants suffer from oxidation and polymerization, and crystalline dispersions may form in sulfurized oils at elevated temperatures. Therefore, indoor storage at 10 to 30 C is most favorable. In heated bulk storage facilities, a minimum flow rate of 1.5 m/s must be maintained to avoid overheating around the heat source. If barrels are stored outdoors, free air circulation around them should be ensured. They should be stored in a horizontal position to avoid entry of rainwater [12].

Lubricant application techniques are specific to processes and will be discussed in more detail in Chapters 6 to 11. However, there are some general principles that are of interest here. Basically, there are only two methods of application:

1. The quantity of lubricant necessary for successful operation is preapplied to the surface of the workpiece. Film thickness can be controlled by several means:

a. The workpiece is dipped into a bath of the lubricant or emulsion, or into a bath of the lubricant in an organic solvent or aqueous dispersion. Coating weight is controlled by bath temperature and concentration, by workpiece temperature, and by the drying cycle following removal from the bath.

b. A film of controlled thickness is deposited by means of rollers, air-atomizing sprays [272], airless sprays, or electrostatic deposition [273], usually in a location adjoining the actual production unit.

Because, in principle, only as much lubricant is provided as is absolutely necessary, no attempt at lubricant recovery is made. Therefore, the system is often described as total-loss lubrication. If cooling is required, it is separately provided either by internal cooling of tools or by application of a coolant, usually water, to the interface. This then creates problems of disposal of the cooling water, which invariably becomes contaminated with the lubricant.

2. An excess of lubricant is delivered to the work zone, and the requisite lubricant film thickness is established by the process itself. Excess lubricant is reclaimed and, because of possible environmental pollution, invariably recirculated. In such systems the lubricant also fulfills the role of a coolant.

4.8.3 Lubricant Treatment

The used lubricant can seldom be returned to the work zone without treatment. The lubricant itself undergoes some changes as a result of its action: soaps, E.P. reaction products, and friction polymers are washed out from the interface, and exposure to air and exposure to the machine tool cause oxidation or corrosion products to form. Wear debris, often coated with lubricant products, gradually builds up. In machining, the lubricant/coolant also serves to remove chips formed in the work zone, and is thus mixed with large quantities of metal particles. In abrasive machining, particles of the abrasive are mixed with metallic fines. In aqueous lubricants, microbial growth not only results in the accumulation of a biomass but can also destabilize the system with consequent separation of the dispersed phase.

A collection of papers on the treatment of metalworking fluids is available [274].

Tramp Oil. One of the most insidious sources of lubricant degradation is tramp oil (i.e., oil, grease, or emulsion that seeps into the metalworking lubricant from bearings, hydraulic components, gear boxes, etc.). Tramp oil increases the viscosity and staining propensity of oil-base lubricants and degrades the stability of emulsions, and oil floating on the emulsion can lead to the growth of anaerobic bacteria. One or more of three approaches may be taken:

a. Special design and maintenance efforts can be made to minimize leakage.

b. The problem can be entirely bypassed by using the same lubricant for the process and for the machine tool. This, however, is often not possible, because the viscosities of oil-base lubricants used in recirculating systems tend to be much lower than those of bearing and hydraulic lubricants, and because emulsions may create fatigue problems in highly stressed components.

c. As a compromise, hydraulic fluids can be formulated to be compatible with the metalworking lubricant, so that in the event of an accidental failure (cut seal or hose) the metalworking lubricant can cope with the contamination. In particular, emulsifiable oils, W/O emulsions, and also O/W emulsions have become available [275].

Treatment Systems. The lubricant system must be designed so that it does not become a source of problems in itself. Piping, tanks, filters, and other components must be laid out without dead corners where sludges deposit and microorganisms thrive. All parts of the system that require periodic cleaning must be readily accessible and cleanable.

A lubricant can be conditioned by one of four basic approaches [276,277]:

1. In the batch process, the lubricant is collected and periodically subjected to treatment. This can be the least expensive solution and is viable for small quantities, but it also results in the greatest variability and shortest life of the lubricant.

2. In the full-flow system, the lubricant is returned to the holding tank, the total flow is subjected to treatment, and the clean fluid is returned to the work zone. The cleanness of the lubricant depends on the efficiency of treatment.

3. In the bypass system, the lubricant is continuously returned to and pumped from the reservoir, while a separate pump moves some of the lubricant through the treatment system. Quality is intermediate between those of the batch and full-flow systems.

4. In the combined system, all lubricant is supplied to the work zone after treatment, as in the full-flow system, but, in addition, the lubricant is kept circulating through the treatment plant even when supply to the work zone is shut off. This gives the best control of quality.

The sizes and complexities of installations vary greatly. The smallest systems deliver 100 to 500 L/min, whereas large ones deliver more than 20,000 L/min. Systems designed for emulsions must be built with particular care. To prevent creaming, piping is usually designed for turbulent flow. In addition, systems for very loose emulsions (mechanical dispersions) sometimes contain baffles, but these can create cleaning problems. For biological control and to retard odor production, Smith and West [278] recommend a minimum flow velocity of 0.7 m/s. Adequate aeration should also be ensured, however, without features such as free drops, which increase foaming and release odors. Piping should have cleaning plugs so that deposits can be not only flushed out but also mechanically removed.

Cleaning of Lubricants. A large variety of techniques are available for removing fines and contaminants from fluids [276,279-283].

1. Separation by Density. According to Stokes' law, a particle suspended in a viscous medium separates at a speed V:

$$V = \frac{d^2 \Delta\rho a}{\eta} \tag{4.7}$$

where d is particle diameter, $\Delta\rho$ is density difference, a is acceleration, and η is viscosity. Thus, larger particles settle more rapidly in a less viscous fluid, especially if the force acting on them is large.

a. Settling or sedimentation of heavier particles is possible when the container or pit is large enough to allow a minimum retention (residence) time, which must increase with decreasing particle size. A 0.5-mm sand particle takes 6 s to settle 30 cm in still water [284]. Typical residence times are 30 min for oil systems, 10 to 20 min for stable emulsions, and only 5 to 6 min for less stable emulsions which would cream on longer holding. The technique is very effective also for settling of scale, sand, and chips. A weir inserted in the tank separates the clean fluid (Fig.4.26a). Sedimented matter is removed by periodic cleaning, or, for larger throughputs, drag chains are used. Sedimentation of ferrous particles is accelerated by magnetic coagulation [285].

b. Lighter particles such as tramp oil and breakdown products, together with entrapped fines, rise to the surface of an aqueous fluid and can be skimmed off with brooms, paddles, or drag conveyors. Surface baffles separate the clean fluid (Fig.4.26a).

c. Gravitational separation is aided by flotation. Air bubbles are introduced either by stirring or by compressed air. The bubbles rise and, especially in the presence of surfactants, fines attached to the bubbles are carried to the surface to be trapped in a foam which is removed (Fig.4.26b).

d. Greater acceleration is imposed on suspended particles in a hydrocyclone (Fig.4.26c). The fluid, pumped at high velocity, is fed tangentially into a conical vessel

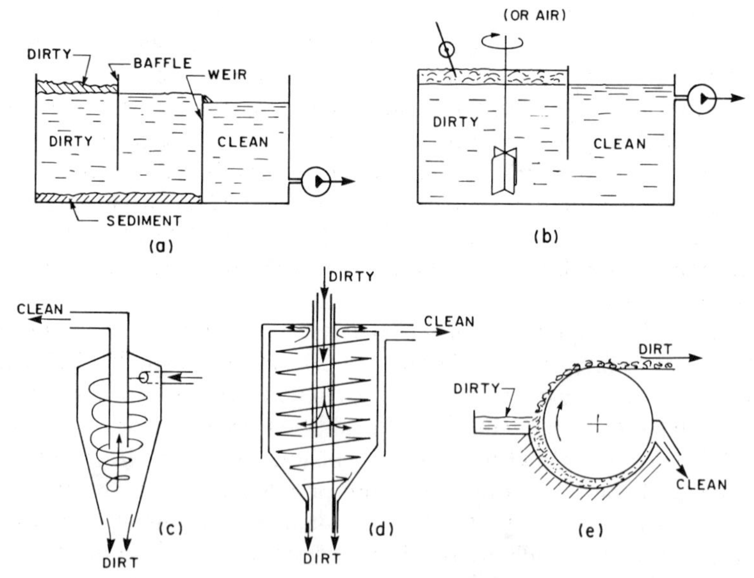

Fig.4.26. Removal of fines from lubricants (a) by settling, (b) by flotation, (c) in a hydrocyclone, (d) in a centrifuge, and (e) in a magnetic drum

so that heavier particles are forced to the wall where they collect, drop and exit at the bottom, while clean fluid leaves at the top. Chemical or magnetic coagulation is again helpful [286]. Some fluids may suffer from excessive foaming.

e. Yet higher accelerations are imposed in a centrifuge (Fig.4.26d). Smaller particles can be removed if they are first coagulated, as is done with a sodium carbonate solution for rolling of aluminum [287]; oil is concentrated in the sludge, which is easy to burn because it does not contain particles of any solid filter medium. Because of the high acceleration, centrifuges can be placed in any position and may be made to operate continuously. Centrifuges are also suitable for reclaiming the 150 to 350 L of oil adhered to every ton of machining swarf (chips) [288]. The chips may be placed in baskets and spun, or fed into continuous centrifuges.

2. Filtration. Depending on size of particles, various filtration media are used. The driving force is simply gravity, or pressure or vacuum is applied to prevent retention of the dispersed phase of emulsions [289]. The filter configuration ranges from flat stationary beds to moving belts, rotating drums, tubes (socks), flattened tubes (leaves), etc. The accumulated solids can be removed by scraping, shaking, or reverse flow.

a. Wire filters of varying mesh sizes are the simplest, but the smallest particle they can retain is around 100 μm. To prevent rapid clogging with coarse particles, a sand bed may be built up.

b. Smaller particles (down to 10 μm) can be removed by the use of woven or nonwoven filter media of controlled pore size. A sediment cake of larger particles develops and then acts as a deep bed capable of removing smaller particles [290]. When the pressure differential increases to a preset value, it triggers the cleaning mechanism or switches to reverse flow to remove (blow off) the cake.

c. Disposable filters are filter media supplied in rolls, to be fed over a rigid wire screen. The choice of medium depends on the fluid [291]. Cellulose fibers and rayon absorb water and swell excessively, whereas polymer (such as polyethylene, polypropylene, and polyester) fibers absorb oil, and could deplete an aqueous lubricant. The sediment is allowed to build up until a pressure drop of 4 to 15 kPa (0.5 to 2 psi) develops. The sediment cake acts as a filter capable of removing particles smaller than 10 μm. Pressures on the order of 50 kPa would extrude gelatinous contaminants through the filter medium; therefore, at a preset pressure the medium is advanced over the filter bed, and thus continuous operation is ensured. Operation is limited to oils and stable emulsions; a loose emulsion would clog the medium and the oily phase would be removed together with the contaminants. Variants of the principle have been developed for such situations: for example, in the system described by Joseph [292], the oil phase is allowed to float to the surface of the tank to be filtered separately and returned to the clean supply after being remixed with the separately filtered emulsion. The importance of filtration is shown by the data of King [124]: one pass in a cup ironer produces 10^3 to 10^4 wear particles 1 to 5 μm in size.

d. The principle of deep-bed filtration can be employed by use of precoated filters in which a layer of permeable inorganic material such as a diatomaceous or Fuller's earth is applied over a perforated or woven metal screen [293]. The filters may be preformed for a single use (although removal of deposits by scraping will allow operation for considerable lengths of time), or precoating may take place in the system itself. Such systems are complex but in the long run economical if particles one micron and less in size are to be removed. The precoating medium is mixed with a measured quantity of clean oil and circulated from the storage tank through the filter bed, often with the addition of a filter aid (powdered earth) which increases the length of the filtration cycle by maintaining porosity in the sediment cake. When the pressure drop increases to

200 kPa, the sludge cake is blown off. The medium attracts polar compounds; therefore, some additives are lost from oil-base lubricants, and emulsions could be broken.

e. Ultrafiltration differs from filtration in that a higher pressure (150 to 200 kPa) is maintained over a filter membrane with very small pores (below 5 nm) [294]. The pores are small enough to reject emulsion and suspension particles (Fig.4.27), and therefore the technique is suitable only for clarification of synthetics.

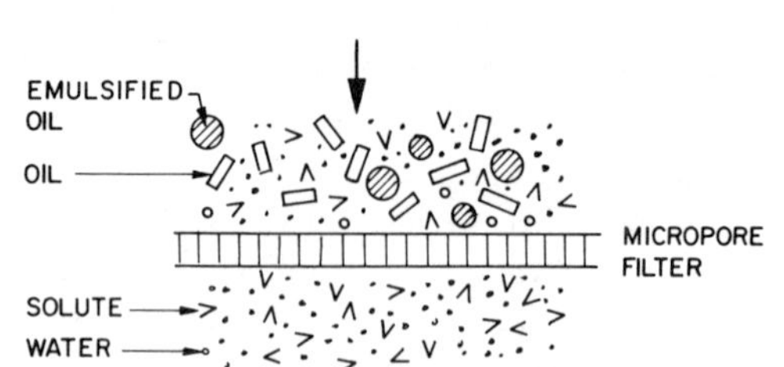

Fig.4.27. Ultrafiltration

3. Electrical Separation. Two electrical effects can be exploited:

a. A magnetic drum removes ferrous fines from fluids. Nonferrous fines stuck by oil films to the ferrous particles are also removed. If the flow is directed between the magnetic drum and a cast iron sleeve, bridging fibers build up which remove further fines (Fig.4.26e) [280].

b. Small (<5 μm) metal fines can be removed by charging them in a 30- to 50-kV dc electrostatic field. In a system described by Fritsche and Sansonetti [295], the electrode is surrounded by ceramic beads on which the fines are deposited. They are removed by back-flushing while the field is switched off. In a 4-h cycle, solids in a rolling oil are reduced from 0.07% to 0.01 or 0.02%.

4. System Configuration. Most lubricant cleaning systems incorporate more than one technique. Coarser fines are often taken out by settling tanks, cyclones, or centrifuges so as to relieve the load on filter media and, especially, on coated filters. The efficiency of filtration should be regularly checked by analysis for residual particulates [296], and, if necessary, make-up to the oil or emulsion has to be made.

Control of lubricant temperature is an important part of the system, calling for the installation of heating and cooling equipment. The temperature at which the bath is operated depends on the lubricant: 60 to 70 C is usual for mechanical dispersions, 50 to 60 C for unstable emulsions, and 35 to 55 C for stable emulsions. Mineral oils are usually held at temperatures around 30 to 35 C, although higher temperatures may be desirable for higher-viscosity lubricants and lower temperatures may be desirable for very light mineral cuts (kerosene) of low flash point. The heat exchanger normally utilizes water or, in areas with limited water supply, air.

There is no single treatment system that would be universally applicable to all lubricants. Therefore, the treatment system must be designed at the same time as the lubricant is chosen, and a total systems approach must be taken [297]. For continuing efficiency, the circulating system must be regularly and thoroughly cleaned, the cleaning chemical neutralized, and the system subsequently maintained [298].

4.8.4 Lubricant Removal

Lubricant residues often interfere with subsequent operations such as welding, adhesive bonding, painting, etc. [299] and are removed. Removal techniques are described in detail in Metals Handbook [157].

a. Aqueous degreasers include emulsions, detergents, and alkaline compounds formulated in various strengths. Many of them are acceptable from an environmental point of view.

b. Solvents include both liquids and vapors of organic materials, very often chlorinated. Trichloroethylene used to be popular but is subject to stringent environmetal (E.P.A.) and health (OSHA) regulations [300]. Other solvents, such as methylene chloride, 1,1,1-trichloroethane, and perchloroethylene [301], are gaining popularity.

Some lubricants, especially those containing solids and layer-lattice compounds, are more difficult to remove, especially if they are burnished onto the surface. Removal is generally aided by mechanical means; therefore, sprays are more effective than immersion. Electrocleaning, commonly applied in the electroplating industry, is also most effective [302].

4.8.5 Disposal of Lubricants

After some time, lubricants in a recirculating system reach a point where irreversible changes in the lubricant, contamination, or maintenance costs force their disposal. Lubricants are often disposed of also when the lubrication system is shut down for cleaning. Lubricant application by the total-loss method leads to accumulation of lubricants on machines, in pits, and in trenches. If cooling water is applied, large quantities of oily waste water are generated. Cleaning solutions also become laden with lubricants. Thus there are large volumes of oils, emulsions, and water-oil mixtures that interfere with sewage treatment plants and are injurious to aquatic life, and these must be dealt with in an environmentally (and legally) acceptable manner.

1. Oils are of value and, if possible, are reclaimed. Vacuum distillation [303] is suitable for well-defined oils such as those used for rolling of aluminum. Oils may be sold to refiners, directly burnt in furnaces, or disposed of by other techniques. Disposal by landfill is not acceptable anymore. A bibliography is given in [304]. Solvents may be reclaimed in special plants [305].

2. Emulsions cannot be discharged into rivers or lakes. Tramp oil is preferably removed first, and then the emulsion is broken to separate the oil and water phases. We have seen what is needed for emulsion stability (Sec.4.6.2); the task is now to destabilize the system [306-310]. The techniques are often suitable also for treating waste waters in which, through the simultaneous presence of surfactants and oils, emulsions are unintentionally formed.

a. Chemical methods aim at destroying the surface films and electrical charges that keep the oil droplets apart. Thus, they are specific to the emulsifier system used. An anionic emulsion can be simply broken by adding acid to offset the HLB ratio. Anionic and cationic emulsifiers are mutually destructive. Nonionic emulsions are more difficult to deal with, and need additions of electrolytes, such as ferric or aluminum sulfates or chlorides, to destroy surface films and thus allow the oily phase to flocculate and coagulate. Similar effects are obtained by use of high-molecular-weight polyelectrolytes with ionizable sites along the chain.

b. Heating accelerates the process and can even be used as the sole means of destabilization.

c. Separation is accelerated by an electrical field.

d. Ultrafiltration, already introduced as a cleaning technique, retains large molecules such as oil molecules on the pressure side of the filter but allows salts to pass through (Fig.4.27) [294].

e. If a semipermeable membrane is placed between a solvent (in this case, water) and a solution, the solvent passes through the membrane to dilute the solution; in doing so, it develops an osmotic pressure. If a high enough pressure (2 to 7 MPa) is applied

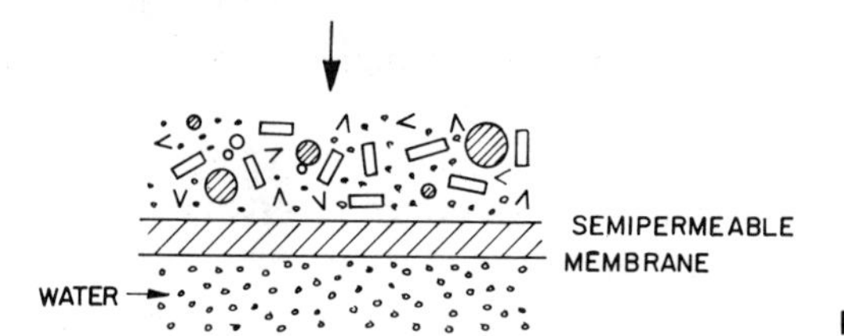

Fig.4.28. Reverse osmosis

on the solution side (Fig.4.28), osmotic flow is reversed and the solvent (pure water) is ejected by reverse osmosis. The concentrated oily phase, which retains all chemicals, is then treated separately. The treatment is suitable even for wastes such as synthetic fluids.

Unless reverse osmosis is used, the aqueous phase may have to be subjected to further treatment before being discharged into rivers, since various salts and possibly also water-soluble organic compounds are left in it.

3. Oily wastewaters present particular problems, not least because of the large volumes generated. Water-pollution control in integrated steel plants, with reference to U.S. rules, is discussed by Kwasnoski [311], and effluent treatment of nonferrous plants, with reference to British rules, is discussed by Stone [312]. Several techniques are employed, often sequentially [313-316]. The basic techniques are similar, irrespective of whether the effluent originates in a forging shop [317], in a rod mill [318], in an aluminum extrusion plant [319], or in general metalworking industries [320,321].

a. Wastewaters from hot rolling mills are often treated by filtration through a sand bed. Particulates such as scale and even some solutes are removed by attachment to the sand-particle surfaces. Variants of this technique have introduced great improvements [322]. Thus, deep-bed filtration through mixed media, in which a coarser medium of lower density is spread on top of a finer medium of higher density, allows filtration to lower particle sizes, and facilitates reconditioning of the filter bed by back washing; afterwards, the particles again occupy their intended positions. Pressure may be applied to increase flow rates [323,324].

b. To relieve later treatment stages of excessive loads, simple settling-tank, flotation, or mechanical filtration [325] techniques are used for separating the bulk of oily contaminants. Dilute mixtures are sometimes disposed of by landfill. Concentrated mixtures can be burnt, if necessary, by adding heavy fuel oil [326].

c. More recently, ultrafiltration [327,328] and reverse osmosis [328] were found to be economical in some applications.

d. Biological processes offer some potential, too [329,330]. Various emulsion constituents represent different pollution hazards but can also become sources of new products [330]. However, water discharged from treatment plants should be low in biological oxygen demand (BOD) and chemical oxygen demand (COD), so as not to place a burden on the environment.

The effect of effluents on microbial ecosystems is still a subject of rapid development [331]. Integrated pollution control is best incorporated at the design stage of a new plant [332]. The aim is usually to concentrate the oily and solid phases to the point where they can be dealt with efficiently. It is not unusual to find then that the cost of treatment can be offset by the value of the recovered product. In general, standards relating to disposal are more readily met with emulsions than with synthetics.

4.9 Health Aspects

In many applications, particularly rolling and machining, the operator is exposed to the liquid lubricant or its mist or vapor for several hours every day. Therefore, even relatively harmless substances such as detergents may become damaging, and more active lubricants can cause problems not encountered outside the work environment. Several general reviews are available [278,333-337]. Legal aspects are considered by Blodgett [338]. There are several possible routes to health damage, but diseases of the skin are by far the most prevalent.

4.9.1 Skin Diseases

The human skin is a complex structure formed of several layers. The outer skin (epidermis) itself is composed of 4 to 5 layers. The outermost horny layer, composed mostly of keratin, is dead, but it is kept supple by natural oils and is replenished by the underlying layers. When the surface layers are deprived of the natural oils, or are damaged by chemical or mechanical action, skin diseases, collectively termed "dermatitis," may occur either by direct action on the horny layer or as a result of allergy. Sensitivity is highly individual, and some people appear to develop it on exposure, while others become accustomed to exposure. Reviews relating to metalworking [339-343] and those of a more general scope [344-346] are available. Various reactions may occur:

a. Light fractions of lubricating oils (typically, with final boiling points below 350 C), solvents, or detergents remove the protective natural oils, and the dry skin develops a rash. Disruption of keratinization allows leakage of fluid.

b. Heavier fractions of mineral oils plug the pores, leading to blackheads and boils (oil acne). Prolonged contact with oils that contain heavy polycyclic aromatic hydrocarbons may cause cancer, especially cancer of the scrotum. It is, therefore, recommended that solvent-refined oils be used where contact with the skin is unavoidable.

c. Additives can be particularly severe irritants. Some chlorinated or sulfochlorinated oils cause chloracne. In the presence of water, they hydrolize to form mineral acids which then accelerate the breakdown of chlorinated additives, and the resultant acidity leads to an attack on the skin. The problem is particularly common in shops where both chlorinated oils and aqueous lubricants are used and where oil is introduced into the emulsion by incomplete cleaning of the machine and by carryover onto parts or on hands [20]. Many of the bactericides are also irritants. Contrary to a common misconception, bacteria themselves do not cause dermatitis.

d. Human skin is slightly acidic (pH 6.7); excessive acidity leads to direct attack of the skin, and excessive alkalinity loosens the horny layer, exposing the skin to further damage. Therefore, where skin contact is unavoidable, a maximum pH of 8.5 is recommended.

Prevention of skin diseases is primarily a matter of avoiding contact and of personal hygiene. Thus, splashguards and enclosures are used to help reduce contact; in addition, removal of oil-soaked clothing, washing of the exposed skin with mild detergents to remove traces of oil, and, if necessary, the use of protective creams are all helpful. Some people, however, are sensitive to the barrier creams themselves. Creams designed to protect in an oily environment must be water soluble, and vice versa. For alternating exposure to oils and emulsions, which is commonly encountered by many tool setters, barrier creams are useless [20]. In general, rules of industrial hygiene [347,348] must be observed.

4.9.2 Inhalation

There is no evidence linking lubricants to respiratory diseases, except perhaps in people predisposed by asthma or emphysema. Nevertheless, oil mists heavier than 5 mg oil per m^3 air are visible, and levels should be kept below this value for general comfort [349]; still lower levels have been set in some countries [350]. Obviously, for a given operation this level is easier to maintain with emulsions or semisynthetic fluids than with neat oils. Fine mist, of a particle size below 10 μm, remains suspended in air and is avoided.

There is no evidence that microorganisms in emulsions would present health hazards [119,351,352].

A new potential problem arises with synthetic fluids in which diethanolamine is the major constituent and sodium nitrite is used as a corrosion inhibitor. When such a fluid is introduced into the acid environment of the stomach through the saliva, a reaction can take place whereby N-nitrosodiethanolamine (NDEA), a known carcinogen, is produced. It appears that, even in the typically alkaline environments of cutting fluids, NDEA may form under storage conditions or as a result of bacterial attack, and that this compound can form also from triethanolamine [87,353-355]. It is possible also that intake may take place through the skin. However, as pointed out by Evans [354], a typical serving of breakfast bacon would introduce ten times as much nitrite as a one-day exposure to a cutting fluid. Although the direct link between cutting fluids and cancer is missing, the potential hazard has, nevertheless, prompted the development of fluids with different anticorrosion additives (Sec.4.6.4).

The misting tendency of oils and aqueous lubricants can be reduced by suitable formulation. Local exhausts reduce both vapor and mist levels.

4.9.3 Other Toxic Agents

Many metals and their oxides are recognized as toxic substances [356,357]. In a few instances, they occur under conditions typical of metalworking lubrication. Chromium compounds found in machining or used in greases as rust inhibitors, as well as cobalt released from tungsten carbide tooling, have been identified as agents for dermatitis. Lead compounds, especially lead naphthenate additive, can cause lead absorption through the skin. Lead poisoning is still a problem in some steel wire factories [358].

In some environments, such as hot forging, several toxic agents (oil, soot, metals) may be present simultaneously [359,360], albeit in small quantities.

Metalworking lubrication practices may be affected also by requirements regarding residues. Thus, food-processing equipment and containers are allowed less than 23 mg/m^2 oily residue [13], and then only harmless substances, such as mineral oil, triethanolamine, methyl esters of fatty acids, etc., are allowed in the U.S.

An extensive compilation of the toxicities of many substances is given by Plunkett [361].

4.10 Summary

1. Lubricants are often thought of as agents that reduce friction and wear. In practice, they must fulfill a number of additional requirements, many of which are difficult to quantify yet may have decisive influences on lubricant selection. Often, a compromise must be struck.

2. Hydrocarbons of widely ranging viscosities, from almost solventlike liquids to semisolid waxes, are used in large quantities as the viscous agents required for the establishment of plastohydrodynamic or mixed-film lubrication. Their lubricating properties and staining tendencies can be closely controlled through control of chain length, structure, and purity.

3. Natural oils and fats are valuable lubricants on account of their viscosity, favorable pressure-viscosity exponents, and early solidification. Derivatives obtained from their processing are important lubricant additives. Soaps formed by reaction with a metal are valued for their favorable rheology.

4. A vast variety of synthetic fluids are available, several of which have found application in metalworking. Synthetic mineral oils are valued for their low staining ten-

dency, synthetic esters for their controlled rheology, silicon compounds for their inertness, and several members of all of the above groups for their high-temperature stability.

5. The prevalance of boundary contact under metalworking conditions dictates the need for endowing lubricants with boundary and E.P. properties. Thus, mineral oils are invariably compounded, and even natural fats, waxes, and soaps benefit from additions of free fatty acids, E.P. agents, or layer-lattice compounds. Greases, composite structures of oil and a thickener, are valued for their ability to form films even under conditions that are not favorable for hydrodynamic lubrication.

6. The great cooling power of water is combined with selected lubricating functions in the various aqueous lubricants. Emulsions contain an oily phase in a dispersed form, whereas synthetic lubricants may be true solutions. The aqueous environment creates special problems, not the least of which is biological attack. Control of aqueous lubricants calls for a much higher level of sophistication than that required for control of oily lubricants.

7. A large variety of substances may be regarded as coatings which, when applied to the workpiece surface, perform the function of a lubricant or endow a separately applied lubricant with better performance. Among these are metal coatings, polymer and polymer-bonded coatings, fusible coatings including glasses, and conversion coatings grown on the workpiece surface. The latter ensure firmest lubricant attachment.

8. A lubricant can function only if properly applied and maintained. Maintenance and control of recirculating systems are of particular importance. Spent lubricants must be finally disposed of in an ecologically acceptable manner.

9. Continued exposure to lubricants can create health hazards, foremost of which are skin diseases. Oil mist is perhaps more of a nuisance, and at this time it is not clear whether there is a potential for cancer as a result of exposure to cutting fluids containing nitrites as corrosion inhibitors.

References

[1] C.H. Riesz, in [3.1], pp.143-239.
[2] E.L.H. Bastian, *Metalworking Lubricants,* McGraw-Hill, New York, 1951.
[3] D. Zintak (ed.), *Improving Production with Coolants and Lubricants,* SME, Dearborn, 1982.
[4] E.L.H. Bastian, I. Rozalski, and K.F. Schiermeier, Lubric. Eng., *17*, 1961, 17-40.
[5] L.B. Sargent, Lubric. Eng., *21*, 1965, 282-286.
[6] E.L.H. Bastian, in [3.32], pp.23-1 to 23-25.
[7] E.L.H. Bastian, Lubric. Eng., 25, 1969, 278-284.
[8] R.E. Savoit, Lubrication, *55*, 1969, 65-75.
[9] G.W. Rowe, in [3.33], Sec.B13.
[10] H. Radtke, Maschinenmarkt, *83*, 1977, 129-133 (German).
[11] T.G. Bryer, Paper 74-LUBS-15, ASME, New York, 1974.
[12] D.B. Dallas (ed.), *Tool and Manufacturing Engineers Handbook,* 3rd Ed., McGraw-Hill, New York, 1976, pp.18.1-18.28.
[13] L.B. Sargent, in [3.31], pp.427-476.
[14] *Kirk-Othmer Encyclopedia of Chemical Technology,* Wiley-Interscience, New York.
[15] E.R. Booser, in [14], 3rd Ed., Vol.14, 1981, pp.477-526.
[16] A.W. Ackerman, Lubric. Eng., *25*, 1969, 285-291.
[17] Th. Mang, in [3.56], Paper No.16 (German).
[18] W.J. Bartz, in [3.55], pp.1-12.
[19] J.A. Schey, Lubric. Eng., *23*, 1967, 193-198.
[20] R.K. Springborn (ed.), *Cutting and Grinding Fluids,* ASTME (SME), Dearborn, 1967.
[21] Th. Mang (ed.), *Wassermischbare Kühlschmierstoffe für die Zerspanung,* Expert Verlag, Grafenau, Württ., 1980.

[22] Th. Mang, in [21], pp.178-199.
[23] H. Jost, Iron Steel Eng., *43* (5), 1966, 88-107.
[24] R.S. Burton, Iron Steel, *43*, 1970, 367-373.
[25] J. Halling, Proc. Inst. Mech. Eng., *192*, 1978, 189-196.
[26] T.J. Edwards, Lubric. Eng., *34*, 1978, 679-687.
[27] K. Ruthemeyer, H. Boer, F. Gros, and E. Gulber, Stahl u. Eisen, *99*, 1979, 90-95 (German).
[28] R.H. Davis, C.E. Hinojosa, W.P. Trautwein, and D.E. Wagner, in [3.55], pp.39-47.
[29] G. Schroeder and J.A. Schey, in [3.56], Paper No.17 (German).
[30] C.J.A. Roelands, J.C. Vlugter, and H.I. Waterman, Trans. ASME, Ser.D, J. Basic. Eng., *85*, 1963, 601-610.
[31] J.K. Appeldoorn and F.F. Tao, Wear, *12*, 1968, 117-130.
[32] M.D. Hersey and R.F. Hopkins, *Viscosity of Lubricants under Pressure: Coordinated Data from Twelve Investigations,* ASME, New York, 1954.
[33] E. Kuss, Schmiertechn. Trib., *25*, 1978, 10-13.
[34] I.B. Goldman et al, Lubric. Eng., *27*, 1971, 334-341.
[35] R.F. Pywell, in [3.263], pp.118-129.
[36] W.G. Johnston, ASLE Trans., *24*, 1981, 232-238.
[37] H. Bennett, *Industrial Waxes* (2 volumes), Chemical Publ. Co., New York, 1975.
[38] E.S. McLoud, in [14], 2nd Ed., Vol.22, 1970, pp.156-173.
[39] D. Swern (ed.), *Bailey's Industrial Oil and Fat Products,* 4th Ed., Vol.1, Wiley, New York, 1979.
[40] T.H. Applewhite, in [14], 3rd Ed., Vol.9, 1980, pp.795-831.
[41] K. Takatsuka, S. Matsuo, T. Matsushita, and K. Matsui, in [3.56], Paper No.53.
[42] N. Ahmed, Lubric. Eng., *34*, 1978, 541-545.
[43] J.W. Hagermann, J.A. Rothfus, and M.A. Taylor, Lubric. Eng., *37*, 1981, 145-152.
[44] Various, in [14], 3rd Ed., Vol.4, 1978, pp.814-871.
[45] K.S. Markley, *Fatty Acids,* 2nd Ed., Interscience, New York, 1960.
[46] R.A. Peters, in [14], 3rd Ed., Vol.1, 1978, pp.716-739.
[47] A.W. Ralston, *Fatty Acids and their Derivatives,* Wiley, New York, 1948.
[48] O. Pawelski, W. Rasp and T. Hirouchi, in *Tribologie,* Springer, Berlin, 1981, pp.479-506 (German).
[49] R.C. Gunderson and A.W. Hart (ed.), *Synthetic Lubricants,* Reinhold, New York, 1962.
[50] D.R. Goddard, in [3.19], pp. 166-196.
[51] R.E. Hatton, in [3.43], pp. 101-135.
[52] D.M. Matthews, J. Am. Oil Chem. Soc., *56*, 1979, 841A-844A.
[53] E.R. Mueller and W.H. Martin, Lubric. Eng., *31*, 1975, 348-356.
[54] W.R. Jones, R.L. Johnson, W.O. Winer, and D.M. Sanborn, ASLE Trans., *18*, 1975, 249-262.
[55] F.C. Brooks and V. Hopkins, ASLE Trans., *20*, 1977, 25-35.
[56] R. Meals, in [14], 2nd Ed., Vol.8, 1969, pp.221-260.
[57] A.E. Jemmett, Wear, *15*, 1970, 143-148.
[58] H.M. Schiefer and R. Holinski, in [3.52], pp.38-43.
[59] W.R. Jones and C.E. Snyder, Jr., ASLE Trans., *23*, 1980, 253-261.
[60] P.H. Molyneux, in [3.19], pp.119-165.
[61] G.W. Rowe, Roy. Inst. Chem. Rev., *1*, 1968, 135-204.
[62] C.V. Smalheer, in [3.43], pp.433-475.
[63] G.W. Rowe, Trans. ASME, Ser.F, J. Lub. Tech., *89*, 1967, 272-282.
[64] B.W. Hotten, Lubric. Eng., *30*, 1974, 398-403.
[65] A.I. Soshko, Sov. Mater. Sci., *16*, 1980, 253-256.
[65a] W.A. Belyi, Ja.M. Solowitzkyi, and Yu.M. Pleskachevskii, in [3.56], Paper No.34.
[66] R.A. Nicholson and C. Hewlett, in [3.56], Paper No.5.
[67] R. Holinski and J. Gänsheimer, VDI-Z, *114* (2), 1972, 131-136 (German).
[68] W.J. Bartz and J. Oppelt, in [3.145], pp.51-58.
[69] J. Gänsheimer, ASLE Trans., *15*, 1972, 201-206.
[70] H. Wochnowski, A. Knappwost and B. Wüstefeld, Schmiertechn. Trib., *23* (1), 1976, 12-14 (German).
[71] E. Vamos and A. Zakar, in [3.56], Paper No.30.
[72] J.M. Palacios, A. Rincon and L. Arizmendi, Wear, *60*, 1980, 393-399.
[73] A.T. Kratschun et al, in [3.56], Paper No.47.
[74] J.H. Harris, in [3.19], pp.197-268.

[75] P.R. McCarthy, in [3.43], pp.137-185.
[76] A.T. Polishuk, in [3.41], pp.67-76.
[77] J.R. Hastings, Lubric. Eng., *37*, 1981, 91-94.
[78] R.B. Batts and A.P. Wenzler, Lubric. Eng., *37*, 1981, 451-456.
[79] J.B. Christian and C. Tamborski, Lubric. Eng., *36*, 1980, 639-642.
[80] M.S. Vukasovich and W.D. Kelly, in [3.144], pp.300-313.
[81] *Synthetic Cutting Fluids*, Machine Tool Ind. Res. Association, Macclesfield, Cheshire, 1975.
[82] I.S. Morton, Ind. Lub. Trib., *23*, 1971, 57-62, 66.
[83] S.J. Barber and W.H. Millett, Am. Mach., *118* (20), 1974, 95-100.
[84] M. Robin, Manuf. Eng., *81* (5), 1978, 53-55.
[85] B.W. Greenwald, in [3.55], pp.55-56.
[86] E.S. Nachtmann, in [3.55], pp.49-54.
[87] T. Mang, in [21], pp.15-41.
[88] P. Becher, *Emulsions: Theory and Practice*, 2nd Ed., Reinhold, New York, 1965.
[89] P. Sherman (ed.), *Emulsion Science*, Academic Press, London, 1968.
[90] K.J. Lissant (ed.), *Emulsions and Emulsion Technology* (2 volumes), Dekker, New York, 1974.
[91] P. Becher (ed.), *Encyclopedia of Emulsion Technology*, Dekker, New York, 1979.
[92] W.C. Griffin, in [14], 3rd Ed., Vol.8, 1979, pp.900-930.
[93] R.W. Mould, H.B. Silver, and R.J. Syrett, Lubric. Eng., *33*, 1977, 291-298.
[94] W.C. Griffin, J. Soc. Cosmetic Chem., *1*, 1949, 311-326.
[95] E.O. Bennett, Lubric. Eng., *30*, 1974, 549-555.
[96] W.A. Sluhan, in [3] , pp.167-170.
[97] L.M. Prince, in [90], Pt.1, pp.125-177.
[98] J.A. Kitchener and P.P. Musselwhite, in [89], pp.77-130.
[99] J. Heidemeyer, Schmiertechn. Trib., *25*, 1978, 167-169 (German).
[100] Yu.G. Proskuryakov and V.I. Isaev, Trenie i Iznos, *1*, 1980, 891-897 (Russian).
[101] Y. Tamai et al, J. Jpn. Petr. Inst., *22*, 1979, 154-156 (Japanese).
[102] G.L. Maurer, in [3.53], pp.203-227.
[103] J. Edwards and E. Jones, Trib. Int., *10*, 1977, 29-31.
[104] M.S. Vukasovich, Lubric. Eng., *36*, 1980, 708-712; in [3.56], Paper No.66.
[105] H.W. Bernhardt, R. Helwerth and H. Hoffmann, Lubric. Eng., *35*, 1979, 36-41.
[106] J.E. Gannon, E.O. Bennett, I.U. Onyekwelu, and I.N. Izzat, Trib. Int., *13*, 1980, 17-20.
[107] S.M. Reshetnikov, Russian Eng. J., *52* (10), 1972, 68-69.
[108] S. Mateera and I. Glavchev, Trib. Int., *3*, 1980, 69-71.
[109] D.A. Hope, Trib. Int., *10*, 1977, 23-27.
[110] E. Drechsel, Schmierungstechnik, *4*, 1973, 298-302 (German).
[111] F. Günther and J. Pocklington, in [3.56], Paper No.86 (German).
[112] R. Mattison, G.I. Lloyd, and J. Schofield, Trib. Int., *8*, 1975, 253-255.
[113] D.W. Thoenes, P. Hermann, and K. Bauer, Schmiertechn. Trib., *25*, 37-39 (German).
[114] E.R. Austin, in [81].
[115] E.C. Hill, in *Microbial Aspects of Metallurgy*, American Elsevier, New York, 1970, pp.129-155.
[116] E.O. Bennett, Lubric. Eng., *28*, 1972, 237-247.
[117] E.C. Hill, Paper No. MF 79-391, SME, Dearborn, 1979.
[118] B.E. Purkiss, Ind. Lub. Trib., *28*, 1976, 44-49.
[119] R.S. Holdom, Trib. Int., *9*, 1976, 165-170, 225-230, 271-280; *10*, 1977, 155-162, 273-280.
[120] P.R. Yust and G.J.P. Becket, Ind. Lub. Trib., *32*, 1980, 220-225.
[121] R. Schweisfurth, in [21], pp.71-78; in [3.56], Paper No.81 (German).
[122] J. Shintaku et al, in [3.55], pp.245-250.
[123] L.C. Daugherty, Lubric. Eng., *36*, 1980, 718-723.
[124] R.D. King in [3.56], Paper No.83.
[125] K.H. Diehl, in [21], pp.79-109; in [3.56], Paper No.88 (German).
[126] E.O. Bennett, in [3.34], Vol.2, Chap.22, *Cutting Fluids—Microbial Action.*
[127] P.J. Mullins, J.J. Obrzut, and R. Shah, Iron Age Metalwork Int., *11* (4), 1972, 19-21.
[128] I.U. Onyekwelu, E.O. Bennett, and J.E. Gannon, Trib. Int., *14*, 1981, 7-9.
[129] E.O. Bennett, I.U. Onyekwelu, D.L. Bennett, and J.E. Gannon, Lubric. Eng., *36*, 1980, 215-218.
[130] E.O. Bennett, Lubric. Eng., *35*, 1979, 137-144.

[131] H.W. Rossmore, J.F. Sieckhaus, L.S. Rossmore, and D. Defonzo, Lubric. Eng., *35*, 1979, 559-563.
[132] H.W. Rossmore, Paper MR 74-169, SME, Dearborn, 1974.
[133] E.C. Hill, in [3.56], Paper No.82.
[134] T.M. Porter and G.W. Rowe, in [3.56], Paper No.91.
[135] R.A. Nicholson, Ind. Lub. Trib., *27*, 1975, 89-90, 92-94.
[136] W. Tempel, Schmierungstechnik, *4* (3), 1973, 65-68 (German).
[137] G.J.P. Beckert, in [3.56], Paper No.105.
[138] R.A. Nicholson, Trib. Int., *10*, 1977, 17-22.
[139] H. Reich, VDI-Z., *120*, 1978, 359-363; in [3.56], Paper No.72 (German).
[140] Anon., *Code of Practice for Metalworking Fluids,* Institute of Petroleum, London, 1978.
[141] I.S. Morton, Ind. Lub. Trib., *24*, 1972, 267-270.
[142] M.D. Smith and J.E. Lieser, Lubric. Eng., *29*, 1973, 315-319.
[143] M. Rollig, R. Lindner, K. Weber, and R. Fischer, Schmierungstechnik, *6*, 1975, 176-180 (German).
[144] J.S. McCoy, Lubric. Eng., *34*, 1978, 180-186.
[145] G.A. Russ, Lubric. Eng., *36*, 1980, 21-24.
[146] W. Neumann, in [21], pp.42-70 (German).
[147] A. Beerbower, Lubric. Eng., *32*, 1976, 285-293.
[148] V.M. Tikhonov and Yu.G. Litvinova, Machines and Tooling, *42* (2), 1971, 63-64.
[149] G.A. Dorsey, Lubric. Eng., *36*, 1980, 713-717.
[150] E.O. Bennett, Lubric. Eng., *30*, 1974, 128-135.
[151] M.R. Rogers, A.M. Kaplan, and E. Beaumont, Lubric. Eng., *31*, 1975, 301-310.
[152] M. Arnold and R. Schweisfurth, Schmiertechn. Trib., *23* (4), 1976, 85-88 (German).
[153] C. Genner and E.C. Hill, Trib. Int., *14*, 1981, 11-13.
[154] J.E. Gannon and E.O. Bennett, Trib. Int., *14*, 1981, 3-6.
[155] E.L. Kane, Lubric. Eng., *29*, 1973, 391-395.
[156] D.R. Gabe, *Principles of Metal Surface Treatment and Protection,* 2nd Ed., Pergamon, Oxford, 1978.
[157] *Metals Handbook,* 9th Ed., Vol.5, *Surface Cleaning, Finishing, and Coating,* ASM, Metals Park, 1982.
[158] H. Safranck, *The Properties of Electrodeposited Metals and Alloys,* Elsevier, New York, 1974.
[159] R.R. Bolt and D.E. Wobbe, U.S. Patent No.3, 360, 157.
[160] R. Duckett, B.T.K. Barry, and D.A. Robins, Sheet Metal Ind., *45*, 1968, 666-670.
[161] W.D. Bingle, J.A. DiCello, A. Saxena, and D.A. Chatfield, Sheet Metal Ind., *55*, 1978, 1076-1087.
[162] H.A. Shepard, Sheet Metal Ind., *21*, 1945, 85-86.
[163] G. Apel and R. Nünninghoff, Stahl u. Eisen, *100*, 1980, 1247-1253 (German).
[164] G.S. Gardner, Sheet Metal Ind., *18*, 1943, 1229.
[165] D. James and J.E. Haynes, in [3.51], pp.55-57.
[166] S. Currie, Metall. Met. Form., *43*, 1976, 376-378.
[167] P. Loewenstein, Paper No. MF 71-139, SME, Dearborn, 1971.
[168] K. Yoshida et al, in *Sheet Metal Forming and Formability,* Portcullis Press, Redhill, 1978, pp.279-285.
[169] G. Stoumpas, Manuf. Eng. Man., *69* (5), 1972, 20-22.
[170] D.S. Newton and P.D. Winchcome, Sheet Metal Ind., *50*, 1973, 506, 508-513, 565-568.
[171] W.J. Wojtowicz, Paper No. MF 72-516, SME, Dearborn, 1972.
[172] M.P. Sidey, Sheet Metal Ind., *52*, 1975, 329-332, 334.
[173] V. Hopkins and M.E. Campbell, Lubric. Eng., *27*, 1971, 389-392.
[174] A. Munster, Maschinenmarkt, *79*, 1973, 1409-1411 (German).
[175] M. Mrozek, MW Interf., *4* (2), 1979, 10-25.
[176] M. Oyane, R. Yoshinaga and Y. Goto, in *Proc. 18th Int. MTDR Conf.,* Macmillan, London, 1977, pp.339-346.
[177] R.M. Davison and T.I. Gilbert, Lubric. Eng., *32*, 1976, 131-138.
[178] R.B. Zimmerli, Metal Progr., *106* (3), 1974, 101-104.
[179] H.M. Schiefer, B.V. Kubczak, and W. Laepple, in [3.55], pp.57-61; in [3.56], Paper No.41.
[180] R.C. Doehring, Draht, *24* (2), 1973, 70-73.
[181] A.J. Groszek and M.R. Milford, in [3.145], pp.237-239.

[182] I.M. Feng, W.L. Perilstein, and M.S. Adams, ASLE Trans., *6*, 1963, 60-66.
[183] A. Di Sapio and J. Maloney, ASLE Trans., *11*, 1968, 56-63.
[184] M. Nishimura, M. Nosaka, M. Suzuki, and Y. Miyakawa, in [3.145], pp.128-138.
[185] D.J. Baldwin and G.W. Rowe, Trans. ASME, Ser.D, J. Basic Eng., *83*, 1961, 133-138.
[186] J.P.G. Farr, A. Franks, and G.W. Rowe, Lubric. Eng., *36*, 1980, 585-591.
[187] G.W. Rowe, in [3.54], pp.9-13.
[188] M.B. Peterson and R.L. Johnson, Lubric. Eng., *13*, 1957, 203-207.
[189] J. Saga and K. Arita, Metall. Met. Form., *39*, 1972, 323-324.
[190] E.N. Klemgard, Lubric. Eng., *20*, 1964, 184-188.
[191] R.C. Haverstraw, Ordnance, *52* (Sept.-Oct.), 1967, 173-178.
[192] C.R. Cook, Lubric. Eng., *28*, 1972, 199-204, 217-218.
[193] J.A. Rogers and G.W. Rowe, J. Inst. Metals, *95*, 1967, 257-263.
[194] V.I. Zaleskii et al, Kuzn. Shtamp. Proizv., *1965* (6), 1-4 (Russian).
[195] G. Graue, W. Lueckerath, and G. Gebauer, Metal Progr., *84* (2), 1963, 81-83.
[196] O. Pawelski, G. Graue and D. Löhr, in [3.51], pp.147-155.
[197] W.L. Clow, Lubric. Eng., *27*, 1971, 20-27.
[198] J. Saga, H. Nojima, and K. Arita, Trans. Iron Steel Inst. Jpn., *17*, 1977, 623-628.
[199] W.D. Kingery, *Introduction to Ceramics*, Wiley, New York, 1960.
[200] F.H. Norton, *Elements of Ceramics*, 2nd Ed., Addison-Wesley, Reading, MA, 1974.
[201] R.H. Deremus, *Glass Science*, Wiley, New York, 1973.
[202] C.L. Babcock, *Silicate Glass Technology Methods*, Wiley, New York, 1977.
[203] D.C. Boyd and D.A. Thompson, in [14], 3rd Ed., Vol.11, 1980, pp.807-880.
[204] Yu. V. Manegin and I.V. Anisimova, *Glass Lubricants and Protective Coatings for Hot Processing of Metals*, Metallurgiya, Moskva, 1978 (Russian).
[205] T. Spittel and A. Kühnert, Neue Hütte, *10*, 1965, 759-760 (German).
[206] J. Nittel and G. Gärtner, Fertigunsgstech. Betr., *7*, 1965, 441-444.
[207] L. Zagar and G. Schneider, Metal Forming, *36*, 1969, 168-171.
[208] Anon., Draht-Fachz., *26*, 1975, 300-301 (German).
[209] J. Buffet and A. Collinet, Glass Technol., *2*, 1961 (Oct.), 199-200.
[210] A.B. Graham, Metal Ind., *97*, 1960, 480-482.
[211] J.P. Pardoe, Metal Ind., *100*, 1962, 426-429, 446-448.
[212] G.W. Morey, *Properties of Glass*, 2nd Ed., Reinhold, New York, 1954.
[213] M.B. Volf, *Technical Glasses*, Pitman, London, 1961.
[214] Yu. V. Manegin and A.B. Lenyashin, MW Interf., *4* (6), 1979, 10-17.
[215] R.F. Huber, J.L. Klein, and P. Loewenstein, Tech. Rep. AFML-TR-67-79, 1967.
[216] P.W. Wallace, K.M. Kulkarni, and J.A. Schey, J. Inst. Metals, *100*, 1972, 78-85.
[217] J.A. Schey, P.W. Wallace, and K.M. Kulkarni, Lubric. Eng., *30*, 1974, 489-497.
[218] W.D. Spiegelberg, Tech. Rep. AFML-TR-77-87, 1977.
[219] M. Leipold, M. Doner, and K. Wang, University of Kentucky Report, 1973.
[220] H. Scheidler, Klepzig Fachber., *80* (2), 1972, 87-89 (German).
[221] H.J. Oel, Ber. Deut. Keram. Ges., *38*, 1961, 258-267 (German).
[222] K.M. Kulkarni, J.A. Schey, P.W. Wallace, and V. DePierre, J. Inst. Metals, *100*, 1972, 33-39.
[223] F.J. Gurney, A.M. Adair, and V. DePierre, Tech. Rep. AFML-TR-77-208, 1977.
[224] G.H.J. Munro, Light Metals, *19*, 1956, 327-328.
[225] A.I. Denisov, Kuzn. Shtamp. Proizv., *1971* (12), 33-34 (Russian).
[226] T. Biestek and J. Weber, *Conversion Coatings*, Portcullis Press, Redhill, 1976.
[227] H.A. Holden and S.J. Scouse, Sheet Metal Ind., *26*, 1949, 123-134, 136.
[228] W.A. Smigel and H.G. Verner, Iron Steel Eng., *28* (4), 1951, 97-102.
[229] W. Lueg and H. Treptow, Stahl u. Eisen, *72*, 1952, 1207-1212 (German).
[230] T.C. Miller, Wire Wire Prod., *29*, 1954, 843-849, 919; *30*, 1955, 1212-1218, 1297.
[231] A.B. Dove (ed.), *Steel Wire Handbook*, Wire Assoc., Guilford, CT, 1965, Vol.1, pp.93-229.
[232] Anon., Wire Ind., *14*, 1947, 688.
[233] E.L.H. Bastian, Iron Age, *165* (March 9), 1950, 65-69, 86.
[234] C. Voigtlander, Wire Wire Prod., *25*, 1950, 656, 684-685.
[235] C.R. Mehl, Wire Wire Prod., *25*, 1950, 657-658, 685.
[236] G.G. Collins, Wire Wire Prod., *34*, 1959, 968-969, 1016.
[237] F.J. Donaghu and R.G. Rhees, Wire Wire Prod., *37*, 1962, 604-608, 610-611, 686-687.
[238] J. Delille, Stahl u. Eisen, *90*, 1970, 850-854 (German).

[239] W. Lueg and H. Treptow, Stahl u. Eisen, *76*, 1956, 1107-1116 (German).
[240] J.L. Marr and A.D. Eckard, Wire J., *11* (10), 1978, 54-56.
[241] E.L. White and P.D. Miller, Lubric. Eng., *14*, 1958, 124-132.
[242] M.A.H. Howes, Sheet Metal Ind., *37*, 1960, 247-252.
[243] V.I. Brabets, Tsvetn. Met., *1968* (6), 78-79 (Russian).
[244] V.I. Pikaeva, V.M. Korbut, and S.Ya. Veiler, Dokl. Akad. Nauk SSSR, *194*, 1970, 566-569 (Russian).
[245] R. Courtel, P. Gilles, M. LeGressus, and E. Sztrygler, in [3.144], pp.93-102.
[246] E.J. Smith, Iron Steel Eng., *44* (7), 1967, 125.
[247] H. Uchida, O. Yanabu, T. Hada, and H. Sato, Iron Steel Eng., *46* (1), 1969, 75-78.
[248] A. Dürer et al, Sheet Metal Ind., *17*, 1943, 1025-1027.
[249] W. Machu, *Die Phosphatierung,* Verlag Chemie, Weinheim, 1950.
[250] W. Wiederholt, *The Chemical Surface Treatment of Metals,* Robert Draper Ltd., Teddington, England, 1965.
[251] W. Rausch, *Die Phosphatierung der Metalle,* Leuze Verlag, Saulgau, Württ., 1974.
[252] R.B. Waterhouse, Wear, *8*, 1965, 421-447.
[253] P.D. Darkins, Trib. Int., *7*, 1974, 107-112.
[254] D. Schmoeckel, Werkstatt u. Betr., *110*, 1977, 673-736 (German).
[255] H.J. Haupt, wt-Z. Ind. Fertig., *69*, 1979, 555-558 (German).
[256] D. James, Metall. Met. Form., *40*, 1973, 317-318.
[257] P.D. Miller, R.A. Jeffreys, and H.A. Pray, Metal Progr., *69* (5), 1956, 61-64.
[258] D. James, Sheet Metal Ind., *49*, 1972, 35-36, 45-46, 108.
[259] R. Lüders, A. Münster, and H. Wild, Z. f. Wirtsch. Fertig., *65*, 1970, 172-174 (German).
[260] G.H. Pimbley, Metal Progr., *80* (8), 1961, 103-107, 116.
[261] R.E. Pettit, Wire J., *11* (11), 1978, 66-67.
[262] K.D. Nittel, in [3.56], Paper No.43 (German).
[263] E. Schlomach, Schmiertechn. Trib., *20*, 1973, 110-115 (German).
[264] Anon., Metall. Met. Form., *42*, 1975, 44-45.
[265] Anon., Metall. Met. Form., *43*, 1976, 83.
[266] Anon., Metal Progr., *104* (1), 1973, 38-60.
[267] G. Juretzek and H. Rohland, Neue Hütte, *16*, 1971, 353-356 (German).
[268] C. Sudo, M. Takatani, T. Masui, and H. Nagai, Wire J., *13* (12), 1980, 74-79.
[269] J.R. Jones and M.N. Gardos, Lubric. Eng., *27*, 1971, 393-400.
[270] W. Herold, Met. Constr. Mecan., *102*, 1970, 389-398 (French).
[271] Anon., Wire Wire Prod., *45* (9), 1970, 67-68.
[272] J.R. Gordon, Lubric. Eng., *35*, 1979, 687-691.
[273] J.H. Netherstreet, Sheet Metal Ind., *48*, 1971, 287-288.
[274] M. Opachak (ed.), *Industrial Fluids: Controls, Concerns, and Costs,* SME, Dearborn, 1982.
[275] R.L. Leslie and H.J. Sculthorpe, Lubric. Eng., *28*, 1972, 165-167.
[276] M.M. Patterson, Lubric. Eng., *26*, 1970, 458-461.
[277] W.A. Sluhan, Cutting Tool Eng., *27* (5/6), 1975, 5-8.
[278] M.D. Smith and C.H. West, Lubric. Eng., *25*, 1969, 321-325.
[279] J.J. Joseph, Am. Mach., *115* (5), 1971, 75-83.
[280] R.L. Quanstrom, Lubric. Eng., *33*, 1977, 14-19.
[281] B.L. Nehls, Lubric. Eng., *33*, 1977, 179-183.
[282] G. Runge, in [3.56], Paper No.35 (German).
[283] P. Wehber, in [21], pp.132-162 (German).
[284] J.P. Wettach, Lubric. Eng., *16*, 1960, 410-413.
[285] L.V. Khudskin and E.M. Bulyzhev, Russian Eng. J., *57* (11), 1977, 40-43.
[286] Yu.V. Polyanskov, Russian Eng. J., *55* (10), 1975, 70-73.
[287] E. Hofling and J. Oertle, Lubric Eng., *37*, 1981, 195-197.
[288] B. Sismey, Ind. Lub. Trib., *23*, 1971, 73-79.
[289] J.P. Wettach, Iron Steel Eng., *43* (8), 1966, 117-123.
[290] A. El Hindi, Iron Steel Eng., *45* (3), 1968, 112-116.
[291] K.A. Brooks, Jr., Lubric. Eng., *30*, 1974, 542-548.
[292] J.J. Joseph, Manuf. Eng. Man., *67* (6), 1971, 23-25.
[293] K.J. Daniells, Sheet Metal Ind., *42*, 1965, 194-196.
[294] S.D. Pinto, in [3.52], pp.129-134; in [3], pp.236-239.

[295] G.R. Fritsche and S.J. Sansonetti, Lubric. Eng., *36*, 1980, 154-159.
[296] R.H. Brandt, Lubric. Eng., *28*, 1972, 254-257.
[297] R.G. Hitchcock, in [3.53], pp.77-83.
[298] F.S. Peterson, Lubric. Eng., *29*, 1973, 534-538.
[299] *Technological Impact of Surfaces,* ASM, Metals Park, 1982.
[300] R.W. Clement, Lubric. Eng., *34*, 1978, 358-363.
[301] W.L. Archer, Metal Progr., *106* (4), 1974, 133-146.
[302] R.G. Bertrand and M.S. Vukasovich, Lubric. Eng., *33*, 1977, 538-543.
[303] A. El Hindi, Lubric. Eng., *34*, 1978, 37-41.
[304] Anon., Lubric. Eng., *30*, 1974, 486-487.
[305] W.O. Heyn, Lubric. Eng., *32*, 1976, 646-650.
[306] E.G. Paulson, Lubric. Eng., *24*, 1968, 508-513.
[307] G.P. Canevari, ASLE Trans., *12*, 1969, 190-198.
[308] H.E. Willoughby, Tribology, *2*, 1969, 221-224.
[309] H.J. Bradke, Metall., *33*, 1979, 267-272 (German).
[310] H. Reich, in [21], pp.163-177 (German).
[311] D. Kwasnoski, Int. Met. Rev., *20*, 1975, 137-145.
[312] E.H.F. Stone, Int. Met. Rev., *7*, 1972, 227-239.
[313] E. Nordell, *Water Treatment for Industrial and Other Uses,* 2nd Ed., Reinhold, New York, 1961.
[314] N.L. Nemerov, *Theories and Practices of Industrial Waste Treatment,* Addison-Wesley, Reading, MA, 1963.
[315] W.K. Mann, H.B. Shortly, and R.M. Skallerup (ed.), *Industrial Oily Waste Control,* Am. Petr. Inst., New York.
[316] S.E. Jorgensen, *Industrial Waste Water Management,* Elsevier, Amsterdam, 1979.
[317] M.R. Hockenbury, J.M. Bower, and A.W. Loven, in *Proc. 31st Ind. Waste Conf.,* Ann Arbor Science Publ., Mich., 1976, pp.982-993.
[318] W.N. Schierven, Jr., Wire J., *4* (6), 1971, 67-71.
[319] J.D. Reynolds, in *2nd Int. Aluminum Extrusion Seminar*, Aluminum Association, Washington, 1977, Vol.1, pp.7-9.
[320] A. Evans, Lubric. Eng., *24*, 1968, 521-524.
[321] C.J. Staebler and B.F. Simpers, in [3.52], pp.135-140.
[322] R. Nebolsine, Iron Steel Eng., *47* (8), 1970, 85-92.
[323] R.S. Patton, F.G. Krikau, and R.J. Wachowiak, Iron Steel Eng., *48* (3), 1971, 98-107.
[324] C. Broman, Blast Furn. Steel Plant, *59* (1), 1971, 19-21.
[325] R.W. Clyne, Lubric. Eng., *21*, 1968, 514-520.
[326] U.J. Moller, U. Boor and G. Runge, Erdöl u. Kohle, *27*, 1974, 70-77 (German).
[327] R. Geutskens, C.P. Heijwegen, and K.J. Smits, Stahl u. Eisen, *99*, 1979, 143-147 (German).
[328] H.R. Hockenberry and J.E. Lieser, Lubric. Eng., *33*, 1977, 257-261.
[329] C. Versino, L. Fogliano, and C. Folonari, in [3.53], pp.189-193.
[330] E.O. Bennett, Lubric. Eng., *29*, 1973, 300-307.
[331] K.W.A. Chater and H.J. Somerville (ed.), *The Oil Industry and Microbial Ecosystems,* Heyden, London, 1978.
[332] H.R. Hockenberry and J.E. Lieser, Lubric. Eng., *29*, 1973, 7-11.
[333] E.G. Ellis, Ind. Lubric., *19*, 1967, 141-145.
[334] T.H.F. Smith, Lubric. Eng., *25*, 1969, 313-320.
[335] D.M.W. Hutcheson, in [81].
[336] M.D. Kipling, Trib. Int., *10*, 1977, 41-46.
[337] F.H. Schneider, G. Seixas, and H.W. Rossmore, in [3.55], pp.241-244.
[338] G.A. Blodgett, in [3.55], pp.251-259.
[339] G.A. Gellin, Lubric. Eng., *25*, 1969, 310-312.
[340] G. Hodgson, J. Inst. Petr., *59*, 1973, 1-8.
[341] O. Schmid, in [21], pp.110-131.
[342] E. Rietschel, in [3.56], Paper No.85 (German).
[343] C.J. Spuij, in [3.56], Paper No.84.
[344] L.B. Bourne *Skin Disease from Oil: Causes and Prevention,* Mason House, London, 1969.
[345] A.A. Fischer, *Contact Dermatitis,* 2nd Ed., Lea and Febiger, Philadelphia, 1973.
[346] V.A. Drill and P. Lazar, *Cutaneous Toxicity,* Academic Press, New York, 1977.

[347] J.B. Olishifski (ed.), *Fundamentals of Industrial Hygiene,* 2nd Ed., National Safety Council, Chicago, 1979.
[348] G.D. Clayton, in [14], 3rd Ed., Vol.13, 1981, pp.253-277.
[349] T. Mang and H. Kräner, Mineralöltechnik, *23* (7/8), 1978, 1-38 (German).
[350] M. Körner, Schmierungstechnik, *7* (1), 1976, 6-7 (German).
[351] G. Lloyd, G.I. Lloyd, and J. Schofield, Trib. Int., *8*, 1975, 27-29.
[352] D. Groenewegen, Schmiertechn. Trib., *25*, 1978, 140-142 (German).
[353] T.Y. Fan et al, Science, *196* (4), 1977, 196-197.
[354] C. Evans, Trib. Int., *10*, 1977, 47.
[355] Anon., MW Interf., *3* (1), 1978, 9-16.
[356] M. Sitting, *Toxic Metals,* Noyes Data Corp., Park Ridge, NJ, 1976.
[357] L. Friberg, G.F. Nordberg, and V.B. Vouk (ed.), *Handbook on the Toxicology of Metals,* Elsevier, Amsterdam, 1979.
[358] Z. Steininger, Wire Ind., *38*, 1971, 798-801.
[359] A.H. Goldsmith et al, Am. Ind. Hygiene Ass. J., *37*, 1976, 217-226.
[360] J.B. Peace, Paper No. MF 73-1193, SME, Dearborn, 1973.
[361] E.R. Plunkett, *Handbook of Industrial Toxicology,* Chemical Publ. Co., New York, 1976.

5

Measurement Techniques

In order for tribological processes to be understood and quantified, a great variety of measurements must be made. Some techniques are quite universal and have been adequately described in other publications [1]; these will be touched upon only briefly. Others, however, are unique to metalworking, and these will be discussed in detail. Many points previously covered by Schey [2] are incorporated here, too.

5.1 Principles of Simulation Testing

The tribology of metalworking is most reliably studied, and lubricants are best evaluated and developed, in an actual production situation. There are, however, powerful incentives to develop laboratory-scale tests. Full-scale trials interrupt production and, especially if a lubricant fails, may carry an exorbitant price tag because the product is ruined. There is also the basic problem that uncontrollable and often unrecognized variables may creep into the evaluation. Therefore, a great deal of effort has been expended in attempts to develop laboratory tests that are, hopefully, less expensive and easier to control and evaluate while still retaining their relevance to production.

Periodically, a test method emerges with some claims to universality, only to fall into disrepute as experience accumulates. Considering the complexity of the system (Fig.3.2), it is clear that no universal test will ever be found. The best that one can hope for is to establish tests that simulate particular aspects of the system, and then select a group of these tests that will provide adequate coverage of all aspects of importance.

As should any other sound engineering decision, the choice of simulation tests should always be based on a consideration of the whole system [2,3]. Indeed, the subject can be treated on the basis of systems analysis [3.22]. For metalworking applications, we shall follow a less formal but no less systematic line of reasoning.

1. The first task is to establish clearly the purpose and scope of testing. Quite often there are several purposes, but they usually can be separated and assigned different weights.

a. The very first step is to create a reference base which also serves as a link to the actual production process. For existing processes this is represented by a well-established lubricant and die, combined with process conditions that are known to result in acceptable quality in the finished product. For new processes, reference points are established in the course of experimentation. The reference base must be quantified—for example, in terms of reduction capacity, coefficient of friction, wear rate, surface roughness and appearance, and other measurable outputs. If possible, conclusions regarding the operative lubrication mechanism are drawn.

b. If the purpose of study is an understanding of the process itself, the reference lubricant and dies are retained, but process variables such as reductions, speeds, temperatures, interface geometry, etc. are systematically varied, and consequent changes in outputs (forces, pressure, friction, wear rate, surface quality of the product) are observed and interpreted in terms of changes in the lubricating mechanism. Indiscriminate changes in variables would lead to a mass of useless data. Therefore, the scope of testing must be limited; the safest procedure is, of course, to change only one variable at a time. If several variables are to be investigated, rules of the statistical design of experiments must be followed, so that the results can be subjected to multiple regression analysis (for examples, see Chapters 9 and 11).

c. If new lubricants or die materials are to be evaluated for purposes unrelated to lubrication mechanisms (e.g., securing new sources of supply, reducing costs, or eliminating undesirable side effects such as smell, toxicity, difficulty of handling, etc.), the established quality of the issuing product must be maintained. If the lubrication mechanism is allowed to change, checks must be made to ensure that the quality of the issuing product is acceptable.

d. If the purpose is an evaluation of lubricants or die materials with the aim of increasing production rates through higher operating speeds or heavier reductions, or of reducing die wear, friction, and power requirements, or of improving the quality of the issuing product, the complexity of the task becomes formidable. The lubrication mechanism will undoubtedly change, and one of the purposes of testing may actually be to establish the operating limits of lubrication. Continuous monitoring of measurable process variables and of the quality of the issuing product is essential.

2. The above reasoning presupposes that something, or rather quite a lot, is already known about the tribology of the production process. Presumably, experience has been systematically built up regarding the process and the influencing variables; observations can be interpreted against existing knowledge (Chapters 6 to 11); dies (tools) and issuing workpieces and, in particular, their surfaces can be examined for evidences of the operative mechanisms; force recordings can be evaluated for changes which point to transients and trends in the lubrication mechanism. Very often trends are more important than absolute values. If, for example, forces rise in the course of a production run, it is likely that the cumulative effect of all variables (e.g., rise in die temperature, accumulation of wear particles, irreversible changes in lubricant composition, incipient pickup on dies, etc.) will lead to poorer lubrication and, possibly, total breakdown. Gradually dropping forces, in contrast, indicate a "running-in" effect, and very long, trouble-free runs are then possible: obviously, the lubrication system is successful. If a lubrication system is to be found for a new process, well-educated assumptions must be made about the lubrication mechanisms to be expected.

3. The purpose of simulation determines what variables must be kept constant; from the preceding discussions it is obvious that, for some variables, no compromise can be made at all:

a. The workpiece material, the die material, and the lubricant are fixed if the effects of process variables are to be explored.

b. The workpiece and die materials are constant if lubricants are to be evaluated or developed, but process conditions must be changed until a surface finish and/or wear rate representative of the reference lubricant of known full-scale performance is obtained. The same reference line is used when die materials are to be evaluated; however, the lubricant may then have to be changed to allow for material interactions (Sec.3.11.6).

c. We have seen (Sec.3.8) that surface roughness greatly affects the formation and breakdown of solid films and determines the degree of boundary versus PHD lubrication in the mixed-film regime. Therefore, the surface finish of the die must be reproduced, and the direction of sliding relative to the direction of surface finish must be maintained. It could be argued that surface roughness should be scaled down in proportion to other dimensions, but this is a sound proposition only if film thickness ratios are maintained. For this, the effects of macrogeometry and other variables on the lubrication mechanism must be known.

d. Because it is relative velocity that establishes fluid lubricant films, the principal directions of movement must be retained. Thus, a process in which normal approach between die and workpiece establishes a squeeze film cannot be simulated by some other test in which sliding motion generates a hydrodynamic film, and vice versa.

4. Other variables will have to be changed if the simulation test is to be different from the process itself, or if certain trends are to be emphasized:

a. In many instances the test setup will be physically smaller than the real process. By proper choice of test geometry and process variables, the operative lubrication mechanism can be reproduced. However, the heat balance of the process is likely to be upset no matter what precautions are taken, and conditions more severe than intended are inadvertently created.

b. Contact geometry determines, to a large degree, the rate at which a liquid lubricant is carried into the interface. If the test geometry is more favorable to lubricant entrapment, the mechanism shifts toward PHD lubrication and the influence of additives may be lost; in contrast, if the geometry is less favorable, a more critical distinction between additives is possible, and tendencies for wear become more evident.

c. Velocity interacts with contact geometry, and higher speed can be used to bring a mixed-film mechanism back to that existing in production. Alternatively, speed can be intentionally changed: slowing down results in shifting toward the boundary regime, thus revealing additive effects and wear, whereas speeding up has several consequences. Initially, lubrication shifts toward predominantly PHD lubrication, but then heating thins the film (Fig.3.38) and, finally, temperatures may be high enough to cause breakdown of the lubricant, with consequent severe pickup, scuffing, and wear. Indeed, the speed at which breakdown occurs can be used as a measure of the critical temperature of the lubricant (Sec.3.5.1).

d. The effect of interface pressure depends on whether enough lubricant is present to fill out the surface roughness features (Fig.3.44). This is, obviously, also connected with the mode of lubricant application, test geometry, and sliding speed, and great care must be taken to ensure that the combination of all these variables results in the right lubrication mechanism.

e. Generation of new surfaces is a crucial aspect of metalworking tribology. In general, the simulation test should give the same surface extension as the process; however, this is a condition one is often forced to violate. Fortunately, extended sliding over the same contact area results in a cumulative exposure of new surfaces, and this is sometimes equivalent to a large surface extension in a single step. However, interpretation of results calls for great caution and substantial experience.

f. Under production conditions, there are long-term changes in die surface finish, in lubricant composition and rheology, and in wear debris accumulation. These are, in general, very difficult to reproduce in the laboratory. Accelerated testing is possible but carries the danger that the lubrication mechanism may become unrepresentative.

g. There is also the danger that surface finish and other process conditions may gradually change in the course of simulative testing; the changes may be sufficient to affect results yet not obvious enough to be detected. Therefore, a cardinal rule of all simulative testing is that a reference line must be established under standardized conditions with pedigreed workpiece materials and lubricants. After every 5 (or, at the most,

10) experimental runs the test must be repeated. If the results cannot be reproduced, two possibilities exist: first, the test may be basically irreproducible and all data must be discarded; second, there may be a gradual and unidirectional drift which can be allowed for by referencing against the drifting base line.

5.2 Reduced-Scale Metalworking Processes

The costs associated with some processes are reasonable enough to allow a virtual replication of a production process to be set up in the laboratory. After proper instrumentation, the ideal test equipment is available. This is particularly true of some metalcutting processes. However, other processes are practiced on too large a scale to permit such an approach. At first sight, it would appear that a straight scaling down of the process should create no problems. Indeed, there are formalized rules of scaling which can be helpful [4], and dimensional analysis can be applied [5]. There are, however, difficulties:

1. As a rule of thumb, it is advisable to scale the geometry of the process linearly. In other words, the ratio of roll diameter to strip thickness should be maintained in rolling, the ratio of diameter to blank thickness should be maintained in deep drawing, etc. This neutralizes process geometry as a variable. However, there are limits to this approach:

a. In cold working, a very thin specimen can create problems that are absent in the full-scale process. Thus, a very thin specimen is less stiff, buckles more readily, and is also difficult to handle; in addition, the required surface roughness may take too large a portion of specimen thickness. The proportion of friction work to deformation work is larger, and performance is not matched when the results are transferred to production [6]. Test conditions must then be adjusted, but this requires full knowledge of the lubrication mechanism.

b. In hot working the problems are greater yet. A small specimen has a much larger surface-area-to-volume ratio than a production workpiece. Consequently, it loses more heat during transfer from the furnace, and is also chilled more rapidly on contact with the colder die. Contact time can be reduced by use of higher deformation velocities, but this undoubtedly changes the lubrication mechanism and also increases the flow strength of the material (Sec.2.1.3). Interface pressures and forces can become completely irrelevant.

2. Even if long-term changes in process variables cannot be reproduced, at least the operating temperatures must be typical of the steady-state operating conditions (unless the effects of start-up are to be studied). In cold working the die must be preheated to the typical production die temperature, often on the order of 150 C. In hot working, much higher temperatures may be necessary. Similarly, the lubricant must be applied at the temperature typical of production conditions.

3. There is often difficulty in reproducing the speeds that prevail in the real process, because the size and cost of drive units may become exorbitant. Therefore, one is often forced to work at lower speeds. This can be permitted only if its consequences are fully understood. In the presence of fluids, lubrication shifts toward the predominantly boundary regime, and heat generation in the workpiece and at the interface, and the distribution of this heat between the die and the workpiece, also change. Appropriate adjustments in process variables will have to be made. If the purpose of experimentation is the evaluation of lubricants, test conditions first must be established with the reference lubricant, and then must be changed until the same surface finish is obtained on the product.

5.3 Transfer from One Process to Another

Scaled-down operations quite often are still too expensive, or demand equipment or measurement techniques that are too complicated, to present an attractive choice for all lubricant research. For this reason, many attempts have been made to perform the majority of tests with a simplified or cheaper metalworking process and then transfer the results to the more expensive production technique of interest. Indeed, some small-scale metalworking techniques that can be performed on inexpensive equipment or that are readily evaluated in terms of the friction component have been suggested as general simulation tests. Success depends on how far the lubrication mechanisms are comparable, as shown by some published examples in the field of bulk deformation:

1. Deformation takes place in a converging gap in both rolling and wiredrawing, and both are steady-state processes, although the speed relations are different (Fig.5.1). Wire can be drawn in the laboratory, even on a tension testing machine. Thus, low-speed wiredrawing has been used for evaluating palm oil substitutes for cold rolling of steel [7] and for brass rolling emulsions [8], and the order of merit was correctly predicted. Obviously, success must have been due to the very limited aim; at the very low speeds used in wiredrawing, only boundary lubrication performance could have been measured. In the work of Williams and Brandt [9] the wire was drawn with heated lubricants; the temperature at which the lubricant broke down gave the same order of merit as the over-all performance in rolling, presumably because the critical temperature of the lubricant was the common criterion. No correlation between coefficient of friction values derived from rolling and from room-temperature wiredrawing could, however, be found.

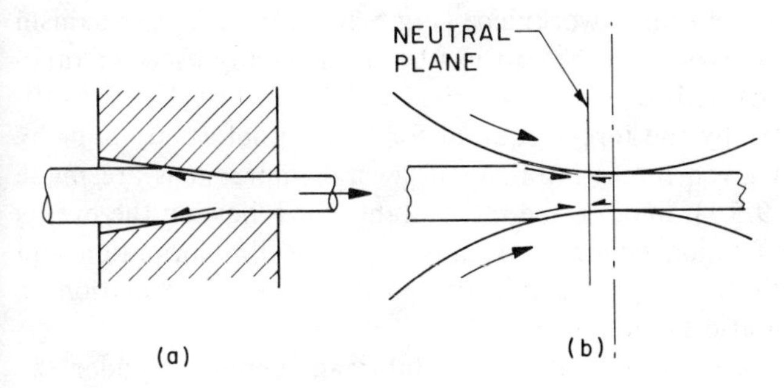

Fig.5.1. Similarities and differences between (a) wiredrawing and (b) rolling

That many of the reported cases of agreement must have been fortuitous was shown by the critical assessment of Lueg and Dahl [10], who tested selected emulsions and dispersions (including palm oil) in carefully controlled experiments. The order of merit in wiredrawing was judged from the level and steadiness of drawing forces; in rolling, the minimum obtainable gauge in eight passes was taken as an indicator of lubricant quality. There was no correlation between the two sets of results, no doubt because lubrication was predominantly boundary in wiredrawing and hydrodynamic in rolling.

2. Forging (upsetting; Fig.5.2a) can be done on any press or compression testing machine, and was among the earliest lubricant evaluation methods; a lower force is taken to indicate a better lubricant.

Much more attractive is ring compression (Fig.5.2b), where a lesser contraction (or greater expansion) of internal diameter gives a clear indication of lower friction (Sec.9.1.2). Because of its simplicity, ring compression has been advocated as a general test method [11]. As in upsetting, the lubricant forms a squeeze film which then becomes entrapped. Strictly speaking, one can expect to simulate only non-steady-state

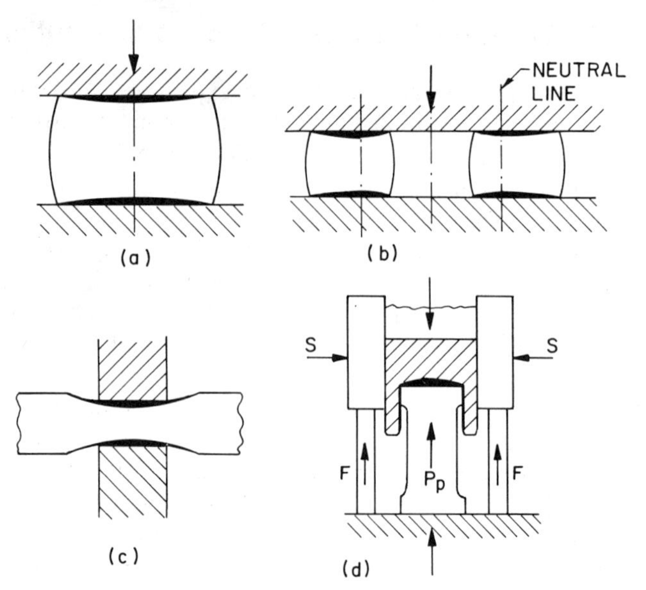

Fig.5.2. Squeeze films in (a) upsetting, (b) ring compression, (c) plane-strain compression, and (d) back-extrusion

processes with normal surface approach, and lubricant breakdown does not necessarily result in severe pickup because sliding is limited.

Plane-strain compression is more severe and has some steady-state characteristics, because the squeeze film established at the time of contact is gradually washed out with the expanding surface of the deforming workpiece (Fig.5.2c). Thus, the mechanism is much closer that in rolling (Fig.5.1b), and the test has been repeatedly used for ranking of rolling lubricants (Sec.6.3.6). It is common experience with this test that the efficiency of lubrication (as measured by the force required for a given deformation, or by the deformation obtained with a given force) improves if several impressions are made without cleaning the anvils (Sec.9.5.1), making the test suitable for following the events that lead to the formation of a friction polymer. Because squeeze-film formation and feed-out of the lubricant are both viscosity- and velocity-sensitive, some information on hydrodynamic characteristics can also be obtained.

Back-extrusion (Sec.9.2) is particularly sensitive to lubricant depletion under the punch and is often used as a small-scale test. In a plane-strain variant, Kapitány and Voith [12] measure the punch force P_p, normal force S, and frictional force F (Fig.5.2d), thus obtaining information on the effects of process variables.

3. Transfer of data from steady-state to non-steady-state processes is even more doubtful. Using lubricants representing different mechanisms, Schey [13] found different orders of merit in different tests. For example, with aluminum alloy 7075 as the workpiece material (Fig.5.3), a high-viscosity mineral oil (M.O.) gave low friction in ring compression, whether on its own or compounded with oleic acid (O.A.), chlorinated paraffin (C.P.), or graphite (GR). The squeeze-film effect and lack of severe sliding gave good results also with water and with a dry graphite film. In contrast, plane-strain compression indicated breakdown of the graphite film and gave very high friction with water. Stresses measured in low-speed wiredrawing showed no direct correlation with either of the tests. The more favorable process geometry allowed the mineral oil plus graphite lubricant to survive, but the dry graphite film broke down as it did in plane-strain compression, and the wire actually fractured during drawing with water. Obviously, these tests cannot be used even for preliminary screening of lubricants unless due

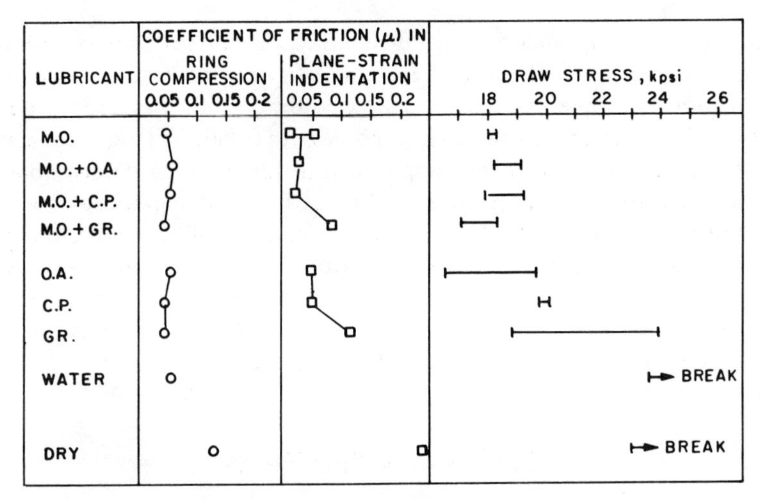

Fig.5.3. Comparison of lubricant performance in ring compression, plane-strain indentation, and wiredrawing of aluminum alloy 7075 [13]

allowance is made for differences in geometry and process conditions, so that the effects of such differences on the anticipated lubrication mechanism can be judged. Similarly, good agreement among ring compression, rolling, and drawing was found by Male [14] with dry lubricants and thick films, but agreement was poor when mixed-film lubrication prevailed. Kapitány and Voith [12] observed only small differences in upsetting, plane-strain compression, and ring compression of aluminum and copper under dry and lanolin-lubricated conditions, even though back-extrusion of aluminum under dry conditions resulted in high forces and die pickup. The difficulty of transferring results from one process to another is shown also in Fig.5.4 [15], where the very large differences in order of merit among the various processes are clearly visible.

4. It might appear that the less severe conditions existing in many sheet metalworking operations would make transfer of data easier, but this expectation is not borne out by results (Sec.5.4).

5. Metalcutting and abrasive machining present interface conditions quite different from the above, and the mechanisms by which certain lubricants are effective in these

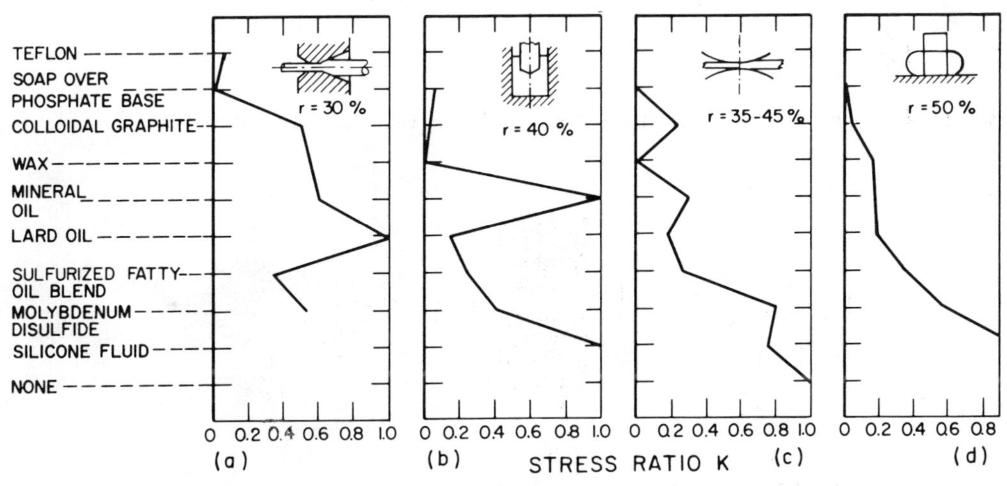

Fig.5.4. Relative performance of lubricants in (a) wiredrawing, (b) piercing, (c) rolling, and (d) axial upsetting [15]

processes are insufficiently understood. Fortunately, cutting tests are often easier to perform than deformation tests. However, reproducibility can be ensured only by the strictest control of tools, workpieces, and process conditions, and in this respect cutting tests are no simpler. Transfer from one cutting process to another is no easy matter either, for very much the same reasons that apply to transfer between deformation processes. Certain limitations are even more drastic for cutting. Cutting speed is a particularly powerful variable: at low speeds the lubricating qualities of the lubricant/coolant prevail, whereas at higher speeds the cooling qualities come to the fore, and extrapolation from one regime to the other is not possible. Furthermore, the importance of lubricating and cooling functions depends also on the process, and transferability of results from one process to another is generally poor. These aspects will be more fully discussed in Chapter 11.

5.4 Simulation Tests

Several techniques have been developed which are not practical deformation processes but are designed to embody specific tribological aspects. The incentive for developing these test methods is the desire to have better control over process variables, and tests are usually designed to facilitate quantitative measurements of friction and wear. Their common feature is that τ_i (or μ) can be obtained without knowing σ_f and without recourse to theory. Partial reviews of this subject have been published [2,16-19].

5.4.1 Tests Involving Bulk Plastic Deformation

A common feature of these tests is that new surfaces are generated by means of deforming the specimen through its entire thickness.

Plane-Strain Drawing Tests. A flat bar or strip of thickness h_0 is drawn between inclined die surfaces to a thickness h_1. The strip is free to spread to the side, and lubricant can escape in that direction, too. Thus, the technique is different from wiredrawing, but offers several advantages:

a. The converging gap may be produced by two rollers (Fig.5.5a). The strip is first drawn with the rolls freely rotating, and then with the rolls clamped. The difference in measured draw forces is approximately equal to the frictional force F. From Eq.2.8,

$$F = \tau_i A_a \tag{5.1}$$

where A_a is the contact area with the die and τ_i is the interface shear strength [20].

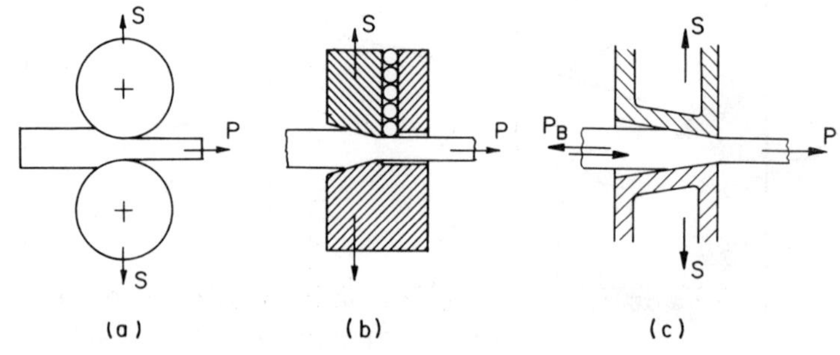

Fig.5.5. Plane-strain drawing (a) between rolls, (b) with low-friction die supports, and (c) with a deflecting die frame and provision for push-pull operation

b. By keeping the rolls clamped, but measuring the separating force S as well as the drawing force P, an average coefficient of friction can be calculated if a number of simplifying assumptions are accepted [21]. The metal between the rollers may be regarded simply as a truncated wedge with an apex half angle α where

$$\tan \alpha = \left(h_0 - h_1\right)/2L \tag{5.2}$$

In calculating the contact length L, elastic deformation must be considered (see Sec.6.1). The coefficient of friction is then

$$\mu = \frac{(P/2S) - \tan \alpha}{1 + (P/2S) \tan \alpha} \tag{5.3}$$

An advantage of this test is that the strip does not need to be pointed; instead, the dies can be closed under sufficient force to indent the strip prior to commencing the draw. If the roll is narrower than the strip, only part of the width is indented [22]. A common problem is that one of the rolls must be allowed to move so that the separating force S can be measured; friction on the guides introduces errors.

c. The problem exists also when wedge-shape dies are used [23], but can be alleviated by various ingenious design solutions. In the design of Pawelski [24], the moving die half is supported against a flat roller bearing (Fig.5.5b). The true included die angle is continuously monitored with an arm bearing on two linear differential transducers, and thus errors due to elastic deflections are eliminated. The designs of Kudo et al [25] and of Wilson and Cazeault [26] utilize the deflection of the die supports for sensing the normal and frictional force components. The main virtue of these devices is that varying reductions can be taken with dies of different angles, and thus both lubricant and material parameters can be evaluated. The test becomes even more versatile when the strip, on entering the die gap, is subjected to pulling or pushing (Fig.5.5c) [25]. This offers a unique opportunity for varying the interface pressure without changing the die angle or reduction. To preserve the surface features of the drawn sheet, the workpiece may be entered at an angle [25]. In the design of Devenpeck and Rigo [27], drawing speeds can be made high enough to allow study of PHD effects and, if a long enough strip is drawn, wear.

d. Tests involving one-sided reduction (Fig.5.6) subject the two faces of the strip to different conditions. When the die set is composed of one flat and one angled surface (Fig.5.6a), an approximate coefficient of friction may be derived, at least for small die angles, simply from the ratio of pull force P to die separating force S: $\mu = P/2S$ (for larger angles, see [27]). The hidden assumption, seldom valid, is that friction is the

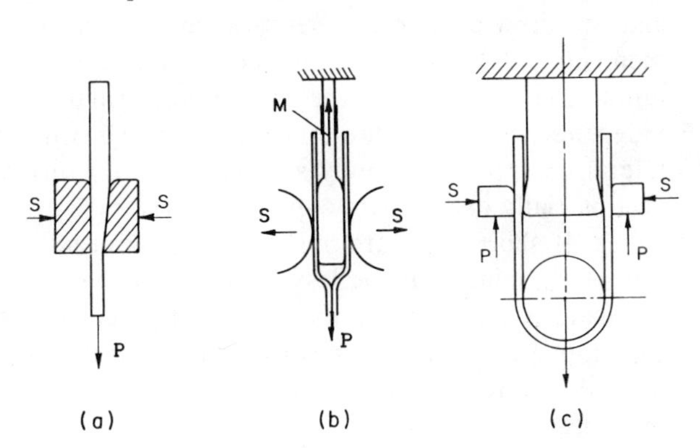

Fig.5.6. Plane-strain drawing with one-sided reduction

same on the parallel and angled die surfaces. Although the intent is often the simulation of ironing, the test actually simulates conditions prevailing in drawing of a tube over a stationary plug. However, for this purpose the die set of Lancaster and Rowe [28] is much more suitable, because the frictional force over the flat die can be separately measured (M in Fig.5.6b) and errors from angular misalignment cannot creep in. The same is true of the version developed by Fukui et al [29], in which a bent strip is drawn by a pin through a pair of dies composed of an inner die, with inclined draw surfaces, and two flat outer dies (Fig.5.6c). In both tests the die- separating force S and draw force P are measured independently; therefore, $\mu = P/S$.

e. In yet other versions of plane-strain drawing, the strip is wrapped around a punch which constitutes the flat die surface (Fig.5.7). It is, therefore, possible for the strip to move at different velocities over the bar and the outer roller or inclined die. Thus, the test closely reproduces the conditions that exist in ironing and in drawing over a bar. The coefficient of friction over the punch is then approximately given by $\mu = P/2S$ (Fig.5.7a). In a modification of the same principle by Veiler and Likhtman [20], one of the cylindrical dies is allowed to rotate freely, and thus the die-side frictional force P_d is transmitted by a flexible band to a load cell (Fig.5.7b). The punch is in two pieces; a well-fitting but freely moving insert in the cutout of the punch transmits the punch-side frictional forces P_p through a pressure pin to another load cell (Fig.5.7c). To avoid the extrusion of material into the gap between punch and insert, the specimen is first partially deformed with a solid punch, and then the draw is completed with the special punch. Friction between the workpiece and the die is $\mu_d = P_d/2S$, and friction between the metal and the punch is $\mu_p = P_p/2S$.

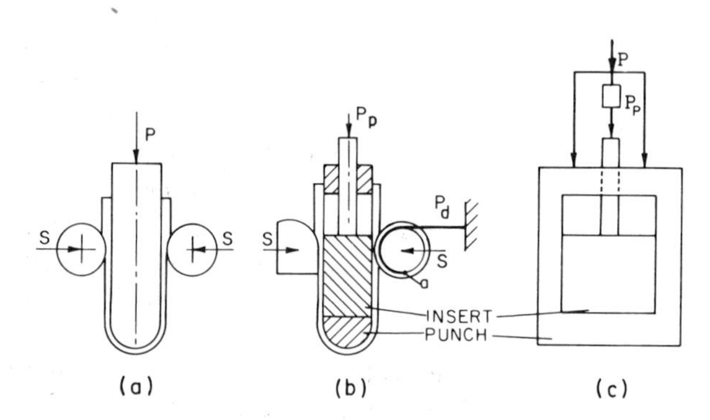

Fig.5.7. Plane-strain drawing over a stationary punch

f. A plane-strain drawing test of special configuration, designed by Sachs [30], is now used extensively. A sheet specimen is cut with a wedge-shape portion which is then drawn through a die consisting of two flat grips (representing the blankholder surfaces) and two draw die surfaces set at a small included angle to reduce the width of the specimen (Fig.5.8a). Greatly improved mechanical designs, such as that of Kawai and Nakamura [31], ensure that the measured normal force Q and die separating force S are independently and accurately measured. Drawing between flat faces with simultaneous lateral compression closely simulates the conditions prevailing in the blankholder zone of a partially deep-drawn cup (Fig.3.5a).

g. If short plugs are pushed rather than pulled through a converging gap, deformation of the plugs will be less constrained than if they were parts of a flat bar, and the lubricant is subjected to more severe duty because of interruptions in contact (Fig.5.8b). Developed by Pawelski et al [4.196], this test can be used also for evaluating hot-working lubricants and, if a sufficiently large number of specimens is pushed through, pickup and wear [32].

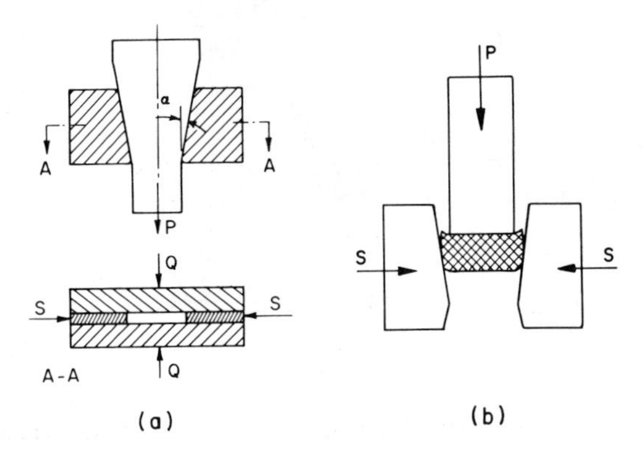

Fig.5.8. Pulling (a) and pushing (b) in plane strain

Compression With Simultaneous Sliding. In the form proposed by Pawlow [33], a cylinder is compressed between parallel die faces while being pulled sideways between the closing dies (Fig.5.9a). Since the frictional force is distributed over two end faces, the average coefficient of friction is $\mu = F/2P$. In the form developed by Saga et al [3.326], a bar representing the die is pulled between workpiece inserts (Fig.5.9b). The device of Nittel [34] moves one of the anvils on a wedge surface, and thus the specimen is compressed while it slides over one of the end faces (Fig.5.9c). The coefficient of friction is simply $\mu = F/P$. This test should be suitable primarily for evaluating dry and solid-film lubricants and the tendency toward die pickup.

All the above simulation tests induce plastic flow over much of the workpiece volume, albeit at widely ranging interface pressures. This aspect has been discussed by Lange and Gräbener [35].

5.4.2 Tests Involving Partial Plastic Deformation

Tests that involve partial plastic deformation are one step farther removed from actual metalworking processes and therefore are generally useful for simulating only selected aspects of the lubrication or wear mechanism.

Plane-strain drawing tests differ from those illustrated in Fig.5.5 to 5.9 in that plastic deformation is intentionally limited and is very often restricted to asperities only. There has been a revival of interest in these techniques for sheet metalworking simulation because the drive for lighter and stronger car bodies necessitates the use of high-strength low-alloy steels, aluminum alloys, and other sheet materials for which, in the absence of shop experience, lubricants have to be developed rapidly and with some assurance that they will perform satisfactorily in production.

a. In the strip drawing test, first used by Sachs [30] and further developed by Wojtowicz [36], a strip is drawn between two opposing flat die surfaces somewhat wider

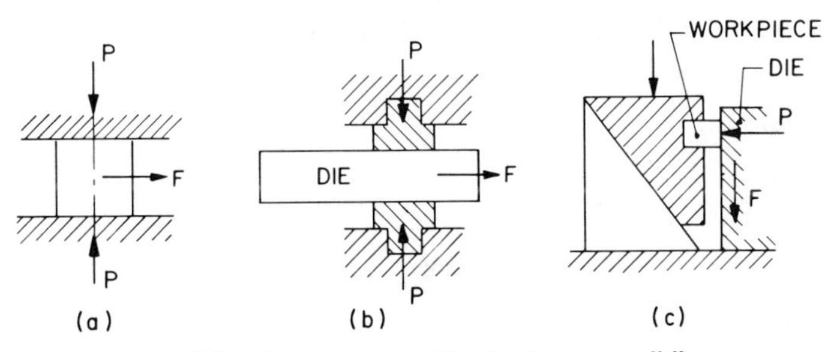

Fig.5.9. Compression with simultaneous sliding

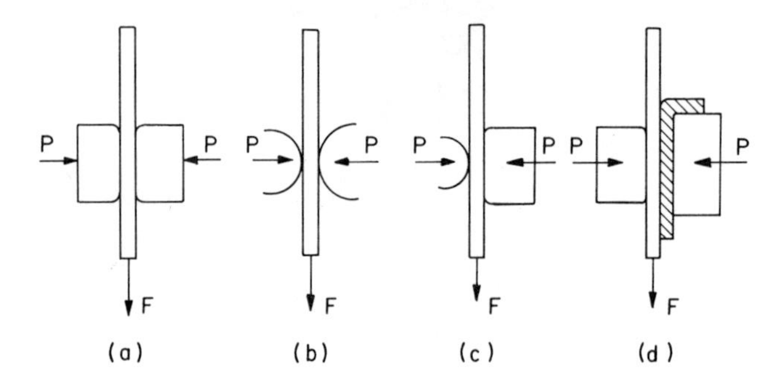

Fig.5.10. Strip drawing tests

than the strip (Fig.5.10a). The coefficient of friction is simply $\mu = F/2P$. Some of the simplicity of the test is deceptive. Sharp entry cannot be tolerated, because the lubricant would be scraped off; any radius must obviously influence the degree of fluid film development, and reproducibility between laboratories is bound to remain poor until geometry is standardized. Several versions of this test have been developed, as reported by Miyauchi et al [37]. When the flat dies are replaced with cylindrical ones (Fig.5.10b), contact pressures become higher but hydrodynamic effects are also more pronounced. Ike et al [38] used this test with either a constant load or constant gap. An intermediate condition is reached when the strip is drawn between a flat and a cylindrical die (Fig.5.10c). For a quick exploration of die composition and surface finish effects, inexpensive and readily interchangeable inserts may be used (Fig.5.10d). In the presence of a liquid, lubrication is of the mixed-film type and trends of the force F are as important as its absolute level, as pointed out by Gibson et al [39]. A good lubricant gives a steady draw force. An increasing trend indicates gradual breakdown, a decreasing trend an activation of additives. A sudden drop at the beginning of drawing is a warning of potential trouble, because high static friction has often been correlated with a tendency toward galling [39,40]. Frequently, these tests are run with incrementally increasing force P to reveal critical die pressures.

Equipment descriptions are often too sketchy to show whether the measurement of normal force P is free from errors due to binding or cocking of dies or to elastic deflections in the die supports. In the equipment of Dehne and Pawelski (reported in [3.64]), the dies are attached to the equivalents of hydrostatic bearings.

b. Because frictional instabilities are often associated with galling, the strip drawing test has been modified to make it more sensitive to stick-slip behavior. Thus, in the solution of Bernick et al [41], the sheet is drawn between a flat and a hemispherical pin, making it a variant of the pin-on-disk principle (discussed below). Tannert [42] fixed the test sheet to an endless belt wrapped around two rollers, with a pendulum attached to one of the rollers (Fig.5.11). The dies are represented by two cylinders which are dragged, under pressure, along the strip surface, thus raising the pendulum. When sticking friction is overcome, the pendulum causes the strip to move; the angle to which the pendulum was raised is taken as a measure of friction.

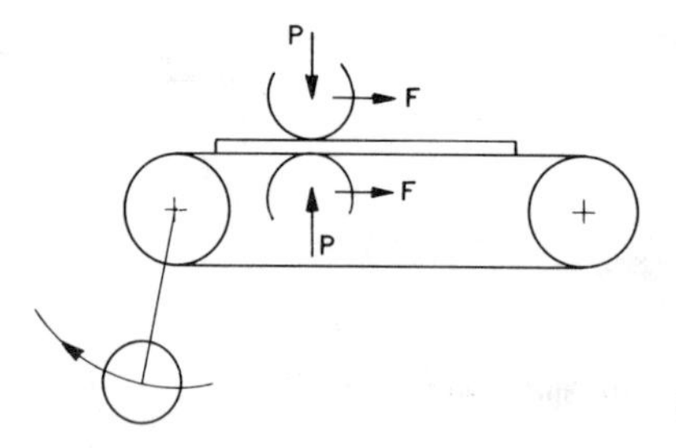

Fig.5.11. Stick-slip tester (after [42])

Draw-Bending Tests. In draw-bending tests, the sheet is bent while being drawn.

a. In many shallow drawings typical of the automotive industry, draw beads are used to restrain the free movement of the blank and thus encourage more stretching (Sec.10.3.4). Sliding with simultaneous bending over the draw bead subjects the lubricant to severe conditions and can be tested in a device designed by Nine [43]. The strip is pulled through a simulated draw-bead die in a tension-testing machine with the beads set to various degrees of interpenetration. The bending component of the draw force is obtained by replacing the bead inserts with rollers. Frictional force (and μ) is then directly available, and pickup and galling are readily observed (Fig.5.12).

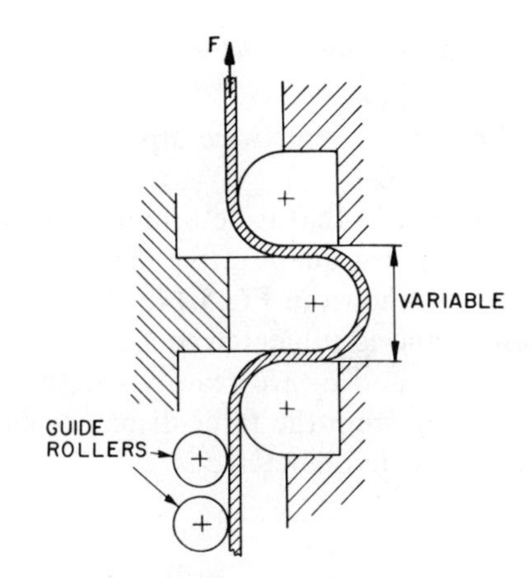

Fig.5.12. Draw-bead tester [43]

b. On the premise that deep drawing involves sliding combined with bending over the draw-die radius, Littlewood and Wallace [44] developed a test wherein the strip is drawn by a force P over a quadrant die against a back tension B (Fig.5.13a), making it possible to test at different interface pressures. Lubricants can be ranked on the basis of the force difference required to maintain the draw, and die pickup is readily observed. With a view toward increasing the sensitivity of the test to pickup and wear phenomena, Woska [45] applied blankholder pressure on the strip (Fig.5.13b). The strip is drawn in 30-mm increments, and the drive allows several thousand strokes per hour. Lubricants are ranked by the number of strokes during which the strip surface deteriorates to a specified roughness. By replacing the quadrant die with a roller (Fig.5.13c), Doege and Witthüser [46] separated the frictional force. The strip is first drawn with the roll freely rotating; from the measured draw force P_1, the force required for bending P_b is:

$$P_b = P_1 - B \tag{5.4}$$

Thereafter, the roll is clamped; the drawing force is now P_2 and the force attributable to friction F is

$$F = P_2 - B - P_b \tag{5.5}$$

If the wraparound angle is $\pi/2$, the coefficient of friction is obtained from

$$\mu = \frac{2}{\pi} \ln \frac{P_2 - P_b}{B} \tag{5.6}$$

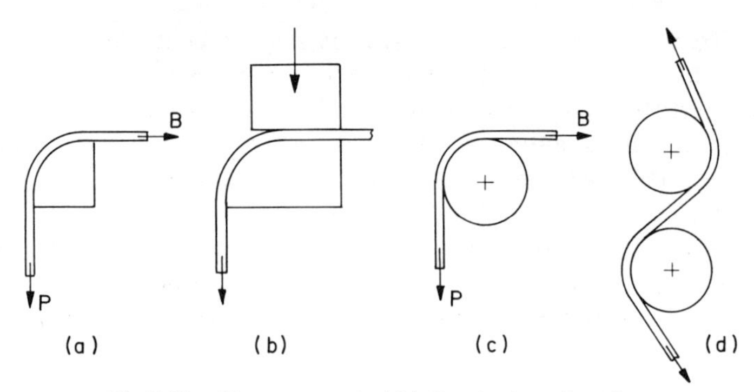

Fig.5.13. Measurement of friction in draw-bending

Somewhat similar conditions obtain when the strip is bent twice around two cylindrical dies (Fig.5.13d) [37].

c. Friction around the punch nose is important in that it determines how much support the sheet receives from the punch surface. For a quantitative measurement of this component, Duncan et al [47] developed the test shown in Fig.5.14. The strip is guided around two pins representing the punch radius, and is subjected to a tensile force P_1 (= P_2 in Eq.5.6 if P_b is ignored). The extensions e_1 and e_2 are readily measured, but not the force P_2 (= B in Eq.5.6); P_2 can be picked up from the force-displacement diagram of the strip material, at a strain indicated by the extensometer e_2.

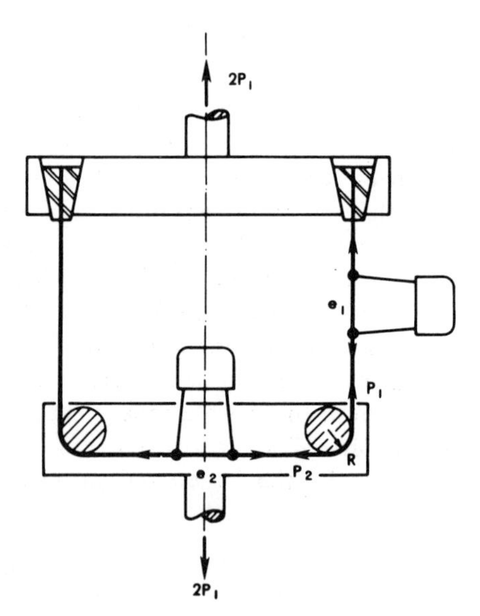

Fig.5.14. Measurement of friction in bend-stretching [47]

Pin-on-disk tests employ essentially the basic "tribometer", which is one of the most frequently used test rigs. The end face of a pin rides on the flat face of a disk or table, or on the cylindrical surface of a disk, sometimes in a reciprocating motion.

a. A flat-ended pin riding on a flat surface (Fig.5.15a) provides an ostensibly constant apparent area of contact, but the slightest cocking of the pin—resulting from elastic deflection in the supporting members—is bound to cause localized contact and plowing. Quiney and Boren [48] tested friction on very thin strip by pulling it past a loaded, flat slider (Fig.5.15b).

b. Much more practical is a pin with a hemispherical end (Fig.5.15c). The pin may

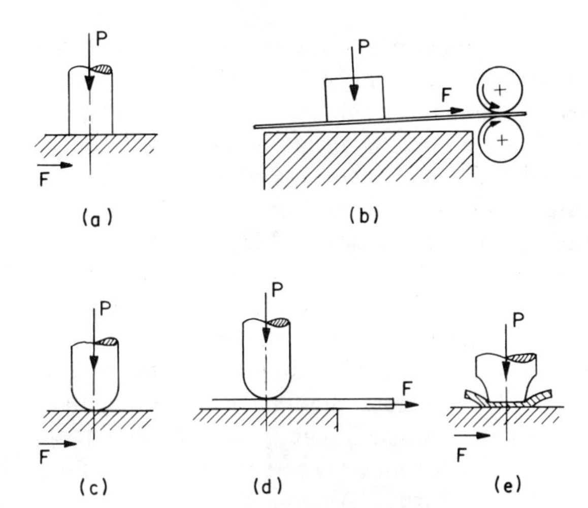

Fig.5.15. Pin-on-disk test and variants

be allowed to penetrate deeply into the specimen surface without basically altering contact conditions. The mode of deformation is between plastic deformation and cutting, since buildup of metal in front of the rider is unavoidable. Extended contact can be simulated by repeated traverses in the same track, provided that interface pressures are kept low enough to avoid deep plowing of the surface. Measurement of normal and lateral forces is relatively simple. However, sturdy construction is essential if stick-slip is to be avoided, although stick-slip may be encouraged by suitable design [49]. The onset of stick-slip in a particular test setup also may be taken as an indication of lubricant quality. A partial review of the subject has been given by Barwell [50]. The tendency toward stick-slip is actually exploited in a variant of this test, developed by Bernick et al [41] for assessing the tendency toward galling in sheet drawing. The sheet is drawn between a flat anvil and the hemispherical end of a piston rod (Fig.5.15d); at the start of the test, high friction is recorded, which drops suddenly on the initiation of sliding. The ratio of the two forces is taken as a quantitative measure of galling tendency.

Difficulties of construction are avoided when the rider is made to rest on three balls. This automatically ensures uniform distribution of the load. Breakaway friction can be measured simply by gradually increasing the inclination of the flat test surface [51]; the frictional force and its variation can be recorded when pulling the rider with a rope and pulley [52].

The pin (usually with a radiused end) may be pressed against the cylindrical surface of a disk, and such geometry has been used for metalworking lubricants by Noordermeer et al [53].

c. In the device developed by Peterson and Ling [54], interface pressures acting on the workpiece are increased by compressing a 1-mm-thick sheet between a flat anvil and punch (Fig.5.15e). Material is extruded from between the dies, and the remaining thickness is able to support a multiple of σ_f. On moving the punch sideways, the coefficient of friction is simply $\mu = F/P$. Cocking of the pin is minimized by sturdy construction, but elastic deflections still make the pressure distribution uncertain.

The contact geometry of pin-on-disk tests makes them most suitable for boundary-lubrication studies, although a limited hydrodynamic contribution is possible when a previously formed groove is traversed at high speeds.

Twist Compression Test. In the various forms of this test, normal pressure is combined with continued sliding over the same surface area.

a. Two cylinders are pressed together on their end faces and then one is rotated against the other (Fig.5.16a). This test has been extensively used in studies of adhesion and of lubricant properties at high normal pressure. However, pressure distribution over the solid end face is uncertain, and the distance traveled (and the tendency toward lubricant breakdown) increases with radius. It is preferable that one specimen be hollow (Fig.5.16b), so as to present an annular surface. Most work has been done in the elastic range; however, Schey [3.123] found that if pressures are high enough, some plastic deformation of the end face can be allowed to occur, and then the test shows remarkably close correlation with lubricant performance in severe industrial operations such as cold extrusion. This is possible only because the limited initial lubricant supply is exposed to continued sliding, and thus its durability and resistance to breakdown are tested. Furthermore, the tendency of the material pair toward die pickup is very quickly evaluated. The test is readily adapted to high-temperature work [3.123],[55]. Local binding of surfaces may cause very high deflecting forces; therefore, construction must be extremely sturdy. Alignment is facilitated by means of a conical contact area (Fig.5.16c) [56]. The frictional force F acts on a mean radius r, and the resultant torque $T = rF$ and normal force P must be carefully separated and measured. The coefficient of friction is then $\mu = T/rP$.

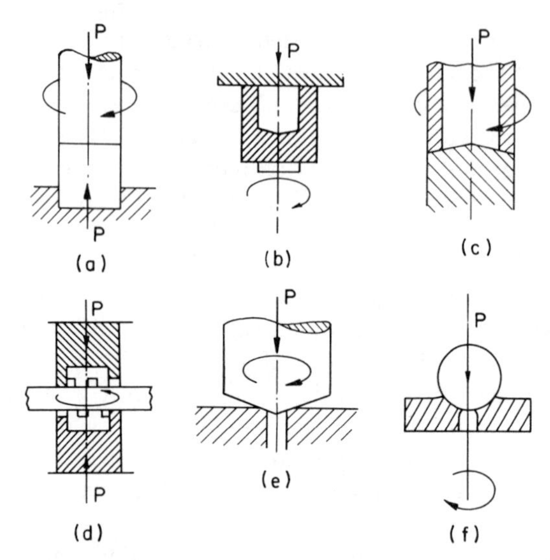

Fig.5.16. Twist-compression test methods

c. In the form developed by Orowan and Los (reported in [57]), two slotted (castellated) specimens are pressed onto the two sides of a flat anvil (Fig.5.16d), which is then rotated by hand. The simplicity of the equipment is appealing, but sliding is limited.

d. A further variant of the twist-compression principle has been used by Williams and McCarthy [58], who pressed and rotated a punch with a 5° conical nose against a drilled-out specimen (Fig.5.16e), and by Shaw et al [3.104], who rotated a steel ball against a predrilled workpiece (Fig.5.16f). An advantage of this version is that more surface extension can be allowed, but the size of the surface area gradually changes and is unknown at any moment; therefore, a coefficient of friction can be determined, but the interface shear strength cannot.

e. In another group of tests the principle of simultaneous twisting and compression is again used but the interface pressure is increased (Fig.5.17a) by confining the workpiece in a container [3.106],[59,60]. Measurements taken on sturdily constructed equipment should be reproducible, although a thin fin of metal is bound to extrude in the

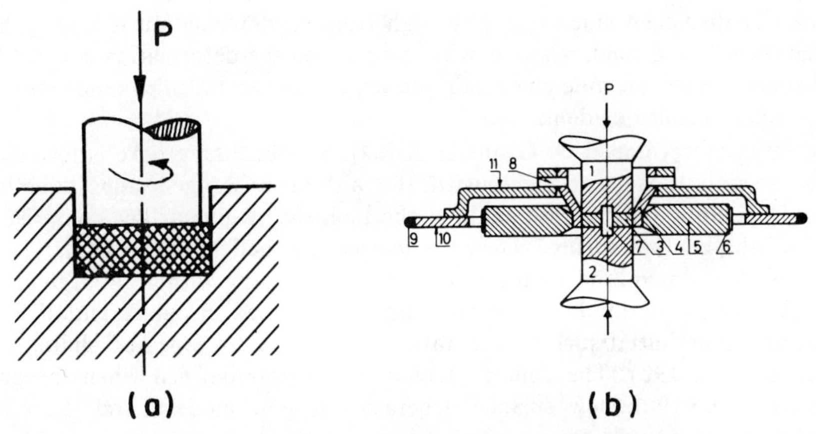

Fig.5.17. Twist-compression with trapped specimens

gap between the rotating punch and the stationary container, and calibration for this effect is difficult. Furthermore, sliding velocities change along the radius. A variant of this test by Haverstraw [4.191] aims at eliminating complex instrumentation. The container is heavy enough to act as a flywheel, and, after it has been brought to a set speed, the specimen is dropped into the cavity of the flywheel, the punch load is applied, and the time required for stopping the flywheel is measured. Thus, speeds vary greatly during the test, and interpretation of the data is difficult. Many objections to this type of test are overcome by the design of Wanheim (Fig.5.17b) [61], in which the specimen, in the form of a hollow disk (3), is compressed between two anvils (1,2) and is radially extruded against an outer die (4) which is rotated by means of a ring (10) and cables (9). Serrated ends on the anvils (1,2) prevent slipping, and all sliding occurs between the container (4) and the outer surface of the cylindrical specimen (3).

5.4.3 Tests Involving Localized Deformation

In another group of tests, plastic deformation is limited to a small part of the workpiece. This process somewhat resembles indentation, and the analysis of Kudo and Tamura [62] shows that the contact pressure is about twice the flow stress, and that large portions of virgin surface are exposed, although at an uneven rate (higher at the tip of the indenter).

a. In one such test (Fig.5.18), a wedge-shape tool is moved in an arc over the surface of a sheet [63], [3.491]. Lubricants are ranked according to the ratio of tangential to normal force (equivalent to a coefficient of friction) and also according to the shape

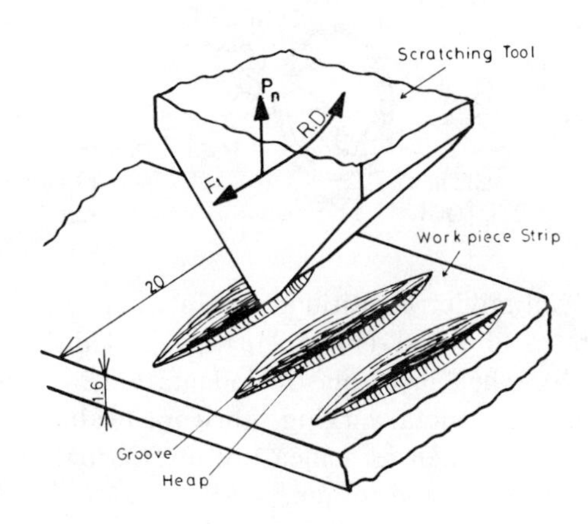

Fig.5.18. Scratch test [3.491]

of the material displaced sideways: with high friction, deformation is highly localized and a ridge (heap) is formed, whereas with low friction the deformation is spread over a greater distance. Because much new surface area is generated, the tendencies toward pickup and wear are also evident.

b. In the tests developed by Oyane et al [64], a threadlike groove is formed on the surface of a cylindrical workpiece with the aid of a wedge-shape indenting tool (Fig.5.19). Depending on the nose angle of the tool, the virgin surface comprises about 30 to 90% of the contact surface when the tool path is helical (Fig.5.19a). The helical groove can be preformed, and then the tool simulates sliding without a large amount of new surface being generated. The groove may be formed circumferentially, without axial feed, and thus initial surface generation is followed by repeated sliding over the same surface (Fig.5.19b). The contact geometry is better defined when the groove is premachined (Fig.5.19c); new surface generation is now modest, and the effects of repeated sliding can be studied.

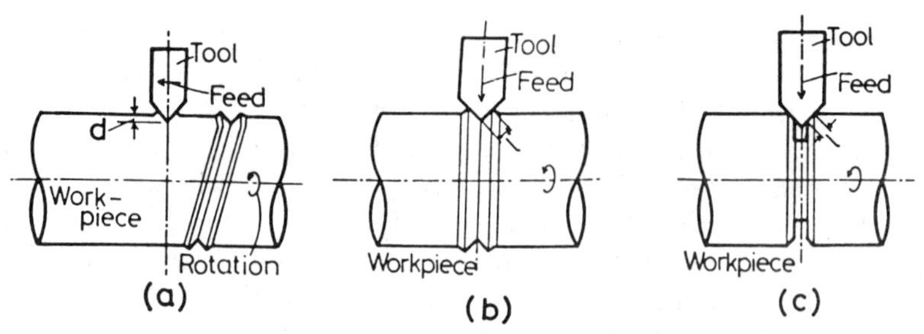

Fig.5.19. Tests involving localized deformation on cylindrical surfaces [64]

c. Somewhat related is the test [64] in which machining of the end face of a thin-wall tube is followed immediately by normal contact with the experimental material (Fig.5.20). Thus, sliding over freshly exposed surfaces reveals the trend toward pickup formation; this trend increases with increasing pressure and sliding distance. The advantages of this test are that it can be carried out on a simple lathe and that it can be used for exploring high sliding velocities and hydrodynamic effects. Under unlubricated conditions, the proportion of tool surface area coated with pickup is taken as a measure of adhesion, and wear can be determined. Brown and Armarego [65] conducted a similar wear test in which a slider was pressed against a tube end face.

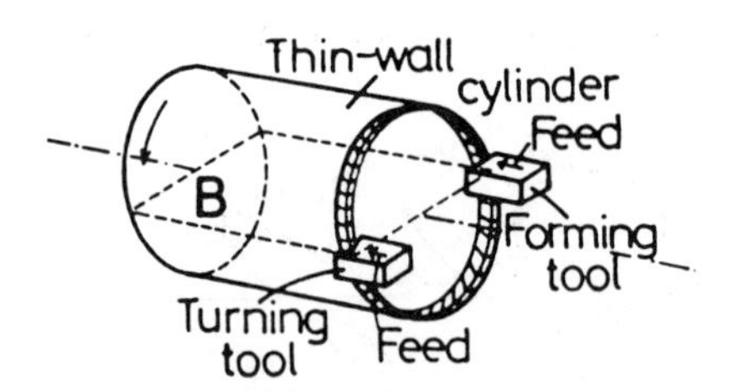

Fig.5.20. Test involving sliding on freshly cut surface [64]

d. Even milder conditions obtain when the workpiece, rotated in the lathe, is contacted by a die rod (Fig.5.21a) [66]. This is similar to the crossed-cylinder test (Fig.5.21b); when local plastic indentation in a helical path is allowed, the technique can be used for metalworking lubricant evaluation [67]. The torque required to rotate the cylinder is taken as a measure of friction, or a coefficient of friction is calculated from normal force and torque.

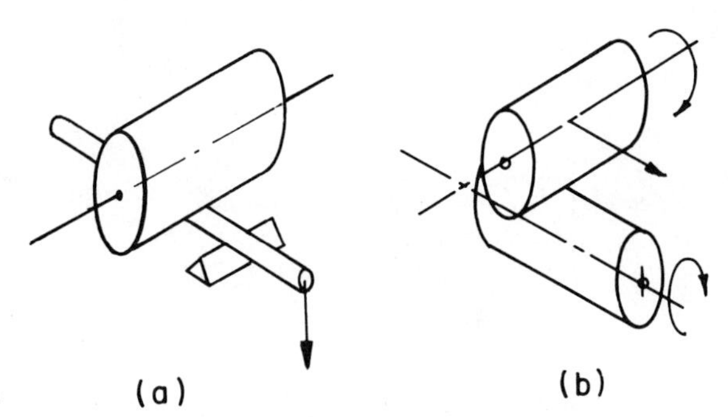

Fig.5.21. Crossed-cylinder tests

e. In hole-expansion tests, interface pressure is generated by driving a harder member (representing the die) into the bore of a hollow cylinder (representing the workpiece).

In the press-fit test (Fig.5.22a), an oversize pin is driven into a bushing. The pressure is calculated from the theory of elastic deformations. Originally developed for the study of the press-fitting process, it was later advocated for general lubricant evaluation [68] but with obviously limited validity.

In the version used by Veiler and Likhtman [69], a ball is repeatedly driven through the bore (Fig.5.22b). The force recorded in pushing the ball through the hole is plotted against the number of passes to show the efficiency of the lubricant in reducing friction and preventing die pickup and scoring.

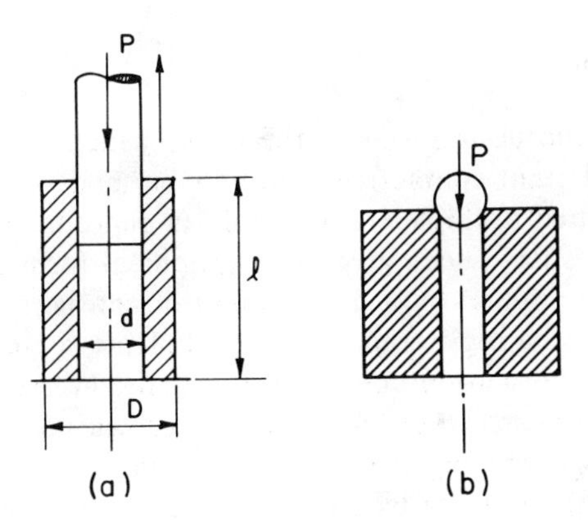

Fig.5.22. Hole-expansion tests

5.4.4 Bench Tests

In the great variety of test machines (tribometers) developed for general friction and wear studies, conditions bear no resemblance to those of actual metalworking processes. Nevertheless, several machines have been repeatedly used for studies of metalworking lubricants. Reviews [70,71] and an extensive compilation of test equipment [72] are available, and the principles of design and construction [1] and the philosophy of development [3] have also been discussed.

Point contact is typical of the four-ball machine (Fig.5.23a), in which a 0.5-in. (12.7-mm) ball is rotated against stationary (clamped) balls. Rotation speed is high (typically 1400 rpm), but for metalworking lubricants the speed is often drastically reduced to as low as 0.3 rpm [73] on the premise that relative sliding speeds are quite low in processes such as rolling. In the pendulum oiliness tester, a steel cylinder attached to a pendulum rubs against four symmetrically arranged steel balls (Fig.5.23b), and μ is derived from the decay of pendulum movement [74].

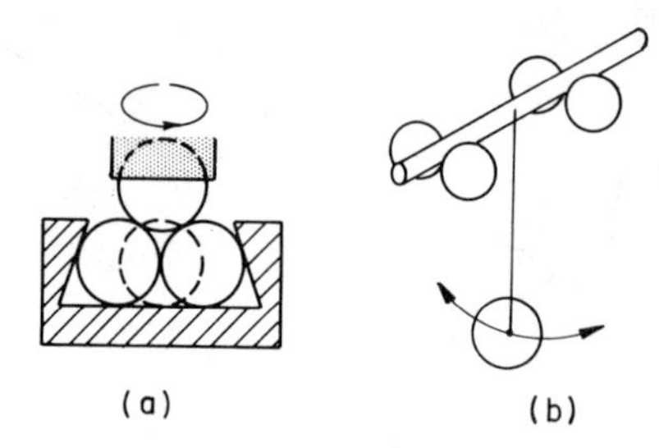

Fig.5.23. Point-contact tests

Line contact prevails in another group of tests (Fig.5.24). In the Timken machine (Fig.5.24a), a steel block is pressed against the rotating outer race of a ball bearing. In the Falex machine (Fig.5.24b) a bar is rotated between two hard V-shape bearing blocks; in the Almen machine (Fig.5.24c), a bar is rotated between split bushings. In the SAE machine and the somewhat similar Amsler tester (Fig.5.24d), two disks are driven independently at different speeds, thus providing a combination of rolling and sliding friction at a variable ratio.

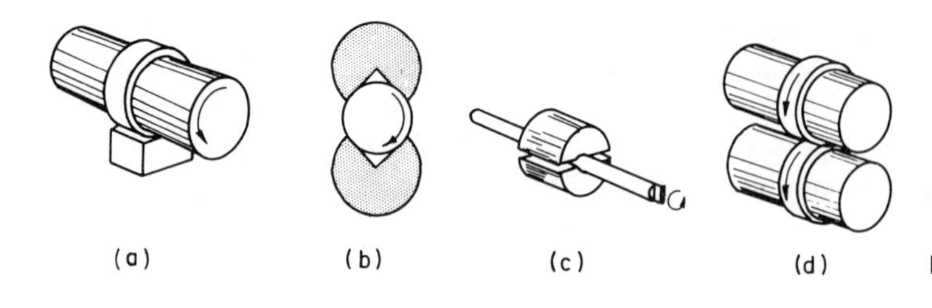

Fig.5.24. Line-contact tests

In the basic forms of the above bench tests, the purely Hertzian nature of contact demands that contacting members be made of hardened steel (very often, rolling bearing elements which are available in reproducible quality and large quantity). Contact pressures have been given in a graphical form by Carroll [75]. In most testers the load either is applied by a dead weight or can be readily measured, and torque or frictional force can be registered or recorded. Thus, a coefficient of friction can be found in addition to information about wear, usually on the basis of the size of the wear scar or the quantity of material lost. The breakdown point of a lubricant is usually found by running for a prolonged time or according to a preprogrammed, incrementally increasing load pattern. Tests based on point contact have been extensively used in boundary and E.P. lubricant development, while the others have found use also in studies of thin-film and hydrodynamic lubrication. For metalworking evaluation, one of the contacting members may be made of the workpiece material, ensuring that at least the right tool and workpiece material combination is used, even though contact is essentially elastic. Because production tools heat up even in cold working, tests must usually be run at an elevated temperature if any correlation is to be expected at all.

Variants of bench tests offer possible applications to metalworking. In the test designed by Armstrong and Lindeman [76], flat pads are pressed, by means of a caliper, against the two flat faces of a disk. In a variant of the four-ball tester, by Bailey and Cameron [77], the stationary balls are replaced with three flat-ended pins or pegs (Fig.5.25a) made of any material; the change to a pin-on-sphere geometry holds promise for boundary-lubricant evaluation. Gonin et al [78] describe a machine in which a cylindrical shaft is rotated against interchangeable flats, split bearings, or pins (Fig.5.25b), thus offering various geometries and differing degrees of sensitivity to hydrodynamic effects. In the device proposed by Kellermann [79], the workpiece is a wire (supported in a V-block) against which a steel disk is rotated (Fig.5.25c). Special machines are needed to simulate special conditions, and thus the impact wear testing machine of Bayer et al [80] may find application for the simulation of cold heading. Sometimes wear is of greater consequence than friction, and therefore Becker [81] rubs a serrated disk of the workpiece material against the cylindrical surface of a piercing punch (Fig.5.26), allowing a rapid evaluation of the wear-reducing qualities of the lubricant.

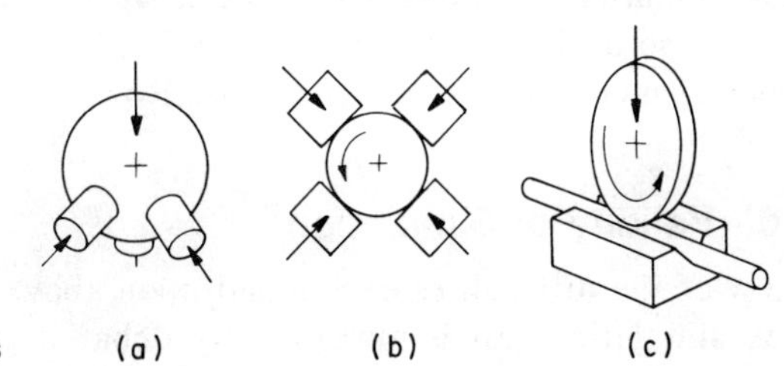

Fig.5.25. Modified bench-test methods

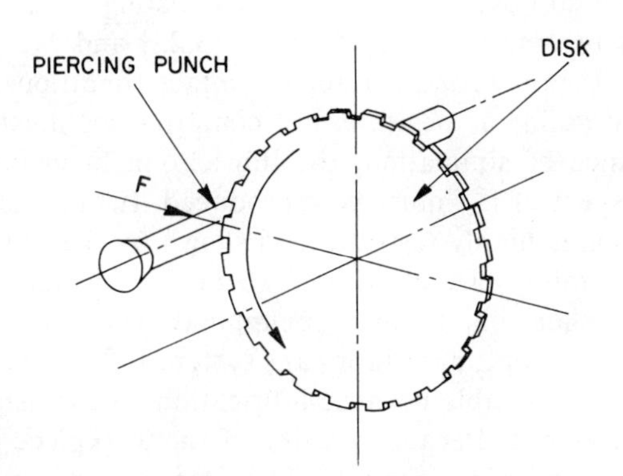

Fig.5.26. Wear test [81]

5.4.5 Assessment of Test Results

From the never-ending proliferation of test methods, it is obvious that no test has yet been devised that could claim general acceptance. Part of the problem is that test results are seldom reproducible: a cooperative effort showed that even ostensibly identical tests can give widely differing results in different laboratories, especially in wear testing [82]. Variations were due not so much to testing procedures or inaccuracies, but rather to the widely different lubrication mechanisms that are obtained under only slightly different conditions. No such cooperative effort has yet been mounted on metalworking lubricants, but it is obvious that any test conducted under ill-defined conditions is practically valueless.

The first step would be to accept some basic standards for selecting and reporting experimental variables, as suggested by DePierre [83]. Such a description should include as a minimum: (1) die and workpiece (material, composition, condition, hardness, dimensions); (2) methods of preparing die and workpiece surfaces (chemical and mechanical, including coatings); (3) surface topography of die and workpiece, including data indicating fullness of profile;(4) heating of die and workpiece (method, temperature, time, atmosphere); (5) lubrication system (lubricant composition, method of preparation and application, thickness); (6) processing conditions (die geometry, amount of deformation, surface and/or deformation speed, working pressures); and (7) surface condition of die and workpiece after testing.

The data will often have to be reported in the briefest possible form, such as coefficient of friction values, but the method of processing the data must also be clearly shown. Unless they can be recalculated from the friction values, it is certainly preferable to present the measured data as well.

Even if reproducible data can be obtained, the question of interpretation still remains. Very often, several tests are conducted and then some judgement must be made regarding the weights that should be attributed to the results. Instead of absolute values, quantitative indeces may be assigned to individual results [84] and then the results (if so desired, with some weighting factors) can be combined to give an over-all number of merit. One such scheme by Rowe and Roy-Chowdhury [85] appears promising.

5.4.6 Validity of Simulation

In view of the difficulties expounded upon above, it is not surprising that the relevance of any simulation test is always hotly debated [86]. Any simulation that fails to take into account the entire system is bound to fail. Because it is easy to overlook an important yet not so obvious point, a formal listing of the tribological characteristics of the system, as recommended by Czichos [3.22] and Molgaard [87], is helpful. As outlined in Sec.5.1, the operating variables, contact conditions, and process environment must be reproduced as far as possible, but compromises must invariably be made. To maintain the relevance of simulation, the lubrication mechanism—or, if this is not possible, a selected aspect of it—must be reproduced. It must then be understood that the validity of simulation is highly restricted, as shown by the following examples.

1. Superalloys are often hot extruded, and lubrication is provided either by a glass pad or, for short billets, by a preapplied glass layer (Sec.8.5). In a program aimed at establishing the optimum lubricant system [88], it was clear from the outset that bench tests were inapplicable because lubrication mechanisms operative in hot extrusion could not be reproduced. Because wetting of the workpiece by the glass and possible reactions with the die and container were of prime importance, the test had to utilize superalloy workpieces and typical extrusion die materials. Consideration of the probable lubrication mechanism indicated that hydrodynamic effects must play a substantial role; plane-strain compression was selected for simulation, because substantial plastic deformation is accompanied by a gradual depletion of a preapplied lubricant film (Fig.5.27). The thickness of the initial squeeze film and its gradual thinning in the course of compres-

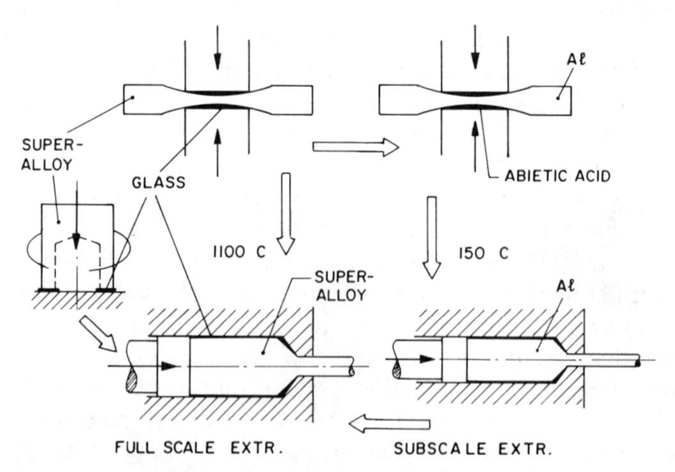

Fig.5.27. Simulation system for glass-lubricated hot extrusion of superalloys [88]

sion were expected to give information on the effects of viscosity, temperature, deforming speed, and, at least to some extent, die pickup. It was also hoped that effects of process geometry could be explored by changing anvil configurations, but interpretation in terms of extrusion parameters proved difficult. Therefore, it became necessary to introduce a second line of investigation, in which extrusion was performed at a reduced scale and a lower temperature, replacing glass with abietic acid and the workpiece material with aluminum. At around 150 C the viscosity of abietic acid and the flow strength of aluminum were in about the same proportion as those of glass and a superalloy at 1100 C. Thus, the effects of die geometry, extrusion speed, and billet-to-container clearance could be explored. When these results were combined with data from the plane-strain compression test, the response of the real system could be correctly predicted, as confirmed by full-scale trials. It is found in practice that, on depletion of the glass film, severe die damage can occur, and to prevent this it is customary to apply a graphite coating to the extrusion die surface. The response of glass to contact with graphite was simulated by elevated-temperature twist-compression tests (Fig.4.24). The entire simulation program was more complex than most, but transfer of the results to the full-scale operation was still successful.

2. There are those who deny the validity of any simulation. There is, of course, some justification for pessimism, as shown by the general lack of correlation even in such seemingly simple cases as the evaluation of lubricants for sheet metalworking. Thus, Gibson et al [39] found no correlation between strip drawing and a cupping (deep drawing) test, nor between laboratory tests and production performance, particularly when the evaluation was based on friction alone. If, however, the trends in forces were observed, correlation became much better, indicating that lubricant breakdown is of much more decisive influence than the magnitude of initial friction. We have already remarked that trends in forces, torques, etc. are very sensitive to changes in lubrication mechanisms and that it is therefore essential for the test run to be long enough to establish steady-state conditions. Details of the surface profile, which determine the load-bearing area and the fullness of the profile (which in turn controls the amount of lubricant entrapped), interact with interface pressure to change friction from low to high values and vice versa (see Fig.3.44). Therefore, more important than the value of friction is whether the lubricant can resist breakdown, metal transfer, and galling. For this, testing at elevated temperatures is essential, as found by Funke and Schlemper [89]. Results are more reproducible and transferable when plastic deformation is induced in the course of testing. Thus, Ebben [90] reported on the evaluation of some 20 lubricants; correlation between a laboratory ironing test and plant performance was satisfactory.

3. The jump from bench-type tests to metalworking is a much larger one, and the results can have only limited validity. Thus, for example, Vojnovic et al [91] found satisfactory correlation between a modified four-ball test and steel rolling performance only if saponification values and free-fatty-acid contents of the lubricants were kept within certain limits. Obviously, only the boundary-lubrication properties were evaluated, and the comparison was necessarily limited.

Thorp [92] used a four-ball tester at two speeds of rotation to evaluate the effects of diluting a drawing soap compound, and then correlated the results with deep drawing performance. Weld load increased and wear scar decreased with dilution, and Thorp surmised that the higher temperatures generated in the high-speed test led to chemisorption. The coefficient of friction remained constant at high speed and decreased at low speed with increasing dilution, yet the deep drawing force increased, indicating that quite different lubrication mechanisms were operating.

No direct transfer of data from bench tests can be hoped for or should indeed be expected; if, however, the lubrication mechanism operative in the real process is known and its contributing factors can be separated, a bench test may be used for limited purposes. This calls for considerable experience and knowledge; some examples will be presented in Chapters 6 to 11.

5.4.7 Choice of Simulation Tests

It is now amply evident that no single test can simulate the great variety of conditions encountered in tribology. This led Akin [3] to suggest that a system of several bench tests be chosen, based on an analysis of lubrication and failure modes, lubricating qualities of the fluids, and test-machine capabilities. Similar recommendations have been made for metalworking purposes, usually with the intention of covering a broad range of lubrication mechanisms from purely boundary to predominantly hydrodynamic ones.

The choice of individual tests depends partly on the mechanisms judged important, and partly on the individual experience and prejudices of the researcher. Thus, Schey [93] suggested a system of three tests:

1. Ring upsetting, to provide information on lubricant entrapment and on the ability of the lubricant to promote free spreading of the workpiece material over the die face;

2. Twist compression, to generate information on boundary and E.P. properties, film durability, and also pickup and die wear;

3. Plane-strain compression, to give a measure of the contribution of hydrodynamic effects.

The ring compression test is chosen as one of the tests by Vámos et al [94], complemented by bar drawing, deep drawing, and a form of plane-strain back-extrusion [12]. Hydrodynamic (as opposed to hydrostatic) contributions are under-represented in this selection. However, the back-extrusion test (Fig.5.2d) is most useful in evaluating the ability of the lubricant to resist thinning.

A variety of tests is needed to predict lubricant performance even in sheet metalworking, as observed by Butler and Pope [95]. Test standards now under development by ASTM should bring some measure of uniformity in test conditions. A broad-ranging and critical round-robin test could then establish the reproducibility and relevance of various test methods.

The lubrication mechanism in metalcutting is predominantly of the boundary and E.P. type; therefore, bench tests are useful. Drilling and tapping have often been used for evaluating E.P. properties, in parallel with or instead of bench tests. Nevertheless, the choice of the best test method is a subject of debate (Chapter 11).

5.5 Evaluation of Friction in Metalworking

Because simulation tests are designed to allow measurement of normal and frictional forces, τ_i and μ are readily extracted. There is, however, also a need to determine frictional forces in actual metalworking processes for two reasons: first, the magnitude of friction provides a clue to the conditions prevailing at the interface and, second, some means of quantifying friction (Sec.2.2) is needed if forces, die pressures, and power requirements are to be calculated.

5.5.1 Total Force or Energy

The force or energy of deformation is often routinely measured, providing useful information:

1. Because friction increases the forces required for deformation or cutting

(Sec.2.3.1), one could, in principle, back-calculate the magnitude of friction from total force if a reliable theory were available. There are two problems:

a. Friction effects are often inextricably interwoven with effects of inhomogeneous deformation (Sec.2.3.4), and many theories, especially those that are easiest to apply, fail to take this into account.

b. Theory can be applied only if the true flow stress of the material σ_f is known. This requires that σ_f be determined on the actual workpiece material, in the direction relevant for the particular process, rather than taken from nominal values. This can be difficult, for example, when a thin strip is rolled and the through-thickness properties have to be determined at high strain rates.

For the above reasons, the calculation of an average μ or m from measured forces must be treated with extreme care; it is likely that many published data based on force measurements are in substantial error.

2. Measured forces are informative if one is willing to set lower aims: a straight comparison of forces allows a ranking of lubricants, if no other variable (reduction, speed, etc.) is allowed to change.

3. The force developed in metalworking also acts on the machinery, and elastic deflections in machine frames or tool holders are proportional to it. Deflections are usually small and need careful techniques for measurement. However, in some instances changes in the dimensions of the issuing product are readily observed. For example, in rolling at a given roll-gap setting, a thicker strip indicates larger elastic opening of the roll gap, thus a higher roll force, and hence higher friction.

4. Through its effect on force, friction also affects torque and power. In some processes only the motor power is conveniently measured, but this tends to be a rather insensitive indicator because of losses in the transmission.

5. The total work of deformation and friction is largely converted into heat (Sec.2.3.3) and, in principle, increase in workpiece temperature can be used as a measure of friction, provided that all other factors are kept constant and the contribution of friction is large enough. Calorimetric techniques have been used in rolling (Sec.6.2.5) and wiredrawing (Sec.7.1.3) but, considering the difficulty of these measurements, the simpler and just as informative force measurements are preferable for lubricant evaluation.

5.5.2 Frictional Force

More direct and relevant information is secured if the frictional force is measured independent of the total or normal forces. The basic problem is the same as in simulation tests: greatest care must be taken to ensure that the measured frictional forces are not affected by normal forces. Examples of appropriate solutions will be given in Chapters 6 to 11.

From the total frictional force only an average external μ or τ_i can be obtained. In discussing lubrication mechanisms (Chapter 3) it became clear that their magnitude may change, even with the same lubricant, over wide ranges as interface pressures, sliding velocities, and lubricant film thicknesses change. In any one metalworking process, these variables are bound to change even within the contact zone, and therefore averages can mask significant changes within the contact zone itself. To reveal such changes, local values of τ_i are needed.

5.5.3 Stress Distribution

For full information on the properties of the interface, both normal and shear stresses must be measured. Appropriate methods have been reviewed by Dohmann [96], Buhler and Lowen [97], and Voelkner [98].

Determination of normal pressure is possible by a number of techniques:

a. A pin can be located in a hole perpendicular to the interface (Fig.5.28a) and the force acting on the pin measured with a suitable load cell. Variants of this technique have been reviewed by Cole and Sansome [99] and by Takahashi [100]. First used by Siebel and Lueg [101] for determining the pressure distribution in rolling, the technique yields information on the shape of the friction hill (as in Fig.2.6).

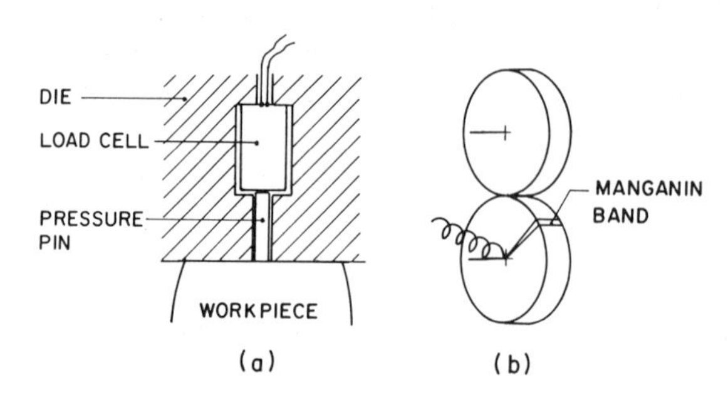

Fig.5.28. Local pressure measurement

b. A pressure-sensitive film may be deposited on the surface of the die. As developed by Kannel and Dow [102], an insulating alumina film is first deposited by vapor sputtering. Then, a very thin (approximately 0.1 μm), narrow (50 μm) manganin alloy strip (which changes its resistivity significantly with applied pressure but not with temperature) is vapor deposited, followed by an alumina protecting film; thus the pressure distribution in the contact zone of plane-strain deformation processes can be measured (Fig.5.28b). Von Laar and Löwen [103] used manganin gages for measuring die pressure in cold extrusion.

c. A good feel for pressure distribution can be obtained by machining a slit in the die so that a thin fin, of a height proportional to the local pressure, is extruded. This technique, apparently first used by Unksov [2.26], has the virtue of simplicity but is not readily interpreted. The lubricant film is locally disturbed, and frictional conditions in the slit make the pressure distribution nontypical at the edges of workpieces. Some improvement is secured by extruding the fin into an outward-tapering slot.

d. Instead of a slit, small holes can be drilled in the die. Placing a thin metallic sheet between workpiece and die, Okhrimenko and Kopyskii [104] established the pressure distribution by measuring the depths to which the sheet was driven into the holes. A somewhat related method, reported by Matsubara and Kudo [105], is more quantitative and has the virtue that the die is not weakened. A sheet of thin metal (such as brass) is finely ridged by a convenient technique such as coining. The sheet is placed in the die with the ridges toward the die surface, and deformation proceeds as usual. The sheet is then removed, and the widths of flattened plateaus on the ridges are measured and converted into pressures with the aid of calibration in simple compression. Pressure variations during the process cannot be followed, unless the process is interrupted at various points and fresh sheets are inserted.

e. More esoteric is the technique of Brouha et al [106], who embedded 40-to 60-μm ruby particles into the surfaces of aluminum specimens and observed, through sapphire anvils, the shift in R-line fluorescence.

All the above techniques yield information only on normal pressure, and the actual value and variation of interface friction can be obtained only through some theory if σ_f is known, with all the attending uncertainties of such procedures.

Measurement of normal and shear stresses calls for more complex techniques:

a. A combination of two pins, one normal to the interface and one at an angle

(Fig.5.29a), offers a direct solution. This method, first used by Van Rooyen and Backofen [107], is not without its problems. Lubricant and even workpiece material may extrude into the space between pin and die; differences in stiffness between tool and pin may cause the pin to protrude and give erroneously high readings [96]; and friction

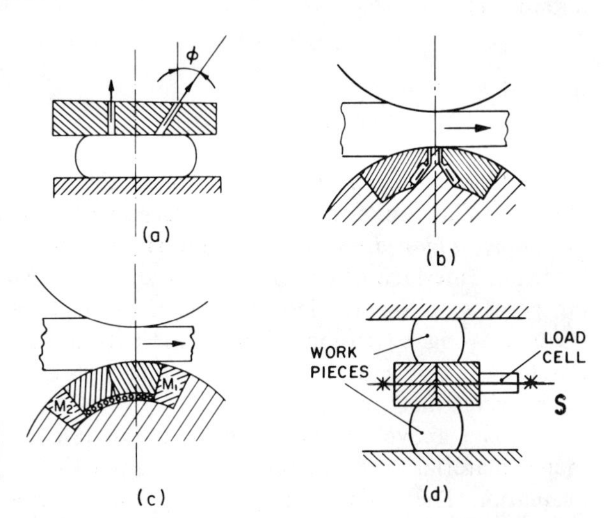

Fig.5.29. Measurement of (a and b) local frictional stresses and (c and d) frictional stresses averaged over entire contact surface

between the pin and hole surfaces can introduce substantial errors (an error of only 1% in pressure can result in a 10% error in the calculated μ [107]). In principle, μ is obtained from the pressures on the oblique pin p_p and on the normal pin p_n:

$$\mu = \left[\frac{p_p}{p_n} - 1\right] \cot \phi \tag{5.7}$$

where ϕ is the angle between the oblique and normal pins. With friction μ_0 between pin and guide hole:

$$\frac{p_p}{p_n} = \frac{1 + \tan \phi(\mu - \mu_0 \mu)}{1 - \mu_0 \mu} \tag{5.8}$$

For all its attractiveness, the technique is laborious, calls for specially constructed (and weakened) dies, and disturbs local frictional conditions. For a mapping of frictional stresses in the contact zone, either a number of pins have to be embedded, or tests must be repeated with the workpiece in different positions.

b. If the gap around a normal-force sensing pin is increased, strain gages attached to the pin give information on both normal and shear forces [108], although metal extruded into the gap is bound to affect the readings. Some improvement is obtained with a steel ball serving as the contact element [109].

c. Some of the objections to the pin technique were overcome by Banerji and Rice [110], who incorporated an integral wedge into a roll (Fig.5.29b). Compression and bending of the wedge is measured by strain gages attached to the wedge, and the roll surface is completed by filler blocks.

d. The technique of Pavlov and De-Yuan, reported in [20], gives information on frictional (but not normal) stresses in plane-strain processes. Closely fitting but freely sliding segments, supported on needle bearings, are built into the die (in this case, roll) surface (Fig.5.29c). The frictional forces are registered on load cells M_1 and M_2.

Extrusion of lubricant and workpiece material must be a problem here too, and averages rather than point-to-point friction values are obtained. In the technique of Skorniakov and Chelesava, reported in [12], a two-piece anvil, held together with instrumented bolts, is placed between two deforming cylinders, thus an average frictional stress is obtained (Fig.5.29d):

$$\tau_{i\,\mathrm{ave}} = \frac{S}{(d/2)^2 \pi} \tag{5.9}$$

e. Photoelastic analysis is possible if the die or tool is made of an optically sensitive material. This limits the choice of workpiece material to soft substances such as lead or plasticine, which allow a reasonable simulation of hot-working but not of cold-working behavior. Interface conditions are also different from those that occur in real metalworking processes. Nevertheless, the technique has been used with some success for plane-strain drawing, extrusion, forging, and cutting. Danckert and Wanheim [111] made the upsetting die of a transparent elastomer whose deformations could be observed by the moire fringe method.

f. The above techniques disturb the lubricant film, or necessitate the use of a model material, and are thus better suited for fundamental studies than for lubricant evaluation. These disadvantages disappear, but evaluation becomes more difficult, when the hoop strain in an axially symmetrical die is measured, as in wiredrawing, extrusion, and sheet metalworking.

In general, the complexity of equipment and the attendant uncertainties of evaluation make the value of the above techniques for lubricant evaluation questionable. They are, nevertheless, important tools for basic investigations into interface conditions.

5.5.4 Deformation Patterns

We have seen in Sec.2.3.4 that friction influences the deformation (strain distribution) of the workpiece.

a. Deformation of a free surface allows a qualitative judgment of the magnitude of friction. With the aid of theories the results can be quantified, but again with some uncertainties. Examples will be seen for spread during rolling, bulging in upsetting, and chip curling in machining.

b. Internal deformations can be observed if the workpiece is made of two halves, with reference lines (such as grid lines) imprinted, engraved, or etched on it. A visual comparison of flow patterns under low and high friction conditions gives qualitative indications; when these are coupled with the theory of visioplasticity [2.2.5], quantitative data can be derived. The accuracy of evaluation decreases when local strains are either very small or very large. Internal markers may also be used; pins were inserted into workpieces in some of the first investigations dating from the 19th century. Second-phase particles, such as MnS inclusions in steel, can serve as built-in markers, as shown by Jones and Walker [112]. Structural features, such as pearlite in a medium-carbon steel, have long been used for following surface deformations in metalcutting. The displacement of markers can be directly observed if the tool is transparent [106].

3. In a few instances, frictional balance actually determines the direction of material flow, and therefore it is possible to obtain a qualitative ranking of lubricants simply by observing changes in the shape of the workpiece. Examples of this are forward slip in rolling (Sec.6.2.6), changes in the internal diameter of a ring in compression (Sec.9.1.2), and a change in the longer dimension of a compressed slab (Sec.9.1.3). If so desired, an average τ_i or μ value can be extracted with the aid of a theory. These

tests are unique in that σ_f need not be known, and errors due to the uncertainty of its value are eliminated.

5.5.5 Metallurgical Changes

In some materials, pressure induces phase changes. Martensite forms in steels [3.366,3.367] and in austenitic stainless steels [113,114]; alpha-plus-beta brasses transform into the beta phase [115]. This technique involves a postmortem examination of the workpiece and needs great care in interpretation, but with proper calibration it can offer important clues.

5.6 Interface Temperature

Interface temperature affects lubricant viscosity, reaction rates, and breakdown of lubricants, and gives an indication of frictional conditions prevailing in the contact zone. Reviews of measurement techniques are given in [116-118] and in [3.22].

5.6.1 Dynamic Thermocouple

The die and the workpiece can be used as the two elements of a thermocouple. Introduced around 1925 for machining, this technique has since been used in various fields. The experimental arrangement of Lueg and Treptow [119] is typical (Fig.5.30). The carbide die insert A, isolated by disk D from die support E, forms one arm of the thermocouple, whereas the drawn wire wound onto the drum H forms the other. A lead, attached at F_1 to the end of the wire, connects to a mercury-wetted contact J attached to the shaft of the drum H. The emf generated in the die throat is recorded on a fast-responding recorder V. In the course of drawing, the distance between the cold junction F_1 and the die throat increases; however, according to Ranger and Wistreich [120], this does not impair the accuracy of measurement. In other applications a rider, made of the workpiece material, may be more convenient. The instantaneous emf is generated by a number of thermocouples formed at asperities making metallic contact with the dies. The high-temperature spots are short-circuited by the low-temperature junctions and a low average is recorded, but this should in no way impair the validity of comparative measurements.

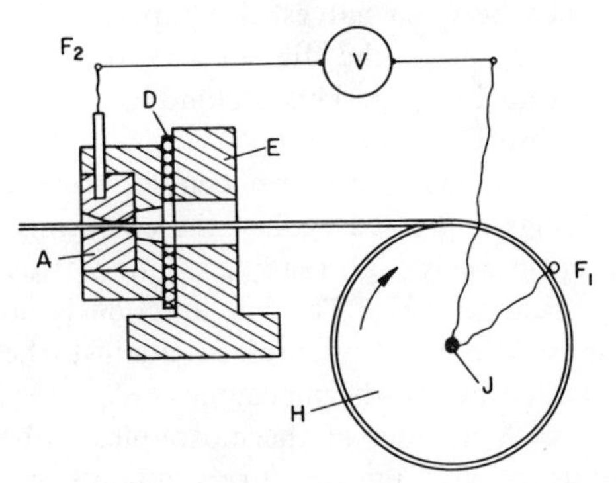

Fig.5.30. Experimental arrangement in which die and workpiece are used as the two elements of a dynamic thermocouple in wiredrawing [119]

The recorded temperature could be influenced by interface films and possible lubricant interactions. However, soap films had no effect in wiredrawing [120]. Similarly, Lueg and Treptow [119] found that a sufficient number of metallic junctions existed (at least in a static calibration test) to allow the formation of couples. Roy and Henry [121] observed no harmful effect of oxides on aluminum. Of course, as soon as full hydrodynamic lubrication is achieved, the method ceases to function.

The basic problem with all dynamic thermocouples is calibration, discussed by Braiden [122]. For a steady-state process the die with a short piece of partly deformed workpiece may be immersed into lubricant baths held at various temperatures. Alternatively, a hollow part may be partially deformed and internally heated by some means. All of these procedures assume that lubrication during static calibration and under dynamic deformation conditions is similar enough not to affect the output.

5.6.2 Thermocouples

Local temperatures can be measured by more or less conventional thermocouples.

1. The hot junction is established by the workpiece or die itself.

a. For wiredrawing the die may be made up of two disks of different compositions [123]. At the plane of contact, the drawn material forms a conducting bridge to complete the thermocouple. Instead of two die halves, two dissimilar dies may be used in tandem, with the drawn wire between them forming the electric contact.

b. Two insulated thermocouple wires may be embedded in the die, close together, in holes perpendicular to the surface, so that the contacting workpiece surface creates the hot junction. Interposed lubricant films invalidate such measurements, but the principle has been used, apparently successfully, under dry friction conditions.

c. Spot checks of surface temperature may be made with prod-type thermocouples; in these, pointed, rigid thermocouple leads protrude and establish contact when stabbed into the hot surface. True temperature readings are obtained if the oxide film is penetrated.

d. Two rollers of different materials pressed against the drawn wire complete the circuit in a temperature meter described by Sturgeon [124]. Readings were reliable, provided that the pressure was high enough to prevent skidding of the rollers and that thick lubricant films were removed with felt wipers before the wire reached the rollers.

e. Belansky and Peck [125] measured the surface temperatures of hot rolling mill rolls with two wire brushes (one iron, the other constantan) pressed against the roll surface. Calibration against a rotating, heated disk eliminated friction as a disturbing factor.

2. Thermocouples of special construction can be designed to give fast response:

a. Flat ribbons of the thermocouple materials are insulated by a mica sheet, sandwiched between halves of a tapered plug, and then press fitted into a hole of the die. On grinding of the die surface, the thermocouple junction is established by smearing and cold welding. This technique gave excellent results for hot rolling [126] and could have broader application.

b. A fine-wire thermocouple, preferably with a flattened junction, gives reasonable readings if pressed against the emerging workpiece surface, although friction can result in erroneously high readings. This difficulty is partially overcome by the rider designed by Zastera [127]. The hot junction is housed in a saddle to insulate it from the wire; the saddle itself is pressed against the wire, and heat radiating from the wire is recorded by the thermocouple.

c. A number of thermocouples embedded in the die at different depths below the interface give temperature gradients from which extrapolation to the surface can be made. In some processes it is possible for the junction to be right at the tool surface. Thus, DeVries et al [128] established the junction between wires fed through the oil hole of a drill with a small ball of silver solder.

d. The thermocouple may be embedded in the workpiece itself. There is no problem if the workpiece is stationary, as in grinding [129], but trailing leads must be fed through a deforming workpiece. Watkins [130] reported on extrusion, and Steindl and Rice [131] on cold rolling. The hot junction must always be at a finite distance from the actual interface and, therefore, measures only subsurface temperatures. Moreover, the thermocouples will take only a limited strain before breaking. Further examples can be found in Chapters 6 to 11.

Lueg and Treptow [119] conclude that the dynamic thermocouple is the most suitable for sensing rapid temperature variations due to friction. A die composed of two disks offers advantages if various materials are to be drawn, since the thermal emf is governed by the choice of the two die components, whereas the output of the dynamic thermocouple is a function of the die-workpiece material combination. A comparison of dynamic and embedded thermocouples was performed by Furey [132].

At very high temperatures, none of these techniques is feasible, and a technique of driving thermocouples into the surfaces of large workpieces by gun action [133] could be considered.

5.6.3 Surface Temperature

Many methods of measuring surface temperature are based on physical phenomena:

1. Infrared detectors can measure local temperatures if one of the contacting members is an optically transparent material such as quartz or sapphire. First used by Bowden and Freitag [134], this technique has been developed for EHD contacts [135,136], for mixed-friction conditions [137], and, with the aid of sapphire windows, also for rolling-element bearings and gears [138]. No application to metalworking could be found. In metalcutting, the interface can be viewed through a hole cut in the tool [139].

2. The surface of the workpiece and/or die may be externally observed either during or immediately after working with radiation pyrometers (see, for example, Cook and Rabinowicz [140]). Pyrometers measuring total radiation or brightness are susceptible to large errors because of variations in surface emissivity. The two-color pyrometer offers substantial improvement, since measurement at two temperatures introduces the ratio of two emissivity values [118]. Therefore, the indicated temperature is dependent only on the slope of the emissivity-versus-wavelength curve. Comparative experiments by Burk [141] favor the two-color pyrometer for steel and stainless steel, even though possible interference from ambient lighting, flames, or loose scale must be considered. An overall improvement is ensured if the product is enclosed. Difficulties increase when both emissivity and temperatures are low. Such is the case in the deformation processing of aluminum and its alloys, and in cold working in general. For these, single-wavelength instruments seem to be preferable [142]. Infrared radiation pyrometers have been used [143,144] and, if calibrated against a surface of known temperature, give high accuracy. Reproducibility of the surface condition is the only limiting factor.

3. Radiant heat can also be registered on infrared-sensitive photographic plates and compared with workpiece surfaces held at a known and controlled temperature, as has been done repeatedly [145] in orthogonal metalcutting. The television camera serves a similar purpose [146]. No record of this technique being used for plastic deformation processes was found.

4. Crayons, lacquers, and inks that change color or change from a solid to a liquid could offer a simple way of measuring surface temperature. Since the phase or color change is a function of time as well as of temperature, these methods can claim only very approximate accuracy unless their application and the cooling rate of the workpiece can be strictly controlled, or if they are used to measure die temperatures. Kato et al [147] placed the powder in the parting line of a split cutting tool.

5. The photoelectric effect was used by Popov and Davydov [148] to measure temperatures developed in grinding less than 1 ms after contact.

6. Resistance change in a conductor can also be used. Thus, Kannel and Dow [102] deposited a very thin, narrow strip of titanium (the resistance of which is little affected by pressure) on a roll surface.

5.6.4 Material Changes

A number of changes (some transient, most permanent) can be utilized.

1. Electromagnetic properties of steels change with temperature, and a recent technique based on phase shift [149] has a high, ±1 C resolution.

2. Surface temperatures may reach high enough values to bring steels into the austenitic range; on emerging from the deformation zone, the surface is quenched back by the colder substrate, and martensite forms [150-153]. Alternatively, a martensitic steel is used which suffers tempering, and the microhardness of surface layers is compared for the same steel held at different temperatures for different times [154]. Some high speed tool steels exhibit microstructural changes which make them particularly well-suited for metalcutting research [155], especially when the examination is made by scanning electron microscopy (SEM) [156]. SEM also reveals reaction layers, especially when the steel is broken at liquid nitrogen temperature and thus a clean fracture surface can be viewed [157]. All of these techniques measure cumulative effects. Furthermore, the underlying assumption—that phase transformation or tempering is not affected by pressure—can be only partly true.

3. A somewhat related technique relies on the radioactive Kr^{85} isotope, introduced into the metal surface by diffusion or ion bombardment. On heating, some of the krypton is lost, thus the maximum temperature to which the surface had been exposed can be judged from the residual radioactivity [158]. The radioactivity is not harmful, calibration is done using test pieces of known thermal history, and the method is useful at temperatures up to 600 C.

5.7 Surface Characterization

Because everything of importance to tribology occurs on the surface, it is no wonder that techniques for revealing and quantifying surface topography and composition have developed at a fast pace, taking advantage of the microcomputer and modern physical analytical techniques.

5.7.1 Surface Topography

Several of the techniques are briefly described in an appendix to the relevant American standard [3.306]. A comprehensive treatment has been given by Thomas et al [3.71].

Stylus Measurement. This method was described in Sec.3.8.1. In addition to books [3.71,3.87], reviews with varying emphases are available [159-163]. For all its convenience and widespread application, stylus measurement does suffer from some drawbacks:

a. The stylus must have a finite radius (typically, 10 μm in industrial instruments, and as small as 1.2 μm in high-resolution research instruments); thus very small details (secondary asperities) are not revealed. The finite radius imposes a limitation also on the sharpest peak than can be measured, and affects the derived roughness parameters [164]. Limitations stem also from stylus kinematics [165].

b. The stylus can neither reach into re-entrant surface details nor follow very steep slopes, yet both features are encountered on deformed surfaces. The stylus may even get stuck in deep crevices such as those found in grinding wheels and coated abrasives; Shah et al [166] solve this problem by oscillating the stylus while the surface below it moves incrementally, so that the surface is stationary when contacted by the stylus.

c. Softer materials are scratched by the stylus, and as a result the measured roughness is less than the actual one [167,168].

Nevertheless, the stylus method is preferred in both industry and research, because it can be readily coupled to computers and thus the trace can be directly processed. The methods of analysis, their limitations, and the meaning of the derived values have been the subjects of intensive research and debate [169-171]. Often, the change in surface topography in the course of contact is of interest; then the profile must be determined before and after contact, and this calls for careful relocation of the specimen [172].

Optical Reflection Techniques. When light hits the surface, at least some of it is reflected. Reflection is called "specular" when the beam is reflected at the same angle as the angle of incidence and is called "diffused" when the reflected energy is distributed. The ratio of the two depends on surface roughness. The human eye is a very critical judge of reflected light and perceives variations that escaped quantification by earlier versions of glossmeters. Gloss is inversely proportional to surface roughness, but the relationship is not simple [173,174]. With the advent of the laser, substantial development has taken place [175,176]. Diffused light reflection gives information on the mean slopes of surfaces including fine details that are missed by the stylus instrument, whereas specular reflection is a function of roughness height distribution. The technique is capable of assessing surfaces that are otherwise difficult to measure, such as diamond-turned finishes [177]. The power spectrum of the surface can be obtained by optical Fourier transformation [178].

Because the technique is noncontacting, it offers the possibility of on-line measurement of surface roughness and of feedback in truly closed-loop automation of machining processes [179]. The signals from fiber-optic transducers can be analyzed [180], or the elliptical polarization of the specular reflection can be measured. An assessment of the applicability of various techniques to manufactured surfaces has been given by Teague et al [181].

The need for inspection at high speeds has also led to the development of surface flaw detectors capable of operating at speeds typical of sheet-processing lines [182,183]. Simple closed-circuit television serves as an extension of the human eye that can be used for visual inspection in inaccessible, dangerous, or hot areas. Inspection of moving surfaces at selected spots is possible if the image is arrested.

There are a number of other techniques, reviewed by Thomas [71,164], suitable for in-plant applications.

Contact Area. In many instances the major concern is not so much the roughness of one surface but the contact area (bearing area) between two surfaces, as reviewed by Woo and Thomas [184]. In metalworking, the contact area is 100% in machining; it is still very high in bulk deformation under boundary or mixed-film conditions, and becomes relatively small only in sheet metalworking. In the latter case optical microscopy is helpful, because asperities become flattened and, if the proper angle of illumination or phase-contrast is used, plateaus show up as reflecting areas. Their proportion to the whole surface can be determined by the usual metallographic techniques (line intercept, Quantimet, etc.).

Interferometry. The surface topography of materials may be directly visualized by interferometry [185-187].

a. In the simplest form, a monochromatic beam of light is split; one beam is reflected from the surface of the specimen, the other from an optically flat reference surface. The two images are superimposed; waves reflected from equal-height locations will be in phase and thus bright, whereas those reflected from different heights will be out of phase, will annul each other, and will be dark. A topographic map with bright and dark contour lines is obtained, in which a level difference between adjacent fringes is equal to half the light wavelength. Multiple-beam interference fringes are sharper. They are formed by repeated reflection between the surface and a reference surface coated with a thin, semireflecting film. In the example shown in Fig.5.31a an electrolytically polished, macroscopically flat, locally rough aluminum surface is further distorted by the passage of a hemispherical slider. A mechanically polished surface (Fig.5.31b) is smooth, as shown by the parallelism of fringes obtained on the slightly tilted surface; the distortion caused by the passage of the slider is now clearly discernible [188].

b. Of particular benefit is the Nomarski technique [3.306]. When white light is used, small differences in height are revealed by changes in color.

c. Some optical microscopes form a light section of the surface which needs no interpretation and can be evaluated for the shape and distribution of asperities, although the resolution is poorer than that obtained in the stylus technique (typically, 1 μm).

d. A powerful aid to the visualization of surfaces is holographic contouring [185,189].

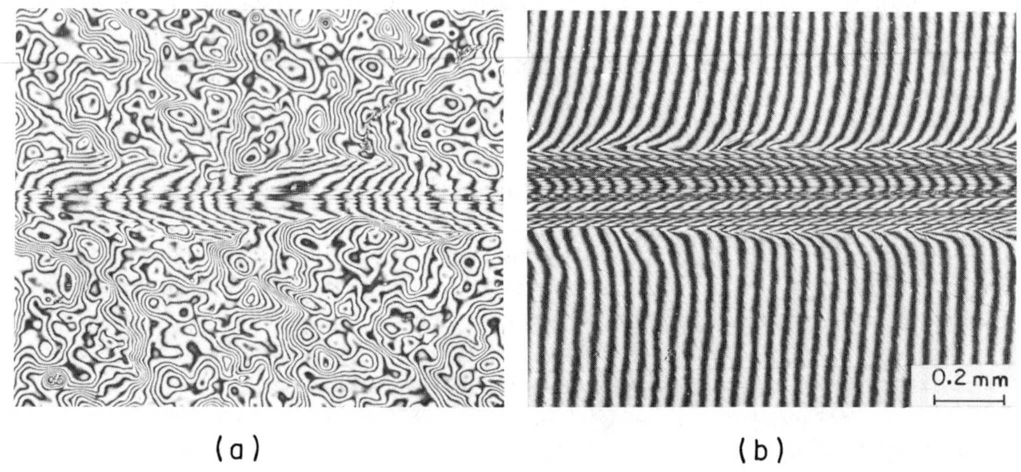

Fig.5.31. Slider tracks on (a) electrolytically and (b) mechanically polished aluminum surfaces [188]. (Courtesy of Dr. D. Scott, National Engineering Laboratory, East Kilbridge, Scotland; British Crown Copyright)

Optical Microscopy.

a. One of the most useful aids for the inspection of surfaces is the magnifying glass or, better yet, the stereo microscope with up to 50× magnification. The over-all view gained at these low powers often reveals features that are lost at higher magnifications.

b. The metallographic microscope is of only limited utility because vertical illumination tends to wash out details unless special techniques are used. Dark-field and, especially, oblique illumination give some feel for three-dimensional features; phase contrast and polarized light are useful in specific instances, as reviewed by Scott [188]. Tapered sections are most informative. The specimen is cut at an angle to the original surface and then carefully polished. To preserve minor surface features, the specimen is first electroplated with a thick film of nickel or coated with solder. Undercuts and jagged features not visible on stylus traces are then revealed (Fig.5.32).

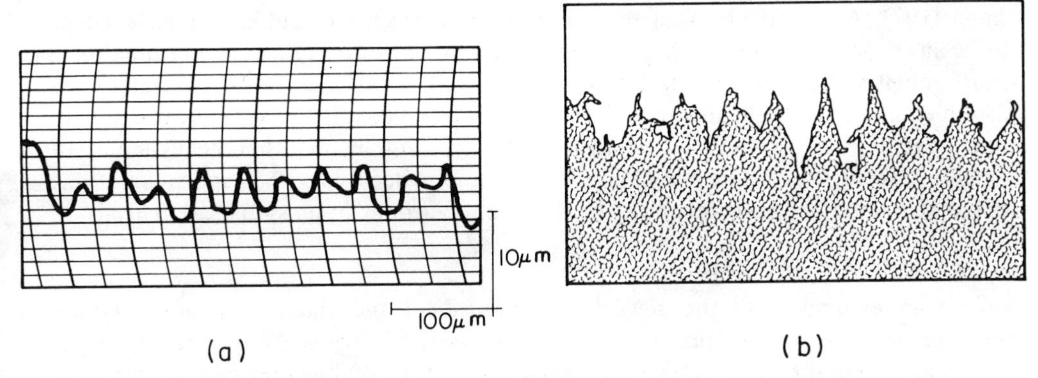

Fig.5.32. Surface features (a) recorded by a stylus instrument and (b) revealed by taper sectioning [188]

Electron Microscopy. Reviews of electron microscopy have been published by Scott and Scott [190], by Quinn [191,192], and by Buckley [193],[3.75]. Originally, transmission electron microscopes (TEM) were used on carefully prepared replicas. Wear debris is picked up by replicas, and therefore they have been found extremely useful in wear studies. The greatest advances in observing qualitative features of surfaces were, however, made with the scanning electron microscope (SEM). It combines high resolution with great depth of focus, creating a three-dimensional effect and revealing previously unobserved details (Fig.5.33) [194]. Modern instruments have quite large chambers to admit tools, dies, or workpieces. Smaller specimens can be cut from the die or workpiece, or may even be designed as inserts to be removed after the forming or cutting process. The taper-sectioning technique reveals surface and subsurface layers [194,195].

In situ observations of friction and wear phenomena are of great value [196], particularly for metalcutting.

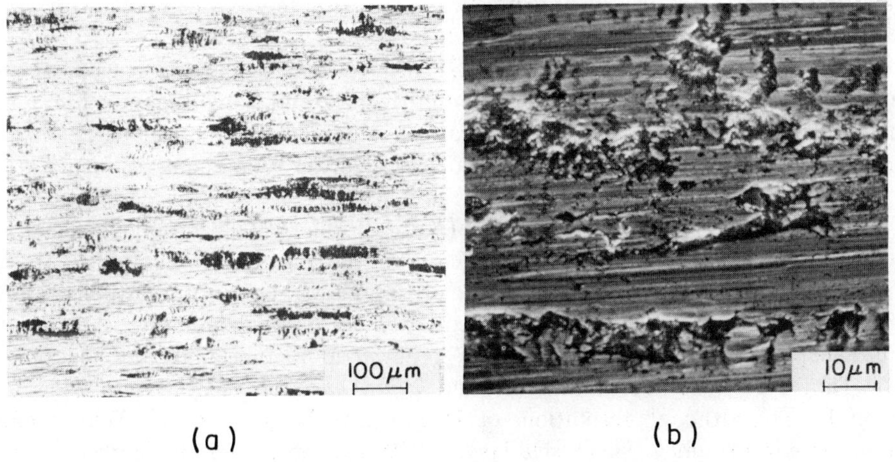

Fig.5.33. Rolled aluminum strip surface as viewed by (a) optical microscopy and (b) scanning electron microscopy (SEM) [194]

Replicas. Portable instruments for on-site roughness measurement seldom yield detailed analyses, and more information can often be obtained from replicas. A number of thermosetting resins produce high-fidelity negative images of surfaces. Details other than undercuts are preserved, but waviness or over-all shape may be distorted by the bowing of quick-setting resin replicas. Epoxies with a minimum of 40% aluminum powder filler set much slower (in days rather than minutes) but retain their macroscopic shape [197]. Alternatively, thin replicas may be prepared by rubbing a softened polymeric sheet on the surface. These replicas are suitable for roughness determination and interferometry, but not for detailed surface analysis [198]. For optical examination, and also to enhance the SEM image and provide electrical conductivity, replicas can be vacuum coated with vapor-deposited gold, silver, chromium, or other heavy-metal films which retain surface detail but ensure opacity and reflectivity. The replica is a negative image of the surface, and this must be taken into account in its statistical analysis.

5.7.2 Surface Analysis

For an understanding of the powerful compositional and chemical factors, there is a need for analytical techniques capable of dealing with very small quantities, very thin layers, and complex compositions. Development in this field has been spectacular in the last decade and has been reviewed from the tribological point of view [3.75],[192,199-203]. Only the briefest outline will be given here.

1. X-ray diffraction relies on the scattering of x-rays by the crystal lattice. It is the oldest technique suitable for studying the crystallinity of fairly thick (approximately 10 μm) surface zones, and in the hands of the knowledgeable practitioner it can identify compounds and analyze wear debris. Glancing x-rays penetrate less deeply and are more suitable for surface films.

2. Electron diffraction has similar applications, but the shorter-wave electron beam penetrates much less and is thus suitable for examination of the uppermost surface layers (about 20 atomic layers).

3. Low-energy electron diffraction (LEED) relies on the scattering and reflection of low-energy electrons and is suitable for examination of the atomic arrangement in the outermost atomic layers, including adsorbed films.

4. Electron microprobe analysis (EMA, EPMA, or EMP) relies on the x-rays emitted by a substrate when bombarded with electrons. The spectrometer gives elemental analysis and, combined with the SEM, it rapidly produces a readily interpreted map of the distribution of elements in a fairly thick (approximately 2 μm) layer, with a resolution (beam area) no greater than 1 μm in diameter.

5. Auger electron spectroscopy (AES) also gives elemental analysis for elements heavier than helium, to very small depths and with a resolution of about 1 μm. Changes in composition as a function of depth can be obtained on removal of layers by ion bombardment (sputtering). By use of a scanning electron beam, compositional variations can be mapped.

6. X-ray photoelectron spectroscopy (XPS), formerly known as electron spectroscopy for chemical analysis (ESCA), measures the energy of electrons emitted on bombardment by x-rays. It has the great advantage of giving information not only on the elements present, but also on their oxidation state and bonding. It detects all elements except hydrogen, and has become a powerful tool in analyzing reaction films [204]. The layer analyzed is thin (around 2 nm), but the spatial resolution is poor (around 1 mm^2, although resolutions of 15 μm have been achieved). When combined with an ion gun to remove successive layers, XPS too can provide in-depth analysis.

7. Energy-dispersive x-ray fluorescence (EDXRF) impinges x-rays on the surface to

detect x-rays characteristic of the elements in the specimen. Penetration depth is rather large, about 10 μm; the method is fast, detects elements heavier than sodium, and has found application in E.P. lubrication.

8. Secondary-ion mass spectroscopy (SIMS) uses a mass spectrometer to analyze ions ejected from the surface bombarded with argon, helium, or neon ions. Both elements and compounds can be identified from a relatively thin surface layer (as thin as 2 to 3 monolayers if light ions are used).

9. Ion-scattering spectroscopy (ISS) relies on measurement of the energy of low-energy, inert ions (such as He or Ne) scattered from atoms of the specimen surface, thus giving elemental analysis of the very surface layer. When combined with SIMS, it has been successful in identifying reaction films [205].

10. The field-ion microscope is capable of resolving individual atom sites. When it is used in conjunction with the atom probe, both structural arrangement and chemical analysis of the surface are possible at the atomic level.

Several of these techniques have been applied to general tribological problems, including the analysis of wear debris, surface films formed by boundary and E.P. lubricants, and solid lubricants. Relatively few applications to metalworking have been reported, mostly in the field of sheet metalworking (Sec.10.4.4). The techniques are not without their problems, especially in the mixed-film lubrication regime. Because techniques suitable for identifying compounds rather than elements tend to have poor resolution, it is difficult to investigate areas (plateaus) on which reaction products could be isolated without the disturbing influence of lubricant residues in the hydrostatic pockets.

5.7.3 Surface Preparation

No work in tribology can be valid if the quality of contacting surfaces is uncontrolled. One of two philosophies is usually followed: first, the surface is prepared by techniques typical of the production environment but according to reproducible, well-defined procedures. Second, surfaces are prepared by laboratory techniques.

Technological Studies. If the experiments are designed to evaluate practical lubricants or the effects of process variables, or are otherwise practically oriented, the main criterion is reproducibility.

a. Dies are given a preparation typical of good work practices. Thus, all lubricant residues from prior operations are removed and, if necessary, the surface is reconditioned to a controlled roughness. This may range from a ground or shot-blasted surface of 1 to 5 μm (40 to 180 μin.) AA roughness for some rolling operations, through a roughness of 0.2 to 0.6 μm (8 to 24 μin.) for most rolling mill rolls, forging dies, and cutting tools, to a mirror finish of sometimes better than 0.05 μm (2 μin.) for bright-finished rolls and draw and cold extrusion dies. The direction of surface finish must be in the correct relation to the direction of material movement (Sec.3.8.6); otherwise, the relevance of the test will be lost.

If oily residues are to be removed, only nonpolar liquids such as mineral spirits, hexane, or heptane should be used. Solvents such as trichlorethylene and alcohols have polar properties, and adsorbed films of lubricating quality are left behind. Industrious rubbing with a cloth immersed in the solvent only helps to build up these films. Because moisture from the air enters into the lubrication mechanisms and because production dies usually run at over 100 C even in cold working, dies should be heated to 110 to 120 C.

The condition of the dies must be closely monitored. Very minor pickup on or scoring of a polished die during testing of one lubricant can interfere with the action of a subsequently tested lubricant, and may render futile even the simplest attempts at ranking. Some materials, such as aluminum, may be removed from the surface of a steel die by etching with a caustic soda solution, but others can be removed only by mechanical means. Surface topography is likely to be changed, and then the only acceptable solution is a complete redressing of the surface.

b. The workpiece, too, must be prepared to a reproducible surface quality. Surface finish must be strictly controlled because of its effect on lubricant trapping and transportation (Sec.3.8.5). The surface should be of a consistent technological quality. Annealed surfaces carry oxide films and are best left alone; if the oxide must be removed by some suitable pickling technique, this should be followed by cold- and hot-water rinsing and rapid drying in blowing air. Rinse water must be of constant quality and, preferably, should be deionized, because deposits due to water hardness may serve as lubricant carriers. Residues from prior processing may resist removal; removal of excess bulk lubricant with a nonpolar solvent may suffice unless the resulting residues could mask the effects of lubricants in subsequent testing. Then alkaline degreasing followed by rinsing in water and drying may be adequate for less critical purposes, or a thin layer of the workpiece material may have to be removed electrolytically or by some chemical etching (pickling) process [4.157]. It is imperative, though, that no polar solvents be used for final rinsing, because they will again leave lubricating residues. Prepared specimens should not be touched by hand.

Specimens for hot working experiments need similarly careful preparation. At preheating temperatures of 300 to 400 C, the lubricant residues oxidize, polymerize, and form tenacious films. At higher temperatures, the residues may influence the course of oxidation. Heating practices themselves may introduce an often undetected variable: oxidation in a small experimental furnace may be substantially different from that occurring in practice, and careful control of heating rates, holding times, and furnace atmospheres is necessary for relevant results.

Fundamental Studies. In fundamental studies all steps of preparation and handling are even more critical. Prepared workpieces must be handled only with tongs or gloved hands, and experiments must be conducted in rooms of controlled temperature and humidity. Storage of workpieces and dies after preparation is usually a source of random variations in results.

Well-reproducible surfaces for basic experiments are often obtained with standard metallographic techniques, by abrasion under water or a flood of methanol. Uniformity and controlled directionality of the surface are important; scratches lead to erroneous results. Electrolytic polishing or sputtering gives cleaner surfaces but a rather nontypical roughness.

It would appear that some of the important precautions have not been observed in many past experiments, making the results suspect. Fortunately, some obvious weaknesses of practice are readily identified and can be taken into consideration when evaluating published information.

Checks for Cleanness. With the development of more sensitive surface-analytical techniques, the concept of a clean surface keeps changing. However, for most metalworking applications, the most important criteria are still reproducibility and technical cleanness. Some very simple checks for freedom of oily residues are sometimes used. Immersion in water followed by observation of the continuity of the water film (water-break test) is not unequivocal, because hydrophilic residues actually improve film continuity. The copper strike test, which is somewhat better, is based on the cementation of copper onto a clean steel surface from an acid copper-sulfate solution [206],[4.157].

Freedom from wear debris and solid lubricant residues is often checked by simply pressing a transparent adhesive tape onto the surface; after the tape has been lifted off, particles can be inspected under the microscope. Schreiber and Ziehm [207] use a strippable polymeric coating. SEM is, in general, helpful too, although the entire battery of chemical and physical techniques is needed for complete characterization of surfaces [208-212].

5.8 Measurement of Film Thickness

The discussion in Sec.3.6.4 showed that, even if the lubricant film is thick enough to be treated by PHD theory, some metallic contact still occurs. Thus, one can speak only of an average film thickness, the magnitude of which can give valuable clues regarding the operative lubrication mechanism.

a. If care is taken in handling the product emerging from the deformation zone (and excess lubricant carried over on the free edges is soaked up), the average film thickness can be obtained by weighing the sample before and after degreasing or, in glass lubrication, by etching or dissolving in molten NaOH.

b. The electrical resistance of an oil film becomes measurable when a fully insulating film develops at least for a fraction of a second [3.174],[213]. Ranger and Wistreich [120] found a resistance of 0.1 to 3 Ω when drawing steel wire with a soap. A high fixed resistance in series with the lubricant film (Fig.5.34a) keeps the current flowing through the film fairly constantly and makes the system more sensitive to variations in film resistance. Even though rapid variations indicate periodic metallic contact, film thickness may be estimated from the average voltage [214]. This technique has been improved by Tonck et al [215] through the provision of a feedback loop so that the voltage drop remains constant. With the aid of electronic circuitry and computer evaluation of data, the formation of boundary films could be followed.

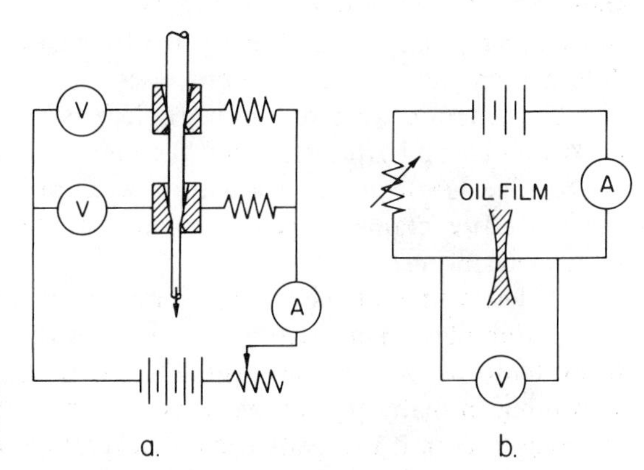

Fig.5.34. Measurement of film thickness by (a) resistance and (b) electrical discharge methods

c. The discharge voltage method used in EHD research [216] could also be applicable. A heavy current, on the order of 1 to 5 A, is passed across the oil film, causing an arc discharge. The voltage V necessary to maintain continuous discharge (Fig.5.34b) is not necessarily directly proportional to film thickness; nevertheless, the technique is useful for relatively thick films (1 μm or thicker) and high sliding speeds.

d. The capacitance of a truly continuous film can be measured with the aid of capacitance bridges [217], although interpretation is difficult when polar molecules change the dielectric constant.

e. The change in magnetic flux across the contact zone has been used in EHD studies [218,219] but demands a special, two-disk machine configuration which does not lend itself to ready adaptation to metalworking.

f. A square, monochromatic beam of x-rays can be directed through the gap between two rollers and the film thickness can be judged from the amount transmitted [220]; this technique would be difficult to apply to metalworking.

g. Somewhat related is the laser-beam diffraction technique [221], again with no obvious application to metalworking.

h. Optical interferometry has been greatly developed for EHD studies [222,223] but requires that at least one of the contacting materials (in our instance, the die) be transparent. Nevertheless, this technique could be used for simulation, although no trace of its application could be found.

i. Radiotracer techniques can be useful, especially with radioactive glasses [224]. The thickness of glass films was measured in hot extrusion [3.442] and that of phosphate coatings in wiredrawing [225].

5.9 Measurement of Wear

There is an increasing awareness that, in many instances, wear is of greater importance than friction itself. Techniques of testing for wear have greatly developed [226]; however, the first hurdle to be overcome is the securing of a sample that can be analyzed and is, at the same time, relevant to the tribological situation.

5.9.1 Sampling for Wear Studies

1. In principle, the tool (die), workpiece, and lubricant all yield valuable information:

a. Wear of the tool or die is most frequently investigated because, in its early stages, it affects the lubrication mechanism; later on it influences the surface quality of the issuing product, and finally it affects the dimensional tolerances in both plastic deformation and metalcutting processes.

b. Moderate wear of the workpiece is accepted and is seldom measured unless it causes problems by transfer onto the die.

c. The lubricant carries with it wear products and also products of tribochemical reactions. It contains, therefore, significant clues regarding the operative lubrication and wear mechanisms.

2. The source of the sample is important, too:

a. Samples taken from production are the most relevant but also the most difficult to evaluate because of the complex, confusing, and sometimes uncontrolled conditions prevailing in many production environments. Furthermore, the length of a run may be too long to permit the evaluation of several die materials or process variables. Testing is accelerated by the use of dies with inserted pins made of several die materials, as has been done for forging dies [227], although the presence of a pin may disturb interface conditions and thus make the results doubtful.

b. Simulation tests, discussed in Sec.5.4, give information on wear, but not all of them are equally suitable. Adhesive wear is detectable in many of them, but the duration of the test is usually insufficient to reveal long-term and cumulative wear tendencies. Acceleration of the wear process by increasing the severity of interface conditions carries with it the danger that the lubrication mechanism may change to one quite unrepresentative of production conditions. The only solution then is to increase the sensitivity of wear detection, for example, by the use of a softened facing applied to a hardened tool backing [228].

5.9.2 Qualitative Wear Analysis

If wear is to be controlled, its sources must be identified, drawing upon all the techniques discussed in Sec.5.7.2 [229], including a variety of metallurgical techniques [230,231].

One of the most valuable tools is the scanning electron microscope (SEM). Coupled with the microprobe or energy-dispersive x-ray analysis, it reveals the extent and composition of metal transfer. Grooves (furrows) are indicative of abrasive wear, and abrasive particles lodged in the surface are also visible. Surface fatigue and delamination wear are identifiable by the characteristic appearance of wear particles and of partially detached surface features. Corrosive wear is evident from the preferential removal of some constituents. Chemical and physical surface-analytical techniques are valuable in the further identification of worn surface features and wear debris. Examples of the application of the SEM and TEM (transmission electron microscope) have been given by Ruff et al [231a].

5.9.3 Quantitative Wear Analysis

Quantification of wear is desirable for predicting the useful life of a tool or die and for designing process models needed for preprogrammed tool changes.

1. Weight loss is readily determined if it is a large enough fraction of the total weight. Plants usually keep records of the quantity (tonnage, number of pieces) produced with any given set of tools (work rolls; drawing, forging, or extrusion dies; punches; deep drawing dies; cutting tools) before regrinding or refinishing of the surface becomes necessary, and before the die has to be discarded or finished to the next larger size because the dimensions of the product no longer fall within the tolerance limits. Wear rates can be expressed as volume, weight, or dimensional change per unit sliding distance (product length), product weight, or time.

2. The more detailed measurements of die dimensions needed for following the course of wear may disrupt production; therefore, most data related to localized die wear are the results of special investigations. Mechanical and optical instruments can be used, and some special devices have been developed for wiredrawing dies and other internal cavities [232-235]. Cast replicas are often preferable, and elastomeric compounds [236] are useful for worn profiles. Ideally, information is obtained in the course of production, and advances have been made in the *in-situ* measurement of tool wear in metalcutting [237,238].

3. Radiactive tracer techniques are among the most powerful tools.

a. A radioisotope is incorporated in the die (or, occasionally, the workpiece) material during preparation (melting, powder compaction, etc.) of the material, or, more frequently, the material is exposed to neutron radiation in a nuclear reactor to activate some of the elements. Transfer of radioactive material onto the nonactive contacting surface may then be followed either by autoradiography or by counting techniques [239,240]. The isotope is chosen for a reasonably short half-life, but the problems of handling and residual radiation remain. Great care must be exercised in evaluating the results, as shown by Golden and Rowe [3.442], and observations are best supplemented by other techniques such as SEM. Nevertheless, the technique is a powerful one when the size of the irradiated die is small enough, as in wiredrawing and metalcutting [241,242].

b. When the isotope of interest has a long half-life and the workpiece or die is so large that total activity would necessitate expensive precautionary measures, an irradiated die or tool insert may be used [243]. Sensing the presence or absence of a very small radioactive particle embedded in the tool flank at a predetermined distance from the cutting edge gives a very rapid indication of whether a prescribed amount of wear has been reached [244]. It could also be applied to other processes as an in-process, go - no go indicator. If wear of the nonradioactive workpiece is of interest, the workpiece can be clad with a layer that contains the radioisotope [3.303].

c. Thin-layer activation [245-249] represents a significant advance. The tool or die surface is made radioactive by exposure to a beam of charged particles from a cyclotron or electrostatic generator, thus limiting the depth of activation to less than 200 μm (and even 50 μm). Direct determination of wear is then possible by measurment of the residual activity of the tool or die. Total activity is low enough to obviate the need for special precautions.

d. The lubricant, too, can be made radioactive; for example, polymers can be doped with a radioactive oxide for the study of transfer [250].

4. Separation of wear particles is greatly facilitated by an instrument (the Ferrograph) developed by Seifert and Westcott [251]. The used oil sample is pumped across an inclined slide which sits on a magnet that has a strength varying along the length of the slide. Magnetic particles ranging in size from 20 μm to several hundred μm are held on the substrate, in a narrow band, graded according to size. The slide is washed and fixed so as to lock the particles in place. The particle distribution is obtained from local opacity, or a bichromatic microscope is used. In the latter, red light is guided to the top of the ferrogram, and thus opaque objects appear red; and green light is projected from the bottom, and thus transparent objects appear green. Translucent material assumes an in-between color. Individual particles can then be analyzed by standard methods. This technique has been used extensively in nonmetalworking studies [252,253], and various improvements have been made, including the use of the Quantimet [254] and a direct reading instrument [255]. The magnet does catch some nonmetallic materials, while missing some of the small magnetic particles; therefore, this method is not a straight replacement for filtering techniques [256], although it could be a powerful aid in metalworking tribology as well.

5. A variety of other techniques, such as flow ultramicroscopy [257] and scintillation suppression [258], may also find application.

6. The acoustic emission technique has found wide application in general fracture-related studies, and has potential also in metalworking tribology [259]. A rather obvious application is the in-process detection of tool wear [260] and breakage [261], but acoustic emission analysis shows promise also for following the variation of friction in such processes as wiredrawing; there appears to be a reasonable relationship between the acoustic energy signal level and the magnitude of the coefficient of friction [262].

7. Another intriguing possibility is the following of the breakdown of lubricant films by monitoring of exoelectron emission [263]. As discussed in Sec.3.1.3, a freshly rubbed surface emits electrons, and therefore a rise in the photostimulated exoelectron emission rate indicates lubricant film failure.

5.10 Staining Propensity and Corrosion

Most lubricants leave a residue on the surface of the workpiece. Three courses of action are possible:

1. The lubricant residue can be removed. This is required when the residue inter-

feres with subsequent machining or finishing, or general use of the workpiece, as would, for example, the glass used in hot extrusion or isothermal forging. Organic-base or aqueous residues may interfere with joining or finishing operations and may have to be removed by degreasing, pickling, or other surface treatment [4.157].

2. The lubricant can be left on the surface—either intentionally, such as when an oil-base lubricant provides protection against corrosion, or simply for economic reasons. It is then the task of lubricant formulation to prevent undesirable side effects such as staining—i.e., the formation of visible marks on prolonged storage, either because of changes (such as polymerization) in the lubrication residue itself, or because of reactions with the metal surface. Reactions may lead also to corrosion of the metal, especially with aqueous lubricant residues.

3. The workpiece can be subjected to heat treatment. The lubricant residue may be converted into a dense, tenacious, and virtually irremovable product which appears as an unsightly stain or may, even if not visible, interfere with subsequent finishing operations.

The formation of stains depends on both lubricant composition and heat treatment conditions. If the surface is freely accessible to the heat treating atmosphere, many lubricants evaporate. However, if small gaps are formed, particularly between laps of a coiled strip or layers of stacked sheets, heating subjects the lubricant to distillation under an increased partial vapor pressure; higher-molecular-weight residues polymerize, and highly viscous or solid products form a brown stain. At higher temperatures most residues oxidize (white stain) and may finally burn off. Because of low heat treatment temperatures, the highly reflective aluminum (and, to a lesser extent, copper and brass) workpieces are most susceptible to this problem. A related defect, known as "snakey edge," forms on steel strips as a result of interactions between lubricant residues and protective atmospheres (Sec.6.3.5).

For the laboratory assessment of staining propensity, a number of tests have been developed, all of which attempt to create a controlled partial vapor pressure while heating a measured quantity of lubricant in contact with the workpiece at a controlled heating rate, designed to simulate production conditions. Vapor pressure can be controlled by placing drops of oil in a shallow can with a perforated lid [264], by clamping together several sheets coated with the lubricant [265], or by coiling a length of strip over a small arbor [266]. Alternatively, the lubricant is placed on a sheet of standardized surface finish and the rate of evaporation is controlled by adjusting access of air to the furnace [267]. Emulsions have insufficient viscosity to stay on flat surfaces and are usually tested on sheet with standard-size depressions [268]. The atmosphere must be the same as in annealing practice [269-271]. White staining was tested [264] by placing the lubricant between two sheets, covering the edge of the sandwich with aluminum foil, and heating first to 350 C (where brown staining occurs) and subsequently to 450 C (where, on holding for 2 h, white staining develops).

In testing for water stains caused by emulsion residues or by moisture condensation from the atmosphere, the essential feature is again that of creating a gap either by stacking sheets or by coiling a strip [272]. Standard corrosion tests exist but are not free of problems [273].

Staining tests become reproducible only if all variables—including the roughness and cleanness of the test sheet, heating rates, and vapor pressure—are strictly controlled. The tests are qualitative, but can be used for lubricant evaluation or production-control purposes if control experiments are carried out on lubricants of known staining propensity. The quantity of carbon left on the surface may be determined by gas-chromatography [271].

A related but different problem is staining or corrosion of workpiece surfaces during storage, usually in the form of corrosion caused or accelerated by the presence of moisture. The task then is to apply a protective oil film, and the purpose of testing is to identify oils and additives which will give protection. Accelerated testing in steam, water fog, and humidity chambers is possible, and gives good correlation with production experience [274].

5.11 Lubricant Composition and Technological Properties

Metalworking lubricants may be rather well-defined, relatively pure substances, naturally occurring mixtures of more or less reproducible quality, or propriety combinations of any of the foregoing groups. Some physical and chemical properties essential for the control of the lubrication process are usually specified even though exact compositions are seldom disclosed.

Standardized methods of determining most physical and chemical properties are available in the publications of the American Society for Testing and Materials, Committee D2 (see Appendix), the American Petroleum Institute, and the (British) Institute of Petroleum. Corresponding standards exist in other countries. The most frequently encountered specifications are defined in the Appendix. Several monographs and reviews are available on rheological techniques [275-277],[3.263]. Great advances have been made in analytical techniques for oils [278] and greases [279], and efforts are being made to analyze used and contaminated oils [280-282], even in the field [283]. Spectrographic analysis has become routine for oils in general [4.147], and infrared monitoring of additive concentrations in aluminum-working emulsions has been reported [284]. Physical tests such as those for air-entraining tendency [285,286] are valuable, particularly for circulating systems. Techniques for monitoring emulsions were discussed in Sec.4.6.6.

There is an almost endless variety of tests specific to particular lubricant applications.

1. The cooling ability of a lubricant can be measured by suspending a thermocouple in a guillotine, and then dropping it into the lubricant. The cooling rate, recorded on a fast-response instrument, is a measure of quenching ability [73]. Heat transfer rates can be studied by cooling a preheated specimen in which thermocouples are buried.

2. Wetting is of considerable concern when complete coverage of the working surface with a preapplied film has to be relied upon. Good wetting is regarded as desirable also in flooding applications. Most wetting tests are qualitative or semiquantitative.

a. In the simplest form of wetting test, the lubricant is placed on a prepared test plate which is then inclined at an angle to allow the fluid to run. More quantitative is the method where a measured quantity of lubricant is placed on a prepared horizontal metal surface and the rate of spread of the drop is measured, if desired, as a function of substrate temperature. This technique is widely used for testing the wetting of metal surfaces by glasses used in hot working. A more accurate measure of wetting is the contact angle, which is, however, affected by glass-metal reactions.

b. In the test described by Snow [268], an aluminum-rolling emulsion sample is placed in a trough, and a roller, partially submerged in the emulsion, is rotated. After rotation has been stopped, the pattern and rapidity of dewetting of the surface is observed.

c. A more quantitative test by Miyagawa et al [287] relies on a thermocapillary effect. Because the surface tension of an oil decreases with temperature, a drop of oil placed on a plate heated at one end will spread toward the cooler end. The rate of spreading was found to be inversely proportional to cold-rolling performance.

d. In some instances the wetting of the workpiece with the lubricant occurs through a surface reaction, which may cause rapid erosion or corrosion of the surfaces. Experiments for the determination of such tendencies have been devoted mostly to hot working lubricants of the glass or ceramic type. Rate of reaction may be determined by weight loss, by dimensional loss (of sheet), or by taking microhardness measurements close to the reacted edge so as to determine the depth to which reaction between lubricant and workpiece material occurred. More sophisticated techniques, such as radioactive tagging, are usually necessary for cold working lubricants, because reaction rates are slower.

5.12 Summary

1. Metalworking tribology can be successfully studied only if the complexity of the system is recognized. The variables to be investigated must be selected with due regard to the operative lubrication mechanism, and, if conditions are varied at all, the important features of the mechanism must still be retained.

2. To satisfy the above criteria, some factors such as temperature must be retained unchanged even in reduced-scale metalworking operations. For preliminary screening, test conditions may be made more severe only if the lubrication mechanism remains basically unchanged.

3. Transfer of data from one process to another is permissible only if the lubrication mechanisms, or at least some aspects of the mechanisms, are common to both processes.

4. Simulation tests are, in general, suitable for reproducing only some selected aspects of the real process, but are useful if their limitations are recognized. The extent of new surface generation, interface geometry, and lubricant resupply are particularly powerful variables. Tests can usually be designed for derivation of quantitative data on friction and, sometimes, also on wear. Bench tests are more suitable for evaluation of particular lubricant functions than for process simulation. Simulation tests must be repetitive or of longer duration if longer-term effects (incipient pickup, lubricant aging) are important.

5. The magnitudes of normal and frictional forces and the distributions of normal and shear stresses can be measured, with more or less confidence, by direct and indirect means. Measuring errors can be substantial unless special machine and die constructions are adopted. Back-calculation of frictional stresses, from data into which flow stress enters as an additional variable, is frought with danger.

6. Interface temperature is perhaps best measured with a dynamic thermocouple, although specially constructed conventional thermocouples suffer less from calibration difficulties. Measurements of radiation and metallurgical changes offer useful alternatives.

7. The stylus instrument is still the dominant technique for measuring surface roughness, especially when quantification is required. However, optical methods hold considerable promise and are suitable for on-line applications. Recent physical methods of surface analysis offer better characterizations of surface films and of surfaces prepared for experiments.

8. Average film thickness can be measured with some confidence by gravimetric and radiotracer techniques; local measurement is difficult because of the prevalence of mixed-film lubrication.

9. Wear is readily quantified by measuring dimensions, weight losses, or changes in radioactivity, provided that the run is long enough. All techniques of wear-particle analysis are applicable to metalworking, too.

10. Staining and corrosion are of special concern and can be measured by standard or special techniques.

11. Physical and chemical analyses are most helpful in identifying changes during prolonged lubricant use. Technological tests of cooling ability and wetting are of particular importance.

12. No single test can possibly yield all the information necessary for lubricant evaluation. Strategically chosen simulation or small-scale metalworking tests, coupled with functional tests (such as tests of staining, corrosion, etc.) are likely to yield the most relevant results. The ultimate judgment on the suitability of a lubricant still has to be made under operating conditions, but preliminary ranking and the knowledge necessary for the interpretation of production performance are best acquired in controlled laboratory experiments. A systematic evaluation of process conditions and lubrication mechanisms (Table 5.1) is helpful in avoiding major pitfalls. Such an approach is useful also for the selection of lubricants.

References

[1] *Experimental Methods in Tribology,* Proc. Inst. Mech. Eng., *182*, Pt.3G, 1967-68.
[2] J.A. Schey, in [3.1], pp.241-332.
[3] L.S. Akin, ASLE Trans., *9*, 1966, 249-256.
[4] O. Pawelski, Arch. Eisenhüttenw., *35*, 1964, 27-35 (German).
[5] D. Kast, Paper No. MF 70-185, SME, Dearborn, 1970.
[6] E. Siebel and E. Kotthaus, Mitt. Forsch.-Ges. Blechverarb., No.15, 1955, 181-185.
[7] W.R. Johnson, H. Schwartzbart, and J.P. Sheehan, Blast Furn. Steel Plant, *43*, 1955, 415-423.
[8] R.S. Barnes and T.H. Cafcas, Lubric. Eng., *10*, 1954, 147-150.
[9] R.C. Williams and R.K. Brandt, Lubric. Eng., *20*, 1964, 52-56.
[10] W. Lueg and W. Dahl, Stahl u. Eisen, *76*, 1956, 1669-1671 (German).
[11] A.T. Male, in [3.41], pp.33-38.
[12] S. Kapitány and M. Voith, Gép, *25*, 1973, 353-357 (Hungarian).
[13] J.A. Schey, in *Metal Forming: Interrelation Between Theory and Practice,* A.L. Hoffmanner (ed.), Plenum, New York, 1971, pp. 275-292.
[14] A.T. Male, in [1], pp.64-67.
[15] *Development of Method for Evaluating Lubricants for Cold Working Metals,* MIT Final Rep. WAL-TR 364/35, July 1959.
[16] O. Pawelski, Schmiertechnik, *13*, 1966, 267-273 (German).
[17] I.S. Morton, Ind. Lub. Trib., *24*, 1972, 163-169.
[18] N. Kawai and T. Nakamura, J. JSLE, *18* (3), 1973, 27-36 (Japanese).
[19] W. Voelkner, Fertigungstech. Betr., *26*, 1976, 678-681 (German).
[20] S.Ya. Veiler and V.I. Likhtman, *The Action of Lubricants in Pressure Working of Metals,* Izd. Akad. Nauk SSSR, Moscow, 1960, (translation: Rep. MCL-1389/1+2), W-PAFB, Ohio, 1961.
[21] N.H. Polakowski and L.B. Schmitt, Trans. AIME, *218*, 1960, 409-416.
[22] H. Wiegand and K.H. Kloos, Werkstatt u. Betrieb, *93*, 1960, 181-187 (German).
[23] P.R. Lancaster and G.W. Rowe, Wear, *2*, 1959, 428-437.
[24] O. Pawelski, Stahl u. Eisen, *84*, 1964, 1233-1243 (German).
[25] H. Kudo, S. Tanaka, K. Imamura, and K. Suzuki, CIRP, *25*, 1976, 179-184.
[26] W.R.D. Wilson and P. Cazeault, in *Proc. 4th NAMRC,* SME, 1976, pp.165-171.
[27] M.L. Devenpeck and J.H. Rigo, in *Proc. 7th NAMRC,* SME, Dearborn, 1979, pp.81-88.
[28] P.R. Lancaster and G.W. Rowe, Proc. Inst. Mech. Eng., *178*, 1963-64, 69-89.
[29] S. Fukui et al, Int. J. Mech. Sci., *4*, 1962, 297-311.
[30] G. Sachs, Metallwirtschaft, *9*, 1930, 213-218 (German).
[31] N. Kawai and T. Nakamura, Bull. JSME, *17*, 1974, 810-818.
[32] E. Doege, R. Melching, and G. Kowallick, J. Mech. Work. Tech., *2*, 1978, 129-143.
[33] I.M. Pawlow and P.S. Kostytschew, *Grundlagen der Metallverformung durch Druck,* Verlag Technik, Berlin, 1954, pp.654-665 (German translation).
[34] J. Nittel, Fertigungstech. Betr., *18*, 1968, 301-304 (German).

Table 5.1. Data sheet for evaluating the process, lubricating mechanism, and simulating test functions
(The process example given is for cold rolling.)

Process: cold rolling with tensions

ISSUING PRODUCT Cold-rolled strip (5000-kg coil) Material: 3003 Al Size: 1.5 mm × 1.5 m Surface: mill finish R_qL 0.4 μm R_qT 1.0 μm free from herringbone and from staining after annealing	PROCESS CHARACTERISTICS Contact: Periodic Lubricant supply: Continuous Process geometry: converging gap; $\alpha = 3$ deg. Speed: sliding, approach } rolling speed 600 m/min Pressure $< \sigma_f$ Atmosphere: air Access: limited Lubricant functions: pickup control ③ 2 1 0 low friction 3 ② 1 0 low wear 3 ② 1 0 heat extraction 3 ② 1 0 heat insulation 3 2 1 ⓪ other: low staining ③ 2 1 0	R = 200 10 m/s 2.0 1.4 α $\sin \alpha = \frac{L}{R}$
WORKPIECE Hot-rolled 3003 Al alloy strip (5000-kg coil) Temperature: room Size: 2.0 mm × 1.5 m Surface: R_qL 3 μm R_qT μm	Surface extension 43% Workpiece/die adhesion ③ 2 1 0 ③ 2 1 0 lubricant reaction 3 2 ① 0 Lubricating mechanism: hydrodynamic } mixed 3 ② 1 0 boundary 3 ② 1 0 E.P. 3 2 1 ⓪ solid-film 3 2 1 ⓪ other 3 2 1 ⓪ Wear: die 3 2 1 ⓪ workpiece 3 2 ① 0 Lubricant: Mineral oil (4 cSt at 40 C) + additive; temperature: 25 C.	DIE Forged steel roll 95 Shore C hardness Temperature: 80 C Size: 400 mm × 1.8 m work 1.2 m × 1.8 m backup Surface: circumferentially ground R_qL 0.2 μm R_qT 0.5 μm
SIMULATING TEST (geometry) laboratory rolling mill (converging gap; $\alpha = 0.66$ to 4.4 deg.)		
Work material: 1.0 mm × 25 mm × 1 m 3003-O strip Temperature: room Surface: R_qL 0.2 μm R_qT 0.5 μm	Contact: Periodic Lubricant supply: Continuous Lubricant temperature: 25 C Speed: sliding, approach } rolling speed 100 m/min Pressure: $\sim \sigma_f (1.0 - 2.0\sigma_f)$ Surface extension: 10–200% Lubricating mechanism: mixed-film Wear:	"Die" material: 52100 steel roll 150 mm diam Surface: R_qL 0.1 μm R_qT 0.3 μm

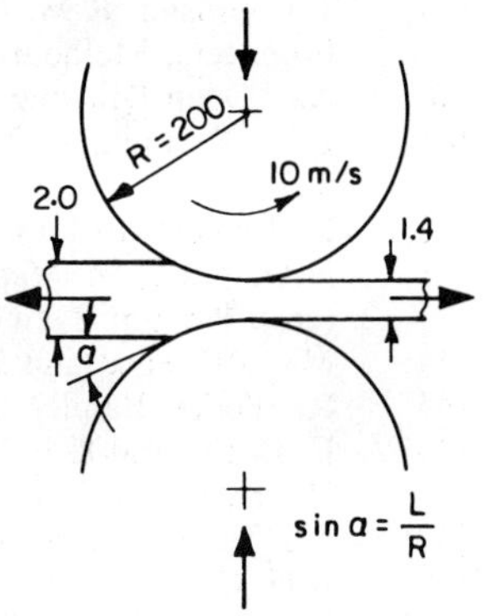

NOTE: 3 2 1 0: The property shown is known (or desired) to be present to a high, medium, or low degree or not at all (or is not important).

[35] K. Lange and Th. Gräbener, in [3.56], Paper No.15.
[36] W.J. Wojtowicz, Lubric. Eng., *11*, 1955, 174-177.
[37] K. Miyauchi, T. Furubayashi, T. Herai, and C. Sudo, in *Sheet Metal Forming and Formability* (Proc. 10th Biennial Congr. IDDRG), Portcullis Press, Redhill, Surrey, 1978, pp.287-301.
[38] H. Ike, K. Yoshida, and M. Murakawa, Wear, *72*, 1981, 143-155.
[39] T.J. Gibson, R.M. Hobbs, and P.D. Stewart, in *Int. Conf. Prod. Technol.*, Institution of Engineers, Melbourne, 1974, pp.328-332.
[40] Am. Deep Drawing Res. Group, Sheet Metal Ind., *54*, 1977, 147-153.
[41] L.M. Bernick, R.R. Hilsen, and C.L. Wandrei, Sheet Metal Ind., *55*, 1978, 827-830, 832-833.
[42] E. Tannert, Erdöl u. Kohle, *14*, 1961, 926-932 (German).
[43] H.D. Nine, in *Mechanics of Sheet Metal Forming*, D.P. Koistinen and N.M. Wang (ed.), Plenum, New York, 1978, pp.179-211.
[44] M. Littlewood and J.F. Wallace, Sheet Metal Ind., *41*, 1964, 925-930.
[45] R. Woska, Bänder Bleche Rohre, *20*, 1979, 562-564 (German).
[46] E. Doege and K.P. Witthüser, *HFF-Bericht 9.UKH*, Gesellschaft fur Produktionstechnik, Hannover, 1977, pp.129-137 (German).
[47] J.L. Duncan, B.S. Shabel, and J.G. Filho, SAE Paper No.780391, 1978.
[48] R.G. Quiney and W.E. Boren, Lubric. Eng., *27*, 1971, 254-258.
[49] P.L. Ko and C.A. Brockley, Trans. ASME, Ser.F, J. Lub. Tech., *92*, 1970, 543-549.
[50] F.T. Barwell, Met. Rev., *4*, 1959, 141-177.
[51] E.A. Smith, Metal Ind., *101*, 1962, 460-462.
[52] C.F. Hinsley, A.T. Male, and G.W. Rowe, Wear, *11*, 1968, 233-238.
[53] L.J. Noordermeer, A.L. Braun, and A.G. Tangena, in *Proc. 7th NAMRC*, SME, Dearborn, 1979, pp.159-163.
[54] M.B. Peterson and F.F. Ling, in [3.50], pp.39-68.
[55] K. Iwata and J. Aihara, Wear, *15*, 1970, 435-448.
[56] J.M. Bradford, K.B. Wear, and M.E. Sikorski, Sci. Instr., *41*, 1970, 1345-1347.
[57] L.R. Underwood, *The Rolling of Metals*, Wiley, New York, 1950, p.154.
[58] J.E. Williams and D.F. McCarthy, Wear, *58*, 1980, 381-385.
[59] P.S. Venkatesan, N. Ahmed, and I.B. Goldman, Wear, *17*, 1971, 245-258.
[60] R. Lauterbach, F. Lira, and E.G. Thomsen, Wear, *10*, 1967, 469-482.
[61] T. Wanheim, Wear, *25*, 1973, 225-244.
[62] H. Kudo and K. Tamura, CIRP, *17*, 1969, 297-305.
[63] H. Kudo and M. Tsubouchi, CIRP, *24*, 1975, 185-189.
[64] M. Oyane, S. Shima, and T. Nakayama, in [3.55], pp.13-21.
[65] R.H. Brown and E.J.A. Armarego, Wear, *6*, 1963, 106-117.
[66] B. Mills and A.H. Redford, CIRP, *28*, 1979, 165-169.
[67] Anon., Iron Steel Eng., *42* (12), 1965, 157.
[68] A. Sonntag, Prod. Eng., *30*, 1959 (June 22), 64-66.
[69] S.Ya. Veiler and V.I. Likhtman, Sov. Phys—Tech. Phys., *2*, 1957, 989-995.
[70] H.T. Azzam, Lubric. Eng., *24*, 1968, 366-376.
[71] R.D. Brown, in [3.28], pp.241-292.
[72] R. Benzing et al (ed.), *Friction and Wear Devices*, 2nd Ed., ASLE, Park Ridge, IL, 1976.
[73] J.F. Griffin, J. Metals, *9*, 1957, 1042-1043.
[74] M. Nagano, J. JSTP, *9*, 1968, 246-251 (Japanese).
[75] J.G. Carroll, Lubric. Eng., *24*, 1968, 359-365.
[76] E.L. Armstrong and M.S. Lindeman, Lubric. Eng., *32*, 1976, 182-189.
[77] M.W. Bailey and A. Cameron, Wear, *21*, 1972, 43-48.
[78] A. Gonin, A. Berger, and J. Neyron, Trib. Int., *13*, 1980, 25-29.
[79] E. Kellermann, Draht-Welt, *48*, 1962, 81-87 (German).
[80] R.G. Bayer, P.A. Engel, and J.L. Sirico, Wear, *19*, 1972, 343-354.
[81] H. Becker, Bänder Bleche Rohre, *19*, 1978, 362-366 (German).
[82] A. Begelinger and W.J. deGee, Lubric. Eng., *26*, 1970, 56-63.
[83] V. DePierre, Air Force Materials Laboratory, Dayton, OH, private communication.
[84] W.J. Bartz, Mineralöl-Technik, *10* (7/8), 1965, 1-34.
[85] G.W. Rowe and S.K. Roy-Chowdhury, Wear, *42*, 1977, 35-47.
[86] Discussion, J. Inst. Metals, *94*, 1966, 429-432.

[87] J. Molgaard, Wear, *41*, 1977, 57-62.
[88] J.A. Schey, P.W. Wallace, and K.M. Kulkarni, Tech. Rep. AFML-TR-68-141, 1968.
[89] P. Funke and C.-A. Schlemper, Bänder Bleche Rohre, *18*, 1977, 70-73 (German).
[90] G.J. Ebben, American Can Co., Chicago, private communication.
[91] S.N. Vojnovic, R.R. Somers, and W.L. Roberts, Iron Steel Eng., *41* (5), 1964, 95-101.
[92] J.M. Thorp, Wear, *37*, 1976, 241-250.
[93] J.A. Schey, Paper No. MF 71-126, SME, Dearborn, 1971.
[94] E. Vámos, I. Valasek, and A. Eleod, Schmierungstechik, *9*, 1978, 141-144.
[95] R.D. Butler and R.J. Pope, in [3.70], pp.162-170.
[96] F. Dohmann, Ind. Anz., *96*, 1974, 1617-1618, 2337-2341 (German).
[97] H. Bühler and J. Löwen, Stahl u. Eisen, *92*, 1972, 698-704 (German).
[98] W. Voelkner, Fertigungstech. Betr., *26*, 1976, 92-96 (German).
[99] I.M. Cole and D.H. Sansome, in *Proc. 9th Int. MTDR Conf.*, Pergamon, Oxford, 1968, pp.271-286.
[100] S. Takahashi, J. JSTP, *8*, 1967, 698-706 (Japanese).
[101] E. Siebel and W. Lueg, Arch. Eisenhüttenw., *15*, 1933, 1-14 (German).
[102] J.W. Kannel and T.A. Dow, Trans. ASME, Ser.F, J. Lub. Tech., *96*, 1974, 611-615.
[103] K. von Laar and J. Löwen, Ind. Anz., *99*, 1977, 1826-1828, 1988-1990 (German).
[104] Ya.M. Okhrimenko and B.D. Kopyskii, Chern. Met., *1961*, 45-53 (Russian).
[105] S. Matsubara and H. Kudo, CIRP, *25*, 1977, 95-99.
[106] M. Brouha, J.E. de Jong, and K.J.A. van der Weide, in *Proc. 7th NAMRC*, SME, Dearborn, 1979, pp.57-64.
[107] G.T. Van Rooyen and W.A. Backofen, Int. J. Mech. Sci., *1*, 1960, 1-27.
[108] C.W. MacGregor and R.B. Palme, J. Appl. Mech., *15*, 1948, 297-302.
[109] P. Davidkov and L. Karagiosov, Neue Hütte, *13*, 1968, 33-38 (German).
[110] A. Banerji and W.B. Rice, CIRP, *21*, 1972, 53-54.
[111] J. Danckert and T. Wanheim, Scand. J. Metall., *6*, 1977, 185-190.
[112] A. Jones and B. Walker, Metals Techn., *1*, 1974, 310-315.
[113] H. Wiegand and K.H. Kloos, Metalloberfläche, *13*, 1959, 229-233 (German).
[114] K.L. Hsu, T.M. Ahn, and D.A. Rigney, in [3.343], pp.12-26.
[115] S. Jahimir, Wear, *58*, 1980, 387-389.
[116] *Proc. Conf. Process Instrumentation in the Metals Industry*, Inst. Measurement and Control, London, 1971.
[117] G. Barrow, CIRP, *22*, 1973, 203-211.
[118] C.D. Coe, in [116], pp.170-176.
[119] W. Lueg and K.H. Treptow, Stahl u. Eisen, *76*, 1956, 1690-1698; *77*, 1957, 859-867 (German).
[120] A.E. Ranger and J.G. Wistreich, J. Inst. Petrol., *40*, 1954, 308-313.
[121] R. Roy and W.G. Henry, Metals Techn., *3*, 1976, 483-485.
[122] P.M. Braiden, in *Proc. 8th Int. MTDR Conf.*, Pergamon, Oxford, 1967, pp.653-666.
[123] W. Reichel, Stahl u. Eisen, *70*, 1950, 1141-1146 (German).
[124] G.Mc.H. Sturgeon, Wire Ind., *27*, 1960, 261-263.
[125] A.M. Belansky and C.F. Peck, Iron Steel Eng., *33* (3), 1956, 62-64.
[126] P.G. Stevens, K.P. Ivens, and P. Harper, J. Iron Steel Inst., *209*, 1971, 1-11.
[127] A. Zastera, Stahl u. Eisen, *74*, 1954, 461-464 (German).
[128] M.F. DeVries, S.M. Wu, and J.W. Mitchell, Microtecnic, *21* (6), 1967, 1-3.
[129] V.V. Tatarenko et al, Russian Eng. J., *49* (1), 1969, 53-54.
[130] M.T. Watkins, J. Inst. Metals, *90*, 1961-62, 209-210.
[131] S. Steindl and W.B. Rice, CIRP, *22*, 1973, 89-90.
[132] M.J. Furey, Trans. ASLE, *7*, 1964, 133-146.
[133] B. Lux, N. Perlhefter, and J. Schniewind, Stahl u. Eisen, *85*, 1965, 875-878 (German).
[134] F.B. Bowden and E.H. Freitag, Proc. Roy. Soc. (London), *A248*, 1958, 350.
[135] W.O. Winer, in [3.253], pp.125-130.
[136] H.S. Nagaraj, D.M. Sanborn, and W.O. Winer, ASLE Trans., *22*, 1979, 277-285.
[137] P. Deyber and M. Godet, Tribology, *4*, 1971, 150-154.
[138] D.G. Wymer and P.B. MacPherson, ASLE Trans., *18*, 1975, 229-238.
[139] E. Lenz, in *1st Int. Cemented Carbide Conf.*, ASME, New York, 1971 (Paper No. MR 71-905).
[140] N.H. Cook and E. Rabinowicz, *Physical Measurement and Analysis*, Addison-Wesley, Reading, MA, 1963.

[141] D.L. Burk, Iron Steel Eng., *44* (9), 1967, 189-206.
[142] R. Barber, Adv. in Instrumention, *33*, 1978, 417-434.
[143] J. Ihlefeldt, H.J. Kopineck, and W. Tappe, Stahl u. Eisen, *100*, 1980, 474-477.
[144] D.D. Beattie, in [116], pp.105-110.
[145] S. Jeelani, Wear, *68*, 1981, 191-202.
[146] G. Spur and H. Beyer, CIRP, *22*, 1973, 3-4 (German).
[147] S. Kato, K. Yamaguchi, Y. Watanabe, and Y. Hiraiwa, Trans. ASME, Ser. B., J. Eng. Ind., *98*, 1976, 607-613.
[148] S.A. Popov and V.M. Davydov, Russian Eng. J., *49*, 1969, 74-77.
[149] P. Keller, Draht, *31*, 1980, 159-161.
[150] E.M. Trent, J. Iron Steel Inst., *143*, 1941, 401-419.
[151] N.C. Welsh, J. Appl. Phys., *29*, 1957, 960-968.
[152] H. Uetz and K. Sommer, Wear, *43*, 1977, 375-388.
[153] D.G. Powell and S.W.E. Earles, in [1], pp.114-117.
[154] P.K. Wright, Trans. ASME, Ser.B, J. Eng. Ind., *100*, 1978, 131-136.
[155] E.F. Smart and E.M. Trent, Int. J. Prod. Res., *13*, 1975, 265-290.
[156] B. Mills, D.W. Wakeman, and A. Aboukhashaba, CIRP, *29*, 1980, 73-77.
[157] J. Kielblock, Schmierungstechnik, *11*, 1980, 231-234 (German).
[158] O. Cucchiara and P. Goodman, Materials Evaluation, *25*, 1967, 97-128.
[159] *Proc. Int. Conf. Surface Technology,* SME, Dearborn, 1973.
[160] J.F. Archard, R.T. Hunt, and R.A. Onions, in [3.89], pp.282-303.
[161] E. Green, in [3.70], pp.330-343.
[162] R.E. Reason, CIRP, *19*, 1971, 559-563.
[163] D.J. Whitehouse, Trib. Int., *7*, 1974, 249-259.
[164] T.R. Thomas, in *Proc. 19th Int. MTDR Conf.*, Macmillan, London, 1979, pp.383-390.
[165] M.N.H. Damie, Wear, *26*, 1973, 219-227.
[166] G.N. Shah, A.V. Bell, and S. Malkin, Wear, *41*, 1977, 315-325.
[167] J.L. Guerrero and J.T. Black, Trans. ASME, Ser.B, J. Eng. Ind., *94*, 1972, 1087-1094.
[168] R.G. Quiney, F.R. Austin, and L.B. Sargent, Jr., ASLE Trans., *10*, 1967, 192-202.
[169] J.A. Greenwood and J.B.P. Williamson, in [3.250], pp.167-177.
[170] E.G. Thwaite, Wear, *51*, 1978, 253-267.
[171] R.S. Sayles and T.R. Thomas, Trans. ASME, Ser.F, J. Lub. Tech., *101*, 1979, 409-418.
[172] K.J. Stout, T.G. King, and D.J. Whitehouse, Wear, *43*, 1977, 99-115.
[173] J. Westberg, in [3.70], pp.260-273.
[174] *Appearance of Metallic Surfaces,* STP 4, ASTM, Philadelphia, 1970.
[175] G.M. Clarke and T.R. Thomas, Wear, *57*, 1979, 107-116.
[176] L.H. Tanner, Wear, *57*, 1979, 81-91.
[177] E.L. Church, Wear, *57*, 1979, 93-105.
[178] E.G. Thwaite, Wear, *57*, 1979, 71-80.
[179] K. Horn, CIRP, *25*, 1977, 257-262.
[180] H. Takeyama, H. Sekiguchi, R. Murata, and H. Matsuzaki, CIRP, *25*, 1976, 467-471.
[181] E.C. Teague, T.V. Vorburger, and D. Maystre, CIRP, *30*, 1981, 563-569.
[182] H. Droscha, Sheet Metal Ind., *55*, 1978, 1230-1231, 1244.
[183] R.A. Brook, in [116], pp.142-145.
[184] K.L. Woo and T.R. Thomas, Wear, *58*, 1980, 331-340.
[185] S. Tolansky, *An Introduction to Interferometry,* 2nd Ed., Longman, London, 1973.
[186] H. Trumpold, in [3.70], pp.241-254.
[187] S. Tolansky, *Multiple-Beam Interference Microscopy of Metals,* Academic Press, London, 1970.
[188] D. Scott, in [3.35], pp.670-673.
[189] M. Lech, I. Mruk, and J. Stupnicki, Wear, *57*, 1979, 263-268.
[190] D. Scott and H.M. Scott, in [3.35], pp.609-612.
[191] T.F.J. Quinn, Tribology, *3*, 1970, 198-205.
[192] T.F.J. Quinn, *The Application of Modern Physical Techniques to Tribology,* Butterworths, London, 1971.
[193] D.H. Buckley, Wear, *46*, 1978, 19-53.
[194] J.A. Schey and O. Johari, unpublished work.
[195] T.S. Eyre and K. Dutta, Lubric. Eng., *31*, 1975, 521-529.
[196] C. Beesley and T.S. Eyre, Trib. Int., *9*, 1976, 63-69.
[197] A.F. George, Wear, *57*, 1979, 51-61.

[198] K. Narayanasamy, V. Radhakrishnan, and R.G. Narayanamurthi, Wear, *57*, 1979, 63-69.
[199] H. Czichos, Schmiertechn. Trib., *21*, 1974, 25-30 (German).
[200] H. Seiler, Chemie-Ing.-Techn., *46*, 1974, 797-804 (German).
[201] D. Godfrey, in [3.49], pp.945-967.
[202] R.S. Carbonara and J.R. Cuthill (ed.), *Surface Analysis Techniques for Metallurgical Applications,* ASTM STP 596, Philadelphia, 1976.
[203] J. Ferrante, Lubric. Eng., *38*, 1982, 223-236.
[204] R.J. Bird and G.D. Galvin, Wear, *37*, 1976, 143-167.
[205] J. Oppelt, K. Müller, and W.J. Bartz, ASLE Trans., *24*, 1981, 71-76.
[206] H.M. Krillic, Iron Steel Eng., *51* (7), 1974, 45-48.
[207] D. Schreiber and H. Ziehm, Stahl u. Eisen, *100*, 1980, 1232-1237 (German).
[208] R.W. Roberts, in [3.400], pp.20-66.
[209] P.F. Kane and G.B. Larrabee (ed.), *Characterization of Solid Surfaces,* Plenum, New York, 1971.
[210] K.L. Mittal (ed.), *Surface Contamination,* Plenum, New York, 1979.
[211] G.R. Sparrow, Manuf. Eng. Trans., *1978*, 1-10.
[212] N. Ohmae and T. Tsukizoe, in *Proc. 4th Int. Conf. Production Engineering,* Jpn. Soc. Prec. Eng., Tokyo, 1980, pp.989-998.
[213] M.J. Furey, ASLE Trans., *4*, 1961, 1-11.
[214] J.A. Leather and P.B. MacPherson, in [3.42], pp.155-162.
[215] A. Tonck, J.M. Martin, Ph. Kapsa and J.M. Georges, Trib. Int., *12*, 1979, 209-213.
[216] W.G. Fiennes and J.C. Anderson, in [3.42], pp.109-119.
[217] A. Dyson, H. Naylor, and A.R. Wilson, Proc. Inst. Mech. Eng., *180*, Pt.3B, 1965-66, 119-134.
[218] F.S. Attia and T.L. Whomes, Trib. Int., *12*, 1979, 215-217.
[219] A. Cameron, in [1], pp.59-61.
[220] J.W. Kannel and J.C. Bell, Trans. ASME, Ser.F, J. Lub. Tech., *93*, 1971, 478-484.
[221] T. Willis and B. Seth, Trans. ASME, Ser.F, J. Lub. Tech., *99*, 1977, 280-293.
[222] F.J. Westlake and A. Cameron, in [1], pp.75-78.
[223] C.A. Foord, L.D. Wedeven, and A. Cameron, ASLE Trans., *11*, 1968, 31-43.
[224] I. Peyches, Glass Ind., *35*, 1954, 309-313, 340.
[225] S. Komura, Wire Wire Prod., *42*, 1967, 429-431, 502-504.
[226] *Selection and Use of Wear Tests for Metals,* STP 615, ASTM, Philadelphia, 1976.
[227] H. Voss, E. Wetter, and F. Netthofel, Arch. Eisenhüttenw., *38*, 1967, 379-386 (German).
[228] B. Mills and A.M. Redford, CIRP, *27*, 1978, 45-48.
[229] D. Godfrey, in [3.344], pp.283-311.
[230] *Metals Handbook,* 8th Ed., Vol.10, *Failure Analysis and Prevention,* ASM, Metals Park, 1975.
[231] T.S. Eyre, Lubric. Eng., *37*, 1981, 603-607.
[231a] A.W. Ruff, L.K. Ives, and W.A. Glaser, in [3.48a], pp.235-289.
[232] R.M.J. Withers, J. Iron Steel Inst., *164*, 1950, 63-66.
[233] R.M.J. Withers, Metal Treat. Drop Forg., *18*, 1951, 191-194.
[234] W. Lueg, Stahl u. Eisen, *71*, 1951, 157-170 (German).
[235] S. Werth, Stahl u. Eisen, *71*, 1951, 66-69 (German).
[236] J.G. Wistreich, J. Iron Steel Inst., *158*, 1948, 496.
[237] G.F. Micheletti, W. Koenig, and H.R. Victor, CIRP, *25*, 1976, 483-496.
[238] W. König, R. Bierlich, and K. Spira, Schmiertechnik Trib., *23*, 1976, 150-156 (German).
[239] J. Kohl, R.D. Zentner, and H.R. Lukens, *Radioisotope Applications in Engineering,* Van Nostrand, Princeton, 1961.
[240] A. Gerve, in [3.42], pp.91-92.
[241] B.W. Rooks and S.A. Tobias, *Proc. 9th Int. MTDR Conf.,* Pergamon, Oxford, 1968, pp.679-698.
[242] R.F. Hill and M. Skunda, Mech. Eng., *94*, 1972, 22-27.
[243] S.K. Bhattacharyya and S.K. Jetley, Tribol. Int., *12*, 1979, 277-280.
[244] N.H. Cook and K. Subramanian, CIRP, *27*, 1978, 73-78.
[245] E. Amini, A.W. Kwiatkowski, and R.H.S. Winterton, in *Proc. 18th Int. MTDR Conf.,* Macmillan, London, 1977, pp.623-628.

[246] T.W. Conlon, Wear, *29*, 1974, 69-80.
[247] S.K. Bhattacharyya, S. Jetley, and B. Ivkovic, Wear, *48*, 1978, 399-407.
[248] R. Evans, Wear, *64*, 1980, 311-325.
[249] B. Ivkovic, Trib. Int., *15*, 1982, 3-8; in [3.56], Paper No.79.
[250] N.S. Eiss, K.C. Wood, J.A. Herold, and K.A. Smyth, Trans ASME, Ser.F, J. Lub. Tech., *101*, 1979, 212-219.
[251] W.W. Seifert and V.C. Westcott, Wear, *21*, 1972, 27-42.
[252] P.B. Senholzi, in [3.48], pp.113-118.
[253] D. Scott and V.C. Westcott, in [3.48], pp.107-112.
[254] S. Odi-Owei, A.L. Price, and B.J. Roylance, Wear, *40*, 1976, 237-253.
[255] D.P. Anderson, Lubric. Eng., *35*, 1979, 203-211.
[256] A.W. Ruff, in [3.343], pp.260-264.
[257] N.M. Mikhin, N.K. Mishkin, V.N. Litvinov, and M.N. Dobytchin, in [3.343], pp.60-63.
[258] I.I. Zemskova, R.M. Matveevsky, and M.M. Khruschov, Wear, *23*, 1973, 225-229.
[259] D. Dornfeld, in *Proc. 8th NAMRC,* SME, 1980, 207-213.
[260] B.A. Glagorskii and I.B. Moskovenko, Russian Eng. J., *60* (6), 1980, 37-39 (Russian).
[261] Y. Kakino, in *Cutting Tool Materials,* ASM, Metals Park, 1981, pp.25-39.
[262] T. Sato et al, J. JSTP, *21*, 1980, 608-613 (Ann. Rep., *2*, 1981, 46).
[263] E. Rabinowicz, D.M. Boyd, and N. Ohmae, in [3.145], pp.9-13.
[264] R.D. Guminski and J. Willis, J. Inst. Metals, *88*, 1959-60, 481-492.
[265] L.B. Zlotin et al, Sov. J. Nonferrous Metals, *7* (7), 1966, 97-100.
[266] F.L. Reynolds, Lubric. Eng., *14*, 1958, 98-103, 120.
[267] L.B. Sargent, Metal Progr., *75* (5), 1959, 113-114.
[268] H.A. Snow, Sheet Metal Ind., *31*, 1954, 601-608.
[269] R.J. Nekervis and R.M. Evans, Iron Steel Eng., *25* (10), 1948, 72-80.
[270] J.F. Griffin, J. Metals, *9*, 1957, 1041-1043.
[271] J.H. Rigo, Lubric. Eng., *33*, 1977, 345-356.
[272] R.W. Belfit and N.E. Shirk, Lubric. Eng., *17*, 1961, 173-178.
[273] I.S. Morton, Ind. Lub. Trib., *24*, 1972, 267-270.
[274] G.P. Koch et al, Lubric. Eng., *33*, 1977, 407-411.
[275] K. Walters, *Rheometry,* Chapman and Hall, London, 1975.
[276] R.W. Whorlow, *Rheological Techniques,* Wiley, New York, 1980.
[277] H.D. Moore, Tribology, *2*, 1969, 19-24.
[278] L.L. Stavinoha, G.E. Fodor, F.M. Newman, and S.J. Lestz, ASLE Trans., *21*, 1978, 217-225.
[279] G.M. Stanton, Ind. Lub. Trib., *28*, 1976, 126-130.
[280] M.L. Bauccio, Lubric. Eng., *38*, 1982, 549-556.
[281] F.E. Lantos and J. Lantos, Lubric. Eng., *27*, 1971, 184-189.
[282] J.E. Steenbergen, Lubric. Eng., *34*, 1978, 625-628.
[283] W.T. Williams, Lubric. Eng., *33*, 1977, 191-194.
[284] F.R. Penney, I.S. Kolarik, and I.P. Hammer, Lubric. Eng., *32*, 1976, 238-241.
[285] H.H. Rowand, Jr., R.J. Patula, and L.B. Sargent, Jr., Lubric. Eng., *29*, 1973, 491-497.
[286] G.A. Volpato, A.G. Manzi, and S. Del Ross, in [3.45], pp.335-342.
[287] H. Miyagawa, S. Kusunoki, T. Sakai, and F. Hirano, J. JSLE (Int. Ed.), *1*, 1980, 97-102.

Rolling

A greater volume of material is worked by rolling than by any other deformation process. The largest quantity is rolled into flat products (plate, sheet, strip, foil) by the process of hot rolling followed by cold rolling. Smaller quantities are rolled into shapes (sections), usually hot. The emphasis in this chapter will be on flat rolling, and material previously covered by Schey [1] will be included.

As befits its economic significance, rolling has long occupied an important position in the literature, and, in addition to its discussion in the references given in Chapter 2, there are entire books devoted to the rolling process [2-7], [5.57]. The proceedings of a conference on rolling technology have been published in a series of articles [8]. A recent review has been published by Sparling [9]. There are books in Russian devoted to the frictional aspects of the process [10,11], but unfortunately they could not be consulted for the present discussion.

6.1 The Rolling Process

All flat products are made in a fairly standard sequence of operations:

a. The starting material is an individually or continuously cast slab which is hot reduced (broken down) on two- or four-high reversing mills. Unless the end product will be wide plate, rolling is continued to form a long, coiled, wide band. If production volumes warrant it, rolling from about 100 mm to about 25 mm may take place on several in-line mills with the slab running out between the stands (semicontinuous line).

b. The thick strip (band) is hot rolled to a thickness of about 1.5 to 5 mm; if quantities are large, a continuous hot strip mill is employed in which the stands are placed close enough together for the strip to run through all stands simultaneously, with controlled tensions between stands. The hot strip may be the finished product in the form of a coil, cut-to-length plate, or sheet.

c. If a thinner gage, a smoother surface finish, or a strain-hardened end product is required, the strip is further rolled cold on reversing or tandem mills, with accurate control of coiler and/or interstand tensions. If the material strain hardens excessively, intermediate process anneals are carried out. The resulting wide strip or sheet is typically 0.2 to 2 mm thick. For small quantities, individual sheets may be rolled by hand.

d. The thinnest strip (less than, say, 0.05 mm) is called foil, and is produced in specialized cold rolling mills. The work rolls are invariably backed up; the number of backup rolls ranges from two (four-high mill) to eighteen (Sendzimir mill).

If it is technically feasible and if the investment can be justified, the strip is rolled wide for economy of production and then is slit into narrower widths or cut into shorter lengths (sheets). Narrow strip with rounded corners is obtained by flattening of wire.

In the rolling of shapes from cast billets of rectangular or round cross section, material flow is much more complicated, but lubrication practices are not greatly different from those used in flat rolling.

As shown in Fig.2.8, rolling is essentially a steady-state process, from both the mechanical and lubrication points of view, thus the lubricant supply at the roll entry (bite) is continuous. There are, however, non-steady-state aspects:

a. Material flow is abnormal and lubrication is interrupted at the beginning and end of the slab.

b. The contact zone between the workpiece and the roll changes continuously, and any part of the roll surface contacts the workpiece surface only once during every roll revolution. Thus, contact is really of a cyclic nature as far as lubricant breakdown, metal transfer (roll pickup), and surface temperature are concerned. In the idle period of the cycle, accidental or intentional changes in the roll surface may occur—for example, through rubbing against scrapers, wipers, or wire brushes.

6.2 Flat Rolling

At first sight, flat rolling appears to be a very simple process. Because the workpiece volume must remain constant, the material displaced by reduction in thickness will go mostly into length. Plane-strain conditions are approximated when the strip is wide (Fig.2.2b) and, for simplicity, rolling will first be viewed as a plane-strain process. Complications arising from inhomogeneous deformation will be introduced later and, finally, shape rolling will be considered.

6.2.1 Entry of Workpiece

In the rolling of a sheet or strip, thickness is reduced from h_0 to h_1 by rolls of radius R (Fig.6.1). The absolute reduction in a single pass is:

$$\Delta h = h_0 - h_1 \tag{6.1}$$

or in relative (or percentage) terms:

$$r = \left(h_0 - h_1\right)/h_0 \tag{6.2a}$$

or

$$r\% = 100\left(h_0 - h_1\right)/h_0 \tag{6.2b}$$

Reduction expressed in terms of natural strain

$$\varepsilon = \ln\left(h_0/h_1\right) \tag{6.3}$$

is of greater value for purposes of calculation and has the advantage that strains imposed in individual passes may be added up.

To initiate rolling, the workpiece must be drawn into the roll gap by friction (Fig.6.2). It is common experience that the rolls refuse to bite if they are very smooth or well lubricated. This condition is reached when the horizontal component of the frictional force F is just equal to the horizontal component of the radial force Pr. From Fig.6.2:

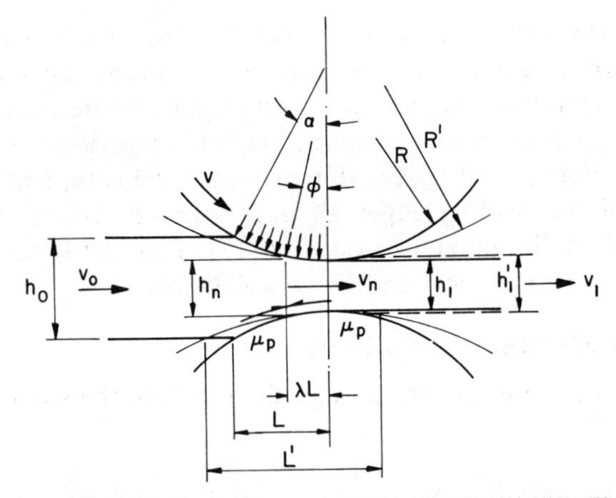

Fig.6.1. Geometry of the roll pass and associated velocities. Elastic deflection is shown in broken lines. Half arrows indicate shear stresses acting on workpiece.

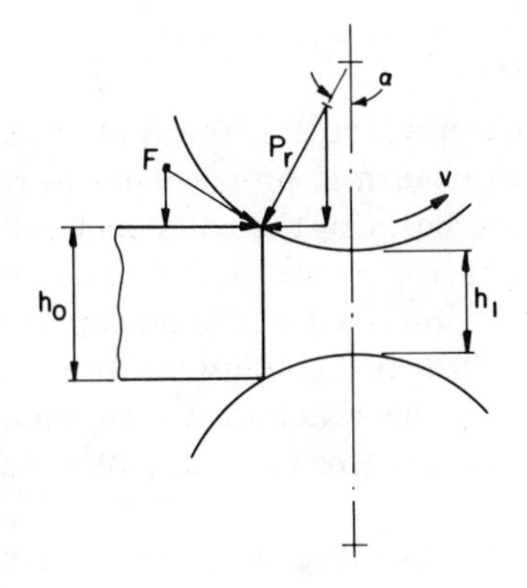

Fig.6.2. Frictional conditions at entry of workpiece

$$F \cos \alpha > P_r \sin \alpha \tag{6.4a}$$

or

$$\frac{F}{P_r} > \tan \alpha \tag{6.4b}$$

From the geometry of the pass, the maximum possible draft is

$$\Delta h_{max} = \mu^2 R \tag{6.5}$$

By definition, $F/P = \mu$, which can be expressed also as $\tan f$ (where f is the friction angle). Therefore, it is usual to state that the bar will enter the rolls unaided only if the friction angle f exceeds the contact angle α. This represents one of the instances where friction is desirable (Fig.2.4b).

This would appear to offer a very simple way of determining the coefficient of friction: the roll gap is increased until the slab enters the rolls without being pushed. However, the value of this test is limited because the frictional force is unidirectional at the point of acceptance, and lubricant films may be scraped off; neither of these conditions is representative of steady-state rolling. Nevertheless, the method is convenient under conditions of dry lubrication and for a very approximate ranking of lubricants. It is suitable also for production-scale application, although local variations of surface roughness, localized pickup, roll speed, slab approach velocity, and intentional or accidental chamfering of the leading edges all have marked effects, as shown by Kortzfleisch [12] and by Kortzfleisch and co-workers [13]. This test is considered in a critical review of test methods by Dobrucki and Odrzywolek [14].

6.2.2 Speed and Stress Distributions

Once the slab is accepted and drawn through the roll gap, the situation changes considerably.

Relative Slip. In the absence of spread, continuity may be maintained only if the products of thickness and velocity are constant at all points along the zone (arc) of contact. Thus (Fig.6.1):

$$v_0 h_0 = v_n h_n = v_1 h_1 \tag{6.6}$$

Accordingly, the exit speed of the slab, v_1, must increase in proportion to the elongation (which in turn is proportional to reduction), while the rolls move at some speed intermediate between the entry and exit speeds. Only at one point in the arc of contact can slab and roll move at the same speed ($v = vn$); because there is no relative slip, the point is described as the no-slip or neutral point or, in a three-dimensional presentation, the neutral plane. Between the entry and neutral points the strip moves more slowly than the rolls (backward slip), and between the neutral and exit points the strip moves more quickly (forward slip). We shall come back to this later in Sec.6.2.6.

The Friction Hill. Relative movements within the arc of contact are opposed by frictional stresses, pointing toward the neutral plane. As shown in the example of forging in Fig.2.6, this gives rise to a friction hill, the shape of which depends on the magnitude of friction and on the pass geometry:

a. With very low friction, the inclination of the roll surfaces causes most of the material to flow backward, thus moving the neutral plane close to the exit. Deformation is almost homogeneous. The friction hill is low and most of the pressure is attributable to the plane-strain flow stress (Fig.2.1, point 4), equal to 1.15 σ_f (or $2k$) if the material yields according to the von Mises criterion. In cold working the material strain hardens during its passage through the rolls and the friction hill appears as in Fig.6.3a. Frictional stresses are low and frequently described by a constant μ, implying the τ_i distribution shown in Fig.6.3a by broken lines. Direct measurements generally show a decay of τ_i toward the neutral plane; even if there is some error in the measurements themselves, a variable μ (or m) within the arc of contact is more realistic.

b. With intermediate friction, the friction hill is steeper, the neutral point is farther from the exit and, if friction is high enough, the condition for sticking (Eq.2.9) is satisfied near the neutral plane. Sliding is arrested, and the neutral plane broadens into a neutral zone. Shear stresses reach a limiting value, and correspondingly the friction hill is rounded (Fig.6.3b). Deformation is inhomogeneous, because a dead-metal zone forms adjacent to that part of the contact arc in which sticking friction prevails. Modeling of the shear stress distribution becomes more difficult.

c. On rough rolls under unlubricated conditions, sticking may exist within the

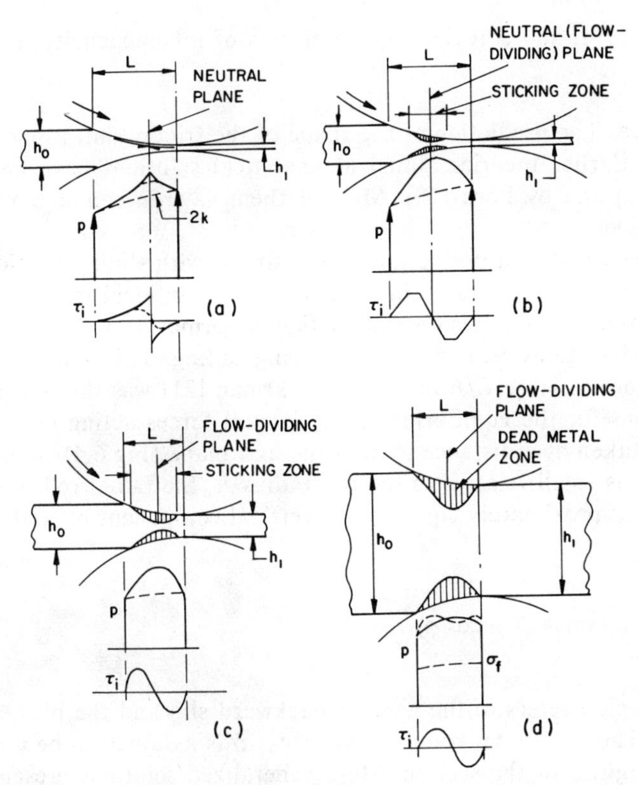

Fig.6.3. Material flow, pressure, and shear stress distributions for various friction conditions and *L/h* ratios

entire contact zone. Deformation becomes highly inhomogeneous, resembling backward and forward extrusion. Instead of a neutral zone, it is more realistic to speak of a flow-dividing zone. Because deformation is concentrated away from the roll surface, shear stresses on the roll decrease and the friction hill is heavily rounded (Fig.6.3c).

d. Rolling a strip into a coil permits the application of tensions. Back tension retards the movement of the strip and thus shifts the neutral plane toward the exit; front tension has the opposite effect. Tensions cause the strip to yield in a combined compressive-tensile state; this results in a lower roll pressure because, in terms of Fig.2.1, the yield criterion is satisfied by moving from point 4 toward point 5.

e. The above discussion assumes that the strip is relatively thin—i.e., $L/h > 2$ (where L is the projected length of the arc of contact and h is the mean strip thickness)—and that inhomogeneity results only from high friction. At lower L/h ratios, and especially when $L/h < 1$ (or $h/L > 1$), inhomogeneity results from the process geometry itself, just as it does in the indentation of a thick slab with narrow anvils (Fig.2.7b). The friction hill may show a double hump (Fig.6.3d). This situation is typical of early passes during the hot rolling of thick slabs [15], and separation of the effects of friction and process geometry becomes extremely difficult.

The effects of friction can be observed experimentally. Roll pressures have been repeatedly measured by the techniques described in Sec.5.5.3, and frictional stresses by techniques discussed in Sec.5.5.2. A recent study by Al-Salehi et al [16] includes a review of earlier work. The double hump is particularly pronounced in the work of Knauschner [17]. The effects of friction on deformation within the workpiece have been followed by various techniques. Extensive studies on thick slabs have been reported by

Tarnovskii et al [18], and a review of the effects of inhomogeneity has been given by Schey [19].

Rolling Theories. The prediction of the shape of the friction hill is one of the principal aims of theory. Earlier theories aimed at analytical solutions have been reviewed by Underwood [5.57] and by Ford [20]. Most of them have in common a number of simplifying assumptions.

Rolling is regarded as a problem of plane strain, with sliding friction along the arc of contact (except at the neutral point), performed on an ideal rigid-plastic material of constant flow strength. It is also assumed that deformation is homogeneous—that is, originally vertical sections remain vertical during rolling. This assumption restricts the validity of solutions to large L/h ratios. Von Karman [21] was the first to write the differential equations for the equilibrium of horizontal forces acting on a vertical section of thickness d_x, taken at a distance x from the exit plane (Fig.6.4). Assuming that friction is low and x is small relative to the roll radius R, the radial roll pressure p may be considered to be approximately equal to the vertical component p_v, and equilibrium can be expressed as

$$\frac{d(\sigma h)}{dx} = p_v(\sin \phi \mp \mu \cos \phi) \tag{6.7}$$

where the minus sign refers to the zone of backward slip and the plus sign to the zone of forward slip. The horizontal compressive stress σ is assumed to be uniformly distributed over the height h of the section. More generalized solutions presented by Orowan [22] and by Geleji [2.18] eliminated some of the unrealistic assumptions; notably, strain hardening of the workpiece material, inhomogeneous deformation, and the presence of a sticking zone are allowed for. The use of the digital computer allows solution of the full Orowan theory as shown by Alexander [23], who also gives a review of rolling theory. Lenard [24] used a variable μ along the arc of contact. Solutions based on slip-line fields [2.27],[25] and on the upper-bound approach [2.12,2.21] are also available. Frequently, m is chosen as the descriptor for interface friction; as noted in Sec.2.2.2, its use offers great computational convenience. Numerical (finite-element) methods, reviewed by Li and Kobayashi [26], offer the greatest freedom of choice, from constant m or μ to mixed sliding-sticking friction. The double hump in the friction hill and the non-steady-state conditions prevailing while the slab enters the rolls can be treated, too. A powerful technique is visioplasticity, which was applied by Thomson and Brown [26a] to study strain distribution in dry rolling of aluminum. It allows the construction of slip-line fields and gives information on pressure distribution and interface conditions.

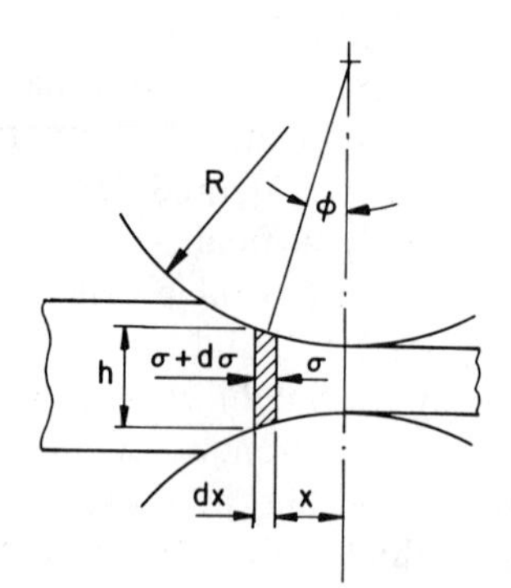

Fig.6.4. Stresses acting on a vertical section of a rolled workpiece

6.2.3 Roll Force

Few theories predict the shape of the friction hill and the distribution of shear stresses in full agreement with actual measurements. To some extent, there is doubt also about the accuracy of the measurements themselves. Fortunately, the average pressure and thus the roll force are less sensitive to the detailed shape of the friction hill. Roll force is thus a reliable indicator of the magnitude of friction: with higher friction, roll force is also higher.

Influencing Factors. For a calculation of roll force, the area under the friction hill curve must be integrated. The roll force P is, in principle, a function of several basic variables:

$$P = (2k - t)(L)(w)\left[f\left(\tau_i ; \frac{L}{h}\right)\right] \tag{6.8}$$

If forces are to be predicted with any accuracy, realistic values of these variables must be inserted:

a. The plane-strain flow stress $2k$ (or, in cold rolling, the average flow stress of the strain-hardening material) must be measured in the through-thickness direction, in a plane-strain compression test (Sec.9.1.4), especially for anisotropic materials [27]. Errors in this value could overwhelm all other factors, including the effects of friction. In well-lubricated cold rolling at reasonable speeds, the friction contribution is only around 5% of the total force, and could be easily lost in the uncertainty of the flow stress value. Therefore, the flow stress must be determined at the relevant strain rate, even in cold rolling [28]. As mentioned earlier, the tension t (taken as the arithmetic average of front and back tensions) is a powerful factor in reducing roll forces.

b. The arc of contact L is readily calculated from

$$L = \left[R\left(h_0 - h_1\right)\right]^{1/2} \tag{6.9}$$

However, in the rolling of thin, hard strip it must be recognized that the rolls, as any other engineering structures, suffer elastic deformation when subjected to stresses. One consequence is that they flatten to present a locally increased radius R' (Fig.6.1) which can be calculated with adequate accuracy from Hitchcock's formula [29]:

$$\frac{R'}{R} = 1 + \frac{16P\left(1 - \nu^2\right)}{\pi E w\left(h_0 - h_1\right)} \tag{6.10}$$

where ν is Poisson's ratio and E is Young's modulus. Some other approaches are reviewed by Atreya and Lenard [30], who use a finite-element technique for calculation of roll shape.

Hitchcock's solution is still the simplest, but its validity has often been questioned and, compared to direct observation, it seems to underestimate the length of arc of contact.

c. The width of the strip w may be taken as the starting width w_0, especially when the strip is wide, but spread must be taken into account for narrower strip. Spread increases with increasing friction (this is discussed later in this section).

d. The function $f[\tau_i;(L/h)]$ signifies the contribution of friction combined with pass geometry. It is in this function that rolling theories differ. For purposes of predicting rolling forces, simple, practical solutions, many of them in graphical form, are acceptable.

For very low friction with no sticking zone (Fig.6.3a), the simple approach of Stone and Greenberger [31], based on compression between parallel surfaces, was often used. Much more satisfactory is the rolling theory of Ford et al [32].

For partial sticking (Fig.6.3b), these solutions are unrealistic, and those of Tamano and Yanagimoto [33], or a plane-strain approximation based on Fig.9.11, give better results.

For hot rolling, solutions based on full sticking (Fig.6.3c) [34] or partial sliding are useful [35].

None of these theories takes into account the inhomogeneity of deformation, which becomes the dominant effect at $L/h < 1$ (Fig.6.3d). Then roll pressures can be estimated by analogy to the indentation of thick slabs (Sec.9.1) from Fig.9.14, which can be approximated by

$$p = 2k\left(0.8 + 0.2\frac{h}{L}\right) \tag{6.11}$$

It is most important to note that friction has only a minor effect on roll force when $h/L > 1$.

The magnitude of friction can be taken into account by a suitable choice of m or μ, as will be discussed in Sec.6.3 and 6.4. The danger is that friction may become an adjustable variable to make experimental results fit a given theory.

With the aid of a powerful digital computer, it is possible to model the entire process of rolling with only a few simplifying assumptions. An example of such an approach and its application to aluminum rolling is described by McPherson [36].

Limiting Gage. Since the roll force P increases with strip hardness, friction, and the L/h ratio (Eq.6.8), the rolls flatten (Eq.6.10) and the length of the arc of contact also increases. Elastic recovery of the strip on release from the roll gap further lengthens the arc of contact, and the exit point can move outside the centerline of the rolls (Fig.6.1). For a given change in roll gap setting, the actual reduction in strip thickness gradually diminishes. Thus, the roll gap efficiency (also called roll-gap transfer coefficient) drops from around 0.5 on thicker strip to 0.01 on the last stand of a cold tandem mill. This is observed not only in the rolling of steel [37-39] but also in the rolling of the much softer aluminum foils [40]. In the limiting case, no further reduction is obtained on closing the roll gap. The strip thickness at this point is commonly referred to as the limiting gage. In general, difficulties can be expected when the strip thickness drops below 1/400 to 1/600 of the roll diameter.

A number of solutions for predicting the minimum gage have been reviewed by Ford and Alexander [41], who also proposed a solution that takes the elastic properties of the strip material and of the roll into account. A number of analyses have appeared since that show that the causes of the phenomenon are still not fully understood. There is no doubt that the minimum thickness may be reduced only by increasing the tensions, by decreasing the roll diameter, by increasing the roll elastic modulus (e.g., with WC rolls) or, most importantly, by reducing friction. The latter effect is clearly visible in Fig.6.5, taken from the work of Pawelski and Käding [42]. When rolling thin gages, the rolls make direct contact outside the strip, but this is only part of the reason for reaching a limiting reduction. An important contributor is roll flattening to the point

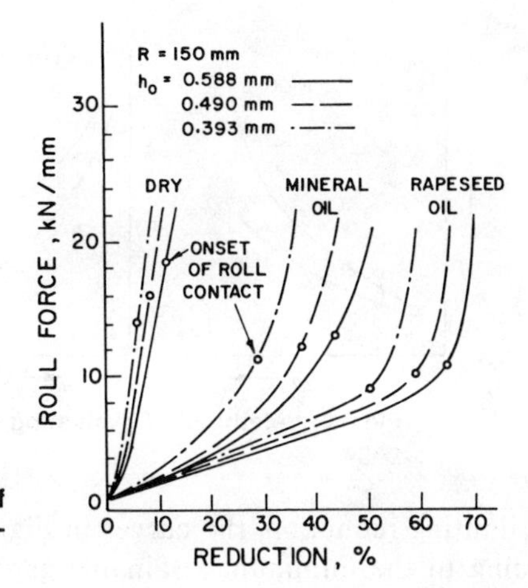

Fig.6.5. Roll forces measured in rolling of steel strip under conditions of limiting reduction [42]

where strip deformation is all elastic. Under these conditions, sticking friction prevails at least over part of the arc of contact even at an apparently low coefficient of friction [43].

Friction From Roll Force. In principle, solutions of Eq.6.8 allow assessment of frictional conditions by several techniques:

a. If the plane-strain flow stress is accurately known and the rolling theory is perfectly valid, one can, in principle, extract τ_i from measured forces. Such an enterprise is extremely hazardous, especially when friction is low, because a small (say, 5%) change in $2k$ could easily result in a 100% change in μ or m. Furthermore, the results are highly dependent on the theory chosen [44]. The situation becomes worse in rolling with tensions. Many friction values quoted in the literature have been back-calculated from roll force and they should be viewed with caution. With higher friction, typical of hot rolling, the contribution of friction to the roll force is large, many errors cancel, and reasonable estimates of μ or m are sometimes obtained. In all instances, the contact length L must be calculated with roll flattening (Eq.6.10). Despite any misgivings one may have about the procedure, it has the advantage that the necessary data can be acquired during production rolling with the aid of instrumentation available on modern mills. Even if the calculated μ or m is in error, it is still usable for modeling (and control) of the rolling process, as long as the same theory is applied for extracting μ and for predicting the effects of friction.

b. If the aim of investigation is simply that of ranking lubricants, roll force serves as a perfectly adequate measure: a better lubricant gives a lower roll force. Many mills are equipped with load cells that allow direct measurement of forces, and better discrimination is ensured if strips of constant starting gage are rolled at different roll gap settings to give a roll force vs reduction curve (Fig.6.6a).

c. All elements of the rolling mill, including the housing, are subject to elastic deflection on loading. Thus, the issuing strip thickness is greater than the preset roll gap. If strips of constant starting thickness are rolled with gradually diminishing roll gap settings, a better lubricant will result in smaller exit gages because lower roll forces give less mill spring (Fig.6.6b).

d. We have seen that friction is an especially powerful factor when the limiting thickness is approached. This can be exploited by rolling a single strip repeatedly with gradually decreasing roll gap settings, and plotting the issuing gage against the number of passes or the actual roll gap settings for each pass. If rolling is taken far enough to

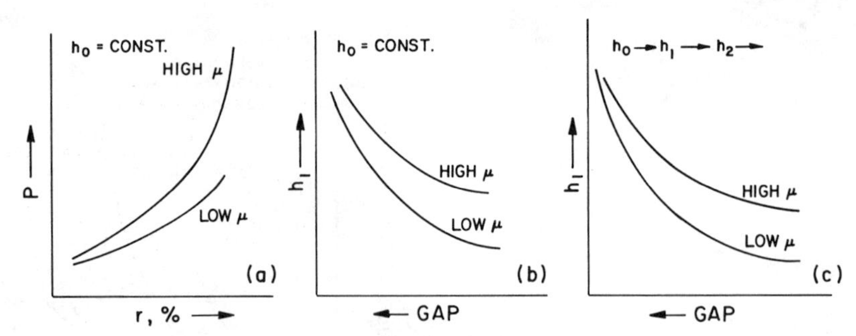

Fig.6.6. Methods of evaluating lubricants from roll force and rolled gage

reach limiting reduction, the curves in Fig.6.6c flatten out and lubricants can be ranked according to the minimum attainable gage. The difficulty of measuring thin strip can be circumvented by measuring the increase in length.

It should be mentioned that with many lubricants, a steady-state, relevant lubrication mechanism is established only after rolling a certain length of strip, and evaluations will be meaningless unless strips of sufficient length are rolled.

6.2.4 Torque and Power

In principle, the torque required for rolling can be readily calculated by assuming the roll force to be concentrated at a distance L from the centerline of the rolls (Fig.6.1). The total torque for both rolls becomes

$$M = 2P\lambda L \tag{6.12}$$

The position of the lever arm is determined by the actual shape of the friction hill. For very approximate calculations, $\lambda = 0.4$ can be taken for cold rolling and $\lambda = 0.5$ for hot rolling.

In more accurate calculations it must be considered that in the neutral plane reversal of interface friction takes place, giving rise to opposing torques. When the rolls flatten, the zone of forward slip extends beyond the centerline of the rolls. Any shift in the position of the neutral plane is reflected in substantial changes in net torque, and theoretical solutions tend to be less accurate for torque than for force.

The net power of deformation can be obtained from

$$\text{Power} = Mv/R \tag{6.13}$$

where v is rolling speed and R is roll radius. The actual motor power must be increased to allow for the inefficiency of transmission elements.

Friction From Torque. Not many mills are equipped with torque meters, and motor power tends to be a rather insensitive measure of frictional conditions.

A special situation exists when, in the rolling of strip, back tension is increased until the neutral plane moves to the exit point and the strip begins to skid (Fig.6.7a). Under these conditions only exists a simple relationship

$$\mu = \frac{M}{PR} \tag{6.14}$$

that was first used by Whitton and Ford [45] for an evaluation of lubricants.

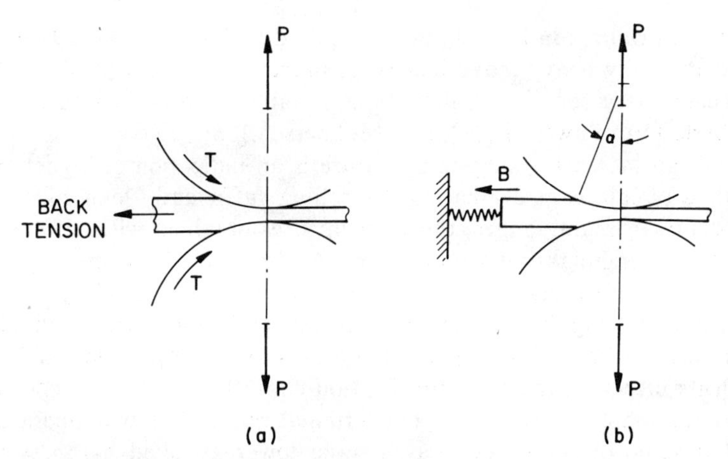

Fig.6.7. Determination of friction by inducement of skidding during rolling

If the mill is not equipped with torque meters, the strip may be stopped by attaching it to a spring dynamometer (Fig.6.7b) according to the technique of Pavlov [46]. Then

$$\mu = \frac{B}{4P} + \tan\frac{\alpha}{2} \tag{6.15}$$

Even though these techniques are attractive because μ is derived without the aid of theory, interface conditions are not necessarily representative of those prevailing under normal rolling conditions. Instead of allowing the balance of frictional forces to establish the position of the neutral plane, skidding is artificially induced and the lubrication mechanism is changed, giving μ values consistently higher than those calculated from forward slip measurements [47]. Otherwise, the two techniques illustrated in Fig.6.7 gave practically identical results in rolling of steel strips with minimum lubrication [48]. The neutral plane is moved back but passage of the slab is maintained when the emerging slab is made to push against a hydraulic cylinder attached to an accumulator, and the resulting pressure is measured [49].

6.2.5 Heat Generation

Most of the work expended in rolling is converted into heat (Sec.2.3.3). Friction contributes to heat generation through at least two sources. First, the work required to maintain interface sliding against frictional resistance is transformed into frictional heat at the interface. Secondly, redundant work from inhomogeneity of deformation is also transformed into heat. There are several consequences:

a. Frictional heat raises the surface temperature of the rolled workpiece. Heat diffusion is rapid enough to equalize temperature throughout the thickness of a thin strip soon after it emerges from the roll gap. Substantial temperature gradients may, however, persist in the hot rolling of thicker slabs, where the cooling effects of the rolls, of the lubricant and/or coolant, and of radiation to the environment cause surface temperatures to drop.

b. The total heat input may be sufficient to maintain a constant or even increasing bulk workpiece temperature if strain rates (which affect flow stress) and rolling speeds (which affect the rate of heat loss) are high enough. Whereas heating is a common occurrence in cold rolling, in hot rolling it is limited to fast continuous lines and to mills that take exceptionally heavy reductions (e.g., planetary mills). Calculation of heat balances can become very complex. Calculations suitable for practical purposes have been given by Roberts [50], Pawelski [51], and McPherson [36].

In general, an increasing strip temperature is an indication of higher friction. This offers a method of lubricant evaluation. By rolling sufficiently long coils with rolls set to produce identical issuing gages, the temperatures obtained with different lubricants can be used as semiquantitative measures of lubricant efficiency [52]. Such an approach offers the advantage of minimum instrumentation, but close control of initial roll temperature and highly reproducible cooling techniques are required. Lubricant evaluation from roll force or rolled gage (Fig.6.6) is more simple and reliable.

Through its effect on temperature, friction can affect also the mechanical properties of the issuing material. Materials that soften at relatively low temperatures (such as aluminum and some of its alloys) may show a lower as-rolled hardness if friction is allowed to rise or cooling is inadequate. In hot rolling, of course, the cooling efficiency of the lubricant is a decisive factor in determining the properties of the as-rolled sheet through its influence on finishing temperatures.

6.2.6 Relative Slip

The importance of relative slip between strip and roll surfaces can hardly be overstated. In a typical rolling situation, forward slip seldom exceeds 15%; the rest of the speed differential is taken up in the backward slip zone. Therefore, the strip enters the roll gap at a high interface sliding velocity which diminishes very rapidly, within a space of typically 20 mm or less, to zero in the neutral zone, only to pick up again as the strip leaves the roll gap. This relative sliding and its reversal, while helpful in many respects, also account for most of the problems encountered with lubrication in strip rolling.

The Neutral Angle. We have seen in Fig.6.3 that with increasing friction the neutral plane moves toward the middle of the arc of contact. Therefore, the angle of the neutral plane ϕ is a sensitive indicator of friction balance and thus of the magnitude of friction. The neutral angle may be readily calculated if some simplifying assumptions are accepted. One of the most widely employed formulae was presented by Ekelund [53], who assumed that deformation is homogeneous, that slipping friction exists everywhere except in the neutral plane, and that radial roll pressure p and interface friction μ are constant along the arc of contact. These two latter assumptions are certainly unjustified, but it appears that many of the errors offset each other. Then, from the horizontal components of acting forces, the position of the neutral plane (Fig.6.1) is

$$\phi = \frac{\alpha}{2} - \frac{1}{\mu}\left(\frac{\alpha}{2}\right)^2 \tag{6.16}$$

where the α angle sustained at entry can be expressed from

$$\sin\alpha = \frac{L}{R} = \left[\frac{h_0 - h_1}{R}\right]^{1/2} \tag{6.17}$$

Substituting into Eq.6.16 we obtain

$$\phi = \left[\frac{h_0 - h_1}{4R}\right] - \frac{1}{\mu}\left[\frac{h_0 - h_1}{4R}\right] \tag{6.18}$$

and thus the angle between the neutral and exit planes is completely determined by pass geometry and the coefficient of friction and, most conveniently, is independent of the flow stress of the material. This offers one of the most reliable methods of friction determination.

For rolling with tensions, the position of the neutral plane can be calculated only if the mean plane-strain flow stress $2k$ is known. According to Ford et al [32]:

$$\phi = \frac{\alpha}{2} - \frac{2k\left(h_0 - h_1\right) + h_0 t_0 - h_1 t_1}{4kR'\mu} \tag{6.19}$$

where t_0 is the back and t_1 the front tension, both expressed in units of stress.

When friction is high enough for a sticking zone to develop, the surface of the strip exhibits forward slip only from the front end of the sticking zone. At a first glance, if sticking prevails over the whole arc of contact, forward slip should reduce to zero, at least on the strip surface. However, the center of the strip is now subjected to severe deformation, resembling forward and back extrusion, as noted in conjunction with Fig.6.3d. At the point of exit, the strip surface is accelerated to keep up with the center, and the position of the flow-dividing plane is still a sensitive indicator of frictional conditions [3.123].

Friction From Forward Slip. Since, by definition, the roll and strip move at the same speed at the neutral plane, any reduction in thickness between the neutral and exit planes must result in a higher exit velocity v_1. Forward slip is defined as

$$S_f = \frac{v_1 - v}{v} \qquad \left(= \frac{h_n - h_1}{h_1} \right) \tag{6.20}$$

It increases with ϕ and thus with increasing friction. From the geometry of the pass, forward slip is approximately equal to

$$S_f = \frac{1}{2}\phi^2 \left[\frac{2R'}{h_1} - 1\right] \tag{6.21}$$

Forward slip can be simply measured by scratching one or more lines on the roll surface (Fig.6.8) parallel with the roll axis; the imprints of the lines will be clearly visible on the rolled strip. At a given roll speed v the rolls take a certain time to travel the carefully measured distance (l_0) between successive scratch lines; during the same time

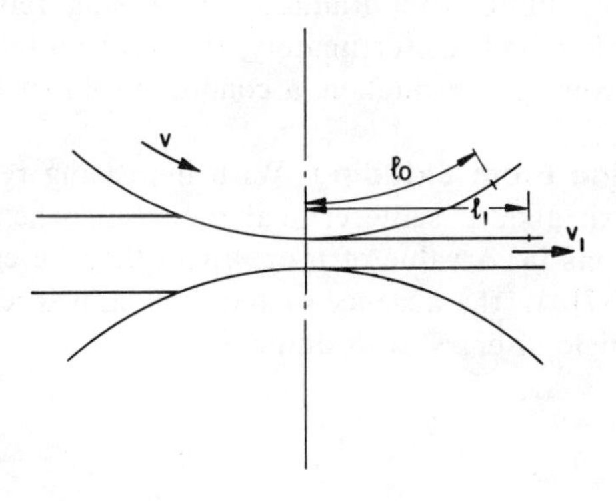

Fig.6.8. Measurement of forward slip

period the strip travels at a higher speed v_1 and the distance between the imprints of the two marks is greater (l_1). Therefore,

$$S_f = \frac{l_1 - l_0}{l_0} \tag{6.22}$$

From Eq.6.21 the position of the neutral plane, and from Eq.6.18 the average external coefficient of friction μ, can be calculated.

Alternatively, forward slip values obtained with different lubricants at a given roll gap setting (or, better yet, over a range of roll gap settings) are directly compared; lower forward slip is indicative of lower friction. Forward slip values observed at different pass reductions are not comparable because, for constant friction, forward slip first rises and then drops as pass reduction increases.

This technique was extensively applied in early rolling research, as reviewed by Underwood [5.57]. Other theories are available, too, but the values of μ extracted from measurements are not greatly different for different theories [44].

Two great advantages of the technique are that $2k$ need not be known and that no special instrumentation is needed. Some precautions are necessary, though. Whenever roll flattening is significant, the flattened roll radius (Eq.6.10) should be substituted. For rolling at elevated temperatures, the shrinkage of the l_1 distance on cooling must be taken into account [54].

The position of the neutral plane is extremely sensitive to the presence of tensions. Friction can be obscured but the method can still be applied if roll and strip velocities are accurately measured, as Kondo [55] did with proximity sensors that gave the work roll and exit deflector roll (billy roll) speeds on an oscillogram. Forward slip values are comparable only for preset tensions, otherwise μ must be extracted from Eq.6.19 or some similar formula.

The techniques proposed by Roberts [7] and by Inhaber [56] derive μ from simultaneously measured roll force, torque, and forward slip values. Their great advantage is that $2k$ need not be known, and the coefficient of friction is obtained even in the presence of tensions [7]:

$$\mu = \frac{M}{PR'\left[1 - \dfrac{2S_f h_1}{h_0 - h_1}\right]} \tag{6.23}$$

For a constant μ, forward slip reaches a maximum when the angle of contact equals the friction angle, as shown by Dahl [47]. This would allow direct determination of μ by simply rolling with gradually increasing reductions until the maximum forward slip value is found. Unfortunately, this procedure is valid only if μ remains unchanged with increasing pass reduction, a condition seldom fulfilled.

Friction From Skidding. With increasing reductions, the neutral angle moves toward the exit and, at some critical reduction when the neutral plane reaches the exit, skidding sets in. A value of μ or m can then be calculated, as proposed by Evans and Avitzur [57]. In the absence of tensions, and when α is small so that in Eq.6.17 $\sin \alpha = \alpha$, the angle of entry at skidding is

$$\alpha_{\text{skid}} = \left(\frac{h_0 - h_1}{R'}\right)^{1/2} = 2\mu \tag{6.24}$$

The onset of skidding can be determined, in principle, also from a single experiment if a wedge-shape specimen is rolled (see, for example, [14]). Once the critical reduction is reached, skidding sets in and the entry angle is twice the friction angle.

Other Methods of Friction Determination. A method related to forward slip measurement has been proposed by Capus and Cockroft [58]. Small irregularities such as pickup on the roll surface cause scratches to appear on the surface of a polished strip. If rolling is carried out slowly enough to stop the rolls with the strip in the gap, the length and direction of the scratches may be observed under a microscope. From the relative lengths of forward and reverse scratches, the position of the neutral plane can be determined and, from Eq.6.18, μ can be derived. The scratches become hooks in the spread zone, allowing fully rolled strips to be examined, although some of the tailing scratches formed in the backward slip zone tend to be obscured by later contact.

The position of the neutral plane can be determined also from backward slip, provided that some accurate means of measuring entry velocity v_0 and roll velocity v is available. Slotted disks pressed onto the entering strip and attached to the roll have been used in conjunction with photocells [59]. Coded disks could also be used, but this technique offers no advantage over the much simpler forward slip measurement.

6.2.7 Strain Distribution

In the idealized case of flat rolling, a rectangular body is homogeneously deformed into another rectangular body of greater length. By definition, all originally vertical sections of the workpiece remain vertical. We have already seen that, in reality, inhomogeneity of deformation develops in the thickness, width, and length directions due to the effects of friction interwoven with the effects of pass geometry [19]. From experimental evidence obtained by techniques described in Sec.5.5.4, the following observations can be made.

Through-Thickness Inhomogeneity. Inhomogeneity of through-thickness deformation has several consequences:

a. The ideal of homogeneous deformation is approached only in rolling with a large L/h ratio and a low enough friction to maintain sliding along the whole arc of contact (Fig.6.3a). As soon as a zone of sticking develops, the formation of a dead-metal zone (Fig.6.3b and c) leads to the extrusion effect already mentioned, causing a backward convexity of originally vertical planes in the entry zone, and a forward curvature in the exit zone. Deformation may be likened to compression between inclined plates, and therefore the angle of entry α, or, more realistically, the $(h_0 - h_1)/R$ ratio, affects the position of the flow-dividing plane. Friction determines, together with the L/h ratio, the extent of the sticking zone; therefore, it also affects inhomogeneity. Furthermore, frictional drag may cause a severe retardation of surface layers, as shown by a reversal of the curvature of grid lines [18,60] and by the distribution of MnS inclusions [5.112].

b. Through-thickness inhomogeneity may cause variations in properties within the rolled workpiece. The interaction of friction and pass geometry is clearly evident from the work of Hundy and Singer [61], who defined an inhomogeneity factor as the difference in hardness between the core and the surface, expressed as a percentage of core hardness. Inhomogeneity decreased with decreasing friction (Fig.6.9) but, at very light reductions, inhomogeneity was evident even with the best lubricant, no doubt because the L/h ratio was small. At high L/h values, deformation became homogeneous irrespective of the lubricant used.

c. Deformation can be inhomogeneous even in the rolling of very thin strip, provided that the pass reduction (or, rather, the L/h ratio) is small. This condition is

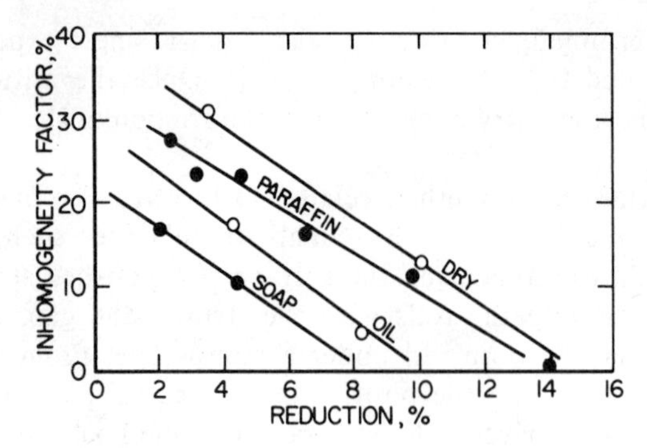

Fig.6.9. Effect of lubrication on inhomogeneity of deformation [61]

intentionally induced in temper (skin pass) rolling of steel. As shown by Bochkov et al [62] in rolling of low-carbon steel with reductions of up to 1% on 500-mm-diameter rolls, the strength, hardness and residual stress levels of the issuing strip were lower when rolling was done with dry rolls than when it was done with lubricated rolls. Rolling at higher reductions (up to 7%) gave the reverse trend.

d. Localized deformation due to friction and pass geometry can affect also the rate at which the crystallographic texture of cold rolled sheet develops [63].

e. The homogeneity of deformation affects also the location at which new surfaces are generated. The absolute increase in surface area is a function only of pass reduction, thus, if a 50% reduction is taken, half of the surface is old (covered with adsorbed films, oxides, contaminants, etc.) and half of it is virgin metal. In well-lubricated rolling where sliding friction predominates, the new surfaces are generated along the arc of contact and, presumably, the lubricant effectively separates the two surfaces. With poor lubrication or in the absence of a lubricant, the surfaces break up just before entering the roll gap; virgin surfaces enter the contact zone and metal-to-metal contact could occur. The consequences of this depend on the prevailing lubrication mechanism.

f. Occasionally, surface cracking is encountered in the hot rolling of difficult-to-work materials. The cause is a combination of low ductility (at less than optimum hot working temperatures) and of secondary tensile stresses generated by cooling of the workpiece surface in contact with the rolls. Localized deformation associated with sticking friction contributes to these stresses. Lubricants alleviate the problem partly by increasing the homogeneity of deformation and partly through their heat-insulating properties.

g. When friction is higher on one roll than on the other, the larger elongation on the well-lubricated surface causes the rolled slab to curl around the roll with higher friction. This happens also when one of the rolls is smaller [64].

Friction and inhomogeneity of deformation are important factors in achieving bonding in roll cladding and in rolling of bimetallic strip [65].

Lateral Inhomogeneity. The assumption of plane-strain deformation implies that material flow is directed only in the longitudinal direction. In practice, it is inevitable that some lateral flow (spread) should also occur.

a. Directions of material flow, as viewed in a direction perpendicular to the arc of contact, were experimentally first established by Siebel and Lueg [5.101]. Lateral flow is clearly evident, as is the resultant increase in width (Fig.6.10). Most of the spread develops in the zone of backward slip, and as such it is sensitive to the position of the neutral plane. Spread increases with increasing friction, but other factors enter, too. For a given pass reduction, spread is greater with a larger roll; for the same roll diameter, spread increases with the L /w ratio, because a longer arc of contact presents greater

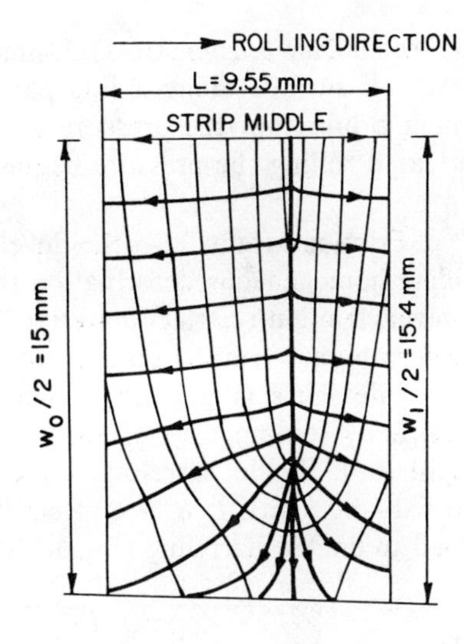

Fig.6.10. Directions of material flow in dry rolling of aluminum strip (plan view of contact zone over half width) [5.101]

frictional resistance in the rolling direction. These observations have been repeatedly confirmed. A partial review has been given by Lahoti et al [66]. Confusion arises only when dry and lubricated experiments are conducted at a constant roll gap setting; lubricated rolling gives heavier reductions and, therefore, greater spread, creating the false impression that spread decreases with friction [67].

b. The friction hill shapes shown in Fig.6.3 apply only to the central portion of a slab where plane strain prevails. Pressures gradually decay toward the spread zones, as already shown by Siebel and Lueg (Fig.6.11) [5.101]. At the edge the normal pressure drops below $2k$ and yielding takes place in a stress state composed of the normal stress and a longitudinal tensile stress induced by the elongation of material adjacent to the

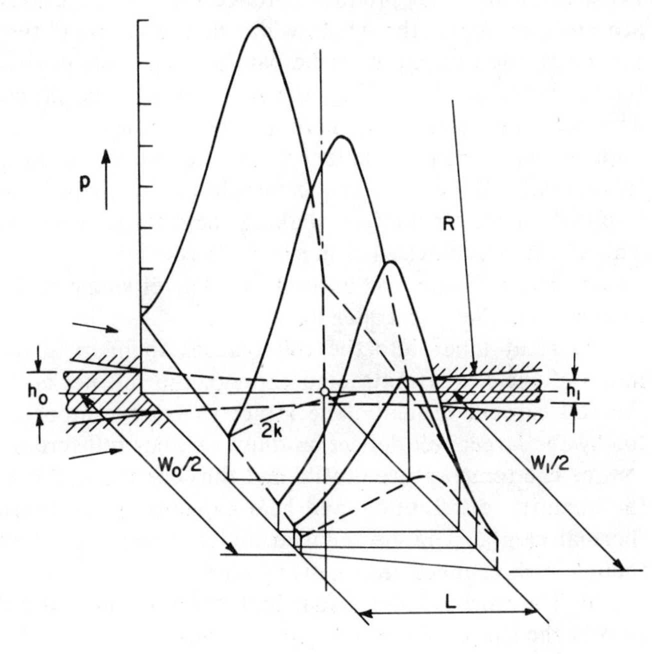

Fig.6.11. Interface pressure distribution in dry rolling of aluminum strip on rough rolls (after [5.101])

edge (secondary tensile stress). Using a large number of pressure transducers embedded in the roll surface along a line parallel to the roll axis, Vater et al [68] showed that without a lubricant the pressure drop-off is steep and limited to the edges, whereas in lubricated rolling the pressure begins to drop off, very gradually, from the center of the strip.

c. Friction affects also the development of the side surfaces of the rolled material. In ideal homogeneous deformation the edge profiles would remain straight (Fig.6.12a). However, friction restricts material flow at the interface in the width direction, too, and therefore higher friction results not only in greater spread but also in greater barreling of the side faces (Fig.6.12b). Under sticking conditions, some of the spread actually develops by folding over of the side surfaces (Fig.6.12c), as shown by Sheppard and Wright [69]. Single barreling is typical of heavy reductions taken on a relatively thin material—i.e., with $L/h > 2$. Relatively light reductions on a thick workpiece ($L/h < 1$) lead to double barreling (Fig.6.12d) which is only weakly dependent on friction.

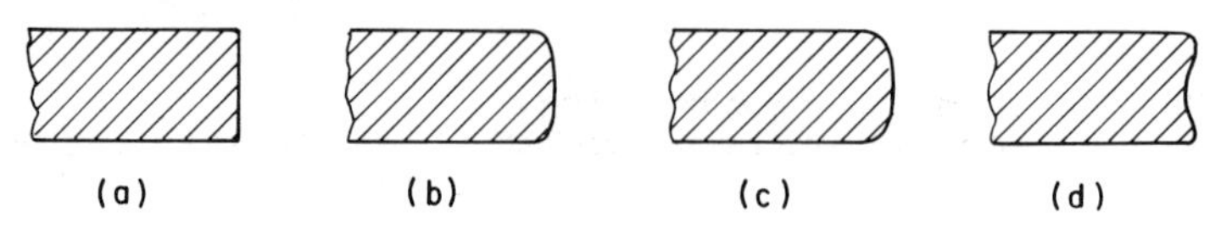

Fig.6.12. Edge profiles of rolled slabs

d. Through its influence on the homogeneity of deformation, friction affects also the occurrence of rolling defects. Edge cracking is a result of longitudinal secondary tensile stresses generated at the edges of the rolled workpiece, where the roll pressure drops (Fig.6.11). Secondary tensile stresses become especially large when the workpiece barrels severely; therefore, the danger of edge cracking increases with increasing friction. Friction, especially sliding friction, has also been suspected [70] of contributing to opening up of the billet in the horizontal center plane (crocodiling or alligatoring), although no direct evidence or detailed analysis is available to support this assumption.

Strip Shape. A flat product is rolled only if reductions and, consequently, elongations are uniform across the whole width of the sheet. If the strip was originally of uniform thickness, the roll gap must be parallel; if the strip is rolled with a slightly thicker center to ensure good tracking, the roll gap must be proportionally thicker in the middle. The roll gap referred to here is that defined by the rolls while the rolling load is applied; since the rolls deflect under the applied loads, cylindrical rolls would present a substantially larger gap in the middle of the strip. Thus, greater elongations would be imposed on the strip edges, making them longer and, therefore, wavy (Fig.6.13) [2.28]. This effect is counteracted in several ways:

a. The rolls may be ground to a barrel shape with the crown (camber) calculated to give a parallel gap under load.

b. Heat input into the rolls causes them to acquire a thermal camber, because much of the heat is extracted through the roll necks. Frictional heating contributes to the roll camber. Luckily, the larger thermal camber is at least partially compensated for by the larger roll deflection due to higher roll forces.

c. The temperature profile and thus the thermal camber of the rolls are affected by the quantity, distribution, and heat capacity of the lubricant. As shown in Fig.6.14, the thermal camber can be reduced and the shape of the roll can be changed [36] even though such changes are relatively slow.

d. The work rolls of a four-high or multiroll mill may be mechanically deflected to control the shape of the roll gap very rapidly.

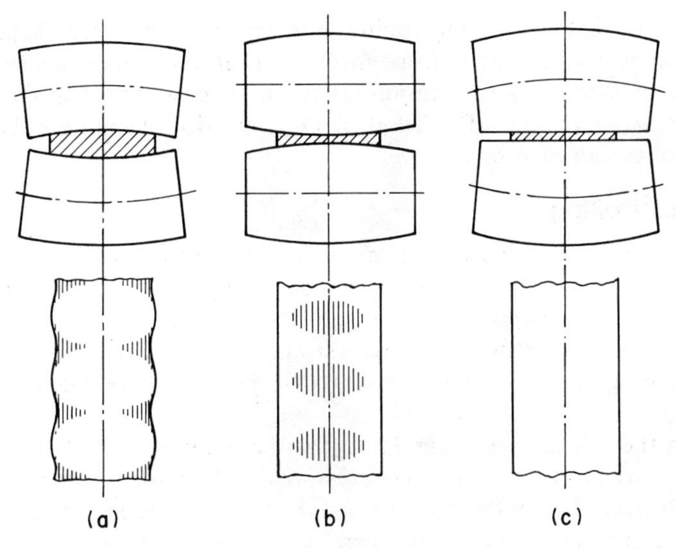

Fig.6.13. Elastic deflection of rolls and its effect on strip shape [2.28]

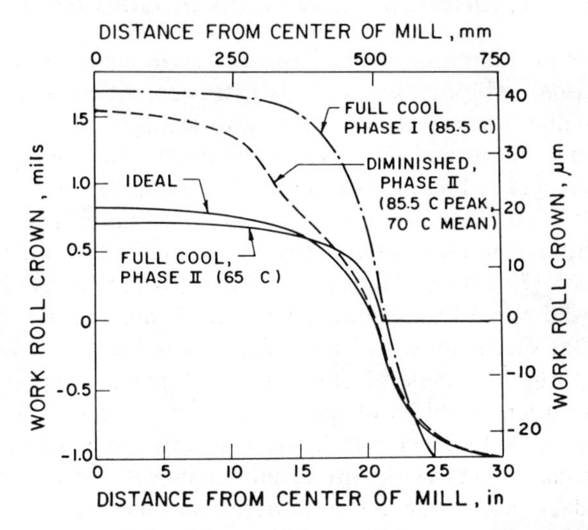

Fig.6.14. Measured variation in roll diameter in cold rolling of aluminum strip compared with calculated ("ideal") diameter distribution necessary to yield a flat product [36]

e. In addition to ground and thermal cambers and roll bending, strip shape is affected also by tensions, because, for a given reduction, roll forces are lower (Eq.6.8) and roll deflections are smaller if higher tensions are applied.

Control of strip shape requires an understanding of the interactions of many variables. Reviews have been given by Sheppard [71] and by Sansome [72], and an entire conference was devoted to the topic [73]. The delicate balance is upset by any change from steady-state conditions. On mill start-up or when cooling by the lubricant is excessive, the proper thermal camber cannot develop and the strip is rolled with a long (wavy) edge (Fig.6.13a). If, on the other hand, insufficient coolant is applied or the lubricant is too warm, excessive thermal camber leads to rolling with a long (wavy) middle (Fig.6.13b). When the waviness shifts off center it is usual to speak of a quarter wave. Highly localized poor shape may develop as a result of a blocked coolant nozzle which allows local heating. During rolling with tensions, the poor shape may be disguised only to appear when the strip is uncoiled. Therefore, shape control increasingly relies on means of measuring the tension distribution across the strip during rolling, as described by Pearson [74], by Davies et al [75], and by Sheppard [71]. Appropriate corrective measures can then be taken.

6.2.8 Powder Rolling

A special case of rolling is the consolidation of metal powders into strip [76,77]. The lubricant may be mixed with the powder to reduce friction between individual particles and thus facilitate densification; however, the lubricant also prevents adhesion between powder particles and thus reduces the green strength of the strip. The lubricant must be chosen to evaporate so as to allow sintering to a product of reasonably high strength and ductility.

Friction on the roll surface is similar to that observed in the rolling of solid bodies, and lubricants can be used to reduce pressures and powder requirements [78]. By inducing sliding over the contact arc, secondary tensile stresses, which may cause the development of transverse cracks on the strip surface, are reduced.

6.3 Lubrication and Wear in Cold Rolling

Interface conditions prevailing in rolling impose relatively mild duties on the lubricant, at least in comparison with processes such as extrusion. At the same time, the process and the issuing product are extremely sensitive to changes in lubrication mechanisms. For these reasons, and because of the economic importance of rolling, great efforts have been made to understand rolling lubrication. In discussing the salient issues, the present discussion proceeds from the better-understood cold rolling to hot rolling, even though the logic of production practice would dictate the opposite sequence. Some general reviews of lubrication [79,80] and lubricant selection [81,82] are available.

Most strip is cold rolled from gages of typically 2 mm and less, on rolls 150 mm or larger in diameter, at speeds up to 2000 m/min, with reductions ranging up to 50% per pass. Consequently, the low angle of entry (Fig.6.1) and high speeds combine to promote fluid film formation. Stability of the operation requires that there should be some positive forward slip; this, together with limitations imposed by surface finish and staining on annealing, usually dictates the use of lubricants with which a full fluid film cannot develop. Boundary lubrication and even occasional metal-to-metal contact are typical, and, if the workpiece material shows adhesion to the roll, pickup ensues. The rate of pickup is relatively low, because the contact zone is only a small fraction of the roll circumference and keeps moving around. Furthermore, in four-high or multiroll mills, the work roll surface and lubricant film are further modified by contact between the work and backup rolls. Consequently, a variety of lubrication mechanisms may come into play, although mixed-film lubrication is predominant.

6.3.1 Full-Fluid-Film Lubrication

The possibility of plastohydrodynamic (PHD) lubrication (Sec.3.6.4) in cold rolling has prompted some of the earliest attempts at the mathematical modeling of friction in rolling. Nadai [83] proposed a solution based on an assumed film thickness. Cheng [84] started from EHD theory; Roberts [85], Bedi and Hillier [86,87], and Avitzur and Grossmann [88] started from minimum energy criteria. Walowit [89] was the first to derive the entrained film thickness by application of Reynold's equation (Eq.3.21) to the inlet zone. Subsequently Wilson and co-workers developed this approach through a succession of papers, reviewed in [3.66] and [90-92], to include more complex effects of heating and of non-Newtonian properties. Heating in the work zone was previously considered by Atkins [93] and by Dow et al [94]. Parallel with this work, extensive investigations were undertaken in Japan (as reported by Azushima [95] and by Ichinoi and Tomizawa [96]), by Soviet workers [97,98], and also by Felder [99]. There has also been mounting experimental evidence to indicate the possibility of at least partial PHD lubrication.

Details of derivations and of observations are to be found in the original publications. The important points that emerge are in agreement with the general discussion given in Sec.3.6.4. The key point is film thickness h as described by Eq.3.29, repeated here as Eq.6.25

$$h = \frac{6\eta v}{2k \tan \theta} \tag{6.25}$$

In various rolling theories, the viscosity η is taken either at room temperature or at the temperature of the roll entry and either at atmospheric pressure or at the pressure prevailing in the entry zone or in the roll gap [4.41], and in some solutions roll gap temperature is considered. The velocity v is taken as the roll velocity or the difference between (or average of) roll and strip entry (or exit) velocities. The flow stress is usually taken in plane strain ($2k$), and some theories allow for strain hardening. The roll entry geometry is expressed by the angle α sustained at entry (Fig.6.1) or by some equivalent geometric factor. Because the equation for film thickness is basically related to journal bearing theory, many authors refer to it as a modified Sommerfeld parameter.

As we shall see, a truly full fluid film represents an abnormal situation and, thus, a number of differences can be observed from the usual rolling behavior previously described:

a. The contribution of friction to the roll force becomes very small. It is doubtful that the pressure distribution can actually be measured without disturbing the full fluid film, but perhaps the experiments of Dow et al [94] with a manganin gage deposited on the roll surface (see Fig.5.28b) come closest to it. The friction hill, such as it is, peaks close to the center of the arc of contact. The friction hill predicted by theory depends on the analysis chosen, and changes greatly when roll flattening is taken into account, as illustrated vividly by the results of Atkins [93] (Fig.6.15). In general, predicted roll

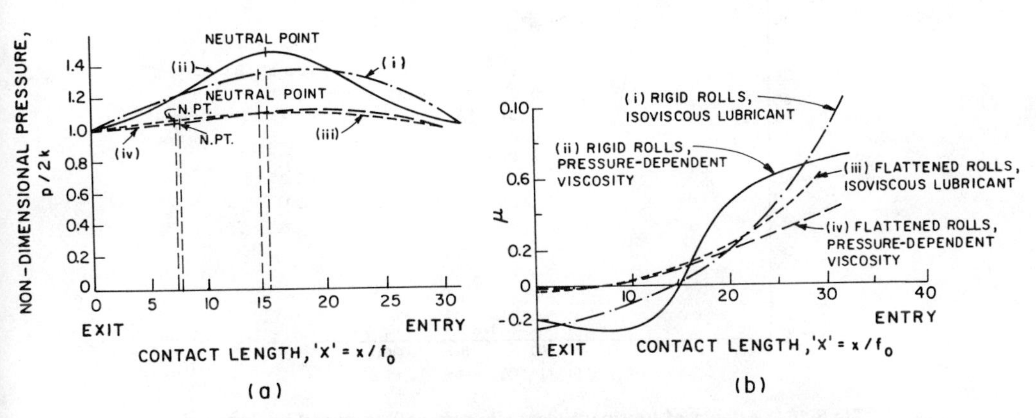

Fig.6.15. Effects of elastic deformations and pressure dependence of viscosity on roll pressure and coefficient of friction variation along the arc of contact, calculated for full-fluid-film lubrication [93]

forces tend to be too high [3.294]. For a check of theory against experiment, several workers back-calculated friction from the measured roll force; however, in view of the uncertainty of flow stress, the results must be viewed with caution.

b. Friction is now due to viscous shearing in the fluid film and should thus be dependent on sliding velocity. The calculated shear-stress distribution depends on the particular theory adopted. Atkins [93] shows that the neutral plane moves close to the exit (which, because of roll flattening, is outside the centerline of the rolls) and that it does not coincide with the peak of the friction hill (Fig.6.15). This is in agreement with measurements [94] but differs from Fig.6.3a where, for Coulomb friction, the neutral plane is at the pressure peak.

c. Friction (traction) is a function of viscosity under pressure. Azushima [100] found in the low-speed rolling of steel (where heating was minimum) with a series of lubricants of identical base viscosity η_0 that friction was highest with a naphthenic oil, lower with a paraffinic oil, and lowest with a synthetic ester, as would be expected from the order of decreasing pressure-viscosity coefficients. Branching in fatty acids and their esters resulted in a high pressure-viscosity coefficient and consequent high friction in the work of Kamita and Yoshida [101].

d. The closeness of the neutral plane to the exit results, at all but very small reductions, in vanishing forward slip, and, if the conditions are appropriate, forward slip becomes negative. This was observed at low reductions by Mizuno et al [102] and was systematically explored by Reid and Schey [3.294]. For a given reduction, forward slip became more negative with higher viscosity or rolling speed (Fig.6.16) and roll forces

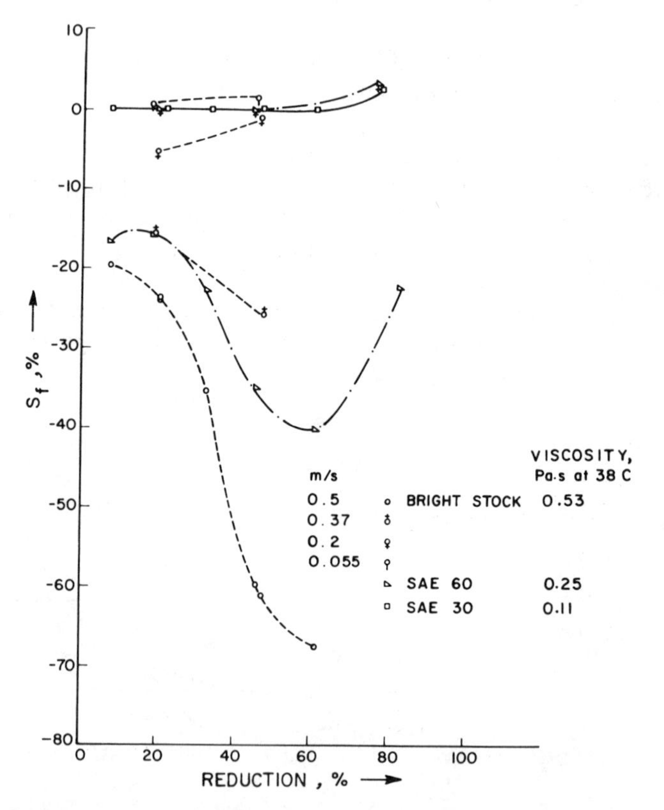

Fig.6.16. Effects of lubricant viscosity and rolling speed on forward slip in rolling of annealed aluminum alloy 6061 [3.294]. Flags attached to symbols indicate rolling speed.

remained low even at high reductions as long as negative forward slip prevailed (Fig.6.17). As indicated by Eq.6.25, speed and viscosity are interchangeable; indeed, negative forward slip sets in at a lower speed with a lubricant of higher viscosity (Fig.6.18). Flow stress was found to be a powerful factor, and negative forward slip was

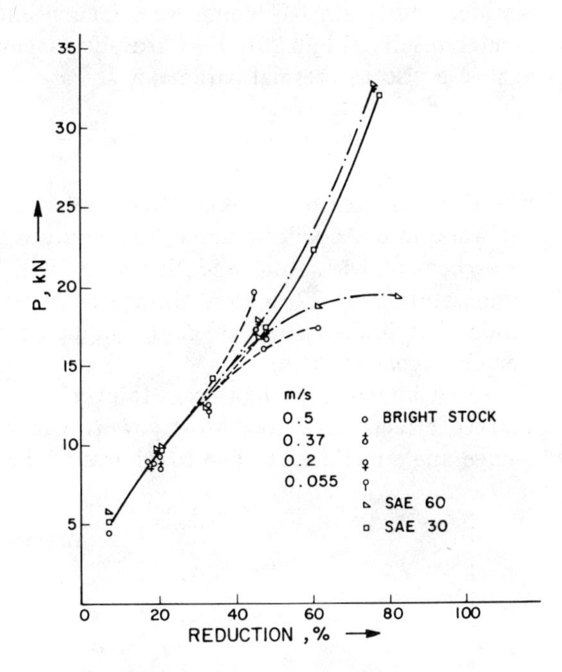

Fig.6.17. Variation of roll force under conditions defined in Fig.6.16 [3.294]

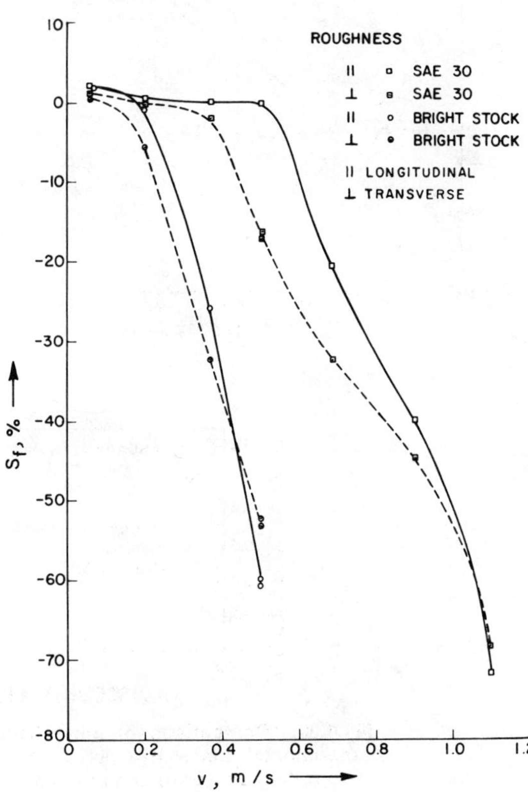

Fig.6.18. Effects of viscosity and workpiece roughness orientation on the onset of negative forward slip [3.294]

difficult to achieve with a workpiece of higher strength (Fig.6.19). Thus, the results are in general agreement with basic hydrodynamic theory but their detailed interpretation poses the greatest challenge. The trends observed in Fig.6.16 are different from those predicted by earlier theories, even in the qualitative sense. Only by considering heat transfer and strain hardening were Chang and Wilson [92] able to match the experimental results (Fig.6.20). Forward slip is plotted as outlet velocity ratio ($Z = v_1/v$) against a viscous thermal parameter L:

$$L = \eta_0\beta\left(v_0 - v\right)^2/4k \tag{6.26}$$

where β is the temperature coefficient of viscosity, and k is the thermal conductivity of the workpiece. As will be noted, temperature is a particularly powerful factor, which is in agreement with Reid and Schey [3.294], who found that an increase of ambient temperature from 22 to 25 C eliminated negative forward slip under a given set of conditions. At lower values of L the higher experimental forward slip is attributed to boundary contact.

e. All factors promoting a thicker film, discussed in relation to Fig.3.36, have marked effects here, too. Most powerful is roll flattening because of its effect on the wedge angle in Eq.6.25. The roughness of the strip is important in that a rougher strip

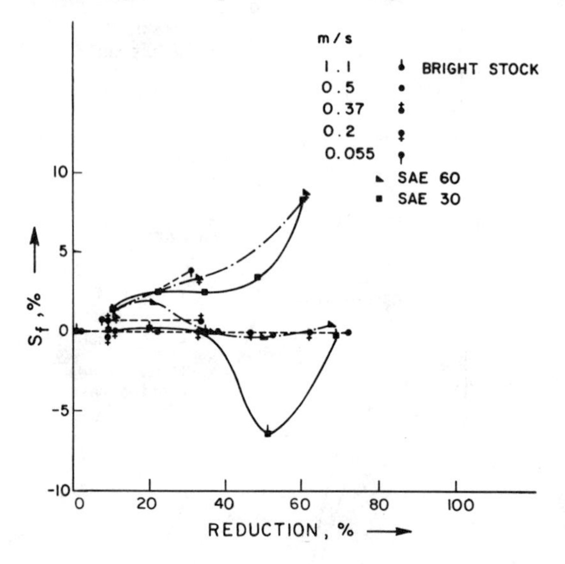

Fig.6.19. Same as Fig.6.16, except that the workpiece material is aluminum alloy 6061-T6 [3.294]

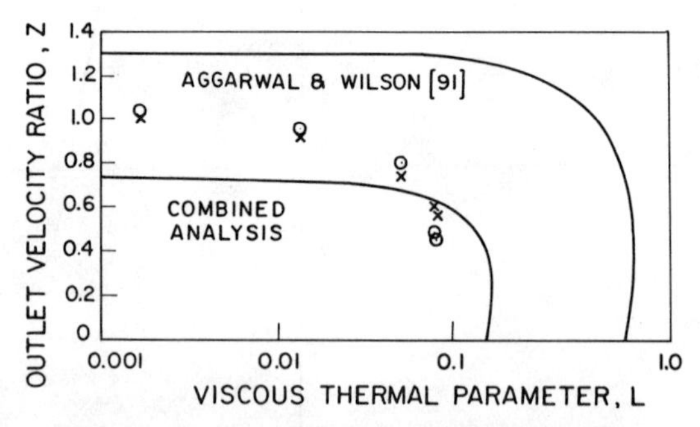

Fig.6.20. Comparison of theoretical predictions (solid lines) with experimental data from [3.294], shown as circles (strip roughness parallel to rolling direction) and crosses (transverse roughness) [92]

carries more lubricant into the roll gap. Roughness is most effective when transversely oriented to the rolling direction (Fig.6.21), an observation that confirms the theory of Sargent and Tsao (Fig.3.50) [3.311]. Deformation of the strip prior to entry into the work zone (rounding off) was found to have an insignificant effect by Tsao and Wilson [103].

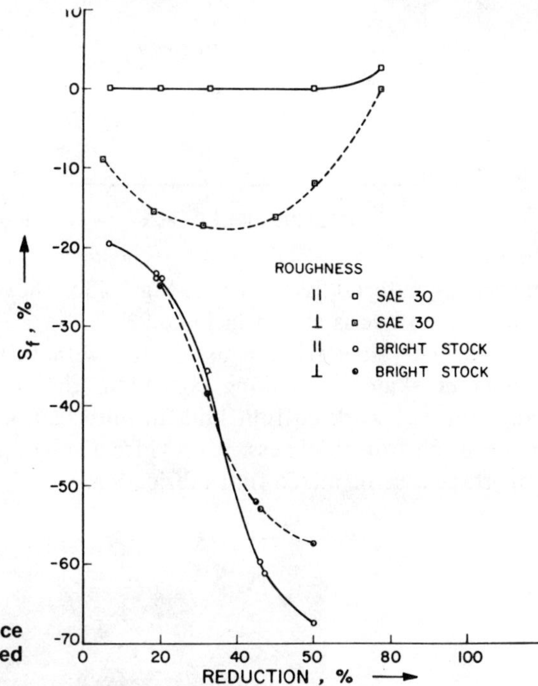

Fig.6.21. Effects of viscosity and workpiece roughness orientation in rolling of annealed aluminum alloy 6061 strip [3.294]

f. In principle, film thickness increases with decreasing flow stress (Eq.6.25) or, rather, roll pressure. For a given metal this can be achieved by applying tensions. The effectiveness of this is difficult to ascertain because the only direct measure of friction—namely, forward slip—is more affected by the balance of tensions than by interface pressure.

g. Comparisons with actual film thicknesses prove unexpectedly difficult. The lubricant bead normally attached to the strip edge causes an overestimation of film thickness from weight determination. By placing a known amount of oil in the middle of a degreased strip and measuring the area covered by the oil on the rolled strip [100], an average film thickness can be estimated, but there are problems with this approach, too. Firstly, the oil film is formed and extended under starvation conditions; for this, Tsao and Wilson [103] developed a correction. Secondly, and more importantly, the lower surface of the strip and the nonlubricated side portions of the top surface are rolled with dry friction, which forces the neutral plane back into the roll gap, imposing more shear on the oil patch and, most likely, reducing the oil film below its equilibrium thickness. Nevertheless, the method is still among the few to give any data at all. Azushima's results [100] are matched quite well by the theories of Dow et al [94] and Aggarwal and Wilson [91], both of which allow for thermal effects (Fig.6.22) [91].

h. By definition, a full fluid film covers the entire interface. A polycrystalline material deformed through such a film suffers substantial surface roughening (Sec.3.8.4) and the issuing strip will be dull. A typical surface obtained on a strain-

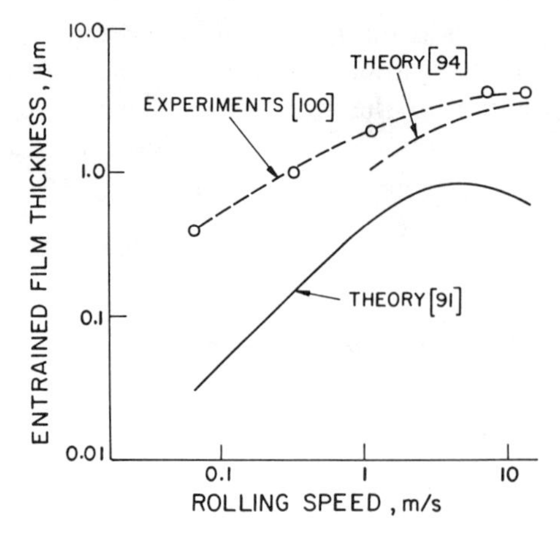

Fig.6.22. Comparison of calculated and experimental film thickness values for rolling of steel at various speeds [91]

hardened material is shown in Fig.3.43c; the nondirectional mottle developing on an annealed surface is shown in Fig.6.23.

The very low friction associated with full-fluid-film lubrication always carries a danger of skidding, making control of the rolling process exceedingly difficult [55]. Nevertheless, work on full-fluid-film lubrication is important because, by allowing estimates of oil film thickness, it provides a starting point for the analysis of the practically important case of mixed-film lubrication.

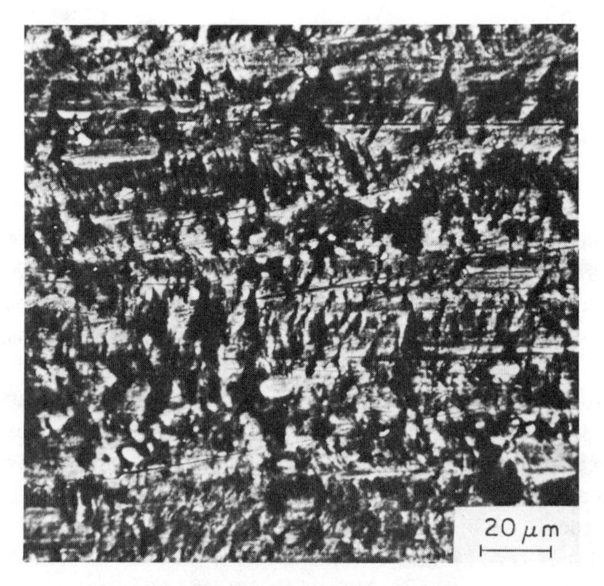

Fig.6.23. SEM photograph of surface of annealed aluminum alloy 6061 strip rolled with bright stock (0.53 Pa·s at 38 C) at 47% reduction and at a speed of 0.37 m/s [3.294]

6.3.2 Mixed-Film Lubrication

All observations made in cold rolling can be fitted into the framework of lubrication regimes discussed in Sec.3.7.3. Indeed, much of the knowledge forming the base of that discussion has been gained in rolling.

Because of the instability introduced by full fluid films, practical rolling is always conducted in the mixed-film regime, on the descending portion of the μ vs v curve in

Fig.3.42. Details of the lubrication mechanism depend on the topography of surfaces as well as on the hydrodynamic and boundary properties of the lubricant. Therefore, an average film thickness has little physical significance, but it does give at least a general indication of the hydrodynamic contribution to lubrication.

Film Thickness. Average film thickness can be measured by various techniques. We have already mentioned dropping a known volume of oil on the surface of a dry strip. Values more typical of equilibrium conditions are obtained by applying a carefully measured quantity of lubricant (from a solvent, by an air gun, or electrostatically) to the strip surface only, and then measuring the quantity of oil carried through the roll gap. It is possible also to calculate an entrapment volume from the combined roll and strip surface roughnesses (Eq.3.23) by assuming zero film thickness in the boundary-lubricated zones [100].

Influencing factors can be rationalized, as in full fluid lubrication, by some form of modified Sommerfeld parameter related to Eq.6.25. A larger value of the parameter signifies that more hydrostatic pockets (and a rougher surface) and, because of the smaller boundary contribution (Eq.3.33), also lower friction and roll force should be expected. The parameters used are roughly equivalent. For example, Japanese workers have widely accepted the expression of Mizuno [104], who used the arithmetic average of roll speed and strip exit speed for v in Eq.6.25. Thinning of the film has been taken into account by the multiplying factor $(1 - 2r/3)$ where $r = (h_0 - h_1)/h_0$.

The use of such parameters allows comparison of data obtained on various mills, as was done by Ogura [105] for rolling of aluminum and stainless steel. Sargent and Stawson [106] also rationalized their results for aluminum rolling against a $\eta v/p$ parameter.

In general, the influencing factors are inlet geometry, viscosity, speed, and interface pressure (some function of $2k$).

Inlet Geometry. Instead of tan θ, the angle of entry α is often used, even though this is somewhat larger than θ in Fig.3.36c because it does not allow for the effect of roll flattening. As can be seen in Eq.6.17, α becomes smaller with increasing roll diameter and decreasing reduction. These trends have been repeatedly confirmed [102,107,108] and are illustrated in Fig.6.24. Kondo [55] found that μ calculated from forward slip in rolling of aluminum strip with tensions decreased roughly as the -1.8 power of roll diameter. In general, this means that low friction is more difficult to attain with small

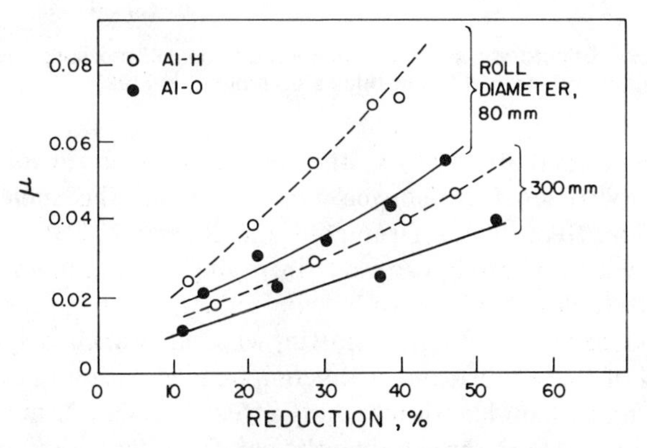

Fig.6.24. Effect of roll diameter on coefficient of friction measured in rolling of 0.4-mm-thick aluminum strip with a light mineral oil [102]

rolls, but at the same time the strip will be brighter and forward slip more positive, thus ensuring greater stability of the rolling process.

Viscosity. There are many observations which show that the average film thickness increases and lubrication becomes more hydrodynamic in character as the lubricant viscosity increases. A typical set of results from Mizuno et al [102] is shown in Fig.6.25. Many of the data can be rationalized by assuming that the coefficient of friction decreases with viscosity according to a power law

$$\mu = f\left(\eta^{a}\right) \tag{6.27}$$

where a might be called a viscosity-sensitivity exponent, the value of which is, unfortunately, dependent also on the method of determining μ. In the work of Thorp [109] on steel strips with three pure paraffins (C_6, C_{12}, and C_{16}), the exponent was roughly -0.2 when μ was back-calculated from rolling force. Kondo [55] found an exponent of -0.5 when calculating μ from forward slip in rolling of aluminum under industrial conditions. Interestingly, the same exponent can be extracted from the results of Mizuno et al [102].

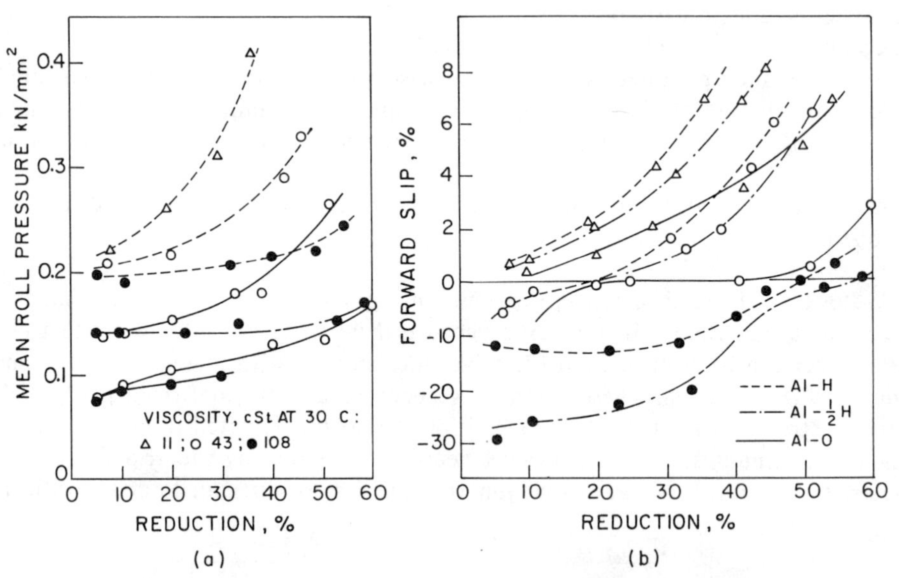

Fig.6.25. Effects of viscosity of mineral oils on roll pressure and forward slip in rolling of 1.2-mm-thick aluminum strip [102]

In judging the effect of viscosity, the composition of the oil and its effect on pressure-viscosity coefficients cannot be ignored. Even the molecular structure of hydrocarbons has an effect. Thus, Drumgold and Rodman [110] obtained, in general, greater reductions with higher-viscosity oils in rolling of aluminum. Unexpectedly, oils of very low viscosity (below 2 cSt) gave heavier reductions than more viscous ones. This was true also of dodecane, a synthetic paraffin, whereas branched synthetics were much poorer. Because all of these lubricants were compounded with a fatty alcohol derivative, it is possible that some chain-length matching effect (Sec.3.5.2) may have occurred. In the work of Iwao et al [111], predominantly paraffinic oils were more effective than predominantly naphthenic oils of the same viscosity. The markedly lower friction observed on adding 3% low-density PE to oil must be, at least in part, due to an increase in effective viscosity [4.65].

The viscosity effect is noticeable also in rolling with fatty lubricants. Whetzel and Rodman [112] found friction to drop in proportion to the logarithm of viscosity. Similar observations were made by Williams and Brandt [113] and by Iwao et al [111] with various fatty substances. Nikolaev and Dolya [114] changed the viscosity of refined cottonseed oil by oxidation and observed a roughly linear drop of forward slip and of roll force with the logarithm of viscosity. There is, however, an uncertainty introduced in these experiments by the simultaneous changes in boundary properties.

Viscosity is very much a function of temperature, and drops steeply for most oils in the temperature range (from 20 to 250 C) normally encountered in cold rolling. This effect was explored in series of papers by Nikolaev [115] and by Grudev et al [116]. With increasing bulk temperature the shift toward boundary lubrication results in increasing friction and, above 200 C, breakdown is likely. This aspect will be discussed later.

Rolling Speed. According to Eq.6.25, viscosity and speed are interchangeable variables. The difference is that, apart from changes induced by temperature, viscosity is fixed in a given rolling mill, whereas speed changes during rolling of a single coil as a matter of necessity.

When a long strip is rolled from a coil, the strip end is entered (threaded) at a low speed, typically 12 to 70 m/min. On gradual acceleration to production speeds (ranging from 300 to 2000 m/min), the thickness of the issuing strip decreases if the roll gap setting is left unchanged. This so-called speed effect was observed in the late 1930's on the then recently installed high-speed tandem mills. If the issuing gage is to be kept within close tolerances, corrective action must be taken. Hence, this phenomenon gave the impetus for much research which contributed to an understanding of lubrication mechanisms in general and the role of the rolling mill response in particular. The latter aspect was well clarified by the early 1960's, as shown by the review of Pawelski [117].

1. Response of the Rolling Mill. Most laboratory mills are relatively small and limited in speed range, and thus they are usually equipped with roller bearings which operate with a very thin EHD lubricant film at all speeds. Thus the roll gap is defined by its setting. If the thickness of the issuing strip drops with increasing speed, only a drop in interface friction can be responsible. This is made clear by the argument presented by Hessenberg and Sims [118]. When the roll separating force is plotted against the rolled strip thickness for a given initial roll gap setting s_0, the actual roll gap increases as a result of elastic distortion in the rolls, bearings, and mill housing (Fig.6.26a). The slope of this so-called elastic curve is the spring constant of the mill, on the order of 4 MN/mm for a production mill. For a strip of h_0 entry thickness, the force rises with increasing reduction according to some "plastic" curve, as determined from rolling theory or experiment. Where this curve intersects the elastic curve, the issuing gage of the strip h_1 is found. If the roll gap is unchanged and all other input variables are kept constant, the issuing gage can decrease to h'_1 with increasing rolling speed only if friction in the roll gap decreases.

The situation is very different in many production mills. Until fairly recently, large roller bearings could not be operated at high speeds. Therefore, many mills are equipped with full-fluid-film back-up roll neck bearings which have a diametral clearance equal to 1/1000 of the roll neck diameter. Thus, the total clearance for the top and bottom rolls may be on the order of 1 or 2 mm, a not negligible value in terms of roll gap settings. At low rolling speeds the roll separating force is carried by a relatively

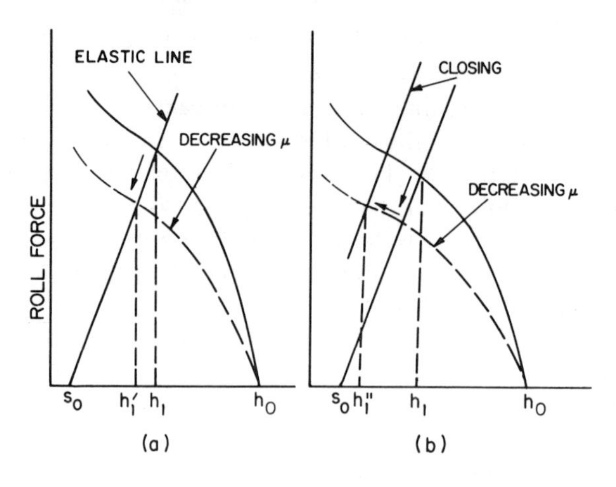

Fig.6.26. Changes in issuing strip thickness [118]

thin hydrostatic oil film between journal and bearing. As the speed of rotation increases, a thicker hydrodynamic film is formed and the roll journals ride up on this film, occupying a new center of rotation. The bottom rolls rise; the top rolls, which have been pushed up by the roll force, descend. Thus, the roll gap closes and the mill elastic line is displaced (Fig. 6.26b). It will be noted that this has the same effect on the issuing gage as that of an increase in friction. For this reason, the individual contributions of journal displacement and interface friction to the speed effect (as expressed by the new gage h''_1) can be separated only if all variables, including roll neck displacement, are simultaneously recorded. Stoltz and Brinks [119] reported a roll gap change on the order of 0.1 mm per stand on a four-stand sheet mill during a slowdown from 1000 to 80 m/min in the midst of rolling a coil. This would readily account for more than the observed speed effect (Fig.6.27). However, the roll gap is attenuated by the decreasing roll gap efficiency in rolling thin sheet. Müller and Lueg [120] found that roll force, tensions, roll journal displacement, and gage variations were intricately interwoven. Unfortunately, they had displacement transducers only on the top roll journal, and thus the total change in roll gap cannot be ascertained. Popov [121] ran the rolls of a four-high mill in contact without metal but with copious application of emulsion. He found the roll separating force to increase with speed steeply up to about 300 m/min and more gradually thereafter. This is also the pattern for the speed effect, and therefore he attributed it entirely to film thickness changes in the journal bearings. However, the thickness of the lubricant film must have increased also between the work rolls and between work and back-up rolls, contributing to the observed effect.

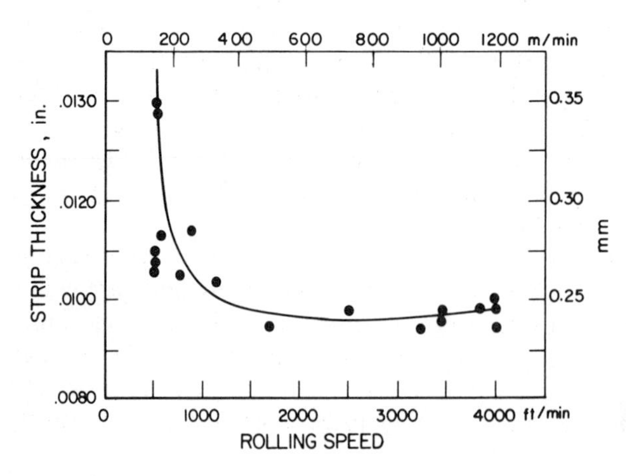

Fig.6.27. Speed effect observed in rolling of low-carbon steel strip on a tandem mill [119]

Bearings constitute a vital element in the rolling process and have been repeatedly discussed [122,123].

2. Change in Lubrication Mechanism. The first undeniable evidence for a speed effect with a truly constant roll gap setting came from laboratory experiments on mills equipped with roller bearings. Ford [107], Sims and Arthur [124], and Billigmann and Pomp [108], all of whom gave reviews of earlier work, invariably found the roll force and issuing gage to decrease with increasing speed. A typical example is given in Fig.6.28 for the rolling of steel in successive passes. It will be noted that the speed effect was marked only in later passes when the strip became hard and thin. Under these conditions the rolls are flattened and the large L/h ratio makes friction a most powerful factor. The L/h ratio necessary for the clear appearance of the speed effect is a function also of roll surface finish and strip material [108], points which will be taken up later.

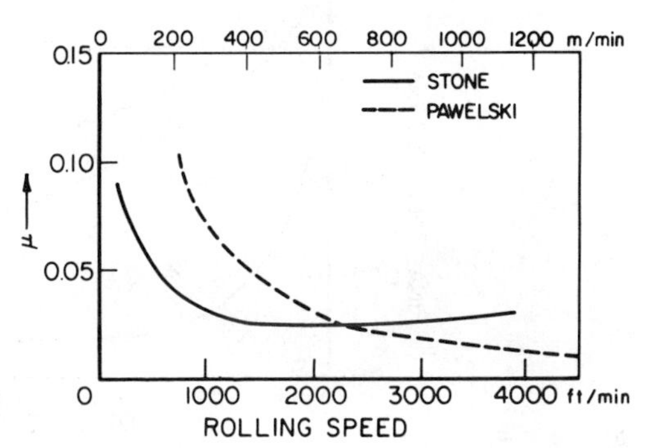

Fig.6.28. Values of coefficient of friction calculated from industrial steel rolling trials [130,134]

The causes of the speed effect were only gradually recognized. Billigmann and Pomp [108] were the first to point out that hydrodynamic effects similar to those of a journal bearing must be active in the roll gap. Sims [125] postulated that at low rolling speeds most of the lubricant is squeezed out, leaving a thinner film in the roll gap. Pawelski [117] considered that the lubricant is subjected to sudden loading on entering the roll gap, causing it to behave in a viscoelastic manner. His calculated stressing times are on the order of a few milliseconds, commensurate with the relaxation times of oils if the effects of high pressure are considered (see Viscoelasticity in Sec.3.6.3). Although such effects may indeed contribute to the phenomenon, the most likely explanation is that lubrication is of the mixed-film type.

As shown in Fig.3.42a for a constant interface pressure, films formed at low speed v and/or low viscosity η are so thin that predominantly boundary lubrication prevails. As the product ηv increases, a thicker film is entrained, and an increasingly larger portion of the surface is supported by hydrodynamic pockets (observed by a number of workers [3.303],[45,126,127]). This leads to a relatively rapid drop in friction as the boundary-lubricated area decreases, and to a relatively modest further drop once most of the load is borne by the hydrodynamic pockets. The possibility of quasihydrodynamic lubrication was suggested by Wistreich [128] by analogy to observations made in wiredrawing, and by Schey [129] from an interpretation of the data of Thorp [109].

The effect is most directly observed by following the drop in issuing strip thickness for a constant roll gap setting. The results may be more readily interpreted if rolling is done at a range of reductions and speeds and if data points corresponding to a constant pass reduction are extracted [109]. A less laborious procedure involves closing of the roll gap during acceleration to maintain a constant exit gage. In either instance, a coefficient of friction may be calculated if so desired.

Examples of the speed effect are abundant [7]. It is shown in Fig.6.27 for issuing gage [119] and in Fig.6.28 for coefficient of friction [117,130]. Whetzel and Wyle [131] found a powerful speed effect with high-viscosity experimental lubricants and with a liquid fat (Fig.6.29). Convincing examples are available in more recent rolling experiments at much higher speeds. Thus, Kondo [55] observed, in the rolling of aluminum, a semilogarithmic relationship between forward slip and rolling speed, and a power-law relationship with an exponent of -0.16 between μ and v (Fig.6.30). Azushima [95] found a similar trend but a somewhat steeper slope in rolling of steel. Generally similar trends were found by Starchenko et al [132] in rolling at speeds up to 3000 m/min and by Kolmagorov and Kolegova [98].

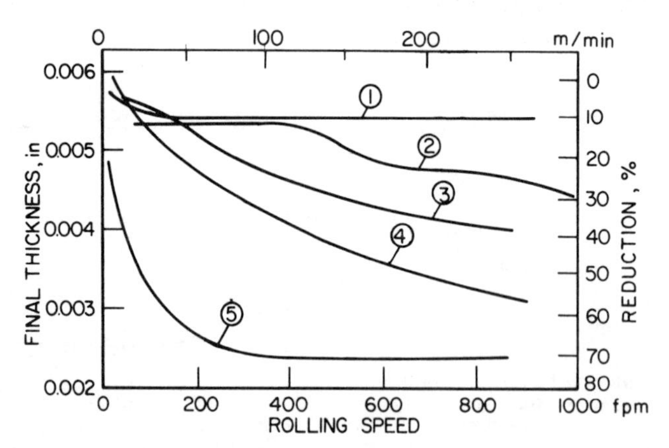

Fig.6.29. Speed effect in experimental rolling of low-carbon steel strip using various lubricants [131]. 1 - soluble oil; 2 - mineral oil; 3 and 5 - experimental lubricants; 4 - liquid fat.

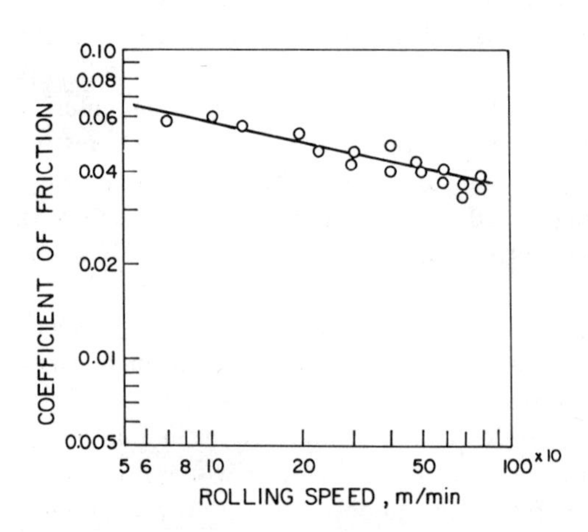

Fig.6.30. Values of coefficient of friction calculated from forward slip in rolling of aluminum with a compounded mineral oil [55]

The influence of the calculating method is evident from earlier evaluations of industrial experience. Stone [130,133] used a simple theory [31] to calculate coefficient of friction values which show a hint of a minimum (Fig.6.28). In contrast, Pawelski [134] calculated a much more gradually changing friction from the tandem mill experiments of Billigmann and Lenze [135] using the more sophisticated theory of Ford et al [32]. Similarly declining values were obtained by Yokote and Nomura [136] from pilot runs on a production aluminum foil rolling mill. The power-law relationship found by other authors also suggests that a minimum should occur at much higher speeds and then only because of lubricant breakdown. Thus, within the range of technical speeds, the mechanism is always that of mixed-film lubrication.

Through its influence on roll force, the speed effect contributes to shape problems. The high forces generated at low speeds deflect the rolls, resulting in a wavy edge, whereas the low friction typical of high speed causes a wavy middle.

That the oil film carried into the roll gap thickens with increasing speed has been repeatedly observed directly. Christopherson and Parsons [137] drew strip through freely rotating rollers (equivalent to Steckel rolling) and found that the surface finish of the strip roughened with increasing speed. Whetzel and Rodman [112] rolled steel strips precoated with controlled films of fatty oils, ranging from very thin, ineffective films to excessively thick ones. After the strip had been rolled to a constant gage at various roll speeds, the thickness of the residual oil film was determined by a solvent method and μ was calculated from roll forces. Allowing for the elongation of the strip, films up to 0.1 μm thick passed through the roll gap without change, and thus, up to this thickness, μ was governed by the available lubricant supply. Thicker oil films could be developed only by providing excess lubricant on the ingoing side, and then rolling velocity became the governing factor. The thickest film obtained with a natural fat was 0.25 μm, and the thickest obtained with a more viscous modified fat was 0.35 μm. It was also noted that the greater viscosity of the modified fat gave markedly lower friction at low speeds whereas both the natural and viscous fats produced identical friction at higher speed, indicating that sufficient lubricant was entrained to make viscosity a secondary factor. Similar conclusions were reached by Shibata et al [138]. In common with other researchers, they found (Fig.6.31) that measured film thicknesses were always greater than those calculated from Eq.6.25. The cause of this discrepancy is to be sought in the effect of surface roughness, a point to which we shall return later. It will be noted that these film thicknesses are almost ten times greater than those measured by Whetzel and Rodman [112]. Fully flooded inlet conditions were obtained with coating weights in excess of 2.0 g/m^2.

The nature of the lubricant also has an influence on the speed effect. Some researchers have noted a rather sudden drop of friction at some critical speed in rolling with mineral oils. Such a drop is visible in Fig.6.29 [131], and an even sharper transition was found by Thorp [109] with pure hydrocarbons. It is not clear whether this effect is an artifact or is real; certainly, with the more usual lubricants, the speed effect is always gradual.

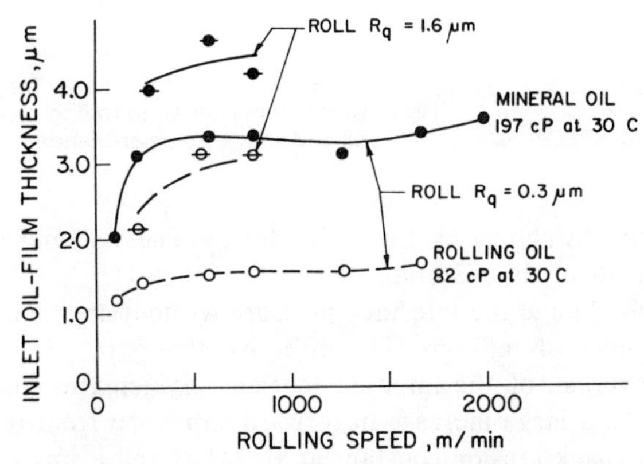

Fig.6.31. Effects of oil viscosity and roll surface roughness on film thickness in rolling of steel strip with a roughness (R_a) of 1.2 μm at 50% reduction [138]

The limiting oil film thickness reached with most oils is controlled by the heat generated at higher rolling speeds. This effect is quite marked in the calculations of Azushima [95] (Fig.6.32), in line with the argument presented in connection with Fig.3.38. The limiting film thickness can be increased only with a lubricant that gives higher traction at higher speeds. Thus, Tamai et al [139] found no limiting film thickness in rolling with 20% modified PMMA (polystearyl-tridecanoyl-methacrylate) polymer blended in a low-viscosity paraffinic mineral oil (Fig.6.33).

An indirect confirmation of the mechanism of the speed effect comes from rolling with ultrasonic vibration of the rolls. Severdenko et al [3.336] found that the small drop in roll force observed on application of vibration during low-speed cold rolling disappeared as rolling speed increased, indicating that the effect of vibration was swamped by the speed effect once a predominantly hydrodynamic film had been established.

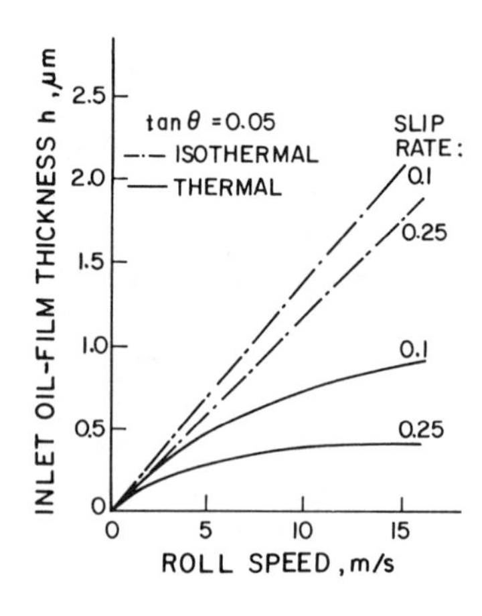

Fig.6.32. Effect of rolling speed and of the resultant temperature rise on calculated oil-film thickness [95]

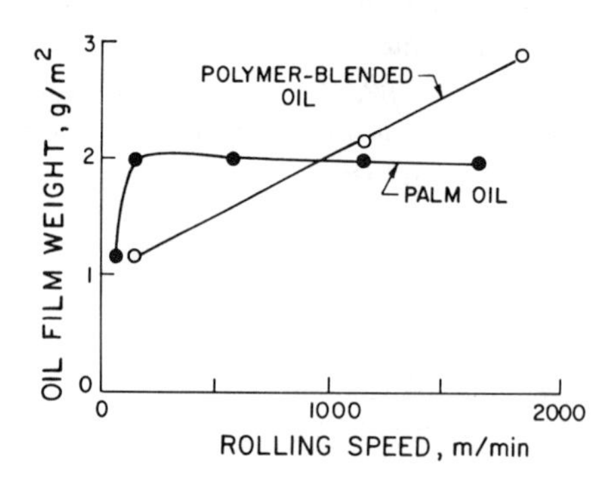

Fig.6.33. Film thickness (expressed as oil-film weights) measured in rolling at high speeds with palm oil and with a polymer-blended oil [139]

Interface Pressure. As shown by Eq.6.25, film thickness should be inversely proportional to the flow stress of the material.

The only way to change the interface pressure without any change in material is by varying the magnitudes of tensions. The effective flow stress is the plane-strain flow stress $2k$ less the average of the imposed tensions. Systematic studies by Kondo [55] revealed, as expected, a large increase in forward slip when front tension was increased from 10 to 80 MPa (back tension constant at 10 MPa) and a small decrease in forward slip when back tension was increased from 2 to 12 MPa (front tension constant at 42 MPa). Somewhat surprisingly, μ remained virtually unchanged; apparently the change in interface pressure was too small relative to the $2k = 200$ MPa of the hard, commercially pure aluminum strip. In principle, however, some change should be expected and was indeed found by Mizuno and Rokushika [140] when back tensions were increased over a larger range. The relative insensitivity of the lubrication mechanism to tensions is beneficial in that a positive forward slip on the order of 2 to 8% can be maintained by varying the front tension, and thus the stability of rolling can be ensured without impairing lubrication.

Interface pressures can be increased without changing the metallurgy of the strip by rolling previously strain-hardened material. Forward slip and μ values are generally higher on hard strips than on soft ones (see Fig.6.24; Fig.6.16 versus Fig.6.19). This was found also by Zhuchin and Pavlov [141] in rolling of a magnetically soft iron alloy. Roberts [142] found that friction was higher on full-hard 0.38-mm-thick mild steel sheet than on annealed material, although the difference disappeared at speeds above 100 m/min; presumably at that speed the limiting film thickness was attained irrespective of interface pressure.

Direct comparison of values of friction measured in rolling of different workpiece materials is not really possible. Because lubrication is of the mixed-film type, boundary lubrication, with its sensitivity to material composition, plays a significant role.

Effect of Pass Reduction. Until now our discussion has focused on the effects of the hydrodynamic contribution to mixed-film lubrication. One must not forget, however, that some boundary contact is always present; it is only to be expected that the boundary lubricating properties of the lubricant and interactions with the workpiece and roll material play important roles.

Boundary effects become highly visible when increasing reductions are taken on strip of constant starting thickness. Unfortunately, this procedure introduces changes in other variables, too:

a. The entry angle α increases, reducing the entrained film thickness. Surface tensions increase, more virgin surface is generated, and the thickness of the entrained film is further reduced. All these effects result in a shift toward predominantly boundary conditions.

b. The increase in the L/h ratio, accentuated by roll flattening, results in higher roll forces, a phenomenon especially evident when limiting reductions are approached.

Conditions are, therefore, complex, but are illustrative of the many effects of changes in the lubrication mechanism. Indeed, experiments of this type led Schey [44] to identify lubrication in rolling as an example of mixed-film lubrication.

It must be emphasized that these phenomena will be observed only if experiments are conducted with the greatest care. Early laboratory experiments often used speeds too low to generate relevant data. The condition of the roll surface was also uncontrolled. Even the most careful cleaning with solvents will not remove incipient pickup which may have been left on the roll surface from prior experiments, especially those involving rolling of high-adhesion materials such as aluminum or titanium. Generally observed in rolling of short strips are random variations in roll force which indicate varying conditions along the roll surface (Fig.6.34a). Neither minimum, nor maximum, nor average forces or forward slip values have any meaning. As shown by Schey [3.303], equilibrium conditions can be established only by running the rolls for a prolonged period of time against wipers or against each other. When a strip long enough to cover several roll circumferences is then rolled, a fairly steady force is recorded during rolling in the predominantly hydrodynamic regime (Fig.6.34c). However, as soon as boundary contact becomes significant, roll force rises in a stepwise fashion from revolution to revolution (Fig.6.34b). The magnitudes of the steps and the final steady-state condition are functions of the efficiency of the boundary lubricant.

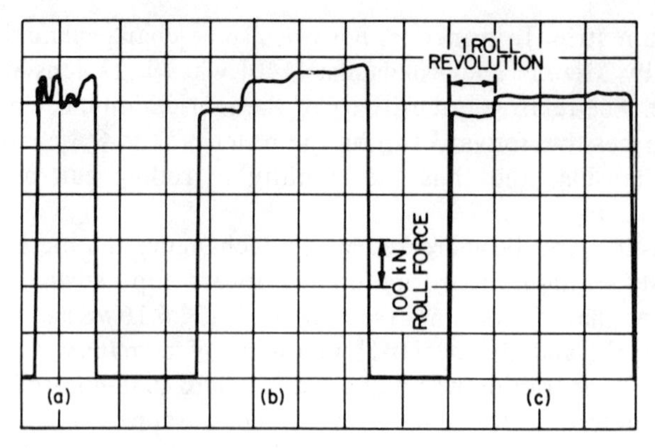

Fig.6.34. Variations in roll force registered in laboratory rolling of aluminum alloy strip [3.303]

While absolute values depend greatly on lubricant and workpiece material, some general trends can be identified:

a. At low pass reductions the L/h ratio is near unity. Because deformation is inhomogeneous, friction plays only a minor role and lubricant variables are obscured. The roll force and the value of μ back-calculated from roll force may be quite high.

b. At relatively low reductions the hydrodynamic element predominates in the mixed-film mechanism. Roll force and forward slip remain sensibly constant or rise only very little from revolution to revolution (Fig.6.35), and limiting values are reached even with a pure hydrodynamic medium. Within a limited range of reductions, the value of μ and the surface roughness of the strip stay constant or rise slightly. Much of the surface is covered with hydrodynamic pockets. If a long enough strip is rolled to reach a

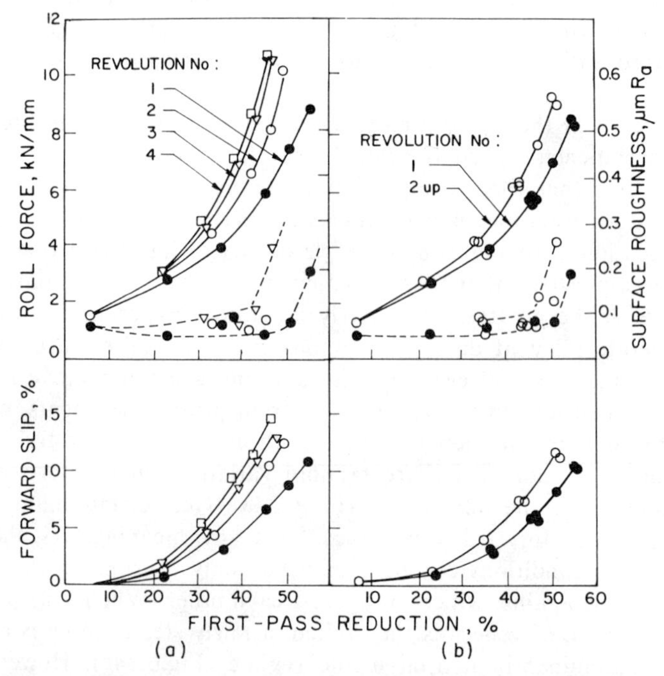

Fig.6.35. Effects of repeated contact with the roll surface on roll force, forward slip, and surface roughness in rolling of hard aluminum alloy 3003 strip with (a) a mineral oil and (b) a compounded oil [3.303]

semblance of equilibrium, there are shifts in the roll force, forward slip, and friction coefficient curves toward lower values with increasing speed or viscosity (Fig.6.36).

c. At some critical reduction the film is thinned to the point where relatively large portions of the surface come into boundary contact. When there is adhesion between roll and workpiece materials, as in rolling of aluminum, steep rises in roll force, forward slip, and coefficient of friction are noted (Fig.6.35). In this regime, the boundary properties of the lubricant come to the fore. Roll forces rise in a stepwise fashion and fail to reach a limiting value in rolling of aluminum with an uncompounded mineral oil; the surface roughens while its specular reflectivity increases. The presence of a powerful boundary lubricant shifts the onset of predominantly boundary lubrication to higher reductions (Fig.6.36) and limits the stepwise increase in roll force to the first one or two

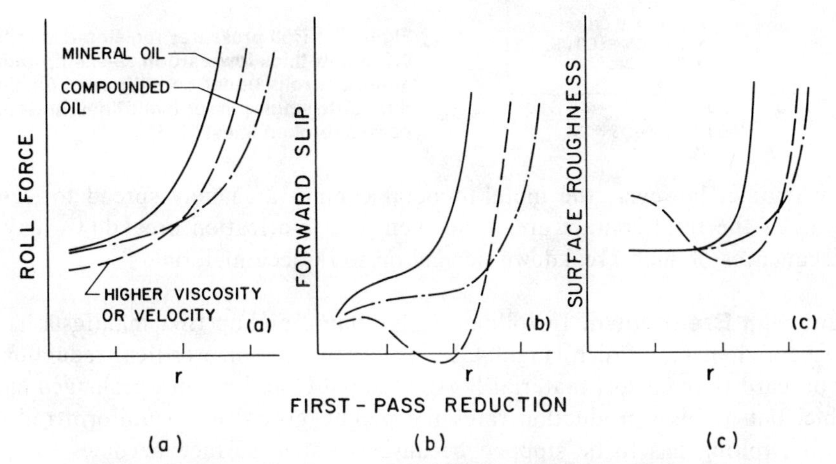

Fig.6.36. General effects of lubricant viscosity and additives on roll force, forward slip, and surface roughness observed in rolling in the mixed-film regime

revolutions (Fig.6.35a), and surface roughening is delayed. If the workpiece material exhibits less adhesion to the roll, the onset of steeply increasing roll force is delayed to heavier reductions. Indeed, unless reductions are taken far enough, the effect may pass unnoticed, as it did in the experiments of Schey [3.303] on Al-5Mg alloy and on mild steel, in part because the rolls (and thus α) were relatively large. Using smaller diameter rolls, LeMay et al [143,144] found a steep rise at 62 to 70% reduction for low-carbon steel. Similar results, indicative of limiting reductions, were reported by Roberts [145] in rolling full-hard mild steel strip with a commercial rolling oil and with acidless tallow, and also by Takahashi et al [146] (Fig.6.37). In this regime, μ is rather high, typically 0.1 or above, indicative of boundary lubrication.

d. Electrical contact resistance measurements [44],[3.294] give evidence of some metal-to-metal contact even in the predominantly hydrodynamic regime. Metal transfer occurs especially with materials of high adhesion, such as aluminum. This point was directly confirmed by measuring radioactivity on the roll surface during rolling of aluminum strip in which a tracer element (Cu^{64}) was activated [3.303]. The consequences of transfer depend on its severity and on material and lubricant. Even under practically acceptable rolling conditions, the rolls acquire a faint white cast in rolling of aluminum and a red cast in rolling of copper. If the lubrication mechanism can prevent cumulative buildup (as in Fig.6.35b), a steady-state condition (dynamic equilibrium) is reached between pickup and removal of pickup in the form of wear debris. In the predominantly

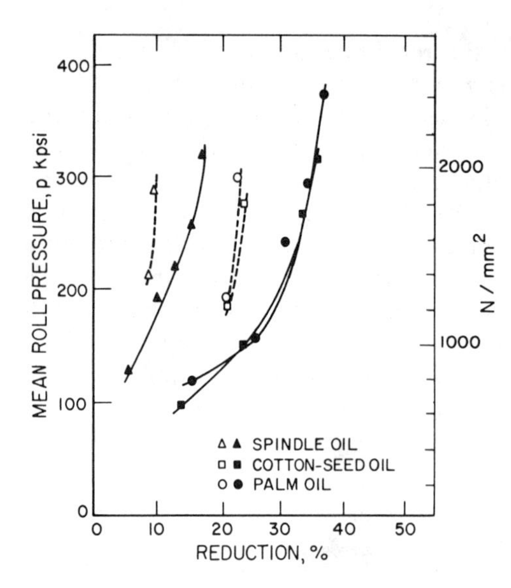

Fig.6.37. Roll pressures registered in rolling of 0.168-mm-thick low-carbon steel strip with 250-mm-diam rolls under conditions of limiting reduction in single passes (solid lines) and multiple passes (broken lines) [146]

boundary regime, however, the metal-to-metal contact area may spread to significant proportions of the total contact area, and then total lubrication breakdown may ensue. The consequences of such breakdown depend on workpiece material.

Lubricant Film Breakdown. In rolling of aluminum, pickup first manifests itself in a heavily smeared surface finish (Fig.3.43b), observable at some critical reduction which is lower on hard than on soft material [44]. At low production rates prolonged operation is possible, but at high production rates the pickup grows into a uniform roll coating and, finally, rolling has to be stopped because the strip surface becomes torn, rough, and coated with debris. Thomson and Hoggart [147] found pickup to be particularly severe when the roll surface was damaged. For a given film thickness, speed, and reduction, the onset of pickup can be delayed by boundary additives.

In rolling of steel and stainless steel, pickup is more localized and becomes damaging only when rolling is done at high enough speeds to cause substantial heating. Therefore, localized, elongated defects visible on the strip surface are often called heat streaks or friction pickup [7,148]. They are primarily responsible for the need to operate mills below their maximum design speeds. In appearance they are somewhat similar to the smeared, damaged surfaces seen on aluminum [147],[3.303] but they tend to be more localized. Heat streaks sometimes show a repetitive pattern corresponding to the roll circumference.

Heat streaks have been the subject of extensive research in Japan, reviewed in part by Azushima [95]. Gokyu et al [149] investigated the occurrence of heat streaks under industrial conditions, but the difficulty of intentionally creating and reproducing these defects led Shintaku et al [150] to perform simulated laboratory tests. A low rolling speed of 30 m/min was combined with heavy reductions and emulsion lubrication. The defects occurred when some critical temperature was reached (Fig.6.38). Kitamura et al [151] found that a local disturbance of the lubricant film caused by a grinding scratch on the roll surface was often the source of such a defect. Sato et al [152] were able to show that heat streaks formed on 18% Cr stainless steel when the mean film thickness fell below a minimum value and the roll surface reached a critical temperature. Because the defect is highly localized, it scarcely affects the average strip surface roughness.

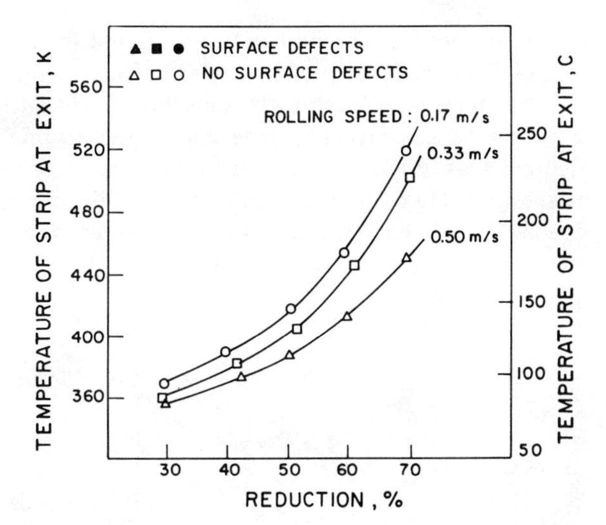

Fig.6.38. Onset of the heat-streak defect in rolling of steel strip [150]

The importance of temperature is shown by observations that the maximum allowable rolling speed increases when E.P. additives (usually of the P type) are added to the lubricant.

Little if any work on the consequences of lubricant breakdown has been reported for other metals. The mechanism should, in general, be the same, and the severity of the problem should depend on adhesion between roll and workpiece materials. Thus, compared to pure aluminum or Al-Mn alloys, an Al-5Mg alloy exhibits less pickup [44].

Aqueous Lubricants. Water is an excellent coolant but a very poor lubricant; it cannot prevent pickup, and roll forces rise with increasing speed even in rolling of steel [7]. Nevertheless, it used to be occasionally used on slow mills for rolling of low-adhesion materials such as steel or copper. As pointed out by Billigmann [153], it also provides a better reference base for lubricant evaluation than does rolling on dry rolls. In present practice, water is always used in combination with a lubricant. Most of the information that has been developed for cold rolling of steel should have fairly general validity. There are basically two methods of rolling:

1. Lubricant With Separate Water.

a. The lubricant may be preapplied to the strip surface in neat form, and water added on the mill. This technique, essentially a total-loss system, was in general use for the first 20 years of rolling steel sheet on tandem mills with palm oil. Judging from the presence of a speed effect and from the surface quality of the rolled strip, water applied on top of the fatty oil modified the lubrication mechanism little or not at all. The same applies, of course, when the strip is coated with a water-insoluble film such as an alkyd amine or epoxy and alkyde amine prior to rolling, and is then rolled using water as a coolant [7].

b. The oil may be mixed with water at the mill and applied in the form of a mechanical dispersion. In the experiments of Lloyd [154], with a preset pass schedule, such mechanical dispersions of 5% mineral oil or 5% oleic acid with water gave the same reductions as the neat lubricants did.

2. Emulsions. The oil or fat can be emulsified in water to yield emulsions of varying degrees of stability. As mentioned in Sec.3.7.1 and 4.6.2, plating out on the metal surface is believed to occur. Evidence for an essentially mixed-film lubrication mechanism includes the following:

a. A speed effect similar to that found in rolling with neat oils is observed, indicative of the contribution of hydrodynamic lubrication. Speed, as opposed to composition, has the decisive effect on the development of surface roughness [155]. An increase in emulsion concentration has the same effect as that of rolling speed, although to a lesser degree. This has been confirmed by a number of researchers [138,156,157]; an example is shown in Fig.6.39 [158]. As with neat lubricants, the speed effect tapers off at higher speeds [159]. The same observations have been made on production mills.

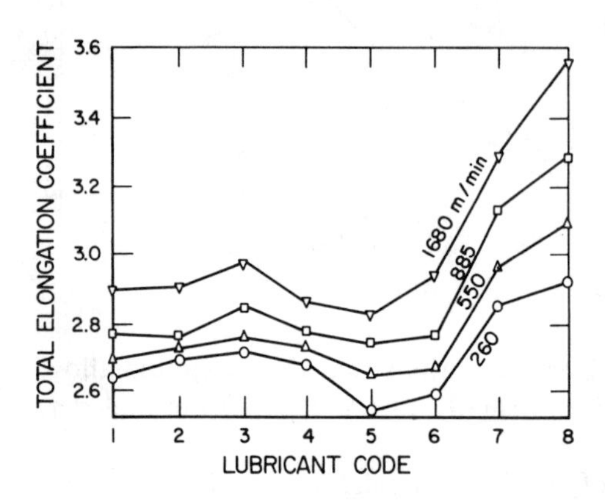

Fig.6.39. Total elongation coefficients achieved in rolling of steel strip at various speeds [158]. 1 - emulsion, 2.5% conc.; 2 - emulsion, 5% conc.; 3 - emulsion, 10% conc.; 4 - water; 5 - mineral oil; 20 cSt at 50 C; 6 - mineral oil containing 1.7% S min; 7 - castor oil applied before first pass; 8 - castor oil applied before each pass.

b. Sensitivity to the composition of the emulsified phase indicates boundary contact and, with the availability of data on emulsions of known composition, it is possible to identify some trends. Emulsions based on mineral oils are poor lubricants and, if of high stability, only little better than water. The addition of fatty compounds improves performance, at least for nonferrous metals, but emulsions based on fats are still better for steel [160-162].

c. Metal-to-metal contact is widespread, as evidenced by roll pickup with aluminum and, to a lesser extent, with copper. Indirect evidence is provided by the observation that the onset of the heat streak defect on steel can be shifted to a higher rolling speed by the addition of E.P. compounds to the oily phase [155]. Of course, the effect of metallic contact becomes obvious only when temperatures are high, and thus a phosphorus additive is effective only on hard strip at high reductions and high rolling speeds (Fig.6.40).

d. Since the inlet is always flooded, film thickness should be controlled only by plating out. Nevertheless, some minimum concentration does seem to be required; with

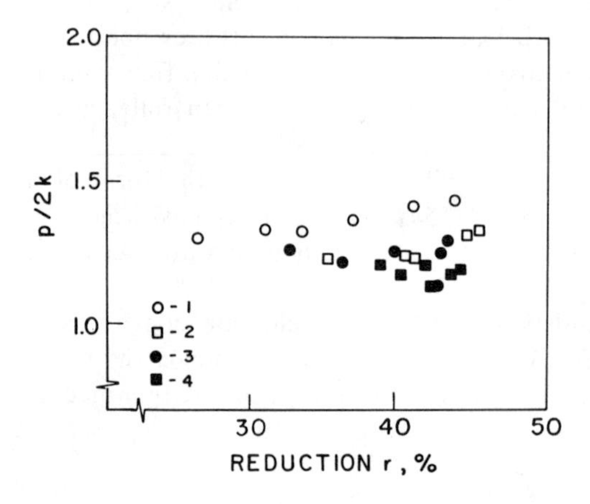

Fig.6.40. Effects of fatty-acid and E.P. additions to tallow in rolling of steel strip at 1000 m/min with stable emulsions [155]. Lubricants: 1 - tallow; 2 - tallow plus E.P. additive; 3 - tallow plus fatty acid; 4 - tallow plus fatty acid plus E.P. additive.

tallow-base steel-rolling emulsions this minimum is around 1.5%. With emulsions based on compounded mineral oils, performance improves with increasing concentration [163,164] at least up to 10%.

e. In line with the plate-out argument, emulsions of low stability (coarse particle size, bordering on mechanical dispersions) are, in general, better than highly stable emulsions. Thus, Whetzel and Rodman [112] observed that film thickness decreased and μ increased on addition of more emulsifier to a natural fat. The point is illustrated by Kihara [155] for low speeds; at high speeds composition becomes more important, obviously because of the higher temperatures developed (Fig.6.41). A translucent emulsion or a true solution is hardly better than water, at least for rolling of steel at low speeds. Lloyd [154] found that glycerin, which is a reasonably good lubricant in its neat form, becomes practically worthless when dissolved in water. Note also the poor performance of soluble oil (mineral emulsion) in Fig.6.29.

f. Since emulsifiers are surface-active agents, they could affect the boundary component of the lubrication mechanism. Funke and Marx [160] report that a cationic emulsifier is best for mineral oil and that an anionic one is best for esters of stearic acid. In general, however, it is difficult to separate the effect of emulsifier composition from the effect of particle size.

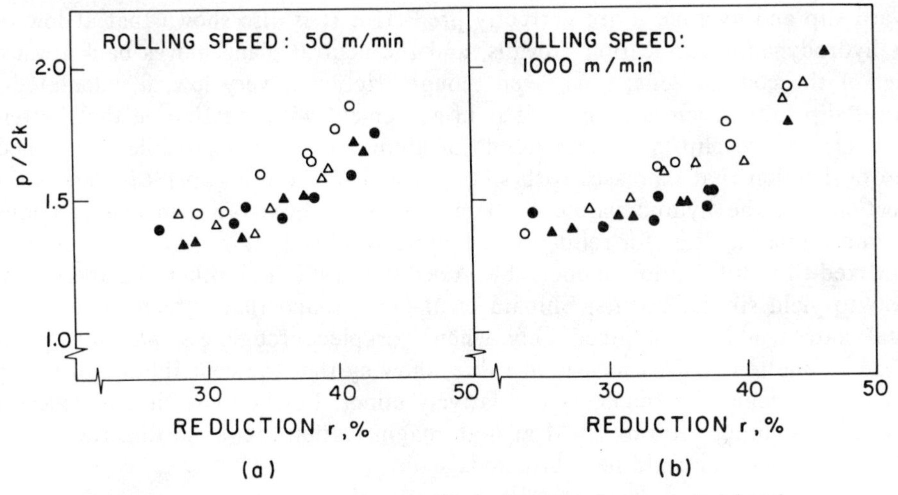

Fig.6.41. Interface pressures in rolling with stable (open symbols) and unstable (closed symbols) emulsions at (a) 50 m/min and (b) 1000 m/min [155]

g. The presence of a mixed-film mechanism is also evident from the surface finish of the rolled strip. Hydrodynamic pockets predominate under conditions that favor the formation of a thick film, and burnishing is visible under conditions giving rise to a thin film. The sensitivity to the lubrication mechanism can be minimized by precoating the strip with a controlled quantity of emulsion. Thus, Starchenko et al [165] preapplied the emulsion, allowed it to age for 45 hours, and then rolled in water only. A smoother (brighter) strip was obtained with no increase in rolling force or mill power.

Modeling of Mixed Films. We discussed in Sec.3.7.2 the modeling of mixed films in general; here we shall investigate the contributions such theories make to our understanding of the rolling process.

All solutions start from an entry film thickness calculated from PHD theory, based on Eq.6.25. Tsao and Sargent [3.301] allow for the effect of roll flattening from Hitchcock's formula (Eq.6.10). The effects of pressure and temperature on lubricant viscosity are assumed to cancel. The material is taken to obey a linear strain-hardening law. In order to obtain a solution for the fraction of the surface in boundary contact, it is necessary to allow both roll and strip to have Gaussian roughness height distributions with asperities of spherical tops; contact of the two rough surfaces is replaced by contact between a flat and a rough surface of composite roughness from Eq.3.23. Thus, the boundary contact area and the frictional stresses acting on it can be obtained, while the frictional stress in the hydrodynamic areas is due to viscous shear. The computation is performed by taking an assumed value of forward slip. The pressure at the exit from the roll gap thus obtained is then compared with that prevailing under the given conditions, and the computation is repeated with new forward slip values until the two pressures converge.

The major assumption relating to Gaussian roughness distribution is certainly invalid, at least as far as the strip is concerned. However, we shall see that, in terms of film thickness, strip roughness is much less significant than roll roughness, and hence it is not surprising that the model yields eminently sensible trends, as shown by a study [166] in which parabolic strain hardening of the workpiece material was assumed. The effects of speed, viscosity, roll radius, tensions, material hardness, and pass reduction on forward slip and average μ are correctly predicted. It is also shown that at low reductions hydrodynamic lubrication prevails, and the neutral plane moves back toward the center of the contact zone; thus, even though friction is very low, μ calculated from forward slip [32] becomes high. This is in agreement with the finding that better prediction of forward slip in tandem rolling of aluminum strip is possible if the model is based on friction that increases with sliding speed in the roll gap [36]. This is a clear indication that the hydrodynamic contribution to the mixed-film mechanism must be quite substantial, at least for rolling with a mineral oil.

Mixed-film lubrication models by Azushima [95] and other Japanese workers appear to yield similar results. Shibata et al [138] found that agreement with experimental data could be obtained only when workpiece roughness was substituted in Eq.3.23 at one tenth of its measured value, showing that the contribution of workpiece roughness to mean film thickness is relatively minor. Furthermore, it was necessary to evaluate the contact ratio by SEM at high magnification (500x) so that the area occupied by small pockets could be subtracted.

6.3.3 Dry Rolling

Rolling of unlubricated, degreased strips with carefully degreased rolls has been a prominent feature of earlier laboratory work conducted with the aim of developing correlations between theory and practice, and also of work with pressure pins built into the rolls. The advantage was that, in the absence of a fluid, the effects of speed and pass reduction, typical of mixed-film lubrication, were avoided, as were the complications arising from entrance of the pressurized lubricant into the gap between pin and roll. These conveniences were bought at the price of generating data unrelated to actual rolling practice.

As mentioned in Sec.5.7.3, vigorous cleaning is apt to form thin residual films of marginal boundary or E.P. lubricating qualities. The μ values of 0.1 to 0.2 often reported are then not unreasonable for rolling of short lengths of strip. However, continuing rolling of adhesion-prone material leads to erratic changes. Friction remains reasonably low and constant for materials such as copper, less so for steel, and not at all for aluminum. If pickup formation is allowed to proceed, friction may rise to $\mu = 0.4$

[44], although average μ has little meaning since the position of the neutral plane may fluctuate violently across the strip width. It also varies along the length of the arc of contact, as shown by Thomson and Brown [26a].

Dry rolling experiments generally confirm the presence of a marginal lubricant. Billigmann and Pomp [108] observed no drop in rolling force on increasing the speed from 5 to 400 m/min; instead, roll force rose gradually with pickup and accumulation of oxides on the roll surface. Dry lubricants such as graphite or MoS_2 are not practical, because a well-burnished coating cannot be maintained. Sims and Arthur [124] rolled steel strip coated with graphite and observed the total absence of a speed effect, confirming the trends shown in Fig.3.42.

6.3.4 Effects of Surface Roughness

We have seen in Sec.3.8 that the roughnesses of the die and workpiece surfaces, combined with the lubrication mechanism, determine the surface finish of the product. This is most clearly evident in rolling, even to the casual observer, because the large and often reflective surfaces of rolled strip make any over-all or local changes in surface quality highly visible.

Effects of Roll Surface Roughness. In general, roll surface roughness can be beneficial or harmful depending on the magnitude of the roughness relative to film thickness, on lubricant rheology, and on the directionality of the roll surface.

Rolls are usually ground under closely controlled conditions on high-precision roll grinding machines to produce a close-to-perfect geometric shape with the desired camber and a specified surface finish. Surface roughness is controlled by selecting an appropriate grinding wheel and by grinding with a prescribed number of passes at preset infeed and traversing rates. The process is very critical, and even minor changes introduced by operators affect both the roughness and its distribution [3.311]. Grinding marks are circumferential; they must be uniformly distributed and free of microscopic variations which may be reproduced as a crosshatch or other pattern on the rolled surface. Roughness is usually specified as R_a, but roughness distribution (e.g., skewness) would be a desirable additional qualifier. An objective means of sensing variations in surface texture, with particular emphasis on detection of crosshatch patterns arising from grinding wheel eccentricity or vibrations in the grinding machine, would be helpful. Various optical techniques (Sec.5.7.1) may find use. An instrument described by May [167] defines a scatter index which, when coupled with profilometric surface roughness determination, gives a better quantitative measure of surface quality than either method alone.

Shot-blasted surfaces are used when a nondirectional surface finish is desired, as in rolling of automotive sheet [168,169]. In this instance, average roughness values are totally inadequate because, depending on the intensity and coverage of shot peening and on the prior ground finish, the actual surface profile may be highly variable for the same R_a roughness. Finishing of rolls by EDM. might offer a greater degree of control [170] but has not found industrial application yet. Wear of rolls (Sec.6.3.8) is inevitable and leads to a gradual change in surface finish; the trend is usually toward smoothing, although roughening may occur under extreme conditions.

The choice of roll surface finish is governed by factors related to the rolling process and to the end use of the product. In early passes a finish is chosen that is rough enough to ensure acceptance of the strip by the rolls, but not so rough as to create pickup problems. Typically a ground finish of R_a = 1 to 2 μm is used. For the finishing passes a ground finish of, say, R_a = 1 μm is desirable for sheet that is to be subsequently painted, enamelled, or otherwise coated, because the rougher finish thus produced provides some mechanical keying for the coating. A smoother ground finish (on the order of R_a = 0.2 to 0.4 μm) is more common for visual appeal, and polished rolls (R_a = 0.02 to 0.1 μm) are used for the production of bright sheet or foil. A fairly coarse (R_a = 1 to 2 μm) shot-blasted finish is preferred for automotive body and appliance sheet (Sec.10.4.3).

The effect of roll roughness depends on the lubrication mechanism:

a. In the mixed-film regime a rougher roll generates more boundary contact, hence roll force is higher. The increase is proportional to the ratio of film thickness to surface roughness (Eq.3.22). This is visible from the results of Shibata et al [138], who controlled film thickness by the deposition of measured amounts of mineral oil (Fig.6.42). Rolls with a roughness of R_a = 1.6 μm needed much more oil to reduce the rolling load to its typical running value which was still substantially higher than on rolls for which R_a = 0.3 μm. The importance of the film-thickness ratio is shown by the results of Sato et al [152], who used rolls with differing Rt (peak-to-valley height) values and found that μ increased with increasing h/Rt (essentially a film-thickness ratio).

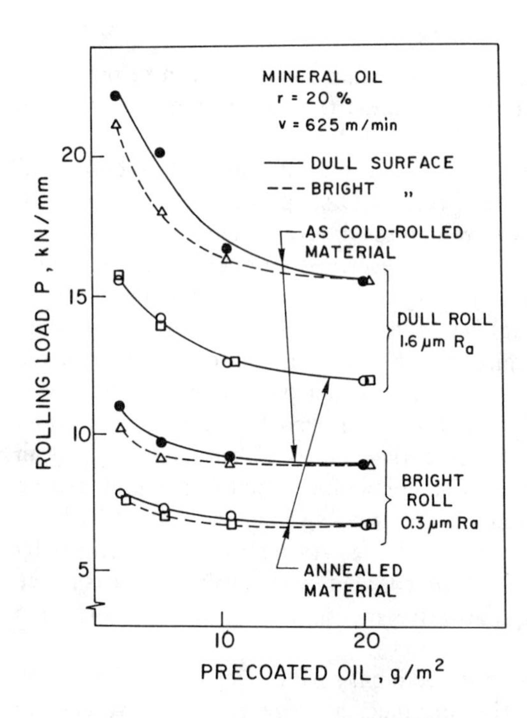

Fig.6.42. Effects of roll and strip surface roughness on quantity of oil required to ensure minimum rolling loads [138]

b. In the course of production rolling with a flood of lubricant, film thickness increases with viscosity and/or velocity; the effect of roll surface finish was observed by Sims and Arthur [124]. In rolling of copper strip, μ dropped from 0.056 to 0.039 during acceleration from 3 to 100 m/min with mirror finish (R_a = 0.1 μm) rolls. A much larger drop—from 0.08 to 0.048—was calculated for matte finish (R_a = 0.5 μm) rolls. Thus, even though the absolute value of friction is higher with rougher rolls, the improvement due to increased lubricant throughput is more marked.

c. When the workpiece material tends to form roll pickup, a small increase in surface roughness causes a large increase in roll forces and forward slip [44,55]. This effect is smaller on steel with its lesser tendency toward adhesion. As noted earlier, copper is unique in that pickup is not cumulative, and therefore the surface roughness effect is also modest in the predominantly boundary regime. Because normal, gradual roll wear causes a general smoothing, the roll force also declines and, for a given roll gap setting, the reductions obtained increase [171].

d. Even though a rougher roll results in higher friction, it can be beneficial when the film is close to breakdown. Thus, a moderately rough roll usually protects against surface damage at reductions at which rolling with very smooth rolls would already lead to roll pickup. This was reported by Ford and Wistreich [172] for copper strip, and has been confirmed by Thomson [3.316] for aluminum and by Sato et al [152] for stainless steel.

Mechanical entrapment of the lubricant is promoted if the rolls are transversely ground. This was already observed by Thorp [109] and has been confirmed by Mazur et al [173].

Surface Finish of Product. As discussed in conjunction with Fig.3.55, the surface finish of a rolled strip depends on the surface finishes of the starting workpiece and roll, and on the lubrication mechanism. The surface roughness of the strip must be closely controlled for both technological and aesthetic reasons; therefore, several investigations and reviews are devoted to this subject in the rolling of steel [174,175], aluminum [176], and copper [177].

The human eye is the most critical judge of the surface quality of a sheet, especially in contrast. Little useful information can be derived from surface roughness measurement itself. Nevertheless, quantification of surface appearance has progressed through the use of optical techniques discussed in Sec.5.7.1. More detailed evaluation calls for microscopic or SEM examination.

a. In the absence of a lubricant, the original surface finish of the strip quickly disappears. Atala and Rowe [178] were able to follow, with the aid of a relocation technique, the fates of individual surface features. Asperities flattened very quickly, and valleys disappeared after some 10% reduction. A dry lubricant such as MoS_2 did not interfere with smoothing yet provided protection against roll pickup, at least under the conditions of these experiments.

b. In the presence of a liquid lubricant, the surface finish of the incoming strip has only a minor effect (Fig.6.42) [138]. Individual features, however, may survive. In the experiments of Atala and Rowe [178], asperities again flattened quickly but valleys persisted to 40% reduction in the presence of a heavy lubricant. This confirms the observation of Thomson [3.316] on the topography of rolled aluminum strip that differences between polished and rough, etched surfaces disappear after some 50% reduction even under predominantly hydrodynamic conditions. Backmann [179] found that the relative improvement in roughness was greater for a softer carbon steel than for a heavily strain-hardening stainless steel.

c. The incoming strip surface finish affects the quantity of oil carried into the roll gap. This leads to a rougher issuing strip, and is of decisive influence when the process is otherwise critical (e.g., at the verge of skidding).

d. The effect of the lubrication mechanism is most powerful. In the predominantly hydrodynamic regime, deformation takes place through a cushion of the lubricant (Sec.3.8.4), leading to a marked roughening of the surface. This was previously described as surface damage [126], but such classification must be avoided because of possible confusion with true surface damage occurring on lubricant breakdown. As lubrication shifts toward the boundary regime, hydrodynamic pockets become smaller and fewer. The surface becomes brighter and the roll surface finish is more faithfully reproduced. Because of forward slip, the issuing strip can be smoother than the roll, as in the work of Thomson [3.316]. Directionality developed in contact with ground rolls gives a directionality in reflectivity, too. This has been explored by Azushima [100], who used the ratio of the intensity of reflection in the rolling direction to that in the transverse direction as a measure of the lubrication mechanism. For hydrodynamic lubrication the ratio is 1; for mixed-film lubrication it becomes greater than 1. Subsequently, Azushima and Miyagawa [180] measured the strip surface quality during rolling from an analysis of the diffraction pattern from a laser beam. In the predominantly boundary lubrication regime the pattern has a high and narrow intensity peak, whereas the peak becomes much lower and broader when hydrodynamic pockets abound at high speeds.

e. As expected, rolling with a less stable emulsion results in greater surface roughness than does rolling with a more stable one, at least at high rolling speeds; at low speeds boundary effects dominate and emulsion stability plays no part [155]. Otherwise, emulsion concentration, speed, and pass reduction have the effects expected from the mixed-film mechanism [181].

f. At low reductions, differential deformation of individual grains governs surface roughening. Thomson [182] found that grain size had little effect above 80% reduction, and that the texture of an aluminum strip had an only minor effect which could not be correlated with any simple measure of orientation.

g. Any variable that affects the lubrication mechanism must also affect surface finish. Thus, Mizuno and Rokushika [140] found that back tensions, which lower the interface pressure, also increase the film thickness and surface roughness.

In summarizing the effects of process conditions on surface finish, it is evident that a bright finish is promoted by large forward slip and boundary lubrication. For a given lubrication mechanism, forward slip can be increased by increasing front tensions and by the use of large-diameter work rolls. If the pass line is set low or high, so that the strip enters the roll gap at an angle, the strip will be dull on the side to which it is deflected and bright on the other side. Unintentional changes in the entry angle can cause periodic and therefore objectionable variations in surface finish. Longitudinal (tiger) stripes are attributed by Roberts [183] to the entering strip being somewhat wrinkled in the lateral direction; localized fluctuation of the entry angle results in a characteristic pattern.

It should be mentioned that powder compacts behave, to some extent, as solid sheet does. Thus, a higher-viscosity lubricant results in a duller finish [184].

Periodic Effects. The large surface area of sheet makes its appearance very sensitive to local variations in surface finish. Some but not all of these variations are accompanied by gage variations. Many are periodic, with a frequency in the audible range, and then they are referred to by the musical octave of vibration [185]. They had already been observed by Trinks [186], but became a nuisance with the increase in aluminum strip production and, later, in rolling of thin-gage steel strip.

a. Herringbone (or chevron) is a periodic marking at some angle to the roll axis (Fig.6.43), sometimes in a V or multiple-V pattern. It is most prominent on aluminum, but Thomson and Hoggart [147] observed it also on copper strip. In rolling with a very viscous lubricant, the alternating bright and dull bands were associated with periodic forward movement of the strip. Under severe rolling conditions with relatively poor lubrication, the bands were attributed to stick-slip motion which was suspected of initiating torsional oscillations in the rolls and drive. At heavier reductions, on the order of 85%, bands occurred at a higher frequency and were ascribed to transverse (vertical) vibrations in the mill. A detailed evaluation of transverse bands has been given by

Fig.6.43. Herringbone defect on rolled aluminum strip

Moller and Hoggart [187]. In rolling aluminum alloy strip on a mill equipped with force and torque instrumentation, they found the ripple to have originated from the torsional vibration of the rolls. It was initiated by any sudden change in torque, such as that caused by the strip entering the rolls, or by sudden gage variations. The vibration was self-sustaining only when μ decreased with increasing speed, and vibrations were damped out if friction rose with increasing speed. The amplitude of vibration depended on the slope of the descending portion of the friction versus speed curve. Since lubricants of lower viscosity developed a more positive speed effect, they were also prone to give ripple more easily. The torsional vibration was in phase for the two spindles and resulted in no variations in roll force. Thus, thickness variations in the strip were minute; the dull part corresponding to hydrodynamic lubrication was slightly thinner and the bright band typical of boundary lubrication slightly thicker. While the findings of Moller and Hoggart [187] are self-consistent, they are not likely to present a full explanation. Torsional vibrations in rolling mills have frequencies of 10 to 35 Hz [188], yet herringbone typically corresponds to frequencies from 140 to 200 Hz. Nevertheless, lubrication does play a critical role, and the onset of this defect seems to be due to a periodic changeover from predominantly hydrodynamic to predominantly boundary lubrication. This is indicated by the usual remedy: the mill is speeded up or, if possible, a higher-viscosity lubricant is used to move into the stable, predominantly hydrodynamic regime.

b. Chatter is observed mostly in cold rolling of very thin (<0.2 mm) strip and is evidenced by alternating bright and dull bars strictly in line with the roll axes. It is attributable to the vibration of the rolling mill rolls in a vertical plane at a frequency of about 200 Hz. It results in violent periodic gage variations of $\pm 25\%$ and can even lead to complete severance of the strip. A detailed experimental and theoretical study of chatter was made by Yarita et al [189], who also reviewed earlier work. They observed that chatter sets in on initiation of rolling with a new emulsion of very fine particle size. As the emulsion ages and the particle size becomes coarser, the effect disappears, only to reappear again when, toward the end of its useful life, the emulsion breaks down.

From the measured roll force Yarita et al deduced, with the use of Bland and Ford's theory [190], that $\mu = 0.04$ in the thick portion and $\mu = 0.02$ in the thin portion of the strip. This leads to a fluctuation in the position of the neutral plane, which then sets up vibrations. Allowing four degrees of freedom, they found one of the frequencies to agree with the observed frequency of 180 Hz. Since chatter is of frictional origin, it was argued that it is best cured by ensuring good lubrication when reductions close to the limiting gage are taken. This they achieved by the development of a new emulsion in which organometallic zinc compounds provided the requisite E.P. lubrication. Roberts [191] comes to similar conclusions and identifies, together with inadequate lubrication, excessively high back tension as the main source of chatter. A computer simulation of chatter is given by Tlusty et al [191a]. It may well be that herringbone and chatter are interrelated. Certainly, bands parallel to the roll axis observed in aluminum rolling, often referred to as gear ripple, must be a form of chatter.

c. A higher-frequency (typically, 600 Hz) chatter observed in strip rolled on wide strip mills has been attributed by Roberts [185] to periodic wear of the backup rolls. This defect, which appears in the form of subtle transverse marks or bands on the strip surface, occurs when the mill is operated for a prolonged period of time at such a speed that the wavelength associated with the vibration is equal to the circumference of the backup roll divided by an integer. External causes, such as local damage to the backup roll, grinding defects (grinding wheel chatter), and mill drive elements, may all excite the vibrations.

d. Vibrations, in general, may be externally excited, as when the strip enters or a weld passes through the roll gap; Misonoh [192] performed an analysis based on five degrees of freedom. Conversely, they may originate in the roll gap; Wunsch et al [193] considered the effect of periodic skidding (slipping) of the strip. There is, at this time, no universally applicable treatment of the problem. It is, of course, quite likely that it involves problems of entirely different kinds.

e. An unrelated cause of gage variations is to be found in the full-fluid-film backup roll bearings. As we have seen, these bearings contribute to the speed effect because the journals climb up on the oil film. Additionally, the journal (and thus the roll) centers may undergo a circular motion as a result of oil whirl which may be externally or self-excited. A review is given by Jain and Srinivasan [194] for general applications. In the rolling process, oil whirl manifests itself in periodic gage variations corresponding to the circumference of the backup rolls. In multipass or tandem rolling, the gage variations become superimposed, sometimes reinforcing and sometimes annihilating each other.

6.3.5 Surface Defects

In addition to surface damage due to lubrication breakdown, there are surface defects that are attributable to frictional or lubrication problems. Reviews of such defeats in steel have been given by Roberts [7], by Thickins and Salmon [195], and by Gregory [196].

Mechanical damage originates from several sources:

a. Adhesion between similar materials is high (Sec.3.11.1), and therefore cold welds or at least scuff marks (tick marks) are commonly encountered when a loose coil is tightened, especially if little residual lubricant film remains between the laps. A scuff mark has a typical head (the adhesive joint) which, in the course of relative sliding, gouges out a tail. The first remedy is rolling with constant tension. If necessary, oil (or oil mist) may be applied at the coiler and, for highest surface quality, interleaving with tissue may be necessary. If welding is allowed in places, the strip may bend sharply and then break away from the coil during uncoiling; such coil breaks (wrench marks) are difficult to straighten out without leaving traces. Scuffing or gouging is possible also on coil slitting lines if tensions are not consistently maintained. Special precautions in handling are needed when the strip is highly reflective, such as bright-rolled stainless steel [197].

b. In four-high mills, the work rolls are often driven, and the backup rolls are dragged by friction at the interface. When friction drops below a minimum (e.g., due to boiling off of the water content of an emulsion during a stoppage), the backup rolls skid on mill start-up. Lines of scuff marks are then formed on both backup and work rolls, and are imprinted on the rolled strip.

c. The strip makes contact with various guide rolls, bridles, stationary side guides, pinch rolls, etc. Whenever adhesion is possible, damage may also occur, which can be particularly severe if slivers of metal are allowed to accumulate. Low-adhesion guide materials (polymers, if possible), good housekeeping, and good tracking minimize this problem.

d. The nose and tail ends of the strip lead to a discontinuity in the lubrication mechanism, and pickup may occur which is then imprinted on the strip surface. With very hard materials, the rolls may be indented. Where such defects cannot be tolerated, the strip end may be cut at an angle and the nose end of the next coil, cut to the same angle, fitted so as to prevent closing of the rolls between two coils.

e. Localized plastic deformations of the rolls by metal slivers, mill wrecks and cobbles, weld joints, etc. are imprinted on the strip surface and, even though they do not really damage the strip, often make it unsaleable.

Defects from all origins mar the surface appearance of the strip. Efforts have, therefore, been made to develop instrumentation for in-process inspection [198] (also, see Sec.5.7.1). On further rolling, the defect may be rolled out or, in the presence of a liquid lubricant, may become more conspicuous. The edges of scratches may fold over to entrap lubricant residues, friction polymers, oxides, etc., which then may be impossible to remove by the usual cleaning techniques and may render the strip unsuitable for many applications.

Problems arise after rolling when, in various high-speed coil treating lines, a full-fluid film builds up between drive rolls and strip, resulting in skidding and surface damage [199]. This can be prevented by the use of patterned pinch rolls, but then this pattern may be imprinted on the strip.

Frictional scratches must not be confused with other, essentially metallurgical defects, often aligned in the rolling direction.

Wear Products. Wear debris formed during rolling is, hopefully, carried away by the lubricant and, in a recirculating system, removed in the course of lubricant treatment. Inevitably, however, some debris remains:

Debris adhering to the rolls can be deposited on or even rolled into the strip. Wipers provide some relief, but felt wipers pressed onto the work and/or backup roll surface are effective only if frequently changed. Wooden wipers, pressed against the roll surface with the end grain, ensure some mild scouring action, which is helpful when the debris adheres only moderately to the roll. They are ineffective against cold-welded junctions.

Some debris remains on the strip surface, often mixed with lubricant residues. Such smut interferes with some finishing operations. Simple wiping with white tissue paper gives a quick appraisal, but only analysis of the surface (Sec.5.7.2) can help in identifying the source. Schreiber and Ziehm [5.207] were able to determine, with the aid of a strippable plastic coating, that the metal content of smut on steel strip was mostly in the form of metallic iron and an iron soap. The amount of metallic iron decreased and the amount of soap increased with a more fatty emulsion, whereas tramp oil and roll roughness increased both. Nonionic emulsifiers reduced soap but increased iron formation; anionic emulsifiers increased both. Similar effects should exist in rolling of other metals.

Oily Residues. Excess lubricant carried out on the strip surface not only represents waste but also creates problems of staining or corrosion. Therefore, every effort is made to reduce the quantity of residual oil. Air jets directed away from the center (air knives) are commonly installed at the exit side of the mill, and, occasionally, wipers or squeeze rolls are used.

With some metals, such as steel, it is possible to take the last pass with a high-detergency emulsion which is a poor lubricant but results in a cleaner strip. Increasing the emulsion temperature to 60 C has a similar but more modest effect [5.207]. The thickness of the residual oil film is a function of lubrication mechanisms; Brinza et al [200] found in studies with C^{14}-tagged lubricant that the residual film decreased with increasing pass reduction and was thinner (about 20 mg/m^2) with coriander oil and thicker (40 mg/m^2) with palm oil (after 60% reduction).

As discussed in Sec.5.10, oily residues interact with the atmosphere on annealing to form stains which may cause rejection of the finished product. The strip surface is exposed to the atmosphere in continuous furnaces and the residues evaporate. However, during annealing of coils (or, less frequently, of sheet stacks), limited access of the atmosphere to the edges results in an irregular wavy pattern called brown stain or, on steel, snakey edge. In general, staining can be reduced by reducing the partial oil vapor pressure by venting the furnace hood and maintaining a gas flow. The causes of snakey edge have been extensively investigated [201-204]. The stain consists mainly of iron oxide, with oxyhydrates, carbon, and iron carbides present, depending on the protective atmosphere used. It is formed by an oxidation reaction between iron and CO [201,202] and can be controlled to some extent by the CO_2/CO ratio. It is promoted by the simultaneous presence of fatty oil and water [204]. Heavier rolling reductions increased the incidence of this defect in the work of Strefford and McCallum [203], leading them to postulate that the higher free surface energy of deformed surfaces had catalyzed the process.

6.3.6 Lubricant Evaluation

There is no doubt that the ultimate test of a lubricant (Sec.5.1) is a prolonged run in a production mill. However, the risks associated with testing of an unknown lubricant are particularly high in rolling. Cleaning of a large recirculating system, filling it with new lubricant, and optimization of the process all require time and money. The cost of poor performance is higher yet in terms of strip scrapped for poor shape, poor surface finish, surface damage, and staining. Thus, there is powerful incentive to perform preliminary testing in the laboratory to aid the selection of promising lubricants [205-209]. All methods discussed in Chapter 5 can be and are used, but some of them have gained prominence.

Lubricating Properties. Closest to actual rolling is testing on a laboratory mill, as evidenced by the investigations that formed the basis of much of the preceding discussion. Benedyk [210] and Sargent [211] discussed methods of evaluation, essentially based on the techniques described in Sec.6.2. Speeds are seldom high enough, and then conditions must be made more severe by increasing the pass reduction and/or temperature; in rolling of aluminum, for example, the reduction may be increased until herringbone appears. In most instances this means rolling beyond the limit set by the angle of acceptance (Eq.6.5). The strip may be entered if the nose is cut (width-wise) at an angle so that a sharp point is presented to the rolls. In rolling of long strip with tensions, reductions may be increased during rolling.

The most severe conditions, suitable for creating heat-streak defects in steel, were obtained by Azushima [212], who rolled a strip on a laboratory mill to part of its length, then arrested it from the back, simultaneously stopping the bottom roll. The top roll continued its rotation to effect a form of grinding action which accentuated the trend toward lubricant breakdown.

The cost of testing can be reduced with simulation tests, but their validity is open to doubt.

The four-ball test is suited mostly for evaluation of E.P. performance and for this reason has limited application. Vojnovic et al [5.91] modified a four-ball tester to obtain lower sliding speed and then found good correlation with rolling performance if saponification value and free fatty acid content of the lubricant were kept within certain limits. Thus, comparison was limited to fatty oils and their derivatives. Surprisingly, Miyagawa et al [213] found some measure of agreement between aluminum-rolling performance and μ and scuffing load in a four-ball tester with steel balls. To improve correlation with rolling, the ratio μ/P_f (where P_f is the load at failure) may be used as a measure of lubricant efficiency [214]. Funke and Marx [214] inserted a thermocouple into the space between the balls to record the temperature rise which was, to some extent, proportional to friction. Operation with lubricant recirculation is sometimes found to give better results [214,215]. Comparison of lubricants of different classes, nevertheless, is difficult. Thus, synthetic solutions (water-soluble chemicals) support a higher failure load in the four-ball tester yet give higher roll forces than emulsions. Marx [216] attributed this to their higher cooling power and thus the higher flow stress at the lower rolling temperature. More likely, however, the four-ball tester revealed only E.P. characteristics, whereas in rolling the rheology of the oil or fatty derivative of the emulsion came to play a larger role. Somewhat related is the pendulum friction tester; because of this, it is perhaps not surprising that Tsuji et al [217] found it suitable for selecting emulsions resistant to the heat-streak defect.

The Amsler tester with a typical surface speed of 60 m/min and a slip ratio of 8 to 20% combines hydrodynamic and boundary effects and should, in principle, provide better simulation [5.91]. It has been used for studies of boundary film development using SEM [218]. Brooks et al [219] reduced the roller face to increase the pressures to values typical of steel rolling, and measured μ over the temperature range from 60 to 180 C. Even though correlation with rolling performance was not quantitative, the order of merit of lubricants was correctly predicted. Billigmann and Fichtl [164] reported some correlation between a modified Timken test and rolling performance.

Various bench tests were critically evaluated by Saeki (reported by Funke [220]). Palm oil, rapeseed oil, and a mineral oil (turbine oil) were tested between 20 and 200 C in a four-ball tester and in two pendulum testers. Coefficients of friction were different even in the two models of pendulum testers, and both friction and wear showed different orders of merit in the three different bench tests. The only common feature was that palm oil was the best and turbine oil the poorest. Not surprisingly, this was confirmed in rolling. This does not mean, however, that the results were directly transferable. For example, palm oil gave μ = 0.11 on the pendulum tester, μ = 0.05 on the four-ball tester, and μ = 0.015 in the rolling mill.

Spreading of oils on a differentially heated surface [5.287] may provide a useful tool. Miyagawa et al [213] observed a sudden increase in spreading rates on adding 0.8% of an emulsifier to a base oil; roll force showed no corresponding change, but the surface quality of the strip, as judged from hydrodynamic pockets, was best when spreading was minimum.

Composition and Stability. Because most rolling lubricants contain some fatty constituent, it is customary to test for composition both indirectly (through saponification, iodine, and acid values) and directly (by various methods of chemical analysis) (Sec.5.11). An application is shown by Hobert et al [221].

The great importance of emulsion stability prompted ASTM to issue a standard (see Appendix) on the measurement of stability of new (unused) rolling emulsions. The emulsion stability index (E.S.I.) is a number showing the proportion of emulsion (total less cream) remaining on standing at 70 C for 4 hours. An emulsion with an E.S.I. of 0.95 would be regarded as very stable, one with an E.S.I. of 0.75 to 0.85 as moderately stable, and one with an E.S.I. below 0.5 as unstable.

Related properties are the plating out, wetting, spreading, and surface tension of the lubricant [7,207,211,213,217], as discussed in Sec.5.11. Some of the techniques are highly technological in nature. For example, McDole [222] suggests checking of plating out on aluminum by etching the strip in 5% caustic soda solution, followed by desmutting in dilute nitric acid and rinsing in distilled water. When some emulsion is then dropped on the dry surface, the oil should plate out and the emulsion should not adhere to the oil; on sheet wetted with water, the oil should displace the water and should resist removal by a stream of water.

Lubricant Changes. Changes occurring during the use of the lubricant are gradual and, if routine examination is not carried out, may be imperceptible until sudden deterioration sets in.

In oil-base lubricants the major problem is contamination with heavier hydraulic and bearing oils which lead to an increase in staining. Bearing seals have been improved and roller bearings are often mist-lubricated with the rolling oil. With the aid of laboratory staining tests (Sec.5.11), the oil can be changed before staining would show up, usually several days after rolling, during annealing. Additives lost by filtration can be made up on the basis of daily analysis.

Emulsions suffer more subtle changes [223]. Their performance often improves for a while after installation of a new batch, for reasons that cannot be explained simply by compositional changes [224]. Aging usually involves some loss of stability and a consequent increase in plating-out rates, which may show up in simulation tests [214,224]. In steel-rolling emulsions Funke et al [225] observed a loss of performance in the four-ball test when the neutral fat portion dropped or the fat oxidized. There is also contamination with fines. Undesirable changes can be limited by good housekeeping and by makeup with concentrate or with selected additives, but finally—usually after a period of several weeks or even months—total replacement becomes necessary. The situation is aggravated in steel rolling where acid residues from the pickling bath contribute to destabilization of the emulsion.

6.3.7 Lubricant Application and Treatment

The best lubricant will fail to serve its purpose unless it is applied in sufficient quantity, uniformly over the entire strip surface, while heat is extracted in a controlled manner to ensure the development of the requisite roll camber. The application system must ensure also that lubricant quality is maintained and that health and environmental effects are considered.

Optimum Lubricant Quantity. The thickness of the lubricant film passing through the roll gap is determined by the lubrication mechanism. If less lubricant is available, the film will be starved, leading to surface defects and, in laboratory evaluation, false rating of lubricants.

In recirculating systems, excess lubricant is applied but is recovered; in direct application, excess lubricant is wasted. Bentz and Sommers [226] applied measured quantities of cottonseed oil by electrostatic deposition to thin (about 0.3 mm) steel strip. The backup rolls were water cooled. In rolling in single passes of 10 to 60% reduction, minimum roll force was obtained on tin plate and soft black plate with 1.25 g of oil per base box (or, since the two sides of the base box represent 40 m^2 of total surface area, 32 mg/m^2). In rolling of full-hard strip, the amount of oil required was 75 mg/m^2, presumably because limiting thickness conditions were approached.

In rolling with emulsions, the weight of the oily phase plating out increases steeply but the film weight passing through the gap rises only slightly for a given set of rolling conditions (Table 6.1) [138]. Since friction in rolling with an emulsion is only slightly higher than in rolling with the neat concentrate, it appears that the emulsion layer lying on top of the plate-out film does not enter the roll gap.

Table 6.1. Oil-film thicknesses in rolling with emulsions [138]

	Emulsion concentration, %				
	1	2	3	4	5
Plated-out oil (g/m^2)	0.12	0.28	0.50	0.67	1.07
Residual oil (g/m^2)	0.16	0.18	0.22	0.22	0.32

Most emulsions are applied in recirculating systems, but in some cases a richer, less stable emulsion is directly applied. Emulsion stability is critical. In a comparison of the two systems, Vucich and Vitellas [227] found that, on double-reduction steel mills (rolling to gages below 0.2 mm), the time required for the emulsion concentration to fall to half of its original value had to be 0.25 to 5.5 min for direct application but twice as long for recirculation.

Application Methods. The prime requirement is always that the lubricant be applied according to a predetermined pattern.

a. In experimental work and small-scale production of sheet it is often adequate to apply the lubricant to the roll and/or strip surface with a brush, cotton swab, or rag.

b. For low-speed strip rolling the lubricant can be evenly distributed by passing the strip through felt pads (press wipers) saturated with the lubricant; the lubricant may be replenished by drip feeding onto the wiper. The press wipers may be replaced with felt or foam rubber (plastic) rollers, particularly if rolling speeds are somewhat higher. Felt, bristle, or wooden wipers applied against the roll surface under a controlled pressure help to spread the lubricant and catch loose debris that might otherwise be rolled into the strip surface.

c. In experiments aimed at lubricant evaluation, uniform and reproducible lubricant application is crucial. A simplified recirculating system (Fig.6.44) incorporating a pump (P), a heater (H), a mixer (M), spraybars (S), and a catch tray (C) is essential, with valving arranged to facilitate circulation prior to rolling. The addition of a small tank (T) allows conversion to total-loss application.

d. In production rolling, the lubricant (whether of oil or water base) is applied through a suitably designed system of jets. Large numbers of jets are normally arranged in banks, parallel with the roll axis [7,196]. Jets can be directed onto the work roll surface, into the roll gap, into the gap between work roll and backup roll, and/or onto the surface of the entering strip. A second bank of jets is frequently located on top of the roll (or, in four-high mills, on top of the backup roll) for cooling. In direct-application systems the water is applied through these jets.

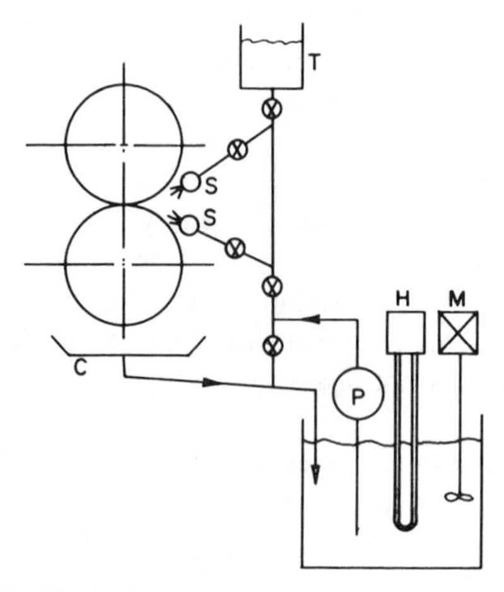

Fig.6.44. Typical experimental lubricant supply system

In total-loss systems special dispensing units are needed for exact control of oil quantities, which may be as low as 10 g/min [196,228]. The cooling water is applied through spraybars.

The lubricant is supplied by pumps at pressures ranging from 50 to 5000 kPa. There is no general agreement regarding the most desirable pressures. Proponents of high-pressure application claim that impingement on the roll and strip surfaces helps to break up a stagnant layer of lubricant or steam and thus increases heat transfer. Adherents of the low-pressure school regard quantity of lubricant and wetting as more important, and point to the benefit of less misting; this is the preferable method.

Cooling. The lubricant (and, if separately applied, the coolant) serves several heat-related functions:

a. A sufficient quantity of lubricant must be available to carry away 75 to 100% of all heat generated, which can be taken simply as the thermal equivalent of the installed motor power. Supply rates of several thousand L/min per mill stand are usual. Tselikov and Smirnov [4] proposed a simple empirical formula for emulsion quantity Q (in L/min) in steel strip mills:

$$Q = vwz \tag{6.28}$$

where v is maximum rolling speed in the last stand (m/s), w is the maximum width of the rolled strip (cm), and z is the number of stands. They also quote experimental work according to which 65 to 75% of the total heat generated is carried away by the emulsion (with a 12 to 15 C temperature rise), 18 to 25% is carried away by the strip, and only small fractions are transmitted to the surrounding atmosphere.

In a Sendzimir mill the heat carried away with the strip is insignificant when thin gages are rolled, and most of the heat must be removed by the rolling lubricant [229,230].

Adequate lubricant quantity is no guarantee of heat removal; the placement and number of spraybars must be optimized, too.

b. The lubricant may be used to heat up the rolls after a shutdown, thus minimizing problems due to insufficient thermal camber.

c. Control of thermal camber by coolant distribution, usually by means of shutting off selected nozzles, is slow-acting but is an indispensable part of normal production control [36]. Automatic control activated by thermocouples or thermistors placed close to the roll surface is possible and has been used in rolling of aluminum foil, where camber control is crucial [231].

Lubricant Maintenance. Rolling mills have some of the largest lubricant systems in operation. The capacity of an oil-base lubricant system is around 20 000 L for single-stand mills and 100 000 L for tandem mills. Emulsion systems can be several times larger. The quantities of cooling water used in total-loss systems place substantial loads on effluent treating facilities. Some mills are equipped with two or three separate systems to deliver different lubricants to various stands of a tandem mill, or to allow rapid changeover of lubricants for different duties [232]. Combination total-loss and recirculating systems also exist [233] for tinplate and double reduced tinplate rolling. Provision is usually made for maintaining circulation during mill shutdowns to preserve homogeneity and control temperatures.

The complexity of the system is a function of the material and strip thickness rolled [234]. In steel and brass mills, a mechanical filter, centrifuge, or magnetic separator is often sufficient even in a 15 to 20% bypass flow. The full flow is passed through mechanical filters in rolling of stainless steel and aluminum, and an earth filter is added in a bypass circuit for mineral-oil-base lubricants. Filtration was actually first applied to aluminum-rolling lubricants [235] but is finding increasing use in steel mills [236]. Special care is exercised in treating lubricants for Sendzimir and other mills in which the lubricant, whether of an oil or water base, must also serve as a bearing lubricant [237]. Solid particles may range from 10 μm to a fraction of a micrometre. It is, of course, impossible to remove all fines economically, but filtration to 0.02 wt % solids is necessary if the reflectivity of aluminum sheet is to be maintained [238]. In steel mills the generation of fines proceeds less rapidly [223], and fines are removed, at least in part, by skimming.

6.3.8 Roll Wear

The discussion in Sec.3.10 showed that most forms of wear are proportional to the sliding distance. In rolling, the relative sliding distance is only a fraction of the total rolled length, and therefore wear is slower than would be expected from the prevailing conditions. Nevertheless, the ground finish is gradually lost and, if the strip finish requirements are not critical enough to force a regrinding of the rolls, rolling may continue until the roll diameter changes sufficiently to make rolling of flat strip difficult. Because any contrast in surface finish shows up readily, it is common practice to start a rolling program with the widest strip and progress to narrower widths (although a narrow strip may be rolled after a shutdown to develop thermal camber).

a. Adhesive and abrasive wear each play a role. With a softer adhesive material (such as aluminum), wear progresses by the abrasive action of the debris; with harder adhesive materials (such as titanium), direct adhesive wear may be possible but does not seem to have been documented. Abrasive wear is evident when the oxide is hard and is supported by a strong substrate; this accounts for the greater wear encountered with the strong aluminum alloys, with stainless steels, and with nickel-base alloys. Corrosive wear may occur with some lubricants and can become localized when the rolls are allowed to stand for a prolonged period of time with a lubricant film between them.

b. The most prominent cause of roll failure is spalling, a form of fatigue failure which may be deep enough to leave little of the hard surface layer typical of cold rolls. Uneven stress distributions due to localized wear, grinding cracks, residual stresses in the roll, and hydrogen embrittlement in contact with an emulsion [239,240] may all contribute to it. Spalling characteristically takes the form of shallow flakes [240,241] or circumferential cracks [239] penetrating to the boundary between the hardened layer and the core of the roll. Spalling in the form of shallow flakes can be minimized by regrinding before excessive stresses can build up. Dean [242] recommends a maximum Scleroscope C hardness increase of 5 to 6 points.

Damage to the roll surface can be more localized and catastrophic. We have already mentioned scuffing on skidding between the work and backup rolls. The rolls may be indented by strip welds, by foreign bodies accidentally entering the bite, or by folding of the strip when shape is exceptionally bad; localized thermal shock may accompany such indentation [243]. Most importantly, cobbles (wrecks) occurring when control is lost in high-speed tandem mills are causes of sudden damage and even breakage. Lucas [241] suggests that cobbles account for 25 to 50% of all roll failures.

The manufacture of rolls is a highly sophisticated industry, and its discussion is beyond the scope of the present treatment. The comprehensive review of Thieme and Ammareller [244] is still valuable. In general, rolls are made by several methods:

a. Forged rolls are produced from steel (0.45 to 1.0% C) that is melted and poured with the greatest care. Dendritic solidification is unavoidable, and microsegregation leads to compositional and hardness differences. More rapid wear of the softer areas results in the development of a surface pattern which is imprinted on the strip surface as orange peel or mottle [241,245,246], not to be confused with surface roughness resulting from the lubrication mechanism. Heat treatment of the steel is a crucial step since the surfaces of cold rolls are often at the highest hardness attainable for steel, with practically no ductility. High residual stresses are induced and, if the treatment has been correctly carried out, surfaces will be in compression. Stresses in the center, where porosity could survive, are relieved by removing some 15% of the diameter prior to heat treatment. A thicker heat treated layer increases roll life by increasing resistance to denting, and minimizes the effects of dendritic structure [241,242].

b. Cast rolls are available in great variety. Both cast steel and cast iron rolls may be produced by double pouring so that the surface (even of white iron) is hard while the core is tough.

c. Composite properties may be ensured by shrinking sleeves onto roll arbors, reducing the expense of replacing an entire roll.

d. Tungsten carbide rolls, with cobalt contents of 8 to 15% for greater ductility, are used when they can be well supported and be of relatively small diameter, as in multiroll (Sendzimir) mills [247].

e. There seem to be few incentives for changing the adhesion or wear resistance of rolls by surface coatings. Spenceley [248] found improvement with electrolytically deposited hard chromium layers less than 25 μm thick. Plating apparently delayed the wear of rough-ground finishes, thus extending useful roll life. In principle, diffusion coatings should be more resistant to the fatigue exposure typical of work rolls. Coating with tungsten carbide is possible but is not widespread.

Wear on tandem mills is most rapid in the middle stands where the heaviest reductions are usually taken, but wear of the finishing rolls is of greatest importance from the point of view of strip surface finish. In cold rolling of tin-plate on typical 5-stand mills [249], the work rolls are changed in the last stand after rolling of 150 tons and in the 4th stand after rolling of 250 tons, but over 1000 tons may be rolled on the first three stands. Backup rolls are reground only after 20 000 tons have been rolled. Sheet rolling is much less demanding and, in rolling with emulsions, quantities ranging from 450 to 1500 tons were quoted for the work rolls of a single-stand reversing mill [250].

6.4 Lubrication and Wear in Hot Rolling

Hot rolling of most industrially important metals is done at temperatures beyond the normal operating range of organic lubricants (Table 2.1). Yet, the problems of lubrication are by no means proportional to rolling temperature; the primary controlling factors are the nature of the metal oxide and the tendency for adhesion between workpiece and roll materials. Accordingly, the first hot rolling lubrication system was developed for aluminum, and the rolling of steel continued until relatively recently without an externally introduced lubricant. The first attempts at steel lubrication were influenced by the difficulties anticipated at the high rolling temperatures, and solutions were sought among practices typical of forging. Only later was a more independent approach adopted, resulting in a gradual convergence of all hot rolling lubrication practices. Indeed, there is now much more in common between aluminum and steel rolling lubrication than between steel rolling and forging practices. The basic reason is that, at least under conditions of rolling, organic substances show rather unexpected benefits at temperatures where only their residues are likely to survive.

6.4.1 Process Conditions

It will be useful at this point to review the factors affecting lubrication, with special emphasis on aspects relating to the hot rolling of relatively thick slabs and strips.

a. The contact zone continuously shifts around the roll circumference, minimizing the time of exposure and making the noncontacting part of the roll accessible to the coolant and to mechanical devices for control of roll pickup. At the same time, periodic heating and cooling (Fig.6.45) [251] aggravates thermal fatigue (heat checking).

b. The entry geometry is favorable for the introduction of a lubricant/coolant.

c. In rolling of a thick slab, heavy deformations are desirable for economy of production. This means that the coefficient of friction must be kept high enough to ensure that the rolls bite (Eq.6.5), and the benefits flowing from lubrication must be weighed against a drop in output due to lower reductions or chronic refusal to bite. Since the running μ (Eq.6.24) is half the μ at acceptance, it is often profitable to cut off the lubricant flow while the workpiece enters the roll gap.

d. Even with heavy absolute reductions, the h/L ratio is large in early passes on a thick slab. Deformation is then inhomogeneous and a sticking zone (dead-metal zone) develops over much of the arc of contact (Fig.6.3d). This is, in itself, not undesirable: the virtual absence of sliding means much reduced wear and, since the rolling force is essentially independent of friction (Eq.6.11), sliding friction would not improve matters. Indeed, at $L/h < 2$ one cannot expect to reduce forces by lubrication, and wear can actually increase. This is the case when wear is due primarily to abrasion by oxides, and

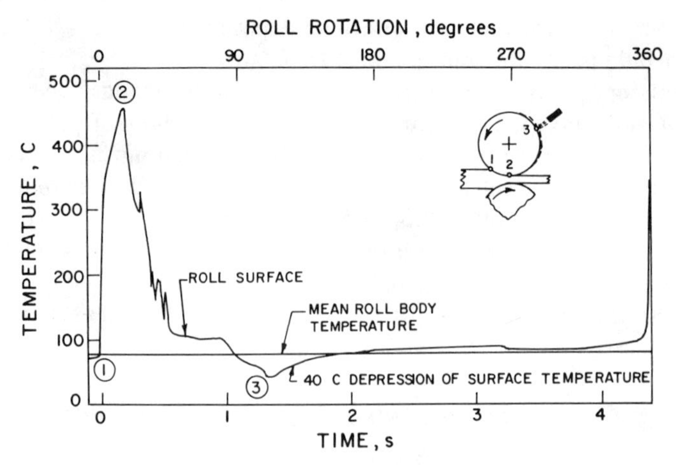

Fig.6.45. Temperature excursions in hot rolling of steel slabs [251]

the application of a lubricant encourages sliding without neutralizing the oxides or without coating the rolls with a lubricant that would fulfill a parting function.

e. At the higher L/h ratios typical of heavy passes on thinner strip (Fig.6.3c), the contribution of friction accounts for a large part of the roll force and torque. Thus it is these later passes that benefit most from lubrication. Forces are reduced and the strip surface is improved. If properly chosen, the lubricant also reduces wear; the sticking zone is reduced anyway, even in the absence of a lubricant, and the effects of increased sliding (rubbing) with a lubricant are easily counterbalanced by the adhesion-reducing and parting effects of the lubricant.

6.4.2 Lubrication Mechanisms

There are many uncertainties about the actual mode of action of hot working lubricants. Much of the existing information has been generated for specific materials; at the danger of oversimplifying complex behaviors, the basic phenomena will be discussed here on the premise that steel can be regarded as a prototype of less adhesive materials (such as stainless steels and nickel-base alloys), whereas aluminum can be taken to be representative of the more adhesive metals (such as titanium and, to a lesser extent, magnesium and copper).

Hydrodynamic Lubrication. Glasses are the only lubricants that can provide truly hydrodynamic lubrication. Their viscosity is high enough to allow a reasonable reduction in finishing passes. In the early work of Chekmarev et al [252] on steel, the angle of acceptance decreased from 13 to 9° with the use of window glass, while forward slip dropped to zero. The absence of surface shear in the work of Jones and Walker [5.112] is also indicative of very low friction. Contrary to cold rolling, full-fluid-film lubrication with glass can actually improve surface finish relative to rolling with natural scale, as found by Chelyshev et al [253]. Even a thin film of glass can reduce heat losses, as shown by the much reduced incidence of surface cracking in edge restraint rolling of titanium- and nickel-base alloys in the work of Kulkarni and Schey [254]. The difficulty of removal and the possible abrasiveness of solidified glass make it an unlikely candidate for widespread industrial use.

Mixed-Film Lubrication. In the vast majority of instances a compounded oil or an emulsion is applied, resulting in mixed-film lubrication. In contrast to cold rolling, the lubricant film operates beyond its breakdown point, and metal-to-metal contact is the rule rather than the exception. The consequences range from relatively high wear rates in rolling of steel to formation of a more or less continuous pickup layer (the roll coating) in rolling of aluminum. That organic substances should remain effective at steel rolling temperatures is astonishing, but perhaps no more so than the ability of rapeseed oil to lubricate in continuous casting of steel.

That lubrication is of the mixed-film type is illustrated by the fact that effectiveness increases with increased rate of application up to a certain point beyond which no further improvement is obtained, indicating that there must be a limiting film thickness. Indeed, attempts have been made, and with some success, to extend film thickness calculations (Eq.6.25) to hot rolling [255,256]. Because of the need for effective cooling, water is invariably present, and one can only assume that the oily phase forms a more or less continuous film on the roll surface. Even though some of this film must burn up, contact times are short and, judging from the varying effectiveness of lubricants of different compositions, the lubricating action must be due to a film which is a mixture of the original lubricant, its reaction products, and residues which range from polymerized gums to carbonaceous residues.

1. Mineral Oils. When applied in the neat form, mineral oils burn readily and pollute the atmosphere [257]. Their efficiency is doubtful; some investigators find little effect on friction and even observe increased wear, presumably because the length of the sticking zone is reduced and thus wear, proportional to sliding distance, increases [258]. Others find mineral oils to be ineffective also in laboratory simulation [259]. The presence of a mixed film is indicated by the fact that increasing viscosity gives better results [260]. In the form of emulsions, mineral oils are relatively ineffective and are suitable only for less demanding tasks such as rolling of copper.

2. Fatty Oils and Synthetics. The importance of the nature of the lubricant residues can be gaged from the marked improvements that are observed when rolling is done using neat fatty oils such as tallow or rapeseed oil or even mineral oils blended with fatty oils or their derivatives. These oils have long been used as the oily phase in emulsions for rolling of aluminum, and have recently become the basis of steel-rolling lubricants. Their effectiveness increases with increasing chain length and saturation [260,261]. In the work of Kihara et al [261], the performance of stearic acid increased with increasing concentration up to 20%, after which no further improvement was obtained. Nishizawa et al [259] found rapeseed oil to be effective in a concentration of 40%; beef tallow gave equal results in industrial tests. Fatty acid soaps present fewer disposal problems [262]. There is general agreement that, at least for steel rolling, synthetic esters are better, presumably because of their higher temperature resistance [258]. Polymer thickeners improve the plating-out tendency [260].

3. E.P. Additives. Phosphorus compounds, especially phosphate esters, have been repeatedly shown to further improve rolling performance when added to mineral oils either on their own or in conjunction with fatty additives. They seem to be effective at quite low concentrations (1%) but give further improvement at concentrations up to 20%. The improvement is greater at higher rolling temperatures, which is indicative of an E.P. mechanism [261]. Neport [263] reports a doubling of roll life.

Solid Lubrication. Because of the high operating temperatures, it would seem logical to try layer-lattice compounds and other high-temperature lubricants. Thus, a lubricant consisting of micronized graphite, resintered boron nitride, and a stabilizer in an amine-amide fluid has been applied to hot steel, thus protecting it from oxidation and reducing friction [264]. Calcium carbonate in an oily suspension and fluorinated mica in an organic base have also been used [3.168]. All these lubricants have little spreading power, and it is questionable if they will find widespread industrial application.

Benefits of Lubrication. There is fairly general agreement on the benefits to be expected. In rolling of aluminum it is a matter of the very feasibility of rolling; in rolling of steel the benefits are less obvious.

Generally, it is reported that the coefficient of friction is reduced, say, from 0.4 to 0.3 or 0.2 [265], although absolute values are uncertain because most friction values are derived from roll forces. The theories used are often based on sliding friction, an obviously unrealistic assumption for hot rolling. Furthermore, the calculated value of μ drops even in the absence of a lubricant with increasing L/h ratio [266,267]. It is not clear whether this is simply an artifact resulting from the method of calculation or whether the more homogeneous deformation that obtains at large L/h ratios indeed results in lower friction. There is, however, no doubt that μ drops with lubrication and that the drop is greater at heavier reductions where L/h (and thus the friction hill contribution) is greater. There is also a drop in roll force and mill power on the order of 10 to 30%. Nikolaev and Zverev [49] measured $\mu = 0.32$ with a cottonseed oil emulsion and $\mu = 0.34$ with a solution of 1.5% $CaCO_3$ plus 1.5% NaCl.

Most importantly, wear is reduced by 20 to 40% and, correspondingly, more steel can be rolled without regrinding, especially in the first three stands of a tandem mill [265,267]. Temperature excursions are also reduced, which could contribute to increased roll life [251], but spalling is not affected. The time saved in roll changing adds to the benefits. The surface finish of the strip also improves, with fewer small pockets and a higher yield in first-quality strips [268]. Williams [269] estimates that over-all savings amount to five times the cost of the oil.

In rolling of aluminum the benefits are dramatic. Without a lubricant a roll coating develops rapidly; there is no opportunity for roll wear, but rolling cannot be continued. With a lubricant, a coating still builds up but at a tolerable rate.

Dry Rolling. As mentioned earlier, hot rolling of steel was, and to some extent still is, practiced without external lubricants.

In rolling with sticking friction (Fig.6.3c and d), the oxide breaks up prior to entering the roll gap, but in rolling in air (and with water) rapid reaction, which provides some measure of protection, is possible. The oxide seems to be sufficient to prevent pressure welding at least with steel, stainless steel, nickel, and copper, but not with aluminum. The thickness of the oxide film remaining on the surface is a function of the adhesion of scale to the workpiece, and increases with scale thickness, pass reduction, and decreasing rolling temperature [270].

The effect of temperature on friction seems to be a function of oxide film thickness and composition. With a thick, preformed oxide film, μ drops with increasing temperature for many metals, at least in the hot working temperature range (Fig.3.16). Nevertheless, conflicting evidence has accumulated in rolling of steel.

In the classic work of Ekelund [53], friction decreased with increasing temperature, and this was true also of the work of Pavlov [271-273] even though the absolute value of μ was lower when rolling was done in vacuum or argon, indicating that the thinner less oxygen-rich film was a better lubricant. The detailed work of El-Kalay and Sparling [274] confirmed this trend with thin oxide films on rough (4 to 5 μm AA) rolls, but interestingly, μ actually increased with temperature on smoother (0.8 μm) rolls for all oxide film thicknesses. Roberts [265] rationalized their results through the ratio of roll roughness to scale thickness. Part of the problem must be the extreme sensitivity of friction to the oxidation conditions [275], the variability of adhesion of the scale to the workpiece [4,270], and the influence of the measurement method. For hot working the most convenient is the technique of the angle of acceptance. Kortzfleisch [12,13] showed that, for ostensibly identical oxidation, the roll speed, approach velocity, and nose shape all had an effect. Friction appeared to decrease with increasing temperature and decreasing rolling speed for specimens with slightly rounded or tapered noses, whereas friction was actually higher at higher temperatures and was independent of roll speed for specimens with sharp leading edges. Under production conditions, these trends may easily counterbalance each other, and temperature effects on friction may be lost. Similar problems exist when friction is calculated from forward slip or back-calculated from roll force.

Pavlov et al [271] found unexpectedly low friction in rolling titanium in vacuum, and only slightly higher friction in rolling titanium in air. As expected, friction dropped dramatically when oxidation of molybdenum and niobium was permitted.

There is thus considerable uncertainty regarding a realistic value of friction for dry hot rolling. Sparling [276] recommends $\mu = 0.23$ for lightly scaled material, $\mu = 0.38$ for lightly scaled material on rough rolls, and $\mu = 0.15$ for lubricated rolling. Denton and Crane [35] recommend $\mu = 0.25$ to 0.4 for unlubricated rolling. The method of pushing against a hydraulic cylinder gave $\mu = 0.455$ [49], perhaps because of the small L/h ratio. Direct measurements by Grishkov [277] of $\mu = 0.17$ to 0.3 are also reasonable in view of the relatively light (20%) reductions taken; for copper, Smith et al [278] deduce $\mu = 0.2$. It is to be expected that friction should also depend on roll material. Ekelund [53] found higher friction on steel rolls than on cast iron rolls, and this was confirmed by Stone [279]. On rough rolls these values could increase substantially, as suggested by Pavlov [5.33].

A number of experiments with pressure-sensing pins built into the roll surface show pressure and shear stress distributions similar to those shown in Fig.6.3c and d. In general, shear stresses diminish toward the neutral plane, indicating sticking. Shear stresses also drop to zero toward the middle of the width of the strip (Fig.6.46) [280].

6.4.3 Lubricant Application and Disposal

The two distinct functions, lubrication and cooling, can be accommodated by several means:

Lubricant Application.

a. In the direct application methods the neat lubricant is applied separately in carefully measured quantities, while the coolant floods the rolls. In the simplest form, a solid slab of tallow, a graphited compact, or some other lubricant block is pressed against the roll surface. Liquid lubricants must be applied in minute quantities, and therefore clogging of orifices is often a problem. Air or steam atomization has led to some viable solutions. The effluent consists of large quantities of water contaminated with small quantities of lubricant, and it can present a disposal problem. The problem is minimized if roll wipers are used to separate the lubricant-rich phase from the essentially pure cooling water phase.

b. The liquid lubricant, either neat oil or a concentrated emulsion, is injected into the cooling water headers. This ensures good mixing but also creates disposal problems.

c. A mechanical dispersion of oil in water is made up and is applied through separate headers to the roll surface. Wipers separate this rich dispersion from the mass of cooling water. Reclamation and recirculation of the dispersion is possible.

d. The lubricant and coolant are combined into an emulsion of controlled stability

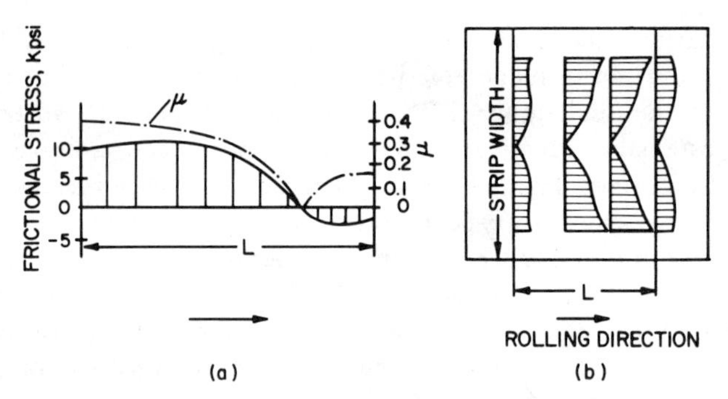

Fig.6.46. Distributions of frictional stresses in rolling of Armco iron at 950 C (after [280]). (a) Longitudinal stresses in middle of strip. (b) Transverse stresses along arc of contact *L*.

and the entire volume is recirculated, using filtration and clarification methods described in Sec.4.8. This method is applied universally in rolling of aluminum and extensively in rolling of copper.

Consumption of oil varies according to the number of stands lubricated and the method of application. In the total-loss system consumption is typically 30 to 40 g per ton of steel [262]; with a premixed dispersion the application rate is typically 10 to 30 L/min/stand, which corresponds to a consumption of 5 g per ton of steel [259], whereas in a full emulsion system consumption is a function of maintenance procedures.

Cooling. The quantity of cooling water, the location of application, and the distribution of flow all have effects on thermal camber (and thus on product shape), on roll wear, and on the properties of the hot rolled plate or band. Therefore, many analyses have been presented for both rolls (e.g., [281]) and runout tables (e.g., [282]).

Treatment and Disposal. Hot rolling mills with their vast cooling water requirements used to be among the major pollutants of water. Today treatment is so thorough that the water is often returned cleaner than when it was taken in. All techniques discussed in Sec.4.8 are employed.

In steel hot rolling mills water serves to remove scale from the ingot, cool the work rolls, flush away secondary scale formed during rolling, control the temperature of the finished product and, incidentally, wash away any lubricant that may be applied. The water flow is directed to move the scale along the pits (underground channels) placed below mill and roller tables, and thus this so-called flume water carries scale ranging from coarse particles to fine sludge, together with oil which may be mechanically dispersed, partially emulsified, or attached to the scale. Treatment takes place in several stages. Scale pits are designed [283] for large capacity to allow sufficient residence time for settling of the scale, which is usually continuously dragged out [4.322]. Free oil floating to the surface can be skimmed off. Deep-bed filtration [4.323] is used to remove fines. It is often more economical to recirculate water, but if it is to be discharged it is subjected to a final treatment.

6.4.4 Surface Finish

As for cold rolled products, the finish of a hot rolled product is a function of several variables.

Roll Surface Finish. In order to aid acceptance of the slab into the roll gap at heavy reductions, roll surfaces are usually rough. Ground surfaces of at least 1 to 3 μm AA roughness are typical. In the early passes of hot rolling of steel on blooming and billet mills, even the maximum reduction afforded by dry, coarse-ground rolls is insufficient for economical production. The surface is then artificially roughened by machining transverse trapezoidal or other systematic patterns into the roll surface [284].

Roll surface roughness has an effect also on lubricant film thickness. Little is published on the subject, but it appears that the effects are the same as in cold rolling, at least as judged from the work of Ivanov and Maksimenko [285], who hot rolled mild steel strip with polymerized cottonseed oil.

Workpiece Surface. The surface of the hot workpiece entering the roll gap is invariably oxidized, and the presence of oxide, together with difficulties of lubrication, result in a hot rolled finish that is less smooth and uniform than that attainable in cold rolling. The surface quality is, nevertheless, acceptable for many applications, provided that localized defects are absent.

In rolling of steel, copper, and similar materials, inadequate removal of scale before the slab enters the first pass results in rolled-in scale pockets which can represent, by the time the hot finishing gage is reached, a substantial portion of the total thickness, and creates a most troublesome form of surface defects [286]. Accumulation of scale on mill accessories can cause secondary defects by scoring the surface of the sheet [195]. One of the benefits of lubrication is minimization of such defects.

Finished Product. When properly conducted, hot rolling produces plate or band with a uniformly rough surface. Roughness is a function of lubrication and of the material (e.g., a copper plate is, generally, smoother than a steel plate).

Hot rolling mills, too, are subject to instabilities resulting from sudden changes which show up as changes in surface appearance. Buckley et al [287] analyzed the consequences of slipping and of the sudden load changes that occur when the slab enters the rolls.

Some defects or potential defects are related to metallurgical causes, primarily to characteristics of the cast structure typical of the starting material. The problem often becomes visible only in the final use of the sheet, usually after cold rolling; however, it is inherited from the casting stage and is carried through hot rolling. Thus, a columnar crystal structure, or the presence of large numbers of similarly oriented grains, may result in clusters of recrystallized grains with a common crystallographic orientation, and become visible to the naked eye. Such roping in stainless steel [288] and directionality in aluminum alloys [289] can be particularly annoying when the sheet is subsequently formed and the part exhibits localized banding or, in deep drawn cups, parabolic markings. Their avoidance calls for strict control of the thermomechanical processing variables and, even though they are sometimes attributed to lubrication problems, they are not influenced by lubrication.

6.4.5 Roll Wear

Hot rolling rolls are exposed to violent fluctuations in temperature, abrasion by hard oxides, and periodic loading. It is no wonder that the resulting rapidity of wear has prompted a wide range of investigations, which are partially reviewed by Williams and Boxall [290] and by Dickinson and Porthouse [291]. Wear is either fairly uniform across the contact surface or localized in deeper wear bands.

Uniform Wear. There is general agreement that uniform wear is caused primarily by abrasion in combination with thermal fatigue, with corrosive wear playing a subordinate role. Abrasion is identified as the major cause also by Shaughnessy [292], and Shiraiwa et al [293] point to the almost constant hardness difference existing between oxide and roll material over the hot rolling temperature range. A regression analysis of industrial wear data by Funke et al [294] showed that wear and roughening of the surface was a function of the flow stress of the workpiece material, of relative sliding velocity, of total rolled length, and of maximum temperature. Wear thus follows the general trends expressed in Eq.3.39. A form proposed by Toda et al [295], often used in process models, is:

$$d \propto \frac{P}{w}\frac{l}{R} \tag{6.29}$$

where d is diametral wear, P is roll force, w is strip width, l is rolled length, and R is roll radius. In view of the lubricating qualities of oxides, one might expect that oxidation of the rolls should offer some protection. However, as pointed out by Ohnuki and Nakajima [296], thermal fatigue still progresses and destroys the surface. The uncertainties associated with investigations based on industrial experience prompted Kihara et al [297] to simulate hot rolling on a one-tenth scale using rolls with readily interchangeable sleeves, and with strips rolled from a coil preheated in a furnace. They were able to reproduce wear patterns found in typical production operations, showed that lubrication with a compounded oil reduces wear, and surmised that the black film often observed on roll surfaces is due to transfer of scale from the strip.

Localized Wear. Wear localized along bands on the roll surface can be more damaging. Several causes contribute to this problem. In the work of Funke et al [298], accumulation of secondary scale appeared to be the primary cause. Such accumulation is more likely with a rougher roll finish, and is especially troublesome in the early stands of a tandem mill where temperatures are still high. It is partly for this reason that the first stands benefit most from lubrication, by prevention of accumulation of scale. However, the primary cause of roughening is linked to the metallurgical structure of cast iron rolls. Double-poured rolls have a wear-resistant white iron surface and a tougher gray iron core. Judd [299], who also gave a review of earlier work, came to the conclusion that banding is a result of the following sequence: thermal fatigue initiates cracks perpendicular to the roll surface; at large eutectic carbides the cracks turn parallel to the surface; oxidation in the cracks causes swelling and spalling of cells; and layers of cells are successively removed to form the bands.

Apart from regular, progressive wear, hot work rolls, too, may be damaged by sudden events such as cobbles [300] and by skidding of the rolls; dynamic effects, including those due to the slab entering the rolls and to skidding of the rolls, have been analyzed by Buckley et al [287].

Because thermal excursions are the primary cause of thermal fatigue, improved roll cooling by properly designed headers can substantially increase roll life, as shown by Stevens et al [5.126].

Roll Materials. Great strides have been made in producing rolls with increased wear resistance [244,301]. Forged alloy steel rolls, lightly forged hypereutectoid (approximately 1.2% C) steel rolls, and double-poured cast iron rolls have been developed. Efforts at improving roll life by surface coating have also become more numerous. In principle, all techniques described in Sec.3.11.5 could be used. In practice, only coatings that resist separation under fatigue loading can be successful. Thus, a hard chromium layer, even though hard and wear resistant [248], is likely to peel off. Diffusion coatings should be better; indeed, a boronized surface gave lower friction in hot rolling of aluminum and nickel alloys [3.416] at least in the laboratory, but durability could not be established. Use of borocyaniding has also been reported [302]. A powder metal fusible coating has been reported successful for aluminum rolling [303], and rebuilding of roll surfaces by electroslag [304] or submerged-arc [305] welding has become accepted practice. Electroslag welding gives, in general, a more uniformly wearing surface because of the absence of inclusions. Precipitation-hardening steel gives longer life in cogging mills with their ragged surfaces [306].

Typical roll lives, expressed in tons of steel per millimetre of diameter removed, are around 20 000 t/mm in roughing stands, 10 000 t/mm in intermediate stands, and 3000 t/mm in finishing stands [244]. Backup rolls range from 40 000 to 200 000 t/mm.

6.5 Rolling of Shapes and Tubes

A large variety of shapes, ranging from bars and wires of circular cross section to complex sections such as rails and beams, are rolled—usually hot but also, increasingly, cold.

6.5.1 The Shape Rolling Process

The process of rolling shapes is considerably more complicated than that of rolling flat products. The reduction in cross-sectional area again results in an increase in length and thus speed; however, the neutral line b becomes a tortuous, three-dimensional curve.

Even in the simplest case, as in rolling of a bar with a diamond or oval cross section, the active roll diameter changes from point to point along the contact surface (Fig.6.47a). Since the section must retain its continuity, it will emerge at some speed that represents a compromise among the various imposed velocities. Thus, sliding over some portions of the contact surface is severe, and wear becomes a major problem. This is particularly true when a leg of a section is rolled between two surfaces of the cooperating rolls (Fig.6.47b) and is thus subjected to a shearing mode of deformation. Such indirect draft is very useful for filling the cavity and is aided by friction; however, wear can become excessive. In contrast, friction hinders development of the profile if the leg is rolled in a closed groove of one of the rolls (Fig.6.47b).

The design of roll passes, with proper regard for material flow, is a highly specialized subject [307,308].

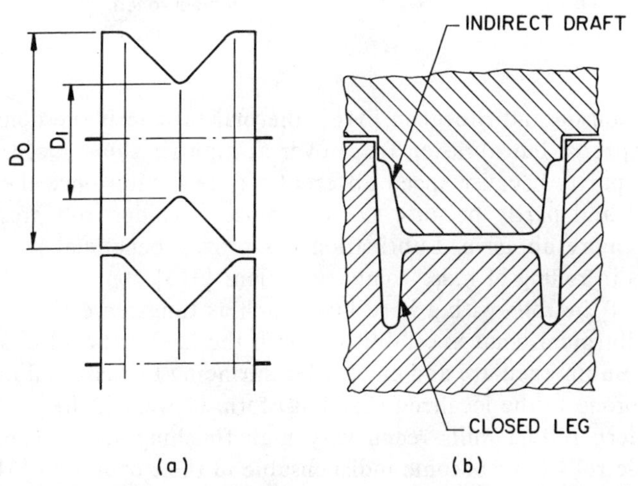

Fig.6.47. Relative roll diameters in rolling of (a) a diamond-shape bar and (b) an I-beam.

6.5.2 Lubrication

Because of the severe sliding conditions, section rolling is a prime candidate for lubrication and, indeed, some of the first hot lubrication attempts were made on section mills. Practices are similar to those followed for flat rolling, except that great care is taken to distribute the lubricant so that it reaches the main wearing surfaces [309]. All techniques and lubricants considered in hot rolling are used here as well [257,258,310-312]. As pointed out by Cichelli [313], the savings in roll life, energy consumption, and improved surface finish vary from plant to plant.

The prime benefit of lubrication is that it reduces nonuniform wear, which makes some parts of the rolled section thicker than allowed by the tolerance limits while other parts of the roll pass are still acceptable. Surfaces that form legs by indirect draft suffer particularly severe exposure. Extensive work has been conducted on these aspects, and Soviet work has been reviewed by Mrozek [314]. In that review, the example taken from the work of Zhadan and Stefanov [315] illustrates the much greater wear encountered in rolling a U channel with straight rather than butterfly pass design (Fig.6.48); this is an example of circumventing rather than combating a tribological problem.

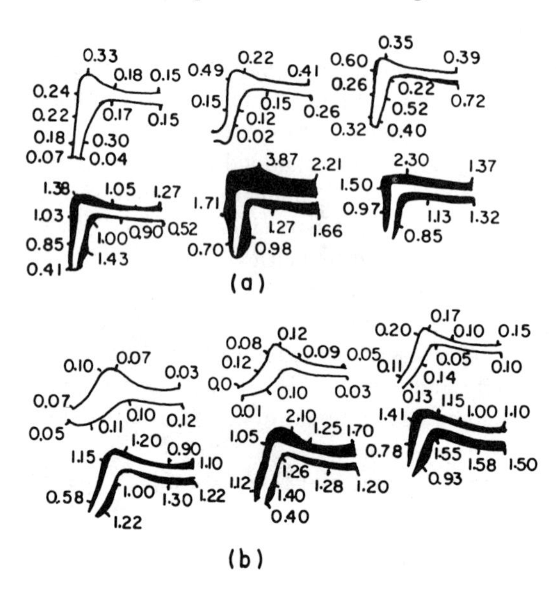

Fig.6.48. Wear of rolls used in rolling of a U-channel with (a) straight and (b) butterfly pass sequences (after [315], quoted in [314]). Numbers indicate wear in units of μm per ton of steel rolled.

6.5.3 Wear

As shown by Johnson and Sturgeon [316], thermal fatigue is the dominant wear mode at low rolling speeds, but abrasion takes over at high finishing speeds. Wear is greater in later passes, partly because speed differentials are greater once the section profile is fully developed and partly because the workpiece is colder, roll pressures are higher, and oxides are more abrasive. Lubrication has proved beneficial in all instances, with greater benefits in rolling of more complex sections [313].

Surfacing of the rolls with a hard layer such as tungsten carbide is successful only if a good metallurgical bond is established with the base; otherwise, spalling is a problem [316,317]. Submerged-arc and electroslag surfacing increase roll life; the electroslag process is less prone to the localized (banding) form of wear [318].

Some modern rolling mills reach very high finishing speeds (up to 75 m/s), and tungsten carbide rolls have become indispensable in their operation [319-323]. The rolls are made in the form of inserts, restrained between steel flanges. Flanges of opposing rolls are usually run hard against each other to ensure dimensional accuracy, and lubrication provides great improvement in the wear life of these flanges, too. Carbide rolls are used also for rolling small sections other than round bars.

6.5.4 Tube Rolling

Tubes are manufactured by a variety of processes, which are reviewed by Blazynski [2.16],[324] and discussed also by Geleji [2.7,2.18]. Extrusion will be discussed in Chapter 8; there are also various rolling processes which present tribological problems similar to those of shape rolling, with the added complication of the relative inaccessibility of the tube bore to the lubricant. These problems have been extensively investigated, particularly in the Soviet Union.

One method of production involves hot rotary piercing of billets between skewed rolls, cones, or disks, with a stationary or rotary plug ensuring a sound inner surface. Because the plug is in contact with virgin metal at high temperatures, adhesion and wear can be severe. A remedy is sought in heat-resistant compositions and surface coatings, and lubrication through a hollow plug has also been attempted. Several techniques have been evaluated by Vedyakin et al [325].

Further rolling of the hollow bloom may be done on a pilger mill, which performs periodic rolling on a mandrel. The mandrel is lubricated by any of the high-temperature lubricants such as graphited grease, phosphate-containing mixtures [4.195], and salt mixtures. The tube may be rolled also on a lubricated mandrel in a continuous mill [326]. In cold pilgering (cold reducing) [327], lubricants appropriate for the metal are used, such as mineral oil with fatty acids for aluminum [328] and with E.P. additives for warm reducing of stainless steel [329].

Many tubes are reduced by hot rolling over a plug (mandrel), and the severe relative sliding leads to scoring of the inner surface unless an effective lubricant is used. Salt is too thin a liquid, and gradually melting substances such as phosphates [4.195], pumice [330], and filled lubricants such as charcoal in salt [331] have been used.

6.6 Lubrication for Iron-Base and Nickel-Base Materials

Industrial lubrication practices follow from the principles discussed in Sec.6.3 and 6.4.

6.6.1 Hot Rolling Lubrication

Rolling temperatures are high, from 1100 to 1200 C at the start, and even finishing temperatures are above 900 C unless metallurgical structure control dictates a lower temperature.

Early European experiences in full-scale trials were described by Beese [332] and by Edmundson [333]. Specific lubricant types, in addition to those discussed in Sec.6.4, have been described by several workers (e.g., [334-337]). Rolling speeds in modern mills exceed 1000 m/min in the finishing stands, but wear is greatest in the first few stands of the finishing train and it is here that lubrication is mostly applied.

Present practice seems to favor the use of lubricants composed of synthetic esters, fortified with a phosphorus-containing E.P. additive. For ease of application, a mineral oil is sometimes used as a carrier. Application is mostly done by the direct technique (using air atomization or positive-displacement metering pumps) or by injection into the header water. The technology is still developing, and no dominant method is likely to emerge for a while yet. To avoid danger of refusal, lubricant flow is initiated only after the strip nose has entered the roll gap, and flow is stopped before the end of the workpiece passes through so that the oil burns off before the next piece enters.

Little has been published about hot rolling lubrication practices for stainless steels and nickel-base alloys but, in principle, differences should be small. The higher hardnesses of oxides and spinels result in greater wear, but production quantities are smaller, and thus the problems are similar to those encountered in rolling of alloy steels.

6.6.2 Cold Rolling Lubrication

Lubrication practices in cold rolling of steel strip are a function of the rolled gage. Wide strip (over 600 mm) of thicker gage (say, above 0.5 mm) is usually called sheet. Thinner strip (down to 0.2 mm) is properly termed blackplate although, if it is subsequently tinned, it is sometimes—and most confusingly—called tinplate. This term should really be reserved for strip that has been tin coated and then further rolled. Strip less than 0.2 mm thick is usually produced by being passed, after annealing, through a tandem mill again, and then it is referred to as double-reduced or extra-thin blackplate (or tinplate). For economy of production, rolling speeds are high, in excess of 1000 m/min, and preferably are at the limit set by the lubricant, typically just below 2000 m/min.

Low-carbon steel exhibits yield-point elongation which would lead, on subsequent stretch forming, to the appearance of Lüders bands (also called stretcher-strain marks). To suppress this, the annealed strip is temper rolled to a very light (about 0.5%) reduction, and this step is often used to imprint or modify the surface finish.

Stainless steels and nickel-base alloys are rolled at lower speeds. Special alloys are often rolled in narrower widths; otherwise, wide strip is split. Wire is often flattened to produce narrow strip with rounded edges.

Experimental Work. Much of the information summarized in Sec.6.3 was actually obtained in experimental rolling of steel. Some of this work was aimed at evaluating the friction-reducing capacities of lubricants. Typical of such work is that of Yamanouchi and Matsuura [338] and Whitton and Ford [45], who used the back-tension method. The coefficient of friction ranged from 0.034 to 0.08, typical of mixed-film lubrication at low speeds. As expected, friction decreased with increasing viscosity of mineral oils [339] and fatty oils. Fatty acids and alcohols added to mineral oil are effective if the mineral oil is of low enough viscosity to give predominantly boundary lubrication. Nevertheless, improvement with steel is relatively small [340].

Fatty oils, as might be expected from their rheology, are generally better. The standard of comparison is always palm oil applied to the strip, with water supplied to the rolls. Much work has been devoted to understanding the reason for its success and to finding suitable alternatives [108,111-113,158,163,164,341-345]. As shown by Takatsuka et al [4.41], the viscosities of various fatty oils increase at the same relative rate, and therefore it is not surprising that oils of higher room-temperature viscosity are generally better lubricants. Composition is more controversial. Nekervis and Evans [5.269] and Kihara et al [162] found that free fatty acid in palm oil gave heavier reductions, but work by Whetzel and Rodman [112] and by Williams and Brandt [113], as well as the detailed investigations of Lueg et al [345] and Grudev et al [346], showed free fatty acid to give only minor improvement. Johnson et al [5.7] came to the same conclusion in wiredrawing simulation. As can be derived from general principles, unsaturated fats are generally less effective than saturated counterparts of the same chain length, and this has led to the general recommendation that rolling oils should have low iodine values and high titers.

Shortages of palm oil during World War II led to the development of so-called synthetic palm oils, usually derived from tallow [5.7,5.269]. They are still acceptable lubricants (Fig.6.49 and 6.50) [5.91],[347,348]. Castor oil, coriander oil, and cottonseed oil can be of comparable performance, while rapeseed oil is somewhat poorer [111,163,345,349], although when mixed with tallow it shows promise [143,144]. Highly unsaturated oils such as linseed oil are undesirable because of their tendency to form gums.

Improving upon the performance of palm oil or synthetic palm oil has proved to be difficult, although some advances have been made. Unfortunately, most are shrouded in proprietary secrecy. Among published attempts are the addition of 50% methyl ester of tallow fatty acid [142]; the use of glyceryl trioleide based entirely on C_{18} chain length [350]; the use of partially hydrogenated lard oil and of triglycerides formed by reacting the hydroxyl group of ricinoleic acid in castor oil with palmitic or stearic acid to obtain a multiple ester linkage [350]; and the copolymerization of animal fats with hydrocarbons to yield modified ethylene polymers [351]. Natural waxes have outstanding properties (Fig.6.49), especially on thin, hard strip (Fig.6.50) [347], but present problems of application. This is true also of lubricants based on the combination of a long-chain fatty acid with a phosphonic acid group [341].

The problems associated with oily waste waters resulting from direct lubricant application led to early attempts at developing emulsions suitable for use in closed

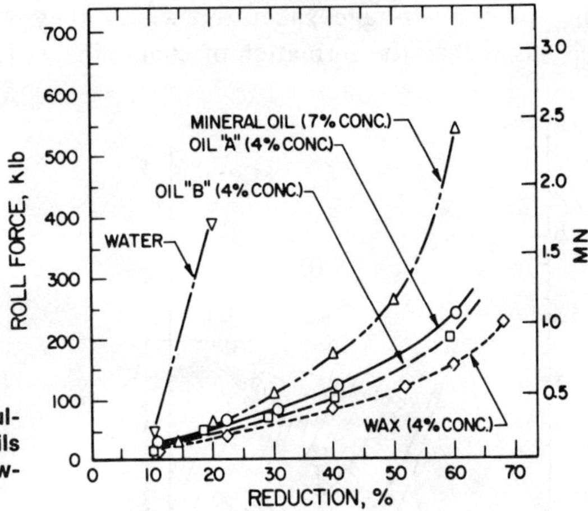

Fig.6.49. Relative efficiencies of water, emulsions based on mineral oil, synthetic palm oils (oils A and B), and wax in rolling of soft low-carbon steel strip [347]

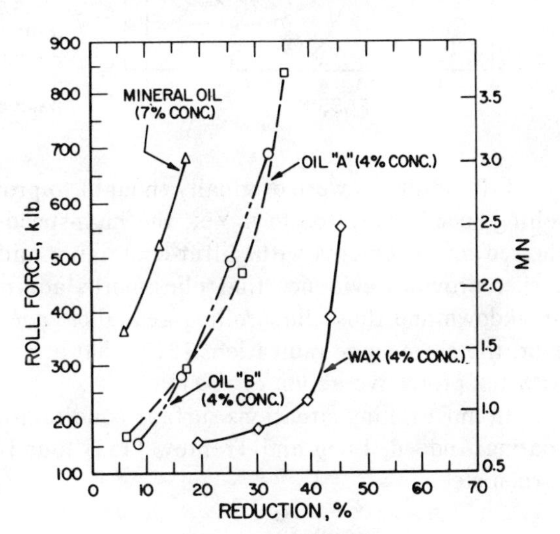

Fig.6.50. Same as Fig.6.49, but for fully hard strip [347]

recirculating systems [153,163,164,342-345,352,353]. The results of Billigmann (Fig.6.51) [153] are typical. Evaluation was based on the minimum strip thickness attained after rolling the same strip in eight consecutive passes; reductions obtainable with water served as the reference line. Not surprisingly, mineral oil emulsions were poor although often better than the mineral oil itself, which is in line with the observation that water on top of oil can be better than oil alone [153,158]. Emulsions of compounded mineral oil were only slightly better, and real improvement came only with emulsified fatty oils [160], none of which could quite match neat palm oil or palm oil substitutes. In general, if a neat oil is superior to palm oil, it also performs better as an emulsion (unless there are undesirable interactions with the emulsifier). Improved performance was reported when high percentages of wax were added to a rapeseed oil emulsion [342]. Funke and Marx [160] found the ethyl ester of stearic acid to be the best. The basic problem is that long-chain, highly saturated esters are difficult to emulsify, and need great care in handling in the recirculating system to avoid trapping of dirt and formation of invert emulsions. In one of the best-documented evaluations of commercially available semisynthetic and chemical rolling fluids, Rigo [5.271] found that all fat-base lubricants were superior to the mineral-oil-base or chemical solution (synthetic) varieties, at least as far as their reduction capacity was concerned. Some synthetics were, however, more readily cleaned off and thus offer advantages in the rolling of heavier-gage sheet for which they were actually designed. High-stability emulsions reduce the formation of contaminants [354].

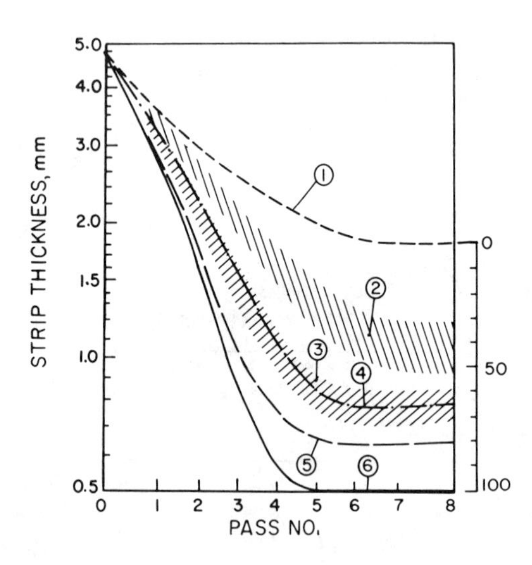

Fig.6.51. Limiting strip thicknesses obtained in multipass rolling of steel strip [153]. 1 - water; 2 - commercial emulsions; 3 - improved emulsions; 4 - undiluted concentrate; 5 - best emulsion; 6 - palm oil.

E.P. additives were originally thought to provide no benefits [111], perhaps because rolling speeds were too low. Yet, the high-speed experiments of Starchenko et al [158] showed improvements with sulfur-containing mineral oil, and the Japanese work quoted earlier provides evidence that phosphorus additives can delay the onset of lubrication breakdown and thus allow rolling at higher speeds without the heat-streak defect. However, there are some indications [212] that a free fatty acid such as oleic acid interferes with the protective action of ZDTP.

In most rolling situations surface conditions are not severe enough to justify surface coating. Indeed, Lueg and Treptow [355] found little benefit from phosphating a high-carbon steel.

Lubrication Practices. Rusting of steel presents special problems and necessitates measures not needed in rolling of other materials. General reviews of lubrication practices have been given repeatedly [7,356-359].

1. Pickler Oil. Hot rolled strip is pickled to remove residual scale which would otherwise lead to rolled-in defects (fleck scale) [196], contribute to the wear of the cold rolls, and present problems in the final application of the sheet. Pickling with sulfuric acid can cause metal loss and overly rough surfaces unless inhibitors are applied, whereas hydrochloric acid removes no steel and leaves the as-rolled roughness. Funke and Mikulla [360] found that if the scale was properly removed, the surface roughness of the hot rolled strip had little effect because most of it disappeared after some 20% cold reduction. Overpickling is dangerous in that the very rough surface may take in excess lubricant, resulting in a loss of control [356]. Careful rinsing after pickling is mandatory because acid residues cause emulsion breakdown [354].

To prevent scuffing during handling and uncoiling, and rusting during storage, a lubricant film (pickler oil or pickle oil) is applied to the strip as it emerges from the rinsing bath. The quantity of oil is controlled through its viscosity, which in turn is governed by the choice of the lubricant and by heating of the pickler oil. If the hot band is fed from a reel, the danger of scuffing is less and a moderate-viscosity oil (200 SUS at 100 F) suffices. However, if the coil is freely rotating on rollers of an open-box uncoiler, with only its edges confined between flanges, a heavy duty, viscous (up to 450 SUS) oil with up to 12% free fatty acid is applied [196,357]. The pickler oil is often the only lubricant present on the first stand of the cold rolling mill, and it may be the only oil used in rolling of heavy-gage sheet, with water applied to the mill stands. An alternative is a thin, dry wax coating deposited from a 2 to 3% aqueous emulsion [196]. The pickler oil represents a source of contamination in rolling with emulsions, and then the emulsion concentrate may be used as the pickler oil. It is imperative, though, to prevent overheating of the concentrate by the strip, because loss of water would prevent re-emulsification and lead to inversion.

Even though the pickler oil is not called upon to perform onerous duties, it cannot be a low-quality product [357]. Some of it remains on the strip and is one of the main sources of staining on annealing. For this reason, a fatty oil is often cut with a mineral oil to reduce the saponification value to 32 to 40 and the free fatty acid content to 1 to 2%.

2. Sheet Rolling Lubricants. Thicker gages of sheet are given only 50 to 60% total reduction distributed over several stands, and therefore lubricant requirements are not severe. Fatty oils blended with mineral oils, in the form of emulsions or dispersions, are often used, and it is here that some of the newer synthetic fluids find application. If staining is to be miminized or a "mill clean" sheet produced, a detergent is applied at the last stand [361].

For higher speeds and heavier reductions, fatty oil emulsions are used in a recirculating system. If the emulsion is formulated so as not to emulsify the pickler oil, it can be kept in service for long periods of time because the pickler oil separates on the tank surface and can be skimmed off. Oil consumption has been quoted as 0.3 to 0.7 kg per ton of steel [250].

3. Lubricants for Thin Strip. Thin-gage materials require reductions ranging from 80 to 90% on a 5- to 7-stand tandem mill. Speeds of 2000 m/min are aimed for. Because at least the last stands operate under conditions approaching limiting reduction, the reduction capacity of the lubricant assumes paramount importance. The typical lubricants are unstable emulsions bordering on mechanical dispersions, and are essentially mineral-oil-free emulsions of fats. Few analyses are available in the open literature; the ones given by Roberts and Somers [347] and by Rigo [5.271] show that all substitute palm oils have high proportions of saturated long-chain fatty oil components. The physical and chemical properties of typical conventional fatty oil lubricants are given in Table 6.2. As already mentioned, phosphorus additives are gaining acceptance for thin-gage, high-speed rolling.

Table 6.2. Typical lubricants used in rolling of steel

Property	Roberts and Somers [347]	Rigo [5.271]
Melting (or pour) point, C	5-45	. . .
Viscosity, SUS:		
At 38 C	50-850	200-260
At 100 C	45-200	55
Viscosity index	130-160	. . .
Saponification value	125-200	165-192
Iodine value	40-75	46-65
Free fatty acid, %	3-20	0-11

Oil consumption is high, approximately 2 to 3 kg per ton of steel [357] in the total-loss system in which 3 to 8 parts of water are added to 1 part of oil. In recirculating systems oil consumption is reduced to 0.5 to 1.5 kg/ton.

4. Double-Reduced Strip. Thin-gage strip is produced by a second pass through the tandem mill, with total reductions ranging from 30 to 50%. A bright finish is often desired to ensure a reflective finish on the strip, and therefore lubricant properties are balanced to ensure low enough friction for rolling under conditions approaching limiting reduction, while the surface is brightened through boundary contact and the burnishing action of the rolls [359]. Unstable emulsions are typical. For example, Drake [357] reported that an oil-in-water dispersion of 8 to 14% concentration, applied at 66 C, gave a better, mottle-free surface than direct application of a more concentrated dispersion at 45 to 50 C. Vucich and Vitellas [227] found that the two techniques are comparable provided that emulsion stability is controlled. Improvements in the lubricant and coolant delivery system are often needed [362].

5. Coated Strip. As mentioned already, occasionally a strip is tin plated and then given a further reduction. From a general point of view, the experiments of Roberts [363] are interesting. In addition to tinplate, aluminum-, lead- (terne-), and zinc-coated sheets were rolled. Somewhat surprisingly, a heavy aluminum coating (366 g/m^2) reduced friction most, followed by unmelted tin at a weight of 17 g/m^2, terne metal at a weight of 27 g/m^2, and remelted tin at a weight of 17 g/m^2, while a heavy zinc coating (287 g/m^2) was the worst. The good performance of the aluminum surface must be attributed to the ease of lubricating aluminum with fatty oils and their derivatives. The much better performance of the unmelted relative to the melted tin coating is in line with the argument that the microgeometry of the electrolytic tin coating encourages entrapment of lubricant, and offers further proof that metal coatings act not simply as friction-reducing agents but also as lubricant carriers (Sec.4.7.1).

6. Temper rolling of steel at very light (0.5 to 1.2%) reductions is sometimes done without a lubricant, frequently on shot-blasted rolls. The transfer of roll roughness is primarily a function of reduction (or, rather, elongation) in the pass [169,364-368]. Rolling speed and cleanness of the strip appear to have little effect. The transfer factor, defined as sheet roughness/roll roughness, is typically 0.25 to 0.5 (and as high as 0.7) at the usual low reductions in dry rolling [367], and less in lubricated rolling [368]. A disadvantage of dry rolling is the generation of metallic fines and a dirty mill [7]. Rolling with a synthetic lubricant eliminates this problem without affecting the transfer of roll roughness onto the strip [62,366-368].

In maintaining a recirculating system, all the measures discussed in Sec.4.8 merit consideration. Laboratory testing can be quite extensive and includes the measurement of such properties as wetting, although their significance is by no means clear. An example of a full testing program is given by Rigo [5.271].

A particular concern is rusting of the rolled sheet. A semiquantitative assessment by Billigmann and Fichtl [164] indicated that mineral oil emulsions, even when compounded with fatty additives, allowed moderate to heavy corrosion, as did also emulsions based on tallow. In contrast, with emulsions based on palm oil, rapeseed oil, or mixtures of tallow and rapeseed oil, rusting was light, even though still somewhat more severe than with neat palm oil or even pickler oil. No relationship between composition and rusting could otherwise be detected. Rusting in standard tests using cast iron chips showed no correlation with actual rusting on rolled strip. Cleanability is another concern and is usually tested by a laboratory version of the full-scale process [5.271].

Despite the difficulties of simulation, most laboratories attached to steel mills employ some kind of a bench test in the hope that correlation with practice can be found [219].

6.6.3 Alloy and Stainless Steels

High-chromium steels, stainless steels, and nickel-base superalloys have in common high flow stresses (thus high roll pressures) and relatively nonreactive surfaces (thus poor response to the presence of boundary additives), and some of them also have a tendency for adhesion. Conditions of limiting reduction are very soon reached, and therefore it is common to roll with small-diameter work rolls, often on Sendzimir mills. This means that the angle of entry for a given reduction is rather high (Eq.6.17) and the hydrodynamic contribution is discouraged. Relatively few avenues of solution are available:

a. The lubricant is chosen to have a high enough viscosity to give a reasonable hydrodynamic contribution without inducing skidding or a deterioration of the surface finish of the strip.

b. Fatty oils or their derivatives are added [131] for their favorable pressure-viscosity coefficients. It is possible also that the wall effect mentioned in Sec.3.6 becomes effective even on the nonreactive surfaces. Neat fats are useful [348].

c. E.P. compounds, particularly chlorinated paraffins, are added. The reactivity of the compound must be controlled in order to prevent corrosion of the rolls and mill furniture.

Information on industrial practice is scarce [369,370], but published data fit within the above framework. The lubricant is usually oil-based, although emulsions have been used on four-high mills. In Sendzimir mills the rolling lubricant has to lubricate also the backup roll bearings [371] and, even though the use of emulsions has been reported, their lubricating power is often inadequate [372], and accelerated fatigue in the aqueous environment may occur. In a more desirable development, the bearing lubricant is applied separately in the form of an oil mist.

6.7 Lubrication of Light Metals

The rapidly increasing demand for aluminum has prompted the introduction of rolling techniques similar to those practiced in rolling of steel. Semicontinuously cast slabs (scalped for best surface quality) 200 to 600 mm thick are hot rolled on reversing mills into blanks 18 to 25 mm thick. The blank (or a continuously cast band of similar thickness) is warm rolled at temperatures of 400 to 200 C into a band 2 to 6 mm thick and then coiled. The hot band is then cold rolled to thinner gages on single-stand mills (which may be reversing) or on tandem mills comprising 2 to 4 four-high mills, at speeds up to 600 m/min. The thinnest conventional strip (can stock) is 0.15 to 0.25 mm thick. Still thinner foil is usually rolled on single-stand, nonreversing mills from annealed, 0.45-mm-thick stock to finished gages of 6 to 12 μm at reported speeds exceeding 2000 m/min.

The technology of magnesium alloys is similar, except that ingot sizes and rolled quantities are much smaller and single-stand mills are usually employed. The hexagonal metal has very little ductility below 220 C, and most rolling is done warm.

6.7.1 Hot Rolling Lubrication

Aluminum has a hard and brittle oxide which offers no protection during hot rolling. Virgin surfaces generated during the elongation of the slab adhere to the roll surface, and, unless a suitable lubricant is used, cumulative pickup will ultimately result in the slab sticking to and wrapping around the roll.

In the early days of low production rates, a smear of mineral oil or animal fat on the roll surface was sufficient. Greater slab weights, increasing outputs, and higher rolling speeds have made oil-in-water emulsions, in concentrations of 2 to 15%, the universal lubricants. They are chosen for specific stages of processing and, even though they cannot prevent the development of a roll coating, they contribute significantly to its control. Lubrication techniques have been repeatedly described [5.264],[370,373-377].

Roll Coating. Early knowledge of coatings was based on industrial observations [5.264,5.268],[70]. More recent laboratory simulations, coupled with advanced surface analytical techniques, have shed light on the details of the mechanism [378-382].

When a freshly ground roll is put into service, the rolled blank is quite smooth (Fig.6.52a) for a very short while. Aluminum transfer is localized to points of discontinuity (nose, tail, and side marks) and to isolated points on the contact surface [378]. With the aid of removable roll inserts, Smith et al [381] were able to follow details of coating buildup by SEM and x-ray spectroscopy. They found that individual pickup formed on first contact streaks out in the rolling direction on further contact; additional pickup points develop while earlier ones gradually wear down and, even though they may persist for hundreds of contacts, they become gradually submerged in a finer, more general transfer film. With a suitable emulsion, a limiting condition is reached; at this point, the rolled blank has a rough but uniform, fish-scale-like, satin-finish surface (Fig.6.52b). When, for any reason, the coating becomes unstable, some patches become detached and, rolled into the surface, result in the black-speck defect (Fig.6.52c and d). The roll coating was found to consist predominantly of aluminum oxide (Al_2O_3) with 10 to 15% metallic aluminum and small amounts of carbonaceous matter attributed to lubricant residues [381]. The free metal content increases when lubricant failure sets in [382].

The rate of coating buildup and the stability of the coating depend on a number of factors. Baba et al [380] found that at least 80 revolutions were needed for the roll force and roll temperature to reach steady-state values on a laboratory mill. From simulation tests Tripathi [379] concluded that rougher rolls were more prone to pickup; it is not clear that this is always the case in practice.

The emulsion has a dominant effect. Plating out can be controlled to reduce friction and metal transfer:

a. For a given emulsion, plating out is prompted by warmer rolls, as found by

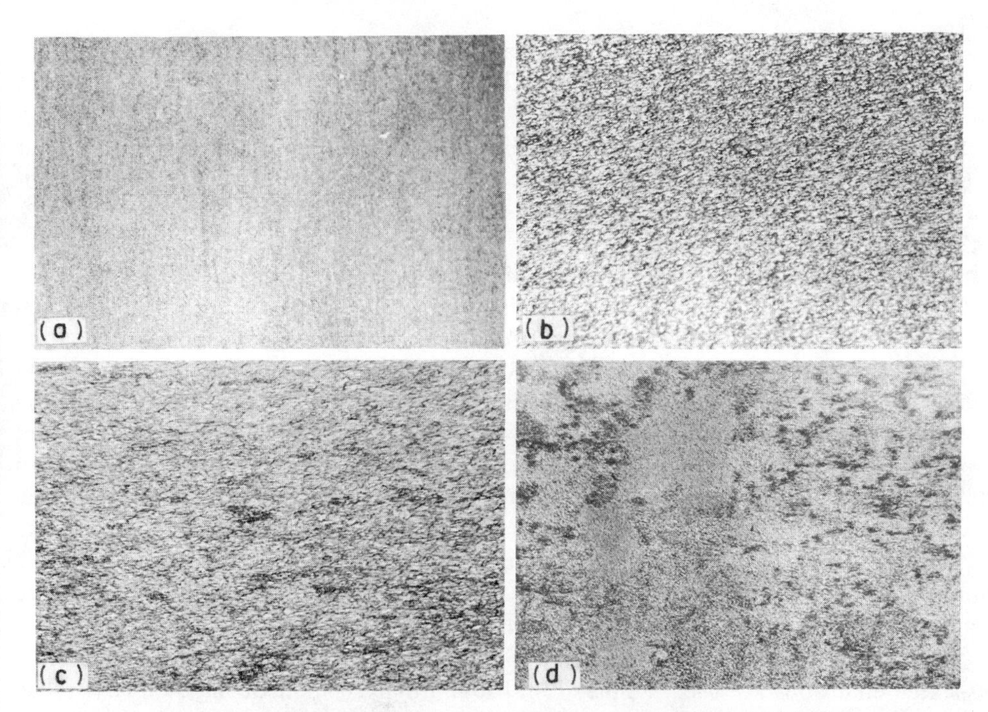

Fig.6.52. Surface appearance of hot rolled aluminum blanks [5.268]. (Courtesy Edgar Vaughan Co., Ltd., Birmingham, England)

Frontini and Guminski [378] with a nonionic emulsion; but hot rolls result in emulsion breakdown and high friction. Indeed, emulsion temperature is one of the factors used to control roll coating in practice. Plating-out tendency is sometimes quantified by weighing the oil film separated on a foil placed in the flow of a warm emulsion stream [378].

b. The other factor is emulsion stability. A stable emulsion allows heavy coating buildup, as found with a very stable anionic emulsion (Fig.6.53, curve A). A semistable anionic emulsion reduced pickup (Fig.6.53, curve B), and a nonionic emulsion that suffered a loss of stability on warm rolls allowed coating buildup, which improved as the rolls heated up (Fig.6.53, curve C) [378].

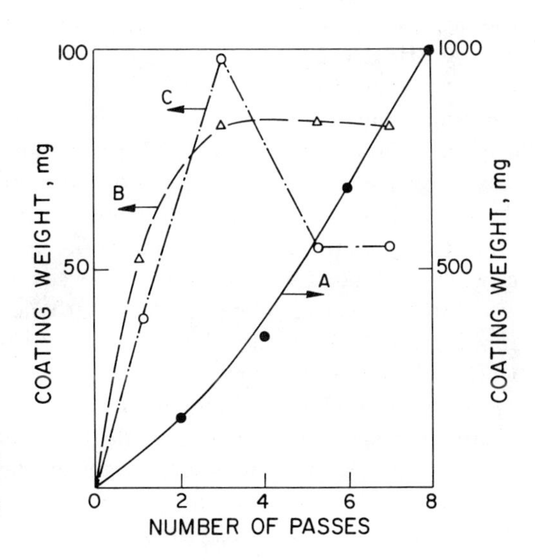

Fig.6.53. Effect of emulsion stability on coating weight [378]. Curve A - very stable anionic emulsion; curve B - semistable anionic emulsion; curve C - nonionic emulsion.

Emulsion stability affects also the stability of the coating. The thick coatings developed with tight emulsions release particles which are rolled into the slab surface and, in the extreme case (Fig.6.52c and d), are visible as black specks; otherwise they become visible only when the blank is anodized [376], whereupon the specks stand out by their dark color. Only when lubrication breaks down are heavily smeared, folded-over features observed [375].

Lubrication Practices. On breakdown mills, drafts of 50 mm per pass are commonly expected, and with the usual roll diameters this is possible only with very high friction. Indeed, Varley [383] observed sticking friction. To ensure high friction, the coating must be allowed to build up through the use of a stable emulsion. Lockwood and Matienzo [384] quoted an average particle size of 1.5 μm for a new emulsion. Aging and contamination, detected by infrared spectroscopy and high-performance liquid chromatography, increased particle size until, at an average size of 3.5 to 3.8 μm, slab refusals occurred. To prevent excessive coating thickness and instability, power-driven rotary and/or oscillatory wire brushes are periodically brought into contact with the surface to remove much of the coating by scouring. More recently, nylon bristles impregnated with abrasive particles, or flap wheels of particle-coated cloth sheets, have been introduced [384a].

On tandem mills, lower friction and a stable coating are desirable, and therefore less-stable emulsions are used. Penney [385] found that good anodizing quality is rolled when about 50% of the emulsion is of a particle size of 3 to 8 μm; more than 50% of such coarse particles cause skidding, whereas less than 45% result in black-speck defects. Tighter emulsions are permissible for general-purpose strip rolling [4.149].

Tripathi [379] emphasized the contribution of friction polymers. While direct evidence for such compounds is difficult to obtain, it is not unreasonable to assume that some tribochemical reactions should take place and that some of the carbon found in the coating should be in the form of polymers. Plating-out of an emulsion implies displacement of water by the emulsified oily phase on the surface of the rolled metal. Experiments of Lockwood et al [385a] showed that a nonemulsified naphthenic oil barely wetted the surface if the oxide was pure Al_2O_3. Wetting improved only slightly with the addition of oleic acid, but improved dramatically as the MgO content of the oxide increased. At the same time, the magnesium ion concentration increased in the water phase. Thus, it appears that adsorption of emulsified oils on the metal surface is greatly affected by surface composition. However, the formation of magnesium soaps leads to a degradation of the rolling emulsion. Aluminum soaps are found in used emulsions but could not be identified on the surfaces of sheets. A black film occasionally observed on roll surfaces [379] is a sign of abnormal operation, linked perhaps to a loss of emulsion stability and a drop in friction. Even though control of coating in tandem mills depends primarily on control of emulsions, some smoothing of the coating is effected by contact between the work and backup rolls and by the wire brushes or cross-grain wooden wipers pressed against the roll surface.

Little is known about material effects. Cast rolls have been suspected of more rapid crazing [374] and of localized wear of the free graphite [375]. Of the various workpiece materials, pure aluminum forms the heaviest coating, while alloys containing magnesium develop coatings that are thinner and harder, but no less difficult to control. Coarse segregations or hot tears in the ingot may give rise to localized unevenness in the coating. The morphology of oxides on the hot rolled band is also a function of composition. Oxides on Al-Mg alloys are enriched in Mg [384] and contain Mg-Al spinels. The surface of a hot band produced under poor lubrication conditions shows transverse tears and a thick surface layer of an oxide-lubricant-residue-metal mixture ("undercast"). Further hot rolling under good lubricating conditions can lead to healing of transverse tears [385b].

For highest-quality sheets, caustic etching of the hot band prior to cold rolling is still sometimes practiced.

All emulsions are based on mineral oils with up to 20% fatty constituents, usually oleic acid and/or fatty alcohols, in addition to esters, etc. The viscosity of the base oil is adjusted to give adequate lubrication without excessive staining. Baba et al [380] conducted well-documented laboratory and field tests exploring lubricant variables. Increasing viscosity (in the range of 14 to 65 cSt at 40 C) gave lower roll force; naphthenic oils were better than paraffinic oils. Oleic acid gave improvement, at least to 12% concentration. Mono-, di-, and triethanolamines added as surfactants (in concentrations of 2 to 6%) did not reduce roll forces, whereas TCP in concentrations up to 2% effected definite reductions; the reason for this is obscure. Wetting agents contribute to emulsion stability and thus reduce coating buildup [378] but may also lead to foaming [5.268]; in some plants they are systematically replaced; in others they are regarded as less important. The additive concentration is usually less in the tighter emulsions used for breakdown rolling. The development of emulsions and optimization of their use call for patient laboratory and field testing [370]. Most compositions are proprietary. An example of development is given by Mysyakin et al [386], who found that an oil phase thickened with polyisobutylene gave better coating control.

As a minimum, the lubrication system includes provisions for separation of tramp oil and filtration over moving filter beds, as well as heating facilities [387,388]. Resin-impregnated roll neck bearings lubricated with the rolling emulsion are used in some reversing roughing mills. Temperature then must be kept between 30 and 50 C. Contamination with tramp oil reduces stability [384], as does microbiological attack [384,389,390].

6.7.2 Cold Rolling Lubrication

Even though the high cooling power of emulsions (Fig.6.54) [36] would make them ideal cold rolling lubricants, water staining during storing of coils prior to annealing or shipment has prevented their use, although occasional claims have been made regarding the development of satisfactory cold rolling emulsions [222,391]. Livanov describes an emulsion, used at concentrations of 10 to 15%, containing complex alkyl esters, some mineral oils, and nonionic emulsifiers [392].

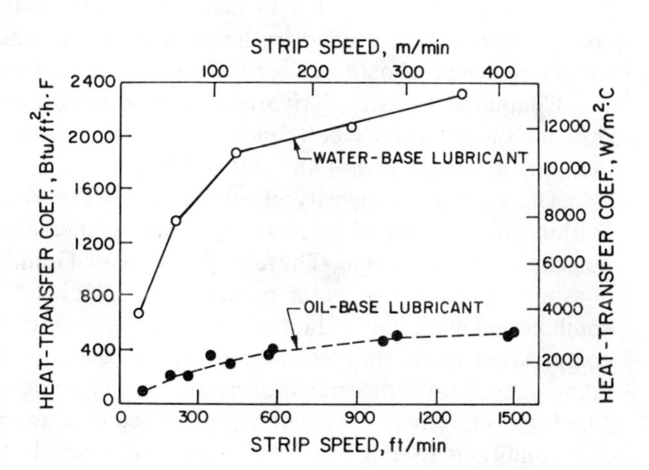

Fig.6.54. Illustration of the cooling powers of emulsions and oil-base lubricants [36]

The vast majority of rolling is performed with mineral-oil-base lubricants. There is no doubt that lubrication is of the mixed-film type and that industrial conditions involve rolling in the predominantly hydrodynamic regime. Nevertheless, the high adhesion of aluminum results in the development of a thin roll coating which is carefully controlled by the addition of boundary additives. Typical lubricants have been discussed by various authors [393-397], and Sargent has published a review of challenges that must be faced in order to develop improved lubricants [398]. Kromykh et al [394] studied the kinetics of contamination.

Lubricants. The base oil is usually chosen at the highest viscosity that is compatible with acceptable staining. Friction appears to be independent of whether the oil is paraffinic or naphthenic [111]. However, for minimum staining, highly refined paraffinic oils with a narrow boiling range (typically 60 C) are better, and a bromine number below 1 and sulfur content below 0.08% have been recommended by Guminski and Willis [5.264]. Because of their low staining tendency, synthetics such as alkylated isobutane are preferred. A C_{14} paraffin that is a byproduct of the production of biodegradable detergent has also been shown to be successful [399].

Because viscosity affects surface finish [400], somewhat heavier oil (typically 20 cSt at 40 C) can be used in the early stages of rolling to heavy gages, but lighter fractions (typically 4 cSt at 40 C) are most widespread for intermediate and finishing passes [374,391,401,402]. If a bright product is to be rolled, finishing speeds are kept low (around 600 m/min) and very light lubricants similar to kerosene are used. However, a flash point below 20 C poses the danger of fire and even of explosion if the oil mix is ignited by a diesel effect [374]; a flash point of 140 C is preferable.

All lubricants are compounded with boundary additives. For a given chain length, saturated fatty acids are more effective than unsaturated ones, and unsaturated fatty acids are better than alcohols, but alcohols stain less [5.264],[403]. A fatty acid ester is slightly less effective than an alcohol [55]. Great improvement is obtained with as little as 0.02% oleic acid, and no further gain is achieved above 1% concentration; fatty alcohols and esters such as butyl stearate are used in higher concentrations. The results of Yoshioka et al (quoted by Kondo [393]) indicate that 5% lauryl alcohol is much better than 1% lauric acid, and that the latter is about equal to 5% methyl laurate. Lanolin, palm oil, coconut oil, and rapeseed oil have all been used but are now bypassed because of brown staining, especially in concentrations required to match the performance of fatty acids or alcohols. For example, Schey [3.303] found that more than 5% lanolin or 2% rapeseed oil was needed to match the performance of 1% oleic acid. Even higher concentrations were needed in the work of Kleinjohann [404]. It is likely that much of the improvement should be ascribed to increased viscosity.

Compared to fatty derivatives, E.P. additives are ineffective, although occasional reference to them may be found—such as, for example, the inclusion of TCP in a formulation given by Vamos and Zakar [405].

The staining propensity of additives is a function of composition and chain length. Within any one homologous series, staining increases with chain length and with the polarity of the molecule. Thus, in the work of Guminski and Willis [5.264] the highest nonstaining molecules were: paraffin C_{14}, alcohol C_{10}, aldehyde C_8, and acid C_6. In combination with the reduction capacity of lubricants (as judged by the plane-strain compression test), they concluded that alcohols give the best compromise. However, many successful industrial lubricants contain acids or esters as the primary additives. The high reflectivity makes the appearance of aluminum strip very sensitive to lubrication conditions (Sec.6.3.5). The surface is generally typical of a mixed-film lubrication mechanism. More extensive transverse tearing close to the edges of the strip has been attributed to excessive oxide film thickness on alloys containing magnesium [385b].

Foil Rolling. A substantial portion of aluminum is rolled into foil. Production methods have been reviewed by Kerth et al [406]. Rolling takes place on four-high mills with 200- to 400-mm-diam rolls, under conditions of limiting reduction. Accurate tension control and excellent lubrication are primary requirements.

The lubricants employed are similar to those used in rolling of thin strips [407,408]. Heavy reductions of typically 50% per pass are taken. Finishing passes are made with a light lubricant (typically 6 cSt at 20 C; final boiling point, 260 C), partly to ensure the desired bright finish and partly to avoid damage to the foil in the last pass. Lubrication is again of the mixed-film type, as shown also by the experiments of Amann [406], who found that μ decreased with increasing oil viscosity and decreasing roll roughness. As in strip rolling, fatty oils (primarily palm-seed oil) were the traditional additives but have been replaced by fatty acids and alcohols.

For the thinnest gages, two strips are rolled together (pair rolling). Only the outer surfaces are bright; the contacting strip surfaces show the free deformation of a polycrystalline material [409]. To minimize waviness of the contact surface, process annealing is limited to low-temperature recovery anneals. Entrapment of lubricant of heavier viscosity can lead to perforation of the foil [182]. Such perforations sometimes extend in a line in the rolling direction, following the breakup of an individual oil drop into smaller droplets. A very thin oil film must, however, be maintained to prevent welding of the paired foil. Perforation could result also from the presence of wear debris in the lubricant.

Because rolling takes place in the mixed-film regime, speed variation is used in combination with tension control to achieve the desired foil thickness. Heat generation may be sufficient to cause some drop in the effective flow stress of the material in the roll gap [136,406], although the resulting strip still shows the consequences of heavy cold working. Because of the extreme thinness and great width (now reaching 2 m) of the product, control of the roll gap is of extreme importance, and particular attention must be paid to the appropriate distribution of the oil over the width of the roll barrels [231].

Lubricant Application and Control. The oil-base lubricant must perform both cooling and lubricating functions, and therefore it is invariably recirculated through a fairly complex system equipped with heaters, coolers, settling tanks, and filters (often coated with active earth). Tramp oils from bearings and hydraulic systems are dissolved in the oil and gradually increase its staining tendency; various efforts have been made to improve oil seals, to change the oils to nonstaining synthetic varieties, or to construct bearings that can operate satisfactorily with the very low-viscosity rolling lubricants. Staining is a particularly severe problem when a low-temperature (typically 180 to 260 C) recovery anneal is given. To reduce the oil carried over on the strip, air jets (air knives) are directed at the exit side of the strip, and sometimes felt wipers or edge rollers are used to soak up excess oil. Checks on viscosity, staining propensity, and additive content are routinely run and appropriate additions are made. Rolling performance is judged from production observations or tests on laboratory mills, although occasionally the use of a simulation test (an Amsler test with an aluminum disk) is reported [410].

Control of oil temperature is of great importance because excessive thinning of the oil can lead to lubricant breakdown, a very rapid rise in rolling load, and severe pickup on the roll surface [411].

Large quantities of fines are produced. According to the data of Ogura and Saeki (quoted by Kondo [393]), fines are formed at a higher rate with fatty acids than with alcohols or esters. Small particles (below 8 μm) cause blackening. Active earth filtration is used to keep the ash content typically below 0.02% to ensure a clean strip finish. For general purposes, 10 to 20% of the flow is treated; for foil rolling, 50 to 60%. Hard alloys are sometimes rolled on Sendzimir mills in which the rolling lubricants also lubricate the bearings [412]. For the protection of bearings, 30 to 60% of the total flow is passed through earth filters.

The oil is applied to the rolls at the rate of 1.5 to 3 m^3/min per metre of width of rolled strip at pressures of 150 to 1000 kPa. Rolling mills are usually fitted with exhaust hoods to remove oil vapors [413]. After the oil has been extracted, the exhaust air is released to the atmosphere. When highly flammable kerosene is used as a coolant, sufficient air must be entrained to ensure a noncombustible mixture. Fire extinguishers capable of quenching a full fire with a flow of CO_2 gas are incorporated and may be operated manually or automatically.

6.7.3 Lubrication of Magnesium and Beryllium Alloys

Magnesium forms a roll coating, although of lesser severity than that formed by aluminum. Hot rolling emulsion lubrication practices are the same as for aluminum. Warm rolling of strip is conducted with emulsions or mineral-oil-base lubricants.

The oxides of beryllium are toxic. The metal is canned in steel for hot rolling above 750 C; at lower temperatures it can be rolled bare. It does not seem to present particular lubrication problems. Presumably it can be rolled with an emulsion, although information on work practices could not be located.

6.8 Lubrication of Copper-Base Alloys

Slabs of copper or its alloys are either semicontinuously cast or cast into permanent or water-cooled molds. Slab thicknesses are as great as 200 mm, and these are broken down, usually but not always, by hot rolling to around 12 mm. Surface imperfections are then removed by milling the whole surface. Further rolling is carried out cold on single-stand or tandem mills. Alternatively, strip is cast to a thickness of 10 to 30 mm and cold rolled to finished gage.

Hot Rolling Lubrication. The oxides of copper are reasonably good lubricants and parting agents, and the oxides of brass are friable. Therefore, hot rolling is usually carried out with mineral-oil emulsions [414]. Only at low production rates will smearing with fat suffice.

Cold Rolling Lubrication. Transfer of copper onto the roll surface results in a red cast, but pickup is, in contrast to that with aluminum, noncumulative, and lubrication requirements are relatively modest. Nevertheless, lubrication is required for lower friction and, especially with the harder alloys, for reduced roll wear. Two techniques are practical:

a. Lubricants based on mineral oils [415,416] provide mixed-film lubrication. Boundary additives are relatively ineffective when the copper soaps are oil-soluble. Aggressive additives such as oleic acid may even increase friction in rolling of brass. Fats are sometimes incorporated and are of modest benefit, most likely because of their rheological properties. E.P. additives, especially sulfur-containing ones, could be effective but are avoided because of the unsightly staining they produce. At higher rolling speeds the film is thick enough to cause skidding, and the oil viscosity must be kept low (below 8 cSt at 40 C [5.272]). Therefore, rolling with emulsions is preferable.

b. Emulsions applied in a recirculating system are now generally used [414]. Emulsions are based on mineral oils, sometimes with fatty additives, and are on the slightly unstable side. The viscosity of the base oil is generally higher (around 400 cSt at 40 C) for initial breakdown passes and lower for finishing passes. Emulsion concentrations are also higher for initial heavier passes than for finishing ones. In the rolling of very thin strip, Shevelkin et al [417] indicate that a 2 to 4% emulsion concentration applied at a high rate to the mill gave the best surface quality and mill output.

Oil staining is of some concern although, because of the higher annealing temperatures, it is by no means as severe as in aluminum rolling. Well-adhering lubricant films are actually of benefit in preventing the multicolored staining that appears on brass as a result of oxidation and dezincification [418].

6.9 Lubrication of Titanium Alloys

Titanium and its alloys present one of the most formidable rolling tasks for two principal reasons. First, titanium transfers readily to the roll surface just as aluminum does, but in contrast to aluminum it does not react with common lubricants. Second, the hexagonal structure of α-Ti makes it very strong in the through-thickness direction and thus results in high cold rolling forces.

As far as possible, rolling is conducted hot at temperatures around 900 C, although the actual temperature depends on the type of alloy and on the metallurgical structure that is to be produced at the end of hot rolling. The oxide serves as a partial parting agent and, even though no reference could be found in the literature, it can be assumed that emulsions would serve as reasonably good lubricants. In dry rolling of titanium, Riegger [419] found that spread increased with temperature in the range from 800 to 1100 C, and that friction was high (from ring-compression tests, $m = 0.8$).

Below a thickness of approximately 2.5 mm the strip is processed cold, with frequent process anneals (typically after only 2 to 4% cold reduction). Mills with small-diameter work rolls, such as Sendzimir mills, are favored, because in these mills roll forces are lower and limiting reductions are not reached so soon. Remarkably little has been published about lubrication in cold rolling. Wilcox and Whitton [420] evaluated various lubricants in cold rolling of 0.9-mm-thick commercially pure titanium strip with light (4 to 5%) reductions. As shown in Fig.6.55, sulfur and chlorine additives (lubricant B) were ineffective relative to a straight mineral oil (A). A soap consisting of

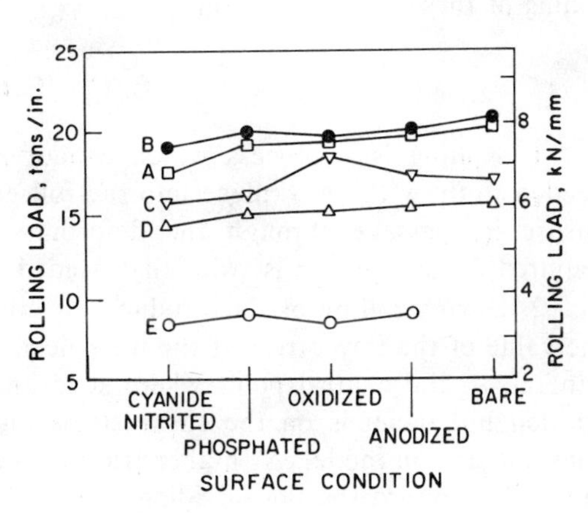

Fig.6.55. Efficiencies of lubricants in cold rolling of titanium at 10% reduction [420]. See text for identification of lubricants.

K-palmitate and stearate (C) was considerably better on all but the oxidized surface. Lanolin (D) was the best conventional lubricant, no doubt because of its high effective viscosity. A graphited chlorofluorinated hydrocarbon grease (E) was best of all but was judged unsuitable for industrial application. The importance of adhesion was shown by the observation that a bare titanium surface produced by pickling gave the highest friction, whereas a surface oxidized by annealing in air for 45 min at 700 C was only slightly worse than a fluoride-phosphate coating. A cyanide-nitrided surface was better yet, but the gains from surface treatment were small compared with improvements resulting from lubrication with high-viscosity substances. Fukuda et al [421] state, however, that oxidizing is effective in reducing cold rolling loads. Pavlov et al [422] found a synthetic compound consisting of esters with amino groups to be best, followed by castor oil, which was judged most suitable for thinner gages. A soap glue containing MoS_2 was also found to be useful [423].

6.10 Lubrication of Other Metals

Refractory metals such as tungsten, molybdenum, niobium, tantalum, and their alloys present a number of rolling difficulties [424], but few of them are associated with friction. Many of these metals show limited ductility in the as-cast condition or are more conveniently cast into round ingots. A sheet bar is then extruded or forged, after which it is further hot rolled into sheet. True hot working temperatures are high, between 1600 and 1900 C, but many alloys can be effectively warm worked between 900 and 1200 C. Periodic recrystallization anneals are then needed. Generally, the working temperature may be gradually lowered as the cast structure is eliminated. The ductile-to-brittle transition temperature of tungsten drops with the amount of prior work. Warm rolling at around 400 C is favorable because neither gas pickup nor oxidation occurs and because conventional lubricants may be used. Organic polymers such as PTFE have been found to be very effective in reducing roll forces in rolling at 400 C and at room temperature. At room temperature even mineral oils with boundary and E.P. additives are useful [425].

Vanadium, vanadium alloys, and zirconium are readily rolled either hot or cold.

Zinc is rolled at 150 to 250 C with an occasional smear of a mineral- or fatty-oil lubricant, with emulsions, or even on dry rolls. Cold rolling to thinner gages benefits from use of a fatty oil - light mineral oil mixture as a coolant and lubricant.

The soft metals tin and lead were the first materials to be rolled and in small quantities require no lubrication. Light mineral oils or dilute emulsions serve well in rolling of these metals into foil.

6.11 Summary

1. Rolling is a process which cannot be conducted without friction. Friction is needed to draw the workpiece into the roll gap and, once the workpiece is accepted, to ensure its passage through the deformation zone. The minimum value of friction required for acceptance is twice that needed for continuous rolling.

2. During rolling of flat products, friction causes interface pressures to rise above the value of the flow stress of the workpiece material. The apex of the friction hill is, in general, at the neutral point where strip and roll speeds are equal. The shape of the friction hill depends on the shear stress distribution, which seldom follows a simple constant μ or m model. At higher friction, a sticking zone develops and the shear stress diminishes toward the flow-dividing plane.

3. The effects of friction are interwoven with the effects of pass geometry, expressed as the L/h ratio, where L is the projected length of the arc of contact and h is the mean strip thickness. At $L/h > 2$, deformation is nearly homogeneous; pressures, roll force, torque, and roll flattening increase with increasing friction, and a limiting strip thickness may be reached. At $L/h < 2$, the inhomogeneity of deformation becomes significant, and at $L/h < 1$ it becomes the dominant factor, relegating friction to a subordinate role. Sticking friction often prevails.

4. The position of the neutral plane is a sensitive measure of the frictional balance in the roll gap; with decreasing friction, the neutral plane moves toward the exit and the difference between strip exit speed and roll speed (forward slip) decreases. Furthermore, when friction is too low, the difference between these two speeds may even become negative; the strip skids, and the stability of the operation is lost.

5. Friction, together with pass geometry, determines the homogeneity of deformation. Friction increases lateral spread and, through its effect on heat generation, contributes also to the development of a thermal camber. This, in turn, affects longitudinal inhomogeneity, which manifests itself in poor strip shape.

6. Lubricants are applied primarily to reduce friction and wear, and to ensure temperature control. The relative importance of these functions depends on the process and the workpiece material.

7. In cold rolling, one of the main purposes of lubrication is that of reducing friction. Because of the need for some minimum friction, hydrodynamic lubrication is impractical, but its theory is of value in predicting the thickness of the film entrained for mixed-film lubrication. The gradually converging entry zone, periodic contact with the roll surface, and limited sliding in the contact zone make rolling responsive to lubrication with fluids. As expected from basic principles, friction in the mixed-film lubrication regime decreases with decreasing roll entry angle (smaller reduction and larger roll diameter), increasing viscosity, and increasing speed. Decreasing friction means lower forces and, for a given roll gap setting, thinner issuing gage. Thus, on acceleration of the mill, the strip thickness decreases; this speed effect is reinforced by changes in oil film thickness in mills equipped with full-fluid-film bearings.

8. With increasing pass reduction, lubrication shifts from predominantly hydrodynamic to predominantly boundary in character. Forward slip increases and roll forces rise; the rise in roll force becomes very steep once conditions of limiting reduction are approached. Very heavy reductions and/or high speeds thin the film to the point where breakdown occurs, resulting in heavy smearing and roll pickup with materials prone to adhesion (such as aluminum) and the development of a localized heat streak defect with other materials (such as steel). Increasing roll surface roughness generally shifts the mechanism toward the boundary regime but can be beneficial in delaying the onset of the heat streak defect.

9. The surface finish of the strip is controlled by the roll surface finish and the prevailing lubrication mechanism. A bright finish is prompted by poor lubrication, a large roll radius, and smooth rolls.

10. Rolling is a dynamic process and periodic changes in the lubrication mechanism, as well as sudden events, can initiate oscillations which result in periodic changes of the strip surface finish, sometimes accompanied by variations in strip thickness.

11. The purpose of lubrication in hot rolling depends on the adhesion between workpiece and roll materials. If adhesion is high, as in rolling of aluminum, lubrication is essential for the control of roll coating buildup. If adhesion is low, as in rolling of steel, the cooling and wear-reducing functions of the lubricant prevail. At high L/h ratios, lubricants aid in reducing roll force and power.

12. Lubricants may be of an oil or a water base. Because of the predominance of the mixed-film mechanism, boundary and/or E.P. additives are essential. Oil-base lubricants are used, in general, only when water staining would develop, as in cold rolling of aluminum, or when an aqueous lubricant would prove inadequate in terms of reduction of friction or prevention of adhesion. The lubricating power of emulsions can be increased by reducing their stability. It is possible also to combine the high lubricating power of a neat fat or oil with the high cooling power of water by applying the lubricant to the strip and the water to the roll, or by mechanically mixing the two at the point of application.

13. The same principles apply to the rolling of sections (shapes) and tubes, except that the spatial configuration of the deformation zone becomes complex, and large speed differentials develop around the circumference of the section. Filling of the shape is generally hindered by friction, but friction is actually helpful in filling a leg formed by indirect draft.

14. Roll wear is a clearly observable phenomenon which leads to changes in surface finish and to a loss of tolerances or product shape. In cold rolling the dominant mechanism is abrasive and adhesive wear accompanied by spalling, a form of fatigue wear. In hot rolling, abrasive and thermal fatigue wear dominate. Lubrication then serves to reduce abrasive and adhesive wear and to minimize the thermal excursions that culminate in thermal fatigue.

15. The most frequently used lubricants are tabulated in Table 6.3. For purposes of preliminary, approximate calculations, typical average coefficient of friction values are also shown, although it must be understood that, because of the prevalence of the mixed-film lubrication mechanism, these values change greatly with process conditions.

Table 6.3. Commonly used rolling lubricants and typical μ values

	Hot rolling		Cold rolling	
Material	Lubricant(a)	μ(b)	Lubricant(a)	μ(c)
Steels	None + cooling water	ST	Emulsion, 3-6% conc., of synthetic palm oil	MF
	Emulsion of fat (+E.P.)	0.4	Synthetic palm oil + water	MF
	Fat (ester) (+E.P.) + cooling water	0.3		
Stainless steels and Ni alloys	As for steel		M.O. (10-20) with Cl additive	MF
Al and Mg alloys	Emulsion, 2-15% conc., of M.O. (20-100) with < 20% fatty acid, alcohol, ester	0.4	M.O. (4-20) with 1-5% fatty acid, alcohol, ester	MF
			As above, but synthetic mineral oil	MF
			Foil: as above, but (1.5-6)	MF
Cu and Cu alloys	Emulsion, 2-8% conc., of M.O. (80-400) with fat	0.3	Emulsion, 2-10% conc., of M.O. (80-400) with fat	MF
			M.O. (8-50) with fat (fatty acid)	MF
Ti alloys	None	ST	Oxidized surface, with:	
	Fat (+water)	ST	esters or soaps	0.2
			castor oil (fatty oil)	0.2
			Compounded M.O. (4-10)	0.2
Refractory metals	Canning + lubricant for can material	0.4	M.O. with boundary and E.P. agents	MF
	Bare, dry	0.3		

(a) M.O. = mineral oil; viscosity in cSt at 40 C in parentheses. (b) ST = sticking friction. (c) MF = mixed-film lubrication; $\mu = 0.10$ at low speeds, dropping to 0.03 at high speeds and viscosities.

References

[1] J.A. Schey, in [3.1], pp.22-37, 247-256, 333-456.
[2] C.W. Starling, *The Theory and Practice of Flat Rolling*, London University Press, 1962.
[3] E.C. Larke, *The Rolling of Strip, Sheet and Plate,* 2nd Ed., Chapman and Hall, London, 1965.
[4] A. Tselikov and V.V. Smirnoff, *Stress and Strain in Metal Rolling,* Pergamon, Oxford, 1965; MIR Publishers, Moscow, 1967.
[5] Z. Wusatowski, *Fundamentals of Rolling*, Pergamon, Oxford, 1969.
[6] W.L. Roberts, *Hot Rolling of Steel*, Dekker, New York, 1983.
[7] W.L. Roberts, *Cold Rolling of Steel*, Dekker, New York, 1978.
[8] Various, Metals Techn., *2*, 1975, 90-152, 282-403, 450-503.
[9] L.G.M. Sparling, Int. Met. Rev., *22*, 1977, 303-313.
[10] A.P. Grudev, *External Friction During Rolling,* Metallurgiya, Moscow, 1973 (Russian).
[11] P.I. Polukhun, *Contact Interaction of Metal and Tool During Rolling,* Metallurgiya, Moscow, 1974 (Russian).
[12] B. von Kortzfleisch, Stahl u. Eisen, *87*, 1967, 424-432 (German).
[13] B. von Kortzfleisch, O. Pawelski, and U. Krause, Stahl u. Eisen, *87*, 1967, 588-597 (German).
[14] W. Dobrucki and E. Odrzywolek, J. Mech. Work. Tech., *4*, 1980, 263-284 (French).
[15] O. Pawelski, Stahl u. Eisen, *83*, 1963, 1440-1451 (German).
[16] F.A.R. Al-Salehi, T.C. Firbank, and P.R. Lancaster, Int. J. Mech. Sci., *15*, 1973, 693-710.
[17] A. Knauschner, Stahl u. Eisen, *88*, 1968, 35-38 (German).
[18] I.Ya. Tarnovskii, A.A. Pozdeyev, and V.B. Lyashkov, *Deformation of Metals During Rolling,* Pergamon, New York, 1965.
[19] J.A. Schey, J. Appl. Metalwork., *1* (2), 1980, 48-59.
[20] H. Ford, Met. Rev., *2*, 1957, 1-28.
[21] Th. von Karman, Z. Angew. Math. Mech., *5*, 1925, 139-141.
[22] E. Orowan, Proc. Inst. Mech. Eng., *150*, 1943, 140-167.
[23] J.M. Alexander, Proc. Roy. Soc. (London), *A326*, 1972, 535-563.
[24] J.G. Lenard, in *Proc. COBEM 81*, Rio de Janeiro, 1981.
[25] U. Stahlberg, Scand. J. Metall., *7*, 1978, 42-48.
[26] G.J. Li and S. Kobayashi, Trans. ASME, Ser.B, J. Eng. Ind., *104*, 1982, 55-64.
[26a] P.F. Thomson and J.H. Brown, Int. J. Mech. Sci., *24*, 1982, 559-576.
[27] R.A. Griffin and J.A. Schey, unpublished work.
[28] T. Arimura, M. Kamata, and J. Kihara, in *Proc. Int. Conf. Sci. Tech. Iron Steel,* Tokyo, 1971, pp.659-662.
[29] J.H. Hitchcock, ASME Research Publication, *Roll Neck Bearings*, 1935.
[30] A. Atreya and J.G. Lenard, Trans. ASME, Ser.H, J. Eng. Mat. Tech., *101*, 1979, 129-134.
[31] M.D. Stone and J.I. Greenberger, Iron Steel Eng., *20* (2), 1943, 61-68, 72.
[32] H. Ford, F. Ellis, and D.R. Bland, J. Iron Steel Inst., *168*, 1951, 57-72.
[33] T. Tamano and S. Yanagimoto, Bull. JSME, *21*, 1978, 592-599.
[34] R.B. Sims, Proc. Inst. Mech. Eng., *168*, 1954, 191-200.
[35] B.K. Denton and F.A.A. Crane, J. Iron Steel Inst., *210*, 1972, 606-617.
[36] D.J. McPherson, Met. Trans., *5*, 1974, 2479-2499.
[37] W.L. Roberts, R.J. Bentz, and D.C. Litz, Iron Steel Eng., *47* (8), 1970, 77-84.
[38] O. Pawelski and G. Käding, Arch. Eisenhüttenw., *41*, 1970, 249-254 (German).
[39] J. Billigmann, K. Kottmann, and P. Funke, Stahl u. Eisen, *91*, 1971, 629-64 (German).
[40] G.K. Cheng, Light Metal Age, *37* (9-10), 1979, 6, 8, 10-12.
[41] H. Ford and J.M. Alexander, J. Inst. Metals, *88*, 1959-60, 193-199.
[42] O. Pawelski and G. Käding, Stahl u. Eisen, *87*, 1967, 1340-1355 (German).
[43] H.A. Kuhn and A.S. Weinstein, Trans. ASME, Ser.F, J. Lub. Tech., *93*, 1971, 331-341.
[44] J.A. Schey, J. Inst. Metals, *89*, 1960-61, 1-6.
[45] P.W. Whitton and H. Ford, Proc. Inst. Mech. Eng., *169*, 1955, 123-140.
[46] I.M. Pavlov, *Theory of Rolling and Fundamentals of Metalworking,* Gonti, Moscow, 1938 (Russian).
[47] T. Dahl, Stahl u. Eisen, *57*, 1937, 205-200.

[48] H.G. Müller and P. Funke, Stahl u. Eisen, *78*, 1958, 1564-1574.
[49] V.A. Nikolaev and V.F. Zverev, Steel USSR, *4*, 1974, 985-986.
[50] W.L. Roberts, Iron Steel Eng., *45* (5), 1968, 128-136.
[51] O. Pawelski, Arch. Eisenhüttenw., *42*, 1971, 713-720 (German).
[52] F.L. Reynolds, Lubric. Eng., *14*, 1958, 98-103, 120.
[53] S. Ekelund, Steel, *93* (Aug.21), 1933, 27-29.
[54] W. Lueg and A. Pomp, Mitt. Kaiser Wilhelm Inst. Eisenforsch., *21*, 1939, 163-170 (German).
[55] S. Kondo, Trans. ASME, Ser.F, J. Lub. Tech., *97*, 1975, 37-43.
[56] H. Inhaber, Trans. ASME, Ser.B, J. Eng. Ind., *88*, 1966, 421-429.
[57] W. Evans and B. Avitzur, Trans. ASME, Ser.F, J. Lub. Tech., *90*, 1968, 72-80.
[58] J.M. Capus and M.G. Cockroft, J. Inst. Metals, *92*, 1963-64, 31-32.
[59] K. Saeki and Y. Hashimoto, J. JSTP, *9*, 1968, 253-257 (Japanese).
[60] F.A.A. Crane and J.M. Alexander, J. Inst. Metals, *91*, 1962-63, 188-189.
[61] B.B. Hundy and A.R.E. Singer, J. Inst. Metals, *83*, 1954-55, 401-407.
[62] N.G. Bochkov, Yu.V. Lipukhin, L.I. Butylkina, and V.V. Ryzkov, Stal', *1975* (9), 825-827 (Russian).
[63] A.K. Grigor'ev et al, Steel USSR, *7*, 1977, 93-94.
[64] C.R. Heiple, Paper 73-MAT-F, ASME, New York, 1973.
[65] V.I. Znamensky, I.Ya. Tarnovsky, and V.B. Lyashkov, Tsvetn. Met., *1970* (4), 162-165 (Russian).
[66] G.D. Lahoti, N. Akgerman, S.I. Oh, and T. Altan, J. Mech. Work. Tech., *4*, 1980, 105-119.
[67] I. Pavlov and Y. Gallai, Metallurg, *12* (9-10), 1937, 107-111 (Brutcher Transl. 1219).
[68] M. Vater, G. Nebe, and J. Petersen, Stahl u. Eisen, *86*, 1966, 710-720 (German).
[69] T. Sheppard and D.S. Wright, Metals Techn., *8*, 1981, 46-57.
[70] F. Kasz and P.C. Varley, J. Inst. Metals, *76*, 1949-50, 407-428.
[71] T. Sheppard, Sheet Metal Ind., *55*, 1978, 1222-1229; *56*, 1979, 1149-1154.
[72] D.H. Sansome, Metals Techn., *2*, 1975, 522-531.
[73] *Shape Control* (Proc. Conf.), Metals Society, London, 1976.
[74] W.K.J. Pearson, in [73], pp.46-54.
[75] W.E. Davies, O.G. Sivilotti, and M.W. Tulett, Metals Techn., *2*, 1975, 494-498.
[76] G.M. Sturgeon and R.L.S. Taylor, *Metal Strip from Powder,* Mills and Boon, London, 1972.
[77] V.P. Katashinskii, Sov. Powder Met. Met. Cer., *18*, 1979, 155-160.
[78] K.M. Radchenko and P.L. Klimenko, Sov. Powder Met., *14*, 1975, 708-712.
[79] W. Gorecki, in [3.45], pp.431-437.
[80] A. Ohm, in [3.51], pp.327-330.
[81] H. Henschen, Bänder Bleche Rohre, *15*, 1974, 66-69 (German).
[82] R.J. Williams and G.W. Williams, Trib. Int., *11*, 1978, 18-19.
[83] A. Nadai, Trans. ASME, J. Appl. Mech., *6*, 1939, 54.
[84] H. Cheng, in [3.50], pp.69-89.
[85] W.L. Roberts, U.S. Steel Corp. Report, 1967.
[86] D.S. Bedi and M.J. Hillier, Proc. Inst. Mech. Eng., *182*, Pt.1, 1967-68, 153-162.
[87] D.S. Bedi and M.J. Hillier, Int. J. Mech. Sci., *12*, 1970, 827-836.
[88] B. Avitzur and G. Grossmann, Trans. ASME, Ser.B, J. Eng. Ind., *94*, 1972, 317-328.
[89] J.A. Walowit, in [3.50], pp.271-274.
[90] W.R.D. Wilson and L.E. Murch, Trans. ASME, Ser.F, J. Lub. Tech., *98*, 1976, 426-432.
[91] B.B. Aggarwal and W.R.D. Wilson, in [3.258], pp.351-359.
[92] C.T. Chang and W.R. Wilson, in [3.56], Paper No.52.
[93] A.G. Atkins, Int. J. Mech. Sci., *16*, 1974, 1-19.
[94] T.A. Dow, J.W. Kannel, and S.S. Bupara, Trans. ASME, Ser.F, J. Lub. Tech., *97*, 1975, 4-13.
[95] A. Azushima, in [3.55], pp.159-166.
[96] J. Ichinoi and G. Tomizawa, in [3.51], pp.331-355.
[97] A.P. Grudev and O.P. Maksimenko, Obr. Met. Davl., *57*, 1971, 209-221 (Russian).
[98] G.L. Kolmogorov and E.D. Kolegova, Izv. VUZ Chern. Met., *1973* (12), 76-79 (BISI Transl. 12219).
[99] E. Felder, Collect. Colloq. Semin., Inst. Franc. Petrol., *29*, 1975, 163-178 (French).

[100] A. Azushima, in [3.52], pp.81-87.
[101] T. Kamita and T. Yoshida, J. JSLE, *21*, 1976, 819-823 (Japanese).
[102] T. Mizuno, K. Matsubara and H. Kimura, Bull. JSME, *12*, 1969, 359-367.
[103] P. Tsao and W.R.D. Wilson, in *Proc. Int. Conf. Steel Rolling,* Iron Steel Inst. Japan, Tokyo, 1980.
[104] T. Mizuno, J. JSTP, *7*, 1966, 383-389 (Japanese).
[105] S. Ogura, J. JSTP, *20*, 1979, 215-221 (Ann. Rep. *1*, 1980, 34).
[106] L.B. Sargent, Jr., and C.J. Stawson, Trans. ASME, Ser.F, J. Lub. Tech., *96*, 1974, 617-630.
[107] H. Ford, J. Iron Steel Inst., *156*, 1947, 380-398.
[108] J. Billigmann and A. Pomp, Stahl u. Eisen, *74*, 1954, 441-461 (German).
[109] J.M. Thorp, Proc. Inst. Mech. Eng., *175*, 1961, 593-603.
[110] L.D. Dromgold and S. Rodman, Lubric. Eng., *26*, 1970, 26-31.
[111] Y. Iwao, H. Hirano, and I. Kokubo, J. JSTP, *8*, 1967, 248-255.
[112] J.C. Whetzel, Jr., and S. Rodman, Iron Steel Eng., *36* (3), 1959, 123-132.
[113] R.C. Williams and R.K. Brandt, Lubric. Eng., *20*, 1964, 52-56.
[114] V.A. Nikolaev and V.S. Dolya, Steel USSR, *5*, 1975, 222-223.
[115] V.A. Nikolaev, Izv. VUZ. Chern. Met., *1978* (1), 117-119 (Russian).
[116] A.P. Grudev, Yu.V. Zilberg, and V.A. Bondarenko, Steel USSR, *1*, 1971, 55.
[117] O. Pawelski, Rheol. Acta, *2*, 1962, 273-280 (German).
[118] W.C.F. Hessenberg and R.B. Sims, Proc. Inst. Mech. Eng., *166*, 1952, 75-90.
[119] G.E. Stoltz and J.W. Brinks, Iron Steel Eng., *27* (10), 1950, 69-88.
[120] H.G. Müller and W. Lueg, Stahl u. Eisen, *79*, 1959, 325-331 (German).
[121] A.N. Popov, Stal, *1966* (4), 296-298.
[122] Various, in [3.51], pp.83-131.
[123] P.L. Charpentier, R.H. Goodenov, and D.J. Miller, Metals Techn., *2*, 1975, 52-61.
[124] R.B. Sims and D.F. Arthur, J. Iron Steel Inst., *172*, 1952, 285-295.
[125] R.B. Sims, J. Inst. Petr., *40*, 1954, 314-318.
[126] P.W. Whitton, Australas. Eng., *46*, 1955, 81-127.
[127] A. Azushima, J. Kihara, and I. Gokyu, J. JSTP, *18*, 1977, 337-343 (Japanese).
[128] J.G. Wistreich, in Discussion to [127].
[129] J.A. Schey, in Discussion to [109].
[130] M.D. Stone, Iron Steel Eng., *30* (2), 1953, 61-74.
[131] J.C. Whetzel and C. Wyle, Metal Progr., *70* (6), 1956, 73-76.
[132] D.I. Starchenko et al, Steel USSR, *2*, 1972, 982-984.
[133] M.D. Stone, Iron Steel Eng., *38* (6), 1961, 67-77.
[134] O. Pawelski, in [3.36], pp.80-92.
[135] J. Billigmann and J. Lenze, Stahl u. Eisen, *82*, 1962, 313-333 (German).
[136] Y. Yokote and S. Nomura, J. Inst. Metals, *88*, 1959-60, 241-254.
[137] D.G. Christopherson and B. Parsons, in *Proc. Conf. Properties of Materials at High Rates of Strain,* Inst. Mech. Eng., London, 1957, pp.115-121.
[138] Y. Shibata, K. Nakajima, and Y. Uehori, in [3.55], pp.171-181.
[139] Y. Tamai, M. Mizoguchi, K. Nakajima, and Y. Shibata, in [3.52], pp.88-91.
[140] T. Mizuno and Y. Rokushika, Bull. JSME, *21*, 1978, 341-348.
[141] V.N. Zhuchin and I.M. Pavlov, Stal in English, *1963* (3), 201-203.
[142] W.L. Roberts, in [3.50], pp.103-121.
[143] I. LeMay and F.R. Vigneron, Lubric. Eng., *21*, 1965, 276-281.
[144] I. LeMay, K.D. Nair, and M.L.J. Hoffman, Lubric. Eng., *23*, 1967, 415-418.
[145] W.L. Roberts, Blast Furn. Steel Plant, *56*, 1968, 382-394.
[146] K. Takahashi, K. Nakajima, and K. Murata, in [3.50], pp.137-146.
[147] P.R. Thomson and J.S. Hoggart, J. Australian Inst. Metals, *12*, 1967, 189-198.
[148] J. Loose et al, Neue Hütte, *24*, 1979, 336-338 (German).
[149] I. Gokyu, J. Kihara, and T. Arimura, J. JSTP, *14*, 1973, 160-167 (Japanese).
[150] J. Shintaku et al, in [3.55], pp.187-192.
[151] K. Kitamura et al, in *Proc. 30th Conf. JSTP,* 1979, pp.81-84 (Japanese).
[152] K. Sato, H. Uesugi, I. Hagiwara, and T. Nagami, J. JSTP, *19*, 1978, 395-402 (Japanese).
[153] J. Billigmann, Stahl u. Eisen, *75*, 1955, 1691-1705 (German).
[154] K.A. Lloyd, in Discussion to [109].
[155] J. Kihara, H. Nagamori, H. Matsuda, and Y. Nakagawa, in [3.55], pp.177-181.

[156] T. Sakai, Y. Saito, A. Okamoto, and K. Kato, J. JSTP, *19*, 1978, 972-979 (Japanese).
[157] V.A. Nikolaev, P.P. Kalashnikov, A.P. Aleshin, and A.F. Primenov, Steel USSR, *5*, 1975, 666-668.
[158] D.I. Starchenko, V.I. Kuzmin, S.A. Obernikhin, and V.I. Kaplanov, Stal in English, *1966* (2), 128-131.
[159] I. Yarita and K. Nakagawa, J. JSLE, *19*, 1974, 45-54 (Japanese).
[160] P. Funke and J. Marx, Arch. Eisenhüttenw., *45*, 1974, 603-608 (German).
[161] J. Marx and P. Funke, Arch. Eisenhüttenw., *45*, 1974, 597-602 (German).
[162] J. Kihara, H. Kawanaka, H. Nagamori, and H. Matsuda, J. JSTP, *20*, 1979, 323-330 (Japanese).
[163] W. Lueg and P. Funke, Stahl u. Eisen, *78*, 1958, 333-343 (German).
[164] J. Billigmann and W. Fichtl, Stahl u. Eisen, *78*, 1958, 344-355 (German).
[165] D.I. Starchenko et al, Steel USSR, *5*, 1975, 439-442.
[166] Y.H. Tsao and L.B. Sargent, Jr., ASLE Trans., *21*, 1978, 20-24.
[167] C.F. May, Sheet Metal Ind., *44*, 1967, 803-816, 841.
[168] R.H. Tressler, in *Mechanical Working of Steel,* Vol.1, Gordon and Breach, New York, 1964, pp.386-404.
[169] W. Panknin and H. Kranenberg, Stahl u. Eisen, *88*, 1968, 484-491 (German).
[170] J. Buza and P. Bartl, Bänder Bleche Rohre, *17*, 1976, 21-23 (German).
[171] M.M. Saf'yan and G.T. Fokin, Obr. Met. Dav., *1971* (56), 176-184 (Russian).
[172] H. Ford and J.G. Wistreich, J. Inst. Metals, *82*, 1953-54, 281-290.
[173] V.L. Mazur, V.I. Timoshenko, and I.E. Varivoda, Izv. VUZ Chern. Met., *1977* (8), 92-96; (12), 72-76 (Russian).
[174] A.P. Grudev and Yu.V. Zil'berg, Obr. Met. Davl., *1971* (56), 184-191 (Russian).
[175] D.I. Starchenko, V.I. Kaplanov, and V.M. Shapko, Izv. VUZ Chern. Met., *1972* (2) 93-95 (Russian).
[176] H. Meisel, Z. Metallk., *61*, 1970, 323-325 (German).
[177] V.P. Severdenko, L.I. Gursky, and N.V. Rumak, Dokl. Akad. Nauk Belor. SSR, *16*, 1972, 798-800 (Russian).
[178] H.F. Atala and G.W. Rowe, Wear, *32*, 1975, 249-268.
[179] A. Backmann, CIRP, *17*, 1969, 427-434.
[180] A. Azushima and M. Miyagawa, J. JSLE, *24*, 1979, 94-97 (Japanese).
[181] W. Gorecki and K. Mniszek, Prace Inst. Hutn., *22*, 1970, 23-29 (BISI Transl. 8983).
[182] P.F. Thomson, J. Australian Inst. Metals, *15*, 1970, 34-46.
[183] W.L. Roberts, in [3.51], pp.320-326.
[184] V.F. Molodychenko et al, Sov. Powder Metall. Met. Cer., *14*, 1976, 617-622.
[185] W.L. Roberts, Iron Steel Eng., *55* (10), 1978, 41-47.
[186] W. Trinks, Blast Furn. Steel Plant, *25*, 1937, 285-288, 713-715.
[187] R.H. Moller and J.S. Hoggart, J. Australian Inst. Metals, *12*, 1967, 155-165.
[188] G. Monaco, AISE Yearbook, 1977, pp.399-410.
[189] I. Yarita et al, Trans. Iron Steel Inst. Jpn., *18*, 1978, 1-10.
[190] D.R. Bland and H. Ford, Proc. Inst. Mech. Eng., *159*, 1948, 144-153.
[191] W.L. Roberts, in *Proc. Conf. Steel Rolling,* Iron Steel Inst. Japan, Tokyo, 1980, pp.1215-1224.
[191a] J. Tlusty, G. Chandra, S. Critchley, and D. Paton, CIRP, *31*, 1982, 195-199.
[192] K. Misonoh, J. JSTP, *21*, 1980, 1006-1010 (Ann. Rep. *2*, 1981, 69).
[193] D. Wünsch, W. Laumann, and H. van de Löcht, Arch. Eisenhüttenw., *49*, 1978, 575-580 (German).
[194] P.C. Jain and V. Srinivasan, Wear, *31*, 1975, 219-225.
[195] J.B. Thickins and C.E. Salmon, in [3.51], pp.306-316.
[196] W. Gregory, in [3.51], pp.297-305.
[197] J.H. Spalton, in [3.51], pp.317-319.
[198] H.J. Kopineck and W. Tappe, Arch. Eisenhüttenw., *43*, 1972, 489-494 (German).
[199] F.J. Lenze and W. Stoll, Stahl u. Eisen, *93*, 1973, 421-429 (German).
[200] V.N. Brinza et al, Stal', *1975* (11), 1013-1016 (Russian).
[201] C.R. Lillie and D.W. Levinson, Iron Steel Eng., *34* (5), 1957, 69-73.
[202] F.V. Schossberger, K. Hattori, and H. Marver, Iron Steel Eng., *34* (5), 1957, 74-81.
[203] R. Strefford and N. McCallum, in [3.51], pp.313-316.
[204] Y. Tamai and M. Sumitomo, Lubric. Eng., *31*, 1975, 81-83.
[205] A.H. Cubitt and A.R. Eyres, in [3.51], pp.356-360.

[206] A.P. Grudev and Yu.V. Zil'berg, Obr. Met. Davl., *1971* (57), 236-242 (Russian).
[207] I.S. Morton, Ind. Lub. Trib., *25*, 1973, 141-148.
[208] J. Petrik and V. Vach, Neue Hütte, *22*, 1977, 547-551 (German).
[209] R.C. Chaturvedi, in [3.56], Paper No.56.
[210] J.C. Benedyk, Light Metal Age, *29* (1-2), 1971, 6-10, 12.
[211] L.B. Sargent, Light Metal Age, *33* (5/6), 1975, 15-17.
[212] A. Azushima, in [3.56], Paper No.54.
[213] H. Miyagawa, S. Kusunoki, T. Sakai, and F. Hirano, in [3.56], Paper No.90.
[214] P. Funke and J. Marx, Arch. Eisenhüttenw., *45*, 1974, 585-590 (German).
[215] G. Fleischer and J. Behringer, Schmierungstechnik, *8* (3), 1977, 77-82 (German).
[216] J. Marx, Arch. Eisenhüttenw., *45*, 1974, 609-610 (German).
[217] K. Tsuji et al, J. JSTP, *21*, 1980, 67-72 (Ann. Rep. *2*, 1981, 19).
[218] A. Sethuramiah, S.D. Phatak, and B.M. Shukla, in [3.56], Paper No.58.
[219] W.R. Brooks, D.S. Harries, and D.B. Jess, in [3.53], pp.123-130.
[220] P. Funke, Stahl u. Eisen, *77*, 1957, 1836-1838 (German).
[221] H. Hobert et al, Schmierungstechnik, *10*, 1979, 368-371 (German).
[222] E.E. McDole, Lubric. Eng., *27*, 1971, 91-94.
[223] J.R. Ludwig, Blast Furn. Steel Plant, *57*, 1969, 641-651.
[224] P. Funke, I. Piorko, and W. Mikulla, Stahl u. Eisen, *94*, 1974, 850-854 (German).
[225] P. Funke, I. Piorko, and R. Wittek, Arch. Eisenhüttenw., *45*, 1974, 591-596 (German).
[226] R.J. Bentz and R.R. Somers, Lubric. Eng., *21*, 1965, 59-64.
[227] M.G. Vucich and M.X. Vitellas, Iron Steel Eng., *53* (12), 1976, 29-38.
[228] V.G. Dodoka et al, Steel USSR, *5*, 1975, 390-391.
[229] H. Reihlen, Stahl u. Eisen, *92*, 1972, 204-209 (German).
[230] L.R. Selig, Proc. Inst. Mech. Eng., *197*, Pt.3D, 1964-65, 151-158.
[231] B. König and H. Nordhoff, Neue Hütte, *17*, 1972, 275-277 (German).
[232] H.E. Riester, Lubric. Eng., *21*, 1965, 513-517.
[233] Anon., Iron Steel Eng., *43* (3), 1966, 157-158.
[234] P. Wehber, in [3.56], Paper No.57 (German).
[235] J.O. McLean, Lubric. Eng., *11*, 1955, 337-339.
[236] J.D. Lykins, Iron Steel Eng., *52* (4), 1975, 72-74.
[237] J.A. Robertson, Metal Ind. (London), *104*, 1964, 568-569.
[238] E.G. Nikolaeva et al, Tekhn. Legk. Splavov N-T Byul. VILSa, *1977* (9), 21-24 (Russian).
[239] D.A. Melford, V.B. Nileshwar, R.E. Royce, and M.E. Giles, J. Iron Steel Inst., *210*, 1972, 163-167.
[240] A.N. Anderson, Iron Steel Eng., *56* (4), 1979, 39-41.
[241] G. Lucas, in [3.51], pp.361-362.
[242] M.S. Dean, Lubric. Eng., *35*, 1979, 383-389.
[243] K. Sakabe and H. Tabe, Trans. Iron Steel Inst. Jpn., *13*, 1973, 38-47.
[244] J.C. Thieme and S. Ammareller, *Rolling Mill Rolls,* Climax Molybdenum Co., New York, 1966.
[245] C.D. Hopkins, R.J. Bassett, and O. Meijer, in *Proc. Mech. Work. Steel Proc. Conf. XVIII,* pp.22-30.
[246] S. Iketaka and K. Nohara, Tetsu-to-Hagane, *57*, 1971, 795-807 (Japanese).
[247] J. Blum, Wire, *22*, 1972 (117), 1-5.
[248] G.S. Spenceley, Sheet Metal Ind., *42*, 1965, 408-416.
[249] H.E. McGannon (ed.), *The Making, Shaping and Treating of Steel*, 10th Ed., U.S. Steel Corp., Pittsburgh, 1982.
[250] H. Pannek, Stahl u. Eisen, *75*, 1955, 767-769 (German).
[251] C.L. Robinson and F.J. Westlake, in [3.45], pp.389-398.
[252] A.P. Chekmarev, A.P. Grudev, and Yu.V. Zil'berg, Izv. VUZ Chern. Met., *1964* (11), 131-136 (Brutcher Transl. 6471).
[253] N.A. Chelyshev et al, Izv. VUZ Chern. Met., *1978* (2), 74-78 (Russian).
[254] K.M. Kulkarni and J.A. Schey, Paper No. MF 69-161, SME, Dearborn, 1969.
[255] J. Ichinoi and G. Tomizawa, in [3.46], pp.538-545.
[256] A.P. Chekmarev et al, Obr. Met. Davl., *1971* (57), 228-231 (Russian).
[257] F. Weber and H. Ahrens, Stahl u. Eisen, *92*, 1972, 519-527 (French, German).
[258] F.J. Westlake, C.L. Robinson, and C. Webb, Schmiertechn. Trib., *21*, 1974, 51-54 (German).

[259] K. Nishizawa, T. Mase, N. Hase, and T. Kono, in [3.46], pp.555-563.
[260] K. Sato, H. Uesugi, T. Hagihara, and T. Nogami, J. JSTP, *19*, 1978, 942-949 (Japanese).
[261] J. Kihara, K. Watabe, K. Doya, and A. Takashima, in [3.55], pp.183-186.
[262] M.M. Molchanov, B.Ya. Shneerov, B.F. Melnikov, and L.M. Perminova, Steel USSR, *5*, 1975, 614-616.
[263] G. Neport, Iron Steel, *44* (2), 1971, 103-104.
[264] A.G. Globus, Iron Steel Eng., *47* (8), 1970, 93-94.
[265] W.L. Roberts, Lubric Eng., *33*, 1977, 575-580.
[266] Ya.S. Gallai, A.A. Zhdanov, and V.P. Volegov, Steel USSR, *1*, 1971, 801-808.
[267] L.M. Bernick and C.L. Wandrei, in [3.52], pp.71-80.
[268] K. Suzuki and K. Ueda, in [3.55], pp.167-169.
[269] K.L. Williams, in [3.54], pp.23-27.
[270] M. Spannenberg, Werkstoffe u. Korr., *23*, 1972, 880-886 (German).
[271] I.M. Pavlov et al, Russian Metall. Mining, *1963*(3), 90-91.
[272] I.M. Pavlov, Yu.M. Sigalov, Ya.B. Gurevich, and A.M. Zubko, Tr. Inst. Met. im. A.A. Baikova, No.9, 1962, 109-114.
[273] I.M. Pavlov and Yu.M. Sigalov, Izv. VUZ Chern. Met., *1961* (8), 195-197 (Brutcher Transl. 6448).
[274] A.K.E.H.A. El-Kalay and L.G.M. Sparling, J. Iron Steel Inst., *206*, 1968, 152-163.
[275] Ya.S. Gallai, A.A. Zhdanov, and V.P. Volegov, Steel USSR, *1*, 1971, 806-808.
[276] L.G.M. Sparling, Metals Techn., *4*, 1977, 301-306.
[277] A.I. Grishkov, Stal in English, *1964* (1), 46-49.
[278] C.L. Smith, F.H. Scott, and W. Sylwestrowicz, J. Iron Steel Inst., *170*, 1952, 347-359.
[279] M.D. Stone, Lubric. Eng., *19*, 1963, 239-241.
[280] V.G. Grosvald and N.I. Svede-Shvets, Izv. VUZ Chern. Met., *1961* (6), 75-86 (Russian).
[281] D.M. Parke and J.L.L. Baker, Iron Steel Eng., *49* (12), 1972, 83-88.
[282] H.F. Izzo, Iron Steel Eng., *49* (6), 1972, 57-61.
[283] C.A. Lee, Iron Steel Eng., *47* (12), 1970, 117-119.
[284] H.E. Muller, Iron Steel Eng., *37* (5), 1960, 107-118.
[285] K.A. Ivanov and O.P. Maksimenko, Izv. VUZ Chern. Met., *1975* (12), 83-86 (Russian).
[286] G.J. McLean, Iron Steel Eng., *42* (12), 1965, 131-138.
[287] G.W. Buckley, M. Lewis, and R.T. Maddison, Metals Techn., *5*, 1978, 228-250.
[288] R.N. Wright, Met. Trans., *7A*, 1976, 1385-1388.
[289] C.M. Jackson and E.H. Welch, Sheet Metal Ind., *52*, 1975, 34-43.
[290] R.V. Williams and G.M. Boxall, J. Iron Steel Inst., *203*, 1965, 367-377.
[291] W.A. Dickinson and D. Porthouse, in [3.48], pp.71-81.
[292] R.N. Shaughnessy, J. Iron Steel Inst., *206*, 1968, 981-986.
[293] T. Shiraiwa, F. Matsuno, and H. Tagashira, Tetsu-to-Hagane, *57*, 1971, 131-145 (Japanese).
[294] P. Funke, J. Holland, and R. Kulbrok, Stahl u. Eisen, *98*, 1978, 403-409 (German).
[295] K. Toda, I. Imai, and R. Inui, in *Proc. Conf. Sci. Tech. Iron Steel,* Iron Steel Inst., Japan, Tokyo, 1971, pp.736-739.
[296] A. Ohnuki and K. Nakajima, in *Proc. 4th Int. Conf. Production Engineering,* Tokyo, 1980, 1041-1046.
[297] J. Kihara, K. Nakamura, K. Doya, and M. Suenaga, in [3.56], Paper No.55.
[298] P. Funke, R. Kulbrok, and H. Wladika, Stahl u. Eisen, *92*, 1972, 1113-1122 (German).
[299] R.R. Judd, Iron Steel Eng., *56* (1), 1979, 51-60.
[300] Y. Sekimoto, Trans. Iron Steel Inst. Jap., *10*, 1970, 341-349.
[301] M. Grounes, Iron Steel Eng., *49* (5), 1972, 73-78.
[302] A.D. Stepanovich, V.E. Nikolaev, and V.F. Shulgin, Mat. Sci. Heat Treat., *17*, 1975, 232-234.
[303] V.P. Gusev et al, Light Metal Age, *36* (7-8), 1978, 13.
[304] W.R. Foley, V. Pres, and W.R. Huber, Iron Steel Eng., *51* (4), 1974, 72-74.
[305] P. Blaskovic, in [3.45], pp.185-190.
[306] I.I. Frumin and A.M. Reznitskii, Automat. Weld., *28* (9), 1975, 44-46.
[307] *Roll Pass Design,* United Steel Co., Sheffield, 1960.
[308] H. Neumann, *Kalibrieren von Walzen,* VEB Deutscher Verlag f. Grundstoffindustrie, Leipzig, 1969.
[309] R.R. Preston, Metall. Mat. Techn., *12*, 1980, 687-691.

[310] D.I. Starchenko et al, Steel USSR, *6*, 1976, 204-206.
[311] J. Behringer et al, Neue Hütte, *21*, 1976, 186-187 (German).
[312] K.A. Ivanov, G.M. Chezhevsky, V.I. Emets, and V.A. Bondarenko, Metallurg, *1972* (5), 32-34 (Russian).
[313] A.E. Cichelli, Iron and Steel Eng., *51* (6), 1974, 56-62.
[314] M. Mrozek, MW Interf., *1* (5), 1976, 20-25.
[315] V.T. Zhadan and V.E. Stefanov, in *14th Conf. Plastic Deformation of Metals,* Metallurgiya, Moscow, 1969, pp.123-126 (Russian).
[316] T.W. Johnson and G.M. Sturgeon, in [3.51], pp.213-215.
[317] J.M. Fein, Iron Steel Eng., *54* (12), 1977, 52-57.
[318] P. Blaskovic, J. Nainar, J. Dorda, and O. Matous, in [3.51], pp.216-222.
[319] G. Hartmann, Stahl u. Eisen, *95*, 1975, 1161-1166 (German).
[320] K. Edsmar, Iron Steel Eng., *52* (4), 1975, 80-88.
[321] A.M. Parkes and S. Giege, Wire Ind., *42*, 1975, 835-837.
[322] F. Hofmann, Draht, *30*, 1979, 255-260; *31*, 1980, 265-267 (German).
[323] F. Grossjean, Bänder Bleche Rohre, *5*, 1980, 175-178 (German).
[324] T.Z. Blazynski, Met. Rev., *15*, 1970, 27-45; Int. Met. Rev., *22*, 1977, 313-322.
[325] N.M. Vedyakin, A.N. Yachmenov, and L.P. Mikhailova, Steel USSR, *4*, 1974, 159-161.
[326] G.I. Golyaev, P.I. Chuiko, and G.A. Evsyukova, Steel USSR, *6*, 1976, 337-340.
[327] H. Yoshida et al, CIRP, *24*, 1975, 191-196.
[328] F. Magyary, Aluminium, *48*, 1972, 614-618.
[329] V.F. Frolov et al, Steel USSR, *9*, 1979, 41-43.
[330] R. Aitken, A. Hamilton, and R.V. Riley, in [3.51], pp.70-74.
[331] V.A. Tarasenko, N.V. Dorbysh, R.E. Uvarova, and L.Y. Beloyusova, Stal, *1977* (5), 440 (Russian).
[332] J.G. Beese, Wear, *23*, 1973, 203-208.
[333] M.R. Edmundson, Iron Steel Eng., *47* (10), 1970, 66-69.
[334] T. Mase, J. JSLE, *24*, 1979, 14-19 (Japanese).
[335] A.P. Grudev et al, Metallurg, *1975* (5), 40-41 (Russian).
[336] V.G. Dodoka et al, Steel USSR, *4*, 1974, 48-49.
[337] R. Terakado, J. JSME, *81*, 1978, 59-64 (Japanese).
[338] H. Yamanouchi and Y. Matsuura, Rep. Castings Res. Lab., Waseda Univ., No.8, 1957.
[339] S.F. Chisholm, Proc. Inst. Mech. Eng., *179*, Pt.3D, 1964-65, 56-64.
[340] V.V. Vainshtok and B.D. Gotovkin, Chem. Techn. Fuels Oils, *10* (1-2), 1974, 18-19.
[341] P. Funke, O. Pawelski, and M. Kühn, Stahl u. Eisen, *85*, 1965, 785-794 (German).
[342] W. Gorecki and J. Madejski, Prace Inst. Hutn., *17*, 1965, 303-314 (Brutcher Transl. 6866).
[343] H. Kraut, Neue Hütte, *7*, 1962, 729-735 (German).
[344] E. Tsuji, W. Lueg, and P. Funke, Trans. Nat. Res. Inst. Metals (Tokyo), *4*, 1962, 135-142.
[345] W. Lueg, P. Funke, and W. Dahl, Stahl u. Eisen, *77*, 1957, 1817-1830 (German).
[346] A.P. Grudev, Yu.B. Sigalov, and L.S. Lega, Met. Koksokhim., *1975* (46), 63-67 (Russian).
[347] W.L. Roberts and R.S. Somers, Lubric. Eng., *18*, 1962, 362-368.
[348] I.K. Tokar and I.A. Chamin, Metallurg, *1960* (4), 28-29; Metallurgist, 1960 (4), 161-163.
[349] A.A. Petrovsky, N.I. Oseyuko, and Ya.S. Shvartsbart, Stal, *1976* (8), 725-730 (Russian).
[350] M. Shamaiengar, Iron Steel Eng., *44* (5), 1967, 135-141.
[351] M. Shamaiengar, Lubric. Eng., *17*, 1961, 83-87.
[352] J. Behringer and H. Mistecki, Neue Hütte, *15*, 1970, 665-672 (German).
[353] J. Loose, V. Lange, W. Knechtel, and R. Steinhauer, Schmierungstechnik, *10*, 1979, 339-341 (German).
[354] V.I. Meleshko et al, Steel USSR, *1*, 1971, 539-542.
[355] W. Lueg and K.H. Treptow, Stahl u. Eisen, *75*, 1955, 1085-1092 (German).
[356] I.A. Proctor, J. Australian Inst. Metals, *11*, 1966, 30-37.
[357] H.J. Drake, Iron Steel Eng., *42* (12), 1965, 110-116.
[358] V. Dedek, Hutn. Listy, *19*, 1964, 27-33.
[359] T.J. Bishop, Iron Steel Eng., *45* (11), 1968, 81-94.
[360] P. Funke and W. Mikulla, Stahl u. Eisen, *89*, 1969, 1333-1338 (German).
[361] R.W. Kenyon, Lubric. Eng., *19*, 1963, 244-245.
[362] H. Nakata, K. Hikino, and T. Jimba, Iron Steel Eng., *48* (11), 1971, 53-59.

[363] W.R. Roberts, in [3.41], pp.45-52.
[364] K. Mietzner, Stahl u. Eisen, *83*, 1963, 336-344 (German).
[365] P.I. Polukhin et al, Stal, *1972* (4), 327-329 (Russian).
[366] J. Loose, M. Galjatin, and J.A. Zalavin, Neue Hütte, *18*, 1973, 307-312 (German).
[367] M. Espenhahn and A. Hunting, Stahl u. Eisen, *95*, 1975, 1166-1172 (German).
[368] M.M. Saf'yan et al, Steel USSR, *4*, 1974, 216-218.
[369] A.J. Stock, Lubric. Eng., *17*, 1961, 580-586.
[370] I.S. Morton, Ind. Lub. Trib., *25*, 1973, 185-189.
[371] H.W. Ward, Sheet Metal Ind., *46*, 1969, 407-411.
[372] K. Gerhardt, Bänder Bleche Rohre, *11*, 1970, 574-583.
[373] L.H. Butler, Sheet Metal Ind., *33*, 1956, 571-577, 647-654, 727-734, 736.
[374] Discussion on "Metal Working Lubricants," Metal Ind. (London), *84*, 1954, 63-66, 83-86, 109-111, 129-131.
[375] J. Herenguel, Sheet Metal Ind., *30*, 1953, 591-596, 598.
[376] H.A. Snow, Sheet Metal Ind., *30*, 1953, 907-908.
[377] M. Suzuki, MW Interf., *5* (2), 1980, 10-22.
[378] G.F. Frontini and R.D. Guminski, Lubric. Eng., *25*, 1969, 60-68.
[379] K.C. Tripathi, Lubric. Eng., *33*, 1977, 630-636.
[380] T. Baba, O. Nagata, S. Kumagai, and S. Yamashita, in [3.46], pp.564-571.
[381] A.M. Smith, E. Barlow, M.P. Amor, and N.C. Davies, ASLE Trans., *21*, 1978, 226-230.
[382] G.A. Dorsey, ASLE Preprint No.81-LC-2B-6.
[383] P.C. Varley, in Discussion to "Friction in Hot Rolling," J. Inst. Metals, *97*, 1962-63, 320.
[384] F.E. Lockwood and L.J. Matienzo, in [3.56], Paper No.50.
[384a] A. Moxon, Lubric. Eng., *38*, 1982, 743-748.
[385] F.R. Penney, Lubric. Eng., *27*, 1971, 87-90.
[385a] F.E. Lockwood, K. Bridger, and M.E. Tadros, ASLE Trans. (to be published).
[385b] P. McNamara, in [4.299], pp.103-126.
[386] B.A. Mysyakin, K.I. Kuznetsov, S.I. Rylskaya, and I.G. Basova, MW Interf., *4* (3), 1979, 11-15; Light Metal Age, *37* (7/8), 1981, 23-24.
[387] E.J. Schaming, Light Metal Age, *27* (7/8), 1969, 8-9.
[388] D.H. Swan, Metallurgia, *83*, 1971, 217-218.
[389] E.L. Kane, Jr., and W. Pfuhl, Lubric. Eng., *32*, 1976, 249-253.
[390] E.C. Hill, O. Gibbon, and P. Davies, Trib. Int., *9*, 1976, 121-130.
[391] Discussion on "Lubricants for Cold Rolling Aluminum," J. Inst. Metals, *89*, 1960-61, 297-303.
[392] V.A. Livanov, Tsvetn. Met., *1975* (4), 68-69 (Russian).
[393] K. Kondo, MW Interf., *5* (5), 1980, 11-19.
[394] V.F. Kromykh, V.A. Livanov, and B.A. Mysyakin, MW Interf., *2* (5), 1977, 12-19.
[395] R.V. Singh, in [3.56], Paper No.51.
[396] K.G. Silvestrov, L.B. Zlotin, S.G. Soltan, and Kh.Z. Uretskaja, Sov. J. Nonf. Metals, *10* (7), 1969, 84-85.
[397] F. Magyary, Aluminium, *50*, 1974, 291-295.
[398] L.B. Sargent, in [3.52], pp.65-70.
[399] P.F. Serna and E. Louis, Aluminium, *52*, 1976, 120-122.
[400] Yu.P. Abdulov and V.A. Kopilenko, Izv. Akad. Nauk SSSR, Metally, *1972* (4), 147-150 (Russian).
[401] S.F. Chisholm, J. Inst. Metals, *78*, 1950-51, 482-500.
[402] Discussion on "Lubricants in Metalworking," Sheet Metal Ind., *31*, 1954, 206-210.
[403] V.L. Derkach et al, Light Metal Age, *35* (11/12), 1978, 32-33.
[404] K. Kleinjohann, Z. Metallk., *56*, 1965, 685-690.
[405] E. Vamos and A. Zakar, in [3.56], Paper No.108.
[406] W. Kerth, E. Amann, X. Räber, and H. Weber, Int. Met. Rev., *20*, 1975, 185-207.
[407] F. Magyary, Neue Hütte, *10*, 1965, 83-87 (German).
[408] R.J. Dean, Sheet Metal Ind., *53*, 1976, 406-413.
[409] M. Brzobahaty, Sheet Metal Ind., *39*, 1962, 341-454.
[410] V.K. Jain, Lubric. Eng., *33*, 1977, 195-197.
[411] J. Hoberg and G. Niebel, Neue Hütte, *22*, 1977, 261-266 (German).
[412] Anon., Sheet Metal Ind., *37*, 1960, 447-454.
[413] R.A. Roos, Lubric. Eng., *38*, 1982, 288-294.

[414] H. Exley, Proc. Inst. Mech. Eng., *180*, Pt.3K, 1965-66, 205-212.
[415] B.J. Meadows and G.W. Drinkwater, Metall. Met. Form., *38*, 1971, 287-289.
[416] J.W. Johnson, Lubric. Eng., *10*, 1954, 80-83.
[417] S.D. Shevelkin et al, Tsvetn. Met., *1975* (3), 64-65 (Russian).
[418] R. Hannesen and U. Heubner, J. Inst. Metals, *98*, 1970, 208-214.
[419] H. Riegger, in Proc. 8th NAMRC, SME, Dearborn, 1980, 187-191.
[420] R.J. Wilcox and P.W. Whitton, J. Inst. Metals, *88*, 1959-60, 200-204.
[421] M. Fukuda, T. Nishimura, and Y. Moriguchi, MW Interf., *5* (3), 1980, 14-21.
[422] I.M. Pavlov et al, Ivz. VUZ Chern. Met., *1964* (9), 88-94.
[423] L.F. Molotkov and Yu.L. Molotkov, Obr. Met. Dav., *57*, 1971, 232-235 (Russian).
[424] W. Rostoker, in *Fundamentals of Deformation Processing,* Syracuse University Press, 1964, pp.375-387.
[425] H.H. Scholefield, J.E. Riley, E.C. Larkman, and D.W. Collins, J. Inst. Metals, *88*, 1959-60, 289-295.

7

Drawing

Significant quantities of wire, bar, and tube are produced in all deformable materials, and tribology plays a vital role in all processes. The subject material discussed here includes the fields previously dealt with by Newnham [1] and Schey [2]. Drawing processes have been reviewed by MacLellan [3], Wistreich [4], Duckfield [5], and Rowe [6]. Extensive coverage is given in books devoted to wiredrawing [7-13] and to drawing of wire, bar, and tube [14].

Drawing is a steady-state process from both the mechanical and tribological points of view (Fig.2.8). The workpiece is long and is continuously drawn into the die. Unless the lubricant breaks down, forces and temperatures attain steady levels. Even though during acceleration non-steady-state conditions exist, they are of significance in only a few instances. Wire and bar drawing are basically the same and will be discussed first, followed by a discussion of the drawing of tube.

7.1 Drawing of Bar and Wire

The starting material is usually a hot rolled or, less frequently, hot extruded rod [6.249],[8,12]. The distinction between bar and wire is ill-defined. Bar is generally of larger diameter (over 6 mm) and is drawn straight, although it is increasingly produced in coils and straightened during or after final drawing. Wire is always produced in coils but is also available in a straightened form. The cross section is most frequently circular but may be square, rectangular, or of some complex shape. Drawing of wide sheet is not practical, but drawing in plane strain (with a strip width-to-thickness ratio of at least 10) is of importance for theory. Much experimental work has been conducted on sheet at low reductions, as a simulation of sheet metalworking, and this research will be discussed in Chapter 10.

The purpose of drawing may be simply to tighten tolerances, improve surface finish, or increase the strength of the product, and then a single draw (pass) at room temperature may suffice. More frequently, the end product is of a smaller cross section than can be produced by hot working, and then a sequence of passes is employed, usually at cold working temperatures, with process anneals if required. Occasionally, material of limited ductility is drawn warm or hot.

Bars are drawn on mechanically or hydraulically driven draw benches. Wires are drawn in long lengths, with a power-driven drum providing the draw force. Single drums (bull blocks) are used for the heaviest gages, but the large number of passes required for thinner gages can be economically taken only on machines in which the wire passes through a succession of dies (holes). Between dies, either the wire is accumulated on drums (capstans) in a machine that may be regarded as a succession of individual bull blocks, or two to three turns of wire are wound on capstans so that friction on the capstans provides the draw force. The capstans either rotate a few percent faster than the speed of the emerging wire (drawing with slip) or are individually driven with dc motors. Typical equipment and practices are described in [9,13,15,16].

The large number of steps could, conceivably, be bypassed by techniques based on a form of extrusion [13,17,18] or cutting [19], reviewed by Tuschy [20]. These processes have not achieved production status and, even though friction plays a substantial role in them, will not be further discussed here.

The end of the workpiece must be reduced (pointed) for threading through the die. Pointing techniques are related to forging (swaging or forge rolling) or containerless extrusion (push pointing), and usually present no special tribological problems. Pointing, threading, and die changing account for most of the time lost in production [21-23], and therefore the process is designed to run without wire breaks as far as possible. Ends of coils can be welded together. Thus, wiredrawing represents the most continuous of all processes, with special demands on the wear resistance of the die.

7.1.1 The Drawing Process

Deformation in drawing, as in rolling, takes place in a converging channel (Fig.7.1). However, the similarity ends here. In rolling, the rolls rotate; in drawing, the workpiece is pulled through the stationary die. The entrance section (bell and entrance) is shaped to encourage lubricant flow into the interface. The wire makes contact in the draw cone (also called the approach), where reduction of the cross-sectional area from A_0 to A_1 takes place. To prevent sudden decompression of the lubricant film and to avoid a rapid increase in diameter on die wear, a parallel land (bearing) is always provided. The back relief ensures a gradual drop of stresses in the die.

Reduction is defined as the engineering compressive strain

$$r = \frac{A_0 - A_1}{A_0} \quad \text{or} \quad r = \frac{A_0 - A_1}{A_0} \times 100\ (\%) \tag{7.1}$$

For purposes of calculation, the natural strain ε is preferred:

$$\varepsilon = \ln \frac{A_0}{A_1} \tag{7.2}$$

Since the volume of the wire must remain constant, the wire elongates. If drawn at a speed v_1, it enters the die at a lower v_0 speed, which can be calculated, just as in rolling (Eq.6.7), from constancy of volume:

$$v_0 A_0 = v_1 A_1 \tag{7.3}$$

However, in contrast to rolling, there is no neutral plane here, and the wire always slides over the whole contact zone. To a close enough approximation, it may be assumed that the interface sliding velocity increases from entry to exit linearly in plane-strain drawing and parabolically in axially symmetrical drawing in a conical die. Sliding generates a frictional stress opposing the movement of the wire (Fig.7.1), and the contribution of this stress to the draw force is of immediate concern. Deformation occurs under the combination of the longitudinal (draw) stress and the indirect compressive stress generated in the die. Friction is equivalent to a back tension and thus lowers the interface pressure, but at the expense of higher draw forces.

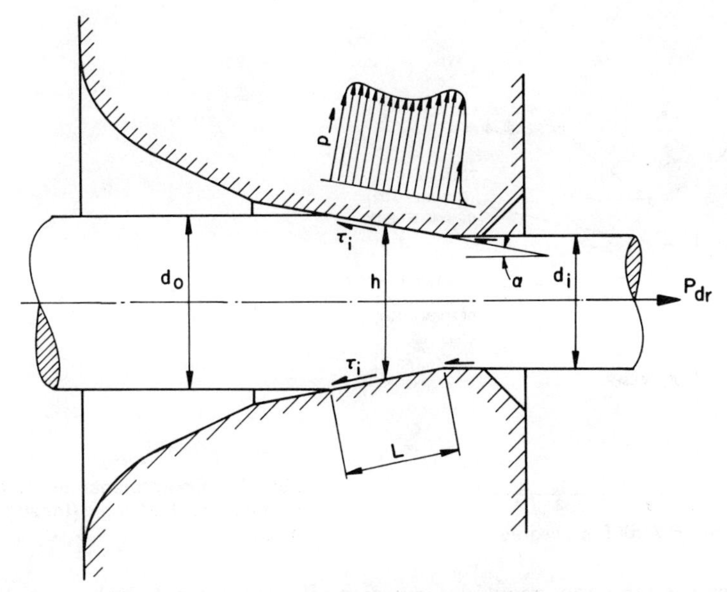

Fig.7.1. Schematic illustration of the drawing process

7.1.2 Stresses and Forces

The force required for drawing can be calculated by a variety of methods, reviewed by a number of researchers [5,8,13,24-26], also for plane-strain drawing [27]. Some more recent solutions are also available [28,29]. By analogy to Eq.2.11, all solutions can be brought to the general form

$$P = \sigma_{fm}\left[f\left(\varepsilon;\frac{\tau_i}{\alpha};\frac{h}{L}\right)\right] = \sigma_{fm}Q \qquad (7.4)$$

The bracketed term (pressure-multiplying factor Q) shows that the draw force can be visualized as having three components:

1. The force required for supplying the pure deformation work, which is proportional to the strain ε and the flow stress σ_f (or, in cold working, the mean flow stress σ_{fm}) of the wire material.

2. The contribution of friction, which increases with increasing interface shear strength τ_i and, for a given reduction, decreases with increasing die half angle α (Fig.7.2) because of a decrease in sliding length. The interface is usually described by a constant μ even though evidence indicates that friction varies along the contact surface with the die.

3. The contribution of redundant work due to inhomogeneous deformation, which increases—as in other processes—with an increasing h/L ratio (see Fig.2.7). The ratio may be calculated from various formulae (see [30]), but, as shown in Fig.7.1, it can be simply taken as the mean diameter (or, in plane-strain drawing, the mean thickness) of the workpiece divided by the contact length L. For a given reduction, the ratio increases with increasing α (Fig.7.2). It causes the interface pressure to rise at entry and exit (Fig.7.1). The effects of redundant work are considered in great detail by Blazynski [2.16].

A simple but useful formula is

$$P = \sigma_{fm}Q = \sigma_{fm}\left(1 + \mu\cot\alpha\right)\phi\varepsilon \qquad (7.5)$$

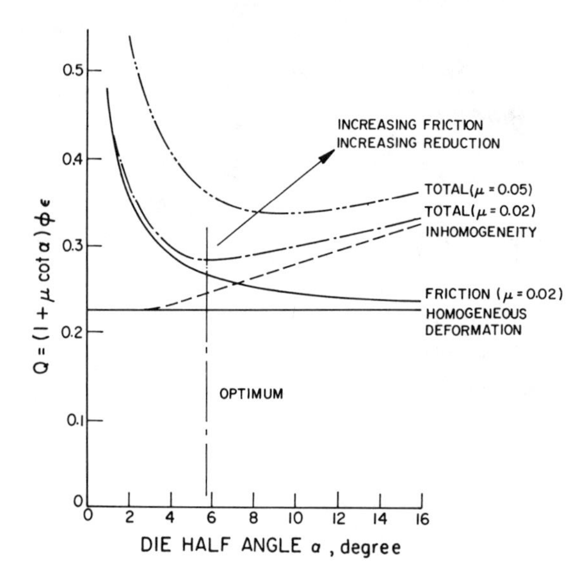

Fig.7.2. Components of draw force and optimum die half angle (for 20% reduction in area)

where ϕ allows for inhomogeneous deformation. Its value can be taken, on the basis of experiments [2.14], for round wire as

$$\phi = 0.88 + 0.12\frac{h}{L} \tag{7.6a}$$

and for plane-strain drawing as

$$\phi = 0.8 + 0.2\frac{h}{L} \tag{7.6b}$$

There is no general agreement on an appropriate formula; the above formulae can give underestimates, and other experiments give better agreement with different formulae [30,31].

Since, for a given reduction, the friction contribution decreases and the contribution of inhomogeneity increases with die angle, an optimum half angle is found, usually between 5 and 7°, where draw force is minimum (Fig.7.2). Lower angles are permissible only with very low friction. With the optimum die angle, the contributions of friction and inhomogeneous deformation are of the same order of magnitude. With increasing reduction, die pressures drop, deformation becomes more homogeneous, and the efficiency of the drawing process (the ratio of total work to pure deformation work) increases. The optimum angle shifts to higher values. Because of this, the trumpet-shape dies used in the past actually gave a better approximation of the optimum die profile.

Friction on the die land also contributes to the draw force [2.12].

Just as in rolling, a back pull (back tension) can be applied to the wire. The resultant drop in interface pressure can be beneficial in terms of the lubrication mechanism and wear, but at the expense of greater draw force.

The draw force cannot exceed the strength of the drawn wire. This sets a limit on the attainable reduction, which decreases with increasing friction and h/L, of some 55% under the best conditions [32]. However, no wire can be perfect, and drawing with critical reductions would lead to frequent wire breaks. Therefore, practical reductions are limited to 20 or 30% per pass.

Friction Evaluation. In principle, the magnitude of friction should be obtainable by a number of techniques. In practice, there are several problems.

1. Friction From Draw Force. Back-calculation of friction from some form of Eq.7.4 is difficult. First, the value of σ_{fm} is usually determined in pure tension, whereas yielding in the die takes place under combined tension and compression. Thus, the effective flow stress may be different for a material of anisotropic properties. A further complication with tensile testing is that the wire may contain residual stresses, and removal of a surface layer to create a standard specimen is not desirable. It is better then to test in axial compression [33]. Since the friction contribution in drawing can be quite small, any error in σ_{fm} can give large errors in calculated μ values. Second, friction and redundant work mutually influence each other. If, to a first approximation, they are assumed to be independent, the contribution of redundant work can be calculated from any appropriate theory, but any error in this term greatly affects the calculated μ, a problem that has been considered by several researchers [31-34]. There are a number of ways to overcome this problem:

a. Basily and Sansome [35] obtained the redundant work term from the difference in strain hardening measured in tension tests on drawn and undrawn material.

b. Evans and Avitzur [6.57] drew identical bars or wires at various reductions with dies of widely varying half angles (from 2.75 to 45°). The optimum die angle was found for each reduction, and μ or m could be calculated from the appropriate theory without the need for knowing the flow stress.

c. If it is assumed that the application of back pull does not significantly affect the lubrication mechanism (a debatable proposition), friction can be more directly determined by measuring the back pull B, the draw force P, and the force S acting on the die support (Fig.7.3). As suggested by Lunt and MacLellan [36], the draw force is

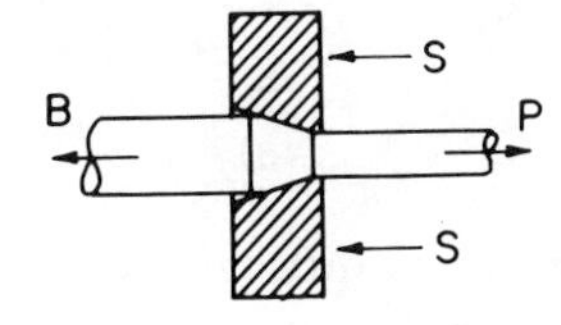

Fig.7.3. Measurement of the contribution of friction with back pull

measured first without back pull ($P_0 = S_0$) and then with increasing back pull. For convenience, a back-pull factor is defined as $b = (P_0 - S)/B$ and then

$$\mu \cot \alpha = \frac{\ln(1 - b)}{\ln(1 - r)} - 1 \qquad (7.7)$$

where r is the reduction from Eq.7.1.

d. Further possibilities exist if drawing of a strip is accepted as a valid simulation of wiredrawing, on the assumption that side leakage of the lubricant does not affect the lubrication mechanism. The simplest is to draw between two rollers (Fig.5.5a), first with the rollers idling and then with the rollers clamped. An approximate τ_i is then determined from Eq.5.1.

2. Friction From Draw and Normal Forces. Simultaneous measurement of draw force and normal force requires more sophisticated equipment and instrumentation, but gives a direct method of measuring frictional forces.

a. In axial symmetry the difficulties are substantial. Strain gages applied to the outer die surface give a measure of hoop stress, but evaluation is affected by the nonuniform deformation of the die. Majors [37] and others calibrated for hoop stress by applying hydrostatic pressure to the die bore sealed with closing plugs.

b. For a direct measurement of radial forces, MacLellan [38] attempted the use of a die split in an axial plane, and measured the force S separating the two halves (Fig.7.4a). Wistreich [39] was successful. He gradually reduced the force H holding the two die halves together until the halves separated, as shown by a break in an electrical circuit. At this point, $H = S$. From a simultaneous recording of the draw force P, the coefficient of friction may be determined—for a straight tapering die—without resorting to theory from

$$\mu = \frac{\cot \alpha - (\pi S/P)}{(\cot \alpha)(\pi S/P) + 1} \tag{7.8}$$

Opening of the die halves along the parting line under load is bound to cause lubricant leakage, thus upsetting the intended lubricating action and generating an additional separating force. Burkin et al [40] constructed a die of four segments.

c. Lubricant leakage is overcome by the split die of Yang [41], which is restrained in a large ring (Fig.7.4b). A small load cell built into a window measures the hoop strain. Sagar and Gupta [42] used a similar arrangement but with strain gages attached to the window.

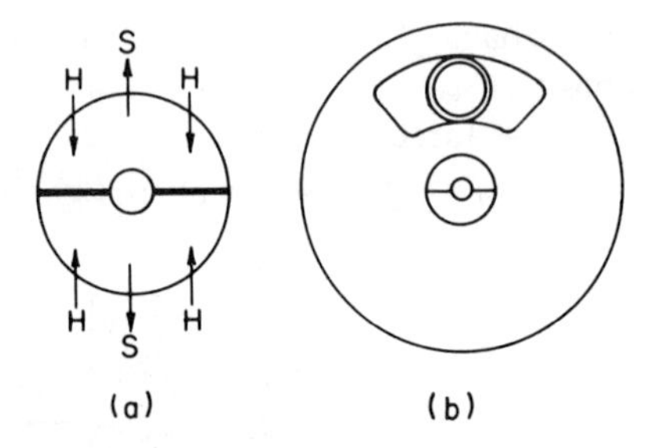

Fig.7.4. Split-die techniques

d. Many difficulties are avoided by drawing in plane strain. When both draw force P and die separating force S are measured, μ is obtained from Eq.5.3 when either fixed rollers (Fig.5.5a) [5.21] or inclined die faces (Fig.5.5b and c) [5.24,6.26],[6.14] are used.

e. Localized pressure distribution could, in principle, be obtained from pressure-pin measurements (Sec.5.5.3), but the interface would be seriously disturbed. The photoelastic method is better but is limited by the strength of the polymer die to lead or similar soft workpiece materials [43]. The results of Ohashi and Nishitani [44] are shown in Fig.7.5.

7.1.3 Heat Generation

Most of the total work expended in wiredrawing is converted into heat. The die is of small mass and is in continuous contact with the wire, and thus heating presents an even greater danger than in rolling. At high drawing speeds most of the heat is retained in the wire; still, the die heats up too and, in multihole drawing, the entry temperature of the wire also rises. The surface temperature of the wire is higher because of concentration of redundant work and frictional heating. From the tribochemical point of view peak temperatures are of importance. On emergence of the wire from the die, temperatures rapidly equalize, and the main concern is the over-all temperature level. Therefore, cooling becomes an important function of the lubricant.

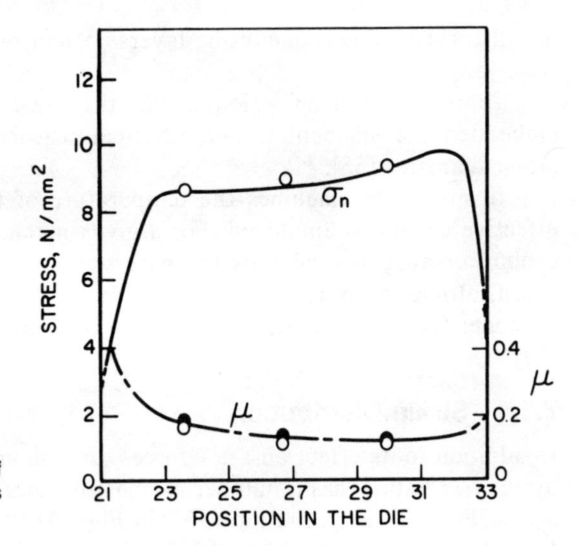

Fig.7.5. Normal pressure and coefficient of friction determined by photoelastic analysis [44]

As befits its importance, heat generation has been studied both theoretically and experimentally. The typical temperature distribution is exemplified by the simplified analysis of Siebel and Kobitzsch [45] (Fig.7.6). More recent and detailed analyses [25,46-49] are available. Reviews can be found in Wistreich [4] and Altan [46].

Various experimental techniques have been used. In the calorimetric method of Rosenhain and Stott [50], the wire emerging from the die was led through an oil bath and the temperature rise of this bath was measured. More information is obtained by the technique of Eichinger and Lueg [51], who measured the temperature of the drawn wire with a thermocouple and the heat content of the thermally insulated draw die by calorimetric methods. Lewis and Godfrey [52] refined the technique by running the wire emerging from the draw die through a heated copper tube. The temperature of the tube was varied until heat exchange between the drawn wire and the heated tube ceased, and thus the true wire temperature was obtained. The friction contribution to total heat generation is, however, so small that it is better to measure interface temperatures by techniques described in Sec.5.6.

The dynamic thermocouple method, introduced to wiredrawing by Thompson and Dyson [53], is certainly the most appropriate [5.119], but much information can be obtained also from thermocouples embedded in the die [54-57]. Measurements invariably show that the temperature increases with drawing speed and peaks at entry and exit [55]. The eddy-current technique described by Keller [5.149] is noncontacting and is

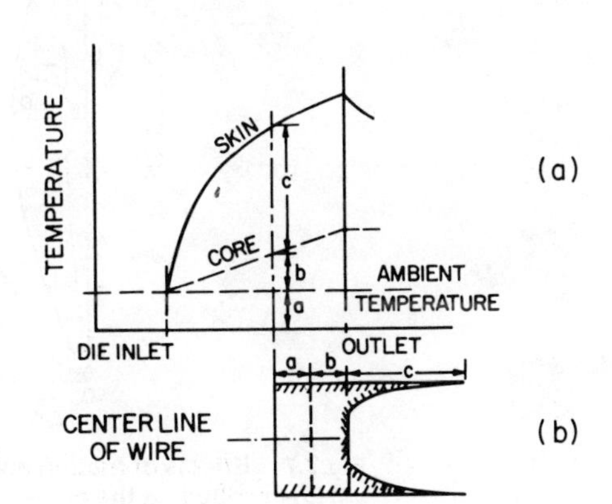

Fig.7.6. Calculated temperature distributions (a) along the wire and (b) across the wire [45]

not disturbed by nonconducting layers. A noncontacting thermometer was used also by Rigo [58].

Temperature distribution within the drawn product can be measured only with embedded thermocouples. For practical reasons, the technique is limited to bars of larger diameter [55].

In multihole machines the temperature of the wire increases progressively unless effective cooling is employed. For a given machine arrangement, the time available for cooling decreases—and thus the wire temperature increases—with increasing drawing speed, although improved lubrication at higher speeds may counterbalance this effect. A model for the prediction of temperatures has been given by Flanders and Alexander [59].

7.1.4 Strain Distribution

In addition to its effect on draw forces, the inhomogeneity of deformation—as expressed by the h/L ratio—has a number of consequences.

a. For a given reduction (36% in Fig.7.7) [2.23], inhomogeneity is governed by the h/L ratio but increases also with friction. The first consequence is that residual stresses are set up in the wire. At low h/L ratios the surface residual stress is tensile. For a given die angle α, stresses decrease with increasing pass reduction and improved lubrication [60-63], both of which serve to make deformation more homogeneous. Consequently, the service properties of the wire also change, as already observed by Linicus and Sachs [64]. In a bar of large enough diameter, inhomogeneity is reflected also in hardness variations. Hundy and Singer [6.61] found this variation to increase with friction for a 20° die angle but to be independent of friction for a 30° angle, presumably because inhomogeneity had an overwhelming effect. Istomin [65] proposed to determine friction from hardness measurements. The surface deformation associated with very high h/L ratios can be beneficial because residual compressive stresses impart greater fatigue resistance.

b. The simple conical die used in practice does not ensure the most homogeneous deformation, and ideal die profiles can be theoretically calculated for more uniform strain or strain-rate distribution while also minimizing friction [2.16],[66-69]. However, the improvement is seldom sufficient to justify the cost of making dies with convex, sigmoidal, or other complex shapes [68,70].

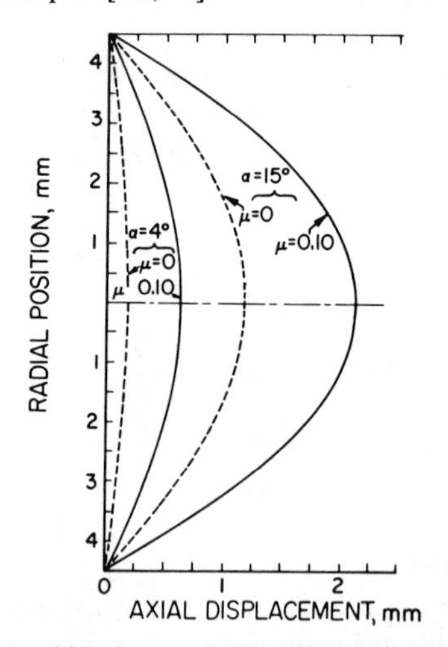

Fig.7.7. Effects of friction and die angle on distortion of a vertical line scribed on the cross section of a round bar [2.23]

c. Surface compressive stresses are balanced by central tensile stresses. At high h/L ratios triaxial tensile stresses are set up in the center [71,72]. The density of wire decreases, and discontinuity in material flow can result in periodic centerburst defects (cuppy break, Fig.2.7c) in wire of moderate ductility, especially if it is subjected to a succession of high-h/L passes. Because friction increases the back tension, it aggravates the situation [72], and the theoretical safe zone for drawing is reduced (Fig.7.8) [2.13],[73]. Of course, this represents only a necessary but not a sufficient criterion for the onset of centerburst: it is also necessary that the material have low ductility. A model based on the volume fraction of voids has been proposed [74].

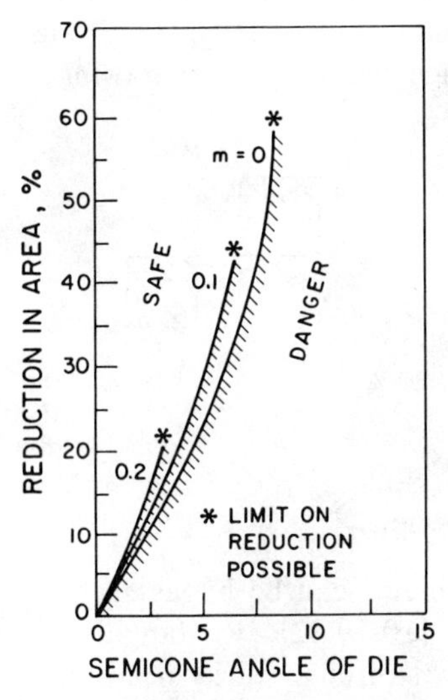

Fig.7.8. Criterion for centerburst defect in drawing of a round wire [73]

d. High h/L ratios affect the external deformation, too. First, a bulge forms at the die entry; for a given die angle, bulging sets in at a lower reduction when friction is high (Fig.7.9) [73,75]. At yet higher angles the surface of the wire is shaved off by a process related to cutting with a negative-rake-angle tool [2.12],[75]. Such shaving is an important step in making wire free of surface defects inherited from the hot processing stages.

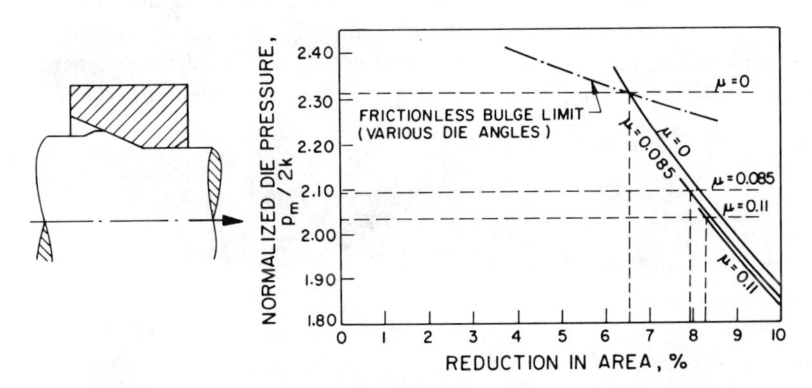

Fig.7.9. Condition for bulge formation in drawing with light reductions [75]

7.1.5 Drawing of Profiles

An increasingly important activity is the drawing of bars with more or less complex cross sections, often to close tolerances, and in difficult-to-draw materials [76,77]. Drawing stresses are somewhat higher, but the major concern is uniformity of material flow, which can be promoted with convex dies [78,79]. Much of the development has been experimental until the recent appearance of analytical treatments [80].

This process presents some special difficulties from the lubrication point of view, as discussed by Schroeder [81]. Material flow is less uniform, local stress concentrations arise at corners and grooves, and surface expansion can be much greater than in drawing of a round to a round. Lubricants may be evaluated by drawing profiles representing varying degrees of difficulty (Fig.7.10) [81]. Results show that even the order of merit of lubricants may change relative to wiredrawing.

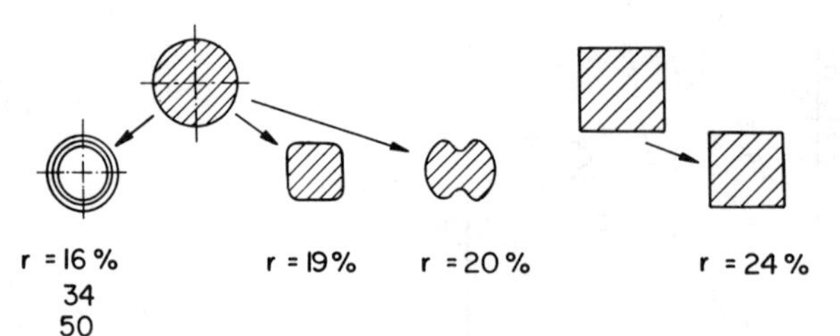

Fig.7.10. Shapes used in evaluation of lubricants for drawing of shaped wire [81]

7.1.6 Roller Dies and Rolling

Many of the lubrication problems can be bypassed by using roller dies or by switching to cold rolling of sections [10,79,82]. Rectangular sections with varying width-to-thickness ratios can be drawn through turks heads in which four rollers are mounted adjustably, with their axes in the same plane.

Roller dies are usually constructed with two or three rollers (Fig.7.11) mounted in a frame with their axes in a single plane [83-87]. A consideration of the three-dimensional geometry will show that the bar is fully enclosed only in the plane of the axes, and that gaps open farther back. This limits the reduction attainable without flash formation, and successive die sets must be set at a different angular position so that any incipient flash is rolled back into the body of the bar. Some spread is unavoidable, and the pass must be slightly opened to accommodate it. The repeated deformation places greater demands on workpiece ductility. However, sliding friction is replaced by the much lower rolling friction, and the tribological demands are much less stringent: the danger of die pickup and die wear are much reduced. Lubrication practices are similar to those of cold rolling (Sec.6.3) [88] and need not be further discussed here.

From a mechanical point of view, the process is equivalent to rolling with a high

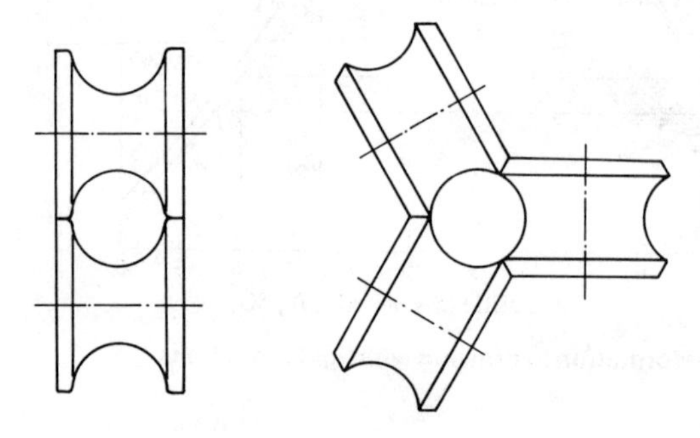

Fig.7.11. Configurations of roller dies

enough front tension to obviate the need for the rolls to be driven. This simplifies construction but limits the attainable reduction. This limit is removed when the rolls are driven [79]. Three-roll arrangements can become quite complex but are feasible even for small-diameter wire [89]. It is not actually necessary to have profiled rolls. As pointed out by Sayer and Moller [90], ductile wires can be reduced in a succession of flat rolls, and then shaped in the last rolls or finished by drawing.

7.2 Tube Drawing

Seamless tubes are mostly produced by one of the hot tube-rolling (Sec.6.5.4) or tube-extrusion (Sec.8.2.3) techniques. Only the Ehrhardt process [2.18],[91] utilizes a technique related to drawing. Hot, heavy-wall, closed-ended tube blanks are pushed through a series of roller dies in mechanical pushbenches to reduce the tube wall. Indeed, it appears that roller dies were first employed in this process.

The surface finish of a tube blank can be improved, and its wall thickness and/or diameter reduced, by cold pilgering (Sec.6.5.4) or, more frequently, by cold drawing. Reviews of all processes are given by Blazynski [2.16],[6.324] and by Geleji [2.7,2.18].

In tube drawing both the diameter and wall thickness change. In calculating strain from Eq.7.1 and 7.2, A_0 and A_1 now are cross-sectional areas of the tube before and after drawing, respectively. Tube drawing may be performed in four basically different forms:

1. Sinking (Fig.7.12a) is closest to wiredrawing in that the tube is drawn through a die. Its diameter is reduced while the wall slightly thickens and the inside surface roughens. Thickening of the wall is a function of friction and of the tube wall-thickness-to-diameter ratio (Fig.7.13) [92]. The effects of friction are the same as in bar drawing, but interface pressures are lower and optimum die angles are larger. Dévényi [93] observed optimum die angles of about 12° at 10% reduction and about 20° at 40% reduction in drawing aluminum tubes.

2. The internal diameter and surface finish are controlled when drawing is done on a plug (also called a mandrel in the U.S.). The plug is held in position by a bar

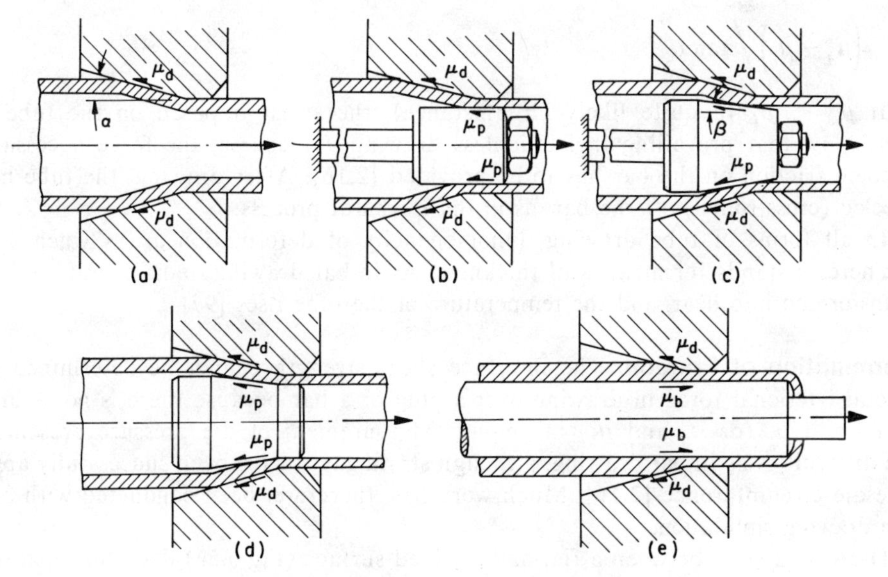

Fig.7.12. Tube drawing processes

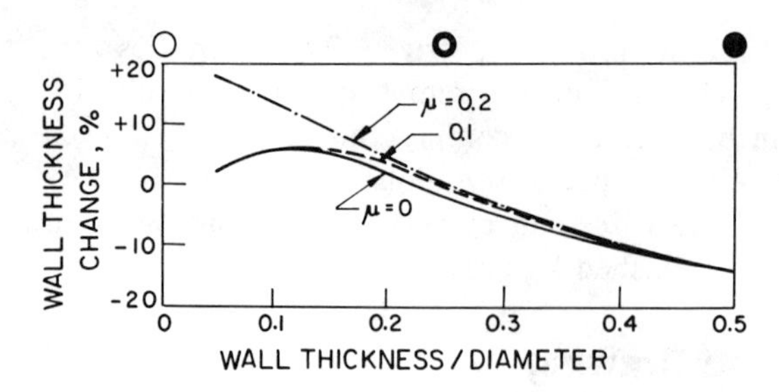

Fig.7.13. Change in wall thickness in tube sinking with 15% reduction in diameter, in dies with 9° half angle [92]

(Fig.7.12b) fixed to the end of the draw bench. In addition to the diameter (sink) being reduced, the wall thickness (draw) is also reduced to a dimension defined by the diameter of a cylindrical plug (Fig.7.12b) or the position of a conical plug Fig.(7.12c). The sink-to-draw ratio affects interface pressures, sliding velocities, and surface extension. The severity of operation increases with increasing draw. The additional friction μ_p on the plug increases the draw force. This effect can be allowed for in Eq.7.5 by replacing $\mu \cot \alpha$ with the term

$$\left(\mu_d + \mu_p\right) \Big/ \left(\tan \alpha - \tan \beta\right) \tag{7.9}$$

The optimum die half angle is usually 12 to 15° in cooperation with 10 to 11° plugs.

3. If the plug is designed so that the frictional force keeps it in the deformation zone, the holding bar can be omitted. Such a floating plug (Fig.7.12d) allows drawing of long lengths on a drum (bull block), and heavier reductions can be taken. Plugs can be designed from basic principles if friction can be characterized [2.16],[94,95]. Here, too, there is an optimum angle. An analysis for hydrodynamic lubrication is also available [96].

4. In drawing on a bar (called a mandrel in Britain), the tube material is forced to slide over the bar. Thus, a frictional stress is set up which transfers the drawing stresses from the tube wall to the bar (Fig.7.12e). Since friction on the bar opposes friction on the die, the stress in the wall of the tube can be approximated by substituting in Eq.7.5 for $\mu \cot \alpha$ the term

$$\left(\mu_d - \mu_b\right) \Big/ \tan \alpha \tag{7.10}$$

If $\mu_d < \mu_b$ (a quite likely circumstance), the stress imposed on the tube wall drops below that prevailing in frictionless drawing. Of course, the force necessary to overcome friction on the bar has to be provided [2.16]. After drawing, the tube has to be reeled (cross rolled) off the bar, as in the Ehrhardt process.

In all forms of tube drawing, inhomogeneity of deformation arises when h/L is high; here, h stands for mean wall thickness. As in bar drawing, most of the total work is transformed into heat, and the temperature of the tube rises [97].

Determination of Friction. The interface shear strength cannot be determined from the total frictional force in drawing over a plug or a bar because there is no assurance that μ_d and μ_p (or μ_d and μ_b) are equal. Measurement of die pressure presents the same difficulties as in bar drawing, although strain gages have been successfully applied to the die circumference [2.16]. Much work has, therefore, been conducted with plane-strain drawing simulation.

Drawing a strip between a flat and inclined surface (Fig.6.5a) does not separate μ_d

from μ_p and is of limited value. The arrangements in Fig.5.6b and c are, however, relevant.

Drawing a strip over a solid bar (Fig.5.7a) cannot separate the two friction components; the more complex arrangement of Fig.5.8e is needed.

7.3 Lubrication and Wear

In contrast to rolling, no friction is needed at all for wire-drawing, tube sinking, and tube drawing on a fixed plug (but some minimum friction is essential for drawing with a floating plug, and friction is helpful on the tube/bar interface in drawing on a bar). Thus, if at all possible, the lubricant is chosen to give lowest friction and minimum wear. It is essential, though, that the heat generated should be extracted, especially in high-speed drawing, because otherwise the lubricant may fail and the properties of the wire may suffer. These requirements can be satisfied by one of two techniques:

1. The lubricant is chosen for its tribological attributes and the wire is cooled while it resides on the internally cooled draw drums (capstans) of single-hole bull blocks and of multihole machines drawing with accumulation. Additionally, external air cooling of the wire coil and water cooling of the die holder are possible. If water is applied to the wire at all, it must be totally removed before the wire enters the next die. The lubricant is usually a dry soap powder, placed in a die box and picked up by the wire surface upon its passage through the box. This technique is used for steel wire greater than 0.5 to 1 mm in diameter, for which the relatively rough surface produced is acceptable. For the most severe draws and for tubes, the soap is often preapplied from a solution, if necessary, over a conversion coating; the soap must be allowed to dry. These techniques are customarily described as dry drawing (a process that has nothing to do with dry working in the sense of an unlubricated interface).

2. The lubricant is chosen both for its tribological attributes and for its cooling power [98,99], and can be either oil-based or aqueous [100]. It can be applied to the die inlet, to the wire, and often also to the capstan, or the entire machine can be submerged in a bath. When the machine operates with slip, the lubricant must reduce wear of the capstan while maintaining some minimum friction. This so-called wet drawing practice is typical of all nonferrous metals and of steel wires less than 0.5 to 1 mm in diameter [98,101].

A transition between the two techniques is sometimes employed, particularly in low-speed drawing of bar and tube; a high-viscosity liquid or semisolid is applied to the workpiece and/or die.

Drawing speeds range from 0.3 to 2 m/s for bar and heavy tube, through 7 m/s for heavy (say, 5-mm) wire, to 40 m/s for fine steel wire and 70 m/s for nonferrous metals. The limit is often set by die life and coil length, which would force a die change or rethreading at unreasonably short intervals.

As in other processes, mixed-film lubrication is often encountered, but the high speeds—coupled with a favorable process geometry—make plastohydrodynamic lubrication possible and even desirable, at least for early draws in which a rougher surface can be tolerated.

7.3.1 Full-Fluid-Film Lubrication

The importance of this topic has inspired several investigations, some of which have been reviewed by Pawelski [102], Lancaster [103], and Avitzur [104,105]. In principle, a thick film can develop with any lubricant if other conditions are conducive to it.

High-Speed Wet Drawing. As given by Eq.3.29a, the entrained film thickness increases with increasing ηv, with diminishing wedge angle θ (= α for a conical die), and with diminishing flow stress σ_f (or, more relevantly, with die pressure p_m, which is governed by reduction, die angle, friction, and back tension).

There is experimental evidence to suggest that a full film can be generated with a high-viscosity lubricant in drawing of soft metals. Felder [106] observed thick (15-μm) films (as judged from weight differential) and no electrical contact in drawing soft aluminum and copper wires with a chlorinated oil of 200 Pa · s viscosity, at speeds from 0.5 to 5 m/s. The presence of a bamboo-like defect (a periodic variation of diameter) also suggested full-fluid-film lubrication on soft copper. However, only short pieces of wire were drawn, and it is not clear whether fluid-film conditions could have been maintained on long lengths of wire after attaining a thermal equilibrium. There is little doubt that film thicknesses increase with velocity (Fig.7.14a) [24] ; this is, however, no proof of a full-fluid-film mechanism in itself.

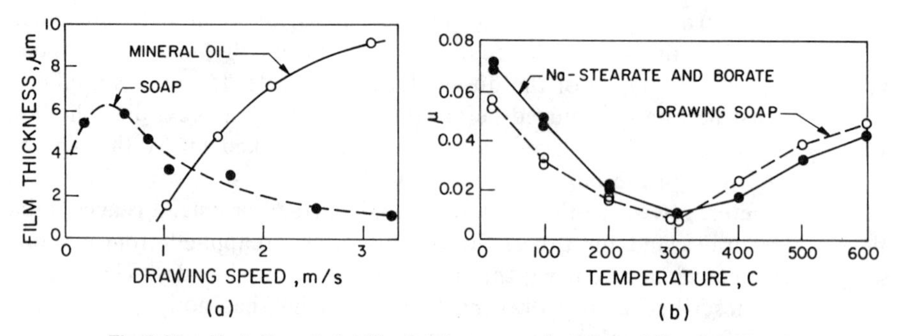

Fig.7.14. Variations in (a) film thickness as a function of drawing speed and (b) coefficient of friction measured in drawing of steel sheet with 20% reduction between dies with 9° half angle [24], [3.149]

There are numerous theories for drawing with PHD lubrication with viscous fluids. Eichinger and Lueg [51] estimated the film thickness from entry geometry. Walowit and Wilson [107,108] gave an isothermal solution, for a fluid of pressure-dependent viscosity, which shows all of the characteristics of Eq.3.29. The analyses of Bloor et al [3.289],[109] for plane-strain drawing and of Dowson et al [110] for wiredrawing showed that heating of the film reduced its thickness whereas elastic deformation had little effect. It will be noted (Fig.7.15) that calculated film thicknesses are too low to ensure full-fluid-film lubrication. Felder [106] accounts for the large measured film thicknesses by assuming that heating generates a quasirigid boundary layer on the wire. Other theories have also been developed [3.292], [111-114].

High-Speed Dry Drawing. There is a much greater opportunity for forming a fully separating film with a solid-like substance. Dodds and Lancaster [115,116] drew short lengths of aluminum and copper, at speeds up to 40 m/s, with lanolin preapplied as a coating. The film thickness was determined by dissolving the lubricant and comparing the transmittance of the solution with those of solutions of known concentration. Film thicknesses of 4 μm on wire with an initial roughness of 1 μm indicated full fluid films, as did the bamboo-like deformation of soft copper. As expected, film thickness decreased with increasing pass reduction. The presence of a thick film was also deduced by Parsons et al [117] from roughening of the wire in drawing at speeds up to 45 m/s. Pawelski [3.149] drew heated strips at low speeds in cold dies and observed that μ

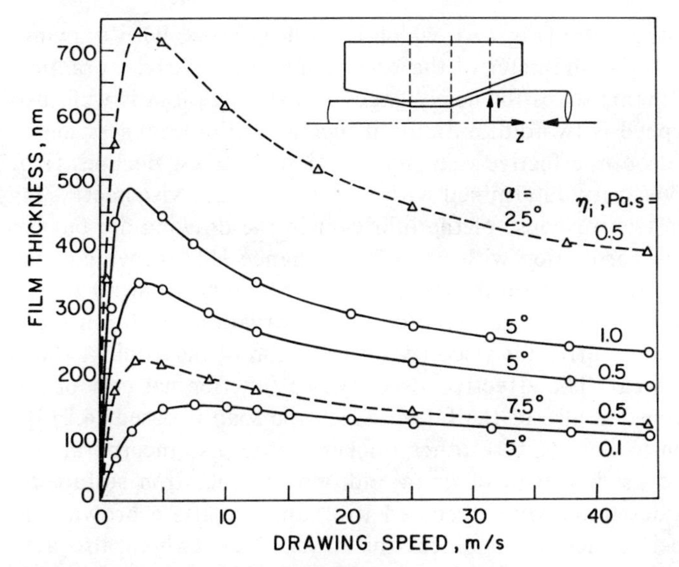

Fig.7.15. Film thicknesses calculated for drawing with mineral oils under hydrodynamic conditions [110]

decreased with increasing temperature (and decreasing viscosity), until the film became too thin and/or the soap degenerated at high temperatures (Fig.7.14b). The evidence is, however, not unequivocal, because in the work of Halliday and Wilson [118], at low die angles and very low speeds (0.015 to 0.14 m/s), the entrained film thickness was independent of speed and proportional to the applied coating thickness, indicating that it was not controlled by a PHD mechanism.

A problem with all these experiments is that thermal equilibrium could not have been established. Rittmann [56] found that temperature rose steeply for the first metre of draw even at the relatively low speed of 1.5 m/s in drawing steel bars 17 mm in diameter. Much higher temperatures and a corresponding drop in film thickness could be expected when drawing long lengths at high speeds. There is some doubt whether true fluid-film conditions can be established in practical circumstances. In the steel drawing experiments of Lancaster and Smith [119], the draw force (and calculated μ) dropped with speeds increasing from 1 to 20 m/s, whereas it should have risen under hydrodynamic conditions (unless heating reduced the viscosity to give lower drag). Similarly, in drawing of steel with lanolin [116], the film thickness was on the order of the surface roughness, even at speeds of 30 m/s. If full fluid films are to be established, soaps, with their readily controlled flow properties, offer the best chance of success.

1. Effects of Soap Composition. Wiredrawing soaps are seldom the pure metal salts of a single fatty acid (Sec.4.3.4); as mentioned in Sec.4.7.7, they are usually compounded with thickeners, primarily lime, in proportions ranging from 25% (for fine wire) to 75% (for heavy wire and bar).

The original rationale for such fillers was that they increase melting point and hardness [120,121]. Yet, as pointed out by Schmidt [122], melting point cannot be the prime criterion since sodium soaps are less suitable for heavy duties than the lower-melting calcium soaps of the same fat [123,124]. It is now generally recognized that viscosity at operating temperature is the controlling factor; hence thicker films are observed with calcium soaps than with sodium soaps.

Higher viscosity is needed for lower drawing speeds, as expected from hydrodynamic theory (Eq.3.29a), and for stronger materials. The viscosity of soap is greatly affected by temperature (Fig.4.9), which in turn increases with drawing speed and with the flow strength and diameter of the wire. Thus, the industrial practice of using leaner (higher-lime-content) soaps for heavy gages and strong alloys is well justified [125,126]. The effect of speed is twofold; initially, it increases film thickness, but then heat generation causes a drop in effective viscosity and film thickness declines (Fig.7.14a).

Effective viscosity determined with the aid of rotary viscometers may not necessarily describe the true response of the lubricant in the drawing die but has been found to give satisfactory correlation with plant performance [11], provided that the test is run at the die operating temperature (typically in the range from 50 to 250 C). A soap of excessive viscosity will not enter the die, and starvation results in incomplete film formation; a soap of insufficient viscosity forms a film of insufficient thickness and allows die pickup and wear. The effective viscosity is a function not only of the thickener type and concentration but also of the fat on which the soap is based [4.239],[127].

In addition to lime (CaO), other thickeners are also incorporated. Borax and soda ash (Na_2O) are used instead of or in addition to lime. Iron sulfate replaces the hydrated iron oxides found on sull-coated wire, and imparts a brown color to the drawn wire. Graphite is added for a black finish, and TiO_2 (which also acts as a polishing agent) for white color. Mica and talc, even though of lamellar structure, act only as parting agents. Often, MoS_2 is added to a sodium soap, sometimes in company with graphite and lime, mostly for difficult-to-draw, strong alloys. Yunusov [128] suggests that reaction with sodium in the soap and/or the borax coating leads to the formation of molybdates of low shear strength. Sulfur added to steel-drawing soaps acts as an E.P. agent. Evidently, experience has demonstrated that film thickness is frequently insufficient to ensure complete die/workpiece separation. It is not clear whether or not such additives affect lubricant rheology to a significant degree. They can, however, interfere with film formation, as found by Funke et al [6.341] on adding organic phosphonates to soap used in drawing limed wire. Occasionally, the additive, such as graphite, can be used as a coating applied prior to drawing [4.229].

In considering the rheology of soap one must remember that the soap undergoes changes in service. Under the high pressures and temperatures prevailing in the die, a sodium soap reacts with the lime coating to form a calcium soap [122,129]. Reactions with the wire, and especially reactions with oxides remaining after descaling [130,131] and with borax [128], are also possible. The quantity of solids gradually increases as the soap passes through several holes in a machine. The pressure and temperature cause changes in the soap structure itself, forming chains both shorter and longer than those originally present [127].

On nonferrous wire, sodium and potassium soaps are used as coatings without carriers, but their rheology is most likely affected by the changes due to pressure, temperature, and reactions with the substrate.

Of course, soap viscosity and composition matter little if the wire coating and/or wire roughness are inadequate to carry soap into the die. The soap must be of controlled particle size; beaded (granulated) soaps help avoid tunneling and reduce dusting. Vibration of the die box or agitation of the soap is necessary when the wire runs free of vibration or whipping which would aid feeding. Melting point is important in controlling feeding [120,122]. Excess soap picked up by the wire is rejected by the die; if it flows back in a molten form, the clumps thus formed deny further access of soap to the die. Sodium soaps are prone to caking because of their hygroscopic nature.

2. Modeling of Soap Films. There are numerous PHD theories of drawing with soaps, based on various models of soap behavior. There is no doubt that soaps are non-Newtonian substances [132], but not all theories take this into account.

Kolmogorov et al [133] assume a pseudoplastic behavior according to a power law

$$\eta = \eta_0 \dot{\gamma}^a \tag{7.11a}$$

where a is a constant. Montmitonnet et al [135], who give also a review of earlier work, use a thermo-piezo-pseudoplastic model to describe their results from capillary viscometry on a commercial, 60% lime soap, by an extension of Eq.3.26:

$$\eta = \eta_0 \dot{\gamma}^m \exp(\alpha p - \beta T) \tag{7.11b}$$

The Bingham behavior observed by Pawelski et al (Fig.4.9) [4.48] has apparently found no application yet to a single-hole die. Montmitonnet and Delamare [132] use a plastic-viscoelastic model based on hardness measurements. Felder and Breinlinger [134] propose simply a temperature-dependent viscosity

$$\eta = \eta_0 \exp(-\beta T) \tag{7.11c}$$

where β is the temperature coefficient of viscosity. Tattersall [136] lumps together the temperature and strain-rate effects by simply taking

$$\eta = \eta_0 \exp(-Bv) \tag{7.11d}$$

where B is a constant and v is drawing speed.

Any of these models can be made to fit the typical experimentally observed speed dependence of film thickness (Fig.7.14a). For very low speeds, the theory of Wilson and Halliday [137], based on a rigid-plastic lubricant model, can fit data [118]— probably because heating and strain-rate effects are negligible and the initial yield strength of the soap predominates. A further development of the theory [138] shows that film thickness is proportional to applied film thickness and increases with a larger τ_i/σ_f ratio, smaller α, and rounding-in at the entry.

An unconventional fluid-film mechanism is obtained when wire (in the reported case, copper) is cooled in liquid nitrogen (-196 C) and then drawn in humid air [3.168]. The observed low friction must be attributed to a form of phase-change lubrication.

In view of the difficulty of ensuring a full fluid film in a standard die, various attempts have been made to increase film thickness. Of the various mechanisms shown in Fig.3.36, rounding off at the die entry (Fig.3.36a) has been observed by Eichinger and Lueg [51]. The sigmoidal and convex dies designed for minimum redundant work provide a lower entry angle (Fig.3.36c). The effect of roughness (Fig.3.36d) is well documented and will be discussed later. At this point, methods of increasing the pressure of the entering lubricant (Fig.3.36b) will be considered.

Inlet Tubes. Christopherson [139,140] proposed that a tube of controlled clearance placed ahead of the die (Fig.7.16) will increase the inlet pressure. On starting up, such a system traverses the entire Stribeck curve (Fig.3.33). Initially, boundary lubrication prevails and there is danger of die pickup; friction is high, but it drops rapidly as the film thickness builds up, and then rises again when speeds become high enough to increase the frictional drag in the true PHD regime. The frictional drag imposes a back pull and sets an upper limit on the attainable inlet pressure. Measured pressures were high enough to cause deformation of the wire prior to entry into the die; this rounding

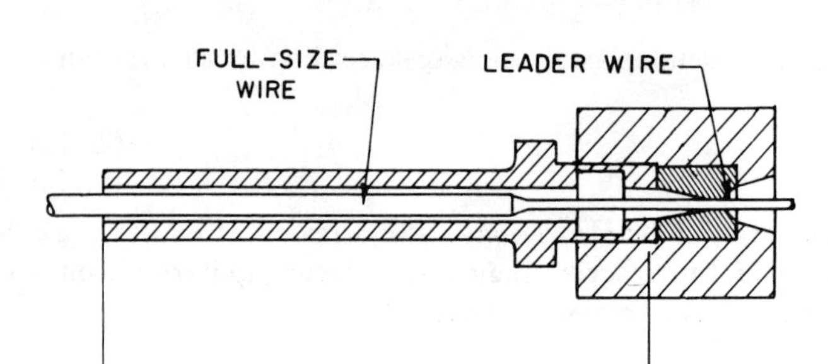

Fig.7.16. Inlet tube for generating hydrodynamic films with viscous fluids [139]

off improves lubrication in the die and also reduces the contact length, thus reducing the frictional force. The die shape becomes noncritical.

Christopherson and Naylor [140] obtained an analytical solution for the tube length required for a given pressure, starting from the premise that the wire occupies an eccentric position in the tube to minimize drag.

Osterle and Dixon [141] extended their numerical analysis to the deformation zone, allowed viscosity to vary with pressure and temperature, and incorporated heating effects, but agreement with experiments [140] was poor. Tattersall [136,142] calculated the flow rates for isothermal conditions allowing for a pressure-dependent oil viscosity. While pressures were in good agreement with experiments [140], calculated flow rates were too high. Calculations for optimum inlet tube dimensions were given also by Chu [143] and by Kartak and Strakhov [144].

There are, however, practical difficulties in executing this concept. The tube must be long (50 to 400 mm), the tube-to-wire clearance must be small (0.05 to 0.01 mm), and the wire must be kept to very close tolerances. Small particles (surface dross) inevitably present on the wire plug up the tube [145] and, even though drawing forces and (more importantly) temperatures are reduced, the economic benefits are generally judged doubtful [146]. Nevertheless, an increased die life is often reported, e.g., a tripling of die life by Nakamura et al [147] in drawing of stainless steel (see Sec.7.3.9).

A potential application of the inlet tube is in the coating of wire with a polymer, where the polymer serves also as the lubricant. Thompson and Symmonds [148] modeled the polymer as a Newtonian fluid characterized by an apparent viscosity. In a follow-up paper, Crampton et al [149] considered non-Newtonian flow with an experimentally determined pressure-sensitive viscosity and with the possibility that, at a critical shear stress, the polymer slips on the wire surface. Film thickness was found to be controlled by a so-called Sommerfeld number

$$S = \frac{6L\eta v\xi}{\sigma_f h^2} \tag{7.12}$$

where L is the length of the inlet tube, h is the radial clearance, and ξ is the slip factor (expressing the velocity of polymer adjacent to the wire). At low speeds and melting temperatures bambooing of the wire occurred; this was attributed to melt fracture (although disappearance of the defect at higher speeds could also have been caused by thinning of the film through heating).

Pressure Die for Soaps. The problem encountered with the original Christopherson tube led BISRA to modify the concept for soap lubrication (Fig.7.17) [23]. As reported by Wistreich [23,24,150] and by Sturgeon and Tattersall [151], the high apparent viscosity of soap permitted shortening of the inlet tube while still generating pressures up

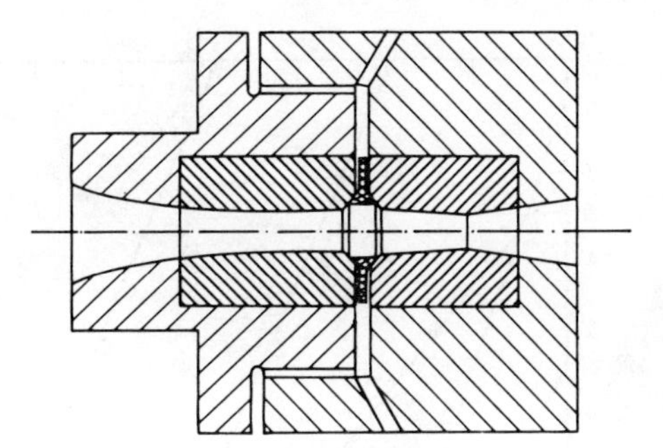

Fig.7.17. Nozzle die for promoting hydrodynamic lubrication with soaps [23]

to 500 N/mm^2, which were sufficient to cause yielding of steel wire. Since the soap is also an effective boundary lubricant and can be compounded with fillers [5.119], difficulties on start-up are mitigated. The inlet is short (12 to 25 mm), and the clearance is more generous (0.13 to 0.5 mm). Die life improved by a factor of two when only the first die had a pressure nozzle in multihole drawing of low-carbon and medium-carbon steel, and by a factor of 20 in single-hole drawing of stainless steel. Film thickness rose to about 6 μm at 0.5 m/s and dropped to 1 μm at 3 m/s (Fig.7.14a). A soap that lost little of its viscosity with temperature (or drawing speed) was best. It fed most uniformly at a particle size of 20 to 40 mesh (0.4 to 0.9 mm), and feeding was prompted by agitation. In the experiments of Tattersall [136], the maximum die pressure and film thickness were observed at different speeds for different soaps. Excessive film thickness led to an oscillatory behavior, resulting in bambooing.

These results could be rationalized through theory [136]. The apparent viscosity derived from measured pressures showed a power-law relation with drawing speed (Eq.7.11d) for any given soap irrespective of inlet geometry (Fig.7.18). Practical drawing speeds are much higher, and the apparent viscosity of soap should be much lower—on the order of the 15 Pa · s suggested by Pawelski [102].

More important than draw force is die wear. Nakamura et al [147] observed that, in drawing of patented, phosphated, 0.6% C steel wire, die life (defined as 0.14 mm wear on 4 mm diameter) increased from 40 to 110 tons of wire with the use of a pressure die.

Double and Multiple Dies. A further, industrially successful variant utilizes a double die. Dies of similar construction were apparently in industrial use in 1933, but were provided with a return passage to circulate the soap and reduce heating. An example is given by Hedman [152]. Double dies, aimed at inducing full-fluid-film lubrication, were developed by Kolmogorov et al, whose book [133] also contains a review of earlier work; an update is also available [153]. In this construction (Fig.7.19) the first die acts as a seal with a positive clearance to allow soap to enter. Tegel-Küppers [154] found that a radial clearance of 0.05 mm was needed in drawing a 3.5-mm-diam medium-carbon steel wire; Zhilkin and Latokhin [155] obtained maximum pressure with a clearance of 0.15 to 0.2 mm. The soap, drawn in on the wire surface, is compressed in the chamber between the two dies. The chamber fills up with compacted soap, and pressure that is sufficient to cause yielding in the wire builds up by the flow of the soap, in a film as thin as in an inlet tube. Pawelski [102] quotes an active layer thickness of 0.15 mm in drawing a 2.28-mm-diam steel wire. The active film thickness h_{act} can be calculated either from the pressure that gives maximum film thickness in the die or from the condition of minimum energy dissipation. Both approaches yield the same relation [102]:

$$h_{act} = \left(\frac{2\eta l v}{\Delta p}\right)^{1/2} \qquad (7.13)$$

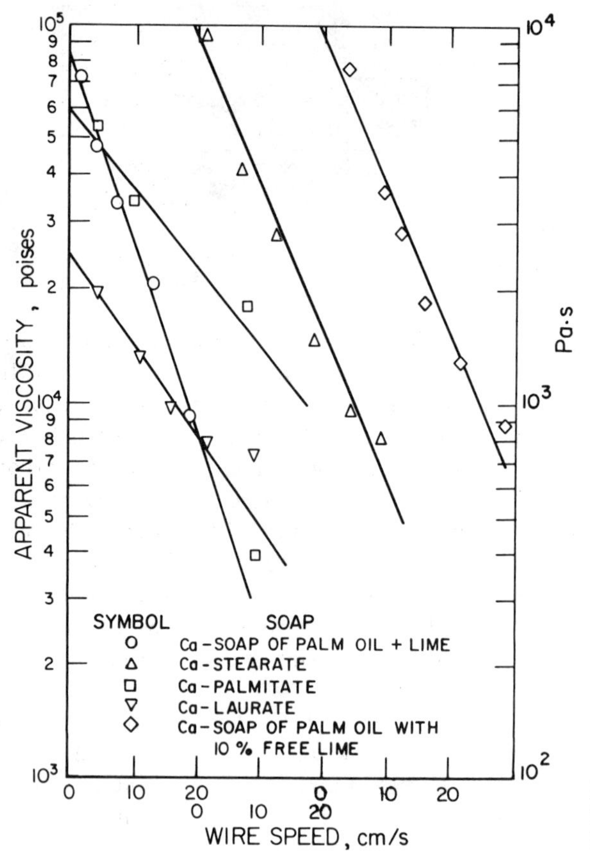

Fig.7.18. Apparent viscosity calculated from drawing experiments with different lubricants [136]

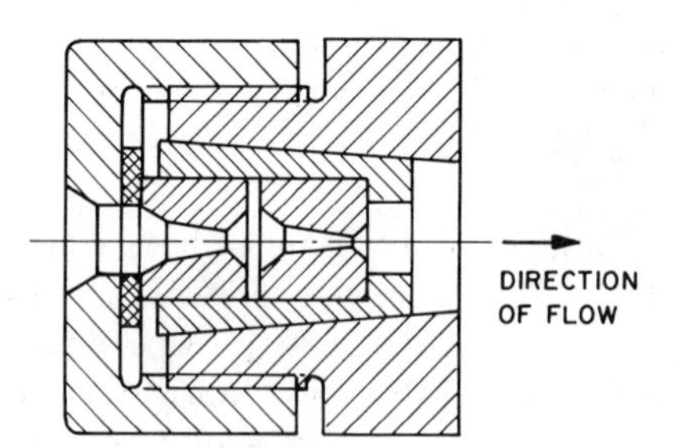

Fig.7.19. Double die for promoting hydrodynamic lubrication [133]

The similarity between Eq.7.12 and 7.13 is evident; Δp is the pressure drop in the tube, and reaches σ_f when the wire deforms prior to entry into the second die. If the film thickness is calculated on the basis of a Bingham model, the theory fits experiments when the initial shear strength is allowed to vary with speed (Fig.7.20) [154].

Draw force reductions of up to 30% have been reported [153,154,156,157]. Double dies have been used also with grease [158]. A form of double die, mentioned by Harper [159], worked with oil and improved lubrication sufficiently to destroy the desired bright finish of nonferrous wire. Wirth and Schatte [160] used three die nibs pressed into a common holder. Because a larger clearance (0.25 to 0.3 mm) was allowed in the first die, the tolerances of the hot rolled wire were less critical, pressure generation was limited, the danger of the die bursting was eliminated, and, even though draw forces were reduced by only 5%, die life increased greatly.

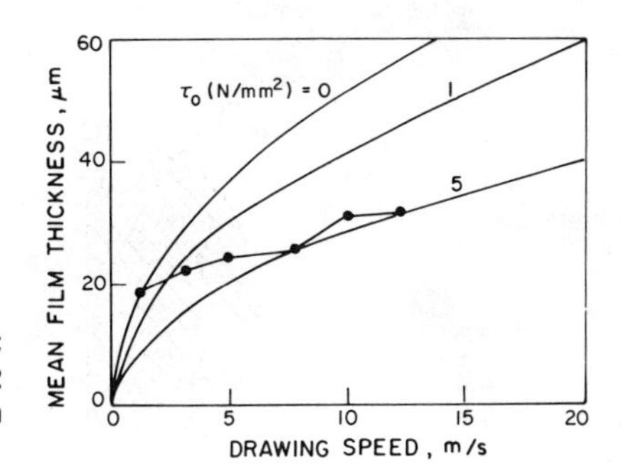

Fig.7.20. Film thicknesses measured in drawing 0.3% C steel wire (nozzle length, 37 mm; reduction from 3.4 to 3.2 mm diam with a Na-Ca-stearate soap) and calculated from a Bingham model (η = 1.5 Pa·s) [154]

Tegel-Küppers [154] found that five dies in tandem gave the lowest draw force when the first two dies acted as seals with a positive clearance and the last three dies performed the reduction. Draw force was then some 10% lower than in a double die, at least at speeds over 4 m/s. However, at low speeds the draw force was actually larger than in a single die, no doubt because the total redundant work was higher in the succession of light passes.

A special form of double die has been developed in Poland [161,162] for liquid lubricants. The dies are externally cooled, and back tension is developed in the wire by taking a reduction in the first die. Lubrication of this first die is ensured by provision of a short inlet tube (Fig.7.21). Reductions of up to 55% can be taken in double dies, with increased die life.

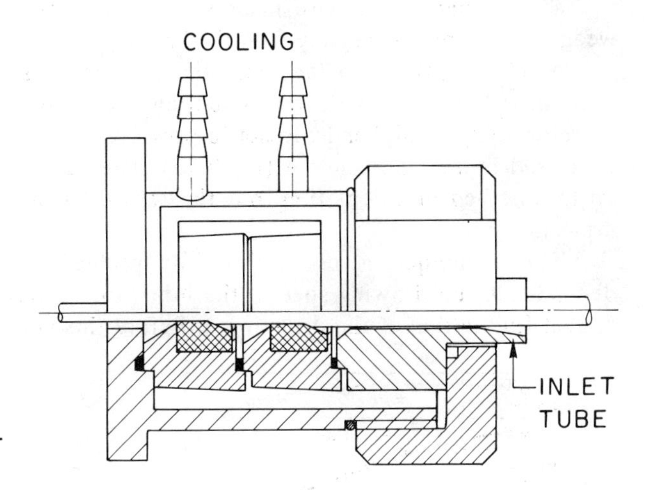

Fig.7.21. Die for hydrodynamic lubrication with viscous fluids [161]

Hydrostatic Die. Attempts at supplying the lubricant under pressure have been repeatedly made. Direct supply to the die/wire interface had to be abandoned because the orifice became blocked with debris [163]. Milliken [164] proposed a porous die material through which oil would be pumped. The idea of supplying oil through radial holes in the die was put forward by Tourret [165].

Some measure of practical success came only when the pressurized fluid was supplied to the chamber of a double die (Fig.7.22) [166]. The first die serves as a seal while the second die performs the actual draw. Kron [167] shows an industrial design dating from 1931. The attraction of this solution lies in its ability to establish a full fluid film at the beginning of the draw. A number of researchers have investigated the variables that affect success in drawing of aluminum [168,169], copper [166,170,171], and steel [166,172-175]. Several theories are also available [106,176].

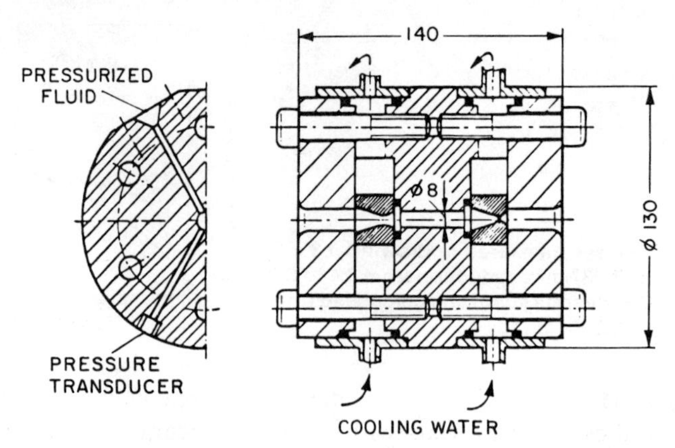

Fig.7.22. Hydrostatic drawing die [166]

Observations clearly show that substantial reductions in draw forces can be expected only when the applied pressures are high enough to cause initial yielding in the chamber. The results of Kopp and Moik [166] are typical. The draw force dropped continuously during drawing of copper, which exhibits gradual yielding, whereas a more sudden drop was noted on steel, with its better-defined yield strength (Fig.7.23). The relative benefits are greater at lower reductions, which are inherently less efficient. These benefits have several sources:

a. Yielding of the wire prior to entry into the working die reduces the effective wedge angle, thus increasing film thickness.

b. The length of contact with the die and thus the draw force are reduced. However, distribution of the same reduction over two dies increases redundant work, and therefore only at higher hydrostatic pressures will the draw force drop below the value measured in a single-hole die (Fig.7.23). The calculated value of μ depends very much on the method of calculation, but the results given in the lower graph in Fig.7.23 are typical.

This technique is not without its problems. At some critical pressure (which depends also on drawing speed), the lubricant gushes from the die [177], resulting in a violent fluctuation of the draw force. Before this point is reached, the pressure becomes

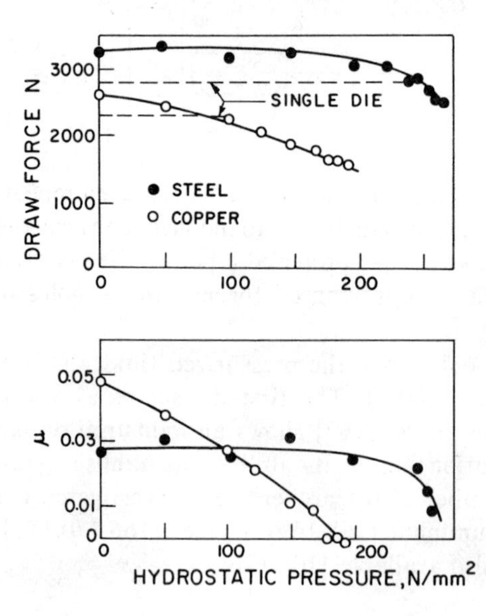

Fig.7.23. Effects of hydrostatic pressure on draw force and calculated coefficient of friction [166]

sufficient to cause necking of the wire. A strain-hardened wire is likely to fracture [169], whereas severe bambooing has been observed by Thomson et al [170] on soft copper. Even in the absence of bambooing, the surface may roughen to an unacceptable degree.

There are practical problems, too: a rather expensive high-pressure pump is needed, and the pressurized fluid must be cut off before the end of the wire passes through the draw die. Most importantly, the sealing (inlet) die is poorly lubricated and may suffer excessive wear.

This led Middlemiss [178,179] to replace the first die with an inlet tube, which created a fluid seal. The resulting configuration is similar to that shown in Fig.7.21. The inlet tube must have generated substantial hydrodynamic pressure, because the pump could be omitted for low-carbon steel but not for stainless steel [179]. Related to this hybrid hydrodynamic-hydrostatic system is the proposal of Thomson and Hoggart [171] to pressurize the fluid entering the inlet tube and thus protect the die during acceleration when the fluid film is not yet formed. Such hybrid systems may have a better over-all chance of practical application than the pure inlet tube or hydrostatic dies [145,180].

In yet another version of the pressure die, the inlet sealing die is replaced with a shaving die (Fig.7.24) [181]. The pressurized fluid lubricates the shaving process and ensures better surface quality.

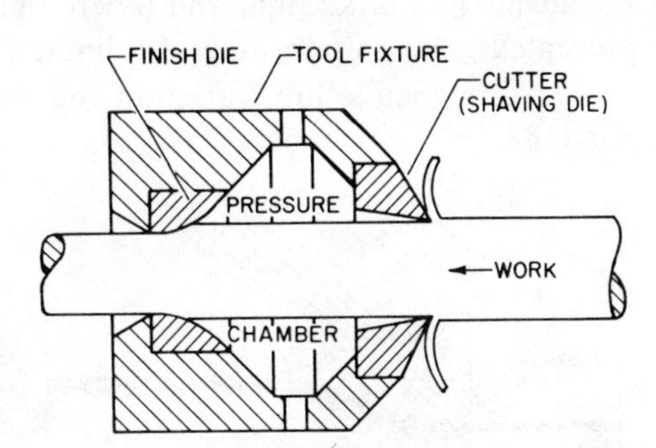

Fig.7.24. Schematic illustration of pressure-lubricated shaving/drawing die [181]

Hydrostatic Extrusion-Drawing. Total submersion of the die and wire in a pressurized medium [3.162],[182] has the primary advantage of increasing the workability of the wire while, incidentally, enhancing the film thickness. In practice such an arrangement would be cumbersome; therefore, hydrostatic extrusion (Sec.8.2.4) is used instead. A combination of extrusion and drawing is also possible, as discussed in Sec.8.2.4.

Tube Drawing. The possibility of increasing the film thickness in tube drawing has also attracted attention.

In tube sinking, a pressure tube or multiple dies can offer benefits [133]. Hydrostatic pressure can be applied to a double die [183], but collapse of the tube sets a limit. A die similar to that shown in Fig.7.19 was used also in drawing of copper and brass capillary tubes with a sodium soap [184].

In drawing on a plug, the critical problem is to increase lubricant supply to the plug/inner tube interface. Pumping of the lubricant through a hollow plug is feasible but in itself does not help to build up a lubricant film. For this, a double-plug arrangement, with supply of pressurized lubricant to the enclosed space, is needed. Matsuura [183] has proposed such a system. Kron [167] reports on results with a hybrid system utilizing a compound plug (Fig.7.25) that is similar to earlier Polish designs [185], but with pressurized fluid applied at the beginning of the draw. Draw force was reduced by

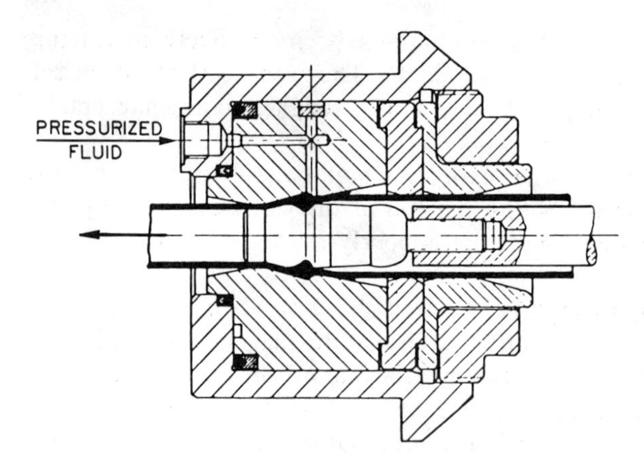

Fig.7.25. Construction for hydrostatic lubrication of both external and internal tube surfaces in drawing on a plug [167]

10 to 15% and reductions could be increased by some 10% on steel tubes, but the economy of operation remained doubtful. Some improvement in plug drawing is achieved simply by use of a suitable plug profile, as reported by Rees [186]. The oscillating floating plug (Fig.7.26a) is designed to trap lubricant; the pressure in the lubricant is assumed to force the shoulder back against the spring, creating an oscillating-pumping action. The adjustable floating plug (Fig.7.26b) creates a lubricant cavity and combines the advantages of straight and tapered plugs. The fluid pressure generated in the cavity prevented galling in drawing of a limited number of stainless steel tubes [186].

Hydrodynamic lubrication of the die/tube interface can be useful in drawing on a bar [187].

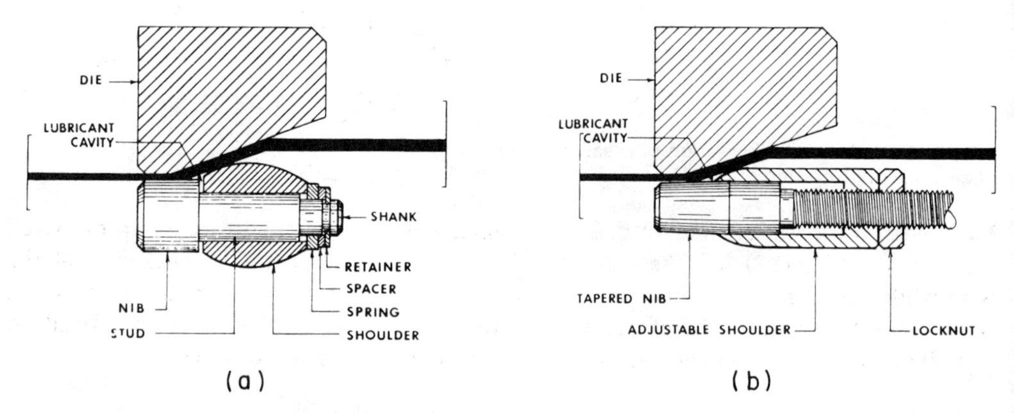

Fig.7.26. Oscillating (a) and adjustable (b) floating plugs [186]

7.3.2 Mixed-Film Lubrication

Even though, in contrast to rolling, there is no neutral plane in drawing, the two processes are similar enough for us to make our observations under the same headings. Lubrication mechanisms have been discussed in general [188-190].

Film Thickness. Some decades ago it was believed that only boundary lubrication could exist under the severe conditions typical of wiredrawing. However, evidence for mixed-film lubrication was gradually forthcoming. Relatively high lubricant throughput was measured with soaps [191,192], and Ranger and Wistreich [5.120] calculated wax film thicknesses ranging from 1 to 30 μm from electrical resistance measurements.

Funke et al [127] estimated 1.5 to 3 μm at speeds of 0.5 to 5 m/s. Lancaster and Rowe [193] used a radioactive sodium stearate soap in strip drawing and found that, compared with the R_a roughness of the strip, the film thickness was inadequate to cover all asperities but was certainly thicker than a boundary film (Fig.7.27). At speeds over 30 m/s, Lancaster [116] measured a film thickness of 2 μm with lanolin, and less than 1 μm with polyglycol, on steel that entered with a roughness of R_a = 3 μm and issued at R_a = 1 μm from the die. The presence of a partial fluid film was shown also by violent fluctuations in electrical resistance [5.120] and draw force [5.119],[194]. It is not clear what governs the magnitudes of these fluctuations. In the work of Flender et al [194], mechanically descaled steel wire was drawn in long enough lengths to establish thermal equilibrium. A soap that minimized fluctuations also reduced wear. Whether fluctuations were related purely to film thickness or also to boundary properties is difficult to decide: for soaps of high ash content (60%), lubricity (as judged from fluctuations) improved when the fat content dropped from 40 to 17%. In contrast, in low-ash (40%) soap, lubricity improved when fat content increased from 40 to 56%.

Hydrostatic pockets scattered on the surface of the wire also constitute evidence of a mixed-film mechanism. The proportion of these pockets or, rather, of the boundary-lubricated area (b in Eq.3.34a) was used by Nakamura et al [147] to rationalize the rate of die wear in both dry and wet drawing (Fig.7.28), thus supporting the existence of mixed-film lubrication.

Average film thickness has little meaning under these conditions. Nevertheless, it is a useful concept for visualizing process conditions. As indicated by Eq.3.29a, film

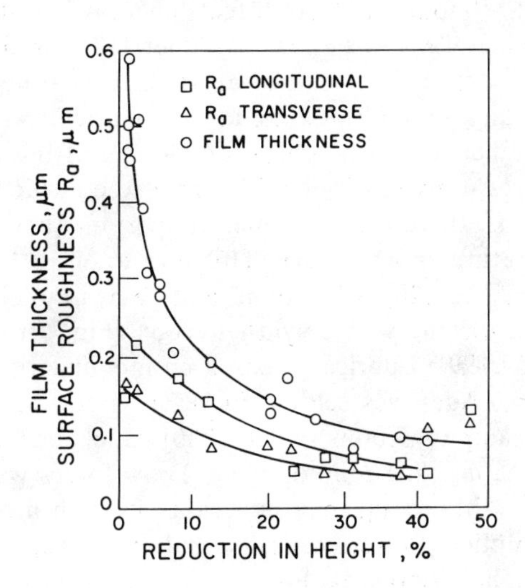

Fig.7.27. Film thickness measured in plane-strain drawing with radioactive sodium stearate [193]

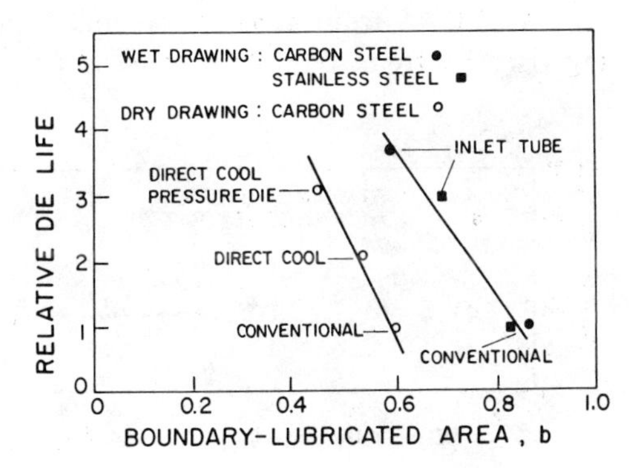

Fig.7.28. Relationship between die life and boundary-lubricated area in drawing of steel wire [147]. Dry drawing: 7.5 m/s, 0.6% C steel, patented, phosphate/soap lubricated. Wet drawing: 1.25 m/s, austenitic stainless steel.

thickness increases with decreasing entry angle. Indeed, it has been repeatedly observed that lubricant flow increases greatly with very low (about 2°) half angles [5.120],[192]. Even in low-speed experiments, Wilson et al [5.26],[118] observed a transition from thick-film to mixed-film lubrication on increasing α to some critical value.

Die geometry is important in industrial practice, too. Schmidt [122] notes that in drawing of steel wire 5 or 6 mm in diameter at speeds greater than 8 m/s, the half angle must be kept at about 6° instead of at the previously used 8 to 10°; obviously, the lower angle is needed to maintain a sufficiently thick film at the higher temperatures generated by the higher speeds. For this reason, cooling must be improved as well. Land length is also of importance. An excessively long land strips the wire of its soap coating, whereas a short land wears rapidly. Schmidt [122] suggests that, for steel wire 2.0 to 6.5 mm in diameter, land length (in mm) = $5/v$, where v is drawing speed in m/s. A similar relation may well hold for other materials and lubricants.

Viscosity. There is ample evidence that increased viscosity increases film thickness and thus reduces draw forces and measured or calculated values of μ.

The effective viscosity of soap increases with the addition of lime (Sec.7.3.1). Correspondingly, Schwier et al [195] found a heavier residual soap film and a rougher surface on the wire. A lime coating, which combined with the sodium soaps used, gave a rougher surface than a borax coating.

With oil-base lubricants, the base-oil viscosity is dominant, as shown by Tourret [196] in drawing aluminum wire with mineral oils (Fig.7.29). Bühler and Rittmann [197] found similar effects in drawing steel. The importance of the structure of the mineral oil was shown by Boor [198] in drawing patented steel wire at 8 m/s. For oils of identical viscosities at 40 C and atmospheric pressure, friction was highest for an aromatic, intermediate for a naphthenic, and lowest for a paraffinic lubricant. Thus, the higher viscosity index of the paraffinic oil more than counterbalanced the higher pressure-viscosity coefficient of the cyclic and aromatic oils (see Table 4.1 and Fig.4.3). It is, therefore, extremely important that viscosity be measured at the operating temperature and pressure. This is true of all lubricants, as illustrated by Ranger and Wistreich [5.120], who found a large drop in electrical resistance (indicating a thinning film) in drawing with a synthetic wax (Fig.7.30). Similar effects have been observed with oils [54,199]. Lubricants based on fats undergo solidification.

Additives seldom increase viscosity, except when added in large proportions. For example, Kotov et al [200] thickened an oil with 10% sodium soap powder for floating-plug tube drawing. Draw forces were reduced by 20%.

Many aqueous lubricants used in drawing are water-diluted soaps, and increasing dilution (lower viscosity and, perhaps, lower boundary lubricating power) results in higher friction both on nonferrous metals [196] and on steel [201].

As follows from general principles, the optimum die angle decreases with increasing viscosity from some 9° with oil to 6° with dry soap [76].

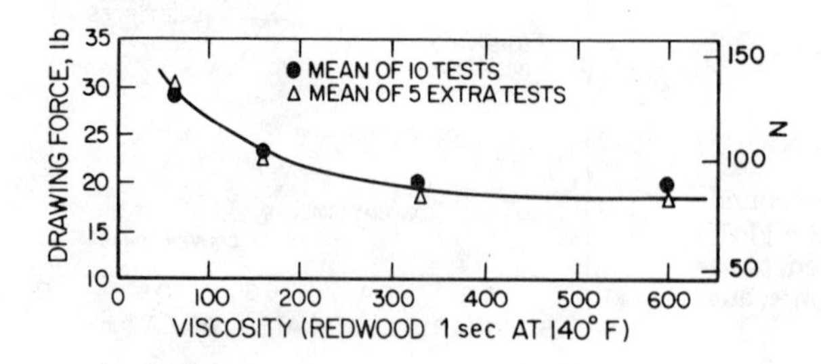

Fig.7.29. Effect of oil viscosity on force required for drawing of aluminum wire [196]

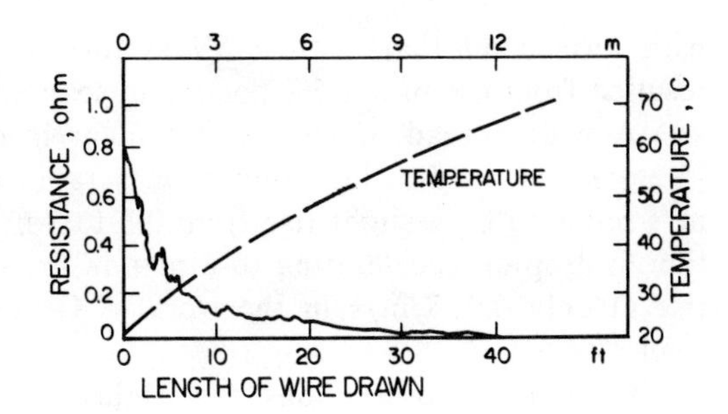

Fig.7.30. Effect of length of wire drawn on temperature and electrical resistance between die and wire [5.120]

Speed. In the mixed-film regime, increasing speed should increase the film thickness and reduce friction (or draw force) (Fig.3.42) until heating causes the film to thin again, with an accompanying rise in friction.

Decreasing draw forces have been noted by Pomp and Becker [202] in drawing of steel with soap at speeds up to 7.6 m/s, by Ranger [203] in drawing of steel with soap at speeds up to 10 m/s, and by Komura [5.225],[204] in drawing of steel with methyl stearate at speeds up to 1 m/s. Increased electrical resistance also gives evidence of increasing film thickness (Fig.7.31) [5.120] in drawing with a wax. Excessive thinning occurs at higher speeds only; Wistreich [39] noted rising friction with soaps at speeds greater than 20 m/s.

With oils the speed effect becomes noticeable at lower speeds, but so does excessive film thinning at high speeds. Friction dropped in drawing of steel bars with speeds increasing to 0.8 m/s in the work of Luksza et al [205]. Bühler and Rittmann [197] noted a slight upturn at only 1 m/s in bar drawing. In wiredrawing, speeds are much higher. A quite dramatic drop in draw force at speeds up to 8 m/s was reported by Soshko [4.65] in drawing of fine stainless steel wire with a short-chain lubricant to which 0.5% low-density PE had been added. In drawing patented wire with a mineral oil/fatty oil lubricant (viscosity, 500 mm^2/s at 40 C), Boor [198] found a rapid drop in friction as speed increased from 1 to 8 m/s, followed by an upturn from 8 to 12 m/s. In contrast, the speed effect was virtually absent when phosphated wire was drawn with either rapeseed oil or a drawing fat, suggesting that film thickness was controlled by impregnation of the conversion coating rather than by speed. The speed effect is usually more noticeable on softer workpiece materials and with lighter lubricants. It is evident also with emulsions. Kiss [54] observed an increase in draw force at speeds above 12 m/s in drawing soft low-carbon steel wire with a potassium soap emulsion, but a much smaller change in drawing strain-hardened wire. In drawing steel, Komura [5.225] noted a drop in μ from 0.1 to 0.07 on accelerating from 0.1 to 1.0 m/s. A soap solution of poor adhesion can give the reverse effect [197]. Olsen and Larkin [206] observed that the draw force peaked at 1 m/s in drawing of fine wire. Surprisingly, the draw force was higher with oil than with an emulsion. In contrast, Kiss [54] found lower forces

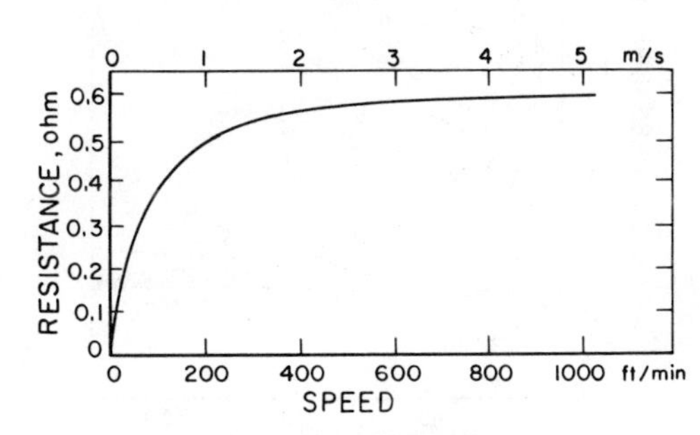

Fig.7.31. Effect of drawing speed on electrical resistance between die and wire [5.120]

with linseed oil than with a 5% sulfurized castor oil emulsion, while highest forces resulted from use of a 2.5% potassium soap emulsion.

A wide spread of results is evident in data on plane-strain drawing. Kudo et al [5.25] noted, in drawing aluminum, a rapid drop in friction as speed increased to 0.5 m/s followed by a slight rise from 0.5 to 3 m/s. Devenpeck and Rigo [5.27] found friction to drop on accelerating to 6 m/s in drawing of steel. In contrast, friction began to rise at only 0.075 m/s in the work of Gräbener and Lange [207] on medium-carbon steel.

Obviously, observations are highly specific according to the conditions of the experiment. The range of drawing speeds which provide minimum friction without the danger of lubricant breakdown can be quite narrow, and speeds are usually kept at a safe level to ensure maximum die life and good product surface [99]. Effective cooling can be more influential in allowing an increase of speed than a change of lubricant, because the limit is set by the rapidly increasing die wear resulting from lubricant breakdown at high temperatures. Since die temperature rises roughly with the square roots of speed and wire diameter [45], die wear also increases with the square root of speed.

Interface Pressure. The mean die pressure p_m is only slightly affected by friction (Fig.7.32) [6.134]. Pressure drops rapidly with increasing reduction and, for a given reduction, is higher for larger die angles. The effect of pressure is often difficult to separate, because other factors that affect film thickness also change simultaneously, and observations can be contradictory. It is important to distinguish clearly between cases where the interface is filled up with lubricant and those where some empty space remains (Fig.3.44).

Kudo et al [5.25] could vary interface pressure while keeping other variables constant in their plane-strain drawing test with imposed push or pull (Fig.5.5c). With a wax, μ = 0.02 to 0.03, independent of p_m, as would be expected for a full film of a semisolid of pressure-dependent shear strength. This agrees also with the constant μ = 0.02 calculated by Boor [198] for drawing of phosphated wire at 1 m/s with a calcium soap; increased die pressure was obtained by successively drawing the wire in six passes.

With liquid lubricants the μ versus p_m plots of Kudo et al [5.25] exhibit a minimum, as shown in Fig.3.44. The pressure was not truly independent of material variables, because different materials had to be used to obtain a wide range of p_m. Pawel-

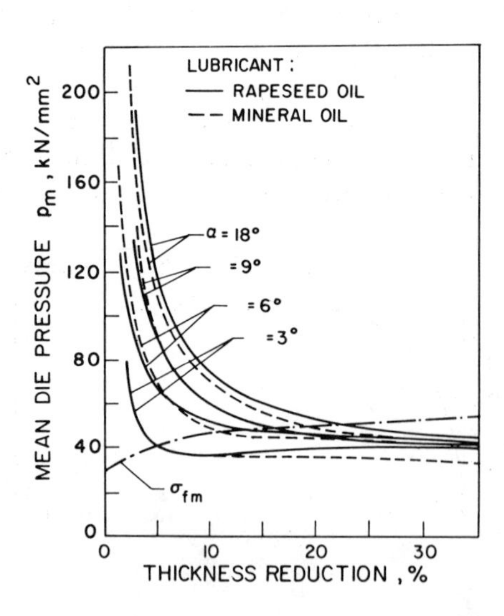

Fig.7.32. Effect of lubricant on mean die pressure in drawing of steel strip [6.134]

ski [6.134] found μ to increase with p_m (Fig.7.33), even when p_m was varied by prerolling the same material to different degrees. This is, indeed, the general trend when the interface is fully flooded but not fully separated, as in the work of Boor [198] with patented but not phosphated wire, and even with phosphated wire when a drawing fat was used as the lubricant. The same relationship was observed by Lueg and Treptow (Fig.7.34) [208] and by others [192,196,209,210],[4.239]. A μ versus p_m curve showing a minimum [5.22],[207] can be explained only by assuming a partially filled interface.

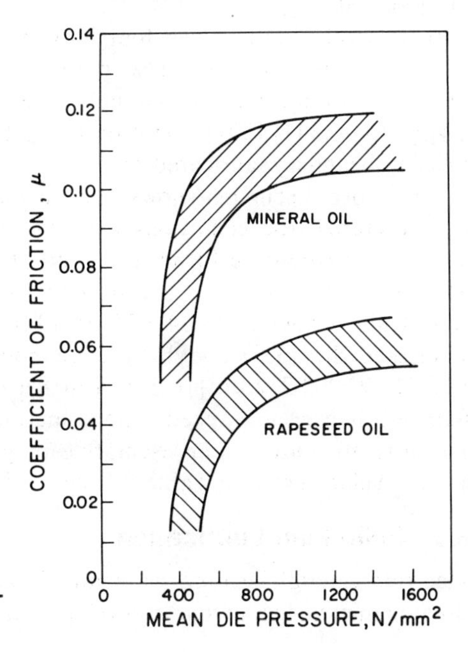

Fig.7.33. Effect of lubricant on measured coefficient of friction [6.134]

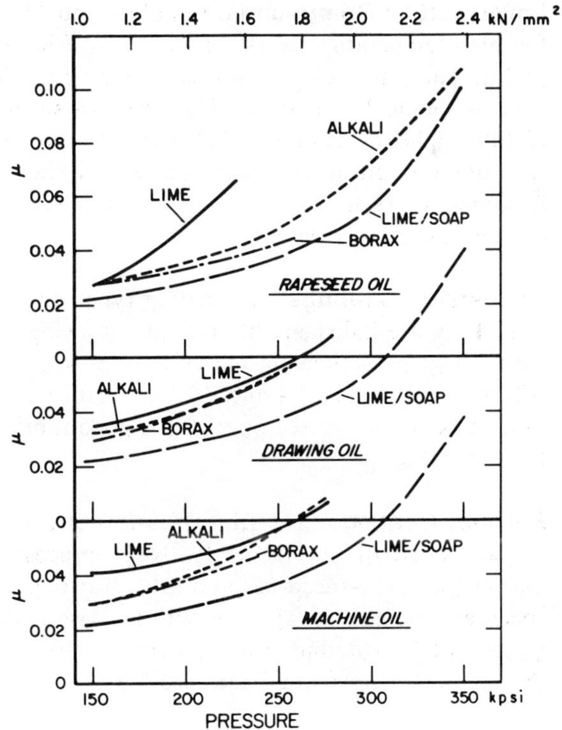

Fig.7.34. Effect of die pressure on coefficient of friction in drawing of 0.53% C steel wire [208]

Effect of Reduction. With increasing reduction the surface area expands more, more virgin surface is exposed or, at least, the boundary-lubricated area is increased, and thus friction rises, too. However, increasing reduction also reduces interface pressures (Fig.7.32), and the over-all effect on μ is variable and difficult to predict.

Kudo et al [5.25] kept p_m constant while increasing the reduction. In drawing of soft aluminum with a compounded oil, μ dropped, whereas in drawing of hard aluminum it rose. With wax, μ increased, presumably because the boundary-lubricated area increased.

In general, friction was found by most workers to increase with pass reduction [35,37,42]. This is consistent with the drop in film thickness found by Golden et al [211] by a radiotracer technique. In experiments where μ dropped with pressure [5.24],[6.134,6.341],[207], the drop in p_m must have outweighed the increase due to the increasing boundary-lubricated area.

The above discussion shows that no single value of μ or other descriptor can possibly characterize the conditions existing for a wide range of speeds, geometries, and reductions. It would be futile to compile lists of μ values; in addition to the effects of process variables, μ values are influenced also by the method of their determination (plane-strain drawing [5.24,5.25],[6.134],[207]; split die [5.120],[38,39,41,42,192]; strain-gaged die [37]) and by the method of calculation from measured draw force [33,35,76,192,203,212,213],[6.57]. Furthermore, many laboratory measurements are conducted at very low speeds and sometimes on lengths as short as 25 mm, and thus the results are quite unrepresentative of production practices. For predictive calculations, the values listed in Table 7.1 should be adequate.

7.3.3 Solid-Film Lubrication

Under this general heading we shall discuss all lubricant systems that do not need a hydrodynamic wedge action to develop a film.

Layer-Lattice Compounds. Graphite and MoS_2 are seldom used on their own, except for high-temperature work [214]. Their friction is found to be independent of speed [5.25]. The effect of pressure is controversial. With a truly continuous, well-developed film, μ should be constant (Fig.3.19). In sheet drawing, μ has been observed to drop [3.149],[5.24] or to rise [5.25] with pressure. It is conceivable that in some instances the film was burnished and improved, while it may have lost its continuity in other experiments. Generally, μ rises with reduction, most likely because surface extension exposes more asperities.

Conversion Coatings. All coating types discussed in Sec.4.7.7 and 4.7.8 find application. They are indispensable for tube drawing, low-speed bar drawing, and wiredrawing with heavy reductions on materials that are otherwise difficult to lubricate. Examples will be found in Sec.7.4 and 7.7. A phosphate coating in itself does not reduce friction; lower forces are ensured only with a good lubricant, as shown by Hantos et al [215] and by Boor [198].

Polymer Coatings. The film thickness of a solid coating deposited from a solvent can be controlled quite independently of process conditions (Sec.4.7.2); this is a great advantage in low-speed bar and tube drawing. The friction of coatings is independent of speed as long as no major temperature rise is encountered. PTFE has often been investigated [207,216], but other polymers, closely related to the coatings used on sheet (Sec.10.4.2), have found wider use [58].

Table 7.1. Commonly used drawing lubricants and typical μ values (lubricants listed according to increasing severity of conditions)

Material	Wiredrawing: Lubricant(a)	μ(b)	Bar and tube drawing: Lubricant(a)	μ(b)
Steels	Over 1 mm: dry (Ca-Na) soap on		Heavy oil, soap-fat paste, grease	MF
	lime or borax	MF	(+E.P.) (+MoS_2, etc.)	MF
	phosphate + soap	MF	Polymer coating + E.P. oil	0.07
	Under 1 mm: EM (M.O. + fat + E.P.)	0.07	Phosphate + soap	0.05
	phosphate + EM	0.1	Metal + M.O. (or EM)	MF
	Metal (Cu, Zn, brass) + EM	MF		
Stainless steels and Ni alloys	EM (M.O. + Cl) (on lime)	0.1	M.O. + Cl additive	0.15
	M.O. + Cl additive (on lime)	0.07	Chlorinated wax	0.07
	Chlorinated paraffin, wax	0.05	Polymer (chlorinated) (+M.O.)	0.07
	Oxalate + soap	0.05	Oxalate + soap	0.05
	Metal (Cu) + EM (or oil)	MF	Metal (Cu) + M.O.	MF
Al and Mg alloys	M.O. + fatty derivatives	MF	M.O. + fatty derivatives	MF
	Synth. M.O. + fatty derivatives	MF	Soap coating	0.07
			Wax (lanolin) coating	0.05
			Polymer coating	0.05
Cu and Cu alloys	EM (M.O. + fat) (+E.P.)	MF	EM (fat)	MF
	Metal (Sn) + E.M. or M.O.	MF	M.O. (+fat) (+E.P.)	MF
			Soap film	0.05
Ti alloys	Oxidized + Cl oil (wax)	0.15	Polymer coating	0.07
	Fluoride-phosphate + soap	0.1	Oxidized + Cl oil (wax)	0.15
	Metal (Cu or Zn) + soap or M.O.	0.07	Fluoride-phosphate + soap	0.1
			Metal + soap	0.07
Refractory metals	Hot: GR coating	0.15	Hot: GR coating	0.15
	Warm (cold): GR or MoS_2	0.1	Warm (cold): GR or MoS_2	0.1
	Oxidized + wax	0.15	Oxidized + wax	0.15
	Metal (Cu) + M.O.	0.1	Metal (Cu) + M.O.	0.1

(a) EM = emulsion of ingredients shown in parentheses. M.O. = mineral oil; of higher viscosity for more severe duties, limited by staining. E.P. = E.P. compounds (S, Cl, and/or P). (b) MF = mixed-film lubrication; $\mu = 0.15$ at low speeds, dropping to 0.03 at high speeds.

Metal coatings are among the most powerful aids to lubrication. In many wire applications they may be left on the surface or, indeed, are needed for service functions. All considerations discussed in Sec.4.7.1 apply.

7.3.4 Lubrication Breakdown

The favorable effects of process geometry and speed on film formation are counteracted by heat generation, severe rubbing, and their consequences.

As we have noted, the temperature rise can be high—on the order of 300 C even in hydrodynamic drawing. As shown by Asada et al [217] by means of finite-element calculations, temperature is most effectively reduced by lowering friction, because cooling of the die (or, in tube drawing, of the plug) is of little avail. The speed limit is set by the failure of the lubricant film; in multihole drawing the entry temperature of the wire must be controlled, too, by interpass cooling.

Indeed, the cooling and lubricating effects can be difficult to separate; in multihole drawing of low-carbon steel wire, Kiss [54] measured a die temperature of 120 C with linseed oil and only 70 to 80 C with emulsions, even though the linseed oil gave lower draw forces. Obviously, the greater cooling power of emulsions outweighed the friction-reducing capacity of linseed oil.

Metal-to-metal contact is a consequence of lubricant breakdown, and, under favorable conditions, leads only to wear of the workpiece. The accumulation of wear debris is readily observed, even visually. Details of the mechanism have been followed by radiotracer techniques. Fischer et al [218] observed a continual interchange of brass between die and wire, and such dynamic equilibrium is likely to be true of all steady-state situations. Accumulation of pickup on the die would terminate the process [219]. A good lubricant should, therefore, minimize pickup by reducing friction and thus temperature, and it should also extend the temperature range over which a protective film is maintained.

Another cause of lubricant breakdown is deterioration of the lubricant. This includes the accumulation of wear debris. The debris enters the interface, becomes lodged in the entry cone, and contributes to film breakdown and die wear. Heating also leads to the gradual deterioration of both oil-base and aqueous lubricants [220]. Particularly in compounded lubricants, formation of insoluble soaps [24,221], reactions with metallic debris, and formation of friction polymers all contribute to gradual changes. In sum total, the effects are likely to be harmful, and one of the purposes of lubricant management is the maintenance of constant quality.

Lubricant breakdown leads to a visible change in surface finish. Under steady-state conditions the large boundary-lubricated and breakdown zones impart a bright finish, and, for this reason, drawing—especially finish drawing—is often conducted under marginal lubrication conditions. However, surface damage in the form of longitudinal scratches or heavily smeared areas is undesirable and is a sign of total lubricant breakdown.

7.3.5 Instabilities and the Application of Vibration

Periodic changes in dimensions and surface finish can arise from several sources.

Chatter Marks. Even though a draw die is much stiffer than a rolling mill, the low stiffnesses of the wire and draw bench (or bull block) combine to allow periodic effects to develop. These are related, at least in the mechanical sense, to stick-slip (Fig.3.56). A prerequisite is that μ should drop with increasing speed. This condition is always fulfilled in mixed-film lubrication, but chatter and consequent periodic changes in surface finish are observed only under marginal lubrication conditions, mostly in difficult-to-lubricate materials. Bambooing, with its very marked periodic change in diameter, is more typical of full-fluid-film conditions.

In tube drawing the presence of the plug introduces a further component of low stiffness into the system. Therefore, chatter is much more frequently observed in drawing with a fixed plug. The frequency of vibration is below the natural frequency of the system. The amplitude of vibration can be high enough to break the bar holding the fixed plug.The stiffness of the bar may be increased by increasing its diameter; hence plugs are often brazed rather than held on a nut (Fig.7.12b and c). A floating plug gives a much stiffer system in which the drawn tube is the least stiff element, and thus, as pointed out by Cook [222], the problem appears only after some length of tube has been drawn. With increasing stiffness the frequency of vibration increases and its amplitude decreases. Excessive vibration may cause the tube to be severed.

Analysis of the problem is not simple. Jones and Harvey [223,224] reported on both experimental and theoretical work. In agreement with general principles (Sec.3.9.1), chatter set in when a high starting μ dropped rapidly with speed, and the damping coefficient of the system was negative. Significantly, no chatter could be induced with a solid soap film. The plug configuration had an effect, too: chatter was more likely to occur when the land was too long or when the die-to-plug angle (α - β in Fig.7.12d) was too large.

Wear reduces the plug angle and thus the restoring force on a floating plug, and die wear could even bring about a divergent gap, resulting in a cut tube.

Rotating Dies. One method of eliminating frictional instabilities has been proposed by Linicus and Sachs [64]. The die is rotated at a relatively low speed (typically 30 to 150 rpm): since there can never be a stick phase, there is no chatter either. Conceivably, lubrication could also be improved. More importantly, incipient pickup points are sheared and, hopefully, the debris leaves the die. The direction of the friction vector changes and its axial component is reduced (Fig.7.35). This could result in a large drop in friction and draw force [86,225] also when a rotating pipe is plunged into a draw die [226]. In practice, the gain is most evident at low drawing speeds. At high speeds the main benefit is the brighter, polished appearance of the wire [227]. The increased complexity of the die can be justified only for difficult-to-lubricate materials such as alloy steels [228,229], although rotation has been applied to aluminum drawing dies to equalize die wear and produce round wire [230].

This technique can be used also for friction determination. The wire is drawn first with the die stationary, then with the die rotating. If the lubrication mechanism is not affected (a doubtful proposition), a μ value can be calculated from the difference in draw force [64].

Nakayama and co-workers [231] explored, in a series of papers, the use of a test in which a rotating tube is plunged into a stationary die. Lubrication is typically of the mixed-film type and lubricant variables can be explored, although interpretation in terms of other processes is difficult.

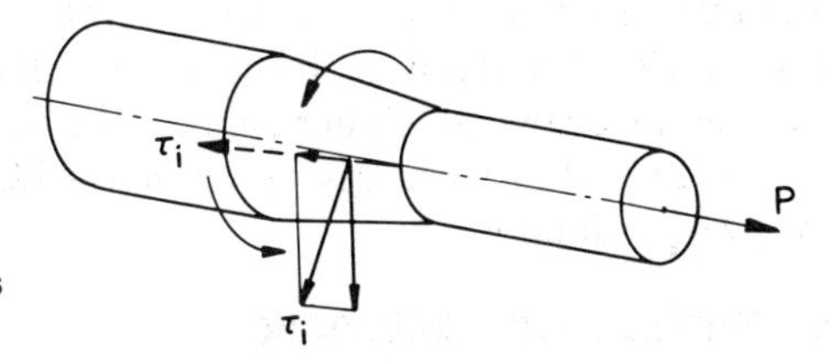

Fig.7.35. Change in frictional stresses during rotation of the die.

Vibrating Dies. As discussed in Sec.3.9.2, vibrations (often of ultrasonic frequencies) can be used to stabilize the drawing process. In addition to the general reviews quoted there, reviews specific to wiredrawing [232,233] and tube drawing [6.324] have been given. The present discussion will cite only some later references.

1. Wire and Bar Drawing. An improvement in friction conditions can be expected when the die is axially vibrated and lubrication is marginal. The effect of vibration decreases with increasing drawing speed (Fig.7.36) [234]. The effect is particularly marked in multihole drawing, where oscillatory back stress seems to result in a genuine reduction of draw stress [233,235]. Radial and tangential vibrations have been found effective, too; the latter can be regarded as a variant of the rotating die [236]. Some ultrasonic cleaning effects also occur [237], and stick-slip is eliminated [238]. The properties of the drawn product are not materially affected, although more uniform hardness and a different texture have been found [236].

It appears that the most successful industrial applications of vibrating dies are

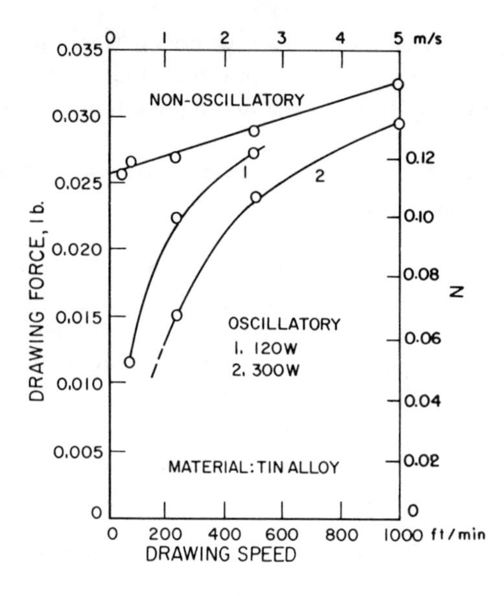

Fig.7.36. Effect of ultrasonic vibration on draw force [234]

going to be in fine wiredrawing [239,240,241] and in drawing of very soft wires such as lead-tin solders [242] and of difficult-to-lubricate materials such as titanium alloys [238,243] and molybdenum alloys [244].

2. Tube Drawing. The difficulty of lubricating the inner surface of a tube and the danger of chatter have made tube drawing on a fixed plug a prime candidate for the application of ultrasonics. Vibration applied to the die is still occasionally reported to give benefits [245], but most applications concentrate on vibration of the fixed plug. There is a measurable reduction in draw force beyond the replacement of static force by ultrasonic force, most likely as a result of reduced friction. The surface finishes of tubes are also superior [246], chatter is suppressed [247,248], and the technology is of considerable practical value in drawing of thin-wall, hard-to-lubricate tubes [249] of stainless steels, superalloys, and titanium alloys. In drawing of aluminum alloys and other materials drawn at higher speeds, the benefits can disappear at production speeds [250]. A theory of the process by Atanasiu [250] incorporates the doubtful assumption of flow-stress reduction.

7.3.6 Surface Characteristics

The surface finishes of both die and workpiece have powerful effects.

a. Sliding over the die surface demands the best possible finish. Therefore, dies are invariably polished and, if at all possible, are finished by polishing in the direction of material flow. There is, however, considerable flexibility in choosing the roughness of the entering wire.

b. The importance of surface roughness in carrying the lubricant into the interface was first demonstrated by Lancaster and Rowe [5.23,5.28] with shot-blasted surfaces. Since then, surface effects have been repeatedly explored, often in plane-strain drawing. A finely roughened surface, such as that produced by etching or fine shot blasting [56,251], transports more lubricant in uniformly distributed micropockets and keeps the boundary-lubricated area small. Therefore, μ is low [3.149],[5.149], and heavier reductions can be taken without pickup. In contrast, a very smooth surface gives high and erratic friction, and may cause die pickup and workpiece scoring. However, a very rough, heavily shot-blasted surface is not desirable either, because the large boundary contact area increases friction [3.149]. Shot blasting followed by etching can be much more favorable [56]. With increasing drawing speed the importance of workpiece roughness diminishes.

c. The surface finish of the end product is determined by the starting roughness and the operative lubrication mechanism (Fig.3.55). The original roughness persists to quite heavy reductions, and an etched bar gives a smoother wire than a shot-blasted bar. Smoothing proceeds more rapidly if there are longitudinal features which allow escape of the lubricant. Examples are sulfide inclusions in steel, which are elongated during hot rolling and which etch faster during pickling [197]. Whether roughening or smoothing occurs depends on the film-thickness ratio (Eq.3.22). Assuming a simple triangular roughness profile, Felder [252] developed a simplified hydrodynamic theory to account for these effects and compared it with the results of Kudo et al [5.25].

d. Roughness changes occurring during drawing and their effects have been repeatedly studied [253-255]. The characteristics of production bar and wire surfaces have been compiled by Springmann [256]. Typical roughness distributions are given in Fig.3.48; the almost Gaussian distribution of pickled bar becomes skewed as more plateaus form. The rate of smoothing increases with diminishing film thickness. Therefore, use of a steeper die angle and a lower-viscosity lubricant results in a smoothing and polishing of asperities; a brighter wire is produced, provided, of course, that no lubricant breakdown or die pickup occurs [6.172], [257-260]. The thick film formed in dry drawing makes the wire dull and rough. In wet drawing, the lowest viscosity that still ensures protection against breakdown is chosen to produce a brighter wire. A solid lubricant such as graphite, which does not flow at all, remains lodged in surface pockets and yields the roughest product [197].

e. One of the problems remaining is linked to the best method of characterizing the surface. Roughness measurements made with stylus instruments are often taken longitudinally, for convenience. When longitudinal features are present, this method gives readings that are too low, and measurement in the circumferential direction is essential [261]. Simple measures such as maximum or mean roughness height are uninformative, and a composite measure such as the fullness of the profile is more meaningful. Statistical analysis of roughness [262] certainly deserves more attention. Plateaus and hydrostatic pockets are most vividly seen under the microscope, especially the SEM. Nakamura et al [147] used photos taken at 500x to detect the boundary-lubricated area by quantitative metallography. The SEM is also a powerful tool in studying surface defects [5.154],[263,264].

f. Surface defects of nonfrictional origin have been described in general by Thompson and Hoggart [6.147], for steel by Kiefer [265], and for aluminum by Wright and Male [264]. Defects include laps, seams, pockets, etc. inherited from prior processing, and defects of material-related origin. These can interfere with lubrication and greatly increase wear, apart from impairing the finish of the product.

g. Most wire is further manufactured into cables, ropes, weaves, fencing, barbed wire, and various bent and formed (upset, etc.) products [10]. To ensure that the surface of the drawn product is of acceptable quality for the subsequent application, the surface composition must also be controlled. The residual lubricant film may be minimized by thinning the film in the last draw or draws, with the use of a steeper die angle, finishing in a wet hole on a dry drawing machine, or using wet drawing only. If necessary, the residue must be removed. Sophisticated surface-analytical techniques are used when characterization of the surface is vital, as for brass-coated tire-cord steel wire [266]. In other applications a controlled amount of lubricant must be available for subsequent working, and dry drawing or wet drawing with a surface coating becomes necessary.

7.3.7 Lubricant Selection and Evaluation

General lubricant recommendations have been given by several authors [3.58],[4.2,4.12],[125,267-273]. Selection on the basis of an empirical data bank was suggested by van Hessche [274], while Rowe [188] proposed to set up a data bank by drawing at 20 and 40% reductions using dies with 4 and 20° half angles, with drawing stress, die pickup and wear, stability, and other criteria taken as the parameters.

Lubricants range from soap solutions, through soap-fat compounds (anionic emulsions of fats) and nonionic emulsions, to mineral oils or compounded mineral oils. The so-called synthetic fluids [275-277] should, in the terminology adopted in Sec.4.6, be described as synthetic emulsions of small particle size.

Lubricant evaluation techniques were reviewed by Boor [270]. Evaluation in the laboratory is often done in low-speed drawing, at speeds insufficient to generate relevant conditions, even though a heavier reduction can go some way toward increasing the severity of the draw. It is better to use high-speed equipment, although repeated draws on a single bull block are not equivalent to drawing in a multihole machine [278]. Tests on short lengths of wire are inadequate [279]; long lengths should be drawn even in strip-drawing tests [5.27].

For friction evaluation, at least the draw force must be measured. Production units can be more difficult to instrument, especially for fine-wire drawing [206].

Simulation tests have been used repeatedly, but correlation is, as might be expected, poor. Only selected aspects of the lubrication mechanism can be explored. Thus, McFarlane and Wilson [280] used a pin-on-disk tester to evaluate boundary lubricants for tube drawing, with only limited success. Nagymányai et al [281] used the four-ball and Timken tests for preliminary selection of additives for aluminum drawing. Oyane et al [5.64] found good correlation between friction in their simulation test (Fig.5.19) and in low-speed drawing of wire. Takahama et al [282] evaluated polybutenes for aluminum drawing by a "keybroach" test in which bars with notches or localized acrylic resin coatings are drawn, and the difference in draw force between coated and uncoated portions is taken as an indicator of lubricant efficiency.

The acoustic emission technique is increasingly used. Sato et al [5.262] found a large increase in signal level after some length of wire had been drawn; this length increased with oil viscosity and drawing speed. Dornfeld [5.259] observed a gradual increase in emission level with increasing speed in low-speed drawing of aluminum alloy wire, and, as might be expected, emission level was higher in unlubricated drawing.

Wiredrawing is unique in that the quantity of lubricant remaining on the wire is undoubtedly equal to that passing through the die. If the wire roughness is well characterized, a film-thickness ratio can be calculated. This opportunity is lost in plane-strain (strip) drawing because, as in rolling, a bead of lubricant is attached to the edges.

All test methods described in Sec.5.10, 5.11, and 4.6 are employed. Some tests have been specifically developed for wiredrawing. For example, accumulation of copper fines and their effect on the lubricant is a serious problem, and Klasky [283] investigated emulsion inhibitors by abrading a copper surface in a modified drill press.

7.3.8 Lubricant Application and Treatment

Any surface oxide or dirt carried into the die greatly accelerates die wear and lubricant deterioration. Therefore, the first step for successful lubrication is preparation of the surface.

Surface Preparation. Both mechanical and chemical techniques (Sec.4.8.1) are described in Metals Handbook [4.157] and the Steel Wire Handbook [4.231]. Here it will suffice to concentrate on recent developments specific to drawing.

In addition to blast cleaning with cast iron or cut-wire shot [284], steel wires can be descaled with the scale itself [285]. Much but not all of the scale can be removed by reverse bending, provided that the sum of the wrap angles exceeds 180° [286] or that a total elongation of 5 to 10% is achieved [4.268]. Elongation of the wire reduces the scale residue to some 0.6% [287]. Normally, the residue must be removed by pickling, but Bernot [288] found wet blasting (with scale in water) to be an ecologically and economically attractive alternative. Abrasive belts have been used on heat treated rod [289]. Continuing efforts are made to increase the efficiency of pickling, not only by control of chemistry but also by application of ultrasonics [290,291]. Recycling of spent liquors is increasingly practiced [292]. Even though these examples refer to steel, similar problems exist, to varying degrees, with other metals.

Prevention of oxidation eliminates the need for descaling. Steel wire rod cooled by water or nitrogen has a thin scale (4.5 μm max) consisting of the less-abrasive FeO and could be drawn, at least experimentally, without descaling, in a lubricant consisting of a stearate soap with 20% inorganic and 12% E.P. additives [293]. Copper annealed in a protective atmosphere can be drawn without descaling.

Lubricant Application. The general considerations of Sec.4.7.2 apply here, but some special aspects need to be discussed also.

1. Application Techniques. Both total-loss and recirculating lubrication are feasible.

a. By definition, dry drawing with a soap is essentially a total-loss technique. Soap films preapplied to the wire or tube, with or without a surface or conversion coating, fall into the same category. Application of lubricant to the inside of a tube is another example. Dipping (submerging of open-ended tubes) is the method used most frequently; the lubricant must completely cover the surface and must dry uniformly prior to drawing. Any excess lubricant that is applied is finally lost. Oil- or paste-type lubricants applied to the plug are also in this category, as are greases applied from a die box. Greases are formulated to a stringy, ropy consistency so that they roll and work in the die box without tunneling. With reference to externally applied oil-type lubricants, the term "total-loss" is used to describe a technique in which a carefully measured quantity of lubricant is delivered, continuously or intermittently, to the interface so that virtually no excess builds up [294].

b. Recirculation calls for continuous supply of lubricant from a well-designed system with controlled residence time. Brandth and Morton [277] recommend a tank capacity of 35 to 50 L per kilowatt of motor power, and a flow rate of 5 L/min/kW, supplied to the die entry.

c. Occasionally these two application techniques are combined. Thus, in multihole wet drawing, the first hole (or first few holes) is run dry [277], and a well-burnished lubricant film is built up which is only gradually worn off in subsequent wet drawing. The problem is contamination of the recirculating system. Calcium soaps form hard deposits and must be settled out. Soluble soaps are better, although they may still upset the stability of the emulsion. If at all suitable, the emulsion concentrate is the best choice. Conversely, the last one or two holes in a dry-drawing machine may be run wet to reduce lubricant residues and improve surface finish.

d. The geometry of the drawing process lends itself to electrical techniques of accelerated lubricant deposition. Electrodes placed in front of the die facilitate deposition of the lubricant on the wire by electrophoresis [295]. With a suitable additive, C_6H_5-O-P($OC_{10}H_{21}$), Kurihara et al [296] found much improved lubrication on steel wire; electron diffraction measurements suggested the formation of metallic compounds. Electric current supplied through sliding contacts on copper wire reduced friction in the work of Spitsyn et al [297]. Perhaps more promising yet is electrostatic deposition of dry lubricants by powder coating techniques; the wire is the grounded target. With soap having a particle size of 20 to 80 μm, coverage is good enough to obviate the need for liming the wire [298]. *In situ* deposition of MoS_2 or other solid lubricant layer by electrochemical deposition [4.186,4.187],[299] may offer a solution for difficult-to-lubricate materials.

2. Cooling. Removal of the heat generated in drawing presents a considerable challenge and is the subject of continuing development (in part reviewed by Thompson and Symmonds [300] and also in [11]). The problem is acute in dry drawing, where lubricant breakdown due to heating sets a limit on the attainable drawing speed. Furthermore, in drawing of steel wire, heating to over 160 C results in strain aging. Even though temperatures can be much higher in the draw die, residence time is too short for aging to take place; the task is, therefore, to effect reductions in temperature between successive holes, where residence times are much longer.

The wire becomes progressively hotter in multihole drawing as the speed increases to accommodate the elongation of the wire. To compensate for this, reductions may be gradually reduced (so-called taper draft) [58,59]. This, however, increases the number of holes needed for a given total reduction. Alternatively, over-all speed can be reduced, but this reduces output.

We have seen that heat generation is reduced by improved lubrication. Full-fluid-film lubrication has the additional benefit that much of the heat generated is removed by the fluid, but the viscosity of oil drops so much in continuous operation that film thickness drops into the mixed-film regime. The most practical approach is, therefore, removal of the heat.

a. Not much heat can be extracted from the die. Wistreich [23] estimates for drawing of 2-mm-diam steel wire that, of the total heat, only 7% is transmitted to the die at a speed of 1 m/s, 3% at 5 m/s, and 1% at 25 m/s. Nevertheless, dies are usually directly cooled with water on their circumferences [301] and, in one system, also on their back faces [302].

b. The thrust of conventional cooling is extraction of heat through internally water-cooled draw drums (capstans). Heat transfer is improved by reducing the gap between the stationary inner cylinder and the rotating outer drum, and by the turbulence created by the stationary supply tube extending into the gap [303]. It has been claimed, however, that a well-designed internal spray system is equally effective [304]. Either way, cooling is limited by the interface films and by the small contact area between wire and capstan and water. The latter is much reduced by the deposits that form from hard water. For maximum effect, the wrap height should be maintained at the largest possible value [59,278]. External cooling with forced air removes some 35% of the heat [278] but creates noise and draft. The contact surface per unit length is increased by replacing the capstan with a single V-groove wheel [305].

c. Most efficient is, of course, direct water cooling of the wire, but the wire must be dried off before it enters the next draw box. One possibility is to cool the wire on the capstan with a water spray and to remove the water with an air knife [306,307]. Alternatively, a very fine mist is applied in a carefully measured quantity so that the water evaporates; because the latent heat of vaporization is also utilized, smaller quantities of water are needed and the wire is dry [308]. The second possibility is to cool the wire directly as it leaves the die. In the version described by Nakamura et al [147,302], drawing speeds for 0.69% C steel wire could be increased from 6 to 8.5 m/s, maximum die temperature dropped from 540 to 430 C, and die life increased (Fig.7.37). In the version described by Wolf and Schubert [304], cooling of the wire is achieved by running it through a long water-filled tunnel attached to the die. Water passing through the

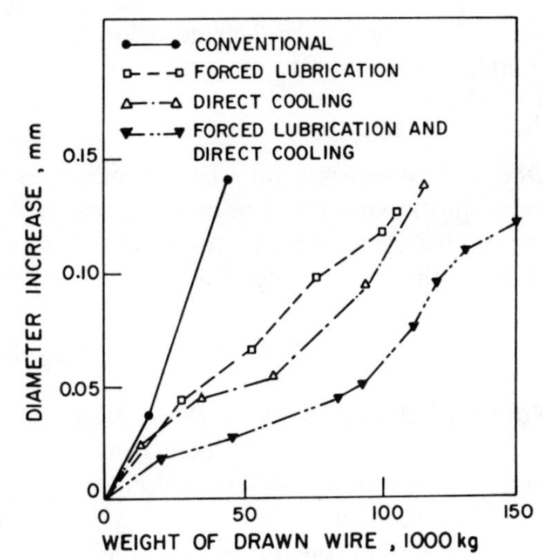

Fig.7.37. Effect of die cooling on wear rate in drawing of 0.6% C steel wire (patented, phosphate/soap lubricated, reduced from 4.5 to 4.0 mm diam at 6 m/s) [147]

exit seal is sucked off by vacuum, which also contributes to cooling. Cooling of the capstan is abandoned.

The difficulties of removing fines from soap-fat emulsions in drawing of copper wire prompted Harper et al [85] to separate the lubrication and cooling functions by applying the neat, viscous oil-base lubricant to the die in a die box resembling a double die, and a coolant containing only a surfactant (polyoxyethylene lanolate) to the capstans (and to roller dies used in early reductions). A fourfold increase in die life (compared with lubrication by emulsion) was obtained.

Maintenance and Disposal of Lubricants. The large quantities of metal debris generated in the course of drawing must be removed. Systems, even though smaller, generally incorporate many of the features found in rolling mills. Their design is discussed by El Hindi [309]. As a minimum, settling tanks are included. Emulsion tanks are often simply skimmed [310], although hydrocyclones are increasingly used [271,311-314]. Positive-pressure, belt-type filters are also employed, especially for drawing of fine wire [315-317]. Soap-fat emulsion particles can be trapped by filters, and flotation is then preferred [310,315,316]. Oil-base lubricants are filtered with positive-pressure or vacuum-belt filters [315,318-320]. Heat exchangers are usually incorporated for cooling of the lubricant [315,321], and, to facilitate start-up with oil-type lubricants, heating is also provided [321].

A valuable aid in neutralizing the effects of wear debris is ultrasonic vibration applied to the fluid in submerged drawing machines [232,322). Vibration apparently prevents entry of debris into the die. There may also be a cleaning effect due to cavitation; the net result is a better finish and longer die life.

It is, of course, essential that routine checks of lubricant quality, particle size distribution, etc., be made [323]. Maintenance follows along the lines discussed in Sec.4.8.3.

Reclamation of spent lubricants is receiving attention for both ecological and economic reasons [324]. Disposal follows the usual routes (Sec.4.8.5). Vacuum distillation has been applied for emulsion splitting [325], and low-pressure steam for the concentration of effluents [326]. Waste-water treatment has been discussed in relation to wire operations [327].

Drawing lubricants present no unique health hazard (Sec.4.9). Dusting is a problem, mostly in mechanical descaling and in liming operations; other coatings are less troublesome [328].

7.3.9 Wear

One can compensate for wear in rolling by resetting the roll gap. No such opportunity exists in drawing through a die. Thus, wear results in an increase in the dimension of the product, and the tolerances set an absolute limit on the allowable wear. The geometry of the converging hole also changes, and thus the lubrication mechanism is affected, lubricant breakdown may occur, and the product finish suffers. Therefore, minimization of wear is a principal preoccupation of tribology in drawing.

Forms of Wear. In a typical die, wear occurs at three major points (Fig.7.38):

a. Severe wear at the point where the wire enters the die results in so-called ringing. The change in entry geometry changes the film thickness and thus the surface finish of the product.

b. Wear of the draw cone changes the geometry of operation and affects film thickness.

c. Wear of the die land is directly responsible for loss of tolerances, and also affects the surface finish of the product. If wear of a wire die is nonuniform, the wire acquires a noncircular profile.

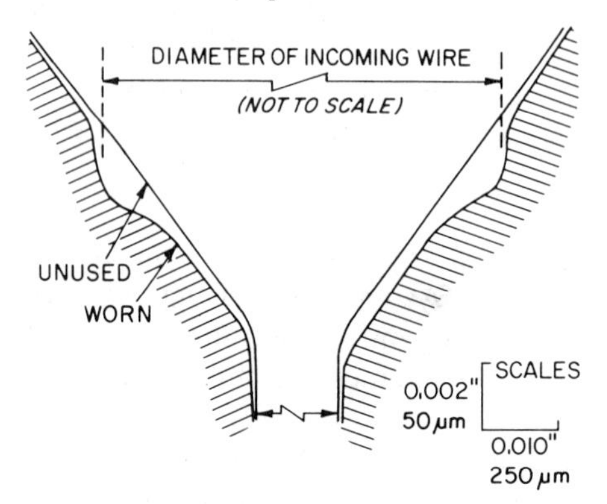

Fig.7.38. Section of a typical worn wiredrawing die [4]

Wear Mechanisms. All wear mechanisms (Sec.3.10) can be active. They have been studied extensively by Wistreich [21], and the applications of recent wear hypotheses have been reviewed by Shatynski and Wright [329]. Wear generally follows Eq.3.39 and increases linearly with the length (or weight) of wire drawn. There is a phase of early rapid wear (Fig.7.39), which Papsdorf [129] attributed to removal of the asperities in a

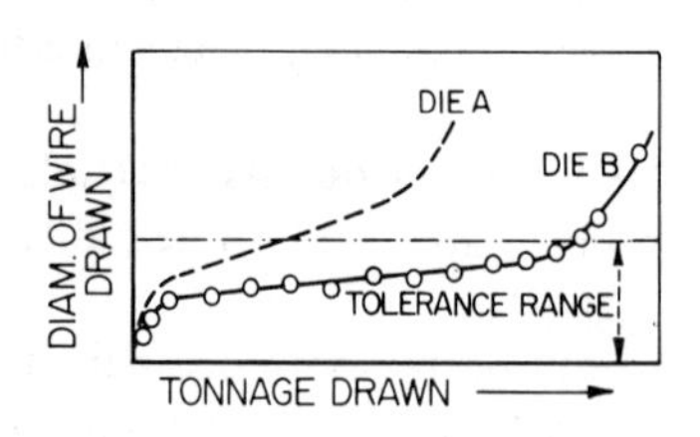

Fig.7.39. Effect of wear rate on die life for a given tolerance range [129]

new die, and an acceleration in the last stage, which is associated with heavy ringing. Details of the wear process, which are often specific to the die material, can be followed by radioactive tracer techniques. In general, the following mechanisms are most active:

a. Abrasive wear is unavoidable. Most metals carry a harder oxide; even if the heavy oxide inherited from hot working is removed, fresh oxide always forms. Some lubricants and coatings are mildly abrasive; others contain abrasive constituents (e.g., SiO_2 or Al_2O_3 in lime) [330]. Wear debris in wet drawing, unless continuously removed, creates a three-body wear condition. In a simulation test with a copper disk rotating against a diamond flat, Harper et al [85] found wear to double when a new lubricant was recirculated and to increase sevenfold when Cu_2O particles were added. A used lubricant, in which fines were entrapped in the soap-fat emulsion, gave similarly high rates of wear.

b. Adhesive wear is likely whenever metal-to-die contact occurs, although the severity of wear is highly dependent on the specific workpiece-die combination.

c. Surface fatigue wear has been suspected as the main culprit in ringing. The high stress gradient, even if loading is continuous, is damaging in itself, and is aggravated by vibration of the wire [331]. Wistreich [332] suggested that cyclic, localized compression of the surface is the cause of ringing. It has also been found [191,332,333] that entering of the wire at an angle greatly increases wear rates; therefore, accurate alignment and elimination of vibration by guide pulleys reduce wear. Hoggart (quoted in [85]) suggested that excess fluid, rejected by the die, causes ringing by cavitation erosion. This hypothesis received some support in the work of Harper et al [85] with viscous oils.

d. Thermal fatigue wear can lead to crazing of steel dies but is, generally, not a major problem.

e. Catastrophic failure (fracture) of dies is usually due to overloading. The danger is greater with lower die angles, and brittle, hard dies must be encased in steel rings.

Dies. Since die wear is proportional only to the length of wire drawn and is not related to wire diameter, a given amount of absolute wear represents a greater percentage increase for a thinner wire. This means that more wear-resistant die materials must be found for finer wires if tolerances are to be held.

Wear rates are expressed as the length of wire that can be drawn before the increase in wire diameter reaches 1 μm. Typical values are: 1 to 2 km for mild steel with die steel dies, 10 km for patented wire [334], and up to 500 km for mild steel wire with tungsten carbide dies. Lives are 10 to 200 times longer for diamond dies (e.g., 160-000 km of copper wire may be drawn before repolishing). All die materials (Sec.3.11.4) have their places; discussions of various types are given in the wire handbooks [8,11,13].

a. Die steels are still used for sections and large-diameter tubes and bars, as they are for roller dies and rolls. They could, no doubt, be improved with coatings, but some of the incentive for doing so has disappeared with the ready availability of large carbide dies. Hard-chromium plating is commonly used for plugs [3.46]. Boronizing has been reported to give better results [335]. Ceramic coatings, particularly ZrO_2, look promising for drawing of stainless steel [207]. Pure zirconia plugs have been used, too [336].

b. Cobalt-bonded sintered tungsten carbide (WC), encased in steel, is the standard die material for drawing of bars, tubes, and wires in all but the smallest gages [337,338]. Wear proceeds fairly uniformly, although radiotracer studies by Button et al [339] and by Dahl and Lueg [340] seem to indicate that, after the cobalt is worn away by adhesive wear, grains of WC are lost from the surface and become embedded in the wire. Golden and Rowe [3.442],[341] concluded that against copper both Co and WC wore simultaneously, except when defects in the copper surface removed WC grains. The Co binder may be attacked by lubricants; the work of Rowe [188] implicates stearate soaps in this respect. Some wear particles enter the lubricant, as was shown by

Holzhey et al [342] in drawing a chromium steel, whereas others become lodged in the wire surface and contribute to abrasion. A die-life model of Witanov [343] postulates fatigue as the major mechanism, although evidence for this seems to be missing. When dies receive inadequate support, they crack through or suffer conchoidal fracture around the die exit [6].

In a series of articles, Linial [344] describes high-temperature boronizing to a depth of 5 to 13 μm. The cobalt is changed to a boride, and the tungsten carbide to another crystal structure, with resultant increases in hardness and die life. Ion implantation with nitrogen or carbon has been reported to give a three- to fivefold increase in die life [3.488], apparently because of reduced adhesion and increased strength of the cobalt matrix.

c. The hardest, most wear-resistant die material is diamond [3.451]. Originally, only natural stones were used, in diameters below 3.0 mm for copper and below 2.0 mm for harder materials [3.426],[345]. They give excellent surface finish and long life but are subject to random sudden early failure. This problem is eliminated by the more recently developed polycrystalline diamond, which is made from powder by synthetic diamond technology and is available in diameters up to 4.6 mm [346]. Monolithic sintered dies or, more frequently, 0.5-to-1.5-mm-thick coatings inside integral tungsten carbide rings [347] give longer life because cracks developing in the easy cleavage direction are arrested by neighboring crystals of hard orientation. Die lives 3 to 10 times that of natural diamond have been reported [346,348,349]. Because wear of diamond is highly dependent on crystallographic orientation [350], natural stones may wear in a noncircular pattern [351]. Polycrystalline bodies are free of these defects; however, they give a somewhat rougher finish. For ease of boring, single crystals are not cut in the highest hardness orientation, and the polycrystalline body presents, on the average, higher-strength directions. Wear is little affected by speed and is perhaps a form of fatigue and adhesive wear. In drawing of ferrous materials, high temperatures induce graphitization and also wear by diffusion.

Of the many articles describing the manufacture of dies, only a few need be quoted [352,353]. The last article in a series by Teller [353] discusses the instruments used for checking the die profile, which has such importance in lubricant film formation. The surface finish obtained on finish polishing is equally important and must be measured. Cast silicone rubber replicas [354] allow inspection without special equipment.

Die life is increased not only by minimizing the presence of abrasive oxides and removing debris from the lubricant but also by early polishing— in particular, removal of incipient ringing.

A drastic improvement is obtained by the use of roller dies—an improvement that is equivalent to a tenfold to 100-fold increase in die life [85].

Wear of Machine Elements. More than any other metalworking process, wiredrawing places exacting demands on the lubricant in controlling the wear of machine elements such as draw cones, capstans, guide pulleys, etc. Capstans in machines drawing with slip are especially vulnerable. Lubrication is, at best, of the mixed-film type in drawing thick wires. Kiss [54] measured μ = 0.14 to 0.2 for steel wire with emulsion lubrication; friction dropped with increasing speed. In drawing fine wires, Valberg [355] found a speed-independent μ, suggesting boundary lubrication. There is thus ample opportunity for reducing wear by the use of coatings, including chromium oxide coatings on tungsten carbide [356], boronizing [357], and alumina and zirconia coatings or components [358,359].

7.4 Lubrication of Iron- and Nickel-Base Alloys

Lubrication practices are dictated by workpiece hardness and adhesion. Many general discussions are available, of which only a few examples are quoted here [5.8],[8,11,79,123,188-190,213,360-362]. The production of steel rod is discussed in the Steel Wire Handbook [8,11].

7.4.1 Steels

Practices for steels are influenced by the diameter, hardness (carbon content and heat treatment), and surface coating of the wire or bar, by drawing speeds, and by surface finish requirements. The general principles of descaling (Sec.7.3.8), surface preparation (Sec.4.8.1), and lubrication apply.

a. Low- to medium-carbon steel wire is drawn dry. The carrier is lime or borax. Lime is chosen for heavy gages, and lower cost has been claimed [363,364], although the same claims have also been made for borax. Borax gives better corrosion protection in a moderately humid atmosphere [365] although it reverts to the crystalline form at high humidity. If the wire need not be cleaned after drawing, the preferred lubricant is calcium stearate, usually compounded with some sodium stearate [4.234],[199] and lime. In an alternative technique, a mixture of calcium stearate and borax is deposited from a very fine dispersion [366] in a single step. Sodium stearate, mixed with some calcium stearate, is used on mild steel and medium-carbon steel if the residues are to be removed prior to annealing. If not removed, soap and carrier residues are converted into a carbonaceous deposit during heat treatment, and this deposit forms a sludge on subsequent pickling, some of which deposits on the wire surface. Schwier et al [195] showed that the quantity of deposit increases with the lime content of a sodium soap, more rapidly with lime contents of over 40%, and is generally higher for a lime than for a borax carrier. For ease of removal with alkaline cleaners, sodium soaps are used with a borax carrier. Because of their tendency for lumping and caking, sodium soaps are often applied by dipping. Additives such as MoS_2 could not be shown to improve die life [259,367] but are sometimes used in low-speed drawing. The use of pressure dies is increasing but is not as widespread as might be thought from their claimed advantages, especially the reduced wire temperature [368]. The reason for this may be that, at high speeds, the lubricant film is thick enough even without a special die.

b. Carbon steel wire containing more than 0.6% C is often patented for maximum strength by quenching from the austenitizing temperature in a lead or salt bath or by forced air. A somewhat similar result is obtained by water quenching directly after hot rolling. The very fine pearlitic structure is capable of heavy deformation (in excess of 90%), while the flow stress of the wire rises above 2 GPa. For thinner gages, patenting is performed after preliminary drawing. After the application of a lime, borax, or phosphate coating, the wire is drawn dry [369,370]. A critical requirement is adequate cooling during drawing, because the properties of the wire—in particular, its ductility—deteriorate through strain aging, and even martensite formation has been suspected. Well-designed capstan cooling combined with die cooling has been found to be successful in drawing of 4-mm-diam reinforced concrete tensioning wire at 8 m/s [371]. A pressure die can reduce temperature by 100 to 150 C [368].

c. In addition to high-carbon steel wire, bars and wires of medium-carbon steel are sometimes phosphated, and all tubes are almost always phosphated [6.330],[272,372]. The experiments of Dahl and Lueg [259] did not show a clear superiority of phosphating over lime coating, but the protection against lubricant breakdown afforded by phosphating certainly justifies its expense in tube drawing. Machu [373] indicates that a coating of 3 to 7 g/m^2 will suffice for a cumulative reduction of 95% in wiredrawing; coating weights of up to 12 g/m^2 are encountered in practice. Developments, including continuous processing lines, have made the application technique compatible with the needs of the wiredrawing process [374-380]. The lubricant is most often a dry soap. A sodium stearate soap is applied, following a borax or lime neutralizing rinse, from a hot solution. The lubricant not only reduces friction but also promotes uniform thinning of the phosphate, as shown by Komura [5.225],[204]. The reactive oil method of application, with the reaction taking place in the die, has greatly simplified the process sequence [381,382] for both wire and tube.

d. Bars, tubes, and heavy wire are drawn, after pickling or shot blasting, at low speeds using a heavy (up to 800 mm^2/s) oil, grease, tallow, or soap-fat drawing paste for a bright finish [123]. At higher drawing speeds the temperatures are high enough to activate E.P. additives, and sulfur compounds (or, sometimes, elemental sulfur [98,383] or sulfo-chlorinated compounds) are added. Friction is reduced only slightly, if at all [198], but lubricant breakdown is prevented [121]. At low speeds the benefits of E.P. additives cannot be shown [197,280], and fatty oils or their derivatives become essential ingredients [56,198,384] in quite high proportions; a minimum of 15% is needed to reduce draw forces and provide protection [197]. Solid fillers—primarily MoS_2, but also graphite, lime, talc, or mica—are also added, in quantities ranging from 2 to 40%, although they increase the difficulty of cleaning [385-387]. In a recent development, E.P. oils or soaps are applied in the die box during drawing of a bar or wire on which a semiplastic coating of undisclosed composition has been deposited [388].

e. Multihole wet drawing of steel is almost invariably performed in an aqueous fluid [389]. Dilute soap-fat emulsions with 2.5 to 3% free fat in the emulsified phase used to be favored [390] but have been largely superseded by nonionic emulsions. The coating/lubricant residue carried over from dry drawing provides additional protection against breakdown. If heat treatment was performed after dry drawing, the wire has to be limed, or it may have to be phosphated, as is done for patented high-carbon steel wire. Compatibility with the emulsion is important, to prevent the formation of solid reaction products which would clog pipes. Soap-fat emulsions and sodium soap solutions, once used almost exclusively in drawing of very fine wire, are particularly prone to formation of hard calcium soaps.

f. Metal coatings are frequently applied for their service properties, but their presence affects lubrication practices if they are applied prior to final drawing. Zinc-coated (galvanized) wire is drawn with soaps, and its response to lubricants has been discussed in conjunction with Fig.4.20 and 4.21. Copper coatings, deposited electrolytically or, more frequently, by cementation, are used for their color [391], and are of great help in wet drawing. In conjunction with dry soaps, they are used for cold heading of wire; the residual coating/lubricant contributes to lubrication in cold heading. A specific application is the manufacture of tire-cord wire of 0.7 to 0.9% C steel. The rod is dry drawn to 2.2 to 2.7 mm, patented, dry drawn to 0.7 to 1.3 mm, patented again, cleaned, brass plated, borax/stearate coated, and wet drawn to the final diameter of 0.12 to 0.35 mm [10,392]. The brass coating ensures better rubber adhesion. The lubricant residues must be compatible with the rubber, and unsaturated fatty oils appeared to be better in tests in which the wire was dipped (but not drawn) in the lubricant [393].

g. High-alloy steels have limited ductility at room temperature. Some can be drawn 85%, and others only 10 to 15% [394], before annealing becomes necessary, followed by pickling. Lueg and Pomp [395] found lime mixed with salt to be a good carrier for dry drawing with a soap.

h. Several efforts have been made to draw both low- and high-alloy steels at elevated temperatures [214,396]. Little is gained in warm drawing of low-alloy steels because those prone to aging pass a ductility minimum. Definite benefits are reaped in drawing of high-alloy steels at temperatures between 400 and 800 C. Fuchs [394] reported experiments with tungsten carbide dies, the casings of which had to be cooled to prevent loss of support by excessive expansion. Coatings are essential but become less effective as smoothing of the surface progresses. After preheating in a lead bath, the thin residual lead film is useful, but only below 500 C; it is also toxic. Copper is acceptable and improves conduction in heating by electrical resistance, but it oxidizes at temperatures above 150 C. In general, MoS_2 breaks down at around 500 C (Fig.3.22), and graphite is the usual lubricant. Flake graphite has been found to be better than other forms of graphite [394]. A general problem is that graphite carried over on the wire makes the wire turns slip upward on the capstan, causing the wire to enter the die at an angle, with resultant wire breaks and die wear [394].

7.4.2 Stainless Steels

Stainless steels, especially those of the austenitic type, form pickup on dies and must be either separated by a thick film or lubricated with a chlorine-containing lubricant [124,209,397-400]. Apart from lower drawing speeds, technologies are similar to those for steel.

a. For severe duties and, particularly, tube drawing, an oxalate coating impregnated with a reactive soap is the standard lubricant [4.258], [397,401-404]. This allows several consecutive draws and has been repeatedly shown to be best [258,395]. Sometimes chlorinated oil is used instead of soap, but its performance is poorer [387]. Removal of the coating in acid or caustic baths causes only slight pitting.

b. Of the metal coatings, only copper is now used. Lead, once in widespread use, is avoided, with the possible exception of instances in which a lead bath is used for preheating prior to warm drawing. Removal in nitric acid can cause severe pitting [402].

c. Polymer films obviate the need for the expensive oxalate treatment. Dispersions in oil or in water have been examined [405], but in most instances a film is deposited from a solvent or an emulsion. The polymer is often chlorinated, or a chlorinated oil is used in conjunction with the coating. Even though it is not a polymer, chlorinated wax falls into the category of film formers.

d. Lime has been used as a carrier for both soaps and chlorinated oils in drawing of bars and heavy wire. Lime with salt has been shown to perform better [258,395], and salt coatings have been mentioned [406], but corrosion is a danger when salt is present. Borax with sodium silicate has been used [404]. A mixture of 30% soap in borax applied at near-boiling temperature has been successful without oxalating [4.258]. Ease of removal and good surface finish are claimed for a semiplastic coating with superimposed E.P. oil [388]; presumably, the oil serves to lubricate the die. This must be the case whenever sulfur is found to be effective [407]. Some formulations contain graphite or MoS_2 [385,408], and a soap with ZnO and MoS_2 [409] has been recommended, but the difficulty of removal usually outweighs whatever advantages there may be.

e. Wet drawing of heavy wire is possible without carriers, using chlorinated waxes or oils, or pure chlorinated paraffin. Fine wires are drawn using chlorinated oils with viscosities of 20 to 60 mm^2/s at 40 C [124], emulsions of mineral oils containing E.P. additives [410], or emulsions of fatty acid distillates with high (40%) free fatty acid contents [411].

7.4.3 Nickel-Base Alloys

Lubrication techniques for nickel-base alloys are similar to those for stainless steels, with an oxalate/soap system being encountered most frequently in drawing of bar and heavy wire [402]. Oxidation of the surface layer during annealing may lead to surface cracking; this defect was prevented in the work of Schroeder [81] by fine bead-blasting prior to oxalating. Reoxalating after each draw in profile drawing was uneconomical, and a ceramic-bonded MoS_2 coating, applied for each draw, eliminated the need for surface treatment. The coating had to be totally removed prior to heat treatment. Lime has also been used as a carrier for oils, chlorinated paraffins, or drawing pastes in drawing of bars, and for oil-base lubricants or emulsions in wet drawing [412]. Wet drawing without surface treatment, using a lubricant consisting of 79% sodium soap, 8% MoS_2, 6 to 10% graphite, and 12 to 20% lime, has been reported [409]. All sulfur-containing residues must be carefully removed prior to heating.

Tube lubrication follows the practices of bar drawing. Additionally, polymer coatings deposited from solvents or emulsions are useful, especially if the fine pitting observed on removal of an oxalate coating is objectionable.

7.5 Lubrication of Light Metals

The lubrication techniques used in producing aluminum, magnesium, and beryllium wires differ greatly from those employed in producing wires of iron- and nickel-base alloys.

7.5.1 Aluminum Alloys

The large quantities of wire produced for electrical conductors justify hot rolling of semicontinuously or continuously cast wire bar [12]. Alloys are often hot extruded. All drawing is done cold. To prevent die pickup, lubricants must always contain fatty derivatives; to avoid corrosion and staining, the base oil is of the lowest viscosity permissible for the drawing speed [413]. General discussions are available [414-417].

a. For low-speed drawing of bar and tube, a highly viscous oil (viscosities to 2000 mm^2/s at 40 C) or polybutene (to 10 000 mm^2/s) is chosen [417]. Alternatively, dry soap (such as aluminum stearate [418]), grease, or wax can be used, although with these lubricants the bar may not pick up a powder uniformly. Therefore, some bars and most tubes are dipped in an aqueous soap solution or an aqueous or organic wax or lanolin solution, and the coating is allowed to dry before drawing [372]. Beeswax performs well [93] but is seldom used in practice. Hydrostatic pockets make for a dull surface which needs considerable further drawing under boundary conditions to become smoother and brighter [264]. This is the problem also with hydrostatic wiredrawing [419]. There is no justification for fillers; solid polymers such as PTFE are effective [405] but also difficult to remove.

b. With decreasing diameter, the wire is drawn at increasing speed. Correspondingly, the viscosity of the oil or synthetic mineral oil is gradually reduced from the 150 to 250 mm^2/s (at 40 C) used in coarse drawing to 20 to 60 mm^2/s for wires up to 0.5 mm in diameter, and to 5 to 8 mm^2/s for fine wire [124,209,315,318,417]. Heavier oils are used for alloy wire [417]. Little systematic work has been published on the effects of additives. Quite large additions of fats (7 to 30%) have been quoted, but these must contribute to increased staining. The fatty acids, alcohols, and esters used in aluminum rolling oils (Sec.6.7.2) should give satisfactory service. A wax dispersion in oil has also been mentioned [420]. Emulsions pose the problem of water staining and are, generally, avoided, even though they are occasionally mentioned in the literature [124,207,421]—particularly for fine-wire drawing with diamond dies [209]. In submerged drawing, ultrasonic energy applied to the fluid allows drawing to finer gages [322]. In the past, debris was allowed to accumulate up to 20% [230,417], but with filtration it is now reduced to 2% [318] or even 0.1% [315], with consequent improvements in surface finish and lubricant life.

A major concern in the production of electrical conductor wire is surface finish, discussed by Wright and Male [264] and by Zini [422]. Some defects are inherited from the hot rolling stage and are modified in drawing. Shaving of the hot rod, to a depth of some 0.1 mm, removes many defects but can introduce defects of its own if done with poor lubrication or worn dies [264]. As in rolling, excessive lubricant viscosity leads to pockmarking (hydrostatic pockets), while lubricant breakdown results in pickup and smearing. Pre-existing defects diminish with an effective mixed-film lubrication bordering on the boundary regime but persist longer with a viscous lubricant [264]. Sharp impressions (which may be true cracks in alloys of limited ductility) close up without healing and trap lubricant residues. Surface defects create stress concentrations which lead to cracking of enamels on bending of the wire.

Magnesium wires are hot extruded. Drawing, if required, is conducted warm, with lubricants similar to those used in sheet metalworking (Sec.10.6.2).

7.5.2 Beryllium

For increased ductility, beryllium wire is warm drawn. Temperatures range from 380 to 450 C in coarse-wire drawing. Bare drawing with a graphite or MoS_2 lubricant is possible. Gross [423] recommended a mixture of the two in a phenolic binder. More frequently, the wire rod is clad. Colloidal graphite served well in drawing with a mild steel sheath [424]. Nickel cladding is used more frequently; to lubricate it, MoS_2 may be deposited from a dispersion with supplementary graphite lubrication in the die box. Alternatively, MoS_2 or graphite (or a mixture of the two) is applied in the die box. The dies are heated to 400 C [425]. Nickel sheaths were used also by Uy et al [426] in drawing 9.5-mm hydrostatically extruded rod to an intermediate gage of 0.5 mm at 370 to 430 C, and a copper coating was used for the finish draw, in three passes, to 0.127 mm at 150 C.

7.6 Lubrication of Copper-Base Alloys

Most copper (for electrical conductors) is hot rolled from continuously cast or chill-cast bar [12]. Alloys are frequently extruded. All drawing is performed cold.

General discussions are given by Harper et al [85], Boon [427], Pepmöller [428], and Tassi and Drotleff [416]. There are two important aspects: the presence of oxide, and reaction of lubricant with the copper.

a. Oxides produced during hot working are thick and discontinuous and can be the source of surface defects such as scale pockets. Thus, they should be removed or their formation prevented (e.g., by extrusion into water). Pops and Hennessy [429] recommend that the oxide film be less than 0.2 μm thick and consist of mixed oxides. This agrees with the finding of Shumilina et al [430] that the mixed Cu_2-CuO film formed at 200 C gave lower draw forces with caprylic acid than the pure Cu_2O formed at 20 C or the CuO formed at 900 C. The difference was particularly noticeable in drawing at 150 C. The effect is specific to the lubricant; stearic acid gave lowest friction with Cu_2O [431], while decyl alcohol gave higher draw forces which peaked with the mixed oxide formed at 200 C [430]. The fines generated were smaller (<1.5 μm) with acid than with the alcohol (<4.5 μm) except in drawing at 150 C, when some stick-slip was also observed with the acid, indicating insufficient lubrication.

b. Copper pickup is not cumulative; instead, a dynamic equilibrium between pickup and removal sets in, resulting in a large amount of debris (about 0.5 g per kilogram of wire [85]), although some of this debris comes from the pickling process, where a copper dust is deposited on the rod. The new surfaces promote reaction with the lubricant, especially with anionic emulsions which form copper soaps [432]. This reduces the stability of the emulsion by reducing its pH, and thus affects the plate-out characteristics of the lubricant. In general, the reduction in draw force with lubrication is greater for annealed than for hard-drawn wire [433], but a lubricant is needed in all drawing, to control wear.

The choice of lubricant is affected by drawing speed:

a. In low-speed drawing of bar and tube, the traditional lubricant is a soap-fat paste which has the advantage of water solubility. This advantage is lost on adding solid fillers [372], which are sometimes regarded as desirable in drawing of hard brasses and bronzes. For these applications, however, dry soaps are preferable, usually as coatings deposited from aqueous solutions. Grease (with rapeseed oil) or tallow, sometimes with added sodium or calcium stearates, picks up well from a draw box to form a coating that, in conjunction with a liquid lubricant, maintains good lubrication through several draws. Mineral oils of high viscosity, always compounded with fatty derivatives and E.P. additives, are readily applied either through a small recirculating system or, preferably, by the total-loss technique. They are also preferred for lubrication of tube plugs [434].

b. High-speed drawing is done almost exclusively in aqueous lubricants. High-soap/low-fat solutions (anionic emulsions) with low ($<1\%$) free fat contents are quite adequate [288], but suffer from soap formation with water hardness and copper fines. Therefore, admixture of a nonionic emulsion (a so-called synthetic fluid) increases stability [435]. Plating out is controlled by the fatty acid content. Some acid is essential for reaction, but excessive concentrations lower the pH and cause rapid attack on copper with the formation of green soaps. The pH is kept between 8 and 9.5; the emulsion breaks down below 7.5 [209,230,427,428]. The concentration is higher (up to 10%) for rod and lower (down to 0.5%) for very fine wire. E.P. additives are incorporated, especially for brass and bronze. Emulsion formulations must take into account possible side effects. Thus, some emulsions attack the cobalt of tungsten carbide dies [427], whereas others contain silicones as foam depressants which can interfere with subsequent processing such as enameling.

Rapid deterioration is prevented by removing the copper fines by filtration [432], and the temperature of the system is kept below 50 C to prevent rapid reaction. When a weir is used in the tank, a minimum residence time of 20 min is ensured, to allow settling. Control of lubricant systems has been discussed by Drotleff [436]. The application of ultrasonics to the bath allows the drawing of finer wire [322]. The surface of the drawn wire is generally good, but for highest-quality trolley and magnet wire the hot rod is shaved [437].

c. Of the metal coatings, tin improves the soldering characteristics of the wire. When applied before drawing, tin facilitates lubrication and allows higher drawing speeds, but it also increases the cost of lubricant system maintenance [438].

d. Roller dies minimize lubrication problems [85] and reduce the quantity of fines generated.

7.7 Lubrication of Titanium and Zirconium Alloys

Both titanium and zirconium alloys are characterized by die pickup and scoring (galling).

Titanium can be lubricated with polymer coatings, but the film must be very uniformly deposited to avoid fracture in drawing of thin wire or thin-wall tubing [439]. Most frequently, a conversion coating (Sec.4.7.8) is used. Fluoride-phosphate coatings [4.241],[77], anodic oxide films [440], an oxide grown around 700 C [441], and copper plating [77,442,443] have all been used. The disadvantage of oxide coatings is that oxygen diffusing into the metal hardens it and reduces the allowable cold work. This is a major drawback in drawing of thin-wall tube [439]. The lubricant can be grease with MoS_2 [4.241], PTFE [77,440], a soap-MoS_2 mixture [441], or a chlorinated wax. An unusual lubricant is a film of asphalt, deposited from a xylene solution [444], in conjunction with lithium stearate soap. Titanium alloys are drawn at 480 to 760 C, with a graphite coating applied prior to drawing. Loewenstein [77] reports the use of additional graphited grease at the die entry.

Films of metals such as copper and silver are mentioned in the early literature. More recently, zinc has been recommended [445]; it is deposited from a $ZnCl_2$ electrolyte, with or without phosphating [446].

Zirconium can be drawn after conversion coating [447] followed by soap application [448]. Among metallic coatings, copper has been mentioned. Asphalt deposited from a xylene solution [444] with superimposed soap has been successful at low speeds.

7.8 Lubrication of Refractory Metals

Refractory metals present a variety of tribological problems.

a. The ductile-to-brittle transition temperature of tungsten drops with accumulated deformation, and therefore drawing temperatures also drop from the initial 1000 C to 550 C [449]. Tungsten carbide dies, heated to around 400 C, are used with graphite lubricant applied to the wire, prior to heating, from an aqueous slurry. Daga [450] showed that the coefficient of friction, calculated from minimum die angle [6.57], increased from 0.12 at 850 C to 0.15 at 950 C, because both graphite and oxide films were thinner at the higher temperature. The optimum die angle was around 12°. Carburization of the wire was avoided by Zhilkin et al [451] by application of a glass lubricant, compounded with some clay and 10 to 15% MoS_2, as a slurry prior to heating. The die had to be heated to 600 C to prevent stiffening of the glass. Heavily worked wire less than 0.25 mm in diameter can be drawn at room temperature in diamond dies [449].

b. Molybdenum technology is similar [408], although thin wire can be cold drawn more readily, with small reductions per pass [444]. For drawing in air, Zhilkin and Zhukov [452] used graphite in an epoxy resin, deposited on top of a borax coating; for drawing in vacuum, MoS_2 in a silicone carrier was used.

c. Chromium wire has been produced experimentally. After plating with nickel and then a thicker layer of copper, small reductions per pass could be taken with both die and wire at 350 C, with a graphite lubricant [453].

d. Tantalum presents problems only because of its adhesion to dies. One solution is to coat heavy wire with copper [454]. At gages below 1 mm, the copper is removed by pickling, and the wire is anodically oxidized and drawn at low speeds (up to 0.12 m/s) [455]. Dry film lubricants on anodic oxide coatings are used also in drawing of shapes [77]. In a departure from this technique, Heier [455] utilized a modified double die, with the first die replaced by a seal with a radial clearance of 0.1 to 0.15 mm. Beeswax, deposited on top of the oxide coating, gave consistently the best quality and provided almost total freedom from pickup. Zinc stearate was marginal; calcium stearate, a soap, and graphite plus MoS_2 were poor. The oxide and the wax film had to be renewed for each draw, but speeds could be increased to 0.5 m/s. Zhukov [456] reported the use of an epoxy-bonded graphite-plus-MoS_2 film for drawing in vacuum or argon. Another technique involves the use of Al-Fe bronze dies with beeswax [457,458].

e. Uranium, too, can be drawn with aluminum bronze dies and beeswax, or with colloidal graphite in water in conjunction with chromium-plated steel dies.

f. A new but rapidly developing field is the drawing of niobium-alloy (Nb-Ti, Nb-Zr, Nb_3Sn) superconducting wire [459]. The wires can be made up into bundles in an aluminum or copper matrix by hydrostatic extrusion, followed by drawing of the bundles [460]. Friction plays an important role in that shear stresses transmitted from the matrix to the wire help to prevent fracture of individual wire strands.

7.9 Summary

1. No friction between workpiece and die is needed in wiredrawing and tube drawing on a fixed plug; some moderate friction is required on the plug in drawing tubes with a floating plug, and friction on the bar is beneficial in drawing on a bar.

2. The frictional force is additional to the force required to perform the deformation (work of homogeneous deformation) and the force arising from the inhomogeneity of deformation (redundant work). The frictional force increases, whereas the force for inhomogeneous deformation decreases, with increasing die angle, and therefore an optimum die angle is found. Since the contributions of friction and inhomogeneous deformation to the total force are of similar magnitude, back-calculation of friction from the total draw force is difficult. However, for a given workpiece material and process geometry, a larger draw force is always indicative of higher friction.

3. Yielding takes place under the combined effects of axial tension and radial compression, and interface pressures are below the flow stress of the material except at low die angles, when they can rise to higher values. Even though pressures are, normally, not excessive, the process presents severe tribological conditions because sliding prevails over the whole contact zone, and heating due to deformation and friction raises the temperature. In multihole drawing the high temperature of the entering wire aggravates the situation.

4. Lubricants are applied to reduce friction, wear, and temperature. A good lubricant reduces temperature by reducing friction. Nevertheless, in high-speed multihole drawing the lubricant must also have adequate cooling capacity to remove heat from the product. The method of coolant application becomes most important. Wear is brought within tolerable limits, in part, by the selection of highly wear-resistant die materials, primarily tungsten carbide and diamond.

5. Drawing without a lubricant would result in immediate pickup. So-called dry drawing is conducted with a nonliquid lubricant, usually a soap, whereas wet drawing is performed with viscous oils or aqueous emulsions. The distinction is practical rather than fundamental: under the conditions prevailing at the interface, soap can be modeled as a viscous fluid of temperature- and strain-rate-sensitive viscosity, or as a Bingham substance.

6. Drawing is a steady-state process with a favorable entry geometry for the entrainment of lubricants. Full-fluid-film lubrication can be encouraged by special die constructions that aid in developing hydrodynamic films or that allow the introduction of externally pressurized fluids. The most practicable of these is the pressure die for soaps; it builds up sufficient pressure by purely hydrodynamic action to cause yielding of the wire prior to entry into the die. The surface finish of the workpiece produced under full-fluid-film conditions is rough.

7. Most practical drawing is conducted under mixed-film conditions. Initial draws may be made with a dry lubricant, but brighter, smoother surfaces can be produced only by deliberately thinning the film to the point where boundary lubrication predominates. This is accomplished by increasing the die angle (although this increases also the inhomogeneity of deformation and could lead to internal arrowhead fracture) and by

drawing with a low-viscosity lubricant. The potential for die pickup necessitates the use of boundary or E.P. additives.

8. Solid film lubrication is of greatest importance in warm and hot drawing, and in cold drawing of adhesion-prone alloys at low speeds, in which an effective fluid film cannot be maintained.

9. Tube drawing on a fixed or floating plug presents special difficulties of lubricant application to ensure effective lubrication of the inner tube surface.

10. Many problems of lubrication and wear can be bypassed by replacing the draw die with a rotary die composed of two to four rollers, particularly in drawing of more ductile materials.

11. Lubricant recommendations are summarized in Table 7.1. For preliminary calculations, coefficient of friction values are given with the understanding that their magnitude is a function of drawing speed, especially for lubricants operating in the mixed-film regime.

References

[1] J.A. Newnham, in [3.1], pp.457-547.
[2] J.A. Schey, in [3.1], pp.47-48, 256-263.
[3] G.D.S. MacLellan, J. Iron Steel Inst., *158*, 1948, 347-356.
[4] J.G. Wistreich, Met. Rev., *3*, 1958, 97-142.
[5] B.J. Duckfield, Wire Ind., *40*, 1973, 618-623, 702-707.
[6] G.W. Rowe, Int. Met. Rev., *22*, 1977, 341-354.
[7] A. Pomp, *The Manufacture and Properties of Steel Wire,* Wire Industries, London, 1954.
[8] A.B. Dove (ed.), *Steel Wire Handbook,* Vol.1, The Wire Association, Guilford, CT, 1965.
[9] Ibid., Vol.2, 1969.
[10] Ibid., Vol.3, 1972.
[11] Ibid., Vol.4, 1980.
[12] L.W. Collins, J.G. Dunleavy, and O.J. Tassi (ed.), *Nonferrous Wire Handbook,* The Wire Association, Guilford, CT, 1977.
[13] O.J. Tassi (ed.), *Nonferrous Wire Handbook,* The Wire Association, Guilford, CT, Vol.2, 1981.
[14] *Ziehen von Drähten, Rohren, Stangen,* Deutsche Ges. f. Metallkde, Oberursel, Germany, 1976.
[15] T. Cahill, Wire Ind., *40*, 1973, 942-946.
[16] W. Teller, Wire World Int., *18* (3), 1976, 131-135.
[17] D. Green, J. Inst. Metals, *99*, 1971, 76-80.
[18] S. Thiruvarudchelvan, J. Mech. Work. Tech., *2*, 1979, 347-356.
[19] M.C. Shaw and T. Hoshi, Wire J., *9* (3), 1976, 74-80.
[20] E. Tuschy, Metall, *36*, 1982, 269-279 (German).
[21] J.G. Wistreich, Wire Ind., *17*, 1950, 889-899.
[22] J.G. Wistreich, J. Inst. Petr., *40*, 1954, 345.
[23] J.G. Wistreich, Wire Wire Prod., *34*, 1959, 1486-1489, 1550-1551.
[24] J.G. Wistreich, in [3.35], pp.505-511.
[25] R.N. Wright, Wire J., *12* (10), 1979, 60-61;Wire Techn., *4* (9/10), 1976, 57-61;*5* (7/8), 1977, 106-109.
[26] R. Zimmermann, Draht, *29*, 1978, 152-163, 231-239 (German).
[27] A.P. Green, Proc. Inst. Mech. Eng., *174*, 1960, 847-864.
[28] E. Felder, CIRP, *25*, 1976, 147-151.
[29] W.R.D. Wilson, in *Proc. 5th NAMRC,* SME, Dearborn, 1977, 80-86.
[30] R.M. Caddell and A.G. Atkins, Trans. ASME, Ser.B, J. Eng. Ind., *90*, 1968, 411-419.
[31] R.W. Johnson and G.W. Rowe, J. Inst. Metals, *96*, 1968, 97-105.
[32] W. Johnson, R. Sowerby, and R.M. Caddell, CIRP, *19*, 1971, 311-315.
[33] R.W. Johnson and G.W. Rowe, in [3.51], pp.43-46.

[34] A.W. Duffill and P.B. Mellor, in *Proc. 9th Int. MTDR Conf.*, Pergamon, Oxford, 1969, pp.427-445.
[35] B.B. Basily and D.H. Sansome, in *Proc. 17th Int. MTDR Conf.*, Macmillan, London, 1977, pp.475-481.
[36] R.W. Lunt and G.D.S. MacLellan, J. Inst. Metals, *72*, 1946, 67-96.
[37] H. Majors, Trans. ASME, *78*, 1956, 79-87.
[38] G.D.S. MacLellan, J. Inst. Metals, *81*, 1952-53, 1-13.
[39] J.G. Wistreich, Proc. Inst. Mech. Eng., *169*, 1955, 654-665.
[40] S.P. Burkin, A.N. Levanov, I.Ya. Tarnovsky, and B.P. Kartak, Izv. VUZ Chern. Met., *1971* (5), 108-111 (Russian).
[41] C.T. Yang, Trans. ASME, Ser.B, J. Eng. Ind., *83*, 1961, 523-530.
[42] R. Sagar and R.K. Gupta, in [3.55], pp.117-126.
[43] P.M. Cook and J.G. Wistreich, Brit. J. Appl. Phys., *3*, 1952, 159-165.
[44] Y. Ohashi and T. Nishitani, Jap. Soc. Mech. Eng., Semi-International Symp. on Exp. Mech., 1967, pp.129-138.
[45] E. Siebel and R. Kobitzsch, Stahl u. Eisen, *63*, 1943, 110-113 (German).
[46] T. Altan, Wire J., *3* (3), 1970, 54-59.
[47] O. Pawelski and K. Reiner, Arch. Eisenhüttenw., *41*, 1970, 201-208 (German).
[48] G. Reisz, Neue Hütte, *19*, 1974, 193-207 (German).
[49] R.W. Snidle, Wear, *44*, 1977, 279-294.
[50] W. Rosenhain and V.H. Stott, Proc. Roy. Soc. (London), *A140*, 1933, 9-25.
[51] A. Eichinger and W. Lueg, Mitt. Kaiser-Wilhelm Inst. Eisenforsch., *23*, 1941, 21-30 (German).
[52] D. Lewis and H.I. Godfrey, Wire Wire Prod., *24*, 1949, 873-877, 880-885, 982-983.
[53] F.C. Thompson and H.G. Dyson, Metallurgia, *6*, 1932, 191-192.
[54] E. Kiss, Neue Hütte, *13*, 1968, 164-169 (German).
[55] R. Pawelski and R. Kopp, Stahl u. Eisen, *89*, 1969, 1345-1353 (German).
[56] K. Rittmann, Arch. Eisenhüttenw., *42*, 1971, 265-271 (German).
[57] E. Felder and A. LeFloch, CIRP, *25*, 1976, 173-178 (French).
[58] J.H. Rigo, Wire J., *5* (6), 1972, 43-52.
[59] N.A. Flanders and E.M. Alexander, Wire J., *12* (3), 1979, 60-68.
[60] H. Bühler and P.J. Kreher, Draht, *19*, 1968, 531-537 (German).
[61] H. Bühler and K. Lueg, Arch. Eisenhüttenw., *41*, 1970, 19-23 (German).
[62] Z.I. Kostina et al, Sov. Mater. Sci., *10*, 1974, 301-302.
[63] J. Luksza, Hutnik, *44* (1), 1977, 45-49 (Polish).
[64] W. Linicus and G. Sachs, Z. Metallk., *23*, 1931, 205-210 (German).
[65] V.N. Istomin, Tsvetn. Met., *1969* (6), 126-129 (Russian).
[66] O. Mahrenholtz, Arch. Eisenhüttenw., *37*, 1966, 847-852 (German).
[67] O. Richmond and M.L. Devenpeck, J. Mech. Phys. Solids, *15*, 1967, 195.
[68] M.L. Devenpeck, in *Metal Forming: Interrelation Between Theory and Practice*, Plenum, New York, 1971, pp.215-234.
[69] G. Zouchar, Neue Hütte, *16*, 1971, 407-411 (German).
[70] J.R. Douglas, Wire J., *4*, 1971, 115-120.
[71] H.C. Rogers, in *Metal Forming: Interrelation Between Theory and Practice*, Plenum, New York, 1971, pp.453-474.
[72] H.C. Rogers and L.F. Coffin, Jr., Int. J. Mech. Sci., *13*, 1971, 141-155.
[73] B. Avitzur, Wire J., *7* (11), 1974, 77-86.
[74] M. Mohamdein and T. Vinh, CIRP, *25*, 1976, 169-172.
[75] R.W. Johnson and G.W. Rowe, Proc. Inst. Mech. Eng., *182*, 1967-68, 521-526.
[76] K. Becker, Wire World Int., *11*, 1969, 147-153.
[77] P. Loewenstein, Paper No. MF70-204, SME, Dearborn, 1970.
[78] D.Q. Cole, Paper No. MF71-140, SME, Dearborn, 1971.
[79] B.J. Duckfield, Wire Ind., *38*, 1971, 43-47, 120-122.
[80] J.S. Gunasekera and S. Hoshino, Trans. ASME, Ser.B, J. Eng. Ind., *104*, 1982, 38-45.
[81] G. Schroeder, in [3.55], pp.103-106.
[82] A.F. Sperduti, Wire J., *13* (7), 1980, 78-83.
[83] A.L. Randazzo, Wire Wire Prod., *39*, 1964, 527-528, 530-531, 620.
[84] I. Gokyo and T. Okubo, J. Iron Steel Inst. Japan (Overseas Ed.), *4*, 1964, 44-52.
[85] S. Harper, A.R. Goreham, and A.A. Marks, Metals Mat., *4*, 1970, 335-339.
[86] H. Marciniak, Hutnik, *39* (1), 1972, 6-10 (Polish).

[87] P.O. Strandell, Scand. J. Metall., *2*, 1973, 7-10.
[88] H. Knobloch, Wire World Int., *17* (2), 1975, 67-71.
[89] G. Properzi, Wire J., *12* (12), 1979, 58-62.
[90] R. Sayer and R. Moller, Wire J., *11* (2), 1978, 68-72.
[91] H. Mühlenweg, Stahl u. Eisen, *79*, 1959, 1792-1800, 1844-1852 (German).
[92] O. Pawelski and V. Rüdiger, Arch. Eisenhüttenw., *47*, 1976, 483-487 (German).
[93] G. Dévényi, Draht, *13*, 1962, 223-231 (German).
[94] A.N. Bramley and D.J. Smith, Metals Tech., *3*, 1976, 322-331.
[95] C.S. Hartley, in *Proc. 6th NAMRC,* SME, Dearborn, 1978, pp.193-198.
[96] G.L. Kolmogorov, V.S. Parshin, V.I. Boyarkin, and V.A. Kondrikov, Tsvetn. Met., *1977* (4), 57-59 (Russian).
[97] B. Zimmermann and R. Zimmermann, Arch. Eisenhüttenw., *49*, 1978, 121-124 (German).
[98] B.W. Siemon and W.B. Bauzenberger, Iron Steel Eng., *33* (11), 1956, 105-108.
[99] G. Verner, Wire Wire Prod., *30*, 1955, 47-50.
[100] W.M. Halliday, Wire Ind., *24*, 1957, 1145-1148; *25*, 1958, 59-61.
[101] J.K. Annandale, Wire Wire Prod., *35*, 1960, 607-610, 654-655.
[102] O. Pawelski, in *9. Umformtechnisches Kolloquium,* Inst. Bildsame Formgebung, Tech. Hochschule Aachen, 1976, Paper No. IX (German).
[103] P.R. Lancaster, Wire Ind., *43*, 1976, 562-564.
[104] B. Avitzur, Wire J., *9* (6), 1976, 75-79.
[105] B. Avitzur, Metal. Odlew., *5*, 1979, 101-125.
[106] E. Felder, in [3.258], pp.365-369.
[107] J.A. Walowit and W.R.D. Wilson, Paper No. MF69-102, SME, Dearborn, 1969.
[108] W.R.D. Wilson and J.A. Walowit, Trans. ASME, Ser.F, J. Lub. Tech., *93*, 1971, 69-74.
[109] S.M. Bloor, D. Dowson, and B. Parsons, J. Mech. Eng. Sci., *12*, 1970, 178-190.
[110] D. Dowson, B. Parsons, and P.J. Lidgitt, in [3.257], pp.97-106.
[111] A. Kneschke and H. Bandemer, Neue Hütte, *16*, 1971, 225-234 (German).
[112] K.H. Weber, Neue Hütte, *16*, 1971, 235-238 (German).
[113] P.F. Thomson, J. Australian Inst. Metals, *16* (3), 1971, 157-166.
[114] J. Zawadzki and H. Orzechowski, Arch. Hutn., *17*, 1972, 73-82 (Polish).
[115] W.A. Dodds and P.R. Lancaster, Wire Ind., *36*, 1969, 333-336.
[116] R.P. Lancaster, Stahl u. Eisen, *91*, 1971, 911-915 (German).
[117] B. Parsons, R. Taylor, and B.N. Cole, Proc. Inst. Mech. Eng., 180, Pt.3I, 1965-66, 230-240.
[118] K.R. Halliday and W.R.D. Wilson, in *Proc. 5th NAMRC,* SME, Dearborn, 1977, pp.165-170.
[119] P.R. Lancaster and B.F. Smith, Wire Ind., *41*, 1974, 933-937.
[120] H.F. Frost, Wire Ind., *21*, 1954, 1199-1201, 1225.
[121] L.H. Butler, Sheet Metal Ind., *33*, 1956, 571-577.
[122] W. Schmidt, Stahl u. Eisen, *91*, 1971, 1374-1381 (German).
[123] C.A. Fischer, Wire Wire Prod., *35*, 1960, 194, 244-245.
[124] L. Salz, Lubric. Eng., *10*, 1954, 190-192.
[125] Anon., Wire Ind., *29*, 1962, 163-165.
[126] S.L. Stalson, in discussion to [4.228].
[127] P. Funke, B. Werdelmann, and W. Lueg, Stahl u. Eisen, *80*, 1960, 918-925 (German).
[128] M.G. Yunusov, Sov. Eng. Res., *1* (5), 1981, 57-60.
[129] W. Papsdorf, Stahl u. Eisen, *72*, 1952, 393-399 (German).
[130] B. Lunn, Wire Wire Prod., *40*, 1965, 1554-1558, 1636-1637.
[131] M. Kühn, Stahl u. Eisen, *72*, 1952, 1212-1216 (German).
[132] P. Montmitonnet and F. Delamare, Trib. Int., *15*, 1982, 133-137.
[133] V.L. Kolmogorov, S.I. Orlov, and K.P. Selischev, *Drawing under Conditions of Hydrodynamic Lubrication,* National Lending Library for Science and Technology, Boston Spa, Yorkshire, England, 1968.
[134] E. Felder and G. Breinlinger, in *Proc. 4th NAMRC,* SME, Dearborn, 1976, pp.158-164.
[135] P. Montmitonnet, M. Brison, and F. Delamare, Wear, *77*, 1982, 315-328.
[136] G.H. Tattersall, J. Mech. Eng. Sci., *3*, 1961, 378-393.
[137] W.R.D. Wilson and K. Halliday, Wear, *42*, 1977, 135-148.
[138] K.F. Kennedy, A. Wadhawan, and W.R.D. Wilson, in [3.56], Paper No.36.
[139] D.G. Christopherson, H. Naylor, and J. Wells, J. Inst. Petr., *40*, 1954, 295-298.

[140] D.G. Christopherson and H. Naylor, Proc. Inst. Mech. Eng., *169*, 1955, 643-665.
[141] J.F. Osterle and J.R. Dixon, ASLE Trans., *5*, 1962, 233-241.
[142] G.H. Tattersall, Wire Ind., *29*, 1962, 975-982, 992, 1083-1087, 1115.
[143] P.S.Y. Chu, in [3.37], pp.104-111.
[144] B.R. Kartak and A.H. Strakhov, Steel USSR, *4*, 1974, 851-853.
[145] J. Zawadzki and H. Orzechowski, Wire Ind., *38*, 1971, 412-415, 417.
[146] K.D. Lietzmann and J. Bittersmann, Neue Hütte, *14*, 1969, 32-37 (German).
[147] Y. Nakamura, H. Kawakami, T. Matsushita, and H. Sawada, Wire J., *13* (6), 1980, 54-58.
[148] P.J. Thompson and G.R. Symmonds, Proc. Inst. Mech. Eng., *191*, 1977, 115-123.
[149] R. Crampton, G.R. Symmonds, and M.S.J. Hashmi, in [3.55], pp.107-115.
[150] J.G. Wistreich, Wire Ind., *24*, 1957, 954-958, 1027-1029, 1046.
[151] G.M. Sturgeon and G.H. Tattersall, Wire Ind., *26*, 1959, 1183-1185, 1192.
[152] A.V. Hedman, Wire Wire Prod., *33*, 1958, 1205, 1271.
[153] V.L. Kolmogorov et al, Tsvetn. Met., *1978* (3), 54-57 (Russian).
[154] L. Tegel-Küppers, Stahl u. Eisen, *101*, 1981, 529-534 (German).
[155] V.Z. Zhilkin and A.D. Latokhin, Steel USSR, *10*, 1974, 849-850.
[156] K.M. Petzold and W. Giehler, Neue Hütte, *14*, 1969, 404 (German).
[157] H. Vollmer and G. Kaiser, Wire, *28* (4), 1978, 137-142.
[158] Ya.S. Shvartsbart, Yu.I. Sinelnikov, S.I. Lutkovsky, and S.I. Orlov, Stal', *1972* (11), 1049-1051 (Russian).
[159] S. Harper, Wire Ind., *43*, 1976, 623-626.
[160] E-E. Wirth and D. Schatte, Neue Hütte, *23*, 1978, 34-36 (German).
[161] Z. Polek, T. Prajsnar, and R. Wusatowski, in *9. Umformtechnisches Kolloquium,* Techn. Hochschule Aachen, 1976 (German).
[162] L. Godecki and T. Prajsnar, Wire, *23* (4), 1973, 171-174.
[163] A. Cameron, J. Inst. Petr., *40*, 1954, 336 (discussion).
[164] M.P. Milliken, Wire Wire Prod., *30*, 1955, 560, 592.
[165] R. Tourret, Wire Ind., *23*, 1956, 41-44, 56.
[166] R. Kopp and M. Moik, Arch. Eisenhüttenw., *48*, 1977, 267-272 (German).
[167] H. Kron, Stahl u. Eisen, *97*, 1977, 332-336 (German).
[168] L.H. Butler, J. Inst. Metals, *93*, 1964-65, 123-125.
[169] V.F. Moseev and A.A. Korostelin, Stal in English, *1962* (3), 237-239.
[170] P.F. Thomson, J.S. Hoggart, and J. Suiter, J. Inst. Metals, *95*, 1967, 152-156.
[171] P.F. Thomson and J.S. Hoggart, J. Inst. Metals, *96*, 1968, 225-228.
[172] Ya.D. Vasilev, Stal in English, *1963* (6), 492-493.
[173] V.P. Volkov, A.V. Nerebov, A.A. Rakcheev, and S.I. Pozdnyakov, Metallurg. J., *1973* (7), 37-39 (Russian).
[174] I.N. Nedovzii, V.D. Chistota, Z.I. Kostina, and A.P. Petrov, Steel USSR, *10*, 1974, 847-848.
[175] S.I. Orlov, V.L. Kolmogorov, V.I. Uralsky, and V.T. Stukalov, Steel USSR, *10*, 1974, 844-846.
[176] S. Kumar and C.B. Misra, J. Inst. Eng. (India), *53*, 1972 (Nov.), 91-94.
[177] N. Tsujimura, Tetsu-to-Hagane, *53*, 1967, 843-846 (Japanese).
[178] A. Middlemiss, in [3.51], pp.47-54.
[179] A. Middlemiss, Wire Ind., *38*, 1971, 188, 199-192.
[180] G. Juretzek, H. Fuhrmann, and K.M. Petzold, Neue Hütte, *10*, 1965, 193-199 (German).
[181] R.W. Gottschlich and N.N. Breyer, J. Metals, *15*, 1963, 364-367.
[182] B.I. Beresnev, L.F. Vereshchagin, and Yu.N. Ryabinin, Tsvetn. Met., *31* (8), 1958, 61-63 (Russian).
[183] Y. Matsuura, Rep. Casting Res. Lab., Waseda Univ. (Tokyo), No.13, 1962, 49-58.
[184] G.L. Kolmogorov, V.I. Boyarkin, and I.M. Mishunin, Tsvetn. Met., *1974* (9), 51-53 (Russian).
[185] T. Prajsnar, J. Rulinski, and E. Zglobicki, Hutnik, *39*, 1972, 301-309 (Polish).
[186] T.W. Rees, J. Appl. Metalwork., *1* (4), 1981, 53-57.
[187] V.V. Shveikin et al, Stal', *1973* (10), 929 (Russian).
[188] G.W. Rowe, Wire Ind., *45*, 1978, 229-231, 236.
[189] B.F. Smith and A. Copper, Wire J., *12* (1), 1979, 76-81.
[190] A.L. Tarnavskii, Steel USSR, *4*, 1974, 502-505.

[191] E.P. Riley-Gledhill, Wire Ind., *21*, 1954, 407-411, 421.
[192] H.G. Baron and F.C. Thompson, J. Inst. Metals, *78*, 1950-1951, 415-462.
[193] P.R. Lancaster and G.W. Rowe, Paper No.61-LUBS-5, ASME, New York, 1961.
[194] H.C. Flender, P. Funke, and A. Jain, Stahl u. Eisen, *95*, 1975, 134-138 (German).
[195] F. Schwier, M. Martin, P. Funke, and J. Pfeiffer, Stahl u. Eisen, *90*, 1970, 120-126 (German).
[196] R. Tourret, Wire Wire Prod., *30*, 1955, 299-303, 347.
[197] H. Bühler and K. Rittmann, Stahl u. Eisen, *91*, 1971, 105-121 (German).
[198] U. Boor, Stahl u. Eisen, *102*, 1982, 13-18 (German).
[199] J.S. Hoggart, Australasian Eng., *1954*, June, 44-50.
[200] V.V. Kotov et al, Tsvetn. Met., *1974* (2), 46-47 (Russian).
[201] K.H. Treptow, Stahl u. Eisen, *76*, 1956, 1133-1134 (German).
[202] A. Pomp and W.L. Becker, Mitt. Kaiser Wilhelm Inst. Eisenforsch., *12*, 1930, 263 (German).
[203] A.E. Ranger, J. Iron Steel Inst., *185*, 1957, 383-388.
[204] S. Komura, Wire Ind., *34*, 1967, 69-73.
[205] J. Luksza, L. Sadok, and A. Skolyszewski, Metal. Odlew., *6*, 1980, 491-502.
[206] K.M. Olsen and C.F. Larkin, Wire Wire Prod., *41*, 1966, 1631-1633, 1707-1709.
[207] Th. Gräbener and K. Lange, in *Proc. 10th NAMRC,* SME, Dearborn, 1982, pp.122-129.
[208] W. Lueg and K.H. Treptow, Stahl u. Eisen, *72*, 1952, 399-416 (German).
[209] A.L.H. Perry, Sci. Lubric., *7* (5), 1955, 14-18.
[210] R.H. Hendrick, Wire Wire Prod., *38*, 1963, 1487-1490, 1574-1576.
[211] J. Golden, P.R. Lancaster, and G.W. Rowe, Int. J. Appl. Radiat. Isotop., *4*, 1958, 30-35.
[212] K.B. Lewis, Wire Wire Prod., *13*, 1938, 441-443, 476-477.
[213] K. Nakamura and J. Kurihara, J. Gov. Mech. Lab (Japan), *15* (July), 1961, 434-441 (Japanese).
[214] R. Froeschmann et al, Wire World Int., *5* (Nov.-Dec.), 1963, 257-262.
[215] R. Hantos, J. Heeringer, and J. Schey, Acta Tech. Acad. Sci. Hung., *15*, 1956, 127-140.
[216] C.B. Miller, Wire J., *3*, 1970, 51-57.
[217] Y. Asada, K. Mori, K. Yoshikawa, and K. Osakada, J. JSTP, *22*, 1981, 488-494 (Ann. Rep. *3*, 1982, 42).
[218] C. Fischer et al, Lubric. Eng., *14*, 1958, 310-315.
[219] M. Oyane, K. Yoshikawa, and T. Sato, Kyoto Rep. No.5, 1980, pp.29-34.
[220] B.A. Kolomiets, V.L. Kolmogorov, I.N. Nedovizii, and B.R. Kartak, Stal', *1979* (5), 376-377 (Russian).
[221] Discussion on "Metal Working Lubricants," Metal Ind. (London), *84*, 1954 (Feb.19), 151-152; (Feb.26), 167-168.
[222] C.S. Cook, J. Appl. Metalwork, *1* (2), 1980, 69-75.
[223] R. Jones and S.J. Harvey, in *Proc. 20th Int. MTDR Conf.,* Macmillan, London, 1980, pp.141-149.
[224] R. Jones and S.J. Harvey, in *Proc. 21st Int. MTDR Conf.,* Macmillan, London, 1981, pp.243-249.
[225] E.L. Francis, H. Greenwood, and F.C. Thompson, Iron Steel Inst. (London), Carnegie Schol. Mem., *22*, 1933, 15.
[226] R. Furuichi and M. Nakayama, Trans. JSME, *41*, 1975, 349 (Japanese).
[227] K.D. Lietzmann and K.R. Eichner, Neue Hütte, *14*, 1969, 458-463 (German).
[228] K. Kramer and J. Steeger, Neue Hütte, *18*, 1973, 756-757 (German).
[229] F. Muntzwyler, Wire Wire Ind., *39*, 1964, 1799-1800.
[230] F.T. Cleaver and H.J. Miller, J. Inst. Metals, *78*, 1950, 537-562.
[231] M. Nakayama, J. JSLE, *24*, 1979, 8-15 (Japanese).
[232] K.D. Lietzmann and G. Kemper, Neue Hütte, *18*, 1973, 79-84 (German).
[233] C.E. Winsper and D.H. Sansome, J. Inst. Metals, *97*, 1969, 274-280.
[234] C.A. Boyd and N. Maropis, DMIC Rep. 187, Battelle Memorial Inst., 1963, 13-22.
[235] K.H. Sämann, Werkstattstechnik, *61*, 1971, 672-676 (German).
[236] J. Bazan and A. Pasierb, Arch. Hutn., *17* (1), 1972, 55-71 (Polish).
[237] H. Maeda, A. Nakagiri, and I. Yoshida, J. Jpn. Inst. Metals, *35*, 1971, 299-307 (Japanese).
[238] N. Maropis and J.C. Clement, Wire Ind., *38*, 1971, 336, 338-341.
[239] J.B. Jones, in *Proc. Int. Conf. Manufacturing Technology,* ASTME (SME), Dearborn, MI, 1967, pp.983-1006.

[240] B. Langenecker and O. Vodep, Wire J. Int., *14*, 1981(9), 246-248.
[241] K. Edelmann, G. Haussler, D. Kandler, and E. Ischeffler, Neue Hütte, *19*, 1974, 33-37 (German).
[242] V.P. Severdenko, V.V. Klubovich, and A.S. Masakovskaya, Dokl. Akad. Nauk BSSR, *19*, 1975, 699-701 (Russian).
[243] V.P. Severdenko, V.V. Klubovich, and L.A. Kononova, Dokl. Akad. Nauk BSSR, *17*, 1973, 325-328 (Russian).
[244] V.P. Severdenko, V.V. Klubovich, R.A. Repin, and L.K. Konyshev, Izv. Akad. Nauk BSSR, *1973* (Fiz.-Tekhn.)(3), 5-8 (Russian).
[245] M.J.R. Young, Ind. Anz., *98* (2), 1976, 25-27 (German).
[246] C.E. Winsper and D.H. Sansome, Metal Forming, *38*, 1971, 71-75.
[247] J.T. Buckley and M.K. Freeman, Ultrasonics, *8*, 1970 (3), 152-158.
[248] T. Sugahara et al, Nippon Kokan Tech. Rep. (Overseas), *12*, 1971 (June), 27-36.
[249] Anon., Machinery U.S., *74*, 1968 (9), 88-89.
[250] N. Atanasiu, Draht, *31*, 1980, 70-73 (German).
[251] R.W. Johnson and G.W. Rowe, Mach. Des. Eng., *1964*, 34-38.
[252] E. Felder, in [3.254], pp.308-312.
[253] W. Lueg and U. Krause, Stahl u. Eisen, *79*, 1959, 1837-1843 (German).
[254] N. Hansen, Draht-Welt, *53*, 1967, 545-551; *54*, 1968, 145-150 (German).
[255] K. Mietzner, Stahl u. Eisen, *82*, 1962, 1423-1432 (German).
[256] K. Springmann, Stahl u. Eisen, *88*, 1968, 1043-1051 (German).
[257] D.B. Woodcock, Proc. Inst. Mech. Eng., *169*, 1955, 668-670.
[258] H. Kuntze and A. Pomp, Wire Ind., *22*, 1955, 58-62, 65 (from Stahl u. Eisen, *74*, 1954, 1325-1334).
[259] E. Dahl and W. Lueg, Stahl u. Eisen, *77*, 1957, 1368-1374 (German).
[260] S.S. Stalson, Wire J., *5*, 1972 (2), 39-43.
[261] H. Kessler and H.J. Bockenhoff, Stahl u. Eisen, *88*, 1968, 1051-1053 (German).
[262] W. Späth, Draht-Welt, *54*, 1968, 797-803 (German).
[263] R.N. Wright, Metal Progr., *114*, 1978 (2), 49-53.
[264] R.N. Wright and A.T. Male, Trans. ASME, Ser.F, J. Lub. Tech., *97*, 1975, 134-140; Wire Ind., *42*, 1975, 522-525.
[265] J.M. Kiefer, Metals Eng. Quart., *15* (1), 1975, 21-31.
[266] G. Hamers, Wire J., *11* (9), 1978, 134-140.
[267] R.F. Morton, Wire Wire Prod., *40*, 1965, 982-985, 1022.
[268] J.B. Bean, Wire Wire Prod., *30*, 1955, 55-57.
[269] L. Salz, Wire Wire Prod., *30*, 1955, 51-55.
[270] U. Boor, Draht Welt, *59*, 1973, 385-391 (German).
[271] B.W. Greenwald, Wire J., *10* (5), 1977, 74-78.
[272] I.S. Morton, Ind. Lub. Trib., *26*, 1974, 205-212.
[273] M.N. Rao, in [3.56], Papers No.44 and 49.
[274] W. van Hessche, Wire Ind., *40*, 1973, 871-873.
[275] A.W. Ackerman and F.V. Allan, Wire Wire Prod., *41*, 1966, 1620-1623, 1705-1707.
[276] R.K. Brandth, Wire Wire Prod., *47* (5), 1972, 54-56.
[277] R.K. Brandth and R.F. Morton, Wire J., *11* (11), 1978, 58-61.
[278] C. Eisenhuth, Stahl u. Eisen, *83*, 1963, 1459-1466 (German).
[279] W. John, G. Gonschior, and H. Hermann, Schmierungstechnik, *11*, 1980, 19-21 (German).
[280] J.S. McFarlane and A.J. Wilson, J. Inst. Petr., *40*, 1954, 324-330.
[281] A. Nagymányai, I. Valasek, and E. Vamos, in [3.56], Paper No.31 (German).
[282] S. Takahama, T. Nakamura, and S. Hironaka, J. JSLE, *24*, 1979, 454-457 (Int. Ed., *1*, 1980, 134).
[283] S.W. Klasky, Wire J., *4* (4), 1971, 54-56.
[284] J.R. Ponteri, Wire J., *9* (8), 1976, 52-58.
[285] J. Bernot, Wire J., *12* (5), 1979, 71-73.
[286] R.W. Helman and J.N. Snyder, Wire J., *10* (10), 1977, 47-53.
[287] M. Baroux, Wire J., *15* (5), 1979, 62-66.
[288] J. Bernot, Wire J., *12* (12), 1979, 80-81.
[289] J.A. Holbrook, Wire J., *7* (12), 1974, 88-94.
[290] A. Valente, Wire J., *9* (5), 1976, 68-73.
[291] M.J. Kelly and R. Phillips, Wire J. Int., *15* (6), 1982, 60-65.
[292] J.N. Stone, Wire J., *13* (8), 1980, 98-100.

[293] O. Miki, S. Fuji, and M. Vemura, Wire J., *12* (5), 1979, 56-60.
[294] M. Donnison, Wire Ind., *41*, 1974, 283-284.
[295] R.C. Williams, Ind. Eng. Chem., *31*, 1939, 725-727.
[296] Z. Kurihara, F. Asuka, and K. Nakamura, J. Mech. Eng. Lab. (Tokyo), *26* (3), 1972, 121-130 (Japanese).
[297] V.I. Spitsyn, O.A. Troitskii, V.G. Ryzhkov, and A.S. Kozyrev, Dokl. Akad. Nauk SSSR, *231*, 1976, 402-404 (Russian).
[298] J.L. Miller, Wire J., *11* (10), 1978, 74-77.
[299] H.F. Atala and G.O.A. Laditan, Wire Ind., *39*, 1972, 455-456.
[300] P.J. Thompson and G.R. Symmonds, Wire Ind., *42*, 1975, 517-520, 596-600.
[301] G. Bader, Wire, *1970* (105), 39-45.
[302] Y. Nakamura, H. Kawakami, T. Fujita, and Y. Yamada, Wire J., *9* (7), 1976, 59-66.
[303] G.M. Sturgeon, Wire Wire Prod., *39*, 1964, 220-225.
[304] J. Wolf and A. Schubert, Draht, *31*, 1980, 61-66 (German).
[305] W.M. Tomlinson, J.W. Pamplin, and R. Shillito, Wire J., *13* (4), 1980, 66-69.
[306] A. Middlemiss, Wire Ind., *39*, 1972, 34-37.
[307] J.H. Rigo, Wire J., *6* (4), 1973, 73-79.
[308] R. Phillips, M.J. Kelly, and M. Hurst, Wire J., *12* (3), 1979, 50-56.
[309] A. El Hindi, in [13], pp.221-248.
[310] B.C. Gentry, Wire Wire Prod., *47* (6), 1972, 63-66.
[311] W. Taller, Wire World Int., *13* (2), 1971, 49.
[312] H.L. Miller, Wire J., *7* (12), 1974, 79-83.
[313] B. Dickerhoff, Wire J., *10* (11), 1977, 87-90.
[314] J. Joseph, Wire J., *9* (8), 1976, 38-41.
[315] D.H. Ward, W.A. Zarbis, and A. El Hindi, Wire J., *6* (6), 1973, 79-87.
[316] J.J. Joseph, Wire J., *7* (10), 1974, 75-84.
[317] H. Knobloch, Draht-Welt, *60* (2), 1974, 53-58 (German).
[318] A.L. Rosener and J.C. Carreiro, Wire Wire Prod., *44* (6), 1969, 41-45.
[319] A. El Hindi, Wire J., *2* (4), 1969, 58-61.
[320] J. Joseph, Wire J., *12* (1), 1979, 88-92.
[321] W. Teller and G. Schönauer, Draht-Welt, *57*, 1971, 626-627 (German).
[322] K.M. Olsen, R.F. Jack, and E.O. Fuchs, Wire Wire Prod., *40*, 1965, 1563-1568, 1637-1638.
[323] F. McDonough, W. Ludwig, and P. Vandenberg, Wire J., *10* (10), 1977, 70-74.
[324] I. Tripp, Jr., Wire J. Int., *14* (5), 1981, 96-99.
[325] Anon., Wire J., *10* (1), 1977, 42-43.
[326] O.G. Reininga, R.H. Wagner, and R.A. Bonewitz, Wire J., *9* (10), 1976, 48-53.
[327] R.J. Devlin, Jr., Wire J., *10* (9), 1977, 182-185.
[328] A.D. Eckard, Wire J. Int., *14* (6), 1981, 76-78.
[329] S.R. Shatynski and R.N. Wright, Wire Technol., *7* (4), 1979, 59-62, 66-69.
[330] J. Vogel, Schmierungstechnik, *2*, 1971, 293-295 (German).
[331] F.T. Cleaver and H.J. Miller, in *The Cold Working of Non-Ferrous Metals and Alloys*, Institute of Metals, London, 1952, p.205.
[332] J.G. Wistreich, J. Iron Steel Inst., *167*, 1951, 162-164.
[333] E.P. Riley-Gledhill, Wire Wire Prod., *29*, 1954, 746-751, 767, 775.
[334] L.C. Shaheen (ed.), in [3.426], pp.521-525.
[335] B. Kastner, L. Sadok, P. Kastner, and Z. Rajwa, Hutnik, *45*, 1978, 290-294 (Polish).
[336] H.H. Sturhahn and P. Schorr, Bänder Bleche Rohre, *16*, 1975, 347-349 (German).
[337] R. Brühl, Wire World Int., *18*, 1976, 195-196.
[338] W. Weith, Draht, *26*, 1975, 607-611 (German).
[339] J.C.E. Button, A.J. Davies, and R. Tourret, Nucleonics, *9* (5), 1951, 34-43.
[340] W. Dahl and W. Lueg, Stahl u. Eisen, *76*, 1956, 257-261 (German).
[341] J. Golden and G.W. Rowe, Brit. J. Appl. Phys., *9*, 1958, 120-121.
[342] J. Holzhey, D. Bechstein, and G. Krüger, Neue Hütte, *12*, 1967, 756-759 (German).
[343] D. Witanov, Wire, *25* (3), 1975, 101-104.
[344] A.V. Linial and H. Sidell, Wire J., *8* (3), 1975, 63-66.
[345] I. Hush, Wire Ind., *38*, 1971, 266-269.
[346] L.C. Carrison, Wire Ind., *44*, 1977, 402-403.
[347] D.G. Flom, R.E. Hanneman, W.A. Rocco, and R.H. Wentorf, Wire J., *8* (3), 1975, 76-80.
[348] R.M. Lefever, Wire J., *9* (11), 1976, 80-83.

[349] L.P. King, Wire J., *10* (4), 1977, 65-68.
[350] O. Schob and W. Vrieze, Draht, *25*, 1974, 288-297 (German).
[351] N. Gane, J. Austral. Inst. Metals, *19*, 1974, 259-265.
[352] Anon., Wire Ind., *45*, 1978, 993-1000.
[353] W. Teller, Wire World Int., *18* (2), 1976, 62-66.
[354] S. Leber and D. White, Wire Wire Prod., *45* (2), 1970, 72-75.
[355] H. Valberg, Wire J., *14* (2), 1981, 78-81.
[356] J.D. Reardon, Wire J., *8* (9), 1975, 123-126.
[357] P.E. Doherty, Wire J., *7* (3), 1974, 62-64.
[358] W. Stannek, Wire Ind., *45*, 1978, 933-935.
[359] H.H. Sturham and H.C. Eichas, Wire J., *11* (9), 1978, 150-153.
[360] P.F. Thomson, J.B. Carroll, and E. Bevitt, J. Iron Steel Inst., *173*, 1953, 36-51.
[361] R. Kellermann and G. Turlach, Werkstattstechnik, *55*, 1965, 309-313 (German).
[362] K. Nakamura and J. Kurihara, J. Gov. Mech. Lab (Japan), *15* (July), 1961, 442-456 (Japanese).
[363] Anon., Wire Ind., *39*, 1972, 678-682.
[364] I. Sasse and W. Frass, Neue Hütte, *14*, 1969, 157-162 (German).
[365] F.J. Donaghu and R.C. Rhees, Wire Wire Prod., *42*, 1967, 604-608, 610-611, 686-687.
[366] D.J. Muse and A.T. Gorton, Wire J., *10* (4), 1977, 43-48.
[367] A.F. Karmadonov et al, Russian Eng. J., *49* (9), 1969, 78-79.
[368] S.I. Orlov, Metallurg, *1972* (1), 35-36 (Russian).
[369] F.J. Harbone, J. Australian Inst. Metals, *16* (2), 1971, 73-87.
[370] A. Robonyi, Wire World Int., *19* (2), 1977, 70-73.
[371] E.M. Alexander, Wire J., *8* (9), 1975, 151-159.
[372] A.L.H. Parry, J. Inst. Petr., *40*, 1954, 319-323.
[373] W. Machu, Wire Ind., *35*, 1968, 151-153.
[374] J.F. Leland, Wire Wire Prod., *29*, 1954, 1440-1443, 1479-1481.
[375] R.C. Rhoades, Wire Wire Prod., *34*, 1959, 949-951, 1016-1019.
[376] W. Machu, Draht-Welt, *55* (1), 1969, 13-17 (German).
[377] D. James, Wire Ind., *38*, 1971, 858-861.
[378] J.F. Richards, Wire J., *5*, 1972, 67-72.
[379] D.C. Rutledge, Wire J., *7*, 1974, 56-59.
[380] J. Donald, Wire Wire Prod., *45*, 1970, 37-41.
[381] K.D. Nittel, in [3.56], Paper No.43 (German).
[382] R.E. Pettit, Wire J., *11* (11), 1978, 66-67.
[383] J.S. Hawkins, Prodn. Eng., *39*, April 1960, 226-234.
[384] W. Lueg and H. Treptow, Wire Ind., *21*, 1954, 1211-1216, 1238 (from Stahl u. Eisen, *74*, 1954, 1334-1342).
[385] P. Magie, Wire Wire Prod., *35*, 1960, 997-999, 1054-1055.
[386] A. Sonntag, Wire Wire Prod., *30*, 1955, 782-786, 816.
[387] W.A. Smigel and W.A. Stillwell, Wire Wire Prod., *36*, 1961, 313-314, 376.
[388] K. Matsumoto, Lubric. Eng., *34*, 1978, 258-264.
[389] H. Panhuis, Wire Prod., *41*, 1974, 57-60.
[390] L. Salz, Wire Wire Prod., *28*, 1953, 1056-1058, 1125.
[391] J.A. Annandale, Wire Ind., *26*, 1959, 741-746.
[392] J. Zouck, Wire J., *5* (4), 1972, 41-48.
[393] J.F. McConnell and J.F. Richards, Wire J., *4* (10), 1971, 41-45.
[394] D. Fuchs, Stahl u. Eisen, *97*, 1977, 154-158 (German).
[395] W. Lueg and A. Pomp, Stahl u. Eisen, *70*, 1950, 977-984 (German).
[396] M. Burgdorf, Form. Trait. Metaux, *1972* (33), 23-26, 73 (French).
[397] A.W. Dahl, Lubric. Eng., *17*, 1961, 570-579.
[398] Z. Halas, Neue Hütte, *8*, 1963, 86-91 (German).
[399] D.R. Wilkinson, Wire Wire Prod., *40*, 1965, 1764-1765, 1800.
[400] Anon., Lubrication, *45*, 1959, 85-100.
[401] M. Buch, M. Hildebrand, L. Eberlein, and H. Göhler, Neue Hütte, *10*, 1965, 338-344 (German).
[402] H.A. Holden, Sheet Metal Ind., *30*, 1953, 775-778, 819.
[403] P.I. Chuiko, V.N. Kolesnikov, and G.A. Savin, Metallurg, *1972* (3), 32-33 (Russian).
[404] R.F. Hood, Wire J., *9* (7), 1976, 37-41.
[405] A.F. Gerds and C.B. Ogle, Wire Wire Prod., *38*, 1963, 1696-1700, 1755-1756.

[406] D.B. Freeman, Wire Ind., *42*, 1975, 526-527.
[407] G.D. Duplii, G.M. Prikhodchenko, and G.I. Khaustov, Stal in English, *1964* (11), 891-893.
[408] Anon., Iron Steel Eng., *33* (11), 1956, 140-145.
[409] V.Z. Zhilkin and A.D. Latokhin, Sov. J. Non-Ferrous Met., *1975* (5), 62.
[410] A.J. Stock, Lubric. Eng., *17*, 1961, 580-586.
[411] Yu.M. Postolov et al, Sov. J. Non-Ferrous Met., *1976* (5), 60-61.
[412] Anon., Wire Ind., *23*, 1956, 223-224, 242.
[413] W.B. Bauzenberger, Wire Wire Prod., *30*, 1955, 888-890.
[414] D.J. Kreml, Wire Wire Prod., *41*, 1966, 308-310.
[415] R.J. Schoerner, Wire Wire Prod., *30*, 1955, 883-885, 938.
[416] O.J. Tassi and J.R. Drotleff, in [13], p.249-255.
[417] W. Torrington, Wire Wire Prod., *47* (8), 1972, 51-52.
[418] C.F. Wickwire, Wire Wire Prod., *30*, 1955, 880, 940.
[419] J. Zawadzki and H. Orzechowski, Arch. Hutn., *17*, 1972, 143-161 (Polish).
[420] J. Werner, Wire Wire Prod., *30*, 1955, 1223-1225.
[421] H.T. Jones, Wire Wire Prod., *30*, 1955, 890-893, 941.
[422] D.L. Zini, Wire J., *12* (8), 1979, 74-79.
[423] A.G. Gross, Wire Wire Prod., *35*, 1960, 1369-1374.
[424] D.B. Wright and G.C. Ellis, in *The Metallurgy of Beryllium,* Monogr. Rep. Ser. No.28, Institute of Metals, London, 1963, pp.418-444.
[425] N.P. Pinto and T.P. Denny, Metal Progr., *91* (6), 1967, 107-118.
[426] J.C. Uy, B.D. Richardson, T.S. Felker, and F.J. Fiorentino, Wire Ind., *39*, 1972, 40-43, 110-112.
[427] C.B. Boon, Wire Ind., *45*, 1978, 1010-1013.
[428] R. Pepmöller, Wire J., *12* (8), 1979, 68-70.
[429] H. Pops and D.R. Hennessy, Wire J., *10* (3), 1977, 50-57.
[430] E.B. Shumilina, V.I. Pikaeva, V.M. Korbut, and S.Ya. Veiler, Sov. Mater. Sci., *12*, 1976, 101-103.
[431] V.M. Korbut, V.I. Pikaeva, and S.Ya. Veiler, Sov. Mater. Sci., *10*, 1974, 227-229.
[432] F. Paparoni, Wire Ind., *33*, 1966, 965-973, 1067-1072.
[433] P. Gregory, Wire, *92*, 1967, 272-274.
[434] I. Tripp and G.E.S. Nicol, Wire Ind., *37*, 1970, 35, 37-39.
[435] Anon., Wire Wire Prod., *44*, 1969, 61, 101-102.
[436] J.R. Drotleff, in [13], pp.257-270.
[437] J.E. Bunner and N.R. Smith, Wire J., *8* (3), 1975, 83-89.
[438] P.E. Lawler and L.N. McKenna, in *Proc. 19th Int. Wire and Cable Symp.,* Atlantic City, NJ, 1970, U.S. Army Electronics Command, Ft. Monmouth, NJ, pp.131-135.
[439] D. Stöckel, Bänder Bleche Rohre, *15*, 1974, 444-448 (German).
[440] G. Schröder, Wire, *30*, 1980, 184-187.
[441] D.H. Wilson, Wire Wire Prod., *30*, 1955, 1246-1247, 1305.
[442] F.R. Larson and W.A. Backofen, Wire Wire Prod., *38*, 1958, 985-987, 1052-1053.
[443] A.D. Merriman, Metal Treat., *19*, 1952, 99-106.
[444] Anon., Iron Age, *193* (Feb. 20), 1964, 130.
[445] V.I. Sokolovskii, U.S. Parshin, and Yu.N. Stepanov, Tsvetn. Met., *1975* (7), 62-64 (Russian).
[446] L.S. Vatrushin and V.G. Osintsev, Tsvetn. Met., *1975* (11), 64-66 (Russian).
[447] S.G. Nelson, Reactor Mater., *10* (2), 1967, 122-124.
[448] P.B. Eyre, J.L. Williamson, and L.R. Williams, in *Beryllium Technology,* Gordon and Breach, New York, 1966, pp.769-806.
[449] D.J. Jones, Metall Mat. Techn., *5*, 1973, 503-512.
[450] R.L. Daga, in [3.145], pp.30-37.
[451] V.Z. Zhilkin et al, Sov. J. Non-Ferrous Met., *1976* (10), 70-71.
[452] V.Z. Zhilkin and A.V. Zhukov, Tsvetn. Met., *1974* (7), 78-81 (Russian).
[453] F. Henderson, F.P. Bullen, and H.L. Wain, J. Inst. Metals, *98*, 1970, 65-70.
[454] R.T. Natapova, Tsvetn. Met., *1964* (1), 66-69 (Russian).
[455] E. Heier, Neue Hütte, *19*, 1974, 207-211 (German).
[456] A.V. Zhukov, Tsvetn. Met., *1977* (1), 72 (Russian).
[457] Anon., Metal Treat., *27*, 1960, 148-150.
[458] R.I. Batista, G.S. Hanks, J.M. Taub, and D.J. Murphy, Lubric. Eng., *17*, 1961, 414-418.

[459] R.N. Wright, in *Manufacture of Superconducting Materials,* R.W. Meyerhoff (ed.), ASM, Metals Park, 1977, pp.175-181.

[460] S. Murase, Y. Koike, and E. Suzuki, J. JSTP, *22*, 1981, 445-451 (Ann. Rep., *3*, 1982, 36).

Extrusion of Semifabricated Products

We saw in Sec.2.4 (Fig.2.8) that extrusion represents a transition from steady-state to non-steady-state processes. Non-steady-state conditions always exist at the beginning and end of extrusion. Steady-state conditions prevail during extrusion of long, semifinished products such as bars, sections, and tubes; end defects may be allowed to develop to a limited degree, because the residue of the billet remaining in the container is discarded. This technique forms the subject of the present chapter.

In contrast, non-steady-state effects predominate in extrusion of short billets for the purpose of making parts; defects cannot be allowed to develop, because the unextruded portion of the billet forms part of the finished component. In many respects, the problems of lubrication are similar to those of lubrication in forging, and therefore the process will be discussed in Chapter 9.

The subject matter of the present chapter incorporates material previously discussed by Kalpakjian [1], Schey [2], and Newnham [3]. In addition to the books referenced in Chapter 2, there are various monographs and reviews that are devoted to extrusion [4-11], of which the most recent and comprehensive is the one by Laue and Stenger [6,7]. The review of extrusion equipment given by Chadwick [12] also deals with several general aspects.

8.1 The Extrusion Process

With few exceptions, extrusion is a batch process. The starting material is usually a cylindrical, cast or previously rolled billet which is placed into a container and pressurized by means of a punch attached to the ram of a hydraulic or mechanical press (Fig.8.1). From the frictional point of view it is of greatest importance whether the billet is actually moved relative to the container (direct extrusion) or is at rest inside the container (indirect extrusion). In either case, the billet is first upset to fill out the container of cross section A_0. The pressure then rises until plastic flow begins, and a product of cross section A_1, which has a short, parallel land, emerges from the die. In principle, it is immaterial whether the extrusion is a round bar, a shaped section, or a tube. Products vary greatly in cross-sectional dimensions and in length, but they typically range from, say, 5-mm-diam wire to 500-mm-diam tubes or sections. Presses range from 2 to 150 MN in capacity [7].

8.1.1 Strain

In common with wiredrawing, extrusion involves flow of metal through a conical channel. In extrusion, however, the indirect compressive stresses are generated by pushing, rather than pulling, the billet through the extrusion die. Deformation in extrusion tends

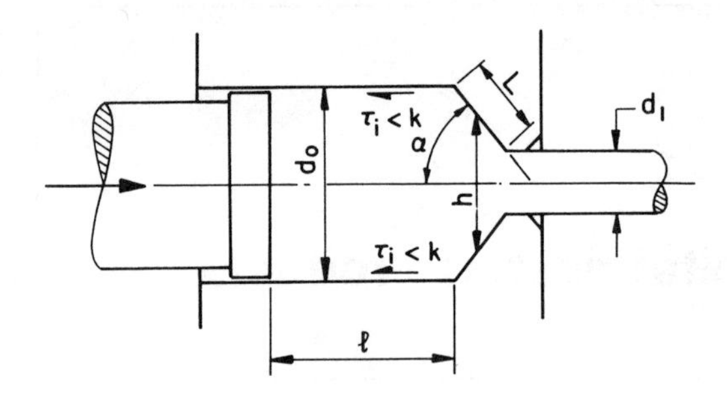

Fig.8.1. Forward extrusion with conical die entry (half arrows indicate interface shear stresses acting on the material)

to be nonuniform, a point to which we shall return later. Nevertheless, it is customary to calculate an average strain as either reduction in area

$$r = \frac{A_0 - A_1}{A_0} \tag{8.1}$$

or extrusion ratio

$$R_E = \frac{A_0}{A_1} \tag{8.2}$$

which conveniently leads to the true strain

$$\varepsilon = \ln R_E = \ln \frac{A_0}{A_1} \tag{8.3}$$

The extrusion ratio can reach very high values, in excess of 400:1.

8.1.2 Extrusion Pressure

Most extrusion theories lead to solutions in which the extrusion pressure p_e at the en of the stroke is proportional to strain

$$p_e = \sigma_{fm}(a + b\varepsilon) \tag{8.4}$$

where σ_{fm} is the mean flow stress of the material, and the parenthetical term allow for the pressure required for uniform deformation, the inhomogeneity of deformatio (redundant work), and die face friction. The constants a and b may be calculated (e.g. [2.12,2.21],[13]) or determined from experiments. For very approximate calculations, i may be assumed that $a = 0.8$ and $b = 1.2$.

As in other processes, the effects of inhomogeneity of deformation and interfac friction interact [2.16]. Deformation becomes more homogeneous with a decreasing $h/$ ratio (Fig.8.1), which implies, for a constant reduction, a smaller die half angle α. Thi results, however, in a longer surface over which friction must be overcome. Conversely a larger die angle reduces the frictional components but leads to increased inhomogene ity; therefore, as in wiredrawing, an optimum die angle is found, the value of whic increases with increasing extrusion ratio and friction (Fig.8.2) [14].

Flow becomes truly streamlined only when friction is very low [15]. It can b shown that die shapes other than simple cones ensure even greater homogeneity of flo

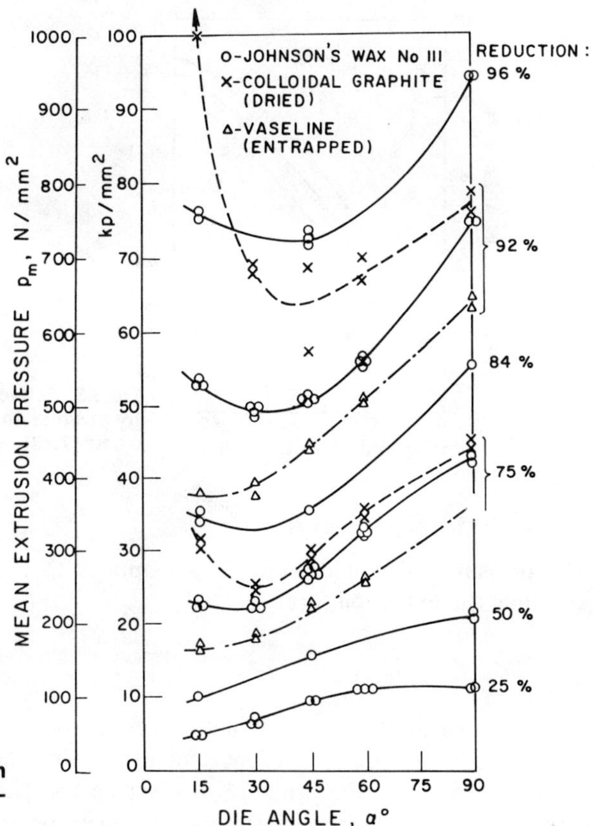

Fig.8.2. Effect of lubrication on optimum die angle in forward extrusion of aluminum [14]

[16,17], especially with high friction. With low friction the improvement does not, in general, justify an extra expense in diemaking, even though with numerically controlled machining this expense may be slight [18].

Stresses in the deformation zone are fully compressive when $h/L > 1$, and, for this reason, extrusion is particularly favorable for working of materials of limited ductility. If, however, the extrusion ratio is low and the die angle is large (and thus h/L is large, too), a hydrostatic tensile stress can develop in the center of the extrusion, and a centerburst defect similar to that found in wiredrawing occurs. Because increasing die friction increases the pressure required for extrusion and thus generates a hydrostatic pressure, higher friction is actually desirable in extending the range over which sound extrusions can be obtained (Fig. 8.3) [7.73],[19,20].

On the basis of pressure considerations, a conical die entry would often be optimum. However, it has the disadvantage that the press stroke must be arrested before the punch can touch the die, thus increasing material lost in the remnant (butt). Furthermore, the extrusion must be pushed back in the die if the remnant is to be cut off. Alternatively, the remnant may be pushed through with a deformable follower block (e.g., a graphite block), or billet must follow billet without removal of the remnants. In the latter case the interfaces between the billets must be clean to facilitate welding and prevent the development of internal defects, and thus the use of a lubricant is precluded. For practical reasons, extrusion is often conducted with a flat-face die (α = 90°), especially at heavier reductions where there is no danger of a centerburst defect.

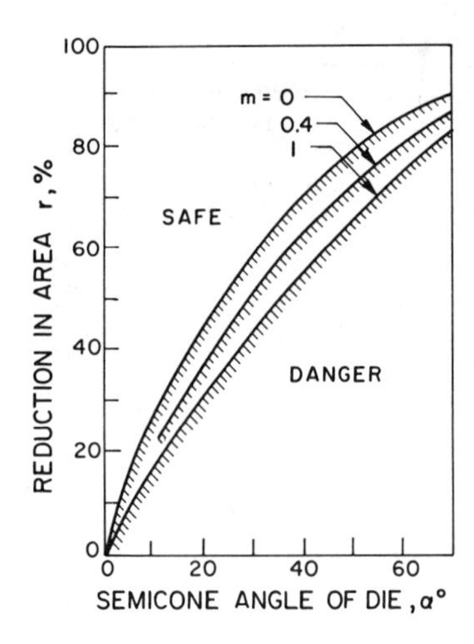

Fig.8.3. Effect of friction on process geometry at which the danger of centerburst defect exists [7.73]

8.1.3 Extrusion Force

The pressure calculated from Eq.8.4 applies to the container cross-sectional area A_0, and thus the extrusion force is

$$P_e = p_e A_0 \tag{8.5}$$

For measurement of this net extrusion force, a billet may be pushed through the extrusion die without the support of a container (Fig.8.4a). To avoid upsetting of the unsupported billet, the pressure p_e must be less than the σ_f of the workpiece material. Thus, the effects of friction and die angle can be explored, but only for a rather limited range of reductions. This technique is, nevertheless, suitable for lubricant ranking. In order for the interface shear strength τ_i (or, if preferred, μ or m) to be derived, the value of σ_{fm} must be accurately known. Alternatively, the axial and normal (radial) forces can be simultaneously measured by the techniques described in Sec.5.5.2. Thus, the split die principle was employed by Townend and Broscomb [21] for axial symmetry and by Polakowski and Schmitt [5.21] for plane strain.

If heavier reductions are to be explored, a conventional extrusion configuration is needed. It is possible, however, to measure separately the forces acting on the extrusion die by leaving a small, well-lubricated clearance between die and container, so that no fin is extruded but free movement of the die is ensured even under load (Fig.8.4b). DePierre [22,23] found such measurements to be reproducible. Lengyel and Mutlu [24] used load cells for punch and die loads, and strain gages on the container for measurement of internal pressure in cold forward and backward extrusion.

The extrusion force is limited by the potential failure of the punch or container. A short punch fails in upsetting, a long one in buckling. The container is an internally

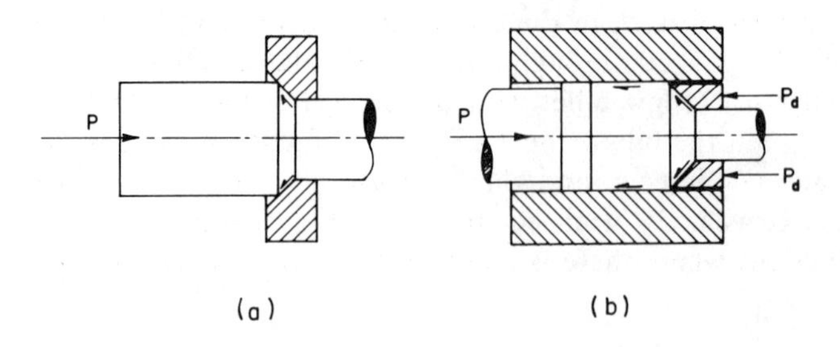

Fig.8.4. Measurement of die friction by (a) containerless extrusion and (b) separation of die force

pressurized vessel which can be reinforced by a shrink-ring construction [7]. Nevertheless, internal pressure is typically limited to 1.2 GPa in hot extrusion and to 2.5 GPa in cold extrusion. This means that for any given material the extrusion ratio is limited according to Eq.8.4 (Fig.8.5) [25].

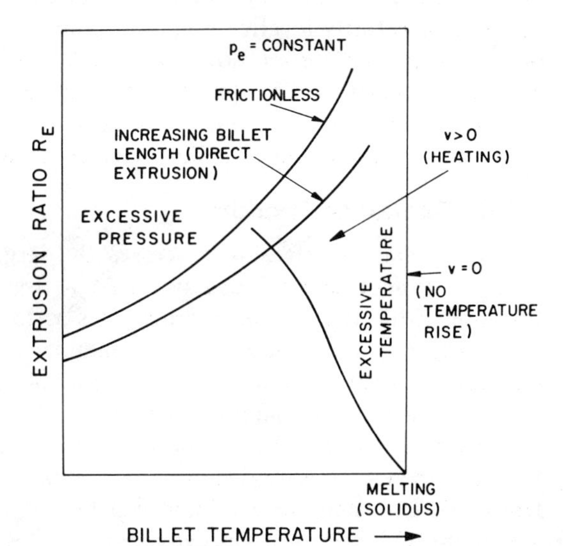

Fig.8.5. Process limitations for hot extrusion arising from container friction and workpiece heating (after [25])

8.1.4 Heat Generation

The work of plastic deformation and friction is converted into heat (Sec.2.3.3). Because strains can be very large and frictional energy is expended in a thin surface layer, heat generation in extrusion is substantial. In a fast stroke most of the heat is carried away by the extruded product, but heat is conducted back into the undeformed part of the billet when extrusion is relatively slow [26,27]. Preheating temperature in hot extrusion is usually high, partly to reduce σ_{fm} and thus allow a large extrusion ratio, and partly to ensure adequate ductility. The temperature rise due to deformation and friction could heat the emerging product above the melting point (or, in alloys, above the solidus temperature). The parallel die land, while contributing only a few percent to the extrusion force, is a major source of heating [28]. Temperature is thus highest on the surface, and grain-boundary melting causes a disintegration of the product in the form of the characteristic fir-tree cracking (speed cracking). Extrusion must be slowed down to minimize the rate of heat generation (Fig.8.5). Alternatively, the billet can be preheated, with a temperature gradient dropping from front to back [29].

8.1.5 Extrusion of Shapes

One of the main advantages of extrusion relative to rolling is that the dimensions or shape of the extruded product can be changed relatively easily and inexpensively by fabricating a new die or die insert. If the extruded product is to be straight and free of internal stresses, the rate of extrusion must be the same in all parts of the extruded section. The rate of flow reflects the resistance that a given part of the section encounters in exiting from the die. Friction on the die face and over the die land hinders free flow and can be used to equalize flow rates. A thicker section will emerge more rapidly than a thinner one, and therefore the land length may be increased or the die half angle reduced to increase frictional retardation. Products of small cross-sectional area usually are extruded through several holes. Flow rates can then be equalized by similar techniques. The difference between the rates of flow through two holes of different diameters can be used for evaluating frictional conditions [30].

Die design is still very much an art, although rational design techniques can be and are employed [7]. Luckily, some internal equalization takes place in the product, and the process is fairly forgiving; in an example of computer-aided die design for extrusion of a T-section, material flow was much more uneven than anticipated and yet the extruded product was sound [31]. The increased frictional resistance along the longer die perimeter and the more complex flow increase the punch pressures relative to extrusion of a round bar of equal A_1 cross-sectional area [7.80],[31-33].

8.1.6 Container Friction

In most practical extrusion processes the large pressures generated demand that the billet be supported by a container. At a first approximation it can be assumed that pressure within the container propagates as in a hydraulic medium and that the extrusion pressure prevails at the billet/container interface (in reality, pressures are lower). Because even the minimum pressure (Eq.8.4) can reach high values at large extrusion ratios, friction at the container-to-billet interface assumes great importance.

From the practical point of view there are only two possibilities: either the interface is arrested (sticking friction), or sliding of the interface is ensured (lubricated flow). No intermediate condition can be allowed, because partial sticking and partial sliding invariably lead to unsteady material flow and a defective extruded product.

8.2 Direct Extrusion

In forward or direct extrusion, the billet is pushed bodily along the container wall and the extrusion emerges in the same direction. The crucial aspect is, however, that the billet must move relative to the container; therefore, in addition to friction on the die, friction on the container wall has to be overcome, too. This has a profound effect on all aspects of extrusion.

8.2.1 Material Flow

The effects of friction on material flow have long been investigated; reviews have been offered by Chadwick [34] and by Dürrschnabel [35]. Some basic types of flow patterns can be identified:

a. In the total absence of friction, grid lines inscribed on the center plane of a split billet remain undistorted until they reach the deformation zone just ahead of the die (Fig.8.6a) [2.28]. Even at the prevailing high interface pressures, this condition can be maintained only as long as $\tau_i << k$.

b. With increasing friction, friction on the container wall causes the grid lines to curve. In seeking minimum energy flow, movement at the transition to the flat die is arrested, and a dead-metal zone is formed (Fig.8.6b). Deformation now occurs in the body of the billet along the boundary of the dead-metal zone. The "internal die angle" thus formed increases with decreasing friction and increasing reduction. Dead-metal zone formation in lubricated extrusion may be avoided if the die is originally made to the appropriate angle, as in Fig.8.1. Sliding then takes place over the entire container/die interface, but only if the lubricant thins out uniformly without suffering local breakdowns. Otherwise, an incipient dead-metal zone forms and localized subsurface defects are generated. Extension of the lubricant is substantial; even at a modest extrusion ratio of 16 (extruding a round bar to one-quarter of the billet diameter), the surface increases fourfold.

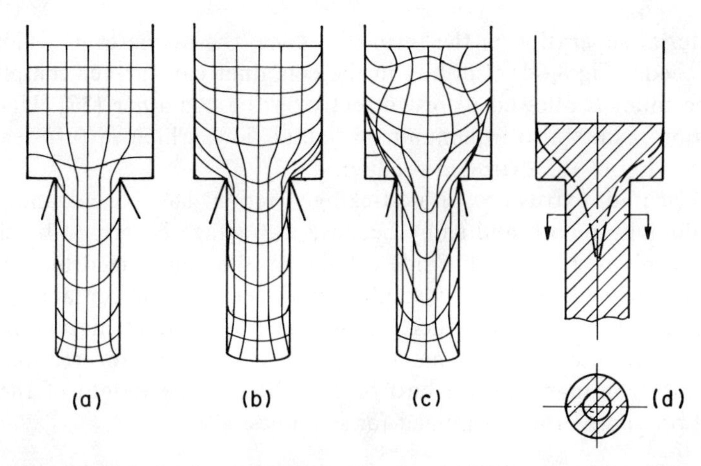

Fig.8.6. Material flow in forward extrusion. (a) Without container friction. (b) With container friction. (c) With cooling. (d) Extrusion defect. (Parts a, b, and c are from [2.28].)

c. If the above conditions cannot be fulfilled, it is much better to extrude without any lubricant whatsoever. Then sticking friction prevails over the entire container/die interface, and extrusion proceeds by shearing along the container wall (Fig.8.7). A dead-metal zone of maximum angle forms, and this angle increases with reduction ratio. It is essential, though, that no trace of a lubricant or contaminant film be present on the billet surface, because otherwise the sticking condition (A-B in Fig.8.8a) will be locally disturbed and the lubricant will be washed in below the product surface (at F-E in Fig.8.8b) [34]. This can lead to surface lamination and, if the product is subjected to subsequent heat treatment, to blistering.

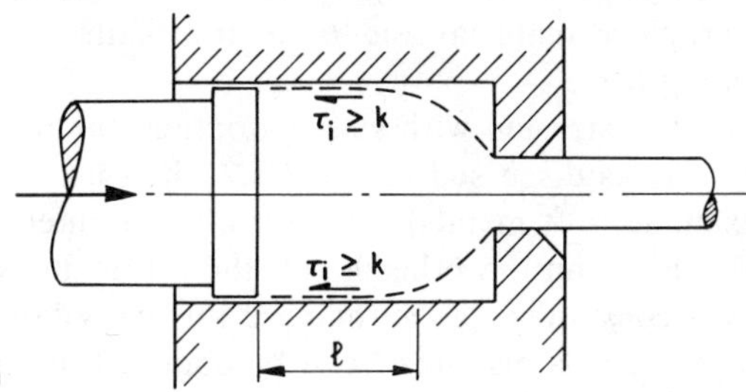

Fig.8.7. Forward extrusion with sticking friction

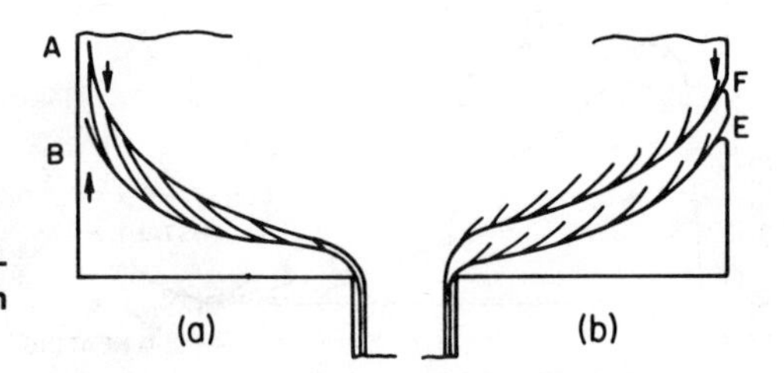

Fig.8.8. Material flow in unlubricated extrusion (a) under normal conditions and (b) with intermittent lubrication [34]

d. When a hot workpiece is extruded in a colder container, chilling contributes to the inhomogeneity of deformation, and a very complex flow pattern develops (Fig.8.6c). In later stages of extrusion the dead-metal zone spreads throughout the entire length of the billet. Material near the punch face moves toward the center and carries with it surface oxide films, as first shown by Genders [36]. This leads to the development of the so-called extrusion defect (also called piping): a ring of oxide inclusions causes complete material separation in the form of a cone, the diameter of which increases as extrusion proceeds (Fig.8.6d). Chilling in the container can further complicate material flow when the billet is allowed to rest directly on the container [37]. Uneven temperature distribution results also in nonuniform metal flow, which in turn leads to internal stresses and curving of the extruded product.

e. Unlubricated extrusion with sticking friction has its advantages, too. The extruded product is smooth and shiny because its surface is formed by shearing inside the body of the billet. Undesirable side effects can be neutralized by extruding with a punch (or follower block) of slightly smaller diameter than the container, thus leaving a skull which traps all oxides and other foreign material. The formation of the extrusion defect can be prevented by leaving a large butt, and yet, for reasons of economy, extrusion usually continues until a butt of 5 to 20% of the weight of the billet is left. The extruded product is then examined for soundness [7].

8.2.2 Effects on Forces

In the direct extrusion of a billet, the pressure required to move the billet against the frictional stress τ_i has to be added to the extrusion pressure p_e. The calculation is simple when τ_i is a constant, and also for sticking friction when $\tau_i = k \simeq 0.5\sigma_f$. The area to be sheared is of circumference $d_0\pi$ and length l, where l is the length of the unextruded billet measured to the die entry (Fig.8.1) or to the beginning of the dead-metal zone (Fig.8.7). Thus, the pressure at any point in the extrusion stroke is

$$p_l = p_e + 4\tau_i\left(l/d_0\right) \tag{8.6}$$

The punch force drops gradually (Fig.8.9) to the minimum p_e measured at the point where the die or dead-metal zone is touched. Sometimes a higher initial breakthrough pressure is registered which can be attributed to the initiation of flow in the workpiece material and/or to transients associated with the development of the lubricating film.

In extrusion with sliding friction, the magnitude of τ_i must be known:

a. As discussed in Sec.3.3.2, the shear strengths of some lubricating substances (for example, soft metals) are insensitive to interface pressure, and then Eq.8.6 applies.

b. If, on the other hand, the lubricant is pressure-sensitive and τ_i can be described by a constant μ (as is the case for many boundary lubricants and polymers), the punch pressure at a distance l can be obtained by integration in the following form:

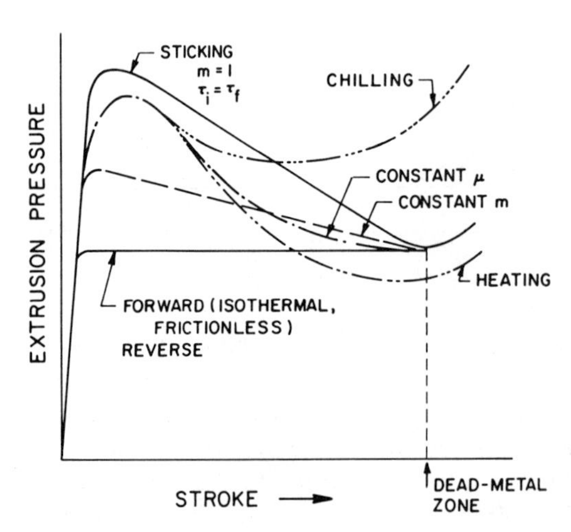

Fig.8.9. Variation of extrusion pressure during direct extrusion under various container-friction and thermal conditions

$$p_l = p_e \exp\left(4\mu l/d_0\right) \tag{8.7}$$

and the extrusion pressure rises exponentially. If, however, the condition $\tau_i = \mu p_l$ is satisfied anywhere along the container wall, sliding at that point is arrested even in the presence of a lubricant (Fig.8.9).

The increased pressure (and particularly, breakthrough pressure) limits the length of billet that can be directly extruded (Fig.8.5).

Evaluation of Friction. In principle, measurement of the punch force at two points in the stroke offers a means of determining the interface shear strength from Eq.8.6, or the coefficient of friction from Eq.8.7 (e.g., [23]). Of course, any such evaluation can be valid only if all other variables remain constant. This condition is seldom satisfied, because σ_{fm} is bound to vary even in cold extrusion and is known to vary greatly in hot extrusion. If the billet heats up in the course of extrusion (as in cold extrusion and isothermal hot extrusion), the extrusion force drops steeply (Fig. 8.9) and a misleadingly high friction value is calculated. In contrast, chilling in nonisothermal hot extrusion raises σ_{fm} and quite unrealistically low friction may be calculated. Sometimes, when the extrusion force actually rises during the stroke (Fig.8.9), a negative friction value is obtained. Therefore, coefficients of friction quoted without further specification should be looked upon with suspicion.

Nevertheless, the extrusion force and, even more so, its variation in the course of the stroke provide important clues regarding the efficiency of lubrication. They can be used for comparative lubricant evaluation provided that other conditions are not allowed to change (see Sec.5.1). More reliable quantitative data could be obtained by direct measurement of the frictional shear stress, by one of the techniques described in Sec.5.5.2. However, because the high pressures and temperatures often encountered present serious experimental difficulties, few attempts have been recorded. For example, the oblique pressure pin technique was used only in room-temperature extrusion of lead [38].

8.2.3 Tube Extrusion

Hollow products with a variety of cross-sectional shapes may be extruded, in principle, by two direct techniques:

1. A hole pierced or machined in a billet can be maintained in the course of extrusion by a fixed or floating mandrel long enough to reach into the die land (Fig.8.10a). A solid billet can be used if a separately actuated piercing ram is available. Lubricated extrusion is possible only with a hollow billet; in working with a piercing ram, sticking friction must be encouraged in order to avoid subsurface defects on the internal tube surface. The effect of friction on material flow is similar to that found in extrusion of

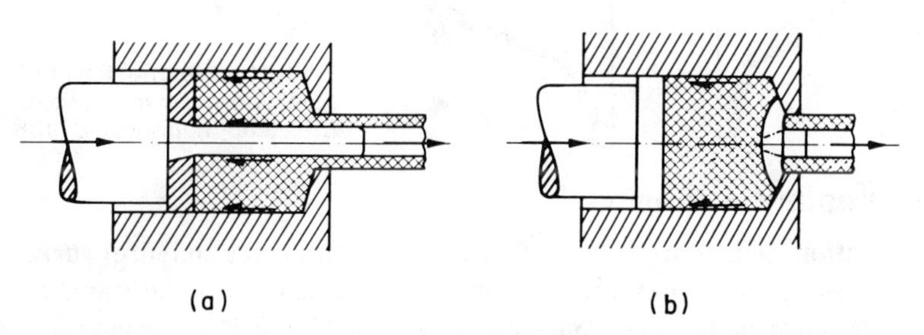

Fig.8.10. Forward extrusion of a tube with (a) a fixed mandrel and (b) a bridge-type die

bars. Marker pins radially inserted into the billet show the development of a dead-metal zone in unlubricated extrusion (Fig.8.11a), whereas pins are successively extruded in lubricated extrusion (Fig.8.11b) [39,40]. The residual skull clearly shows that the original skin remains in the container in the absence of a lubricant but is drawn in during lubricated extrusion.

2. Alternatively, a solid billet is extruded through a bridge, spider, or porthole die (Fig.8.10b) which divides and then reunites the flowing material to form a hollow product with several internal pressure welds. All traces of lubricant must be carefully kept out: otherwise, the weld quality will be impaired. This technique is suitable only for materials that can be extruded at low enough temperatures to retain the strength of the die (in practice, to lead and the more readily weldable aluminum alloys). Die design calls for great skill [7].

The additional frictional surface created on the mandrel or around the bridge adds considerably to the extrusion pressure. Toward the back of the billet, sticking may prevail between mandrel and billet, but toward the deformation zone the inner surface of the tube slides on the mandrel, subjecting it to large frictional forces. Because the punch is exposed to high temperatures in hot extrusion, it may not be able to withstand these forces and may be torn off or thinned out by plastic flow. Therefore, lubrication of the mandrel surface is usually mandatory, at least above aluminum extrusion temperatures.

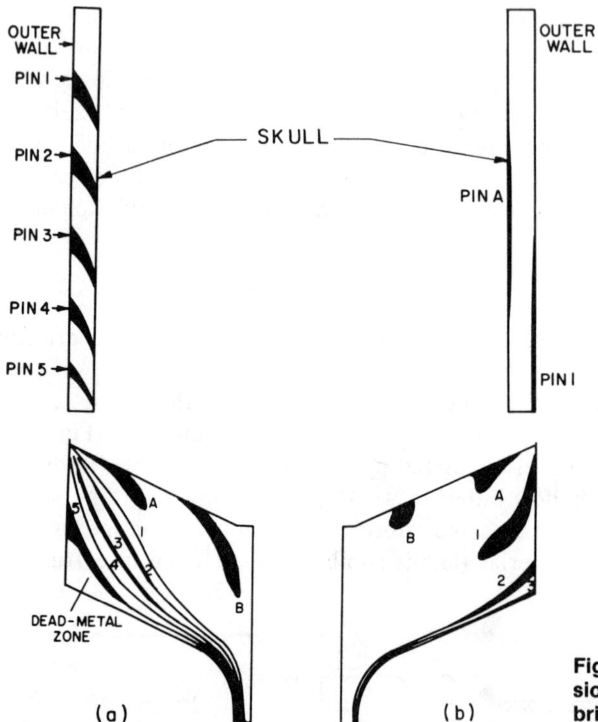

Fig.8.11. Material flow in tube extrusion, shown by marker pins: (a) unlubricated; (b) lubricated [40]

8.2.4 Hydrostatic Extrusion

Container friction and, in extrusion of tubes, friction on the mandrel surface are major contributors to extrusion pressure and thus limit both billet length and extrusion ratio for given maximum punch and container pressures (Fig.8.5). Envelopment of the billet in a fluid medium could practically eliminate these sources of friction and could even affect the lubrication mechanism in the die. Therefore, hydrostatic extrusion has attracted a great amount of interest, as attested to by the enormous volume of literature that it has generated in the course of some thirty years [7.20],[41-47]. Despite its great theoretical advantages and the vast experimental and theoretical efforts that have been devoted to its development, hydrostatic extrusion has been slow in making a major impact on production practices, mostly because of practical difficulties [48,49]. Various attempts at solving these difficulties have led to a number of propositions, all with different tribological implications:

a. In the basic process, the pressure is built up by the penetration of a punch which traps the fluid medium with a high-pressure seal (Fig.8.12a). Elimination of contact between container and billet surfaces is particularly advantageous when long billets are to be extruded into semifinished products. Billet length is limited only by container construction, and, if desired, a spool of wire may be placed inside the container and long lengths of fine wire extruded. If the pressure medium has lubricating qualities, friction is much reduced over the die face, permitting the use of lower die angles which then ensure greater homogeneity of deformation.

b. The hydrostatic pressure acting on the workpiece can be increased by taking advantage of low die friction and extruding with a low die half angle (very low h/L ratio) or by extruding into a pressurized medium (Fig.8.12b). With the latter technique even brittle materials can be deformed, although at the expense of high container pressures.

c. A basic problem of all the above techniques is low production rate; the container has to be pressurized for each billet, and large quantities of fluid have to be supplied and drained. Fiorentino et al [50] apply a thick film of semisolid lubricant to the surface of the billet, which can thus be loaded as in usual cold extrusion but into a preheated container in which the lubricant melts to become the pressurized medium (Fig.8.12c).

d. If the clearance around the punch is small enough to act as a dynamic seal with a controlled leakage rate (Fig.8.12d), sufficient pressure for extrusion may be developed [51,52]. Less practically, lubricant may be trapped also with a billet that has a flange at the punch end [14]. Extrusion at high velocities is also helpful in minimizing leakage [53,54].

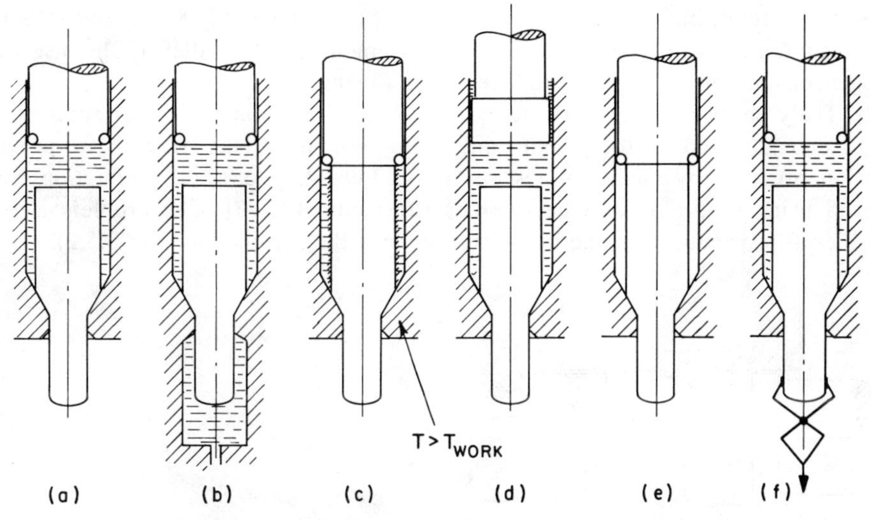

Fig.8.12. Basic processes of hydrostatic extrusion (after [47]). See text for identification.

e. A common problem is that an excessively thick lubricant film not only roughens the product but may also generate periodic instability, leading to "bambooing" which manifests itself in periodic variations in the surface finish and diameter of the extruded product (see Sec.8.5). To suppress this effect, the billet-augmented process, in which the punch, with a high-pressure seal, bears on the back face of the billet (Fig.8.12e), has been proposed [46,55].

f. Alternatively, the extruded product is subjected to a tensile force (Fig.8.12f), and thus a combined extrusion-drawing process is created [46,56,57], with the advantage of lower die pressures and the possibility of heavier reductions. In extrusion of tubes, drawing reduces also the frictional stresses on the mandrel [58].

g. In extrusion through two dies in tandem, the first reduction is equivalent to drawing under hydrostatic pressure with fluid-film lubrication, and the second reduction, where most of the work is done, corresponds to hydrostatic extrusion [48,59].

h. The stability of extrusion increases also when an elastomer is used for the punch and a separate lubricant is applied to the billet [60].

i. Various ingenious schemes have been proposed to extrude bars continuously and semicontinuously, as reviewed by Alexander [61] and by Tuschy [7.20]. Continuous billet augmenting has also been proposed [62]. The large reduction capability of hydrostatic extrusion has prompted much work into wire production [63].

j. There is also the potential for hot extrusion [48,64-66] with grease, polymer, or glass media which melt at the extrusion temperature and thus transmit the extrusion pressure. Extrusion with gas has also been proposed [67] but requires separate lubrication and appears to offer no advantage.

The first commercial applications of hydrostatic extrusion have been in fields where conventional processes have shortcomings. The greater homogeneity of deformation makes hydrostatic extrusion particularly favorable for cladding, and the much reduced mandrel friction makes it advantageous for extrusion of tubes [68].

8.2.5 Friction-Assisted Extrusion

There are several ingenious schemes, reviewed by Avitzur [69], in which friction is actually exploited to aid extrusion.

a. In friction-assisted forward extrusion, the container is moved at a somewhat higher speed than, but in the same direction as, the punch (Fig.8.13), and thus friction on the container wall actually aids the movement of the billet [70], ensures more homogeneous material flow, and reduces the remnant.

b. Truly continuous hydrostatic extrusion could be achieved by circulating a fluid of high enough viscosity (or a semisolid of high enough shear strength) to provide the extrusion pressure by means of viscous drag. Indeed, pilot-scale production has been achieved using castor oil and beeswax as the media [71-73]. The problem is that the very thick film produced in the die zone roughens the surface (see Sec.8.5.2).

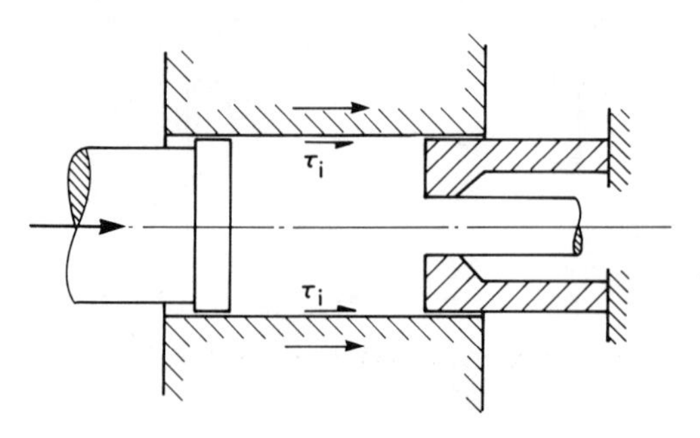

Fig.8.13. Friction-assisted forward extrusion

8.3 Indirect Extrusion

Indirect extrusion is also called inverse, reverse, or backward extrusion, because the billet is stationary in the container cavity, and a punch moving against the billet forces the extrusion to emerge in a direction opposite to the punch movement (Fig.8.14a). The crucial point is that the billet remains at rest in the container and therefore container friction does not come into play.

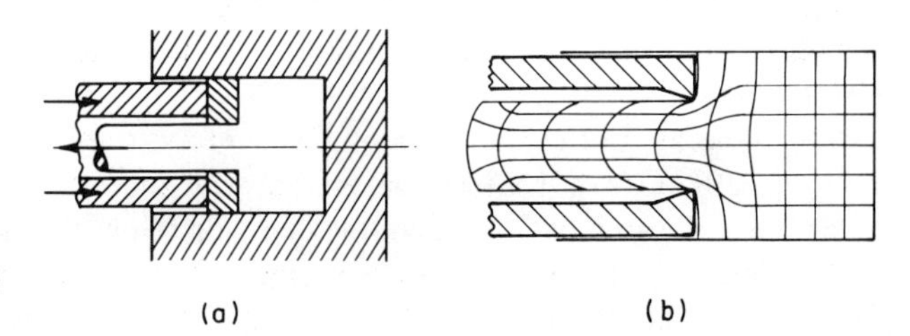

Fig.8.14. Backward extrusion (a) and the resultant material flow (b)

8.3.1 Material Flow and Forces

In indirect extrusion, as in direct extrusion, material flow in the vicinity of the die depends on die geometry, reduction, and the presence of a lubricant. With a good lubricant and a tapered die entry, a homogeneity of flow comparable to that of hydrostatic extrusion can be obtained [74]. In lubricated flow, the dead-metal zone is limited to a thin layer on the die face (Fig.8.15) [7]. The nonextruded part of the billet shows no effect of deformation (Fig.8.14b). Since only the extrusion pressure p_e needs to be developed, heat generation is also reduced. As can be deduced from Fig.8.5, for a given punch and container pressure either a larger extrusion ratio can be taken or, more importantly for difficult-to-extrude materials, workpiece temperature can be lowered and extrusion speeds increased without running the danger of incipient melting. In direct extrusion, the temperature in the unextruded portion of the billet keeps rising, whereas in indirect extrusion it remains essentially constant. This is of particular importance in extrusion with sticking friction.

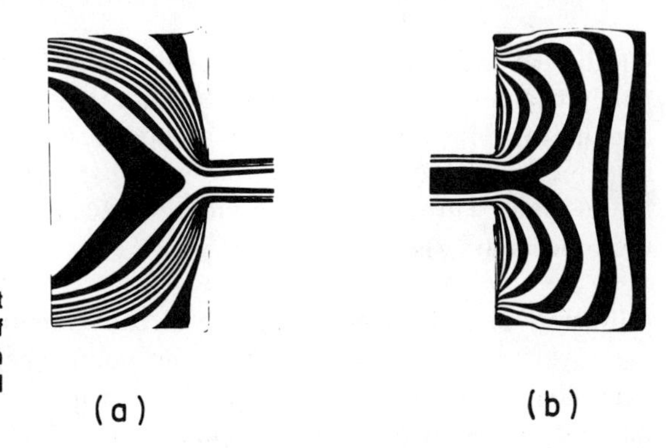

Fig.8.15. Material flow in unlubricated hot extrusion of a composite billet made up of disks of two aluminum alloys (Lang, quoted in [7]). (a) Forward extrusion. (b) Backward extrusion.

A major problem in indirect extrusion is the construction of the hollow ram and the die; recent advances in the design of these components have contributed to the extended application of the process [7,12]. An entirely different form of back extrusion is obtained when a solid punch penetrates the billet; this process will be dealt with in Sec.9.2.

8.3.2 Friction-Assisted Extrusion

In recent years a number of attempts have been made to take advantage of container friction [69]. Remembering that the movement of the billet relative to the container determines whether extrusion is direct or indirect, it will be recognized that many of these processes are combinations of the two techniques or are, essentially, reverse extrusion processes even though the workpiece moves toward the die.

a. In the technique described by Ziehm [75], the die is fixed in a hollow stem which at the beginning of extrusion penetrates into the billet. Thus extrusion begins without a high breakthrough pressure; the punch then advances and the process continues as direct extrusion.

b. Several methods of making the process essentially continuous have been proposed [7.20]. Green [76] suggests that the material be wedged into the groove of a roller with the aid of a stationary shoe. Friction on the groove walls moves the billet while friction on the shoe hinders its movement, and the net extrusion force available is governed by the balance of the two. This technique has been used for extrusion of both aluminum and copper [77]. In the technique of Avitzur [78], a closed pass is formed between a grooved roll and a roll provided with a flange, and the frictional forces developed on the roll surface and on the groove sides are used for extrusion through a die.

c. A longer contact length and thus greater frictional force can be obtained in a linear arrangement. In one of the processes described by Black and Voorhes [79], articulated grips move a square bar from two sides while lubricated stationary dies prevent its spread in the lateral direction. In the version proposed by Fuchs and Schmehl [80], the frictional force between a tightly fitting elastically deforming clamp and a precision billet is used to push the material through the extrusion die. Lengyel and Kamyab-Tehrani [81] propose to remove the need for close-fitting clamps by using segmented grips moving around all four sides of the billet.

None of the above-mentioned processes has achieved production status, partly because of difficulties of construction and product quality. Nevertheless, they provide examples of utilizing rather than combating friction.

8.4 Non-Steady-State Effects and Defects

Of all bulk deformation processes, extrusion is perhaps the most sensitive to lubrication, partly because the non-steady-state conditions at the beginning and end of extrusion are conducive to instabilities, and partly because any unwanted change in lubrication leads to defects which affect not only the appearance but also the properties and integrity of the product. Through this, the yield of acceptable material and the over-all economy of the process are also influenced.

Initiation of Extrusion. We have already mentioned the pressure peak (breakthrough pressure) that is observed just at the point where extrusion is initiated. In lubricated extrusion it indicates that the lubrication mechanism typical of steady-state conditions has not yet been established. Thus, it is most marked when a square-ended billet is extruded through a tapered die; the sharp edge of the billet may actually scrape off whatever lubricant film is present on the die, and if this occurs a new, low-friction layer can develop only after the die throat has been filled.

End of Extrusion. When extrusion proceeds to the point where the end face touches the deformation zone, flow becomes nonuniform even in lubricated flow. The flow rate from the sides to the center is insufficient, and a crater (pipe) forms. The depth of the pipe increases with increasing container and die friction [5,82] but is reduced by friction on the back face of the billet. In extrusion of semifabricated products, the pipe increases the discard, especially if it fills up with oxides and lubricant residues; it can form a difficult-to-detect defect in the rear end of the product.

Extrusion Defects. We have already discussed the origins of internal defects such as centerburst and the extrusion defect proper (Fig.8.6d). Differences in grain size and metallurgical structure, resulting from recrystallization following inhomogeneous deformation, are also regarded as defects in some materials [7].

There are also external, friction-related defects in both hot and cold extrusion that, when severe, make a product unsuitable for use.

a. We have already mentioned the fir-tree effect (speed effect) in hot-short materials with friction on the die land as a contributory cause. Wilcox and Whitton [83] attribute a cold extrusion defect of similar appearance to the simultaneous occurrence of slip on the die face and sticking on the die land.

b. Subsurface defects caused by intermittent stick-slip on the container wall (Fig.8.8) may emerge at the surface in the form of laminations or flakes.

c. Pickup on the die or on the die land leads to deep scoring of the extruded surface (Fig.8.16a) [84-86]. Pickup tends to be cumulative and may force a die change and clean-up, often after extrusion of each billet in hot extrusion.

d. In lubricated extrusion, the surface roughens in the presence of a thick lubricant film (Fig.8.16b) [21,34,37,51,58,85,86], a phenomenon often described as orange peel. An excessively thick, unstable film leads to bambooing (Fig.8.16c) and, in some less-ductile materials, even to shear cracks (Fig.8.16d). If the billet has been turned in preparation for extrusion, the machining marks are stabilized by the lubricant (Fig.8.16e).

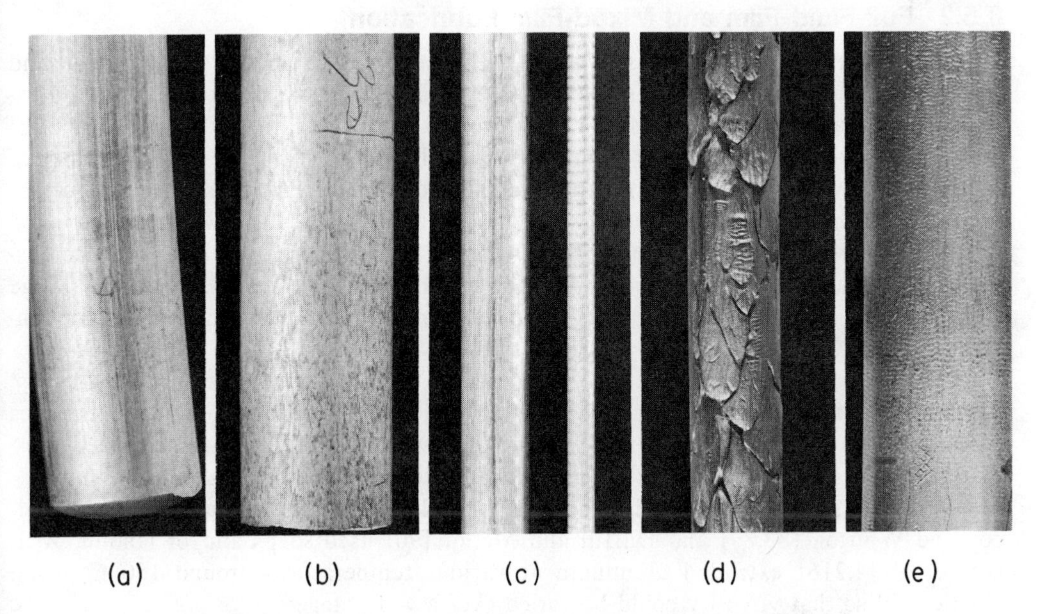

Fig.8.16. Typical defects observed in cold extrusion of aluminum alloys [85]. See text for identification. (Courtesy Dr. R. Akeret, Swiss Aluminium Ltd., Neuhausen am Rheinfall, Switzerland)

8.5 Lubrication and Wear

Lubrication, when desired at all, can be effected by any of a number of mechanisms provided that some basic criteria are met. Reviews have been given by Male [87] and Chadwick [88].

8.5.1 Attributes of Extrusion Lubricants

An extrusion lubricant must have all the attributes listed in Sec.4.1, but some requirements assume special significance.

An uninterrupted separating film between billet, die, and container, and its low shear strength, are important not only because they determine the allowable extrusion ratio and billet length in direct extrusion but also because they control material flow and thus the properties of the product and the length of the discard.

Heat insulation is imperative in nonisothermal extrusion when the extrusion temperature is high enough to cause softening of the die, mandrel, or container on prolonged exposure and thus when, in the absence of such insulation, the die would close in or open up (die wash), the mandrel would thin out or tear off, and the container would open up.

Protection against oxidation is vital when the workpiece material forms an abrasive oxide, or when the lubricant will not wet as a result of oxidation. Because oxidation can be controlled not only by a preapplied lubricant film but also by the preheating atmosphere, the method of heating must be chosen with the entire tribosystem in mind. Induction heating, resistance heating, and glass or salt baths can all be considered in addition to gas-fired or electric muffle furnaces, with or without a protective atmosphere. If scaling of steel is allowed, the scale may be removed with high-pressure water, but great care must be taken to avoid scale pockets which would break through any lubricant film. The relative advantages of various methods have been discussed by Naden [89] and by Laue and Stenger [7].

8.5.2 Full-Fluid-Film and Mixed-Film Lubrication

Extrusion is one of the processes in which a thick lubricant film can be developed and maintained. The similarity to wiredrawing is evident [7.34]; the difference is that lubricant pressure at the die inlet can be higher even in nonhydrostatic extrusion (Fig.3.36b) and thus thick-film formation is promoted. Film thickness is governed by a number of factors, as shown by the results of experiments conducted on various extrusion processes with a great variety of lubricants and at various temperatures. As would be expected from PHD theory (Eq.3.29), film thickness increases with increasing $\eta v/p$ and decreasing die half angle α, and by the factors shown in Fig.3.36. Film thicknesses may range from excessively thick full-fluid films to mixed films with predominantly boundary contact.

Isothermal Extrusion. Even though most extrusion of semifabricated products takes place at elevated temperatures, many of the observations that have been reported were made at room temperature or at slightly elevated temperatures with the aid of simulating materials, so that the complicating effects of chilling could be excluded. Thus, Wilcox and Whitton [83,89] and Duffill and Mellor [90] used soaps and/or lanolin. Wallace et al [4.216] extruded aluminum at various temperatures around 150 C, using abietic acid so that viscosity could be varied over a wide range:

a. At higher viscosities (lower temperatures) and a lower half angle α, pressure builds up fairly gradually until extrusion begins (143 C in Fig.8.17). The first part of

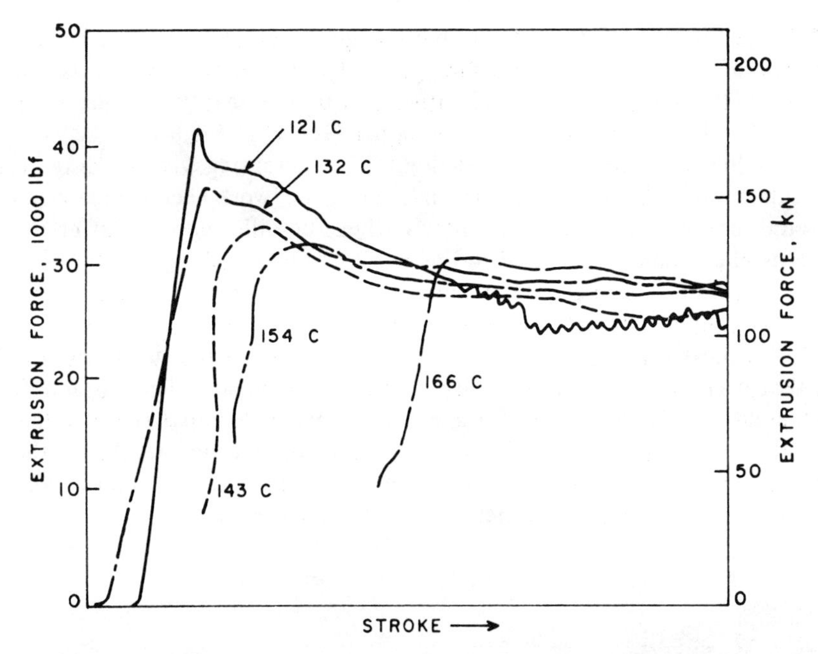

Fig.8.17. Effect of lubricant viscosity (extrusion temperature) on forces registered in isothermal extrusion of aluminum using abietic acid as the lubricant [4.216]

the section to emerge is fairly bright, typical of boundary contact, but then the surface becomes duller, indicating the presence of a thicker film. If extrusion is interrupted, the remaining billet shows deformation prior to entry into the die zone (Fig.8.18), proving that the pressures that developed were high enough to initiate yielding. The extrusion pressure drops and the surface roughens with decreasing α, again showing evidence of improved lubrication. The same observations apply to hot extrusion under isothermal conditions [91]. The die configuration also has an effect, even in hot extrusion of aluminum, where a concave die proved to be more efficient than a conical one [92].

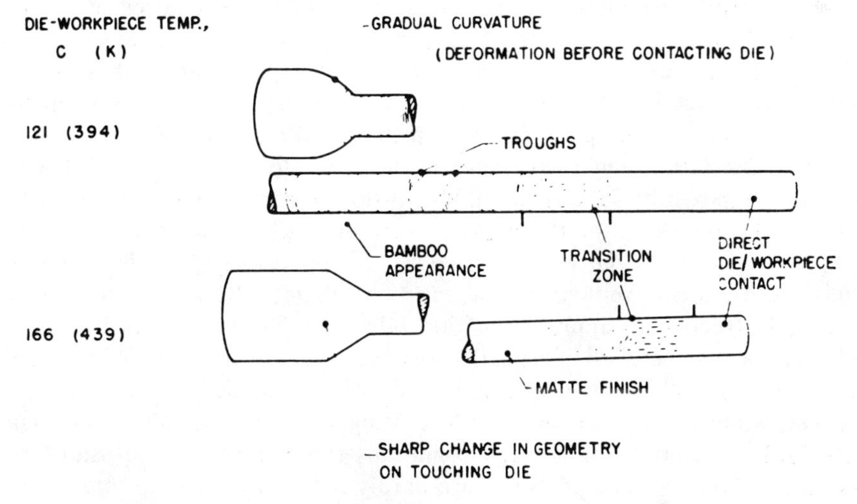

Fig.8.18. Deformation prior to die entry and appearance of extruded product in isothermal extrusion with fluid lubrication [4.216]

b. When conditions are not favorable enough to create a thick film, deformation begins in the die throat, and the surface is much brighter with only occasional hydrostatic pockets (Fig.8.18), which is indicative of a predominantly boundary-type, mixed-film lubrication. The extrusion pressure is higher (166 C in Fig.8.17).

c. At higher viscosities (132 C in Fig.8.17), increasing viscous drag results in a slightly higher force. With certain combinations of die/workpiece geometry and $\eta v/p$, a breakthrough pressure is noted, particularly when the billet nose is flat or just matches the die. This phenomenon is related to the process of establishing a PHD film, as shown by the disappearance of the peak when the billet nose angle is smaller than the die angle, so that a tapered lubricant wedge is preformed [4.216].

d. With excessively high viscosity (121 C in Fig.8.17) the lubricant film becomes so thick that it cannot maintain stability, and periodic collapse leads to development of the bamboo defect (Fig.8.19). Duffill and Mellor [90] noted that the wavelength of the bamboo defect was equal to the length of the die land, and they attributed the defect to intermittent lubricant failure. Most information on this defect comes from research related to hydrostatic extrusion, which will be discussed next.

Fig.8.19. An example of severe bambooing [4.216]

Hydrostatic Extrusion. As shown in Fig.3.36b, pressure applied to the inlet zone aids in developing a hydrodynamic film. Thus, hydrostatic extrusion promotes thick-film lubrication, although there are some very practical limitations:

a. The hydraulic medium must be chosen to develop the required pressure without solidification. Thus, a medium with good lubricating properties, such as castor oil, may be used if pressures are not too high, as in model experiments with wax [93] or in extrusion of soft metals. Otherwise, favorite substances are low-viscosity mineral oils, castor oil with methylated spirits (useful up to 1.4 GPa), glycerine with 25% ethylene glycol (2.0 to 2.8 GPa), and isopentane and gasoline (to 3.0 GPa) [41-48]. Because of the limitation imposed by solidification, lubrication is frequently of the mixed-film type, as evidenced by the need for the pressure medium to be compounded with boundary and E.P. additives (and even with layer-lattice compounds) and by the response of surface finish to die angle, reduction, and $\eta v/p$ [41-48,94]. Often it is found to be more practical and effective to apply a separate lubricant film such as a grease or, in the case of steel, a conventional phosphate/soap coating to the billet, and to utilize the hydrostatic feature only to reduce friction and improve the homogeneity of deformation [45,95]. Oxidation of the surface prior to coating with soap has also been found to be effective [96]. Of course, under such conditions the benefits of hydrostatic operation become rather dubious since pressures are hardly reduced relative to those encountered in well-lubricated conventional extrusion [97].

b. Because pressures in hydrostatic extrusion are sufficient to initiate yielding of the billet, the lubricant film may easily thicken to the point where it becomes unstable, and bambooing is often observed [4.216],[98,99]. In experiments on wax as a model material, Wuerscher and Rice [100] associated the defect with fluctuations between PHD and boundary conditions, noting that the calculated coefficient of friction dropped with increasing viscosity. Bambooing occurred only with the stiffer, less compressible castor oil and not with higher-viscosity silicone oils. Egerton and Rice [101] showed that the extrusion actually comes to a momentary rest but then accelerates sufficiently to establish, even though for only a brief time, a fluid film. Support for these observations comes from the work of Krhanek and Jakes [102], who noted that the bamboo defect is suppressed with increasing extrusion speed. Stick-slip combined with poor die lubrication can even result in fir-tree-like cracking [83,103], and therefore lubrication reduces the defect.

c. The origin of the bamboo defect lies in the lubrication mechanism, but its development, magnitude, and damping out depend on the entire system. Thus, billet augmentation suppresses fluctuations not only by physically restraining the billet but also by reducing the film thickness. This, incidentally, increases die friction [55]. Because the billet is now pushed (Fig.8.12e), it will first upset if the billet-to-container clearance is too large for the prevailing lubricant viscosity [4.216]; in contrast, lubricant starvation sets in when the clearance is too small. Bambooing may still occur if $\eta v/p$ is excessive, α is small, or the die land is slightly tapered inward [4.216]. If the augmenting pressure is too high, the billet is upset into the die throat, lubricant excess is reduced, and the entry geometry is impaired, resulting in a rise in friction and a change to boundary lubrication [55,104].

d. The same effects were noted by Kulkarni and Schey [51] in extrusion with a controlled follower block clearance (Fig.8.12d). The whole billet is extruded if loss of fluid around the follower block is limited. The leakage rate Q can be calculated [105] from

$$Q = \frac{\pi a c^3}{6\eta} \frac{dp}{dl} \tag{8.8}$$

where a is the radius of the container, c is the radial clearance, and dp/dl is the rate of pressure drop.

e. Mixed-film lubrication prevails also in hydrostatic extrusion/drawing (Fig.8.12f), as shown by the roughening of the product surface with increasing speed in the experiments of Dunn and Lengyel [106]. Deformation of the bar prior to entry into the die and a rougher emerging surface indicate a shift toward a more hydrodynamic contribution at higher extrusion/draw stress ratios, at least for a strain-hardening material such as copper.

Nonisothermal Extrusion. Lubricated nonisothermal hot extrusion is usually conducted with a phase-change mechanism (discussed later), but when, at low extrusion ratios, the required glass pad cannot be retained in the container, a technique relying on preapplied glass films is preferred.

a. The glass powder is usually applied to the surface from a spray or brush mixture with a polymeric binder [22,107], although better coverage is ensured when the billet is rolled in glass powder after preheating to an intermediate temperature and then rerolled on reaching the extrusion temperature [4.222]. The results of Kulkarni et al [4.222] emphasize the need for a hot, well-fitting follower block, indicating that controlled or even zero leakage is necessary to supply lubricant to the die. The process operates in the mixed-film regime, as shown by its response to viscosity, speed, and die angle changes and by the need for a graphite coating on steel dies (or, alternatively, protection of the dies by a zirconia or alumina coating). Under optimum conditions, uniform glass coverage and minimum extrusion pressure can be established. However, because velocity not only increases $\eta v/p$ but also reduces cooling (and thus η), optimum operation is determined by the entire system. In extrusion of superalloys, successful glasses had viscosities of 10^2 to 10^4 Pa · s at the extrusion temperature [4.222].

b. Tight sealing of the container is important in the technique described by Sauve [108] in which a polymeric material of undisclosed composition is introduced with the hot billet into the container (Fig.8.20). Seals at the die and follower block prevent escape of the polymer while it melts and envelops the billet, and thus extrusion proceeds by a fluid-film mechanism. Somewhat similar techniques, but with a grease, were used by Nishihara et al [109] for extruding copper at temperatures up to 1000 C, and by Fiorentino et al [48] for extruding titanium.

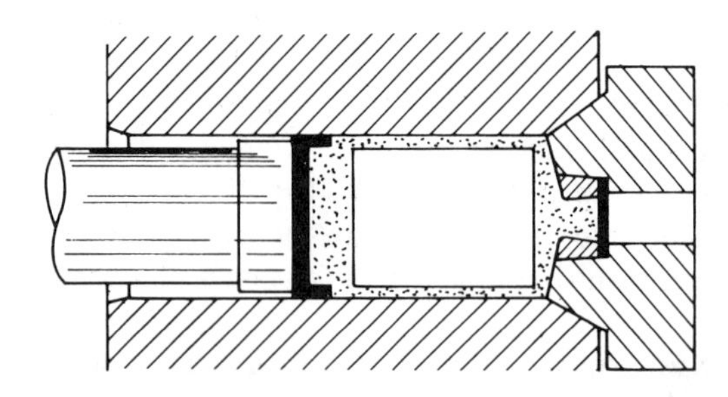

Fig.8.20. Hot extrusion with gradually melting polymeric lubricant [108]

Theories of Lubrication. Analysis of mixed-film lubrication for extrusion is practically nonexistent. Most theoretical work is related to hydrostatic extrusion, but the conclusions are applicable to other cases of liquid-lubricated extrusion, at least in terms of the effects of process variables and geometry on film thickness.

Even simple hydrodynamic analysis will show that pressurization of the fluid at entry should reduce friction on the die [110]. Early theories of hydrostatic extrusion were based on isothermal, isoviscous lubricant behavior [111,112], or neglected flow at the inlet zone [2.12],[113]. In the first attempts at considering the inlet zone, isothermal calculations [7.108],[114] indicated an unrealistically large friction contribution. Numerical techniques [115,116,117] and the upperbound approach [118] allow consideration of work hardening and, most importantly, of heating due to the plastic deformation of the workpiece and shearing in the lubricant film. These theories now correctly show that the contribution of friction to the extrusion pressure is almost negligible, and calculated film thicknesses are in good agreement with experimental values [116]. An analytical solution is possible only if one accepts simplifying assumptions, as did Snidle et al [119] in ignoring viscous heating in the lubricant film. Unfortunately, this limits the solution to the extrusion of soft materials at low reductions and low speeds. That heating effects due to friction are significant is also shown by the analysis of Guha and Lengyel [120] of high-speed hydrostatic extrusion/drawing. The results of Hsu [121] for high-temperature isothermal extrusion with small die angles are questionable because comparison was made with nonisothermal experimental data of DePierre [23].

Wilson et al [122] analyzed the transient conditions existing during the nosing of a billet. Wilson and Mahdavian [99] suggested that the sudden pressure drop occurring when the full fluid film is established is responsible for setting up the oscillations accompanying bambooing. The onset of bambooing later in the stroke (e.g., curve for 121 C in Fig. 8.17) indicates that other factors must have influences, too.

8.5.3 Phase-Change Lubrication

The technique of phase-change lubrication is associated with the name of Séjournet [37,123-125]. It actually does not involve a true phase change, but only a gradual loss of viscosity.

In the search for a lubricant that would allow hot extrusion of steel at large extrusion ratios and in long lengths, substances that liquify at the extrusion temperature were lost through the die, and those that were solid (e.g., talc) caused die wear. This convinced Séjournet and his co-workers to consider substances that soften gradually [124]. Since about 1950 the technique of glass lubrication (the Ugine-Séjournet process and its further developments) has spread to become dominant not only for extrusion of steel tubes but also for extrusion of sections and bars in steels and other materials when quantities are insufficient or shapes too complex to allow rolling. The process sequence is well established.

Lubricant Application. To protect the die with a uniform, heat-insulating, low-friction glass coating over the entire length of a long extrusion, a pad of glass is placed in front of the billet. Plate glass, originally used, cracked on contact with the billet, and pads made of about 100-mesh powder became standard [124]. They are self-sintered or are made with binders such as sodium silicate or bentonite clay [126]. Other forms of glass such as wool, cloth, and foam were also tried but with less success or economy [4.207]. The viscosity of glass used for making commercial foams is too high [127], and the graphite used as a foaming agent interferes with wetting of the metal; however, foam is successful as a carrier for powder.

To provide heat insulation and lubrication at the billet/container interface, the billet is preheated with minimum oxidation and then rolled down an inclined table covered with glass powder or fiber, so that it becomes coated with a layer of molten lubricant. For tube extrusion, a woven glass sock is pulled over the mandrel, or glass powder is sprayed into the billet bore. The references cited in Sec.4.7.6 provide information on composition and properties.

Extrusion. The billet thus protected against heat loss is loaded into the container where it touches the front pad (Fig.8.21). As extrusion begins, a thin film of glass melts off the pad. If melting characteristics are properly chosen and the press speed is adequate (typically, 0.15 to 0.5 m/s), the pad supplies glass to the end of the stroke. Glass on the billet surface only serves to supplement lubrication toward the end, as shown by radiotracer studies by Rowe et al [4.193],[128,129]. The desirable glass property is usually

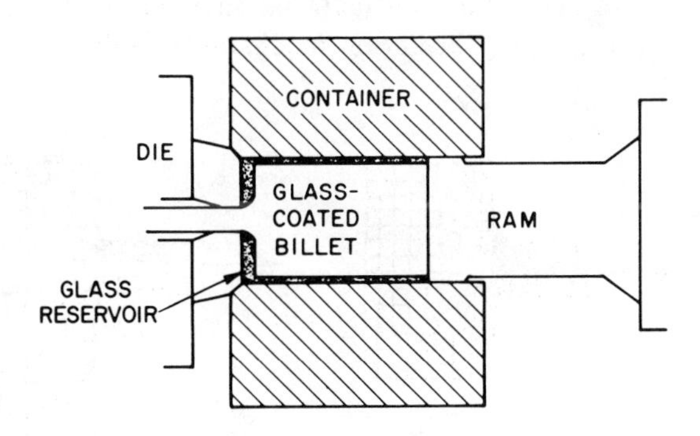

Fig.8.21. Glass-lubricated hot extrusion

defined as a viscosity of 10^1 to 10^2 Pa · s at the billet temperature (Sec.4.7.6), although a viscosity-temperature curve of low slope would appear to be needed for optimum melt-away performance, and thermal properties such as heat conduction and specific heat should be just as important. Several improvements have been suggested, usually in the form of patents. For example, it has been suggested that a desirable viscosity gradient could be established by the use of glass layers of differing viscosities, either between the billet and container or between the front and back of the billet bore.

Even though controlled melting provides the glass supply, flow itself is controlled by hydrodynamic factors.

a. Experiments conclusively show that film thickness increases with $\eta v/p$ and decreases with increasing die angle [4.193],[129,130]. A tapered nose is essential with a conical die, lest the lubricant be scraped off by a square billet. Insufficient die angles lead to nonuniform glass distribution, no doubt for the same causes that result in bambooing, and included angles of 120 to 150° are regarded as optimum [131]. The glass pad actually forms its own optimum die shape, the angle of which increases with decreasing viscosity and increasing pad thickness [132,133]. There is seldom need for a pad thicker than half the container diameter. At heavy reductions a flat die is the simplest, eliminating the need for a conical billet end and ensuring optimum conditions, as shown by the remarkably smooth material flow pattern even at extrusion ratios that would lead to severe inhomogeneity in unlubricated extrusion (Fig.8.22) [132,133,134].

b. That lubrication is of the thick-film type is also shown by the almost steady extrusion force usually registered and by the relative insensitivity of this force to glass viscosity [129,131], indicating that operation is in the ascending part of the Stribeck curve (Fig.3.42). Larger variations in extrusion force occasionally reported [135] must be attributed to insufficient lubrication. At the same time, the very low friction coefficients (μ = 0.002 to 0.05) calculated by several authors [37,123,136,137] are not prima facie evidence of hydrodynamic lubrication, because interface pressures are multiples of the flow stress; indeed, this is one of the instances where μ is not very informative.

c. The main problem with this process is the difficulty of avoiding excessive surface roughness with overly thick films. Films 10 to 30 μm thick are desirable; thinner films are prone to breakdown, and somewhat heavier films are often used. The original surface of the billet is extruded, and machining marks and defects are carried onto the surface of the extruded product.

d. The lubrication mechanism is not appreciably changed by relatively small additions to the glass, provided that they do not drastically change its flow properties. Graphite, as mentioned, causes foaming and thus aids removal of the glass, as does up to 20% potassium palmitate [4.193]. The difficulties encountered in extrusion of high-temperature materials have prompted various suggestions for compounding the glass, but without substantial improvement.

In principle, any material that melts at a controlled rate could be used in a true phase-change lubrication mechanism. Indeed, common salt has been used at extrusion

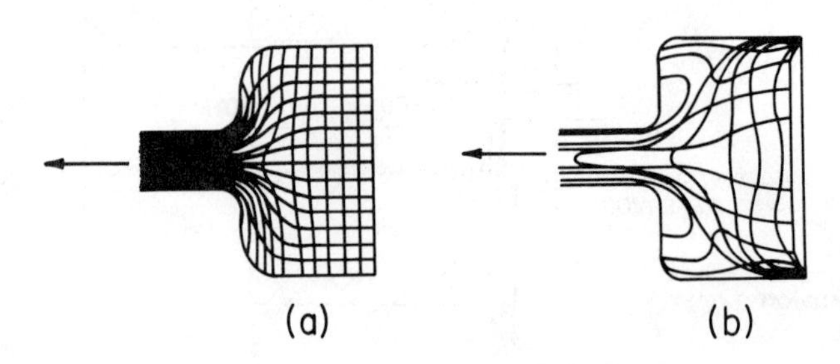

Fig.8.22. Grid deformation in remnants after (a) unlubricated and (b) glass-lubricated hot extrusion [132]

temperatures of 900 to 1100 C [37],[4.193], as has lithium fluoride. However, salts lack the gradual melting character of glass, and the addition of a solid filler such as slate [4.193] or silica or magnesia [138] cannot entirely compensate for this. Nevertheless, filled salts (such as NaCl, K_2SO_2, or K_2CrO_4) have shown some promise [138], although the threat of rapid corrosion of the product necessitates immediate cleaning. Basalt, a natural Fe-Mg-silicate, can be treated and then used at higher temperatures (1100 to 1200 C) [139].

Theory. Few analyses of the phase-change lubrication mechanism are available. It was early recognized [123,140] that the rate of melting is controlled by thermal diffusivity K

$$K = \frac{\lambda}{\rho c} \tag{8.9}$$

where λ is thermal conductivity, c is specific heat, and ρ is density of the glass. As shown by Baque et al [141], flow is proportional to the slope (γ) of the temperature profile at the point of contact, and this in turn is inversely proportional to K. A low K is typical of all glasses [124], and the temperature profile is very steep. The viscous flow rate is, however, inversely proportional to the slope γ; where melting flow and viscous flow rates balance, the steady-state flow rate is obtained [141]. Film formation is thus found to be insensitive to viscosity (as long as it is not so high that the glass would be unable to follow the movement of the steel), and the flow rate Q is

$$Q = \left[\frac{KvL}{\left(T_s - T_i\right)\beta}\right]^{1/2} \tag{8.10}$$

where v is the velocity of the steel, L is the length over which rubbing occurs, T_s and T_i are the working and initial temperatures of the glass pad, respectively, and β is the temperature-sensitivity exponent of viscosity for a glass obeying a power law. The film thus formed behaves like a viscous fluid; the film is more stable at exit if $\eta v/p$ is high and the initial film thickness is moderate. The calculated thicknesses of around 16 μm were in good agreement with values measured on steel tubes [141]. Viscous drag is very low, and this accounts for the uniformity of metal flow [142].

8.5.4 Solid-Film Lubrication

Because of the high interface pressures, solids with pressure-insensitive shear strengths can be very useful as lubricants.

Oxides. Among the oxides grown on metals, only some (e.g., on copper) can serve as lubricants. Others (e.g., on titanium) serve merely as parting agents. Externally introduced oxides never repaid the effort put into investigating them [4.191,4.192],[143] although Rogers and Rowe [4.193] found that solids such as lime or slate gave adequate lubrication in the form of a friable pad trapped between die and billet. Lubrication must be attributed to a controlled form of wear (and perhaps melting) of the pad.

Metals. Metals are used for two distinct purposes in extrusion. In the role of a lubricant, either they are applied as a coating (Sec.4.7.1) or, if a thicker coating is required or a powder is to be extruded, a can may be fabricated.

The typical canning technique is rather simple (Fig.8.23). To ensure smooth,

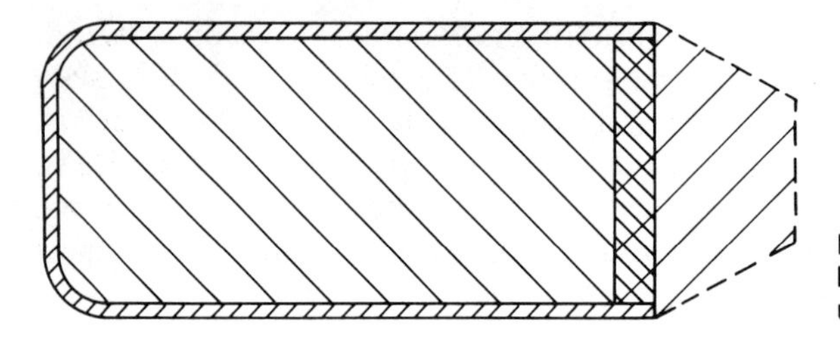

Fig.8.23. Typical canning technique for hot extrusion; broken lines indicate plug used for conical die entry

defect-free material flow, extrusion is invariably carried out with lubrication, and then it is preferable to have a well-fitting billet with a conical nose. The can is often evacuated through a tube to prevent oxidation of the powder and puffing up of the can. Because the can material will be removed after extrusion, preferably by mechanical means, a parting agent is often placed on the billet/can interface. In some instances a thin sheet of diffusion barrier is also used. By the proper choice of can material, extrusion pressures can be reduced, as found by Unckel [144] in extrusion of a high-strength Al-Cu-Mg alloy canned in pure aluminum. However, the can material must not be too soft or too thick [145,146], lest it be extruded ahead of the billet or behave like a fluid film with a typical unsteady bambooing flow. A can material of 1/2 to 1/3 the strength of the billet material gives satisfactory results [147]. An example of a complex canning arrangement for extrusion of a tungsten tube is shown in Fig.8.24 [148]. The can metal, because of its lower strength, allows roughening of the extrusion surface. Therefore, coarse-grain castings should be grain refined first by forging [148].

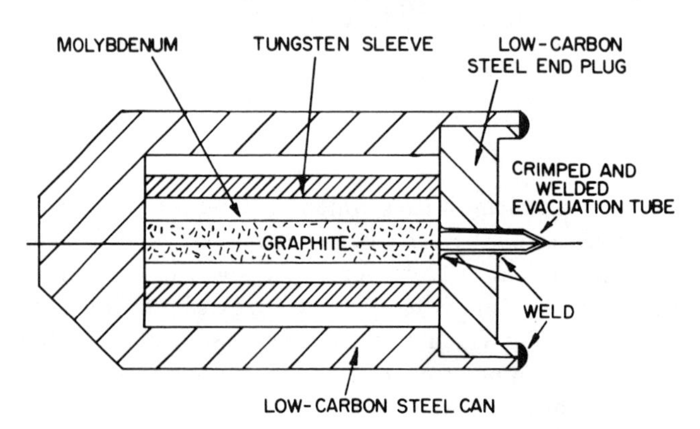

Fig.8.24. Complex canning arrangement, including composite billet for extruding tungsten tubes at 1210 C [148]

Another, entirely different purpose of extruding dissimilar materials is the making of bimetallic products, usually in the form of a relatively thin cladding on a bar or on one or two sides of a tube. The aim then is to ensure uniform material flow and to establish a good mechanical and metallurgical bond. An incidental side benefit may be easier lubrication of the cladding.

Other Solid Lubricants. Layer-lattice compounds are widely employed. When they are the only means of lubrication, uniform coverage of the billet is critical because localized film failure would lead to die pickup and workpiece damage. The shear stress of the interface increases with temperature; lower friction is ensured with grease-base lubricants filled with copper and lead powder [149]. To minimize breakthrough pressure, the billet should be nosed [148]. Because of their limited spreading ability, these lubricants are more suitable for short billets.

Graphitic lubricants are often used for selective lubrication of die and mandrel, either alone or in conjunction with another lubricant.

Polymers are used in cold extrusion as solid-film lubricants. In hot extrusion they may be gradually melted to provide a form of phase-change lubrication.

Soap and boundary lubricants can be very effective in cold extrusion.

With all forms of solid-film lubrication, pressures in direct extrusion of long billets may reach high enough values to satisfy the condition of sticking friction (Eq.2.9), and then the billet length has to be reduced or the process changed to reverse extrusion. Solid lubricants, as already mentioned, may be applied to the billet when the pressure medium lacks sufficient lubricating properties. Johnson and Wilson [150] continued development of a model for solid lubricant entrainment in hydrostatic extrusion, allowing for pressure-sensitive shear strength.

8.5.5 Application of Ultrasonics

As pointed out in Sec.3.9.2, the imposition of ultrasonic vibration aids in establishing or maintaining a lubricant film if otherwise the film would be marginal. Indeed, Petukhow [151] found that friction dropped and deformation became more homogeneous with increasing vibration amplitude, especially if vibration was applied to both die and container. Jones [7.239] reported similar improvements in direct extrusion of lead and aluminum. This technique is, however, likely to remain a laboratory curiosity, because adequate lubricant systems can be found for most situations, and the application of ultrasonics would be impractical in instances where difficulties in lubrication are encountered (such as in the hot extrusion of refractory alloys).

8.5.6 Die Wear

The extrusion of long bars and tubes imposes some of the most severe demands on the various tooling elements [7]. High pressures and sliding velocities combine with adhesive and abrasive action and, in hot extrusion, with sudden temperature fluctuations and prolonged exposure to high temperatures. It is, therefore, not surprising that some of the improvements in tool materials (Sec.3.11) were first applied to extrusion dies.

All containers are exposed to internal pressure. In addition, there is thermal fatigue in nonisothermal extrusion and abrasive wear in forward extrusion (whether dry, solid- or glass-lubricated). Wear is usually concentrated toward the die end of the container because the cumulative sliding length (Eq.3.39) is greater there. Hot working die steels are generally adequate except in isothermal extrusion of titanium, for which superalloy dies are more satisfactory.

Follower blocks suffer mostly from thermal fatigue and, if overloaded, from plastic deformation. The materials employed are similar to those used for containers.

Mandrels are subject to high heat and frequently fail by plastic deformation. Therefore, they are often cooled internally or, between strokes, externally [7,152].

Dies are the most severely loaded. In extrusion of a long section the die softening temperature may be reached: tongues deform, the orifice may open up, and dimensions are lost. Abrasive wear is more gradual but is, again, much accelerated at elevated temperatures. Such die wash is sometimes aggravated by adhesive wear. Much work has gone into increasing wear resistance by selection of appropriate steels [7],[153,154], by use of surface treatments such as nitriding [155] and W_2C and W_3C coating [156], and by use of tungsten carbide inserts [157]. Ceramic coatings, particularly ZrO_2 and Al_2O_3, have come into general use in extrusion of iron- and nickel-base alloys and have found application as solid die inserts, too [158]. The low thermal conductivity of many high-temperature materials is a disadvantage in that high tool temperatures can build up. The molybdenum-base alloy TZM has been shown to give a better compromise in properties [7].

8.6 Lubrication of Iron- and Nickel-Base Alloys

There are basically only three practical techniques for hot extrusion of steels, stainless steels, and nickel-base semifabricated products:

Solid-Film Lubrication. Hot nonisothermal extrusion on mechanical presses at high speeds and with graphitic lubricants was the only method available before the introduction of glass lubrication, and is still practiced for shorter, smaller products, including tubes. The lubricant used is graphite in oil or fat (such as tallow): the oil burns off, and the graphite provides the lubricating function but without heat insulation. The addition of salt has been reported to be beneficial [159]—perhaps by providing a liquid carrier at extrusion temperature.

The search for alternative or better lubricants has proved to be essentially fruitless. Shaw et al [160] found in extruding 25-mm-diam 4340 steel billets (at 1180 C and a 4:1 extrusion ratio using dies with 130° included angles) that only lubricants containing graphite gave reasonably low and uniform extrusion pressures. Solids such as BN, vermiculite, and talc added to a conventional graphitic lubricant did not reduce pressures, and MoS_2 failed at the high temperatures. Glasses were unsuccessful, too.

The very short contact time on high-energy-rate machines prompted some investigations in the 1960's into the potential use of these machines for hot extrusion. As in forging (Sec.9.1.1), die life problems prevented the widespread industrial use of the technology.

Preapplied Glass Film. Hot nonisothermal extrusion at moderately high ram speeds (about 0.1 to 0.2 m/s) on hydraulic presses, at low extrusion ratios, relies on glass preapplied to the billet. Dies are tapered, typically at an included angle of 120°, the die is protected by an alumina or zirconia coating, and a graphite lubricant is applied to the die and sometimes also to the container. Interface shear strength values on the order of 35 MPa (or approximately 1/3 of the values measured in unlubricated extrusion) were reported by DePierre [23]. The drop in shear stress with extrusion speed indicated the presence of hydrodynamic effects. Several alternative lubricants have been explored [4.192] but appear to have made little impact on practice. The billet remnant is pushed through the die with a deformable graphite follower block, and thus high yields are ensured even with tapered dies.

Glass Lubrication. By far the largest quantities are extruded by phase-change lubrication. This technique allows longer lengths, thinner walls, and more complex shapes than graphite lubrication [89,161-163] and eliminates carbon pickup, too. Numerous articles deal with the technology of the process [123,126, 135,161-167]. For economy of extrusion, flat dies (180° included angle) are used except when the high strength of the workpiece material necessitates smaller extrusion ratios, and then a 120° included angle is preferred.

Tool materials have a vital influence on the economy of the process. Temperature changes during extrusion are primarily a function of extrusion speed [168] and glass viscosity [133]. Die temperatures exceed the tempering temperature of steel dies in extrusion of long products, especially of complex shapes. Apart from use of a heat-resistant die material (Sec.3.11.4), alumina or zirconia coatings 0.25 to 1 mm thick are routinely deposited by the plasma-arc technique onto a rough-machined die surface. Dies are often preheated to 250 to 430 C to reduce heat shock.

The glass is chosen for minimum cost compatible with technological performance.

Window glass or its equivalent, in powder or fiber form, mixed with a binder, is satisfactory for many steels and even for stainless steels. Lubrication practices influence surface defects [169]. Some producers replace borosilicate glass with the cheaper basalt for Ni-Cr alloys and with basalt mixed with borosilicate glass for Ni-Cu alloys extruded at lower temperatures [139]. The extrusion temperature range of nickel-base superalloys is so narrow that chilling on contact with the container would cool the billet below the acceptable minimum temperature, and sheathing made of a Ni-Cr alloy is then used as a ductile heat barrier [139]. Extrusion at the usual high speeds would lead to overheating and speed cracking, and therefore the press speed is controlled by the provision of several pumps, which can be successively switched off as extrusion proceeds.

Hydrostatic Extrusion. Other techniques of lubrication have found only limited application. Variants of hot hydrostatic extrusion are occasionally reported in industrial use. Lipinsky et al [170] describe the extrusion of milling cutters with preformed flutes. Kostava and Muraviev [65] report on extrusion on bars with a mixture of pitch and graphite which melts at 200 to 250 C to act as a hydrostatic fluid. In the process described by Gurdus and Zheltogirko [171], a hot graphite follower block is pressed around the extrusion to become a parting agent/lubricant. These methods are, by the nature of their products, more closely related to forging.

Hydrostatic extrusion at room temperature has not proven to be economical, although hydrostatic extrusion/drawing of stainless steel wire offers the advantage of large reductions (up to 86%) in a single pass, and the issuing wire has higher strength and greater ductility than conventionally drawn wire [172].

8.7 Lubrication of Light Metals

Extrusion of light metals is characterized by intermediate temperatures that allow near-isothermal operation.

8.7.1 Aluminum Alloys

Extrusion of aluminum alloys has been critically reviewed by Akeret [84,85] and by Chadwick [88]. Basically, there are two options for extruding aluminum and its alloys.

Unlubricated Extrusion. By far the largest quantity of aluminum alloys is hot extruded without any lubricant whatsoever. A breakthrough pressure of up to 20% is sometimes found and is related to the thermal activation of the flow process [173]. Once material flow begins, it is characterized by sticking friction with a large dead-metal zone (Fig.8.25) in forward extrusion, although Sheppard et al [174] found the interface shear factor m to depend on pressure, suggesting (at least in extrusion of aluminum alloy 7075) that some form of stick-slip, rather than full sticking, prevails. As long as the billet is of good, uniform quality, the die is free of major pickup, and no lubricant or contaminant finds it way into the container, the quality of the extrusion will be excellent [88].

For best surface quality, a follower block of smaller diameter is used to leave a skull, which is removed by specially designed follower blocks [47] or in a separate stroke. The skull entraps imperfections of the billet surface. When extrusion is done without a skull, the billet is scalped for highest quality. Shearing along the container wall contributes to heat generation; in direct extrusion it may reach 50% of the total force [12] and thus greatly adds to the temperature rise. Because the heat rise is proportional to σ_f (Eq.2.15), the permissible exit speed is found to be inversely propor-

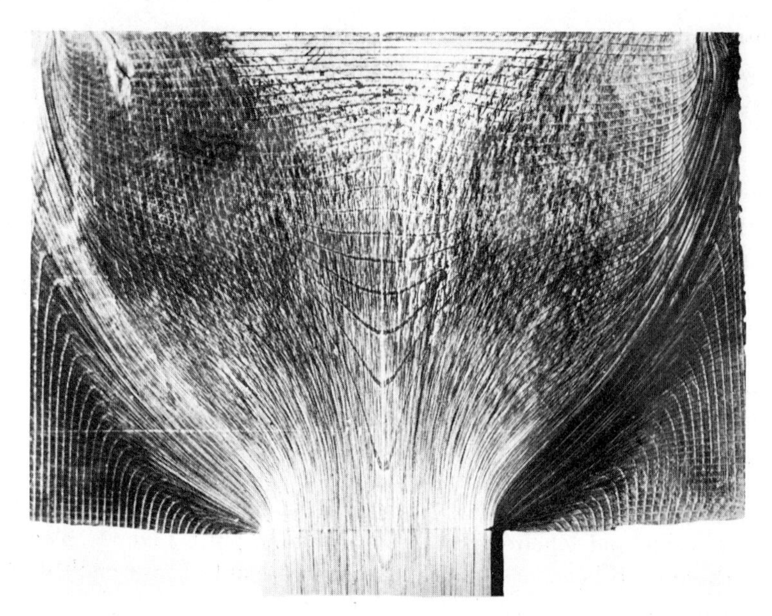

Fig.8.25. Material flow in hot extrusion of an aluminum alloy billet without lubrication [132]. (Courtesy Cefilac, Paris, France)

tional to the square of σ_f [84], and drops to 1.5 mm/s for 2024 and 7075 alloys. In these alloys localized severe deformation may also lead to an undesirable, recrystallized coarse grain structure of inferior properties on the periphery of the extrusion [7].

Reverse extrusion imposes much smaller strains (Fig.8.14b). The work of shearing along the container wall does not contribute to an increase in the temperature of the extruded product, and a threefold increase in speed is possible [84,88].

The main problem in extrusion with sticking friction is the die. The die is designed and made [175-177] so that optimum material flow is ensured and an extrusion free of defects is produced. The die usually has a flat face, which is protected by the dead-metal zone. However, the high adhesion of aluminum to steel soon results in the buildup of a coating on the die and die land. Local lubrication of the die land or of the billet face with a graphitic lubricant is sometimes practiced, even though the lubricant is lost soon after the extrusion emerges [178] and extrusion proceeds over the bare die. No lubricant is allowed in some press shops, because it may encourage attempts at more extensive lubrication, with consequent material defects, as shown in Fig.8.8.

The die is usually polished, although in the experiments of Tokizawa et al [179] transversely ground dies that were moderately rough (R_a = 0.5 to 0.75 μm) gave better extrusion quality than those polished or ground parallel to the extrusion direction, probably because reaction products filled up the troughs and then prevented severe adhesion. This explanation is supported by the observation that product quality, as judged from surface roughness, became poorer with increasing adhesion at higher temperatures.

Die coatings can be helpful. In experiments of Tokizawa et al [180], sliding friction could be maintained and pickup limited to a very thin, stable film with nitrided tool steel and cemented carbide dies, whereas sticking and heavy coating were observed with bare tool steel and with chromium carbide and TiC coated dies. Indeed, nitrided dies have found extensive practical application, with a threefold to sevenfold increase in die life [7].

Pickup is typical of bare tool steel. Kemppinen [181] and Merk and Naess [178] observed that debris containing Fe_2O_3 is often present when pickup damage occurs;

damage diminishes when oxidation-resistant dies are used but increases in extrusion of strong alloys where die temperatures are higher. There are several sources of oxygen [88,182]; one of them is thought to be Al_2O_3 in or on the billet, and another is atmospheric oxygen. Thus, any measures that eliminate oxygen are helpful, including filtration of the aluminum melt to remove oxides, admission of an inert gas to the die (Altwicker, reported in [12], and Heffron and Hull [183]), and, in extrusion of tubes, the use of an oversize mandrel which shears off the oxide formed on the tube bore [88].

Lubricant applied to the tool or billet leading face may be helpful in preventing oxidation of the tool until the billet face passes through. Still, most of the time, uniform die pickup is accepted as a side effect of extruding with sticking friction, and emphasis is placed on die cleaning and maintenance with the aid of wire brushing and caustic soda etching. In another practice, a spray of a graphite and caustic soda mixture is applied to remove remnants of the previous coating and provide lubrication for the next push [184].

Lubricated Extrusion. Lubricated hot extrusion of aluminum alloys has found only limited application. It is not easy to identify an obviously best lubricant. There are few glasses of the right viscosity for phase-change lubrication [4.209], and the rough product surface typical of the PHD mechanism is inferior to the accustomed surface. Graphite cannot maintain sliding friction over the entire course of extrusion of a long billet, although both graphite and MoS_2 have been used for mandrel lubrication in tube extrusion [185]. PTFE has been found to be useful in cable extrusion [182] and in experimental hot extrusion of alloy 2024 [160]. Despite repeated investigations of lubricated hot extrusion [185-187], especially for strong alloys, the unlubricated process remains dominant. In extrusion of high-magnesium alloys, a short cylinder of soft aluminum helps to initiate more homogeneous flow [88].

The low cold flow stress of pure aluminum and its low alloys permits their extrusion at room temperature [86,189] using fats or waxes, such as lanolin, as a lubricant. Even though the temperature increase at high extrusion speeds is sufficient to cause annealing and other undesirable metallurgical changes [189], the lubricant provides effective mixed-film lubrication, and surface quality is good [86]. Hydrostatic extrusion was once also thought of as a competitive process, especially for electrical conductor wire, but the resulting surface quality is inferior because of surface roughening and defects related to the starting material [190,191]. Lubrication is essentially of the mixed-film type, but local entrapment causes roughening, and Lugosi and Male [190] suggest the use of a separate lubricant applied to the billet surface to avoid stick-slip. Experimental warm hydrostatic extrusion of alloys has been reported to give freedom from die pickup coupled with good surface quality [192].

8.7.2 Magnesium and Beryllium Alloys

The oxide of magnesium is powdery and friable, and ensures lubricated flow [193,194] without the addition of extra lubricant. In experiments on extrusion of magnesium alloy AZ80A, PTFE reduced extrusion pressures with conical dies, but inhomogeneous flow resulted in surface cracking with flat dies [195].

Beryllium presents problems chiefly because of the toxicity of its oxide. Therefore, it can be extruded bare only at temperatures below 750 C, using graphite or MoS_2 as a lubricant [196,197]. Both bars and tubes are extruded, with good surface quality, in the temperature range between 400 and 600 C. Silver coating has also been successful [4.211].

Hot extrusion of bare beryllium at temperatures above 750 C is dangerous and also difficult because, in the event of glass lubricant breakdown [197,198], rapid die wear ensues. Canning in mild steel eliminates both problems [198-200]. When short lengths are extruded, graphitic lubricants suffice, although glass could be used equally well at temperatures up to 1065 C, beyond which alloying between can and billet takes place [198].

8.8 Lubrication of Copper-Base Alloys

As pointed out in Sec.3.3.1, the nature of an oxide has a profound influence on its lubricating characteristics. Nowhere is this more clearly illustrated than in hot extrusion of copper and its alloys. Cuprous oxide has a low enough shear strength to serve as a lubricant and give essentially lubricated flow even with a flat-face die (Fig.8.26a) [201]. In contrast, the oxide of brass is a poor lubricant and causes sticking friction which, compounded by chilling, results in a very complex, nonlubricated flow (Fig.8.26b), which then leads to development of the extrusion defect [7,202]. Vater and Koltzenburg [203], who gave also a review of the subject, found even more complex flow with bronzes, as did Hesse et al [204] in extrusion of Cu-Cr alloys. For production of high-quality extrusions, several options exist [7]:

1. Extrude with a skull, leaving all oxides behind. This is common practice for $\alpha + \beta$ brasses and aluminum bronzes. For other alloys, it is best to avoid the formation of oxides by heating in a protective atmosphere, or to remove the oxide by descaling [203], and then even copper will flow like brass. Ram speeds can also be increased to as high as 15 mm/s [203]. Any oxide present carries with it the danger of intermittent lubricated flow (Fig.8.8), which would allow oxide to enter into a subsurface location. If the product is subsequently annealed in a reducing atmosphere, the oxide is reduced by hydrogen to form copper and H_2O, the latter of which then causes blistering.

2. Extrude copper alloys without a skull, relying on the consistently poor lubrication afforded by their oxides, and limit the extent of the extrusion defect by leaving about a 20% butt, coupled with an inspection of the product for the defect. This is the technique often followed in mass production of brass.

3. Extrude copper with a consistently well-lubricated flow. Cairns and Such [205]

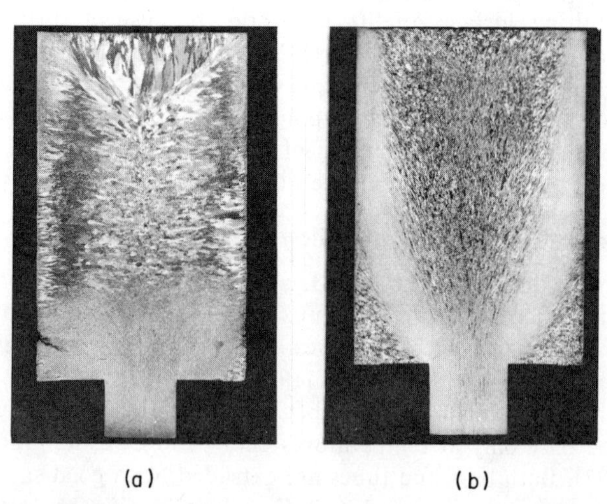

Fig.8.26. Microstructures of billets of (a) copper and (b) 70/30 brass partially extruded at 800 C [201]. (Courtesy Dr. A.T. Male, Westinghouse Corp., Pittsburgh, PA)

describe a sequence in which the billet is preheated, scalped, and then further heated in an oxidizing atmosphere to produce a strongly adherent cuprous oxide film which, in conjunction with an oil-graphite mixture applied to the container, die, and mandrel, ensures reliable sliding friction throughout the push.

4. Extrude copper, and especially its hard alloys, in conical dies using glass lubrication.

In all instances, the proper choice of die material is important, because temperatures are high enough to cause softening on prolonged exposure. Dies of high-temperature die steel materials [7], nickel-base superalloys [205], or ZrO_2, Al_2O_3 , or Mo-ZrO_2 [158] are justified in many instances.

Experimental cold and warm hydrostatic extrusion of copper and its alloys have shown promise [48,206-207] and have led to the first industrial applications [12,68]. The low flow stress of copper and the large demand for small-diameter seamless tubing make the process competitive with conventional manufacturing sequences. If the extrusion temperature is increased to around 350 to 500 C, much higher extrusion rates are possible [208] and operation is still practical. The fluid used is castor oil, which comes into contact with the hot billet at atmospheric pressure only for a short period of time, and boiling ceases as soon as the pressure is applied. Very efficient mandrel lubrication is ensured. Extrusion of fine (0.1-mm-diam) wire is more difficult [209] because, at the high speeds desirable, viscous drag increases the required pressure and, if a very low-viscosity fluid such as kerosene is used, E.P. and boundary additives are needed even with a diamond die. For general hydrostatic extrusion silicones are among the best lubricants [210].

In cold hydrostatic extrusion of copper-clad aluminum, Lugosi et al [211] found 20% MoS_2 in stearyl stearate to be effective. Friction on the Cu-Al interface affected the process: intermittent thickening of copper was observed when the shear strength of the billet/sheath interface was low, and a good bond could not be produced even with a clean interface at the 19:1 extrusion ratio employed. This observation confirms the need for severe interface sliding (shearing) if a metal-to-metal bond is to be established.

8.9 Lubrication of Titanium, Zirconium, and Uranium Alloys

Titanium presents two special problems. One is that, even though titanium oxide serves a useful function as a parting agent, oxide access must usually be denied to prevent the formation of an alpha case. The second problem is that titanium adheres badly to steel, and lubricant breakdown results in immediate pickup and scoring. Extrusion temperatures depend on the alloy and, in the case of $\alpha + \beta$ alloys, on whether extrusion is conducted in the α or the β range. In general, temperatures are between 800 and 1000 C, and therefore the techniques typical of steel extrusion are applicable.

Small billets, short lengths, and small extrusion ratios favor the nonisothermal extrusion process with graphite-type lubricants. Even though graphite is the prime lubricating agent, the carrier fluid or grease also has an effect [160]. Various additives have also been incorporated into greases, including solid parting agents such as talc, mica, soapstone, and heavy semisolids such as asphalt. Alternatively, preapplied glass films can be used [23,107]. Preapplication of the glass film has the advantage that the billet is protected from oxidation during preheating. Use of a borosilicate glass has been reported for isothermal extrusion [212].

Long lengths are extruded by the phase-change lubricating technique [4.167],[167,213,214]. Because of the lower extrusion temperature, lower-viscosity glasses than those used for steel are required. Die wear could be severe, and plasma coating with zirconia or alumina is routinely done, with the coating often being replaced after every push. Canning in copper has the benefit of protecting the billet as well as facilitating lubrication with a graphite-grease mixture [215]. Recent work practices are described by Markworth and Ribbecke [216].

Zirconium is in many ways similar to titanium and is dealt with by the same techniques [217]. Cladding in copper (for temperatures up to 800 C) [218] and in steel (to 900 C) are also practiced [4.211].

Uranium also suffers from the problems of oxidation and contamination, and therefore cladding with copper is advisable [4.211], although bare extrusion is also possible [219]. When the uranium is preheated in a salt bath, the salt can serve as a lubricant [3.426]. Because uranium has several allotropic transformations, extrusion temperature greatly affects extrusion pressures. For atomic energy applications, uranium/zirconium composites were extruded at 680 C in copper cans [220] using a graphitic lubricant. In hot extrusion of UO_2 fuel elements, steel or copper has been used for the outer can, and stainless steel, tantalum, or molybdenum for the inner sheathing [221].

8.10 Lubrication of Refractory Metals

Alloys of refractory metals (vanadium, molybdenum, tungsten, niobium) share a relatively high extrusion temperature (1000 to 2200 C; see Table 2.1), rapid oxidation, and sensitivity to contamination.

Depending on the extrusion ratio and the danger of contamination, refractory metal alloys have been extruded either bare or canned, using either graphitic or glass-type lubrication. Cans are usually made of mild steel or stainless steel for temperatures up to 1250 C [222], or of molybdenum. It is often difficult to ensure uniform coating of the billet by a glass because of oxidation, but the oxides themselves have some lubricating qualities, and thus the difficulties are less severe than might be anticipated. Coating with a glass is effective if the billets are preheated in a protective (argon) atmosphere. The molybdenum alloy TZM exhibited good surface finish after being extruded with glass [4.222]. Low alloys of vanadium and molybdenum are freely extruded, using either the nonisothermal or the phase-change lubrication technique [223]. Cold hydrostatic extrusion is also feasible for some vanadium alloys; the low inhomogeneity of deformation minimizes strain hardening [224].

8.11 Lubrication of Lead

Extrusion of lead is of interest for two reasons. First, lead is an important industrial metal in its own right, and once was extruded in large quantities into cable sheathing with either conventional or screw-type presses [225]. Extrusion of lead still finds some application for bar and wire products as well as for tubing. Secondly, lead is an excellent simulating material. Because of its low melting point, pure lead hot works at room temperature and has often been used for simulation of hot extrusion processes.

The low flow stress of lead allows very large extrusion ratios and, even though the calculated coefficient of friction may be low, in reality extrusion proceeds by sticking friction, as shown by Krysko and Fenton [226]. To avoid surface and internal defects, it is again important that no lubricant or contaminant should be allowed. This applies also to extrusion of cable sheathing, where two streams of metal must be pressure welded. Screw extrusion is feasible, and the use of sodium palmitate has been reported for lubrication of screws [227], but presumably it is feasible for solid products only.

Extrusion ratios tend to be smaller in simulation studies, and then roughening of the container is necessary to ensure sticking friction. Otherwise, any wax or grease-type lubricant will facilitate sliding friction in extrusion through conical dies.

8.12 Summary

1. In extrusion of long, semifabricated products, friction is generally unnecessary and undesirable. Friction on the die increases extrusion pressures and, for a given die geometry, impairs the homogeneity of deformation. If under the prevailing pressures the shear strength of the interface exceeds the shear flow stress of the workpiece material, a dead-metal zone forms.

2. In direct (forward) extrusion, friction on the container wall increases the extrusion pressure. In the limit, sticking friction develops and extrusion proceeds by shearing through the billet, leaving behind a skull. For a given press capacity, the length of billet that can be extruded is limited by container friction.

3. Friction on the container wall is immaterial in indirect (reverse) extrusion, because there is no relative movement.

4. Friction contributes to heat generation and limits attainable reductions and speeds in hot extrusion. Friction on the die land promotes surface cracking (speed cracking, fir-tree effect).

5. Friction is beneficial only when there is danger of internal (arrowhead) cracking or when the container is moved so that the frictional force contributes to the extrusion force. Friction on the die and die land may be utilized to equalize material flow in extrusion of complex shapes.

6. Extrusion is a process that can only be conducted either totally unlubricated or fully lubricated. Intermediate conditions are undesirable, because changeover from unlubricated to lubricated flow during extrusion of a billet results in formation of subsurface defects.

7. Unlubricated extrusion is essential for extruding hollow products with bridge, spider, or porthole dies, because use of a lubricant would prevent rewelding of the separated material streams. Unlubricated extrusion is desirable for hot extrusion of aluminum alloys with flat dies, because the surface of the extruded product is freshly formed by internal shear in the billet, and is feasible also for nonisothermal hot extrusion of copper and its alloys.

8. Lubricated extrusion is essential for hot extrusion of steel, nickel- and titanium-base alloys, refractory metal alloys, and high-strength copper alloys, as well as for all cold extrusion. The geometry of the process makes it particularly favorable for lubrication with glasses that melt gradually to form a lubricating film. The billet surface becomes the surface of the extrusion, modified by surface extension and by the presence of the lubricant.

9. Friction is reduced, and the homogeneity of flow is improved, by lubrication. Lubricant film thickness increases with viscosity, extrusion speed, and decreasing die angle. Tapering of the billet eliminates breakthrough pressures. Film thickness increases also with the pressure applied to the lubricant, and thus hydrostatic extrusion not only eliminates container friction but can also reduce die friction. It is particularly effective in reducing friction on the mandrel in tube extrusion. Excessive film thickness leads to rough surfaces and, possibly, to periodic collapse of the film (bamboo defect).

10. Severe sliding and prolonged exposure to high temperatures and high pressures combine to make hot extrusion extremely demanding on die materials. Adhesive and abrasive wear, thermal fatigue, and plastic deformation may all occur. Surface treat-

ments such as nitriding are beneficial even in unlubricated extrusion, and ceramic die coatings are often essential in high-temperature glass-lubricated extrusion.

11. The most commonly used lubricants are given in Table 8.1. For purposes of preliminary calculation, typical μ or m values are also shown, although it must be understood that under hydrodynamic or mixed-film conditions the coefficient of friction is greatly affected by process conditions. At the high pressures developed with high extrusion ratios, sticking may be attained even with relatively good lubricants.

Table 8.1. Commonly used extrusion lubricants and typical friction values

Material	Hot extrusion		Cold extrusion	
	Lubricant(a)(b)	μ(c)	Lubricant	μ
Steels	GR (for short pieces)	0.2	NA	
	Glass (10-100 Pa·s) pad and coating	0.02		
Stainless steels and Ni alloys	Glass (20-200 Pa·s) pad and coating	0.03	NA	
	Melting solids	0.05		
Al and Mg alloys	None	ST	Lanolin	0.07
	None (GR on die face)	ST		
Cu and Cu alloys	GR (for short pieces)	0.2	Castor oil (for hydrostatic extrusion)	0.03
	None (with or without skull	ST		
	Glass (10-100 Pa·s) coating	0.05		
Ti alloys	GR (for short pieces)	0.2	NA	
	Glass (10-100 Pa·s) coating + GR on die	0.05		
	Glass (10-100 Pa·s) pad and coating	0.03		
Refractory metals	Glass (10-200 Pa·s) coating + GR on die	0.05	NA	
	Glass (10-200 Pa·s) pad and coating	0.03		
	Canning + lubrication for can			

(a) GR = graphite, often with a polymeric binder. (b) Viscosities of glasses are given for typical extrusion temperatures. (c) When $\mu p = k$ is satisfied, sticking friction sets in.

References

[1] S. Kalpakjian, in [3.1], pp.549-601.
[2] J.A. Schey, in [3.1], pp.47-58 and 263-266.
[3] J.A. Newnham, in [3.1], pp.672-675.
[4] C.E. Pearson and R.N. Parkins, *The Extrusion of Metals,* 2nd Ed., Chapman & Hall, London, 1960.
[5] W. Johnson and H. Kudo, *The Mechanics of Metal Extrusion,* Manchester Univ. Press, 1962.
[6] K. Laue and H. Stenger, *Strangpressen,* Aluminium-Verlag, Düsseldorf, 1976.
[7] K. Laue and H. Stenger, *Extrusion,* ASM, Metals Park, OH, 1981.
[8] *Strangpressen,* Deutsche Gesellschaft f. Metallkunde, Frankfurt, 1976.
[9] L.V. Prozorov, *Extrusion of Steels and Refractory Alloys,* Mashgiz, Moscow, 1969.

[10] J.F.W. Bishop, Met. Rev., *2*, 1957, 361-390.
[11] J.M. Alexander, J. Inst. Metals, *90*, 1961-62, 193-200.
[12] R. Chadwick, Int. Met. Rev., *25*, 1980, 94-136.
[13] E.R. Lambert and S. Kobayashi, in *Proc. 9th Int. MTDR Conf.*, Pergamon, Oxford, 1969, 253-269.
[14] H. Kudo and H. Takahashi, CIRP, *13*, 1965, 73-78.
[15] S. Sohrabpour, A.H. Shabaik and E.G. Thomsen, Trans. ASME, Ser.B., J. Eng. Ind., *92*, 1970, 461-467.
[16] B.I. Bachrach and S.K. Samanta, in *Proc. 2nd NAMRC,* SME, Dearborn, 1974, pp.179-193.
[17] P.V. Vaidyanathan and T.Z. Blazynski, J. Inst. Metals, *101*, 1973, 79-84.
[18] J. Frisch and E. Mata-Pieri, in *Proc. 18th Int. MTDR Conf.*, Macmillan, London, 1978, pp.55-60.
[19] F.J. Gurney and V. DePierre, Trans. ASME, Ser.B, J. Eng. Ind., *96*, 1974, 912-916.
[20] W.B. Rice and J.N. Garner, in *Proc. 6th NAMRC,* SME, Dearborn, 1978, pp.138-141.
[21] G.H. Townend and D.G. Broscomb, Proc. Inst. Mech. Eng., *169*, 1955, 671-676.
[22] I. Perlmutter, V. DePierre, and C.M. Pierce, in [3.50], pp.147-161.
[23] V. DePierre, Trans. ASME, Ser.F, J. Lub. Tech., *92*, 1970, 398-405.
[24] B. Lengyel and D. Mutlu, in *Proc. 19th Int. MTDR Conf.*, Macmillan, London, 1979, pp.265-269.
[25] T. Sheppard and A.F. Castle, in *Proc. 16th Int. MTDR Conf.*, Macmillan, London, 1976, pp. 535-545.
[26] R. Akeret, J. Inst. Metals, *95*, 1967, 204-211.
[27] T. Altan and S. Kobayashi, Trans. ASME, Ser.B, J. Eng. Ind., *90*, 1968, 107-118.
[28] H.P. Stüwe, Metall, *22*, 1968, 1197-1200 (German).
[29] G. Lange, Z. Metallk., *62*, 1971, 571-584 (German).
[30] L.F. Mondolfo, A.R. Peel, and J.A. Marcantonio, Metals Techn., *2*, 1975, 433-437.
[31] V. Nagpal, C.F. Billhardt, and T. Altan, Trans. ASME, Ser.B, J. Eng. Ind., *101*, 1979, 319-325.
[32] M.H. Ahmed and M.M. Farag, in *Proc. 4th NAMRC,* SME, Dearborn, 1976, pp.180-187.
[33] T. Sheppard and E.P. Wood, in *Proc. 17th Int. MTDR Conf.*, Macmillan, London, 1977, pp.411-421.
[34] R. Chadwick, Met. Rev., *4*, 1959, 189-255.
[35] W. Dürrschnabel, Metall, *22*, 1968, 426-437, 995-998, 1215-1219 (German).
[36] R. Genders, J. Inst. Metals, *26*, 1921, 237-245.
[37] J. Séjournet, Lubric. Eng., *18*, 1962, 324-329.
[38] A.M. El-Behery, J.H. Lamble, and W. Johnson, in *Proc. 4th Int. MTDR Conf.*, Pergamon, Oxford, 1964, pp.319-335.
[39] C. Blazey and D. Stead, Metal Ind., *103*, 1963, 88-89.
[40] C. Blazey, L. Broad, W.S. Gummer, and D.B. Thompson, J. Inst. Metals, *75*, 1948-49, 163-184.
[41] H.Ll.D. Pugh (ed.), *Hydrostatic Extrusion (Proc.),* Inst. Mech. Eng. London, 1974.
[42] J.M. Alexander and B. Lengyel, *Hydrostatic Extrusion,* Mills and Boon, London, 1971.
[43] L.V. Prozorov, A.A. Kostava, and V.D. Revtov, *Extrusion of Metals by Liquids Under Pressure,* Mashinostroenie, Moscow, 1972 (Russian).
[44] H.Ll.D. Pugh (ed.), *Engineering Solids Under Pressure,* Inst. Mech. Eng., London, 1971.
[45] A. Bobrowsky and E.A. Stack, in [3.50], pp.122-136.
[46] H.Ll.D. Pugh and C.J.H. Donaldson, CIRP, *21*, 1972, 167-186.
[47] R. Akeret, in [8], pp.138-155.
[48] R.J. Fiorentino, G.E. Meyer, and T.G. Byrer, Metall. Met. Form., *41*, 1974, 193-197, 210-213.
[49] J.A. Schey, Metals Eng. Quart., *10* (2), 1970, 16-18.
[50] R.J. Fiorentino, G.E. Meyer, and T.G. Byrer, Metall. Met. Form., *39*, 1972, 200-203.
[51] K.M. Kulkarni and J.A. Schey, Trans. ASME, Ser.F, J. Lub. Tech., *97*, 1975, 25-31.
[52] D. Ruppin and K. Müller, Aluminium, *56*, 1980, 263-268, 329-331, 403-406 (German).
[53] M. Nishihara, T. Fujita, T. Matsushita, and M. Noguchi, J. Soc. Mat. Sci. Jpn, *20*, 1971, 918-923.
[54] C.E.N. Sturgess and T.A. Dean, in *Proc. 13th Int. MTDR Conf.*, Macmillan, London, 1973, pp.389-399.

[55] J.M. Alexander and S. Thiruvarudchelvan, CIRP, *19*, 1971, 39-52.
[56] B. Parsons, D. Bretherton, and B.N. Cole, in *Proc. 11th Int. MTDR Conf.*, Pergamon, Oxford, 1971, pp.1049-1074.
[57] B. Lengyel and J.M. Alexander, Wire Ind., *39*, 1972, 978-982.
[58] B. Lengyel and M.J.M.B. Marques, in *Proc. 20th Int. MTDR Conf.*, Macmillan, London, 1980, pp.113-117.
[59] S. Thiruvarudchelvan and J.M. Alexander, Int. J. Mach. Tool Des. Res., *11*, 1971, 251-268.
[60] D.B. Marreco and H.A. Al-Quareshi, in *Proc. 4th NAMRC*, SME, Dearborn, 1976, pp.213-218.
[61] J.M. Alexander, Mat. Sci. Eng., *10* (2), 1972, 70-74.
[62] H.K. Slater and D. Green, Proc. Inst. Mech. Eng., *182*, Pt.3C, 1967-68.
[63] H.Ll.D. Pugh, T. Wilkinson, and M.H. Hodge, Wire Ind., *40*, 1973, 49-52.
[64] S. Takahashi et al, in [44], pp.166-177.
[65] A.A. Kostava and V.K. Muraviev, Kuzn. Shtamp. Proizv., *1977* (1), 9-11 (Russian).
[66] M. Nishihara et al, in *Proc. 20th Int. MTDR Conf.*, Macmillan, London, 1980, pp.87-92.
[67] L.F. Vereshchagin, Yu.S. Konyaev, and E.V. Polyakov, in [44], pp.131-132.
[68] J.O.H. Nilsson, High Temp.-High Pressures, *8*, 1976, 691-693.
[69] B. Avitzur, Wire J., *11* (3), 1978, 78-84.
[70] Ya.M. Okhrimenko, V.L. Berezhnoi, V.N. Shcherba, and G.S. Sharikov, Light Metal Age, *31* (3/4), 1973, 27, 29-30.
[71] F.J. Fuchs, Wire Ind., *38*, 1971, 258-264.
[72] J.S. Cartwright, B. Lynch, and J.R.S. Shaffer, Wire Wire Prod., *47* (2), 1972, 38-48; (3), 1972, 45-52.
[73] J. Tirosh and G. Grossman, Trans. ASME, Ser.H, J. Eng. Mat. Tech., *99*, 1977, 52-58.
[74] W. Ziegler and K. Siegert, Metall, *31*, 1977, 845-851 (German).
[75] K.F. Ziehm, Light Metal Age, *28* (11/12), 1970, 6-10.
[76] D. Green, J. Inst. Metals, *100*, 1972, 295-300.
[77] H.K. Slater, Wire J., *12* (2), 1979, 76-82.
[78] B. Avitzur, Paper No. MF 75-140, SME, Dearborn, 1975.
[79] J.T. Black and W.G. Voorhes, Trans. ASME, Ser.B, J. Eng. Ind., *100*, 1978, 37-42.
[80] F.J. Fuchs and G.L. Schmehl, Wire J., *6* (11), 1973, 53-57.
[81] B. Lengyel and S. Kamyab-Tehrani, in *Proc. 17th Int. MTDR Conf.*, Macmillan, London, 1977, pp.423-428.
[82] J. Wantuchowski and J. Sloniowski, Arch. Hutn., *1*, 1975, 135-156 (Polish).
[83] R.J. Wilcox and P.W. Whitton, J. Inst. Metals, *88*, 1959-60, 145-149.
[84] R. Akeret, Z. Metallk., *64*, 1973, 311-319 (German).
[85] R. Akeret and P.M. Stratman, Light Metal Age, *31* (3/4), 1973, 6, 8-10; (5/6), 15-18.
[86] J. Willis and A.J. Bryant, Z. Metallk., *61*, 1970, 683-692.
[87] A.T. Male, Amer. Mach., *114* (16), 1970, 107-108; (17), 1970, 81-82.
[88] R. Chadwick, Metals Mat., *4*, 1970, 201-207.
[89] J.W.R. Naden, J. Iron Steel Inst., *193*, 1959, 278-284.
[90] A.W. Duffill and P.B. Mellor, Proc. Inst. Mech. Eng., *180*, Pt.3I, 1965-66, 260-269.
[91] V.V. Boitsov, Yu.G. Kalpin, and S.Z. Figlin, Kuzn. Shtamp. Proizv., *1971* (11), 3-5 (Russian).
[92] P.I. Polukhin, Yu.P. Glebov, and V.S. Gorakhov, Tsvetn. Met., *1976* (11), 58-60 (Russian).
[93] W.B. Rice and H.S.R. Iyengar, CIRP, *18*, 1970, 193-198.
[94] Yu.F. Cherny, V.M. Ershov, and V.S. Koviko, Kuzn. Shtamp. Proizv., *1976* (5), 19-21 (Russian).
[95] Yu.F. Cherny and Zh.N. Ognetova, Fiz. Met. Metallov., *35*, 1973, 1322-1325 (Russian).
[96] Yu.F. Cherny et al, Russian Eng. J., *54* (9), 1974, 67-68.
[97] A. Knauschner and R. Skrzypek, Neue Hütte, *17*, 1972, 283-287 (German).
[98] J.M. Alexander and B. Lengyel, Proc. Inst. Mech. Eng., *180*, Pt.3I, 1965-66, 317-327.
[99] W.R.D. Wilson and S.M. Mahdavian, in *Proc. 3rd NAMRC*, SME, Dearborn, 1975, pp.52-71.
[100] A. Wuerscher and W.B. Rice, Trans. ASME, Ser.B, J. Eng. Ind., *94*, 1972, 795-799.
[101] D. Egerton and W.B. Rice, Trans. ASME, Ser.B, J. Eng. Ind., *98*, 1976, 795-799.
[102] P. Krhanek and J. Jakes, in [44], pp.117-121.

[103] T. Yamada, T. Abe, M. Noguchi, and M. Oyane, Bull. JSME, *15*, 1972, 672-680.
[104] P.W. Wallace, K.M. Kulkarni, and J.A. Schey, J. Inst. Metals, *100*, 1972, 78-85.
[105] M.M. Kemal, Trans. ASME, Ser.F, J. Lub. Tech., *90*, 1968, 412-416.
[106] P. Dunn and B. Lengyel, J. Inst. Metals, *100*, 1972, 317-321.
[107] I. Perlmutter and V. DePierre, Metal Progr., *84* (5), 1963, 90-95, 128, 130, 132, 134, 136.
[108] C. Sauve, J. Inst. Metals, *93*, 1964-65, 553-559.
[109] M. Nishihara, M. Noguchi, T. Matsushita, and Y. Yamaguchi, in *Proc. 18th Int. MTDR Conf.*, Macmillan, London, 1978, pp.91-102.
[110] W.R.D. Wilson, Trans. ASME, Ser.F, J. Lub. Tech., *93*, 1971, 75-78.
[111] C.H.T. Pan, Tech. Rep. MTI-67-TR10, Mech. Techn. Inc., 1967.
[112] H.S. Iyengar and W.B. Rice, CIRP, *17*, 1969, 117-122.
[113] M.J. Hillier, Int. J. Prod. Res., *5*, 1967, 171-181.
[114] R.W. Snidle, B. Parsons, and D. Dowson, in [3.257], pp.107-117.
[115] R.W. Snidle, D. Dowson, and B. Parsons, Trans. ASME, Ser.F, J. Lub. Tech., *95*, 1973, 113-122.
[116] W.R.D. Wilson and S.M. Mahdavian, Trans. ASME, Ser.F, J. Lub. Tech., *98*, 1976, 27-31.
[117] S. Thiruvarudchelvan, Wear, *72*, 1981, 325-333.
[118] B. Avitzur, in *Proc. 17th Int. MTDR Conf.*, Macmillan, London, 1977, 445-451.
[119] R.W. Snidle, B. Parsons, and D. Dowson, Trans. ASME, Ser.F, J. Lub. Tech., *98*, 1976, 335-343.
[120] R.M. Guha and B. Lengyel, J. Inst. Eng. (India), *54*, 1974, 117-124.
[121] Y.C. Hsu, Trans. ASME, Ser.F, J. Lub. Tech., *92*, 1970, 228-242.
[122] W.R.D. Wilson, B.B. Aggarwal, A.B. Norelius, and H. Quist, in *Proc. 7th NAMRC*, SME, Dearborn, 1979, 129-136.
[123] J. Séjournet, in [3.50], pp.162-184.
[124] J. Séjournet and J. Delcroix, Lubric. Eng., *11*, 1955, 389-398.
[125] J. Séjournet, Rev. Met., *58*, 1961, 1029-1037 (French).
[126] E.K.L. Haffner and J. Séjournet, J. Iron Steel Inst., *195*, 1960, 145-162.
[127] J. Nittel, G. Gärtner, and J. Möckel, Neue Hütte, *10*, 1965, 647-649 (German).
[128] G.W. Rowe and J. Golden, Sci. Lubric., *13* (4), 1961, 18.
[129] J.A. Rogers and G.W. Rowe, Proc. Inst. Mech. Eng., *179*, Pt.3D, 1964-65, 93-102.
[130] C.M. Pierce, Tech. Rep. AFML-TR-67-83, 1967.
[131] V.F. Vdovin et al, Met. i Gorn. Prom., *1968* (3-4), 39-41 (Russian) (Brutcher Transl. No.7541).
[132] J. Séjournet, Mem. Soc. Ing. Civils Fr., Sect.V, *109* (5), 1956, 361-379 (French).
[133] Yu.V. Manegin, Kuzn. Shtamp. Proizv., *1976* (4), 13-16 (Russian).
[134] T.H.C. Childs, Metals Techn., *1*, 1974, 305-309.
[135] Yu.F. Cherny, N.D. Zolotukhina, and I.D. Pisarenko, Kuzn. Shtamp. Proizv., *1966* (4), 11-13 (Russian).
[136] B. Pocta, Hutn. Listy, *21*, 1966, 325-328 (Brutcher Transl. No.6916).
[137] K.E. Hughes, K.D. Nair, and C.M. Sellars, Metals Techn., *1*, 1974, 161-169.
[138] N.J. Parratt, in [3.51], pp.159-162.
[139] R.J. Courtney, J. Inst. Metals, *99*, 1971, 261-266.
[140] B. Jaoul and J. Séjournet, Metaux, No.406, June 1959, 221-235 (French).
[141] P. Baque, J. Pantin, and G. Jacob, Trans. ASME, Ser.F, J. Lub. Tech., *97*, 1975, 18-23.
[142] J. de Charsonville, E. Felder, and P. Baque, Rev. Met., *70*, 1973, 497-505 (French).
[143] W.D. Spiegelberg, C.R. Cook, and A.L. Hoffmanner, in [3.41], pp.53-65.
[144] H. Unckel, Aluminium, *25*, 1943, 342-346.
[145] L.N. Moguchin, Tsvetn. Met., *34* (3), 1961, 75-81 (Russian).
[146] C. Panseri, G. Bedeschi, and G. Bracale, Alluminio, *31*, 1962, 3-20 (Italian).
[147] L.N. Moguchin, Kuzn. Shtamp. Proizv., *1967* (9), 11-13 (Russian).
[148] P. Loewenstein, J.G. Hung, and R.G. Jenkins, *Refractory Metals and Alloys*, III, The Metallurgical Society AIME, Gordon and Breach, New York, 1965, pp.151-167.
[149] C.E.N. Sturgess and T.A. Dean, J. Mech. Work. Tech., *3*, 1979, 119-135.
[150] J.R. Johnson and W.R.D. Wilson, ASLE Trans., *24*, 1981, 307-316.
[151] V.I. Petukhow, O.V. Abramov, A.M. Zubko, and Yu.V. Manegin, Light Metal Age, *36* (6), 1973, 6-9.
[152] W. Kaczmar and H. Drzeniek, Rudy i Metale Niez., *15*, 1970, 376-381 (Polish).

[153] J.W. Pridgeon, R.S. Cremisio, and D.W. Mills, Light Metal Age, *36* (3/4), 1978, 20, 22.
[154] A.A. Nagaitsev, D.Kh. Piguzova, and L.V. Vainpres, Tsvetn. Met., *1975* (9), 56-60 (Russian).
[155] S. Raft and F. Kaszás, Kohaszat, *107*, 1974, 91-95 (Hungarian).
[156] N.J. Archer and K.K. Yee, Wear, *48*, 1978, 237-250.
[157] Anon., Stahl u. Eisen, *98*, 1978, 905.
[158] H. Leibold and B. Rönigt, Z. Metallk., *62*, 1971, 270-273 (German).
[159] A. Kolsch, Z. Metallk., *62*, 1971, 649-652 (German).
[160] H.L. Shaw, F.W. Boulger, and C.H. Lorig, First Eng. Summary Rep. Contr. AF 33(600)-26272, 1957.
[161] R.M.L. Elkan (Pt.I), and R. Cox (Pt.II), J. Iron Steel Inst., *202*, 1964, 236-245, 246-260.
[162] P.J. Sukolski and G. Hoyle, J. Iron Steel Inst., *193*, 1959, 270-277.
[163] J. Burggraf, Bänder Bleche Rohre, *11*, 1970, 559-564 (German).
[164] R. Cox, T. McHugh, and F.A. Kirk, J. Iron Steel Inst., *194*, 1960, 423-434.
[165] R. Hubert and H. Thannberger, Metal Progr., *84* (2), 1963, 74-76.
[166] Anon., SEAISI Quart., *1972* (1-2), 40-45.
[167] A.D. Roubloff, Iron Steel Eng., *40* (3), 1963, 79-88.
[168] K.E. Hughes and C.M. Sellars, J. Iron Steel Eng., *210*, 1972, 661-669.
[169] R. Buhlig, Neue Hütte, *22*, 1977, 189-192 (German).
[170] V.V. Lipinsky, N.V. Grishanin, and E.G. Tolmatskaya, Kuzn. Shtamp. Proizv., *1978* (9), 7-9 (Russian).
[171] I.I. Gurdus and F.P. Zheltogirko, Kuzn. Shtamp. Proizv., *1978* (9), 4-7 (Russian).
[172] E.R. Lambert, N.V. Bekaert, and B. Lengyel, CIRP, *22*, 1973, 75-76.
[173] A.F. Castle and T. Sheppard, Metals Techn., *3*, 1976, 465-475.
[174] T. Sheppard, P.J. Tunnicliffe, and S.J. Patterson, J. Mech. Work. Tech., *6*, 1982, 313-331.
[175] *Aluminum Extrusion Die Design,* Kaiser Aluminum, Oakland, CA, 1963.
[176] I.E. Boczor, Proc. Hung. Res. Inst. Non-Ferrous Metals, *1971*, 349-357.
[177] P.J. Thompson, J. Inst. Metals, *96*, 1969, 257-261.
[178] G. Merk and S.E. Naess, Z. Metallk., *68*, 1977, 683-687.
[179] M. Tokizawa, K. Dohda, and K. Murotani, Bull. Jpn. Soc. Prec. Eng., *10*, 1976, 145-149 (Japanese).
[180] M. Tokizawa, K. Murotani, T. Oyobe, and K. Dohda, J. Jpn. Soc. Prec. Eng., *8*, 1978, 55-61 (Japanese).
[181] A.I. Kemppinen, Metal Progr., *93* (6), 1967, 147-150.
[182] G.M. Bouton et al, Bell Syst. Tech. J., *34*, 1955, 529-561.
[183] J.F. Heffron and R.L. Hull, Light Metal Age, *34* (3/4), 1979, 31-32.
[184] Anon., Iron Age, *180* (Aug.29), 1957, 70-71.
[185] R. Akeret, Z. Metallk., *55*, 1964, 570-573 (German).
[186] Yu.P. Glebov and U.S. Gorokhov, Tsvetn. Met., *1975* (2), 67-69 (Russian).
[187] R. Chadwick, Metal Ind., *100*, 1962, 202-205.
[188] N.S. Rakhmanov et al, Tsvetn. Met., *1975* (10), 65-69 (Russian).
[189] C.J. Dangerfield and L. Gwyther, J. Inst. Metals, *100*, 1972, 233-238.
[190] R. Lugosi and A.T. Male, in *Proc. 6th NAMRC,* SME, 1978, pp.127-132.
[191] R. Lugosi and A.T. Male, ibid, pp.133-137.
[192] M. Seido, S. Mitsugi, and D. Oelschlagel, in *Proc. 2nd Int. Aluminum Extrusion Techn. Sem.,* Vol.1, Aluminum Assoc., Washington, 1977, pp.133-141.
[193] C. Blazey, Australasian Eng., Aug. *1953*, 53-56.
[194] T.F. McCormick, Trans. ASME, *75*, 1953, 1525.
[195] H.L. Shaw, F.W. Boulger, and C.H. Lorig, Summary Rep. Contract AF 33(600)-26272, 1955.
[196] J. Williams, Met. Rev., *3* (9), 1958, 1-44.
[197] W.W. Beaver, in *The Metallurgy of Beryllium,* Monogr. Rep. Ser. No.28, Institute of Metals, London, 1963, pp.781-807.
[198] J.M. Siergiej, in *Symposium on Newer Structural Materials for Aerospace Vehicles,* ASTM STP-379, 1965, 106-118.
[199] W.J. Wright and J.M. Silver, in *The Metallurgy of Beryllium,* Monogr. Rep Ser. No.28, Institute of Metals, London, 1963, pp.734-742.
[200] G. Griedman, Tech. Paper No. MF 67-204, ASTME (SME), Dearborn, 1967.

[201] A.T. Male, J. Inst. Metals, *93*, 1964-65, 288.
[202] A. Sandin, J. Inst. Metals, *90*, 1961-62, 439-447.
[203] M. Vater and K. Koltzenburg, Bänder Bleche Rohre, *11*, 1970, 587-595 (German).
[204] H. Hesse, J. Broichhausen, and P. Beiss, Metall, *36*, 1982, 363-368 (German).
[205] J.H. Cairns and D.B. Such, J. Inst. Metals, *98*, 1970, 289-298.
[206] B.W.H. Lowe, Wire, *1970* (110), 320-326.
[207] W. Svedberg and R. Erbe, Erzmetall, *24*, 1971, 534-537 (German).
[208] H.T. Larker and J.O.H. Nilsson, in [44], pp.161-165.
[209] K. Osakada and R. Asada, J. Mech. Work. Tech., *2*, 1977-78, 277-290.
[210] Yu.F. Shevakin, L.N. Orlov, and L.M. Grabarnik, Tsvetn. Met., *1972* (7), 79-81 (Russian).
[211] R. Lugosi, C.S. Hartley, and A.T. Male, in *Proc. 5th NAMRC,* SME, Dearborn, 1977, pp.105-113.
[212] V.V. Boyutsov et al, Kuzn. Shtamp. Proizv., *1975* (12), 1-3 (Russian).
[213] A.M. Sabroff, Metals Eng. Quart., *3* (2), 1963, 31-35.
[214] Anon., Steel, *155* (Dec.21), 1964, 46-48.
[215] A.M. Sabroff and P.D. Frost, Modern Metals, *13* (5), 1957, 50 ff.; *13* (6), July 1957, 52 ff.
[216] M. Markworth and G. Ribbecke, Metall, *28*, 1974, 777-796 (German).
[217] J. Thevenet and J. Buffet, Rev. Met., *59*, 1962, 553-561 (French).
[218] N.N. Folomin, V.G. Osintsev, and A.G. Egorov, Sov. J. Non-Ferrous Met., *1976* (10), 71.
[219] F.R. Lorenz, W.B. Haynes, and E.S. Foster, Trans. AIME, *206*, 1956, 1076-1080.
[220] C. Sauve, Mem. Sci. Rev. Met., *59*, 1962, 208-224 (French).
[221] A. Kaufman, P. Lowenstein, and J. Hunt, in *Fuel Element Fabrication with Special Emphasis on Cladding Materials,* Academic Press, London/New York, 1961, pp.325-341.
[222] R.W. Buckman, Int. Met. Rev., *25*, 1980, 158-162.
[223] C.E. Lacy and C.J. Beck, Trans. ASM, *48*, 1956, 579-594.
[224] V.P. Buryak, A.N. Gridnev, N.I. Matrosov, and Yu.M. Ivanov, Sov. J. Non-Ferrous Met., *1976* (10), 70.
[225] W. Hofmann, *Lead and Lead Alloys, Properties and Technology,* Springer, Berlin, 1970.
[226] W.W. Krysko and R.G. Fenton, Acta Techn. Acad. Sci. Hung., *57*, 1957, 345-372.
[227] W.W. Krysko and M.W. Lui, Z. Werkstofftech./J. Mat. Techn., *1*, 1970, 83-89 (German).

Forging

Forging, in all of its forms, is a batch process from both the mechanical and tribological points of view. Steady-state conditions are never achieved; during the stroke of the press or hammer the lubricant is always exposed to continuously changing pressures and velocities, and the initial lubricant supply must suffice for the duration of the operation. Lubricant residues and wear or pickup products are also continually changing, although a dynamic equilibrium may be attained through careful control. The absence of steady-state conditions accounts for the great difficulties encountered in the systematic analysis of the forging processes despite their often deceptive simplicity.

The subject matter of the present chapter includes all forms of forging, whether cold, warm, or hot, and whether conducted in an open die, an impression die, or a closed die. Forging often involves elements of extrusion; indeed, processes such as hot forging on an upsetter or cold forging on a press are often examples of direct and indirect extrusion in which non-steady-state effects associated with the initiation and end of extrusion dominate the situation.

In many ways, the various forging processes are competitive with one another, and the competitive position of each is greatly influenced by its lubrication system. Thus, hot forging followed by finish machining may be replaced by cold forging, with all its advantages (Sec.2.4), provided that a suitable lubricant can be found. Indeed, economy of production has often been the major impetus for the development of new forging processes and associated lubrication techniques.

Subject areas previously covered by Kalpakjian [1], Newnham [2], and Schey [3] are included here. Books [4-8], reviews [9], and conferences [10] dealing with hot forging and with cold forging [11-14] are available.

Because the variety of processes is so great, basic principles will first be discussed in terms of the frequently encountered and ostensibly simple problem of the axial upsetting of a cylinder, and then these principles will be applied to other processes.

9.1 Upsetting

Upsetting is the prototype of open-die forging operations in which the workpiece is deformed between two die faces that allow free deformation in some directions. Frequently, the end of a bar is upset while the bar is held in a die; the process is then equivalent to upsetting between one platen with sticking friction and another platen with end-face expansion. Upsetting is practiced cold or hot, from the miniature to the huge scale, which are exemplified, respectively, by the upsetting of a pin head and the compression of an ingot weighing in excess of 100 tons.

9.1.1 Upsetting of a Cylinder

In Sec.2.3.2 we used the upsetting of a cylinder to demonstrate the effects of friction (Fig.2.6) and shall now investigate it more closely.

Effects of Friction. Because the side surfaces can deform freely, the effects of friction are much more clearly visible in upsetting than in rolling. The workpiece deforms in a manner that minimizes the energy required, and this leads to complex interactions between material flow and friction.

The average strain, at any point of the stroke, is simply (Fig.9.1)

$$\varepsilon = \ln \frac{h_0}{h_1} = \ln \frac{A_1}{A_0} \tag{9.1}$$

where the new cross-sectional area A_1 is obtained from constancy of volume.

The average pressure is

$$p_a = \frac{P}{A_1} \tag{9.2}$$

where P is the force at any point in the stroke.

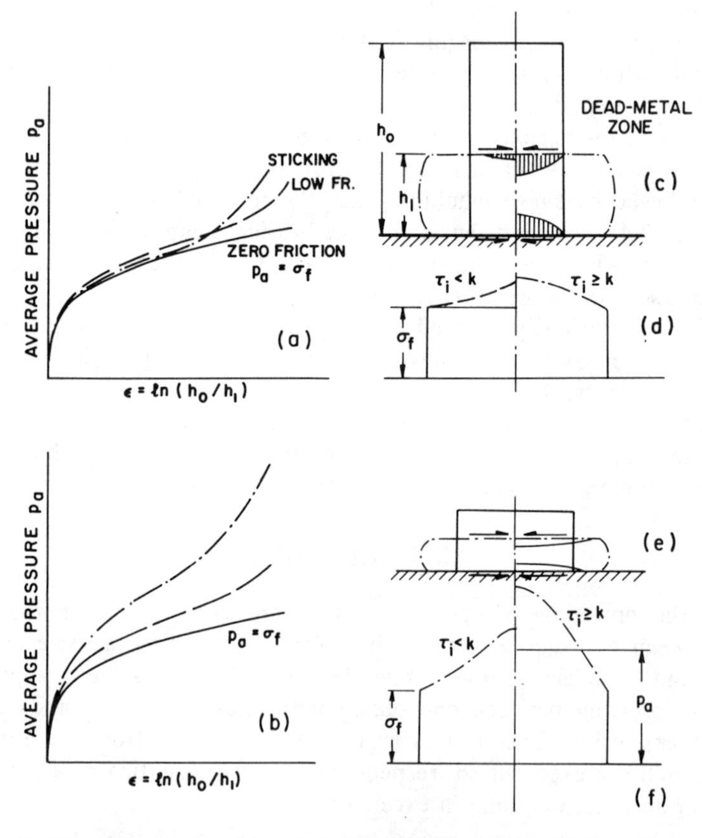

Fig.9.1. Increase in average die pressure with strain, workpiece deformation, shear stresses, and friction hill in upsetting of billets with small (a, c, d) and large (b, e, f) diameter-to-height ratios

As in rolling, pressures depend on both friction and process geometry (Fig.9.1a and b) and can be rationalized against the instantaneous d/h ratio. Several possibilities exist:

1. In the absence of friction, the end face expands freely, the workpiece remains cylindrical, and $p_a = \sigma_f$ (Fig.9.1a and d).

2. With low friction, expansion of the end face is opposed by the frictional shear stress τ_i, which is much less than k; therefore, even though some barreling is noticeable, the end face slides on the die from the center point (which can be regarded as a neutral point) outward (Fig.9.1c), and a low friction hill (Fig.9.1d) develops. The average pressure is only slightly (say, 5%) higher than the flow stress (Fig.9.1a). If friction is expressed by μ, the calculated friction hill is exponential in shape.

3. When the d/h ratio is large enough, the condition for sticking, $k = \mu p$, is satisfied at some point and a small dead-metal zone forms (left-hand side of Fig.9.1c) even with low friction [2.18]. Sliding is now limited to an annulus, and, according to Shutt [15], the width of this zone can be used to calculate μ at the onset of yielding. The test must be repeated at several reductions and the results extrapolated to zero reduction. Of course, the extrapolation is invalidated if the lubrication mechanism changes during compression. The frictional drag can be modeled also by m or a mixture of m and μ (Unksov [2.26]; Wanheim and Bay [3.102]). The friction hill becomes rounded and p rises more steeply.

4. When $\tau_i > k$ at all points—for example, in upsetting with rough, unlubricated dies—sticking prevails from the edge in, and a workpiece with a low initial d/h ratio deforms highly inhomogeneously (Fig.9.1a and right-hand side of Fig.9.1c), with most of the end-face growth resulting from a folding over of the side surfaces. This is evident from the lack of growth of the original end face in hot forging (Fig.9.2) [16]. The friction hill might be expected to rise steeply, but in practice it is observed to remain quite low, because deformation of the outer fibers of the workpiece occurs by simultaneous compression and bending [17]. Average pressures can even dip below those observed in upsetting with lubricated end faces (Fig.9.1a) [17,18], although not below σ_f. At larger d/h ratios (right-hand side of Fig.9.1e) the dead-metal zones interpenetrate and the end face begins to expand, irrespective of the initial d/h ratio, when the instantaneous d/h reaches about 4 [16],[2.26]. Interface pressures now rise steeply (Fig.9.1f), and so does the average pressure (Fig.9.1b). Simple slab or upper bound solutions indicate that pressure in the friction hill should rise linearly, but all direct measurements show a rounding off of the pressure peak (Fig.9.1f), indicating that analyses based on decreasing interface shear stress toward the neutral point are more relevant. Recent numerical solutions also confirm this point.

The above observations have a number of important practical implications.

Pressures and Forces. The distribution of normal pressure over the contact surface is of substantial theoretical interest, because it reflects the effects of friction. Measurements have been made by all of the techniques discussed in Sec.5.5.2 and 5.5.3, and the summary given above is based on measurements utilizing various forms of the pressure pin [19,20],[2.33],[5.97], photoelastic [2.26], deforming rubber tool [3.102], grooved sheet [5.105], and ruby insert [5.106] techniques. Pressure distribution is of practical interest also when elastic deformation of the tooling and its effect on dimensional accuracy of the forging are of sufficient concern to be compensated for in the course of die design.

Of great practical importance is the average die pressure p_a (Eq.9.2). Conveniently, it is expressed as a Q_a multiple of the flow stress σ_f prevailing at the point of interest during the stroke

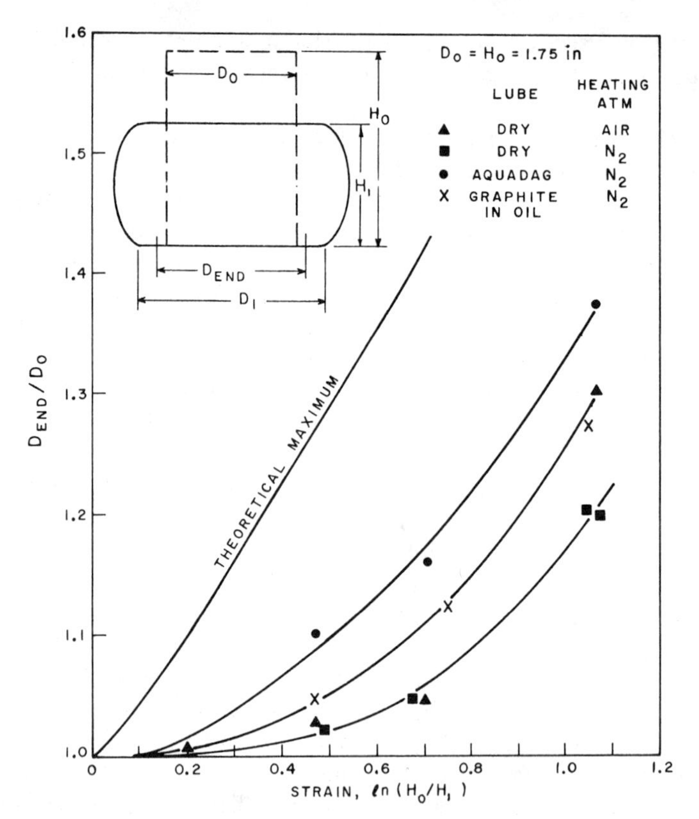

Fig.9.2. Effects of atmosphere and lubrication on end-face expansion in hot upsetting of type 1045 steel cylinders (press speed, 0.76 m/min) [183]

$$p_a = Q_a \sigma_f \tag{9.3}$$

There are two important applications.

1. Prediction of Forces. Pressure-multiplying factors Q_a can be obtained from various theories, based either on the assumption of a constant μ in the sliding zone (with due allowance for the dead-metal zone), or on a constant m (which gives simply a straight-sided friction hill and does not allow for partial sticking). Both approaches ignore inhomogeneity of deformation, and thus grossly overestimate Q_a at low d/h ratios; therefore, the empirical maxima given by Schey et al [17] should be taken as limits (Fig.9.3). At high d/h ratios deformation is more homogeneous (Fig.9.1e), theoretical solutions are more valid, and average pressures indeed reach high values, which accounts for the difficulty of forging thin shapes. The upsetting force P_a also becomes very high because of large Q_a and A_1

$$P_a = p_a A_1 = Q_a \sigma_f A_1 \tag{9.4}$$

Lubrication is then essential, although a marked improvement can be achieved only when μ is very low (Fig.9.3).

In principle, the calculation could be made in reverse, and friction could be evaluated on the basis of measured upsetting forces. One can upset with a constant force and measure the resulting deformation, or measure the force required for upsetting to a predetermined height. Either method is suitable for an over-all ranking of lubricants, but the extraction of μ or m values is fraught with danger. As can be seen in Fig.9.3, simple theories are unreliable, and the more complex ones (such as finite element analyses) are not available in a convenient form—especially for mixed sliding/sticking friction and for folding over. Furthermore, even if the theory were reliable, σ_f (and its change from point to point in the billet) would have to be known with sufficient accuracy. This is seldom the case and, especially in upsetting with good lubrication, the uncertainty of σ_f is sufficient to swamp the effects of friction. Therefore, μ and m values reported on this basis should be regarded with reservations, and should be set against the known lubrication mechanism.

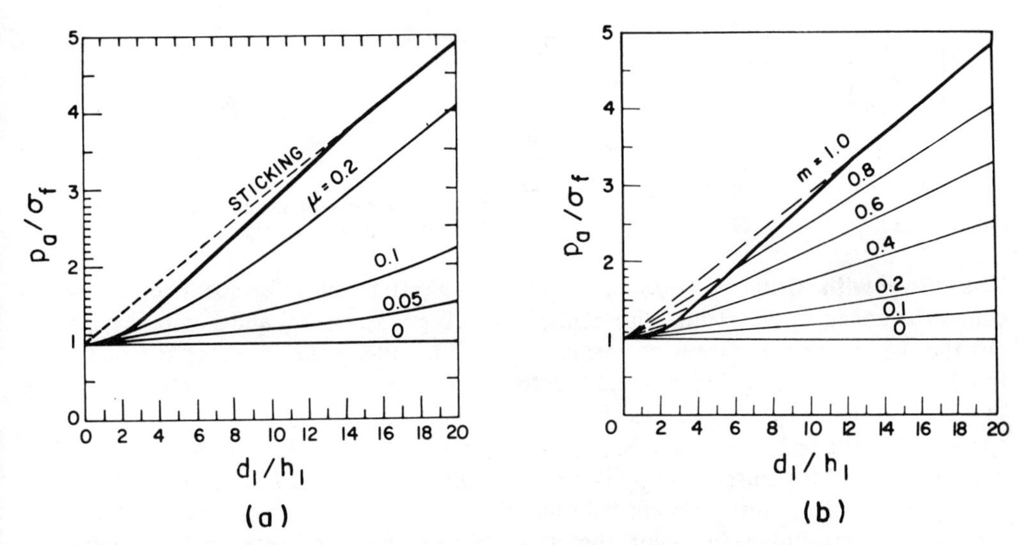

Fig.9.3. Pressure-multiplying factors for predicting average die pressures in upsetting of cylindrical billets [17]. (a) Using a coefficient of friction. (b) Using an interface shear factor. Empirical maxima are shown by heavy lines.

2. Determination of Flow Stress. The uniaxial compressive flow stress σ_f is a basic property that is needed for all analyses of deformation processes, and its unequivocal determination is of fundamental significance. Friction at the interface is inevitable and must be accounted for if a valid σ_f is to be extracted from measured p_a values.

a. If the magnitude of friction is known (e.g., from a ring-compression test or from pressure-pin measurements), the measured p_a can be reduced by Q_a. Simple theories greatly overestimate Q_a at the low d/h ratios typical of flow-stress determination, and σ_f will be unrealistically low unless empirical Q_a values (Fig.9.3) are used.

b. As recognized originally by Polakowski [21], and recently by Herbertz and Wiegels [18] and by Schey et al [17], the effect of friction can be practically ignored (Fig.9.1a) if upsetting with sticking friction is limited to a final d/h ratio of 2 or less. This allows σ_f determination to a sufficient reduction if the initial d/h was 0.5 or, if buckling is of concern, 0.67.

c. While still keeping the final d/h ratio small, one can attempt to ensure close-to-perfect lubrication. This can be done by replenishing the lubricant in the course of incremental upsetting (which is rather a nuisance), by grooving the end face of the billet to entrap a liquid lubricant (Fig.9.4a), by shot blasting the billet surface for the same purpose, or by providing an ample supply of entrapped lubricant. A hydrostatic lubricant cushion can be maintained by means of a rimmed specimen (Fig.9.4b) as originally proposed by Rastagaev and further investigated by Pöhlandt [22]. A conical impression (Fig.9.4c) may be used on more brittle materials where the rim would split [18]. In any event, the die surface should be smooth and polished, and the friction contribution will then be low enough (Fig.9.1a) to be ignored.

d. Other methods rely on extrapolation or on special specimen geometries but suffer from disadvantages (reviewed in [17,21]) and can be quite laborious; they are justified only for basic studies. Real problems are encountered at strains $\varepsilon > 1$, and new specimen geometries have been sought [23] which, however, still cannot ensure homogeneous deformation.

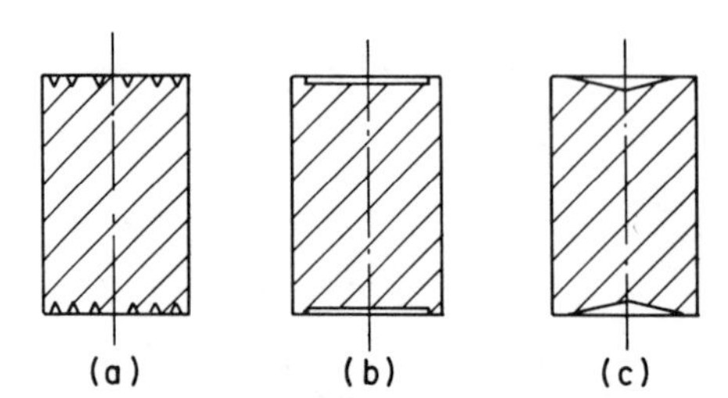

Fig.9.4. Means of promoting full-fluid-film lubrication in upsetting

Upsetting with Conical Dies. The restraining effect of interface friction can be counterbalanced by upsetting with conical dies (Fig.9.5). Siebel and Pomp [24] pointed out that when the component of normal force acting along the interface is equal to the frictional force, the cylinder should deform uniformly; when the frictional force is higher, barreling occurs, whereas if the radial force component is higher, the end faces spread. However, the artificial geometry could severely interfere with the lubrication mechanism, and measurements by Dobrucki and Odrzywolek [19,25] show that even the basic premise of the test does not hold up. Thus, the technique is of limited validity but highlights a very important point: the effect of friction can be neutralized and material flow in a radial direction encouraged by the use of inclined surfaces. This principle is the basis of much impression-die forging design.

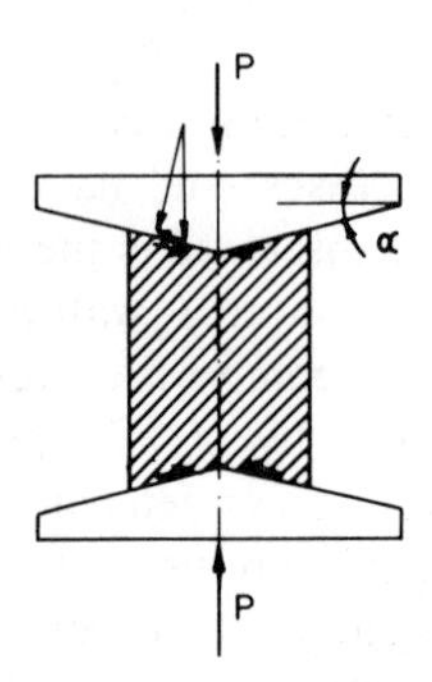

Fig.9.5. Upsetting between conical platens (after [24])

Barreling is important for a number of reasons.

a. It can be used as a quantitative indicator of friction [26], although folding over creates complications. Kulkarni and Kalpakjian [27], who presented also a review of the subject, showed that the profile of barreled specimens could be closely approximated by an arc of a circle R_c (Fig.9.6). The relative curvature $H_{ur} = H_f/R_c$ was found to increase with increasing reduction and friction, whereas aspect ratio had little influence.

b. Barreling is of interest for predicting the length of billet needed for producing a cylinder of given size; data are available for sticking friction [28].

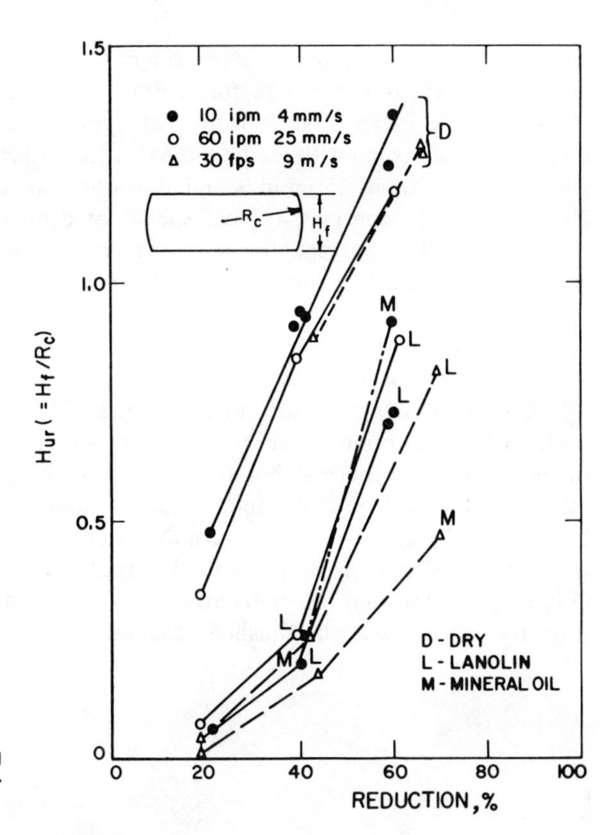

Fig.9.6. Effect of lubrication on barreling in upsetting of specimens of aluminum alloy 7075 at room temperature [27]

Formation of Defects. The inside of an upset cylinder is in a compressive stress state, and therefore friction-related defects are limited to the surfaces.

a. Barreling generates tangential (circumferential) secondary tensile stresses which may lead to cracking on the free surfaces of upset cylinders if the workpiece has only moderate ductility. Combined with the temperature rise due to the work of deformation (Sec.2.3.3), cracking may be particularly severe in hot forging of materials with narrow hot working temperature ranges. Cracking may then be avoided by the use of lubricants that lower friction sufficiently to ensure near-homogeneous deformation.

b. Cracks on the interface are rare, but can occur if the center of the workpiece is constrained by a rough die while the edges can move out freely on a well-lubricated surface. Discontinuity of material flow then creates tensile surface stresses which lead to radial cracks [29,30].

Effects of Temperature. As in other plastic deformation processes, the temperature of the workpiece rises because of plastic deformation (Eq.2.15) and interface friction. Thus, in cold working the workpiece is warmer than the die. However, in hot working under nonisothermal conditions the hot workpiece loses some of its heat to the die, while the die heats up, at a rate determined by a number of factors:

a. The rate of heat transfer increases rapidly as the pressure is applied, because conformity is established and insulating surface films are broken through; the effects of various lubricants will be discussed in Sec.9.5.

b. Heat loss increases with increasing temperature differential between workpiece and die. Is is greatest in conventional hot forging and is absent in isothermal forging.

c. A most important factor is contact time. This factor depends on die approach velocities, which vary over a wide range depending on the forging equipment. Speeds range from 25 to 150 mm/s for hydraulic presses, but the hydraulic system usually incorporates a valve to shift from fast approach to forging speed, and the associated dwell increases contact times to values on the order of 0.5 s. Mechanical presses hit the workpiece at around 250 mm/s and slow down to zero speed, but contact times are only 20 to 30 ms. In contrast, typical speeds of drop hammers and power hammers are 3 m/s and 6 to 9 m/s, respectively, giving contact times of only a few microseconds. Yet, the speeds attained in high-energy-rate forging (HERF) units, which are 3 to 5 times higher, have not brought the anticipated benefits of even shorter contact times, mostly because of die wear problems. The wide range in contact time has an effect not only on cooling but also on lubrication mechanisms.

Chilling strengthens the surface layers of the billet and thus has the same effect as an increase in friction; therefore, separation of friction and cooling effects is extremely difficult in nonisothermal hot forging. From the practical point of view, of course, it matters little whether movement of the workpiece over the die surface is arrested purely by friction, by chilling, or by a combination of the two; it is the end result in terms of material flow that is important. Internal deformations become highly nonuniform (Fig.9.7) [3.123], and the concentration of strain in shear bands can lead to undesirable property variations in the finished product.

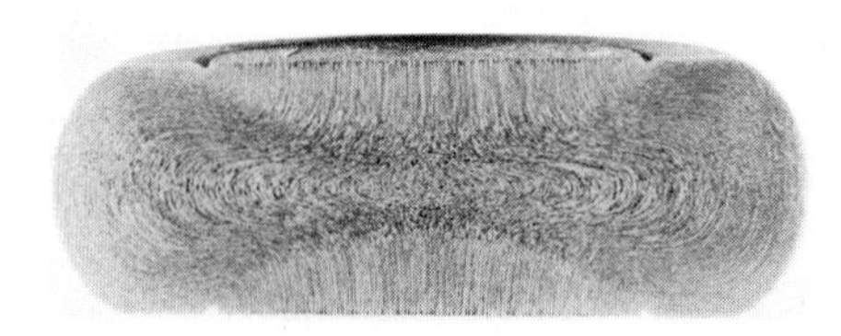

Fig.9.7. Folding over observed in unlubricated nonisothermal hot upsetting of steel [3.123]

9.1.2 Ring Compression

Compression of rings offers the great advantage that frictional conditions can be judged from deformation alone, and a knowledge of σ_f is not needed.

Ring Deformation. With zero friction at the interface, the ring expands as though it were part of a solid disk, and velocities increase radially over the entire surface. With increasing friction, it takes less energy for some of the material to flow toward the center, the diameter of the hole grows less rapidly, and a neutral line (neutral circle) develops (Fig.9.8a). With yet higher friction the internal diameter decreases, and both internal and external surfaces barrel (Fig.9.8b). This test is thus the forging equivalent of forward slip measurement in rolling, and its attraction for lubricant evaluation is

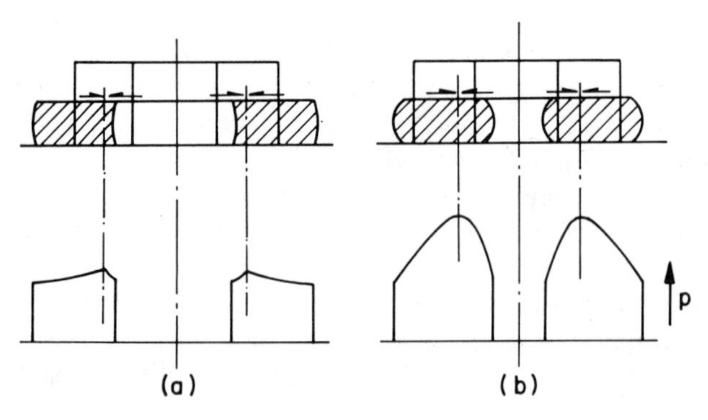

Fig.9.8. Deformation of a ring and pressure distribution in upsetting with (a) low friction and (b) high friction

obvious. If the specimen geometry is kept constant and the reduction in height is exactly reproduced, observation of the change in internal diameter is sufficient for ranking of lubricants; a lesser decrease in diameter indicates a lesser resistance to shear and thus a lower friction value. Therefore, this technique, proposed by Kunogi [31] for cold working and further developed and adapted for hot working by Male and Cockroft [32], has been used extensively and has spawned a voluminous literature.

Friction can be quantified if one of the theories linking the position of the neutral circle to the frictional balance is adopted. Available solutions are based on an average external μ [33], on upper bound analyses (reviewed by Avitzur et al [34]), and on finite element techniques [35], some of which allow analysis of hot working problems [36]. The solutions are generally based on the change of internal diameter, and two examples are illustrated in Fig.9.9. Some solutions allow for barreling of the surfaces, but only finite element techniques can admit folding over [36]. The effect of chilling on the die can also be considered [36,37]. Because there is no easy method of calibration, it is difficult to choose the most valid theory, and derived average μ or m values can easily vary by ±50% in response to just the choice of theory.

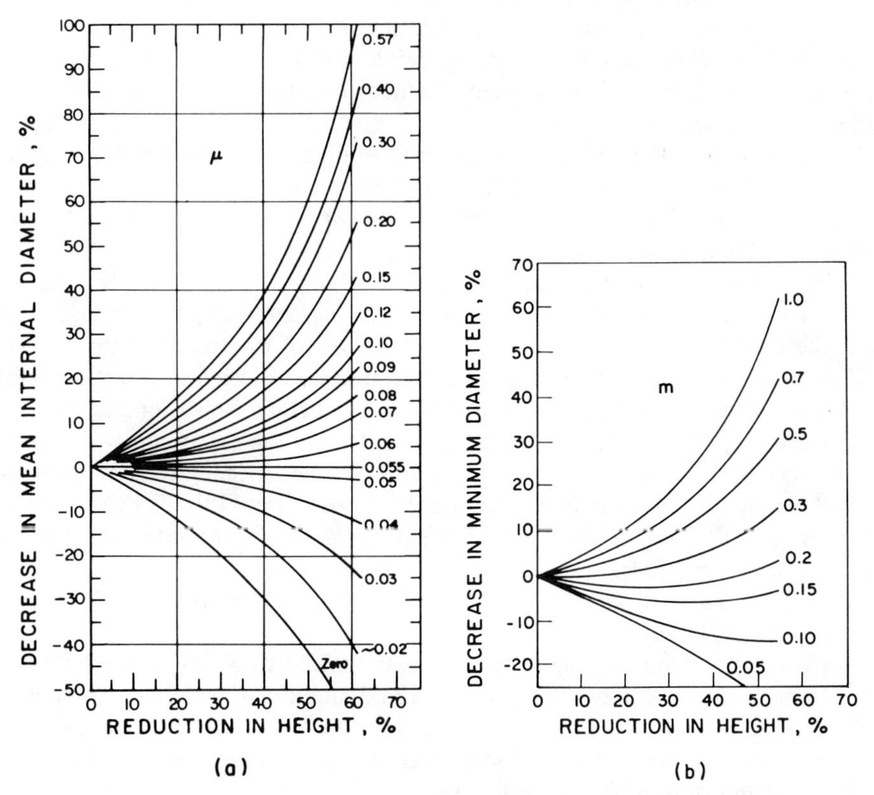

Fig.9.9. Calibration curves for extracting (a) coefficient of friction (from experimental calibration [32]) and (b) interface shear factor (from theory [37])

Determination of Flow Stress. Because the deformation of the ring provides its own calibration for friction, ring compression would appear extremely attractive for flow-stress determination with compensation for friction effects (see, for example, [38]). For this, a theory linking σ_f and friction (or, more directly, hole contraction) is needed. Data points calculated by Saul et al [39] on the basis of one of the theories of Avitzur

[34] are convenient for plotting into curves from which the pressure-multiplying factor can be read directly [2.12],[40]. Unfortunately, all but the numerical solutions appear to overestimate greatly the effect of friction on interface pressures, as shown by experiments with artificially roughened anvils and complete sticking. Numerical methods require considerable calculating effort and, until more accurate theories become available, it is safer either to minimize friction to the point where an expansion of the hole is secured or, instead, to rely on simple upsetting for flow-stress determination.

Test Methods. Results obtained by various authors are directly comparable as long as relative specimen geometries are identical. The inside diameter is usually chosen as one-half the outside diameter, while the thickness of the specimen influences the homogeneity of deformation and thus interface sliding. A favorite geometry employs a height equal to one-third the outside diameter (a ring of 6:3:2 proportions). Male and Cockroft [32] used this geometry for an empirical calibration with the aid of cylinder upsetting from which mean coefficients of friction were deduced. At high friction and reduction the ring closes up, and therefore Löwen [41] utilized a ring with a ratio of 3.5:2:1.

The effect of the height-to-diameter ratio on the deformation of the ring led Avitzur and Kosher [42] to recommend that friction be evaluated by conducting tests with rings of identical diameters but differing heights; if the internal diameter of the ring remains unchanged, friction is uniquely defined, because the neutral circle is right at the edge of the inner bore.

The expansion of the end faces often leads to changes in the lubrication mechanism in the course of compression, and following these changes the test may be repeated at different reductions. Evaluation in terms of μ or m is difficult, because the history of deformation affects these values.

In hot working with nonisothermal dies, chilling of the relatively thin ring may overwhelm effects of friction. However, this may be a blessing in disguise, because it reflects the conditions that are encountered in forging of workpieces with thin webs.

Very often, the ring becomes oval either because friction is directional (and this offers a method of studying the effect of, say, die surface finish) or because the workpiece material has an oriented structure. Differences in the expansion of top and bottom faces can usually be traced to differences in lubrication, although materials with a negative strain-rate sensitivity may exhibit similar asymmetry [43]. The very sensitivity of the test makes it suitable for quick evaluation of many variables by the statistical design of experiments [44].

Taller rings suffer a variety of instabilities, many of which are influenced also by interface friction [45,46].

In a variant of the ring compression test, described by Veiler and Likhtman [5.20], the specimen is upset between two punches the diameters of which are smaller than the outside diameter of the specimen but larger than the hole. Lubricant efficiency is again judged from the change in hole diameter, but the restraint provided by the overhanging material must interfere with such evaluation.

9.1.3 Upsetting of Slabs

When a rectangular slab is upset in plane strain between flat anvils that overhang the entire slab (Fig.9.10), flow is directed away from the centerline (the neutral line) in two opposing directions. Just as in rolling, the neutral line widens into a neutral zone when sticking sets in over part of the surface. In practice, plane strain is imposed on some parts of a forging (say, the shank of a connecting rod) when adjacent parts prevent a change in dimension (as in Fig.2.2, where the die provides the restraint).

Frictional resistance again results in a friction hill (Fig. 9.10), and the pressure

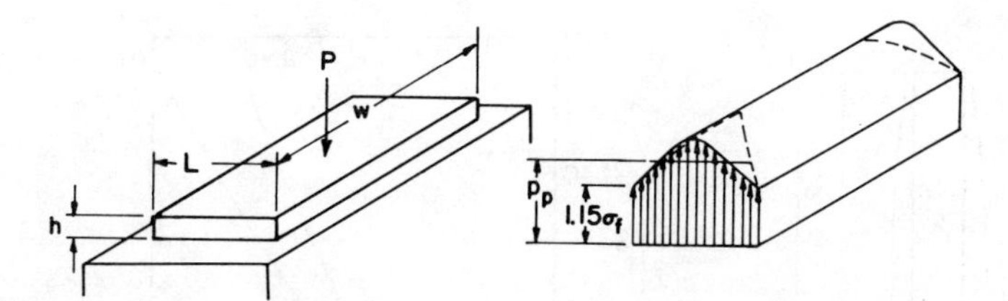

Fig.9.10. Upsetting of a flat slab with overhanging anvils [2.11]

distribution and average pressure p_p can be found as in the upsetting of cylinders. The difference is that for the same μ or m value, average pressures will be higher (Fig.9.11) because the hill is now a ridge rather than a cone [47].

It is important to recognize that the major material flow always takes place in the direction of lesser resistance—i.e., along the shorter dimension—and that it is this contact length L that is the basis of analysis.

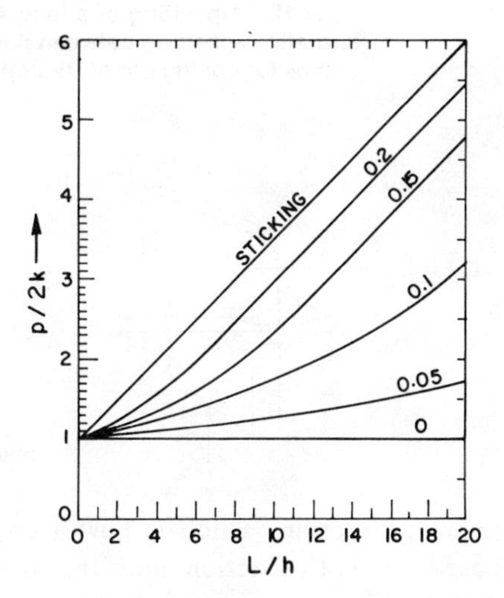

Fig.9.11. Pressure-multiplying factor for predicting average die pressure in upsetting of a slab (after [47])

In the absence of restraint, deformation takes place in the width direction, too. Because of the much longer sliding distance, such spread is very sensitive to low values of friction, and this led Hill [48] to propose its use for lubricant evaluation. The workpiece must be at least 10 times wider (longer) than the L dimension; its height should be similar to L ($h_0 = L/2$ in Fig.9.12). The middle of the specimen deforms most, resulting in the characteristic cigar shape which gives its name to this test (Fig.9.12a). Analysis by Fricker and Wanheim [49] confirms the sensitivity of the test to low friction values (Fig.9.12b). For w/L values below 10 an analysis was done by Wilson and Whitworth [50].

This test loses its sensitivity at high friction values, and this led Kosher and Avitzur [51] to use the degree of folding over as a measure of friction in strip (and also disk) forging.

The importance of frictional sliding distance is visibly demonstrated when a square or triangular prism is compressed. Because of the greater frictional resistance in the diagonal directions, the specimen gradually assumes a circular shape.

9.1.4 Plane-Strain Compression

Plane-strain deformation is approximated also when a large workpiece is compressed between narrow anvils (Fig.2.2b and Fig.9.13). Major material flow is again in the

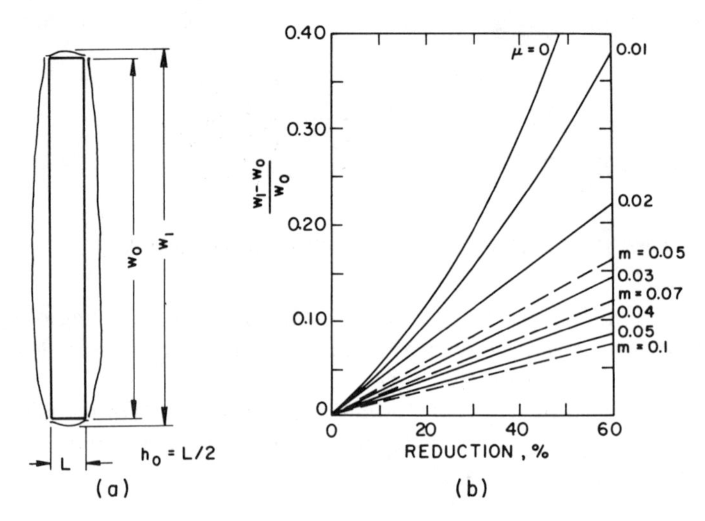

Fig.9.12. Upsetting of a long slab (lamina) [49]. (a) Plan view before and after upsetting between parallel anvils. (b) Theoretical calibration curves for coefficient of friction μ and interface shear factor *m*.

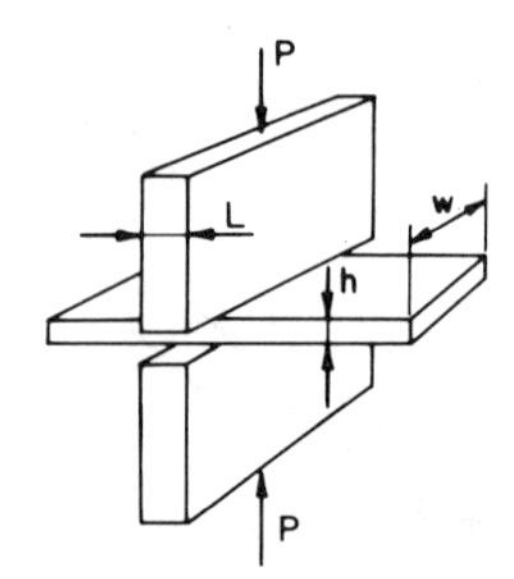

Fig.9.13. Schematic illustration of the plane-strain compression test (after [52])

shorter L direction, which is now a constant, given by the anvil dimension. Spread is opposed by both friction and the physical restraint imposed by the nondeforming material adjacent to the deformation zone.

Interface Pressures. A complication arises when the workpiece is relatively thick and $h/L > 1$. As discussed in conjunction with Fig.2.7, the deformation zones fail to interpenetrate, and a pressure higher than σ_f (or $2k$) must be applied. Because this pressure increase has nothing to do with friction, all situations must be carefully analyzed, the direction of major material flow found, and, from this, L determined. If $h/L > 1$, the pressure increase is due to geometrical factors and is affected little, if at all, by friction (Fig.9.14). If, on the other hand, $L/h > 1$, friction becomes dominant, and the curves in Fig.9.11 can be used to predict average pressures.

The plane-strain compression test is suitable for determination of the plane-strain compressive flow stress if, following Watts and Ford [52], $w > 8L$ to ensure plane-strain conditions, and the L/h ratio is kept between 2 and 4 so that friction effects remain minor in the presence of a very good lubricant.

Determination of Friction. Measured average pressures can be used to back-calculate an average μ or m provided that the unknown σ_f can be eliminated. This can be achieved, following the method of Alexander [53], by compressing the same specimen between two anvils chosen to give L/h ratios of 3 and 7 at the same strain. Difficulties are encountered because deformation is not quite homogeneous in compression with the

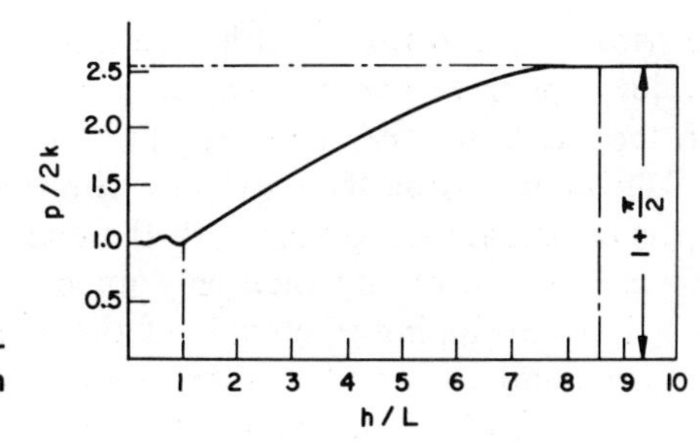

Fig.9.14. Pressure-multiplying factor for predicting average interface pressure in plane-strain indentation with or without lubrication [2.19]

narrower anvil; furthermore, lubricant entrapment is also less efficient and thus friction tends to be higher, leading sometimes to a crossover of the two pressure curves from which friction is to be found. The same problem surfaces in the technique of Rowe [2.23] in which a linear approximation to the exponential friction hill is utilized for a direct derivation of the coefficient of friction from

$$\mu = \frac{2h\left(p_2 - p_1\right)}{p_1 L_2 - p_2 L_1} \tag{9.5}$$

where h is the common thickness to which a plate is compressed with anvils of different dimensions L_1 and L_2, generating the corresponding pressures p_1 and p_2.

Direct measurement of the average external friction can be done by a technique similar to that shown in Fig.5.29d, but with rectangular workpieces [54].

A relative ranking of lubricants may be obtained by presetting the compression force and measuring the issuing thickness or, conversely, by compressing to a preset thickness and measuring the maximum force developed. If average pressures are to be calculated, spread should be taken into account.

Deformation. Sectioned specimens show that almost laminar flow is possible with very good lubricants, but sticking zones develop when high friction is combined with cooling. The profiles of well-lubricated specimens reveal internal shear along velocity discontinuities; Thomson [55] found this effect to be greatest at integral L/h ratios.

Spread in plane-strain compression reflects the ease with which the workpiece can slide in the width direction on the anvil; in this sense, spread is similar to that found in the cigar test but is reduced by the restraint of adjacent material. Lonn and Schey [56] observed that spread and, in particular, the shape of the spread zone are indicative of friction, with a gradual spread pattern developing under conditions of low friction and very localized spread with high friction (Fig.9.15).

Spread assumes significant proportions if the workpiece is too narrow relative to its thickness to ensure close enough approximation to plane-strain conditions. This is typical of the process of cogging or drawing out, where the width and height of the workpiece are similar. As in rolling, spread increases with increasing friction.

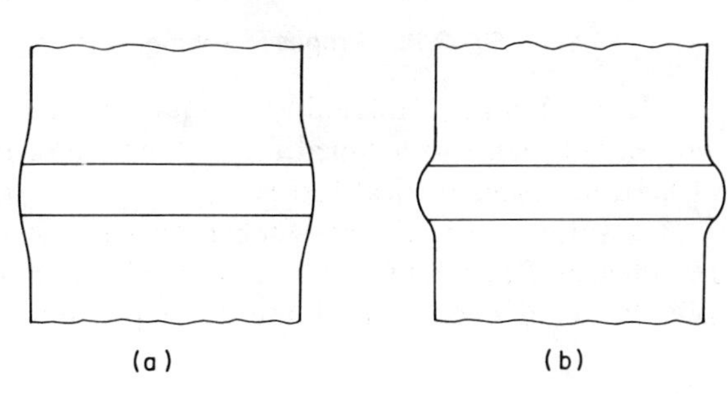

Fig.9.15. Spread in plane-strain compression with (a) low friction and (b) high friction (after [56])

Interface Conditions. In both upsetting and plane-strain compression a liquid lubricant film develops in response to velocity, viscosity, and pressure, as given in Eq.3.30. However, the similarity ends here (see Fig.5.2).

During upsetting the entrapped film remains entrapped and, depending on its shear properties, expands together with the end face to a greater or lesser degree (Fig.5.2a). It covers the entire end face only under exceptional circumstances, such as when the original or developing roughness of the end face helps in transporting the lubricant. In most instances, a boundary-lubricated or unlubricated annulus forms and one cannot speak of uniform lubricating conditions over the entire interface.

In contrast, in plane-strain compression the deforming material flows out from between the anvils and in doing so carries with it a lubricant film (Fig.5.2c). Thus, full-fluid or mixed-film lubrication can be maintained right up to the edge of the anvil, where decompression causes a collapse of the film. At this point, pickup is also observed if the workpiece/die combination is prone to it, and thus additional information can be obtained without jeopardizing the lubrication mechanism operative over much of the surface. On continued compression, a boundary-lubricated zone develops near the edges, and the hydrostatic zone is limited to the center.

Axial upsetting of a cylinder, ring compression, and plane-strain compression are suitable for experiments in the cold, warm, and hot working temperature ranges, as long as the limitations imposed on the lubrication mechanism by the mode of contact and deformation are recognized (Sec.9.5). Special die construction is needed for heated-die and isothermal studies.

9.1.5 Inclined Die Faces

Forging of a rectangular slab between inclined die surfaces results in a shift of the neutral plane toward the converging side and an increased material flow toward the expanding side. This principle is widely utilized when material movement in a certain direction is desired. It is also suitable for lubricant evaluation, because the position of the neutral plane depends not only on the angle of inclination but also on frictional resistance (Fig.9.16a), just as it does in rolling. A better lubricant allows more material flow in the widening direction, and lubricants can be ranked by simply observing flow, or by determining average external μ or m values from an appropriate theory.

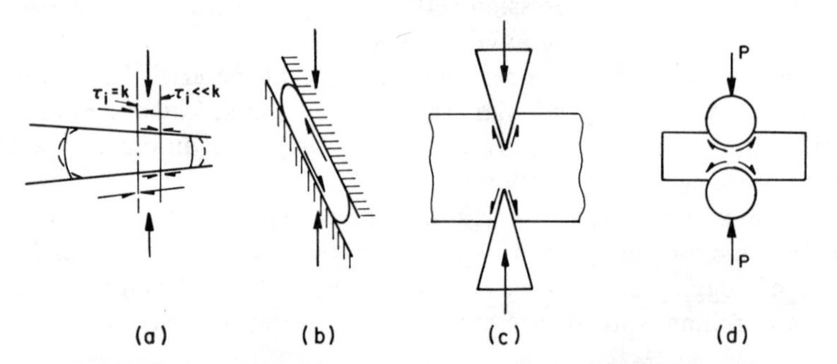

Fig. 9.16. Proposed test methods for evaluating forging lubricants

In a somewhat related technique introduced by El-Magd [57], a sheet is sandwiched between two inclined faces (Fig.9.16b); μ can be back-calculated from the ratio of horizontal and vertical forces.

A variant of the cone-indentation test, developed by Tamura and Kudo, and also reported by Kudo [58] in a review article, employs two wedges with included angles of, say, 30° (Fig.9.16c). The force developed during penetration may be calculated from slip-line field theory, and this test seems to retain sufficient sensitivity even for higher friction values.

The technique of Kravcenko [59] uses two cylindrical anvils (Fig.9.16d); since the indentation force is dependent on penetration, a constant penetration is standardized. The indentation pressure is first determined for sticking friction with a roughened indenter, and then with a smooth indenter and the experimental lubricant. A coefficient of friction may be calculated. Alternatively, the depth of penetration can be measured for a constant load [4.185]. Because the lubricant thins out around the nose, this test has been found to be useful for fast and simple ranking of lubricants, particularly at elevated temperatures.

9.2 Back Extrusion and Piercing

In a very important group of processes a punch is used to create a hollow workpiece.

9.2.1 Back Extrusion

When the workpiece is supported in a container and the punch is relatively large in diameter, the process is a variant of back or reverse extrusion (Fig.9.17a) and is used to produce tubular, canlike products such as tube blanks, pressure vessels, toothpaste tubes, and similar shapes. The extrusion pressure (acting on the container cross-sectional area A_0) is obtainable from Eq.8.4, but care must be taken to use the tube (and not the punch) cross section as A_1. As shown by measurements [8.24],[60-61a], friction between the extruded tube and the container could become substantial, and therefore the container is kept as short as possible; similarly, the punch diameter is reduced after a short land. This solution is, however, not practicable for hot working, because the product would shrink on the punch; instead, a detachable punch nose of larger diameter can be used if the billet is pierced right through. Alternatively, friction on the container can be made to aid extrusion by moving the container in the direction of can motion but at a higher speed (Fig.9.17b). Reduced friction makes for lesser inhomogeneity of material flow [61a].

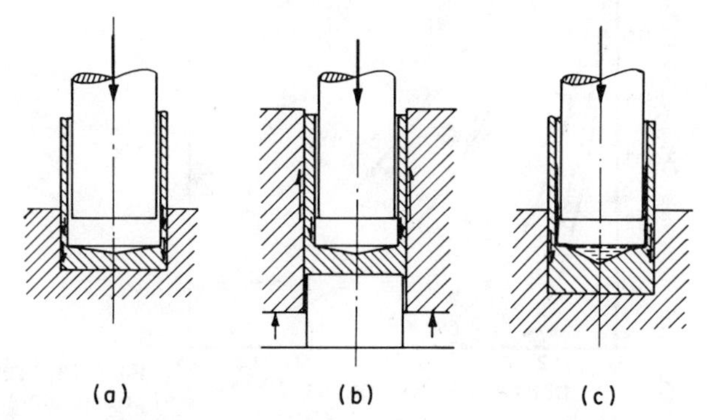

Fig.9.17. Variants of back extrusion (arrows show shear stresses imposed on material)

Penetration of the punch into the workpiece creates extremely severe interface conditions because surface extension is large and the lubricant trapped under the punch must extend, under high pressures, to follow the growth of new surface. Surface extension is a function of punch profile, as can be seen by following the deformation of a ring scribed on the end face of the workpiece (Fig.9.18) [62]. Correspondingly, lubricant thinning is also a function of punch profile (Fig.9.19). Ideally, the lubricant should be metered out gradually to provide uniform coverage of the freshly developing surfaces while protecting the punch face until the end of the stroke. It will be noted that for a

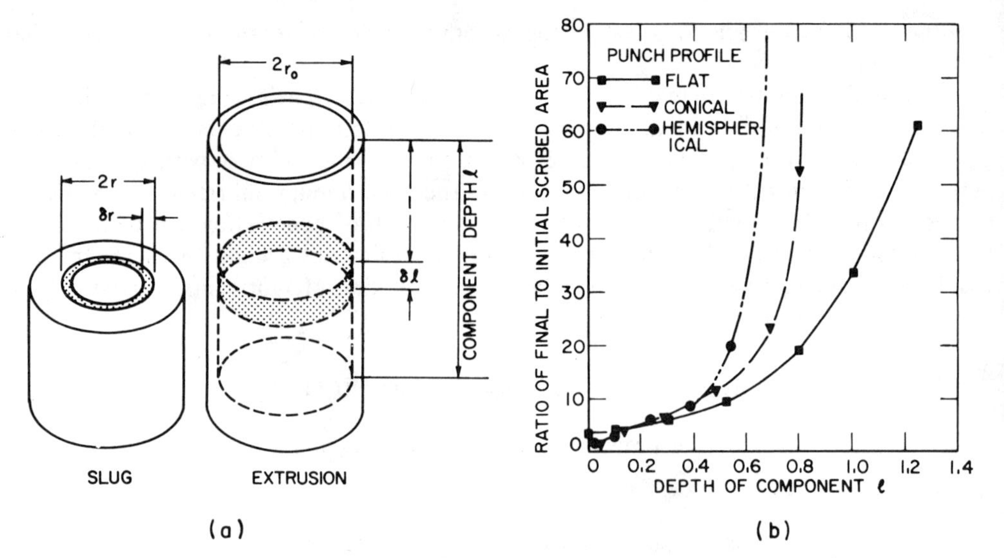

Fig.9.18. Expansion of surface area in back extrusion of a cup (a) and the effect of punch profile on the rate of expansion (b) in extrusion of a steel can with 76.5% reduction [62]

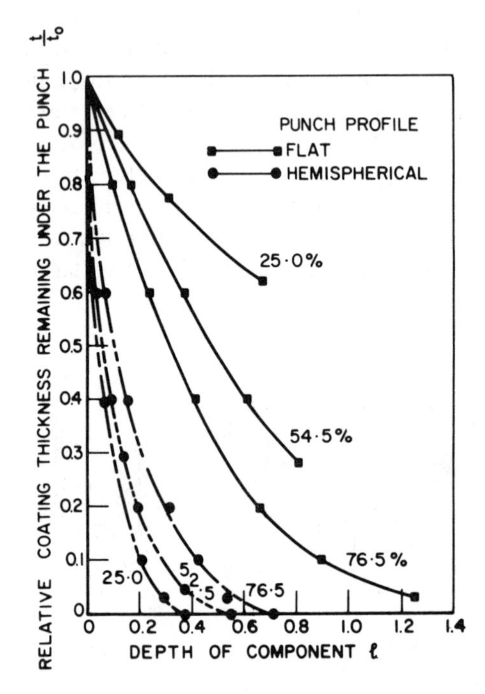

Fig.9.19. Effect of punch profile on thinning of a phosphate-soap lubricant under the punch face [62]

given depth of penetration a hemispherical punch retains more lubricant at heavy reductions, whereas the reverse is true of a flat-nose punch. Over-all, however, lubricant thinning is much more rapid with the hemispherical punch, and this shape would be preferred only for a shallow cup. A good compromise is the slightly tapered punch adopted in practice. The principal concern is breakdown of lubrication and die pickup around the punch corner, and therefore a slight radius is usually helpful.

Because of the severity of conditions, can extrusion has often been advocated as a lubricant evaluation technique [5.12,5.15,5.94], and several examples of its application will be discussed later in this chapter.

Lubricant supply can be improved by trapping an excess quantity between a flat-nose punch and a conical recess machined into the workpiece end face (Fig.9.17c), as proposed by Kudo et al [63] and further developed by Kaupilla et al [64].

9.2.2 Piercing

As the punch becomes smaller, the cross-sectional area A_1 of the extruded product (tube) becomes larger (Fig.9.20a) and the extrusion ratio is reduced, as is the extrusion pressure from Eq.8.4. However, pressure on the punch cannot drop below $3\sigma_f$ because the container behaves like the nondeforming part of a semi-infinite workpiece (as in Fig.2.7a). Because pressure is now controlled by the inhomogeneity of deformation, lubrication can do little to reduce it. It, may, however, be necessary to minimize friction on the surface of the punch if the hole is deep. In cold working, the length of contact surface can be reduced as in Fig.9.17a. If the workpiece is long, the material displaced by the punch must flow against friction on the container wall (Fig.9.20b) and pressures can reach very high values. Therefore, effective lubrication of the outer surface becomes critical. When this is not possible, the problem may be circumvented by piercing a round-cornered square billet, so that the material flows only radially.

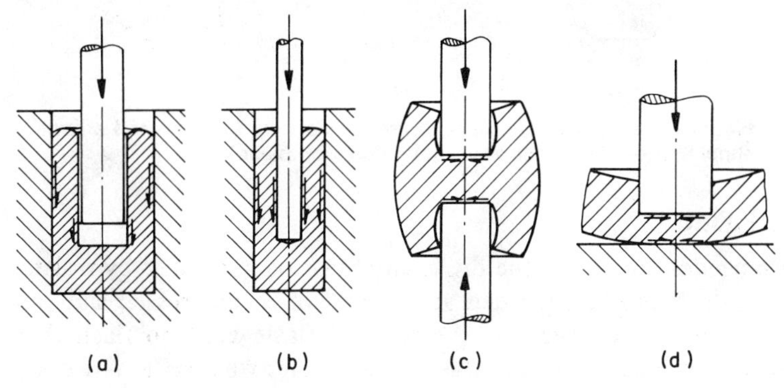

Fig.9.20. Piercing processes

From a theoretical point of view, similar solutions apply to extrusion and piercing [8.14], and the die geometry has the same effects. The important point to note is that punch pressures can never drop below $3\sigma_f$; indeed, they begin to rise again when the extrusion ratio drops below 2 [60]. Analysis is usually based on steady-state deformation, although non-steady-state conditions prevail at the beginning of punch penetration. Even in these early stages, friction increases punch pressures [65].

9.2.3 Indentation

When the workpiece is unsupported by a container and is indented from one or both end faces with a punch, deformation depends on the ratio of workpiece diameter to punch diameter. When this ratio is greater than 6, the workpiece behaves as a semi-infinite body, and punch pressure cannot be less than $3\sigma_f$ and is largely independent of friction. As in back extrusion, thinning out of a lubricant depends on the punch nose configuration. As the workpiece-to-punch-diameter ratio drops, pressure on the punch also drops gradually. The workpiece is free to deform, the outer diameter increases, and a complex shape develops (Fig.9.20c), little affected by lubrication [16]. However, reduced friction on the punch side surface helps in retracting the punch.

When the punch penetrates from one side into a billet (sheet, plate) supported on a flat die surface, the edges of the workpiece lift away from the die (Fig.9.20d). Male [66] found that flatness can be maintained to greater punch penetration if friction is reduced on the punch/workpiece interface while high friction is maintained between the lower surface of the workpiece and the flat die. Punch shape has an effect here, too; with a rounded punch, a lubricant reduces pressures provided that it does not thin out excessively. A well-adhering solid lubricant is effective, but breakdown of a liquid film may allow pressures to rise above their unlubricated values [66].

9.3 Impression-Die Forging

The technique that used to be called closed-die forging and also drop forging is now more accurately termed impression-die forging, because the two cooperating dies define only part of the forging geometry, and material is free to flow out along the parting line into a flash (Fig.9.21).

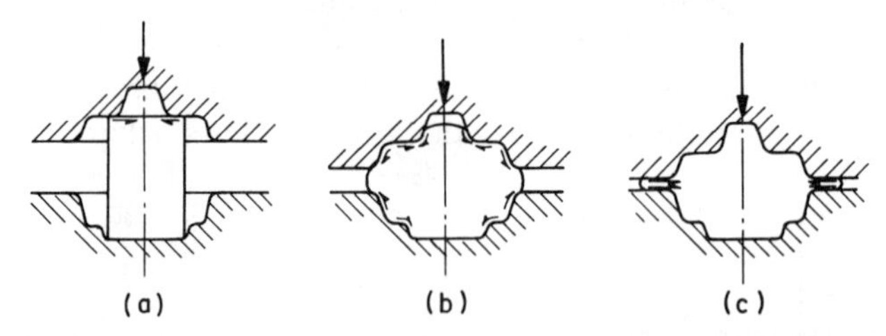

Fig.9.21. Stages of die filling in impression-die forging (half arrows show frictional shear stresses acting on workpiece)

9.3.1 Die Filling

The pressure needed for filling the die cavity must be generated by resistance to flow in the flash. The flash can be regarded as one half of a compressed slab, and pressure at the die entry point is governed by the ratio of flash width to flash thickness and by friction. Thus, high friction in the flash is desirable, whereas low friction is needed in the cavity. These contradictory requirements are not easily reconciled, although the situation is somewhat helped by the breakdown of the lubricant that occurs in the flash land because of the intense rubbing that takes place there.

The resistance developed in the flash is most clearly seen when the die cavity is simply an outwardly tapering slot (Fig.9.22a). The height of the extruded rib increases

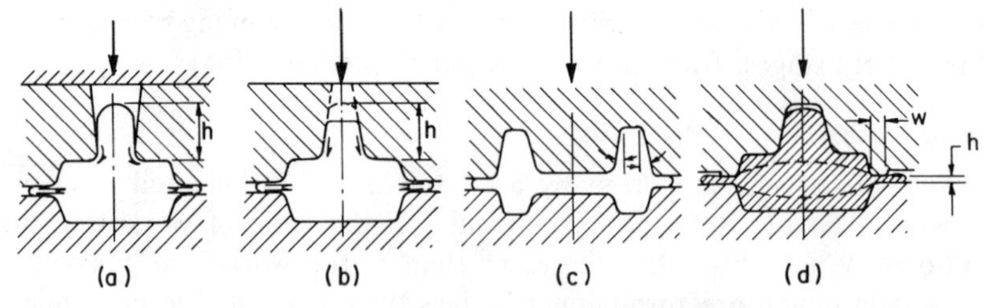

Fig.9.22. Simplified forging processes used in the study of forging lubricants

with increasing friction on the flash surfaces, although it is influenced also by the geometry of the operation: a billet of greater height-to-diameter ratio gives deeper fill [67-70]. This process has also been suggested for determination of friction [68].

Conventional impression dies are of a different shape. In order to facilitate release of the forging from the cavity, it is customary to forge with a draft (Fig.9.22b), the angle of which depends on workpiece material, equipment, and workpiece configuration. The greater the temperature difference between workpiece and die, and the longer removal from the die is delayed, the greater is the danger of the workpiece shrinking onto bosses (protrusions) of the die; therefore, larger internal draft angles must be applied (Fig.9.22c). Mechanically actuated, rapidly operating ejectors allow smaller drafts, and draft can be minimized under isothermal forging conditions.

The commonly held view that draft helps die filling was shown by Wallace and Schey [16] to be fallacious. Forging with draft is equivalent to forward extrusion into a narrowing gap. Frictional forces become high and, especially under conditions of nonisothermal hot working, sticking friction is soon attained. This is true even of the forging of simple shapes (Fig.9.21); barreling causes the workpiece to touch the die and become immobile away from the corners. Cavity filling then has to proceed by internal flow in the bulk of the material—a very difficult proposition, especially if the surface zones are chilled. Flash formation is of little help, at least initially, and pressures sufficient for filling the last die details are built up only when the flash is thin. In hot forging the effect of friction is augmented by chilling of the thin flash. This results, however, in very large forging forces, high flash pressures [71], and limited die life. Therefore, excess material is accommodated in a flash gutter (Fig.9.22d), with the flash land limited to a w/t ratio of less than 5.

It is hardly possible to have a lubricant that creates high friction in the flash land and low friction in the cavity; therefore, lubrication is of secondary importance to die temperature as a means of promoting die filling [72,73]. Lubrication becomes truly helpful only when the billet is smaller than the width (or diameter) of the cavity, so that the greater end-face expansion obtained with lubrication facilitates filling [16,71].

The configurations of many practical forgings are too complex for systematic studies, and therefore simplified shapes similar to those shown in Fig.9.22b and d have often been used. With a deeper hole (broken line in Fig.9.22b), the height h of the boss formed is a measure of the balance between the friction on the wall of the die and the friction in the flash gutter. While the test is by no means readily analyzed, it is acceptable as a scaled-down version of real forging and has taken a prominent place in both German work (e.g., Tolkien [74] and Stoter [75]) and U.S. work. The benefit of lubrication is clearly visible in the results of Altan et al [76]. In a two-dimensional simulation of filling of a rib, Regazonni et al [76a] found that over-all lubrication prevented filling because the material escaped in the flash.

9.3.2 Forging and Ejection Forces

The forces to be expected in forging can be relatively simply calculated if a number of simplifying assumptions are accepted. Flow can be visualized as being concentrated in a zone around the parting plane (broken lines in Fig.9.22d). Some constant friction factor m or μ can be assumed for the flash, and, from the pressure developed in the flash gap at entry to the die, an operative cavity pressure can be obtained [77]. Various upper bound solutions are also available [78]. Only numerical techniques can follow the details of the process [36,79]. These techniques allow also the calculation of deformations, and in some respects this is even more important than calculation of forces, because die filling is the major concern.

Frictional effects may be evaluated on the basis of measurements of the total force required to forge a given (usually simplified, axially symmetrical) shape (Fig.9.23a). This technique was used by Tolkien [74], who forged a flat disk somewhat similar to a gear blank and thus achieved die filling mostly by spread over the end faces. The total forging force P was therefore regarded as an indication of sliding friction.

Easy removal of the workpiece from the die cavity is a prerequisite of a smooth production flow. Forces required for ejecting a forged workpiece from the die cavity reflect friction and adhesion between die and workpiece after conformity between the two surfaces has been achieved through plastic deformation. Several techniques have been proposed for simulating this effect.

Tolkien [74] took the force P_E required to eject the workpiece as a measure of sticking (Fig.9.23a). Sakharov, quoted by Veiler and Likhtman [5.20], utilized a ring die with a conical hole (Fig.9.23b). A cylindrical billet was upset forged to fill the hole; after the die was turned over, the force P_E required for ejection was measured. Breznyak and Wallace [80] used a conical punch coated with the experimental lubricant to penetrate the workpiece material to a predetermined depth (Fig.9.23c). The force required to extract the punch after a specified holding time (5 s) was taken as a measure of sticking.

Cavity walls tend to roughen with increasing die wear, and therefore the ejection force can be taken as a measure of die wear, too [81].

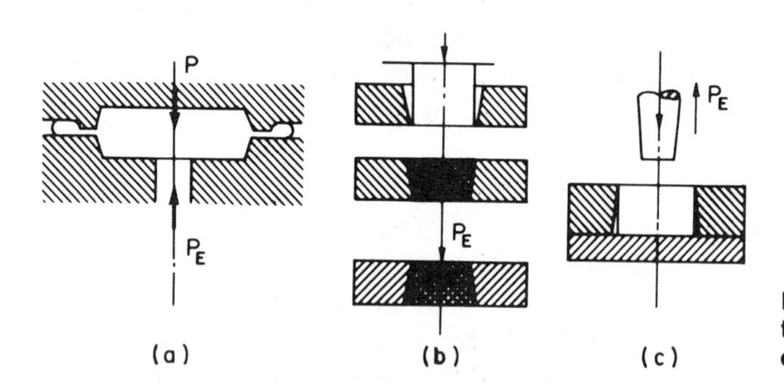

Fig.9.23. Techniques for testing the parting action of forging lubricants

9.4 Closed-Die Forging and Extrusion

In true closed-die forging (also called trapped-die forging), the die elements completely enclose the workpiece at the end of the stroke. Other operations also fall within the broad meaning of the term, and in some of these part of the workpiece may be unconfined to provide a safety vent in case of overfill.

9.4.1 Processes

In true closed-die forging, the flash, if formed at all, degenerates into a fin, usually in the direction of punch motion. Basic forging operations such as upsetting and piercing are usually combined with forward and back extrusion, either in a sequence of steps or simultaneously.

Closed-die forging is, in reality, often an example of forward (Fig.9.24a) or forward and back (Fig.9.24b) extrusion. It should be remembered that forward extrusion always entails sliding of the billet against the container wall, even when extruding a hollow part (Fig.9.24c), and friction hinders material flow. In contrast, in combined forward/back extrusion, friction acting in the direction of material flow aids extrusion (Fig.9.24b and d), as discussed in Sec.8.2.5. Consequently, the force required for such combined operations is less than for either stage separately, and the balance between material flow in the forward and reverse directions is very sensitive to friction [82,83].

In contrast to the extrusion of semimanufactured products (Chapter 8), steady-state conditions are seldom attained and the development and exhaustion of the lubricant film during the transition stages assume critical importance. Thinning of the lubricant film entrapped under the punch in extrusion of tubes (Fig.9.24c and d) presents the greatest challenge, and therefore scaled-down versions of such processes are useful for

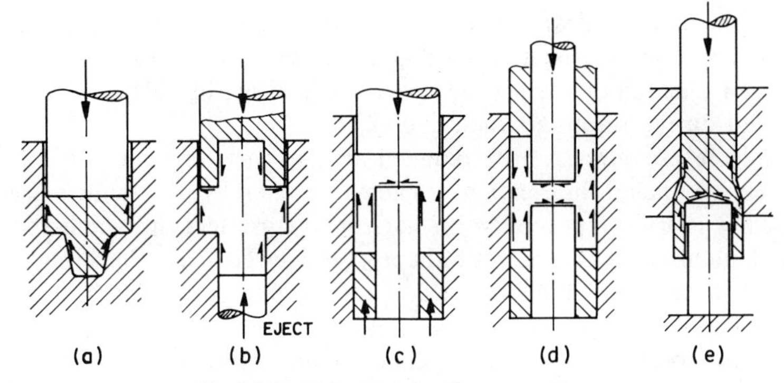

Fig.9.24. Extrusion-forging operations

evaluation of lubricants. The punch nose is often made slightly conical to facilitate material flow without the excessive lubricant thinning characteristic of rounded punches (Fig.9.19). Whenever possible, the container is made short (as in Fig.9.17a), and the punch is relieved after a short land. In an alternative proposed by Kunogi [31], container friction is reduced, at least in the deformation zone, by extruding out of the container (Fig.9.24e), and thus the process becomes a form of forward extrusion. In a special form of forward extrusion the billet is pushed through the die entirely without the support—and thus the friction—of a container (Fig.8.4a); the extrusion ratio must be kept small enough to avoid upsetting or buckling of the billet. It is obviously desirable that die friction be kept low.

It is sometimes more favorable to make flanged parts by upsetting (Fig.9.25a) rather than by extrusion. Friction governs the deformation mode [84]; high friction is now preferable because expansion of the end face could lead to the development of a circumferential fold. High friction is desirable also to avoid piping toward the end of forward extrusion (Fig.9.25b) [85] or lateral extrusion (Fig.9.25c) [86].

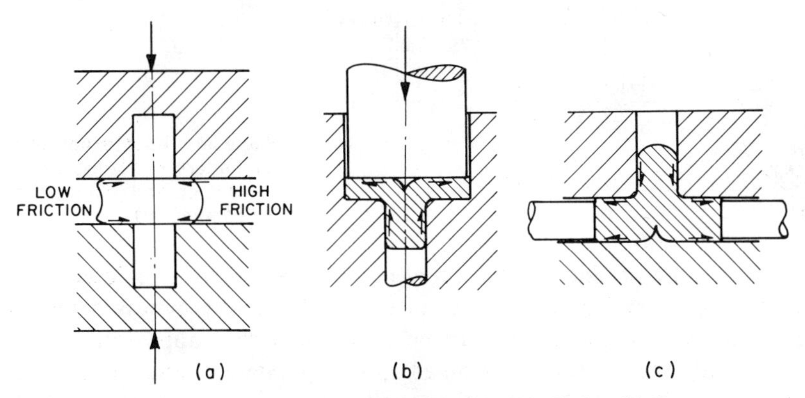

Fig.9.25. Development of forging defects with low friction

9.4.2 Practices

Workpieces may be cut to length and, if required, prelubricated on all surfaces. Alternatively, the part is forged at the end of a bar (wire), and then the freshly cut end face must be effectively lubricated. Frequently, the feed stock is cold drawn in a die attached to the upsetter, and then the same lubricant is used for drawing and forging.

The processes conducted cold are referred to as cold forging, cold forming, cold extrusion, or, in the back extrusion of thin wall cans, impact extrusion [12]. Parts are usually formed without a draft, and ejectors are routinely employed. An important requirement is that the lubricant should protect the already formed surfaces during reverse sliding or ejection of the finished part. Very often, lubricant breakdown and die pickup occur not during forming but during ejection.

Punch and die pressures can be reduced and/or parts of greater complexity made by warm working, although lubrication problems are thereby greatly increased.

Hot working is widely practiced on horizontal upsetters, in which the die is split along an axial plane so that undercut shapes (as viewed from the direction of punch motion) can be formed. Combined extrusion and forging is possible also on mechanical presses, provided that part geometries are similar to those shown in Fig.9.24.

A specialized form of closed-die forging is compaction of metal powders and cold and hot forging of sintered preforms. Generally, lower friction between the particles aids consolidation, and lower friction on punch and container surfaces is desirable for the same reasons as in forging of solid bodies [87-89].

In compaction on a single-action press (Fig.9.26a), the pressure exerted on the lower punch is reduced by friction on the container wall, and the ratio of upper to lower punch pressures can be used for lubricant evaluation [90]. If friction cannot be sufficiently reduced, it is preferable to allow the container to float and thus seek its lowest energy position, as dictated by the frictional balance (Fig.9.26b). Occasionally, friction may be helpful, as in hot upsetting of sintered preforms, where barreling results in support of the side surface and thus suppresses cracking (Fig.9.26c) [91].

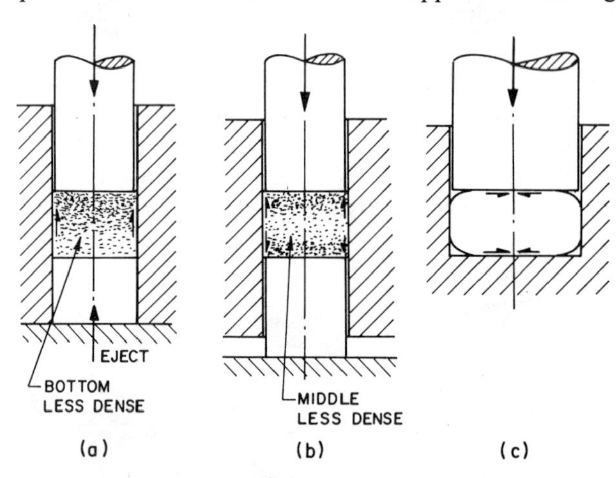

Fig.9.26. Powder compaction with (a) stationary container and (b) floating container, and (c) upsetting of a cylindrical preform

On a final note, it should be pointed out that problems associated with high friction and consequent high die pressures often can be totally bypassed by changing the deformation mode. This usually entails replacing the normal approach of the dies with a rotary motion. Sliding friction is changed to rolling contact, and the process geometry can become more favorable, too, although the danger of inhomogeneous deformation may increase and dies may be of more complex shape. This topic is discussed by Marciniak [92] and is now the subject of conferences [93].

9.5 Lubrication and Wear

The great variety of processes calls for a yet larger variety of lubricants, and their operative mechanisms are best discussed according to workpiece temperature, starting with the simpler cold forging mechanisms and progressing to the complexities introduced by higher temperatures.

9.5.1 Cold Forging and Extrusion

Lubricants are chosen according to the severity of the operation, as determined by interface pressures, process geometry, and the extent of sliding and surface expansion. Thus, back extrusion of a thin-wall deep cup in a strong alloy would create the most severe conditions, especially on the punch end face where the preapplied lubricant thins out, whereas forging of a shallow recess or upsetting to small diameter-to-height ratios would not call for special lubricant duties. Lubrication mechanisms will be discussed in a sequence reflecting increasing severity of duties. Many results come from experimental work; in judging the suitability of lubricants, the magnitudes of forces are often of less importance than the variation of forces and the extent of die pickup (Sec.5.1).

Mixed-Film Lubrication. In the presence of a liquid lubricant, PHD lubrication could develop at the beginning of deformation, but evidence indicates that some boundary contact soon occurs; therefore, it is more relevant to consider the situation as a case of mixed-film lubrication.

1. Effects of Deformation and Speed in Upsetting. A large mass of experimental results shows that, in cold upsetting between flat platens, friction drops while average film thickness increases when lubricant viscosity and approach speed increase. This is typical of lubrication in the mixed-film regime, corresponding to the descending part of the Stribeck curve (Fig.3.42).

The initially entrapped lubricant cushion is sealed at the edges, where a zone of pure boundary contact develops [94,95]. Thus, lubrication is mixed in the macroscopic sense, with PHD conditions in the center and boundary conditions at the edges (Fig.3.37). As deformation proceeds, the film thins out as it follows the expansion of the end face, and roughening of the workpiece results in a shift toward conventional (microscopic) mixed-film lubrication in the original PHD zone. Roughening has been extensively studied by Butler [96], by Oyane and Osakada [97], and by workers cited in Sec.3.8.4.

If desired, an average μ can be derived from ring compression or, less reliably, back-calculated from mean pressures in cylinder upsetting. However, the magnitude of μ in itself conveys little about the operating mechanism; it is more useful to observe surface deformation and to assess the area of boundary-lubricated surface. As discussed in conjunction with Eq.3.32 to 3.34, most of the frictional stress is generated on the boundary-lubricated surfaces, with hydrodynamic pockets contributing only negligible amounts. Experimental observations indicate a very wide range of μ according to process conditions:

a. At sufficiently high impact velocities, such as are attained in high-energy-rate forging equipment or in conventional hammers (impact velocities in excess of 5 m/s), even water will generate PHD films [27,98,99]. At the lower speeds more typical of mechanical presses, upsetters, and hydraulic presses, oil-base lubricants are needed. High viscosity alone can ensure film formation, but a high speed is also needed when viscosity is low [5.13]. Whenever ηv is high enough, the squeeze film reduces friction to typically $\mu = 0.05$ (Fig.9.27) and the presence of boundary or E.P. additives passes unnoticed, irrespective of the die/workpiece material combination. Therefore, adequate viscosity is crucial in forming of nonreactive metals, and in lubricating the freshly formed end faces in forging from a bar.

b. In many practical applications the surface roughening resulting from the presence of a PHD film is objectionable and viscosity is reduced so that most of the interface is in boundary contact. Therefore, friction cannot drop below $\mu = 0.05$ and is typically on the order of $\mu = 0.1$ (see Fig.3.19) [100-104]. Additives are then very

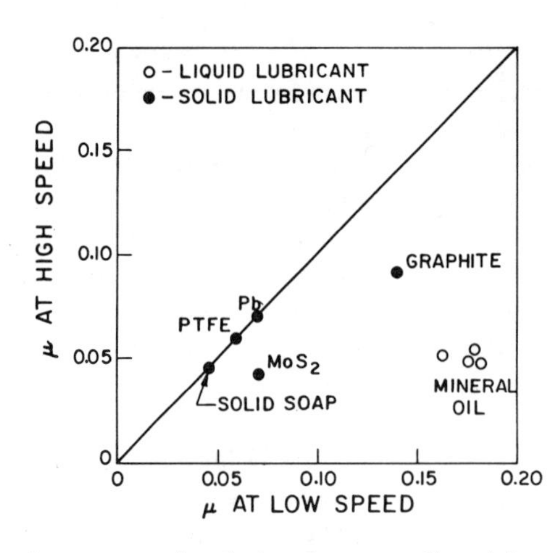

Fig.9.27. Coefficient of friction measured in low- and high-speed compression of 18-8 stainless steel to 30% reduction in height [98]

important; in their absence, die pickup occurs toward the edges of the workpiece if adhesion between die and workpiece is high, as in upsetting of aluminum and, to a lesser extent, stainless steels and titanium. Friction could then gradually rise to values as high as $\mu = 0.3$. Less adhesive combinations maintain lower friction, on the order of $\mu = 0.1$ to 0.15, as in upsetting of brass or copper. Nevertheless, additives are required even here, because die wear could otherwise become too severe. Contact in a single stroke is too brief to activate E.P. lubricants, and it is likely that the necessary cumulative time and temperature combination can be ensured only on the die surface in the course of several hundreds or thousands of contacts. Protection of the die during the first few contacts is particularly important in forging from a bar with unprotected, freshly sheared end faces, until reactions are initiated and friction polymers are formed. As expected, the effectiveness of boundary additives increases with increasing chain length, and saturated fatty acids reduce friction more than their unsaturated counterparts [100]. However, it is not always clear whether this is purely a boundary effect or is also attributable to the higher viscosities of the more effective agents. Fatty oils find use, too, for both their viscosities and their boundary properties.

c. At the very low speeds typical of many laboratory experiments, boundary lubrication effects become more pronounced and hydrodynamic effects are suppressed. Therefore, many of the μ values reported in the literature are higher than the most likely values attained under production conditions.

d. The effect of speed depends also on lubricant rheology. Thus, a highly viscous or semisolid lubricant such as lanolin may give lower friction at low speeds than a viscous oil (Fig.9.6), yet result in higher friction at high speeds where its shear strength is higher than that of an oil film.

e. The squeeze film is more readily established, more uniform, and less prone to breakdown during upsetting in a highly pressurized oil bath [105]. Even though this is not a very practical proposition, it illustrates the benefit of an increase in viscosity due to increased pressure.

2. Effects of Process and Surface Geometry in Upsetting. As in all PHD lubrication, geometry plays an important role. The shape of the tool is dictated by the configuration of the finished part, but, nevertheless, elastic deformation of a flat platen can help in entrapping the lubricant and may be a major source of squeeze-film formation in low-speed compression [106].

A squeeze film is formed even under a spherical ball impacting a flat. Miyagawa and Hirano [107] found that surface roughening was proportional to viscosity. For the same base viscosity, naphthenic oils, with their higher pressure-viscosity coefficients, gave better developed films than paraffinic oils, and the films were better able to resist breakdown on subsequent sliding. Narrow-cut fractions were more resistant to breakdown than oils of wider molecular weight distribution, an observation that justifies the preference shown in practice for narrow-cut mineral oils.

Entrapment of a viscous lubricant by mechanical means is particularly effective. Premachined pockets (Fig.9.4b) are not practical for production purposes, but grooving of the end faces (Fig.9.4a) is acceptable when other measures prove inadequate. The latter is a valuable method for reducing friction in determinations of flow stress [108]. As shown by Schey and Hoba [109], the volume of the grooves gradually diminishes in the course of compression, and the displaced lubricant flows over the ridges to provide micro-PHD lubrication. Grooves are most effective if they are wider and deeper than the natural roughness that develops with the given lubricant, and thus grooves should be smaller for a lighter lubricant and a finer-grain material. Pyramidal hollows created in the surface produced a similar effect in the work of Wilson and Rowe [110]. Small, closely spaced hollows were better in reducing friction and preventing pickup. In general, workpiece roughness is helpful, whether produced by prior working, pickling (etching), shot blasting, or machining, as long as the finish is nondirectional or, if it is directional, as long as the grooves are perpendicular to the major sliding direction (Sec.3.8.5).

Die surface roughness is, as mentioned in Sec.3.8.6, generally harmful in the mixed-film regime, but moderate roughness is helpful if the lubricant forms friction polymers and wear products which then fill up transversely oriented valleys, provided that no pickup occurs during the first few contacts. In most laboratory work the dies are cleaned after every test and too few repeats are made to attain steady-state conditions; for this reason, too, the condition $\mu = 0.05$ should occur in production much more frequently than published laboratory results would indicate.

3. Replenishing Squeeze Films in Upsetting. In the course of compression the growth of the entrapped squeeze film seldom keeps up with the expansion of the end face, and therefore the boundary-lubricated edge zone broadens, the average μ rises, and pickup may occur.

Friction can be kept low and pickup minimized if the squeeze film is periodically re-established. This can be achieved in several ways:

a. Compression can be interrupted and the die and workpiece surfaces relubricated. Such incremental compression is acceptable for laboratory flow-stress determinations but is impracticable for production.

b. Vibration may be superimposed on the platen. As discussed in Sec.3.9.2, the squeeze film is re-established if the tool lifts away from the workpiece. This effect is particularly noticeable when the workpiece is reactive with the lubricant and adheres to the die (Fig.9.28) [111]. Vibration at both low (10 to 50 Hz [112]) and higher frequencies has been found to be effective in reducing friction and barreling [113-115], although equally good results can usually be obtained by choosing a better lubrication system. Much of the force reduction reported by various workers can be attributed to a replacement of the static compression force by a vibrational force [3.338].

4. Plane-Strain Compression. As discussed in Sec.9.1.4, a squeeze film is established also in plane-strain compression, but this film is then gradually washed out from the work zone. The similarity to rolling is evident, and mixed-film lubrication in plane-strain compression has been studied not only for its own sake [55,116-118] but also as a simulation test for rolling [5.264].

Viscosity and additive effects are the same as in upsetting of a cylinder. A zone of hydrodynamic squeeze film forms around the neutral line, and the width of this zone

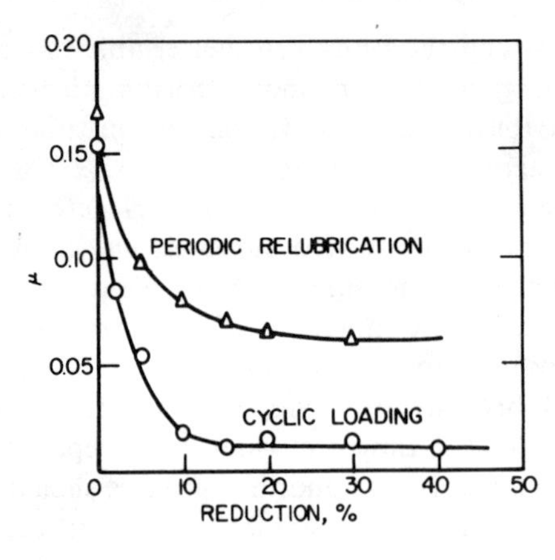

Fig.9.28. Reduction in friction obtained by periodic relubrication and cyclic loading during compression [111]

increases with increasing speed and viscosity [117]. Compared to cylinder upsetting, the effect of additives is much more clearly revealed because the freshly formed metal surfaces slide over the edges of the anvil. Tool pickup becomes immediately visible as scoring of the deformed surface. The test geometry allows one to make several indentations in succession; in general, the force required for compressing to a constant thickness (or the thickness obtained with a constant force) decreases with the number of indentations to reach a steady-state minimum (Fig.9.29a) [117],[5.264]. The effect is marked only when the anvils are ground perpendicular to the direction of material flow and it must be attributed to the accumulation of wear debris and the formation of a friction polymer. In the work of Delamare et al [117], equilibrium was reached earlier with a smoother die (Fig.9.29b) but the film broke down sooner on subsequent unlubricated indentations, whereas the film developed more gradually on rougher dies but was then resistant to breakdown. After the bulk of the friction polymer had been removed the transfer film on the die was studied with Auger electron spectroscopy, which showed elements transferred from the workpiece material.

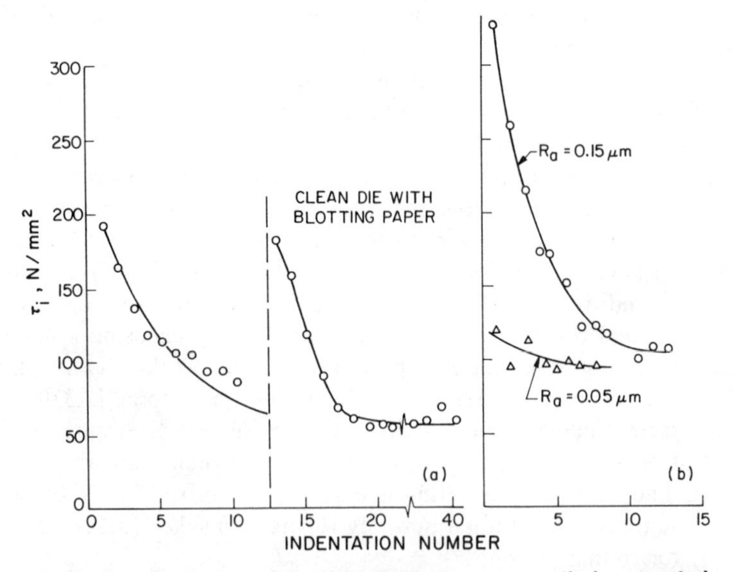

Fig.9.29. Drop in calculated interface shear strength in repeated plane-strain compression, showing the effects of (a) cleaning of the die and (b) die surface roughness [117]

5. Extrusion. Under relatively mild conditions, cold extrusion can be performed with liquid lubricants. Such practice is typical of working from a long bar or wire, where the freshly cut end faces must be protected with a liquid lubricant applied in the machine, and this liquid then provides lubrication for relatively minor forward extrusion (e.g., for reducing the shank of a bolt in preparation for thread rolling) and back extrusion (e.g., for creating the depressions in socket-head bolts). Lubrication in cold heading of steel and stainless steel was investigated by Tevaarwerk et al, and the results were summarized by Plumtree and Sowerby [119]. A viscous mineral oil (SAE 30 or heavier) gave low friction under the relatively high-speed conditions imposed by a cold header, but tool pickup and welding were observed with stainless steel wire against steel tooling, indicating that lubrication must have been in the mixed-film regime. This is shown also by the results of Noda et al [120], who found that ZDTP was needed to improve surface roughness in forging of socket-head bolts. The use of additives, so widespread in practice, is thus justified. Additives range from boundary to E.P. The effectiveness of commercial lubricants was shown by Abdul [121], who found also that sawdust served as a good carrier for liquid lubricants. A film of excessive thickness leads to surface roughening and, in the extreme case, to bambooing. The choice of die material is important in preventing adhesion, as demonstrated by the absence of pickup in extrusion of stainless steel with tungsten carbide tooling [119]. Cold closed-die forging is feasible with MoS_2 or phosphate/soap lubrication, depending on the extent of material flow [122].

6. Theory. Early analyses of upsetting [3.322] were restricted to calculation of the thickness of the squeeze film developed with an isoviscous lubricant (Eq.3.30). Later analyses [3.66],[99,123] allow for the effect of pressure on viscosity (Eq.3.30a). Kumar et al [124] consider non-Newtonian effects in plane-strain compression. Thompson and Symmons [125] include the effect of plastic wave propagation on the lubricant film developed in high-speed upsetting.

Analyses of surface roughening at the end face of the impacted cylinder are also available. Oyane and Osakada [97] used the roughening of a free surface and the thickness of the entrapped squeeze film as inputs in their analysis, and found reasonable agreement between calculated and experimentally observed fractions of the boundary-lubricated contact area. Experiments of Wilson and Silletto [126] generally confirmed an earlier analysis of Wilson, which linked roughening to a time-of-exposure parameter. They found that roughness was equal either to the natural (free) roughening or to one-quarter of the entrapped film thickness, whichever was smaller. The theory as extended by Wilson and his co-workers (summarized in [3.66]) shows that viscous heating is harmful while workpiece roughness is beneficial because it ensures transport of the lubricant film and prevents or postpones film breakdown, thus confirming the effects discussed in Sec.3.8.5.

In the analysis of cold extrusion, non-steady-state effects (Sec.8.4 and 8.5) become most important. Blancon and Felder [127] gave an analysis of the following sequence: squeeze film formation, then upsetting within the container, followed by extrusion.

Solid-Film Lubrication. The high pressures and large surface expansions typical of extrusion would lead to pickup and tool wear with liquid lubricants, and therefore solid or bonded lubricants are used extensively.

For maximum benefit, all surfaces of a precut billet or slug must be coated. In forging from bar or wire, precoating of the bar or wire is beneficial if severe cold extrusion is combined with moderate back extrusion or upsetting for which a liquid lubricant is supplied at the cold header. In judging the severity of the operation it should be remembered that die temperatures reach 100 to 200 C under steady-state conditions if workpieces follow in rapid succession and heating due to plastic deformation is substantial. All kinds of solid and semisolid lubricants (Sec.3.3) have been applied, in forms tailored to individual metals and to specific severities of operation.

a. The major advantage of metal films deposited on the billet is that they follow the expansion of the end face in upsetting and flow with the workpiece in extrusion. Separately interposed metal or polymer films are cut through at the edges (Fig.3.17a) in upsetting [128], and are only moderately effective even though their shear strength is low, because friction on the unlubricated annulus raises the average coefficient of friction. Metal as well as polymer films can be modeled as strain-rate-sensitive substances [3.127]. The main danger with polymer and layer-lattice films is localized breakdown, which then allows metal-to-metal contact and die pickup. Nevertheless, several commercial formulations have found use and are effective [129].

Transport of the film and thus resistance to breakdown are improved with appropriate workpiece surface roughness, and moderate die surface roughness oriented perpendicular to the sliding direction can also be beneficial in providing lubricant reservoirs (Sec.3.8.6). For this to occur, however, the film must be thick enough to provide full separation. In the compression of phosphate-soap coated steel rings, Hart and Modlen [130] found that friction increased as the roughness of spark-eroded dies increased from 0.1 to 7 μm R_a, no doubt because some asperity contact occurred. Analysis of solid lubricant transport is more difficult than analysis of liquid lubricant transport [3.66],[8.150] because no single rheological model can encompass the wide variety of lubricant behavior.

b. Semisolids such as soaps, waxes, lanolin, tallow, and other fats can be deposited in a thin film from a solvent, although they are not necessarily much better able to resist wiping off than are some of the more viscous liquids. Some improvement can be ensured by imparting E.P. properties (e.g., sulfonated tallow). For heavier duties these lubricants must be keyed to the surface by a lubricant carrier or, for severe conditions, by a conversion coating (Sec.4.7.7). The most severe conditions may justify use of MoS_2, deposited on top of the soap. As pointed out in Sec.4.7.8, the phosphate film itself is not a lubricant and ensures low friction only when impregnated with a semisolid. For example, Wiegand and Kloos [3.321] found that friction and thus surface temperatures were much higher with oleic acid than with lithium stearate.

The film thickness must be kept at the minimum value that still ensures adequate lubrication. Excess lubricant builds up in recesses in the die, resulting in underfilled forgings; it can also lead to undesirable roughening of the surface.

Lubricant throughput and thus roughening of the surface in extrusion are functions of many variables. Dannenmann [131] found, in cold extrusion of steel with a commercial phosphate system (and also with an MoS_2 grease), that the peak-to-valley roughness R_t increased when the billet was ground but decreased when the billet was turned (Fig.9.30), which is in line with the principles illustrated in Fig.3.55. The effect of die angle was complex; in general, one would expect more roughening (or less smoothing) with a smaller included angle, but this was true only at certain extrusion ratios. The reasons for this deviation are obscure. Certainly, thick-film effects are undesirable, as found by Suh and Bachrach [132] in extrusion of a tubular component. In order to cool the mandrel and prevent pickup, a liquid lubricant was applied to the bore surface; a bamboo defect developed with a stepped mandrel but could be eliminated by the use of a straight mandrel, which discouraged the development of a thick film.

For studies of surface roughness, the lubricant film (including conversion coating) must be removed. However, the residual lubricant film itself gives valuable clues

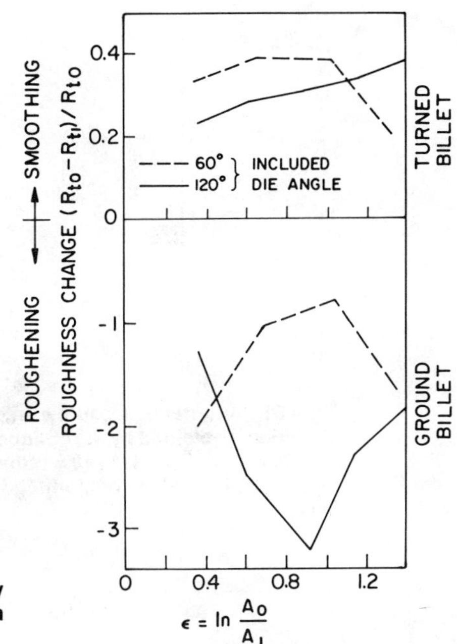

Fig.9.30. Change in average peak-to-valley roughness in cold extrusion of steel with a phosphate-soap lubricant [131]

regarding the lubrication mechanism and the adequacy of protection against metallic contact. The coating thickness can be determined by various techniques (Sec.3.8), such as gravimetry [133] or, preferably, a radioactive tracer technique [134,135]. This is particularly suitable for phosphate coatings. Gilbert et al [134] showed that close to 90% of the original coating was retained in extrusion of artillery shells. Stylus measurement has limited application [136], because the stylus could plow through even the residual, hardened phosphate-soap film.

Distribution of Frictional Stress. Some additional information on lubrication mechanisms may be obtained from studies of the shear stress distribution along the interface. Many of the techniques discussed in Sec.5.5.3 have been utilized, but the most detailed information comes from experiments with the oblique pin technique (Fig.5.29a). The results that van Royen and Backofen [5.107] obtained in compressing commercially pure aluminum are typical.

A technically clean, unlubricated surface resulted in high interface pressures (Fig.9.31) and an almost constant τ_i, and thus gave a rising μ from center to edge (adhesion must have prevailed, and modeling with m would have been better). A solid metal (lead) film gave such a low friction hill that both τ_i and μ remained reasonably constant. With oleic acid in mineral oil, the rise in μ toward the edge was indicative of the low friction of the entrapped film in the center and of the higher friction in the boundary-lubricated annulus.

As the variations of τ_i and μ were followed during the course of compression, a variety of behaviors was noted (Fig.9.32). The decrease in μ on the unlubricated interface again suggests that modeling with m would have been better. Friction rose rapidly with an uncompounded mineral oil as the initially entrapped film thinned and boundary contact increased. Both oleic and lauric acids initially reduced the shear stress, as would be expected from their boundary properties; on further compression, the rapid increase in τ_i and μ suggest that the lubricant film must have been broken through. The constancy of μ with MoS_2 (after initial lubricant alignment had been achieved) is typical

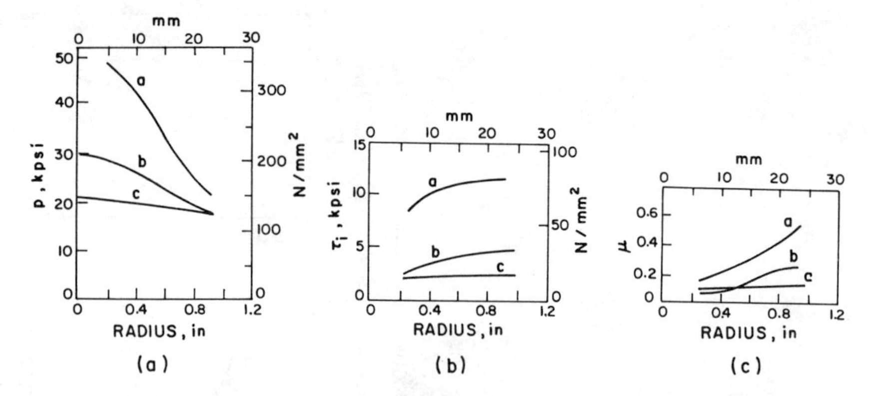

Fig.9.31. Interface pressure p, interface shear stress τ_i, and coefficient of friction μ obtained by the oblique pin technique in upsetting of aluminum billets ($d_0/h_0 = 4$) to 10% reduction (after [5.107]). (a) Dry platens. (b) Lubricated with a compounded mineral oil. (c) Lubricated with lead foil.

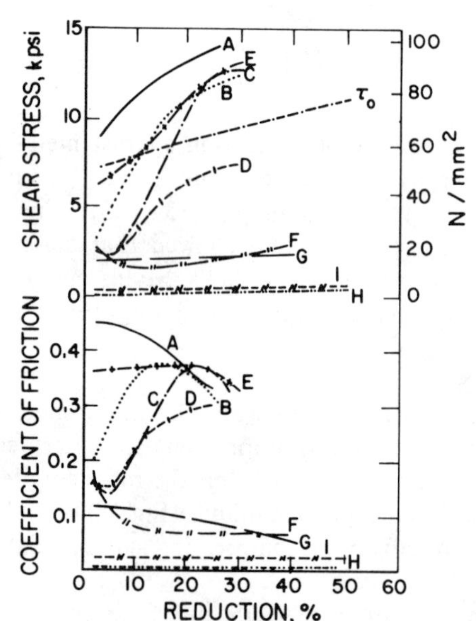

Fig.9.32. Shear stress and coefficient of friction measured in upsetting of aluminum with various lubricants [5.107]. A - Dry. B - Oleic acid in mineral oil; abraded surface. C - Same as B, but etched. D - Lauric acid in mineral oil; etched. E - Mineral oil; etched. F - MoS_2. G - Lead foil. H - PTFE sheet. I - Soap.

of layer-lattice compounds. The lead foil gave a constant τ_i whereas μ dropped because of increasing interface pressures with reduction; modeling with m would have been no better because of strain hardening of the workpiece material. The very low friction obtained with PTFE over the entire reduction range is somewhat surprising. Breakdown at the edges should have given a rising trend, and the coefficient of friction is also too low. The very low values obtained when soap was applied on top of a chromate-phosphate coating are perhaps more realistic, and reflect the low shear strength of soaps in general.

Shear stresses and μ values tend to be more constant with materials that are less prone to adhesion. Thus, Armco iron (Fig.9.33a) and copper (Fig.9.33b) gave reasonably constant values of μ over the limited reduction range explored. The improvement that was observed with an oxidized iron surface was most likely due to physical entrapment of the lubricant in the oxide film, although reaction with the boundary additives may

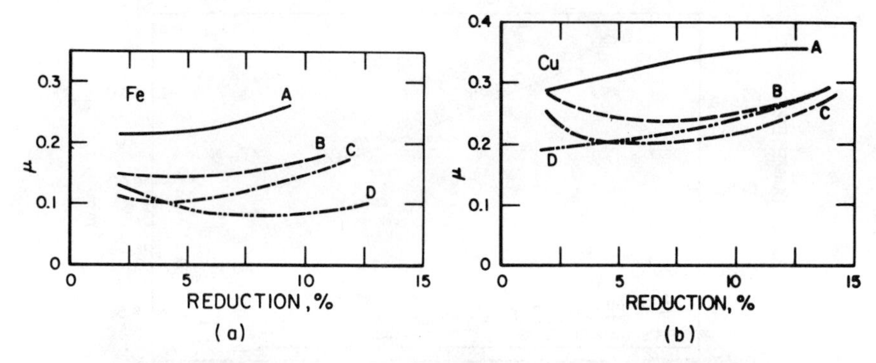

Fig.9.33. Coefficient of friction in upsetting of low-carbon steel (a) and copper (b) with various lubricants [5.107]. A - Mineral oil. B - Lauric acid in mineral oil. C - Oleic acid in mineral oil. D - Same as C, but oxidized surface.

also have added to the effectiveness of the treatment. In general, τ_i and μ were too high for the liquid lubricant, perhaps because the pin disturbed the lubricant film.

Die Wear. Cold forging production runs are usually long (tens of thousands of parts). Die lives of 100 000 parts are sought in cold extrusion with steel dies, and lives 3 to 5 times longer with tungsten carbide dies [137]. Thus, the die material must be hard enough to resist abrasion and reactive enough to allow E.P. agents to function but without causing excessive chemical wear (Sec.3.11). The ultimate tool selection, as always, is made on the basis of long runs, but simulation has shown considerable promise [5.63].

Some developments in die steels include fabrication by powder metallurgy techniques [138] and application of various coatings [139]. Carbide tool inserts or dies are used widely in cold extrusion [140,141].

Several studies have shown the effect of lubrication on die wear. James [133] reported on back extrusion of mild steel at 50% reduction. Wear, judged from the change in punch diameter, was greatest with chlorinated paraffin in a light kerosene base, much less severe with a more viscous commercial cold heading lubricant containing an unspecified E.P. agent, and negligible with a phosphate/soap system (Fig.9.34a). Surface finish in the hole, specified as peak-to-valley average (PVA), reflected the roughening that normally accompanies die wear (Fig.9.34b). Radioisotope studies are particularly helpful and indicate that wear generally increases with severity of operation [142] and can be minimized by use of suitable lubrication [143].

9.5.2 Warm Forging and Extrusion

As defined in Sec.2.4, the term "warm forming" is loosely used to denote deformation above room temperature but below hot working temperatures. Typical temperatures are 200 to 300 C for aluminum alloys, 650 to 800 C for steels, and 500 to 700 C for nickel-base alloys. In everyday usage the term "warm forming," unless further specified, implies the working of steel. The main incentive is that die pressures can be approximately halved while oxidation is still limited, and thus surface quality and tolerances are much improved relative to hot working. This subject has been reviewed by Mamalis et al [144].

In order to produce a smooth surface, PHD effects are discouraged, and most warm working processes rely on solid-film lubrication. Lubricant effects are specific to the workpiece material; there are, however, some observations of a general nature.

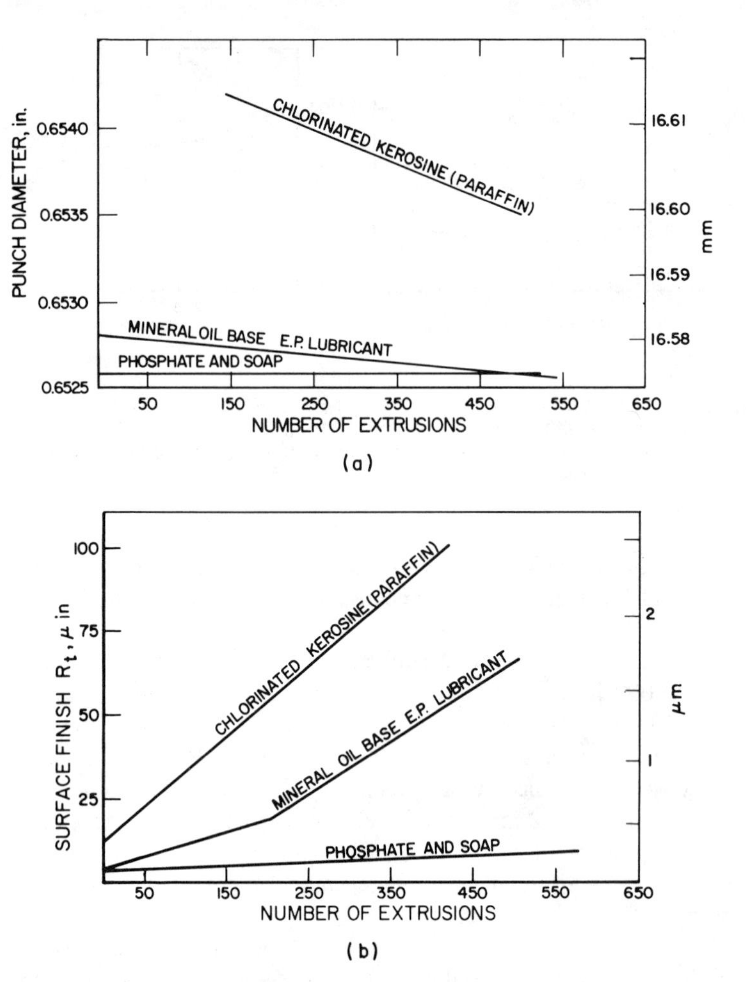

Fig.9.34. Effects of lubrication on punch wear (a) and bore finish (b) in back extrusion of steel cans [133]

Effect of Process Geometry. Because warm working is employed primarily for the forging and extrusion of more complex parts, surface expansion and interface sliding at high pressures tend to break down the lubricant. To retain relevance, lubricant evaluation must also take place under appropriately severe conditions. Because of the potential for film breakdown, tool pickup and wear become extremely important; they also tend to make for poor reproducibility of results unless application methods are thoroughly controlled, multiple tests are conducted, and, if necessary, statistical techniques are used for evaluation.

The ring-compression test has been popular because of its convenience [145-150]. A viscous lubricant develops squeeze films and gives lower friction than dry films [148], even though the viscous lubricant may fail under more severe conditions. Radial score marks, which are indicative of lubricant breakdown, are often observed with lubricants that otherwise give quite low friction, at least in the initial stages of compression. As pointed out by Dean [147], the greater surface expansion obtained with a lubricant giving lower friction can actually result in greater die wear; sticking friction, which limits relative sliding, is beneficial in this respect. Nevertheless, the test is useful in judging friction for operations involving surface expansion, and, if several specimens are forged without renewing the lubricant coating on the die surface, wear of the coating can be followed [3.328].

Plane-strain compression [151,152] permits heavy reductions (up to 90%), and, if the lubricant is prone to breakdown, pickup is immediately evident on the surface of the deformed strip [3.328]. Resistance to breakdown is evaluated even more readily by pushing short specimens through a converging die (Fig.5.8b). Doege et al [5.32] found that wear data are also obtained fairly readily; their results show that low friction is not a guarantee of low wear rate.

The most demanding conditions are created in back extrusion of thick-wall cans. There is usually too much scatter in measured extrusion forces to allow lubricant ranking on this basis alone. Trends in extrusion force, during one stroke or between successive billets, are more sensitive indicators of breakdown. Most directly, scoring of the bore surface can be observed [153-157]. Kaiser [153] quantified results by determining the bore length-to-diameter ratio at which pickup set in, but this procedure is laborious. Dean [147] rationalized his results by plotting against the lubricant spread ratio, defined as the sum of punch nose area and internal area of bore divided by punch nose area (rather like in Fig.9.18a).

The differences between the process conditions prevailing in ring compression and those in back extrusion are best illustrated by the observation that if testing is done before a volatile carrier can evaporate, friction in ring compression is usually less than friction after the coating has dried. In contrast, a partially wet, imperfectly formed coating breaks down in back extrusion. Another general observation is that in back extrusion best results are obtained when both billet and die are lubricated, whereas in ring compression coating of the die alone is sufficient. Because back extrusion really tests resistance to pressure welding, a low-adhesion coating on the workpiece (such as copper on steel) reduces pickup when a graphite coating is applied only to the die [147].

Forward extrusion of complex parts under industrial conditions [158] or in the laboratory [159] is also suitable for exploring friction and, particularly, wear phenomena.

Effects of Temperature and Time. Working at elevated temperatures creates two problems. First, the die becomes quite hot (and is most often intentionally preheated to 200 C or above), making wetting of the die by the lubricant difficult. Second, any coating preapplied to the workpiece may be rendered useless during preheating to the working temperature. Thus, the rate of heating should generally be high, and the lubricant should be formulated to protect the layer-lattice compound (which, at temperatures over 400 C, is usually graphite) from rapid loss of adsorbed films and from oxidation. This need has accelerated the trend toward induction heating, and has also resulted in development of lubricants with glassy [156] or polymer bonds [160]. Some industrial lubricants retain sufficient film thickness on heating for a prolonged time, but the residual film has a somewhat higher friction than a nonheated one.

The temperature of lubricant breakdown thus depends on both the active substance and the carrier. In ring tests, Geiger [146] found a very rapid breakdown of phosphate/soap at interface temperatures above 200 C; aqueous MoS_2 and graphite both failed above 500 C, but graphite resulted in much lower friction before breakdown set in. Commercial graphitic lubricants with synthetic binders yielded $\mu = 0.05$ even above 700 C.

Warm working lubricants bonded with polymers reduce heat transfer, whereas others increase it [5.32]. All reduce convective cooling of the dies. Wetting of the usually quite hot (200 to 400 C) dies is a prime concern and depends also on the die material.

Temperatures are high enough to drive off volatile components of the lubricant, building a gas pressure in the die cavity. When a sample was heated in a closed container, gas pressure was found to vary greatly for various commercial lubricants [5.32].

Die wear has been the primary limiting factor in the spread of warm working and, in addition to the obvious hot working die steels, high speed steels for temperatures from 600 to 800 C [5.32], cast carbides, and nitrided die steels have been examined. In warm extrusion of difficult materials (e.g., high speed steels), wear is still too rapid even with a nitrided maraging steel [158]. Sintered carbides hold promise [161,162] but survive only if supported well enough to prevent their fracture on sudden temperature change; therefore, they are likely to succeed only as die inserts. Interestingly, Rao and Biswas [163] found in unlubricated forging that wear was minimum at warm working temperatures, but it is doubtful that the same would hold true for lubricated extrusion.

General reviews of warm forging lubrication have been published by Gordon [164], Saga [156], and Matthes [165].

9.5.3 Hot Forging

We have seen in Sec.1.1.1 that hot forging is the most ancient of all metalworking processes, yet it is also the one for which lubrication studies have been neglected until fairly recently. The reason is, of course, that in drop (hammer) forging of steel the new surfaces generated during deformation are exposed to air between successive blows, and thus they reoxidize before making the next die contact. We have also seen (Sec.9.3.1) that die filling is not necessarily aided by lubrication. Therefore, it was only with the spread of hydraulic and, particularly, mechanical press working, and with the introduction of light metals and difficult-to-forge materials, that lubrication became a concern of first importance.

Workpieces are heated to typical hot working temperatures (Table 2.1) but dies are usually substantially colder. Therefore, a lubricant not only should ensure die/workpiece separation and lower friction, but preferably should also act as a heat insulator. Excessive heating of the dies would lead to die damage and wear, and the lubricant—or its carrier—should also perform a cooling function. High temperatures at the interface severely limit the choice of lubricant; because of their importance, solid films will be discussed first.

Solid-Film Lubrication. With the exception of ferrous materials, oxides formed on the workpiece material seldom serve as true lubricants (Sec.3.3.1). Oxides of other metals externally introduced have been repeatedly studied [166],[4.191,4.192]. However, in forging they are generally not competitive with layer-lattice substances, and most forging lubricants contain graphite (or, for lower temperatures, MoS_2) in various carriers [167-169].

With occasional exceptions, such as in forging of aluminum or magnesium, the lubricant is applied only to the die because it would be destroyed in the course of preheating the workpiece. The die itself is usually preheated or reaches some steady-state temperature during forging, and therefore the carrier is chosen with wetting of the die surface in mind. Oils and greases used to be preferred, but occupational health and ecological considerations favor aqueous lubricants [4.58]. Furthermore, capillary forces that aid alignment of platelets are less effective while the oil decomposes on heating. The attributes of an ideal nonpolluting lubricant were discussed by Jain [170]. Only exceptionally are other carriers considered. Thus, potassium iodide was used on 1100 C dies; when the powder was sprayed onto the die surface, the KI flashed off, leaving a uniform, dry graphite coating [73].

1. Effect of Temperature. Lubricant breakdown and oxidation are functions of contact time, die and workpiece temperature, and surface deformation and sliding. For a given contact time, various lubricants break down at different temperatures. Ring-compression tests by Löwen [41] showed an early breakdown of ZnS, relatively high friction for BN, and the superiority of graphite (Fig.9.35a). Upsetting to higher strains (Fig.9.35b) increased friction because of surface expansion and sliding. It is remarkable that BN retained its value as a parting agent. Löwen attributed this to oxidation; the resulting boron oxide, which softens at 550 C, actually gave lower friction at heavy reductions than a damaged graphite film could have given. Similar observations on the effects of deformation were made by Löhr [4.196]

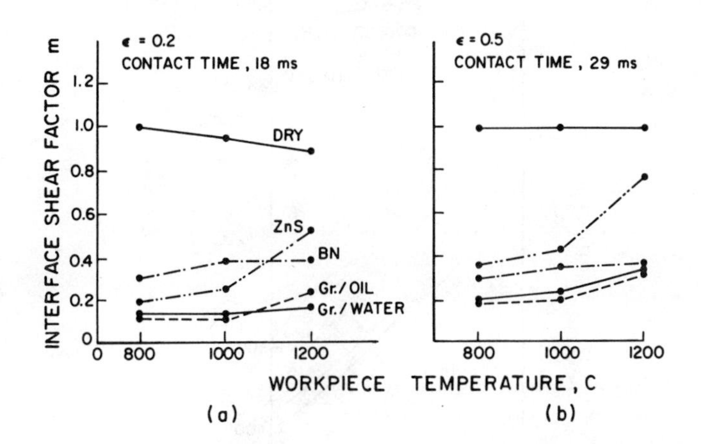

Fig.9.35. Friction in hot upsetting of steel rings with dry lubricants [41]

2. Effect of Heat Transfer. A potentially confusing factor is chilling of the workpiece on the die surface; this restricts deformation just as much as friction does, and also affects measured interface stresses [171]. The relationships are not straightforward, because oxides, which are heat insulators, interact with the lubricant to determine the rate of heat transfer, as reviewed by Mrozek [172]. These effects have been most extensively studied in forging of steel, although they are equally important in forging of other metals.

A thicker oxide is, in general, a better heat insulator [173-176], and the reduced heat flow reduces the temperature gradient in the surface layer of the die [177].

Breakup of the oxide film greatly increases heat transfer, and therefore the rate of heat loss increases suddenly when pressure is applied and plastic flow begins, irrespective of whether the workpiece is of steel [178,179] or aluminum [180]. This effect is quite dramatic when a workpiece with a rough surface is loaded and deformation of asperities creates oxide-free junctions [178,180]. Heat transfer is then the same with or without a lubricant, as observed for graphite on steel [176,178] and MoS_2 on aluminum [180]. Heat transfer rates approach values of completely uninsulated surfaces, as pointed out by Reynolds [178]. With increasing strain more fresh surface is exposed, and heat transfer rates increase further (if possible).

Heat transfer is also a function of contact time and thus of forging speed. Temperature gradients are greater at higher speeds [175,177], and Dean and Silva [181] estimated from measured surface temperatures that the very surface of the die reaches 900 C on a hammer and 800 C on a mechanical press during forging of steel at 1040 C.

Cumulative contact time is, however, less on the hammer, and this keeps bulk temperatures lower.

An important lubricant function is cooling of the die. Aqueous lubricants are the most effective and reduce the bulk temperatures of hammer dies—with their steeper temperature gradients—more than those of press dies [181].

It is not unusual to find that heat transfer rates increase with lubrication (Fig.9.36) [104], because the insulating properties of the oxide are destroyed when it is impregnated with a more conductive substance—especially with a liquid carrier such as water, oil [182], or grease [104]. Only when the carrier forms a heat-insulating film is heat transfer reduced.

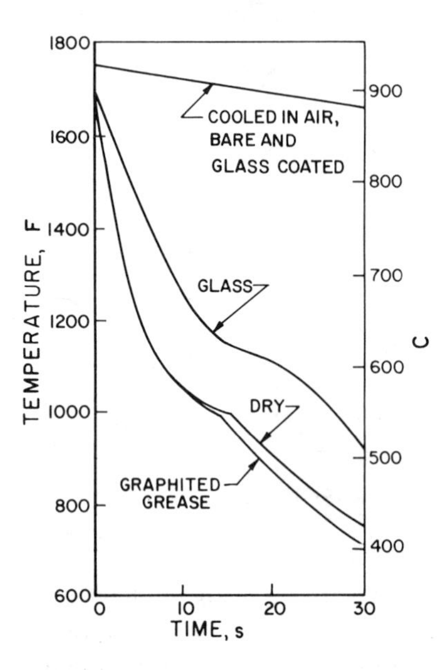

Fig.9.36. Cooling of mild steel specimens as measured with a thermocouple located 4 mm below the surface [104]

Lower heat transfer rates do not necessarily mean lower friction, and conflicting data on the effects of oxide film thickness on friction have been reported. These results were reconciled to some extent by the work of Keung et al [173], who found friction to reach a minimum at an oxide film thickness of about 50 μm, climb to an intermediate maximum at 150 μm, and decline further at 250 μm. Lower friction was associated with a higher FeO content in the scale.

Similar effects must apply also to oxides of other workpiece materials, although the relative degrees of hardness and ductility (or, rather, brittleness) of various oxides are quite different (Sec.3.3.1). Oxides of other metals may also be interposed as lubricants but are not capable of following the surface extension of the workpiece. They were useful only to low strains in the ring-compression tests of Male [166].

3. Effect of Forging Speed. With higher forging speeds, several effects take place simultaneously:

a. Flow stress σ_f increases because of the higher strain rate, but decreases if much of the heat generated is retained.

b. Contact time is reduced and cooling minimized, counterbalancing to some extent the increase in σ_f due to higher strain rates.

c. The time of lubricant exposure to high temperature is reduced.

d. In the presence of a liquid carrier or carrier residue, squeeze-film effects develop.

The combination of the above effects generally reduces friction at higher speeds. This is observed even with unlubricated but oxidized steel workpieces. The reason for this is not clear, although Wallace and Schey [183] were able to show that it is not due simply to lesser cooling. The squeeze-film effect dominates with liquid carriers: Jain and Bramley [184] observed a marked drop in μ with Copaslip, an antiseize compound consisting of fine lead and copper particles in a bentone grease. With colloidal graphite in water (Aquadag), Wallace and Schey [183] found lower friction even with water residues retained on a cold hammer die; Keung et al [173] attribute this to the development of a steam cushion. A speed effect was found at all speeds when oil was the carrier for graphite, MoS_2, or PbO. When the graphite film applied to the die is allowed to dry completely, the squeeze-film effect is absent and μ is virtually constant [173,183].

Competing with the squeeze-film effect is lubricant breakdown due to oxidation and thermal decomposition. At given forging and die temperatures, lower speed results in longer contact time and more damage to the continuity of the film. Depending on the nature of the viscous carrier, friction may even become higher than with a dry film. Copaslip was virtually useless at a low press speed [184], and the breakdown of a mineral oil appeared to interfere with the lubricating action of graphite [183], presumably by preventing alignment of platelets. Die temperature and the time elapsed between lubricant application and contact with the workpiece play decisive roles in the evaporation of the carrier. With increasing interface temperature, film damage and thus the increase in μ are greater [5.32]. These effects are summarized schematically in Fig.9.37.

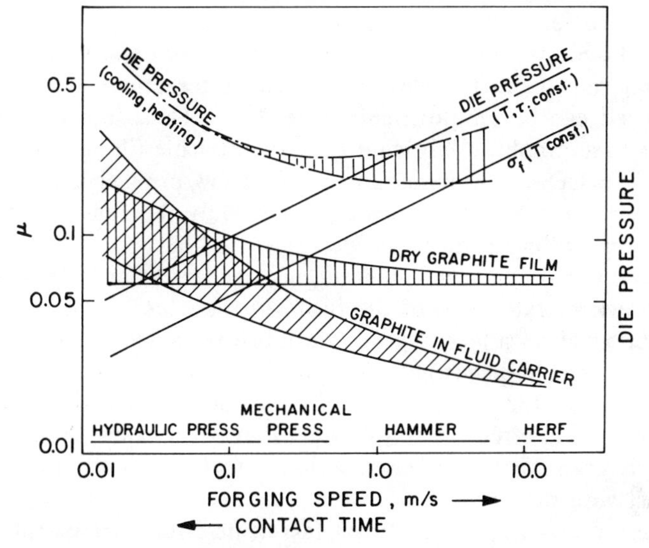

Fig.9.37. Schematic illustration of the combined effects of deformation speed (contact time) and lubrication on friction and die pressure in hot forging

Interactions of oxides with the lubricant also enter into the speed effect. A heavy oxide film interferes with a dry colloidal graphite lubricant [171,173,174,183]. A thin (say, 25-μm) oxide film gives lower friction with dry graphite at both hammer and press speeds. A thick oxide film can, however, negate the squeeze-film effect; in the work of Keung et al [173], the carrier receded on the oxide film and moved the suspended graphite with it. Thus, even though a heavy oxide film is a good heat insulator [185], it is generally objectionable in forging.

The effect of graphite particle size is still debated. It has been claimed that semi-colloidal graphite is more effective than colloidal graphite [186], and it has been suggested that flake graphite results in less oxide penetration and thus cleaner aluminum forgings [6.374]. Schlomach [187] found no effect of particle size at low strains. At higher strains and die temperatures a commercial lubricant with finer graphite gave lower friction; Schlomach attributed this to the interference of the coarser graphite with the additives present in this formulation. At lower die temperatures this effect could not be observed, because the additives had not yet been activated. It would certainly appear that colloidal graphite is less tolerant of a thicker oxide film [183].

4. Effects of Die Geometry and Temperature. Lower friction at higher forging speeds does not necessarily mean lower forging forces, because the increase in σ_f often outweighs the effect of the lower pressure-multiplying factor that accompanies a lower μ. Thus, for a given lubricant, forging on a high-energy-rate hammer results in higher forces than does forging on a press [174]. A lubricant film of better heat insulation brings the situation closer to adiabatic conditions, and the greater temperature rise may counterbalance the strain-rate effect. Results are, therefore, highly dependent on experimental conditions [188,189].

An optimum speed in terms of minimum forging load and best die filling can sometimes be found. This optimum was about 0.05 m/s for a 20-mm-thick web in the work of Parikh et al [190] and increased with decreasing web thickness.

In judging the effectiveness of a lubricant, one must keep in mind that entirely different criteria apply to various forging geometries.

a. In upsetting and ring compression the predominant variable is end-face expansion, and this is promoted by lubrication (Fig.9.2) [16].

b. In true closed-die (trapped-die) forging the major deformation mode is extrusion, into a narrowing gap if draft angles are applied. Interactions among oxides, lubricant, and forging speed can become difficult to separate. Thus, in the work of Parikh et al [190], a grease-base graphitic lubricant gave the best die filling in closed-die forging of billets heated in a CO atmosphere and forged at low press speeds. However, the same lubricant was the worst for billets heated in air. Only the aqueous graphitic lubricants resulted in good die filling at all press speeds.

c. In conventional impression-die forging the extrusion effect is combined with upsetting (and lateral extrusion) of the flash. In the experiments of Sheikh et al [174] the higher friction of an aqueous graphitic lubricant aided die filling at press speeds, whereas at high hammer speeds both the aqueous lubricant and Copaslip gave the same result. It appears that the short contact time in the hammer aided die filling and also neutralized the friction effect. A simple steel shape forged by Sharan and Prasad [191] filled best with graphite in oil, next with Na_2CO_3 in water, then with graphite in water, and least with sawdust.

Die temperature is a most significant factor; its effects are complex:

a. Higher die temperatures result in less cooling and thus facilitate material flow, especially in impression- and closed-die forging.

b. If increasing interface temperature results in an earlier breakdown of the lubricant, interface sliding decreases [192] and less outward expansion is found in ring compression [193]. Some lubricants fail to wet a hotter die, and friction increases [5.32].

5. Effect of Application Method. Even the best lubricant will fail if it is deposited discontinuously; at the same time, excessive coating thickness can lead to lubricant accumulation, unfilled forgings, and poor surface quality. Therefore, controlled deposition is essential.

Hand application by swabbing is still practiced but is not satisfactory, particularly with aqueous lubricants [194]. Several mechanical methods of application have been developed [194-196]. A good application system must prevent settling out of solids in the holding tank, ensure reliable and uniform atomization (breaking up of the liquid droplets) by mechanical means or air pressure, and deliver the fine droplets to the die in a controlled manner. Hand-held spray heads suffice for production at lower rates, but mechanically operated stationary or oscillating spray bars are essential for high production rates. Spray heads with sufficiently large orifices can be kept open if air is blown through them after each lubricant application [197].

Schlomach [198] investigated the effects of spray variables on the thickness and continuity of graphite films. In forging steel rings with a 3.5:2:1 ratio of dimensions at 1150 C, he found with two commercial lubricants that, at 2% and 10% solid concentration, friction dropped with increased spraying time because a squeeze film developed [187]. On prolonged spraying (over 2 s) the previously deposited film was washed off. With die temperatures increasing from 200 to 400 C, the maximum lubricant weight that could be applied dropped from 0.4 to 0.2 mg/cm^2. No improvement was obtained by using coatings heavier than 0.5 mg/cm^2. Doege et al [5.32] observed a linear increase in friction with increasing dilution.

Most application is by the total-loss technique, but reclamation of graphite-base lubricants by selective filtration is possible [199].

6. Effect of Ejection Force. Pressures generated by gaseous components of the lubricant aid the removal of forgings from the cavity and have been repeatedly studied in specially constructed devices and dies. Tolkien [74] and Sheikh et al [174] incorporated a pressure transducer communicating with the die cavity through a small hole. Schlomach [179] countersunk the die to create a thin membrane at the bottom of the cavity, and calibrated the deflection of this membrane in terms of internal pressure.

When the shape of the preform fits the cavity surface tightly, entrapped air is adiabatically compressed in the course of die filling. In the presence of a lubricant, evaporation and combustion of lubricant components can have various effects:

a. If the cavity is sealed early in the stroke, gas pressures rise further and pressures may become high enough to cause deformation of the workpiece [174]. This pressure aids ejection, but ejection forces will be low only if the lubricant also serves as a good parting agent. Thus, in the work of Tolkien [74], the ejector force dropped to 1% of the dry value when an aqueous graphitic lubricant was used. In contrast, sawdust (which increased the pressure by 110% over the unlubricated value) was much poorer as a parting agent, and the resulting ejection force was 15% of the dry value.

b. If the preform does not seal the die early in the stroke, the combustion products create a gas cushion around the forging which then facilitates ejection even though the gas pressure at the end of the stroke may be below the adiabatic value [179].

c. If the lubricant does not contain volatile components (e.g., a dry graphite film), the gas pressure is the same as in dry forging but ejection forces drop to much lower values [174].

d. Excessive gas pressures could eject the workpiece too violently. It is not clear, however, whether they can actually prevent die filling.

Thick-Film Lubrication. At the temperature of hot forging, only glass or similar inorganic substances can be used. The forging process imposes some special requirements:

a. In both isothermal and nonisothermal forging, any accumulation of lubricant residues in the die cavity results in underfilled forgings. Therefore, the lubricant must be applied to the workpiece only, in the form of a thin coating.

b. The glass should wet the workpiece in order to follow surface deformation, but it should do so without attacking (corroding) the die or the workpiece. It should adhere to the forging sufficiently to be lifted out with it.

c. If glass adheres to the die, it should allow ejection of the workpiece without excessive force and without long, strong stringers.

d. In nonisothermal forging the heat-insulating capacity of the lubricant should be high.

e. Because glass is not capable of preventing pickup if the film is locally damaged, additional protection must be secured by adding a compatible parting agent such as BN to the glass and/or by applying a thin, dry or grease-base [200] graphite coating to a colder die. Isothermal forging temperatures are too high for graphite to survive, and BN serves as a useful parting agent.

1. Effect of Process Geometry. Forging configurations range from very flat (fan and turbine blades) to deeply recessed (airframe components). Lubricant evaluation and selection must take this into account.

a. Ring compression is useful for flat configurations. The thickness of the squeeze film is a function of forging speed and of glass viscosity at the interface temperature. A combination corresponding to the minimum in the Stribeck curve (Fig.3.42), which results in minimum friction and upsetting force, can usually be found. With thicker films, the squeeze film tapers from the neutral line outward and causes concavity of the inner surface of the ring (Fig.9.38) [183]. If a film is too thin at the beginning of the stroke, it loses continuity on deformation, friction rises, and pickup may occur.

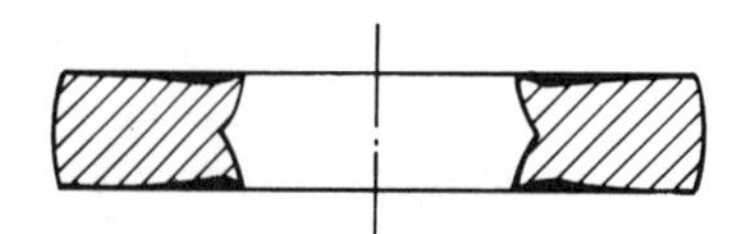

Fig.9.38. Deformation of a ring upset using a thick glass film as a lubricant [183]

b. Hydrodynamic effects in plane-strain compression reveal the ability of the lubricant to follow surface deformation [4.217], and this test deserves more attention than it has received. It is particularly suitable for selecting lubricants for forgings in which the metal flows over the edge.

c. For more severe deformation, a back-extrusion operation would appear to be suitable.

d. Pure sliding tests, such as pressing of a slug through a converging channel (Fig.5.8b) [5.32] and the twist-compression test (Fig.5.16b) [4.217], have also been used.

2. Lubricant Variables. There is no general agreement regarding the optimum viscosity, although most investigators recommend 10 to 100 Pa · s at the workpiece temperature (Sec.4.7.6). More relevant is the viscosity at the average of the die and workpiece temperatures [4.196,4.217]. In nonisothermal forging the workpiece surface temperature tends to drop with a good heat-insulant liquid film, but it can actually rise when the film breaks down and high friction generates heat.

For isothermal forging, it is more meaningful to relate glass viscosity to the flow stress of the material [4.219].

There is no definite minimum film thickness, but typical values are around 0.05 to 0.1 mm in forging and ranged from 0.1 to 1 mm in pushing of slugs [4.196]. Excessively thick films lead to surface roughening and glass buildup in the die.

Some lubricants other than glass can serve as viscous fluids. As shown by the diminishing film thickness on increasing workpiece temperature in Fig.9.39, Phosphatherm behaved as a viscous fluid when applied to a nonreactive stainless steel workpiece, but reacted with the scale on lightly oxidized carbon steel in a manner that reduced

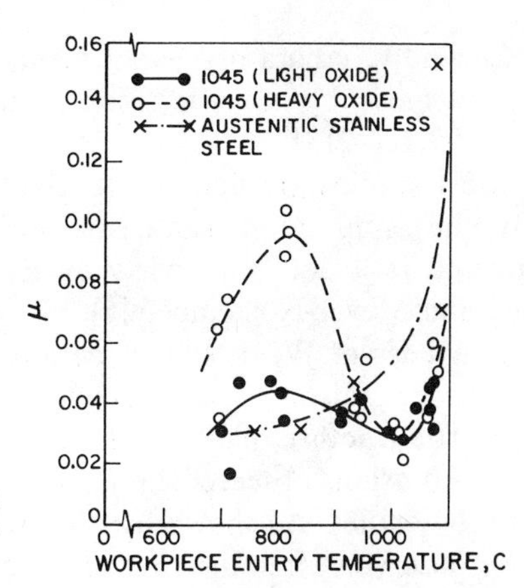

Fig.9.39. Effects of temperature and oxide film thickness on friction in pressing of Phosphatherm-coated slugs through a converging gap [4.196]

friction [4.196]. However, a heavy oxide layer could not be dissolved in the lubricant and resulted in increased friction at lower temperatures.

As occurs in PHD lubrication in general, one would expect appropriate surface roughness of the workpiece to be helpful in retaining the lubricant. Surprisingly, no such effect was noted by Leipold et al [4.219].

Wetting of the workpiece surface by the glass and protection during preheating are important (Sec.4.7.6), but the glass with the best protective properties is not necessarily the best lubricant on steel [201], and this might be expected also with other metals. Best protection is ensured by preheating in a glass bath, the glass then serving also as a lubricant [202]. Glass is a good heat insulator and reduces cooling rates during transfer from the furnace (Fig.9.36). There is conflicting evidence on cooling during forging: Cook and Spretnak [104] observed slower cooling (Fig.9.36), whereas Boer and Schröder [203] detected no difference in their more detailed study.

Lubricants of a different class are based on nonpigmented eutectic salt mixtures applied from an aqueous solution, with or without other, proprietary additives [204]. A salt such as NaCl provides a melt, and a carbonate and/or a nitrate provides the lift. They are nonpolluting, are claimed to be noncorrosive, and combine excellent cooling with the advantages of melt lubrication. However, no protection is available in the event of film breakdown. Common salt (NaCl) or BaCl employed in heat treating baths, have also been used, but corrosion can be a problem. This difficulty is avoided with lubricants based on salts of carboxylic acids. Even though they cannot match the lubricating power of layer-lattice substances, they help to maintain a clean work environment.

A rather unusual approach relies on a Pb-Sn-Sb alloy as the lubricant in ausforging of steel [205].

Die Wear. The life of hot forging dies ranges from a few hundred to some tens of thousands of parts. It is short enough to have prompted serious investigations into causes of die wear, especially because die costs account for some 15% of total production costs.

1. Wear Test Methods. As do all other wear studies (Sec.5.9), studies of die wear suffer from a lack of reproducibility. Basically, two approaches are possible:

First, a large number of observations from production practice may be subjected to statistical treatment [191,206-210]. There is no doubt about the relevance of the results, but uncontrolled variables make for a wide scatter of data. Thomas [211] estimated that, after systematic effects have been allowed for, there still remains a variability of 75% that is beyond the control of the forger.

Secondly, laboratory experiments can be carried out under well-controlled conditions but must, by necessity, be limited to relatively few (at the most, a few thousand) parts [192,212-218].

Most studies are done on forging of steel but have more general implications; certainly, they apply to nonisothermal forging of all but the light-metal alloys. Earlier work is covered in a detailed review by Kannappen [219], and a full treatment is given by Lange and Meyer-Nolkemper [5].

2. Modes of Wear. It is generally agreed that most die wear can be accounted for by four mechanisms (Sec.3.10) which, in order of decreasing incidence [210], are abrasion, brittle fracture, plastic deformation, and thermal fatigue.

a. Abrasion, observed in 70% of all worn dies, is a gradual form of wear caused mostly by oxides. As shown by Sharma and Arrowsmith [220], oxide transfer to the die can be observed after the first few contacts, and abrasion proceeds with the number of blows. Wear is proportional to distance [221], which leads to a relationship such as Eq.3.39. The surface has a typical washed-out appearance [5]. In forging of a flat configuration with end-face expansion, wear is concentrated in zones of sliding rather than of sticking. Similarly, in impression dies it is concentrated in parts of the cavity where heavy material flow occurs (Fig.9.40) [213]. Initial surface roughness seems to have little effect, and roughness reaches a maximum for a given part of the cavity [221].

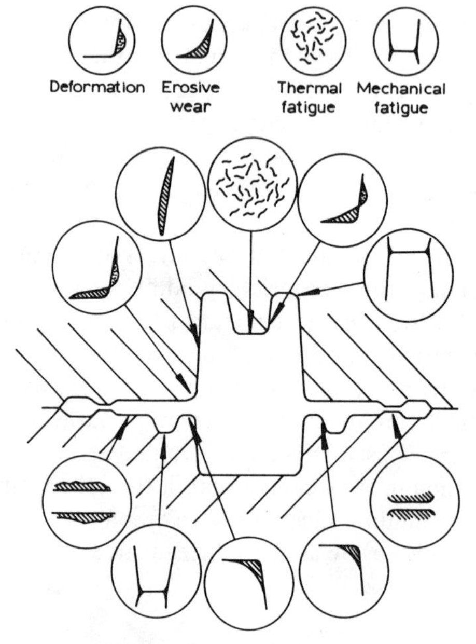

Fig.9.40. Modes and locations of die wear encountered in nonisothermal hot forging [213]

b. Brittle fracture, found in some 25% of failed dies, is attributable to mechanical fatigue. Corner cracks are most frequent, but spalling and surface cracks are also observed [210]. The main cause of brittle fracture is the development of high local stresses around stress raisers such as tight radii [207], combined with inadequate ductility of the die material. It may lead to premature destruction of the die by cracking.

c. Plastic deformation results from overloading of a ductile die at points of high pressure. It leads to indentation of flat die surfaces, upsetting of protrusions, and opening up of cavities, and repeated loading may exhaust the ductility of the die material.

d. Thermal fatigue exhibits a typical grainy appearance on the die surface and is found mostly on the flat parts of the die.

3. Influencing Variables. Most dies fail by several wear mechanisms; the effects of process conditions are closely interwoven—sometimes reinforcing and at other times opposing each other. Nevertheless, some general trends are discernible:

a. High die temperature is destructive because the die surface loses strength in a thin layer [222], which makes it less resistant to abrasion and plastic deformation. Therefore, wear under unlubricated conditions was found to be inversely proportional to die temperature in ring-upsetting tests by Rooks [192]. Conversely, rapid heating above the transition temperature on contact, followed by quenching by the cold backing, leads to the formation of a brittle, hard (martensitic) white layer, which is prone to fatigue but more resistant to abrasion [223].

b. Oxides are helpful as heat insulators but harmful as the major sources of abrasion [214]. Their effects depend on part geometry, which controls sticking and sliding, and on the entrapment of oxides. Uncontrolled oxidation is one of the main causes of scatter of results [211].

c. Lubrication has perhaps the most complex effect. In forging of flat parts it promotes end-face expansion; therefore, wear progresses over a greater part of the surface, but to a shallower depth, leading to a general reduction of wear [213]. In forging into a cavity, wear is reduced [191,215,218], unless flow past a die feature becomes so intense that the lubricant breaks down. Dean [215] found wear to increase in the warmer bottom die but to decrease in the cooler top die. Cooling on application of an aqueous lubricant is helpful in reducing bulk temperature [215a] but harmful if quenching is excessive; therefore, an aqueous lubricant can increase thermal fatigue. It is suspected that high gas pressure contributes to pitting, and directional marks are sometimes attributed to erosion by combustion gases [215]; however, direct evidence of this is missing, and the observations could be interpreted in terms of sliding (abrasion) and quenching (thermal fatigue). The most powerful factor is relative sliding; if end-face expansion is the dominant deformation mode, wear is often inversely proportional to friction, as in the work of Tolkien [74].

4. Forging Variables. The effects discussed above are sufficient to explain most practical observations:

a. The main variable is die temperature [192]. Die life decreases with increasing weight of the forging, because the higher heat content of a larger piece results in higher die temperatures [208,210].

b. Interruptions of the smooth flow of production are harmful [210], because they increase temperature excursions.

c. Contact (dwell) and cycle times affect bulk die temperatures and surface temperature gradients. In one study, an increase in contact time from 5 to 1000 ms reduced lubricated wear because of white layer formation, but increased dry wear in the absence of a white layer. Thus, it appears that a white layer forms only when dwell time is long enough for the austenitic temperature to be reached and when a water-base lubricant accelerates quenching to a degree sufficient for martensite to form [224]. Contact time increases with the number of blows on a hammer, leading to more severe wear if the hammer is too small for the part [207,210]. Wear is especially severe in the finishing blows on strong alloys [225]. Stickers are most harmful and result in much-reduced die lives, especially in press forging without an ejector [210].

d. Excessive temperature is harmful, but so are excessive temperature excursions. Thermal shock and thus thermal fatigue are minimized by appropriate preheating of the die, preferably with an evenly distributed heat source (gas or electric) rather than with a localized high-temperature source (such as a hot workpiece). The optimum preheating temperature is a function of both die and lubricant composition [5.32].

e. Die configuration, together with lubrication, determine material flow. More complex parts need higher pressure and more blows to fill in a hammer, are more likely to stick, and throw more flash, resulting in increased wear. This is also true of presses and HERF equipment [214]. A wider flash land wears less but at the expense of higher cavity pressure.

Fully valid mathematical modeling of these complex interactions is not yet feasible, but Storen et al [226] made a beginning by considering the thermal and mechanical properties of the die and workpiece, the fracture toughness of the die, the weight and complexity of the forging, and the mode of loading. Rooks [192] modeled the wear profile in upsetting.

5. Selection of Die Materials. The metallurgical aspects of die materials are beyond the scope of this monograph. In general, alloy steels are required [227],[3.426]; chromium and molybdenum are very beneficial, especially at higher concentrations. A 5% Cr steel (H11) has been found to be more wear resistant than a 0.55 C, 1.5 Ni, 0.6 Cr, 0.25 Mo steel [216]. It should be possible to reduce the large variety of die steels to only a few [228]. Major improvements have been achieved by reducing the inclusion contents of steels. Newer die materials have been reviewed by Assmann [162].

Many forms of surface coatings (Sec.3.11.5) have been employed:

a. Surfacing by weld deposition is routinely done for repairs and also for building up new dies; die lives 2 to 10 times longer have been reported [229].

b. Nitriding has proven to be useful for die inserts [230]; it is effective in reducing stickers [217] and, in various forms of treatments, is one of the better-established techniques [231-233]. It improves abrasion resistance but impairs impact wear resistance [234]. Selective nitriding is often preferable [235].

c. Diffusion coating techniques—especially boronizing [236-238] and chromizing [234]—result in die lives 2 to 3 times longer. Boronized dies are especially resistant to abrasion of die protrusions but are less effective in reducing wear in upsetting (Fig.9.41) [239].

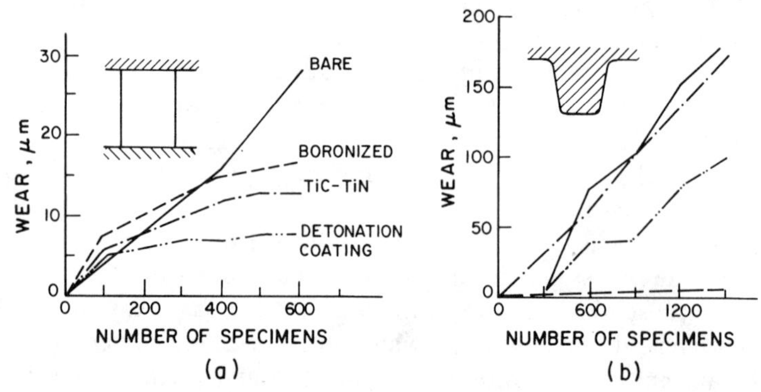

Fig.9.41. Effect of die surface coating on wear in (a) upsetting and (b) forging around a boss [239]

d. Variable success has been recorded with surface coatings such as TiC-TiN and detonation coatings of WC bonded with cobalt and chromium (Fig.9.41). A WC coating deposited by spark hardening is useful only when the substrate steel is free of inclusions that could initiate fatigue [240]. A cermet (70 Ni, 20 Cr, 5 Si, 5 B), applied as a slip and then sintered, gave improved abrasion resistance [241]. Electrodeposited Co-W and Co-Mo coatings reduced adhesive transfer [220] and improved wear resistance [242,243].

e. Because the first few contacts are decisive in the life of a die [220], pretreatment of die surfaces is quite widespread. The blue oxide film formed on heating in the presence of water vapor adsorbs and bonds the lubricant. A degreased die surface may be coated with graphite, which is then burnished to ensure alignment of platelets. The controlled roughness of a shot blasted or finely finished EDM surface helps in retaining the graphite.

9.6 Lubrication of Iron- and Nickel-Base Alloys

Apart from sheet metalworking, forging is the most important manufacturing process for making finished parts. It is then not surprising that lubrication practices continue to be developed and improved. A general review has been given by Busby and Fogg [244].

9.6.1 Cold Forging and Extrusion

Cold forging of steel encompasses operations that vary widely in severity, with an appropriately broad range of lubricants for processes variously described as cold heading [119,245-247], cold forming [156, 248-250], four-slide forming [251], and cold extrusion [11-14,252,253]. In general, lubricants are either tailored for specific duties or formulated as general-purpose lubricants with additive contents sufficient for the most severe duty anticipated.

Cold Heading. Nails and bolts are made by upsetting heads on segments of wire or bar. This imposes fairly mild conditions, and an emulsion or a noncompounded mineral oil should, in principle, suffice [119,254]. However, freedom from die pickup, good product finish, and long die life make additives desirable. This is especially true when the wire is given a light sizing draw prior to entering the cold header, in preparation for more severe working. Fatty oils are adequate for lighter duties, but E.P. additives are mandatory in more severe upsetting [246,255]. Tanaka et al [100] found chlorine to be the most effective, but only at slightly elevated temperatures. More frequently, sulfur or phosphorus is added [120].

The higher adhesion of stainless steel results in sticking friction under dry conditions [100]. Their lack of reactivity with fatty acids renders these lubricants ineffective unless they are of high viscosity (such as palm oil or tallow) or are semisolid at the forging temperature (stearic acid below 60 C). The high viscosities of waxes and the reactivities of highly chlorinated oils and chlorinated resins make them also suitable. Sulfur has been found to be effective [100], but residues can cause grain-boundary corrosion during heat treatment.

The repeated contact typical of cold heading should create ideal conditions for the formation of a friction polymer. Such formation is affected also by the metallurgical characteristics of the die and workpiece. Delamare and Kubie [118] found in plane-strain compression with a compounded cold rolling lubricant that the transfer film developed more rapidly with tooling that had more and harder carbides, and surmised that these carbides had rubbed the surface of the workpiece and thus had prompted the development of a transfer film. Friction was initially lower and reached its steady-state value more rapidly with M4 high speed steel than with D2 tool steel (Fig.9.42). When various batches of austenitic 302 stainless steel were tested, friction was found to be higher for those batches that formed martensite on deformation. This may partially explain the higher friction that is associated with the ferritic type 430 stainless steel. For this material, surface condition also appeared to be important; small quantities of carbide particles found on the surfaces of some sheets contributed to increased friction.

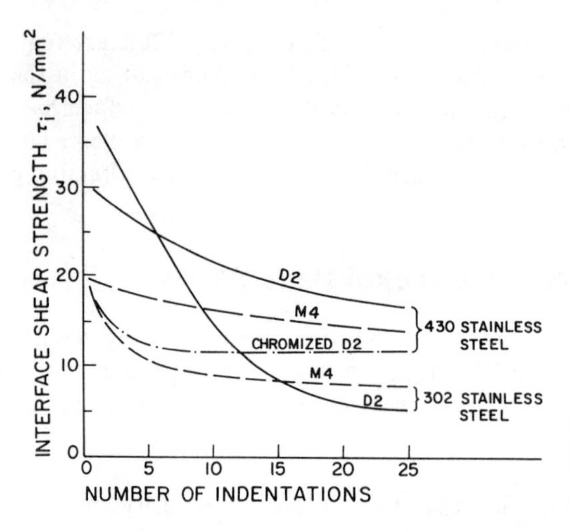

Fig.9.42. Formation of well-lubricating films in repeated indentation of stainless steel plates with anvils of different compositions and with a compounded mineral oil lubricant [118]

Occasionally graphite or MoS_2 is added to the lubricant, but the removal of the resulting residues creates a problem.

Upsetting Combined with Modest Extrusion. Intermediate lubricant duties are typical of operations such as forward extrusion of a bolt shank or indentation or back extrusion of a hollow in a bolt head. Lubricant residues on the wire surface help in feeding the wire, provide lubrication in the course of upsetting, and suffice for light forward extrusion. The stearate soap in a lime carrier, which remains after dry drawing (Sec.7.3), is very effective, and metallic coatings such as copper are even more so, especially on stainless steel. However, the removal of copper may create pollution problems, and the use of MoS_2 has been advocated [256]. A phosphate coating on steel or an oxalate coating on stainless steel, together with the superimposed soap layer, is most effective [257,258].

A problem common to all prelubricated wire is that the freshly cut end face is totally unprotected, and therefore supplementary lubrication is essential [254]. More severe deformation must be distributed over several blows [259] to allow access of fresh lubricant. The oil-base lubricant frequently contains sulfonated fatty oil in conjunction with soaps such as stearates or palmitates [260]. Lead, copper, and tin soaps seem to be equally effective [261].

Cold Extrusion. The terminology is diffuse but the term "cold extrusion" is usually reserved for operations involving forward or back extrusion or combinations of the two. The severity of lubricant duty depends on the reduction (and surface expansion). Basically, two approaches are feasible:

a. The advantage of working from a bar can be retained if a cold header with several die stations is used in conjunction with a flood of heavily compounded oils of which some contain graphite [260,262].

b. More typically, slugs are cut from the bar and lubricated over their entire surfaces. Often the sheared ends of the slugs are smoothed and reshaped in an upsetting operation for which liquid lubricants may suffice. Slugs of higher-carbon steels (say, in excess of 0.4% carbon) are spheroidized for maximum ductility. Prior to extrusion, the slug or preform is coated with the lubricant over its entire surface.

1. Lubrication of Slugs. Lubricants are of several classes:

a. Metal coatings in conjunction with liquid lubricants were the first to be used.

Tin is a possibility [263], especially when the part would have to be tinned anyway for corrosion resistance. Zinc has been found to be effective although it is not clear whether or not it really ensures lower forces than does the more common phosphate coating [264,265]. In general, differences between various metals, such as copper, cadmium, and zinc [266], are small, and the choice must depend on factors other than lubrication. Lead was commonly used on stainless steel but fell into disfavor because of its toxicity. Copper is still occasionally used [260,267]. Recommendations on film thickness range from 2 μm [253] to 20 μm [267]. The optimum thickness must depend on the severity of operation and on the application method: excessively thin coatings fail, and those that are too thick flake off.

b. Solid lubricants in a great variety have been tested and recommended at one time or another. Under laboratory conditions, with a limited number of extrusions, many lubricants are successful [129]. These include graphite and MoS_2 applied as pastes, applied as dry coatings [268] or bonded films, or dispersed in fatty bases such as lanolin, rapeseed oil, or tallow [265].

c. Polymer coatings have been repeatedly tested because some of them are readily removed with organic solvents. However, protection against breakdown and tool pickup cannot be guaranteed. This is true even of polysulfides, which could be expected to react with the metal [269], and of PTFE [268,270], which actually gives quite low extrusion forces. Somewhat better resistance to breakdown can be obtained with heavier coatings, but rippling (bambooing) of the extruded surface may then set in.

d. Liquid lubricants in general cannot prevent breakdown in severe extrusion. Occasional successes are reported, but only at relatively light duties [121]. A high viscosity is helpful when the process allows a fluid film to develop. Hydrostatic extrusion would appear attractive, too, but the improvement relative to a phosphate/soap system is so small that the loss in productivity (and the greater expense) cannot be justified.

e. The vast majority of industrial cold extrusion is conducted with coatings. Coatings range from methacrylic resin through lime, borax, rust, intentionally grown oxides, and sulfurized surfaces [271]. Although extrusion pressures are often quite acceptable, no other coating can match the performance of conversion coatings such as phosphate coatings on steels [133,156,272-279] and oxalate coatings on stainless steels [267,272,280,281] (there are innumerable articles on this subject, and only a selection is quoted here). Details of such treatments are discussed in Sec.4.7.8. These coatings often make cold extrusion competitive with warm and hot forging, as illustrated for a socket wrench by Sieber [282]. Basically, two variables can be controlled [4.254],[249,250]:

First, the weight of the phosphate coating is varied according to the severity of operation, and ranges from 10 to 20 g/m^2 for lighter duties and from 20 to 35 g/m^2 for heaviest back extrusion [283,284]. Laboratory experiments and even the forces measured during production [285] often fail to reveal any difference between much lighter and much heavier coatings [272,273], but the benefits of a heavier coating for severe duties are evident in long production runs; scoring, die pickup, and die wear are all reduced. However, excessive coating thicknesses can cause buildup of excess lubricant [133,272].

The second variable is the lubricant applied on top of the phosphate film. Fats, waxes, E.P. compounds, and layer-lattice compounds all show reasonable performance [133,268,286], although extrusion forces are often higher with layer-lattice compounds [263,271]. Heavy E.P. oils are best when lubricant trapping occurs. All of the above lubricants have been reported in industrial use at one time or another [249,250,252,258,284]. Direct comparison of performance is difficult: for example, graphite has sometimes been found to be poor [287], but at other times very good [288].

Obviously, process conditions must influence lubricant ranking. The particle size may also have an effect: Georgescu-Cocos [289] found MoS_2 of coarser particle size to give higher friction in ring compression than finer-grain ($<30\ \mu m$) varieties. The standard lubricant is still a reacted sodium stearate soap [290], which represents the bench mark against which all other lubricants must be measured. While extrusion forces may not vary greatly, differences are immediately evident from punch wear (Fig.9.34a) and from the roughness of the inside surface of a back-extruded can (Fig.9.34b) [133].

A significant and fairly recent development is the industrial application of an MoS_2 film on top of the phosphate/soap system [4.180,4.254,4.259]. The film is deposited by barreling in dry MoS_2 powder or from proprietary aqueous suspensions [4.179],[291-293]. The additional lubrication and surface protection afforded by the MoS_2 film allow very severe operations to be performed in a single stroke, as exemplified by the combined forward and back extrusion of spark plug bodies. Lanolin and tallow cannot match MoS_2 [294].

2. Effect of Workpiece Material. The composition of the steel and the associated flow stress variations should have an effect on the phosphate film, but such effects are not evident from forces measured in back extrusion [295].

Steels containing more than 5% Cr are difficult to phosphate and thus are oxalated. This is true also of nickel-base alloys.

For severe duties, phosphating of the wearing surfaces of the tool set has been attempted [283], but it is then more difficult to strip the extrusion off the punch.

3. Lubricant Evaluation. Lubricant development often utilizes the ring-compression test (in fact, this test was originally developed for evaluation of cold extrusion lubricants) [31]. The test correctly predicts such features as the breakdown of the phosphate/soap system above 200 C [154,296] and the improvement in lubricating performance with increasing coating weight. Ashar and Weinmann [297] found best performance to occur at weights above 24 g/m^2, where $\mu = 0.04$.

There is no doubt that back extrusion is the most discriminating test for judging the resistance of the lubricant to breakdown [153,154,295]. Kunogi [31] used the configuration shown in Fig.9.24e.

It should be noted that the absolute magnitude of the extrusion force is seldom a reliable indicator. More important are the trends observed in the course of repeated experiments, from which extrapolation to industrial conditions can be more reliably made. With the best lubricants, long laboratory runs are needed for establishing resistance to lubricant breakdown.

Powder Compaction. In powder compaction the lubricant may be mixed into the powder prior to compaction. It must not impair the flow of the powder, and therefore a small quantity (0.5 to 1%) of a very finely distributed semisolid lubricant such as stearic acid, zinc stearate, or lithium stearate is added [87-89]. Larger additions reduce the density of the compact. The lubricant must be burnt off prior to sintering, and some commercial waxes have been developed to have lower burn-off temperatures than zinc and, particularly, lithium stearates. Evaporation of the lubricant may cause swelling or spalling of the part and contamination of the furnace atmosphere, and residues may remain in the compact; therefore, efforts are directed toward compaction without admixed lubricant.

It is essential, however, to lubricate the die wall and thus facilitate compaction and ejection. When an admixed lubricant is used, lubricant exuding from between particles serves as the die-wall lubricant. In its absence, the lubricant must be applied to the die surface between successive parts.

9.6.2 Warm Forging and Extrusion

Few organic lubricants can survive temperatures in excess of 300 C. In upsetting, PTFE can be used at up to 400 C [298]. The industrially important warm forging of steel in the temperature range from 600 to 800 C eliminates all organic lubricants as possibilities. Only two choices remain:

1. Practical lubrication relies almost exlusively on MoS_2 or graphite deposited from an aqueous dispersion. Most experimental investigations have been based on commercial formulations of undisclosed composition, and thus few generalized conclusions can be drawn. Molybdenum disulfide is useful only at low temperatures and short heating times [153-156]. Graphitic formulations, especially those with oxidation-protection additives [299], can survive higher temperatures. There are obvious differences in friction and resistance to breakdown, but these differences cannot be meaningfully correlated with the incomplete information that has been given on particle size, nature and quantity of binders, additives, and fillers. A ZnS paste, applied intermittently to the die, has been reported to match the performance of graphite [165].

The lubricant is normally applied to the die, although at least one commercial lubricant can be applied to the workpiece [291]. Graphite bonded with the glass-forming B_2O_3 serves well up to 800 C, but MoS_2 gives excessively high friction [156]. It has been shown that aqueous colloidal graphite dispersions containing small quantities of cadmium oxide can develop firmly bonded films on steel and reduce friction in plane-strain compression at around 500 C [152], but toxicity would be an objection to commercial use of this technique.

2. Essentially thick-film lubrication is ensured by use of lubricants such as the Bi_2O_3 + $Na_2B_4O_7$ mixture formulated by Saga [156]. Although this lubricant is specifically recommended for stainless steel, its industrial usage could not be ascertained.

Increased ductility makes warm heading of steel (at about 600 C) and of stainless steel (at 300 to 600 C) attractive [300]. At relatively low temperatures (up to 400 C), liquid lubricants suffice if the wire is copper coated [119]. A high-chlorine oil as well as colloidal graphite are also satisfactory [300], as is a fat-base lubricant applied by an airless spray method [301]. Graphite in oil was the best lubricant in ring upsetting of stainless steel [302]. Warm working on multistage presses is also practiced [303].

In ring-compression tests on nickel-base superalloys, Berry et al [148,149] found BN and MoO_3 to perform essentially a parting rather than a lubricating function (μ = 0.3). Mixtures of MoO_3 with a graphited oil were better, but were surpassed in the temperature range from 150 to 500 C by 30% MoS_2 added to a graphited oil.

9.6.3 Hot Forging

Hot forging temperatures in excess of 1000 C severely limit lubricant choices. Differences in practice arise from the different oxidation characteristics of the various workpiece materials.

1. In a recent study, Luong and Heijkoop [304] chose atmospheres so as to obtain scale of controlled thickness, consisting entirely of one type of oxide. Thicker scale, up to a limit, reduced friction. Contrary to earlier suggestions that FeO is responsible for reduced friction, they found mixed oxides to be more effective. Porosity of scale and the composition of the workpiece were also important. Thus, in unlubricated forging the situation is still not entirely clear. Some well-distributed oxide is beneficial, but heavy scale contributes to die wear and results in forged-in scale pockets. Scale blow-off is, therefore, important. On stainless steel the scale is more tenacious and less effective, as shown by the ring-compression results of Male (Fig.9.43) [305].

2. The most widely used lubricant is graphite in an aqueous carrier, which is

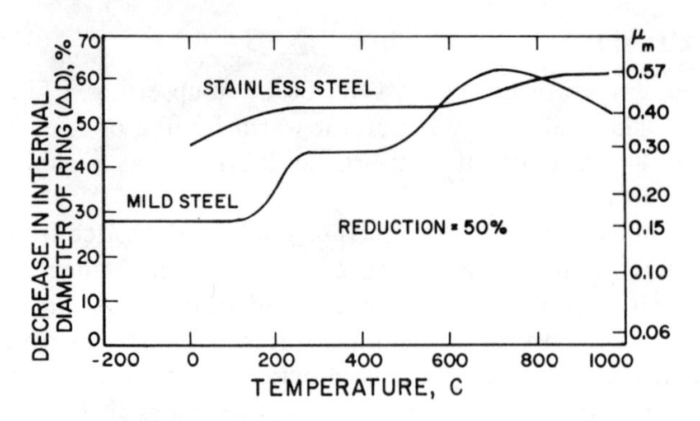

Fig.9.43. Friction measured in hot ring upsetting [305]

applied to the die between strokes by appropriate equipment [194-196,306-308]. Oily carriers, once popular, are now shunned because of the pollution caused by the burning oil [307,308]. In forging of steel, the superiority of graphite over other lubricants has been repeatedly shown in upsetting [80,104,183,309], ring upsetting [104,150,183,217], simplified forging [70,74,183], and production forging on hammers [80,217,310] and high-energy-rate equipment [311,312]. Aqueous suspensions have proven to be equal or superior to oily mixtures, at least from the point of view of end-face expansion [37,183] and ejection force (Table 9.1), although gas pressure is higher with an oil-base lubricant [74,309]. In general, a lubricant that ensures the greatest end-face expansion does not necessarily give the lowest ejection forces [74] or the least roughening of the forging die. Tolkien [310] found an oil-base lubricant to be better than an aqueous one (although this may reflect the state of the art in 1960). A heavy oil-base lubricant performs better when it is trapped, as in an extrusion-type operation [313]. None of the solid lubricants can prevent metal-to-metal contact, and if the workpiece is prone to adhesion some tool pickup is inevitable.

Table 9.1. Comparative performance of some forging lubricants [310]

	Performance(a)			
Lubricant	Sliding friction	Sticking friction	Gas pressure	Die wear (dimensional change)
Sawdust	4	4	1	1
3.6% colloidal graphite in water	1	1	5	3
4% colloidal graphite in oil	3	3	2	2
Polyalkylene glycol	5	5	4	4
17% Na_2CO_3 in water	2	2	3	5

(a) Best performance is marked by 1, poorest by 5.

3. Various fillers, carriers, and bonding agents have been suggested for specific applications. Binders may either lessen or intensify wear [215a]; unfortunately, most of the relevant information is proprietary. A graphited oil benefited from the addition of MoO_3 in ring-compression of superalloys [148]. Addition of clay to a graphite suspension was reported to improve lubrication and reduce pollution [314]. In an extensive study, a suspension of grease with waterglass as a binder formed an effective heat-insulating film in hammer forging [315]. Grease-base lubricants once were popular and proved successful in laboratory experiments on stainless steel [8.195]. Some commercial lubricants consist of particles of metals such as aluminum, copper, or lead in a grease

base. They usually give quite low friction [104,174], although in the work of Bühler and Löwen [5.97] the shear stresses were no lower than those obtained with graphite in water (Fig.9.44).

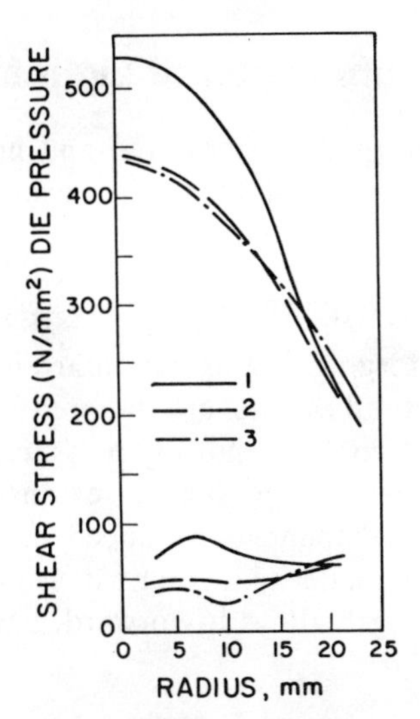

Fig.9.44. Die pressures and shear stresses measured in upsetting of mild steel at 1000 C to 50% reduction in height (d_0 = 35 mm; h_0 = 21 mm) (after [171]). 1 - Dry. 2 - Graphite in water. 3 - Copper in bentone grease.

4. Sawdust was one of the earliest lubricants, thrown by hand at critical points of the die cavity. Its effects are perhaps manifold [310,316]. Gases developed on contact with the hot workpiece form a gas cushion and lead to high ejection pressures (Table 9.1), and combustion residues serve as parting agents. Sawdust reduces wear and, occasionally, is still found in practice.

5. Glasses, preapplied to the workpiece from an aqueous slurry or to a preheated workpiece in the form of powder, form a smooth coating only if no thick oxide layer develops. Therefore, glasses are used primarily in forging of stainless steels and nickel-base alloys [203,317]. To protect against pickup in the event of film failure, a thin graphite coating is usually applied to the die. The glass film reduces heat transfer only when it is thick enough, as was found in upsetting in a container in preparation for piercing [318], but the thin films preferred for good surface finish in forging of shapes have very little effect on cooling of the billet or heating of the die [203]. A heavier glass coating which feeds out gradually is beneficial in piercing of billets [319].

6. More recently, salts have again been advocated, especially where cleanness is important. Various proprietary, nongraphited (nonpigmented) lubricants are also available. No systematic study of the performance of these lubricants relative to graphite has been published, but they are known to perform satisfactorily in production of all but the most difficult, complex forgings.

7. In forging of superalloys, prevention of cooling becomes essential when the ductility of the workpiece drops steeply below the hot forging temperature range. An insulating layer of glass or an asbestos mat reduces chilling. In the most critical cases canning may be used, in much the same manner in which it is used in extrusion (Sec.8.5.4), although at the expense of detail in the finished forging. Sprayed or electrodeposited metal coatings are better in this respect, and the lubricant can be chosen to fit the coating. Afterward, the coating must be removed by chemical means.

8. One should not forget that the tribological system includes the die surface. For example, Felder and Montagut [217], showed that nitriding reduces the number of stickers and leads to lower friction in forging without a lubricant.

9.7 Lubrication of Light Metals

Aluminum and its alloys are forged by both cold and hot forging techniques, whereas the limited number of slip systems available in the hexagonal magnesium and beryllium alloys necessitates deformation at elevated temperatures.

9.7.1 Cold Forging and Extrusion

Lubrication practices in cold forging and extrusion are influenced by the high adhesion of aluminum to the die, and lubricants are tailored to specific duties. Light duty involves predominantly normal contact and low pressures and is exemplified by upsetting, whereas heavy duty involves large surface extension and high pressures and is exemplified by extrusion. The components produced in the largest quantities are commercially pure aluminum tubes made by impact extrusion; smaller quantities of components are extruded in high-strength alloys by forward, backward, and combination techniques.

Cold Forging. All experimental evidence supports the thesis that lubrication in cold forging is of the mixed-film type. In all forms of compression, lower friction is invariably observed with higher lubricant viscosity and speed [100,116,320,321],[5.20,5.264], but it is highly dependent also on the nature and concentration of additives. As is true for aluminum in general, E.P. additives are not useful [100],[3.492], but fatty oils and their derivatives are. Guminski and Willis [5.264] found in plane-strain compression that the reduction attained at a constant applied load increased with the chain length of the additive (Fig.9.45). For a given chain length, branched molecules were less effective than straight-chain molecules, and saturated acids were more effective than unsaturated ones. The formation of a friction polymer was important, as shown by the fact that reduction increased with the number of impressions taken without cleaning the anvils. Johnston and Atkinson [322] observed that, at constant load, reduction gradually increased in response to addition of boundary lubricants, reaching a maximum at concentrations over 1%. As might be expected, fats such as tallow are very effective [323] because they combine high viscosity with boundary activity. Solid lubricants such as

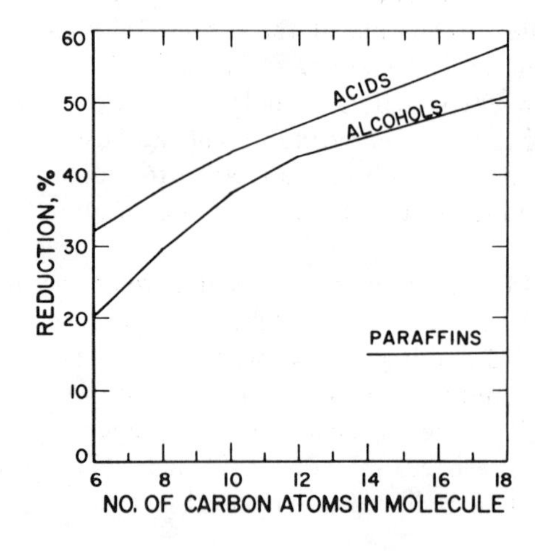

Fig.9.45. Reductions obtained in plane-strain compression of Al-1.25%Mn alloy at a constant load (2% additive in light mineral oil base) [5.264]

layer-lattice compounds are of debatable value. Some experiments have shown them to be effective [321], but equally good lubrication can be attained without them, and their removal creates difficulties. Ultrasonic vibration improves lubrication if a poor lubricant is used [324], but its use cannot be justified.

The presence of a liquid roughens the product surface; this subject is explored in substantial detail by Butler (for a summary of his work, see [96]). In practice, a smooth surface is usually desired, and the thinnest lubricant film that still ensures good lubrication is used. Film thickness is controlled by viscosity and velocity or, in the case of highly viscous substances, by the quantity applied. The extent of the matte center observed in upsetting and plane-strain compression is governed by hydrodynamic factors and is not affected by boundary additives [325].

Die surface roughness is important even in compression under ostensibly dry conditions. Dellavia et al [326] found decreasing friction with diminishing roughness, no doubt because the asperities of the rougher tool broke through the very thin surface film existing under unlubricated conditions. With liquid lubricants, as in mixed-film lubrication in general, the die should be smooth. This also ensures maximum smoothing of the workpiece, as shown by Vater et al [327] in a statistically designed study of the cigar test.

Cold Extrusion. Semisolids are the standard lubricants for cold extrusion [328]. The high adhesion of aluminum to the die makes the effect of lubricant clearly evident, both in surface finish and in forces, as shown by the statistically designed experiments of Farmer [329]. Pickup on the punch and scoring of the extruded surface are particularly sensitive to lubrication in back extrusion.

It appears that any lubricant of sufficiently high effective viscosity that is capable of expanding with the surface and providing boundary lubrication can be used. Sulfonated tallow is frequently mentioned [330-332]; lanolin is most common [333,334],[5.12], as are soaps—particularly zinc stearate [259,334,335] and zinc arachinate [335]. Waxes of medium hardness (melting at 33 to 43 C) combined with fats or fatty acids, and other specially developed lubricants, are sometimes used [334].

Vinyl stearate was found to be effective by Owens and Roberts [336]. Polymers such as PTFE provide low friction but cannot prevent scoring on breakdown. A clear silicone wax, deposited from a chlorinated solvent, is reported to function well [4.58].

Layer-lattice compounds are unable to repair breaks in the lubricant film and are thus of little use on their own, but MoS_2 added to wax [337] or tallow has been advocated for more severe duties. Kemppinen [338] found a layer of MoS_2 superimposed on a soap film to be effective in extrusion of harder alloys. PTFE in tetrahydronaphthalene performed as well as MoS_2 in oil [99.338a]. Only for the most severe extrusion duties is a conversion (phosphate) coating with soap lubrication required [122,272].

Success of lubrication and quality of the finished product depend very much on control of lubricant application. Deposition of the lubricant is often done by tumbling, which also aids lubrication by producing a matte, ideally roughened surface [334]. Lubricant film thickness usually is controlled by dilution in an organic volatile solvent such as a light mineral fraction or, if proper precautions are taken, carbon tetrachloride. A film approximately 0.02 mm thick seems to suffice for severe duties. Bastian [4.2] recommends lubricant quantities ranging from 0.8 to 1% of metal weight.

9.7.2 Warm and Hot Working

Typical forging temperatures are above 300 C for aluminum and magnesium and between 650 and 1100 C for beryllium (Table 2.1).

a. With increasing temperature, organic lubricants gradually fail, although aluminum stearate has been shown to be effective up to 400 C in compression [100]. Water-soluble soaps used to be employed as hammer-die lubricants. A coating of PTFE applied to cold dies before the dies are heated can produce good die filling [8.195], but toxic fumes are generated on contact with the hot workpiece.

b. Because of their stability, layer-lattice compounds are generally used. There is little evidence to recommend the application of MoS_2, even though in the experiments of Schey and Newnham [3.492] it provided remarkably low friction and very good resistance to breakdown, especially when applied to a boronized die (Fig.3.67). However, no evidence could be found of the practical application of such a die-lubricant combination.

c. Graphite is the most generally employed lubricant. For many years, standard practice called for dispersion of the graphite in an oily carrier, and the value of this system has been repeatedly demonstrated in laboratory upsetting and trapped-die forging [8.195] and also in tests based on ejection force [5.20]. Much of the forging of aluminum and magnesium is near-isothermal forging at die temperatures between 315 and 425 C, and wetting of the die by the oily carrier is a major reason for the good performance that is observed. The high gas pressures and low ejection forces are also helpful. Critical experiments of Shah [339], ranging from simulation to full-scale forging, indicated that oil-base lubricants still provide lower friction and better surface finish than aqueous ones, although acceptable performance can be ensured if some oil is emulsified in the water carrier.

Schiefer and Holinski [4.58] reported greatly improved performance with a formulation consisting of fairly coarse graphite in a silicone emulsion: very fine graphite (of unspecified particle size) was not able to form a film. Some earlier work also indicated that coarser graphite should give better results, but Shah [339] and Lockwood et al [339a] found colloidal graphite (of 1 to 5 μm particle size) to be invariably best. Continuing laboratory and plant experiments of Lockwood and co-workers [339a] have shown that water-base formulations perform well if they contain both graphite and high-molecular-weight hydrocarbons or polymers. Application was critical in their work. An air-atomizing system gave poor coverage and became clogged rapidly; a non-air-atomizing system (originally used for oil-base lubricants) was useless; a system incorporating a sonic nozzle was best.

Graphite applied to the workpiece as a dry, rubbed-on coating is ineffective [3.328], but improved lubrication is ensured if the workpiece is etched in a caustic solution (10% NaOH) and then dipped in an aqueous graphite dispersion which is allowed to dry before heating. The fresh, roughened aluminum surface gives better bonding than an old, heavily oxidized one, but the forging must be cleaned to forestall corrosion; residual graphite leads to pitting corrosion. A caustic soda etch at 40 C, followed by water rinsing, nitric acid desmutting, and water rinsing, is effective.

The importance of die surface finish is illustrated by the results of Schey and Lonn [3.328]. A graphite film applied to polished dies broke down in the third or fourth indentation, as shown by rapidly rising forces (Fig.9.46) and pickup at the anvil edges. In contrast, the coating survived when it was applied to bead-blasted anvils (R_a = approx. 1 μm).

d. Several attempts aimed at improving the graphite lubrication system have met with little commercial success. Bailey and Singer [151] found that a small quantity of CdO admixed with graphite served as a parting agent, prevented die pickup in plane-strain compression, and reduced friction at heavy reductions. However, the toxicity of CdO is a problem.

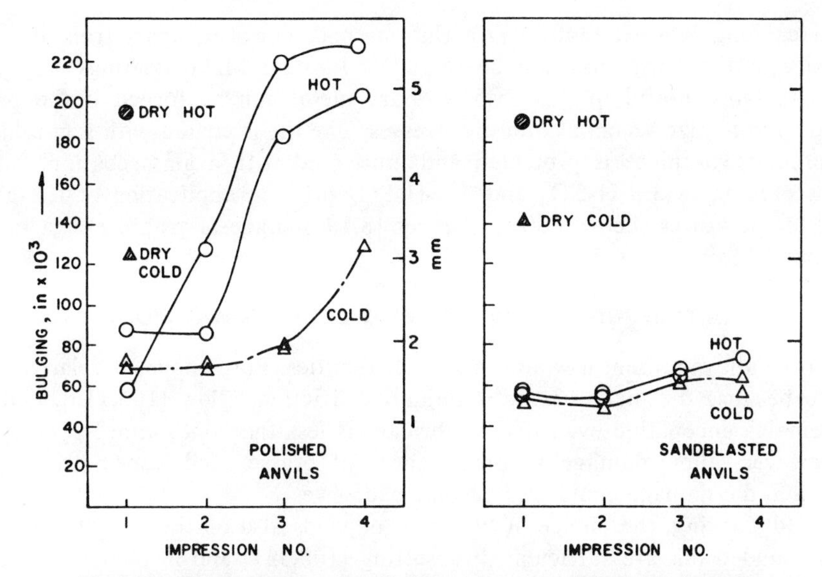

Fig.9.46. Effect of tool surface finish on wear of a predeposited graphitic lubricant during repeated plane-strain compression of aluminum at 400 C (hot) and at room temperature (cold), as judged from bulging of the edge profile [3.328]

e. Of the polymers, PTFE applied to the workpiece improved die filling in laboratory tests [8.195] but generates toxic fumes on heating.

f. Commercial oil-base lubricants with lead naphthanate additive have given satisfactory performance [339b] but present—in addition to the smoke problem common to other oil-base lubricants—possible hazards of lead poisoning.

g. Salt-base aqueous formulations, recommended for steel, are unable to prevent aluminum pickup [339],[4.58].

Magnesium lubrication technology follows aluminum practice, with graphite the dominant lubricant. The friability of magnesium oxide is helpful and reduces friction in dry upsetting (Fig.9.47) [305], but lubrication is still essential. In laboratory experiments a mixture of graphite and MoS_2 gave the deepest die filling [8.195], but industrial use could not be ascertained. Pitting corrosion is a danger here, too, and therefore parts are cleaned by sandblasting and then etched in an 8% nitric, 2% sulfuric acid solution and rinsed.

Beryllium fractures easily if tensile stresses are generated during forging, and the toxicity of beryllium oxide is a problem at high temperatures. Therefore, early practice

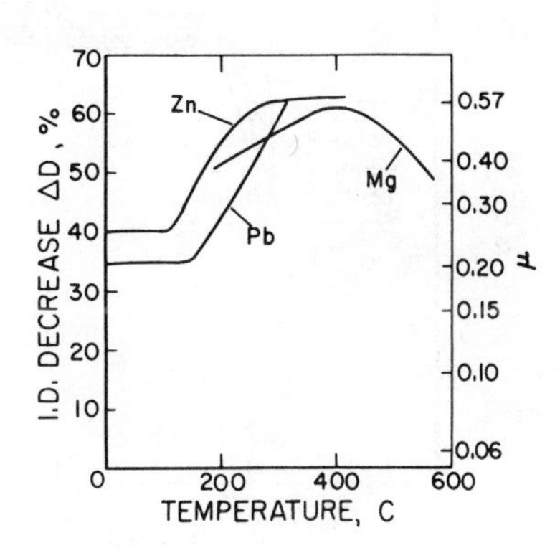

Fig.9.47. Friction in unlubricated ring upsetting of zinc, lead, and magnesium [305]

involved canning in steel [340]. With this method, however, apart from its cost, the complexity of the parts that can be forged is limited. Metal coatings (Ag, Ni, Cu) [8.197] are more useful in this respect. Bare metal can be forged if the process is designed to minimize secondary tensile stresses. The die is coated with a graphite lubricant, and the workpiece is protected and lubricated with a glass coating. Jain [308] indicates etching in 3% H_2SO_4 and 3% H_3PO_4 prior to application of the glass frit. For low temperatures (below 500 C) Beaver [8.197] suggests graphite or a mixture of graphite and MoS_2.

9.8 Lubrication of Copper and Other Nonferrous Metals

Copper-base alloys present a wide range of difficulties. Pure copper is relatively easy to lubricate because die pickup is not cumulative; friction is low [100,116], and surface finish remains acceptable even if the lubricant is less than optimum. A poor lubricant does, however, allow damage to the surfaces of brasses and, especially, the harder bronzes, and die damage could also become excessive.

In cold working, the choice of lubricant depends also on the severity of operation. Noncompounded oils are sufficient in upsetting [100,321] and in plane-strain compression [116]. Unsaturated fatty acids such as oleic acid are ineffective or may even increase friction and pickup because the soap is soluble in oil, whereas saturated acids such as the semisolid stearic acid are effective [100]. Many E.P. additives are useless, and lubricants that contain sulfur are avoided because of the unsightly dark staining that accompanies formation of CuS. For this reason, sulfonated tallow is undesirable, even though it has been frequently used in experimental work [331,332] and has been recommended by some authors. Much better are the nonsulfurized semisolids such as lanolin, zinc stearate, soaps, tallow, and waxes. Graphite greases are also effective [104] but introduce unnecessary cleaning problems. Polymers, and particularly PTFE, can be very satisfactory; however, in view of the sufficiency of viscous and semisolid lubricants, they are seldom used.

When workpiece temperatures are increased to 400 C, organic lubricants gradually fail, but without catastrophic consequences for copper [100]. At yet higher temperatures, the copper oxide acts as a lubricant (Fig.9.48) [305]. Under unlubricated conditions, friction is lower with a smoother die. As discussed in Sec.3.3.1, the oxide of brass is abrasive and thus increases friction (Fig.9.48). Hot forging dies used to be lubricated with graphite in oil, but this has gradually been replaced with graphite in water [308].

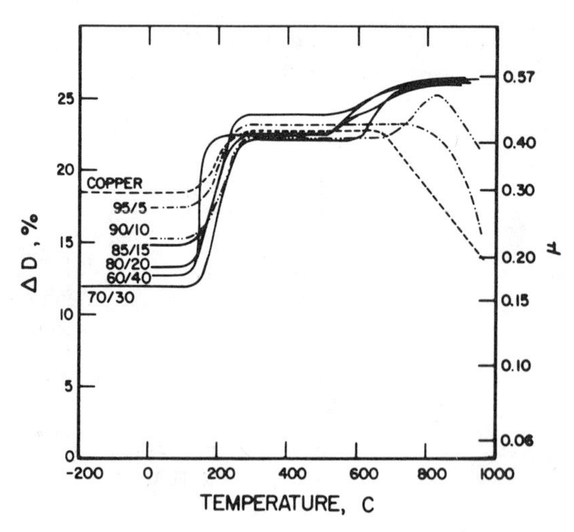

Fig.9.48. Friction in unlubricated ring upsetting of copper and several brasses [305]

Lead and tin were the first materials to be impact extruded. Their low flow strength, combined with favorable adhesion properties, make lubrication of these metals a relatively easy task. In upsetting with a viscous lubricant, the speed effect is clearly observable [108]. Viscous or semisolid lubricants, such as cottonseed oil, hydrogenated cottonseed oil, zinc stearate, wax, and fatty acid mixtures, have been employed for extrusion [4.2]. In experiments, tallow was used for tin and sulfonated tallow for lead [332]. Because of the softness of the workpiece, excessively hard waxes may damage the part. If the part is to be subsequently decorated, removal is a major concern and zinc stearate is avoided. Zinc, even though of hexagonal structure, has a low enough flow strength to be classified with lead and tin, and lanolin seems to serve well for impact extrusion of zinc products [341].

9.9 Lubrication of Titanium and Zirconium Alloys

Titanium and zirconium share severe die/workpiece adhesion problems and require similar lubrication practices.

1. Cold working is quite limited. High adhesion results in sticking friction in dry upsetting (Fig.9.49) [342]. Liquid lubricants are only moderately effective and then only in simple upsetting. The heavier oxide film that is formed when air is admitted on cooling from the annealing temperature is useful only when surface extension is not large. Otherwise, a more effective parting/lubricant carrier film must be applied. This can be a metal (usually copper) but more frequently is a fluoride-phosphate coating (Sec.4.7.8). The lubricant may be a soap or, for heavier duties, graphite in a gum resin, which provided the best compromise with regard to finish and friction in the work of Sabroff et al [343,344]. Wax reduced friction more but did not protect the extrusion against damage on ejection.

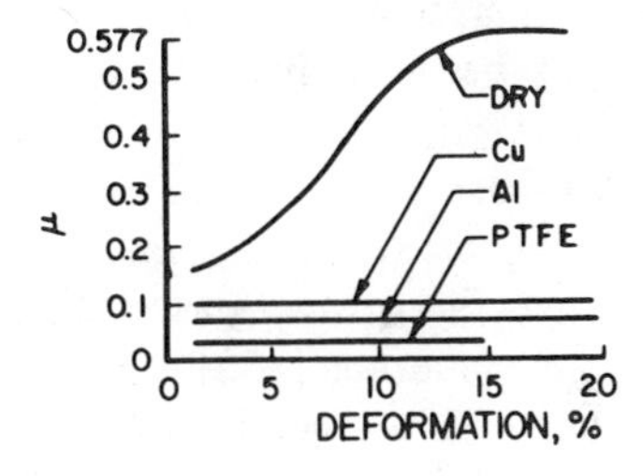

Fig.9.49. Friction measured in cold compression of titanium rings with metal and PTFE films [342]

2. Warm working at intermediate temperatures is aided by the chloride coating of Farr et al [4.186]. In forming below the typical hot forging temperatures, solid-film lubrication is common, typically with commercial compounds based on graphite, MoS_2, or a mixture of the two. Berry et al [345] found a mixture of commercial graphite-in-oil (HDL) and MoS_2 to be particularly effective in isothermal ring forging (Fig.9.50). Titanium oxidizes in air, but the oxide is hard, breaks up, and fails to prevent adhesion. Thus, friction increases with increasing temperature even when forging is carried out under isothermal conditions to avoid cooling effects. In a practical sense, sticking friction is typical of temperatures above 400 C [149]. At such temperatures the choice of die material becomes critical to prevention of adhesion; in forging of Ti-6Al-6V-2Sn at 675 C, Meehanite was best [346].

3. The choice of hot working temperatures is influenced by the beta (BCC) to alpha (HCP) transformation that takes place in titanium at 960 C; the properties of the finished forging are greatly affected by the temperature of deformation. Alloys that

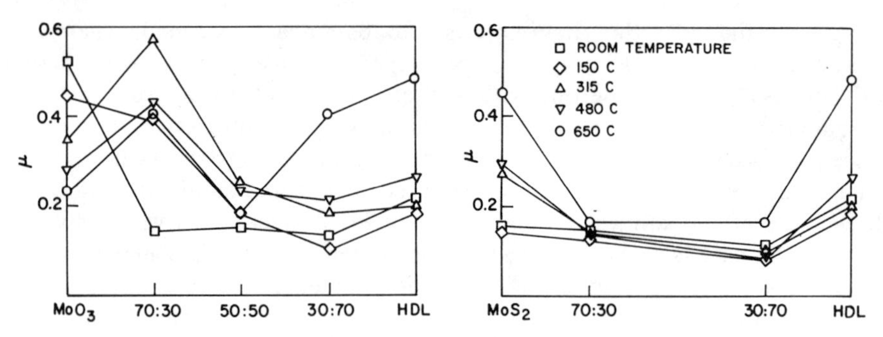

Fig.9.50. Friction measured in compression of alloy Ti-6Al-4V rings (HDL is a commercial oil-base graphite preparation) [345]

possess $\alpha + \beta$ structures at room temperature are usually forged either above or below their transformation temperatures, and thus forging temperatures range from 1060 to 840 C, whereas the beta-stabilized alloys can be forged at lower temperatures (920 to 800 C). Lubrication practices are influenced by forging temperature, die temperature, and severity of deformation.

a. Nonisothermal forging with dies heated to 250 to 425 C is possible with solid die lubricants [347,348], such as commercial products or graphite and MoS_2 with a polymeric binder, or lubricants containing particles of lead and tin in a volatile organic base. A viscous mixture, such as graphite in a silane, was best in the work of Gurney and Male [347] provided that forging was done fast enough to prevent degradation of the carrier. For higher forging temperatures, a glass coating is almost universally used (e.g., [349,350]). The workpiece is sandblasted, and then an aqueous dispersion of glass and a polymeric binder is deposited by dipping, spraying, or painting. Thus, the coating prevents oxidation during heating and ensures low friction during forging. A graphitic lubricant is applied to the die. Heat transfer is extremely important, and lubricant effects may be neutralized by excessive cooling [8.143].

b. Isothermal forging is conducted in air with superalloy dies, or in vacuum with TZM (Mo-base alloy) dies. If deformation is not severe and surface finish is critical, as in restriking or sizing of finished forgings, solid lubricants are sufficient. At higher temperatures, however, glass lubrication is the normal practice [350,351]. Borosilicate glasses are favored [352] although other glasses are also suitable (Fig.4.23). The requirements of a good isothermal hot forging lubricant have been systematically explored by Spiegelberg [4.218],[353] and by Gurney et al [4.223]. Buildup in recesses of the die is possible if the lubricant is overly viscous or, conversely, is of insufficient viscosity and is squeezed out from other contact surfaces. A parting agent is also required [4.29]. At the usual high temperatures, graphite does not survive and ceramic materials such as BN are needed. Chen [354] reports that the best combination of lubricity, adhesion, and surface finish was obtained for $\alpha + \beta$ titanium alloys with a lubricant consisting of 300 g of borosilicate glass, 24 g of TiC or BN, 450 g of xylene, and 78 g of acrylic binder. At the lower temperatures used in forging beta titanium alloys, 4.5% CrC was added to a 49% PbO glass. Surprisingly, Price and Alexander [355] judged glass to be inferior to graphite in isothermal forging at 800 to 980 C, because surface finish was poorer and material flow was not aided in upsetting of rings and in filling of a simple die shape. Most probably, the glass was not of optimum viscosity and the die cavity was not of sufficient complexity to reveal the advantages of glass.

9.10 Lubrication of Refractory Metals

As we have noted in discussing the extrusion of these metals (Sec.8.10), the main problems are oxidation during heating and the high forging temperatures.

Both molybdenum and tungsten form lubricating oxides. The melting point of MoO_3 is 795 C, and that of WO_3 is 1473 C; at forging temperatures these oxides volatilize. For preheating in protective atmospheres, glasses provide effective coatings for all refractory metals, including tantalum [356]. The dies are protected by MoS_2 or a mixture of MoS_2 with graphite, although carbon contamination is usually undesirable because of carbide formation. Powder preforms can be forged, too [150]. Polymeric carriers, used in forging of steel, are effective here also. To reduce cooling, glass cloth, glass wool, mica, asbestos, or sawdust is used.

In a period of intensive research in the 1960's, several metallic coatings were explored [1], but it is not clear whether any of them gained widespread application. Certainly, either coating or cladding is feasible, provided that a selective etchant can be found.

9.11 Summary

1. In simple open-die operations such as cylinder, ring, or slab upsetting, friction induces inhomogeneity of deformation and raises the interface pressures and forces. Fortunately, sticking friction sets an absolute limit on the pressure increase, but, even so, pressures can become excessive when friction is high and the diameter-to-height (or L/h) ratio is large. The very feasibility of forging then depends on the application of a lubricant that reduces the interface shear strength. Most conveniently, the effectiveness of a lubricant can be judged directly from the change in the internal diameter of a compressed ring. Low friction results in hole expansion, but this does not necessarily mean that the lubricant will also resist breakdown when there is intense relative sliding between the die and workpiece surfaces. Paradoxically, sliding induced by lubrication may be a cause of increased die wear.

2. The material may be made to flow against frictional resistance by forging it between inclined faces, and this is the basis of design for preforging. Even then, interface sliding is arrested at high interface pressures, and this accounts for the difficulty of filling converging gaps that are formed in an impression- or closed-die cavity in forging with draft.

3. An entirely unrelated source of high die (punch) pressure is the restraint imposed by nondeforming parts of the workpiece; deformation is highly inhomogeneous and little affected by friction. Nevertheless, lubrication is beneficial in reducing wear.

4. Filling of the die cavity for complex forgings involves elements of open-die forging, indentation, and forward and back extrusion. Friction controls the direction of material flow. To prevent escape of material into the flash during impression-die forging, it is desirable that low friction in the cavity be combined with high friction on the flash land.

5. In nonisothermal hot forging, chilling of the surface zone of the workpiece combines with the effects of friction to retard material flow and make deformation more inhomogeneous. Therefore, increased die temperatures may be as effective as or even more effective than lubrication in promoting die filling.

6. The duties of lubricants vary over an enormous range, from the relatively mild normal contact in upsetting to the severe conditions created by the surface extension,

high local pressures, and intense relative movement typical of back extrusion or piercing. Correspondingly, lubricants are available in a wide variety.

7. The process geometry is seldom favorable for the formation of thick films, although plastohydrodynamic squeeze films form in upsetting and plane-strain compression. However, at the edges of the contact zone boundary contact is inevitable, and lubrication is generally of the mixed-film or solid type.

8. Lubrication in cold working relies on compounded oils, on semisolids such as fats, soaps, and waxes, or, for severe duties, on systems based on a conversion coating with a superimposed reactive soap. For the most demanding applications, MoS_2 may be applied as the outermost layer of the system.

9. At the usual warm working temperatures most organic lubricants fail, and layer-lattice compounds are applied to the die and/or to the workpiece, often bonded with a polymer. Lubrication is still one of the stumbling blocks to the wider application of warm working technology.

10. Hot working is performed at temperatures at which workpiece materials are highly reactive. The oxide formed on heating and forging in air can be helpful, as in forging of steel or copper; it can also interact with the separately applied die lubricant. Environmental considerations have led to significant changes, first in a move away from oil-base graphitic lubricants to aqueous formulations and more recently to graphite-free lubricants which do not quite match the performance of graphitic ones but have the virtues of white color and water solubility. For titanium, nickel, and refractory metal alloys glass coatings are often preferred, partly because they provide protection against oxidation. Viscosity must be carefully chosen to avoid unacceptably rough surfaces on the forging. Additional protection of the die in the event of film breakdown is required in both isothermal and nonisothermal forging.

11. In hot working, high temperatures, pressures, and interface sliding make die wear a major problem. Lubrication is often carried out with the primary purpose of wear reduction. Die surface coatings are becoming an important part of the lubricant system.

12. The choice of lubricant depends, as in other processes, on many considerations, but some typical practices are given in Table 9.2, together with typical μ values. The latter should be taken as very rough guides only.

References

[1] S. Kalpakjian, in [3.1], pp.603-659.

[2] J.A. Newnham, in [3.1], pp.661-702.

[3] J.A. Schey, in [3.1], pp.58-68, 266-275.

[4] J.E. Jenson (ed.), *Forging Industry Handbook,* Forging Industry Association, Cleveland, 1966.

[5] K. Lange and H. Meyer-Nolkemper, *Gesenkschmieden,* 2nd Ed., Springer, Berlin, 1977.

[6] A.N. Bruchanow and A.W. Rebelski, *Gesenkschmieden und Warmpressen,* Verlag Technik, Berlin, 1955.

[7] A.M. Sabroff, F.W. Boulger, and H.J. Henning, *Forging Materials and Practices,* Reinhold, New York, 1968.

[8] T. Altan et al, *Forging Equipment, Materials and Practices,* Metals and Ceramics Information Center, Columbus, OH, 1973.

[9] T. Altan and V. Nagpal, Int. Met. Rev., *22,* 1977, 322-341.

[10] *HFF-Bericht 9. Umformtechnisches Kolloquium Hannover,* Hannoversches Forschungsinst. f. Fertigungsfragen, 1977.

[11] H.D. Feldmann, *Cold Forging of Steel,* Chemical Publ. Co., New York, 1962, pp.185-198.

[12] J.J. Everhart, *Impact and Cold Extrusion of Metals,* Chemical Publ. Co., New York, 1964, pp.80-90.

Table 9.2. Commonly used forging lubricants and typical friction values

(Lubricants are listed according to increasing severity of conditions.)

Material	Hot forging lubricant(a)	μ	m	Cold forging/extrusion lubricant(b)	μ
Steels	None	ST	1.0	Soap solution	0.2
	Salt solution (on die)	0.4	0.7	EM (M.O. + fat)	0.2
	Soap (on die)	0.3	0.5	EM (M.O. + fat + E.P.)	0.2
	GR in water (on die)	0.2	0.4	M.O. (20-800) + fat + E.P.	0.15
	with binder (on die)	0.2	0.4	Compounded M.O. + GR or MoS_2	0.15
				Sulfonated fatty oil	0.1
				Lime + compounded oil	0.1
				Copper + compounded oil	0.1
				Phosphate + soap	0.05
				Phosphate + soap + MoS_2	0.05
Stainless steels and Ni alloys	GR in water (on die)	0.2	0.4	M.O. (20-800) + Cl additive	0.2
	Glass (10-100 Pa·s)			Lime + compounded oil	0.15
	+ GR (on die)	0.05	. .	Copper + compounded oil	0.1
				Polymer coat	0.05
				Oxalate + soap	0.05
Al and Mg alloys	Soap (on die)	ST	1.0	M.O. (10-100) + fatty	
	GR in water (on die)	0.3	0.5	derivatives	0.15
	GR with binder (on die)	0.2	0.4	Lanolin; dry soap film	0.07
				Phosphate + soap	0.05
Cu and Cu alloys	Soap (on die)	0.3	0.5	Soap solution	0.1
	GR in water (on die)	0.15	0.3	EM (M.O. + fat)	0.1
				EM (fat)	0.1
				M.O. (20-400) + fat	
				(+ Cl additive)	0.1
				Fat; wax (lanolin)	0.07
				Soap (Zn-stearate)	0.05
				GR or MoS_2 in grease	0.07
Ti alloys	MoS_2 on die	0.2	0.5	M.O. (20-800) + Cl additive	0.2
	Glass (10-100 Pa·s)			Polymer coating	0.05
	+ GR (on die)	0.05	. .	Oxidize + lubricant	0.15
				Cu or Zn coat + lubricant	0.1
				Fluoride-phosphate + soap	0.05
Refractory metals	Canning + lubricant	0.2	. .		
	GR in water (on die)	0.2	. .		
	Glass + GR (on die)	0.05	. .		

(a) ST = sticking friction. GR = graphite. Glass viscosity is given for forging temperature. (b) EM = emulsion. M.O. = mineral oil; viscosity in cSt at 40 C is given in parentheses. E.P. = E.P. additive (S, Cl, sometimes P; also sulfochlorinated fats). GR = graphite.

[13] H.Ll.D. Pugh and A.H. Low, J. Inst. Metals, *93*, 1964-65, 201-217.

[14] M.T. Watkins, Int. Met. Rev., *18*, 1973, 123-174.

[15] A. Shutt, Int. J. Mech. Sci., *8*, 1966, 509-511.

[16] P.W. Wallace and J.A. Schey, in *Proc. 8th Int. MTDR Conf.*, Pergamon, Oxford, 1968, pp.1361-1380.

[17] J.A. Schey, T.R. Venner, and S.L. Takomana, J. Mech. Work. Tech., *6*, 1982, 23-33.

[18] R. Herbertz and H. Wiegels, Stahl u. Eisen, *101*, 1981, 89-92 (German).

[19] W. Dobrucki and E. Odrzywolek, J. Mech. Work. Tech., *4*, 1980, 155-176 (French).

[20] N.M. San'ko, Yu.V. Andreev, and S.S. Seleznev, Obr. Met. Davl., *1971* (57), 172-174 (Russian).

[21] N.H. Polakowski, J. Iron Steel Inst., *163*, 1949, 250-276.

[22] K. Pöhlandt, Ind. Anz., *101* (48), 1979, 28-29 (German).

[23] K. Pöhlandt and K. Roll, in *Formability 2000 A.D.*, ASTM Symp., Chicago, 1980.

[24] E. Siebel and A. Pomp, Mitt. Kaiser-Wilh. Inst. Eisenforsch., *10*, 1928, 55-69 (German).
[25] W. Dobrucki and E. Odrzywolek, J. Mech. Work. Tech., *4*, 1981, 351-368 (French).
[26] J.C. Gelin, J. Oudin, and Y. Ravalard, J. Mech. Work. Tech., *5*, 1981, 297-308.
[27] K.M. Kulkarni and S. Kalpakjian, Trans. ASME, Ser.B, J. Eng. Ind., *91*, 1969, 743-754.
[28] J.A. Schey, T.R. Venner, and S.L. Takomana, Trans. ASME, Ser.B, J. Eng. Ind., *104*, 1981, 79-83.
[29] S.K. Samanta, Trans. ASME, Ser.H, J. Eng. Mat. Tech., *97*, 1975, 14-20.
[30] H.A. Kuhn and S.K. Suh, in *Proc. 4th NAMRC,* SME, Dearborn, MI, 1976, pp.29-33.
[31] M. Kunogi, J. Sci. Res. Inst. (Tokyo), *50*, 1956, 215-246.
[32] A.T. Male and M.G. Cockroft, J. Inst. Metals, *93*, 1964-65, 38-46.
[33] M. Burgdorf, Ind. Anz., *89*, 1967, 799-804 (German).
[34] B. Avitzur, C.J. van Tyne, and C. Umana, in *Proc. 18th Int. MTDR Conf.,* Macmillan, London, 1978, pp.305-313.
[35] S.I. Oh, G.D. Lahoti, and T. Altan, J. Mech. Work. Tech., *6*, 1982, 277-290.
[36] N. Rebelo and S. Kobayashi, Int. J. Mech. Sci., *22*, 1980, 699-705, 707-718.
[37] G.D. Lahoti, V. Nagpal, and T. Altan, in [3.52], pp.52-59; Trans. ASME, Ser.B, J. Eng. Ind., *100*, 1978, 413-420.
[38] J.R. Douglas and T. Altan, Trans. ASME, J. Eng. Ind., *97*, 1975, 66-76.
[39] G. Saul, A.T. Male, and V. DePierre, Tech. Rep. AFML TR-70-19, 1970.
[40] A.N. Bramley and N.A. Abdul, in *Proc. 15th Int. MTDR Conf.,* Macmillan, London, 1975, pp.431-436.
[41] J. Löwen, Ind. Anz., *94*, 1972, 238-241 (German).
[42] B. Avitzur and R. Kosher, Paper 76-WA/PROD-7, ASME, New York, 1976.
[43] D. Bhattacharyya and R.H. Brown, in [3.55], pp.23-30.
[44] A. Mulc and S. Kalpakjian, Trans. ASME, Ser.B, J. Eng. Ind., *94*, 1972, 1189-1192.
[45] F.W. Travis, Int. J. Prod. Res., *8*, 1970, 133-148.
[46] M.N. Janardhana and S.K. Biswas, Int. J. Mech. Sci., *21*, 1979, 699-712.
[47] J.F.W. Bishop, J. Mech. Phys. Solids, *6*, 1958, 132-144.
[48] R. Hill, Phil. Mag., *41*, 1950, 733-744.
[49] D.C. Fricker and T. Wanheim, Wear, *27*, 1974, 303-317.
[50] W.R.D. Wilson and H. Whitworth, in *The Production System* (Proc. 3rd Int. Conf. Prod. Res.), R.H. Hollier and J.M. Moore (ed.), Taylor and Francis, London, 1977, pp.169-181.
[51] R.A. Kosher and B. Avitzur, Trans. ASME, Ser.B, J. Eng. Ind., *100*, 1978, 428-433.
[52] A.B. Watts and H. Ford, Proc. Inst. Mech. Eng., *1B*, 1952-53, 448-453.
[53] J.M. Alexander, J. Mech. Phys. Solids, *3*, 1955, 233-245.
[54] I.Ya. Tarnovskii, A.N. Levanov, V.B. Skornyakov, and B.D. Marants, Izv. VUZ Chern. Met., *1961* (6), 53-59 (Russian).
[55] P.F. Thomson, J. Inst. Metals, *97*, 1969, 225-231.
[56] A.H. Lonn and J.A. Schey, in *Proc. 2nd NAMRC,* SME, Dearborn, MI, 1974, pp.165-178.
[57] E. El-Magd, Metall, *32*, 1978, 781-786 (German).
[58] H. Kudo, J. JSTP, *9*, 1968, 268-275 (Japanese).
[59] N.A. Kravcenko, Kuzn. Shtamp. Proizv., *1961* (9), 12-13 (Russian).
[60] J.B. Haddow and J.M. Chudobiak, Can. Met. Quart., *2*, 1963, 325-340.
[61] E.P. Unksov and Yu.S. Safarov, Int. J. Mech. Sci., *17*, 1975, 597-602.
[61a] M. Plancak, in [3.56], Paper No.38.
[62] PERA Rep. No.102, Production Engineering Research Association, U.K., 1962.
[63] H. Kudo, H. Takahashi, and K. Shinozaki, CIRP, *14*, 1967, 465-471.
[64] R.W. Kaupilla, K.J. Weinmann, and D.J. Agnew, in *Proc. 10th NAMRC,* SME, Dearborn, 1982, pp. 213-217.
[65] B. Avitzur, E.D. Bishop, and W.C. Hahn, Trans. ASME, Ser.B, J. Eng. Ind., *94*, 1972, 1079-1086.
[66] A.T. Male, in *Proc. 7th NAMRC,* SME, Dearborn, MI, 1979, pp.145-150.
[67] M. Burgdorf, Ind. Anz., *89*, 1967, 1406-1411 (German).
[68] K. Herold, Fertigunstech. Betr., *18*, 1968, 440-443 (German).
[69] J.A. Newnham and G.W. Rowe, J. Inst. Metals, *101*, 1973, 1-9.
[70] S.C. Jain, A.N. Bramley, C.H. Lee, and S. Kobayashi, *Proc. 11th Int. MTDR Conf.,* Pergamon, Oxford, 1971, pp.1097-1115.
[71] P.W. Wallace and J.A. Schey, in *Proc. 10th Int. MTDR Conf.,* pp.525-535.
[72] K.T. Chang and I.J. Choi, CIRP, *24*, 1975, 163-165.

[73] P.R. Gouwens, Am. Mach, *103*, 1959, 137-144.
[74] H. Tolkien, Werkstattstechnik, *51*, 1961, 102-105 (German).
[75] H.J. Stöter, Werkstattstechnik, *49*, 1959, 775-779 (German).
[76] T. Altan, G.D. Lahoti, and V. Nagpal, J. Appl. Metalwork., *1* (1), 1979, 29-40.
[76a] R. Regazzoni et al, J. Mech. Work. Tech., *5*, 1981, 281-296.
[77] T.L. Subramanian and T. Altan, J. Appl. Metalwork, *1* (2), 1980, 60-68.
[78] A.N. Bramley and J.T. Thornton, in *Proc. 6th NAMRC,* SME, Dearborn, MI, 1978, pp.96-102.
[79] L.G. Cser, CIRP, *25*, 1976, 153-157.
[80] E.J. Breznyak and J.F. Wallace, *Lubrication during Hot Forming of Steel,* Forging Industry Educational and Research Foundation, Cleveland, 1965.
[81] D. Heinemeyer, wt-Z. Ind. Fertig., *67*, 1977, 79-82 (German).
[82] R. Geiger, in *Proc. 1st NAMRC,* McMaster Univ., Hamilton, Ont., 1973 (Suppl.).
[83] B. Avitzur and C.E. Umana, in *Applications of Numerical Methods to Forming Processes,* ASME, New York, 1978, pp.175-181.
[84] K. Dieterle, in *Proc. 3rd NAMRC,* SME, Dearborn, MI, 1975, 179-190.
[85] K. Lenik, C. Kajdas, and K. Lojek, in [3.56], Paper No.109.
[86] G.P. Masura, K.J. Weinmann, and R.W. Kauppila, in *Proc. 3rd NAMRC,* SME, Dearborn, MI, 1975, pp.191-205.
[87] G. Bockstiegel and O. Svensson, in *Modern Developments in Powder Metallurgy,* Plenum, New York, *4*, 1971, 87-114.
[88] G. Dowson, Met. Powder Rep., *33* (3), 1978, 115-116, 118.
[89] E.K. Weaver, in [3.55], pp.145-149.
[90] T. Tabata, S. Masaki, and K. Kamata, J. JSTP, *21*, 1980, 773-776 (Ann. Rep. *2*, 1981, 53).
[91] H.A. Kuhn and C.L. Downey, in *Proc. 1st NAMRC,* McMaster Univ., Hamilton, Ont., 1973, Vol.1, pp.205-220.
[92] Z. Marciniak, in *Proc. 4th Int. Conf. Production Engineering,* Jpn. Soc. Prec. Eng., Tokyo, 1980, pp.759-768.
[93] *Proc. 1st Int. Conf. Rotary Metalworking Processes,* IFS Ltd., London, 1979 (see also P.M. Standing, J. Mech. Work. Tech., *4*, 1980, 177-190).
[94] G.W. Pearsall and W.A. Backofen, Trans. ASME, Ser.B, J. Eng. Ind., *85*, 1963, 329-338.
[95] T. Sata, D. Lee, and W.A. Backofen, in [3.50] pp.1-19.
[96] L.H. Butler, in [3.51], pp.63-69.
[97] M. Oyane and K. Osakada, Bull. JSME, *12*, 1969, 368-375, 1555-1561.
[98] K. Osakada and M. Oyane, Bull. JSME, *13*, 1970, 1504-1512.
[99] M. Oyane and K. Osakada, Bull. JSME, *12*, 1969, 149-155.
[100] E. Tanaka, S. Semoto, and Y. Suzuki, Sci. Rep. Res. Inst. Tokyo Univ., Ser.A, *17* (4), 1965, 193-207.
[101] J. Stöter, Werkstattstechnik, *49*, 1959, 223-230 (German).
[102] H.Ll.D. Pugh, Metal Treat., *27*, 1960, 189-195.
[103] G. Gartner, Freiberger Forschungsh., Reihe B, No.27, 1958, 1-140.
[104] C.S. Cook and J.W. Spretnak, Tech. Rep. AFML-TR-67-280, 1967.
[105] M. Shinohara and M. Miyagawa, J. JSTP, *19*, 1978, 926-933 (Japanese).
[106] K. Osakada, Int. J. Mech. Sci., *19*, 1977, 413-421.
[107] H. Miyagawa and F. Hirano, J. JSLE (Int. Ed.), *1*, 1980, 55-64.
[108] N. Loizou and R.B. Sims, J. Mech. Phys. Solids, *1*, 1953, 234-243.
[109] J.A. Schey and R. Hoba, unpublished work.
[110] D.V. Wilson and G.W. Rowe, J. Inst. Metals, *95*, 1967, 25-26.
[111] D. Lee, T. Sata, and W.A. Backofen, J. Inst. Metals, *93*, 1964-65, 418-422.
[112] V.P. Severdenko, I.G. Dobrovol'ski, and V.N. Bulakh, Izv. VUZ Chern. Met., *1971* (7), 67-69 (Russian).
[113] M.Y. Karnov, Kuzn. Shtamp. Proizv., *1961* (3), 16-18 (Russian).
[114] Y. Bocharov, S. Kobayashi, and E.G. Thomsen, Trans. ASME, Ser.B, J. Eng. Ind., *84*, 1962, 502-508.
[115] V.P. Severdenko and V.V. Petrenko, Sov. Mater. Sci., *6*, 1970, 655-656.
[116] H. Takahashi and J.M. Alexander, J. Inst. Metals, *90*, 1961-62, 72-79.
[117] F. Delamare, M. DeVathaire, and J. Kubie, Trans. ASME, Ser.F, J. Lub. Tech., *104*, 1982, 545-551.
[118] F. Delamare and J. Kubie, Trans. ASME, Ser.F, J. Lub. Tech., *104*, 1982, 538-544.

[119] A. Plumtree and R. Sowerby, Lubric. Eng., *32*, 1976, 585-595.
[120] M. Noda et al, J. JSLE, *23*, 1978, 291-295 (Japanese).
[121] N.A. Abdul, in *Proc. 21st Int. MTDR Conf.,* Macmillan, London, 1981, pp.389-396.
[122] Kh. Hoang-Vu and K. Lange, in *Proc. 10th NAMRC,* SME, Dearborn, MI, 1982, pp.196-204.
[123] G.L. Kolmogorov, Izv. VUZ Chern. Met., *1975* (4), 101-104 (Russian).
[124] S. Kumar, U. Chandra, and T.V. Balasubramanian, Wear, *71*, 1981, 293-305.
[125] P.J. Thompson and G.R. Symmons, in *Proc. 17th MTDR Conf.,* Macmillan, London, 1977, pp.587-595.
[126] W.R.D. Wilson and J.G. Silletto, in [3.55], pp.87-94.
[127] R. Blancon and E. Felder, in [3.258], pp.360-364.
[128] A.J. Holzer, CIRP, *29*, 1980, 135-139.
[129] N.A. Abdul, in *Proc. 18th Int. MTDR Conf.,* Macmillan, London, 1978, pp.331-337.
[130] D.B. Hart and G.F. Modlen, Prod. Eng., *53*, 1974, 153-156.
[131] E. Dannenmann, CIRP, *17*, 1969, 363-365 (German).
[132] S.K. Suh and B.I. Bachrach, in *Proc. 7th NAMRC,* SME, Dearborn, MI, 1979, pp.52-56.
[133] D. James, Sheet Metal Ind., *43*, 1966, 193-204.
[134] L.O. Gilbert, S.L. Eiser, J. Doss, and W.D. McHenry, Metal Finishing, *33* (Apr.), 1955, 56-68, 61.
[135] Anon., Metal Forming, *35*, 1968, 34-40.
[136] J. Stefanakis, Schmiertechn. Trib., *21*, 1974, 137-139 (German).
[137] *Metals Handbook,* 8th Ed., Vol.4, *Forming,* ASM, Metals Park, OH, 1969.
[138] J. Gustafson, Draht, *31*, 1980, 84-88 (German).
[139] L.N. Mironov, Kuzn. Shtamp. Proizv., *1975* (6), 16-18 (Russian).
[140] W.L. Kennicott, Metals Eng. Quart., *13* (2), 1973, 1-18.
[141] J. Blum, Draht, *27* (1), 1976, 1-7 (German).
[142] R. Szyndler, Sbornik V.S.B. Ostrave (Hutn.), *15*, 1969, 209-215 (Czech).
[143] E. Schlowag and J. Quaas, Umformtechnik, *1975* (4), 14-20 (German).
[144] A.G. Mamalis, W. Johnson, and H.J. Marezniski, in *Proc. 18th Int. MTDR Conf.,* Macmillan, London, 1978, pp.173-182.
[145] R. Haverberg and J. Throop, Paper No. MF70-140, SME, Dearborn, MI, 1970.
[146] R. Geiger, Inz. Anz., *92*, 1970, 623-629, 1553-1554 (German).
[147] T.A. Dean, in *Proc. 16th Int. MTDR Conf.,* Macmillan, London, 1976, pp.483-488.
[148] N. Misra, M.H. Pope, and J.T. Berry, in [3.41], pp.39-43.
[149] J. Berry, Trans. ASME, Ser.F, J. Lub. Tech., *97*, 1975, 33-36, 51.
[150] E. Geldner, Metall, *29*, 1975, 1209-1212 (German).
[151] J.A. Bailey and A.R.E. Singer, J. Inst. Metals, *92*, 1963-64, 378-380.
[152] P.F. Thomason and B. Fogg, in [3.51], pp.142-146.
[153] H. Kaiser, in *13th Int. MTDR Conf.,* Macmillan, London, 1973, pp.555-558.
[154] R. Geiger and J. Stefanakis, Ind. Anz., *96*, 1974, 2245-2246 (German).
[155] E. Dannenmann and J. Stefanakis, Ind. Anz., *97*, 1975, 1743-1746 (German).
[156] J. Saga, MW Interf., *5* (1), 1980, 9-23.
[157] J.T. Berry and M.H. Pope, in *Metal Forming - Interrelation Between Theory and Practice,* A.L. Hoffmanner (ed.), Plenum, New York, 1971, pp.307-324.
[158] T.A. Dean and C.E.N. Sturgess, J. Mech. Work. Tech., *2*, 1978, 255-265.
[159] N. Dycke, in *Cold Forming,* VDI-Ber. Nr.266, 1976, pp.101-109 (German).
[160] R. Doehring, Paper No. MF70-161, SME, Dearborn, MI, 1970.
[161] S.M. Chmara et al, Kuzn. Shtamp. Proizv., *1970* (2), 41-42 (Russian).
[162] R. Assmann, Ind. Anz., *99*, 1977, 2072-2076 (German).
[163] T.R.R. Rao and S.K. Biswas, J. Mech. Work. Tech., *3*, 1979, 137-150.
[164] I.O. Gordon, Metal Forming, *37*, 1970, 127-132.
[165] H. Matthes, Wire, *29*, 1979, 271-273; in [3.56], Paper No.28.
[166] A.T. Male, in [3.50], pp.200-208.
[167] R.C. Doehring, Werkstatt u. Betr., *104*, 1971, 749-753 (German).
[168] G. Scholz, Wire, *22*, 1972, 120-121.
[169] R. Carr, Can. Mach. Metalwork., *82* (11), 1971, 74-75.
[170] S.C. Jain, Paper 77-LUBS-33, ASME, New York, 1977; in [3.56], Paper No.29.
[171] H. Bühler and J. Löwen, Arch. Eisenhüttenw., *43*, 1972, 603-607 (German).
[172] M. Mrozek, MW Interf., *1* (4), 1976, 14-20.

[173] W.C. Keung, T.A. Dean, and L.F. Jesch, in *Proc. 3rd NAMRC,* SME, Dearborn, MI, 1975, pp.72-84.
[174] A.D. Sheikh, T.A. Dean, M.K. Das, and S.A. Tobias, in *Proc. 13th Int. MTDR Conf.,* Macmillan, London, 1973, pp.342-346, 347-350.
[175] M.A. Kellow, T.A. Dean, and F.K. Bannister, in *Proc. 17th Int. MTDR Conf.,* Macmillan, London, 1977, 355-361.
[176] Yu.G. Burov and B.M. Pozdneev, Kuzn. Shtamp. Proizv., *1979* (9), 3-6 (Russian).
[177] H.A. Kellow, A.N. Bramley, and F.K. Bannister, Int. J. Mach. Tool Des. Res., *9*, 1969, 239-260.
[178] C.C. Reynolds, in [3.55], pp.83-86.
[179] E. Schlomach, VDI-Z., *115*, 1973, 1151-1154 (German).
[180] R. Dahlheimer and W. Schmädeke, Ind. Anz., *92*, 1970, 37-39 (German).
[181] T.A. Dean and T.M. Silva, Trans. ASME, Ser.B, J. Eng. Ind., *101*, 1979, 385-390.
[182] V.I. Zelesskii et al, Kuzn. Shtamp. Proizv., *1976* (5), 12-13 (Russian).
[183] P.W. Wallace and J.A. Schey, Trans. ASME, Ser.F, J. Lub. Tech., *93*, 1971, 317-323.
[184] S.C. Jain and A.N. Bramley, Proc. Inst. Mech. Eng., *182*, 1967-68, 783-798.
[185] H.R. Nichols and W.H. Graft, Tech. Rep. AMC TR 59-7-579, 1959.
[186] W.E.J. Cross, Metal Treat., *16*, 1949, 141-145.
[187] E. Schlomach, Werkstatt u. Betr., *107*, 1974, 145-148 (German).
[188] H.-J. Metzler, Ind. Anz., *92*, 1970, 1995-2000 (German).
[189] G.K. Sanakoev, V.A. Matyazh, and A.P. Produkin, Kuzn. Shtamp. Proizv., *1971* (9), 11-12 (Russian).
[190] N.M. Parikh, K. Kulkarni, J. Benedyk, and F.C. Bock, in [3.53], pp.137-146.
[191] R. Sharan and S.N. Prasad, in [3.55], pp.77-82.
[192] B.W. Rooks, in *Proc. 15th Int. MTDR Conf.,* Macmillan, London, 1975, pp.487-494.
[193] A.B. Gercikov et al, Kuzn. Shtamp. Proizv., *1977* (4), 21-31 (Russian).
[194] M.L. Surmanian, Paper No. MS77-340, SME, Dearborn, MI, 1977.
[195] H. Glusing and H. Vetter, TZ f.prakt.Metallbearb., *68*, 1974, 412-415 (German).
[196] I.A. Noritsyn, Russian Eng. J., *55* (6), 1975, 69-73.
[197] D. Doehring, private communication, Acheson Colloids Co.
[198] E. Schlomach, Schmiertechn. Trib., *20*, 1973, 110-115 (German).
[199] R. Carroll and J.R. Tucker, Paper No. MF73-195, SME, Dearborn, MI, 1973.
[200] J.J. Gurney and A.T. Male, Paper No. MF74-626, SME, Dearborn, MI, 1974.
[201] V.I. Zalessky and A.V. Slezovsky, Izv. VUZ Chern. Met., *1973* (1), 122-125 (Russian).
[202] A.I. Demisov, Kuzn. Shtamp. Proizv., *1971* (12), 33-34 (Russian).
[203] C.R. Boer and G. Schröder, in *Proc. 21st Int. MTDR Conf.,* Macmillan, London, 1981, pp.209-215.
[204] P.D. Darkins, Metallurgia, *46*, 1979, 451-452.
[205] H.P. Liebig, Ind. Anz., *91* (11), 1969, 47-51 (German).
[206] H. Lindner, Werkstattstechnik, *53*, 1963, 913-919 (German).
[207] J.L. Aston and A.R. Muir, J. Iron Steel Inst., *207*, 1969, 167-176.
[208] J.L. Aston and E.A. Barry, J. Iron Steel Inst., *210*, 1972, 520-526.
[209] S.E.A.E. Kandil, Metal Forming, *42*, 1975, 145-146.
[210] H. Meyer-Nolkemper and D. Heinemeyer, Ind. Anz., *99*, 1977, 599-602 (German).
[211] A. Thomas, Metal Forming, *38*, 1971, 41-45.
[212] J.L. Aston, A.D. Hopkins, and K.E. Kirkham, Metall. Met. Form., 39, 1972, 46-49.
[213] A. Thomas, in [3.51], pp.135-141.
[214] B.W. Rooks, S.A. Tobias, and S.M.J. Ali, in *Proc. 11th Int. MTDR Conf.,* Pergamon, Oxford, 1971, pp.745-760.
[215] T.A. Dean, in [3.55], pp.69-75.
[215a] E. Doege and R. Schneider, in [3.56], Paper No.39.
[216] E. Felder and P. Bauduin, Mec. Mater. & Electr., *1981* (Jan), 4-15 (French).
[217] E. Felder and J.L. Montagut, Trib. Int., *13*, 1980, 61-68.
[218] Y.V. Smirnova et al, Kuzn. Shtamp. Proizv., *1978* (10), 41-49 (Russian).
[219] A. Kannappen, Metal Forming, *36*, 1969, 335-343; *37*, 1970, 6-14, 21.
[220] R. Sharma and D.J. Arrowsmith, Wear, *74*, 1981, 1-10.
[221] T.M. Silva and T.A. Dean, in *Proc. 15th Int. MTDR Conf.,* Macmillan, London, 1975, pp.479-486.
[222] V.K. Lobanov and V.M. Pilipenko, Kuzn. Shtamp. Proizv., *1973* (7), 5-8 (Russian).
[223] A.K. Singh, B.W. Rooks, and S.A. Tobias, Wear, *25*, 1973, 271-279.

[224] B.W. Rooks, A.K. Singh, and S.A. Tobias, Metals Tech., *1*, 1974, 449-455.
[225] Yu.M. Belov and V.A. Krasavchikov, Kuzn. Shtamp. Proizv., *1970* (3), 7-9 (Russian).
[226] S. Storen, J. Ebbesen, J. Slutas, and I. Saetre, in *Proc. 15th Int. MTDR Conf.*, Macmillan, London, 1975, pp.473-478.
[227] *Metals Handbook,* 8th Ed., Vol.5, *Forging and Casting,* ASM, Metals Park, OH, 1970.
[228] H.J. Bevker and K. Rasche, J. Mech. Work. Tech., *2*, 1978, 267-278.
[229] L.A. Solntsev et al, Kuzn. Shtamp. Proizv., *1979* (2), 36 (Russian).
[230] N.H. McBrown and T.B. Smith, Iron Steel, Sp. Issue 1970, 21-26.
[231] R.N. Bayliss and A.D. Hopkins, Form. Trait. Metaux, *1971* (29), 29-35 (French).
[232] D. Heinemeyer and H. Meyer-Nolkemper, Ind. Anz., *98*, 1976, 424-426 (German).
[233] S.A. Kadnikov, Zh.M. Urbanek, and V.E. Koloskov, Kuzn. Shtamp. Proizv., *1979* (3), 9-10 (Russian).
[234] M. Gierzynska-Dolna, J. Mech. Work. Tech., *6*, 1982, 193-204.
[235] C. Miland and W. Panasiuk, J. Mech. Work. Tech., *6*, 1982, 183-191.
[236] S.I. Gorelik, E.M. Fayunshmidt, G.I. Belyaeva, and M.A. Zmanovskaya, Kuzn. Shtamp. Proizv., *1976* (7), 39-41 (Russian).
[237] L.M. Sorkin, Kuzn. Shtamp. Proizv., *1966* (8), 8-10 (Russian).
[238] E.I. Bel'skii, Kuzn. Shtamp. Proizv., *1973* (3), 8-11 (Russian).
[239] H.G. Joost, in [10], pp.99-104 (German).
[240] P.H. Thornton, Metals Techn., *7*, 1980, 26-31.
[241] A.N. Babyn'kin et al, Kuzn. Shtamp. Proizv., *1976* (12), 38-39 (Russian).
[242] K.J. Lodge, F.A. Still, J.K. Dennis, and D. Jones, J. Mech. Work. Tech., *3*, 1979, 63-75.
[243] F.A. Still and J.K. Dennis, Metall. Met. Form., *44*, 1977, 10-21.
[244] W.H. Busby and B. Fogg, Metall. Met. Form., *43*, 1976, 370-372, 374-376.
[245] N. Bowers, Wire J., *6* (5), 1973, 54-61.
[246] N. Bowers, Wire J., *12* (9), 1979, 156-159.
[247] N. Yamagoshi, T. Minami, and S. Aihara, Wire J., *6*, 1973, 68-81.
[248] G.K. Schwartz, Bänder Bleche Rohre, *10*, 1969, 349-355 (German).
[249] D. Schmoeckel, wt-Z. Ind. Fertig., *63*, 1973, 401-404 (German).
[250] D. Schmoeckel, Schmiertechn. Trib., *22*, 1975, 12-16 (German).
[251] J. Ivaska, Jr., Tool. Prod., *45* (7), 1979, 62-65.
[252] T.C. Atkiss, Paper No. MF73-562, SME, Dearborn, MI, 1973.
[253] D.F. Baxter, Jr., Metal Progr., *102* (6), 1972, 59-63.
[254] G.H. Townend and F.B. Wilson, Metal Treat., *25*, 1958, 353-359.
[255] J.C. McMurray, Wire Wire Prod., *35*, 1960, 1649-1652.
[256] B.H. Beverly, W.L. Karpen, and D.F. Snook, Wire J., *9*, 1976, 57-63.
[257] W. Machu, Wire World Int., *9* (4), 1967, 125-131.
[258] D.C. Ruthledge, Wire J., *7* (7), 1974, 56-59.
[259] E. Faust, Ind. Anz., *87* (57), 1965, 135-139 (German).
[260] A.M. Cooper, Production Eng., *40*, 1961, 283-296.
[261] M. Noda et al, J. JSLE, *23*, 1978, 296-300 (Japanese).
[262] K.W. Hards, Am. Mach., *106* (Apr.30), 1962, 111-113.
[263] J. Pomey, A. Royez, P. Mathonu, and M. Mouflard, CIRP, *11*, 1962-63, 185-193.
[264] H. Hauttmann, Iron Age, *167* (Mar.15), 1951, 99-104.
[265] F. Tychowski, CIRP, *11*, 1962-63, 237-242.
[266] J. Saga and H. Iwasaki, J. JSTP, *9*, 1968, 93-99 (Japanese).
[267] C.V. Sciullo, Autom. Machining, *26* (Feb.), 1965, 63-64.
[268] J. Saga, A. Takiguchi, and H. Iwasaki, J. JSTP, *8*, 1967, 131-137 (Japanese).
[269] D. Blake, H.P. Bailey, J. Bonnar, and A. Rakison, Metall. Met. Form., *39*, 1972, 30-31.
[270] Anon., Ind. Lub. Trib., *22*, 1970, 26-27.
[271] H. Ll.D. Pugh, M.T. Watkins, and J. McKenzie, Sheet Metal Ind., *38*, 1961, 253-279.
[272] S. Samanta, Jernkontorets Ann., *149*, 1965, 712-733.
[273] F. Howard, H.A.J. Dennison, and N. Angus, Sheet Metal Ind., *38*, 1961, 403-421.
[274] R.G. Dermott, Metal Progr., *86* (4), 1964, 118-123.
[275] R.E. Okell, Sheet Metal Ind., *38*, 1961, 14-20.
[276] M.D. Verson, Wire Ind., *27*, 1960, 1091-1095.
[277] C.R. Weymueller, Metal Progr., *82* (4), 1962, 67-77.
[278] C.H. Wick, *Chipless Machining,* Industrial Press, New York, 1960, pp.347-366.
[279] M.D. Verson, Sheet Metal Ind., *38*, 1961, 493-499.
[280] R.W. Perry, Metals Eng. Quart., *2* (1), 1962, 56-61.

[281] Anon., Metal Forming, *38* (3), 1971, 80-81.
[282] K. Sieber, Draht, *29*, 1978, 317-321 (German).
[283] R.A.P. Morgan, Sheet Metal Ind., *38*, 1961, 99-122.
[284] H. Fischer, Sheet Metal Ind., *30*, 1953, 447-463.
[285] H.D. Witte, Ind. Anz., *90*, 1968, 203-209 (German).
[286] S.C. Poulsen, Machinery (London), *97*, 1960, 980-995.
[287] G. Murari, E. Rebola, and V. Vullo, in *Proc. 20th Int. MTDR Conf.,* Macmillan, London, 1980, pp.133-139.
[288] M.G.L. Austin and C.J. Essom, Metall. Met. Form., *38*, 1971, 317-321.
[289] S. Georgescu-Cocos, Constructia de Masini, *28*, 1976, 513-519 (Rumanian).
[290] G.K. Schwartz, Wire World Int., *12* (6), 1970, 125-126.
[291] R.C. Doehring, Paper No. MF72-526, SME, Dearborn, MI, 1972.
[292] H. Vetter, Ind. Anz., *93*, 1971, 1681-1684; *99*, 1977, 409-411 (German).
[293] Anon., Metall. Met. Form., *39*, 1972, 22.
[294] J.C. Hendry and D.J. MacArthur, Nat. Eng. Lab. Rep. No.499, Jan., 1972.
[295] G. Schmitt, quoted in [3.61], p.430.
[296] R. Geiger, Ind. Anz., *96*, 1974, 2245-2246 (German).
[297] V.J. Ashar and K.J. Weinmann, in *Proc. 4th NAMRC,* SME, Dearborn, MI, 1976, pp.115-122.
[298] S.M. Doraivelu and V. Gopinathan, Trib. Int., *12*, 1979, 123-126.
[299] J. Hinte, in *Proc. 10th NAMRC,* SME, Dearborn, MI, 1982, pp. 205-212.
[300] P.F. Thomason, Proc. Inst. Mech. Eng., *184*, 1969-70, 885-895.
[301] N. Yamokoshi and T. Minami, Wire, *25* (2), 1975, 39-43.
[302] W. Shichun, J. Mech. Work. Tech., *6*, 1982, 333-345.
[303] S. Kusada, T. Kaneko, and K. Yuasa, Ind. Anz., *99*, 1977, 1247-1249 (German).
[304] L.H.S. Luong and T. Heijkoop, Wear, *71*, 1981, 93-102.
[305] A.T. Male, J. Inst. Metals, *93*, 1964-65, 489-494.
[306] H. Tolkien, Werkstattstechnik, *50*, 1960, 224-226 (German).
[307] S.C. Jain, in *Proc. 1st NAMRC,* MacMaster Univ., Hamilton, Ontario, 1973, 143-163.
[308] S.C. Jain, Ind. Lub. Trib., *25*, 1973, 98-99.
[309] E.F. Khayuretdinov, V.M. Greshnov, A.B. Kashintsev, and V.D. Yakovlev, Izv. VUZ Chern. Met., *1975* (11), 97-99 (Russian).
[310] H. Tolkien, Werkstattstechnik, *51*, 1961, 431-435 (German).
[311] C.E. Cruff and C.C. Reynolds, Trans. ASME, Ser.F, J. Lub. Tech., *97*, 1975, 14-17.
[312] J. Crawley and G. Wills, in *Proc. 7th Int. MTDR Conf.,* Pergamon, Oxford, 1967, pp.5-24.
[313] H. Grotz, Ind. Anz., *92*, 1970, 1104-1105 (German).
[314] V.N. Krashevich et al, Kuzn. Shtamp. Proizv., *1969* (11), 46 (Russian).
[315] L.D. Demidov and V.P. Petrunin, Izhevsk. Mekhanich. Inst. Trudy, *2*, 1967, 151-157.
[316] A. Knauschner, Fertigungstech. Betr., *10*, 1960, 139-145 (German).
[317] I.V. Anisimova et al, Kuzn. Shtamp. Proizv., *1971* (7), 17-19 (Russian).
[318] M. Vater and W. Elfgen, Bänder Bleche Rohre, *13*, 1972, 443-453 (German).
[319] P.I. Chuiko et al, Stal in English, *1967* (3), 1967, 240-241.
[320] E.S. Nachtman, Paper No. MF72-527, SME, Dearborn, MI, 1972.
[321] H.L.D. Pugh, Metal Treat., *27*, 1960, 189-195, 231-236.
[322] C. Johnston and R.E. Atkinson, Lubric. Eng., *32*, 1976, 242-248.
[323] V. Gopinathan and P. Venugopal, Trib. Int., *10*, 1977, 152-154.
[324] G. Leps, Neue Hütte, *19*, 1974, 659-663 (German).
[325] W.R.D. Wilson and E.A. Kokoska, Wear, *32*, 1975, 25-32.
[326] A. Dellavia, A. Bugini, E. Gentili, and R. Pacagnella, Metall. Met. Form., *44*, 1977, 442, 444, 446.
[327] M. Vater, W. Heller, K. Yoshii, and M. Aoki, Bänder Blecher Rohre, *11*, 1970, 289-297 (German).
[328] G. Brix, Draht, *26*, 1975, 216-219, 292-297, 481-483 (German).
[329] L.E. Farmer, CIRP, *27*, 1978, 129-134.
[330] C. Belvedere, Alluminio, *35* (4), 1966, 163-172.
[331] R.J. Dower, J. Inst. Metals, *95*, 1967, 1-7.
[332] M.T. Watkins, K. Ashcroft, and J. McKenzie, in *Proc. Conf. Technology of Engineering Manufacture,* Inst. Mech. Eng., London, 1958, pp.155-167.
[333] D.F. Galloway, ibid, pp.141-148.

[334] E. Elliott, Metal Ind., *101* (July 27), 1962, 62-64.
[335] W. Gmöhling, Z. Metallk., *55*, 1964, 567-569 (German).
[336] R.S. Owens and R.W. Roberts, in *Proc. Int. Conf. Manuf. Technol.*, ASTME (SME), Dearborn, MI, 1967, pp.1193-1205.
[337] G.H. Kitchen, Lubric. Eng., *23*, 1967, 181-186.
[338] A.I. Kemppinen, Metal Treat., *32*, 1965, 245-249.
[338a] S. Suresh and S. Bahadur, Trans. ASME, Ser.F, J. Lub. Tech., *104*, 1982, 552-558.
[339] S. Shah, in [3.55], pp.95-101.
[339a] F.E. Lockwood, J.P. Faunce, W. Boswell, and C. Shelton, Lubric. Eng. (to be published).
[340] J.P. Denny and J.D. McKeogh, J. Metals, *13*, 1961, 432-433.
[341] B. Parsons, D.B. Laycock, and B.N. Cole, in *Proc. 8th Int. MTDR Conf.*, Pergamon, Oxford, 1968, pp.1003-1020.
[342] A.T. Male, referred to in [2.23], p.300.
[343] A.M. Sabroff and P.D. Frost, Trans. ASLE, *3*, 1960, 61-68.
[344] A.M. Sabroff, O.J. Huber, and P.D. Frost, Trans. ASME, *80*, 1958, 124-132.
[345] J.T. Berry, M.H. Pope, and N. Misra, in *Titanium Science and Technology*, R.I. Jaffe and H.M. Burte (ed.), Plenum, New York, 1973, Vol.1, pp.419-429.
[346] A.M. Bubbico, Paper No. MF70-121, SME, Dearborn, MI, 1970.
[347] F.J. Gurney and A.T. Male, Paper No. MF74-627, SME, Dearborn, MI, 1974.
[348] G.O. Eccles, Metal Treat., *28*, 1961, 50-52.
[349] E.G. Shastin et al, Kuzn. Shtamp. Proizv., *1978* (5), 6-8 (Russian).
[350] M. Mrozek, MW Interf., *1* (2), 1976, 21-26.
[351] N.A. Agapov, V.V. Boitsov, V.N. Starchenkov, and S.Z. Figlin, Kuzn. Shtamp. Proizv., 1977 (6), 20-22 (Russian).
[352] A.V. Gerasimov et al, Kuzn. Shtamp. Proizv., 1976 (5), 14-16 (Russian).
[353] W.D. Spiegelberg, Tech. Rep. AFML-TR-77-86, 1977.
[354] C.C. Chen, in [3.53], pp.147-155.
[355] J.W.H. Price and J.M. Alexander, in *Proc. 4th NAMRC*, SME, Dearborn, MI, 1976, pp.46-53.
[356] A.A. Saprykin, V.A. Malafeev, and I.V. Anisimova, Kuzn. Shtamp. Proizv., *1977* (1), 11-13 (Russian).

10

Sheet Metalworking

In Chapter 6 we mentioned that by far the largest quantity (typically some 70 to 80%) of all metals is rolled into sheet and plate. Heavy plate is directly used after various manufacturing steps in heavy construction, but most of the thinner material—say, below 5 mm—is further worked by a great variety of techniques, collectively termed sheet metalworking processes. They place widely variable demands on lubricants, and the lubricants often determine the success of the process itself. The discussion given here encompasses subject areas previously covered by Newnham [1] and Schey [2].

In addition to more general discussions in the books referenced in Chapter 2, there are several books devoted to sheet metalworking [9.137],[3-8]. A review has recently been published by Mellor [9]. Conferences have been devoted to the subject [10], and the International Deep Drawing Research Group holds biannual congresses, of which some have particular relevance to the present topic [11,12].

The first step in all processing is some form of shearing, a process more related to machining, whereas the subsequent bending, stretching, drawing, and ironing operations share many of the lubrication problems of plastic deformation processes. Therefore, in the following discussion the tribology of shearing will be discussed first, followed by the tribologies of the other sheet metalworking operations.

10.1 Shearing

Shearing involves the separation of sheet material with the aid of cooperating dies. Shearing may be performed along a straight line with shear blades or rotary (slitting) cutters, or along a contour with the aid of a blanking or punching die. Cropping of bars is also related.

10.1.1 The Shearing Process

The basic shearing process can be illustrated by an example of hole punching (Fig.10.1).

When the tool first contacts the workpiece, elastic deformation occurs, the magnitude of which is a function of the elastic properties of the material, the cutting zone geometry, and the restraint provided by the adjacent material and die elements such as holddowns and strippers. Then plastic deformation (a slight extrusion) begins and, after some critical penetration, which is a function of the die clearance and the ductility of the material, cracks are generated at the die and punch edges (Fig.10.1a). On further penetration the cracks propagate to complete the cut, while the cutting force reaches its maximum, which is only slightly dependent on clearance and not at all dependent on friction. The part must be pushed farther to remove it from the cutting zone; in this

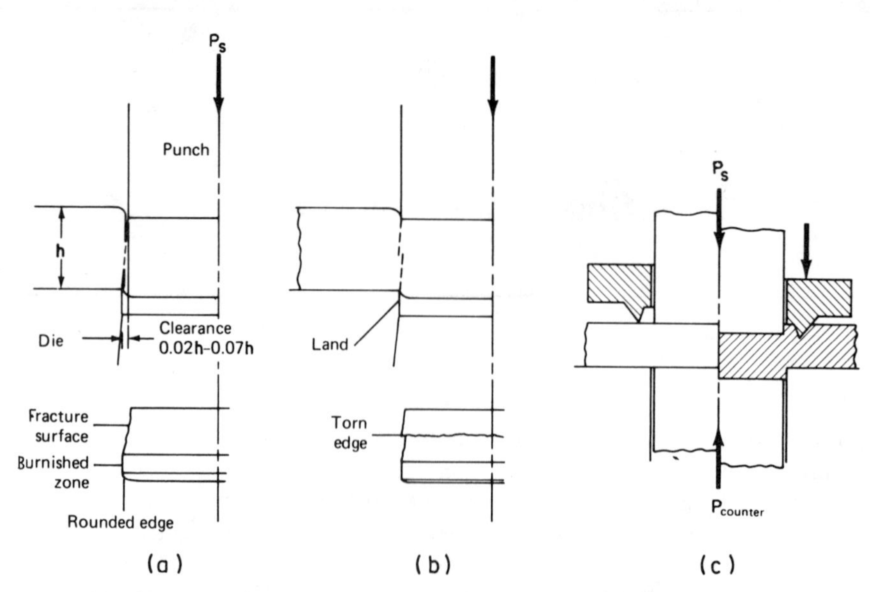

Fig.10.1. Schematic illustration of the shearing process [2.11]

phase the force drops steeply and is partly attributable to friction on the die land and punch side surfaces. When cutting is done with a less-than-optimum clearance, secondary shearing takes place (Fig.10.1b), resulting in a second peak in the force-displacement curve, the magnitude of which is slightly affected by friction on the punch side surface.

The resulting cut surfaces are neither perpendicular nor very smooth—at best, R_t = 3 to 6 μm transverse to the punch movement and 0.1 to 2 μm parallel to it [3.314].

Much-improved accuracy, perpendicular edges, and a smooth finish may be ensured by various special processes such as fine blanking (precision blanking), in which a specially shaped blankholder (Fig.10.1c) imposes compressive stresses on the cutting zone, delays crack initiation, and ensures that the whole thickness is plastically sheared.

Crack generation in itself is not affected by lubrication [13]. Meyer and Kienzle [14] observed only a slight increase in the smoothly deformed edge surface on application of a lubricant. This does not mean, however, that lubrication is not important. Indeed, in many instances it is essential if tool wear is to be held to tolerable rates and the surface quality of the cut is to be acceptable.

10.1.2 Wear

Shearing creates severe wear conditions which have been a subject of study for some forty years. A partial review is given by Hogmark et al [15].

1. Die and tool edges and the die land and punch side surfaces (the flanks) are exposed to the virgin surfaces generated in the course of shearing. Relative sliding creates ideal conditions for adhesive wear of the edges and flanks (Fig.10.2).

2. The workpiece material suffers severe strain hardening. Even though the depth of this heavily deformed zone is shallow (typically, 30 to 50% of the sheet thickness) [3.314], it increases local tool pressures and gives increased support to any abrasive particles that may be present, thus making them more damaging.

3. Shearing forces and, particularly, localized pressures can be high. Forces can be reduced by selection of an adequate clearance (typically around 8% [16]), but only slightly. High production rates involve impact conditions [17] imposed at high rates of

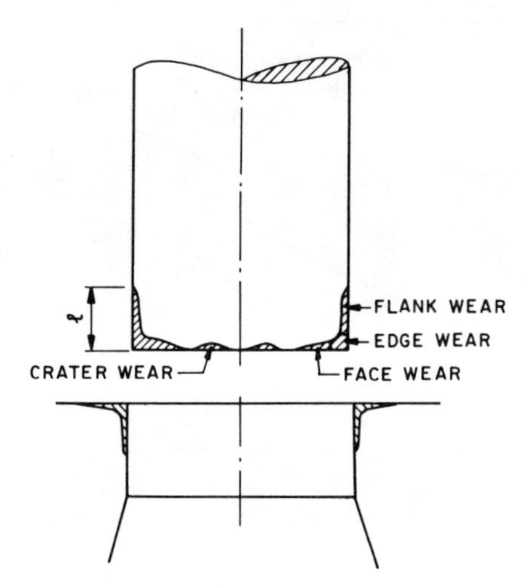

Fig.10.2. Wear of shearing punch and die

repetition. Repeated loading, especially in the presence of adhesive bonds, leads to spalling (fatigue) of the edges [15], which terminates the run. Crater wear is also possible on the punch end.

4. Elastic deformation of the workpiece results in relative movement along the punch end face, first radially outward and then, after crack initiation, inward [18]. This leads to abrasive wear on the punch face. By limiting elastic deflections, a stripper reduces this form of wear [19]. In combination with high normal loads, crater wear is observed.

5. Elastic springback of the workpiece material increases the pressure acting on the punch during retraction and thus increases flank wear, particularly in piercing of a hole [20].

6. High production rates contribute to a temperature increase. In punching of stainless steel, the temperature rise was 80 C under dry and 55 C under lubricated conditions [21], values sufficient to increase adhesive and oxidative wear.

7. Thermoelectric currents generated in shearing promote wear. This source of wear can be neutralized with a compensating circuit [22].

The course of wear follows some specific patterns and may be quantified by various measurements:

a. Flank wear can be characterized by length (l in Fig.10.2) or area [15,19]. Flank wear is important because it determines the length which is lost in regrinding. Its origin is in adhesion and abrasion and, according to Eq.3.39, increases with the number of strokes—at a higher rate when the clearance is too small (Fig.10.3) [17]. The wear length increases asymptotically to the maximum given by the punch penetration. A semiquantitative assessment is possible by inspecting the flank for scoring and pickup.

b. Edge (tip) wear, even though difficult to separate from flank wear (Fig.10.2), is important in that it determines burr height [23-25]. This, too, increases with the number of parts [16] and is at a minimum at some fairly generous clearance (12% in Fig.10.4) [17]. Excessive clearance (17% in Fig.10.4) leads to a large burr also, but only because the part is finally separated by tensile fracture, and not because of wear.

c. Face wear is mostly abrasive in origin and as such increases linearly with the number of parts [18,19].

Punch wear can be conveniently followed by radioactive tracer techniques [27] or by dimensional measurements. After a stage of rapid initial wear, wear progresses at a lower, steady rate (Fig.10.3) [16,17,25,26] until a phase of rapid wear sets in again. At

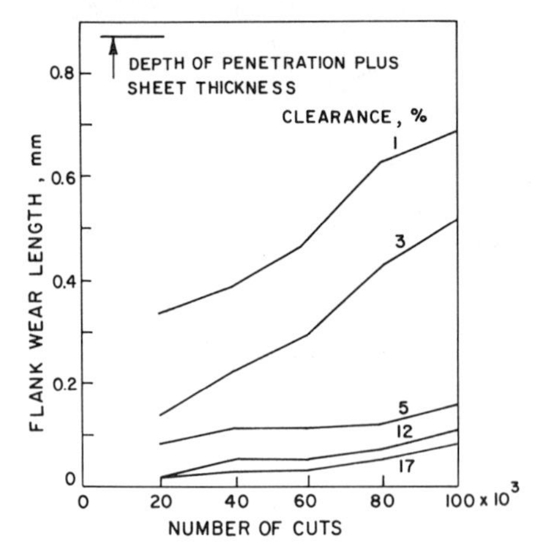

Fig.10.3. Dependence of flank wear on punch-to-die clearance in cutting of steel blanks [17]

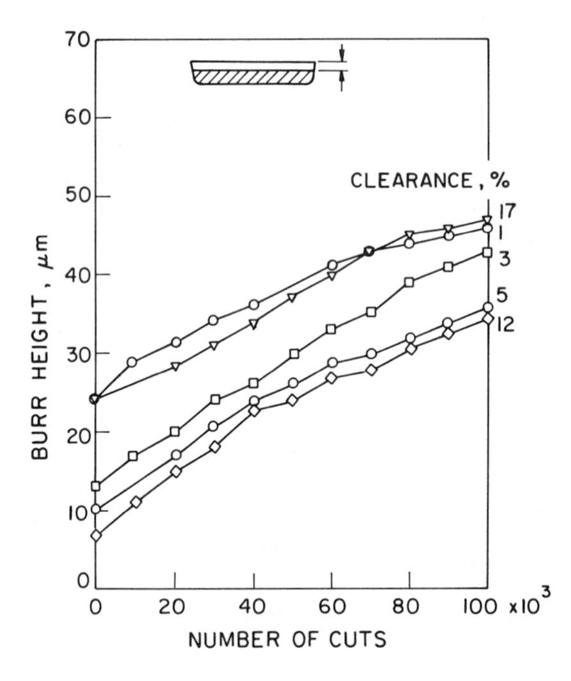

Fig.10.4. Effect of clearance on burr development in blanking of steel sheet [17]

this point the product becomes unacceptable: the part conforms to the worn punch [25], and a burr develops. Therefore, die life is usually defined by the maximum tolerable burr height. However, wear cannot be judged from absolute burr height alone, because this is a function also of material properties. Burr height increases with increasing ductility and is thus generally less on cold rolled than on annealed material. Similarly, alloying that reduces ductility also reduces burr height; on silicon steels burr height increased with decreasing silicon content and decreasing *r* value [17]. Because of these effects, Kienzle and Buchman [26] suggested the use of face wear as an indicator. Tool life generally follows a Weibull distribution [28]—i.e., it has a normal distribution but with a finite lower limit. Excessive face wear causes plastic deformation in the form of dishing of the part due to contact prior to the initiation of fracture. The deformed blank shape can then be taken as a criterion for critical tool wear [29].

10.1.3 Wear Control

Since lubrication does not directly affect the shearing process, it is more meaningful to speak of wear control, which encompasses a number of aspects.

Tool Materials. Because of the predominance of adhesive and abrasive wear, both tool material composition and hardness are important, but the danger of spalling demands also some measure of ductility [3.426],[9.137].

The wear resistance of tool steels is not simply a matter of hardness. Maeda and Aoki [19] heat treated a 12% Cr steel (similar to AISI D2 steel), a high speed steel (M2), and a low-alloy steel (02) to a common hardness of 61 HRC. Bainitic steel sheets in hardnesses ranging from 290 to 440 VHN were punched. Face wear, which is due mostly to abrasion, was least with D2, greater with M2, and greatest with O2. Obviously, Rockwell hardness is an inadequate measure of wear resistance, essentially because it measures only the matrix hardness H_m, whereas resistance to wear is affected also by the hardness of the dispersed (carbide) phase H_c, which occupies an area α. An effective hardness H_{ve}, calculated from

$$H_{ve} = H_c + (1 - \alpha)H_m \qquad (10.1)$$

gave the correct relationship with wear resistance. This is in agreement with Hogmark et al [15] and Bühler et al [25], who found wear to be inversely proportional to the amount of carbide, and at a minimum with the 12% Cr steel. Flank wear, which is greatly affected by adhesion, was also highest with the O2 steel in the work of Maeda and Aoki [19]. Excessive punch hardness results, however, in increased spalling [15,26].

Carbide tooling is common for long runs. Heat treatable, steel-bonded titanium carbide (Ferro-TiC) has the advantage that it can be machined in the annealed state [30].

As might be expected, tools respond extremely well to surface treatments. All forms of treatment have been tried, including hard facing with nickel-base alloys [31], boronizing, VC, hard-chromium coating, and nitriding [32,33]. At least a doubling and as much as a quadrupling of tool life can be expected. In general, CVD coatings with TiC [31,34,35,36] are the most effective. The benefits increase with increasing sheet thickness [35], and up to a hundredfold increase in die life is found [36]. Tools coated with TiC also greatly improve the surface roughness of fine-blanked parts [37].

Workpiece Materials. The workpiece material is of importance from the points of view of both adhesive and abrasive wear.

The presence of abrasive oxides or coatings is, obviously, damaging. Thus, a softer bainitic steel covered with a thicker oxide film suffers greater wear than a harder strip covered with a thinner blue oxide film, whereas a polished (bright) strip minimizes abrasive wear [19]. Similarly, the SiO_2 powder found in some insulating coatings on transformer steel contributes to increased wear [38].

The absence of an oxide is, however, not always desirable. When adhesion is a problem, some oxide is preferable. Thus, a bright soft steel adhered to the tool, removed the matrix, broke carbide particles loose, and thus increased flank wear in the work of Maeda and Aoki [19]. Adhesion becomes more important with smaller clearances, and below 5% clearance the harder strip suffered greater wear because its thin oxide could not prevent adhesion. Chipping then became a danger, as it did also with stainless steel [18]. As expected (Sec.3.11), adhesion is governed by the tool/workpiece material combination, as evidenced by the weight loss measured in a simulation test with a castellated disk (Fig.5.26). With an austenitic stainless steel, TiC reduced wear substantially but suffered from spalling. In contrast, a TiN coating virtually eliminated wear (Fig.10.5). However, with a ferritic stainless steel it was TiC that eliminated wear while TiN only reduced it [32].

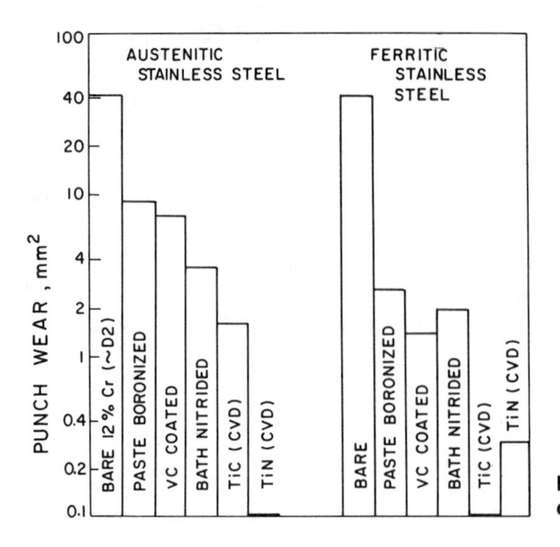

Fig.10.5. Effect of punch surface treatments on wear measured in a simulation test [32]

Lubrication. Various lubricants are effective in reducing wear, especially wear that is due primarily to adhesion [13].

Liquid lubricants form a squeeze film when the punch hits the sheet. Most of the film squeezes out, but in doing so it also lubricates the end face and reduces face wear. The squeezed-out oil is available to seep into the cutting zone, reducing adhesion on the punch flank. As the punch retracts, the oil lubricates the flank, reduces the retracting force, and minimizes scoring and adhesive wear, but also causes the part to stick to the punch face [39].

Because of the need to reduce adhesion, most lubricants are compounded. Repeated contact and elevated temperatures activate E.P. additives, and these are found in all formulations except those destined to be used with aluminum alloys. Heavier sheet imposes more onerous duties, and therefore both base oil viscosity and additive levels are higher. Aqueous lubricants can be used, although oil-base ones are more suitable for severe duties. Liquid lubricants are helpful also in removing wear debris, which becomes a major cause of wear especially when clearances are large, and thus adhesive wear is less severe.

Of the most frequently used sheet metals, austenitic stainless steel is prone to adhesion [18] and requires a minimum of 7.5% Cl in the oil [32]. Ferritic stainless steel is next in severity. Carbon steel, especially in thicker gages and at higher hardness levels, creates severe conditions, and both chlorine [40] and sulfur are effective [41,42]. In work at PERA [41], MoS_2 and/or PTFE powder were also effective as additives. Very low-carbon steel, with its tendency to smear, necessitates frequent regrinding of carbide tooling; the use of a lubricant with active sulfur dramatically increases die life [43]. Oils with boundary additives are usual for aluminum alloys, while both boundary and E.P. additives are often employed for copper alloys [13]. If the part is cleaned soon after forming, sulfur compounds are allowable [43]. Various other substances, such as caprolactam-base lubricants, have been reported [4.73].

Coatings can be effective, too. In blanking of austenitic stainless steel, a PVC coating was found to be excellent [42]. Similarly, some of the insulating coatings deposited on transformer steel, such as an isophthalic-acid-base alkyl film or a fused Mg-phosphate-chromate-borate coating, reduce wear [38]. Inorganic coatings, especially those containing silica, increase wear and spalling [16].

Lubricants are effective also in fine blanking [37], especially in reducing flank wear. Transverse roughness of the part improves with chlorine and sulfur additives and with higher oil viscosity. By appropriate shaping of the die elements, lubricant reservoirs can be formed (Fig.10.6) and the quality of cut improved.

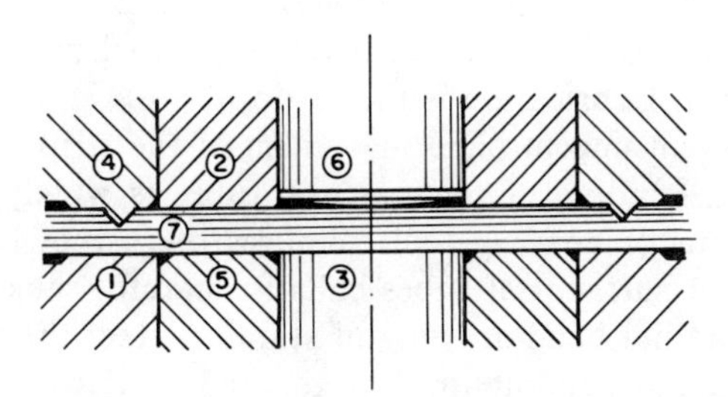

Fig.10.6. Lubricant pockets developed in fine blanking by use of appropriately shaped tools [37]

Thick plate is sometimes punched hot. Hot forging lubricants are generally suitable, ranging from graphitic formulations to glass cloth. An example for stainless steel has been given [44].

10.2 Bending and Spinning

Parts blanked from sheet can be further formed by processes that create localized deformation.

10.2.1 Bending

In the basic process of bending, a part is deformed between a punch and a die (Fig.10.7). The sheet gradually conforms to the punch nose, at low pressures, and with very little sliding: friction effects are negligible here. Meanwhile the sheet slides over the die radius, and a frictional force is generated. Because interface pressures are low—well below the flow stress of the workpiece material—contact is limited to asperities, and the coefficient of friction can be quite high (Fig.2.3a). If the die/workpiece combination is prone to adhesion, pickup develops. If the workpiece is covered with an abrasive oxide, the die radius wears. In either case, the surface of the part is scored, die wear can become a problem, and lubrication is then essential.

Bending along a straight line is often performed with a steel punch and an elas-

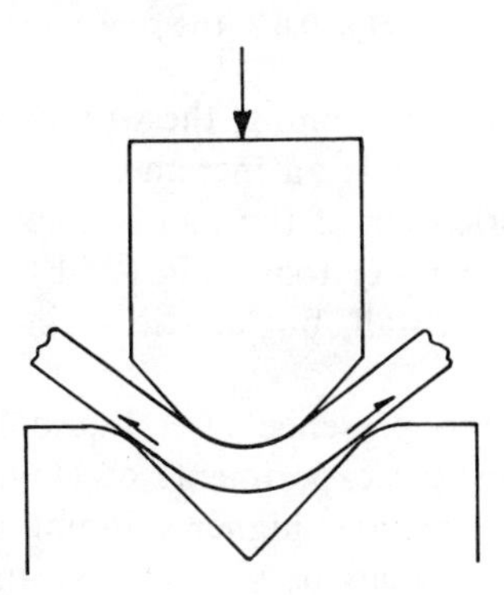

Fig.10.7. Bending of sheet (half arrows indicate frictional stresses acting on material)

tomeric (polyurethane) pad. The elastomer imposes tensile stresses on the sheet, somewhat as in stretch forming. The stresses are transmitted by a frictional force that can be modeled by a coefficient of friction the value of which depends on the surface condition of the blank and on the properties of the elastomer. Higher friction reduces springback [45].

Long lengths of shapes such as tubes, corrugated sheet, etc., are produced in large quantities by roll forming. Just as in shape rolling (Fig.6.47), the workpiece moves at a given constant speed while local velocities around the die profile vary, giving rise to relative sliding. Even though pressures are only fractions of those that occur in shape rolling, adhesive or abrasive wear, tool pickup, and workpiece scoring may be encountered. Lubrication is then helpful [46,47], although some friction must be maintained to prevent uncontrolled movement of the strip.

Bending of sections and tubes is widely practiced, and friction on the wiping elements of bending tools requires the use of lubricants. The difficulty of lubricating the inner surfaces of tubes calls for careful selection of the components of the tribosystem. In radial bend-drawing of stainless steel the wiper die and the flexible ball mandrel are made of aluminum bronze, and a viscous, compounded lubricant is applied which reduces mandrel tension and minimizes thinning on the tension side [48].

10.2.2 Spinning

Spinning is an incremental forming process in which a circular blank is rotated and held between a male die (spinning block) and the tail stock in a machine tool resembling a lathe. The blank is pressed with a hand- or power-actuated tool so that it gradually conforms to the shape of the die (Fig.10.8a). Localized pressures on the tool are high and there is intensive sliding, and thus lubrication of the outer blank surface is essential.

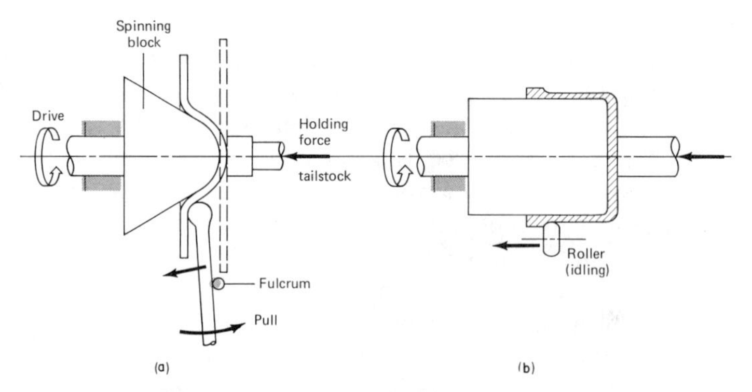

Fig.10.8. The processes of (a) spinning and (b) shear spinning [2.11]

In shear spinning, the wall of a previously formed cup or vessel is reduced. Strictly speaking, this is an incremental bulk deformation process with high surface pressures, and lubrication of the tool is essential. Friction problems are partially circumvented by the use of roller tools (Fig.10.8b). Friction on the die (mandrel) hinders the free extension of the wall, but elastic springback helps to relieve the pressures and reduces friction.

In the presence of a liquid lubricant, lubrication is of the mixed-film type, as shown by the experiments of Hayama [49]. A rougher surface is formed with a more viscous lubricant, higher spinning speeds, or larger rollers. However, it is essential that boundary agents be present, too. In shear spinning of steel, the workpiece is coated with fat, and an emulsion is applied during spinning to extract the generated heat. In the experiments of Gütlbauer [50], spinning was possible without preapplied fat if an emulsion based on rapeseed oil, compounded with a solid lubricant (ZnS), was used. Mineral-oil-base emulsions were less successful in that ovality and wall thickness variations were greater. Wear of the mandrel was significant, even when it was hard-chromium coated.

10.3 Stretching, Deep Drawing, and Ironing

The largest quantity and variety of parts are made by the processes of stretching, deep drawing, ironing, and their combinations. To ensure success, interactions among the process, material properties, and friction effects must be understood. Those interested in the more general aspects will benefit from recent reviews by Mellor [9] and Hasek [51]. The tribological aspects have been discussed by Fogg [52].

10.3.1 Material Properties

An important point to note is that in most sheet metalworking processes yielding occurs in a predominantly tensile stress state. Typical conditions range from balanced biaxial tension (point 3 in Fig.2.1), through plane-strain tension (point 4 in Fig.2.1), to a combination of tension and compression (moving toward point 5 in Fig.2.1). The compressive stress often originates in the geometry of the process and, with the exception of ironing, die pressures remain low, well below the flow stress σ_f. This means that contact is limited to asperities, and the Coulomb law of friction (Eq.2.8) holds. As noted in Sec.3.2.5, asperities yield at relatively low pressures when the substrate deforms in tension [53,54]; thus the real contact area and, with it, frictional stresses increase rapidly, resulting in a relatively high coefficient of friction (Fig.2.3a). Correspondingly, friction plays an often decisive role in the success of the operation, which is measured by the absence of unacceptable necking or fracture of the workpiece. In contrast to bulk deformation, the magnitudes of pressures and forces, and of the effects exerted on them by friction, are of lesser concern.

The response of the material to tensile stresses is often inseparable from the effects of friction, as reviewed by Mellor [9], Gibson et al [55], Havranek et al [56], and others. Properties of importance include the following:

a. The *strain-hardening rate* is often expressed by the n value, which is the exponent in a flow stress relation of the type

$$\sigma_f = K\varepsilon^n \tag{10.2}$$

where K is a material constant and ε is natural strain; on the basis of Eq.2.5

$$\varepsilon = \ln \frac{t_0}{t_1} \tag{10.3}$$

where t_0 and t_1 are the sheet thicknesses before and after deformation. The strain sustained prior to necking is proportional to n. Elastic springback is often important, and then a flow stress relation of the form

$$\sigma_f = K\left(\varepsilon_0 + \varepsilon\right)^n \tag{10.4}$$

is more suitable; ε_0 is the strain at which plastic deformation commences. Because deformation is often nonuniform, a mean flow stress can be difficult to calculate from Eq.10.3 or 10.4 and then the ultimate tensile strength (UTS) is often used for simplified calculations.

b. *Strain-rate sensitivity* is expressed by the exponent m in a flow stress relation of the form

$$\sigma_f = C\dot{\varepsilon}^m \tag{10.5}$$

where C is a material constant and $\dot{\varepsilon}$ is defined by Eq.2.6. Strain (elongation) after necking is proportional to m. The total elongation in the tension test is a measure of combined strain hardening and strain-rate sensitivity and has, therefore, fundamental value.

c. The *plastic anisotropy* of the sheet is expressed by the r value

$$r = \frac{\varepsilon_w}{\varepsilon_t} \tag{10.6}$$

where the width strain $\varepsilon_w = \ln(w_0/w_1)$. For an isotropic material, $r = 1$. A material of high r value resists thinning, whereas one of low r value thins down readily while resisting width reduction. Sheets having planar anisotropy show a variation in r value measured in different directions relative to the rolling direction, and form ears in deep drawing.

10.3.2 Stretching

In pure stretching the sheet is firmly clamped along its circumference while a male die (punch) deforms it (Fig.10.9). The shape is developed entirely at the expense of sheet thickness. Therefore, necking and, finally, fracture of the sheet must eventually occur. The depth of draw before fracture and the location of the fracture point are functions of friction and of material properties.

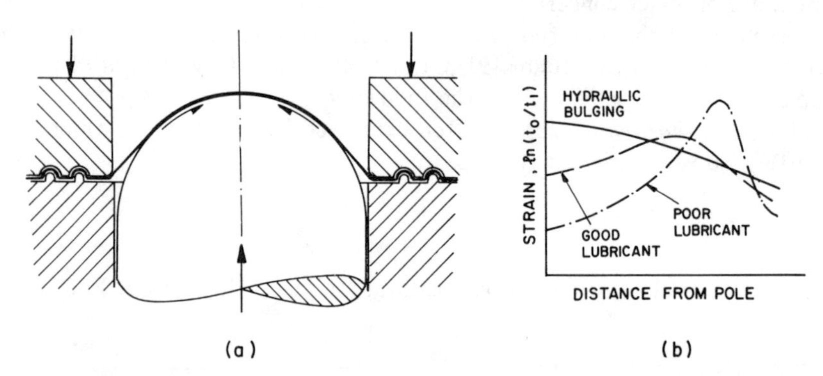

Fig.10.9. Schematic illustration of the limiting dome height (LDH) test (a) and strain distributions obtained under different lubrication conditions (b)

Thinning of the sheet takes place against frictional restraint over the punch surface. In the absence of friction, a condition that can be obtained by use of a hydraulic fluid as the punch, the sheet thins out gradually toward the apex, and fracture finally occurs in balanced biaxial tension [57,58]. In the presence of friction, free thinning on the punch nose is hindered, the strain distribution becomes more localized (Fig.10.9b) [59], and the fracture point moves toward the die radius where plane-strain conditions prevail (for reasons, see Fig.2.2). The total depth, obtainable before fracture sets in, diminishes. A higher *n* value has the same effect as reduced friction, whereas an increasing *r* value has the opposite effect [60,61]. These effects can be exploited for a number of purposes:

1. Pure stretching over a hemispherical punch can be used to evaluate the stretching capacity (stretchability) of a metal. The sheet must be firmly clamped to prevent draw-in from the flange; in practice, it is necessary to incorporate beads in the die and blankholder (Fig.10.9). The limiting dome height (LDH) increases with decreasing friction, and sensitivity to material properties such as the *n* value is reduced (Fig.10.10) [62]. Similarly, the effect of surface roughness, an important factor in practical applications, is also neutralized [63]. Therefore, it is now accepted that the test should be conducted with an unlubricated punch [64].

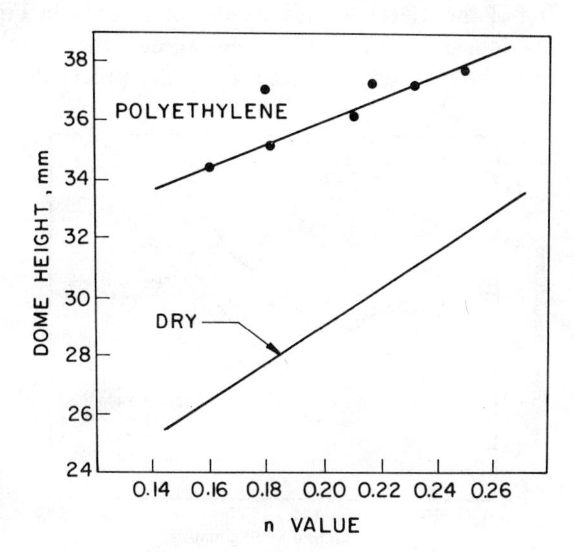

Fig.10.10. Effects of strain hardening (*n* value) and lubrication on dome height in the LDH test [62]

2. The formability of a material under strain conditions ranging from balanced biaxial tension to plane-strain tension can be explored with the aid of tests in which the strip width is varied [52]. From the onset of necking (or fracture), a forming limit diagram (FLD) (Fig.10.11) is obtained which, following the work of Keeler [65] and

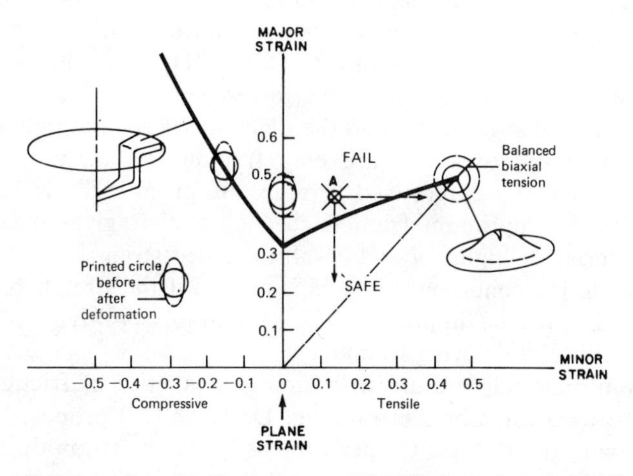

Fig.10.11. Forming limit diagram (FLD) typical of steels [2.11]

Goodwin [66], has become a widely used diagnostic tool in industry. Prior to deformation, a suitable grid which incorporates circles is etched on the surface of the sheet, so that from the distortion of these circles the magnitudes of major and minor strains can be ascertained. For an introduction to this technique, see [67,68]. The FLD is lower with higher friction [51]. Thus, if a part fails in production (point A in Fig.10.11), one possible remedy may be lubrication of the punch to raise the FLD and redistribute the strain toward the safe biaxial region (horizontal arrow from point A), provided of course that stretchability is greater in balanced biaxial tension (a situation that holds for steels). In a variant of the LDH test, rectangular strips of varying width are drawn so that the resulting draw ("dome") height can be plotted against transverse (minor) strain or, in the absence of gridding, against sample width. The results are affected by a number of test variables, as discussed by Story [69]. Most significantly, even the ranking of the metals varies, as shown by data in Fig.10.12. In this instance, the results for the lubricated samples did not agree with production experience. In general, correlation must be established before a specific procedure is adopted.

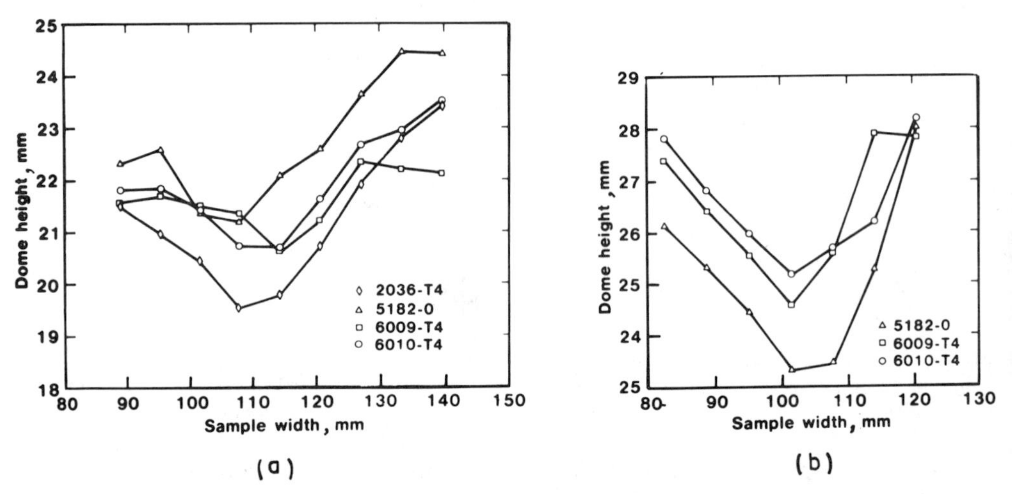

Fig.10.12. Dome heights obtained with (a) dry and (b) lubricated punches [69]

3. Because of the interactions between material properties and friction, independent friction evaluation is difficult. If local strains are known, stress ratios can be calculated and, if the r value is determined in tension, a coefficient of friction can be derived according to Kaftanoglu [70]. According to Gosh [71], μ can be obtained, without a knowledge of material parameters, from the force-displacement curve in the LDH test by relating the rate of change of force to the diameter of contact radius.

In general, interface pressures are only fractions of σ_f, and μ becomes high. Examples of experimentally determined μ values are given in Table 10.1. Kobayashi et al [72] found that the optimum friction coefficient that gives maximum stretch is smaller for a material of higher n and r values. Since strain hardening has the same effect as lubrication, the condition $\mu = 0.05$ is needed if the stretchability of a material for which $n = 0.2$ is to be improved. The strain-rate sensitivity of a material also increases stretchability by delaying fracture.

4. For a given material, strain distribution is a function of friction (Fig.10.9b and 10.13) and can be used for lubricant ranking. Devedzic [73] proposed that a statistical parameter expressing the flatness (or peak intensity) of the strain distribution curve be used. Much simpler and perhaps just as satisfactory is the strain at the pole. The

Table 10.1. Experimentally determined coefficient of friction values

Sheet material	Lubricant	Strain	μ	Ref. No.
Al-killed steel	Dry (rough punch)	0.06	0.6	[70]
	Grease	0.04	0.42	
		0.09	0.29	
		0.13	0.24	
	Dry (smooth punch)	. . .	0.2	[71]
	Mill oil	. . .	0.16	
	PTFE spray + PE sheet	. . .	0.12	
Stainless steel (409)	Dry	. . .	0.14	[71]
	Mill oil	. . .	0.13	
	PTFE spray + PE sheet	. . .	0.09	
Soft copper	Grease	0.1	0.32	[70]
	PTFE film	0.16	0.135	
		0.33	0.07	
Soft brass (70/30)	Grease	0.1	0.31	[70]
	Dry	. . .	0.36	[71]
	PTFE + PE film	. . .	0.29	
Soft aluminum	Grease	0.12	0.38	[70]
3003-O Al	Dry	. . .	0.22	[71]
	PTFE + PE film	. . .	0.07	
2024-T4 Al	Dry	. . .	0.31	
	E.P. oil	. . .	0.19	
	PTFE + PE film	. . .	0.17	
	Polymer coat + oil	. . .	0.045	

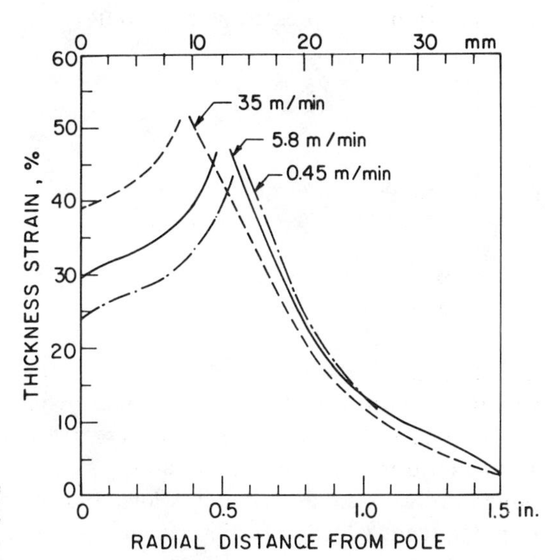

Fig. 10.13. Speed effect in stretch forming of steel sheet with chlorinated oil (sheet thickness, 0.9 mm; punch diameter, 50 mm) [57]

effects of material properties can be neutralized by comparing the pole strain obtained in lubricated stretching with that obtained in hydraulic bulging, in which friction is totally absent [52,57,58]. The distance of fracture from the pole is another possible measure, as is the distance of the fracture point from the limiting curve on the FLD [74]. A theory of thinning in stretching is given by Chakrabarty [75].

At large ratios of sheet thickness to punch diameter, (t/d_p), the bending component becomes significant. This can be explored by stretching in plane strain with an angular punch. As shown by Demeri [76], lower friction again contributes to greater stretchability and reduced strain gradients.

10.3.3 Deep Drawing

In ideal deep drawing, in contrast to stretching, it is actually desired that the blank be drawn into the gap between punch and die to form a cup-shape part without over-all thinning (Fig.10.14). Several reviews are available [9,52,77,78].

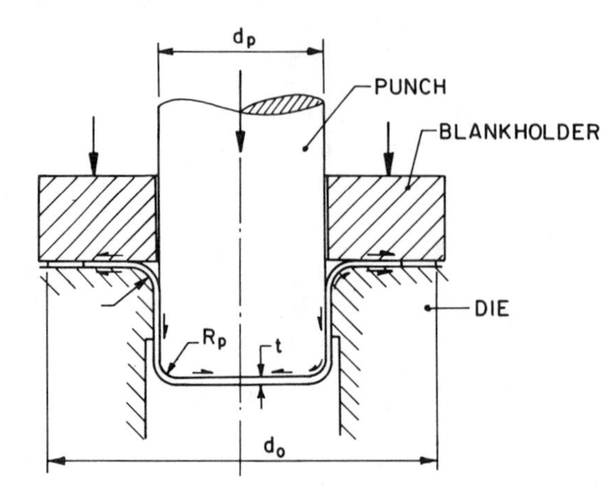

Fig.10.14. Deep drawing of a cylindrical cup (half arrows indicate frictional shear stresses acting on partially formed cup)

Drawing a Flat-Bottom Cup. When a circular blank is drawn with a cylindrical punch, a blank of larger d_0 diameter must be fitted over a punch of smaller d_p diameter. The strain is expressed as reduction r_d

$$r_d = \frac{d_0 - d_p}{d_0} \tag{10.7}$$

or draw ratio

$$\mathrm{DR} = \frac{d_0}{d_p} \tag{10.8}$$

Difficulties arise with increasing draw ratios because of increasing draw stresses. The situation is best examined at some intermediate stage of drawing (Fig.10.14):

a. The end face of the punch creates biaxial tension in the base of the cup. The sheet conforms to the punch radius R_p by combined bending and stretching. If the punch radius is small, friction is low, and the material has low r and n values, excessive thinning leads to fracture [79]. Friction on the punch base and radius is helpful because it transfers stresses onto the punch and minimizes thinning, an important point with strain-hardened sheet which has a low n value. Wallace [63] suggested that fracture moves to the wall when $R_p/d_p < \mu\pi/4$.

b. The wall of the partly formed cup is in tension and the natural tendency would be toward a reduced diameter. However, the punch prevents this, creating a plane-strain condition. High friction on the punch is again helpful, because tensile stresses which could lead to fracture are transferred from the wall onto the punch. The fracture point then moves toward the die exit, where uniaxial tension prevails. With excessive punch and die radii the unsupported sheet puckers, especially if the r value is low.

c. The sheet is bent and unbent around the die radius. Friction on the die imposes higher drawing stresses on the cup wall and is, therefore, harmful.

d. The remaining annulus of the blank is drawn into the gap by radial tension while the reduction in diameter creates circumferential compression, corresponding to the left-hand side of the FLD (Fig.10.11). This could lead to buckling (wrinkling) of a relatively thin blank, as can be deduced from elastic-plastic considerations [80,81]. Thus, while it is possible and usual to draw relatively thick blanks without further restraint, blanks of larger d_0/t ratios must be drawn with a blankholder (holddown) if the draw ratio exceeds 1.2. This means that further frictional contact is established between the blank and the blankholder and die surfaces. This friction increases the draw stress and is, again, harmful. However, a thick lubricant film may also be undesirable because it allows wrinkling.

e. Circumferential compression thickens the outer edges of the blank so that, when the gap between punch and die is equal to the original sheet thickness, some wall reduction (ironing) takes place. This is essentially a bulk deformation process similar to drawing of a tube on a bar, and thus creates more severe frictional conditions.

Limiting Draw Ratio. In the absence of friction on the punch wall, the partly drawn cup must carry the drawing force. When the force exceeds the strength of the partly drawn cup, fracture ensues. The location of fracture depends, among other factors, on friction [9,79,82-84].

As the blank diameter increases for a given sheet material and thickness, the minimum blankholder pressure (BHP) necessary to prevent wrinkling (Fig.10.15) must be increased. For a given lubricant, the frictional force increases with increasing blank diameter, resulting in a diminishing upper limit of blankholder force that can be tolerated without fracture. At a critical blank diameter even the minimum blankholder force is sufficient to cause fracture; this is expressed as the limiting draw ratio

$$\mathrm{LDR} = \frac{d_{0\,\max}}{d_p} \tag{10.9}$$

Lower blankholder friction results in reduced tensile stresses in the wall, allowing larger-diameter blanks to be drawn and opening the window of BHP within which a good draw can be achieved (Fig.10.15), thus making the operation less critical. This is particularly important in drawing of thin sheet, where the slightest misalignment of the blankholder can create a significant increase in BHP. The LDR is affected by a number of other factors, too:

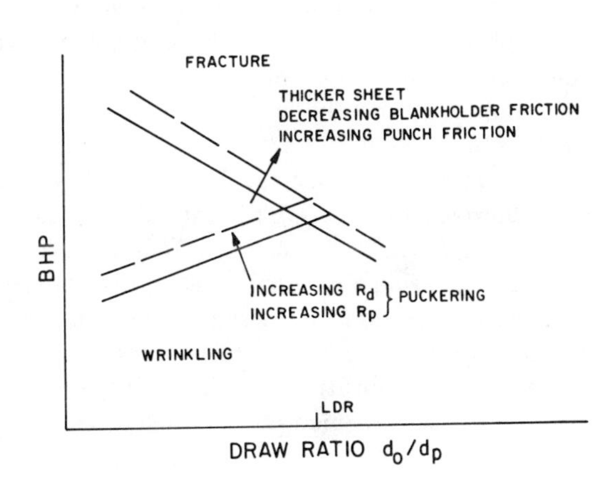

Fig.10.15. Effects of friction and process geometry on permissible blankholder pressure in deep drawing

a. The above discussion assumes a certain level of friction on the punch radius; Fogg [52] suggests shot blasting of a punch with $R_p = 2.5\ t$ to eliminate the effects of friction. Friction on the punch surface does not affect the drawing force but reduces the stress acting in the cup wall, thus increasing the LDR.

b. With a thinner sheet, the frictional force under the blankholder constitutes a larger proportion of the total drawing force; at the same time the load-carrying capacity of the cup diminishes, and thus the LDR drops also (Fig.10.16) [85,86]. This creates difficulties in conducting reduced-scale tests. Siebel and Kotthaus [5.6] suggested the use of increased blankholder pressure, but this does not overcome the problem entirely [87], because friction is a function also of pressure [53].

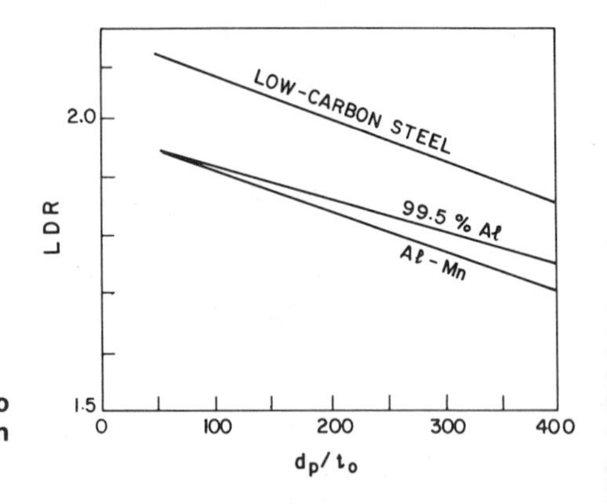

Fig.10.16. Effect of ratio of punch diameter to sheet thickness on limiting draw ratio attainable in deep drawing [85]

c. Because fracture occurs in tension, all factors that influence tensile properties also affect the LDR and thus modify the effects of friction. Most powerful is plastic anisotropy: materials with high r values resist thinning, are stronger in plane strain, and thus result in higher LDR values. The n and m values are of lesser importance [60]. Typical LDR values are 2.0 for aluminum and 2.2 for steel and brass (for $d_p/t < 100$).

1. Lubricant Evaluation. The influence of friction on the LDR offers a method of lubricant evaluation. Lubricants can be ranked simply by the LDR, provided that the blankholder pressure is high enough to prevent wrinkling (i.e., it must be in the upper half of the safe zone in Fig.10.15). However, the test is laborious: at least ten blanks must be drawn at each increment of diameter, and increments must be small. The test is also of limited sensitivity. As pointed out by Fogg [52], sensitivity is improved by increasing friction on the punch radius, but even so the variation between a very good and a poor lubricant is only 4%. A quick evaluation is possible by drawing larger-than-maximum blanks and noting the depths of fractured cups [5.46].

For a given LDR, the high blankholder force permissible with a good lubricant and low punch wall friction promote thinning of the cup. Therefore, the volume of the cup also increases, and this could be taken as an additional measure of lubricant performance [88].

2. Drawability Testing. The suitability of a sheet metal for deep drawing may be characterized by a standard test such as the one by Swift (reported by Willis [4]), in which the LDR is determined under preset conditions and die geometries. To avoid interference from frictional effects, a standardized lubricant must be found. In principle, the test becomes most sensitive to material variables if friction is minimized on all contact surfaces. This used to be approximated [89] with a viscous, compounded drawing oil, but the sensitivity of this oil to surface conditions and drawing speed has led to the introduction of oiled polyethylene sheet as a standard lubricant [90-92].

Forces. A recording of draw force versus stroke (Fig.10.17) shows a gradual rise while the sheet conforms to the punch nose. The force rises more rapidly once the wall begins to form; the rise is steeper with higher friction in the blankholder zone, and the maximum force attained is higher also. Punch friction increases the draw force only slightly; more importantly, it transfers stresses onto the punch and increases the LDR (Fig.10.15 and 10.18a).

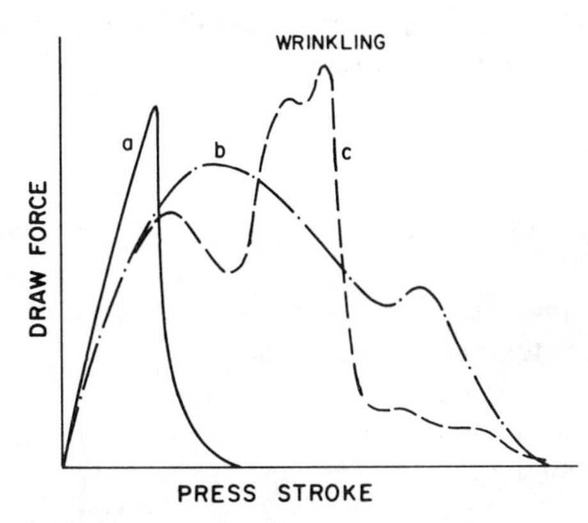

Fig.10.17. Typical draw force diagrams registered in drawing with (a) excessive, (b) correct, and (c) insufficient blankholder pressures

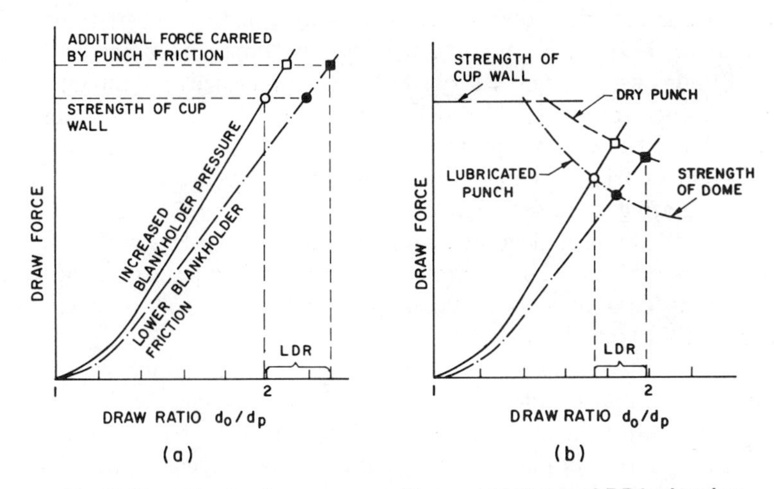

Fig.10.18. Effects of process variables and friction on LDR in drawing of cylindrical cups with (a) flat and (b) hemispherical bottoms

1. Lubricant Evaluation. The radial drawing stress accounts for 50 to 70% of the total force, and bending and unbending account for 8 to 14% [93,94]; thus, the contribution of friction in the blankholder zone and on the draw radius is large enough to offer the possibility of ranking lubricants, at a constant draw ratio, by the magnitude of the maximum draw force. The number of tests required is small, and discrimination is better than by the LDR [52,95], but only if the d_0/t ratio is large. The spread between a good and a poor lubricant can be 20%. A larger apparent spread of data is obtained, according to Reitzle et al [96], from the difference D between the force required to induce incipient fracture P_{fr} and the actual maximum draw force P_{dr} measured with the particular lubricant

$$D = \frac{P_{fr} - P_{dr}}{P_{fr}} \times 100 \tag{10.10}$$

A better lubricant gives a higher value of D. Gibson et al [5.39] define a lubricant efficiency η as the ratio of a calculated frictionless draw force to the actual, measured draw force; a higher value of η indicates a better lubricant.

An indirect measure of draw force is the thinning of the cup in the bottom half; greater thinning is an indication of higher friction [97].

As the blank is drawn into the die gap, the deforming area under the blankholder diminishes and, even though for a constant blankholder force the pressure on the flange increases, the draw force declines. When the edge is pulled through the die gap, a secondary force maximum is noted (Fig.10.17, curve b), the magnitude of which depends, for a given gap and sheet thickness, on friction. Therefore, this too can be used for judging lubricants, as pointed out by Kretschmer [97], who published a series of papers on draw forces. A potentially confusing factor is the incipient wrinkling that can develop with good, thick lubricant films, leading to a rise in the secondary force maximum. An indirect measure of friction is given by the amount of martensite formed in austenitic stainless steels [98].

The area under the force-displacement curve represents the energy of the draw, and can be used for lubricant ranking. It has the advantage that the contribution of ironing at the edge is included [88].

2. Theory. Interactions of material parameters with friction effects in the various deformation zones make the theory of deep drawing complex. Reviews have been published by Alexander [99], by Kaftanoglu [81], and by Pope and Berry [94], all of whom also developed their own theories. Blazynski [2.16] emphasizes redundant work. Avitzur [100] and Hasek and Krämer [101] provide upper-bound solutions. Numerical solutions can allow variable μ in various parts of the contact zone [81,102]. In principle, no average coefficient of friction can have true meaning. Nevertheless, a μ value is often assumed for calculating purposes. Thus, Chung and Swift [93] assumed $\mu = 0.06$ for drawing of steel with graphited tallow, while Fukui [103] took $\mu = 0.16$ for drawing of aluminum and $\mu = 0.19$ to 0.26 for drawing of iron with rapeseed oil lubrication. Values of μ as low as 0.05 have been taken [86] and could well be even lower in the blankholder zone [70] (see also Sec.10.4).

The effects of friction on draw force and LDR are summarized in Fig.10.18a. Lower friction on the blankholder increases LDR by reducing the draw force, whereas increasing punch friction raises it by transferring the load from the cup wall to the punch. The over-all effect is that all-around lubrication increases the LDR, but not as much as selective lubrication does.

Friction due to blankholder contact can be readily separated experimentally by using the technique of Kawai et al [104]. The blankholder is split into two halves, and the frictional force FR is measured along with the punch force P and blankholder force P_B (Fig.10.19). Otherwise, the contribution of blankholder friction to the draw force can be separated in tests with incrementally increasing blankholder force [70]. Doege and Witthüser [5.46] separated the frictional component by measuring the force for bending P_b (Eq.5.4) and frictional force F (Eq.5.5) for sliding in strip drawing tests (Fig.5.13c and 5.10a). These forces, divided by strip width and multiplied by cup circumference, give the bending and frictional components P_{bd} and F_d. Then, starting from the premise that draw force is directly proportional to sheet thickness, a cup is drawn from a single blank, and then from two blanks placed on top of each other. Thus the force for deformation in the blankholder zone (P_{bh}) and the frictional force in the blankholder zone (F_{bh}) are found, and from these an average μ is calculated according to Siebel and Panknin [85] from

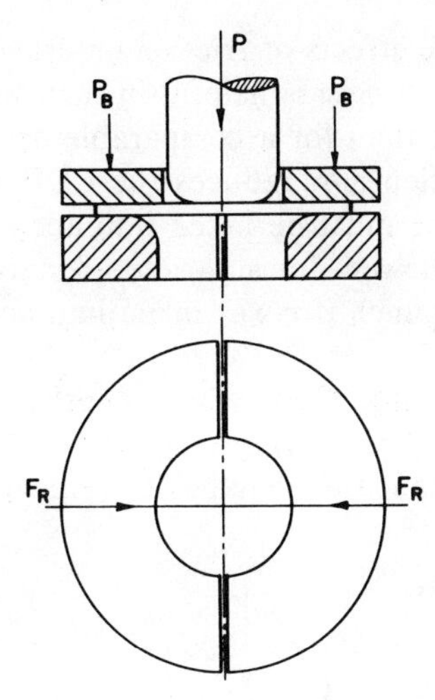

Fig.10.19. Die set with split blankholder for measuring blankholder friction [104]

$$\mu_{bh} = F_{bh} \frac{2}{\pi} \frac{\beta}{\beta_0^2 - 1} \frac{1}{d_0^2} \frac{1}{\text{BHP}} \tag{10.11}$$

where β_0 is the initial draw ratio, β is the draw ratio when the maximum force is reached, and BHP is the blankholder pressure. Some results are given in Table 10.2.

Rajagopal [105] proposed a method of extracting μ for the cup wall/punch interface from measured draw forces and draw depths. The values thus obtained were quite high: for tinplate, $\mu = 0.32$ dry and 0.25 with an emulsion; for aluminum alloy 3004, $\mu = 0.47$ dry and 0.3 with an emulsion. Of course, the frictional force depends on whether or not the cup lifts away from the wall by springback.

Drawing of a Hemispherical Cup. When the punch has a hemispherical end, drawing begins with stretching over the punch nose [106]. Higher blankholder force is needed to provide frictional restraint and prevent puckering. This could, however, lead to excessive thinning and fracture on the punch. High friction on the punch is therefore beneficial but narrows the window for good draws (Fig.10.15).

Table 10.2. Comparison of coefficient of friction values obtained in simulation tests and in deep drawing
(Material: low-carbon steel) (from data in [5.46])

	μ			
Lubricant	Strip drawing (Fig.6.10a)	Draw bending (Fig.5.13c)	Deep drawing	Maximum draw force, kN
Spray PTFE	0.150	0.116	0.210	126
Talcum	0.161	0.120	0.158	116
Phosphate + soap	0.174	0.140	0.140	116
Graphited grease	0.197	0.172	0.154	124
Dry	0.203	0.196	0.168	134

The effects of friction on draw force and LDR are shown in Fig.10.18b. Friction on the punch nose is helpful in that it prevents thinning and thus premature fracture. LDR is lower than for a comparable draw with a flat-ended punch (Fig.10.18a), and over-all lubrication now reduces the LDR [57,107]. High n, m, and r values are helpful. Their influence must be taken into account in evaluating lubricants; Nakagawa and Okazaki [107] showed, for a variety of materials, that LDR increased linearly with increasing μ on the punch if r was taken into account in deriving the value of μ.

Redrawing. Cups whose depths are greater than that allowed by the LDR are produced by redrawing (Fig.10.20). Frictional conditions are similar to those in drawing, except that the blankholder (punch sleeve) is now tubular.

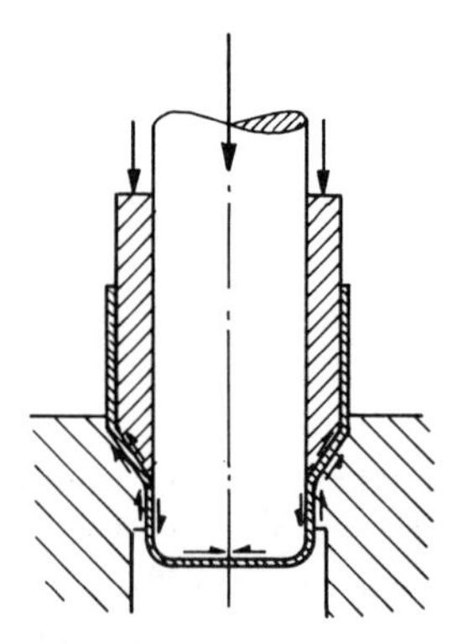

Fig.10.20. Redrawing of a cylindrical cup (half arrows indicate frictional stresses acting on workpiece)

10.3.4 Combined Stretching and Drawing

Many parts are made by pressworking processes which incorporate elements of both stretching and drawing. The shapes of these parts range from ostensibly simple rectangular containers [108,109] to the complex, irregular shapes of automotive components [110,111]. At the corners the stress and strain states are similar to those observed in deep drawing, and the wall height is built up by circumferential compression (Fig.3.5a), for which the high r-value direction in a sheet with planar anisotropy is favorable. At the sides the wall is formed by draw-in of the sheet in plane strain (Fig.3.4b), and a high n value is favorable. The base of the part is frequently domed (Fig.10.21), and therefore contact between punch and sheet is localized and a large area of the sheet is unsupported, resulting in puckering. Another problem is springback which distorts the shape of gently curving parts. The process is controlled by several means:

a. Higher pressure on a flat blankholder increases the resistance to draw-in. However, more positive control is obtained with draw beads (Fig.10.21a), which function by

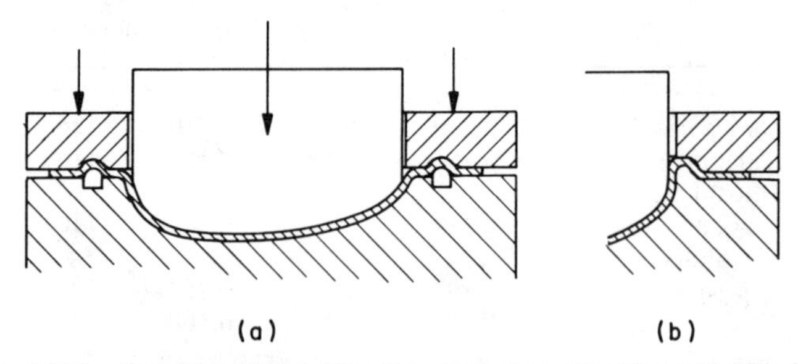

Fig.10.21. Combined stretching/drawing of an irregular part with (a) draw bead and (b) edge bead

repeated bending and unbending of the sheet. In principle, they are effective even without friction. The restraining force increases with decreasing bead radius [109,112], and a round-cornered square is more effective than a fully radiused bead [113]. Draw beads are usually made as inserts and are more readily adjustable than edge beads (Fig.10.21b). Friction on the bend radii, on the blankholder surface, and on the die radius are important contributors to the draw force, as shown by Weidemann [114] (Fig.10.22) and by Nine [5.43] in the draw-bead simulator (Fig.5.12). With proper lubrication, the draw beads change the strain history in various parts of the drawn part [115,116] and through this contribute to an increased limiting depth, as shown by Hasek and Lange [109]. Even though the interface pressures are low by bulk deformation standards, relative sliding is severe. The coefficient of friction is a function of lubricant: Nine [5.43] measured $\mu = 0.2$ with a mill oil and $\mu = 0.05$ with a soap, on both steel and 2036 aluminum. Draw beads present the dangers of pickup and sheet damage, a point to which we shall return (Sec.10.4.4). Draw beads made of polyurethane are less effective but minimize scoring [117].

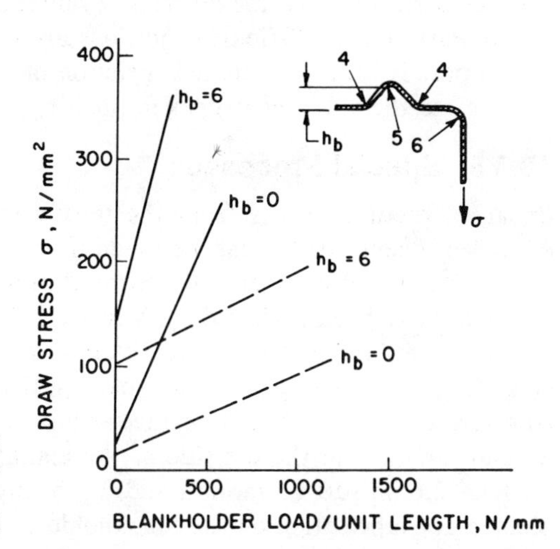

Fig.10.22. Draw stresses developed in drawing of 0.9-mm steel sheet with a low-viscosity rust-prevention oil ($\mu = 0.15$, solid lines) and with a high-viscosity drawing oil ($\mu = 0.035$, broken lines) [114]

b. Combined stretching and drawing offers a wide scope for differential lubrication [55]. When the drawing component dominates, use of a dry punch and a lubricated blankholder zone (combined with a high r value) results in the deepest parts. Over-all lubrication provides only a small improvement when stretching is dominant; a high n value is much more helpful [55]. Punch friction is still important, though. High friction minimizes thinning whereas low friction promotes deformation in a curved bottom and thus minimizes springback. The punch radius must be lubricated sufficiently to allow plastic deformation over the entire base (for which the stress must exceed the initial flow stress or yield strength, YS) without exceeding the strength of the wall (for which stresses must remain below the UTS). This led Duncan and Bird [111] to state for a punch of 90° radius that

$$\text{UTS} > \text{YS} \exp(\mu\pi/2) \tag{10.12}$$

c. Friction is important in tests devised for evaluation of the stretch/drawing properties of sheet metal. The Erichsen and Olsen tests are performed by clamping an oversize blank between flat, polished blankholder and die surfaces. On advancing a ball-ended punch, smaller than the die diameter, a cone is stretched. The depth of the cone at incipient fracture is taken as the Erichsen or Olsen value. Unlike the LDH test (Fig.10.9), where draw beads eliminate the effects of draw-in [118], the Erichsen and Olsen tests allow some material to be fed in. Therefore, these tests provide fair simulations of combined stretching and drawing, and correctly show the benefits of higher *n*, *m*, and *r* values. Unfortunately, the results are affected also by friction and its distribution. When only the punch is lubricated, the limiting value increases, more for materials of low *n* value (Fig.10.10). When the lubricant is applied also to the blankholder zone, draw-in increases the limiting depth. This means that Erichsen values determined under different frictional conditions are not comparable. Much work has gone into understanding these relationships, as reported by Fogg [52]. Attempts have been made to standardize the lubricant; PTFE has been recommended [119,120] but has been found to have inadequate ductility [121]. Polyethylene sheet [119,120,122] and, more recently, PE sheet with oil applied to both contact surfaces have been proven to give the most reproducible results [123].

d. Drawing of square boxes from cylindrical blanks is highly sensitive to lubrication and has been proposed for lubricant evaluation by Hasek [124].

e. Particularly difficult conditions are created when the bottom of an already deep drawn part is locally stretched. Friction on the punch shifts the point of thinning from the apex to the sides, as it does in pressing of ridges into flat panels [125].

10.3.5 Special Processes

Because fracture results from the tensile strength of the partially formed wall being exceeded, deeper draws can be made if the tensile stresses are either transmitted to the punch or reduced in magnitude. Several practical techniques have been developed, some of which have been reviewed by Mrozek [126]:

a. When the draw die is fitted into a container filled with a liquid lubricant (Fig.10.23a), punch penetration causes the fluid pressure to build up. Stresses are transferred onto the punch surface, while the fluid displaced also improves lubrication around the die and the die side of the blankholder. Kasuga et al [127,128] achieved an LDR of 2.6 on soft aluminum and 2.7 on steel using mineral oil as the fluid, with supplementary lubrication on the blankholder. Continuing work by Chauzov and co-workers [129,130] with oils, and by Granzow [131] with water, indicates that transfer of stresses to the punch is most important. Indeed, El-Sabaie and Mellor [132] showed that fracture moves from the plane-strain location around the punch radius to the uniaxial location at the die exit. As a result of this, the stress supported by the cup is greatly

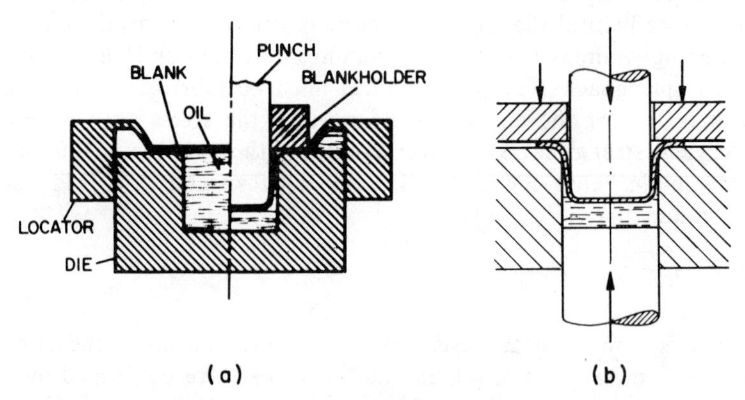

Fig.10.23. Deep drawing into a fluid-filled die cavity (a) without and (b) with pressure relief

increased. In their work, an LDR of over 3 was obtained on brass. Disadvantages are that nonuniform sheet thickness and irregular punch configuration can lead to nonuniform material flow and fracture, and that the force requirements are high. This latter disadvantage is reduced by incorporating a retractable counterpunch (Fig.10.23b) [130].

b. In the process of hydromechanical forming—also known under a variety of proprietary names—escape of the fluid is prevented by a seal in the die face (Fig.10.24). A controlled pressure is built up prior to commencement of drawing, and the increase in pressure is limited in a programmed fashion through a relief valve. Therefore, the sheet is laid up against the punch, and a part presenting a large punch-to-die clearance can be drawn by forming a reverse, stable bulge which prevents puckering. Thus, conical shapes and other difficult-to-form configurations can be made in a single draw. Deep draws are achieved because high friction transfers the draw stresses onto the punch [133-136]; for materials of low *n* values the pressure must rise more steeply to provide early support [132]. In a variant of the process, the fluid is retained behind a rubber membrane [137]; there is some change in the stress distribution.

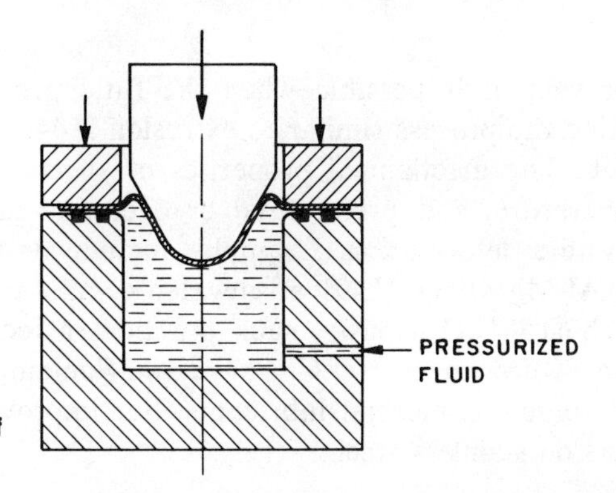

Fig.10.24. Schematic illustration of hydromechanical drawing

c. The stress distribution is markedly different when the die is replaced with a rubber cushion. The sheet is deformed as in hydroforming, but friction on the rubber is more important than friction on the punch. Fracture occurs at the punch radius, but draws deeper than those of conventional drawing can still be achieved [138,139]. If a lubricant is used, it should be compatible with the rubber. Usually, a neutral soap or a mineral or synthetic oil compounded with neutral fatty oils is acceptable.

d. A circumferentially grooved, expanding punch serves to transfer stresses and increases the LDR [140].

e. The cup wall is formed under compressive stresses in the technique proposed by Maslennikov [141]. The punch is replaced with a rubber ring (Fig.10.25); on application of load, the rubber flows into the die. An LDR increase from 2 to 6 was claimed for aluminum. Yamaguchi et al [142] found that when rosin was applied to the inner part of the blankholder/blank contact area, and wax to the die/blank interface, a draw ratio of 5 could be attained in 7 strokes. A rigid punch, used with a double die to allow penetration of the polyurethane ring, produced parts of better gage uniformity but necessitated too many strokes.

f. Smaller improvements are found in conventional deep drawing with a solid punch when rubber pads are placed on both the die and blankholder surfaces and thus partial radial pressurization is achieved [143].

g. Hydrostatic pressure does not delay the onset of necking, but the application of all-around hydrostatic pressure does improve lubrication, and the radial pressure imposed on the edges of the blank slightly increases the LDR [143-146]. A large

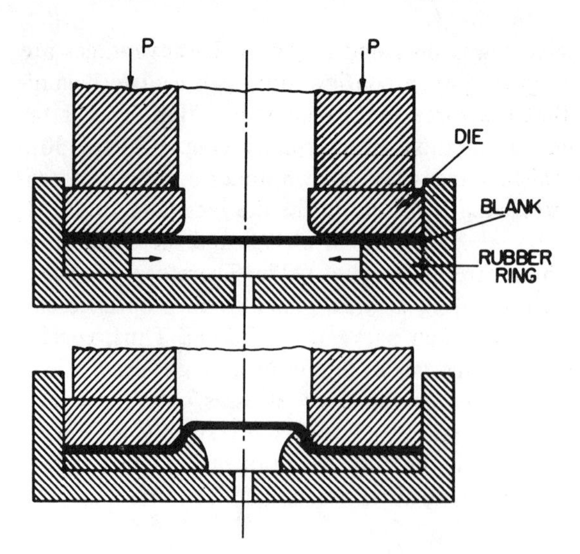

Fig.10.25. Drawing with an elastomeric ring in place of the punch [141]

improvement is possible when the fluid pressure is applied only to the blank edges making the process similar to extrusion [144].

h. The mechanical properties of sheet materials often change quite rapidly with temperature, and preferential or over-all heating or cooling can be used to increase allowable deformation. Examples include drawing of steel [147], stainless steel [148] and Al-Mg alloys [149] at elevated temperatures, and drawing of steel with a cooled punch [150]. Obviously, such practices affect lubrication. Even at modestly elevated temperatures up to 100 C, Funke and Schlemper [5.89] observed marked drops in LDR with some commercial lubricants, but improvements with others—some on mild steel others on stainless steel.

10.3.6 Ironing and Necking

In ironing, the wall of a cup is reduced over its entire height without materially changing its inner diameter (Fig.10.26). The similarity between this and drawing of a tube or a bar (Fig.7.12e) is evident. Ironing can be applied simultaneously with drawing [151] but is generally a separate operation. Long employed for the manufacture of cartridge cases and similar products, it has acquired great significance with the spread of two-piece beverage cans which are drawn and then ironed, usually in three stages, to a much-reduced wall thickness [152]. Substantial expansion of the surface demands a lubricant capable of following surface extension.

Friction transfers stresses to the punch and increases the attainable reduction, as

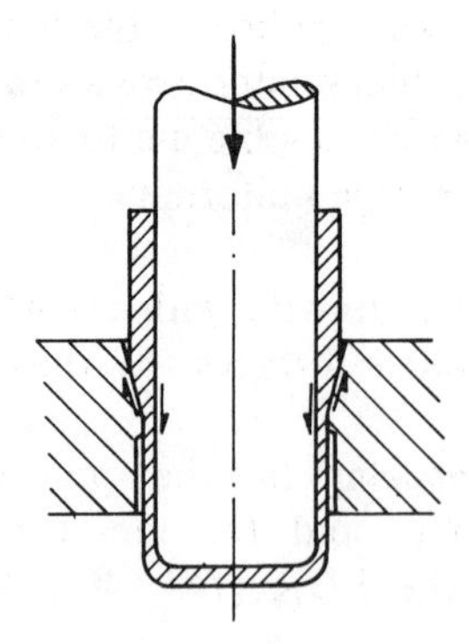

Fig.10.26. Ironing of a cup (half arrows indicate directions of frictional stresses acting on cup)

shown by the results of Osakada et al [153] (Fig.10.27), who used a split die to separate die friction with the aid of Eq.5.3. Shawki [154] applied strain gages to the draw die for the same purpose and calculated μ = 0.058 for a mineral oil, 0.041 for cottonseed oil, and 0.342 for a 10% soap solution. With the aid of strain gages attached to the end surfaces of partially drawn cups, he concluded that friction between the punch base and the cup was 0.13, 0.114, and 0.101, respectively, for the above lubricants. Rajagopal and Misra [155] measured the ironing load at reductions ranging from zero to the limiting value (where fracture occurred). An approximate value of die friction could be derived from the initial slope of the load-vs.-reduction curve, and friction on the punch was obtained from the difference between the ironing load at failure and the load-carrying capacity of the wall (as found by tensile testing). The technique could not be applied to aluminum cans, because the heavily strain-hardened material fractured at loads lower than those predicted from tensile strength. An indirect measure of friction is obtained from a strain gage measurement of draw-die expansion [156].

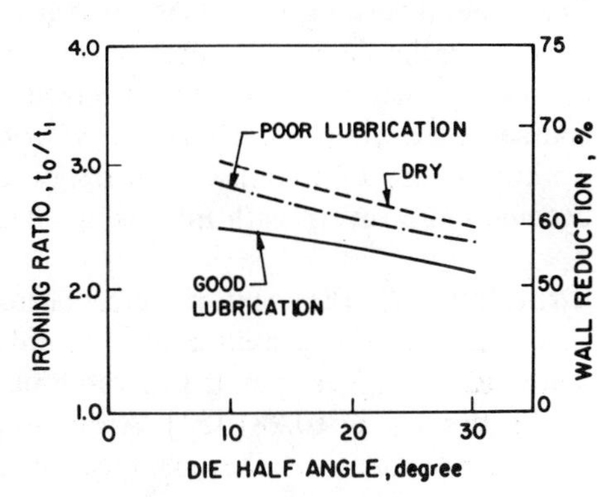

Fig.10.27. Effect of lubrication on reduction obtainable in ironing with dies of different half angles [153]

We have seen (Sec.7.2) that, theoretically, very high reductions can be taken if friction on the punch is high (Eq.7.10). This conclusion is drawn also in theories of ironing [78,153,157,158] and has been confirmed in practice [159]. Reductions close to 60% are feasible with a lubricated punch, and reductions up to 70% can be attained with a dry punch [153], provided that the die angle is small enough. Even though the frictional force on the die increases with a decrease in half angle, so does the frictional force on the punch. Reductions in excess of 90% are possible in a single pass when the die half angle is less than 6°. A neutral point develops inside the ironing zone, and the material can emerge at a speed higher than that of the punch [160]. A limit is set by the difficulty of stripping the ironed can, especially when it is a thin-wall can, as in two-piece canmaking. To prevent collapse of the can, the punch is usually smooth and is lubricated together with the die. Rajagopal and Misra [155] concluded from their indirect method that punch friction is actually lower than die friction. However, for homogeneous material flow, friction on the two surfaces should be equal, and from practical observations Panknin [159] suggests that μ should be from 0.02 to 0.03 on the die and twice that on the punch. If friction on the punch were lower, the maximum attainable reduction would drop rather than increase with a decrease in die angle.

Processes of great practical importance are necking (nosing) and expansion of can or tube ends [161-163]. The maximum necking ratio (ratio of diameters before and after necking) is limited by development of circumferential folds, or by axial collapse or upsetting of the wall. The latter two failure modes are especially sensitive to both friction and die angle. For a given angle, the tube may actually become longer with lower friction, as shown by Lahoti and Altan [162]. The maximum necking ratio is greater for a 30° than for a 60° included die angle. In the work of Ebertshauser [161], the low friction of a polymer film resulted in high reduction ratios on thick-wall products whereas oil-type lubricants were just as good on thin-wall ones, apparently because the polymer film could not follow the deformation. A dry MoS_2 film was not quite as good as a drawing paste. This is essentially an instance of squeeze-film formation followed by moderate sliding, and thus entrapment of liquids in surface roughness features is beneficial.

10.4 Lubrication and Wear

The principles of lubrication in pressworking have been repeatedly discussed [43,52,57,164-168].

10.4.1 Process Conditions

Sheet metalworking processes present a wide range of demands on lubricants, from the almost negligible to the most severe. In these processes, as in other non-steady-state processes, whatever lubricant is introduced at the beginning must suffice for the entire course of deformation. Therefore, lubricant breakdown, tool pickup, and scoring of the workpiece are of major concern. For a given process geometry, conditions generally become more severe with increasing sheet thickness.

Stretching. At the start of deformation there is normal approach between punch and sheet, and, in the presence of a liquid lubricant, a squeeze film develops. Contact is made at the highest point; in stretching over a hemispherical punch, there is no sliding at this point (Fig.10.28) [54]. Away from it, sliding distance and velocity increase at a rate proportional to the distance, if friction is low. With high friction, sliding is arrested. Interface pressures are low, because the membrane stress is typically $4(t/d_p)$UTS and is on the order of 0.01 to 0.1 UTS. Nevertheless, surface deformation is substantial.

A surface element of an originally smooth blank roughens prior to making contact with the punch, generally in line with the discussion in Sec.3.8.3 (Fig.3.49), although details of the roughening process are affected by anistropy and the strain ratio [169,170]. Some sheet materials—notably, low-carbon steels and Al-Mg alloys—are prone to discontinuous yielding, and at low strains the surface is covered with highly visible, mutually intersecting families of slip bands. No analysis of the effects of such yield point phenomena on lubrication appears to have been made, and the remarks made here apply only to sheet that begins to deform plastically over the whole surface. The rate of roughening is lower in balanced biaxial tension than in plane strain, but

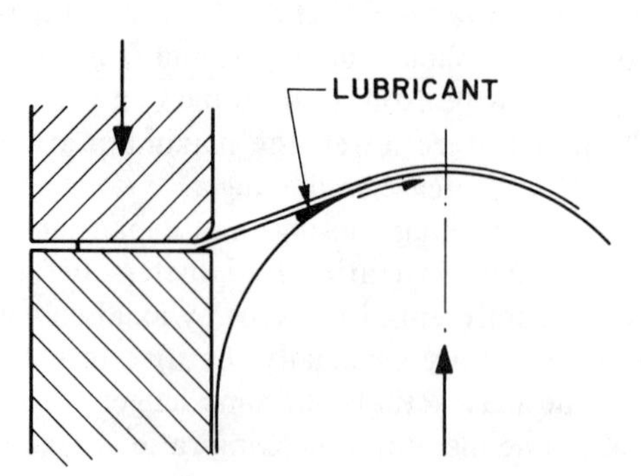

Fig.10.28. Negative wedge developing in stretch drawing (after [54])

roughness is proportional to the equivalent strain. Asperities deform and flatten, and the real (or boundary-lubricated) contact area (b in Eq.3.34a) increases rapidly to 60 to 70% of the total area [85,171]. On a dry sheet, the b ratio increases in proportion to pressure, giving a constant μ (Fig.3.34, left). In the presence of a liquid lubricant, the b ratio is stabilized and μ decreases with interface pressure (Table 10.1). The liquid film is squeezed out in a wedge which advances with the line of contact [52]; the divergent gap is unfavorable for hydrodynamic lubrication (Fig.10.28). Nevertheless, some lubricant usually remains trapped in surface features and prevents metal-to-metal contact. Indeed, when the film is very thick (as it is when an oiled PE sheet is used), the surface roughens as in free stretching or hydraulic bulging [54].

Deep Drawing. Conditions vary greatly at various points of a partially drawn cup (Fig.10.29):

a. Conditions on the punch bottom (zone 1) are the same as in stretching, except that friction around the radius (zone 2) limits the transmittal of stresses (Eq.10.12), and interface pressures may drop close to zero. The sheet roughness is virtually unchanged [172]. On the punch radius the stresses are higher, some relative sliding takes place, and the lubricant thins out.

b. Between the punch flank and the partially drawn cup (zone 3), the lubricant is trapped in a squeeze film, whereas it is drawn in under favorable conditions at the point of transition (zone 4). Pressures are low and can easily drop to zero if elastic springback occurs, or if the clearance is large enough to allow the cup to separate from the punch in the early stages of drawing. The surface roughens as in free deformation, in combined tension/compression, and the presence of a lubricant makes little difference [172-174].

c. Around the die radius (zone 5), pressures are somewhat higher—by analogy to a frictionless pulley, on the order of (t/R_d)UTS. Since R_d = 5t to 8t, the pressure is around 0.2 UTS, which is still a quite small value. Sliding conditions are, however, severe. The speed is equal to the drawing speed and the surface, deformed by circumferential compression and bending and unbending, becomes progressively rougher [173].

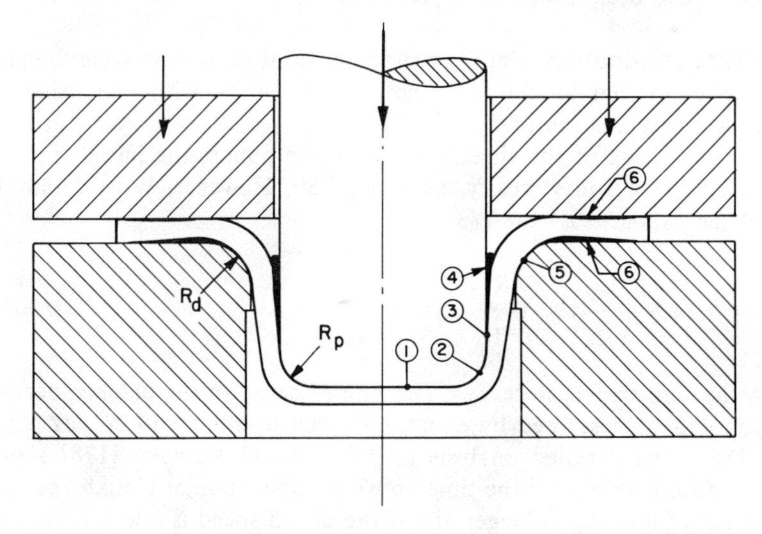

Fig.10.29. Zones of differing lubrication conditions in deep drawing (see text for identification)

There is danger of exposing virgin surfaces, and tool pickup is observed only on the die radius, if at all.

d. In the blankholder zone (zone 6), pressures are low—typically 0.002 to 0.02 UTS. A liquid lubricant is trapped by a squeeze-film effect. Thickening of the flange creates a wedge which contributes to lubrication by hydrodynamic action, as shown by pressure-transducer measurements taken by Witthüser [175]. In materials of planar anisotropy there is more thickening in the low r-value direction [102], and thus the film is thicker in the high r-value direction. The surface roughens under the imposed tensile/compressive stresses at a higher rate than in pure tension [170], but roughness is still proportional to strain if the effects of anisotropy are considered [169].

e. When the clearance is such that ironing takes place at the rim, the surface becomes smoother [173] and roughness depends on the thickness and nature of the lubricant film.

f. A sheet of directional surface finish or metallurgical structure exhibits visible parabolic markings [6.289], which are accentuated in the presence of a liquid lubricant.

Combined Stretching and Drawing. Conditions in combined stretching (on the punch) and drawing (around the die radius and under the blankholder), with the added demands set by the draw beads, have the effect of drawing over several die radii. A constant μ is suitable for modeling friction except when heavier sheet is drawn over tight radii; then μ increases and tool pickup can occur [176]. If this is a major concern, the R/t ratio is increased.

Ironing. Problems in ironing are similar to those encountered in tube drawing on a bar, except that the part is shorter and the lubricant finds relatively easy access to the inside. Interface pressures are on the order of σ_f, and sliding velocity on the die is equal to the punch velocity. Sliding velocities on the punch are much lower.

10.4.2 Lubrication Mechanisms

There is a vast variety of lubricants that are marketed for pressworking, but their modes of action can be fitted into the general framework of lubrication mechanisms. Some of these have been discussed by Wilson [3.288].

Full-Fluid-Film Lubrication. Under favorable conditions it is possible to maintain a full fluid film, as evidenced by the increase of draw forces with increasing speed when highly viscous substances are used.

a. Upon impacting of the sheet surface by the punch, the initial film thickness h_0 is governed by speed and viscosity, as in Eq.3.30. However, Wilson and Wang [177] showed that the exponent is now 2/3

$$h_0 = 2.24\, R_p \left(\frac{\eta v}{\sigma_f}\right)^{2/3} \left(\frac{R_p}{t}\right)^{2/3} \tag{10.13}$$

Thereafter, the film thickness decays. The rate of decay in the blankholder zone (zone 6 in Fig.10.29) can be described by equations given by Moore [3.29], as pointed out by Wilson [3.288]. The detailed analysis of Mizuno and Kataoka [178] shows that the remaining oil film is thinner if the time between application of blankholder pressure and commencement of drawing is longer and if the punch speed is lower.

b. The situation around the die radius (zone 5 in Fig.10.29) is more complicated. Mizuno and Kataoka [178] begin their analysis with the assumption that, initially, the sheet is wrapped around the die radius, creating a squeeze-film effect which is partially counterbalanced by the negative wedge effect at the die exit. Thus a fluid film is more difficult to maintain. Wilson [3.288] suggests that, at least in steady state, this zone could be regarded as a foil bearing, for which Blok and Van Rossum [179] have shown that the film thickness h should increase with increasing die radius R_d, increasing ηv, and decreasing tension T:

$$h = 1.405 R_d \left(\frac{\eta v}{T} \right)^{2/3} \qquad (10.14)$$

For the entry to the punch/flank interface (zone 4, Fig.10.29), the relevant tension is circumferential and is difficult to determine.

c. Application and removal of the highly viscous lubricant required for generation of a full fluid film can present difficulties. A lighter oil can maintain a film only if externally pressurized. Through a series of publications, Fogg [180] showed the development of the technique to the point of practical application. The lubricant is supplied to orifices inserted into the die or punch at points where the lubricant would thin out most, and where good lubrication is desired to ensure free sliding and stretching. Thus, in pressing of a heat exchanger panel formed over a series of hemispherical punches, orifices were placed in each punch and in the adjacent die and blankholder surfaces [181]. Capillary flow restrictors regulate the oil film. Oil pressure depends on the prevailing local pressures and ranges from 4 to 40 MPa in forming of 0.75-mm-thick sheets of aluminum and stainless steel. The required pressures and flow rates can be calculated by finite-element techniques [180].

d. In a version of phase-change lubrication, Wallace [182] coated the surface of a stretch-forming punch with a 2-mm-thick layer of ice kept at a temperature of -3.3 C. Full-fluid-film lubrication was ensured when the sheet was kept at 15 C. Lower-temperature operation is possible with solid CO_2 [183] or solid krypton [184].

Mixed-Film Lubrication. Most liquid lubricants operate in the mixed-film regime. The coefficient of friction diminishes with increasing ηv, corresponding to the declining portion of the curve in Fig.3.42, no doubt due to trapping of a thicker film (Eq.10.13).

a. In stretch forming this leads to the movement of the fracture site toward the apex, a larger pole strain, and a higher forming limit [57,185,186]. Viscosity alone is not sufficient, as shown by Fogg [52]. For the same viscosity, silicones were much less effective than polyglycols, indicating that Eq.10.13 is not a sufficient descriptor of reality. Tackiness is also needed and is imparted by additives such as polybutenes, polyisobutylenes, bentone clays, silica, and carbon black.

b. In deep drawing with a flat-nose punch, lower friction in the blankholder zone and around the die radius results (Fig.10.18a) in a lower draw force [176,187,188] and a higher LDR (irrespective of whether lower friction is obtained through higher speed or higher viscosity; see Fig.10.30 and 10.31). Isachenkov [189] traversed a wide viscosity range by varying the sugar content in water. Draw force first decreased but, at high concentrations, increased again, presumably because full-fluid-film lubrication had been attained. A similar trend can be seen in the results of Ferron [168] in deep drawing of stainless steel with various oils, and in those of Mizuno and Kataoka [178], who separated out blankholder friction (Fig.10.32). The mechanism does not remain unchanged over the entire draw; initially the lubricant film builds up and blankholder friction drops, only to rise again as more boundary contact develops [178,190]. With increasing viscosity, the optimum blankholder pressure rises, partly because of the need for pre-

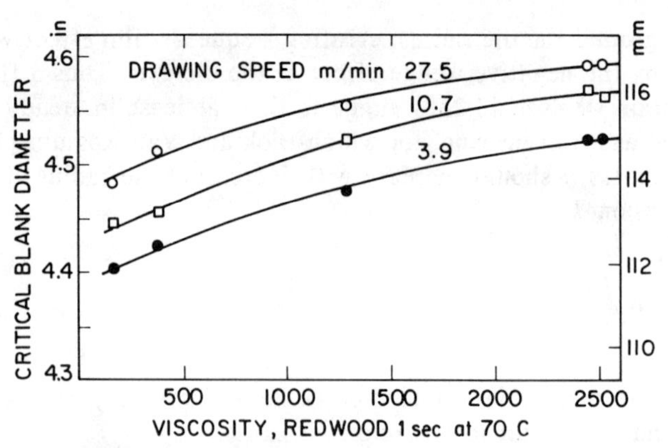

Fig.10.30. Effects of viscosity and speed on the maximum blank diameter that could be drawn from 1-mm-thick rimming steel sheet into 50.8-mm-diam cups [57]

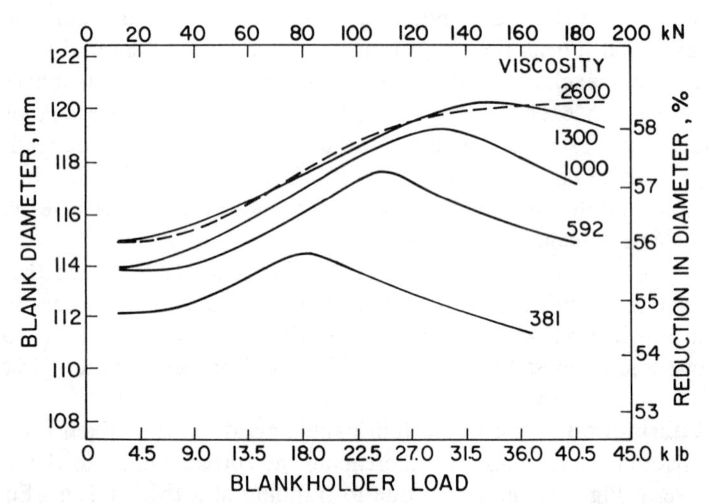

Fig.10.31. Effects of lubricant viscosity (measured in Redwood 1 seconds) and blankholder load on maximum blank diameter (0.9-mm-thick mild steel sheet; 50.8-mm-diam cup) [191]

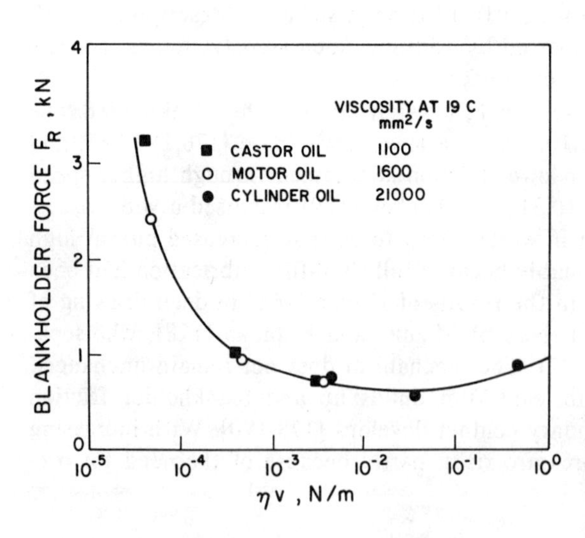

Fig.10.32. Variation of blankholder force with speed and viscosity (72-mm-diam, 0.5-mm-thick mild steel blanks drawn with 40-mm-diam punch) [178]

venting wrinkling, which would then result in higher draw forces (Fig.10.31) [191]. The reason for the drop in LDR at intermediate blankholder forces in drawing of brass with low-viscosity lubricants is obscure (Fig.10.33) [192]. The draw force drops with increasing speed also in redrawing [193].

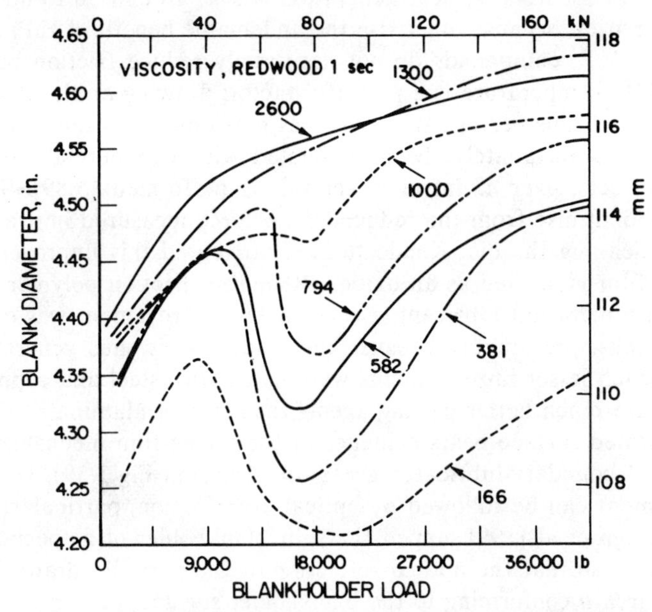

Fig.10.33. Effects of lubricant viscosity on maximum blank diameter and required blankholder force (viscosity in Redwood 1 seconds; 70/30 annealed brass blank 0.96 mm thick; 50.8-mm-diam cup) [192]

c. With a hemispherical punch the reduced support given by the punch (Fig.10.18b) leads to a lower LDR [57,106]. The mechanism shifts toward the boundary end of the regime around the draw radius [178], although the ηv effect is still noticeable.

d. Increased ηv results in reduced ironing loads [156,194].

e. The ηv effect is observed also in simulation tests [54]. It was noted in drawing over a quadrant die (Fig.5.12a) [5.44],[102] and in the strip drawing test (Fig.5.10a) [5.36].

f. In bulk deformation processes, increased reduction and/or pressure shift the mechanism toward the boundary end, with a concurrent rise in friction. In sheet metalworking at low interface pressures, μ drops with pressure because yielding of the substrate in a tensile stress state allows the real contact area to increase and real contact pressure to decrease rapidly [53,175,195]. Only when the pressure reaches the compressive flow stress of the material will μ become constant or even increase (Fig.3.44). The simplicity of the strip drawing test lends itself to an exploration of pressure effects, but a direct relation to pressworking is difficult to establish [5.39].

g. Some of the surface is in boundary contact, as evidenced by the improved performance obtained with compounded oils [191,192]. The magnitude of the boundary-lubricated zone (the b ratio in Eq.3.34a) depends only on ηv, and not on composition [196]. However, in the b zone even the base oil has an effect. In the work of Kawai et al [195,197], the aromatic constituents found in naphthenic oils gave lower values of blankholder friction than paraffinic oils of the same viscosity. However, the aromatic constituents also interfered with the functions of boundary and E.P. additives [197]. Very light oils (below 20 cSt at 20 C) dissolved large quantities of oxygen and provided much lower values of friction than could be expected on the basis of viscosity alone. In agreement with general principles, additives were more effective in highly refined, low-viscosity oils. The widely used emulsions and soap solutions [198] also operate in the mixed-film regime, as shown by the reduced draw forces observed at higher concentrations [5.92]. Except for ironing, heat generation is low, around 30 to 80 C, but it must be higher on asperities, because otherwise the undeniable benefit of E.P. additives could not be explained. E.P. compounds do not necessarily reduce friction but limit pickup and scoring [5.44]. Temperature is a powerful factor; drawing at elevated temperatures reduces viscosity and thus performance, whereas performance is improved by activation of E.P. additives. Unfortunately, tests conducted with commercial lubricants cannot separate these effects, even if infrared analysis is performed [5.89]. The mixed-film mechanism is evident also from the reduced draw force measured in drawing of several blanks without cleaning the die. Kondo [6.393] attributed this improvement on aluminum to a black film identified as an oxide, although a friction polymer may also have formed with the mineral oil lubricant. Even greater improvement was obtained by dry polishing the blanks, presumably because the oxide debris thus generated acted as a parting agent. Much lesser improvements were noted with steel and copper, the natural oxides of which are much better parting agents than that of aluminum.

h. The deformed surface bears evidence of the mixed-film mechanism, with hydrostatic pockets and boundary-lubricated areas (of b ratio in Fig.3.39). Details of surface contact development can be followed by optical examination, particularly in the blankholder zone. Kasuga et al [199] lapped the split blankholder of a special die to optical flatness (Fig.10.19) so that the optical reflection of the partially drawn blank was proportional to the area b conforming to the blankholder surface.
Kawai et al [196] used phase contrast to make the contact area more clearly visible. The b ratio increases with increasing blankholder pressure and decreasing viscosity, corresponding to a shift toward a greater boundary contribution. Frictional force is directly proportional to b, indicating that friction in the hydrostatic pockets is negligible. Typical boundary contact ratios are shown in Fig.10.34 [178], around the die radius and under the blankholder, for low and high ηv values, as a function of punch penetration. It will be noted that b under the blankholder increases monotonically during the stroke but changes around the die profile in a more complex manner. Initially, conformance is found only around the die entrance (zero position), and then conformance

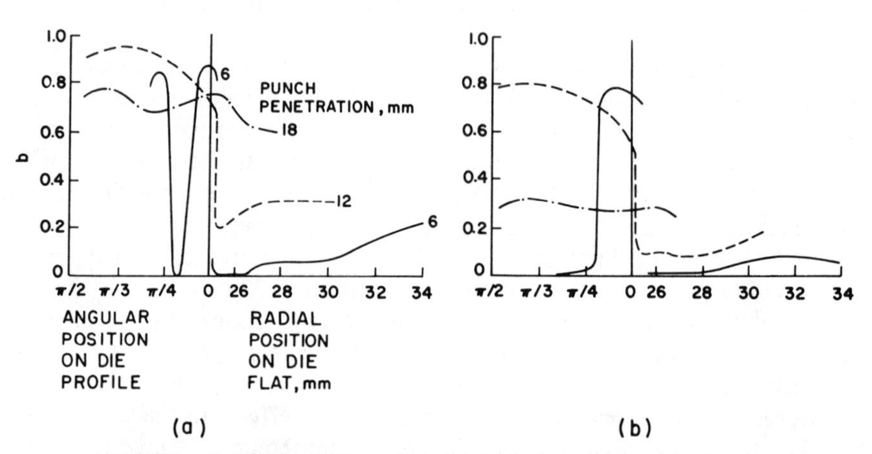

Fig.10.34. Proportion of boundary-lubricated contact area in drawing of steel sheet with (a) castor oil at 0.2 mm/s and (b) cylinder oil at 2 mm/s [178]

becomes more general around the whole radius, and finally a lower b ratio develops as the circumferentially compressed part is drawn through. Thickening under the blank-holder is greater with high ηv. The b ratio decreased roughly linearly with the logarithm of the relative oil film thickness h/R_{max}, where R_{max} is the surface roughness developed in free stretching. This is in agreement with Kasuga et al [199], who found that the contact ratio and frictional force decreased with decreasing grain size, and that the shear stress over the contact area could be described by a constant μ, the value of which is characteristic of the lubricant.

Solid-Film Lubrication. In most pressworking, relative sliding speeds are too low and the process geometry is not favorable enough to maintain a full fluid film, and lubrication with dry films is preferred so that the film thickness can be controlled by application rather than by process conditions alone. Such films also offer the possibility of differential or selective lubrication.

1. Dry Soaps. Even though dry soaps may be modeled as viscous substances and have speed-dependent actions (Sec.7.3), their fields of application place them in the dry-film category [4.172],[200]. For good adhesion, the surface must be clean, preferably freshly pickled. Soap-film thickness is controlled by the concentration and/or temperature of the aqueous solution. Subsequently, the water is driven off by heating in ovens, making the whole process most suitable for application in rolling mills [201]. Sidey [4.172] quotes coating thicknesses of 1.5 g/m^2 increasing to 3.0 g/m^2 on rough finished sheet. These coatings are related to the soaps used in wiredrawing (Sec.7.3.1) and are based on mixed soaps of selected fatty acids. Borax is often added as a filler and presumably to improve attachment to the surface. Sensitivity to moisture can be reduced by replacing borax with a polymer. The melting point is adjusted according to the severity of operation, so as to avoid melting. Soaps have been most successful on steel [202]. On stainless steel they are adequate only with coated tools [3.490]. They are seldom used on nonferrous metals because the slightly alkaline coating etches the metal and develops stiffer, higher-melting soaps on storage [200],[4.172]. The coating does not protect against corrosion or rough handling, but it has the advantage of water solubility and does not seriously interfere with welding. Proprietary formulations are claimed to be easier to apply [201].

Soaps applied to conversion-coated surfaces (Sec.4.7.7 and 4.7.8) ensure a high lubricating power that is needed only in severe ironing or in drawing of stainless steel. Their use may be justified also when the part has to be phosphated anyway [203]. The application and use of such coatings have been repeatedly discussed [204,205]. The drawing and ironing loads decline slightly with speed, indicating mixed-film lubrication. However, when a reactive organic phosphate-oil system is applied, forces increase with speed, indicating that some breakdown must occur [194].

Waxes have excellent lubricating properties for all metals but have to be applied from and removed by organic solvents [206]. Commercial water-soluble formulations exist. Friction is generally low; Fukui et al [5.29] found $\mu = 0.032$, virtually independent of speed.

2. Polymer Coatings. All varieties discussed in Sec.4.7.2 find application.

a. Loose films are practical only in prototype production or in the laboratory. After some early partial success with PE and polypropylene films [207-209], oiled PE became a standard for severe stretching [210] and for stretchability and stretch-bend testing [5.47]. PVC is also useful [148,210]. The initial promise of PTFE [208] was not fulfilled because of the limited stretching capacity of the film. An unusual product, suitable only for mild draws, is paper approximately 20 μm thick [202,211].

b. Most polymers are now used as coatings. Among the earliest were methacrylate copolymers deposited from trichloroethylene solutions or from aqueous dispersions [212]. The value of PTFE, especially in the baked-on form, was discovered early on [213], and PTFE has potential application in drawing of aluminum hollowware. For better adhesion, these films are often deposited onto a ceramic substrate, fired onto the sheet surface. The pressworking of strip precoated in a special facility, often at the rolling mill, has received much attention [214-220],[4.171]. These coatings ensure clean shop conditions and protect the part during handling.

c. Polymer coatings impart greatly increased stretchability, approaching that obtained in hydraulic bulging [216,220]. In deep drawing and incidental ironing the forces are substantially reduced, and the LDR is increased to 2.58 on steel [219]. In a series of articles, Franks and Plevy [88,221] observed that PTFE deposited on a ceramic substrate reduced draw forces and, in particular, the force peak associated with incidental ironing in drawing with a minimum or negative clearance. Force reduction was greater with a shot-blasted substrate. Surprisingly, the mean asperity height was also reduced relative to oil lubrication [88].

d. An important requirement is that the film should not wrinkle or stick to the dies. Duke and Gage [215] observed that coatings containing cross-linked polymers suffered damage through hairline cracking (crazing) under the influence of compressive stresses in the blankholder zone; the defect was similar to that found on the inside surface in a bend test. Flaking of a polyester coating was found by von Finckenstein and Lawrenz [219]. To avoid cracking and peeling in tensile regions, the polymer must have high elongation and good adhesion to the sheet. Coatings that soften at low temperatures may be damaged when worked on high-speed mechanical presses [219].

e. Under mild conditions, coated sheets can be formed dry. In more severe situations, additional lubrication is needed. This may be part of a system, as discussed in Sec.4.7.2. Among mill-applied lubricants are esters 2 μm thick deposited on top of a 5-μm polymer film [222]. Lubricants applied at the pressworking plant must be chosen for compatibility. For example, PVC is attacked by chlorine-containing lubricants [223]. In general, water-base lubricants, mainly waxes, are preferred [224]; often they can be left on the finished part. It is necessary, however, to test them for possible delayed damage to the coating on the formed part [219].

3. Layer-Lattice Compounds. Until recently, layer-lattice compounds were used mostly in pastes, greases, and high-viscosity oils designed for use on heavy-gage material or in pressing at elevated temperatures. For demanding applications, particularly for metals that do not react readily with lubricants, dry films deposited from liquid or volatile carriers have been used. Gurney and Adair [225] found, in agreement with experience in forging (Sec.9.5), that the remnant of the carrier greatly affected friction and the response of the film to imposed pressure and sliding speed. In principle, a dry film shows no speed effect, and this has indeed been found by Coupland and Wilson [106] in cupping tests and by Fukui et al [5.29] in simulation tests. Only when the liquid carrier remains, or the layer-lattice compound is added to an oil or grease, can there be a speed effect. Surprisingly, Kozlov [226] found in tests on deep drawing of steel that the optimum graphite content in oil, as judged from the level of the draw force maximum, increased with increased drawing speed. Proprietary or locally made lubricants often contain a variety of other ingredients, sometimes as tackifiers or parting agents and at other times with little clear justification.

Some bonded films can survive severe ironing. A lubricant containing 0.35-μm MoS_2 particles was reported to allow total reductions of 60%, typical of the three-stage ironing customary in two-piece beverage can production [227],[4.177]. The ironing loads decreased linearly with increasing MoS_2 content to 20%, and with increasing coating weight to 5 g/m^2. Coating weights of 1 to 2 g/m^2 were found most cost effective. A proprietary system relying on MoS_2 in acrylic waxes for the outsides and thinner wax coatings for the insides of the cans has been claimed to be readily removable by standard cleaning techniques [4.179]. In general, however, a common problem with layer-lattice compounds and nonorganic fillers is the relative difficulty of their removal.

4. Metal Coatings. Metal coatings find their most extensive use on sheet products. Most applications involve steel, but the principles are universal (Sec.4.7.1).

The dual function of coatings is evident from observations made on tinplate [78]. Under mild conditions, tin alone can facilitate drawing; draw force and LDR increased as coating thickness increased from 0.058 to 1.5 μm in the work of Duckett et al [4.160]. In practice, a lubricant is invariably applied which then masks the effects of tinplate variables so that no improvement is found with coatings over 0.4 μm thick [228]: blankholder friction keeps decreasing but so does the load-carrying capacity of the cup. The severe conditions of ironing bring out more subtle effects. A nonreflowed coating gives better lubricant entrapment [4.159],[229]. A thinner coating (say, 0.1 to 0.2 μm) on the punch side versus a 0.7-μm coating on the die side ensured differential lubrication and kept punch friction high yet allowed stripping of the can in the work of Misra and Rajagopal [228]. Once a coating of minimum thickness (>0.2 μm) had been applied, no further drop in ironing force was observed by Bingle et al [4.161]. However, thicker coatings are needed to protect against localized breakdown at asperities. Not surprisingly, a superimposed layer of a polymer such as PTFE masks the effect of the tin coating [230]. Chemical treatments applied to the tin coating do not appear to interfere with lubrication even though they may prevent wetting of the surface by water [231].

With increasing attention on the corrosion resistance of car bodies, galvanized (zinc-coated) sheet finds increasing use. LDR is improved in deep drawing, and higher blankholder forces are needed in press forming [231a]. Cracking or flaking of the film may occur, particularly in the compressive zone, and die pickup can be a problem. Polished, hard-chromium coated tools give best results [232].

Dry Pressworking. In light bending, stretching, and drawing it is quite possible to operate without an intentionally applied lubricant; surface oxides and incidental lubricant residues are sufficient, and often give lower friction than a light mineral oil which can dissolve surface films. Friction depends greatly on composition. For example, Duncan et al [5.47] found in their stretch-bend test (Fig.5.14) that friction was highest with 2036 (Al-Cu) alloy, lower with 6010 and 6009 (Al-Mg-Si), and lowest with 5182 (Al-Mg).

Application of Vibration. As pointed out in Sec.3.9.2, vibration of the tool can have beneficial effects when the process geometry is not particularly favorable for film formation. Vibration can increase the film thickness and break incipient pickup points [3.334].

Benefits have been reported in sheet drawing [233], dimpling [234], deep drawing [235], and ironing [236], but improvements seldom justify the expense. Kristoffy [3.338] found no significant improvement on application of 20-Hz or 20-kHz vibration to a deep drawing or ironing punch when a good commercial lubricant was used. Möllers and Fischer [233] reported a substantial drop in friction in drawing of strip and in deep drawing, but it is not clear what lubricant, if any, was used. Low-frequency vibrations in the range from 10 to 45 Hz have been applied by mechanical means such as cams. A vibrated blankholder reduced draw force and increased LDR in grease-lubricated deep drawing of aluminum blanks [237]. Extensive Soviet work is summarized in a book by Severdenko et al [238].

An important factor is heating due to vibration. Temperatures can reach 200 C and impair lubrication by liquids or waxes, whereas layer-lattice compounds and polymers respond well. The application of ultrasound prevents rupture of films of limited stretchability, such as PTFE. High-amplitude vibration (in excess of 12 μm) induces galling, irrespective of whether the oscillation is axial, radial, or tangential (torsional).

10.4.3 Effects of Surface Roughness

As a result of the low pressures prevailing in pressworking, contact between die and workpiece is most often limited to asperities. Thus, details of the surface topography assume an even greater importance than in bulk deformation processes [187,239-243] for two principal reasons:

First, many sheet products have large surface areas, and therefore roughness affects the appearance of surfaces, whether painted or unfinished. For a pleasing appearance, painted surfaces should usually have a controlled roughness, neither burnished nor excessively rough. Local variations in surface topography are to be avoided because of their high visibility.

Second, details of the surface topography determine the distribution of contact points between die and sheet, and the quantity and distribution of entrapped lubricants. Thus, topography affects press performance and the onset and consequences of lubricant breakdown. In general, rougher sheet is more tolerant: it accommodates slight unevenness such as that encountered with mechanical blankholders [240]. Also, it is capable of entrapping more wear debris before surface damage occurs.

The problem of topography is most evident in automotive production [244] but is also of importance in other applications.

Surface Characterization. A basic problem at the present state of development is that no truly relevant measure of surface quality has yet been found. As discussed in Sec.6.3.4 and 6.4.4, the sheet surface carries the imprints of the last rolling passes. For good surface quality the sheet is cold rolled and the roll roughness, modified by the lubrication mechanism, is imprinted to give the finish of mill-hard or annealed-last sheet. Sheets subjected to temper rolling exhibit the surface finish of the temper rolls superimposed on the prior finish, to an extent determined by the temper reduction. Thus, a wide variety of finishes is encountered:

1. Easiest to characterize are finishes obtained with ground rolls. The height distribution is approximately Gaussian and R_a (CLA) or R_q (RMS) values are sufficient if taken in the rolling and transverse directions. When hydrodynamic pockets are numerous, the Gaussian distribution is lost, and differences between the grinding (rolling) and transverse directions diminish.

2. In principle, shot-blasted surfaces should be easy to characterize if blasting took place with 100% coverage. Lesser coverage results in localized dimples in the as-ground surface of the roll. Of course, the strip carries the imprint (negative) of the roll surface profile. Therefore, the temper rolls produce generous asperity radii which, when superimposed on the roughness carried over from previous rolling, result in surfaces exhibiting plateaus and irregular valleys (Fig.10.35). Characterization has been attempted in various ways:

a. Early recommendations were based on statistical averages such as R_a or R_q. Thus, Littlewood and Wallace [5.44] suggested R_a = 1.0 μm for general pressworking

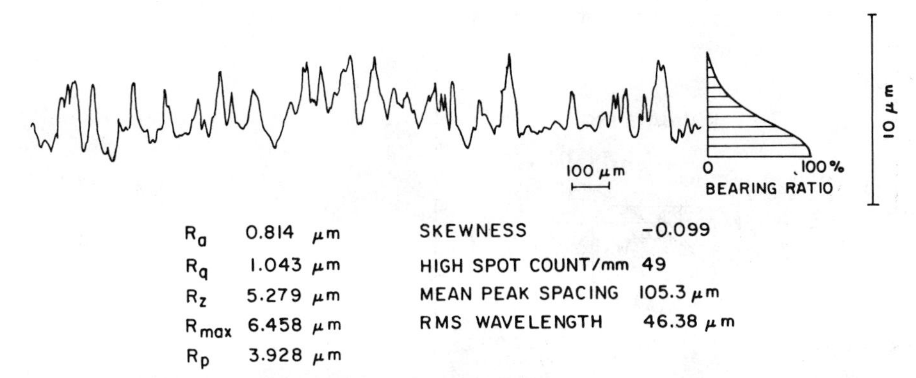

Fig.10.35. Surface roughness profile of typical steel sheet used in automotive body construction

and 1.4 to 2 μm for automotive production, a range found suitable also by Butler and Pope [5.95],[245].

b. It was soon recognized that statistical averages are quite inadequate in themselves. The surface can be visually characterized by stylus instrument recordings, but correlation with press performance is difficult [240].

c. Statistical averages combined with a count of the number of peaks per unit length give a better description of the surface. A difficulty is, however, that the number of peaks measured on a given surface depends on the cutoff wavelength, the height above the centerline at which the peaks are counted, and, ultimately, the definition of what constitutes a peak. Commercial instruments are available that give peak counts (see Table 10.3). Within these limitations, recommendations given for automotive sheet by various authors [5.40,5.44],[245-248] are combined in Fig.10.36. Leroy et al [248a] developed a criterion based on the mean peak (or valley) length on the centerline of the profile, ignoring peaks or valleys less than 5 μm wide (long). For the sake of uniformity, this has been converted into the equivalent number of peaks per millimetre in Fig.10.36, even though such conversion depends on the definition of a peak.

d. As shown in Fig.3.46, surfaces of identical maximum roughness can have highly different real bearing areas. Fischer et al [249] showed in 1960 that sheets with peaked profiles allow relatively easy squashing of asperities. Contact points are closely spaced and well distributed, yet the real contact area is limited to a small fraction of the

Table 10.3. Effect of measurement method on roughness parameters [5.40]

		Peaks/mm						
		Cutoff (mm): 0.75			2.5			
Sample	R_a	Height(a) (μm): 0.25	1.25	6.25	0.25	1.25	6.25	Remarks
A:								
Top	0.95	197	137	10	187	118	13	Non-galling
Bottom	1.00	210	136	11	173	116	14	
B:								
Top	0.68	191	106	2	160	86	2	Galling
Bottom	0.55	192	87	0	156	68	0	

(a) Measured from centerline.

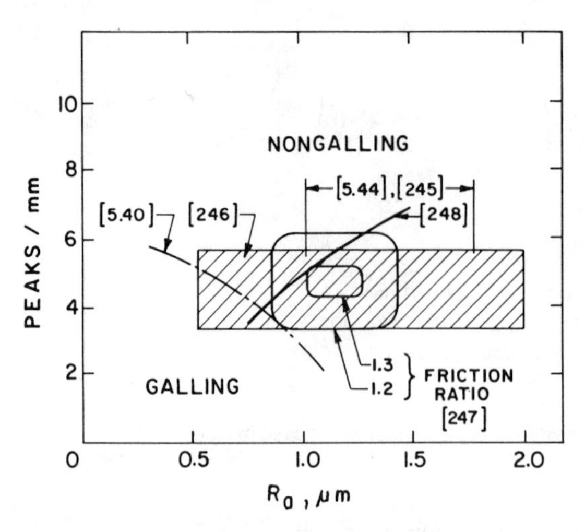

Fig.10.36. Recommended surface roughness parameters for automotive presswork-ing

apparent area of contact. Therefore, sliding is well controlled without excessive frictional forces. The bearing area measured at 1.5 μm below the reference line gave the best correlation with press performance (Fig.10.37). A further development, by Hastings and Gagnon [250], introduces a parameter p that combines peak density and bearing area:

$$p = \sum_{-3R_a}^{+3R_a} \left(N_{hT} \cdot A_{hT} \cdot N_{hL} \cdot N_{hL}\right) \Delta h \tag{10.15}$$

where N_h is the number of peaks at a given height, A_h is the bearing area at the same height, Δh is the height increment, and the subscripts L and T refer to rolling direction and transverse direction, respectively.

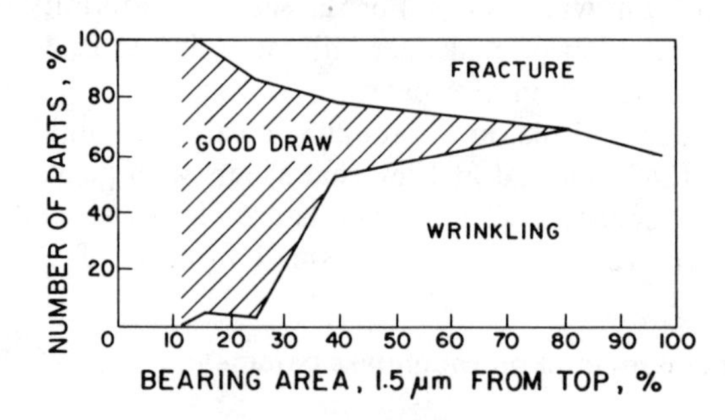

Fig.10.37. Success in drawing of an automotive part as a function of bearing area [249]

e. Statistical measures such as the bearing area curve [242] or, more likely, the skewness of the profile could give valuable information but do not appear to have been used extensively. Computer evaluation of surface topography (CEST), introduced by Shuaib and Wu [251], offers great promise. Butler [240] reports a lack of correlation between statistical measures and press performance, but does not specify which measures were taken into account. Leroy et al [248a] found that analysis of the autocorrelation function yields little new information.

3. The roughness of the sheet is not constant even if the sheet was procured from a single source. With wear of the rolls, peaks are worn off, and both R_a (or R_t) and peak count diminish [247]. The roughness changes also when surface treatments such as phosphating are applied [251].

4. Sheet roughness changes during deformation, and, if the sheet makes contact

with the tool only after it has been stretched, the as-rolled roughness is modified by the straining history (Fig.3.49). The real contact area increases much more rapidly if the sheet is subjected to bulk deformation (as in the blankholder zone in deep drawing) than if it is only stretched over a punch [54] or bent in U-forming [252].

Effect on Lubrication. The effect of roughness on lubrication is as would be expected from general principles (Sec.3.8.5). Increased roughness is beneficial in that more lubricant is entrained and wear debris is accommodated. The effect is more pronounced when the features of a directional surface finish are perpendicular to the sliding direction. A very rough surface can, however, be objectionable in appearance, and the lubricant may leak out in the direction of valleys. The desirability of roughness also depends on the process and, more directly, on interface pressure (Fig.3.44).

a. At low interface pressures, increased roughness develops more asperity contact and μ increases [185,253]. Because of the low pressures prevailing in the blankholder zone, a smooth sheet gives inadequate control, allows materials to run in, and may allow puckering to occur in stretch drawing [5.36],[114].

b. At higher interface pressures, the effect of roughness depends on the lubricant. With a viscous fluid, more lubricant is entrained and μ drops, as observed in strip drawing at heavier reductions [5.29,5.89],[53,185,187,242]. The effect is more pronounced with higher viscosities and/or higher speeds [242]. Because of the more severe conditions created at the draw radius, LDR generally increases [254] and draw force decreases [96]. This is true also of ironing [56]. Little or no effect of roughness can be expected when the lubricant is solid [3.304],[53,255] or when a marginal lubricant such as a corrosion-protection oil is used [187]. When the lubricant is of low viscosity or is aqueous, and does not fill out surface features, rougher sheet develops higher friction [5.37],[187] because of increased boundary contact.

c. Excess oil film thickness allows incipient wrinkling and thus gives higher draw forces; the effect is more marked on rougher sheet. In deep drawing with an emulsion, the optimum coating thickness corresponded to 1 g/m^2 of emulsified phase (dry weight) on top of 2 g/m^2 of corrosion-protection oil [96].

10.4.4 Lubricant Breakdown and Galling

Even under processing conditions that appear mild in comparison with bulk deformation processes, lubricant breakdown is often encountered because the lubricant is wiped off in the course of sliding and asperities are exposed to direct contact with the die. Tool pickup then follows, which in turn leads to scoring of the workpiece surface. These phenomena are usually described by the term "galling;" even though the OECD glossary [2.1] recommends that this term be avoided, it has gained such wide acceptance that we shall use it, too. Because galling is an event that follows local lubricant failure, it is greatly affected by workpiece and die topography and by adhesion between the contacting materials.

Incidence of Galling. Galling is typical of all processes in which the workpiece slides over the die. In their review of Japanese work, Miyauchi et al [5.37] indicate that, of all galling problems observed, 40% are associated with draw beads, 22% with die radii, and less than 10% with ironing. Galling is responsible for some 70% of all rework (polishing) of automotive dies. Several points are to be noted:

a. Galling must still be judged visually, even though attempts at quantifying it have been made. Ike and Yoshida [256] offer the following grading on the basis of SEM observations:

Phenomenon	R_q (μm)
Smoothing (with no fragments on the deformed surface)	Original
Whitening	2 to 12
Scratching(a)	4 to 20
Light galling(a)	10 to 40
Severe galling(a)	20 and over

(a) Fragments visible.

More generally, a scratched surface for which $R_q > 10$ to 15 μm is regarded as galled [5.37], although the limit depends also on visibility.

b. As are all wear phenomena, galling is a function of pressure and total sliding distance (Eq.3.39). Pickup occurs after some length of sliding, and then proceeds in a manner that depends on the sheet/die material combination. The sliding distance increases with the depth of cup or pressing, and with the width of the blankholder or draw bead. In wedge drawing (Fig.5.8a) of strip, Kawai et al [195] observed an initial pickup-free sliding with low, essentially boundary friction. On further sliding of steel, increases in draw force and scoring set in. Draw force was proportional to the welded area of contact, which in turn increased with decreasing lubricant viscosity. Finally, a steady-state condition corresponding to dry lubrication was established. On aluminum workpieces [257], metal transfer was affected also by lubricant composition, with 5% sulfurized fatty oil or 5% stearic acid in paraffin giving the lowest friction and pickup. On both metals the shear strength of the welded zones was roughly equal to the shear strength of the strain-hardened workpiece material, indicating that adhesion had indeed taken place. Similar observations were made by Oyane et al [5.64] in their simulation test.

c. At low interface pressures, conformance of the sliding surfaces depends also on the elastic deformation of the sheet, and thus the effect of blankholder pressure [5.37],[256] can be normalized through

$$\mathrm{BHP}_{\mathrm{norm}} = \left(\frac{t}{0.7}\right)^2 \mathrm{BHP}_{\mathrm{actual}} \tag{10.16}$$

where t is sheet thickness in mm.

d. On sliding over a draw bead, galling is observed on the opposing surface, where reverse bending takes place over two radii. In deep drawing and ironing, galling is limited to the die side of the cup.

e. Systematic studies of galling in production are difficult, and simulation tests abound. All tests are based on sliding at low interface pressure and include sheet drawing tests (Fig.5.10), draw-bead simulators (Fig.5.12 and 10.38) [248], tests with combined drawing and bending (Fig.5.13), and tests in which breakaway friction is measured (Fig.5.11 and 5.15d). Because galling is a function of sliding distance, several specimens are usually tested in sequence and, in addition to visual inspection, the trend in draw forces is noted [248,256]. Tests based on breakaway friction have been quantified [5.41] by a friction ratio which is the breakaway force divided by the running force. High values (over 1.3) are often accompanied by chattering and scoring of the sheet [258].

Mechanism of Galling. Galling is a result of localized adhesion and is, therefore, affected by a number of process variables:

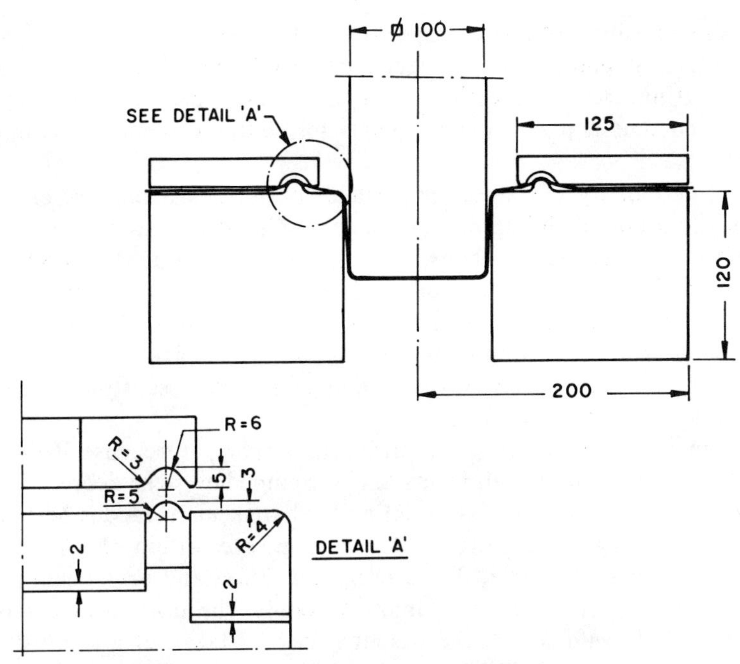

Fig.10.38. Double-bead test for drawing of steel strip [248]

a. The effects of asperity shape may be rationalized by analogy to bulge formation in ironing, as analyzed by Hill [2.19], Avitzur [2.12], and others. An asperity of low slope deforms and is pushed into the bulk. As the asperity slope increases, a point is reached where a bulge forms [256], which is then removed by a shearing (cutting) action; the critical slope is less for higher friction, as shown by Rault and Entringer [259] in simulation tests on plasticine and also on steel. The shearing action is related to prow formation (Fig.3.57) and to the cutting mode of wear (Fig.3.12). The sheared material is usually piled up at the ends of individual scratches.

b. Debris formed in the course of sliding may be trapped and rendered harmless in valleys of the sheet surface, may become embedded in other sheet asperities, or may lodge against die scratches. Repeated encounters at many points result in deformation, thinning, and lamination of particles; some of them may even roll up [259]. If adhesion to the die is high, new large asperities form which lead to scratching, formation of more thick particles, and galling.

c. Galling is affected by the macroscopic tool geometry. It usually originates at transition points such as die radii, but its severity greatly increases if the already damaged sheet, with embedded particles, is dragged along a parallel surface such as a die land. Ike and Yoshida [256] classify these phenomena as primary and secondary galling.

d. The incidence of galling is a function also of the stiffness of the system. Damage is localized in a soft (constant-force) system, whereas it is widespread and uniformly distributed in a stiff (constant-gap) setup [5.38].

Effects of Surface Roughness. Both workpiece and die roughness have significant influences on the incidence and severity of galling.

a. A smoother sheet surface presents a larger bearing area and is more prone to galling. For a given reduction in thickness, smoother sheet galls after shorter sliding or fewer repeat tests [5.37]. This effect can be quite dramatic; a small change in roughness is sufficient to initiate galling in drawing of materials prone to pickup.

b. The size and distribution of asperities is most important. For a given average roughness, larger flat plateaus give longer contact (sliding) distance and more galling. For a given bearing area, a smoother sheet has smaller plateaus but a much greater peak density, and thus it presents more points for sliding contact and is more prone to galling [248].

c. The directionality of the sheet surface profile has a marked effect. Plateaus aligned in the direction of sliding present longer sliding distances and are more likely to allow embedment of the debris. Thus, they are more harmful than transverse ridges, which trap debris in valleys and may even act as rasps to remove incipient pickup from the tooling [256,259]. In apparent contradiction are the results of Story and Weinmann [260], who found in drawing aluminum that the weight of pickup increased with asperity width rather than length; however, this may simply represent the development of a thicker but no less stable surface coating.

d. The depth profile of the sheet surface is important because it determines the ability of the surface to entrap lubricants and accommodate wear debris.

e. There has been some success in defining what sheet surface is best, even though there is no general agreement. Recommendations are within the ranges shown in Fig.10.36. It is doubtful, however, that average roughness and peak count are sufficient descriptors of an antigalling profile. Almost certainly, the bearing area should also be specified. To exclude wild peaks, the baseline may be taken at a small distance from the highest peaks—e.g., by excluding the first 2% of the bearing area [5.40]. Invariably, steel sheets with bearing areas of less than 20% at 1.5 μm from the surface prove to be nongalling, whereas those with larger bearing areas (and those whose bearing areas increase more rapidly with depth) are prone to galling [3.320],[5.40],[247]. A large bearing area also yields a high friction ratio in the galling test [258]. The bearing area may be calculated from flattened spots observed on sheet that has been slid between parallel jaws (Fig.5.10a) at different interface pressures [248]; the bearing area increases more rapidly on sheet that is prone to galling. The mean peak length at the centerline, combined with R_a, has also been suggested as a measure [248]. It is likely, though, that a combined value such as that in Eq.11.15 will be more generally useful. In view of the effect of asperity shape, Rault and Entringer [259] define a nongalling finish for low carbon steel as one that has a mean plateau height of 10 to 25 μm, a plateau length of 60 to 120 μm, and a mean ratio of valley to plateau width of 3:2 to 5:2.

f. A rough die surface is desirable only on the punch because it then increases the LDR [192,255,261] and the depth of draw in automotive pressworking [262]. If the sheet material has to slide over die elements, roughness leads to higher friction [242], lubricant breakdown, and galling, although the effect depends also on die composition and hardness. The harder the tool, the more damaging its roughness. Thus a cast iron die [242] or a nonhardened steel die [260] can be rougher than a heat treated steel or WC die.

Material Effects. Because of their importance in adhesion, material composition and hardness have decisive influences. Adhesion between die and workpiece governs whether die pickup will occur and, if it occurs, whether it will be cumulative. Details of the mechanism are still obscure (Sec.3.11.2), but the effects can be marked:

a. Compatibility of die and workpiece materials will be discussed in Sec.10.4.5. Workpiece material effects can be quite subtle. Thus, increasing the magnesium content of aluminum alloy 2036-T6 reduces adhesion [260]. In steels, segregation of manganese, chromium, and other alloying elements has been suspected to reduce adhesion, perhaps by strengthening the surface layer [248]. Steels precipitation hardened with niobium or titanium additions are more resistant to galling than solid-solution (Si, Mn) hardened steels [5.37] for the same reason. These effects should be quite general. Martensitic transformation induced in austenitic stainless steel increases friction and draw forces but reduces adhesion [156].

b. Surface composition plays an important role. A sheet subjected to sliding contact galls more readily on subsequent contact [259]. An untouched surface has surface layers which may or may not protect it. For example, steel pickled in HCl was found to be more prone to galling than that pickled in H_2SO_4 [5.37]. The oxides and carbonaceous deposits formed on annealing steel in a DX (exothermic) atmosphere prevented galling whereas the cleaner surface produced in an HN atmosphere (mixture of N_2 and H_2) required good lubrication and more closely controlled surface topography [247]. Hilsen et al [258] found that galling tendency, as expressed by the friction ratio from the galling test (Fig.5.15d), decreased from 1.4 with a 10-nm-thick contaminant film to 1.1 or 1.2 with a 50-nm-thick film. The reactivity of a surface changes on cleaning and deformation, as shown by measurement of electron emission [263], and this too must affect adhesion. Comprehensive reviews of work on surface composition have been published by Leroy et al [248a] and Coduti [263a] for steel, and by McNamara [263b] for aluminum.

c. The hardness of the workpiece material plays a role, too. High-strength steels are not much more adhesive than low-carbon steels but generate higher pressures and are more prone to galling [264], especially on draw beads [264a]. Previously strain-hardened material undergoes less hardening during working, and this is beneficial because it reduces the thickness of wear debris [259]. However, the sheet hardness should be not more than half the tool hardness, otherwise strain hardening of the asperities makes them strong enough to cause abrasive wear of the die [5.37]. Some researchers suggest that asperities should be softer than the bulk of the sheet [3.320] to allow better conformance, while others claim that this leads to more galling [259]. Perhaps no generalization can be made because of the influence of material variables.

d. The elastic properties of the die material affect conformance. Kasuga et al [265] investigated this effect by removing partially drawn stainless steel cups at the point of reaching the force maximum, and by taking roughness measurements on the remaining flange and over the surface that contacted the die radius. A roughness parameter, formed by dividing the change in average roughness height by the original roughness, increased in direct proportion to maximum draw force (Fig.10.39). This, in turn, increased with the elastic modulus of the die material, indicating that a die that con-

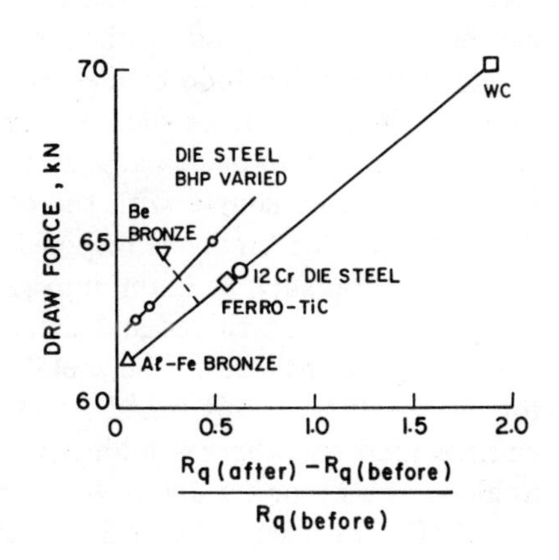

Fig.10.39. Effect of die material on smoothing or roughening of type 304 stainless steel cups drawn with a graphite/stearic acid mixture [265]

forms less to the sheet generates a rougher surface and higher forces. A beryllium bronze was the only exception, because it wore by metal transfer.

e. The presence of a lubricant is important. If the film is thin and boundary contact dominates, the properties of the tool material influence draw force and galling. This is the case when only corrosion-protection oil is present [5.37]. In simulation tests, Bragard et al [248] applied a lubricant only to the first strip and then observed the increase in force and the onset of galling in the double-bead tester (Fig.10.38). As the film thickens through higher ηv, galling diminishes also at any given pressure [5.37],[242]. On a very smooth surface, a liquid lubricant cannot prevent galling [53]; a solid lubricant protects only when surface reaction is also possible. Thus, MoS_2 prevented galling on smooth steel and especially on copper-plated steel, but not on brass [53]. Hilsen and Bernick [247] propose that friction ratios of 1.2 to 1.3 call for controlled lubrication, higher ratios are indicative of galling with any lubricant, and lower ratios signify resistance to galling without regard to the type of lubricant. With a given steel, Bernick et al [266] found friction ratios of 1.02 to 1.05 with soap/borax, wax-base, and acrylic film lubricants. A synthetic ester/oil mixture and an emulsion gave ratios of 1.20 to 1.25, and rust-preventive oils and light paraffins allowed a ratio of 1.30, which is beyond the galling limit.

10.4.5 Die Materials

The high visibility of surface scratches dictates the choice of pressworking dies. Of course, wear is also of concern; however, if galling sets in, production has to be stopped long before wear could become a problem. The choice of die material depends greatly on workpiece material and on interactions with lubricants, as discussed in Sec.3.11.6.

a. Die steels are still the most extensively used die materials [3.426],[9.137]. Graphitic steels combine adequate hardness with some built-in lubricity. Cast iron is advantageous for large dies and gives low friction, but cast iron dies allow more galling on low-carbon steel than do steel dies [242]. To reduce the cost of dies, areas subject to wear (such as draw radii) can be made as inserts [267]. Steel dies are adequate for stainless steel and other high-strength, high-adhesion materials only if the sheet is fully protected by an oxalate/soap or polymer film [268].

b. Of the nonferrous metals, zinc-base alloys are suitable for low-volume production. Of specific importance are bronzes typically containing 14% Al, 4% Fe, and 1% Ni, which are unique in that stainless steel can be drawn with them without pickup and scoring. Their high hardness keeps wear to a reasonable rate [3.418,3.490],[269-271]. Fats, soaps, and even water-soluble lubricants can be used with them [268,272]. They are, however, brittle, and, apart from encasing them in steel, the problem of support can be solved by depositing them on a steel die by ion plating [273].

c. Tungsten carbide bonded with cobalt is the staple die material for large production runs in applications such as deep drawing and ironing—for example, in production of tin boxes or two-piece cans. In the presence of an aqueous lubricant, the tin coating forms a galvanic couple with the cobalt binder. Komanduri [274] suggests that anodic dissolution of the binder is responsible for punch wear. Tool steel can be less damaging to a soft metal such as aluminum [256]. TiC is reported to outperform WC in canmaking [275]. Other carbides and ceramics have been explored experimentally. Their success is highly specific to the metal and lubricant, and results are sometimes contradictory. Oberländer [276] found Al_2O_3 to be suitable for stainless steel when used with a compounded oil, whereas high friction and die pickup were noted by Schlosser [3.490]. Schlosser also found a die of 40% zirconia and 60% Mo to be better.

d. Die coatings offer advantages for some of the difficult-to-draw metals. Most prominent of these is stainless steel, and several experiments have been aimed at finding coatings that would allow pressworking without oxalation. For austenitic steels, coatings of TiC appear to be best [3.490],[34,277,278] when used with waxes, soaps, or other well-adhering lubricants. For ferritic steels, Woska [279] found TiC, TiN, and nitrided surfaces to be best, followed by WC-coated and boronized dies. Hard chromium was little better than D2 steel, perhaps because of the observed poor wetting of lubricants. For mild steel, Furubayashi et al [264a] found no galling with carbide diffusion or vapor coating, but increasingly more galling with chromium-plated, nitrided, and sulfurized dies. Little information on industrial experience has been published for other metals.

e. Polymers such as epoxy resins are of relatively low strength and need metal inserts for tight radii [280]. Polyurethane allows polished sheet to be formed without the danger of scratching [281].

10.4.6 Lubricant Selection, Application, and Testing

General principles discussed in Chapters 4 and 5 apply here, but some specific concerns need to be highlighted.

Lubricant Selection. Several general reviews of the topic are available [4.2,4.12],[9.137],[8,181,203,256,282-300].

The major difficulty of a systematic selection of lubricants [52,299,301] is the wide variety of conditions that is encountered. Classifications based on processes are not fully adequate because conditions may change greatly in, say, combined stretching/drawing, depending on the degree of drawing. Production rates affect equilibrium temperatures of the tooling and thus the potential breakdown or activation of lubricants. At this time, selection is best based on a knowledge of lubrication mechanisms, whereas those less familiar with the subject must rely on commercial recommendations (and on Table 10.4).

It is often economical to select a better-quality lubricant which then allows the use of a less-expensive grade of sheet metal. Beyond the lubricating function, compatibility with later treatment of the part is important. This may necessitate removal of residues or use of lubricants that do not interfere with processes such as welding, either by impairing weld quality or by generating excessive fumes [32,167,302].

Lubricant Application. All techniques considered in Sec.4.8.2 apply here [167,303-305].

a. Manual application with brushes, swabs, or rollers is still acceptable at low production rates [121]. Drip application is the first step in mechanization and is suitable for lubricants of moderate viscosity.

b. Roller coating is practiced mostly for coiled stock but is suitable also for precut parts [13,306].

c. Mist application can be economical and versatile and is suitable for deposition of thin films of lubricants ranging from emulsions to waxes [307]. Application to the die is also possible. Electrostatic deposition is attractive but more costly.

d. Mill-applied coatings remove the burden from the press shop but may be too expensive. Some systems, such as a combined phosphating-oiling treatment with a lubricating oil that contains an organic phosphate [194], are suitable for press shop application [308].

Lubricants are often made in forms that allow several methods of application [289,293,297,309]. Thus, polymeric additives endow the lubricant with thixotropic

Table 10.4. Commonly used sheet metalworking lubricants and typical μ values
(Lubricants are listed according to increasing severity of conditions)

Material	Shearing, bending(a)	Pressworking(a)	μ(b)
Steels	Dry (pickle oil)	Dry (pickle oil)	0.2
	EM (M.O. + E.P.)	Soap solution	0.15
	M.O. + E.P.	EM (M.O. + fat) (+E.P.)	MF
		M.O. + fat (+E.P.)	MF
		Fat (tallow)	0.07
		Soap (water-soluble) film	0.05
		Wax (chlorinated)	0.05
		MoS_2 or GR in grease	0.05
		Polymer coating (+MoS_2)	0.05
		Phosphate + soap	0.05
		Metal (Sn) + EM	0.05
Stainless steels and Ni alloys	EM (M.O. + Cl)	EM (M.O. + Cl) (or fat)	0.2
	M.O. + Cl additive	M.O. + Cl additive	MF
		Chlorinated wax	0.07
		Polymer coating	0.07
		Oxalate + soap	0.07
		Metal (Cu) + M.O.	0.05
Al and Mg alloys	EM (M.O. + fatty derivatives)	EM (M.O. + fatty derivatives)	0.15
	M.O. + fatty derivatives	M.O. + fatty derivatives	MF
		Soap or wax (lanolin) coating	0.05
		Polymer coating	0.05
		Warm: GR film	0.1
Cu and Cu alloys	Soap solution	Soap solution	0.1
	EM (M.O. + fat)	EM (M.O. + fat) (or fat)	0.1
	M.O. (+fat) (+E.P.)	Drawing paste (EM of fat)	0.1
		Fat (tallow)	0.07
		Pigmented tallow (MoS_2, etc.)	0.05
Ti alloys	M.O. + E.P.	Wax coat	0.07
		Oxide + soap coating	0.07
		Polymer (+MoS_2)	0.05
		Fluoride-phosphate + soap	0.05
		Metal (Cu or Zn) + M.O. (or soap)	0.1
Refractory metals	M.O. + E.P.	Warm: MoS_2 or graphite	0.2
		Cold: Al-Fe-bronze dies with wax	0.07

(a) EM = emulsion of ingredients shown in parentheses. M.O. = mineral oil; of higher viscosity for more severe duties, limited by staining. E.P. = E.P. additive (S, Cl, and/or P; also sulfochlorinated fats). (b) MF = mixed-film lubrication: μ = 0.15 to 0.05.

properties, which prevent dripping. Pastes and synthetic lubricants are often made to be water-extendable so that they can have consistencies ranging from semisolid to liquid.

We have seen that deeper draws are often possible when friction is low in the blankholder zone but high on the punch. This can be accomplished by differential lubrication of the blank [310,311]. It is difficult to keep a liquid lubricant away from the punch, and roughening of the punch is then more practicable (provided that stripping is still easy enough). Differential lubrication can be better controlled with dry or solid lubricants. Spray phosphating and electrochemical deposition [4.187] allow lubrication in a pattern. Rowe [4.187] showed that in deep drawing the critical zone is close to, but does not extend to, the outer edge of the blank. Otherwise, good lubrication on the die side and poorer lubrication on the punch side can be remarkably effective (Fig.10.40) [57]. Similar benefits can be reaped in ironing [59].

Lubricant Evaluation. All techniques discussed in Chapter 5 are used, but the major effort is usually directed toward evaluating lubricating qualities. Part of the problem is that laboratory tests fail to show long-term trends unless run in large enough numbers to stabilize surface conditions and temperatures [312]. Die and lubricants should be heated to operating temperatures so that lubricant breakdown or activation becomes evident [5.89],[313]. As mentioned in Sec.5.4, trends are often more important than absolute forces. As discussed in conjunction with galling, this is particularly true when adhesion is a problem. Lubricant evaluation can take different approaches:

a. Scaled-down production tests are feasible, especially in deep drawing, provided that the d_0/t ratio is kept large enough (usually in excess of 100). If enough cups (in excess of 50) are drawn with any one lubricant, the maximum draw force, its change, and pickup on the die give reasonable indications. The same applies to ironing tests. Correlation with production experience is usually adequate [314-316], although it can be poor when temperatures or d_0/t ratios are too low [52].

b. The large number of simulation tests advocated indicates that none of them can give full evaluation. Indeed, it has been repeatedly shown that several tests are needed for a reasonable prediction of plant performance [5.39,5.95],[32,317]. In general, the relevance of tests diminishes as they become further removed from the process condi-

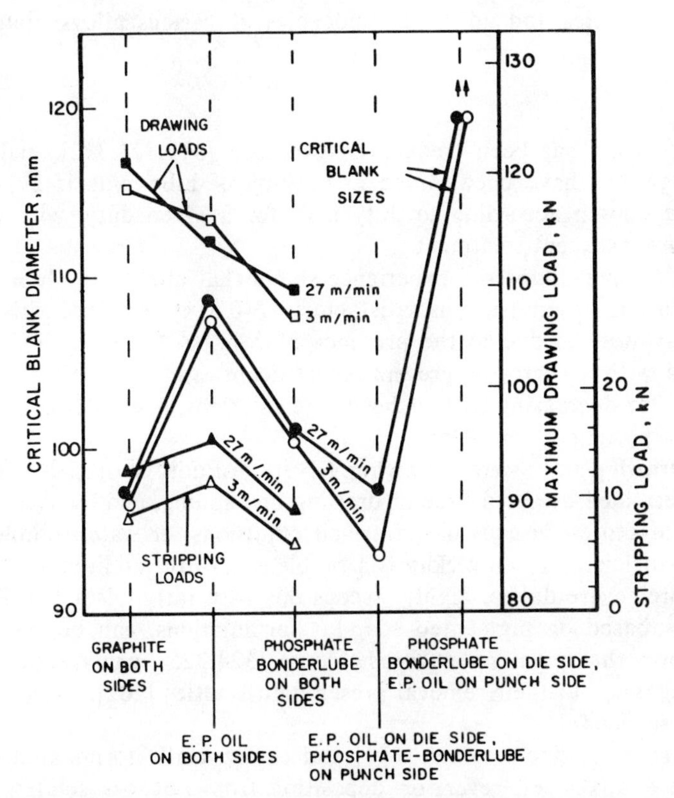

Fig.10.40. Effects of selective lubrication on limiting draw ratio, draw force, and stripping force in deep drawing of 1.6-mm-thick annealed low-carbon steel blanks into 50.8-mm-diam cups with simultaneous wall ironing [57]

tions they are meant to simulate. For deep drawing, the wedge-drawing test (Fig.5.8a) does not really simulate the blankholder zone from the material point of view [318] but is adequate for lubricant exploration. This is less true of the strip drawing test (Fig.5.10a), where lateral compression is absent [5.89]. Because of the importance of friction around the draw radius, bend-draw tests (Fig.5.13 and 5.14) are more relevant [175,319], although correlation with actual deep drawing can be poor (Table 10.2) [5.46]. Draw bead tests can be conducted with variable penetration (Fig.5.12) or with contact on the flat binder areas. Again, trends are more important than absolute values [290]. The actual draw force versus drawing distance curve can be numerically represented by reading the force at the beginning of the draw and at, say, two points during the draw. By multiplying the three values by each other, a composite, more sensitive number is obtained.

c. Bench tests (Fig.5.23 and 5.24) can show only limited correlation with some specific aspect of lubricant performance. The four-ball test gives adequate correlation with deep drawing when oil viscosity is dominant, but the correlation is nonexistent when boundary properties dominate with soap solutions [5.92]. In general, poor correlation is found [199,315].

d. Functional tests for cleanability, corrosion due to the residual oil film, etc., are routinely performed, as discussed in Sec.5.10 and 5.11 [5.206],[320].

10.5 Lubrication of Iron- and Nickel-Base Alloys

The differing activities and adhesion tendencies of various alloys dictate lubrication practices.

10.5.1 Steels

Lubrication of steels has been frequently discussed [317,321,322], and many experimental investigations have been aimed at ranking of lubricants [5.29],[174,323,324]. Lubricants are chosen according to duty and, for a given duty, with a view toward optimization of subsequent treatment.

1. Light Pressing. Industrial experience shows that almost any lubricant (light oil, dilute soap solution, or emulsion) is satisfactory [240,314]. The insensitivity of steels to lubrication may well be due to the presence of residual films. Annealed and pickled sheet is coated with a corrosion-prevention lubricant, and cold rolled sheet carries an oily residue. After degreasing with orthosilicate solutions, a silica film remains which is likely to have a parting action [248].

2. Deep Drawing and Stretch Drawing. Severity of duty in deep drawing and stretch drawing is determined by the degree of drawing and sliding, and by sheet thickness.

a. For light duties, viscous oils, fatty-oil emulsions, or water-soluble polymer formulations are sufficient. When pickup is a problem, a sulfur additive is needed [314].

b. For more severe duties, highly viscous oils with fatty oils and E.P. additives, or drawing pastes based on pigmented soap-fat formulations, can be used. It has been repeatedly shown that pigments such as talcum [324,325], chalk, and ZnO [191] are good parting agents, but their removal presents difficulties [202]. Fats, particularly tallow [323], are successful.

c. Soaps, such as zinc stearate [174] and the specially formulated dry soaps, are excellent, as are waxes. However, on deposition from aqueous solutions, the heating required to drive off the water causes aging in rimming steels [321]. Chlorinated wax is one of the heaviest-duty lubricants [324].

d. Graphite and MoS_2 do not necessarily ensure low friction [5.29],[174] but when added to a fat such as tallow [323] or to a drawing paste they give good results [326]. They are effective under severe conditions such as drawing of high speed steel [327]. Removal is again a problem, and this can be minimized by applying the layer-lattice compound locally to the die radii.

e. Polymers such as PE and PTFE have often been found to be useful [326] and, as discussed in conjunction with mill-applied coatings, have found their specific fields of application.

3. Heavy Drawing and Ironing. High surface pressures and large surface expansions demand a phosphate/soap system [121,285]. Reactive oils do not give quite the same performance [194] but are suitable for moderately heavy duties. Among the layer-lattice lubricants, only the recently developed bonded MoS_2 coatings can survive [4.177]. The alternative is to coat with tin 0.25 to 0.75 μm thick [4.161], as in production of drawn and ironed cans, and to lubricate and cool with the aid of a compounded emulsion.

10.5.2 Stainless Steels and Nickel-Base Alloys

The lack of reactivity coupled with the frequent demand for very high surface quality presents substantial problems which have inspired many investigations [108,328-332]. Plant practices have been repeatedly discussed [200,253,282,283]. As discussed in Sec.10.4.5, an Al-Fe bronze prevents pickup and allows the use of rather indifferent lubricants [268]. Much work is done with steel dies, however, and the following remarks concern such practices:

For lighter duties, a viscous oil with chlorine additives suffices [330]. Performance improves with increasing viscosity, and also when polybutenes are used [329]. Neat chlorinated paraffins give the best performance. Graphited tallow and pigmented oils and pastes are sometimes used, but the graphite causes carburization on annealing and presents difficulties of removal.

Of the separating films, dry soaps [121] and waxes such as beeswax [328] are adequate for medium duties. Chlorinated polymers deposited from aqueous or organic vehicles are more satisfactory, as are polymers in general [121,212,253]. Peelable films are indispensable in making parts with high surface quality. However, for the most severe duties, including ironing, an oxalate/soap coating is best. If necessary, MoS_2 may be applied on top of it. When the die is coated with TiC, chlorinated oil suffices [156]. Alternatively, the workpiece can be coated with a metal such as lead, but since the advent of the oxalate coating this is seldom practiced.

Martensite transformation induced by deformation can lead to cracking of deep drawn, metastable austenitic stainless steel cups. Lubrication seems to play no role in this [331].

The few available publications on working of nickel-base alloys [333,334] indicate that practices similar to those used for stainless steel apply. At room temperature, PE, and MoS_2 in acrylic resin, yielded the lowest forces [334]. At elevated temperatures only graphite or MoS_2 has a chance of survival and, as shown by Berry et al [9.148,9.149],[335], a mixture of the two in oil may be better than either alone.

10.6 Lubrication of Light Metals

Lubrication practices for light metals are governed by the need to avoid tool pickup.

10.6.1 Aluminum Alloys

Numerous experimental investigations [323,336,337] and a review specifically directed toward aluminum [338] are available.

a. For light pressworking, low-viscosity mineral oils, synthetic oils, or oil-base emulsions are suitable, invariably with boundary additives.

b. For more severe drawing and stretch drawing, the oil viscosity is higher, the emulsion is less diluted, and the boundary additive concentration is higher. A specific formulation introduced by Kravtsov et al [339] consists of 35% fatty acid esters, 7% diethylene glycol, and 0.6% copper dioleate, remainder water. Oils with E.P. additives are also used but are difficult to justify unless it is expected that the tool can develop protective films. When pickup is concentrated at specific areas of the die (such as draw beads), a concentrated lubricant may be applied locally. Interest in this field has grown with the introduction of aluminum automotive components such as bumpers [340]. Nine [341] found good correlation between the results of the draw bead test (Fig.5.12) and plant performance.

c. Solid-film lubricants such as soaps and waxes are useful—especially if they possess boundary lubricating qualities. Graphite is less favorable because its success depends very much on application [323] and because it leads to corrosion problems. Better is MoS_2, which yielded higher friction but also larger draw ratios in rectangular deep drawing [326]. Except for working at elevated temperatures [342], there should be no need for layer-lattice lubricants. Polymers are extensively used in the form of precoated (painted) sheet. Neumann [338] quotes a total draw and redraw ratio of about 2.5. Peelable films of PVC or PE protect parts with critical surface requirements while providing excellent lubrication when oil is superimposed [5.47],[326,336]. Loose PTFE films are useful [221,343] but are limited by their lack of ductility. Fired-on films can withstand substantial amounts of deformation [88].

d. An application of great importance is the manufacture of two-piece drawn and ironed beverage cans, which reached an annual production rate of 47 x 10^9 cans in 1981 in the USA [344]. Reviews have been published by Mastrovich [345], by Knepp and Sargent [346], and by Knepp [344]. Interestingly, the deep drawing (cupping) step requires the heavier lubricant, usually an emulsion of controlled stability. To allow plate-out, the roller applicator is placed at some distance from the cupper. The base mineral oil or polyisobutylene typically has a viscosity of 80 mm^2/s and contains 20 to 30% boundary additives, some esters, and nonionic (less frequently, anionic) emulsifiers. Emulsions are used at concentrations of 2 to 30% in water [344]. Water loss on passage from the applicator to the cupper increases the apparent viscosity, whereas residues of the much less viscous rolling oil (about 1/20th as viscous) could prevent wetting and cause dilution, with consequent excessive friction and cup failure. Therefore, residual oil is kept to 30 mg/m^2 [344]. The ironer (body maker) emulsion is more dilute (4 to 20% in water) and is sometimes replaced with synthetics based on polyglycols. Kelly [347] discusses the virtues of synthetics of undisclosed composition, including lower (experimental) cupping forces. Emulsions are applied from fully controlled and filtered recirculating systems. To prevent contamination from tramp oils, they are used also as the press hydraulic fluid. Aging, especially of anionic emulsions, can increase friction and metal pickup. In view of the vast quantities produced, lubricant carry-off, washability, and reclaiming from the washing solution are of concern [344,347].

e. All aluminum surfaces are covered with a natural oxide. The thickness of this oxide can be increased by electrolytic techniques. If the oxide formed is relatively soft and noncrystalline, it can serve as a lubricant carrier for stearic acid or a fatty oil, as shown by Petronio [348]. Detailed studies by Overfelt et al [349] on automotive alloys have shown that friction drops slowly as film thickness increases from 5 to 10 nm, and that it is lower at higher interface pressures and in the presence of a lubricant. Friction drops more steeply when film thickness reaches 15 nm on alloys 6009 and 6010 and when it reaches 25 nm on alloy 2036. It remains constant at thicknesses above 50 nm, irrespective of interface pressure and lubrication. There is evidence that the presence of MgO in the oxides of Al-Mg alloys (visible by the golden tint of the oxide) leads to more rapid wear of forming dies [263b].

f. Die materials are typically steels or, for very high production, tungsten carbide. Among coatings, nitriding has proved to be successful. Ceramics such as Al_2O_3 suffer excessive wear, perhaps because of the oxidation of aluminum [338].

10.6.2 Magnesium Alloys

Only moderate deformation of magnesium alloys is possible at low temperatures, and then soaps, waxes, or (in more difficult cases) graphited tallow will suffice. More severe stretching and deep drawing must be done at temperatures from 220 to 370 C. The lubricant is invariably colloidal graphite in a volatile (organic or water) base [350,351]. It is applied to the blank prior to heating by spraying or roller coating. To facilitate removal, the blank should not be etched prior to coating, so as to retain the oxide which is then removed together with the graphite coating. Caustic etching is followed by chromic acid - sodium nitrate pickling.

10.7 Lubrication of Copper-Base Alloys

Lubrication difficulties increase from copper (which does not form cumulative pickup) to brass to bronze.

a. For light pressworking, medium-viscosity mineral oils, oil emulsions, or soap solutions are used. For heavier duties, fats are added—sometimes in concentrations up to 50%.

b. For deep drawing and stretch drawing, straight fatty oils or compounded mineral oils, emulsions, or soap-fat compounds are needed. The LDR increases with increasing viscosity [192]. Chlorine or sulfur E.P. additives are often encountered, although sulfur staining is a problem. Dry soaps are effective [323] as are, of course, polymers, including methacrylate resins [208]. Layer-lattice compounds have been found to be effective [323,333] but should not be needed except in working at elevated temperatures. Fillers such as kaolin have been used [323] but, again, should not be needed.

10.8 Lubrication of Titanium Alloys

Because of their limited ductility, titanium alloys are usually warm formed, although limited forming and drawing at room temperature are possible [352].

a. In cold working, the high adhesion of titanium to die materials necessitates the use of a continuous separating film, such as a polymer (Fig.10.41) [208,353] or, for stretch forming, even a dry wax. Only for light duties, such as roll forming, will an E.P. oil suffice. For heavier work, various surface treatments combined with wax or dry soap lubricants can replace the polymers. Oxidizing, fluoride/phosphate coating [354], and sulfidizing have been found to be effective [4.242]. Nitriding provides only marginal improvement (Fig.10.41).

b. At elevated temperatures the options are few. Low-melting glasses [355,356] protect also during heating but are difficult to remove and may require additional graphite lubrication of the die elements. Graphitic lubricants are simpler to apply to the die and blanks but require subsequent removal. They have been used in hot hammer forging [357]. The synergistic effects noted between graphite and MoS_2 could be exploited also [335]. At relatively low forming temperatures (380 C) a fluoride/

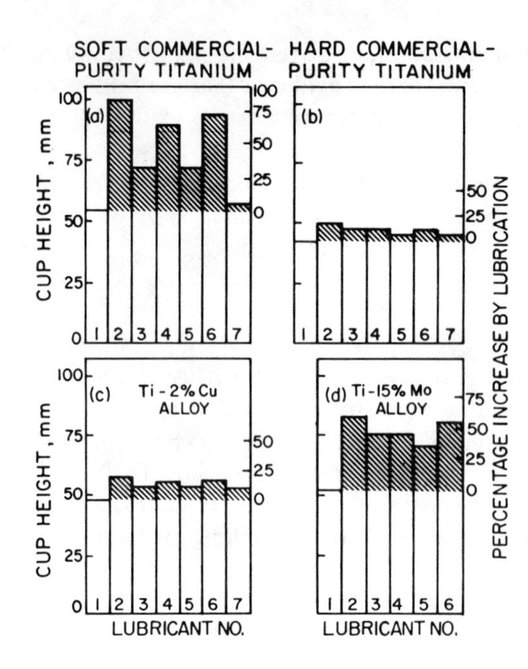

Fig.10.41. Effects of lubricants and surface treatments in deep drawing of titanium sheet [353]. 1 - Dry. 2 - 0.05-mm PE sheet. 3 - 0.25-mm PE sheet. 4 - PTFE dispersion. 5 - MoS_2 dispersed in a resin. 6 - Methacrylic resin. 7 - Nitrided surface.

phosphate coating with resin-bonded MoS_2 survived long enough to allow deep drawing [358] when some graphite was applied to the dies.

c. Little has been reported on the selection of die materials or coatings for minimum adhesion. Dies of beryllium bronze, lubricated with graphite, eliminated galling at 375 C (359).

10.9 Lubrication of Refractory Metals

The main problem in lubrication of refractory metals usually is brittleness or adhesion.

a. The ductile-to-brittle transition temperature of tungsten sheet is between 100 and 400 C. Forming is done at elevated temperatures. On heating of sheets with infrared lamps, a graphite coating increased heating rates and allowed forming [360]. Molybdenum and its alloys (TZM) are also too brittle to be formed at room temperature. At around 400 C, MoS_2 is the preferred lubricant, although graphite has also been used [361,362]. Dies of aluminum bronze have been used with castor oil as the lubricant [363]. Bronze dies with beeswax, or chromium-plated steel dies with graphite, are satisfactory in forming of uranium [364].

b. The basic problem with tantalum is its severe galling on steel dies. One solution relies on bronze dies with beeswax [364], another on surface treatment. In one study [365] it was found that, after anodic oxidation in dilute phosphoric acid at a current density of 100 to 150 A/m^2, tantalum sheet could be drawn and redrawn with a saponified castor oil lubricant. Die life was longest with tungsten carbide tooling, less with tool steel tooling, and least with an Al-Fe bronze, presumably because of the abrasive action of the oxide. Use of sulfonated tallow has also been reported [366].

10.10 Summary

1. Friction does not affect the process of shearing (blanking, punching) itself, and therefore the prime purpose of lubrication is to reduce die wear. Adhesive, abrasive, fatigue, and chemical wear mechanisms contribute to a loss of punch and die profile

and thus to an increase in clearance, with the consequent formation of a burr on the sheared part. Wear is reduced by appropriate choice of die materials and lubricants. Die materials must be hard and yet be of adequate ductility and also low adhesion. Die surface coatings play an important role. Lubricants are invariably compounded to minimize adhesive and abrasive wear through formation of boundary and E.P. layers.

2. Bending places minor demands on lubricants, whereas spinning and, especially, flow turning involve high local pressures and require good lubrication.

3. In stretching, the shape is developed by thinning out the sheet, with only relatively minor sliding on the punch. Friction governs, together with the tensile properties of the sheet material, the strain distribution, the attainable depth of stretching, and the location of fracture. Therefore, these features can be used for ranking lubricants. Upon initial approach of the punch to the sheet, squeeze films are formed; however, in the course of the sheet being wrapped around the punch, the divergent shape of the tool/workpiece gap discourages hydrodynamic lubrication, and adherence of the lubricant to the surfaces becomes important. The original roughness of the sheet and its roughening prior to punch contact are important in determining lubricant entrapment.

4. In deep drawing the sheet is drawn into the die gap, and simultaneous circumferential compression contributes to the formation of a cup of substantial depth without over-all thinning. Friction on the blankholder and die radius increases the draw force and reduces the limiting draw ratio; lubricants can be ranked on these bases. Friction on the punch allows deeper draws, making differential lubrication an attractive but seldom practicable proposition. Even though some squeeze-film formation occurs between the flange and the die and blankholder faces, lubrication around the draw radius is almost always of the mixed-film type, which results in lower draw forces and higher LDR values with higher-viscosity fluids and/or higher drawing speeds.

5. Pressworking often involves a combination of stretching and drawing, with the degree of draw-in being controlled by draw beads. The draw beads create severe conditions, sometimes more severe than those found around the die radius. Pickup on the tools and subsequent scratching (galling) of the workpiece are of major concern.

6. Ironing is similar to drawing of a tube on a bar. Interface pressures can exceed the flow stress of the material, and sliding conditions are severe but are mitigated by the favorable die geometry. Friction on the punch allows drawing at heavier reductions. The process has acquired new significance with the advent of the two-piece drawn and ironed beverage can, with aqueous lubricants supplied from central recirculating systems.

7. The frictional balance can be changed by special techniques that either ensure preferential lubrication of critical areas such as draw radii (hydrostatic and externally pressurized lubrication), transfer the tensile stresses to the punch (roughened or expanding punch; hydraulic or rubber cushion drawing), or impart compressive stresses to the workpiece.

8. Interface pressures are often much below the flow strength of the sheet material, and surface features are not always filled out by the lubricant. The topography of the sheet surface assumes paramount importance. Since the sheet yields in a tensile or tensile/compressive stress system, asperities deform and the real area of contact increases rapidly. Best control of pressworking is ensured when the real contact area is relatively small (to minimize friction) yet well distributed (to avoid localized wrinkling). Such a surface is also best for the avoidance of galling. Some progress has been made toward defining the properties of nongalling low-carbon steel sheet surfaces, with implications for other sheet materials.

9. Even under conditions that appear mild relative to those of other processes, the

use of lubricants is essential because of the need to prevent die pickup and subsequent surface damage (galling). The lubricant has to operate in an often unfavorable process geometry at relatively low sliding velocities. Therefore, highly viscous, pasty, or solid-film lubricants are often employed, together with surface coatings. Typical lubricant recommendations are summarized in Table 10.4. The final choice is often dictated by ease of removal, interference with subsequent finishing processes, and cost.

References

[1] J.A. Newnham, in [3.1], pp.703-770.
[2] J.A. Schey, in [3.1], pp.68-74, 276-277.
[3] E. Siebel and H. Beisswänger, *Tiefziehen,* Carl Hanser Verlag, München, 1955 (German).
[4] J. Willis, *Deep Drawing,* Butterworths, London, 1954.
[5] G. Sachs and H.E. Voegeli, *Principles and Methods of Sheet Metal Forming,* Reinhold, New York, 1960.
[6] F.W. Wilson (ed.), *Die Design Handbook*, McGraw Hill, New York, 1965.
[7] G. Oehler, *Schnitt-, Stanz- und Ziehwerkzeuge*, 5th Ed., Springer, Berlin/New York, 1966.
[8] D.F. Eary and E.A. Read, *Techniques of Pressworking Sheet Metal,* 2nd Ed., Prentice-Hall, Englewood Cliffs, NJ, 1974.
[9] P.B. Mellor, Int. Met. Rev., *26*, 1981, 1-20.
[10] D.P. Koistinen and N.-M. Wang (ed.), *Mechanics of Sheet Metal Forming,* Plenum, New York, 1978.
[11] *Sheet Metal Forming and Energy Conservation* (Proc. 9th IDDRG Congress), ASM, Metals Park, 1976.
[12] *Sheet Metal Forming and Formability* (Proc. 10th IDDRG Congress), Portcullis Press, Redhill, Surrey, 1978.
[13] Th. Mang, H. Becker, D. Schmoeckel, and K.H. Schubert, Blech Rohre Profile, *28*, 1981, 277-280 (German).
[14] M. Meyer and O. Kienzle, CIRP, *11*, 1962-63, 111-116.
[15] S. Hogmark, O. Vingsbo, and S. Fridström, Wear, *51*, 1978, 85-104.
[16] P. Funke, R. Wittek, and S. Schubert, Bänder Bleche Rohre, *19*, 1978, 472-475 (German).
[17] E. Doege, C.P. Neumann, K.H. Schmidt, and B. Fugger, in *HFF Bericht No.6,* Hannover, 1980, Paper No.20 (German).
[18] T. Maeda and I. Aoki, J. Fac. Eng., Univ. Tokyo (B), *32*, 1974, 443-475.
[19] T. Maeda and I. Aoki, J. Fac. Eng., Univ. Tokyo (B), *36*, 1978, 537-577.
[20] Ch. H. Kim, Ind. Anz., *97*, 1975, 2141-2142 (German).
[21] H. Becker, in [3.56], Paper No.48.
[22] F.P. Mikhalenko et al, Vestn. Machinostr., *1974* (2), 76-80 (Russian).
[23] F.W. Timmerbeil, Werkstattstech. u. Maschinenb., *46*, 1956, 58-66; *47*, 1957, 231-239, 350-356 (German).
[24] O. Kienzle and W. Kienzle, Stahl u. Eisen, *78*, 1958, 820-829 (German).
[25] H. Bühler, F. Pollmar, and A. Rose, Arch. Eisenhüttenw., *41*, 1970, 989-996 (German).
[26] O. Kienzle and K. Buchman, Arch. Eisenhüttenw., *34*, 1963, 443-451 (German).
[27] H. Peter and R. Neider, Mitt. Forsch.-Ges. Blechverarb., *1959* (6/7), 57-66 (German).
[28] A.G.M. Buiteman, F. Doorschot, and P.C. Veenstra, CIRP, *24*, 1975, 203-205.
[29] P.C. Veenstra and J.A.H. Ramaekers, CIRP, *27*, 1978, 157-158.
[30] J.J. Marklew, Machinery (London), *118*, 1971, 531-533.
[31] A.F. Astashov and A.S. Kryzhanovsky, Kuzn. Shtamp. Proizv., *1973* (1), 45-46 (Russian).
[32] H. Becker and R. Woska, Bänder Bleche Rohre, *22*, 1981, 176-178 (German).
[33] B. Edenhofer, Werkstatt u. Betr., *109*, 1976, 289-293 (German).
[34] J.G. Sauer, Metal Progr., *105* (4), 1974, 80-82.
[35] A. Oldewürtel, Bänder Bleche Rohre, *19*, 1978, 137-141 (German).
[36] H. Benninghoff and H. Zickler, Metalloberfläche, *32*, 118-129 (German).
[37] F. Birzer, in [3.56], Paper No.46 (German).

[38] K.G. Brownlee and T.W. Smythe, J. Iron Steel Inst., *208*, 1970, 806-812.
[39] E. Dannenmann and S. Sugondo, CIRP, *30*, 1981, 167-170 (German).
[40] G.R. King, Sheet Metal Ind., *54*, 1977, 453-462.
[41] PERA Report No.97, Melton Mowbray, England, 1961.
[42] T. Maeda and I. Aoki, J. JSTP, *21*, 1980, 241-249 (Int. Ed., *2*, 1981, 30).
[43] D.J. Kraan, Trib. Int., *14*, 1981, 29-32.
[44] P.K. Teterin and V.P. Lukyanov, Kuzn. Shtamp. Proizv., *1965* (6), 4-7 (Russian).
[45] H.A. Al-Qureshi, J. Mech. Work. Tech., *1*, 1977-78, 261-275.
[46] J. Ivaska, Precis. Met., *34* (2), 1976, 26-28.
[47] D.I. Starchenko and M.I. Chelovan, Stal', *1968* (2), 153-155 (Russian).
[48] K. Inoue and P.B. Mellor, J. Mech. Work. Tech., *3*, 1979, 151-166.
[49] M. Hayama, Bull. Fac. Eng., Yokohama Nat. Univ., *20*, 1971(Mar.), 61-69 (Japanese).
[50] F. Gütlbauer, in [3.56], Paper No.59 (German).
[51] V. Hasek, Bänder Bleche Rohre, *25*, 1978, 213-220, 285-292, 493-499, 619-627 (German).
[52] B. Fogg, Sheet Metal Ind., *53*, 1976, 294-304.
[53] W. Panknin and M. Reihle, Stahl u. Eisen, *82*, 1962, 470-479 (German).
[54] A.K. Sengupta, B. Fogg, and S.K. Ghosh, J. Mech. Work. Tech., *5*, 1981, 181-210.
[55] T.J. Gibson, J. Havranek, and R.M. Hobbs, in *Int. Conf. Prod. Techn.*, Inst. of Engineers, Melbourne, 1974, pp.333-339.
[56] J. Havranek, R.M. Hobbs, and I.S. Brammar, in [12], pp.121-126.
[57] D.V. Wilson, Sheet Metal Ind., *43*, 1966, 929-944.
[58] R. Pierce and P.G. Joshi, Trans. ASM, *57*, 1964, 399-416.
[59] E.M. Loxley and P. Freeman, J. Inst. Petr., *40*, 1954, 299-307.
[60] K. Yoshida and Y. Hayashi, Sheet Metal Ind. Int., *55*, 1978(Dec.), 7-14.
[61] K. Suzuki, N. Ohgoshi, and I. Gokyu, *Proc. Int. Conf. Sci. Techn. Iron Steel,* Iron Steel Inst., Tokyo, 1971, pp.951-953.
[62] S.S. Hecker, Metals Eng. Quart., *14* (4), 1974, 30-36.
[63] J.F. Wallace, J. Inst. Metals, *91*, 1962-63, 19-22.
[64] R.A. Ayres, W.G. Brazier, and V.F. Sajewski, J. Appl. Metalwork., *1* (1), 1979, 41-49.
[65] S.P. Keeler, SAE Paper No.650535, 1965.
[66] G.M. Goodwin, SAE Paper No.680092, 1968.
[67] S. Dinda, K.F. James, S.P. Keeler, and P.A. Stine, *How to Use Circle Grid Analysis for Die Tryout,* ASM, Metals Park, 1981.
[68] R.M. Hobbs, Sheet Metal Ind., *55*, 1978, 451-464.
[69] J.M. Story, J. Appl. Metalwork., *2* (2), 1982, 119-125.
[70] B. Kaftanoglu, Wear, *25*, 1973, 177-188.
[71] A. Ghosh, Int. J. Mech. Sci., *19*, 1977, 457-470.
[72] M. Kobayashi, Y. Kurosaki, and N. Kawai, Trans. ASME, Ser.B, J. Eng. Ind., *102*, 1980, 142-150.
[73] B. Devedzic, in [3.56], Paper No.21.
[74] B. Devedzic and M. Stefanovic, in *Proc. 21st Int. MTDR Conf.*, Macmillan, London, 1981, pp.397-403.
[75] J. Chakrabarty, Int. J. Mech. Sci., *12*, 1970, 315-325.
[76] M.Y. Demeri, J. Appl. Metalwork., *2* (1), 1981, 3-10.
[77] V. Hasek, Werkstattstechnik, *66*, 1976, 649-653 (German).
[78] S. Rajagopal, in [3.55], pp.135-144.
[79] P. Ng. J. Chakrabarty and P.B. Mellor, in *Proc. 17th Int. MTDR Conf.*, Macmillan, London, 1977, pp.579-585.
[80] J.S. Gunasekera and P.F. Thomson, in *Proc. 10th NAMRC*, SME, Dearborn, 1982, pp.173-179.
[81] B. Kaftanoglu and A.E. Tekkaya, Trans. ASME, Ser.B, J. Eng. Mat. Tech., *103*, 1981, 326-332
[82] N. Kawai, M. Hiraiwa, and M. Arakawa, Bull. JSME, *13*, 1970, 1513-1521.
[83] N. Kawai and M. Maeda, Bull. JSME, *13*, 1970, 1522-1530.
[84] E. Doege, wt-Z. Ind. Fertig, *66*, 1976, 615-619 (German).
[85] E. Siebel and W. Panknin, Z. Metallk., *47*, 1956, 207-212 (German).
[86] R. D'Haeyer, J. Gouzou, and A. Bragard, in [11], pp.220-234.
[87] W. Panknin and W. Eychmüller, Mitt. Forsch.-Ges. Blechverarb., No.17, 1955, 205-209 (German).

[88] T.A.H. Plevy, Wear, *58*, 1980, 359-380.
[89] O.H. Kemmis, Sheet Metal Ind., *34*, 1957, 203-208, 251-255.
[90] D.V. Wilson and R.D. Butler, J. Inst. Metals, *90*, 1961-62, 473-483.
[91] M. Atkinson and I.M. MacLean, Sheet Metal Ind., *42*, 1965, 290-298.
[92] D.V. Wilson, B.J. Sunter, and D.F. Martin, Sheet Metal Ind., *43*, 1966, 465-476.
[93] S.Y. Chung and H.W. Swift, Proc. Inst. Mech. Eng., *165*, 1951, 199-223.
[94] M.H. Pope and J.T. Berry, Trans. ASME, Ser.B, J. Eng. Ind., *95*, 1973, 895-903.
[95] D.F. Eary, Paper No. MF69-518, SME, Dearborn, 1969.
[96] W. Reitzle, H. Drecker, and F. Fischer, Bänder Bleche Rohre, *21*, 1980, 8-13 (German).
[97] G. Kretschmer, wt-Z. Ind. Fertig., *60*, 1970, 71-73 (German).
[98] R. Zeller, Ind. Anz., *100* (5), 1978, 26-27 (German).
[99] J.M. Alexander, Met. Rev., *5*, 1960, 349-411.
[100] B. Avitzur, in *Proc. 10th NAMRC*, SME, Dearborn, 1982, pp.157-164.
[101] V. Hasek and G. Krämer, Ind. Anz., *97*, 1975, 2139-2140 (German).
[102] R. Reissner, in [3.56], Paper No.19 (German).
[103] S. Fukui, Sci. Papers Inst. Phys. Chem. Res. (Tokyo), *34*, 1938, 1422-1527.
[104] N. Kawai, M. Hiraiwa, K. Suzuki, and T. Ryuzin, J. JSTP, *8*, 1967, 125-130.
[105] S. Rajagopal, Trans. ASME, Ser.B, J. Eng. Ind., *103*, 1981, 197-202.
[106] H.T. Coupland and D.V. Wilson, Sheet Metal Ind., *35*, 1958, 85-103.
[107] K. Nakagawa and S. Okazaki, in *Proc. Int. Conf. Sci. Techn. Iron Steel,* Iron Steel Inst., Tokyo, 1971, 908-910.
[108] P. Funke, C. Pavlidis, and E. Lange, Bänder Bleche Rohre, *16*, 1975, 315-318 (German).
[109] V. Hasek and K. Lange, wt-Z. Ind. Fertig, *68*, 1978, 545-554 (German).
[110] K. Siegert, in [3.56], Paper No.23 (German).
[111] J.L. Duncan and J.E. Bird, Sheet Metal Ind., *55*, 1978, 1015-1025.
[112] J. Painter and R. Pierce, Sheet Metal Ind., *53* (7), 1976, 12-15, 18-19.
[113] S. Fukui and K. Yoshida, in *Proc. 2nd Int. MTDR Conf.*, Pergamon, Oxford, 1962, pp.19-26.
[114] C. Weidemann, Sheet Metal Ind., *55*, 1978, 984-986, 989.
[115] D.H. Lloyd, J. Australian Inst. Metals, *12*, 1967, 106-118.
[116] K. Yoshida, K. Miyauchi, and H. Komorida, Bull. JSME, *10*, 1967, 188-196.
[117] H.D. Nine and T.R. Smith, J. Appl. Metalwork., *2* (1), 1981, 19-27.
[118] A.K. Ghosh, Metals Eng. Quart., *15* (3), 1975, 53-61, 64.
[119] B. Kaftanoglu and J.M. Alexander, J. Inst. Metals, *90*, 1961-62, 457-469.
[120] M. Yokai and J. M. Alexander, Sheet Metal Ind., *44*, 1967, 466-475.
[121] D.H. Lloyd, Sheet Metal Ind., *43*, 1966, 220-235, 307-315, 393-400, 432-447, 530-539.
[122] D.H. Lloyd, Sheet Metal Ind., *39*, 1962, 6-19, 82-91, 158-166, 236-245, 306-315.
[123] L.R. Hawtin and G.M. Parkes, Sheet Metal Ind., *47*, 1970, 533-540, 543.
[124] V.V. Hasek, Ind. Anz., *101* (93), 1979, 44-45 (German).
[125] I. Adak and F.W. Travis, in [12], pp.67-77.
[126] M. Mrozek, MW Interf., *1* (3), 1976, 15-23.
[127] Y. Kasuga and S. Tsutsumi, Bull. JSME, *8*, 1965, 120-131.
[128] Y. Kasuga, N. Noaki, and K. Kondo, Bull. JSME, *4*, 1961, 394-405.
[129] V.I. Kazachenok and A.S. Chauzov, Kuzn. Shtamp. Proizv., *1965* (1), 25-26 (Brutcher Transl. No.6470).
[130] A.S. Chauzov and V.I. Kopylov, Kuzn. Shtamp. Proizv., *1978* (11), 28-30 (Russian).
[131] W. Granzow, Manuf. Eng. Man., *72* (1), 1974, 38-39.
[132] M.G. El-Sabaie and P.B. Mellor, Int. J. Mech. Sci., *15*, 1973, 485-501.
[133] G. Oehler, Werkstattstechnik, *62*, 1972, 284-288 (German).
[134] J. Tirosh, S. Yosifon, R. Eshel, and A.A. Betser, Trans. ASME, Ser.B, J. Eng. Ind., *99*, 1977, 685-691.
[135] U. Herold, Bänder Bleche Rohre, *22*, 1981, 45-49 (German).
[136] J. Reissner and P. Hora, CIRP, *30*, 1981, 207-210.
[137] O.P. Lay, Sheet Metal Ind., *48*, 1971, 658, 660, 665-672.
[138] W. Panknin, Werkstattstech. u. Maschinenb., *47*, 1957, 295-303 (German).
[139] M. Fukuda and K. Yamaguchi, Bull. JSME, *14*, 1971, 504-511.
[140] A. Mayer, F. Fischer, and G. Sonntag, Bänder Bleche Rohre, *15*, 1974, 395-399 (German).
[141] N.A. Maslennikov, Eng. Digest, *17*, Sept. 1956, 366-368.

[142] K. Yamaguchi, N. Takakura, and M. Fukuda, J. Mech. Work. Tech., *3*, 1979, 357-366.
[143] R. Narayanaswamy, S.M. Doraivelu, V. Gopinathan, and V.C. Venkatesh, J. Mech. Work. Tech., *6*, 1982, 227-234.
[144] F.J. Fuchs and J.W. Archer, Paper No. MF69-147, SME, Dearborn, 1969.
[145] H.Ll.D. Pugh, in *Proc. Conf. Manufacturing Technology*, ASTME (SME), Detroit, 1967, pp.1053-1082.
[146] W.A. Mir and M.J. Hillier, Trans. ASME, Ser.B, J. Eng. Ind., *91*, 1969, 766-771.
[147] N.K. Wong, B.H. Arkun, W.T. Roberts, and D.V. Wilson, Rev. Met., *77*, 1980, 413-422.
[148] E. Lange, Bänder Bleche Rohre, *16*, 1975, 511-514; *17*, 1976, 9-12 (German).
[149] R.A. Ayres and M.L. Wenner, Met. Trans., *10A*, 1979, 41-46.
[150] W.G. Granzow, Sheet Metal Ind., *56*, 1979, 561-564.
[151] V. Nagpal, in *Proc. 6th NAMRC*, SME, Dearborn, 1978, pp.158-165.
[152] W. Panknin, Ch. Schneider, and M. Sodeik, Sheet Metal Ind., *53* (8), 1976, 137-142.
[153] K. Osakada, S. Fujii, R. Narutaki, and S. Sakakura, in *Proc. 18th Int. MTDR Conf.*, Macmillan, London, 1978, pp.137-144.
[154] G.S.A. Shawki, Acta Techn. CSAV, *13*, 1968, 605-625.
[155] S. Rajagopal and S. Misra, in *Proc. 7th NAMRC*, SME, Dearborn, 1979, pp.89-95.
[156] J.H. Kerspe, in [3.55], pp.127-133; in [3.56], Paper No.42.
[157] B. Avitzur, in *Proc. 18th Int. MTDR Conf.*, Macmillan, London, 1978, pp.145-150.
[158] B. Dodd, J. Strain Analysis Eng. Des., *14* (2), 1979, 43-47.
[159] W. Panknin, in *Proc. 1st Int. Tinplate Conf.*, Int. Tin Res. Inst., London, 1976, pp.200-214.
[160] R.K. Busch, Report No.10, Inst. f. Umformtechnik, Univ. Stuttgart (Girardet, Essen), 1969 (German).
[161] H. Ebertshauser, Blech Rohre Profile, *27*, 1980, 86-93 (German).
[162] G.D. Lahoti and T. Altan, in *Proc. 6th NAMRC*, SME, Dearborn, 1978, pp.151-157.
[163] M.N. Gorbunov, Russian Eng. J., *9*, 1969, 57-60.
[164] H.A.H. Crowther, P.D. Liddiard, and K.I. Marword, Sheet Met. Ind., *18*, 1943, 1733-1738, 1915-1920, 2099-2105; *19*, 1944, 81-88.
[165] P. Funke, Blech Rohre Profile, *26*, 1979, 1-7 (German).
[166] P. Funke, Blech Rohre Profile, *27*, 1980, 394-398 (German).
[167] T. Mang, Blech Rohre Profile, *27*, 1980, 175-179 (German).
[168] A. Ferron, in [12], pp.303-316 (French).
[169] H. Kaga, CIRP, *20*, 1971, 55-56.
[170] P.F. Thompson and P.U. Nayak, Int. J. Mach. Tool Des. Res., *20*, 1980, 73-86.
[171] K. Yamaguchi and P.B. Mellor, Int. J. Mech. Sci., *18*, 1975, 85-90.
[172] E. Dannenmann, Ind. Anz., *91*, 1969, 2253-2254 (German).
[173] E. Dannenmann and D. Schlosser, Ind. Anz., *98*, 1976, 1838-1841 (German).
[174] K. Lange and R. Dalheimer, in *Proc. 8th Int. MTDR Conf.*, Pergamon, Oxford, 1968.
[175] K.P. Witthüser, Dr. Ing. Diss., Technical University, Hannover, 1980 (German).
[176] H.D. Nine, J. Appl. Metalwork., *2* (3), 1982, 200-210.
[177] W.R.D. Wilson and J.J. Wang, Trans. ASME, Ser.F, J. Lub. Tech. (to be published).
[178] T. Mizuno and H. Kataoka, Bull. JSME, *23*, 1980, 1016-1023.
[179] H. Blok and J.J. Van Rossum, Lubric. Eng., *9*, 1953, 316.
[180] B. Fogg and M.K.A. Owais, CIRP, *27*, 1978, 183-188.
[181] B. Fogg, Sheet Metal Ind., *52*, 1975, 297-298, 300-302.
[182] J.F. Wallace, Metal Ind. (London), *97*, 1960, 395-398, 415-418, 424.
[183] E.R. Braithwaite and G.W. Rowe, Sci. Lubric., *15* (3), 1963, 92-110.
[184] G.W. Rowe, Proc. Roy. Soc. (London), *A228*, 1955, 1-9.
[185] B. Fogg, Sheet Metal Ind., *44*, 1967, 95-112.
[186] D.V. Wilson, B.B. Moreton, and R.D. Butler, Sheet Metal Ind., *38*, 1961, 25-36, 42.
[187] C. Weidemann, in [12], pp.245-251.
[188] K. Hanaki and K. Kato, J. JSTP, *21*, 1980, 765-772 (Ann. Rep., *2*, 1981, 52).
[189] Ye.I. Isachenkov, *Technological Lubricants for Metalworking under Pressure*, Mashgiz, Moscow, 1960, pp.3-14 (Russian).
[190] N. Kasik and J. Reissner, Blech Rohre Profile, *5*, 1980, 303-308 (German).
[191] PERA Report No.86, Melton Mowbray, England, 1961.
[192] PERA Report No.60, Melton Mowbray, England, 1958.
[193] S.R. Gurumukhi and P.B. Mellor, in *Proc. 16th Int. MTDR Conf.*, 1976, pp.507-515.
[194] H. Khamedy-Zadeh, H.T. Coupland, and P.B. Mellor, Rev. Met., *77*, 1980, 363-369.

[195] N. Kawai, T. Nakamura, and M. Iwata, Trans. ASME, Ser.B, J. Eng. Ind., *99*, 1977, 242-249.
[196] N. Kawai, K. Kondo, I. Shimizu, and T. Ryuzin, Bull. JSME, *15*, 1972, 628-634.
[197] N. Kawai, K. Kondo, I. Shimizu, and T. Nakamura, Bull. JSME, *15*, 1972, 635-641.
[198] M. Robin, Manuf. Eng., *81*, 1978, 53-55.
[199] Y. Kasuga, K. Yamaguchi, and K. Kato, Bull. JSME, *11*, 1968, 354-360, 361-365.
[200] J. Clarke, Sheet Metal Ind., *38*, 1961, 865-880.
[201] D. Daniels, Metal Stamp., *12* (11), 1978, 10-12.
[202] H.T. Coupland and W. Holyman, Sheet Metal Ind., *42*, 1965, 7-14.
[203] R.W. Perry, Metals Eng. Quart., *2* (1), 1962, 56-61.
[204] D. James, Sheet Metal Ind., *41*, 1964, 691-697.
[205] D. Schmoeckel, Werkstattstechnik, *63*, 1973, 401-404 (German).
[206] J. Werner, Wire Wire Prod., *30*, 1955, 1223-1225.
[207] A. Ovreset and J.L. Duncan, J. Inst. Metals, *93*, 1964-65, 403-405.
[208] D.R. Mear, H.H. Topper, and D.A. Ford, Sheet Metal Ind., *40*, 1963, 477-485.
[209] U.S. Rao, Sheet Metal Ind., *44*, 1967, 673-680.
[210] M. Blaich, Report No.61, Inst. f. Umformtechnik, Univ. Stuttgart, (Springer, Berlin), 1981 (German).
[211] W. Panknin and K. Oberländer, Mitt. Forsch.-Ges. Blechverarb., No.17, 1955, 232-235 (German).
[212] D.R. Mear and H.H. Topper, Sheet Metal Ind., *40*, 1963, 567-570.
[213] K.G. Adams, W. Davison, and M.J. Hillier, Can. Mach. Metalw., *79* (9), 1968, 81, 99.
[214] J.E. Cook, Sheet Metal Ind., *45*, 1968, 461-472.
[215] A.J. Duke and A. Gage, J. Oil Col. Chem. Assoc., *53*, 1970, 774-791.
[216] K.J. Lawrenz, Ind. Anz., *99*, 1977, 515-516, 1521-1522 (German).
[217] A. Neubauer, A. Eichorn, and U. Fischer, Fertigungstech. Betr., *25*, 1975, 24-26 (German).
[218] G. Oehler, wt-Z. Ind. Fertig., *66*, 1976, 621-625 (German).
[219] E. von Finckenstein and K.-J. Lawrenz, Bänder Bleche Rohre, *19*, 1978, 130-136 (German).
[220] C.C. Veerman, M. Degroot, J.J. Peels, and L. Hartman, Hoesch Ber., *9* (1), 1974, 46-52.
[221] R. Franks and T.A.H. Plevy, Sheet Metal Ind., *56*, 1979, 500-511, 524.
[222] Anon., Nippon Steel Tech. Rep. No.12, 1978, 137-138.
[223] P.L. Furminger, Sheet Metal Ind., *45*, 1968, 422-425.
[224] Ch. Weidemann, Rev. Met., *77*, 1980, 343-352.
[225] F.J. Gurney and A.M. Adair, AFML-TR-78-66, 1978.
[226] Yu.I. Kozlov, Kuzn. Shtamp. Proizv., *1975* (5), 24-25 (Russian).
[227] R.M. Davison and T.I. Gilbert, Wear, *31*, 1975, 173-178.
[228] S. Misra and S. Rajagopal, in [3.53], pp.85-94.
[229] R. Duckett and C.J. Thwaites, Sheet Metal Ind., *48*, 1971, 274-276.
[230] P.E. Thomas and J.M. Thorp, Sheet Metal Ind., *51*, 1974, 685-687, 700.
[231] E.J. Helwig and L.E. Helwig, ASM Mat. Conf., Paper No.75, 1977.
[231a] I. Aoki, T. Horita, and T. Herai, in [4.299], pp.165-176.
[232] A. Klotzki, Bänder Bleche Rohre, *12*, 1971, 119-121 (German).
[233] J. Möllers and F. Fischer, Bänder Bleche Rohre, *16*, 1975, 457-460 (German).
[234] J. Peacock, Am. Mach., *105* (Nov.27), 1961, 83-85.
[235] J. Mortimer, Engineer, *231*, 1970(5976), 24-25.
[236] B. Langenecker, C.W. Fountain, and V.O. Jones, Metal Progr., *84* (4), 1964, 97-101.
[237] T. Mori and Y. Uchida, in *Proc. 21st Int. MTDR Conf.*, Macmillan, London, 1981, pp.237-242.
[238] V.P. Severdenko, V.S. Pashchenko, and B.S. Kosobutsky, *Deep Drawing of Sheet Metal with Ultrasonics,* Nauka i Tekhnika, Minsk, 1975 (Russian).
[239] F. Shinji, Y. Kiyota, A. Kunio, and O. Koji, Sheet Metal Ind., *40*, 1963, 739-744, 754.
[240] R.D. Butler, in [3.51], pp.58-62.
[241] H. Ziegler, F. Fischer, and K.H. Schmitt-Thomas, Metallweiterverarbeitung, *14*, 1976, 207-208, 210-211 (German).
[242] K.W. Blümel, Sheet Metal Ind., *57*, 1980, 152-159, 171.
[243] H. Hoffmann, in [3.56], Paper No.20 (German).
[244] R.D. Butler and J.F. Wallace, in *Recent Developments in Annealing,* Spec. Rep. No.79, Iron Steel Inst., London, 1963, pp.131-141.

[245] R.D. Butler and R.J. Pope, Sheet Metal Ind., *44*, 1967, 579-592, 597.
[246] C.R. Weymueller, Metal Progr., *88* (4), 1965, 213-233.
[247] R.R. Hilsen and L.M. Bernick, in *Formability Topics - Metallic Materials,* STP 647, ASTM, Philadelphia, pp.220-237.
[248] A. Bragard et al, in [12], pp.253-278.
[248a] V. Leroy, A. Bragard, L. Renard, and P. Flament, in [4.299], pp.3-56.
[249] F. Fischer, K.H. Schmitt-Thomas, and V. Seul, Stahl u. Eisen, *80*, 1960, 1524-1531 (German).
[250] P.R. Hastings and G. Gagnon, Paper, Fall Meeting, ADDRG, 1975.
[251] A. Shuaib and S.M. Wu, in *Proc. 7th NAMRC*, SME, Dearborn, 1979, pp.294-299.
[252] N. Kawai, K. Kondo, and T. Nakamura, Bull. JSME, *17*, 1974, 803-810.
[253] C.J. Taylor and F.B. Drake, Sheet Metal Ind., *42*, 1965, 533-537.
[254] P.W. Whitton and D.R. Mear, Sheet Metal Ind., *37*, 1960, 743-751.
[255] S. Fukui, K. Yoshida, K. Abe, and K. Ozaki, Sheet Metal Ind., *40*, 1963, 739-744, 754.
[256] H. Ike and K. Yoshida, in [3.56], Paper No.18.
[257] N. Kawai, T. Nakamura, and K. Dohda, Trans. ASME, Ser.B, J. Eng. Ind., *104*, 1982, 375-382.
[258] R.R. Hilsen, I.F. Hughes, and D.T. Quinto, in [12], pp.XXXV-XLIV.
[259] D. Rault and M. Entringer, in [11], pp.97-114.
[260] J.M. Story and K.J. Weinmann, in *Proc. 8th NAMRC*, SME, Dearborn, 1980, pp.245-251.
[261] J.F. Wallace, Sheet Metal Ind., *37*, 1960, 901-904.
[262] S.P. Keeler, Sheet Metal Ind., *42*, 1965, 683-691.
[263] J. Buza and J. Buda, CIRP, *26*, 1977, 367-370.
[263a] P.L. Coduti, in [4.299], pp.57-101.
[263b] P. McNamara, in [4.299], pp.103-126.
[264] T. Horita, I. Aoki, and T. Sato, J. JSTP, *19*, 1978, 17-24 (Japanese).
[264a] T. Furubayashi, S. Sato, M. Hirasaka, and N. Yoshihara, in [4.299], pp.147-164.
[265] Y. Kasuga, M. Shirayama, and H. Asano, in [3.46], pp.116-121.
[266] L.M. Bernick, R.R. Hilsen, and C.L. Wandrei, Wear, *48*, 1978, 323-346.
[267] P.G. Nelson, Metal Progr., *91* (1), 1967, 83-89.
[268] D. Schulz, Blech Rohre Profile, *18*, 1971, 135-140 (German).
[269] L. Ashton, Sheet Metal Ind., *54*, 1977, 640-642.
[270] G. Oehler, Maschinenmarkt, *81*, 1975, 1524-1527 (German).
[271] K. Keller, VDI-Ber., *39*, 1959, 55-59.
[272] K. Oberländer, in [3.56], Paper No.22 (German).
[273] H.A. Sundquist, A. Matthews, and D.G. Teer, Thin Solid Films, *73*, 1980, 309-314.
[274] R. Komanduri, Carbide J., *1976* (Nov./Dec.), 3-5.
[275] Anon., Mod. Met., *32* (12), 1977, 81-82.
[276] K. Oberländer, Blech Rohre Profile, *20*, 1973, 222-223 (German).
[277] J.H. Kerspe, Ind. Anz., *102* (91), 1980, 60-62 (German).
[278] J.H. Kerspe, Bänder Bleche Rohre, *21*, 1980, 68-73 (German).
[279] R. Woska, Blech Rohre Profile, *28*, 1981, 202-206 (German).
[280] Anon., Machinery (London), *121*, 1972, 834-839.
[281] K. Farwell, Mod. Mach. Shop, *44* (11), 1972, 80-83.
[282] L.H. Salz, Metal Progr., *92* (3), 1967, 133-140.
[283] L.H. Salz, Light Metal Age, *25* (April), 1967, 8-12.
[284] S.J. Barber, Lubric. Eng., *15*, 1959, 497-498.
[285] J.T. O'Reilly, Lubric. Eng., *8*, 1952, 14-17.
[286] A.A. Brown, Iron Steel Eng., *30* (6), 1953, 96-100.
[287] F.M. Giordano, Tool Mfg. Eng., *50* (Jan.), 1963, 63-66.
[288] H.T. Coupland, Sheet Metal Ind. Int., *55* (Dec.), 1978, 20-24.
[289] G.M. Davies, Trib. Int., *11*, 1978, 17.
[290] W.J. Wojtowicz, Paper No. MF 70-502, SME, Dearborn, 1970.
[291] R.I. Hamilton, Paper No. MF 74-625, SME, Dearborn, 1974.
[292] J. Ivaska, Paper No. TE 77-499, SME, Dearborn, 1977.
[293] E.H. Lee, Sheet Metal Ind., *52*, 1975, 293-296.
[294] P.D. Liddiard, Sheet Metal Ind., *25*, 1948, 1167-1173.
[295] J.J. Obrzut, Iron Age, *202*, 1968(July 25), 105-112.
[296] H.R. Oswell, Sheet Metal Ind., *52*, 1975, 454-456.

[297] T.A.H. Plevy, Sheet Metal Ind., *57*, 1980, 136-149.
[298] A. Salvi, Riv. Mecc., *26* (604), 1975, 29-32 (Italian).
[299] *Sheet Metal Industries Year Book*, London, 1979, pp.79-81.
[300] C. Sudoo, Y. Hayashi, and M. Nishihara, Rev. Met., *77*, 1980, 353-362.
[301] E.M. Spallina, Paper No. MF 75-180, SME, Dearborn, 1975.
[302] L. Berggren, Sheet Metal Ind., *52*, 1975, 458-459.
[303] H.L. Hilbert, Schmiertechn. Trib., *20*, 1973, 16-21 (German).
[304] R.I. Hamilton, Metalwork. Econ., *27* (6), 1971, 24-27.
[305] Anon., Sheet Metal Ind., *47*, 1970, 517-518, 592.
[306] R. Clark, Sheet Metal Ind., *52*, 1975, 335-337.
[307] D. Daniels, Metal Stamp., *12* (10), 1978, 14, 16.
[308] H.T. Coupland, Sheet Metal Ind., *57*, 1980, 180-182.
[309] K.C. Swindell and P. Wainwright, Sheet Metal Ind., *58*, 1981, 290-295.
[310] H.W. Swift, Sheet Metal Ind., *31*, 1954, 817-828.
[311] R. Pearce, Sheet Metal Ind., *41*, 1964, 567-570.
[312] A.H. Cole, Sheet Metal Ind., *52*, 1975, 452-453, 465.
[313] Yu.I. Artemov, T.N. Veretenova, A.I. Balin, and M.M. Savchenko, Kuzn. Shtamp. Proizv., *1980* (2), 17-19 (Russian).
[314] H.T. Coupland, Sheet Metal Ind., *40*, 1963, 49-56, 66.
[315] B. Roncin and M. Thebault, in [3.56], Paper No.25.
[316] W. Krämer, Ind. Anz., *86*, 1964, 2167-2170 (German).
[317] P. Timossi, in [3.53], pp.95-106.
[318] P. Ng, J. Chakrabarty, and P.B. Mellor, Int. J. Mech. Sci., *18*, 1976, 249-259.
[319] P.H. Theimert, Ind. Anz., *100* (25), 1978, 36-37 (German).
[320] W. Katzenstein, Paper No. MF 71-128, SME, Dearborn, 1971.
[321] F.L. Ewald, Metal Progr., *89* (1), 1966, 72-75.
[322] W. Holyman, Sheet Metal Ind., *52*, 1975, 368-371, 386.
[323] E.A. Evans, H. Silman, and H.W. Swift, Sheet Metal Ind., *24*, 1947, 1995-2002, 2209-2213, 2216.
[324] D.F. Eary, Tool Eng., *43* (Aug.), 1959, 83-90.
[325] A.V. Korolev and I.V. Podluzhnaya, Sheet Metal Ind., *39*, 1962, 279-282.
[326] V. Gopinathan and N.K. Murali, Trib. Int., *9*, 1976, 231-235.
[327] B. Fogg and W.H. Busby, Sheet Metal Ind., *47*, 1970, 289-299.
[328] V.B. Kaujalgi and T.R. Sud, Sheet Metal Ind., *46*, 1969, 891-893.
[329] E. Takeuchi and M. Tanaka, J. Jpn. Inst. Metals, *35*, 1971, 989-997 (Japanese).
[330] P. Funke, C. Pavlidis, and E. Korri, Materialprüfung, *15*, 1973, 261-267 (German).
[331] F.J. Dirks, H. Fritz, and J. Walling, Blech Rohre Profile, *22*, 1975, 444-450 (German).
[332] M. Arakawa and H. Sumitomo, Tetsu-To-Hagane, *63*, 1977, 824-831 (Japanese).
[333] J.A. Grainger, Sheet Metal Ind., *33*, 1956, 461-473.
[334] R. Raj and J.F. Berry, Trans. ASME, Ser.B, J. Eng. Ind., *92*, 1970, 412-418.
[335] M.H. Pope, L. Robins, and J.T. Berry, ASLE Trans., *13*, 1970, 148-158.
[336] A.A. Boccaccio, Modern Metals, *18* (July), 1962, 50-53.
[337] R.S. Barnes and T.H. Cafcas, Lubric. Eng., *10*, 1954, 147-150.
[338] W.-D. Neumann, Bänder Bleche Rohre, *17*, 1976, 184-187 (German).
[339] V.P. Kravtsov, V.S. Shipilov, and L.V. Zotova, Tsvetn. Met., *1980* (8), 77-78 (Russian).
[340] L.C. Blayden and J.E. Parnell, Automot. Eng., *86* (3), 1978, 36-40.
[341] H.D. Nine, SAE Paper No.780394, 1978.
[342] V.Z. Kutsova, D.G. Borshchevskaya, T.Ya. Evina, and O.P. Drobich, Sov. Mater. Sci., *16*, 1980, 481-482.
[343] T.A.H. Plevy, in [3.56], Paper No.24.
[344] J.E. Knepp, in [3.56], Paper No.26.
[345] J.D. Mastrovich, Lubrication, *61* (2), 1975, 17-36.
[346] J.E. Knepp and L.B. Sargent, Jr., Lubric. Eng., *34*, 1978, 198-201.
[347] R. Kelly, in [3.56], Paper No.27.
[348] M. Petronio, Lubric. Eng., *14*, 1958, 510-522.
[349] R.A. Overfelt, J.J. Wert, and W.H. Hunt, in [3.53], pp.115-122.
[350] F.L. Coenen, Modern Metals, *13* (Sept.), 1958, 88-90.
[351] Anon., Sheet Metal Ind., *44*, 1967, 476-477, 481.
[352] R.B. Waterhouse and M.H. Wharton, Ind. Lub. Trib., *26*, 1974, 20-23, 56-59.
[353] E. Mitchell and P.J. Brotherton, J. Inst. Metals, *93*, 1964-65, 278-279.

[354] P.G. Patten, Sheet Metal Ind., *34*, 1957, 741-744.
[355] J. Grindrod, Sci. Lubric., *11*, 1959, 13-14.
[356] W.M. Sterry, Metal Progr., *77* (6), 1960, 110-116.
[357] M. Lorant, Sheet Metal Ind., *32*, 1955, 96-100.
[358] PERA Report No.100, Melton Mowbray, England, Dec. 1961.
[359] D.E. Strohecker, R.J. Carlson, S.W. Porembka, and F.W. Boulger, DMIC Report No.203, Battelle Memorial Inst., 1964.
[360] H.W. Babel and R.M. Bonesteel, Metal Progr., *87* (4), 1965, 66-71.
[361] R.E. Jackson, SAMPE Journal, April/May 1967, 20-25.
[362] V.N. Korolev and A.I. Galakhov, Kuzn. Shtamp. Proizv., *1972* (5), 15-18 (Russian).
[363] J.M. White, Tool. Prod., *23* (June), 1954, 106, 110.
[364] R.I. Batista, G.S. Hanks, J.M. Taub, and D.J. Murphy, Lubric. Eng., *17*, 1961, 414-418.
[365] L.S. Blinova and G.A. Gubanova, Kuzn. Shtamp. Proizv., *1980* (4), 22-23 (Russian).
[366] J.L. Everhart, Mater. Methods, *34* (Dec.), 1951, 89-104.

11

Metal Removal

Chapters 6 to 10 were devoted to processes in which fracture is—with the exception of the process of shearing—undesirable, and the shape of the part is produced by plastic flow. Machining differs in that the material is intentionally separated and the shape of the component is developed by removing unwanted metal.

In contrast to most plastic deformation processes, the geometry of cutting is not uniquely defined by the tooling. Instead, it is affected by complex interactions between tool geometry, material properties, and frictional stress distribution, and is further influenced by tool wear. Indeed, many general wear mechanisms are also related to cutting; therefore, one could classify all knowledge pertaining to machining processes as a branch of tribology. Consequently, many articles appearing in tribological publications deal with the mechanics of cutting and with wear mechanisms that, under certain conditions, can be used also for metal removal. Such a broad interpretation of the field would easily result in a treatise of the size of this entire monograph, and the number of references would be in the thousands. Therefore, a narrower view is taken here.

First, the mechanics of metal cutting will be reviewed, with emphasis on the effects of friction and wear. Then the mechanisms of wear and lubrication will be explored, followed by a discussion of testing techniques. The same aspects will then be reviewed for grinding and, finally, the more specific aspects of working individual metal groups will be covered. Whenever possible, background material will be discussed with reference to readily accessible English-language reviews or monographs from which further references (including those pertaining to much excellent work published in foreign languages) may be obtained. Critical reviews authored by subgroups of the International Institution for Production Engineering Research are drawn upon whenever possible; the title of this journal is abbreviated as CIRP.

Manufacturing parts by removing material is essentially wasteful and, whenever possible, parts are made by plastic deformation or other processes (casting, powder metallurgy, joining, etc.). Nevertheless, the requisite tolerances, surface finish, or geometry can often be ensured only by metal removal, and it has been estimated that this activity alone accounts for some 3% of the GNP in an industrially developed economy. There are thus powerful incentives to understand the process, so that metal can be removed at a fast rate, at low cost, and with adequate surface integrity. Accordingly, the literature is vast: in addition to books cited in Chapter 2, metal removal processes are discussed in books [1-12] and in conference proceedings [13-19]. Several reviews are devoted to the tribological aspects of metal cutting [3.29],[20-25].

11.1 Cutting

In many metal-removing processes the excess material is separated from the surface in more or less continuous, well-identifiable chips, with the aid of a tool that has, at the beginning of the cut, a geometrically well-defined cutting edge (see Fig.2.5c). More than one cutting edge may be at work, but the essentials of the process are best discussed with reference to a single-edge tool.

11.1.1 Ideal Orthogonal Cutting

The simplest presentation of cutting becomes possible when the cutting edge of the tool is straight and perpendicular to the direction of motion (Fig.11.1) and when the width of the cut w (perpendicular to the plane of the page) is infinite. Such plane-strain cutting is approximated when a plate is cut on its edge; in order to keep side flow to the minimum, the thickness of the plate (which becomes w) is at least 10 times the depth to which the cutting edge penetrates the workpiece (the width must be even greater for cutting with negative-rake tools). To avoid confusion, the depth dimension will always be referred to as the undeformed chip thickness t. According to the shear model of cutting proposed by Merchant [26] and by Piispanen [27], the chip is formed by shearing closely spaced lamellae. The chip flows up along the rake face of the tool but acquires a curvature and lifts away gradually. Unnecessary friction on the freshly cut surface is prevented by relieving the tool on its clearance or flank face.

Process Geometry and Kinematics. In principle, it makes no difference whether the tool or workpiece is moved. It is more convenient to visualize a stationary tool, with the workpiece moving from left to right at a velocity v (Fig.11.1), as in shaping. The work-

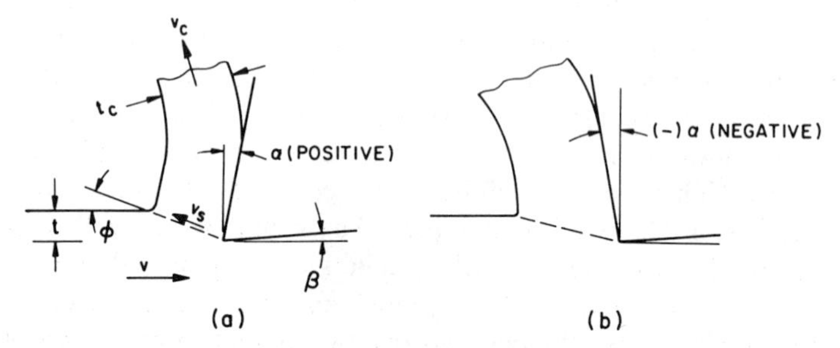

Fig.11.1. Geometry of orthogonal cutting and designation of rake angles

ing (rake) face of the tool is inclined at a rake angle α from the perpendicular to the direction of motion; the angle is taken positive or negative according to convention (Fig.11.1a and b). The flank face is relieved by a clearance angle β. Deformation may be imagined to take place entirely by shear concentrated in a plane, at a shear angle ϕ measured from the cutting direction. This angle determines the chip thickness t_c, and consequently, the chip-thickness ratio or cutting ratio r_c

$$r_c = \frac{t}{t_c} = \frac{l_c}{l} \tag{11.1}$$

is most informative regarding the processes taking place. In some respects, the reciprocal value, called the chip-compression factor, is more convenient and informative [4,28]. Both can be obtained from measured chip thickness or, because the chip tends to be ragged and uneven, from the measured length l_c, utilizing the principle of constancy of volume, and assuming that width change is negligible. If spread cannot be neglected, the average cross-sectional area can be found by weighing a chip of measured length. From the geometry of the process, the shear angle is defined by

$$\tan \phi = \frac{r_c \cos \alpha}{1 - r_c \sin \alpha} \tag{11.2}$$

Because of constancy of volume, the chip-thickness ratio can be expressed also from the chip velocity v_c and cutting velocity v:

$$r_c = \frac{V_c}{V} = \frac{\sin \phi}{\cos (\phi - \alpha)} \tag{11.3}$$

Thus, with increasing shear angle ϕ, the chip is thinner and comes off at a higher speed v_c, and the chip-tool contact length L is shorter. It is usually observed that the chip is also more tightly curled (has a smaller radius of curvature).

Since in the ideal case all shear is concentrated in an infinitely thin shear zone, the shear strain rate would reach infinity; therefore, it is more realistic to regard the shear plane as a shear zone of small thickness.

Forces. It is not difficult to measure the forces acting on the tool with the aid of a two-axis dynamometer (Fig.11.2). The cutting force P_c is needed to keep the tool in the cut. Most importantly from the tribological point of view, the resultant P_R can be

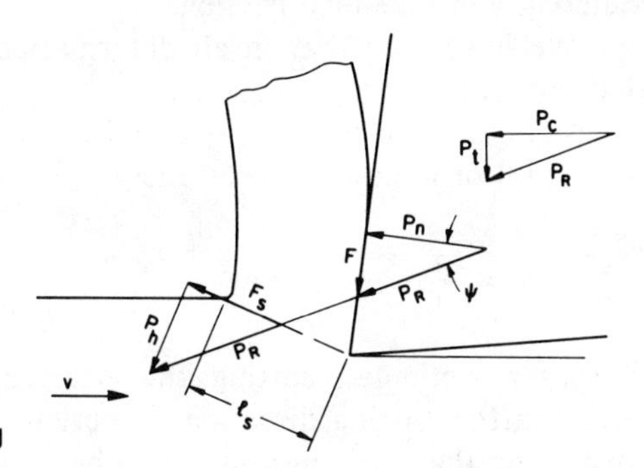

Fig.11.2. Balance of forces in orthogonal cutting

regarded as the vector sum of the normal force P_n acting perpendicular to the rake face and of the friction force F acting parallel to it. For a sharp tool with adequate clearance, friction on the clearance face can be neglected. Reaching back to the definition of the coefficient of friction (Eq.2.8), we can express

$$\mu = \frac{F}{P_n} = \frac{P_c \sin \alpha + P_t \cos \alpha}{P_c \cos \alpha - P_t \sin \alpha} \tag{11.4}$$

For reasons soon to be expanded upon, we shall call this the apparent mean coefficient of friction. From the geometry of the force vector diagram, a friction angle ψ can be defined as

$$\mu = \tan \psi = \frac{F}{P_n} \tag{11.5a}$$

or, from the measured forces,

$$\tan (\psi - \alpha) = \frac{P_t}{P_c} \tag{11.5b}$$

Even more than in other processes, the mutual influences of friction and deformation will be of greatest significance. Therefore, the forces acting on the shear plane are of interest: the material is being sheared by the shear force F_s acting in the shear plane, while the normal force P_h exerts a hydrostatic pressure on the intensely sheared material. We have seen (Sec.2.1.3) that hydrostatic pressure does not affect the flow stress σ_f (or, more relevantly for chip formation, the shear flow stress k) of a material, but it does delay fracture. Thus, in a ductile material, the chip can form in a continuous fashion, without gross fracture, even though the strains are high (typically, on the order of 1 to 3).

The frictional force F hinders the passage of the chip over the rake face of the tool and slows down the chip. Accordingly (Eq.11.3), the shear angle ϕ is reduced. In contrast, cutting with a more positive rake angle α causes ϕ to increase and the chip to thin out. At the same time, the normal force P_n drops, and thus μ calculated from Eq.11.5a often increases with increasing α, even though the frictional force F stays the same or may even drop. Kronenberg [4] drew attention to the confusion that often results, and in many ways it is better not to use the concept of μ at all. If μ is used, it must be understood that it is only an apparent mean coefficient of friction. Thus, the geometry of the chip-forming process is determined, even in the simplest view, by the combined effects of tool geometry, the flow and fracture characteristics of the work material, and rake-face friction.

We have seen that in all deformation processes the material seeks a flow pattern that minimizes energy. This is true also of cutting, and was recognized in early theories, reviewed by Kronenberg [4] and by Finnie [29]. Theories based on maximum shear stress [30] or minimum shear energy [26] lead to

$$\phi = \frac{\pi}{4} - \frac{1}{2}(\psi - \alpha) \tag{11.6}$$

Thus, for frictionless cutting, the shear angle should be half the angle between the rake face and the cutting direction. Experience has shown that Eq.11.6 is actually too optimistic and that the constant has to be smaller than $\pi/4$, as recognized independently by Merchant [26] and by Piispanen [27], and emphasized by Zorev [3] in his book on chip formation. The important point is that, with decreasing α or increasing friction, the shear angle decreases, the shear plane l_s lengthens, the cutting force rises, and the energy expended increases.

In cutting with increasingly negative rake angles (Fig.11.1b), the shear plane lengthens, the cutting and thrust forces increase, and, ultimately, cutting gives way to bulge formation. At this point we are back to bulge formation in plane-strain drawing (Fig.7.9); the situation is analogous to drawing a semi-infinite body between inclined die surfaces, but the die half angle is now $(90 + \alpha)$ degrees, where α has a negative value.

11.1.2 Realistic Orthogonal Cutting

We have seen in discussing drawing with a steep die angle (Sec.7.1.4) that the inhomogeneity of deformation leads to complex material flow. We have also found that, whenever the condition $\mu p = k$ is satisfied (Eq.2.9), sticking friction is established. It will be recalled from Sec.2.2.1 that this does not necessarily imply that the workpiece actually sticks or adheres to the die, but only that relative movement is arrested. Conditions are no different in cutting, and, in view of the high pressures that must act on the rake face of the tool, it is obvious that the simple shear plane model with sliding friction on the rake face can describe reality only under exceptional circumstances. The situation on the rake face is complicated by the fact that, even more than in other processes, the inhomogeneity of deformation interacts with the effects of friction (Sec.2.3.4), and no meaningful discussion of the tribology of cutting is possible without consideration of realistic material flow.

Material Flow. For an observation of material flow, all techniques described in Sec.5.5.4 can be used. Ideally, the process should be observed at realistic cutting speeds, for which high-speed photography is needed. However, the chip edge is not representative of what goes on in the middle of the chip width, and only recently were high-speed films made with the workpiece photographed through a glass plate that ensured plane strain up to the edge, as reported by Warnecke [31]. More frequently, the specimen is parted in the middle, and the deformation pattern developed at cutting speeds is preserved by quick-stop devices, usually by explosive shearing of the tool holder. To reveal the deformation pattern, the parting plane may be gridded: techniques for fine gridding have been developed by Oxley and co-workers [32,32a]. Structural features in steel and nonferrous metals have also been used. The question remains whether the mechanical shock associated with deceleration may affect the final phase of chip formation.

Experience confirms that material flow is indeed affected by both frictional conditions and material properties. In particular, strain distribution is a function of the localized changes in flow stress and friction, as affected by strain hardening, strain-rate sensitivity, temperature distribution, and adhesion between tool and workpiece material. For purposes of discussion, it is useful to separate deformation into several zones (Fig.11.3).

1. Most metals strain harden on deformation (Eq.10.2). The rate of strain hardening (the n value) decreases with increasing strain but still remains positive, even at high temperatures, if strain rates are high enough to prevent dynamic recrystallization and if there are no second-phase particles that could initiate microcracks [3.223]. This means that shear cannot take place in an infinitely narrow plane; instead, a more diffused primary shear zone is found (zone 1 in Fig.11.3). The chip becomes thicker and the cutting energy increases (thus a soft material takes higher energy than a previously strain-hardened one). At higher cutting speeds, shear heating weakens the material in a well-defined zone, approaching adiabatic shear, and thus keeps the shear zone narrow (typically, about 2.5 μm thick). At the usual cutting speeds this corresponds to very high strain rates (on the order of 10^6/s).

2. The deformed chip must move out of the shear zone by flowing up against the tool rake face. Additional energy is thus expended in a zone adjacent to the rake face (zone 2 in Fig.11.3), irrespective of whether sliding or sticking friction prevails.

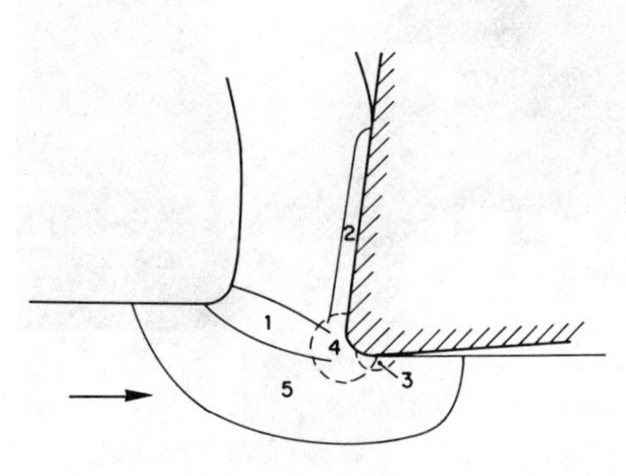

Fig.11.3. Zones of deformation in chip formation [31]

3. The tool moves over the freshly formed surface and, unless the tool is truly sharp, a rubbing surface of finite length generates further friction and deformation against the flank face (zone 3).

4. The tool edge, unless of zero radius, creates some local plastic deformation, the nature of which is affected also by friction (zone 4).

5. Strain hardening and flank-face contact cause plastic deformation to spread below the depth of undeformed chip thickness, and leaves a plastically worked surface behind (zone 5).

11.1.3 Chip Formation

The chip usually formed is not the ideal chip of Fig.11.1. Reviews of chip formation have been given by Shaw [33] and by Armarego and Brown [5]. Williams et al [34] made a comprehensive study on many workpiece materials. A review with emphasis on chip control and chip breaking has been given by Kluft et al [35]. In general, the following chip forms are observed:

Continuous Chip. Under ideal conditions, primary shear can be assumed to take place in closely spaced shear planes (Fig.11.4). The spacing of lamellae (on the order of 2 to 10 μm) and deformation within the chip depend on material properties and have been observed directly in the SEM. Reviews are given by von Turkovich [36] and by Doyle and Samuels [37]. In ductile materials the shear strain can be accommodated without gross cracks. However, in less-ductile materials, and in those containing a second phase, microscopic cracks are likely to develop. The possibility was first suggested by Shaw and also by Kohn [3.224], and the theme has been further developed again by Shaw [3.223]. He postulates that cracks generated in the shear plane reweld when sticking friction on the rake face results in high hydrostatic pressure. Cracks could survive in the presence of sliding friction and then lubricant access would be furthered. Cracks could also exist on the outer surface of the chip where the normal stress is low. Experi-

Fig.11.4. Continuous chip formation in cutting of 60/40 brass at a speed of 100 m/min (courtesy of Dr. P.K. Wright, Carnegie-Mellon University)

mental confirmation comes from the work of Komanduri and Brown [38], followed by a series of papers by Brown and co-workers [39]. With increasing speed the number and size of microcracks decreases in the primary shear zone, thus reducing lubricant access.

1. Continuous Chip Formation with Sliding Friction. Sliding friction can be maintained over the rake face only when $\mu p < k$. Since p is generally larger than k, μ must be kept quite low. This can be achieved only under limited conditions:

a. Access of a lubricant to the chip/tool interface is very difficult (Sec.11.4.1).

b. Surface films preformed on the tool are helpful but can usually survive only limited sliding. Therefore, sliding friction is more frequently observed in the laboratory where experiments with short cutting lengths are often conducted.

c. Lubricant breakdown is accelerated and adhesion is promoted by increasing temperatures. Therefore, sliding friction is usually limited to low-speed cutting.

d. Sliding friction can be maintained over a broader range of conditions if tool/workpiece adhesion is minimized by choosing an appropriate tool material or surface coating (Sec.11.3 and 11.6.5). Alternatively, or additionally, the workpiece can be modified to reduce adhesion, typically by additions that form a readily melting or smearing second phase (Sec.11.7).

e. For a tool of more positive rake angle α, the shear angle ϕ is larger, the shear plane and tool/chip contact zone are shorter, and the tool pressure (or P_n) is lower. Therefore, sliding friction is more easily maintained.

2. Continuous Chip Formation with Sticking Friction. Sliding friction is difficult to maintain in practice for several reasons:

a. Many cutting processes such as turning, boring, facing, and drilling and planing involve sustained contact. Any surface films present on the tooling are worn off, and lubricant access is difficult if not impossible.

b. Cutting generates 100% new surface. In the absence of contaminant and lubricant films on the tooling, and with the substantially increased temperatures, perfect conditions for adhesion are set up. After the initial contaminant films are worn off, metal transfer (tool pickup) and adhesion are likely to develop.

c. High positive rake angles, necessary to maintain sliding friction even under the best conditions, result in a weaker tool. Even ductile or very tough tools (HSS) may have to be made with modest (say 5 to 10°) positive rake angles. The more brittle cemented carbides, and, especially, the ceramics, must be used with a zero or negative rake angle.

It is not surprising that sticking friction (in the metal cutting literature, also referred to as "seizure" [8]) should occur, greatly affecting material flow. The consequences of sticking friction have not been difficult to visualize in other processes. In extrusion, for example, an excessively steep die angle leads to the development of a dead-metal zone (Fig.8.6b); in indenting a body, the punch grows a dead-metal nose (Fig.2.7b). In cutting, the situation is more complex. The tool wedge penetrates the work nonsymmetrically and the hotter chip flows up the rake face at some (typically lower) speed v_c while the cooler new surface passes the clearance face at the cutting speed v. Also, the normal pressure acting on the rake face is large whereas the pressure acting on the clearance face is low. The result is that the dead-metal zone assumes the nonsymmetrical shape of the so-called built-up edge (BUE), which acts as an extension of the tool edge (Fig.11.5). The chip is formed by intense shearing along the boundaries of this zone and flows up freely until it touches the rake face of the tool. The problem is that the BUE is seldom stable, but at least part of it is lost and rebuilt at more or less regular intervals, and its size and shape depend on process conditions and on material properties.

Fig.11.5. Example of BUE formation in cutting of 60/40 brass at a speed of 30 m/min (courtesy of Dr. P.K. Wright, Carnegie-Mellon University)

The essential features of the BUE formed at low speeds have long been known, as pointed out by Pekelharing [40] in his critical review. Much of the book of Kuznetsov [41] is devoted to the BUE, and Ernst and Martelotti [42], Nakayama et al [43], Zorev [3], and Williams et al [34] studied it in detail. Some features were clarified in 1965 by Trent [44], but details are still debated [40] and are the subject of intensive research. Some of the difficulty arises from the tendency to deal with the BUE and the secondary shear zone as two distinct phenomena. Here we will take the view that they are manifestations of the same phenomenon—namely, sticking friction—and that the transition from one to the other is simply a consequence of changes in flow pattern resulting from changes in strain hardening, strain-rate sensitivity, flow stress, and temperature distribution in the shear zone and the chip.

An important aspect should be the homologous temperature which determines the response of metallic materials to stresses. It will be recalled from Sec.2.1.3 that pure metals generally show low flow stress, high ductility, and great strain-rate sensitivity in the hot working temperature range, typically above one-half the melting point T_m on the homologous temperature scale. Alloying shifts the onset of hot working to higher temperatures, and in heavily alloyed materials, such as superalloys, the typical hot working behavior is not observed until close to the melting point—above, say, 0.8 T_m. The following remarks apply to materials for which room temperature is in the cold working range.

a. In low-speed cutting, chip formation is usually discontinuous and will be discussed later.

b. At somewhat higher (but still low) speeds the heat generated is sufficient to localize shearing. Shear strains are very high in the vicinity of the edge where the material flow separates, and a stagnation point develops, as observed by Horne et al [45]. Very pure materials of high ductility may be able to sustain such high strains, but most technical materials fracture; as shown by direct observations in the SEM [46], cracks are initiated away from the edge, at the boundaries of the sticking zone. Layer by layer, a BUE develops, the dimensions of which increase with speed (Fig.11.6) to some maximum value [31]. Cutting now proceeds by intense plastic shearing along the boundaries of the BUE. The effective rake angle becomes more positive, thus the chip thins out, the energy consumption drops, and the apparent mean μ also diminishes. In view of full sticking on the rake face, the latter result is contradictory, and the contradiction is removed if the apparent mean μ is recalculated with the new rake angle (Fig.11.7) [3]. The tool/chip contact length decreases, and the tool edge is protected by the BUE. Thus, a stable BUE would be desirable for a number of reasons. Unfortunately, there are undesirable consequences, too. The undeformed chip thickness is greater than intended, and the part is undersize (a cylinder turned on a lathe may be undersize by a millimetre or so). The surface finish is poor also: intense shearing at the surface of the BUE causes the material of the dead-metal zone to soften, and parts of it are periodically carried away. Pekelharing [40] estimates that some 5 to 10% of the cut surface (and of the underside of the chip) is covered with such particles (scale) (Fig.11.5). Hard particles contribute to flank wear of the tool. The BUE may build up

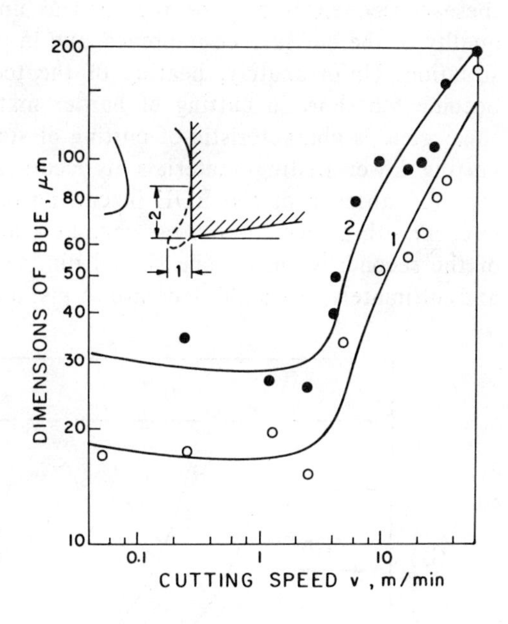

Fig.11.6. Dimensions of BUE in cutting at very low speeds (1045 steel; rake angle, 0°; feed, 10 to 30 μm; dry) [31]

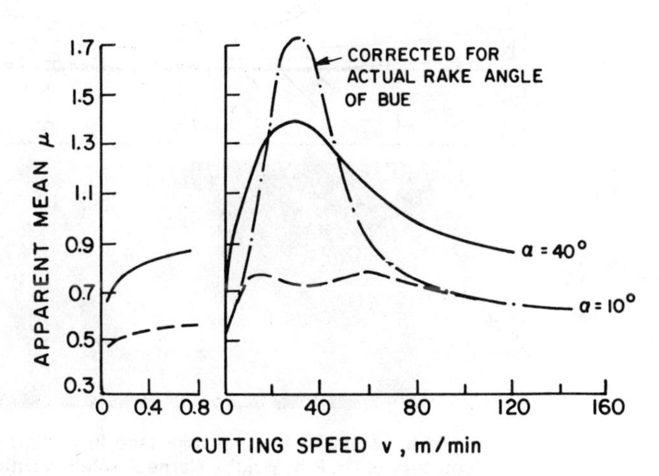

Fig.11.7. Apparent mean coefficient of friction in cutting of free-machining steel (HSS tool, dry, f = 0.15 mm) [3]

to such a large size that the forces acting on it tear it off. When there is no actual adhesion, the built-up material is totally lost, and the cycle of building up the BUE begins anew. However, when there is adhesion, the periodic loss of the BUE contributes to tool wear. Ohgo and co-workers [47] studied these effects in detail. Collapse of the BUE leaves large fragments embedded in the workpiece (and chip) surface, resulting in a very poor finish indeed. Furthermore, fluctuations in cutting force may result in vibrations, with a further decrease in accuracy and surface quality. Thus, a study of the BUE is important, because many processes such as drilling, reaming, tapping, and broaching are conducted in this speed range.

c. At yet higher speeds, heating in the shear zone further weakens the material, causing the BUE to shrink. Ultimately, at sufficiently high homologous temperatures, the BUE degenerates into an apparently uniform, thin layer that is sticking or almost sticking along the rake face (Fig.11.8a) [48]. The effective rake angle becomes the same as the tool rake angle, the apparent mean μ drops slightly (Fig.11.7), but the shear angle decreases and the cutting ratio and cutting energy increase. Even though these consequences may be regarded as undesirable, a significant benefit is that the quality of the cut surface improves, and in many ways this would be the most desirable situation. Unfortunately, heating of the tool becomes substantial, and tool life may become too short in cutting of harder materials. The degenerate BUE or secondary shear zone is characteristic of cutting of steel at speeds over, say, 50 to 70 m/min. In cutting lower-melting materials it occurs at lower speeds, and it is likely that the reported absence of the BUE in cutting of lead, tin, or zinc actually corresponds to cutting with a secondary shear zone at a high homologous temperature. The thickness of the secondary shear zone should diminish with increasing homologous temperature and, ultimately, it should decrease to just a few atom layers. One could surmise that, if

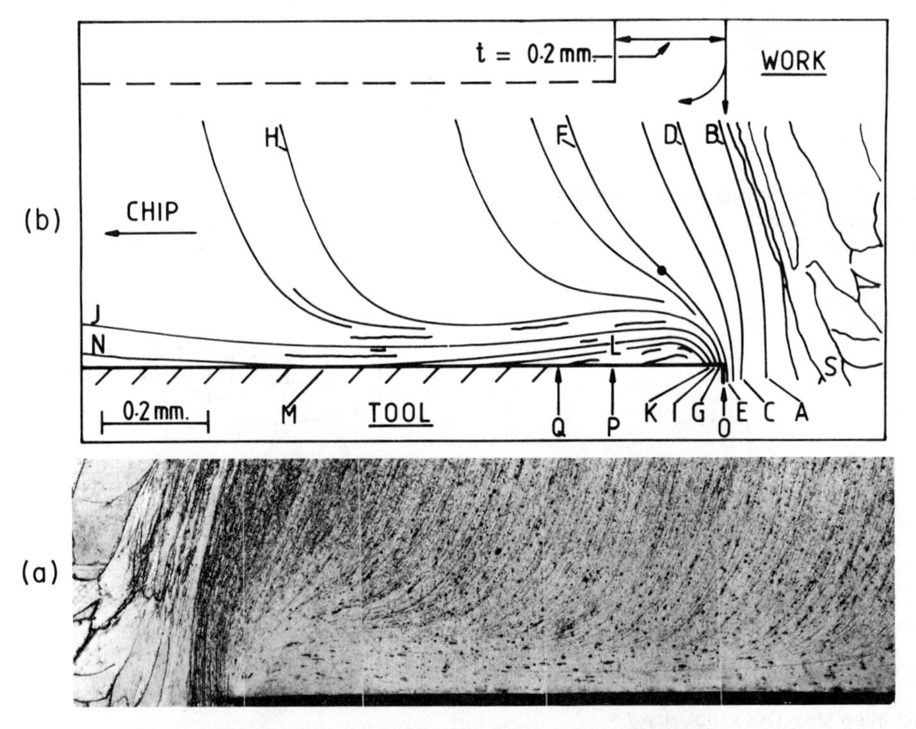

Fig.11.8. Material flow on rake face in cutting of low-carbon steel (courtesy of Dr. P.K. Wright, Carnegie-Mellon University) [48]

sufficiently temperature-resistant tools could be found, all reasonably ductile materials could be cut under such conditions.

Not all of the chip/tool contact zone is sticking (Fig.11.9). In cutting of a chip of finite width *w*, the sticking zone (zone 1) is surrounded by a zone of sliding friction (zone 2) in which the chip and tool are physically separated before the chip lifts away. Whether transition from sticking to sliding sets in suddenly, as suggested by Wallace and Boothroyd [49], or gradually, as found by Finnie and Shaw [50], depends perhaps on the particular material combination. It is in this zone that the atmosphere and the lubricant/coolant can have the greatest influence.

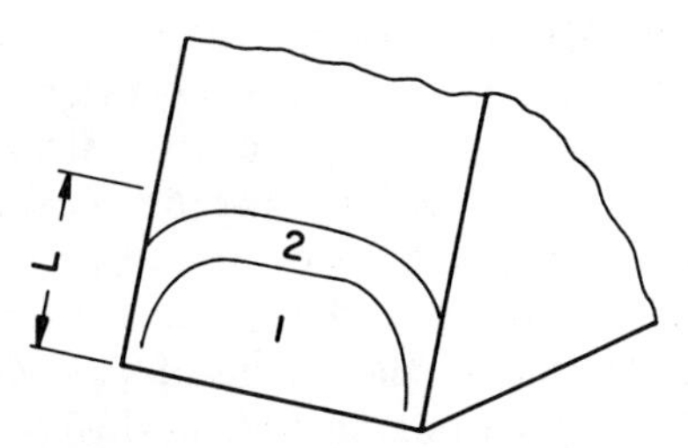

Fig.11.9. Zones of sticking (1) and sliding (2) contact

The physical conditions at the interface and, particularly, in the sticking zone have not yet been fully clarified. The earliest evidence for sticking came from quick-stop tests. The underside of the chip shows the imprint of tool grinding marks in zone 1, indicating the absence of sliding [51]. More evidence comes from direct observations of the interface through transparent sapphire tools, reported in a series of papers by co-workers of Tabor [52]. In the initial stages of cutting, metal transfer and seizure first occur some distance away from the cutting edge, where sliding speeds are greatest. However, as cutting continues and oxide and contaminant films wear off, the sticking zone extends to the tool edge, as found by Wright [48]. Detailed inspection of flow lines in the chip also shows that the material flows around the edge, leaving a stagnant zone (which can be regarded as the remnant of the BUE) beyond which the flow lines turn parallel to the rake face (Fig.11.8b) [48].

If there is strong adhesion between tool and workpiece, the layer on the rake face is physically arrested and the full chip speed is attained in a narrow secondary shear zone. This is the case particularly with aluminum against a sapphire (Al_2O_3) tool [52], which is in agreement with the observation made in Sec.3.11.2 that aluminum adheres strongly to its oxide. This is typical also of all ferrous materials. With less-adhesive combinations, it is conceivable that full seizure is not attained over the entire chip/tool contact area. From experiments with steel tools, Childs et al [53] concluded that, at lower speeds, not all of the contact zone is sticking in machining of brass.

Wright et al [54] suggested that the interface should be viewed as a network of microregions in dynamic equilibrium. At any one moment some regions are fully bonded whereas others experience interfacial sliding; a moment later the bonded regions may be sheared and the sliding ones bonded. This image is thus similar to that of dry friction (Fig.3.13) with a fraction of the surface actually bonded, forming strong junctions (Eq.3.12). The magnitude of a in Eq.3.12 is controlled by adhesion and thus depends on the nature of oxide films and contaminant films on the rake face, the cutting speed (or, rather, the temperature generated), and the compositions and phase distributions of tool and workpiece materials. If oxides and contaminant films cannot be replaced in the course of intermittent cutting, they will gradually wear off. Therefore, the bonded fraction a increases with cumulative sliding distance (or time at a given speed). After cutting of a sufficient length, welded regions are found over the entire zone 1. This should be true of all metals and alloys when cut at sufficiently high speeds to bring the shear zone into the strain-rate-sensitive (hot working) temperature range. Such temperatures can be readily attained because the chip is already heated in the primary shear zone and receives substantial further heat input in the secondary shear zone.

3. Factors Affecting BUE Formation. Quantification of the above influences is not yet possible, but general guidelines seem to emerge:

a. There is experimental evidence linking the formation of a sticking zone to adhesion. Thus, Takeyama and Ono [55] observed BUE with silver, tin, lead, and zinc, all of which have low solubility in the cobalt matrix of the WC tool. Of course, other factors that affect adhesion (Sec.3.11.2) should have effects also, particularly the presence of surface films and the temperature of contact.

b. Evidently, a negative stress gradient must exist perpendicular to the tool face, i.e., the material must be weaker away from the face [56]. Thus, a high strain-hardening rate (as expressed by a large n value) should mean that, because the heavily deformed layer adjacent to the tool will be strong, the BUE should grow. This was one of the major conclusions of Takeyama and Ono [55] and also of Rowe and Wolstencroft [57]. In cutting of brass at different temperatures, the presence of BUE was linked to a high n value once surface films had been worn off (Fig.11.10).

c. A temperature gradient plays the same role as that of strain-hardening rate. A tool of good heat conductivity chills and thus strengthens the root of the BUE and shifts the intense shear zone away from the tool, making the BUE larger. Thus, a larger BUE is found with HSS tools than with WC tools [58]. A tool of good heat conductivity (or a heated tool) weakens the BUE and diminishes it [40].

d. A high workpiece temperature shifts the shear zones into high homologous temperatures even in high-melting materials. Thus, the BUE is reduced [49,57,59], surface finish improves, and the cutting energy is decreased—although at the expense of the energy and equipment required to heat the workpiece [60]. To reduce the total energy requirement, the workpiece can be heated locally, just ahead of the shear zone, by a plasma torch, induction [61], or laser heating [62].

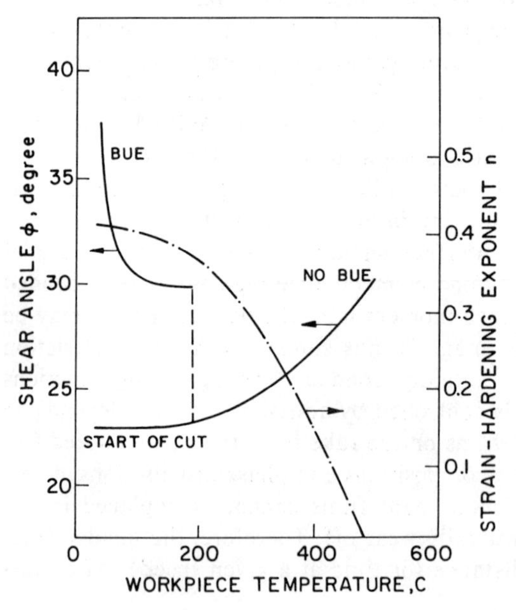

Fig.11.10. Occurrence of BUE in cutting of brass (HSS tool; rake angle, 15°) [57]

e. High localized temperatures are attained also in high-speed machining [63]. In practical applications, problems are encountered mostly with the requisite equipment, and the benefits are lost when the noncutting time is not proportionally reduced [64,65].

f. Heat generation increases with undeformed chip thickness, and thus the BUE develops at a lower speed and to a larger extent but also transforms into the secondary shear zone at a lower speed (Fig.11.11) [57,66,67]. For this reason, it is often said that the BUE disappears on taking a heavier cut.

g. It is frequently stated that only two-phase materials show a BUE. Williams and Rollason [59] found no BUE with vacuum-melted and cast brass and Cu-Ni alloy. Williams et al [34] suggested that commercial-purity metals contain sufficient second-phase particles to qualify as two-phase materials. Yet, pure metals of high adhesion to the tool (typically, iron, aluminum, and molybdenum) also form a large BUE [55], and this point needs clarification. The presence of a second phase does affect, however, the strain that a material can sustain in shear [45]. Inclusions in free-machining materials promote early fracture on the shear plane and reduce the size of the BUE.

h. Variations in mechanical properties with temperature have effects, too. If the flow stress drops steeply with temperature, the BUE is likely to shrink. The special case of low-carbon steels that suffer dynamic strain aging was investigated in great detail by Shaw and co-workers [68,69]. Around 300 C, the ductilities of these steels are low, whereas their strain-hardening exponents and flow stresses are high (this is called also the blue-brittleness temperature because of the color of the oxide formed). On machining at speeds where the primary shear zone is still below this temperature, additional shearing around the tool rake face raises the temperature into the critical range, strengthening the material and thus promoting BUE formation just as a high n value does [56].

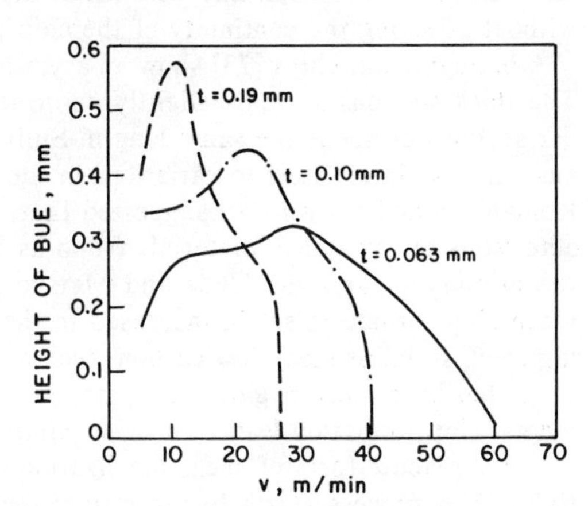

Fig.11.11. Height of BUE as a function of cutting speed and undeformed chip thickness in cutting of steel (rake angle, 0°) [66]

Discontinuous Chip and Chatter. The workpiece, tool, tool holder, and machine tool all suffer elastic deflections when subjected to cutting forces. If these forces undergo variations for any reason, vibrations (chatter) may be initiated (as discussed in Sec.3.9.1), which result in a visible and measurable waviness of the machined surface and in the periodic variation of chip shape.

Several kinds of uneven, serrated chips are encountered, as reviewed by Komanduri et al [70,71].

a. A wavy chip (Fig.11.12a) is continuous but shows periodic, roughly sinusoidal variations in thickness. The regenerative chatter originates in self-induced vibrations due to cyclic variations in cutting speed (e.g., from a gear box or coupling), undeformed chip thickness (from waviness produced by regenerative chatter in prior cutting opera-

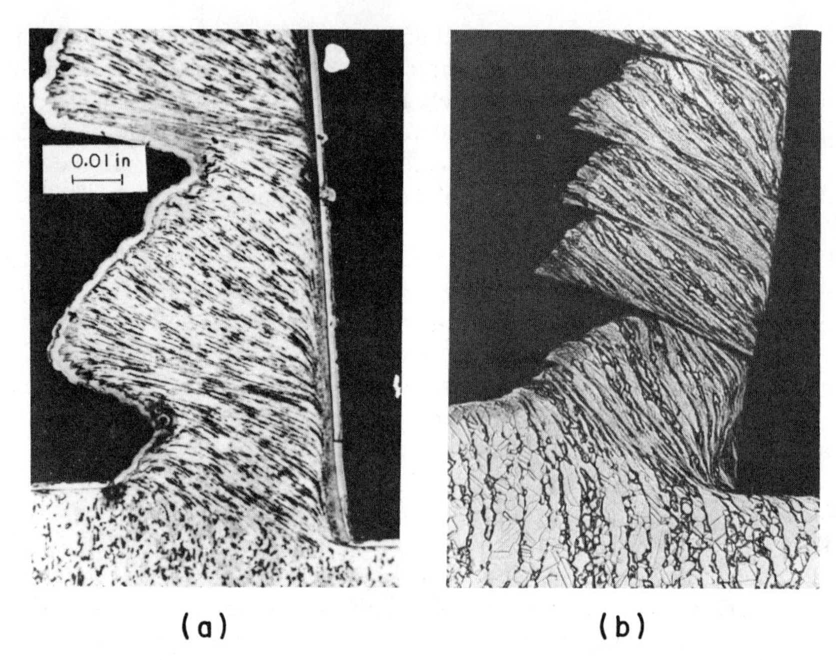

Fig.11.12. Examples of (a) wavy chip (courtesy of Dr. R. Komanduri, General Electric Co.) and (b) segmented chip (courtesy of Dr. P.K. Wright, Carnegie-Mellon University)

tions), and cutting force (from decreasing force with increasing speed) [5,72]. Loss of material from the BUE may be sufficiently constant in frequency to set up vibrations without affecting the continuity of the chip [41].

b. Segmental chips [73] show an asymmetrical, sawtooth-like waviness (Fig.11.12b). The thick sections are only slightly deformed and are joined by severely sheared, thinner sections of about the same length. Sullivan et al [74] attributed their occurrence in austenitic stainless steel to variations in the secondary shear zone (similar to stick-slip). Komanduri and Brown [70] suggested that, additionally, the negative stress-strain characteristics of two-phase materials (such as low-carbon steels) should have an effect on the primary shear zone. Ueda and Matsuo [75] observed that the speed at which sawtooth chip formation set in increased in the sequence: titanium alloy, titanium, maraging steel, stainless steel, low-carbon steel; at a given speed, sawtooth chip formation was promoted by a more negative rake angle. Large cutting force variations typical of segmental chip formation lead to fatigue failure of the tool.

c. An acute form of shear localization is observed in materials of low heat conductivity. The process starts by upsetting ahead of the tool, resulting in localization of shear. Heat generated in the shear plane cannot dissipate, the material of the shear plane weakens, and shearing continues until a chip segment is removed. The process then starts again by upsetting (Fig.11.13) [76]. At higher speeds the segments become smaller and finally fall apart. The average chip-thickness ratio is close to unity, giving the false impression of low friction and a large shear angle, whereas in reality the shear angle is very low, the chip thins out by flattening, and friction (or the secondary shear zone) plays a secondary role, as pointed out by Komanduri [76] for titanium.

d. The classical form of discontinuous chip, studied extensively by Field and Merchant [77], is typical of cutting of ductile materials at very low speeds that generate low homologous temperatures. Severe strain hardening of the material ahead of the tool edge coupled with high friction on the rake race and under the clearance face lead to

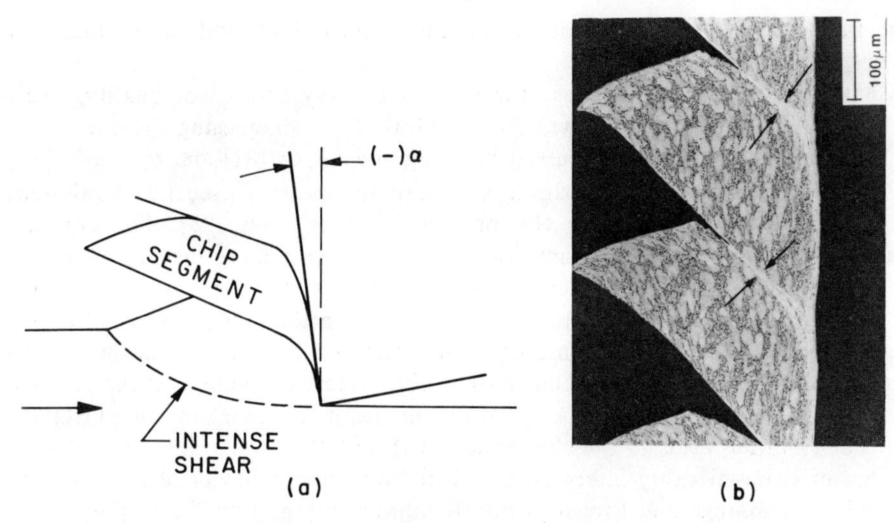

Fig.11.13. Schematic illustration of (a) shear-localized chip formation and (b) chip produced in machining of alloy Ti-6Al-4V (arrows point to intense shear bands) (courtesy of Dr. R. Komanduri, General Electric Co.) [76]

periodic cutting. Again, material is piled up in front of the tool by a process resembling upsetting, until sufficient strain is accumulated to initiate shear. Elastic members in the system (e.g., the tool holder) allow sudden acceleration and the separation of a chip, to be followed by a new upsetting cycle. Thus the chip is ragged, the undeformed chip thickness is variable, and cutting forces fluctuate violently. The new surface is highly uneven. Furthermore, the new surface is formed not by shear but under the influence of high surface tensile stresses. As pointed out by Williams [3.111],[78], a severe deformation zone develops in which shear-tear fracture and tensile separation alternate. This additional (third) zone of deformation accounts for the large cutting energies observed under these conditions. The surface finish of the workpiece is poor, being full of cracks and embedded particles detached from the chip. The speed at which such behavior is observed is dependent on temperature and adhesion. With ductile steels the third zone is typical of cutting below 15 m/min. Shouckry [79] found it necessary to cut aluminum at -190 C to obtain embedded particles; Williams [78] observed the ragged surface in room-temperature cutting of aluminum at speeds below 7 m/min. Similar observations were made by Sullivan et al [74] and by Makino et al [80] in cutting of 60/40 brass. Copper, although it has a higher melting point, does not form cumulative pickup, and it is perhaps for this reason that the ragged third zone disappeared at speeds in excess of only 5 m/min [78]. Of course, a previously strain-hardened material takes less deformation before fracture and thus may not show this kind of discontinuous chip formation.

e. Discontinuous chip formation is intentionally induced in some free-machining alloys by the incorporation of inclusions or second-phase particles that serve as stress intensifiers and cause total separation of tightly spaced chip fragments. The second-phase particles or inclusions often reduce the shear strength in the secondary shear zone and thus serve as built-in lubricants. Materials of very low ductility (such as the hexagonal zinc and magnesium alloys) and two-phase materials with brittle second phases (such as gray cast irons) also form discontinuous chips. The differences with this class of materials are that the spacing of shear zones is very close, frictional contact length on the rake face is short, and thus the shear angle is high and the surface finish is excellent.

Chatter is undesirable because the surface is wavy or of poor quality, tolerances may be lost, and the tool may even break. Methods of diagnosing the sources of and controlling chatter have been discussed by Kegg and Sisson [81] and by Hoshi [82].

As in other processes, application of ultrasonic vibration (Sec.3.9.2) can eliminate chatter, in addition to affecting the process in other ways [83]. The cutting ratio increases, indicating lower rake-face friction. The effect is more marked at low cutting speeds [84]. Conceivably, the interface becomes accessible to the atmosphere or lubricant, although flank wear seems to decrease only when the tool is not lifted clear of all contact [85]. Tanaka et al [84] interpret their results on the atomic scale: dislocations emerging at the surface of the chip reduce the effective dynamic shear stress in the shear zone, and thus ultrasonic energy results in strain softening. In some instances it is found that vibration leads to the elimination of the BUE.

Chatter is inextricably interwoven with the dynamic behavior of machine tools, as discussed by Armarego and Brown [5], by Boothroyd [9], and by Tlusty [86].

11.1.4 Effects of Process Geometry

It will now be necessary to consider the interactions of process geometry with frictional conditions.

Effect of Rake Angle. In principle, the relationship expressed in Eq.11.6 holds under realistic cutting conditions: the shear angle ϕ decreases, the cutting energy increases, and the chip thickens with decreasing (or negative) rake angle α and increasing friction angle ψ. In view of the preponderance of sticking on the rake face, the friction angle is best regarded as a measure of the sticking contact length [28], which in turn is governed by the rake angle and by material properties.

For a given material, the tool/chip contact length increases with a diminishing rake angle. When the rake angle becomes highly negative, cutting gives way to bulging. The critical angle of transition is greatly dependent on material properties, the stress state, and the stiffness of the tool/workpiece system. Abdelmoneim and Scrutton [87] found the transition at -65° in cutting, while Komanduri [88] was still able to obtain a chip at -75°. At the point of transition, the rake-face friction force reverses because of the reversed material flow (Fig.11.14) [88]. The critical angle can be calculated from theory [2.12],[3.112]. Slip-line fields for negative rake angles have been given and reviewed by Abebe and Appl [89].

The higher normal pressures generated with a tool of smaller rake angle promote the formation of a BUE. Thus, at a given cutting speed, the chip form may change greatly according to rake angle, as shown in Fig.11.15, taken from the work of Makino et al [80]. With increasing speed, the boundaries move to lower rake angles.

Effect of Flank-Face Contact. Zone 3 in Fig.11.3 symbolizes the frictional contact between the clearance (flank face) and the newly formed workpiece surfaces. In cutting an ideal rigid-plastic material with a perfectly sharp tool, contact would be limited to a line. In reality, the workpiece material springs back elastically, the tool edge has a finite radius, and rubbing along the new workpiece surface soon develops a small wear land. Therefore, the effect of flank-face friction can be systematically studied with tools having controlled-length artificial wear lands, as surveyed by Bailey [21].

When sliding prevails under the high normal stresses encountered over the rake face, the likelihood is that sliding can be maintained also over the flank face. In cutting

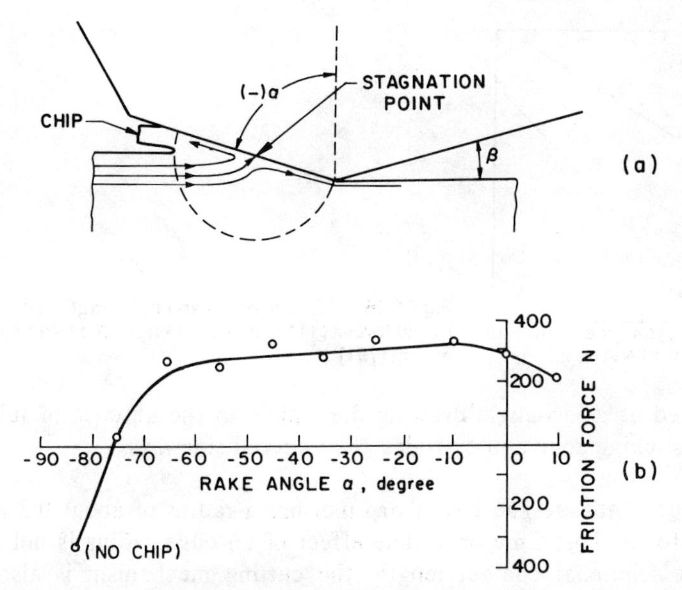

Fig.11.14. Effects of rake angle on chip formation in orthogonal cutting: (a) material flow with negative rake angle and (b) magnitude of frictional force (carbide tool, 0.14% C steel, $t = 0.01$ mm, $w = 3.7$ mm, $v = 1080$ m/min) [88]

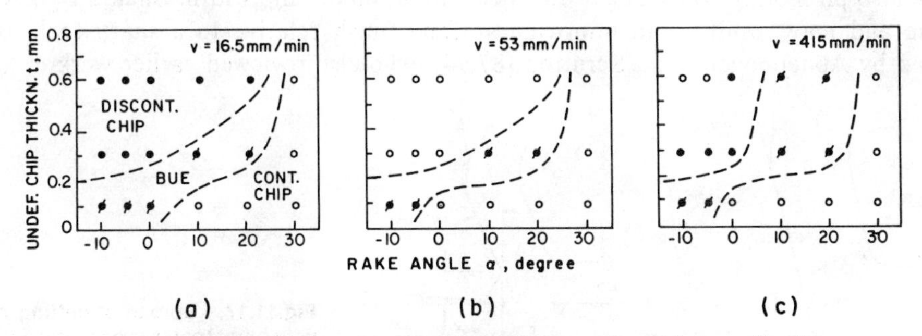

Fig.11.15. Effects of rake angle and cutting speed on chip formation in cutting of 60/40 brass with HSS tools [80]

in the presence of a BUE, the freshly formed workpiece surface is in contact with the bottom surface of the BUE and the tool flank is protected; only when the BUE is reduced is flank contact established. Even then, sticking friction is likely to prevail, and shearing takes place in what is sometimes described as the tertiary shear zone, the length of which is given by the land length. A stable shear zone may develop at higher speeds (temperatures), and there is evidence that sliding and subsurface flow can occur simultaneously [90]. Thus, in general, it is found that both cutting and thrust force components increase with increasing land length (Fig.11.16) [91]. Indeed, photoelastic studies showed a higher average μ on the wear land than on the rake face [92].

Flank profile (and its change through wear) is most important. However, compared with the rake face, the flank face is easier to lubricate. It has a positive clearance angle, it is relatively accessible, and, at least at low cutting speeds, the real area of contact and adhesion can be reduced [93]. The importance of positive clearance was demonstrated by Thomsen et al [91]; a negative clearance of only 1° destroyed the

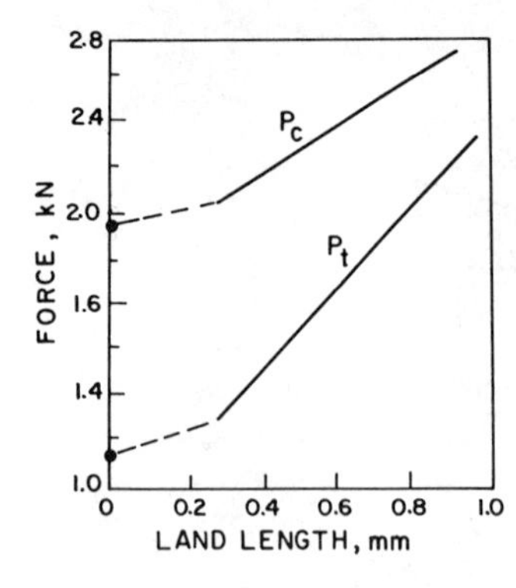

Fig.11.16. Effects of wear land length on cutting and thrust forces (1112 steel; rake angle, 0°; f=0.25 mm; v=117 m/min) [91]

relief and acted as a low-angle drawing die which, in the absence of lubricant access, established a sticking zone and extruded the material sideways.

Effect of Edge Radius. The best sharp tool has a radius of about 0.3 μm but retains its sharpness for a short time only. The effect of an edge radius is not simply that of increasing the frictional contact length: the cutting mechanism is also affected. On progressing along the rake face toward the edge, the effective rake angle diminishes and finally becomes negative (Fig.11.17a). Continuing along the radius, we first find a negative clearance angle, and only later is the necessary positive clearance established. Both negative rake and negative clearance angles promote the formation of a BUE while also promoting transverse flow in the direction of chip width. Such a BUE can be stable and may result in an improved surface finish relative to a sharp tool, as confirmed by Abdelmoneim and Scrutton [87,94], who also reviewed earlier work.

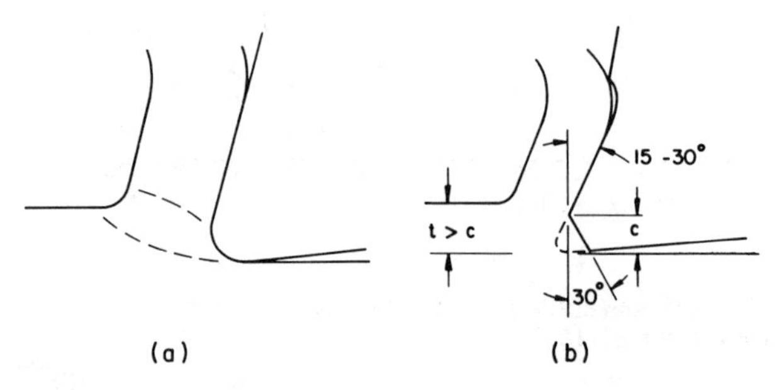

Fig.11.17. Spread of cutting zone with blunt tool (a) and cutting with a tool having a chamfered edge (b) [95]

The stable BUE is exploited in cutting of steels with hard (WC or harder), double-rake tools. The stable BUE becomes the cutting edge (Fig.11.17b), and some of the heat is carried away by a secondary chip that flows out sideways, as discussed by Hoshi [95]. The effective rake angle is large and the chip becomes brilliant white in the absence of high-temperature oxidation, and hence the name "silver-white-chip cutting" is given to this process.

Size Effect. It has long been known that thinner cuts take proportionately more energy and larger cutting forces. This phenomenon has been studied in great detail by a number of workers [3,49,94], and has been reviewed also by Bailey [21].

In considering the effects of undeformed chip thickness, it is best to visualize the events that occur when a tool of finite radius is made to cut gradually increasing undeformed chip thicknesses. First, contact takes place on the flank face between asperities (Fig.11.18a), corresponding to sliding between rough surfaces. Then contact becomes equivalent to superficial ironing of a semi-infinite body with a tool of low half-angle

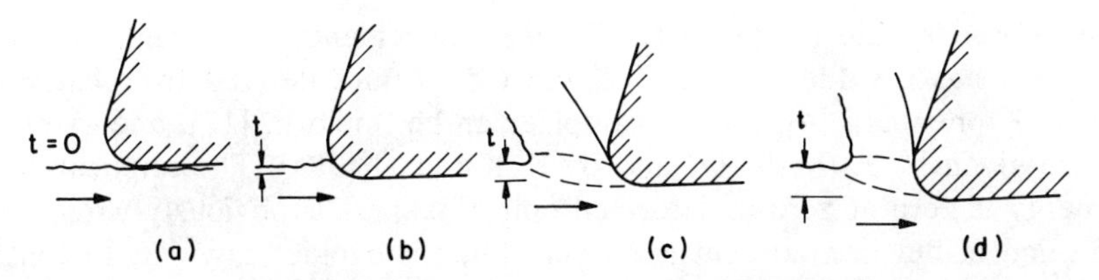

Fig.11.18. Transition from rubbing to cutting on increase of the undeformed chip thickness in cutting with a rounded-edge tool

(highly negative rake angle) (Fig.11.18b). A bulge forms, and energy is expended (more for a heavily strain-hardened material), but no material is removed. At some critical penetration, cutting begins with a negative rake angle (Fig.11.18c). It is in this regime that theories of metal cutting overlap with theories of friction; the shear plane is long, friction losses on the flank face are high, and the energy expended is high. A stationary metal cap may form on the tool edge, and the thrust force P_t increases; often $P_c/P_t \leq 1$, as shown by Sarwar and Thompson [96]. When the undeformed chip thickness becomes equal to the tool radius, the effective rake angle is zero at the surface (Fig.11.18d). Finally, the normal cutting mechanism is established.

In cutting with small undeformed chip thicknesses, plastic deformation of the newly formed surface absorbs a substantial proportion of the total energy consumed, particularly in a strain-hardening material (zone 5 in Fig.11.3). As shown by Larsen-Basse and Oxley [97], strain rate in the primary shear zone increases with decreasing t, and this contributes to an increased cutting force in strain-rate-sensitive materials.

Forces and Energy Requirements. For practical, approximate calculations of forces and energy requirements, it is sufficient to assume that the nominal cutting stress (specific cutting pressure) p_c is largely unaffected by tool geometry and friction, and can be estimated from an experimentally determined cutting force P_c acting on the undeformed chip of tw cross-sectional area:

$$p_c = \frac{P_c}{tw} \qquad \left(\frac{\mathrm{N}}{\mathrm{m}^2}\right) \tag{11.7}$$

The normal (thrust or feed) force P_t may be taken, to a first approximation, as one-half of P_c [25]. Neither groove- nor obstruction-type chip breakers affect P_c greatly [98].

Since work (or energy) is force (P_c) multiplied by the distance (l) over which it acts (Eq.2.13), and the volume of material removed is $V = twl$, the energy consumed in removing a unit volume of material—called the specific cutting energy E_1—is

$$E_1 = \frac{P_c l}{twl} \qquad \left(\frac{\mathrm{J}}{\mathrm{m}^3} \text{ or } \frac{\mathrm{W \cdot s}}{\mathrm{m}^3} \text{ or } \frac{\mathrm{N}}{\mathrm{m}^2}\right) \tag{11.8}$$

It will be noted that, when expressed in consistent units, the numerical values of p_c and E_1 are interchangeable.

The effect of undeformed chip thickness can usually be represented by a simple relation of the type

$$E = E_1 \left(t/t_{\mathrm{ref}}\right)^{-c} \tag{11.9}$$

where the reference chip thickness t_{ref} is some convenient value (such as 1 mm), and the exponent c has a value between 0.1 and 0.5. Values derived from early work are discussed by Kronenberg [4]; a few examples can be found in [12], and extensive compilations are given by Althoff [99] and by König et al [99a]. The implication that the cutting energy is zero at zero undeformed chip thickness is obviously wrong in light of Fig.11.18a and b, but in metal cutting a very low t is avoided anyway. In contrast, this aspect becomes very important in grinding.

A further complication, ignored in the above treatment, is spread of the material which results in the formation of a collar. This collar increases the effective depth (width) of cut in a subsequent pass and thus increases the cutting force and energy, especially with materials of high strain-hardening rate (such as annealed materials).

11.1.5 Temperature

It is evident from the preceding discussion that temperature and, particularly, its distribution have an even greater influence in machining than in bulk deformation processes. Through their effect on the shear properties of the workpiece material, they affect the chip-forming process itself and, through their effect on the tool, they determine the limits of the process and the mode of wear. This topic has been the subject of intensive theoretical and experimental studies, reviewed by Lenz [100], by Barrow [5.117], and by Redford et al [101]. Experimental techniques will be discussed in Sec.11.5.1. There is continuing activity, including approaches based on finite-element [32,58] and finite-difference [102] techniques and on electrical analogs [103].

A rough estimate of temperatures can be obtained by dimensional analysis [4,104], assuming that all energy (Eq.11.9) is converted into heat. Then the maximum temperature rise is

$$\Delta T = E\left(\frac{vt}{k\rho c}\right)^{1/2} \tag{11.10}$$

where k is heat conductivity, ρ is density, and c is specific heat of the workpiece material.

Because the cutting zone keeps moving into the workpiece, little heating ahead of the tool is to be expected and, at least at high cutting speeds, most of the heat (over 80%) is carried away by the chip. However, the tool is in continuous contact with the chip, and, in the absence of an effective heat-insulating layer, the rake face of the tool heats up. The higher rate of heat generation attained in cutting at higher speeds results in a higher temperature, which becomes the most significant factor in limiting the rate of material removal.

An important conclusion from both theory and experiment is that rubbing on the rake face or deformation in the secondary shear zone is a substantial source of heating, and therefore the maximum temperature is developed at the rake face, some distance away from the tool nose but before the chip lifts away (Fig.11.19) [105,105a]. Softening of the workpiece material in this zone and a reduced tool/chip contact length result in a drop of the apparent mean μ with increasing cutting speed, yet the higher over-all temperatures make conditions more demanding on the tool. Thus, wear may drastically increase while apparent friction drops. The importance of heat generation on the rake face is demonstrated in experiments by Chao and Trigger [106] and by Wright et al [107] with tools of restricted contact length. In these tools, the usual, slightly positive rake face is retained for only a short length (typically 0.5 mm), and thereafter the rake face is relieved at a much steeper angle. The tool is thus stronger than one

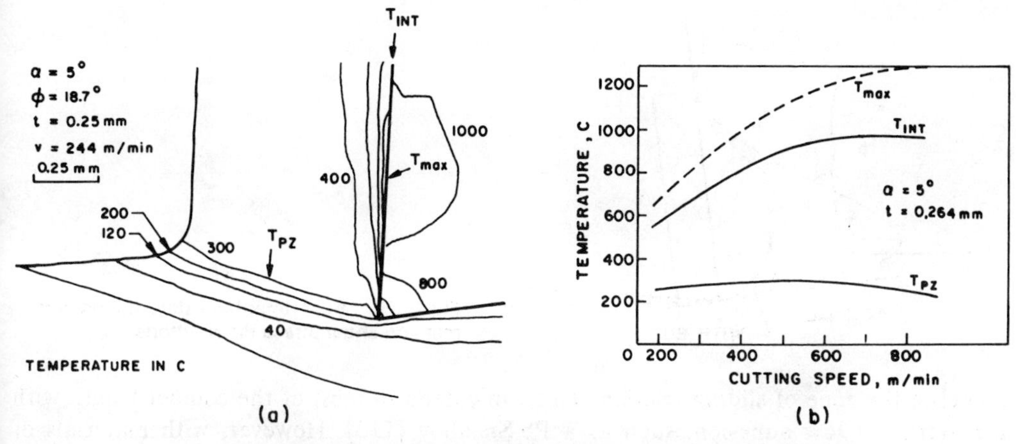

Fig.11.19. Calculated temperature distribution in chip and tool (a) and variation of temperature with cutting speed (b) (carbide tool, 1016 steel) [105]

with a large over-all positive rake angle, yet contact length is reduced and temperatures are some 30% lower in low-speed cutting.

The interaction of frictional conditions with material properties is again evident. The hottest point in cutting of nickel alloys is at the edge [8,108] because of adhesion and because the chip loses more heat to the tool than it gains by rubbing against it. With other materials (particularly steels), the hottest point is away from the edge.

A yet further source of heat is friction on the flank face [109].

Both frictional conditions and workpiece and tool thermal properties affect the partitioning of heat among chip, workpiece, and tool. In general, the greatest proportion of heat is carried away by the chip. A tool of low heat conductivity localizes heat in the interface or secondary shear zone [110]. The effects of tool geometry, process conditions, and metallurgical variables on interface temperature have been reviewed by Levy and Thompson [111].

11.1.6 Stress Distribution

In view of the prevalence of sticking or partially sticking friction, the simple sliding friction model illustrated in Fig.11.2 is seldom valid. It has been known and confirmed by the detailed studies of Childs [112] that the shear stress on the rake face affects chip thickness, contact length, and chip curvature. These, in turn, are affected by the presence of lubricants or coolants, and they affect the wear of the tool. Thus, knowledge of the stress distribution is of great importance from the tribological point of view.

Of the techniques discussed in Sec.5.5.3, only a few can be applied here. Photoelastic techniques are obviously useful [112a,113] but are limited to cutting of soft metals such as lead, and contact with the photoelastic polymer tool is hardly representative. Cutting of a polymeric workpiece with a steel tool allows photoelastic analysis of stress distributions in the workpiece [114]. Segmented split tools disturb the rake face but can be used under more realistic conditions [115].

Earlier measurements are in essential agreement with the stress distribution proposed by Zorev [51]. The normal stress rises from zero at the point of chip separation roughly according to a power law (Fig.11.20). Extrapolation of measured values gives a normal stress maximum at the tool edge; more recent measurements [115,116] indicate that there should be a decline (broken lines in Fig.11.20). The shear stress, too, rises more or less according to a power law, and thus a constant μ is a quite reasonable

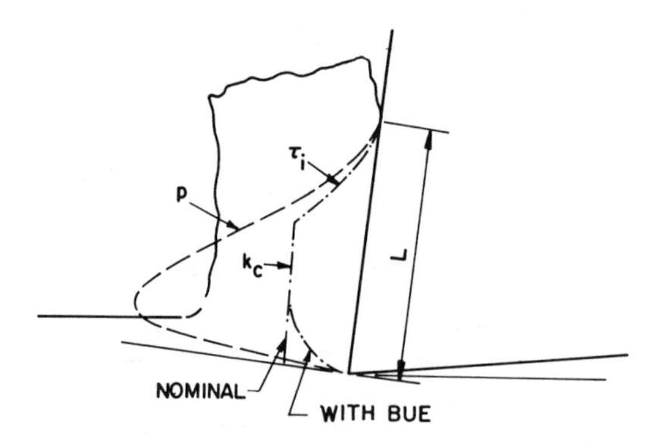

Fig.11.20. Experimentally determined normal and shear stress distributions.

model in the zone of sliding friction. This can extend to most of the contact length with a material of low adhesion, such as a Pb-Sn alloy [115]. However, with materials of higher adhesion (e.g., copper and aluminum), the shear stress cannot rise further once the shear flow stress of the chip material k_c is reached in the zone of sticking friction (although, surprisingly, the shear stress exceeded the normal stress in cutting of aluminum in the work of Kato et al [115]). Close to the tool edge, in the stagnant zone, the shear stress should actually drop below k_c, just as in deformation processes (e.g., Fig.6.3b). This has indeed been found in photoelastic studies [116], but not with split tools [115]. In the presence of a BUE, the shear stress should drop even further.

Obviously, neither an average μ nor a constant m value can describe the interface in its entirety, and it is for this reason that we called the ratio of frictional and normal forces (Eq.11.4) the apparent mean μ. Since the sliding zone is typically only 30 to 40% of the total contact length, a reasonable description of the interface is possible by some value of k_c, which must be about 30% lower than the true k_c.

There is much greater uncertainty regarding the stress distribution in the shear zone, which must affect the chip-formation mechanism and thus possibly also the actions of lubricants. Shaw [3.223],[116a] proposed that the normal stress should reach a limiting k value (which is now the shear flow stress of the essentially undeformed material) at some distance in from the surface.

11.1.7 Machining Theories

The simple shear theory (Sec.11.1.1) cannot describe cutting with sticking friction on the rake face.

Rowe and co-workers [57,117] extended the minimum-energy theory of Merchant [26] by including the work expended in the secondary shear zone. The contact length L is taken as a χ multiple of the undeformed chip thickness, and has to be determined experimentally. To allow for the presence of a sliding zone, the average value of k_c is taken as βk_c, where β is zero for perfect lubrication and unity for sticking friction. For a given rake angle α, the shear plane angle ϕ is thus obtained from

$$k \cos \alpha \cos (2\phi - \alpha) = \beta \chi k_c \sin^2 \phi \tag{11.11}$$

This shows clearly the importance of friction. When the product $\beta\chi$ is low, the primary shear plane energy dominates events and the shear angle is high (Fig.11.21) [117]. When $\beta\chi$ is large (broken lines), the shear angle is low and the total energy dissipated is high. In an extension of this theory, strain hardening is included but strain rate is ignored because, to a first approximation, its effect is cancelled by the effect of temperature. Wright [118] took into account strain hardening, and Wright et al [118a] con-

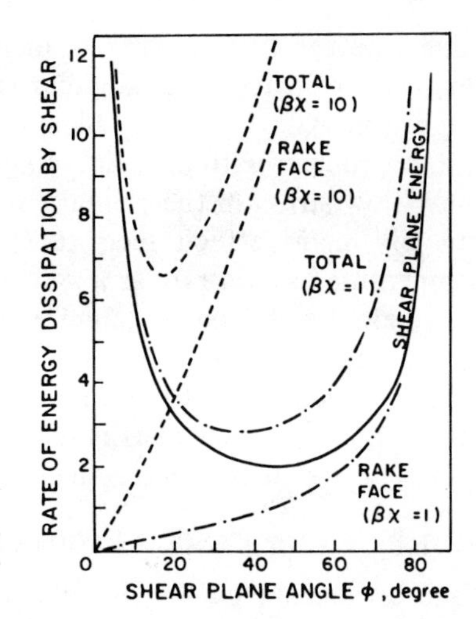

Fig.11.21. Effects of tool/chip contact length ($L = \chi t$) and interface shear strength ($\tau_i = \beta k_c$) on energy consumed in machining [117]

sidered the effects of friction on the shear angle. Theories based on the minimum energy principle have been formulated also by DeChiffre [28] and by Dautzenberg et al [119].

In a series of papers, Oxley and his co-workers [3.113],[32] developed a theory in which the shear flow stress k_c is allowed to vary with strain hardening, strain rate, and temperature. The thickness of the secondary shear zone is assumed to take a value that results in a strain rate and temperature that minimize k_c and thus also frictional work and total work. The shear angle is then found where the calculated interfacial shear stress $\tau_i = k_c$ (Fig.11.22)[32a]. The calculated thickness of the secondary shear zone (expressed as a fraction of the chip thickness t_c) is found to increase with strain hardening, strain-rate sensitivity, and undeformed chip thickness, and to decrease with increasing cutting speed (or temperature), with rake angle, and, in steels, with increasing carbon content, all in agreement with expectations. A material whose k_c drops greatly with temperature will exhibit a thin secondary shear zone. In an application of this theory, the conditions for the formation and disappearance of the BUE could be predicted. Oxley [3.113] suggests that near-seizure conditions, rather than complete sticking, are more likely to occur at the interface, and Challen and Oxley [3.112] constructed a slip-line field showing that a soft asperity is retarded on riding against a hard asperity; thus, full seizure on the rake face would not be necessary. Most likely, adhesion between tool and workpiece and the fracture strength of the chip material deter-

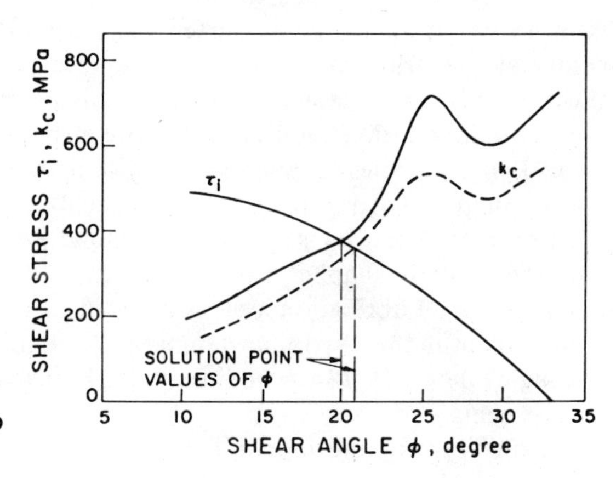

Fig.11.22. Effect of strain hardening of chip material on calculated shear angle [32a]

mine whether full sticking is indeed attained. A rougher tool, especially one with a roughness oriented perpendicular to chip flow, arrests flow, and adhesion to the material trapped in valleys is highly likely.

A problem with all machining theories is the difficulty of determining the relevant shear flow stress in the primary shear zone (k) and the secondary shear zone (k_c). Fortunately, many effects seem to cancel. For more accurate calculations, a temperature-compensated strain rate or a velocity-modified temperature can be used. The flow stress back-calculated from machining data is often in good agreement with extrapolation from compression tests. Indeed, orthogonal machining has been suggested for flow stress determination at high strain rates. The problem of reconciling flow stress data from machining and other tests is discussed by Finnie [120], Childs [121], Wright and Robinson [122], and by Holzer and Wright [122a].

11.1.8 Other Cutting Processes

True orthogonal cutting is encountered only in facing of the end of a tube. In most other processes oblique cutting is practiced: the cutting edge is at some angle of inclination to the direction of motion, and the chip flows up in a helix rather than a spiral. The normal (oblique) rake angle is more positive and is defined by the cutting-velocity vector and the chip-flow direction [1]. Feed is the distance between successive engagements of the cutting edge (or edges) and is somewhat larger than the undeformed chip thickness.

From the tribological point of view, the most important factor is the length or duration of engagement between tool edge and workpiece.

a. Turning and boring of cylindrical surfaces can be regarded as steady-state processes involving continuous engagement at a constant speed, characterized by the development of equilibrium temperatures and very limited access of lubricant. The undeformed chip thickness is only a little less than the feed per revolution. The depth of cut corresponds to the chip width.

b. Facing with a single-point tool, form turning, and parting off are similar to turning except that the velocity diminishes as the tool moves toward the center of the workpiece. The undeformed chip thickness is approximately equal to the feed.

c. In gun drilling and trepanning there is continuous engagement at a constant speed, but contact on the guide and side-relief surfaces creates substantial sliding friction for which lubrication is most beneficial.

d. Drilling with a twist drill or spade drill represents a highly complex situation. Conditions are similar to orthogonal cutting along the cutting edges, although Law et al [123] found from photoelastic studies that shear stresses and temperatures reach maximum values at about three-quarter distance to the outer corner. The chisel point in the center of the drill penetrates the workpiece by indentation and accounts for the high thrust forces encountered. Rubbing at this point and along the margins (side surfaces) creates a lubrication requirement quite distinct from the cutting process. The undeformed chip thickness is approximately half the feed per revolution.

e. Shaping (with a linearly moving tool) and planing (with a linearly moving workpiece) present conditions similar to turning, except that tool contact is now interrupted and, especially in shaping, can be short enough to prevent the development of maximum temperatures. Lubricant access is allowed to the rake face between successive engagements. Even if the maximum temperature builds up, the tool is cooled during the noncutting cycle [124]. If a nonadhesive BUE develops, it may drop off between successive engagements.

f. Fly cutting and face milling are similar to shaping except that the length of cut is usually shorter and the undeformed chip thickness keeps increasing from the beginning to the end of each engagement.

g. Slab milling and end milling with a peripheral cut differ in several aspects from the processes previously discussed. The length of engagement for each tooth is short. The undeformed chip thickness changes from zero to some maximum value determined by the chosen value of feed per tooth. In down or climb milling the chip is thickest at engagement and tapers out to nothing; in up milling the cut begins at zero chip thickness and ends at maximum chip thickness (Fig.11.23). Thus, during these operations the entire range of conditions shown in Fig.11.18 is traversed; cutting begins only at some critical undeformed chip thickness which decreases with increasing cutting speed, as shown by Okamura and Nakajima [125].

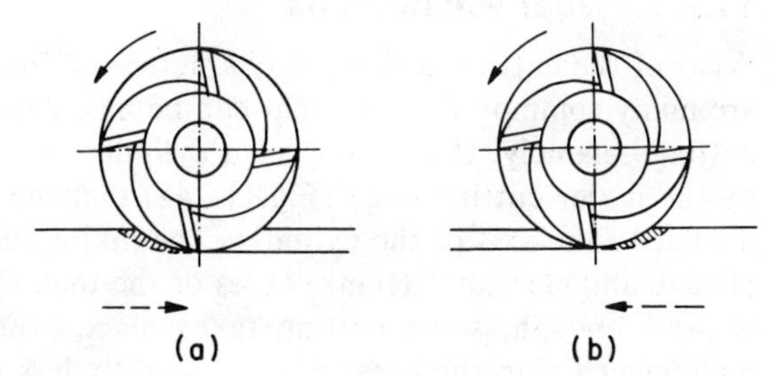

Fig.11.23. Schematic illustrations of slab milling: (a) down (climb) and (b) up (conventional) milling (after [12])

h. The advantages of periodic tool engagement can be secured in turning by the use of a rotary cutter [1]. The tool is disk-shaped and is mounted with its axis inclined to the direction of workpiece rotation. It is rotated by a motor [126] or is self-propelled [127] (driven by virtue of the kinematics of the process). Because the point of engagement keeps moving around the circumference, the tool is allowed to cool and lubrication becomes more feasible.

i. External thread cutting is essentially a form-turning process. Internal thread cutting with a tap is similar in characteristics except that several tool edges are simultaneously engaged, and undeformed chip thickness is determined by tap design. A convenient feature is that the tapping torque is easily measured and is a direct measure of the sum of cutting forces. Threads may be formed also by plastic deformation. Thread rolling on an external surface may be regarded as an example of shape rolling, and lubrication practices are similar to those described in Chapter 6. Cold form-tapping of a hole is a progressive indentation process in which sliding of the tool creates severe interface conditions, and effective lubrication is essential both for reducing the torque and for ensuring adequate surface quality [128].

j. Broaching is similar to tapping in that the undeformed chip thickness is determined by tool design. However, the motion is linear, and speeds are usually low enough to allow (and necessitate) efficient lubrication. In sawing, t is determined by the feed rate, but, in common with broaching, t may be low enough to bring the process into the range where $P_c/P_t \leq 1$ [96].

11.2 Wear in Cutting Processes

In plastic deformation processes, concern over wear is often overshadowed by considerations of forces or material flow (Chapters 6 to 10). Except for hot extrusion, die lives are measured in hours or days, or in thousands of parts. In contrast, tool wear is the dominant concern in metal cutting. Process conditions are chosen to give maximum economy or productivity, often resulting in tool lives measured in minutes or, at the most, hours. Central to the problem are: high tool temperatures, which lead to softening of the tool material; intense sliding between tool and virgin workpiece material, which promotes adhesive and abrasive wear; intimate contact that allows diffusion of elements from the workpiece into the tool, and vice versa; greatly enhanced tribochemical reactions; high tool forces and pressures, which can cause plastic deformation or fracture; and thermal cycling in interrupted cuts, with consequent cracking due to thermal fatigue. Thus, all mechanisms discussed in Sec.3.10 may be active alone or, more frequently, in combination. Identification of the dominant mechanism is far from simple, and most interpretations are subject to controversy. Reviews have been given by a number of workers [3.348,3.354],[5,8,129-134].

11.2.1 Wear Phenomena

Wear of the cutting tool takes characteristic forms in various processes [131], but there are many common features that can be best described by reference to a form of oblique cutting—namely, the turning of a cylinder. The task of cutting is performed primarily by the major cutting edge (Fig.11.24a) inclined at an angle C_s to the direction perpendicular to the axis of the cylinder. The major cutting edge is formed by the intersection of rake and clearance (flank) faces of the tool; the tool nose is radiused, and, if the feed is large enough, some cutting takes place over part of the minor cutting edge. The undeformed chip thickness t is just slightly less than the feed f, and the chip width w is slightly greater than the depth of cut d (Fig.11.24b). Some workers use the chip equivalent q (the length of the engaged cutting edge divided by the area of cut: $q = L/A$ or $q = L/fd$) to normalize observations with respect to the geometry of the cut. In the following discussion, undeformed chip thickness will be equated with feed f, and the symbol t will be reserved for tool life.

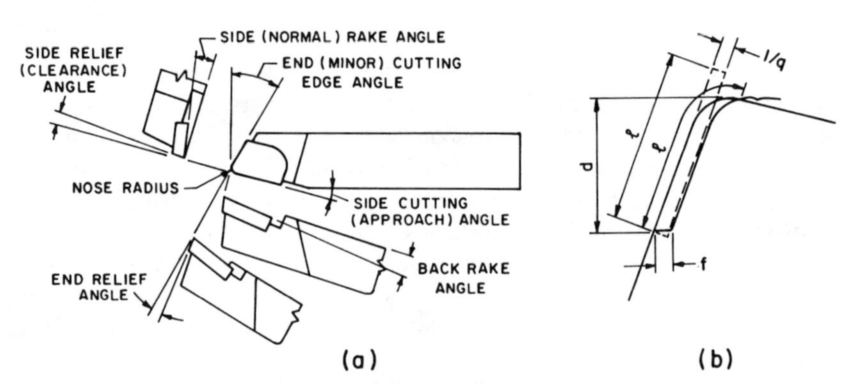

Fig.11.24. Turning: (a) terminology adopted for describing the geometries of turning tools (all rake and relief angles measured in normal directions; British terminology shown in parentheses) [after 12]; and (b) chip equivalent [after 135]

Flank Wear. The clearance face of the tool is subjected to intense rubbing against the freshly formed surface. Very quickly, a wear land forms which, in the simplest case, has a well-measurable average length WB (Fig.11.25) [135]. It tends to progress rapidly in the first few seconds of cut, then settles down to a lower, steady-state rate over the useful life of the tool, and accelerates again after some critical time has elapsed (Fig.11.26a). With many tool materials the rate of wear increases with removal rate (speed and feed), indicating that temperature (Eq.11.10) is a powerful factor, most

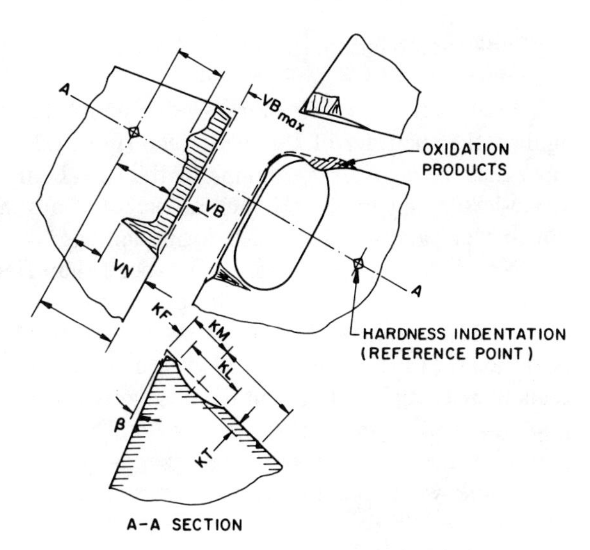

Fig.11.25. Characterization of flank and crater wear (after [135])

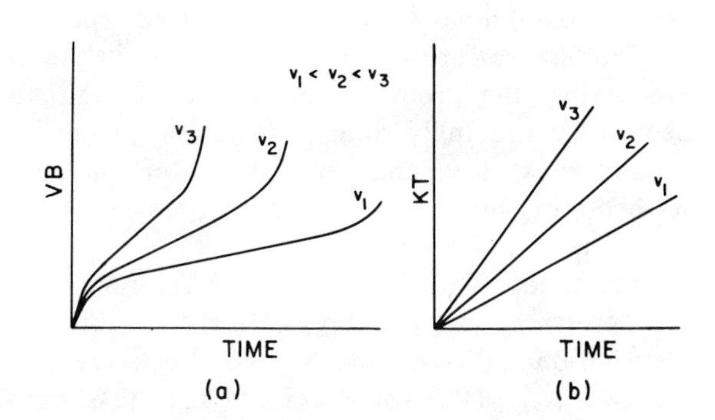

Fig.11.26. Characteristic progression of (a) flank wear and (b) crater wear

likely through weakening of the tool, but perhaps also through promotion of adhesive wear. Nevertheless, temperatures and pressures are relatively low, and the dominant wear mode is likely to be abrasion by fragments of the BUE or by hard constituents in the workpiece material. The latter was shown by Ramalingam and Wright [136] in experiments with workpieces of controlled inclusion content produced by powder-metallurgy techniques. The microcutting mechanism is similar to lapping, with hard inclusions acting as abrasive grit embedded in the softer matrix. Some flank wear models assume thermally activated and purely mechanical components [137]. A model based on adhesive wear has been proposed by Rubenstein [138], who has also reviewed much of the prior experimental work. Other workers (e.g., Trent [8]) suggest that diffusion could also be a significant contributor. No general model is likely to exist since interactions between tool and workpiece are specific. Thus, a titanium-stabilized stainless steel (AISI 321) gave higher wear (most probably through abrasion) than one (AISI 304) without hard particles; adhesive wear may have dominated with the latter.

The steady-state wear rate is usually quite reproducible but the initial rate is variable, especially with cemented carbide tools. Ber and Kaldor [139] have attributed this to microchipping of the edge, and have found that honing a radius of 0.01 to 0.02 mm reduced scatter. In interrupted cutting (as in milling), flank wear appears to be a function of the total number of cutting cycles [140] rather than of total engagement time.

Flank wear must be limited, because it results in a loss of dimensional tolerances (unless, of course, the tool is repositioned), in a poorer surface finish [141], in increasing flank face friction, in the development of a negative clearance angle, and, finally, in edge destruction.

Frequently, a notch or groove of depth *VN* forms at the line where the tool rubs against the shoulder of the workpiece (Fig.11.25). Hard scale on the surface is an obvious cause, as is strain hardening of the surface in a previous cut, especially with heavily strain-hardening materials such as nickel alloys. Side flow [142] and burr formation in the first cut and sawtooth chip formation aggravate notch wear [143,144].

Notch wear is less sensitive to speed than flank wear, suggesting that factors other than temperature must come into play, too. Indeed, this is the zone where the atmosphere and cutting fluid have access to the interface and reactions can take place. Ishibashi et al [145] observed heavy groove wear when cutting low-carbon steel with HSS tools or with an oil that contained an active E.P. additive. When oxygen is excluded by applying a plain mineral oil, a plant oil, or pure water, notching is eliminated. With an emulsion, the extent of grooving depends on the presence of reactive additives.

Flank wear and grooving wear at the minor cutting edge (the trailing edge) are primarily oxidative in nature. They are accelerated when the tool material oxidizes readily and are reduced when the workpiece forms stable oxides (and acts as a getter), or when oxygen is excluded by cutting in argon [146].

The above forms of notch wear are, in themselves, not particularly harmful. However, when the groove forms at the tool tip, cutting geometry is lost, the surface becomes exceedingly rough, machining proceeds with a negative clearance angle, and rapidly rising temperatures soon soften the tool to such an extent that the nose is washed away by plastic flow or even melting, and the tool is totally destroyed.

Crater Wear. The chip flowing up the rake face subjects the tool to intense temperature, pressure, and, possibly, sliding. Consequences depend on pressure and temperature distribution and on adhesion [147]. Tools are normally chosen to have a relatively high hardness (e.g., HSS for aluminum and brass, carbide for steel), so that the edge retains its shape. Since the maximum temperature usually develops farther back along the contact length (Fig.11.19), a characteristic crater forms (Fig.11.25). Crater depth *KT* increases linearly with time, at a faster rate for higher cutting speeds (Fig.11.26b), while the position of the deepest point moves farther away from the edge.

Several mechanisms have been proposed for crater wear; the review of Ramalingam [132] contains an extensive reference list.

a. The chip material, which is strengthened by high strain rates in the secondary shear zone, may drag out a thin film of the tool (especially HSS) which is weaker because it is subject to very low strain rates [48].

b. A sticking secondary shear zone or a stable BUE protects the tool, and cratering begins only where the chip slides (zone 2 in Fig.11.9); in this sense, adhesion is desirable. However, the BUE is usually unstable, and on collapsing it takes with it fragments of the tool if adhesion is high [8,134]. Such attritive wear can be rapid.

c. Adhesion can facilitate the movement of atoms across the interface, leading to diffusion. Stability of the tool material and rates of solution become important. Relative wear rates based on the solution of carbides in the chip have been calculated by Kramer and Suh [148]. Crater wear in machining of nickel can also be predicted from solubility considerations [149].

d. Abrasion of the thermally softened tool material by the chip is also feasible, because carbides of the tool material reside at high temperatures and thus lose much of their strength (Fig.11.27) [150,151], whereas hard inclusions in the chip are exposed to temperature for such a short time that their hardness is not diminished [136].

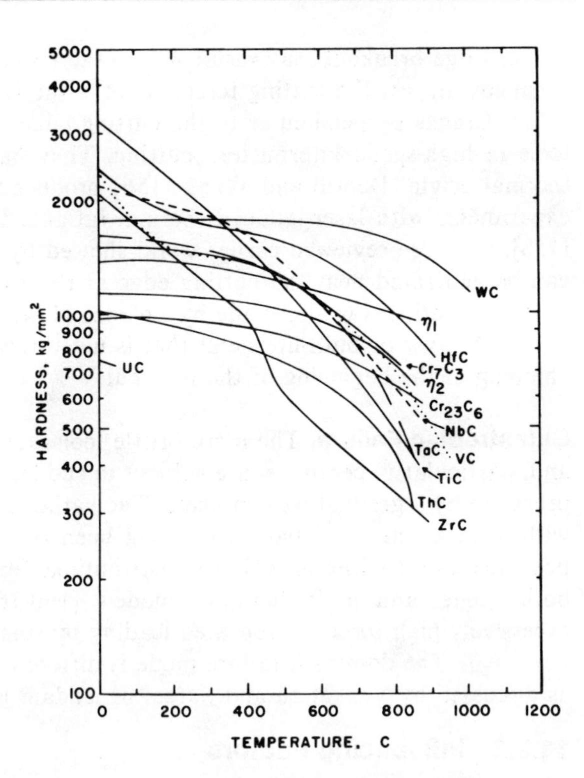

Fig.11.27. Decrease in hardness of carbides with increasing temperature (from [150], after Westbrook and Stover)

Again, no general mechanism is likely to exist. Whatever the mechanism, temperature and its distribution are controlling factors, and mathematical models of wear have been proposed purely on the basis of temperature, irrespective of the metallurgical processes [32,152].

Crater wear in itself is not damaging; indeed, it may even contribute to improved surface finish, and the BUE may disappear once the crater matches the chip curvature [129]. The shear angle may also increase as though the tool had a more positive rake angle. Since crater wear is greatest where sliding sets in on the rake face, tool life may be increased by grinding away that part of the tool [8].

A notch or groove may appear also on the rake face (Fig.11.25) as a result of thermal loading, abrasion, and impact by the side surface of the chip [144].

Edge Damage. The cutting edge may suffer damage from a number of causes; Höglund [153] enumerates 13 mechanisms.

a. If no BUE or secondary shear zone forms, the major cutting edge may become rounded as a result of abrasion in low-speed cutting.

b. The edge may be plastically deformed if pressures and/or temperatures are excessive. This damage is observed mostly at high cutting speeds and is controlled by the shear flow stress of the material relative to the hot strength of the tool, expressed as the plastic factor of safety by Loladze [154]. Edge deformation can create a negative clearance or interference.

c. Periodic break-off of the BUE may cause chipping of the edge, especially when adhesion is high and the tool is brittle.

d. Brittle tool materials of limited transverse rupture strength may suffer microchipping as a result of the tensile stresses generated. In continuous cutting the edge is normally compressively loaded. As discussed by Shaw [155], damage is most likely when the tool emerges from the cut and the edge is suddenly unloaded.

e. Edge breakout may result when crater wear weakens the tool to the point where it cannot support the cutting force, or when the BUE grows to an excessive size.

f. Cracks perpendicular to the cutting edge (comb cracks) are observed in carbide tools in high-speed intermittent cutting. They have long been recognized as having a thermal origin. Deotto and Wang [156] produced a simplified analysis on the basis of experiments with laser-induced thermal fatigue. Pekelharing [157] and Wu and Mayer [158], who also reviewed earlier work, showed by FEM calculations that tensile stresses can be generated near the cutting edge at the end of the noncutting period of a cycle. This observation was also made by Loladze [154] with photoelastic tools.

g. A layer of built-up metal that is not thrown off at the end of one cut can cause chipping at the beginning of the next cut.

Catastrophic Failure. The more brittle tools, such as those made of cemented carbides and, particularly, ceramics, are subject to sudden, catastrophic fracture, not necessarily preceded by a gradual wear process. The variability of tool life and the need for dealing with it on a statistical basis have long been recognized [159]; substantial efforts have been spent on finding probability distributions for the onset of fracture, on the basis of both single- and multiple-injury models [160-162], corresponding to failure due to excessively high forces or repeated loading by smaller forces, respectively.

The dominant failure mode is different for different classes of tool materials, as discussed by Venkatesh and Satchithanandam [163].

11.2.2 Influencing Factors

In addition to the metallurgical aspects of tool and workpiece materials, both tool geometry and process conditions affect wear.

Tool Geometry. Tool geometry affects not only the cutting process (Sec.11.1.4) but also wear.

a. A more positive rake angle reduces cutting forces and thus temperatures, increasing life. However, the mass of the tool decreases, the heat path between rake and flank faces shortens, and the clearance face heats up more, resulting in faster flank wear. Thus an optimum α is found (Fig.11.28a). A brittle tool is weakened and suffers catastrophic failure. Therefore, cemented carbide tools are made with rake angles no greater than $\alpha = 5°$, and ceramic tools with rake angles of $\alpha = 0$ to $-5°$.

b. A large side-cutting angle gives a larger effective rake angle and moves the two main points of heat generation farther apart, thus increasing life. However, for a given feed, it also reduces the undeformed chip thickness, increases the length of tool contact, and increases forces and energy requirements. Thus, an optimum angle may be found for a given set of conditions (Fig.11.28b).

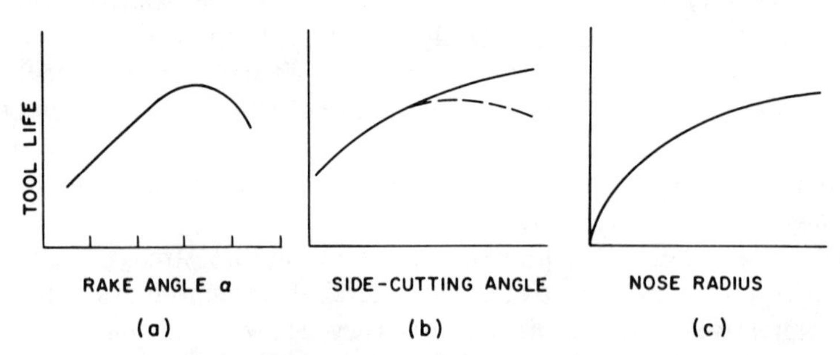

Fig.11.28. Schematic illustration of effects of tool geometry on tool life

c. A larger clearance angle gives less rubbing contact for a given flank wear, reduces temperatures, and increases life. However, an excessive clearance angle weakens the edge and promotes chipping. In dry cutting of low-carbon steel with HSS, Abdel-moneim [164] found tool life to increase with clearance angle, according to a power law, up to 8°.

d. A larger nose radius increases the heat capacity and thus the life of the tool (Fig.11.28c), and improves surface finish.

e. Edge hone or chamfer, while reducing chipping, increases cutting forces and temperatures, and thus reduces the life of carbide tools at high speeds but has less effect at low speeds [165]. It is essential when brittle tools are used.

f. As mentioned, relieving the tool to give a shorter chip contact length (restricted rake-face contact tool) reduces heating and increases tool life [106].

g. Groove- or obstruction-type chip-control devices ensure much easier chip disposal [98]. Their effect on forces, temperatures and wear is minor, unless they also change the effective rake angle.

Process Variables. In addition to the effects of lubrication and tool and workpiece materials, which will be discussed separately, there are a number of process variables that affect wear:

a. The geometry of the process is often different from that encountered in turning. Thus, edge and nose wear may be dominant on drills and milling cutters. Cutoff tools make contact at the corners as well as on the flank and rake faces, and corners of HSS tools fail by catastrophic softening [166]. A somewhat similar situation exists in form turning [167].

b. For a given lathe rotational frequency (rpm), the cutting speed changes in processes such as cutting off, form turning, taper turning, and facing. When a facing cut is started at the minimum diameter, cutting speed and thus temperature increase rapidly, making the process attractive for accelerated testing. Flank wear increases exponentially with time [168].

c. Intermittent engagement, as in milling, has conflicting effects [140] which depend on the dominant wear mechanism. At a given speed, a shorter engagement time and a longer out-of-cut time, that is, a lower duty cycle, result in lower temperatures and thus longer tool life. However, a lower duty cycle also increases temperature fluctuations and reduces tool life by subjecting the tool to thermal fatigue [169]. Thus, a higher duty cycle is beneficial, and higher cutting speeds can give longer tool life as long as the duty cycle is also raised [170]. In general, there is an optimum speed for a given cutter. Cutting fluids increase temperature excursions and are avoided with more brittle tools [155].

d. Increasing workpiece temperatures reduce the flow stress and cutting force but also increase the temperature of the tooling, and thus they are beneficial only if the tool has a high temperature resistance (such as a ceramic). For a HSS tool the optimum temperature may be below room temperature for some workpiece materials.

e. Vibrations are generally harmful, and longer tool life is to be expected with a rigid machine tool. Ceramics are particularly vulnerable. Cracks similar to those observed on ceramics develop also on CBN tool edges if the machine has insufficient stiffness, whereas wear is more like that of carbides in a stiff system [171]. Even polycrystalline diamond suffers a rounding of cutting edges in low-stiffness milling [171].

f. The electric current generated at the tool/workpiece contact can accelerate wear, although the effect depends also on the construction and condition of the machine tool which completes the current path. Thus, insulation of the workpiece from the machine, or the imposition of an external current to make the tool the cathode, is effective, as found by Axer and reported by Opitz [172], but not in all instances [173,174]. An extensive treatment of this subject is given by Postnikov [3.84]. On imposing a magnetic field, improved tool life was reported in hot machining [175] but accelerated diffusive wear was registered in cold machining [3.368].

g. The thermal conductivity of the tool has a controversial effect. Friedman and Lenz [176] found in cutting that forces and crater wear rate were lower with carbide tools of lower conductivity because the tool/chip contact length L was shorter. They surmised that the increased L values found with high-conductivity tools shifted the maximum temperature to the sliding zone where the tool suffers rapid wear. This is in agreement with the finding of Ber [177] that, in cutting of steel with carbide tools, the speed at which severe wear commenced decreased with increasing conductivity. In contrast, Matta et al [178] found that higher conductivity reduced chip curl and increased wear of ceramic tools. The effect seems to be highly specific. Kotval and Barash [179], starting from the premise that a cooler tool would increase life, plated some carbide tools with copper or silver for better cooling but found that wear improved much less than anticipated, for reasons not understood.

11.2.3 Prediction of Tool Life

Tool life affects the choice of tool, process conditions, economy of operation, and the possibility of automatic or computer control. Thus, beginning with the monumental work of F.W. Taylor [180], continuing efforts have been devoted to developing methods of predicting tool life. Full success has proved elusive for several reasons.

Tool life is not an absolute concept but depends on what is regarded as the end point at which the tool has to be reground or replaced.

a. In finishing operations, surface finish, surface integrity, and dimensional accuracy are of concern.

b. In roughing operations a poorer surface can be tolerated and the tool can be used until (or, preferably, until just before) catastrophic failure occurs, or until wear results in an unacceptable dimensional change.

c. The character of the chip changes in the course of tool wear, making the disposal of the chip difficult, or, if segmented chip formation occurs, vibrations may set in.

d. Cutting forces may reach unacceptably high values, resulting in tool fracture.

e. The rate of material removal may drop to uneconomically low levels.

f. Wear and/or metallurgical changes in the tool may lead to unacceptably high regrinding costs.

These criteria, while important from the operational point of view, reveal little about the condition of the tool, and, more conventionally, some value describing tool wear is taken as the end point of tool life [135,163].

a. Most frequently, the value of flank wear VB (Fig.11.25) is specified. Generally suggested values [181,182] are VB = 0.4 mm (or VN or VB_{max} = 0.8 mm) for carbide tools, and VB_{max} = 0.6 mm for ceramic tools. For HSS steel the values are typically VB = 0.75 mm for finishing and VB_{max} = 1.5 mm for roughing. The time elapsed (in minutes) to reach the specified wear land at a given speed, or the speed allowable for reaching the wear value in a certain time, is specified.

b. Crater wear is described as KT, or as KT/KM, or by the crater index $KI = KT/(KT + KL/2)$ [163]. Alternatively, the cross-sectional area perpendicular to the cutting edge is taken [183]. With carbide tools, crater wear tends to dominate over flank wear in the long term, and leads to edge failure. With coated tools the relationship between flank wear and crater wear may correlate very well, and then the time-consuming crater wear determination can be eliminated [184].

c. The mechanism of total tool destruction is different for different tool materials [163]. The noses of HSS tools are lost through overheating; carbides fail when the crater breaks into the end-cutting edge groove or the rake face notch breaks into the flank groove (Fig.11.25). Time to failure depends on the criterion adopted (Fig.11.29) [163].

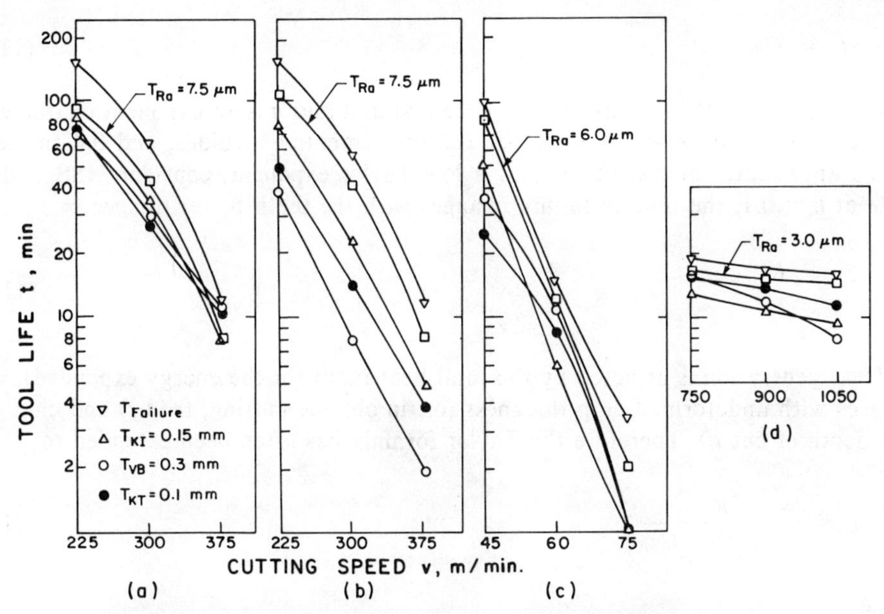

Fig.11.29. Effects of cutting speed on tool life according to various tool-life criteria in cutting of 0.25% C steel (UTS = 410 N/mm²) with (a) cemented TIC tool, (b) carbide tool of steel-cutting grade (ISO P10), (c) HSS tool, and (d) oxide ceramic tool (rake angles: +6° for HSS tool, −5° for other tools) [163]

d. The specific tool wear rate is sometimes used [185]:

$$V_s = \frac{1}{A}\frac{dV}{dl} \tag{11.12}$$

where A is area of wear, V is volume worn, and l is distance rubbed.

e. The total (flank and crater) wear volume has also been suggested as a sufficient measure of wear [186], at least for carbides for which volumetric wear life appears to correspond to the mean of the VB and KT lives. A metal removal ratio (similar to a grinding ratio; Sec.11.6.2) can then be specified as

$$\eta = \frac{\text{volume of metal removed}}{\text{wear volume}} \tag{11.13}$$

Oxidative wear of the minor cutting edge results in a deterioration of the surface, sometimes before flank wear reaches objectionable values [135], and then surface finish is a better indicator.

Opitz [187] suggested that all modes of wear can be expected to occur as a function of $vt^{0.6}$. Shaw [3.354] showed that temperature varies according to the square root of this group, and thus the classification is of thermal origin, a not unreasonable premise since most wear processes described in Sec.11.2.1 are accelerated by temperature.

There can, however, also be sudden changes, as when diffusion wear is activated at some critical temperature.

For practical purposes it is more convenient to retain speed as the primary variable. Then, tool life can be expressed in the form of the classical, simplified Taylor equation

$$vt^n = C \tag{11.14a}$$

where C is a constant for a given workpiece material and n is an exponent characteristic of the tool (typically $n = 0.1$ for HSS, 0.2 for cemented carbides, and 0.4 for ceramics). To appreciate the significance of a low Taylor exponent, consider HSS tools, for which, at $n = 0.1$, the time to failure changes with the tenth power of speed:

$$t = \frac{K}{v^{1/n}} \tag{11.14b}$$

Heat generation is affected by the total heat input (or the energy expended), which increases with undeformed chip thickness (or, in oblique cutting, feed f) and chip width w (or depth of cut d). Therefore the Taylor formula has often been extended to

$$t = \frac{K}{v^{1/n_1} f^{1/n_2} d^{1/n_3}} \tag{11.15}$$

where, usually, $1/n_1 > 1/n_2 > 1/n_3$. For example, typical values are $n_1 = 0.1$, $n_2 = 0.17$, and $n_3 = 0.25$ for HSS tools, showing that, for increased material removal rates, it is preferable to increase feed rather than speed. The exact values of the exponents depend also on workpiece material and cutting conditions, including lubrication.

This relationship is not an absolute one, as shown in the review and critique of Colding and König [135]. First, it depends on the criterion adopted (Fig.11.29) [163]. Second, the log-log plot of tool life against speed (and feed) is seldom truly linear but, if generated over a wide enough speed range, shows a curvature (Fig.11.29 and 11.30). Tool life may show a maximum where a BUE protects the edge, and may drop at yet lower speeds where sliding of the chip sets in [135,188]. Often, a break is observed in the plot where the dominant wear mechanism changes; the break is particularly noticeable with HSS that suffers very rapid wear when the secondary shear zone becomes too small to protect the rake face at high temperatures [188,189]. A slight break is sometimes seen with carbides [190].

As mentioned earlier, brittle tools, particularly ceramics, suffer catastrophic failure for which a Taylor plot is useless. Even with HSS and carbide tools, the Taylor constants must be regarded as statistical means [191] and are better developed by the method of weighted least squares [192].

The tool life relations are of vital importance for determining the choice of the most productive or most economical cutting conditions. Because $n_1 < n_2$, it is usually recommended that the heaviest cut be taken (within the limits of cutting forces, surface roughness, surface integrity, chatter, etc.). However, Colding and König [135] showed that the Taylor relation does not always lead to an optimum speed, necessitating other, analytical approaches. The important conclusion for our purpose is that, when a tool is pushed to the high end of its speed range, small increases in speed may result in an extremely rapid drop in economy.

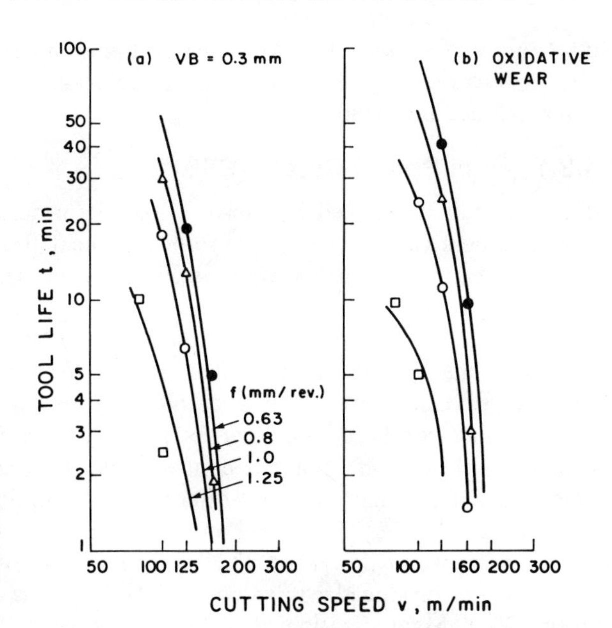

Fig.11.30. Effects of feed and cutting speed on (a) flank wear and (b) oxidative wear limit in cutting of normalized 1055 steel with carbide tools (ISO P15) [135]

11.3 Materials for Cutting Tools

Tool materials were discussed in a general sense in Sec.3.11.4; it now remains to explore their application to cutting, with emphasis on their wear behavior. In view of the various failure mechanisms, the following factors have a bearing:

a. Hardness (strength in compression) and its change with temperature

b. Strength in tension (or transverse rupture strength) at the temperature of operation

c. Fatigue and impact strength (toughness)

d. Thermal properties (heat conductivity, specific heat, coefficient of thermal expansion) and, in general, thermal shock resistance

e. Potential for adhesion and its dependence on temperature

f. Rate of mutual dissolution of tool and workpiece material (solubility and diffusion coefficient)

g. Chemical stability

h. Resistance to oxidation and tribochemical reactions.

These properties must be judged in relation to the strength (high-temperature, high-strain-rate shear flow strength), thermal properties, and microstructure of the workpiece material. Loladze [154] gives examples for systematic selection based on Soviet tool grades. In a broader sense, tool selection is affected also by workpiece- and process-related constraints, as discussed by Komanduri [198]. Specific recommendations are given in the Machining Data Handbook [12] and in Metals Handbook [11]. In addition to references cited in Sec.3.11.4, cutting tool materials are discussed in books [8,19,193,194], collections of papers [195], and reviews [196-198].

Importantly, the tool choice represents a compromise between hardness and toughness. Hardness is particularly important when the workpiece material contains hard second-phase particles. In many ways, machining is a process of mutual wear, and much can be accomplished by bringing the workpiece material—with a suitable heat treatment—into a condition that minimizes tool wear. The finished part can then be heat treated to give it the desired service properties. Refractory inclusions, such as Al_2O_3,

SiO_2, and various spinels, have greater high-temperature hardnesses than many carbides, and then only removal of these inclusions can increase tool life (unless a yet harder tool can be found).

11.3.1 High Speed Steels (HSS)

HSS tools can be shaped by conventional deformation and machining processes, and their toughness can be readily controlled by heat treatment. Hence, they are manufactured into complex configurations such as form-turning tools, drills, milling cutters, reamers, taps, etc. Their physical metallurgy is discussed by Mukherjee [199] and by Wilson [200]. They rely on high hardness (63 to 68 HRC), which is attributable to the martensitic matrix further hardened by very fine precipitates of carbides (of V, W, and/or Mo). Chromium, also a carbide former, increases hardenability, and cobalt ensures higher red hardness. Coarser, primary carbides may contribute to cutting performance, although this point is controversial. Commercially available grades (M: molybdenum and T: tungsten types) offer various combinations of hardness (hot hardness) and toughness [3.426],[12,201]. Powder metallurgy products are reported to give longer and more consistent life [196,202,203], although in some tests they were no better than conventionally produced equivalents [204]. Tools have been made also by liquid infiltration [205]. Magnetization of tools was observed to increase tool life [206], perhaps because of magnetostrictive strengthening [3.84].

A variety of surface treatments can be applied. Some of the more traditional ones are reviewed by Opitz and König [207]. Steam tempering (blueing, oxide coating) relies on the formation of a porous, hard Fe_3O_4 layer on the finish-ground tool, which increases tool life by 150 to 200% when impregnated with oil. Nitriding is particularly effective (20- to 30-fold increase in tool life) for drills and reamers used in working of aluminum alloys, but a 2- to 4-fold increase in working of steel has also been reported. Spark hardening, spark deposition of WC (spark alloying), boronizing, and chromizing have also been used. Recent developments have concentrated on chemical and physical vapor deposition (CVD and PVD) [208]. Thread-cutting tools coated with Ti(C,N) by CVD [209,210] have longer life but require heat treatment after coating. Coating with TiN by radio-frequency sputtering [211], with TiC by activated reactive evaporation [212], and with Ti(C,N) by PVD [213] have the advantage of lower temperatures, allowing treatment of the finished tool. Taps, milling cutters, drills, etc. typically show 2- to 6-fold increases in life.

Several wear mechanisms are active in cutting with HSS; the dominance of any one of them depends on operating conditions and is not yet fully understood [8]. It is certain, however, that insufficient hardness of the tool leads to rapid failure at all cutting speeds by plastic deformation or abrasion. Thus, overtempering accelerates both flank wear and crater wear [214]. Overheating (burning) in grinding is particularly damaging [202,214]. Cooler grinding conditions attained with CBN wheels are attractive in this respect, even though the resulting surface finish is rougher [215]. Hardness (or yield strength) is of paramount importance also when failure is by chipping of the edges, as in tapping or drilling [216].

Wear and tool life are greatly dependent on cutting speed (Fig.11.31) [207]. At relatively low speeds, abrasion by hard constituents of the workpiece damages the flank and leads to the formation of a shallow crater beyond the BUE. Periodic detachment of the BUE leads to attritive wear. Once speeds (and temperatures) are high enough to reduce the BUE to a secondary shear zone, several mechanisms may come into play, all of which are highly dependent on temperature.

Diffusion of iron atoms from the tool into the chip was postulated by Cook [129].

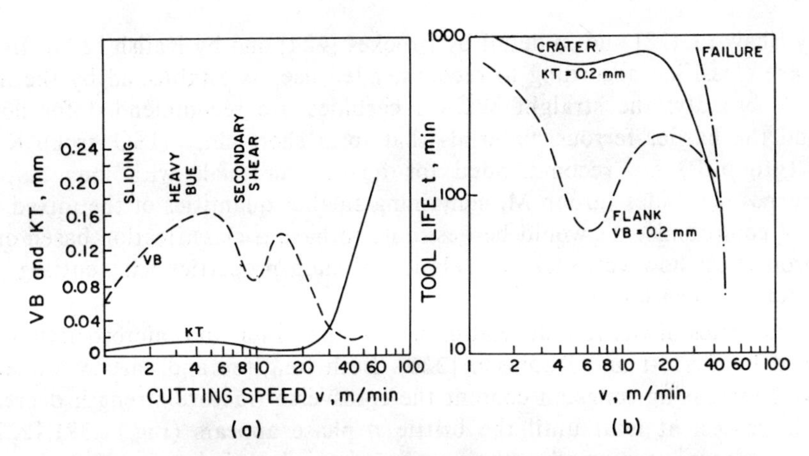

Fig.11.31. Effects of cutting speed on flank and crater wear and on tool life in cutting of 1050 steel with HSS tools [207]

Wright and Trent [217] regarded diffusion as a probable but not dominant factor. Evidence for diffusion comes from the white-etching layer found at the boundary between the BUE and the tool surface. This layer was identified by Wright [218] and by Philip [219] as cementite (Fe_3C) with chromium, vanadium, and molybdenum taken into solution from the tool. Since it is thicker toward the end of the BUE where temperatures are higher, it is likely to form also in the shallow crater but is washed away. Venkatesh [220] pursued, through a series of publications, the theme of diffusion wear. However, some of the doubts raised by Ramalingam [132] regarding diffusion gradients are difficult to answer, and it is likely that the softened tool is washed away by abrasion and/or plastic shearing of a thin surface layer [217].

Oxidative wear is of secondary significance, except in very low-speed cutting. Doyle [214] observed oxide which appeared to have flowed in a viscous manner during cutting and thus may have served as a lubricant. This would perhaps explain the observation of Opitz and König [207,221] that, at certain speeds, dry cutting can give longer life than lubricated cutting (Fig.11.31).

The wear mechanism depends also on the workpiece material. For example, Söderberg and Vingsbo [222] concluded that abrasive wear prevailed in drilling of plain carbon steels but that adhesion was responsible for most of the flank and margin wear in drilling of quenched and tempered steels.

The low value of n_1 in Eq.11.15 indicates that HSS tools are extremely sensitive to speed (temperature). Because tool life is a system output, it can be greatly affected not only by lubrication but also by tool geometry, which can often reduce temperatures more than any other measure.

11.3.2 Cemented Carbides

The higher hot hardness of cemented carbides is coupled with lower toughness. However, in the form of brazed or mechanically clamped (disposable) inserts, they find extensive use in processes previously preserved for HSS tools. The classic straight carbide is WC bonded with 4% (for high hardness) to 12% (for higher toughness) cobalt, which serves also as a crack arrestor. Later developments have involved the use of micrograin carbides and the addition of TiC, TaC, HfC, and NbC [223]; these increase the hot hardness of the mixed carbide even though they themselves may have lower strength at high temperatures than WC (Fig.11.27). Carbide grades are classified according to codes developed in various countries and by the ISO; they are described briefly by Shaw [3.428] and in detail by Brookes [224] and by Kalish [225]. In general, carbides are classified according to recommended use, as established by the manufacturer. Very broadly, the straight WC-Co carbides are recommended for nonferrous metals and the harder ferrous materials that form short chips (ISO group K), mixed carbides (group P) are recommended for ferrous materials with long chips, while general-purpose carbides (group M, containing smaller quantities of the mixed carbide) represent a compromise. It would be desirable to have a classification based on fundamental properties; however, clear correlation of such properties with cutting performance has not yet been found.

The properties of the tool depend on both composition and microstructural factors, as reviewed by Perrott and Robinson [226]. Both tungsten and carbon are soluble in cobalt; with increasing tungsten content the transverse fracture strength decreases but tool life increases, at least until the brittle η phase appears (Fig.11.32) [227]. Both flank and crater wear are speed- (temperature-) dependent, indicating that softening and diffusion must be important. Experience has shown that a straight WC tool fails rapidly by crater wear in machining of steel and many other ferrous alloys, most likely by diffusion. As pointed out by Naerheim and Trent [228], results of static diffusion experiments are not applicable to metal cutting where reaction layers are swept away by the moving chip. From their work it appears that the straight WC grades wear simply by diffusion of cobalt, tungsten, and carbon into the chip material. In the mixed carbides the cubic (Ti,Ta,W)C carbides are chemically more stable and dissolve much more slowly in steel than does WC [3.443]. Thus, the binder and WC wear preferentially by diffusion, leaving the cubic carbides to stand out by some 2 to 3 μm [228]. This slows down material flow and protects the binder and WC from further rapid dissolution. Diffusion of iron from the chip promotes crater wear, increasingly so at higher temperatures, as shown by Okamoto and Doi [229] with internally cooled tooling. According to these mechanisms the formation of protective oxides, considered by others [148,230,231], would play no role. The relative rates of solubility have been invoked by Kramer and Suh [148] to show why TiC is poor and WC is good for machining nickel, and why the reverse holds true for steel. The puzzle of the success of WC in cutting gray cast iron remains [132]; mixed carbides are needed for cutting nodular iron which

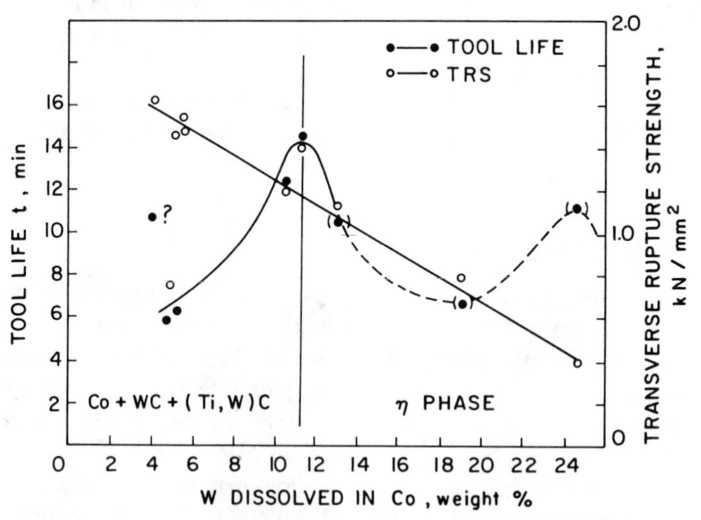

Fig.11.32. Effects of cobalt content on tool life and transverse rupture strength [227]

also contains free carbon. Adhesion is stressed by Komanduri and Shaw [232], and Katz and Rubenstein [233] built an adhesive wear model on the finding that tool life increased when the change in composition of the carbide reduced the tool/chip contact length.

The dominant wear mode depends on stress level and temperature, as governed primarily by cutting speed and feed rate. This led Trent [8,234] to construct machining charts (Fig.11.33). At low temperatures the BUE protects the rake face, and the flank and edge wear mostly by abrasion and attrition [235,236]; the latter is accelerated by adhesion and brittle fracture [237]. Edge destruction is particularly rapid in interrupted cutting, but Negishi et al [238] found that cracks are repaired by diffusion of cobalt within a certain range of feeds (0.5 to 0.8 mm/rev under their conditions). Once speeds are high enough for the BUE to disappear, crater wear by diffusion sets in (although plastic ploughing by hard inclusions [136] and plastic deformation [236] may also play roles). At yet higher temperatures, softening of the carbide results in plastic deformation of the tool edge, followed by rapid destruction of the tool. There are thus preferred operating rates that correspond also to the highest metal removal ratio [186]. The benefit of mixed carbides in cutting of steel can be clearly seen in Fig.11.33 [234].

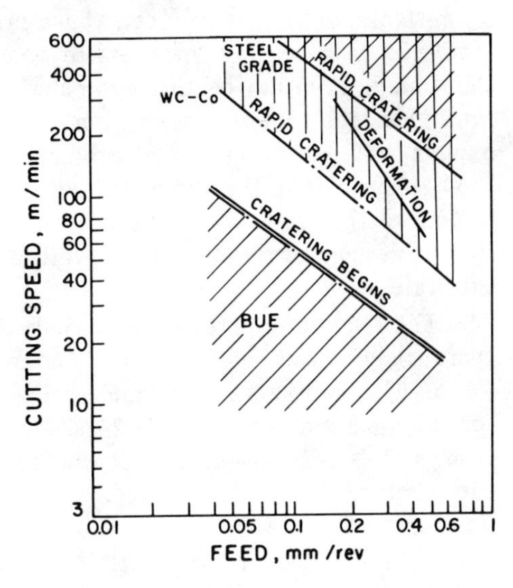

Fig.11.33. Tool-wear map for cutting of 0.4% C steel with a hardness of 200 HV (after [234])

Wear of TiC cermets occurs by diffusion, but at a lower rate [239], and thus they can be used at higher speeds. Diffusion was found to be the dominant wear mechanism above 1000 C [240] with different nitride- and carbide-type cermets; under cutting conditions, diffusion of carbon appeared to be more important than that of the binder. Static diffusion tests indicated the possibility of reaction with nonmetallic inclusions in steel. Thus, MnO reacted with TiC to form a (Ti,Mn)O solid solution, while MnS formed a low-melting eutectic with the nickel binder. On varying the TiN content (from 0 to 53%) in TiC cermets (with 5 Mo_2C and 5 to 20% Ni) in cutting of steel, flank wear decreased with an increasing proportion of the hard TiC-TiN phase, with a concurrent increase in thermal conductivity [241]. Higher TiN additions increased life to fracture in line with the increased thermal shock resistance.

11.3.3 Coated Carbides

As indicated in Sec.3.11.5, an improvement in cutting tool performance can be ensured by applying a more temperature- and wear-resistant coating to the surface of a tougher base. The production of coated carbide tools and, especially, of indexable inserts has grown phenomenally [208,242]. Success was achieved through an understanding of the manufacturing conditions that ensure good bonding of the coating without the formation of a brittle intermediate layer, and by tailoring of the substrate so that heat conductivity, thermal expansion, and chemical affinities were matched. Some substrates have a cobalt-enriched rake face and edge for greater toughness and a cobalt-depleted flank for wear resistance [198]. Both CVD and PVD techniques are used, and the coating is selected on the general premise that a good tool material should be hard, chemically stable, and resistant to diffusion and oxidation.

Normally, both rake and flank faces are coated. Ramalingam and Van Wyk [243] conducted experiments with only the flank or rake face coated with TiN. Flank coating reduced flank wear but allowed crater wear to proceed at the normal rate. Rake-face coating reduced crater wear but also somewhat reduced flank wear. It also prevented a rise in cutting forces; perhaps by keeping friction lower, it reduced temperatures and thus also flank wear. Hale and Graham [244] considered flank wear in more detail. They found that a TiC flank coating afforded protection even after it had been worn through, but that uncoated or only rake-coated tools wore rapidly. They concluded that the critical area was the lower edge of the flank wear land: as long as it had at least a 5-μm-thick coating, it reduced the wear rate. Both flank and crater wear resistance increased with coating thickness to about 6 μm; beyond this no further improvement was found. The thin coating protected the rake face even though crater wear was 50 μm. This is a general observation, and several hypotheses have been put forward to explain why a coating should protect even after its disappearance. Hale and Graham [244] believe that the remaining edge slows down cratering; transport of TiC to the bottom of the crater was observed by Snell [245]; and Venkatesh et al [246] conclude that some coating remains even while the crater forms by plastic deformation of the substrate.

The relative importance of various wear mechanisms is subject to debate, although there is agreement that more than one mechanism may be active.

a. Most workers agree that coatings, particularly TiN, reduce the cutting force P_c (or the apparent mean μ) [25,243,247], although others deny this effect [248]. In this respect, TiN is better than TiC; whether this is due to strengthening of the chip by diffusion of carbon from the TiC coating [249], or to lower adhesion, is still open to debate.

b. Flank wear occurs mostly by abrasion. Hardness is certainly important in abrasive wear, although Suh [133] points out that, once the tool hardness is at least 4.5 times the workpiece hardness, no further improvement can be expected. Hardnesses are to be compared at operating temperatures, and then many components of steel may become damaging. The hard TiC is better than Al_2O_3 or TiN in protecting the flank [244], even though abrasion followed by diffusion is postulated by Chubb et al [250,251], and Dearnley and Trent [248] suggest diffusion as the primary cause of flank wear over 875 C.

c. Crater wear, too, is open to various interpretations. Dearnley and Trent [248] find diffusion and plastic deformation (smearing) to be important, with the former dominating the wear of TiN and the latter the wear of Al_2O_3, with TiC in between. Others find TiN to be more "antidiffusive" [247]. The mechanical properties should dominate when the coating is chemically stable (e.g., alumina). Some workers have observed thermal cracking, which accelerates diffusion of the substrate [250]. Suh and co-workers developed the theme (summarized in [133]) that wear resistance is directly linked to chemical stability, as expressed by the free energy of formation. After chemical dissociation, the velocity component perpendicular to the tool surface would provide the mass transfer necessary to maintain mass continuity while the chip slows down on the tool face; the tool constituents then dissolve in the chip. Even though the end result is the same as with diffusion, the chemical stability argument is important because it suggests that little further improvement would be possible with carbides, and future developments would have to concentrate on oxides or, to avoid their low toughness and hot hardness, on oxycarbides, carbonitrides, and alloyed carbides, oxides, or nitrides. The problem often lies in finding an appropriate treatment or interface that ensures bonding to the substrate. Thus, the deposition technique is also of importance, but the relative ranking is not unequivocal. For example, under the conditions of their milling experiments, Doi et al [213] found PVD to be better than CVD.

d. Grooving is observed on many tools and is related to oxidative stability. Dearnley and Trent [248] found no grooving with Al_2O_3, little with TiN, and much with TiC. The latter seems to suffer fretting fatigue as well.

e. Thermal conductivity is important in protecting the substrate and in influencing chip formation. Wertheim et al [252] find that carbides and nitrides (whose conductivities increase with temperature) offer advantages at low speeds, whereas oxides (whose conductivities decrease with temperature) are better for high speeds. A thin, 5- to 10-μm coating of a low-conductivity material makes the tool behave as though all of it were of low conductivity [110].

f. A combination of desirable properties may be ensured by depositing two or more layers on top of each other. Thus, TiN on TiC improves crater resistance but not necessarily flank wear, while Al_2O_3 on TiC improves both [247,253]. The results are highly process-dependent, and results vary according to geometry, speed, and machine tool, even for the same workpiece material. The operative wear mechanism appears to be similar to that of single-layer coatings [251].

g. The number of compositions and combinations that have already been tried is large, and the number of possibilities is enormous. For example, CVD-deposited Ti(B,N) gave better crater and flank wear resistance than TiN [3.479], and so did Ti(N,C,O) [3.478]. Flank wear of commercial multiphase coatings occurred primarily by plastic deformation [184]. While TiC and TiN coatings were satisfactory for steel, a Mo_2N coating was better for titanium alloys and Cr-Ni steels [254]. The η-carbide Co_6W_6C improved life in continuous cutting as a substrate for or admixture with TiC [255]. A 2- to 5-μm coating of HfC gave superior deformation resistance on a cemented TiC substrate [256]. A diffusion coating of Ti(C,N) forms when mixed carbides [257] and TiC cermets [258] are nitrided. Some commercial tools have nitrogen-implanted alumina coatings.

h. Coating performance is affected also by changes in the composition of workpiece materials of similar mechanical properties. Thus, Schintlmeister et al [259], in turning Ca-Si deoxidized steel, found that the liquid layer formed by the steel was patchy on straight carbide tools, more cohesive on mixed carbides, and continuous on TiC- or Al_2O_3-coated tools. This layer has a wear reducing value, as shown by Narutaki [260], who deposited a sintered oxide layer of appropriate composition ($2CaO \cdot Al_2O_3 \cdot SiO_2$) on the tool surface.

The improvement obtainable is shown in Fig.11.34 [261]; most notable is the increase in total quantity of material removed.

11.3.4 Ceramics

Ceramic cutting tools became practical around 1940; the book by King and Wheildon [262] reviews much of this development.

As mentioned in Sec.3.9.4, most tools are based on alumina. Natural or synthetic

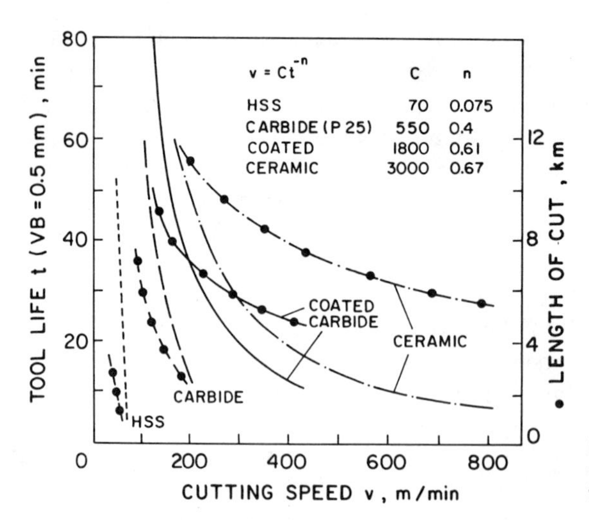

Fig.11.34. Changes in tool life and length of cut for different tool materials as functions of cutting speed [261]

ruby and sapphire have special applications; most industrial ceramics are made of fine-grain (mean grain size much below 10 μm) electrocorundum compacts [263]. Because they are self-sintered, without a metallic binder, they are higher in hardness than cemented carbides (Fig.3.66), especially at the high temperatures encountered in cutting of materials such as steel and cast iron at high speeds. Thus, they ensure shorter production times and much increased material removal capability (Fig.11.34). They have, however, low toughness, and only one-half to one-third of the transverse rupture strength of cemented carbides. A great advantage is their chemical stability.

As do brittle materials in general, ceramic tools suffer from chipping and from variability in tool life, although the latter has been largely solved through improved manufacturing techniques and by the addition of grain-growth inhibitors such as MgO. The high compressive but low tensile strength dictates the use of negative (typically, -5°) rake angles; the edge is chamfered (typically, 0.2 mm at 45°) and polished to prevent excessive early failures and microchipping.

Wear occurs by several mechanisms.

a. Notching at the depth-of-cut line can be severe, but life is usually limited by flank wear (since most applications are in finishing of gray cast iron, where tolerances must be kept, or in cutting of hardened steels, where the high cutting forces developing with increased flank wear lead to catastrophic failure). Crater wear is present but is not life-limiting.

b. Wear of ruby in cutting of nonferrous metals has been attributed to adhesion and spalling [264]. Loladze et al [151] find no affinity of alumina for steel but moderate to high affinity of alumina for titanium.

c. One of the dominant mechanisms, especially in continuous cuts, is a consequence of the oxidation of the workpiece material. Spinels formed with the tool material (with steels and cast irons, $FeO \cdot Al_2O_3$ [265]) are plastically deformed and washed away by the chip. Thus, wear is by plastic deformation, but the rate-controlling process is oxidation [266]. The depth to which oxygen gains access to the interface decreases with increasing cutting speed, and thus wear rates are lower at higher speeds [267].

d. The other major mechanism is fatigue wear on both the microscale and the macroscale [268]. Vibrations set up on entry into the workpiece must be damped quickly, and hence very stiff lathes are needed. In the less frequent application to milling, flank wear was observed to begin with microcracks in a honeycomb pattern some distance away from the edge; when cracks reached the edge, rapid flank wear set in [269].

e. Alumina is capable of substantial plastic deformation at elevated temperatures, and deformation of tools has been observed [268].

f. The low toughness of ceramics makes tools susceptible to thermal shocks, and therefore coolants—if used at all—are applied in a flood. Fluids are known to affect the strength and fracture behavior of ceramics, and surface effects discussed by Westwood [3.225] should have relevance.

Many relatively recent developments have been aimed at increasing the toughness of ceramic inserts [198,261,270]. Additions include other oxides (e.g., 10% TiO) and carbides (up to 30% TiC, and some WC). Commercial tools with TiC additions perform well even in lathes of medium rigidity [247] because of their greater fatigue resistance [268]. Other sintered products include Al_2O_3-TiB_2 and Si-Al-O-N compounds. Another approach relies on the dendritic solidification structure obtained on cooling an Al_2O_3-ZrO_2-W melt containing 10 to 14% ZrO_2 [198]. Zirconia-base precipitation-hardened ceramics showed low flank and crater wear but suffered from severe edge breakdown [271]. Some of the newer products allow machining of nickel-base superalloys; temperatures can approach the melting point of the superalloy [268]. Silicon nitride (Si_3N_4) base tools have higher toughness and promise good cutting performance in machining of cast iron [272]. Cemented ceramics were produced by reaction sintering of Al_2O_3 with 70Ni-30Zr alloy, resulting in tough yet nonreactive tools [133].

11.3.5 Cubic Boron Nitride (CBN)

Originally, tools were produced only by sintering of a 0.5-mm-thick layer onto a cemented carbide base. More recently, inserts have become available, some with a binder phase [198].

The higher hardness of CBN and the greater toughness of the tool structure make these tools suitable also for intermittent cuts, and the lesser reactivity compared with that of carbides allows cutting of hardened steel and of nickel- and cobalt-base superalloys. In agreement with ceramic practice, a negative rake angle is used, and the edge is radiused or provided with a small chamfer (negative land). A stiff lathe, adequately powered, is again necessary.

Several wear mechanisms appear to be active. Edge chipping is a brittle fracture phenomenon and is improved by edge preparation. Excessive loading or thermal cycling can lead to spalling of the CBN layer. Excessive loading can also collapse the end face in machining of hard materials (such as a fully heat treated, 60 HRC, D3 tool steel) at high feed rates [273]. Noncatastrophic wear proceeds both on the flank and on the crater and is speed-dependent [273,274], indicative of a thermally activated process. Bhattacharya et al measured temperatures of 1300 C in cutting D3 tool steel [273] and estimated temperatures in excess of 850 C for nickel-base superalloys [275]. Oxidation is a factor, as shown by the reduction in wear that results when CO_2 is admitted to the atmosphere in cutting at low speeds; boron oxide formed at high speeds increases wear rates [276]. Depth-of-cut line wear is attributed to oxidation and attrition, and is more severe when chip formation is discontinuous, as in dry cutting of D3 steel; the continuous chip formed in wet cutting is much less damaging [273].

More recently, tool inserts have been introduced in which the sintered CBN layer contains up to 40% ceramic bond. They are suitable for cutting all but the hardest steels [277]. The main wear mechanism appears to be attrition. Small lumps of grains, swept along the rake face, result in grooving or striation [278]. The limited ductility of hardened steel keeps temperatures fairly low, but the higher toughness of nickel-base superalloys requires higher CBN content. The presence of a cobalt binder makes diffusion possible, as in cemented carbides [279].

11.3.6 Diamond

Single-crystal diamonds have been used for decades for finish turning of aluminum alloys and other nonferrous metals at very high speeds (30 to 100 m/s). As do all brittle materials, diamonds suffer from unpredictable early failure. Apart from faulty setting, failure seems to be connected with internal stresses [7.351]. Wong [280] observed performance in industrial cutting and noted that N-O bonds showed up in the infrared spectra of those crystals that failed by early fracture. Some 60% of crystals had lives of around 4 h, irrespective of orientation (110 and 100; 111 is a cleavage plane and is not used). In contrast, laboratory experiments show an orientation dependence of wear [7.351]. Normal flank wear is independent of speed, suggesting an athermal abrasive wear mechanism. It is possible also that microfatigue of surface asperities contributes to wear. Diamond is capable of taking an extremely sharp edge, but edge chipping affects some 10% of all crystals. The tool nose is usually rounded, and a slightly positive rake angle is permissible.

Despite its high hardness, diamond deforms plastically at very high temperatures, but this does not seem to be a limiting factor. Oxidation begins at 600 C, but in high-speed machining little oxygen finds access. Diamond does not dissolve in metals, but graphite does. Graphitization is spontaneous at 1500 to 2000 C. Under cutting conditions, the combination of temperature and shear stress promotes graphitization [281,282] in the presence of a Group IV B to Group VIII metal (the metals that are used as catalytic solvents in diamond synthesis). Indeed, graphitization was observed by Tanaka et al [282] on holding diamond in iron and manganese, and thermal erosion was evident also with nickel, cobalt, chromium, and titanium. The graphite diffuses into the hot chip in machining of iron, leading to catastrophic wear. Sawtooth markings observed on the wear face were attributed by Komanduri and Shaw [281] to preferential transformation on 110 planes. Carbon content in steel (or cast iron) slows wear considerably [283]. Graphitization may proceed even without an apparent rise in temperature, as found by Wilks and co-workers [284], implying that the high reactivity of virgin surfaces accelerates transformation. With some metals, such as molybdenum, the interaction is strong enough to fracture the crystal.

Much diamond turning is now done with polycrystalline compacts made by sintering 1- to 30-μm powder in a layer approximately 0.5 mm thick onto a cemented carbide base [198]. The brittleness of the compact makes a slightly negative rake angle necessary for machining hard materials such as WC compacts, but a positive rake angle is commonly used for nonferrous metals. Polycrystalline compacts may be separately sintered and then brazed to steel or WC holders.

11.4 Cutting Fluids

The benefits accruing from the application of cutting fluids were noted long ago [1.23]. In 1881, Mallock [285] conducted some experiments in which soap reduced rake-face friction. Later, Taylor [180] noted the advantages of water as a coolant. Since then, vast efforts have been expended in attempts to understand the action of cutting fluids, yet without arriving at fully satisfactory explanations. To understand the difficulty, we must return to Fig.3.2, which shows the variables of the tribological system. Any change in one of the metal cutting system inputs results in complex changes in outputs, and these changes are greater in magnitude than those observed in most plastic deformation processes. The preceding discussion has already identified the major variables, but it will be useful here to recapitulate them with emphasis on their effects on lubricants.

General reviews [20,286-288], reviews of earlier work [2,289], and reviews of German research [290-292] are available. To aid the reader, readily accessible publications of permanent value will be reviewed here irrespective of age.

11.4.1 Process Conditions

A cutting fluid must possess all the attributes listed in Sec.4.1, but some attributes acquire special significance.

a. In the course of cutting, 100% new, highly reactive surfaces are generated. Pressures on the rake face are high; they range from, say, 200 N/mm^2 for nonferrous metals and low-carbon steels to 1500 N/mm^2 for difficult-to-machine materials, and are even higher for hardened steels. The interface is generally inaccessible to the cutting fluid; direct access is limited to the crevice where the chip leaves the rake face of the tool. Interface temperatures are high enough to promote reaction; however, contact time is brief, on the order of milliseconds, and it is questionable whether the secondary shear zone can be directly affected at all. Thus, reactions on the freshly formed chip surface can take place only after the chip has lifted away from the rake face. Therefore, such reactions are likely to have little effect, even though they were central to earlier concepts of cutting lubrication [293-295].

b. Temperature increases rapidly with increasing cutting speed and may reach 1300 C. Such temperatures are destructive to all lubricants and to many tool materials. Thus, cooling becomes the primary function of cutting fluids at higher speeds. Fluids are likely to affect wear mechanisms, too, but not necessarily in a positive manner; lubrication, cooling, and wear reduction do not always go hand in hand.

c. The flank clearance face and end face of the tool are subjected to severe sliding contact on the freshly formed workpiece surface. The wear land develops rapidly, making access of cutting fluids to the clearance face questionable.

d. Secondary contact surfaces are important in some processes. The margins of twist drills make elastic contact while the points rub in a plastic indentation mode. In tapping, reaming, and broaching, the last contacts are essentially elastic; heating and expansion of the tool intensify interface pressures.

e. Access of fluid to the rake face is highly variable. In turning, the tool is inaccessible for a prolonged period of time, but the chip surface and clearance face can be readily flooded with a coolant. Pekelharing [296] suggested that the tool be vibrated in the feed direction at twice the amplitude of the feed; the cut is interrupted periodically and fluid access to the rake face is allowed. In milling, the rake face is periodically accessible and flooding is possible. However, interrupted cuts expose the tool to wide temperature fluctuations which may be further accentuated by a fluid. In drilling and tapping there is difficulty of supplying lubricant to the point of action; furthermore, the work zone may become clogged, denying fluid access. Thus, the severities of different operations vary and cannot be specified by a single descriptor. Not even a simple ranking can be truly meaningful, even though attempts at defining relative operational severity have long been made [297].

f. The large plastic deformation and intense sliding imposed in thread form tapping set this process in an entirely different category [128,298], more akin to bulk deformation processes.

g. Severity of conditions is not constant for a given process, but increases with increasing tool wear.

11.4.2 Effects of Fluid Application

Very broadly, the possible effects of cutting fluid application can be categorized under the headings of lubrication and cooling.

Lubrication. Some observations can be interpreted in terms of classical lubrication effects.

a. In low-speed cutting with sliding friction, forces—primarily rake face friction and the cutting force P_c—are reduced; for constant material removal rates, the power consumption drops; the shear angle increases; the chip becomes thinner and curls more tightly; and the calculated friction angle drops. These effects have been repeatedly observed in turning [294,299] and in tapping [3.183]. Chip spread and thus collar formation are reduced, too. If, without a lubricant, a discontinuous chip would form, a lubricant may cause a shift to continuous chip formation, especially with a tool of small rake angle.

b. In cutting with sticking friction, the length of the sticking zone is reduced, with the same results as those indicated in the previous paragraph [24,34,152,300-303]. Consequently, normal pressures drop, and interface shear stresses drop more steeply [301]. The length of the sticking zone may be regarded as a measure of lubricant effectiveness, and this led De Chiffre [303] to evaluate cutting fluids using tools of restricted contact length (Fig.11.35). The cutting force is measured while the feed is gradually increased. At a critical feed f_0, sticking persists over the whole restricted contact length L_0, and P_c does not rise any more. With better lubricants, f_0 is larger, and thus the ratio L_0/f_0 is smaller.

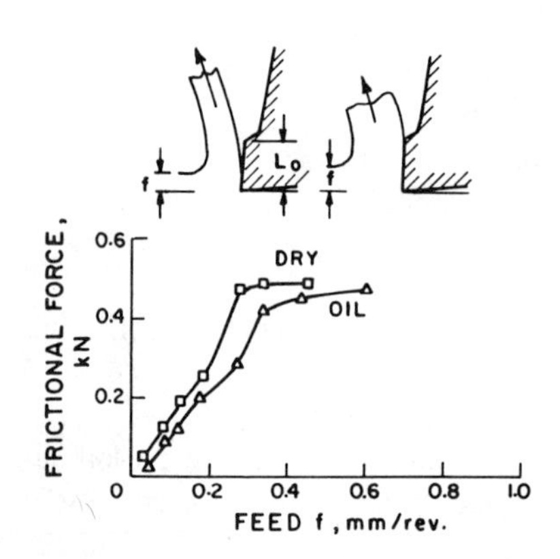

Fig.11.35. Principle of evaluating friction with restricted-contact-length tools (low-carbon steel, $L_0 = 1$ mm, $\alpha = 6°$, $v = 25$ m/min) [303]

c. If a BUE forms in the absence of a lubricant, the onset of its formation is shifted to higher speeds [3.292]. The incipient BUE is more frequently lost [47], permitting more lubricant access. The formation of the BUE may even be totally prevented.

d. Avoidance of the BUE and less frictional drag on the flank face result in better surface finish [304].

e. Lubricant effects are particularly noticeable when secondary friction contacts are significant. Therefore, lubricant ranking can totally change on going from, say, turning to drilling or tapping.

f. Lubricants interposed between tool and chip (or workpiece) reduce adhesion and thus adhesive wear, and also may affect abrasive wear by promoting burnishing without pickup. This, too, contributes to a better surface finish.

g. No significant lubricant effect can be expected when the material is brittle and breaks into small chips, thus keeping the chip contact length short (typically, gray cast iron, magnesium, free-machining brass, etc.).

h. Reduced friction results in less heat generation, which, in general, should lead to less wear or, for the same wear, allow higher speeds. However, this aspect cannot be discussed without reference to the cooling function of the fluid.

Cooling. The fluid may exert its cooling function by reducing the temperature of the workpiece as it arrives at the shear zone, by reducing the temperature of the chip as it leaves the secondary shear zone, and possibly by reducing the bulk temperature of the tool as well.

a. A coolant that reduces the temperature of the workpiece also increases its shear strength in the primary and secondary shear zones, thus increasing the cutting force—albeit usually only slightly [305]. The relationship is seldom straightforward, because lubricating effects confuse the picture [299]. Since the heat source is some distance away from the surface to which fluid has access, the cooling effect is not very marked, and becomes even less so at higher cutting speeds. Shaw et al [306] found that water or emulsion reduced interface temperatures only at speeds below 120 m/min at $f = 0.13$ mm, and below 60 m/min at $f = 0.26$ mm.

b. Cooling with water did not change the chip-thickness ratio in the work of Shaw et al [306], which indicates that cooling did not affect the chip-forming mechanism. However, cooling of the back face of the chip tightens chip curl and reduces L.

c. When wear is due to a temperature-dependent mechanism, cooling should, in principle, increase tool life. Iyengar et al [307] found on using an internally cooled tool that the the cooling effect became significant at 40 m/min in cutting of steel. Zorev [189] noted that in cutting with HSS the heating effect was insignificant below 500 to 550 C, but that above this temperature a coolant could prolong tool life to destruction, even though flank wear rates may have been greater. This was found also by Muraka et al [58], who measured a 5 to 10% reduction in rake-face temperature.

d. Effects of cooling on wear are complex because, even though rake-face temperatures may be unaffected, temperature distribution within the tool is changed [8]. For example, Feng et al [308] observed that water cooling increased the rate of flank wear by shifting the point of maximum temperature closer to the cutting edge. Smart and Trent [5.155],[309] estimated temperatures from metallurgical changes in HSS tools. In cutting of steel, a stagnant zone was found to keep the edge cooler, and therefore the greatest benefit was obtained by directing the coolant jet at the end clearance face; in cutting of nickel, sticking extended to the edge, raising its temperature, and the jet had to be directed at the flank clearance face. Similarly, Hollis [310] found that liquid CO_2 had to be directed to the base of a carbide tip to retard crater wear. Thus, the location of coolant application is also critical.

e. The cooling effect depends on the thermal properties of fluids. An example frequently quoted is due to Beaubien and Cattaneo [311], who found that tool life increased linearly with specific heat in the order: air - mineral oil - 25% NaCl in water - water. However, in view of the highly different characteristics of these lubricants, the relationship must have been fortuitous. Correlation between specific heat and the cooling effect was found also by Eugene [312]. Merchant [294] observed that a chemical solution cooled better than an emulsion, although Shaw et al [306] found little differences among emulsions, chemical solutions, and water in cutting of steel with HSS tools.

f. Improvements observed with refrigerated fluids offer further evidence of the cooling effect [313]. Pahlitzsch [314] found a two-fold increase in the number of holes drilled on chilling an emulsion to 4 C. In milling with chilled oil, life increased some 40%. A blend of 75% petroleum naphtha and 25% CH_3CCl_3 chilled to -45 C gave yet greater improvement [315]. The cooling effect may, however, be complicated by changes in fluid viscosity.

g. Yet higher cooling rates may be achieved with high-volatility coolants that evaporate on impinging on the workpiece, extracting heat equivalent to the latent heat of evaporation. Besides nitrogen and carbon dioxide, fluorocarbons and halogenated hydrocarbons have been applied from a pressurized system through air-mix nozzles [316]. Some of these fluids may combine cooling with lubrication.

h. Higher fluid viscosity causes progressive retardation of flow in layers near the surface and hence reduces the rate of heat removal. Thus, the viscosity decrease resulting from higher temperatures may improve tool life until the reduced cooling effect impairs tool life (Fig.11.36). Boston et al [317] found optimum temperatures of 21 C for a sulfurized mineral oil and 15 C for an emulsion.

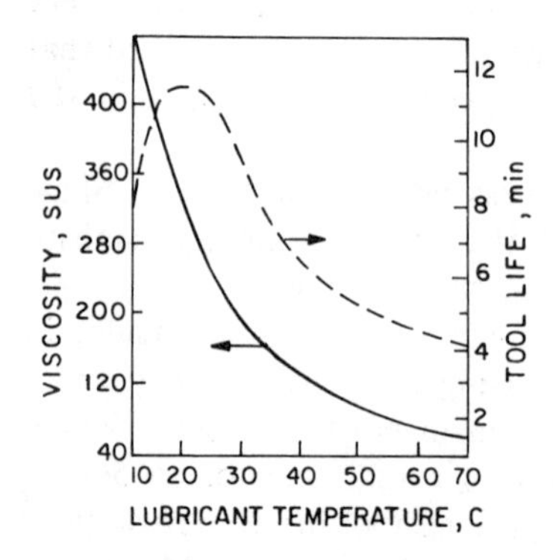

Fig.11.36. Effects of lubricant temperature on viscosity and tool life (annealed 3140 steel, HSS tool, *f* = 0.32 mm, *d* = 2.5 mm, *v* = 41 m/min, sulfurized mineral oil) [317]

i. Temperatures are not necessarily lowered by the application of a fluid. Eugene [312] measured lower temperatures in dry cutting during operation in the range of BUE formation, presumably because shrinkage or periodic break-off of the BUE in the presence of a fluid allowed tool temperatures to rise. A similar observation was made by Shaw et al [306] for heavier (f = 0.26 mm) cuts.

j. The coolant and its distribution affect thermal distortions of machine tools and thus affect accuracy indirectly. In modern, completely enclosed machine tools, removal of the splash guards may significantly change the pattern of distortions [318].

Even though simple relationships between tool life and specific heat of fluid have sometimes been claimed, the ability of the fluid to extract heat depends on many more factors. Heat transfer through a metal/fluid interface is involved, and several attempts have been made to derive over-all heat-transfer coefficients experimentally. Ernst [293] and Merchant [294] measured heat transfer into a fluid passing through a heated tube. The heat-transfer coefficients of emulsions increased with decreasing concentration, and water was the best. Hain [319] studied both laminar and turbulent flow and found good correlation between heat-transfer coefficients and temperatures in cutting in the 60-to-160-m/min range. Closer to actual cutting conditions was the simulation of Shaw et al [308], who directed a fluid jet onto the surface of a steel plate heated from below.

A thermocouple inserted 25 μm below the plate surface showed that at low temperatures a liquid stream was effective because cooling was by conduction; at intermediate temperatures (around 300 C), the liquid formed a steamed blanket, and mist application was better because the latent heat of evaporation helped cooling; at the highest temperatures even mist formed a blanket. Ueno et al [320] highlighted the importance of process geometry. When the fluid was applied over a heated bar, heat-transfer coefficients were 6145 kcal/m$^2\cdot$h$\cdot$C for a 1:20 emulsion and only 943 for oil. However, when a welded junction similar to a tool/workpiece contact was cooled, the difference almost disappeared, because the equivalent heat-transfer surface became greater with the oils.

In practice, total flooding is customary, but the possibility of the formation of a mist blanket cannot be exluded. Small quantities of surfactant in water improve wetting and thus cooling, whereas higher concentrations may reduce cooling by forming an adsorbed insulating film.

11.4.3 Effects of Speed and Feed

We have seen that higher speeds and, to a lesser extent, larger feeds generate higher temperatures. Furthermore, access to the interfaces becomes more restricted. The tool passes through the cutting zone faster, and thus the time available for reactions and cooling is also shortened. Therefore, a number of general effects are observed, typified by the data quoted by König and Vits [292].

a. At very low speeds and modest feeds, a clear lubricating effect is observable: both wear (Fig.11.37) and cutting forces (Fig.11.38) are reduced. The chip-thickness ratio also increases (Fig.11.39), indicating that the lubricant may somehow penetrate to the rake face. How this happens is still controversial, just as it was in 1958 when Tourret [2] wrote on this subject.

If chip formation is discontinuous, the fluid has periodic access to the interface and could act simply by reducing adhesion on the rake face. When chip formation is continuous, access of lubricant from the sides of the chip is possible, as observed directly

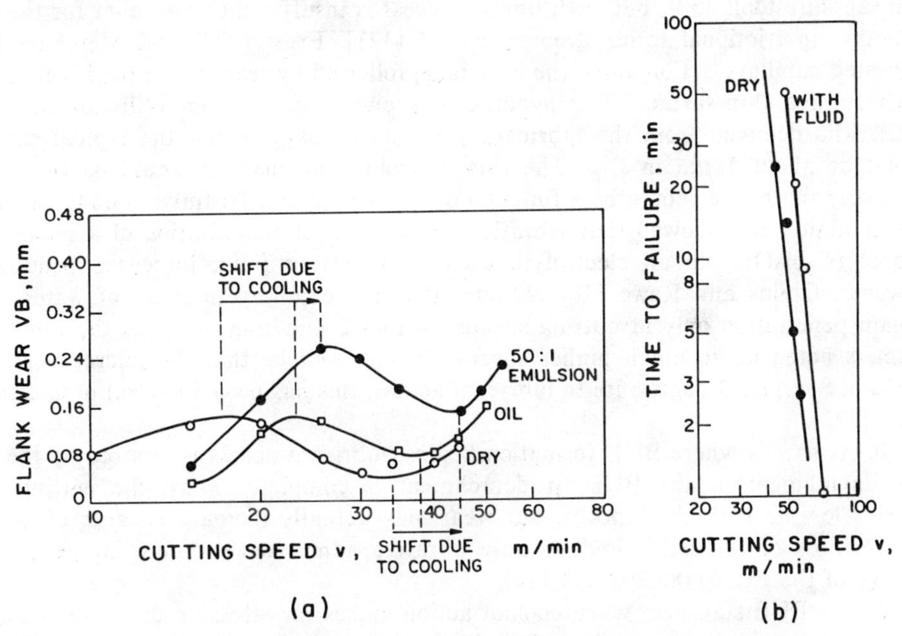

Fig.11.37. Flank wear at 30 min (a) and time to failure (b) in cutting of normalized 1055 steel with HSS tools (f = 0.25 mm, d = 2 mm) [292]

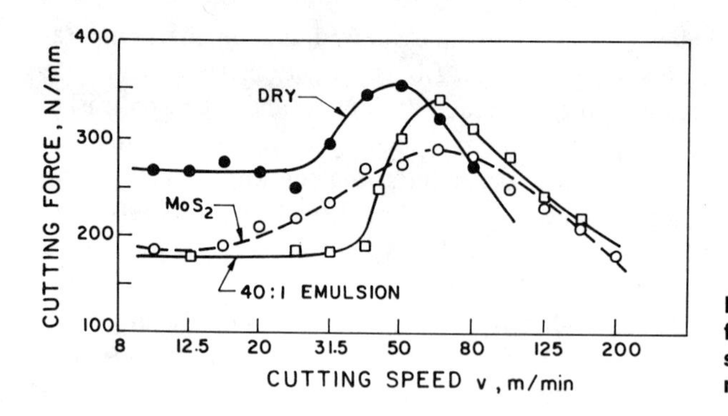

Fig.11.38. Effects of lubrication on force in cutting of normalized 1055 steel with carbide (P30) tools (f = 0.25 mm, d = 3 mm) [292]

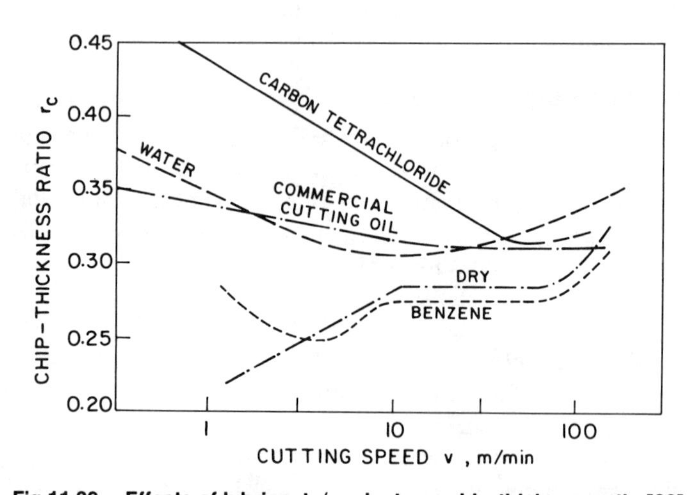

Fig.11.39. Effects of lubricants/coolants on chip-thickness ratio [30]

with sapphire tools [52], but such limited access is insufficient to account for the large reduction in frictional force often measured [321]. Ernst [293] and Merchant [294] suggested capillary action along the rake face, followed by reaction on the highly active, newly formed chip surface. This hypothesis received support from Williams and Tabor [302], who deduced from the lubrication of iron by oxygen that the typical capillary should be about 1 μm in size. The rate-controlling mechanism would be the rate of transport, and hence should be a function of vapor pressure. Postnikov [3.84] assumed a similar action and showed that vibration of the tool, or the addition of a monohydric alcohol (C_3H_7OH) to an electrolyte used as a cutting fluid, increased penetration. However, Childs and Rowe [20] consider that the capillary pressure of water would explain penetration only at cutting speeds up to 1.2 mm/min, whereas the lubrication effect is noted up to much higher speeds. It may well be that the microcracks mentioned in Sec.11.1.3 contribute to lubricant access; this has been a central proposition of Shaw [322].

b. At speeds where BUE formation begins, lubricant access is ensured by the periodic detachment of the BUE, if detachment is complete. Thus, the cutting force remains low up to higher speeds, but wear may actually increase because of attritive wear on loss of the BUE, and because abrasive/adhesive wear may intensify in the absence of the BUE [189,292,323,324].

c. The BUE stabilizes when coolant action makes the stress gradient steep enough; at this point, wear begins to drop. It is likely that there is no fluid access at all to the cutting zone; however, the fluid may still seep into the gap between BUE and chip back face, shortening the base of the BUE.

d. At yet higher speeds, the BUE degenerates into the secondary shear zone and no fluid access is possible; this conclusion was reached also by Kurimoto and Barrow [325]. However, the crevice between chip and rake face is open to the fluid, and reactions are possible at least with the tool. The contact length L (or the sticking zone within the contact length) is reduced; this causes the shear angle to increase and the cutting force and energy to drop [286,303,326]. In cutting with a carbide tool in a 4% caustic soda solution, Opitz et al [221,327] concluded from the length of the etched zone that about one-third of L was accessible. It is doubtful that, in the short time available, reactions on the chip surface could progress far enough to influence L.

It is in this speed regime that Taylor's equation most clearly holds. Even though wear rate (Fig.11.37a) and cutting force (Fig.11.38) may be high because of higher shear flow stresses, tool life to destruction is increased by the cooling effect of cutting fluids (Fig.11.37b), and higher maximum speeds are permissible. Since cooling is the dominant effect, the composition of the fluid is of lesser importance, but aqueous fluids are clearly better than oils, and both give better results than those obtained in dry cutting. This is in agreement with early observations that, for short tool lives, water, water plus borax, or emulsions are best [328-330]. All measures that intensify cooling also increase tool life. However, when life is limited by end grooving due to corrosion, the effect may be reversed. A plain oil that excludes oxygen is then better than a compounded oil or an emulsion [325]. One of the most critical and often overlooked factors is flow rate (Fig.11.40) [292]. Insufficient flow rate reduces tool life, particularly with emulsions. Interruption of fluid flow may lead to catastrophic failure.

e. With increasing feed, higher temperatures are generated, the onset of BUE formation shifts to lower speeds, and the BUE becomes larger (Fig.11.11). Thus, the speed at which lubricating action ceases is also lower. However, a lubricant can prevent BUE formation to heavier feeds; for example, Zorev [3] found a BUE at f = 0.1 mm in air, at f = 0.15 mm with spindle oil, and at f = 0.3 mm with water or CCl_4. With increasing feed, cooling becomes less effective. Shaw et al [306] measured a temperature drop of 156 C with a water-base coolant at f = 0.058 mm, and none at all at f = 0.26 mm. Very small feeds promote flank wear; the increasing wear land prevents lubricant access, and allows temperatures to rise, and this in turn leads to more rapid crater wear [325].

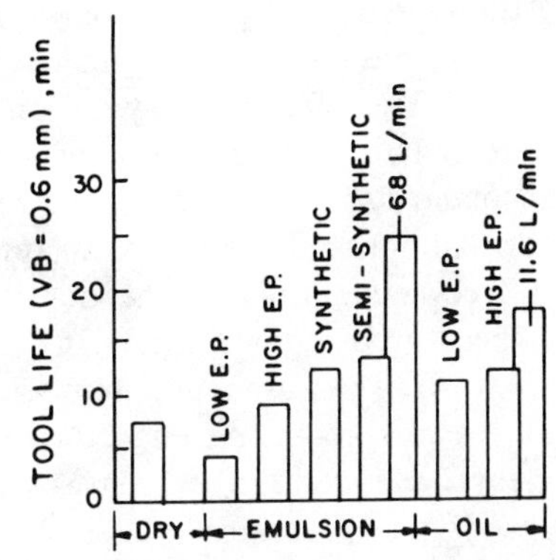

Fig.11.40. Effects of cutting-fluid composition and delivery rate on tool life in cutting of alloy Ti-6Al-4V with carbide tools (K20 carbide, f = 0.25 mm, d = 1.5 mm, v = 60 m/min, fluid flow = 1.7 L/min, unless otherwise indicated) [292]

11.4.4 Lubrication/Cooling Mechanisms

Even under the most favorable low-speed cutting conditions, it is unlikely that thick fluid films could form with externally imposed lubricants. Some early calculations by Bisshopp et al [331] suggested that films would be only 10 times thicker than asperity height. Beaubien and Cattaneo [311] included pressure-viscosity effects in their calculations without arriving at a firm conclusion. It is possible, however, that liquids formed from internal lubricants (inclusions) may develop a wedge, as pointed out by De Salvo and Shaw [332]; the relative movement between chip and rake face creates the correct wedge configuration.

Since full fluid films cannot form, externally introduced fluids can function only by solid-film, E.P., boundary, and, in special situations, mixed-film mechanisms.

Gas and Vapor. Metal cutting is always carried out in a fluid medium. In dry cutting the medium is air. Extensive research has been directed at understanding the influences of air, of its constituent gases, and of other gases on the cutting process.

a. Air consists chiefly of oxygen and nitrogen. Experimentation in vacuum [326], and especially in high vacuum [333], eliminates the formation of oxides. Resulting changes in chip formation and in cutting forces depend on the particular tool/workpiece combination.

In cutting of iron or low-carbon steel with HSS tools in vacuum, adhesion causes the shear angle to drop to very low values, chip formation becomes discontinuous, and cutting forces are high [326]. A very low partial oxygen vapor pressure (a few pascals) was sufficient to reduce adhesion in the work of Horne et al [333], and low cutting forces and continuous chip formation were restored in the experiments of Rowe and Smart [326]. Conditions similar to cutting in vacuum are achieved by admitting nitrogen to the vacuum chamber or by excluding oxygen—in the course of normal machining—with a nitrogen jet [8]. The benefits can be attributed to oxygen access to the crevice, with consequent oxidation of the tool surface and reduction of L.

In cutting of copper or aluminum, the effects are more complex. These metals formed no BUE under the conditions employed by Rowe and Smart [334], and the cutting force was no higher in vacuum than in air. Rowe and Smart postulated that oxygen adsorbed on the free chip surface impeded dislocation movement on the shear plane and counteracted the effect of reduced adhesion. It appears that cutting in vacuum results in a lower cutting force if cutting in air is accompanied by the formation of a strong deposit in the sliding zone (zone 2 in Fig.11.9). This was observed by Doyle and Horne [335], at low speeds, in cutting of aluminum and copper with HSS tools; with carbon steel tools it was evident only in cutting of aluminum. At higher speeds, where oxygen access is limited, the difference between vacuum and dry cutting disappears [48]. Adhesion over the faces of sapphire (Al_2O_3) tools creates sticking [336], and metal transfer ensues on cutting in air; hence, vacuum gives lower cutting forces [333].

b. The effect of oxygen on tool life is also complex. Zorev and Tashlitsky [189] note that, in cutting of steel with HSS tools, oxidation decreases the stability of the secondary shear zone and thus reduces tool life; at high speeds there is insufficient time for reaction and the effect disappears. Trent [8] confirms this point and notes that there is free access only at the depth-of-cut line and at the trailing end of the nose radius (where grooving wear sets in); elsewhere, the workpiece material acts as a getter, as shown by the absence of temper colors on the clearance face of the tool in the vicinity of the minor cutting edge. Excess oxygen blown at the flank face [8,172] or the end-clearance face [296] increases wear; blowing at the rake face increases wear only at low speeds, in the region where the BUE forms, but makes no difference at high speeds. However, Kasyan et al [337] observed that attritive wear of carbide tools was reduced by increased partial oxygen pressure. Exlusion of oxygen by argon increases tool life at higher speeds in machining of steel [172] or titanium alloys [338], and CO_2 increases life by allowing a larger, protective BUE to form on HSS [189].

c. It is generally thought that oxidation is too slow to form protective oxides on the chip; however, protective films may still form on the tool. Opitz et al [327] reported that FeO-MnO-SiO_2 films formed on carbide tools at an increasing rate when cutting was carried out in a more oxygen-rich medium (Fig.11.41), indicating that oxidation must be the rate-controlling mechanism. All aqueous fluids are sources of oxygen, and tool life is increased by introducing excess oxygen into oils or emulsions [339]. Similar improvements were noted by Latyshev [340] with H_2O_2 and $NaMnO_4$.

d. Gases not only lubricate but also cool. This point is demonstrated by the improvements observed with cooled gases. Tool life increases when air or CO_2 is cooled to -40 to -60 C [221,341], even though cutting forces rise [315].

e. Gases known to form lubricating/parting films (such as chlorine, sulfur, and iodine) adsorb readily on virgin surfaces [342] and reduce cutting forces when admitted to a vacuum chamber [343]. Chlorine was effective even for Al_2O_3 sliding against copper, nickel, and iron in the work of Pepper [336].

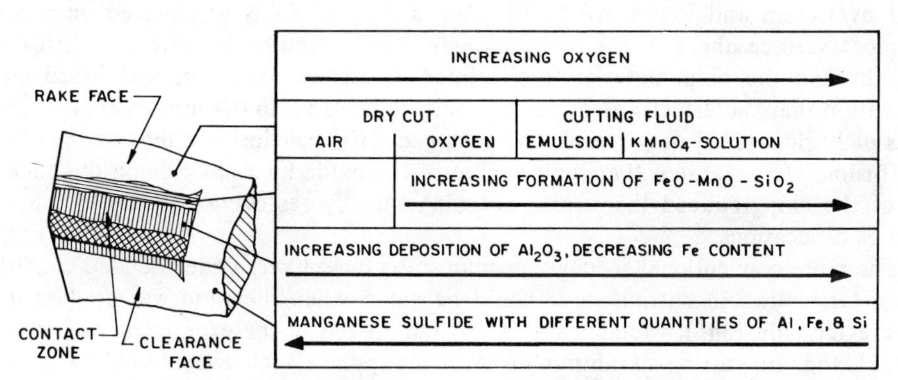

Fig.11.41. Effects of oxygen supply on formation of interfacial layers in cutting of aluminum-deoxidized, normalized 1055 steel with carbide tools [327]

Carbon tetrachloride (CCl_4) is an effective lubricant under some conditions and continues to stimulate research interest because its behavior possibly could shed light on the basic mechanisms of lubrication. It is, however, a toxic substance that has been banned from use under production conditions.

a. Liquid CCl_4 dramatically improves lubrication in low-speed cutting of iron, aluminum, copper, single-phase alloys, and even gold. The shear angle and cutting ratio are increased (Fig.11.39) [30,294]; the cutting force is reduced, often by 50% [34]; the chip is continuous, smooth, and tightly curled; and the machined surface is of good quality. This effect tapers off at speeds above 30 m/min in cutting of steel [294,326]. Improvements are noted only with liquid or mist application; vapor is hardly better than air, perhaps because insufficient quantities of either are available. However, Merchant [295] reported low friction with superheated vapor on aluminum. Liquid CCl_4 shifts the onset of BUE formation to higher speeds [34] and reduces the size of the BUE. At higher speeds (over 30 m/min), Williams et al [34] noted some cooling effects. These observations are consistent with the explanation that the fluid acts as a classical low-viscosity E.P. agent that finds access to the rake/chip crevice, reduces L, creates a steeper shear stress drop, and reduces friction in the sliding zone, as found by Childs [301] on iron. Liquid CCl_4 is a poor lubricant under low-pressure, low-temperature sliding conditions [344] and in the absence of air. Thus, its effectiveness is reduced in argon [333]. This is true also when it is used with carbide tools; it appears that tungsten oxides must form first so that they can be transformed into chlorides [345]. Shaw [322] observed high friction on drawing the tool back over a freshly formed aluminum surface under flooded conditions, whereas low friction was found by Lakhwara and Rice [346] on applying a small amount of lubricant to the chip crevice only. The freshly formed workpiece is smooth, and Williams [347] attributed this to embrittlement of the workpiece material in the tertiary shear zone.

b. Less clear is the mechanism by which CCl_4 arrives at the interface. Usui et al [348] noted a lower P_c even when only the surface of the workpiece was coated, and a lesser but still significant lubricant action was evident when the CCl_4 was allowed to evaporate; this effect was noted also by Rowe and Smart [349]. Only when the workpiece was heated above the boiling point of CCl_4 did the effect disappear. The interpretation given is that the previously machined surface contains microcracks into which the fluid diffuses. This view is supported by the absence of the effect in cutting of a carefully ground surface [3.223]. This might explain also the sudden decrease in force found by Cassin and Boothroyd [350] when a drop of CCl_4 was placed on a copper workpiece surface ahead of the tool, although they attributed the effect to diffusion of CCl_4 through the chip material to the interface. Their argument was based on the observation that the effect was noticeable only at feeds up to 0.4 mm. However, experiments of Barlow [351] with radioactively tagged CCl_4 conclusively showed the absence of diffusion. He suggested that replacement of the oxide by a chloride on the back surface of the chip reduced the strain-hardening rate by removing barriers to the emergence of dislocations.

The more conventional seepage mechanism is most likely under flooded conditions, since a large drop in cutting force could be noted when the fluid was applied to the rake crevice only [346]. Metal transfer was eliminated in the experiments of Horne et al [333] and, in cutting of aluminum with a sapphire tool, even bubbling could be observed, indicating a possible reaction. Other chlorinated compounds provided lubrication only if they were volatile (had a high saturated vapor pressure), reinforcing the argument that access through microcapillaries might be possible. Good results have been obtained with chloroform, with bromine [295], and with the low-toxicity trichloroethane (CH_3CCl_3) [350]. The latter reduced friction of nickel and iron against Al_2O_3 [336].

c. It has been noted that CCl_4 is ineffective or even increases friction in cutting of lead. Possible explanations are that the shear strength of $PbCl_2$ is higher than that of lead [295,333], and that the brittleness of the chloride allows metallic contact [344].

Cutting Oils. A large class of cutting fluids is based on mineral oils with various boundary and, particularly, EP additives, used mostly in conjunction with HSS tooling. The primary function of these fluids is lubrication, although at higher cutting speeds the cooling aspect becomes dominant (Fig.11.37a). Typically, they eliminate or reduce the BUE and shift its onset to higher speeds [290]. While under some conditions a plain oil may be sufficient [45] or even preferable [325], a compounded lubricant is definitely needed when secondary rubbing surfaces are involved. Even though at higher speeds friction (cutting force) may increase, a better surface finish is generally maintained.

a. In principle, the viscosity of the base oil should be low when fluid access is limited, as in drilling, and the data of Tourret [352] on drill life seem to confirm this (Fig.11.42). In contrast, Beaubien and Cattaneo [311] found drill life to increase with viscosity up to about 300 cSt at 40 C. Some minimum viscosity may well be needed to keep the oil along the wall, and there may be an optimum viscosity for a given application. Certainly, higher viscosity is desirable when a light oil would flow out of the contact zone, as in broaching. In machining at very low speeds and with unfavorable geometries, even a pasty consistency may be required. The existence of an optimum

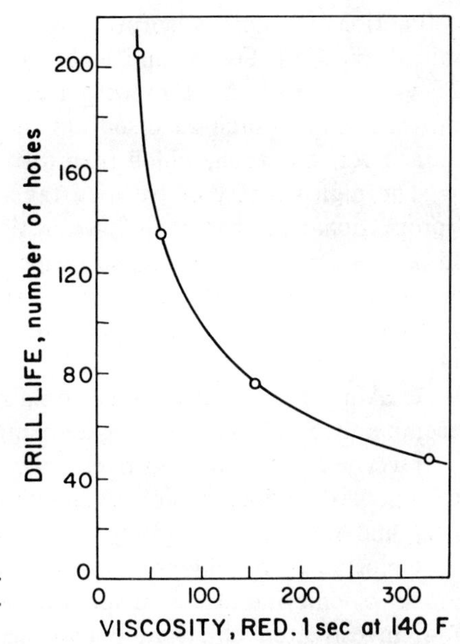

Fig.11.42. Decrease in drill life with increasing lubricant viscosity (active sulfurized oil, 1 L/min, 6.35-mm-diam drill at 4000 rpm) [352]

viscosity is suggested by the results of Boston et al [317] in turning (Fig.11.36); however, the viscosity change was obtained by raising the temperature, and cooling effects and activation of sulfur complicate the picture. It is similarly difficult to judge the observation that addition of a light oil to a heavier cutting oil increased tool life [353], or that addition of a light oil to a heavy E.P. oil first increased, but at higher dilutions decreased, tool life [352]; it is impossible to separate the additive and viscosity effects. Certainly, a lighter, more freely spreading oil should offer better protection against oxidation and thus reduce notch wear [354] and, especially, end-face notch wear, which can be the life-limiting factor [325].

b. Boundary additives (Sec.3.5) are ineffective in the cutting zone in cutting of steel because the physically adsorbed films cannot resist wiping off (Fig.11.43) [24]. They are, however, valuable in the rubbing zones, and perhaps even on the flank face, but certainly on drill margins and in elastic contact zones in tapping and internal broaching. Zaslavsky et al (quoted by Mrozek [355]) found greatly increased drill life on adding 1% "tribopolymer." Long-chain fatty acid derivatives dissolved in a mineral

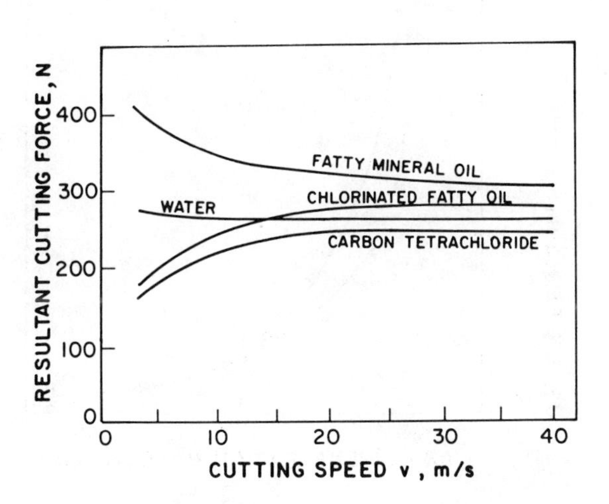

Fig.11.43. Effects of lubricants on forces measured in cutting of mild steel [24]

oil function by rapid adsorption on aluminum and on all metals more electropositive than silver [356]. Owens and Roberts [357] found a reduced feed force in cutting with 20% vinyl stearate in oil, or with 1-cetane. Entelis et al [358] make the unusual suggestion that soaps—such as disodium maleate—function by interacting with the steel to produce active carbon, which then diffuses into the tool surface.

The high activity of fresh surfaces accelerates adhesion, and the rate of formation is proportional to the rate of generation of virgin surfaces. In the work of Mori et al [359], activity decreased in the series: alkyl halide > alcohol > acid >> alkane, alkene. Surprisingly, minor amounts of surfactants (amino acid) were observed to reduce the cutting force even if applied to the workpiece surface ahead of the cutting tool [346].

c. Almost all cutting oils contain E.P. additives (Sec.3.4). They are chosen to become active only at some higher temperatures (say, 200 C).

Overly aggressive additives may actually increase tool wear [360]. This is the problem with iodine, which would otherwise be very good on steel [3.180], titanium [361], and stainless steel [362].

Chlorine is widely used. Wolfe et al [363] found drill life to increase with the number of chlorine atoms in substituted ethane and methane (but not ethylene). Barker [364] investigated chlorinated hydrocarbons in oil. There was no correlation with the number of chlorine atoms in the molecule, but CCl_4 and CH_3CCl_3 were again effective.

Sulfur is used even more frequently. In thread cutting, Davis et al [4.28] found a package of elemental sulfur with a boundary additive to be best, followed by elemental sulfur, di-tert-octylpolysulfide, dibenzyl-disulfide, and lauryl mercaptan, all added to give 1% S in the base oil.

The optimum additive package depends on application. The results of Holmes [365] are instructive (Fig.11.44): at lower drill feed rates the chlorinated additive was better, whereas at high penetration rates (and thus higher temperatures) sulfur was better. Most commercial products contain both chlorine and sulfur [366,367] because of their synergism [3.183]; their presence on the rake face could be proven by analysis [47], indicating that some fluid access must be possible. The severe conditions existing during cutting can even lead to the formation of new compounds [368], but the effec-

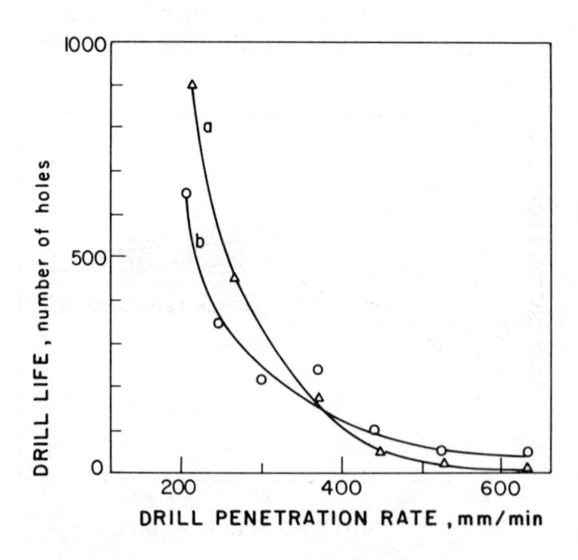

Fig.11.44. Effect of drill penetration rate on drill life with compounded mineral oils [365]. a - 1.5% chlorine. b - 1.5% sulfur.

tiveness of reactions can be judged only by testing under conditions typical of the intended application.

In general, the more intensive the chemical reaction, the lower the speed at which the additive ceases to be effective. High reactivity is undesirable also because it accelerates grooving wear (Fig.11.45) [354]. Merging of the end-face groove with the depth-of-cut line groove results in collapse of the tool [163]. Kurimoto and Barrow [325] observed that neither straight nor compounded oils had any influence on the flank wear and crater wear of HSS tools; ultimate tool failure was a consequence of corrosive groove wear at the end-clearance face, accelerated by E.P. additives but slowed down by the air-exclusion effect of a plain oil. Thus, plain oil was best for both roughing and finishing.

More is not necessarily better, and beyond a critical additive concentration performance is not improved [352,367,369] and may even worsen. Zakar et al [370], using sulfurized and sulfochlorinated oleic acids, chlorinated paraffin, and TCP as their ingredients, found that the different packages were most effective at different concentrations; a S-Cl-P package was best over-all.

Few attempts have been made to calculate optimum composition from fundamentals, partly because of the difficulty of identifying the compounds formed *in situ* under the conditions of cutting. Therefore, elemental or molecular analysis of the cutting fluid provides little guidance. On the premise that fracture is promoted by a medium whose heat of chemical reaction equals the surface energy of the material, Shchukin et al [371] identified mercury chloride and iodine as the most suitable additives for nickel.

The benefits of additives are most clearly demonstrable if rubbing is encountered in addition to cutting, as in drilling [372].

d. As discussed in Sec.3.4.2, interactions with other lubricant constituents are important.

Oxygen can be regarded as an E.P. lubricant in its own right, as shown also by the high friction (low chip-thickness ratio) observed when benzene denied access of air in the work reported by Merchant (Fig.11.39) [294]. Oxygen is also an important contrib-

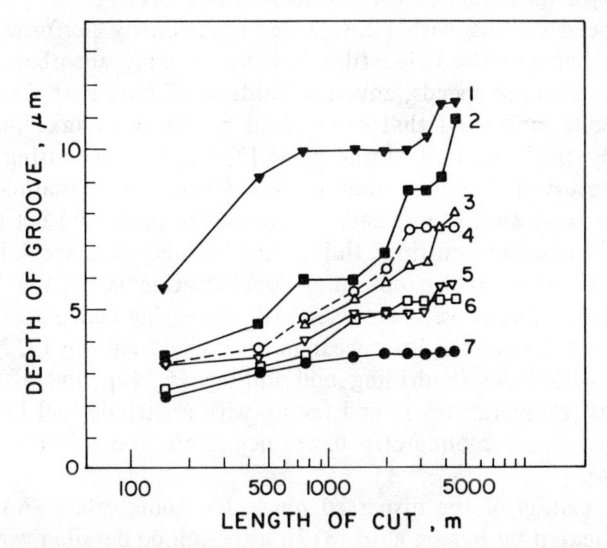

Fig.11.45. Progression of grooving wear in cutting of 1055 steel with carbide tools and different lubricants (v = 150 m/min, f = 0.05 mm, d = 0.2 mm) [354]. 1 - Dry. 2 - Emulsion. 3 - Argon gas. 4 - Highly chlorinated oil. 5 - Standard cutting oil. 6 - Light oil. 7 - Spindle oil.

utor to the action of E.P. additives. For example, in the work of Mould et al [3.183], the number of holes that could be drilled dropped by one-half to two-thirds on displacement of air by argon.

Interactions with other additives are complex. Even the minor constituents of less-refined mineral oils mask some of the E.P. activity. Serov et al [373] found that addition of oleic acid to nonactive sulfur and chlorine additives increased wear, whereas active additives were not disturbed. The performance of dibenzyl disulfide also suffered from addition of oleic acid, but addition of pentachlorobutyric acid resulted in an improvement [374]. Drill life was reduced on addition of MoS_2 to a sulfochlorinated oil [375].

The presence of water in a cutting oil increased wear in the work of Ishibashi and Katsuki [360], except in oils with non-surface-active phosphorus additives; with surface-active types, wear increased to the level found with plain water. In some instances, no benefit of additives could be shown.

e. An unusual additive is sodium triborate in gear cutting and broaching oils. It is believed to form a thick film of tenacious glass when the tool heats up [376].

Aqueous Fluids. Apart from the cooling effect already discussed, water-base (aqueous) fluids fulfill a number of other functions in cutting with HSS or carbide tools. Some lubricating function is evident from the delay in the onset of the BUE (Fig.11.37 and 11.38) and from the improved surface finish usually observed. Even though wear may actually increase, tool life to destruction almost always increases (Fig.11.37b). Aqueous fluids are more effective at higher feeds, which initiate the BUE at lower speeds [290].

a. Plain water is a carrier of oxygen and has, in cutting with carbides, the same effect as a stream of oxygen [8]. Drobysheva and Latyshev [345] found that it leads to the formation of tungsten and cobalt oxides. Because water cannot prevent oxygen access, it is also ineffective in reducing notch wear (Fig.11.45) [354] if notch wear is due to oxidation. That aqueous fluids are more effective agents of oxygen transfer than air or even oxygen gas was indicated in the work of Diederich [327] by the rates of complex oxide layer formation on carbide tools (Fig.11.41).

b. In low-speed cutting with HSS, water occasionally performs better than fatty oils, presumably because the oxide film is more strongly adsorbed than are oil films [349]. However, at higher speeds, aqueous fluids accelerate both flank wear and crater wear of HSS tools, indicating that some fluid access must take place [325]. This is suggested also by the results of Horne et al [333], who, in cutting aluminum with a sapphire tool, observed bubbling due to gas formation; oxidation of the interface resulted in severe zone 2 transfer. Lead, which did not react, did not transfer.

c. Practical aqueous cutting fluids are emulsions (Sec.4.6.2) or synthetics (Sec.4.6.3). With emulsions, an optimum concentration is usually found for a given application. Since cooling power decreases with increasing concentration [325], a leaner emulsion (around 2%) gave the least wear in high-speed cutting [377], a 6% concentration was better with HSS in drilling and milling [5.249], and a 10% concentration ensured the lowest temperatures in end facing with a carbide tool [378]. At low speeds (15 to 20 m/min), where cooling effects are negligible, the effect of concentration was insignificant [324].

d. The composition of the dispersed phase has some effect. An optimum base-oil viscosity was indicated by Boston et al [317], although no detailed work appears to have been published on this aspect. Emulsions of fatty oils and derivatives eliminated the BUE in cutting of zinc, cadmium, lead, aluminum, and copper with HSS tools at very low speeds (1.4 m/min) in the work of Latyshev [379]. Frequently, E.P. additives are incorporated in the oily phase. Tests by Barker [364] confirmed the need for such additives in drilling of steel and stainless steel, although no direct correlation with sulfur (active or total) and/or chlorine content could be found. Chlorine may accelerate corrosive wear of the minor cutting edge [360].

e. Less frequently, polymers are emulsified. For example, a low concentration (0.01%) of polyacrylamide was found to be effective [4.100],[380]. A water-soluble polymer, in combination with ammonium salts of carboxylic acids, was reported on by Kajdas et al [381]. Shestopalov et al [382] suggested that a 5 to 10% emulsion of polyvinyl acetate or a 2 to 3% suspension of LDPE increases HSS drill life because the surface is protected from oxidation by hydrogen released from the decomposition of the polymer, with subsequent diffusion of carbon into the cutting edge.

f. Corrosion inhibitors such as sodium nitrite frequently increase wear [361,383], whereas bromine and iodine derivatives of benzoic acid imparted both E.P. and antioxidant properties [384]. Interactions with the water may change performance: Yang [385] found that hard water improved performance by forming calcium salts of tall-oil fatty acids and petroleum sulfonates, but depleted oleic acid soaps by forming scum.

g. Aqueous electrolytes, obtained on incorporating a halogen in the fluid, were shown by Postnikov [3.84] to promote adsorption of linear polar molecules. Electrochemical effects are discussed by Kuznetsov et al [386]. The electrical double layer (Fig.3.41) is significantly affected by an externally imposed potential. For example, Westwood [3.225] quotes the work of Latanision et al: on imposition of a +1.6 V potential, material removal rates doubled and surface finish improved in cutting of a nickel-base superalloy because the metal was brought into the passivation (oxide formation) range and because the harder, less ductile surface improved the machinability of the otherwise tough, highly strain-hardening material. In contrast, a hard, tempered 4140 steel was made more machinable by imposition of a -0.5 V potential, which brought it into the active dissolution range.

h. It should be noted that aqueous media can have significant effects on the properties of tool materials, especially ceramics [3.220,3.225], but little work seems to have been done on this aspect.

Solid Lubricants. We have already discussed films formed *in situ*; there are a few instances where a solid lubricant is externally introduced.

a. Graphite is seldom used, because deposition in the cutting zone is difficult and because cleanness is a problem. Boston et al [329] found that colloidal graphite improved tool life when added to the oily phase of the emulsion during manufacture, but was ineffective when added to the diluted emulsion.

b. Molybdenum disulfide added to mineral oil improved surface finish in turning of steel with HSS in the work of Banerjee [387]. Increased production rates were noted in milling and sawing with aqueous MoS_2 suspensions [388]. Bagchi [389] attributed the improvement obtained in finish machining to a reduction of the thermal emf. However, addition of MoS_2 to sulfochlorinated oil reduced drill life [375]. A layer of MoS_2 deposited on the tool surface reduced the cutting force for the first 10 seconds of cutting in the work of Meyer, quoted by König [290]. More recently, efforts have been made to coat the tool with a bonded MoS_2 film. Much depends on the strength of the bond. Rowe [24] noted that the film was quickly removed from the nose region but continued to inhibit the growth of the contact length L. One coating was bonded with mono-aluminum phosphate and chromium oxide applied in the form of an aqueous slurry, cured at 200 to 250 C [390]. It reduced friction for the first 2 metres of sliding during tapping, but continued to limit rake-face adhesion and thus delayed clogging of the tap [391]. It was effective also in milling [390]. Success in the laboratory is seldom matched in production, partly because of decomposition of the coating in humid air [392]. The benefits decrease or disappear when an E.P. oil is used [390]. Particularly in turning, an uncoated tool used in conjunction with a cutting oil was as good as a coated tool [391].

c. Soft metals have occasionally been investigated as solid lubricants [393]. A copper film electrodeposited on a milling cutter improved life in cutting of nickel-base and titanium alloys [394]. A continuing supply of copper is ensured by cutting in a copper sulfate electrolyte [394a]. A soft metal serves not only as a film of low shear strength (as does a tin coating on HSS in cutting of aluminum) but may also affect adhesion. For example, it increased adhesion in cutting of lead with a tinned tool. As suggested by Rowe and Smart [395], cutting is an excellent method for studying adhesion because of the large proportion of virgin surface generated.

d. Several papers by Shchukin and other Soviet workers [371,396] show that low-melting metals can greatly increase material removal rates in cutting of hard materials. The mechanism is interpreted in terms of the classical Rehbinder effect and is an example of liquid metal embrittlement. Different alloys are needed for different workpiece materials. A potential problem is that metal tools are also attacked, and the workpiece surface may be damaged [371]. More practical may be the proposition to use a suspension of small (below 80 μm) particles of a metal such as a Sn-Zn eutectic. This allowed a steel with a hardness of 36 to 40 HRC to be drilled with a HSS tool [396].

11.4.5 Application of Cutting Fluids

The inaccessibility of the cutting zone makes the function of cutting fluids difficult enough; poor application methods can render them useless. The fluid must be applied so that it can control friction or contact length on the rake face, reduce friction on the flank face and secondary rubbing surfaces, cool the critical parts of the tool, and, not least, help in removing chips from the contact zone. Several general discussions are available [4.20,4.81].

Manual Application. The application of a fluid from a squirt can or in paste form to a drill or tap is still acceptable in job-shop situations. In turning, manual application is inadequate. For example, lubricant applied with a brush allowed twice as much wear as did lubricant applied in a flood [354].

Application by Flooding. Most machines are equipped with at least a small recirculating system. For convenience, lathe tools are usually flooded from above (Fig.11.46a), even though tool life increases when the lubricant is directed into the clearance crevice (Fig.11.46b) [397]. A combination of the two is sometimes practical. The quantity applied is critical: as a rule of thumb, 5 L of coolant per minute for each kilowatt of motor power has been recommended [398]. The Machining Data Handbook [12] recommends 15 L/min for each simultaneously engaged cutting edge in turning and milling, 3 to 5 L/min for each millimetre of bar diameter on screw machines, and 0.3 to 0.5 L/min for each millimetre of drill diameter.

Worthwhile improvements were noted by Pigott and Colwell [399] when the fluid was directed under a pressure of 200 to 4000 kPa into the clearance crevice (Fig.11.47). In this so-called Hi-jet process, heat extraction increased and tool life improved as fluid pressure was increased to 280 kPa, but further increases reduced tool life; it was presumed that the jet had such a high speed that it reduced boiling and heat transfer. Nagpal and Sharma [400] found emulsions and straight mineral oils to be similar in

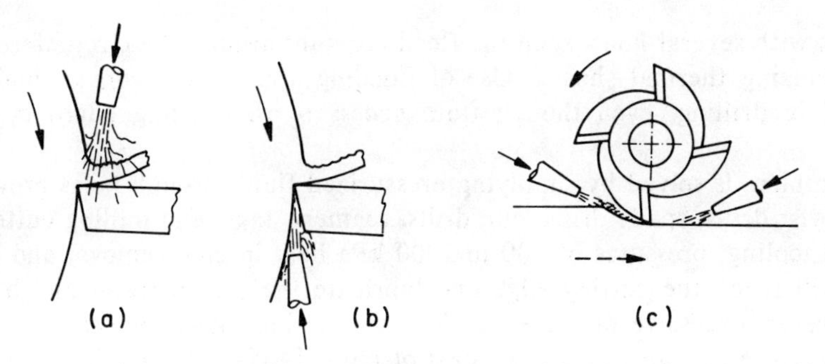

Fig.11.46. Application of cutting fluid

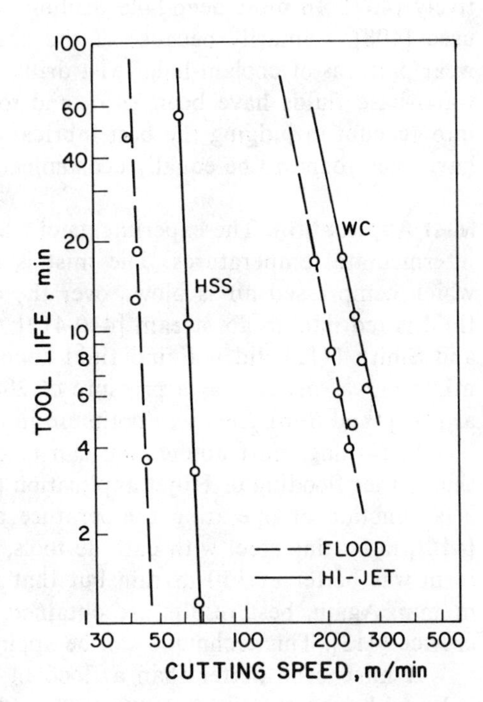

Fig.11.47. Effects of fluid application on tool life in cutting of annealed 1045 steel (fluids: oil with HSS, 40:1 emulsion with WC) [399]

effectiveness except that the optimum pressure—as judged from tool temperature, cutting force, and chip curvature—was a function of feed with emulsions but not with oil. Misting is a problem, and, for operator safety, automatic shut off at the end of operation is desirable.

In the technique of jet infusion, Kishi et al [401] directed a stream of emulsion at pressures of 500 to 2000 kPa into the chip/rake-face crevice in cutting with a carbide tool. Cutting force dropped 10 to 20% at the optimum pressure of 1500 kPa; the effect was even more beneficial in a parting-off operation.

The most direct application was attempted by Sharma et al [402], who injected a high-pressure fluid (up to 70 MPa) into a hole placed at a distance of one to three times the feed from the tool edge. Friction was reduced but the hole became blocked in cutting with a straight or chlorinated oil, but not with a sulfurized oil, indicating that the higher temperature resistance of sulfur was important.

In principle, the lubricant should be directed into the clearance crevice, but a second nozzle may be necessary to clear away chips from milling cutters (Fig.11.46c). Flattened, fan-shape nozzles help distribute the fluid to wide-face slab mills, and ring distributors with several holes keep the flood constant around the circumference of face mills, minimizing thermal shocks. Use of flooding application from a single nozzle is widespread in drilling, even though fluid access to the cutting edges is difficult to ensure.

The problem is solved by supplying pressurized fluid through holes provided in the bodies of twist drills, spade drills, gun drills, reamers, taps, and milling cutters. In such coolant-fed tooling, pressures of 300 to 2000 kPa help in chip removal and ensure that the fluid will reach the cutting edge and lubricate the drill margins and bearing pads [403]. However, excessive pressure can lead to floating [404]. Pulsating fluid pressure increases tool life and aids in the removal of chips [405,406]. Small, measured quantities of fluid delivered to the backs of tool inserts evaporate and cool the inserts effectively [407]. In most deep-hole drilling and deep-bore finishing, oil-base lubricants are used [408], primarily because of the need to minimize friction on bearing pads. The wear patterns of coolant-hole twist drills vary according to the fluid used; both oil- and water-base fluids have been used, and tool, fluid, and regrinding costs must be taken into account in judging the best lubricant. Sulfochlorinated oils and chemical solutions have been found to be equally economical [409].

Mist Application. The experiments of Shaw et al [308] showed that mist is effective at intermediate temperatures. The mist is usually generated by a syphon (aspirator) in which compressed air is blown over the open end of a tube; alternatively, pressurized fluid is fed into an air stream [410-412]. Fluid consumption is low, below 1 L/h. Shaw and Smith [412] did not find fluid concentration to be critical, but a minimum of 6 mL/min was needed (at a pressure of 200 to 500 kPa) in the work of Proskuryakov et al [413], and 6 mL/min was optimum in the work of Rao [414].

In turning, mist application can lead to less flank wear and better surface finish than either flooding or Hi-jet application [414]. Performance relative to that of flooding is a function of operating temperature and thus of speed. Kececioglu and Sorensen [415], in cutting steel with carbide tools, found that application of an emulsion in mist form was better at 150 m/min but that flooding with the emulsion was better at 240 m/min. Again, best results are obtained when the mist is directed into the clearance crevice [413]. This technique can be applied also to end mills and slab mills [411,413].

A mist is no better than a flood in reaching the cutting zone in drilling of deep holes with a twist drill, but it is preferable for drilling on large panels. Rao et al [416] applied the mist from below, through a pilot hole drilled in a plate, and found optimum mist concentration to increase with speed.

Either oil- or water-base fluids can be used for mist application. Special oil formulations and automatic shutoff are helpful in keeping airborn mist at acceptable levels. Demisters can be installed to reduce clouding of the cutting region. A separate flooding system may be necessary to cool the workpiece.

11.4.6 Selection and Treatment of Fluids

Of the large number of publications on this topic, only a few representative ones, of various ages, are quoted here. Some treatments—[3.56],[4.3,4.20,4.21,4.274],[5.17], [417-431], Sec. 16 of [12]—are general. Other publications emphasize additives [432], aqueous fluids [433], synthetics [4.81], and fluids for difficult-to-machine materials [434,435]. Commercial sources of cutting fluids within the U.S. are listed in [436] and in Sec. 16 of [12].

Selection is based on the severity of duty [437]. No firm ranking can be made, but the sequence shown in Table 11.2 (at the end of this chapter) represents increasing severity of conditions. Of the lubricant classes discussed in Chapter 4, oils and aqueous (water-base) fluids have special significance.

Cutting Oils. Fluids based on mineral oil are less effective coolants and are more expensive, at least in initial outlay, although their long life often makes them competitive with aqueous fluids. They are necessary when aqueous fluids would cause staining (as on some nonferrous metals), would not stay in the cutting zone (as in broaching, gear cutting [420], and deep-hole drilling [408]), would contaminate the hydraulic fluid, or would wash the lubricant off of machine elements (as on automatics [438]). In order of increasing severity of duty, they are usually classified as follows:

a. Straight oils (Sec.4.2) typically are naphthenic, and have viscosities of 5 to 50 mm^2/s at 40 C. Sometimes synthetic oils are used. Neat fatty oils have been replaced by compounded mineral oils.

b. The term "compounded oil" (Sec.4.5.1) is often used in the narrower sense to denote mineral oil to which fatty additives (up to 40% lard oil or castor oil) have been added to reduce friction and improve surface finish, especially for cutting of nonferrous metals in automatic screw machines.

c. Compounded oils containing sulfurized fat do not stain copper immersed at 100 C; hence, they are often called nonactive E.P. oils.

d. Chlorinated oils contain chlorinated paraffin. They are relatively ineffective for steel but are indispensible for stainless steels, for soft, "draggy" metals, and for high-silicon aluminum alloys.

e. Sulfurized oils contain either elemental sulfur or additives of varying activity. The composition is usually described by total S (typically 0.5 to 2.0%) and active S (0.3 to 1.0%).

f. Sulfochlorinated oils contain sulfur and chlorine additives or sulfochlorinated fats. The composition is usually 1 to 4% total S, 0.75 to 3% active S, and 0.1 to 10% Cl.

g. Sulfochlorinated compounded oils with fatty additives offer the broadest temperature range of protection (Fig.4.15).

Multipurpose oils are formulated to serve also in hydraulic systems of machine tools, thus totally eliminating the danger of contamination. Heavy pastes are still useful for hand reaming and tapping.

Aqueous Fluids. Water-base fluids are higher in cooling capacity, but generally lower in lubricant efficiency, than oil-base fluids. The terminology given in Sec.4.6 is fairly widely accepted.

Often the same fluid can be used, at different dilutions (ranging from 10:1 to 40:1 water to concentrate), in various processes. Water-base hydraulic fluids have been formulated that can be diluted for service as cutting fluids.

Economics. Cutting fluids may be expensive on a unit volume basis, but, when total production costs and the contributions of fluids to productivity and quality are considered, the costs become minor. In judging the economies of alternative fluids, not only their initial costs but also the costs of makeup, cleaning, handling, and disposal, the effects on the working environment, the actions on the machine tool, and the possibility of reclamation from the swarf must also be considered. The economics of selection is discussed in [4.20,4.21], and the cost of treatment in [439]. Suggestions on trouble-shooting can be found in [4.20].

Treatment. The techniques described in Sec.4.8 find increasing application in the treatment of cutting fluids. General discussions have been given [4.138],[439-444]. Recirculating systems for individual machine tools tend to be relatively simple, but larger, centralized systems have the same features as those used in rolling. Only examples of articles dealing with magnetic separation [445], cooling [446], and filtration can be quoted here. Filtration of coarse particles with a 50-to-100-mesh screen is usually adequate, but tighter requirements are set in processes where frictional surfaces operate. Thus, in deep-hole drilling, filtration to 40 to 60 μm is needed for a surface finish of R_a = 4 to 6 μm, and to 10 μm for finish boring [408]. A diatomaceous earth positive-pressure filter removes fines to 1 μm [447]. Ultrasonic activation of grinding fluids improved both G ratio and surface finish [448].

Reclamation of oil from the chips (swarf) may be economical; more importantly, cleaner material is more acceptable for recycling [449]. Centrifuging is well established for this purpose [4.288],[450].

11.5 Testing

Tribological testing in metal cutting follows the principles discussed in Sec.5.1. Difficulties are greater than those in deformation processes, because steady-state conditions seldom exist for a long enough time to allow a quick exploration of variables.

11.5.1 Principles of Testing

We have seen (Sec.3.10) that wear rates are subject to much larger variability than friction. This creates one of the inherent difficulties of testing, and strategies must be devised to minimize variations. In doing so, the entire system (Fig.3.2) must be kept in mind: workpiece material, tool, process conditions, and fluids are equally important. Outputs (friction and wear) serve as immediate feedbacks into the system. Hence, there is no possibility of conducting, say, a simple wear test on a tool material; the machinability of the workpiece material, the effectiveness of the cutting fluid, the wear resistance of the tool, and the influences of process variables are always tested simultaneously. Indeed, a given technique may be used to evaluate any of these factors. This is true, of course, of all tribological testing; unique in this case are the rates of changes and the extreme sensitivity of the system to the variables. The major inputs have been discussed by several researchers [167,451-453].

Workpiece Material. In plastic deformation processes the relevant properties, such as flow stress and ductility, can be determined, even though some difficulties may be encountered. Variations in properties are not excessive (typically ±20% of the nominal value). In metal cutting the relevant property is machinability, a topic of great importance, as shown by the number of conferences devoted to it [14,16,17,454]. Unfortunately, this term defies precise, quantitative definition [455,456]. A material is highly machinable when satisfactory parts can be made from it at low cost, with minimum difficulty. Machinability is treated as an input to the cutting system, but it is also one of the outputs:

a. A satisfactory part must have an acceptable surface finish and acceptable tolerances. The basic surface finish is defined by process geometry (e.g., in turning, tool-nose radius and feed) but is greatly modified by the process of chip formation, by the presence of BUE, by the cutting fluid, and by the dynamic response of the system. Superimposition of process-related roughness on the geometric (or, rather, kinematic) roughness has been discussed by Lonardo [457], and problems of surface evaluation have been reviewed by Dinichert et al [458] and by Whitehouse et al [459]. Tolerances are affected by tool wear, by temperature, and, again, by dynamic response.

b. A surface of good appearance may actually be damaged by microscopic tears, defects, structural changes, or internal stresses that impair the performance of the part. Indeed, the high demands placed on components subjected to severe service conditions have given rise to much research on surface integrity and related measurement techniques, as reviewed by Field et al [460,461]. This topic is discussed in great depth in Section 18 of the Machining Data Handbook [12]. Effects of machining on microhardness and residual stresses have been reviewed by Tönshoff and Brinksmeier [462].

c. Metal-removal rates must be high for economy of operation, but increasing feed and speed result in higher forces, higher temperatures, and more rapid tool wear. There are usually two optimum material-removal rates—one for greatest economy, and a higher one for highest production rate. The problem is closely linked to the mathematical modeling of tool life and has been the subject of extensive research [135,463]. Techniques for the simultaneous evaluation of several variables with the aid of cutting rate/tool life (R/T) functions have also been developed [464].

d. The chip produced must be easily removable from the chip-forming zone. Discontinuous or tightly curled, easily broken chips give the best machinability from this point of view.

e. A material that can be cut with a lower force (specific energy) is regarded as more machinable, even though this is a lesser consideration with modern machine tools. A low strain-hardening rate is helpful from the point of view of chip formation.

f. Low thermal conductivity and low specific heat result in higher cutting temperatures (Eq.11.10), discontinuous chip formation, poorer surface finish, and severe tool exposure.

g. The relative ranking of materials depends also on the process. Even for the same process, the characteristics (primarily, stiffness and dynamic response) of individual machine tools can elicit quite different responses.

h. A material of a given composition may have quite different machinabilities under different metallurgical conditions. In general, low conductivity and strength are desirable for chip formation, and, if a compromise has to be made, it is preferable to have low ductility and higher strength. However, very high strengths place severe demands on the tool, and then a lower-strength, higher-ductility condition is preferable.

i. The machinabilities of nominally identical materials can vary substantially. The coefficient of variation is about 1/5 of the mean for pedigree materials and may be 1/2 of the mean under production conditions [167]. One of the most powerful variables is hardness, which is directly related to shear flow stress and thus to energy expenditure. We have seen that tool life is greatly dependent on temperature, and thus minor variations in machinability can have serious consequences on repeatability [167,465]. Therefore, the first requirement of tribological testing is a workpiece material of uniform and constant machinability, as determined by mechanical properties and metallurgical structure. Since uniform properties are more difficult to attain in large cross sections, the size of the test pieces is usually limited.

Tool. In most tests the tool is used until it has to be reground or until it is destroyed. Thus, testing involves a number of tools; a reground tool must be regarded as a new one because of possible incipient damage or changes in material properties. The first task is to ensure that the tool lot is homogeneous enough [466]. We have already referred to the variability of tool life with carbide and ceramic tools; less-pronounced but still significant differences can be found with HSS tools. For example, Lorenz [188,467] found by a statistical evaluation of twist drills that drill life may vary by a factor of 5 to 8 in homogeneous production batches and by a factor of 130 in off-the-shelf lots.

Tool preparation is of extreme importance and must include control of surface finish (roughness and directionality) and edge sharpness. If tools are reground, the entire heat-affected zone must be removed. Damage in grinding must be avoided, and at least the finishing passes must be gentle. If possible, new tools or inserts should be used.

Cutting fluid, the third key element of the tribological system, is perhaps more readily available in a reproducible form but is subject to gradual changes during use. The application method and fluid flow rate must be reproducible and representative of production methods. Therefore, a recirculating system equipped with at least a filter is a prerequisite of meaningful testing, with some provision for measuring and, if possible, controlling the temperature and flow rate of the fluid.

Equipment. The equipment must have adequate power and stiffness to allow testing at relevant speeds and feeds, without dynamic instabilities. The stiffness of the tool holder and of its support are particularly important. For many tests, a continuously variable drive is desirable to maintain constant cutting speeds even when the workpiece dimension changes. If the test involves a series of speeds, closely spaced, discreet steps may be acceptable. A wide range of feeds, typical of production practices, should be available.

Testing Strategy. The principles discussed in Sec.5.1 apply, but the variabilities of workpiece and tool materials create a more complicated situation. Reviews of problems have been given by Micheletti [455] in a report on machinability testing and by Tipnis et al [167,182]. Several approaches may be taken:

a. If the purpose is a direct comparison of cutting fluids or tool materials, short-time or accelerated tests can be run under laboratory conditions. Workpiece and tool materials are carefully chosen for minimum variability, and, if necessary, preliminary experiments are carried out to determine the statistical significance of the results [467]. A baseline is established using a reference tool/fluid combination, and experiments with this combination are periodically repeated to compensate for any possible drifts. Tools are selected from a single production lot, preferably of known and documented history, subjected to careful quality control. A workpiece material is chosen that is available in uniform quality, with minimum variations, in a hardness that gives a reasonable tool life. If uniformity from workpiece to workpiece cannot be ensured, each workpiece is referenced to a standard condition (cutting dry or with a reference fluid).

b. If the purpose is generation of data for production planning, experiments are statistically designed. Both workpiece and tool materials may be selected by statistical sampling techniques. To improve reproducibility, tests are preferably conducted under laboratory conditions. For example, steel companies often possess automatic screw machines and conduct tests involving several tons of steel.

c. Field tests can yield meaningful results only if the data are taken in a long-run mass-production situation. Examples are given by Morton [468]. With the spread of numerical-control (NC) machining centers, on-line measurement of variables is increasing in importance [5.238], and, with suitable sensors, collection of data under production conditions is becoming more practicable.

There is basically no difference between cutting-fluid and wear tests. Since wear tests can seldom be accelerated without losing relevance (Sec.5.9), the test must be run under conditions typical of the intended application; extrapolation and transfer to other processes can lead to misleading conclusions [453].

Selection and Measurement of Variables. Scattered throughout our discussion have been references to the variables that are affected by tribological conditions; it will be useful to recapitulate them together with methods for their measurement. Sensors for in-process measurement have been reviewed by Micheletti et al [5.237] and by Groover [469].

1. Forces and Energy. Cutting force and energy change with friction and wear, but not necessarily in a clear-cut manner.

In turning, Micheletti et al [470] found that the increase in cutting force due to flank wear was balanced by the decrease due to crater wear; however, feed force increased with flank wear and was little affected by crater wear. Hence, feed force is a suitable indicator, as found also by König et al [471], whereas cutting force is relatively insensitive to friction changes [308]. Force is also one of the variables suitable for on-line control. For small feeds, the variation of cutting force with undeformed chip thickness is approximately linear. The slope of the straight line and its intercept extrapolated to $t = 0$ increase with friction, and Kalaszi [183] used this for ranking of lubricants in orthogonal cutting. Colwell [472] observed that the signal was most sensitive to tool wear during the start of a cut or during a dwell period, and on-line analysis of these changes can give early warning of excessive wear [473]. In continuous cutting, the cutting force may be adequate for the control of roughing operations [474].

In drilling, feed force and torque change sufficiently to serve as indicators of tool wear. Tapping torque is sensitive to friction and is more suitable for lubricant than for coolant evaluation.

Wear of the tool and, particularly, edge chipping or tool fracture result in a change of vibration modes, and vibration analysis offers an on-line sensing method.

2. Chip Formation. The chip-thickness ratio is an indicator of the shear angle but gives false information if a BUE is present or if chip formation is discontinuous. The tool/chip contact length L may be found by inspecting the tool for deposits or wear patterns, and can be explored with restricted-contact-length tools (Fig.11.35) [303]. In oblique cutting, the direction of chip flow changes and thus the length-to-width ratio of the chip increases with lower friction. This was used by Bezer and Oates [475] for lubricant evaluation in a shaping operation, with the tool oriented at 45° to the cutting direction. Sensors for chip-form detection have also been used [476].

3. Temperature. All techniques discussed in Sec.5.6 have been utilized. Several reviews are available [100,101,477,478].

Information on temperature distribution within a tool has been obtained with embedded thermocouples by Trigger and Chao [478] and from changes in the structures of HSS [5.154,5.155] and iron-bonded carbide tools [248] by Trent and co-workers. Temperature distribution can be studied, but only on the side surface, by infrared photography. Originally, infrared photography required that the workpiece be heated to 600 C [9], but recent developments have removed this restriction [5.145,5.146]. Temperatures on the flank face have been measured by observation through holes drilled in the tool [478] or in that part of the workpiece material that is to be converted into chip [479]. For in-process measurement, two techniques are potentially attractive:

One is the dynamic thermocouple that was shown to be largely unaffected by lubricants, at least under some conditions [306]. The tool and workpiece are insulated from the machine tool, and a mercury-wetted slip ring or a chip of the workpiece is used to make contact. The sensitivity of this method is open to question. Colwell [472] and Kretinin [480] found it to be adequate, and Redford [481] observed that temperature showed less scatter than did cutting force. However, the results of Zakaria and El-Gomayel [482] indicate that the change in emf can be as low as 4% during the life of the tool, and much of the increase is made up by noise.

The other method relies on a small thermocuple embedded in a HSS tool [101] or under a carbide insert [483]. The simplicity of this method is appealing, even though only a bulk temperature—away from the hottest zone—is obtained.

4. Electrical Resistance. Despite the intimate contact between tool and workpiece, a small resistance (approximately 1 mΩ) exists and was shown by Wilkinson to offer on-line capability [484,485].

Lubricant films in bearings develop a resistance of approximately 10 Ω between the rotating shaft and the machine tool frame. When a low voltage is passed through a simple ball contact placed at the end of the spindle, the resistance in the tool/workpiece contact zone presents a current path parallel to the machine frame. Changes in contact conditions result in resistance changes large enough for in-process measurement [486].

5. Acoustic Emission. The cutting process provides several sources of acoustic emission, and analysis of the signal may provide insight into details of the process. Dornfeld and his co-workers [487] linked energy levels to tool wear. In-process monitoring of wear is possible [488], and sudden bursts of energy are a sign of chipping, cracking [489], or total failure [5.261].

6. Wear. There are numerous ways of determining wear:

a. Flank and trailing-edge wear are normally measured under a microscope. Several proposals for in-process sensing have been made, including measurement of the workpiece dimension and of the gap between tool holder and workpiece. Uehara [490] coated the clearance face with a graphite or metal film to form a resistor, the length of which changed with flank wear. Measurement of the tool/chip contact length can be included in the evaluation.

b. Crater depth is usually measured with a light-section microscope. More detailed information can be obtained from plastic replicas [491]. Loladze et al [492] have shown that inspection of the rake surface under a microscope, after only a short cutting time, yields information in agreement with long-term tests. De Chiffre [378] found a 5-min test to give sufficient differentiations in size and shape of crater and flank wear on carbide tools to allow lubricant ranking. In another short-time test, Uehara et al [493] separated WC particles from the chip surface by pickling and, after conversion of the tungsten into thiocyanate, measured the absorbency of the solution. No scheme for in-process measurement of crater wear seems to have been proposed.

c. Radioactive tools have long been used in wear studies. A critique has been given by Cook and Lang [494]. Flank and crater wear can be separated by measuring beta radiation on the rough and smooth sides of the chip, respectively. Alternatively, crater wear can be judged from the activity of the chip produced by the radioactive tool, and flank wear by the activity of the chip produced by a second, nonactivated tool [495]. A problem is that groove wear—which is seldom a life-limiting factor—confounds the results.

When the remaining activity of the tool is measured, flank and crater wear can be separated only if a wear model is assumed. Thus, Chawla and co-workers [496] proceeded on the basis that flank wear was proportional to the square of time, whereas crater wear progressed linearly. The sensitivity of this method is sufficient to allow short-term testing, with samples taken over 15-s periods. Total activity is decreased and sensitivity increased by the use of an irradiated tool insert. With a single tool, 3 to 4 different cutting conditions can be explored in tests of 1-min duration [5.243].

Thin-layer activation [5.245-5.249] has the advantage that radiation intensity is low and tool wear gives large changes in activity. Activation can be limited to the tool area of interest, such as the flank in turning [5.249] and the cutting edge in drilling [5.245]. Various evaluation techniques have been adopted. Amini et al [5.245] simply measured the remaining activities of twist drills after drilling groups of holes. Vamos [497] used a new tool for each 12-min turning test, conducting three tests per oil. A short-term test technique has been reported on, in a series of papers, by Ivcovic [5.249]. Since activity decreases exponentially with flank wear, wear rates are measured in tests of only 0.1- to 1-min duration. At least 6 tests are needed to define the wear function, but, with a possible 24 to 30 tests on a single edge, 4 to 5 cutting fluids can be evaluated. Further examples of the use of this technique are given in [3.56]. Of course, all such approaches are based on the premise that the history of the tool does not affect its wear; this assumption is acceptable in the steady-state wear regime for limited lengths of time. When the tool is accessible to radiation counting between engagements, the technique can be considered for on-line measurement [5.145].

A micro-isotope implanted into the tool [5.244] offers an on-line indication of when limiting wear is reached.

7. Surface finish is a sensitive indicator of changes in chip formation, lubrication, and wear mechanisms [30,308]. Even though quantitative relationships cannot be defined at this time, measurement of surface finish—and of its change relative to a standard (starting) condition—yields useful information in research, and may be used as the principal tool-life criterion in finish turning [5.17]. There are also opportunities for on-line evaluation by various noncontacting techniques (Sec.5.7.1), including optical evaluation by use of a laser beam [498].

11.5.2 Test Methods

Some techniques have gained widespread application.

Turning of a cylindrical bar on a rigid lathe yields results applicable to many cutting operations, and is the subject of standards (Appendix). A reasonable tool life is 60 min for HSS and 30 min for carbides. One of the commonly used workpiece materials is a low-alloy steel such as AISI 4340, because of its good through-hardening properties. Zlatin and Christopher [453] recommend a hardness of 300 BHN, a bar diameter of 100 mm, and a length of not more than 500 mm, so that adequate stiffness is maintained when the diameter is reduced. Micheletti [455] recommended a length-to-diameter ratio not exceeding 10:1. Scale from rolling or heat treatment must be removed. Tool geometry is taken from practice. Feed is typically 0.2 to 0.25 mm, and depth of cut is at least 5 and preferably 10 times the feed and is deep enough to engage the major cutting edge beyond the nose radius. Fluid flow is 15 L/min.

1. Turning of a Cylinder. In the basic form of the test, turning at a constant speed is interrupted at predetermined intervals, the tool is inspected, and *VB*, *VN*, *KT*, etc. are measured. The test is continued until some wear criterion (Sec.11.2.3) is satisfied or total tool failure is reached. Failure may be judged from the sudden change in force or power, although visual observation of a bright band on the uncut shoulder and a change in noise level give early warning [499]. The remaining shoulder must be machined off with a carbide tool to remove burr and embedded remnants of the tool. When experiments are conducted over a wide range of speeds, results can be plotted as force versus cutting speed (as in Fig.11.38 and 11.43) or, more frequently, as tool life (time to reach a given *VB*, *KT*, total failure, etc.) versus cutting speed (as in Fig.11.31, 11.34, and 11.37a). A log-log plot (as in Fig.11.30, 11.37b, and 11.47) immediately reveals whether the Taylor relation holds, and then n and C may be extracted. Other numerical values are speeds for a 20- or 60-min tool life (VB_{20}, VB_{60}, KT_{20}, etc.). Since there is no assurance that the ranking of lubricants remains the same, two speeds are often quoted.

Tests must be conducted at a minimum of four speeds, all within a range characterized by a single dominant wear mechanism. This is usually the temperature-sensitive range before catastrophic failure. Sensitivity is lost if speeds fall within the BUE range. Similarly, cutting at low speeds (e.g., at 14 m/min) [500] may not be demanding enough to reveal differences in additive content. Excessive speeds, while giving fast results, are not representative, and cannot be extrapolated to normal speeds because of the nonlinearity of the Taylor plot at high speeds. Thus, it is doubtful that meaningful constants can be extracted from speed to failure at 1 and 10 min.

Tool bits are usually reground to ensure correct angles and surface finish, but inserts are best used as-received. Even if it is assumed that the tool is of constant quality, it is unlikely that the large quantities of workpiece material required will have constant properties. Oathout et al [499] calibrated the entire experimental lot (Fig.11.48) by running some 20 tools to failure, dry, at varying speeds, on 4 to 5 bars. The reference level for each bar was then established by a dry cut at 25-mm-diam intervals, and bars that gave reference speeds more than 4 m/min outside the standard were rejected. A speed range of 30 to 40 m/min took HSS tools through a 1- to 20-min range on annealed 4340 bars.

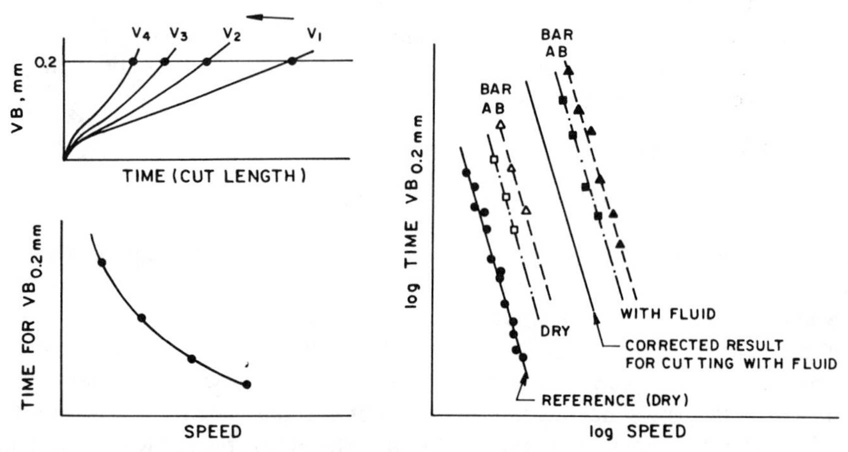

Fig.11.48. Principle of procedure for normalizing tool wear for variability of workpiece material [after 499]

The material quantity required for the evaluation of a single lubricant is large (on the order 100 kg), and the required time is long (on the order of one week).

2. Short-Term Tests. The time and expenditure required by full testing programs have resulted in the adoption of various short-term methods.

In the simplest scheme, direct comparison of lubricants is made at a constant cutting speed, by taking half the length of the cut dry and the other half with the test lubricant, and by correcting all data for the dry level [501]. To allow for speed changes on a fixed-speed machine, weighted averages may be taken; notching of the bar allows the cut to run out, avoiding speed changes while the tool is engaged [502]. There is, however, a danger that at some different speed the lubricant ranking may change.

The two test conditions (e.g., dry and lubricated) may be alternated at short intervals [3.84]. This is the basis of the yet shorter tests utilizing thin-layer activation. With conventional radioactive tools, the reference point can be established by increasing and then decreasing the speed in three increments; tool life calculated from the medium speed was found to be in good agreement with results of conventional tests [496].

In a variant of this test, Boulger et al [503] applied a constant feed force by dead-weight loading and took the resultant feed rates as the dependent variable.

In boring of several holes in a thick plate, tool wear can be directly measured from the change of hole diameter [378], but the plate material must be homogeneous; in this respect, the test is similar to drilling. Instead of wear, the mean feed rate and torque for a given feed force can be registered [504].

3. Variable-Speed Tests. Several approaches are feasible:

a. When a cylinder is faced, starting from the edge of a hole drilled in the center, the tool is subjected to continuously increasing cutting speeds [505,506]. On a large-diameter cylinder the tool wears out in a single cut, but it is doubtful that material properties can be kept uniform. The results can be put into the Taylor form [507]. When several cuts must be taken to obtain one datum point, the tests are repeated at different spindle speeds, and the equivalent speeds are calculated for each [506]. Instead of wear, temperature can be measured by the dynamic thermocouple technique; for reproducibility, the tool is first run in and recordings of the fourth cut are taken as representative [378]. The reference line is again established by dry cutting. To allow a complete run in a single cut, Bäcker et al [508] used an AISI 52100 tool at 60 ± 1 HRC, which had only 30% of the life of a HSS tool. Measurement of the cut profile gives tool wear (Fig.11.49). With testing at 4 spindle speeds, 16 to 20 measurements provide an evaluation.

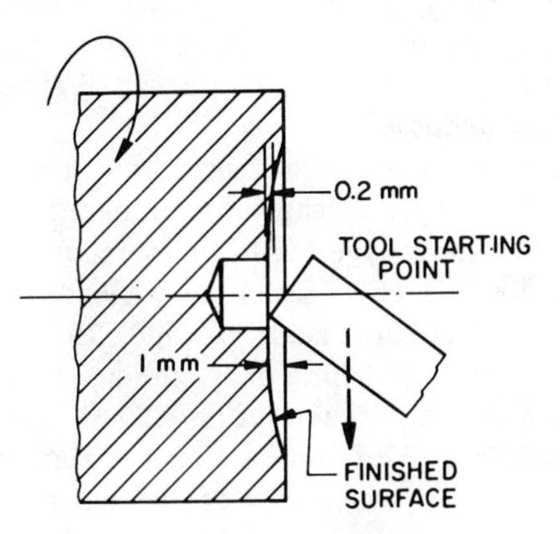

Fig.11.49. Wear testing by facing [508]

b. The need for a large-diameter workpiece is eliminated by turning a taper. The results can be put into the Taylor form, and, as shown by Heginbotham and Pandey [509], agreement with n and C values from conventional tests is good.

c. A continuously variable-speed drive allows acceleration in a single cut at a constant diameter. Good correlation with conventional tests is sometimes found [510]. The length of chip cut to failure of a HSS tool, or the wear of a carbide tool for a given period of cutting, may be used as a wear criterion [190]. Alternatively, speed at failure is compared with speed in a conventional test, on the premise that the speed for tool failure is independent of past history [207].

d. Stepwise acceleration was found suitable by Opitz and König [207] if correction was made for the length of cut at the failure speed (Fig.11.50). For HSS tools, a starting speed of 20 to 30 m/min is used; speed is increased, in a geometrical progression of 1.12 for each 25 m of cut, until the tool fails. The initial speed is immaterial if failure occurs after at least 6 steps.

Good agreement among conventional, shortened conventional (1 to 8 min), facing, and taper-turning tests was found by Thomas and Lambert [511]. Agreement was further improved when a wear land was prehoned. Analyses of variable-speed testing are available [512,513].

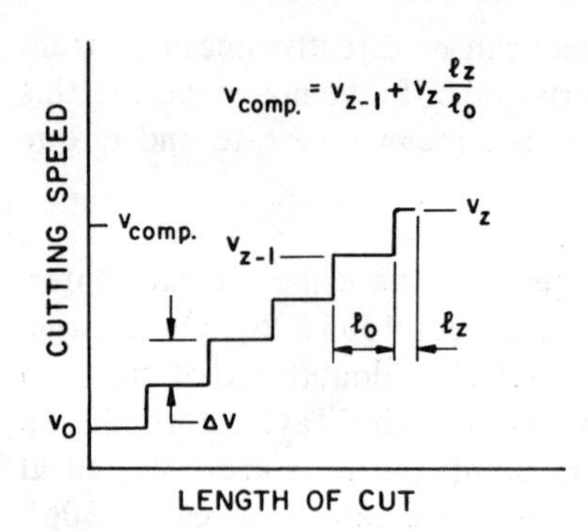

Fig.11.50. Calculation of compensated cutting speed [207]

Other Tests. While turning is typical of many chip-forming operations, the results are not necessarily transferable. Partly for this reason and partly for convenience, several other tests have been used:

a. Parting off is similar to taper turning except that the tool edge and corner are exposed to conditions more typical of screw-machine work [514].

b. Threading of a bar exposes the tool to a temperature distribution different from that in turning. Lubricants can be ranked according to measured torque or maximum cutting speed at which a satisfactory thread is produced [515,516]. In field tests, visual inspection for torn threads allowed Webb et al [517] to differentiate between oils by statistical analysis.

c. Tapping is internal thread cutting. Plug taps with 3 or 4 flutes are readily available. If the hole is carefully prepared, wear of the tap can be followed simply by checking with a gage [518], or by recording torque and temperature [365]. The typical torque diagram (Fig.11.51) has a steady-state portion which can be used as a measure of lubricant effectiveness [365,500,519,520], provided that chip interference (clogging) is prevented by flushing or by the use of a spiral-point tap [521]. A commercial tester, designed at the initiative of an ASTM subcommittee, is now available [518,522]. Webb and Holodnik [523] reported on the statistical analysis of data obtained in two laboratories. The reproducibility of the mean steady-state torque was very good (standard

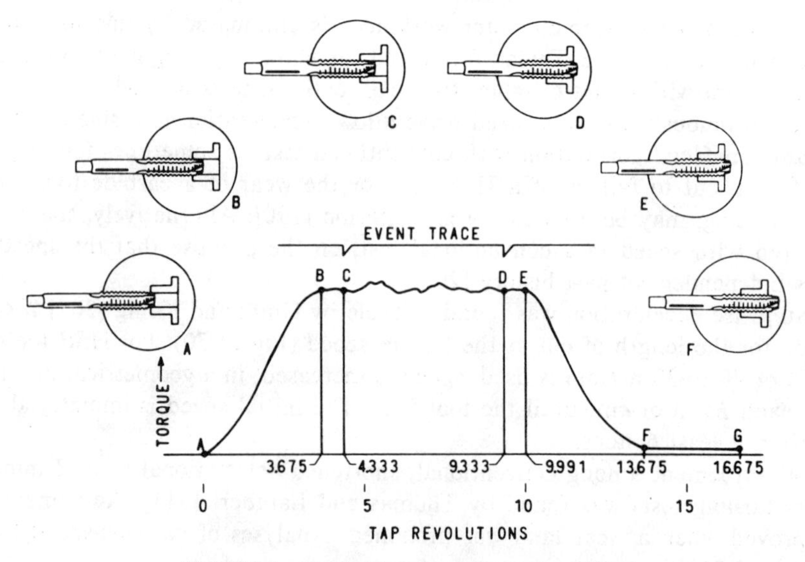

Fig. 11.51. Typical tapping torque trace [523]

deviation less than 2% of mean), and lubricants were ranked correctly when results of the two laboratories were normalized against a reference oil. Critical are the hole tolerance and, to a lesser extent, surface finish. The standard procedure is to follow drilling with reaming and honing; the latter is needed only when form tapping is also planned. A BUE may form and, when lubricants are tested intermixed, has to be removed. A matrix for the testing sequence of ten oils is based on the premise that wear can be ignored. The test is sensitive enough to discriminate among fluids of different field performance [523], although some workers find it less suitable for aqueous fluids [500].

d. Shaping gives periodic engagement, and several thousand cuts must be made if similar oils are to be discriminated [475].

e. Milling is employed relatively infrequently, but end milling has been recommended by De Chiffre [520] as one test in a series. Kitagawa et al [524] used a fly-cutting operation to evaluate flank wear in milling the end face of a cylinder, and thus tested the entire cross section of the bar.

Drilling is of great practical importance and, because it presents lubrication conditions much different from those of turning, has also been developed as a cutting-fluid testing method. Standard HSS twist drills are used, typically with a 118° point angle, a 29° helix angle, and a 12° relief angle [453]. All drills should be reground prior to testing. Skells and Cohen [525] ground a four-facet prismatic point because it allowed more accurate evaluation of edge (flank) wear. A drill of minimum diameter $D = 6$ mm (or 0.25 in.) prevents wandering. Blind or through holes are drilled to a depth of 2 to 3 D, with a 3 L/min fluid supply. The workpiece material is again chosen for uniformity. Type 4340 steel with a thickness of $2D$ and a hardness of 300 BHN has been recommended as a representative steel [453]. Several workers have observed that results are highly sensitive to workpiece hardness.

To ensure reproducible results, drilling also must be performed on a rigid machine kept in good condition [526]. Basically, two techniques are available:

a. The drill is fed at a preset value per revolution, using a power feed; feed force (thrust force) and torque are measured.

b. The drill is fed at a preset, constant force, and the rate of drill penetration and torque are measured [527]. Since the rate of feed decreases with flank wear, axial retardation can also be used as a wear index [528].

In both techniques, drill wear is also assessed, either after frequent stoppages or after a given number (say 100) [501] of holes have been drilled. Edge-wear limit (near the corner) is usually taken at 0.38 mm. Total failure is indicated by noise, smoke, and a change in chip formation [529]. To limit the number of holes to be drilled (say, from 75 to 120 [520]), Sutcliffe et al [530] increased the feed rate, although this may cause the ranking of lubricants to change (Fig.11.44). Testing at a minimum of three speeds is necessary, because tool life becomes very short at excessive speeds and the Taylor plot cannot be extrapolated. Compensation for changes in material hardness [530] is ensured by calibrating against a reference lubricant (dry drilling may be too demanding and may show greater differences than would be found with a lubricant, thus leading to overcompensation).

For drilling, time in the Taylor equation is replaced by the number of holes drilled H. From a log-log plot of number of holes versus speed, n and H can be extracted. Lorenz [188] recommended regression analysis of statistically designed experiments and pointed out that the Taylor plot is nonlinear. Kanai et al [526] found that outer corner wear, drill life, and cutting forces followed a log-normal distribution.

The minimum test program involves three speeds, two drills per speed (if the number of holes is within 10%; if not, more drills have to be used), and a time expenditure of 12 h for a single fluid [525]. Automatic (NC) drilling machines improve reproducibility and decrease the time requirement [500,529].

There is no agreement on the discrimination obtainable by drilling tests. Blanchard and Syrett [500] found that emulsions could be evaluated on the basis of the number of holes but that drilling force did not distinguish among various E.P. fluids. When holes were drilled in rapid succession to allow temperature to increase, there was a sudden, transient rise of about 15% in thrust force, which correlated with E.P. performance. In contrast, Skells and Cohen [525] found drilling force to be a suitable indicator. Drill life has been repeatedly used for E.P. additive evaluation [3.179]. Leep [501] found drilling to be more discriminating than turning or milling.

Drilling with a pilot hole approximates a pure cutting process. With a constant feed force, Henschen [516] recorded the mean feed rate on six holes. Dagnell [528] took flank wear on a flat drill as a measure of lubricant effectiveness.

Transfer of Results. There is reasonable agreement that the order of merit of lubricants is similar in pure chip-forming operations such as turning and milling [453,499], tapping and broaching [453], or tapping, threading, and broaching [519]. Actual values of tool life are, however, different, because engagement times and access of fluid (or air) to the rake face are different. An example, from the work of Kitagawa et al [524], is shown in Fig.11.52.

Correlation between the wear land in laboratory drilling tests and the quality of threads cut in the field was found by Webb et al [517] by statistical evaluation. The relationship between bench tests and drill life is more tenuous. Reasonable correlation for sulfur and chlorine was found by some [3.179], but only modest correlation by others [451]. Yet others found no correlation at all [517,527]. Similarly, tapping torque gave good correlation with welding load in 4-ball tests [3.183] but not with torque measured in bench tests [523]. Similarly, 4-ball test results did not correlate with low-speed broaching performance [475].

Because of the difficulty of transferring results, test sequences have also been recommended. Lindgren [531] used drilling, reaming, and tapping, and also measured cooling rates with a wire anemometer. De Chiffre [520] developed a sequence of drill-

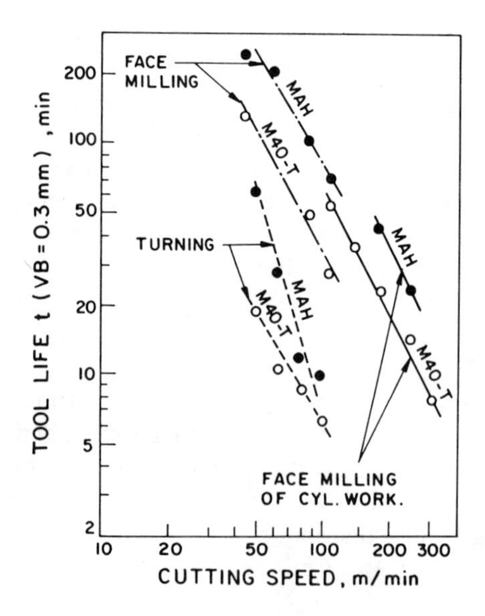

Fig.11.52. Effects of machining process on tool life (normalized 1055 steel, $f = 0.2$ mm/rev, $d = 2$ mm, tools: M40-T = ISO M40; MAH = micrograin carbide) [524]

ing, boring, reaming, and tapping, under specified conditions, and gave examples of results obtained in two laboratories. A total evaluation took some 150 kg of steel, 150 L of fluid, and 68 h. In one version, end milling was also included.

Morton [468] has developed a test sheet for in-plant observation of lubricant performance. The data sheet in Table 5.1 can be readily simplified for cutting-fluid evaluation.

General Tests. Cutting fluids are routinely subjected to tests touched upon in Sec.4.6 and 5.11. There are numerous discussions oriented toward cutting fluids, of which only a few can be referenced here [4.3,4.20,4.21],[532,533]. Health and pollution aspects are important (Sec.4.8 and 4.9) and have been discussed, with special reference to legislative aspects, by Barker [534].

11.6 Abrasive Machining

A large group of machining processes relies on metal removal by a multitude of hard, angular, abrasive particles or grains (also called grits) which may or may not be bonded to form a tool of some definite geometric form. The basic differences relative to cutting are that individual grains are more or less randomly oriented to the cutting direction, and that the undeformed chip thickness is generally small. Cutting speeds may range from a fraction of a metre per second to 200 m/s, and a single grit may be engaged with the workpiece for a short distance. Thus, the range of conditions varies enormously, as do results of encounters between grit and workpiece.

Some common features of metal removal will be discussed first, followed by a review of the tribological aspects of various processes. Wear mechanisms discussed in Sec.3.10 are of direct relevance and should be reviewed. Several books and collections of papers [535,536] and reviews [537,538] are devoted to the subject; practical aspects have been discussed by Farago [539].

11.6.1 The Abrasion Process

Abrasive grit of irregular shape is generally manufactured to present sharp cutting edges. Sharpness is lost by wear, but is restored by fracture of the grit or by release of the worn grit from the bond, either in the course of machining or as a result of dressing. Cleavage of a grit not only restores sharp edges but also creates serrated surfaces (facets) which serve as secondary cutting edges. Thus, grits have complex geometries which can, however, be approximated by regarding the grit as a single-point cutting tool on which a thrust (normal) force P_t acts while it is moved against the cutting (tangential) force P_c (Fig.11.53a). Rake angles range from occasionally positive to generally negative values. The value often quoted is the attack angle ($\gamma = 90 + \alpha$). The depth of engagement (the undeformed chip thickness t) is variable but generally small. Thus, the considerations of Sec.11.1.4 apply, and the conditions shown in Fig.11.18 are encountered. There are, however, substantial differences relative to orthogonal cutting:

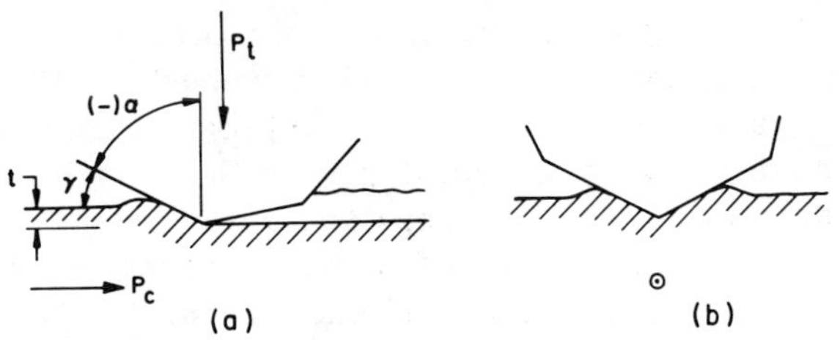

Fig.11.53. Plowing with abrasive grain of highly negative rake angle: (a) side view; (b) front view

a. Grits have finite widths and can displace the material sideways. Such plowing does not result in material removal (Fig.11.53b).

b. Rake angles can be highly negative. Material flow is then directed into both the chip and the workpiece (Fig.11.14) [88], and work expended to overcome friction may become a substantial part of the total work. Values calculated by Kita and Ido [540] show (Fig.11.54) that, with a highly negative rake angle, the friction contribution dominates even when friction is low. Further work is lost in the deformation of the workpiece surface, penetrating to about ten times the depth of indentation.

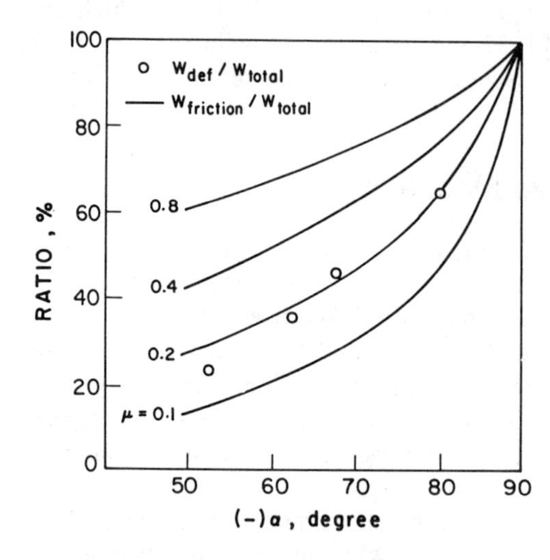

Fig.11.54. Ratios of deformation and frictional work to total work as functions of rake angle [540]

c. The clearance angle is often zero or even negative. We have seen in Sec.11.1.4 under the heading Flank Friction that this greatly increases frictional forces and contributes to heating.

Material Removal. The consequences of grit/workpiece encounters are varied:

a. Deformation may be purely elastic, thus creating high frictional forces and temperatures without metal removal.

b. Plowing of the workpiece surface creates a frontal bulge ahead of the grit, while most material is moved sideways. The frontal bulge (sometimes termed a "prow") either remains on the surface or is lifted out with the grit.

c. Chip formation characteristic of cutting with a highly negative rake angle may occur. The transition from the rubbing-plowing to the cutting mode is usually accompanied by a drop in the thrust force P_t, and the ratio P_c/P_t reaches values greater than unity, as in cutting, whereas it is typically 0.4 to 0.5 in grinding [541].

To a first approximation, these consequences are a function of process geometry. Thus, Kragelski [3.26] regards grits as spheres of R radius and regards t/R as the controlling variable. Real events are much more complex.

Influencing Variables. Effects of process variables have been reviewed by Hahn [542], by Doyle and Dean [543], and by Samuels [544].

a. Speed has a marked effect. Sasaki and Okamura [545] observed in continuous microcutting in a helical path with a single-point tool that, when the speed was increased while a constant t and α were retained, short, broken segments changed to a continuous chip (Fig.11.55). As in cutting, the speed effect must be attributed to a change in temperature. Indeed, the critical rake angle for chip formation in low-speed abrasion of steel changes from 0 to 40° when the steel is heated to 400 C [543].

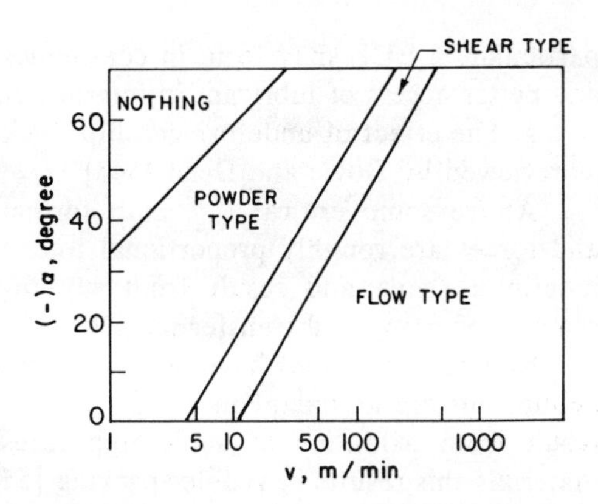

Fig.11.55. Type of chip formed in cutting with a ceramic tool having a plane rake surface [545]

b. The critical rake angle at which chip formation begins [546] became less negative with increasing friction in the work of Kita et al [547]. With a grit of $\alpha = -50°$, a continuous, smooth chip was produced with spindle oil whereas a sawtooth-type chip was formed in dry cutting. As in cutting, sticking friction is likely to prevail in many situations. Then, a more negative rake angle can be tolerated with increasing speed (Fig.11.55), presumably because the low effective shear flow stress within the hot chip reduces the apparent mean coefficient of friction. Sauer and Shaw [548] calculated the value of the critical attack angle as a function of the shear stress of the workpiece material.

c. Adhesion between grit and workpiece material can lead to the formation of a stagnant region or prow (essentially, a BUE), which then acts as a grit of more positive rake angle. This favorable effect is, however, offset by the poor surface produced [549]. Thus, it is generally found that a better finish can be obtained on a strain-hardened than on an annealed material.

d. The shape of the grit affects metal removal. A prismatic grit with the flat face perpendicular to the sliding direction builds up a larger prow and promotes chip formation (Fig.11.56) [544]. A pyramidal chip, with the edge in front, pushes more material sideways, as does a conical tip [550]. Serrations on freshly dressed grit contribute to chip formation, especially at small depths of cut. When favorably oriented, a single grit may cut at more than one edge [551].

e. The structure and properties of the grit determine the efficiency of operation. In general, a harder grit is needed for harder metals. The grit must not break under the imposed static and dynamic forces, yet it must be friable enough to develop new cutting edges by fracture [552].

f. The geometry of the encounter determines the undeformed chip thickness and length of contact [553]. Thus, single-point microcutting tests with continuous engagement do not necessarily reproduce the conditions that exist in grinding, and tests with a grit attached to the circumference of a wheel give more relevant information [552]. In

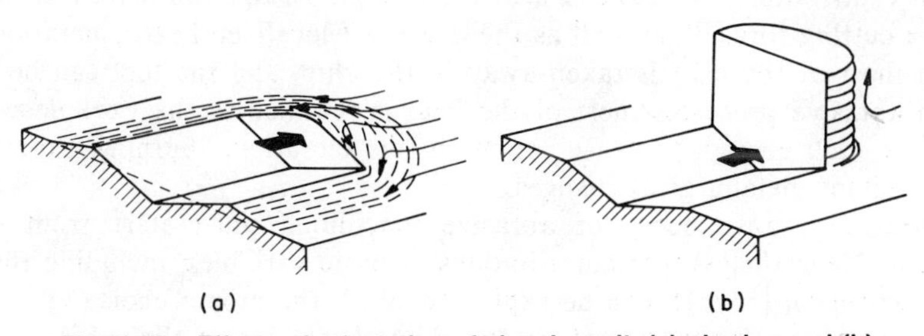

Fig.11.56. Effects of orientation of abrasive grit: (a) plowing and (b) cutting [544]

particular, a BUE may form in continuous cutting, whereas the shorter contact length and better access of lubricant in intermittent cutting may prevent its formation.

g. The effect of undeformed chip thickness depends also on the workpiece material, as reviewed by Doyle and Dean [543].

Above some critical t_{crit}, chip formation dominates, and material-removal rates and forces are roughly proportional to t. Some of the particles removed are of rather indefinite shape and result from rubbing or fatigue-type encounters and from the removal of prows and transfer particles.

Spherical particles have not been observed in grinding of steel in pure nitrogen or helium but are abundant in grinding in air; therefore, it has been postulated that they result from oxidation at such high rates that combustion takes place. With some materials this results in visible sparking [554]. Bowden and Freitag [555] suggested that surface melting can take place. Many of the spherical particles are hollow and may form by curling and agglomeration of thin platelets, as suggested for fine grinding (small t) by Komanduri and Shaw [556].

The proportion of chips decreases with decreasing t, but not as steeply as might be expected. Doyle and Dean [543] attribute this to the very fine subgrain structure developed in the surface layer of the workpiece during previous encounters; when the grit fails to penetrate this fragmented layer, chips are formed even at highly negative rake angles. Doyle and Aghan [557] suggest that plastic instability associated with the fine grain size contributes to chip formation in polishing. Shaw [557a] likens chip formation to a plastic indentation/extrusion process.

h. Grit size is of little importance in grinding as long as it is above 100 μm. Below this size, grits rapidly become less efficient. In contrast to Doyle and Dean [543], Misra and Finnie [558] consider that material is removed only if the grit penetrates the thin (10-μm), heavily deformed surface layer; since the depth of engagement is typically about one-tenth of grit diameter, the observed drop in efficiency would be accounted for. Of course, chip formation could still take place at lower penetration, but at a lower efficiency.

i. Adhesion between workpiece and abrasive and its modification by the presence of a grinding fluid are most powerful influences. A BUE formed in high-adhesion situations changes the effective rake angle and, more importantly, leads to a poorer workpiece finish.

j. Chip removal from brittle materials such as untempered tool steels is essentially a fracture process governed by the equivalent or effective elastic modulus (Eq.3.3) and by the fracture properties of the workpiece material [559]. Below a certain critical grit size, material removal again occurs by ductile fracture.

It is clear from the above review that, as in cutting, material effects, geometry, process conditions, and friction, as well as adhesion, interact in ways that seldom allow separation of friction effects. However, in contrast to cutting, sliding friction becomes a significant contributor to forces and heat generation. Thus, a lubricant can substantially reduce the cutting force P_c as well as the degree of localized heat generation. Whereas in cutting most of the heat is taken away in the chip, and the tool can be effectively cooled, in abrasive processes most of the heat is retained in the workpiece. Temperatures can be high enough to cause oxidation (discoloration, "burning") of the surface, with concomitant metallurgical changes.

By necessity, any theory of abrasive machining must start from simplifying assumptions. Nevertheless, the contributions of basic variables, including that of plowing (or of extrusion [557a]), can be explained. With the proper choice of material constants, agreement with experiments is found. A discussion of the subject, including a review of earlier work, is given by Lortz [560]; more recent treatments have been presented by Spurr [561] and by Suh et al [562].

Much research on abrasive machining has been carried out with single grains. Direct transfer to practical processes is not possible [563], but ranking of grains and the exploration of process variables is feasible [544].

11.6.2 Bonded Abrasives

Several abrasive machining processes utilize a tool of well-defined geometry.

Grinding. In the most extensive application of abrasion, grits are bonded with a glassy (vitreous), polymeric (resinoid), or metallic binder into grinding wheels. Because of its importance, grinding is discussed in many sources, including books [1,4,5,9,10,12,535,536] and reviews [564-566]. Recent development trends are discussed by Colding et al [567].

1. The Bond. Metal removal is affected not only by the grit but also by the binder.

a. The strength of the bond is determined by the quality and quantity of binder. Larger quantities of the same binder form more and larger bond posts which hold the grits more firmly against the forces developed in grinding. Wear of the grits develops flat surfaces and dulls the edges; the plowing and rubbing associated with worn grits result in reduced material-removal rates, higher forces, higher temperatures, and poorer surface finish. Indeed, Nakayama et al [568] judged the sharpness of a wheel by running it against a PTFE plate; when $\mu < 0.2$, there were too many worn grits for effective grinding. Thus, the bond must release worn grits when the force acting on them becomes high. As a result, the wheel is worn, and the bond strength is chosen to ensure restoration of cutting action without excessive wheel loss. Thus, the bond is selected to give a controlled grinding ratio G. The definition of G is:

$$G = \frac{V_w}{V_s} = \frac{Z_w}{Z_s} \tag{11.16}$$

where V is volume removed, Z is volume removed per unit time, and the subscripts w and s refer to workpiece and wheel, respectively.

Since grit wears more rapidly in grinding a harder metal, a general rule states that a softer (less strongly bonded) wheel should be used for harder materials [569]. Wheel hardness is a measure of the force required to dislodge grains. Therefore, it is not a constant but is affected also by all variables that affect the grinding force. Typically, a wheel will appear softer in grinding at a lower speed, without a lubricant, and when the contact length is greater.

b. The openness of the wheel structure is closely controlled. A more open structure has slimmer bond posts and, most importantly, allows lubricant/coolant to move through the wheel.

A grinding wheel undergoes gradual changes in the course of use. Wear of the grit can lead to glazing, while blocking of the surface pores results in loading. Werner and Lauer-Schmaltz [570] distinguish three kinds of loading. Adhesion of low-carbon steel, high-chromium steels, and ductile nonferrous metals results in the formation of large built-up edges which may fill out the interstices. Accumulation and pressure welding of chips of medium- and high-carbon steels and of ductile copper and aluminum alloys result in the formation of layered, entrapped, bridging particles. Adhesion and accumulation without pressure welding is reported with some stainless steels, titanium alloys, and aluminum-silicon alloys. Apart from adhesion between grit and workpiece, loading is promoted also by a harder (denser) wheel structure and a vitrified bond. Much less loading is evident with resinoid bonds. The presence of a coolant (including water) may actually promote loading (perhaps by excluding oxygen), whereas a good lubricant usually reduces it. Grinding forces rapidly increase, metal-removal rates drop, and surface finish deteriorates with loading because some of the adhered material is redeposited on the surface [571]. The situation improves only at the expense of wheel wear when high forces dislodge the ineffectual grains. The nonadhesive type of loading can be reduced by dislodging the accumulated fines with a fluid applied under high pressure.

Because grains are randomly distributed in the wheel, they not only are randomly oriented but also vary in height distribution. The extremely rugged surfaces of grinding wheels present considerable difficulties of characterization, even in the static state, and much effort has been expended in studying them under the SEM [572] and with stylus instruments [573,574]. The difficulty increases when the effective geometry during dynamic (cutting) conditions is to be assessed. Techniques have been developed by, among others, Shaw and his co-workers [575]. Verkerk [576] reported on a round-robin investigation initiated by CIRP.

2. Kinematics. There are two basic forms of grinding. In plunge grinding the only feed motion (the infeed) serves to reduce the distance between grinding wheel axis and workpiece surface, thus setting the depth of wheel engagement a. Wheel wear is more or less uniform across the width (Fig.11.57a and b). When a wide slab or long cylinder or tube is to be ground with a narrower wheel, cross feed must be applied, and then the wheel develops a tapered surface (Fig.11.57c), at an angle governed by the wear mechanism [577].

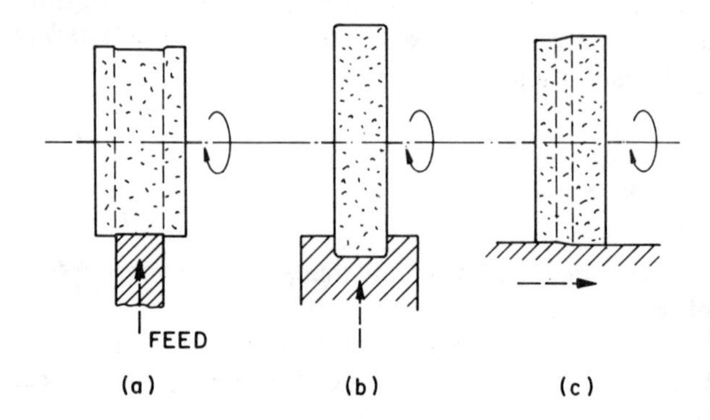

Fig.11.57. Plunge grinding of surface (a) and of slot (b), and surface grinding with cross feed (c)

The undeformed chip thickness t and length of cut l are functions of depth of wheel engagement a, grinding geometry, wheel speed v_s, and workpiece feed rate v_w (Fig.11.58). According to geometry, processes may be classified as follows:

a. In surface grinding (Fig.11.58a) the workpiece is fed in the direction of grinding at a speed v_w and the depth of cut is set between successive passes. The mean length of cut l is then

$$l = \left(ad_s\right)^{1/2} \tag{11.17}$$

Elastic deformations of the wheel and workpiece increase this length (as in roll flattening; Sec.6.2.3).

b. In external grinding (Fig.11.58b) the workpiece is rotated at a surface speed v_w and the infeed determines the depth of cut. The contact length is shorter than in surface grinding.

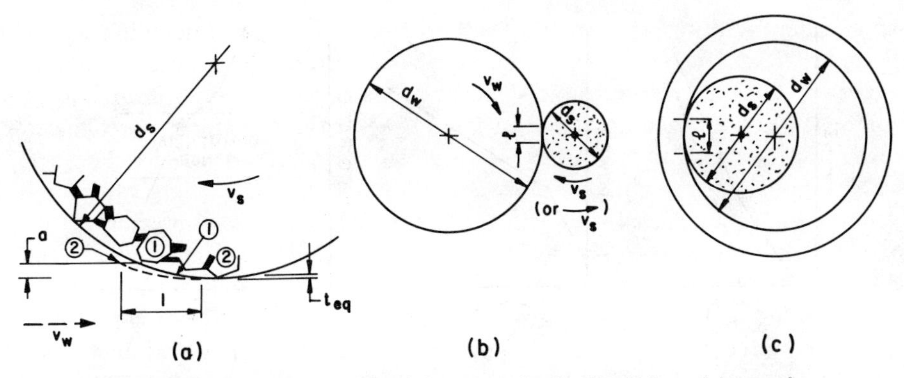

Fig.11.58. Surface grinding (a), external grinding (b), and internal grinding (c)

c. In internal grinding (Fig.11.58c) the contact length is longer and can be calculated from the equivalent diameter d_e:

$$d_e = \frac{d_w d_s}{d_w \pm d_s} \tag{11.18}$$

where the plus sign is for external grinding and the minus sign for internal grinding.

As in milling (Fig.11.23), the workpiece may move against or in the direction of wheel rotation, resulting in up or down grinding, respectively. When the nominal dimension is reached, grinding is sometimes continued without infeed. Such "spark-out" improves the surface finish and releases the elastic stresses in the wheel and machine, with a gradual decay of forces.

3. Grinding Methods. Grinding was originally a finishing process. However, its scope has been expanded considerably, and processes can be meaningfully categorized according to undeformed chip thickness. Because of the random distribution of grits, t varies both in the width and length of the contact zone. Fortunately, it is found that the important parameters such as grinding force, surface finish, and grinding ratio are related to an equivalent grinding thickness t_{eq}, as confirmed by a CIRP study reported on by Snoeys et al [578], who also reviewed earlier work. The equivalent grinding thickness

$$t_{eq} = a\left(v_w/v_s\right) \tag{11.19}$$

can be regarded as the sum of all instantaneous chip thicknesses in a longitudinal section of the contact area, or, as shown by Shaw [564], as the thickness of a continuous ribbon peeled from the workpiece at the wheel speed v_s (Fig.11.58). An important point is that, as the wheel speed v_s decreases, the undeformed chip thickness increases. This results in a higher load on the grains and may result in their premature dislodgement. If, however, the speed increases, temperatures increase also, and then wear can become severe. Thus, there is usually an optimum speed at which the G ratio is highest.

a. *Precision Grinding.* Originally the purpose of grinding was to produce close tolerances and a low-roughness surface. At a very low depth of engagement, only rubbing takes place, and the minimum penetration at which chip formation begins is connected with some minimum force per unit width [579,580]; this force is lower for easy-to-grind materials. The rate of removal Z'_w (volume per unit time and unit width) increases when, at a critical force, more of the material is removed in chip form (Fig.11.59a).

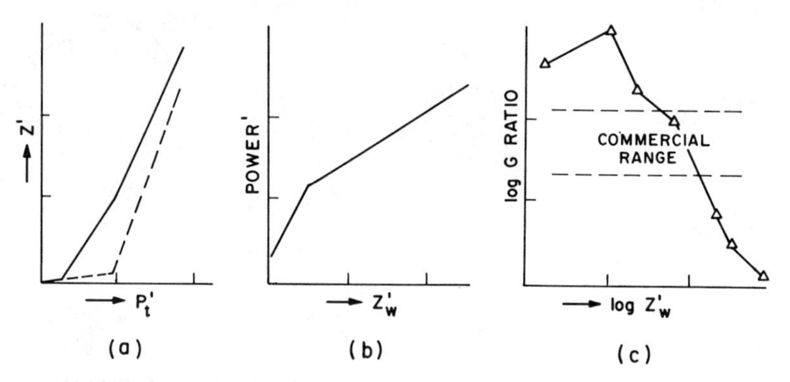

Fig.11.59. Rate of removal (a), power (b), and grinding ratio (c) in precision grinding [579]

The greater efficiency of chip formation results in a more moderate increase in power (Fig.11.59b), and the "metal-removal parameter" (volume of material removed per unit time and unit normal force) increases. However, above some critical force, the rate of wheel wear Z'_s (volume per unit time and unit width) increases rapidly (Fig.11.59a) and the G ratio drops (Fig.11.59c). Thus, practical conditions are chosen to give a compromise between reasonable metal-removal rates and G ratios (Fig.11.59c). Specific energy requirements are between 30 and 100 $W \cdot s/mm^3$.

Wear of the wheel results in the development of undulations which induce chatter and cause the emission of a howling sound. As do all vibration phenomena, this depends on machine tool stiffness [581], but, for a given machine, the chatter-free volume removed is a sensitive indicator of process conditions [582].

The great sensitivity of this process to the force level prompted Hahn [583] to advocate grinding with the application of constant force rather than constant feed, especially in internal grinding where spindle deflections are troublesome.

b. *Coarse Grinding.* When the purpose of operation is fast material removal, the undeformed chip thickness is chosen to be as large as possible. Chip formation dominates the process, and, correspondingly, the specific energy drops to 5 to 10 $W \cdot s/mm^3$. Speeds can be increased to take advantage of the much-increased metal-removal rate, and high G values are observed when temperatures become high enough to soften the material. Wheels must be made to release the worn grit without dressing, yet without excessive wear. Constant-force grinding is preferable. Applications include not only processes of low dimensional tolerance requirements, such as cut-off, surface grinding, and fettling, but also the so-called high-efficiency grinding process based on high surface speeds (for fast material removal prior to finish grinding).

c. *Creep-Feed Grinding.* This process combines some advantages of precision grinding and coarse grinding. The full depth is removed in a single pass but at a low workpiece speed [584,585]. The specific energy is still high—close to that of precision grinding—but, as in coarse grinding, heat buildup ahead of the wheel is sufficient to accelerate metal removal. The surface produced is less damaged, partly because the heated layer is ground away and partly because intense cooling and high-porosity wheels are used [586]. The maximum specific energy allowable without burning is discussed by Ohishi et al [587].

d. *Forces and Energy Requirements.* It is relatively easy to measure the thrust (normal) force P_t and cutting (tangential) force P_c. The latter could be—and often has been—regarded as a frictional force, and then an apparent coefficient of friction can be calculated as P_c/P_t. From the mechanics of grinding it is obvious that this number is, in fact, not a friction coefficient but a force coefficient (grinding coefficient), which can be taken as a measure of process efficiency. Because of a shift toward chip formation, the grinding coefficient increases with larger t, with higher grinding speed, and with strain-hardened material [588]. Discussions of forces in grinding have been given [577,589]. Lindsay [590] proposed the use of maps showing changes in specific power (slope of the power versus metal-removal rate curve) as a function of the metal-removal parameter (slope of the metal-removal rate versus normal force curve). Threshold values are excluded; hence, the main purpose of these maps is to show the effects of changes in system variables on changes in force and power, for both cutting and grinding operations. In grinding with resin-bonded wheels, the grit receives viscoelastic support, and thus the up-grinding force is 2 to 3 times larger than the down-grinding force.

The specific grinding energy (synonymous with specific cutting energy) increases with decreasing t, as it does in cutting (Eq.11.9). This is observed in the range in which the cutting mode predominates [557a]. There is, however, a limiting energy value at approximately t = 700 nm. Here the plowing and rubbing modes predominate, and the high strength of the surface layer further increases the energy required. Malkin [591] equated the minimum energy with the specific melting energy.

e. *Thermal Aspects.* A comprehensive review of this topic by Snoeys et al [586] deals with both theoretical and experimental investigations. As mentioned, much of the heat that is generated remains in the workpiece, and thus surface damage can occur through overheating, and undesirable residual tensile stresses can be generated.

Temperatures can be very high. Sauer [592] calculated peak grain temperatures of 2100 C and wheel and workpiece bulk temperatures of 330 C in surface grinding of steel. In external cylindrical grinding, the surface temperature increases with d_w (larger contact zone) and with lower work speed (lower relative velocity) but is always lower than in surface grinding [593]. Rapid oxidation of the chip becomes a source of further heat input if the stream of chips is carried around the wheel [594].

In general, temperatures are lower at a higher wheel speed v_s; a lower work speed v_w; a lower equivalent wheel diameter d_e (Eq.11.18); and a higher undeformed chip thickness t. The heat of chip formation can be reduced only by changing speeds and by dressing. However, because friction is such a large contributor to grinding energy, lubricants are effective in reducing temperatures [567,586]. Interrupted cuts with segmented wheels also help. Coolants, when properly applied, can further reduce temperatures by removing debris and preventing wheel loading which would increase the frictional component.

Experimental measurement of temperatures presents a challenge, because several grains are active at any one time, and because the contact zone is not directly observable. The dynamic thermocouple method is discussed by Kaiser [595], infrared measurements by Kops and Shaw [594], and a photoelectric method by Popov and Davydov [5.148].

Coated Abrasives. Traditionally, abrasive grains attached to a flexible backing such as paper or cloth have been used for low-speed finishing of surfaces. However, with the development of stronger adhesives and backings, coated belts operating at high speeds (typically up to 70 m/s) have become important production tools, capable of high metal-removal rates [596].

Grains are deposited on a layer of adhesive applied to the backing (make coat), and are held in place by a second layer (size coat). Bond strength must be balanced to prevent stripping of new grains while allowing release of worn grains.

Grains with sharp edges and of elongated shapes can be electrostatically aligned to give cutting edges of low negative rake angles. Furthermore, grains can be spaced some ten times farther apart than in grinding wheels. Therefore, chip formation is the dominant metal-removal mode, chips can be stored without clogging, specific energy requirements and surface temperatures are lower than in grinding, and damage (burning) of the workpiece can be avoided. Grit size is important here, too. Date and Malkin [597] observed that finer grit gives more rubbing contact and lower initial removal rates. Removal rates are further reduced by wear particles that adhere to the grit and become trapped between grains ("loading" of the belt). Coarser grit is more effective initially and retains its cutting ability longer because grain fracture creates new cutting edges; clogging is also less rapid.

Wear leads to rapid development of flat tops on grains, with a simultaneous decrease in the metal-removal rate and in the P_t/P_n ratio [598]. The cutting function is prolonged by one of two mechanisms: first, coarser grit fractures; second, grit is detached at a later stage of wear. Finer grit, containing fewer defects, is stronger, and is detached from the backing by plastic failure (fatigue) of the bonding resin. The rate of detachment decreases with increasing cutting speed (because of the greater efficiency of cutting and hence lower forces), and increases when the cutting direction is reversed [599]. Techniques for determining topography [600] and for following the course of wear [601,602] have been developed.

Grinding fluids serve the important functions of lubrication, cooling, flushing away of debris, and reduction of clogging.

An important element of the process is the slotted contact wheel or platen used behind the belt. Cutting is intermittent and heat can flow into the workpiece during the noncutting period, allowing higher removal rates without burn. By setting the belt tension and choosing a wheel of appropriate hardness (usually a hard rubber) and grooving pattern, vibrations can be induced which ensure a so-called "aggressiveness:" cutting action is enhanced and worn grit is shed more readily.

Grinding with abrasive disks has been considered in detail by Mohun [602a].

Honing. The process of honing is related to grinding, but the grit is bonded into sticks and the motion is not unidirectional. In the most frequent application, an internal cylindrical surface is honed with a number of honing sticks carried in an expanding, axially oscillated head, while the workpiece is rotated. Thus, a cross-hatch pattern is produced; cutting across ridges of the pattern produced in an earlier pass results in short chips and facilitates cutting along a long path. The cross-hatched pattern is, in itself, of tribological interest: it is a significant factor in maintaining an effective lubricant film between piston ring and cylinder wall in internal-combustion engines. The pressure applied on the honing sticks must be large enough to maintain a cutting action. A fluid, usually an oil, serves a critical function in washing out the chips. Diamond and CBN have gained importance as abrasives [603].

In superfinishing, the stone is large, pressure is low, and oscillatory motion is of a shorter stroke and a higher frequency [604].

Friction Cutting. Steel mills have long used a frictional process for cutting of ingots. A toothless, water-cooled soft steel disk (sometimes notched at its periphery), rotated at high speeds (36 to 100 m/s), is pressed against the steel surface. The steel is worn away in the form of a metal oxide debris, with relatively little wear of the disk. This process has been investigated by Childs and co-workers [605].

11.6.3 Loose Abrasives

Several processes rely on unbonded abrasive grit [10] and can be regarded as examples of abrasive wear and erosion, reviewed by Misra and Finnie [3.361].

Lapping is a form of three-body abrasive wear. The abrasive is introduced in the form of an oil-base slurry between the workpiece and a counterformal surface, called the lap. The lap may be a reusable tool—usually a flat-surface disk made of cast iron or brass—or may be another component, as in the lapping of mating gears. Slots in the lap are sometimes used to introduce the slurry and to accept debris. The mechanism involves both metal removal and plastic deformation. With angular, sharp particles, the cutting action is produced by the rolling and sliding of the grit, aided by embedment of grit in the softer lap. To accelerate wear, grit is sometimes intentionally embedded with the aid of a tool. A more spherical grit produces wear mostly by repeated plastic deformation [606].

Metal-removal rates are functions of grit size and applied pressure. Miller [607] found in diamond lapping of molybdenum balls that metal-removal rate increased with increasing load applied to the lap, but only to a critical value; this critical value was higher for larger grit sizes. Fine abrasives needed only very small loads to be effective. Undoubtedly, the fracture properties of the abrasive, as affected by the lapping oil, are of importance in maintaining cutting edges. Softer, more friable, nonembedding abrasives can be used for *in situ* lapping of mating machine components; the abrasive gradually breaks into very fine, nonabrasive particles which need not be removed [608]. Longitudinal ultrasonic vibration increases metal-removal rates [609].

Buffing and Polishing. The fine-grain abrasive may be applied to a soft surface (usually felt or another fabric). This process is of great significance for metallography [610]. In buffing, the abrasive is in a semisoft binder, whereas in polishing it is carried in an oily or aqueous slurry. Material removal is minimal, and the surface is smoothed partly by plastic flow.

Erosive Machining. Various finishing processes rely on wear produced by sliding or by the kinetic energy of impacting abrasive particles, and are thus related to erosion [3.369]. Essentially, the mechanism is repeated deformation of the surface followed by removal of the hardened layer, as discussed by Finnie [3.361]. Several variants exist, all with different applications and different (sometimes proprietary) names.

In abrasive flow machining [12], also called extrude-hone abrasive flow machining [611], the abrasive is made up in a polymer base into a medium of putty-like consistency. The medium is moved, under pressure, over or through the workpiece. In abrasive jet machining [12,612], abrasive particles, fed into a high-velocity gas stream, impinge on the surface.

In ultrasonic machining (also called impact machining or grinding) the tool is attached to an ultrasonic generator, and a slurry of abrasive grit in water (occasionally oil) is circulated to the interface between tool and workpiece [12]. The tool is made of some erosion-resistant material (mild steel, stainless steel, etc.), but, even so, wear of the tool results in a gradual loss of resonance and a drop in efficiency. Similarly, gradual fragmentation of the grit (usually boron carbide) reduces the rate of material removal. Research on this process has been reported in a series of papers by Adithan and Venkatesh [613]. The process is most effective on brittle materials. Drilling is also assisted by ultrasonic vibration.

Electrochemical Processes. Metal-removal rates can be increased by the application given by Pahlitzsch [614]; grinding and the effects of magnetic fields were discussed also by Kuppuswamy [615,616]. Kops [617] noted improved material-removal rates in electrolytic ultrasonic machining.

There are also a number of specialty processes that involve material removal by chemical, electrolytic, or electrodischarge methods. In the broadest sense, these could be regarded as special forms of wear, but are beyond of the scope of this treatment. Descriptions of them will be found in the Machining Data Handbook [12].

11.6.4 Surface Quality

Surface finish in abrasive machining is the cumulative result of multiple encounters with large numbers of grains. A nominal roughness based on the geometry of bonded abrasives can be predicted, but is modified by the details of the chip-forming/plowing mechanisms and is made worse by the redeposition of adhesive fragments [618]. It is in this respect that lubricants can be most helpful. Characterization of ground surfaces by statistical evaluation of roughness profiles has progressed [619,620].

High temperatures generated in grinding result in very fast heating of a thin surface layer, which is then rapidly quenched by the cold substrate. Residual tensile stresses of thermal origin are set up in combination with residual compressive stresses originating from local deformation due to plowing and cutting with a highly negative rake angle; microscopic cracks may be generated perpendicular to the grinding direction in less ductile materials, and heat treatable materials may undergo transformations. Thus, surface damage is an important aspect; it is discussed by Torrance [621] and in Section 18 of [12]. Evaluation of grinding parameters combined with in-process measurement of size, form error, and surface roughness can lead to adaptive control of the process [5.238],[622]. Techniques for in-process detection of wheel wear, loading, and other variables have also been developed [623,624].

11.6.5 Abrasives

As do all other tool materials, abrasives represent a compromise between sometimes conflicting requirements. Hardness helps resist abrasion by hard particles, while friability allows fracture to occur under imposed mechanical and thermal stresses, generating new cutting edges but resulting in loss of abrasive material. Adhesion to the workpiece material controls the BUE, clogging, redeposition of debris on the workpiece, and pull-out from the bonded structure; adhesion to the bond ensures the strength of a bonded abrasive; chemical stability increases wear resistance; and resistance to corrosion reduces wear in the presence of oxygen and cutting fluids and affects adhesion to the workpiece.

Abrasive Grains. Other than natural diamond, all abrasives are man-made. A history of their development is given by Ueltz [625]. In addition to the book by Coes [626], there are articles dealing with the thermodynamic foundations of the properties of abrasives [627] and with the properties of unbonded grains [552,628].

1. Alumina. Emery is a natural rock, only half of which is Al_2O_3; it is of insufficient hardness for most applications. Present abrasives are varieties of the hexagonal alpha alumina, produced from bauxite or purified alumina in electric furnaces, and cooled to provide controlled crystallization. The range of properties obtainable is shown in Fig.11.60a [552,625,629]. Surface morphology was studied by Komanduri and Shaw [551]. White aluminum oxide (type 1) is pure alumina made friable by internal voids and soft β-Al_2O_3 inclusions; when crushed to medium to small sizes, it has sharp

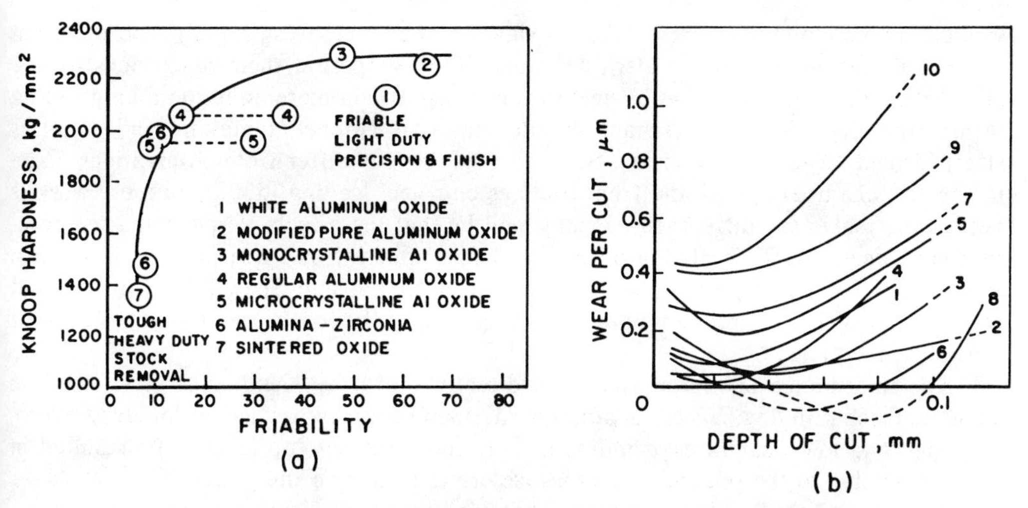

Fig.11.60. Abrasive grains: (a) hardness [629] and (b) wear in fly cutting [552]. 1 to 7 - Alumina grains as shown in (a). 8 - Sintered Al_2O_3. 9 - Green SiC. 10 - Black SiC.

cleavage facets. Small additions of chromium oxide (type 2) increase strength, and grains show pyramidal projections. Monocrystals (type 3) are grown rather than crushed, are less friable, and have sharp edges. Regular aluminum oxides (type 4) are made from bauxite and are strengthened by a fine dispersion of a titanium-containing phase; cutting edges are produced by crushing the strong body. On fast cooling, microcrystalline alumina (type 5) is obtained. The addition of zirconia (type 6 and 7) allows alumina dendrites to grow in a soft Al_2O_3-ZrO_2 eutectic, with properties depending on ZrO_2 concentration and cooling rate [629]. The sintered oxide (type 8) has good impact strength and can be made in various shapes to control cutting action.

Attritious wear was studied in fly milling by Brecker et al [552]; it is less for the harder, more friable varieties (Fig.11.60b), which are thus more suitable for precision grinding. The tough grades and, particularly, the zirconia grades are better for heavy stock removal and snagging [630]. They removed some ten times more steel than alumina grits in belt grinding under mild, constant-load conditions, but the difference almost disappeared under severe conditions in the work of Visser and Lokken [602].

The role and sources of adhesion are the subject of debate. Komanduri [265] suggested that oxides formed on steel enter into a solid-phase reaction with Al_2O_3, thus increasing adhesion. This is in agreement with the high adhesion with alumina observed by Pepper [336] on admitting oxygen to a high-vacuum chamber, and would account for the apparent negative wear rates with some grits (Fig.11.60b). Shirakashi et al [344] extended the argument to other metals by emphasizing the importance of matching the oxide lattices. Indeed, the dense, well-matched oxides of aluminum and titanium gave high friction and adhesion in twist-compression experiments. The slower oxidation of type 304 stainless steel necessitated more sliding to reach equilibrium friction, whereas with low-carbon steel, with its porous oxide, metal-to-metal contact developed and friction rose further. In contrast, the nonmatching oxide of copper kept friction low. Yossifon and Rubenstein [631] argued, on the basis of grinding experiments on type 304 stainless steel, that oxides fulfill the traditional role of lubricants in protecting the tool and reducing adhesion. Conditions that are known to create low grinding temperatures resulted in large-scale adhesion and bridging between grains; this was attributed to slow oxidation. At intermediate temperatures, adhesion was limited to isolated points, whereas at high temperatures—where oxidation would progress rapidly—no adhesion was noted, and grits wore by attritious wear. They supported their conclusions by the observation that grinding of mild steel in argon resulted in more adhesion than grinding in air, and that oxidation-resistant ductile materials (Monel, tantalum, and stainless steel) formed large-scale adhesion. Some of these results differ from observations made in vacuum chambers [333] and from findings on wheel loading [570], and more work is needed to resolve the differences. It may well be that the effects of speed and temperature have been insufficiently considered, and that adhesion and wear aspects have been intermixed. This is suggested also by the work of Kumar and Shaw [631a], who found that wear was due to microchipping resulting from adhesion, without any evidence of solid-state reactions.

The importance of temperature was directly shown by Komanduri et al [632] in simulated hot grinding. Whereas adhesion at room temperature led to buildup, microchipping, and low wear rates, grinding at high temperatures (up to 1000 C) resulted in heavy wear due to the release of micron-size crystallites from the grains.

2. Silicon Carbide. This man-made abrasive is produced in an electric furnace from silica sand and coke. It is harder than alumina but is subject to oxidation, which is accelerated by the presence of water. Black SiC has more impurities than the green variety, but their hardnesses are similar. The grit has many sharp edges, and cleavage planes impart a friability comparable to that of white alumina [552].

An extensive treatment of the tribological properties of SiC has been given by Miyoshi and Buckley [3.447], including the effects of crystal orientation [633]. All metals transfer to silicon carbide, but to different degrees. Reaction with workpiece materials is a significant factor and can be rationalized from chemical affinity. Attritious wear is a result of several mechanisms [634]. Surface atoms are removed layer by layer under the influence of oxidation and shear stresses; cleavage fracture occurs along densely packed crystal planes; dissolution at high temperatures is possible, with the carbon diffusing into the workpiece material to form unstable carbides that decompose on cooling. Since the solubility of carbon in iron-base materials depends on the carbon already in the material, wear rates are high and the grit flattens rapidly on steel [552] but wears less on tool steel [631a] or cast iron. The excess of carbide-forming elements found in HSS also leads to intensive interactions [151]. Some whiskers of unknown composition were observed to form in hot grinding of a cobalt-base superalloy [632].

3. CBN and Diamond. The performances of both CBN and diamond rely on hardness. Their morphologies were studied by Komanduri and Shaw [635] in both the uncoated and metal-coated forms.

Single crystals of CBN have well-developed planes and sharp edges. They are friable and thus are self-sharpening to some extent [636]. Even though their hardness is lower than that of diamond (Table 3.1), their temperature resistance is higher and they react less with ferrous materials. Diamonds range from well-developed single crystals to weakly bonded polycrystalline aggregates. Reaction with carbide-forming elements leads to rapid wear, as on low-carbon steel or HSS.

Grinding Wheels. Grinding wheel characteristics—such as the kind and size of abrasive, the quantity of bonding agent (wheel grade and structure), and the bond type—are specified by a standardized designation system (ANSI B74.13) described in handbooks [11,12].

Wear occurs by adhesion, attrition, thermal shock loading, thermal fatigue [586], chipping, fracture of grains, and loss of whole grains or grain clusters [3.348]. In profile grinding, the corners (edges) wear more rapidly [637]. Wheel life is longer on a grinder of greater stiffness [567]. Models of wheel wear have been produced for control purposes [638].

As a general rule, alumina abrasive is chosen for ferrous materials except austenitic stainless steels. Silicon carbide is best for austenitic stainless steels, cast irons, nonferrous materials, metal carbides, glass, and ceramics. In a job-shop environment, mixed-grit wheels are often used. The bond is chosen for its structure, flexibility, resistance to adhesion (clogging), and compatibility with the grinding fluid, according to the severity of the grinding operation and the geometry of the wheel. Various designs aim at increasing the burst strengths of wheels [567].

Higher efficiencies have been obtained experimentally with wheels containing oriented SiC single crystals [639], and commercial wheels containing sintered alumina sticks are available.

Diamond and CBN wheels may be metal-bonded, and then they rely entirely on the resistance of the grits to wear. Resin-bonded wheels have some self-sharpening ability. Frictional heat is less than with metal-bonded wheels, but the high thermal conductivity of the grit causes softening of the bond. Therefore, the introduction of metal-clad grit represented a great advance. The metal coating improves adhesion to the resin, serves as a heat sink, protects the resin by reducing heat concentration, and keeps abrasive fragments in place [283]. The best cladding metals are nickel for wet grinding and copper for dry grinding—perhaps because of the noncumulative nature of metal transfer with copper. The high thermal conductivities of these abrasives keep workpiece temperatures lower, an important advantage in grinding of heat treated steels. However, the affinity of diamond for iron makes it vulnerable to rapid wear. An optimum speed exists in grinding (Fig.11.61) [283]: below this optimum speed, large chip thicknesses and hence large forces dislodge the grains; above it, temperatures are too high and diffusive wear sets in. Steels with excess carbide-forming element content require lower speeds. CBN wheels do not suffer from diffusive wear and, if properly chosen, can give grinding ratios 20 to 100 times higher than those of alumina wheels [636,640-642], provided that a limiting metal-removal rate is not exceeded [643].

A most important aspect of grinding is dressing of the wheel, which is reviewed by Verkerk [644], by Buttery et al [645], and by Bhateya [646]. In a typical precision grinding operation, only a small fraction of the wheel is worn away; most of it is lost in dressing. The results of dressing depend on the method employed, on the relative friabilities of abrasives and the bond, and on the grinding fluid used.

In general, dressing may take two forms. First, a surface layer is cut away with a single-crystal diamond or is worn away by a diamond-impregnated stick pressed over the rotating wheel surface. Vitreous sticks or diamond in a metal matrix are used for resin-bonded CBN and diamond wheels [640]. A novel method was proposed by

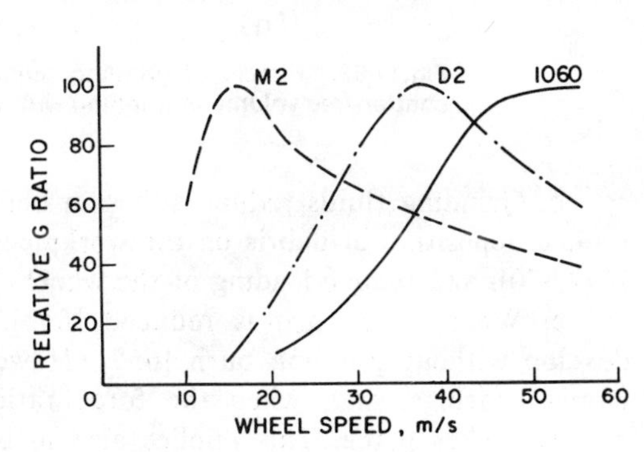

Fig.11.61. Effects of carbide-forming elements on G ratio in grinding of various steels with diamond wheels [283]

Komanduri and Reed [647], who ground a hot-pressed SiC or SiN ceramic alongside the workpiece in the presence of a lubricant; the ceramic chips thus produced adhered to the resin bond, protecting it and increasing wheel life.

Secondly, vitreous bonds can be crushed by running the wheel against rollers. The surface thus produced is rougher, and cleavage of grits produces sharp cutting edges with only slightly negative rake angles [648].

11.6.6 Grinding Fluids

Grinding in air (dry grinding) is possible and often desirable on some materials, but fluids can fulfill a number of important functions. General discussions have been given by Shaw [649], by Tripathi [650], and—with an extensive review of earlier work—by Sluhan [651].

Attributes of Grinding Fluids. The large friction contribution makes the role of grinding fluids different from that of cutting fluids.

a. As a lubricant, a grinding fluid reduces friction on the rake face and on the wear flats, and possibly also on the binder and on metal transferred to the abrasive. Thus, power consumption is reduced, sometimes dramatically (Fig.11.62a) [582,652,653]. The grinding force is usually reduced, but may actually rise with strain-hardening materials [292]. Grinding with a larger equivalent chip thickness becomes permissible, and thus efficiency improves. Most importantly, the reduced energy input results in lower temperatures (Fig.11.63a) (Dederichs, quoted in [570]).

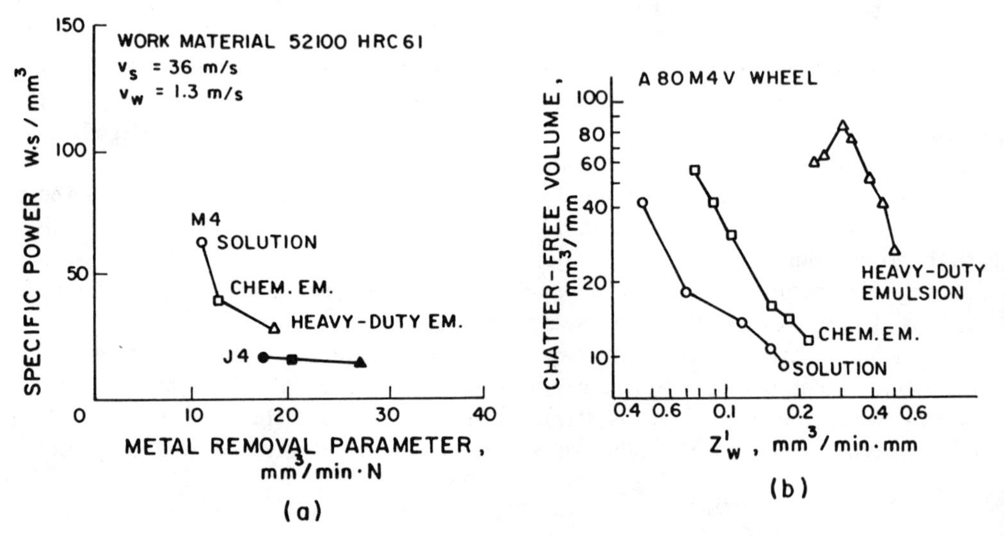

Fig.11.62. Effects of grinding fluids on (a) specific power and (b) chatter-free volume in grinding with vitrified alumina wheels [582]

b. Grinding fluids reduce adhesion between workpiece and grit. In doing so, they reduce deposition of debris on the workpiece surface [571,654], improve surface finish [292,570], and reduce loading of the wheel.

c. Wear of the grit is reduced [653], and larger wear flats can be allowed to develop without grinding burn [655]. However, wear flats may act as microhydrodynamic bearings, and, when the force ratio (apparent μ) drops to 0.2, no material removal takes place. This applies also to belt grinding [598]. When the geometry is favorable for hydrodynamic lubrication (as in internal plunge grinding), oversupply of a

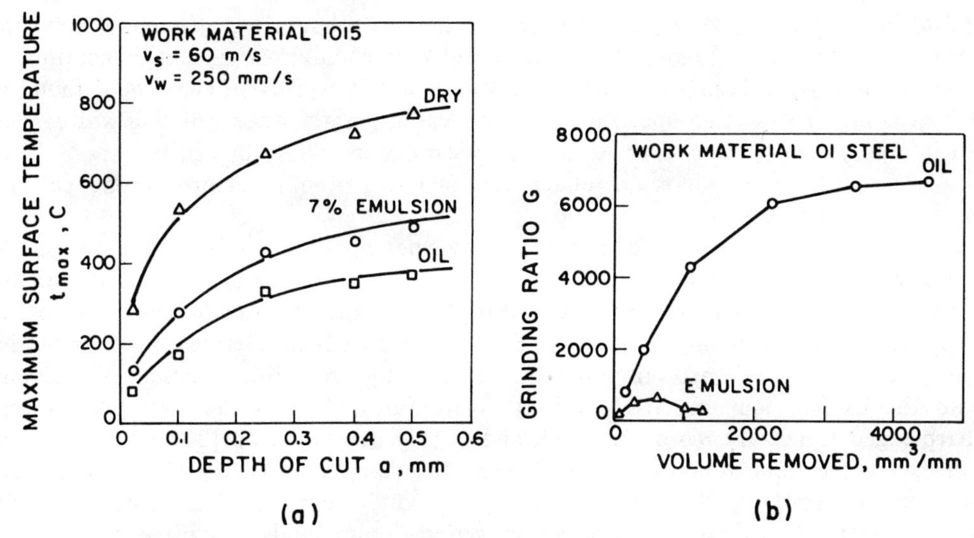

Fig.11.63. Effects of grinding fluids on (a) surface temperature in grinding with resin-bonded alumina wheels [570] and (b) grinding ratio in grinding with CBN wheels [657]

lubricant may stop cutting earlier [434]. If the fluid attacks the binder, severe wheel loss by grit pull-out sets in. The lubricant affects the character of the wheel: a high-lubricity oil makes it soft-acting and a high-E.P. oil makes it hard-acting.

d. Metal-removal rates are increased and/or maintained at high rates for a longer time [570,656]. The chatter-free volume (Fig.11.62b) [582] and the grinding ratio (Fig.11.63b) [657] are larger.

e. The lubricating function relies, with most materials, on E.P. reactions. Engagement time is short, on the order of microseconds; however, engagement is repeating rather than continuous, and the surface is accessible to the fluid between engagements [24]. High temperatures aid the E.P. action but cause chlorine to be ineffective in high-speed steel grinding. Broader coverage is obtained by the incorporation of sulfur; Auger spectroscopy shows sulfur penetration to 50 or more atomic layers in ground surfaces [650]. Studies of wear particles indicate the possibility of extensive interactions leading to the formation of organic-inorganic fibers and a kind of friction polymer in grinding with oils [658]. Oxygen-saturated oils behave as E.P. oils do [659]. Boundary additives are effective only on some nonferrous metals.

f. Cooling of the actual contact zone is not possible, but cooling of the workpiece is feasible and often desirable. Calculations show that, even at maximum convective cooling, only a 20 to 40% reduction in temperature is possible [586]. Cooling reduces bulk temperature, delays grinding burn, and minimizes thermal distortion. Equally important, especially in rough grinding, is cooling of the wheel for protection of the bond.

g. Proper application of fluid aids in chip evacuation and, if high pressures are applied, in physical cleaning of the wheel. Loading is delayed or even totally prevented [567,586].

h. Operating conditions generally improve. Damping of wheel vibrations has been claimed [660]. Rapidly oxidizing materials, the debris of which ignites by exothermic reaction, can be ground safely with a fluid.

Classes of grinding fluids are generally similar to cutting fluids (Sec.11.4) except that their lubricating functions are more important. General recommendations can be found in sources cited in Sec.11.4.6 and in [661,662].

a. In dry grinding it is again oxygen that acts as the lubricant. Its importance has been demonstrated by Shaw [1], who observed greater adhesion and agglomeration of wear debris when grinding was done in argon or nitrogen. Duwell et al [663] found in belt grinding of steel that specific energy increased greatly when grinding was carried out in argon, with either SiC or Al_2O_3 grit. Wear occurs mostly by attrition.

b. Gases, such as chlorine, reduce cutting energy [663] but are impractical and have little cooling power.

c. Water often reduces the specific energy when applied in a flood, but spraying of water onto a grinding belt actually increased energy requirements [663]. The primary function of water is cooling, although adsorption on the grit should reduce adhesion. The effect is, however, highly specific. Pure water greatly accelerates the wear of alumina wheels in grinding of titanium, presumably by promoting adhesion of titanium onto the oxygen ions in Al_2O_3 [663a]. Water also changes the wear mechanism: sharp-edged ridges are more prevalent, and G may actually drop [24]. Chemical solutions serve the same function while reducing the dangers of rusting of the equipment and corrosion of the workpiece. In the work of Chandrasekar and Shaw [663b], sodium nitrite was most effective at 10% concentration. Some fluids contain E.P. additives, although their effectiveness does not seem to be well-documented.

d. Lubrication and cooling functions are provided by emulsions, ranging from semisynthetic fluids to heavily compounded emulsions (soluble oils). Effectiveness measured by the G ratio is generally higher at higher concentrations [561]. The difference is more noticeable on annealed than on hardened workpieces of the same material [650]. There is no doubt that sulfur and chlorine additives improve performance, at least on ferrous materials [663], although hydrogen may be introduced into high-strength steels for grinding with sulfur additives [664]. There are suspicions that sulfur and chlorine may cause stress corrosion of titanium and nickel alloys and high-strength steels, but these suspicions could not be confirmed [665]. The high cooling power and lesser misting of emulsions are attractive, but water may cause corrosion and, in high-strength steels [665], an acceleration of fatigue (crack-tip propagation). To prevent loading, the wheel may be packed with soap. Polymers such as polyacrylonitrile are also effective [666]. Aqueous fluids may attack the binder; water attack on vitreous bonds is greatly accelerated by the presence of soda; alkaline fluids attack resinoid and shellac bonds; and all fluids attack oxychloride bonds.

e. The friction-reducing capabilities of oils make them superior to aqueous fluids in terms of all grinding parameters (Fig.11.62 and 11.63), and even in over-all cost. The results of Peters and Aerens [667] are typical (Table 11.1) and have been repeatedly

Table 11.1. Relative performance of grinding fluids [667]
(O1 steel, HRC 62; EK60L7VX wheel)

Fluid	Speed (v_s), m/s	Metal-removal rate (Z'_w), $mm^3/s \cdot mm$	Specific energy(a) ($E_{0.1}$), $W \cdot s/mm^3$	Grinding ratio, G	Chatter-free volume(b), mm^3	Surface roughness (R_a), μm
Emulsion ...	30	1.78	48	18	230	0.83
(3% conc.) ..	60	4.87	48	18	700	0.83
Oil	30	7.0	28	36	3 600	0.66
(synthetic) ...	60	22.6	28	36	20 000	0.66

(a) At an equivalent chip thickness of 0.1 μm. (b) At an equivalent chip thickness of 0.2 μm.

confirmed [570,653,659]. Even though they are poorer coolants, oils reduce heat generation and keep contact and surface temperatures (Fig.11.63a) lower than aqueous fluids do [292], and even bulk temperatures are only slightly higher [567]. Their main disadvantage is misting; mist extractors and covers need to be well-designed to ensure clean, safe shop conditions [166,570]. They also can present a fire hazard, and may be avoided for this reason alone.

f. Solid lubricants added to fluids are of variable efficiency. Improved surface finish relative to that obtained with an emulsion has been reported [387] for MoS_2 added to oil, although no oil reference data were generated. A powder of PbS fed into an air stream was more effective in reducing grinding energy than was MoS_2 [668], although its toxicity precludes industrial use. Strips of solid lubricant applied to the surface provided intermittent lubrication without wheel clogging [669]. Application of BN in an unspecified base of controlled melting point was effective in reducing wear of the ceramic bonds of CBN wheels [670]. Graphite in an aqueous dispersion was effective in reducing the rate of diamond dissolution in grinding of tool steels with a diamond wheel [671].

g. Solid lubricants may be incorporated into the wheel or stick. Honing sticks have long been infiltrated with sulfur [672]. A mixture of sulfur, stearine, and aluminum stearate has also been tried [673]. Replacements for sulfur have been sought, in graphite, graphite fluoride, MoS_2, and PTFE, with mixed success: the surface usually improved, but material-removal rates dropped [672]. Sulfur added to grinding wheels improved surface finish but at the expense of metal-removal rates [542]; an objectionable odor may also be generated. Removal rates were increased by addition of 10% WS_2 to the resin bonds of diamond wheels; other solids impaired the bonds [674]. Low-melting alloys added to the binders of diamond and CBN disks increased removal rates and G ratios [571].

h. Inorganic fluids have some special applications. In an example of the development of fluids from basic principles, Shaw and Yang [675] concluded, from considerations of surface adhesion based on atom size and charge, that barium should be most effective for adsorption on alumina and that the phosphate radical should be best for adsorption on titanium. Since barium phosphate is not water-soluble, they used $Ba(OH)_2$. Sayutin et al [676] made their choice on the basis of the free energies of formation between titanium and chlorides. The compound with the highest free energy, $NiCl_2$, also proved to be the best grinding fluid in the series.

The choice of fluid depends also on the type and geometry of the operation. Cup grinding and related processes (Blanchard, etc.) have the largest engagement lengths and generate much turbulence, and therefore nonfoaming transparent chemical solutions are preferred. Surface grinding is more demanding and calls for relatively soft wheels and modest lubricating power. Cylindrical and centerless grinding, with their essentially line contact, and additional contact on the wear plate if one is used, demand harder wheels and better lubrication. In internal grinding, the wheel is of small diameter, cooling is difficult, and high conformance results in high friction; therefore, heavy-duty emulsions or oils are used to reduce friction. Thread, profile, and gear grinding almost always require the friction-reducing capacities of oils [677]. The choice of fluid is, however, a function also of abrasive and binder and of workpiece material.

Belt grinding imposes both lubricating and cooling requirements. Low-viscosity oils (4 mm^2/s at 40 C) are suitable for mirror finishing or satining, and heavier oils or emulsions for heavy stock removal and centerless grinding [678]. In honing, removal of the wear debris is most important. Low-viscosity oil, which can be filtered effectively, is again the lubricant of choice [603,679]. Emulsions are used for less-demanding finishes [603]. In lapping, the primary functions are prevention of adhesion and reduction of lap wear, which are fulfilled normally by an oil-base carrier.

Testing. The principles discussed in Sec.11.5 apply here also. The number of repeat tests is, by necessity, limited, and variability in wheel life may be a problem. Strachan [680] discusses statistical techniques suitable for coping with this situation, in conjunction with test methods for diamond wheels.

Most fluids are evaluated in grinding. Forces (P_t and P_c), motor power, wheel wear (diameter change), workpiece weight loss, and surface roughness are usually measured, and wheel loading is observed. When the diameter change is very small, as in diamond grinding, the wheel must be allowed to settle down for 12 hours before measurement [680]. From the data obtained the values of G, Z, and E are calculated. The chatter-free grinding volume (howling limit) is a measure of total material removed [582]. Peters and Aerens [667] have urged the use of grinding charts. Rowe [681] has suggested computer-based ranking in which measured parameters as well as physiological properties, stability, handling, cost, etc. are weighted according to the constraints imposed by the process.

Simulation testing is not easy. Kirk et al [682], after reviewing single-grain abrasion tests, concluded that better simulation is obtained by running a hemispherical wheel segment against the periphery of a heat treated steel disk. Lubricant ranking on the basis of height loss agreed well with wheel breakdown—indicated by a sudden drop in force—in cylindrical plunge grinding. An advantage is that the morphology of wear can be followed in the SEM. Kumar et al [682a] developed a similar technique, named "cluster overcut fly grinding." Either a small cluster of grains (cut from a wheel with a diamond saw) is attached to a metal disk, or a wheel is dressed so that only a small, square patch stands out from the surface.

Bench tests have limited validity, although correlation can be good when E.P. activity is important. For example, Röllig et al [683] ranked fluids according to welding load, critical load for rapid wear, and coefficient of friction measured in the four-ball test.

Application. Proper application of fluids is critical, and the particular problems presented by grinding have received much attention [561,570,677,684].

Fast rotation of the grinding wheel leads to the formation of an air film. Measurements of Kaliszer and Trmal [684] indicated low pressures (around 1 kPa) but high air-stream velocities; velocities increased with wheel speed and grit size and in the presence of a wheel guard. It is often thought that this air film prevents the fluid from reaching the wheel surface. However, Werner and Lauer-Schmaltz [570] point out that the fluid stream bounces off the wheel only when its speed is not matched to the surface speed of the wheel. Fortunately, the misconception about the role of the air film still led to the correct solutions: techniques designed to break the air film allow acceleration of the fluid to the wheel speed. Some devices incorporate an air scraper [570]. Several techniques have found acceptance:

a. The simplest fan-shape nozzle covers the width of the wheel (Fig.11.64a) and is ground to shape to "break the air film" [4.20]. If the fluid flows fast enough, it will follow the wheel. A pressure of 430 to 640 kPa is sufficient for a wheel speed of 30 m/s. Increased fluid-delivery rates greatly reduce workpiece temperatures [685], and feed rates can be increased [686].

b. A nozzle with a large orifice extending over the sides of the wheel (Fig.11.64b) allows gradual acceleration of the fluid.

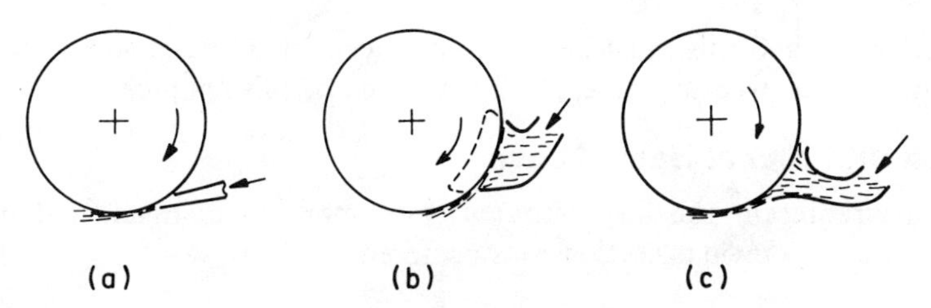

Fig.11.64. Methods of applying grinding fluids

c. A nozzle, again with a large orifice, directs the fluid almost perpendicular to the wheel surface (Fig.11.64c) [684], where the fluid is accelerated.

d. The surface of the wheel can be cleaned by applying fluid at high pressure (up to 2 MPa) through nozzles placed close to the wheel surface, outside the working zone [567,652]. In one variant, pressure jets oscillate across the wheel width to ensure complete coverage [687]; in another, several banks are placed around the wheel circumference [570,677]. It is believed that the fluid, in addition to dislodging debris, is driven into the pores, to be released again in the work zone. Grinding burn is observed wherever the jets fail to overlap. This technique is of particular importance in creep-feed grinding.

e. The fluid may be supplied from the center of the wheel, to be hurled through the pores [586,688]. Clogging is a danger, and fine filtration (below 3 μm) is necessary. Misting is also a problem. Alternatively, small quantities of an oil are applied through the center of the wheel, while large quantities of an aqueous fluid flood the contact zone [688a].

f. The contact zone is totally immersed in a chamber through which pressurized oil streams [689]. This allows creep-feed grinding of large drill flutes.

g. A fluid film is deposited on the workpiece surface from a contacting porous element or a roller, while the contact zone is flooded [690]. Alternatively, bulk metal removal is done by flooding the contact zone, and finish grinding proceeds with a thin oil film applied to the workpiece by a roller [691].

h. In electrolytic grinding, a metal-bonded, alumina, or diamond wheel is operated in an alkaline metal electrolyte (typically, $NaNO_2$, Na_3CO_3), and most metal removal actually takes place by anodic dissolution [12,614].

i. Even distribution of the fluid, at the rate of 5 L per minute per centimetre of width (or 15 L per minute per kilowatt of motor power), is critical in belt grinding [678].

Treatment. All grinding fluids are applied from a recirculating system. In the simplest case, flotation of debris is possible in aqueous fluids, and centrifuges are adequate for removing tramp oil. Filtration, reviewed by Mrozek [692], is essential for high surface quality and for improved removal rates. Polyanskov [693] has discussed the principles of designing a system for diamond grinding. Grinding with CBN also requires filtration, to below 5 μm with oil and to below 8 μm with emulsions [694]. Filtration is critical also for honing [679].

11.7 Cutting and Grinding Fluids for Specific Metals

Detailed treatments of metal-removal processes are given in handbooks [11,12]. Here we shall limit ourselves to a brief summary of the choice of fluids for different metals. As mentioned in Sec.11.5.1, the machinability of a given metal depends much on its metallurgical condition, and this must be taken into account in the choice of fluid. General recommendations are given in Table 11.2, at the end of this chapter.

11.7.1 Iron and Nickel-Base Alloys

Ferrous metals represent the largest quantity of metals machined, and much of the research discussed in previous sections has centered on them.

Steels. Iron is seldom used; as a pure metal of high ductility, it is difficult to machine [34]. A low-carbon steel is stronger, yet it can be cut at lower forces because the contact length L is less.

Inclusions are present from deoxidation practice. Hard ones, such as Al_2O_3 and SiO_2, cause abrasive wear [136,695] but also help by reducing ductility in the shear zone. When of small size, they are not objectionable [695]. Complex oxides that develop on the rake face in cutting of Ca-Fe-Si deoxidized steels reduce wear of carbide tools [25,327,696] but may soften ceramic tools [697]. Reviews have been given by Joseph and Tipnis [698] and by Ramalingam et al [699], who also propose a model [700]. Several workers have explored these effects by testing steels with controlled inclusion populations, produced by powder metallurgy [699,701]. An unusual proposition is temporary embrittlement by cathodic charging of steel with hydrogen [702].

A BUE generally forms; its size depends on the availability of oxygen or cutting fluid. Both aqueous and oil-base fluids are used. With increasing severity of operation, more sulfur is generally required. Chlorine is relatively ineffective in drilling and tapping but has sometimes been found useful in turning [500], and is often incorporated in commercial lubricants. Semisynthetic fluids are adequate at lower metal-removal rates, but emulsions are preferred for high removal rates [703,704]. Oils are generally better for tapping and drilling [500,531], although emulsions and synthetics have been used, too [501,520,525].

High-Strength Steels. Many steels are machined in the fully heat treated condition, at hardnesses over 300 HB. In a series of articles, Bailey [705] explored surface damage and found that fluids extended the range over which steady-state conditions could be established. Aqueous fluids can be used although they may promote fatigue-crack propagation [665], and oils are then preferable [435,704].

Tool steels are ground at hardnesses up to 65 HRC. Diamond suffers rapid wear, and alumina or CBN wheels are used. The latter offer high removal rates, keep the surfaces cool [279,640,657,694,706], and, with vitrified bonds, are excellent also for internal grinding. Oil is the preferred fluid, as it is with alumina wheels [663a], although emulsions can be used at lower removal rates [694]. Dressing is critical [640]. Single-stroke finish honing also calls for oil (or kerosene) with sulfur and chlorine additions.

Free-machining steels owe their desirable properties to controlled inclusions of MnS (resulfurized steels), lead (leaded steels), or both (resulfurized-leaded steels) [37,698,699]. The presence of MnS inclusions, in combination with lead, MnSe, or MnTe [707], reduces the shear strength in the primary shear zone. Lead is insoluble in iron and prevents rewelding of cracks [708].

The sulfur- and/or lead-enriched chip/tool interface reduces the shear strength in the secondary shear zone, limits L and the BUE, reduces cutting forces (especially steeply with resulfurized-leaded steels [34]), and makes the chip easily disposable. Adhesive wear of HSS and carbide tools is minimized [47], and wear of ceramic tools is reduced [697]. Surface analysis shows thin layers (2 to 4 nm) of lead on both tool and chip surfaces [230], and thus lead acts as a typical low-shear-strength lubricant. Theories of the action of inclusions have been proposed [709,710].

Cast Iron. Gray cast iron contains free graphite in the form of curved plates (flakes) which ensure early chip separation and impart free-machining properties. Dry cutting is easily performed with all tools (it is usually done with carbide tools or, for finishing, with ceramic tools [64]), but a suction system is necessary to draw off the dust that is generated. With HSS tools and, particularly, drills, life is highly dependent on hardness (to the 16th power) [711]. If a fluid is needed, as in tapping, it must be a specially formulated detergent emulsion, to prevent caking of the slurry formed with the powdery chips. A small amount of emulsion mist prevented tool buildup in reaming [712]. Filtration is essential. A triethanolamine solution was found to be best for diamond honing [713]. Nodular and malleable irons, in which the graphite is in a globular form, are intermediate in behavior between gray cast iron and steel. White iron (which contains primary Fe_3C) is extremely abrasive.

Stainless Steels, Maraging Steels, and Heat-Resistant Alloys. Slow oxidation, high adhesion, and, in the case of austenitic steels, high strain hardening make these materials difficult to machine. Low heat conductivity and specific heat lead to shear localization and sawtooth chip formation [34,75]. Hard inclusions such as TiC in titanium-stabilized steels cause abrasive (plastic plowing) wear [136]. Invariably, chlorine additives are necessary; many commercial emulsions and oils also contain sulfur [714], especially for heat-resistant alloys. Oils are preferred for HSS tooling in drilling [525], broaching, etc. It is believed that chlorine can lead to corrosion, and lubricant residues are generally removed.

Nickel-Base Alloys. Pure nickel behaves like iron [34], but the adhesion of nickel-base superalloys, combined with their high specific cutting energies and low specific heats and heat conductivities, places great demands on the tooling and allows only low cutting speeds [142]. Since adhesion persists to the cutting edge, the crater moves close to the edge (more so at higher speeds), and resistance to flank wear becomes important [715]. Of the carbides, WC is best, but a binder of greater temperature resistance would be desirable [715,716]. This is the conclusion reached also from thermochemical considerations. The other candidate would be HfC, but experimental HfC coatings suffered delamination [149]. Ceramics are too brittle, but an Al_2O_3-30%TiC cermet and a silicon nitride showed promise, especially when the edge chamfer was confined to the nose radius [717]. Temperatures are reduced and life of carbide tools is increased by changing from a -5° to a +10° rake angle [718].

No general recommendations can be made regarding the best lubricant [435]. Sulfochlorinated oils outperformed emulsions in drilling, whereas chlorinated oils were better in tapping [519]. As a matter of precaution, residues are removed prior to heating. Sulfochlorinated oils gave good results in belt grinding [596], while 5% concentrations of chlorine or stearic acid in oil were satisfactory in lapping [719]. Sulfur stains can be removed by a sodium cyanide etch [11].

11.7.2 Light Metals

The low temperatures generated in machining of aluminum alloys allow higher cutting speeds. Even though dry cutting is possible, flooding with a fluid is preferable. Chemical solutions offer transparency, but emulsions are more effective in preventing pickup.

Oils, when used, are of low viscosity (4 mm^2/s at 40 C) for good cooling, and contain 5 to 10% fatty additives for screw-machine work. Even lighter kerosene is used for milling. Stick grease finds application in sawing and polishing [11].

Pure aluminum and ductile aluminum alloys are more difficult to cut because of adhesion and the formation of a stringy chip. Therefore, strain-hardened or fully heat treated conditions are preferred [720]. Cutting speeds can be high—up to 2400 m/min for carbide and 3600 m/min for diamond [64] tooling. Aluminum-silicon alloys present special difficulties, particularly at silicon contents above 10%, because of the presence of very hard, primary silicon crystals. Diamond (originally single crystals, now mostly polycrystalline tools [721]) is the natural choice for turning at high speeds [722]. Minimum-cost speed was found at 350 m/min (in comparison with 150 m/min for carbides) [723]. Tool life is 3 to 5 times longer than that of carbide tools, and 10 times more total metal is removed [182]. Face milling also can be done with polycrystalline diamond [724]. Wet cutting with carbide tools is a reasonable alternative.

Because of adhesion and redeposition, aluminum alloys are ground and lapped with oil, often with a boundary additive [719], although aqueous fluids have been found to be adequate for wrought materials [725]. In belt grinding, any of various fluids or stick wax is used, depending on duty.

Magnesium has low shear ductility and behaves as a free-machining material. It can be cut dry at high feeds, but tools must be sharp to minimize rubbing and heating. At small chip thicknesses (below 25 μm) the thin chip presents a fire hazard. Oil of low viscosity must be supplied in a generous flow (20 L/min), and must contain less than 0.2% free fatty acid to prevent corrosion. Emulsions attack the metal and make scrap recycling inefficient [11]. Grinding, too, must be done wet; if an emulsion is used, the sump must be vented to allow hydrogen to escape. The sludge must be reacted with a 5% solution of $FeCl_2 \cdot 2H_2O$.

Beryllium is machined dry. Only in deep hole drilling are emulsions used; afterward the workpiece must be washed and dried.

11.7.3 Copper-Base Alloys

Pure copper and its ductile alloys form stringy chips. They are cut with some fatty oil (typically lard oil) in a mineral-oil base. Because of staining by active sulfur, E.P. oils are avoided, except for harder alloys and alloys containing nickel. Any stains that develop can be removed with a sodium cyanide etch. Recommendations for various alloys were made by Shirk [726]. Because the contact length L is shorter, a harder material—such as 70/30 brass—develops a lower cutting force than does pure copper [34].

Several free-machining brasses have long been used. A lead content of about 1% results in the separation of a monatomic lead layer on the interface [727], reducing its shear strength. When lead content exceeds 2.75%, thick striations of lead cover 30 to 50% of the chip surface [727]. Shear ductility and fracture energy are reduced [728], and short chips are formed by a void-sheet mechanism [729]; thus, lead imparts free-machining properties. The cutting force for leaded 60/40 brass is only 1/4 of that measured in cutting of 70/30 brass [34]. A free-machining copper with 0.6% Te gives a continuous chip but a very low cutting force [34].

11.7.4 Titanium Alloys

The high reactivity of titanium, combined with its low heat conductivity and heat capacity, make chip formation discontinuous [76], and machining is very difficult. The best tool protection is given by a stable secondary shear zone [730]. At low speeds,

HSS is the best tool material [731] when used with a highly chlorinated oil or emulsion. At higher speeds, WC and HfC are expected to perform well [716]. All tool materials react, but diffusion rates are reasonably low with tungsten carbide and diamond [730]. Even so, feeds and speeds must be kept low. The danger of stress-corrosion cracking with chlorinated fluids could not be substantiated [182,665]. Surface films containing chlorine were detected on titanium even when water was used as a cutting fluid, indicating that no more harm should result from the use of chlorinated fluids [732].

Grinding speeds are kept relatively low, and then wheel wear is low also [631a]. *G* ratios can be very poor with emulsions but are higher with oils [292]. For grinding at normal speeds, a SiC wheel was found to be best with an E.P. emulsion [663a]. Following the development of inorganic fluids [675], a 10% aqueous solution of K_3PO_4 (buffered with equimolar NaH_2PO_4) was found to be better than emulsions in centerless grinding [663]. At high removal rates, SiC was better than Al_2O_3 [733]. Titanium must never be ground dry, because of fire hazard [11].

11.7.5 Other Metals

Most refractory metals can be cut, with some difficulty, with sulfochlorinated oils. Trichloroethylene has been used for tungsten [734] and molybdenum [735]; an aqueous fluid containing 6% KI, 0.5% I, 1% triethanolamine, and 8% ethyl alcohol has been used for molybdenum [735]; and an emulsion has been used for niobium. Zorev [323] studied adhesive wear with molybdenum. Annealed niobium and tantalum behave like copper, and their alloys are similar to stainless steel.

Except for tungsten, refractory metals load the grinding wheels, and chemical solutions or sulfochlorinated emulsions are needed. An "active" water-base fluid has been described for molybdenum [736]. Most metals are ground with vitrified alumina wheels; tungsten is usually ground with SiC wheels. For depleted uranium, chlorinated emulsions were best with SiC, and sulfochlorinated emulsions with Al_2O_3; sulfur was apparently detrimental with SiC [737].

Zinc alloys are readily machined dry, with emulsions, or with light oils (sometimes compounded with lard oil).

11.8 Summary

1. The process of chip formation is not uniquely defined by process geometry; material properties and interface conditions interact, and frictional and shear effects can be difficult to separate. In general, it can be said that lower friction on the rake face increases the shear angle, thins out the chip, reduces cutting forces and energy requirements, and, most importantly, reduces temperatures generated in the contact zone.

2. Normal pressure on the rake face is high, and sticking friction is often attained over most of the contact length *L* between chip and rake face. Virgin surfaces generated in the course of cutting may adhere to the tool; consequences of adhesion depend on material, speed, temperature, and presence of parting agents. Ductile, heavily strain-hardening materials—especially if they also have low heat conductivity—form segmented chips at low cutting speeds. The surface finish is poor and is made worse by high friction on the flank surface. At intermediate speeds, sticking friction leads to development of a dead-metal zone over the tool edge. This built-up edge (BUE) becomes particularly prominent with two-phase materials. Instability of the BUE results in a surface roughened by fragments of the BUE. The resultant increase in effective rake angle reduces forces, and the BUE may actually protect the tool. At high speeds, thermal softening shrinks the BUE until it is reduced to a secondary shear zone. At this

point, a thin layer of material is stationary (or almost stationary) along the rake face, and the material of the chip must accelerate in a thin zone. Frictional forces along the rake face are governed by the magnitude of L and by the shear flow strength of the chip material.

3. In cutting with negative-rake-angle tools, the thrust (normal) force increases and sticking friction sets in earlier. At some critical angle, which depends on the state of stress and also on friction, the surface bulges without chip formation. This condition arises in cutting of small undeformed chip thicknesses with a tool of finite edge radius, and when the tool has a negative clearance angle. The contribution of friction to total energy (and also to temperature rise) becomes substantial.

4. In drilling with twist drills, friction develops also at the drill point, which performs plastic indentation. Significant friction arises also on secondary rubbing surfaces, such as those present in drilling, gun drilling, trepanning, and tapping. Contact over the wear land in cutting and on worn grit surfaces in abrasive processes contributes substantial frictional resistance and heat.

5. In abrasive machining, the width of the cutting edge is limited (often to a single point), and rake angles range from zero to highly negative. Multiple encounters with the workpiece surface may lead to cutting, plowing, rubbing, or plastic fatigue.

6. Most of the heat generated is carried away with the chip when the undeformed chip thickness is large; nevertheless, tool temperatures become very high when engagement is prolonged. Much of the heat is retained in the workpiece when the undeformed chip thickness is small; further heating occurs when rubbing is significant, as in abrasive machining.

7. The severity of conditions leads to rapid tool wear, with total tool life being expressed in minutes rather than hours. All wear mechanisms may operate. Abrasive wear, microfracture, and (if adhesion is high) adhesive wear (which may lead to attrition) dominate at low temperatures. Thermally activated processes, including creep (plastic flow), diffusion, dissociation, and chemical reactions, become important at high temperatures. Because temperature is proportional to speed, tool life can often be described by a power law based on speed (the Taylor equation), although the exponent tends to be also a function of other variables and is seldom truly constant over a wide speed range.

8. Tool materials are among the most important elements of the tribological system. Hardness and toughness at operating temperature, stability, and resistance to chemical reactions (with the environment and the workpiece material) are desirable attributes. High adhesion is harmful only when it is coupled with rapid diffusion or with high rates of reaction with the workpiece material.

9. At low cutting speeds, fluids fulfill primarily the role of a lubricant, reducing rake-face friction (if sliding prevails) or the contact length L (under sticking conditions) as well as reducing friction on the flank face and on secondary rubbing surfaces. Fluids may also shift the onset of BUE formation to higher speeds. Access to the cutting edge is difficult if not impossible, but benefits are derived from limiting the extent of sticking within the contact length L, especially when the rake face is intermittently exposed.

10. At high cutting speeds the primary function is cooling. The temperature in the zone of chip formation cannot be substantially influenced. However, the bulk temperatures of workpiece and tool can be reduced. Thus, the cumulative effects of heating can be minimized, and tool life can be prolonged. Even though speeds are high in grinding, the friction-reducing effect of fluids remains important, because the heat generated in plowing and rubbing contacts can be substantially reduced.

11. Cutting and grinding fluids may be either oil-base or water-base. The friction-

reducing effect decreases and cooling power increases in the series: oil, emulsion, semi-chemical solution, true solution. The lower heat-transfer coefficient of oils is counter-balanced to a substantial degree by the reduced heat generation and the larger heat-transfer area. Application of fluids is as important as their selection. Best results are obtained by supplying the fluid as close to the cutting zone as possible.

12. The severity of an operation increases, in general, with the difficulty of access to the contact zone and with the presence of a secondary rubbing zone. The atmosphere serves as a powerful cutting and grinding fluid when oxygen suffices as the active medium. The greater cooling power of oils and aqueous fluids is combined with lubricating functions derived from oil viscosity and from the presence of more active species, especially chlorine and sulfur additives. The low temperature resistance of fatty derivatives limits their use to some nonferrous metals or to processes where secondary rubbing is significant. The recommendations given in Table 11.2 should be regarded as guidelines only.

Table 11.2. Commonly used machining fluids (compiled from [12], [420], and Sec.11.7)

Process	Tool	Steel (<275 BHN)	Steel (>275 BHN)	Stainless steel, nickel alloy	Cast iron	Aluminum alloy	Magnesium alloy	Copper alloy
Grinding		O1, E1, C1	O2, E1, C1	O2, E2, C2	E1, C1	O1, Sp	O1, Sp	O1, Sp
Turning	HSS	O1, E1, C1	O2, E2, C2	O2, E2, C2	E1, C1	E1, C1, Sp	O1, Sp	E1, C1, Sp
	Carbide	D, E1, C1	D, E1, C1	D, E1, C1	D, E1, C1	D, E1, C1	O1, Sp	E1, C1
Milling / Drilling	HSS	O1, E1, C1	O2, E2, C2	O2, E2, C2	E1, C1	D, O1, Sp	O1, Sp	E1, C1, Sp
	Carbide	D, E1, C1	O1, E1, C1	O2, E1, C1	D, E1, C1	D, O1, Sp	O1, Sp	E1, C1
Form turning	HSS	O2, E2, C2	O2, E2, C2	O2, E2, C2	E1, C1	E1, C1, Sp	O1, Sp	E1, C1, Sp
	Carbide	D,E1, C1	E2, C2	O2, E2, C2	D, E1, C1	D, E1, C1	O1, Sp	E1, C1
Gear shaping	HSS	O2, E2, C2	O2, E2, C2	O2, O3	E1, C1	O1, Sp	D, O1, Sp	O1, Sp
Tapping	HSS	O1, E2, C2	O2, E2, C2	O2, O3	E1, C1	D, O1, Sp	D, O1, Sp	O1, Sp
Broaching	HSS	O2, E2, C2	O2, E2, C2	O2, E2, C2	E2, C2	E1, C1, Sp	D, O1, Sp	E1, C1, Sp
	Carbide	O1, E1, C1	O1, E1, C1	O1, E1, C1	D, E1, C1	D, E1, C1	D, O1, Sp	E1, C1, Sp

Code: D – Dry. O1 – Mineral oil or synthetic oil. O2 – Compounded oil. O3 – Heavy-duty compounded oil. E1 – Mineral-oil emulsion. E2 – Heavy-duty (compounded) emulsion. C1 – Chemical fluid or synthetic fluid. C2 – Heavy-duty (compounded) chemical or synthetic fluid. Sp – Specially formulated fluid, with boundary and/or E.P. additives.

References

[1] M.C. Shaw, *Metal Cutting Principles*, 3rd Ed., MIT Press, Cambridge, MA, 1957.

[2] R. Tourret, *Performance of Metal Cutting Tools*, Butterworths, London, 1958.

[3] N.N. Zorev, *Metal Cutting Mechanics*, Pergamon, Oxford, 1966.

[4] M. Kronenberg, *Machining Science and Application*, Pergamon, Oxford, 1966.

[5] E.J.A. Armarego and R.H. Brown, *The Machining of Metals*, Prentice Hall, Englewood Cliffs, NJ, 1969.

[6] G. Vieregge, *Zerspanung der Eisenwerkstoffe*, 2nd Ed., Verlag Stahleisen, Düsseldorf, 1970.

[7] V. Arshinov and G. Alekseev, *Metal Cutting Theory and Cutting Tool Design*, Mir Publishers, Moscow, 1970.

[8] E.M. Trent, *Metal Cutting*, Butterworths, London, 1977.

[9] G. Boothroyd, *Fundamentals of Metal Machining and Machine Tools*, McGraw-Hill, New York, 1975.

[10] J. Kaczmarek, *Principles of Machining*, Peter Peregrinus, Stevenage, Wydawmictwa Naukow-Techniczne, Warsaw, 1976.

[11] *Metals Handbook*, 8th Ed., Vol.3. *Machining*, ASM, Metals Park, 1967.
[12] *Machining Data Handbook*, 3rd Ed. (2 volumes), Machinability Data Center, Metcut Research Associates, Cincinnati, 1980.
[12a] T.J. Drozda and C. Wick (ed.), *Tool and Manufacturing Engineers Handbook*, Vol.1, Machining, SME, Dearborn, 1983.
[13] M.C. Shaw et al (ed.), *Proc. Int. Production Engineering Research Conf.*, ASME, New York, 1963.
[14] *Machinability*, ISI Spec. Rep. 94, Iron Steel Institute, London, 1967.
[15] L.V. Colwell et al (ed.), *Proc. Int. Conf. Manufacturing Technology*, ASTME (SME), Dearborn, 1967.
[16] *Influence of Metallurgy on Machinability*, ASM, Metals Park, 1975.
[17] *Machinability Testing and the Utilization of Machining Data*, ASM, Metals Park, 1979.
[18] *Proc. 4th Int. Conf. Production Engineering*, Jpn. Soc. Precision Eng. and Jpn. Soc. Techn. Plasticity, Tokyo, 1980.
[19] *Cutting Tool Materials*, ASM, Metals Park, 1981.
[20] T.H.C. Childs and G.W. Rowe, Rep. Progr. Phys., *36*, 1973, 223-288.
[21] J.A. Bailey, Wear, *31*, 1975, 243-275.
[22] P. Dewhurst, Proc. Roy. Soc. (London), *A360*, 1978, 587-610.
[23] M.Es. Abdelmoneim, Wear, *63*, 1980, 303-318.
[24] G.W. Rowe, Phil. Mag., *A43*, 1981, 567-585.
[25] M.C. Shaw, in [3.34], Vol.2, Chap.20, *Metal Removal*.
[26] M.E. Merchant, J. Appl. Phys., *16*, 1945, 267-275, 318-324.
[27] V. Piispanen, J. Appl. Phys., *19*, 1948, 876-881.
[28] L. De Chiffre and T. Wanheim, in *Proc. 9th NAMRC*, SME, Dearborn, 1981, pp.231-234.
[29] I. Finnie, Mech. Eng., *78*, 1956, 715-721.
[30] H. Ernst and M.E. Merchant, in *Surface Treatment of Metals*, ASM, Metals Park, pp.299-335.
[31] G. Warnecke, in *Proc. 5th NAMRC*, SME, Dearborn, 1977, pp.229-236; Ind. Anz., *96*, 1974, 858-859 (German).
[32] P.L.B. Oxley, in *Proc. 22nd Int. MTDR Conf.*, Macmillan, London, 1982, pp.279-287.
[32a] P. Matthew and P.L.B. Oxley, Wear, *69*, 1981, 219-234.
[33] M.C. Shaw, in [14], pp.1-9.
[34] J.E. Williams, E.F. Smart, and D.R. Milner, Metallurgia, *81*, 1970, 3-10, 51-59, 89-93.
[35] W. Kluft et al, CIRP, *28*, 1979, 441-455.
[36] B.F. von Turkovich, CIRP, *30*, 1981, 533-540.
[37] E.D. Doyle and L.E. Samuels, J. Austral. Inst. Metals, *21* (1), 1976, 2-15.
[38] R. Komanduri and R.H. Brown, Metals Mat., *6*, 1972, 531-533.
[39] L.H.S. Luong and R.H. Brown, Trans. ASME, Ser.B, J. Eng. Ind., *103*, 1981, 431-436.
[40] A.J. Pekelharing, CIRP, *23*, 1974, 207-212.
[41] V.D. Kuznetsov, *Metal Transfer and Build-up in Friction and Cutting*, Pergamon, Oxford, 1966.
[42] H. Ernst and M. Martelotti, Mech. Eng., *57*, 1938, 487-498.
[43] K. Nakayama, M.C. Shaw, and M.C. Brewer, CIRP, *14*, 1966, 211-223.
[44] E.M. Trent, in [14], pp.11-18.
[45] J.G. Horne, P.K. Wright, and A. Bagchi, in *Proc. 9th NAMRC*, SME, Dearborn, 1981, pp.223-230.
[46] K. Iwata and K. Ueda, Wear, *60*, 1980, 329-337.
[47] K. Ohgo, Wear, *51*, 1978, 116-126.
[48] P.K. Wright, Metals Techn., *8*, 1981, 150-160.
[49] P.W. Wallace and G. Boothroyd, J. Mech. Eng. Sci., *6*, 1964, 74-87, 306-308, 422-423; *7*, 1965, 118-123.
[50] I. Finnie and M.C. Shaw, Trans. ASME, *78*, 1956, 1649-1657.
[51] N.N. Zorev, in [13], pp.42-49.
[52] E.D. Doyle, J.G. Horne, and D. Tabor, Proc. Roy. Soc. (London), *A366*, 1979, 173-183.
[53] T.H. Childs, D. Richings, and A.B. Wilcox, Int. J. Mech. Sci., *14*, 1972, 359-375.
[54] P.K. Wright, J.G. Horne, and D. Tabor, Wear, *54*, 1979, 371-390.
[55] H. Takeyama and T. Ono, Trans. ASME, Ser.B, J. Eng. Ind., *90*, 1968, 335-343.
[56] M. Bao and M.G. Stevenson, CIRP, *25*, 1976, 53-57.
[57] G.W. Rowe and F. Wolstencroft, J. Inst. Metals, *98*, 1970, 33-41.

[58] P.D. Muraka, G. Barrow, and S. Hinduja, Int. J. Mech. Sci., *21*, 1979, 445-456.
[59] J.E. Williams and E.C. Rollason, J. Inst. Metals, *98*, 1970, 144-153.
[60] W.B. Rice, R. Salmon, and A.G. Advani, Int. J. Mach. Tool Des. Res., *6*, 1966, 143-152.
[61] W. Pentland, C.L. Mehl, and J.L. Wennberg, Am. Mach., *104* (July 11), 1960, 117-132.
[62] S. Rajagopal, D.J. Plankenhorn, and V.L. Hill, J. Appl. Metalwork., *2* (3), 1982, 170-184.
[63] B.F. von Turkovich, in *Proc. 7th NAMRC*, SME, Dearborn, 1979, pp.241-247.
[64] J.F. Kahles, M. Field, and S.M. Harvey, CIRP, *27*, 1978, 551-560.
[65] J. Chaplin and J.A. Miller, in *Proc. 9th NAMRC*, SME, Dearborn, 1981, pp.311-317.
[66] Z. Ze-Hua, in *Proc. 21st Int. MTDR Conf.*, Macmillan, London, 1981, pp.275-282.
[67] S. Abeyama and S. Nakamura, in [17], pp.184-198.
[68] M.C. Shaw, E. Usui, N.H. Cook, and P.A. Smith, Trans. ASME, Ser.B, J. Eng. Ind., *83*, 1961, 163-192.
[69] E. Usui and M.C. Shaw, Trans. ASME, Ser.B, J. Eng. Ind., *84*, 1962, 89-102.
[70] R. Komanduri and R.H. Brown, Trans. ASME, Ser.B, J. Eng. Ind., *103*, 1981, 33-51.
[71] R. Komanduri et al, Trans. ASME, Ser.B, J. Eng. Ind., *104*, 1982, 121-131.
[72] V.A. Stewart and R.H. Brown, in *Proc. 13th Int. MTDR Conf.*, Macmillan, London, 1973, pp.13-18.
[73] W.C. Rice, Engineering J., *44* (2), 1961, 41-45.
[74] K.F. Sullivan, P.K. Wright, and P.D. Smith, Metals Techn., *5*, 1978, 181-189.
[75] N. Ueda and T. Matsuo, CIRP, *31*, 1982, 81-84.
[76] R. Komanduri, Wear, *76*, 1982, 15-34.
[77] M. Field and M.E. Merchant, Trans. ASME, *71*, 1949, 421-430.
[78] J.E. Williams, Wear, *48*, 1978, 55-71.
[79] A.S. Shouckry, Wear, *55*, 1979, 313-329.
[80] R. Makino, K. Kishi, and K. Hoshi, CIRP, *24*, 1975, 47-51.
[81] R.L. Kegg and T.R. Sisson, Paper No. MR68-615, SME, Dearborn, 1968.
[82] T. Hoshi, in *Proc. 3rd NAMRC*, SME, Dearborn, 1975, pp.788-803.
[83] V.F. Kazantsev, in [3.337], pp.3-98.
[84] Y. Tanaka, M. Ido, and K. Yoshida, in [18], pp.440-445.
[85] T. Stöferle and H. Müller-Gerbes, Werkstatt u. Betr., *106*, 1973, 1025-1030 (German).
[86] J. Tlusty, CIRP, *27*, 1978, 583-589.
[87] M.Es. Abdelmoneim and R.F. Scrutton, Wear, *24*, 1973, 1-13.
[88] R. Komanduri, Int. J. Mach. Tool Des. Res., *11*, 1971, 223-233.
[89] M. Abebe and F.C. Appl, in *Proc. 9th NAMRC*, SME, Dearborn, 1981, pp.341-348.
[90] M.Es. Abdelmoneim and R.F. Scrutton, Wear, *27*, 1974, 35-46.
[91] E.G. Thomsen, A.G. Macdonald, and S. Kobayashi, Trans. ASME, Ser.B, J. Eng. Ind., *84*, 1962, 53-62.
[92] H. Chandrasekaran, Wear, *36*, 1976, 133-145.
[93] K. Hitomi and G.L. Thuering, Trans. ASME, Ser.B, J. Eng. Ind., *84*, 1962, 282-288.
[94] M.Es. Abdelmoneim, Wear, *58*, 1980, 173-192.
[95] T. Hoshi, in [19], pp.413-426.
[96] M. Sarwar and P.J. Thompson, in *Proc. 22nd Int. MTDR Conf.*, Macmillan, London, 1982, pp.295-304.
[97] J. Larsen-Basse and P.L.B. Oxley, in *Proc. 13th Int. MTDR Conf.*, Macmillan, London, 1973, pp.209-216.
[98] A.H. Redford, CIRP, *29*, 1980, 67-71.
[99] K.F. Althoff, Stahl u. Eisen, *95*, 1975, 221-234 (German).
[99a] W. König, I.K. Essel, and I.L. Witte, *Specific Cutting Force Data for Metal Cutting*, Verlag Stahleisen, Düsseldorf, 1982.
[100] E. Lenz, in [15], pp.553-562.
[101] A.H. Redford, B. Mills, and S. Skhtar, CIRP, *25*, 1976, 89-92.
[102] A.J.R. Smith and E.J.A. Armarego, CIRP, *30*, 1981, 9-13.
[103] D. Lenk, J. Nittel, and W. Zscherpel, Schmierungstechnik, *10*, 1979, 267-269 (German).
[104] M.C. Shaw, Techn. Mitt., *5*, 1958, 211-216 (German).
[105] A.O. Tay, M.G. Stevenson, G. DeVahl Davis, and P.L.B. Oxley, Int. J. Mach. Tool Des. Res., *16*, 1976, 335-349.
[105a] P.K. Wright, in *Metallurgical Effects at High Strain Rates*, R.W. Rhode et al (ed.), Plenum, New York, 1973, pp. 547-558.
[106] B.T. Chao and K.J. Trigger, Trans. ASME, Ser.B, J. Eng. Ind., *81*, 1959, 139-151.

[107] P.K. Wright, S.P. McCormick, and T.R. Miller, Trans. ASME, Ser.B, *102*, 1980, 123-128.
[108] A.E. Focke et al, in [3.421], pp.309-413.
[109] Y. Lee, K. Hiramoto, N. Fujita, and T. Sata, Bull. Jpn. Soc. Prec. Eng., *11*, 1977, 147-148.
[110] E. Lenz, D. Pnueli, and L. Rozeanu, Wear, *53*, 1979, 337-344.
[111] B.S. Levy and R.W. Thompson, in *Proc. 3rd NAMRC*, SME, Dearborn, 1975, pp.334-345.
[112] T.H.C. Childs, Int. J. Mech. Sci., *13*, 1971, 373-387.
[112a] E. Usui and H. Takeyama, Trans. ASME, Ser.B, J. Eng. Ind., *82*, 1960, 303-308.
[113] T.N. Loladze, CIRP, *24*, 1975, 13-16.
[114] S. Ramalingam, Trans. ASME, Ser.B, J. Eng. Ind., *93*, 1971, 527-537, 538-544.
[115] S. Kato, K. Yamaguchi, and M. Yamada, Trans. ASME, Ser.B, J. Eng. Ind., *94*, 1972, 683-689.
[116] G. Montag and J. Höppner, Fertigungstech. Betr., *24*, 1974, 407-412 (German).
[116a] W.S. Sampath and M.C. Shaw, in *Proc. 11th NAMRC*, SME, Dearborn, 1983.
[117] G.W. Rowe, E.F. Smart, and K.C. Tripathi, ASLE Trans., *20*, 1977, 347-353.
[118] P.K. Wright, Trans. ASME, Ser.B, J. Eng. Ind., *104*, 1982, 285-292.
[118a] P.K. Wright, A. Bagchi, and J.G. Chow, in *Proc. 10th NAMRC*, SME, Dearborn, 1982, pp.255-262.
[119] J.H. Dautzenberg, J.A.W. Hijink, and A.C.H. van der Wolf, CIRP, *31*, 1982, 91-96.
[120] I. Finnie, in [13], pp.76-88.
[121] T.H.C. Childs, in *Mechanical Properties at High Rates of Strain*, Inst. of Physics, London, 1974, pp.382-392.
[122] P.K. Wright and J.L. Robinson, Metals Techn., *4*, 1977, 240-248.
[122a] A.J. Holzer and P.K. Wright, Mater. Sci. Eng., *51*, 1981, 81-92.
[123] S.S. Law, M.F. DeVries, and S.M. Wu, Trans. ASME, Ser.B, J. Eng. Ind., *94*, 1972, 965-970.
[124] K. Okushima and T. Hoshi, Bull. JSME, *10*, 1967, 206-215.
[125] K. Okamura and T. Nakajima, in [535], pp.305-321.
[126] P.K. Vennvinod, W.S. Lau, and P.N. Reddy, Trans. ASME, Ser.B, J. Eng. Ind., *103*, 1981, 469-477.
[127] R.L. Thomas and R.L.J. Lawson, in *Proc. 16th Int. MTDR Conf.*, Macmillan, London, 1976, pp.125-131.
[128] W.E. Henderer and B.F. von Turkovich, in *Proc. 3rd NAMRC*, SME, 1975, pp.589-602.
[129] N.H. Cook, Trans. ASME, Ser.B, J. Eng. Ind., *95*, 1973, 931-938.
[130] E.M. Trent, in [3.342], pp.443-489.
[131] V.A. Tipnis, in [3.344], pp.891-930.
[132] S. Ramalingam, in [3.49], pp.663-677.
[133] N.P. Suh, Wear, *62*, 1980, 1-20.
[134] P.K. Wright and A. Bagchi, J. Appl. Metalwork., *1* (4), 1981, 15-23.
[135] B. Colding and W. König, CIRP, *19*, 1971, 793-812.
[136] S. Ramalingam and P.K. Wright, Trans. ASME, Ser.H, J. Eng. Mat. Tech., *103*, 1981, 151-156.
[137] Y. Koren, Trans. ASME, Ser.B, J. Eng. Ind., *100*, 1978, 103-109.
[138] C. Rubenstein, Trans. ASME, Ser.B, J. Eng. Ind., *98*, 1976, 221-232.
[139] A. Ber and S. Kaldor, CIRP, *31*, 1982, 13-17.
[140] E. Orady and J. Tlusty, in *Proc. 9th NAMRC*, SME, Dearborn, 1981, pp.250-255.
[141] K. Taraman and D. Troup, Paper No. MR75-705, SME, Dearborn, 1975.
[142] M.C. Shaw, A.L. Thurman, and H.J. Ahlgren, Trans. ASME, Ser.B, J. Eng. Ind., *88*, 1966, 142-146.
[143] L.V. Colwell, D.V. Doane, and I. Tanaka, Metals Eng. Quart., *14* (1), 1974, 32-40.
[144] N. Ueda and T. Matsuo, in [18], pp.512-517.
[145] A. Ishibashi, A. Katsuki, and J. Mikajiri, Bull. JSME, *17*, 1974, 502-510.
[146] T. Ono and H. Takeyama, in *Proc. 5th NAMRC*, SME, Dearborn, 1977, pp.253-257.
[147] P.K. Wright and A. Thangaraj, Wear, *75*, 1982, 105-122.
[148] B.M. Kramer and N.P. Suh, Trans. ASME, Ser.B, J. Eng. Ind., *102*, 1980, 303-309.
[149] B.M. Kramer and P.O. Hartung, in [19], pp.57-74.
[150] L.E. Toth, *Transition Metal Carbides and Nitrides*, Academic Press, New York, 1971.

[151] T.N. Loladze, G.V. Bokuchava, and G.E. Davidova, in *The Science of Hardness Testing,* J.H. Westbrook and H. Condrad (ed.), ASM, Metals Park, 1973, pp.251-257, 495-502.
[152] E. Usui et al, Trans. ASME, Ser.B, J. Eng. Ind., *100*, 1978, 222-243.
[153] V. Höglund, CIRP, *25*, 1976, 99-103.
[154] T.N. Loladze, CIRP, *27*, 1978, 535-539.
[155] M.C. Shaw, CIRP, *28*, 1979, 19-21.
[156] D.R. Deotto and K.K. Wang, in *Proc. 2nd NAMRC*, SME, Dearborn, 1974, pp.88-104.
[157] A.J. Pekelharing, CIRP, *27*, 1978, 5-10.
[158] H. Wu and J.E. Mayer, Trans. ASME, Ser.B, J. Eng. Ind., *101*, 1979, 159-164.
[159] J. Taylor, in *Proc. 8th Int. MTDR Conf.*, Pergamon, Oxford, 1968, pp.487-504.
[160] E. Usui, T. Shirakashi, and T. Ihara, in [18], pp.474-479.
[161] S. Ramalingam and J.D. Watson, Trans. ASME, Ser.B, J. Eng. Ind., *99*, 1977, 522-532; *100*, 1978, 193-200.
[162] R. Levi, Y. Koren, S. Malkin, and O. Masory, CIRP, *31*, 1982, 41-44.
[163] V.C. Venkatesh and M. Satchithanandam, CIRP, *29*, 1980, 19-22.
[164] M.Es. Abdelmoneim, Wear, *72*, 1981, 1-11.
[165] J.E. Mayer, Jr., and D.J. Stauffer, Paper No. MR73-907, SME, Dearborn, 1973.
[166] R.W. Thompson, R.B. Dixon, and B.S. Levy, in *Proc. 3rd NAMRC*, SME, Dearborn, 1975, pp.385-400.
[167] V.A. Tipnis and R.A. Joseph, Trans. ASME, Ser.B, J. Eng. Ind., *93*, 1971, 571-585.
[168] A.S. Jhita, V.K. Jain, and P.C. Pandey, in *Proc. 22nd Int. MTDR Conf.*, Macmillan, London, 1982, pp.247-253.
[169] S.M. Bhatia, P.C. Pandey, and H.S. Shan, Wear, *61*, 1980, 21-30.
[170] S. Ramalingam, in [18], pp.452-461.
[171] G. Chrysolouris, CIRP, *31*, 1982, 65-69.
[172] H. Opitz, in [3.35], pp.664-669.
[173] C.H. Khang and G.L. Viegelahn, in *Proc. 2nd NAMRC*, SME, Dearborn, 1974, pp.105-116.
[174] H.S. Shan and P.C. Pandey, in [18], pp.535-540.
[175] V. Raghuram and M.K. Muju, Int. J. Mach. Tool Des. Res., *20*, 1980, 87-96.
[176] M.Y. Friedman and E. Lenz, Wear, *25*, 1973, 39-44.
[177] A. Ber, CIRP, *21*, 1972, 21-22.
[178] J.E. Matta, W.L. Roper, D.P.H. Hasselman, and G.E. Kane, Wear, *37*, 1976, 323-331.
[179] C.J. Kotval and M.M. Barash, in *Proc. 5th NAMRC*, SME, Dearborn, 1977, pp.247-252.
[180] F.W. Taylor, Trans. ASME, *28*, 1907, 31-279.
[181] G. Barrow, Tribology, *5*, 1972, 22-30.
[182] V.A. Tipnis and J.D. Christopher, in [17], pp.3-35.
[183] I. Kalaszi, Int. J. Prod. Res., *9*, 1971, 37-51.
[184] N.G. Odrey and A.A. Zaman, in *Proc. 10th NAMRC*, SME, Dearborn, 1982, pp.312-319.
[185] R.A. Etheridge and J.C. Hsu, CIRP, *18*, 1970, 107-117.
[186] K. Uehara, CIRP, *24*, 1975, 59-64.
[187] H. Opitz, Werkstattstech. u. Maschinenb., *46*, 1956, 210-217 (German).
[188] G. Lorenz, CIRP, *26*, 1977, 39-43.
[189] N.N. Zorev and N.I. Tashlitsky, in [14], pp.31-34.
[190] T. Ohno, M. Takahashi, T. Sakai, and S. Hamahata, in [17], pp.164-175.
[191] J.G. Wager and M.M. Barash, Trans. ASME, Ser.B, J. Eng. Ind., *93*, 1971, 1044-1050.
[192] R.E. DeVor, D.L. Anderson, and W.J. Zdeblick, Trans. ASME, Ser.B, J. Eng. Ind., *99*, 1977, 578-584.
[193] H.G. Swinehart (ed.), *Cutting Tool Materials Selection*, ASTME (SME), Dearborn, 1968.
[194] *Guidebook for Selecting and Specifying Cutting Tools,* McGraw-Hill, New York, 1981.
[195] *Modern Trends in Cutting Tools*, SME, Dearborn, 1982.
[196] W. König and K. Essel, CIRP, *24*, 1975, 1-5.
[197] N. Gane, J. Austral. Inst. Metals, *21* (1), 1976, 24-31.
[198] R. Komanduri and J.D. Desai, Rep. No. 82CRD220, General Electric Co., Schenectady, NY, 1982.
[199] T. Mukherjee, in [3.424], pp.6-14.

[200] R. Wilson, *Metallurgy and Heat Treatment of Tool Steels,* McGraw-Hill, New York, 1975.
[201] H.D. Weckener, wt-Z. Ind. Fertig, *70*, 1980, 625-630.
[202] R.W. Bratt, in [19], pp.133-158.
[203] K.D. Bouzakis, W. König, and K. Vossen, CIRP, *31*, 1982, 25-29.
[204] W.E. Henderer and B.F. von Turkovich, in *Proc. 6th NAMRC*, SME, Dearborn, 1978, pp.283-289.
[205] G. Langford, Mater. Sci. Eng., *28*, 1977, 275-284.
[206] M.L. Neema and P.C. Pandey, Wear, *59*, 1980, 355-362.
[207] H. Opitz and W. König, in [3.424], pp.6-14.
[208] B.N. Colding, Paper No. MR80-901, SME, Dearborn, 1980.
[209] T. Akasawa et al, in *Proc. 21st Int. MTDR Conf.*, Macmillan, London, 1981, 405-410.
[210] M. Podob, Metal Progr., *121* (6), 1982, 50-53.
[211] K.Y. Su and N.H. Cook, in *Proc. 5th NAMRC*, SME, Dearborn, 1977, pp.297-302.
[212] A.H. Shabaik, in *Proc. 21st Int. MTDR Conf.*, Macmillan, London, 1981, pp.421-430.
[213] Y. Doi, A. Doi, M. Kobayashi, and Y. Mori, in [19], pp.193-206.
[214] E.D. Doyle, Wear, *27*, 1974, 295-301.
[215] G.J. Trmal, in *Proc. 22nd Int. MTDR Conf.*, Macmillan, London, 1982, pp.335-341.
[216] B.F. von Turkovich and W.E. Henderer, CIRP, *27*, 1978, 35-38.
[217] P.K. Wright and E.M. Trent, Metals Techn., *1*, 1974, 13-23.
[218] P.K. Wright, J. Austral. Inst. Metals, *21* (1), 1976, 34-40.
[219] P.K. Philip, Microtecnic, *27*, 1973, 47-50.
[220] V.C. Venkatesh, Trans. ASME, Ser.F, J. Lub. Tech., *102*, 1980, 556-559.
[221] H. Opitz and W. König, in [14], pp.35-41.
[222] S. Söderberg and O. Vingsbo, Wear, *75*, 1982, 123-143.
[223] C. Bonjour, Wear, *62*, 1980, 93-122 (French).
[224] K.J.A. Brookes, *World Directory and Handbook of Hard Metals,* Engineer's Digest, London, 1976.
[225] H.S. Kalish, in [3.421], pp.28-53.
[226] C.M. Perrott and P.M. Robinson, J. Austral. Inst. Metals, *19*, 1974, 241-252.
[227] E. Pärnama and H. Johnsson, in [3.424], pp.166-169.
[228] Y. Naerheim and E.M. Trent, Metals Techn., *4*, 1977, 548-555.
[229] S. Okamoto and M. Doi, Bull. Jpn. Soc. Prec. Eng., *10*, 1976, 89-94.
[230] R.W. Thompson, D.T. Quinto, and B.S. Levy, in *Proc. 2nd NAMRC*, SME, Dearborn, 1974, pp.545-559.
[231] I. Ham, K. Hitomi, and G.L. Thuering, Trans. ASME, *83*, 1961, 142-154.
[232] R. Komanduri and M.C. Shaw, Wear, *33*, 1975, 283-292.
[233] Z. Katz and C. Rubenstein, CIRP, *25*, 1976, 39-44.
[234] E.M. Trent, Inst. Prod. Eng. J., *38*, 1959, 105-130.
[235] D.M. Gurevich, Russian Eng. J., *60* (11), 1980, 41-43.
[236] Yu.G. Kabaldin, Sov. Eng. Res., *1* (2), 1981, 85-87.
[237] K. Uehara and Y. Kanda, CIRP, *26*, 1977, 11-16.
[238] H. Negishi, K. Aoki, and T. Sata, in [18], pp.480-485.
[239] B.J. Ranganath and V.C. Venkatesh, in [18], pp.518-523.
[240] N. Narutaki and Y. Yamane, in [19], pp.319-333.
[241] H. Tanaka, in [19], pp.349-361.
[242] C. Hauser, Wear, *62*, 1980, 59-82 (French).
[243] S. Ramalingam and R. Van Wyk, in *Proc. 7th NAMRC*, SME, Dearborn, 1979, pp.228-231.
[244] T.E. Hale and D.E. Graham, in [19], pp.175-191.
[245] P.O. Snell, Jernkont. Ann., *154*, 1970, 413-421.
[246] V.C. Venkatesh, A.S. Raju, and K. Srinivasan, CIRP, *26*, 1977, 5-9.
[247] A.K. Chattopadhyay and A.B. Chattopadhyay, Wear, *80*, 1982, 239-258.
[248] P.A. Dearnley and E.M. Trent, Metals Techn., *9*, 1982, 60-75.
[249] S.B. Rao, K.V. Kumar, and M.C. Shaw, Wear, *49*, 1978, 353-357.
[250] J.P. Chubb and J. Billingham, Wear, *61*, 1980, 283-293.
[251] J.P. Chubb, J. Billingham, D.D. Hall, and J.M. Walls, Metals Techn., *7*, 1980, 293-299.
[252] R. Wertheim, R. Sivan, R. Porat, and A. Ber, CIRP, *31*, 1982, 7-11.
[253] R. Krishnamurthy, in [3.56], Paper No.110.
[254] I.P. Tretyakov et al, Russian Eng. J., *56* (12), 1976, 36-39.
[255] W.D. Sproul and M.H. Richman, Metals Techn., *3*, 1976, 489-493.

[256] P.D. Hartung and B.M. Kramer, in *Proc. 9th NAMRC*, SME, Dearborn, 1981, pp.424-431.
[257] O. Rüdiger, H. Grewe, and J. Kolaska, Wear, *48*, 1978, 267-282.
[258] S. Takatsu, K. Shibuki, and H. Kiso, in [19], pp.207-224.
[259] W. Schintlmeister, O. Pacher, and T. Raine, Wear, *48*, 1978, 251-266.
[260] N. Narutaki, CIRP, *25*, 1976, 93-98.
[261] R. Abel and V. Gomoll, Ind. Anz., *102* (46), 1980, 27-30 (German).
[262] A.G. King and W.M. Wheildon, *Ceramics in Machining Processes*, Academic Press, New York, 1966.
[263] E.D. Whitney, Powder Met. Int., *6* (2), 1974, 73-76.
[264] A.A. Avakov and G.G. Khachatryan, Russian Eng. J., *54* (1), 1974, 61-64.
[265] R. Komanduri, CIRP, *25*, 1976, 191-196.
[266] A.G. King, in [3.424], pp.43-48, 162-165.
[267] R.C. Brewer, Eng. Digest, *18*, 1957, 381-387.
[268] J.F. Huet and B.M. Kramer, in *Proc. 10th NAMRC*, SME, Dearborn, 1982, pp.297-304.
[269] M. Sakamoto, T. Nakamura, and S. Motoyoshi, J. Jpn. Soc. Prec. Eng., *45*, 1979, 1082-1087 (Japanese).
[270] E.D. Whitney, Paper No. MR76-15, SME, Dearborn, 1976.
[271] R. Van Wyk and S. Ramalingam, in *Proc. 8th NAMRC*, SME, Dearborn, 1980, pp.285-291.
[272] S.T. Buljan and V.K. Sarin, in [19], pp.335-348.
[273] S.K. Bhattacharya, D.K. Aspinwall, and A.W. Nicol, in *Proc. 19th Int. MTDR Conf.*, Macmillan, London, 1979, pp.425-434.
[274] G. Chryssolouris, J. Appl. Metalwork., *2* (2), 1982, 100-106.
[275] S.K. Bhattacharya and D.K. Aspinwall, in [19], pp.249-264.
[276] V.N. Ponduraev, Russian Eng. J., *59* (3), 1979, 42-44.
[277] H. Eda, K. Kishi, and H. Hashimoto, in *Proc. 21st Int. MTDR Conf.*, Macmillan, London, 1981, pp.253-258; in [19], pp.265-280.
[278] Y. Kono et al, in [19], pp.281-295.
[279] N. Narutaki, Y. Yamane, and M. Takeuchi, J. Jpn. Prec. Eng., *45*, 1979, 201-207 (Japanese).
[280] C.J. Wong, Trans. ASME, Ser.H, J. Eng. Mat. Tech., *103*, 1981, 341-345.
[281] R. Komanduri and M.C. Shaw, Phil. Mag., *34*, 1976, 195-204.
[282] T. Tanaka, N. Ikawa, and H. Tsuwa, CIRP, *30*, 1981, 242-245.
[283] D.M. Busch, in [535], pp.906-931.
[284] A.G. Thornton and J. Wilks, Wear, *53*, 1979, 165-187.
[285] A. Mallock, Proc. Roy. Soc. (London), *33*, 1881, 127-139.
[286] G.W. Rowe, in [3.19], pp.269-309.
[287] H. Mistecki, Schmierungstechnik, *4*, 1973, 208-212, 239-243 (German).
[288] J.O. Cookson, Trib. Int., *10*, 1977, 55-59.
[289] C.A. Sluhan, Paper No.399, Vol.62, Book 1, ASTME (SME), Detroit, 1963.
[290] W. König, Schmiertechn. Trib., *19*, 1972, 7-12 (German).
[291] W. König, in *Proc. 3rd NAMRC*, SME, Dearborn, 1975, pp.306-321.
[292] W. König and R. Vits, in [3.56], Paper No.62 (German).
[293] H. Ernst, Ann. New York Acad. Sci., *53*, 1951, 936-961.
[294] M.E. Merchant, Lubric. Eng., *6*, 1950, 163-167.
[295] M.E. Merchant, in [3.35], pp.566-574.
[296] A.J. Pekelharing, in [13], pp.114-119.
[297] J. Beaton, J.M. Tims, and R. Tourret, in [3.36], pp.155-176.
[298] V.M. Men'shakov et al, Machines Tooling, *42* (2), 1971, 59-61.
[299] E. Bickel, Microtecnic, *9*, 1955, 53-58.
[300] E. Usui and K. Takada, J. Jpn. Soc. Prec. Eng., *33*, 1967, 175.
[301] T.H.C. Childs, Proc. Inst. Mech. Eng., *186*, 1972, 717-727.
[302] J.A. Williams and D. Tabor, Wear, *43*, 1977, 275-292.
[303] L. De Chiffre, ASLE Trans., *24*, 1981, 340-344.
[304] M.S. Selvam and V. Radhakrishnan, Wear, *30*, 1974, 179-188.
[305] L. Fersing, Trans. ASME, *73*, 1951, 359-374.
[306] M.C. Shaw, J.D. Pigott, and L.P. Richardson, Trans. ASME, *73*, 1951, 45-56.
[307] H.S. Iyengar, R. Salmon, and W.B. Rice, Trans. ASME, Ser.B, J. Eng. Ind., *87*, 1965, 36-38.
[308] I.M. Feng, A. Gujral, and M.C. Shaw, Lubric Eng., *17*, 1961, 324-329.

[309] E.F. Smart and E.M. Trent, in *Proc. 15th Int. MTDR Conf.*, Macmillan, London, 1975, pp.187-196.
[310] W.S. Hollis, Int. J. Mach. Tool. Des. Res., *1*, 1961, 59-78.
[311] S.J. Beaubien and A.G. Cattaneo, Lubric. Eng., *10*, 1954, 74-79.
[312] F. Eugene, Microtecnic, *9*, 1955, 70-80.
[313] R.J. Delaney, Machinery (London), *91*, 1957, 548-551.
[314] G. Pahlitzsch, Microtecnic, *8*, 1954, 327-331; *9*, 1955, 65-69.
[315] F.A. Monahan et al, Am. Mach., *104* (May 16), 1960, 109-124.
[316] M.B. Parker, Paper No. MR70-278, SME, Dearborn, 1970.
[317] O.W. Boston, W.W. Gilbert, and R.E. McKee, Trans. ASME, *67*, 1945, 217-224.
[318] H. Wiele and M. McPherson, Maschinenbautechnik, *27*, 1978, 151-155 (German).
[319] G.H. Hain, Trans. ASME, *74*, 1952, 1077-1082.
[320] T. Ueno, A. Ishibashi, and A. Katsuki, Bull. JSME, *13*, 1970, 729-736.
[321] W.C. Lauterbach and G.A. Ratzel, Lubric. Eng., *7*, 1951, 15-19.
[322] M.C. Shaw, Wear, *2*, 1958/59, 217-227.
[323] N.N. Zorev, in [15], pp.493-498.
[324] K. Sakuma and T. Fujita, J. JSLE (Int. Ed.), *1*, 1980, 91-96.
[325] T. Kurimoto and G. Barrow, CIRP, *31*, 1982, 19-23.
[326] G.W. Rowe and E.F. Smart, Brit. J. Appl. Phys., *14*, 1963, 924-926.
[327] H. Opitz, W. König, and N. Diederich, Stahl u. Eisen, *88*, 1968, 978-986 (German).
[328] H.J. French, N.J. Bayonne, and T.G. Digges, Trans. ASME, *52*, 1930, 55-86.
[329] O.W. Boston, W.W. Gilbert, and C.E. Kraus, Trans. ASME, *58*, 1936, 371-378.
[330] H.L. Hall, Werkstattstechnik, *49*, 1959, 247-254 (German).
[331] K.E. Bisshopp, E.E. Lype, and S. Raynor, Lubric. Eng., *6*, 1950, 70-73.
[332] G.J. De Salvo and M.C. Shaw, in *Proc. 9th Int. MTDR Conf.*, Pergamon, Oxford, 1969, pp.961-971.
[333] J.G. Horne, D. Tabor, and J.A. Williams, in [3.55], pp.193-200.
[334] G.W. Rowe and E.F. Smart, in [3.37], pp.48-57.
[335] E.D. Doyle and J.G. Horne, Wear, *60*, 1980, 383-391.
[336] S.V. Pepper, J. Appl. Phys., *47*, 1976, 2578-2583.
[337] M. Kasyan, F. Parikyan, and I. Ivanov, Machinostroene, *24*, 1975, 518-521 (Russian).
[338] V.J. Haden, The Prod. Eng., *39*, 1960, 529-536.
[339] N.J. Sereda and V.P. Belyanskii, Sov. Eng. Res., *1* (4), 1981, 89-90.
[340] V.N. Latyshev, Russian Eng. J., *51* (7), 1971, 75-76.
[341] L. Walter, Can. Mach. Metalwork., *76* (8), 1965, 94-97.
[342] V.N. Latyshev, Mech. Sci., *1973* (3), 47-49.
[343] V.N. Latyshev, Sov. Mater. Sci., *7*, 1971, 296-299.
[344] T. Shirakashi, R. Komanduri, and M.C. Shaw, Trans. ASME, Ser.B, J. Eng. Ind., *100*, 1978, 244-248.
[345] O.A. Drobysheva and V.N. Latyshev, Sov. Mater. Sci., *8*, 1972, 288-290.
[346] A.K. Lakhwara and W.B. Rice, in *Proc. 7th NAMRC*, SME, Dearborn, 1979, pp.248-254.
[347] J.E. Williams, in *Proc. 4th NAMRC*, SME, Dearborn, 1976, pp.377-382.
[348] E. Usui, A. Gujral, and M.C. Shaw, Int. J. Mach. Tool Des. Res., *1*, 1961, 187-197.
[349] G.W. Rowe and E.F. Smart, in [3.36], pp.83-95.
[350] C.A. Cassin and G. Boothroyd, J. Mech. Eng. Sci., *7*, 1965, 67-81.
[351] P.L. Barlow, Proc. Inst. Mech. Eng., *181*, Pt.1, 1966-67, 687-705.
[352] R. Tourret, Proc. Inst. Mech. Eng., *1B*, 1952-53, 530-531.
[353] K.J.B. Wolfe, Proc. Inst. Mech. Eng., *1B*, 1952-53, 517-535.
[354] K. Sakima and T. Fujita, in [3.56], Paper No.98.
[355] M. Mrozek, MW Interf., *3* (6), 1978, 12-20.
[356] H.A. Smyth and R.M. McGill, J. Phys. Chem., *61*, 1957, 1025-1036.
[357] R.S. Owens and R.W. Roberts, in [15], pp.1193-1205.
[358] S.G. Entelis et al, in [3.56], Paper No.70.
[359] S. Mori, M. Suginoya, and Y. Tamai, ASLE Trans., *25*, 1982, 261-266.
[360] A. Ishibashi and A. Katsuki, CIRP, *23*, 1974, 19-20, 318-319.
[361] V.N. Latyshev, Machines Tooling, *41* (5), 1970, 60-62.
[362] R.W. Roberts and R.S. Owen, in [3.50], pp.244-255.
[363] K. Wolfe, M.D. Kinman, and G. Lennard, J. Inst. Petr., *40*, 1954, 253.
[364] G.E. Barker, Paper No. MR66-124, SME, Dearborn, 1966.

[365] P.M. Holmes, Prod. Engineer, *44*, 1965, 130-141.
[366] A. Dorinson, ASLE Trans., *6*, 1963, 270-275.
[367] I.S. Morton, in [14], pp.185-192.
[368] M.C. Shaw, J. Appl. Mech., *15*, 1948, 37-44.
[369] A. Dorinson, Lubric. Eng., *12*, 1956, 387-391.
[370] A. Zakar, L. Gyöngyössy, and E. Vamos, in [3.56], Paper No.76 (German).
[371] E.D. Shchukin, L.S. Bryukhanova, Z.M. Polukarova, and N.V. Pertsov, Sov. Mater. Sci., *12*, 1976, 375-384.
[372] I.S. Morton and R. Tourret, J. Inst. Petr., *40*, 1954, 261.
[373] V.A. Serov et al, MW Interf., *5* (6), 1980, 10-15.
[374] J. Pomey, C.M. Prevost, and P. Quantin, CIRP, *11*, 1963, 104-110 (French).
[375] E.K. Henriksend and P.R. Arzt, in [3.50], pp.234-243.
[376] R.A. Stayner, F.E. Barnes, J.J.M. Senden, and H.K. Adamczak, in [3.56], Paper No.96.
[377] B. Jeremic and M. Lazic, in [3.56], Paper No.77.
[378] L. De Chiffre, in [3.56], Paper No.74.
[379] V.N. Latyshev, Sov. Mater. Sci., *4*, 1968, 541-543.
[380] Yu.G. Proskuryakov and V.M. Isaev, Russian Eng. J., *60* (3), 1980, 37-39.
[381] C. Kajdas, B. Misterkiewicz, and M. Dominiak, in [3.56], Paper No.67.
[382] V.E. Shestopalov et al, in *Polim. Tekhnol. Prosessakh Obrab. Met.*, Naukova Dumka, Kiev, 1977, pp.107-110 (Russian).
[383] A. Ishibashi and A. Katsuki, Bull. Jpn. Soc. Prec. Eng., *5*, 1971, 81-82.
[384] S. Watanabe, T. Fujita, K. Suga, and K. Kasahara, Lubric. Eng., *38*, 1982, 412-415.
[385] C.C. Yang, Lubric. Eng., *35*, 1979, 133-136.
[386] V. Kuznetsov, Yu.M. Korobov, and Yu.G. Kotlov, Sov. Mater. Sci., *11*, 1975, 407-409.
[387] J.K. Banerjee, in [3.345], pp.496-500.
[388] J. Vercon, J. Stanic, and B. Popovic, Schmiertechn. Trib., *18*, 1971, 13-17.
[389] H. Bagchi, Ind. Lub. Trib., *32*, 1980, 124-125.
[390] D. Hartley and P. Wainwright, in [3.345], pp.489-495.
[391] T.H.C. Childs and D. Hartley, Trans. ASME, Ser.F, J. Lub. Tech. (to be published).
[392] A.J. Koury, M.K. Gabel, and A.P. Wijenayake, Lubric. Eng., *35*, 1979, 309-314.
[393] E. Rabinowicz and S.P. Loutrel, in [3.46], pp.84-89.
[394] R.P. Walduck and S.J. Harvey, in *Proc. 1st Joint Polytechnic Symp. Manuf. Eng.*, Leicester Polytech., 1977, Paper No.E2.
[394a] P.J. Amaria and R.J. Goosney, in *Proc. 1st NAMRC*, McMaster Univ., Hamilton, Ont., 1973, Vol.2, pp.49-66.
[395] G.W. Rowe and E.F. Smart, Lubric. Eng., *29*, 1973, 12-16.
[396] I.A. Andreeva, L.S. Bryukhanova, V.M. Miroshnichenko, and E.D. Shchukin, Sov. Mater. Sci., *16*, 1980, 250-252.
[397] W.E. Lauterbach, Lubric. Eng., *8*, 1952, 135-136.
[398] R.B. Miebusch and E.H. Strieder, Sci. Lubric., *3* (2), 1954, 74.
[399] R.J.S. Pigott and A.T. Colwell, SAE Quart. Trans., *6*, 1952, 547-566; SAE J., *60*, 1952, 45-48.
[400] B.K. Nagpal and C.S. Sharma, Trans. ASME, Ser.B, J. Eng. Ind., *95*, 1973, 881-889.
[401] K. Kishi, H. Eda, T. Furusawa, and Y. Ichida, in [3.46], pp.579-587.
[402] C.S. Sharma, W.B. Rice, and R. Salmon, Trans. ASME, Ser.B, J. Eng. Ind., *93*, 1971, 441-444.
[403] C. Wick, Manuf. Eng., *81* (5), 1978, 44-51.
[404] F. Pfleghar, Ind. Anz., *98*, 1976, 1220-1221 (German).
[405] Anon., Tool Manuf. Eng., *56* (4), 1966, 52-54.
[406] E.A. Rich, Paper No. MR66-186, SME, Dearborn, 1966.
[407] N.P. Jeffries, Ind. Lub. Trib., *24*, 1972, 179-181.
[408] R. Nicholson, in [3.56], Paper No.94.
[409] J.E. Repass and R.W. Brockman, Am. Mach., *115* (Feb.22), 1971, 84-85.
[410] S.R. Alger, Am. Mach., *119* (May 15), 1975, 69-70.
[411] E. Klein, Schweiz. Aluminium Rundschau, *22*, 1972, 360-371 (French and German).
[412] M.C. Shaw and P.A. Smith, Am. Mach., *100* (Apr.9), 1956, 178-179.
[413] Yu.G. Proskuryakov, N.F. Belov, and V.N. Petrov, Machines Tooling, *32* (6), 1961, 29-33.
[414] P.N. Rao, in [3.56], Paper No.101.
[415] D. Kececioglu and A. Sorensen, Trans. ASME, Ser.B, J. Eng. Ind., *84*, 1962, 49-52.

[416] P.N. Rao, K.V. Siddaraj, and R.P. Arora, in [3.55], pp.201-204.
[417] A. Niedzwiedzki, Machinery (London), *83* (Oct.9), 1953, 717-721, 741; *84* (Feb.6), 1954, 337-343; *85* (Aug.6), 1954, 280-286.
[418] J.J. Dwyer, Jr., Am. Mach., *108* (Mar.16), 1964, 105-120.
[419] F. Croxon, The Prod. Eng., *49*, 1970, 237-246.
[420] P.M. Holmes, Ind. Lub. Trib., *23*, 1971, 47-55.
[421] J. Tomko, Paper No. MF74-170, SME, Dearborn, 1974.
[422] V.A. Serov et al, Machines Tooling, *45* (7), 1974, 53-55.
[423] J.O. Cookson, Trib. Int., *10*, 1977, 5-11.
[424] L.H. Haygreen, Trib. Int., *10*, 1977, 13-16.
[425] G. Zwingmann, Werkstatt u. Betr., *112*, 1979, 409-414, 483-487 (German).
[426] W.J. Bartz, Werkstattstechnik, *68*, 1978, 471-476, 621-624 (German).
[427] M. Albert, Mod. Mach. Shop., *54* (1), 1981, 102-112.
[428] R.A. Holstedt, in [3.56], Paper No.106.
[429] K. Tuffentsammer, in [3.56], Paper No.60 (German).
[430] G.H. Geiger, in [3.56], Paper No.61 (German).
[431] R. Kelly and G. Foltz, in [3.34], Vol.2, Chap.21, *Cutting Fluids*.
[432] W. Schaper, in [3.56], Paper No.71 (German).
[433] T. Sutcliffe and S.J. Barber, Am. Mach., *121* (4), 1977, 138-141.
[434] P.R. Arzt and I.J. Stewart, Lubric. Eng., *19*, 1963, 283-291.
[435] N. Zlatin and E.R. Snider, Lubric. Eng., *7*, 1965, 287-292.
[436] Anon., Manuf. Eng., *75* (5), 1975, 31-38.
[437] H.F. Weindel, in [4.3], pp.34-39.
[438] R.A. Nicholson, Ind. Lub. Trib., *23*, 1971, 63-66.
[439] Various, Am. Mach., *115* (Mar.8), 1971, 75-83.
[440] A.R. Eyres, J. Inst. Petr., *59*, 1973, 9-14.
[441] W.A. Sluhan and C.A. Sluhan, in [3.46], pp.605-609.
[442] C. Evans, Trib. Int., *10*, 1977, 33-37.
[443] E.O. Morgan, Paper No. MF77-207, SME, Dearborn, 1977.
[444] C. Pilletteri, Lubric. Eng., *38*, 1982, 622-625.
[445] A.M. Tukhontsov, Russian Eng. J., *60* (8), 1980, 49-50.
[446] J. Otsuka, Bull. Jpn. Soc. Prec. Eng., *11*, 1977, 109-114.
[447] L.G. Treat and W.C. White, in [3.53], pp.163-170.
[448] L.V. Khudobin, V.I. Kostelnikov, and Yu.V. Polyanokov, Russian Eng. J., *55* (4), 1975, 52-54.
[449] P.J.C. Gough (ed.), *Swarf and Machine Tools*, Hutchinson, London, 1970.
[450] A.A. Afanes'ev, Russian Eng. J., *60*, 1980, 52-54.
[451] L.H. Sudholz, Lubric. Eng., *13*, 1957, 509-515.
[452] M.D. Smith and J. Lieser, Lubric. Eng., *29*, 1973, 315-319.
[453] N. Zlatin and J. Christopher, Lubric. Eng., *29*, 1973, 402-405.
[454] Various, Metals Techn., *3*, 1976, 248-287.
[455] G.F. Micheletti, CIRP, *18*, 1970, 13-30.
[456] N.H. Cook, in [16], pp.1-10.
[457] P.M. Lonardo, CIRP, *25*, 1976, 455-459.
[458] P. Dinichert, R. van Hasselt, E. Matthias, and A. Wirtz, CIRP, *22*, 1973, 249-261.
[459] D.J. Whitehouse, P. Vanherck, W. deBruin, and C.A. van Luttervelt, CIRP, *23*, 1974, 265-282.
[460] M. Field and J.F. Kahles, CIRP, *20*, 1971, 153-163.
[461] M. Field, J.F. Kahles, and J.T. Cammet, CIRP, *21*, 1972, 219-238.
[462] H.K. Tönshoff and E. Brinksmeier, CIRP, *29*, 1980, 519-530.
[463] J.E. Mayer, Jr., and D.G. Lee, in [16], pp.31-54.
[464] M.Y. Friedman, V.A. Tipnis, and M. Field, in *Proc. 16th Int. MTDR Conf.*, Macmillan, London, 1976, pp.579-583; in *Proc. 6th NAMRC*, SME, Dearborn, 1978, pp.395-401.
[465] D.W. Murphy, Lubric. Eng., *30*, 1974, 10-16.
[466] W.R. Russell, Lubric. Eng., *30*, 1974, 252-254.
[467] G. Lorenz, in [17], pp.147-163.
[468] I.S. Morton, Ind. Lub. Trib., *24*, 1972, 221-225.
[469] M.P. Groover, Paper No. MR75-146, SME, Dearborn, 1975.
[470] G.F. Micheletti, A. De Filippi, and R. Ippolito, CIRP, *16*, 1968, 353-360.
[471] W. König, K. Langhammer, and H.V. Schemmel, CIRP, *21*, 1972, 19-20.

[472] L.V. Colwell, CIRP, *24*, 1975, 73-76.
[473] L.V. Colwell, J.C. Mazur, and W.R. DeVries, in *Proc. 6th NAMRC*, SME, Dearborn, 1978, pp.276-282.
[474] A. Kinnander, in *Proc. 22nd Int. MTDR Conf.*, Macmillan, London, 1982, pp.255-259.
[475] H.J. Bezer and P.D. Oates, Lubric. Eng., *29*, 1973, 336-340.
[476] F. Otto and W. Kluft, Ind. Anz., *99*, 1977, 961-962 (German).
[477] E. Bickel, in [13], pp.89-94.
[478] K.J. Trigger, in [13], pp.95-101.
[479] O.D. Prins, CIRP, *19*, 1971, 579-584.
[480] O.V. Kretinin, Sov. Eng. Res., *1* (2), 1981, 88-89.
[481] A.H. Redford, in [17], pp.367-379.
[482] A.A. Zakaria and J.I. El-Gomayel, Int. J. Mach. Tool Des. Res., *15*, 1975, 195-208.
[483] V. Solaja and D. Vukelja, CIRP, *22*, 1973, 5-6.
[484] A.J. Wilkinson, Proc. Inst. Elect. Eng., *118* (2), 1971, 381-386.
[485] J.I. El-Gomayel and A.A. Zakaria, in *Proc. 5th NAMRC*, SME, Dearborn, 1977, pp.328-336.
[486] K. Nakazawa, in [18], pp.486-491.
[487] M.S. Lan and D.A. Dornfeld, in *Proc. 10th NAMRC*, SME, Dearborn, 1982, pp.305-311.
[488] K. Iwata and T. Moriwaki, CIRP, *26*, 1977, 21-26.
[489] I. Inasaki and S. Yonetsu, in *Proc. 22nd Int. MTDR Conf.*, Macmillan, London, 1982, pp.261-268.
[490] K. Uehara, CIRP, *22*, 1973, 23-24.
[491] J. Samuels, M. Tani, C. Beiswenger, and I. Ham, CIRP, *25*, 1976, 77-81.
[492] T.N. Loladze, G.R. Tkemaladze, O.V. Kochiashvili, and A.I. Mikanadze, CIRP, *31*, 1982, 45-48.
[493] K. Uehara, S. Kumagai, H. Mitsui, and H. Takeyama, CIRP, *23*, 1974, 13-14.
[494] N.H. Cook and A.B. Lang, Trans. ASME, Ser.B, J. Eng. Ind., *85*, 1963, 381-387.
[495] H. Opitz and O. Hake, Microtecnic, *10*, 1956, 5-9.
[496] R. Chawla and S.B. Datar, Wear, *58*, 1980, 213-222.
[497] E. Vamos, in [3.56], Paper No.73 (German).
[498] M. Shiraishi, Trans. ASME, Ser.B, J. Eng. Ind., *103*, 1981, 203-209.
[499] J.D. Oathout, W.C. Howell, J.P. Hamer, and H.L. Leland, Trans. ASME, *75*, 1953, 1081-1086.
[500] P.M. Blanchard and R.J. Syrett, Lubric. Eng., *30*, 1974, 62-71.
[501] H.R. Leep, Lubric. Eng., *37*, 1981, 715-721.
[502] A. De Filippi, in [16], pp.396-412.
[503] F.W. Boulger, H.L. Shaw, and H.E. Johnson, Trans. ASME, *71*, 1949, 431-445.
[504] H. Henschen and W.C. Howell, Jr., Lubric. Eng., *30*, 1974, 315-322.
[505] J.R.J. Van Dongen and J.G.C. Stegwee, Stahl u. Eisen, *56*, 1936, 1185-1187 (German).
[506] V. Solaja, Wear, *1*, 1957-68, 512.
[507] G. Lorenz, CIRP, *12*, 1963, 217-222.
[508] L. Bäcker, R. E. Haik, and M. Luciani, in [17], pp.239-252.
[509] W.B. Heginbotham and P.C. Pandey, in *Proc. 7th Int. MTDR Conf.*, Pergamon, Oxford, 1967, pp.515-528.
[510] W.B. Heginbotham and P.C. Pandey, in *Proc. 8th Int. MTDR Conf.*, Pergamon, Oxford, 1968, pp.163-171.
[511] J.L. Thomas and B.K. Lambert, in *Proc. 2nd NAMRC*, SME, Dearborn, 1974, pp.139-153.
[512] V.K. Jain and P.C. Pandey, in [3.345], pp.447-455.
[513] C. Bedrin and B. Roumèsy, Mecan. Mater. Electr., *1976* (313/314), 31-37 (French).
[514] J.P. Richter, ASLE Trans., *22*, 1979, 171-177.
[515] H.A. Hartung, J.W. Johnson, and A.C. Smith, Jr., Lubric. Eng., *13*, 1957, 538-542.
[516] H. Henschen, Lubric. Eng., *29*, 1973, 151-156.
[517] T.H. Webb, W.J. Kelly, S.M. Darling, and L.R. Woodhill, Lubric. Eng., *30*, 1974, 231-240.
[518] C.D. Flemming and L.H. Sudholz, Lubric. Eng., *12*, 1956, 199-203.
[519] E.N. Ladov, Lubric. Eng., *30*, 1974, 5-9.
[520] L. De Chiffre, Lubric. Eng., *36*, 1980, 33-39.
[521] D. Hartley, in [3.56], Paper No.97.

[522] W.A. Faville and R.M. Voitik, Lubric. Eng., *34*, 1978, 193-197.
[523] T.H. Webb and E. Holodnik, Lubric. Eng., *36*, 1980, 513-529.
[524] R. Kitagawa, C.H. Kahng, T. Akasawa, and K. Okusa, in *Proc. 9th NAMRC*, SME, Dearborn, 1981, pp.305-310.
[525] G.W. Skells, Jr., and S.C. Cohen, Lubric. Eng., *33*, 1977, 401-406.
[526] M. Kanai, S. Fujii, and Y. Kanda, CIRP, *27*, 1978, 61-66.
[527] M. Streuber, G. Menzel, and G. Naumann, Schmierungstechnik, *10*, 1979, 277-279 (German).
[528] J. Dagnell, CIRP, *15*, 1967, 301-308.
[529] E. Holodnik and L.M. Edwards, Lubric. Eng., *30*, 1974, 195-200.
[530] T. Sutcliffe, S.J. Barber, W. Dykan, and C.G. Wall, Lubric. Eng., *35*, 1979, 145-152.
[531] B. Lindgren, in [3.53], pp.61-66.
[532] O. Svahn and B. Steen, CIRP, *21*, 1972, 37-38.
[533] J. Müller, in [3.56], Paper No.63 (German).
[534] G.E. Barker, in [4.3], pp.213-226.
[535] M.C. Shaw (ed.), *New Developments in Grinding*, Carnegie Press, Pittsburgh, 1972.
[536] C.P. Bhateya and R.P. Lindsay (ed.), *Grinding Theory, Techniques, and Troubleshooting*, SME, Dearborn, 1982.
[537] A.G. Wetton, J. Mech. Eng. Sci., *11*, 1969, 412-425.
[538] C.P. Bhateja, in [3.49], pp.705-718.
[539] F.T. Farago, *Abrasive Methods Engineering*, Industrial Press, New York, 1976.
[540] Y. Kita and M. Ido, Int. J. Mach. Tool Des. Res., *16*, 1976, 209-219.
[541] E.R. Marshall and M.C. Shaw, Trans. ASME, *74*, 1952, 51-59.
[542] R.S. Hahn, in [3.50], pp.209-233.
[543] E.D. Doyle and S.K. Dean, CIRP, *29*, 1980, 571-575.
[544] L.E. Samuels, in [535], pp.283-304.
[545] T. Sasaki and K. Okamura, Bull. JSME, *3*, 1960, 547-555.
[546] A.J. Sedriks and T.O. Mulhearn, Wear, *7*, 1964, 451.
[547] Y. Kita, M. Ido, and S. Hata, Wear, *47*, 1978, 185-193.
[548] W. Sauer and M.C. Shaw, CIRP, *21*, 1972, 85-86.
[549] M. Field, CIRP, *23*, 1974, 191-192.
[550] Y. Kita, M. Ido, and Y. Tuji, Wear, *71*, 1981, 55-63.
[551] R. Komanduri and M.C. Shaw, Trans. ASME, Ser.H, J. Eng. Mat. Tech., *96*, 1974, 145-156.
[552] J.N. Brecker, R. Komanduri, and M.C. Shaw, CIRP, *22*, 1973, 219-225.
[553] G.S. Reichenbach, J.E. Mayer, S. Kalpakcioglu, and M.C. Shaw, Trans. ASME, *78*, 1956, 847-859.
[554] J.O. Outwater and M.C. Shaw, Trans. ASME, *74*, 1952, 73-86.
[555] F.P. Bowden and E.H. Freitag, Proc. Roy. Soc. (London), *A248*, 1968, 350-367.
[556] R. Komanduri and M.C. Shaw, Phil. Mag., *32*, 1975, 711-724.
[557] E.D. Doyle and R.L. Aghan, Met. Trans., *6B*, 1975, 143-155.
[557a] M.C. Shaw, Mech. Chem. Eng. Trans. Inst. Eng., Australia, *MC 8* (1), 1972, 73-78.
[558] A. Misra and I. Finnie, Wear, *65*, 1981, 359-373.
[559] S. Vaidyanathan and I. Finnie, in [535], pp.813-831.
[560] W. Lortz, Wear, *53*, 1979, 115-128.
[561] R.T. Spurr, Wear, *65*, 1981, 315-324.
[562] N.P. Suh, H.C. Sin, and N. Saka, in [3.49], pp.493-518.
[563] T. Matsuo, CIRP, *23*, 1974, 85-86.
[564] M.C. Shaw, in [535], pp.220-258.
[565] R.S. Hahn and R.P. Lindsay, in [536], pp.3-41.
[566] R.P. Lindsay, in [536], pp.42-60.
[567] B. Colding et al, CIRP, *21*, 1972, 157-166.
[568] K. Nakayama et al, in [18], pp.606-611.
[569] S.J. Pandey, S.M. Halder, and G.K. Lal, Wear, *58*, 1980, 237-248.
[570] P.G. Werner and H. Lauer-Schmaltz, in [3.55], pp.225-232.
[571] E.D. Doyle and D.M. Turley, in *Proc. 4th NAMRC*, SME, Dearborn, 1976, pp.346-350.
[572] Y. Matsuno, H. Yamada, M. Harada, and A. Kobayashi, CIRP, *24*, 1975, 237-242.
[573] P. Sathiamoorthy, V. Radhakrishnan, and J.F. Rehman, Wear, *54*, 1979, 303-313.
[574] W. Scott and R.M. Baul, Wear, *57*, 1979, 247-254.
[575] K.V. Kumar and M.C. Shaw, CIRP, *28*, 1979, 205-207.

[576] J. Verkerk, CIRP, *26*, 1977, 385-395.
[577] J.K. Banerjee, Trans. ASME, Ser.B, J. Eng. Ind., *101*, 1979, 135-146.
[578] R. Snoeys, J. Peters, and A. Decneut, CIRP, *23*, 1974, 227-237.
[579] R.P. Lindsay, in [17], pp.338-351.
[580] R.S. Hahn, in [17], pp.282-296.
[581] M. Weck and K.H. Schiefer, in [18], pp.644-648.
[582] R.P. Lindsay, Paper No. MR74-120, SME, Dearborn, 1974.
[583] R.S. Hahn, Trans. ASME, Ser.B, J. Eng. Ind., *86*, 1964, 287-293.
[584] G.J. Trmal, in *Proc. 21st Int. MTDR Conf.*, Macmillan, London, 1981, pp.323-328.
[585] Q.Z. Zhou and M.C. Shaw, in *Proc. 9th NAMRC*, SME, Dearborn, 1981, pp.267-274.
[586] R. Snoeys, M. Maris, and J. Peters, CIRP, *27*, 1978, 571-581.
[587] S. Ohishi, Y. Furukawa, S. Shiozaki, and S. Okada, in *Proc. 20th Int. MTDR Conf.*, Macmillan, London, 1980, pp.375-382.
[588] T.C. Buttery and M.S. Hamed, Wear, *44*, 1977, 231-245.
[589] H. Opitz, W. König, and G. Werner, in [535], pp.259-282.
[590] R.P. Lindsay, in *Proc. 10th NAMRC*, SME, Dearborn, 1982, pp.277-281.
[591] S. Malkin, in *Proc. 9th NAMRC*, SME, Dearborn, 1981, pp.235-239.
[592] W.J. Sauer, in [535], pp.391-411.
[593] M.Es. Abdelmoneim, Wear, *56*, 1979, 265-296.
[594] L. Kops and M.C. Shaw, CIRP, *31*, 1982, 211-214.
[595] M. Kaiser, Ind. Anz., *97*, 1975, 549-550 (German).
[596] S.E. Amundson, E.J. Duwell, and R.C. Lokken, Wear, *36*, 1976, 99-110.
[597] S.W. Date and S. Malkin, Wear, *40*, 1976, 223-235.
[598] J. Shibata, I. Inasaki, and S. Yonetsu, Wear, *55*, 1979, 331-344.
[599] J. Sugishita, S. Fujiyoshi, and T. Imura, Wear, *52*, 1979, 219-234.
[600] K. Harrison, J.P. Byron, and S.K. Bhattacharya, Wear, *39*, 1976, 335-343.
[601] F.A. Burney and S.M. Wu, Wear, *36*, 1976, 225-234.
[602] R.G. Visser and R.C. Lokken, Wear, *65*, 1981, 325-350.
[602a] W.A. Mohun, Trans. ASME, Ser.B, J. Eng. Ind., *84*, 1962, 431-482.
[603] S.G. Babaev, N.K. Mamedkhanov, and R.F. Gasanov, MW Interf., *5* (4), 1980, 10-17.
[604] H. Frank, in [535], pp.832-849.
[605] T.H.C. Childs and F.H. Kinsella, Wear, *66*, 1981, 241-255.
[606] A. Szuder, P. Kapsa, T. Mathia, and J.M. Georges, CIRP, *26*, 1977, 155-160.
[607] N.E. Miller, Wear, *58*, 1980, 249-259.
[608] G.A. Watt, in [3.54], pp.35-41.
[609] M. Kiselev et al, Bull. Res. Lab. Precis. Mach. Electron., Tokyo Inst. Techn., No.40, Sept. 1977, pp.17-26.
[610] L.E. Samuels, *Metallographic Polishing by Mechanical Methods*, 2nd Ed., Pitman, London, 1971.
[611] L.J. Rhoades, Paper No. MR80-340, SME, Dearborn, 1980.
[612] J.W. Callahan. Paper No. MR76-122, SME, Dearborn, 1976.
[613] M. Adithan and V.C. Venkatesh, CIRP, *27*, 1978, 119-121.
[614] G. Pahlitzsch, in [13], pp.242-256.
[615] G. Kuppuswamy, Trib. Int., *9*, 1976, 29-32.
[616] G. Kuppuswamy, Trib. Int., *10*, 1977, 184-188.
[617] L. Kops, in *Proc. 3rd NAMRC*, SME, Dearborn, 1975, pp.955-967.
[618] D.M. Turley and E.D. Doyle, Wear, *57*, 1979, 237-246.
[619] S.M. Pandit, P.T. Suratkar, and S.M. Wu, Wear, *39*, 1976, 205-217.
[620] P. Ramachandran and S. Vaidynathan, Wear, *36*, 1976, 119-125.
[621] A.A. Torrance, in *Proc. 19th Int. MTDR Conf.*, Macmillan, London, 1979, pp.637-644.
[622] H. Kaliszer, in [18], pp.579-593.
[623] T. Sata, T. Suto, T. Waida, and H. Noguchi, in [535], pp.752-770.
[624] J. Shibata, T. Goto, and M. Yamamoto, CIRP, *31*, 1982, 233-238.
[625] H.F.G. Ueltz, in [535], pp.1-52.
[626] L. Coes, *Abrasives*, Springer, New York, 1971.
[627] E.D. Whitney, in [535], pp.53-74.
[628] K. Takazawa, in [535], pp.75-90.
[629] E.D. Whitney and R.E. Shepler, in *Proc. 5th NAMRC*, SME, Dearborn, 1977, pp.303-308.
[630] T. Matsuo, N. Ueda, S. Sonoda, and E. Oshima, in [18], pp.667-672.

[631] S. Yossifon and C. Rubenstein, Trans. ASME, Ser.B, J. Eng. Ind., *103*, 1981, 144-155, 156-164.
[631a] K.V. Kumar, Wear, *82*, 1982, 257-270.
[632] R. Komanduri, W.H. Laverty, and M.C. Shaw, Trans. ASME, Ser.F, J. Lub. Tech., *99*, 1977, 415-420.
[633] K. Miyoshi and D.H. Buckley, Wear, *75*, 1982, 253-268.
[634] R. Komanduri and M.C. Shaw, Trans. ASME, Ser.B, J. Eng. Ind., *98*, 1976, 1125-1134.
[635] R. Komanduri and M.C. Shaw, Int. J. Mach. Tool Des. Res., *14*, 1974, 63-84.
[636] H.K. Tönshoff et al, Manuf. Eng. Trans., *1977*, 270-277.
[637] Y. Kita, H.H. Damlos, and E. Salje, in [18], pp.612-619.
[638] G. Werner, in [3.49], pp.691-703.
[639] J. Borkowski, in [18], pp.685-690.
[640] H.J. Tönshoff, P.G. Althaus, and H.H. Nölke, in [3.56], Paper No.103 (German).
[641] R.P. Lindsay, CIRP, *23*, 1974, 87-88, 322.
[642] S.K. Bhattacharya and K.K. Hon, in *Proc. 19th Int. MTDR Conf.*, Macmillan, London, 1979, p.645.
[643] E. Saljé and H. Heidenfelder, in *Proc. 22nd Int. MTDR Conf.*, Macmillan, London, 1982, pp.307-314.
[644] J. Verkerk, CIRP, *28*, 1979, 487-495.
[645] T.C. Buttery, A. Statham, J.B. Percival, and M.S. Hamed, Wear, *55*, 1979, 195-219.
[646] C.P. Bhateya, in [17], pp.325-337.
[647] R. Komanduri and W.R. Reed, Jr., CIRP, *29*, 1980, 239-243.
[648] N.N.Z. Gindy and T.J. Vickerstaff, in *Proc. 22nd Int. MTDR Conf.*, Macmillan, London, 1982, pp.315-322.
[649] M.C. Shaw, in [4.3], pp.65-76.
[650] K.C. Tripathi, in [3.56], Paper No.1.
[651] C.A. Sluhan, Lubric. Eng., *26*, 1970, 352-374.
[652] L.V. Khudobin, Russian Eng. J., *50* (6), 1970, 66-70.
[653] M. Osman and S. Malkin, ASLE Trans., *15*, 1972, 261-268.
[654] K.C. Tripathi and G.W. Rowe, in *Proc. 17th MTDR Conf.*, Macmillan, London, 1977, pp.181-187.
[655] J.A. Kirk and J.F. Cardenas-Garcia, in *Proc. 3rd NAMRC*, SME, Dearborn, 1975, pp.426-440.
[656] M. Harada and N. Shinozaki, in [13], pp.218-224.
[657] H.K. Tönshoff, P.G. Althaus, and H.H. Nölke, in [3.55], pp.217-223.
[658] K.C. Tripathi, A.W. Nicol, and G.W. Rowe, ASLE Trans., *20*, 1977, 249-256.
[659] S.A. Popov et al, Sov. Eng. Res., *1* (5), 1981, 84-86.
[660] I.L. Khudobin, Sov. Eng. Res., *1* (5), 1981, 51-53.
[661] H.W. Wagner, Mech. Eng., *73*, 1951, 128-132.
[662] L.P. Tarasov, Tool Manuf. Eng., *46* (7), 1961, 67-73; *47* (1), 1961, 60-67; *47* (2), 1961, 57-63.
[663] E.J. Duwell, I.S. Hong, and W.J. McDonald, ASLE Trans., *12*, 1969, 86-93.
[663a] M.C. Shaw, personal communication.
[663b] S. Chandrasekar and M.C. Shaw, Wear (to be published).
[664] K.B. Das, in [3.52], pp.92-98.
[665] N. Zlatin et al, Techn. Rep. AFML-TR-73-165, 1973.
[666] S. Mateeva and I. Glavchev, Trib. Int., *13*, 1980, 69-71.
[667] J. Peters and R. Aerens, CIRP, *25*, 1976, 247-250.
[668] G.W. Rowe and A.G. Wetton, Wear, *14*, 1969, 455-457.
[669] K. Barker, H. Kaliszer, and G.W. Rowe, in [3.70], pp.177-184.
[670] S.A. Popov and V.A. Panaioti, Machines Tooling, *50* (10), 24-25.
[671] A.Y.C. Nee, Int. J. Mach. Tool Des. Res., *19*, 1979, 21-31.
[672] A. Yamamoto and T. Ueda, Bull. Jpn. Soc. Prec. Eng., *13*, 1979, 177-182.
[673] A.V. Yakimov and R.M. Mubarakshin, Russian Eng. J., *58* (6), 1978, 23-25.
[674] Y. Ueno, S. Miyake, H. Kuroda, and J. Watanabe, CIRP, *28*, 1979, 219-224.
[675] M.C. Shaw, and C.T. Yang, Trans. ASME, *78*, 1956, 861-868.
[676] G.I. Sayutin, V.A. Nosenko, and N.I. Bogomolov, Sov. Mater. Sci., *11*, 1975, 688-690.
[677] G. Spur, in [3.55], pp.233-239.
[678] F. Ghio, Ind. Lub. Trib., *26*, 1974, 99-102.
[679] W. Keyser, in [3.56], Paper No.100 (German).

[680] P.J. Strachan, Trib. Int., *8*, 1975, 153-160.
[681] G.W. Rowe, Wear, *77*, 1982, 73-80.
[682] J.A. Kirk, J.F. Cardenas-Garcia, and C.R. Allison, ASLE Trans., *20*, 1977, 333-339.
[682a] K.V. Kumar, M. Cozminca, Y. Tanaka, and M.C. Shaw, Trans. ASME, Ser.B, J. Eng. Ind., *102*, 1980, 80-84.
[683] M. Röllig, R. Lindner, K. Weber, and R. Fischer, Schmierungstechnik, *6*, 1975, 176-180 (German).
[684] H. Kaliszer and G. Trmal, in [4.3], pp.90-105; Chart. Mech. Eng., *23* (9), 1976, 95-100.
[685] V.V. Efimov et al, Russian Eng. J., *60* (11), 1980, 50-51.
[686] L.V. Khudobin et al, Russian Eng. J., *54* (2), 1974, 65-68.
[687] C. Wick, Machinery (New York), *73* (3), 1966, 92-95.
[688] W. Graham and M.G. Whiston, Int. J. Mach. Tool Des. Res., *18*, 1978, 9-18.
[688a] G. Pahlitzsch, Microtecnic, *8*, 1954, 199-205.
[689] G. Werner, Paper No. MR79-319, SME, Dearborn, 1979.
[690] L.V. Khudobin and V.F. Guryanikhin, Machines Tooling, *43* (10), 1972, 43-47.
[691] I.C. Khudobin, Russian Eng. J., *60* (11), 1980, 46-49.
[692] M. Mrozek, MW Interf., *3* (3), 1978, 11-17.
[693] Yu.V. Polyanskov, Sov. Eng. Res., *1* (2), 1981, 49-52.
[694] P.D. Oates, H.J. Bezer, and A.M. Balfour, in [4.3], pp.158-162.
[695] Y. Sakai, R. Kitagawa, and K. Ohkusa, in *Proc. 21th Int. MTDR Conf.*, Macmillan, London, 1981, pp.431-437.
[696] K. Ohgo, K. Nakajima, and T. Awano, Wear, *40*, 1976, 85-92.
[697] N. Narutaki and A. Murakoshi, Bull. Jpn. Soc. Prec. Eng., *11*, 1977, 121-126.
[698] R.A. Joseph and V.A. Tipnis, in [16], pp.55-72.
[699] S. Ramalingam, B. Thomann, K. Basu, and J. Hazra, in [16], pp.111-129.
[700] S. Ramalingam, in [17], pp.394-403.
[701] J.D. Byrd and B.L. Ferguson, in *Proc. 6th NAMRC*, SME, Dearborn, 1978, pp.310-316.
[702] K. Kishi, H. Eda, and H. Ueno, in *Proc. 19th Int. MTDR Conf.*, Macmillan, London, 1979, pp.469-476.
[703] C.A. Smits, in [3.56], Paper No.104.
[704] V. LaFarina, in [3.56], Paper No.102.
[705] J.A. Bailey, Wear, *44*, 1977, 361-370, 371-376.
[706] S.A. Popov and Y.G. Kravchenko, Russian Eng. J., *57* (9), 1977, 59-60.
[707] K. Kishi and H. Eda, Wear, *38*, 29-42.
[708] K. Iwata and K. Ueda, in *Proc. 4th NAMRC*, SME, Dearborn, 1976, pp.326-333.
[709] K. Yamagouchi and T. Kato, Trans. ASME, Ser.B, J. Eng. Ind., *102*, 1980, 221-228.
[710] E. Usui, T. Obikawa, and T. Shirakashi, in *Proc. 10th NAMRC*, SME, Dearborn, 1982, pp.247-254.
[711] K. Subramanian and N.H. Cook, Trans. ASME, Ser.B, J. Eng. Ind., *99*, 1977, 295-301.
[712] K. Ohgo, A. Satoh, T. Mizuno, and T. Itoh, Wear, *52*, 1979, 79-88.
[713] K.I. Sadykhov et al, Russian Eng. J., *55* (12), 1975, 55-56.
[714] P. Wainwright, Stainless Steel Ind., *4* (22), 1976, 13-15.
[715] A.E. Focke et al, Wear, *46*, 1978, 65-79.
[716] A.I. Bezykornov, N.I. Bogomolov, and M.S. Koval'chenko, Poroshkovaya Met., *1970* (11), 77-83 (Russian).
[717] J. Vigneau and J.J. Boulanger, CIRP, *31*, 1982, 35-39.
[718] T.A. Schroeder and J. Hazra, in *Proc. 9th NAMRC*, SME, Dearborn, 1981, pp.374-381.
[719] N.F. Karpov et al, Machines Tooling, *55* (12), 1974, 52-53.
[720] S. Zaima and W. Okazaki, Bull. JSME, *13*, 1970, 719-723, 724-728.
[721] D. Keen, Wear, *28*, 1974, 319-330.
[722] R.E. DeVor and J.C. Miller, in *Proc. 9th NAMRC*, SME, Dearborn, 1981, pp.296-304.
[723] G.F. Wilson and D.K. Bruschek, in [19], pp.297-316.
[724] H.K. Toenshoff and G. Chryssolouris, in *Proc. 9th NAMRC*, SME, Dearborn, 1981, pp.410-415.
[725] Yu.V. Polyanskov and Yu.M. Prabikov, Russian Eng. J., *57* (6), 1977, 52-54.
[726] N.E. Shirk, CDA-ASM Conf., Paper No.071/2, Copper Dev. Assn., New York, 1972.
[727] C.T.H. Stoddart et al, Metals Techn., *6*, 1979, 176-184.
[728] N. Gane, Phil. Mag., *43A*, 1981, 545-566.
[729] A. Wolfenden and P.K. Wright, Metals Techn., *6*, 1979, 297-302.
[730] P.D. Hartung and B.M. Kramer, CIRP, *31*, 1982, 75-80.

[731] K. Sakuma and T. Fujito, Bull. Jpn. Soc. Prec. Eng., *10*, 1976, 159-160.
[732] H. Simon, M. Thoma, and K. Maier, wt-Z. Ind. Fertig., *69*, 1979, 79-82 (German).
[733] I.S. Hong, in [535], pp.860-869.
[734] P.J. Ziegelmeier, Cutting Tool Eng., *23* (5), 1970, 11-13.
[735] V.N. Latyshev and V.V. Semenov, Machines Tooling, *44* (6), 1973, 47-49.
[736] N.N. Zorev and D.N. Klaustch, CIRP, *16*, 1968, 267-275 (German).
[737] A.C. Hoffmanner and G.E. Meyer, in [17], pp.297-324.

Appendix

List of Standards Applicable to Metalworking

(ASTM Standards unless otherwise shown)

Physical Properties
Density, specific gravity D 1298-80; API gravity D 287-67(1977)
Viscosity, kinematic or dynamic D 445-79, D 446-74(1979); efflux viscosity (Saybolt Universal) D 88-81; viscosity-temperature charts for petroleum products D 341-77
Distillation range D 86-82; vapor pressure D 323-82; evaporation loss D 972-56(1981)
Flash and fire points (Cleveland Open Cup) D 92-78; (Tag Open Cup) D 1310-80
Melting point D 127-63(1977)
Color D 156-82

Chemical Properties
Saponification value (a measure of the mean molecular weight of fatty acids, free and esterified) D 94-80, D 460-78, D 1962-67(1979)
Acid value (a measure of free fatty acids) D 1980-67(1979)
Neutralization number (a measure of acidic or basic constituents): by potentiometric titration D 664-81; by color indicator titration D 974-80
Iodine value (a measure of unsaturated fatty acids) D 1541-60(1979)
Sulfur, total and active D 129-64(1978)
Chlorine D 808-81, D 1317-64(1978)
Unsaponifiable (but oil-soluble) impurities D 1965-67(1979)
Carbon residue: Conradson D 189-81; Ramsbottom D 524-81
Ash D 482-80

Oil Properties
Oxidation characteristics of inhibited mineral oils D 943-81
Foaming characteristics D 892-74(1979)
Chemical analysis for metals in new and used oils D 811-82
Water in petroleum products D 95-70(1980)
Resistance to water spray D 4049-81

Grease Properties
Cone penetration D 217-82(1879), D 1403-69(1980)
Apparent viscosity D 1092-76(1981)
Dropping point D 566-76(1982)
Water washout characteristics D 1264-73(1978)

Aqueous Fluids

Active sulfur in cutting fluids D 1662-69(1979)

Emulsion stability: of soluble cutting oils D 1479-64(1978); of new (unused) rolling oils D 3342-74(1979)

Bioresistance of water-soluble metalworking fluids D 3946-80

Foam in aqueous media (bottle test) D 3601-77(1982)

Ash and total iron in steel rolling dispersions of rolling oils O 4042-81

Surface Properties

Interfacial tension by platinum wire ring test D 971-82

Hydrophobic surface films: by the water break test F 22-65(1976); by the atomizer test F 21-65(1976)

Copper corrosion by the copper strip tarnish test D 130-80

Tool Life Tests

Single-point tools of sintered carbide ANSI B 94.36-1956 (R1971)

Single-point tools of materials other than sintered carbides B 94.34-1946(R1971)

Classification

Classification of metalworking fluids and related materials D 2881-73(1978)

Approximate Conversion Factors

Quantity	SI unit	Multiply by	To obtain
Length	m	39.4	in.
	mm	39.4	0.001 in.
	μm	39.4	μin.
Volume	L	0.264	U.S. gallon
Speed	m/s	3.28	ft/sec
	m/s	196.8	ft/min
	m/min	3.28	ft/min
Force	N	0.225	lb (force)
	kN	225	lb (force)
	MN	112.4	ton force (2000 lb)
Pressure	kPa	0.145	psi
	MPa (= N/mm^2)	0.145	kpsi (1000 psi)
	MPa	9.8	bar (atmosphere)
Work (torque)	N · m	0.74	ft · lb
Power	kW	1.34	horsepower
Temperature	C	32 + (1.8 C)	F
	K	K - 273.15	C
Viscosity	Pa · s (= $N \cdot s/m^2$)	10	P (poise)
	mPa · s	1	cP (centipoise)
	mm^2/s	1	cSt (centistoke)

List of Recurrent Symbols

(Note: some symbols listed below may have other meanings, too, as specified in the text)

A_a apparent area of contact
A_r real area of contact
A_0 initial cross-sectional area
A_1 final cross-sectional area
C constant; Taylor constant
D diameter of tool
E elastic modulus; specific energy
F frictional force
G shear modulus; grinding ratio
H hardness (HB = Brinell hardness number kg/mm^2; HRC = Rockwell C hardness number; HV = Vickers hardness kg/mm^2)
K constant
L length of contact between tool and workpiece
M torque
P force
R radius of tool
R_a surface roughness (arithmetic average)
R_q surface roughness (RMS value; σ)
R_t surface roughness (peak-to-valley height)
S die-separating force
T temperature, absolute (Kelvin)
T_m melting point on Kelvin scale
V volume
W work
Z rate of volume removal
a fraction of contact area in metallic contact; depth of cut in grinding
b fraction of contact area in boundary contact
c fraction of contact area separated by contaminant films
d diameter of workpiece; depth of cut in turning
e engineering strain
f function; feed in turning
h height; thickness; film thickness
h_0 height before deformation; film thickness at entry
h_1 height after deformation; film thickness at exit
k shear flow stress according to von Mises (= 0.557 σ_f)
l length, distance
l_0 length before deformation
l_1 length after deformation
m frictional shear factor; strain-rate-sensitivity exponent (*m* value)
n strain-hardening exponent (*n* value)
p interface pressure
r reduction in height or area; plastic anisotropy (*r* value)
r_c chip-thickness ratio in cutting
s distance
t time; tool life; undeformed chip thickness

v velocity
v_0 velocity at entry to work zone
w width of workpiece
α pressure-viscosity coefficient; die half-angle; entry angle in rolling; rake angle
β temperature coefficient of viscosity; plug half-angle; clearance angle
γ shear strain; attack angle
γ shear strain rate
ε natural strain (logarithmic strain, true strain)
ε strain rate
η dynamic viscosity
θ angle of converging gap
ψ friction angle in cutting
λ film-thickness ratio
μ coefficient of friction
ν Poisson's ratio
ρ density
σ normal stress; standard deviation of asperity-height distribution
σ_f flow stress in uniaxial tension or compression
τ shear stress
τ_f shear flow stress
τ_i interface shear stress
τ_s shear strength interposed film
ϕ shear angle in cutting

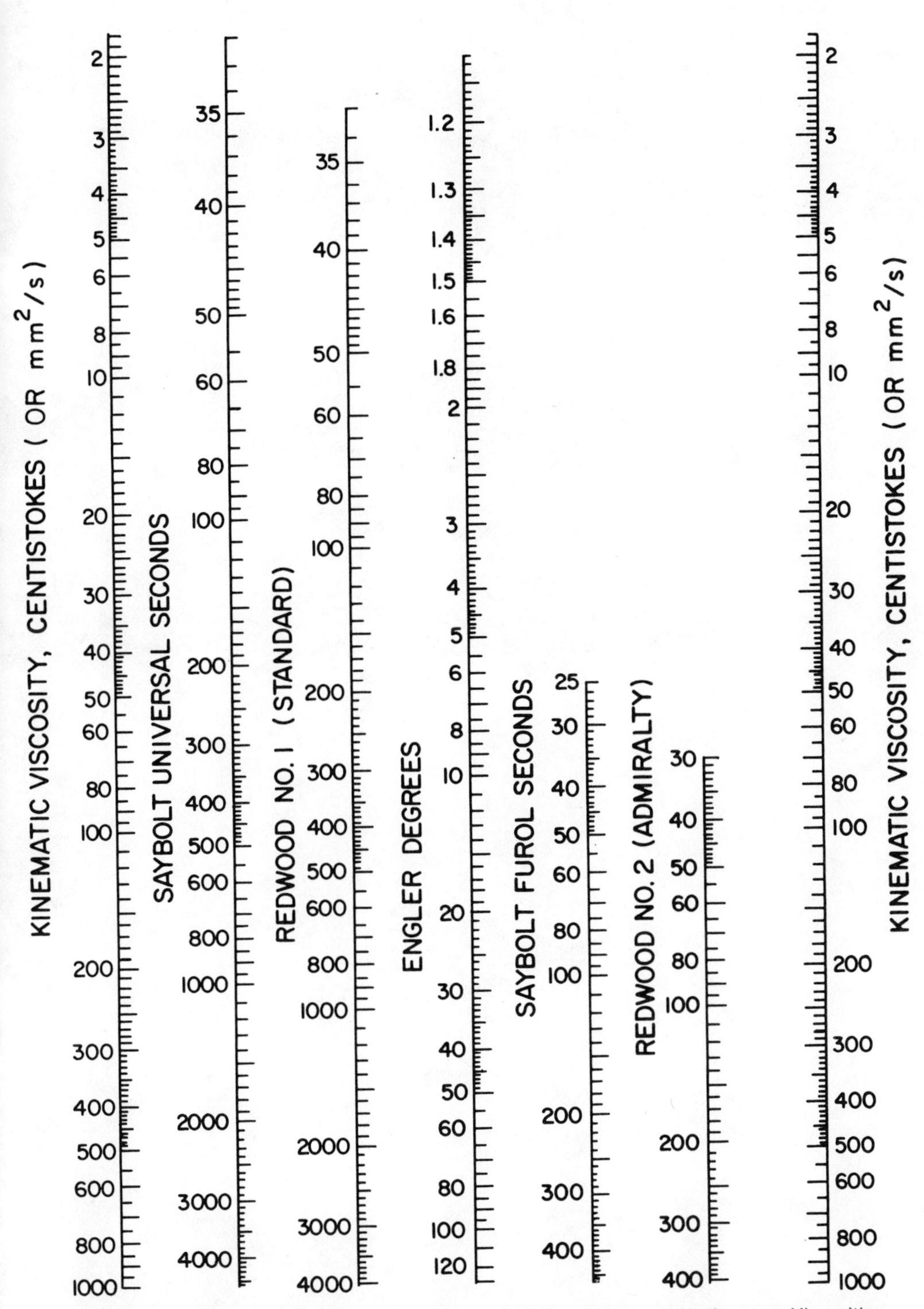

Line up straightedge so centistoke (mm^2/s) value on both kinematic scales is the same. Viscosities at the same temperature on all scales are then equivalent. To extend range of only the kinematic, Saybolt Universal, Redwood No. 1, and Engler scales: multiply by 10 the viscosities on these scales between 100 and 1000 cSt on the kinematic scale and the corresponding viscosities on the other three scales. For further extension, multiply these scales as above by 100 or a higher power of 10. (Example: 1500 cSt = 150 $\times$ 10 cSt = 695 $\times$ 10 SUS = 6950 SUS.)

Fig.A. Viscosity Conversion Nomograph. (Courtesy of Texaco's Magazine LUBRICATION)

Index of Authors Named in References in This Book

Numbers following authors' names refer to numbered entries in bibliographies; e.g., 10.187 is reference 187 in Chapter 10. A few apparent duplications result from variations in spelling or use of initials from chapter to chapter.

A

B

H

I

J

K

L

M

N

O

P

S

T

U

V

W

Y

Z

Subject Index

NOTE: Page numbers ending in F or T indicate that information concerning the entry is presented in a Figure or a Table.

D

M

N

Q

R

U